S0-BLM-136

Active Close Binaries

NATO ASI Series

Advanced Science Institutes Series

A Series presenting the results of activities sponsored by the NATO Science Committee, which aims at the dissemination of advanced scientific and technological knowledge, with a view to strengthening links between scientific communities.

The Series is published by an international board of publishers in conjunction with the NATO Scientific Affairs Division

A Life Sciences **B Physics**	Plenum Publishing Corporation London and New York
C Mathematical and Physical Sciences **D Behavioural and Social Sciences** **E Applied Sciences**	Kluwer Academic Publishers Dordrecht, Boston and London
F Computer and Systems Sciences **G Ecological Sciences** **H Cell Biology**	Springer-Verlag Berlin, Heidelberg, New York, London, Paris and Tokyo

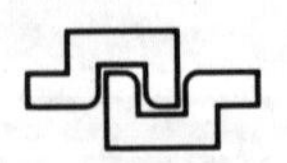

Series C: Mathematical and Physical Sciences - Vol. 319

Active Close Binaries

edited by

Cafer İbanoğlu

Ege University Science Faculty,
Astronomy and Space Sciences Department,
Bornova, İzmir, Turkey

Kluwer Academic Publishers

Dordrecht / Boston / London

Published in cooperation with NATO Scientific Affairs Division

Proceedings of the NATO Advanced Study Institute on
Active Close Binaries
Kuşadası, Turkey
September 11–22, 1989

Library of Congress Cataloging-in-Publication Data

NATO Advanced Study Institute on Active Close Binaries (1989 :
Kuşadası, Turkey)
Active close binaries / edited by Cafer İbanoğlu.
p. cm. -- (NATO ASI series. Series C, Mathematical and
physical sciences ; vol. 319)
"Proceedings of the NATO Advanced Study Institute on Active Close
Binaries, Kuşadası, Turkey, September 11-22, 1989."
"Published in cooperation with NATO Scientific Affairs Division."
Includes index.
ISBN 0-7923-0907-3 (acid free paper)
1. Stars, Double--Congresses. 2. Stars, Variable--Congresses.
I. İbanoğlu, Cafer, 1946- . II. North Atlantic Treaty
Organization. Scientific Affairs Division. III. Title.
IV. Series: NATO ASI series. Series C, Mathematical and physical
sciences ; no. 319.
QB821.N375 1989
523.8'41--dc20 90-44193

ISBN 0-7923-0907-3

Published by Kluwer Academic Publishers,
P.O. Box 17, 3300 AA Dordrecht, The Netherlands.

Kluwer Academic Publishers incorporates the publishing programmes of
D. Reidel, Martinus Nijhoff, Dr W. Junk and MTP Press.

Sold and distributed in the U.S.A. and Canada
by Kluwer Academic Publishers,
101 Philip Drive, Norwell, MA 02061, U.S.A.

In all other countries, sold and distributed
by Kluwer Academic Publishers Group,
P.O. Box 322, 3300 AH Dordrecht, The Netherlands.

Printed on acid-free paper

Printed in the Netherlands

TABLE OF CONTENTS

2. BINARY EVOLUTION AND THE CONTACT QUESTION

3. HIGH ENERGY BINARIES AND EXOTIC OBJECTS

4. COOL CLOSE BINARIES AND RS CVn STARS

PREFACE

Since the 1970s symposia or colloquia devoted to recent research on close binaries have been held around the world almost annually. At meetings of the General Assembly of the International Astronomical Union this topic has also been discussed in detail at presentations in various commission meetings and also as invited talks by leading astronomers in the field. In recent years, fundamental changes have taken place in the study of close binaries due to the improvements in observational techniques, extension of observations from X-ray to radio regions of the electromagnetic spectrum, and advances in theoretical studies.

For more than a decade, a group of astronomers at Ege University Observatory has been concentrating on active close binaries with particular emphasis on the behaviour of the light curves of chromospherically active systems. Thus, we decided to organize an international meeting in Western Anatolia, where this part of Turkey had been the cradle for great developments in science during antiquity. Kuşadasi, located only minutes away from Ephesus, one of the seven wonders of the world, was selected to be the meeting site.

Close binary systems constitute a very rich source of information about the physical properties of the component stars. Some systems are eclipsing variables, where periodic recurrences of eclipses are observed as comparatively brief decreases in the total brightness of the binary system. Precise methods of photometric observations make it possible to obtain the light variations of these systems because of eclipses and other phenomena. Therefore, close binaries may be considered as astrophysical laboratories. The subject matter encompasses a number of broad areas, related to different classes of binaries. The specific areas that were discussed at this meeting were broad, extending from systems such as Algols and W Serpentis stars which are non-degenerate binaries with circumstellar disk and wind structures; to near-contact short period binaries with starspots; chromospheric and coronal activity indicators; and cataclysmic variables and X-ray emitting neutron stars with close binary components which are in the advanced stages of evolution of close binary phenomena.

The light variations of close binaries, excluding the changes from eclipses, were addressed by two different models: i) Photometric effects by oscillating components, and ii) Star-spot and other surface inhomogeneities on close binary companions. The data, gathered with different techniques, indicated that such binaries are commonly chromospheric-active and have large, hot coronae and play an important role in understanding the evolution of the stellar components and the structure of their atmospheres and coronae. Studies of the outer atmospheres may in fact lead to better understanding of solar activity. Although the meeting concentrated on the phenomena associated with the outer atmospheres of the late-type components of these systems, other related topics of interest such as mass-exchange, mass-transfer, and mass loss were also discussed.

Recent advances, such as space-borne instruments and new generation ground-based equipment are having profound effects in the study of active close binaries. The data obtained from X-ray to radio wavelengths allow us to model the coronae and chromospheres of chromospherically active RS CVn systems. The relations between activity and evolution in the RS CVn systems were discussed in detail. Theoretical implications of magnetic flares, tidal torques, and magnetic stellar winds in active binaries were compared with observational results. Period changes, particularly in chromospherically active binaries and synthronization of starspot lifetimes with activity cycles were also considered.

Many phenomena observed in active close binaries could not be explained by the starspot hypothesis alone. Observations of active binaries planned for the next decade from both ground- and space-based observatories, will hopefully lead to further advances in the field.

The meeting was supported financially by NATO and partly by the Scientific and Research Councils of Greece and Turkey. I want to express my very sincere thanks to these organizations, whose financial support made this meeting possible to organize. The Scientific Committee made every effort to assure the success of the meeting. All the local arrangements, including very successful social programmes were on the shoulders of the Local Organizing Committee. Their dedicated work and help throughout the meeting will be remembered by all. Special thanks to Professor Zdeněk Kopal and Dr. Janet Akyüz Mattei, who did not spare their help at every aspect of the meeting. Last, but not least, I would like to thank all the speakers for their scientific contributions and for sending me their camera-ready manuscripts which helped to make my editorial responsibilities much easier. Finally, my sincere thanks to Kluwer Academic Publishers, especially Mrs. Nel de Boer, for their excellent cooperation.

22 January 1990 C. Ibanoğlu

SCIENTIFIC ORGANIZING COMMITTEE

C. İbanoğlu
E.F. Guinan
Z. Kopal
J. Akyüz-Mattei
H. Mauder
M.J. Plavec

LOCAL ORGANIZING COMMITTEE

M.C. Akan
A. Alpar
O. Demircan
S. Evren
V. Keskin
Ü. Kızıloğlu
Z. Tunca

SESSION CHAIRMEN

C. İbanoğlu	September	11, Monday morning
J. Akyüz-Mattei		11, Monday afternoon
R.H. Koch		12, Tuesday morning
M.J. Plavec		12, Tuesday afternoon
H. Mauder		13, Wednesday morning
F. Scaltriti		13, Wednesday afternoon
R. Webbink		14, Thursday morning
J. Kuijpers		14, Thursday afternoon
Z. Kopal		15, Friday morning
D.S. Hall		15, Friday afternoon
H.B. Ögelman		18, Monday morning
O. Demircan		18, Monday afternoon
J. Linsky		19, Tuesday morning
S. Catalano		19, Tuesday afternoon
C. la Dous		20, Wednesday morning
T. Mazeh		20, Wednesday afternoon
M. Kitamura		21, Thursday morning
K.C. Leung		21, Thursday afternoon
E.F. Guinan		22, Friday morning

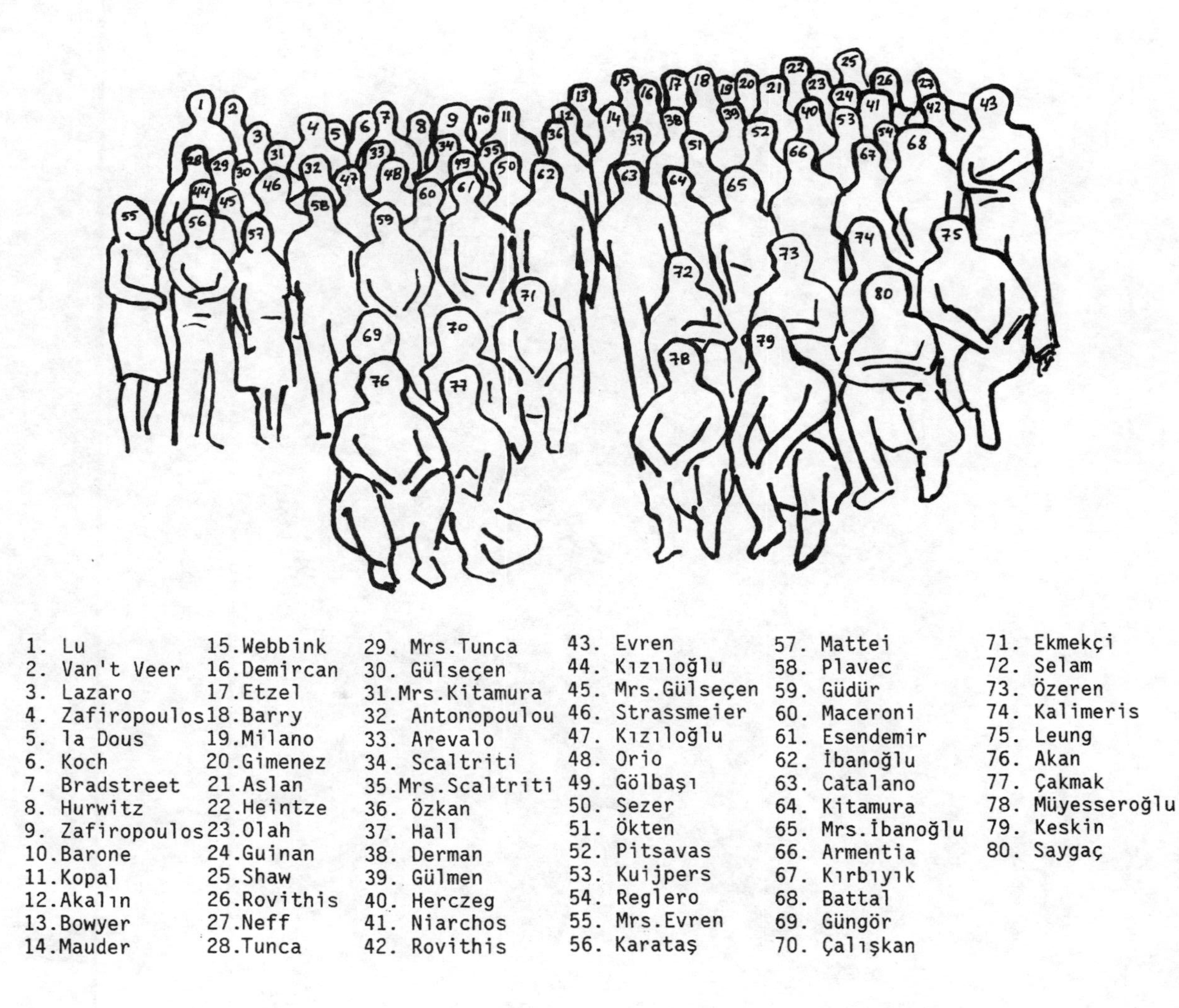

1. Lu
2. Van't Veer
3. Lazaro
4. Zafiropoulos
5. la Dous
6. Koch
7. Bradstreet
8. Hurwitz
9. Zafiropoulos
10. Barone
11. Kopal
12. Akalın
13. Bowyer
14. Mauder
15. Webbink
16. Demircan
17. Etzel
18. Barry
19. Milano
20. Gimenez
21. Aslan
22. Heintze
23. Olah
24. Guinan
25. Shaw
26. Rovithis
27. Neff
28. Tunca
29. Mrs.Tunca
30. Gülseçen
31. Mrs.Kitamura
32. Antonopoulou
33. Arevalo
34. Scaltriti
35. Mrs.Scaltriti
36. Özkan
37. Hall
38. Derman
39. Gülmen
40. Herczeg
41. Niarchos
42. Rovithis
43. Evren
44. Kızıloğlu
45. Mrs.Gülseçen
46. Strassmeier
47. Kızıloğlu
48. Orio
49. Gölbaşı
50. Sezer
51. Ökten
52. Pitsavas
53. Kuijpers
54. Reglero
55. Mrs.Evren
56. Karataş
57. Mattei
58. Plavec
59. Güdür
60. Maceroni
61. Esendemir
62. İbanoğlu
63. Catalano
64. Kitamura
65. Mrs.İbanoğlu
66. Armentia
67. Kırbıyık
68. Battal
69. Güngör
70. Çalışkan
71. Ekmekçi
72. Selam
73. Özeren
74. Kalimeris
75. Leung
76. Akan
77. Çakmak
78. Müyesseroğlu
79. Keskin
80. Saygaç

LIST OF PARTICIPANTS

Akalın, A. Ankara Univ. Dept. of Astronomy, Ankara, Turkey
Akan, M.C. Ege Univ. Dept. of Astronomy, İzmir, Turkey
Akyüz-Mattei, J. AAVSO, Cambridge, MA, USA
Antonopoulou, E. Univ. Athens, Dept. of Physics, Athens, Greece
Arevalo, M.J. Ins. Astrofisica de Canarias, Canary Islands, Spain
Armentia, J.E. Univ. Complutense, Dept. Astrofisica, Madrid, Spain
Aslan, Z. İnönü Univ. Dept. of Physics, Malatya, Turkey
Barone, F. Univ. Napoli, Dept. of Physics, Napoli, Italy
Barry, D.J. Georgia State Univ. CHARA, Georgia, USA
Başal, M. İstanbul Univ. Dept. of Astronomy, İstanbul, Turkey
Battal, C. Ege Univ. Dept. of Astronomy, İzmir, Turkey
Baysal, H. Ege Univ. Dept. of Astronomy, İzmir, Turkey
Bowyer, S. Univ. California, Space Sci. Lab., CA, USA
Bradstreet, D. Eastern College, St. Davids, PA, USA
Çakmak, H. İstanbul Univ. Dept. of Astronomy, İstanbul, Turkey
Çalışkan, H. İstanbul Univ. Dept. of Astronomy, İstanbul, Turkey
Catalano, S. Univ. Catania, Dept. of Astronomy, Catania, Italy
Demircan, O. Ankara Univ. Dept. of Astronomy, Ankara, Turkey
Derman, E. Ankara Univ. Dept. of Astronomy, Ankara, Turkey
Edalati, M.T. Univ. Nebraska, Physics and Astronomy Dept., NE,USA
Eker, Z. Ege Univ. Dept. of Astronomy, İzmir, Turkey
Ekmekçi, F. Ankara Univ. Dept. of Astronomy, Ankara, Turkey
Esendemir, A. Orta Doğu Teknik Univ. Dept.Physics, Ankara, Turkey
Esenoğlu, H. İstanbul Univ. Dept. of Astronomy, İstanbul, Turkey
Etzel, P.B. San Diego State Univ. Dept. Astronomy, CA, USA
Evren, S. Ege Univ. Dept. of Astronomy, İzmir, Turkey
Foing, B.H. ESA, Noordwijk, The Netherlands
Gimenez, A. Inst. Astrofisica de Andalucia, Granada, Spain
Gölbaşı, O. İnönü Univ. Dept. of Physics, Malatya, Turkey
Guinan, E.F. Villanova Univ. Dept. Astron. Astrophys., PA, USA
Güdür, N. Ege Univ. Dept. of Astronomy, İzmir, Turkey
Gülmen, Ö. Ege Univ. Dept. of Astronomy, İzmir, Turkey
Gülseçen, H. İstanbul Univ. Dept. of Astronomy, İstanbul, Turkey
Güngör, S. İstanbul Univ. Dept. of Astronomy, İstanbul, Turkey
Hall, D.S. Vanderbilt Univ., Nashville, USA
Heintze, J.R.W. Sterrekundig Inst. State Un.Utrecht, The Netherlands
Herczeg, T.J. Univ. Oklahoma, Norman, USA
Hurwitz, M. Univ. California, Space Sci. Lab., CA, USA
İbanoğlu, C. Ege Univ. Dept. of Astronomy, İzmir, Turkey
Kalimeris, T. Univ. Athens, Dept. of Physics, Athens, Greece
Karaali, S. İstanbul Univ. Dept. of Astronomy, İstanbul, Turkey
Karataş, Y. İstanbul Univ. Dept. of Astronomy, İstanbul, Turkey
Keskin, V. Ege Univ. Dept. of Astronomy, İzmir, Turkey
Kırbıyık, H. Orta Doğu Teknik Univ. Dept.Physics, Ankara, Turkey
Kitamura, M. National Astronomical Observatory, Tokyo, Japan
Kızıloğlu, N. Orta Doğu Teknik Univ. Dept.Physics, Ankara, Turkey
Kızıloğlu, Ü. Orta Doğu Teknik Univ. Dept.Physics, Ankara, Turkey
Koch, R.H. Univ. Pennsylvania, Dept. Astron. Astrophys.,PA,USA

Kopal, Z.	Univ. Manchester, Dept. Astron., Manchester,England
Kuijpers, J.	Sterrekundig Inst.State Uni.Utrecht,The Netherlands
la Dous, C.	Univ. Cambridge, Inst. Astronomy, Cambridge,England
Leung, K.-C.	Univ. Nebraska, Physics and Astronomy Dept., NE,USA
Linsky, J.	JILA, Univ. Colorado, Boulder, CO, USA
Lu, J.F.	Univ. Sci. and Techn. of China, Rep. of China
Maceroni, C.	Osservatorio Astronomico di Roma, Rome, Italy
Mauder, H.	Univ. Tübingen, Tübingen, Federal Rep. Germany
Mazeh, T.	Tel Aviv University, Ramat Aviv, Israel
Milano, L.	Univ. Napoli, Dept. of Physics, Napoli, Italy
Mutel, R.	Univ. Iowa, Dept. Physics and Astronomy, Iowa, USA
Müyesseroğlu, Z.	Ankara Univ. Dept. of Astronomy, Ankara, Turkey
Neff, J.	NASA Goddard Space Flight Center, Greenbelt,MD, USA
Niarchos, P.	Univ. Athens, Dept. of Physics, Athens, Greece
Ögelman, H.B.	MPI für Extraterr. Physik.,Munich, Fed. Rep.Germany
Ökten, A.	İstanbul Univ. Dept. of Astronomy, İstanbul, Turkey
Olah, K.	Konkoly Observatory, Budapest, Hungary
Orio, M.	Osservatorio Astronomico, Pino Torinese, Italy
Özeren, F.F.	Ankara Univ. Dept. of Astronomy, Ankara, Turkey
Özkan, T.	İstanbul Univ. Dept. of Astronomy, İstanbul, Turkey
Pitsavas, T.	Univ. Athens, Dept. of Physics, Athens, Greece
Plavec, M. J.	Univ. California, Dept. Astronomy, Los Angeles, USA
Reglero, V.	Dept. Matematica Aplicada Astronomia,Valencia,Spain
Rovithis, H.	Univ. Athens, Dept. of Physics, Athens, Greece
Rovithis, P.	National Observatory of Athens, Athens, Greece
Saygaç, T.	İstanbul Univ. Dept. of Astronomy, İstanbul, Turkey
Scaltriti, F.	Oss. Astronomico di Torino, Pino Torinese, Italy
Selam, S.	Ankara Univ. Dept. of Astronomy, Ankara, Turkey
Sezer, C.	Ege Univ. Dept. of Astronomy, İzmir, Turkey
Shaw, S.	Univ. Georgia, Dept.Physics and Astron.,Georgia,USA
Strassmeier, K.G.	Vanderbilt Univ.Dept.Physics and Astr.Nashville,USA
Tunca, Z.	Ege Univ. Dept. of Astronomy, İzmir, Turkey
Van't Veer, F.	Institut d'Astrophysique, Paris, France
Webbink, R.F.	Univ. Illinois, Dept. of Astronomy, Urbana, USA
Yerli, S.K.	Orta Doğu Tek. Univ. Dept. Physics, Ankara, Turkey
Zafiropoulos, B.	Univ. Patras Dept. of Physics, Patras, Greece
Zafiropoulos, F.	Univ. Patras Dept. of Physics, Patras, Greece

OPENING REMARKS

On Physical Causes of the Light Variations of Active Binary Systems and on their Interpretation

ZDENĚK KOPAL
Department of Astronomy
University of Manchester
England

In embarking on the present NATO Advanced Study Institute on Active Close Binaries it is proper for us to define first the scope of its subject and the range of its concern. This can be done in more than one way; but perhaps the simplest and most general approach to our task is to consider the occurrence of transient evolutionary phenomena in close binary systems, taking place on the time-scale of our observations, generally faster than those which would unroll if the stars were single. And if we adopt this statement as our working definition, the task facing us in our discussion will be to account for the observed facts in terms of theories governing stellar physics, and to document our conclusions - tentative as they may be - by observations which are either already available, or could be secured by existing means.

In this sense, we are indeed not short of such data, nor are these likely to be exhausted in the future. For the binary stars in our Galaxy - and no doubt in other galaxies as well - represent one of the most common types of stellar populations. Up to the distances of a few hundred parsecs, physical binaries can be detected astrometrically by effects of their orbital motions (or, in the limit, by common proper motions of their components or interferometric techniques). By virtue of periodic displacements of spectral lines, closer binaries beyond the reach of the preceding methods can be detected spectroscopically with the aid of modern large reflectors up to distances of several thousand parsecs. Beyond this range close binaries can, however, be detected if (and only if) the particular orientations of their orbital planes render them to become "eclipsing variables", whose characteristic light variations can disclose their existence in galactic or extragalactic formations resolvable in individual stars, at distances of the order of a million parsecs or more.

To give a few more quantitative estimates of this situation, within (approximately) 30 parsecs of the Sun we now know of (at least) 10 eclipsing variables in a sample of some 10^4 individual stars; and if the same proportion extends also beyond this distance, the total number of eclipsing pairs of stars in our Galaxy alone should be of the order of 10^9 - quite beyond the means of individual discovery.

But the significance of eclipsing variables in astronomy is based not only

C. İbanoğlu (ed.), Active Close Binaries, 1–6.

on their enormous numbers (disclosing them to be standard handiwork of nature), but also on the unique nature of information which they alone can provide. Spectroscopic observations alone can furnish only the minimum values of the masses of their components, and lower limits for the size of their orbits. The "missing clue" - necessary to convert these lower bounds into actual values - is the inclination of the orbital plane to the line of sight; and its value can be ascertained (from an analysis of the light changes) only if the respective binary happens also to be an eclipsing variable. Moreover, the range of astrophysical data which can be deduced from such an analysis transcends by far the masses and absolute dimensions of their components, or the characteristics of their orbits; for even the internal structure of the constituent components may transpire from some of their observable characteristics. Although the interiors of the stars are concealed from direct view by the enormous opacity of the overlying material, a gravitational field emanates from them which the overlying layers (opaque as they may be) cannot appreciably modify. This field is, to be sure, invisible; but is bound to affect both shape and motion of any masses that may be situated within it - just as the distribution of brightness over the exposed surfaces of the components (influencing the light changes within eclipses) is governed by the energy flux originating in the deep interior of the respective stars.

These, and many other possibilities opening up by the studies of eclipsing variables, have long attracted due attention on the part of the observers. Largely because of the recurrent nature of the phenomena exhibited by such systems, eclipsing variables have always been favourites with pioneers of accurate photometry of any kind - visual, photographic, or photoelectric - and the total number of individual observations made in this field must by now run into many millions. A mere inspection of such data cannot, however, exhaust - or even scratch - the tremendous wealth of information which they contain. To develop this information calls, in turn, for a construction of the *models* of such systems, based on the coupling of the three great conservation principles known to physical sciences - those of *energy*, *mass*, and *momentum* - which apply indeed to all stellar structures, be these single or multiple; though only in close binaries their coupling assumes proportions capable of matching fully theory with the observations.

A conservation of momentum will manifest itself mainly through phenomena belonging to the domain of *hydrodynamics*; some of which can affect the observed phenomena to a very vital extent. Thus their hydrostatic limit requires that the *shape* of each component (spherical if the star were single, and spheroidal if it rotates about one axis) be one appropriate for the prevalent field of force. If, moreover, the stellar material can be regarded as an inviscid fluid, the tides raised by the components on each other will not affect their axial rotation or the position of its axes.

Inviscid fluid represents, however, only a mathematical abstraction which does not exist in nature. In the outer fringe of a star consisting of neutral gas, the viscosity phenomena can as a rule be neglected and neutral gas regarded as an ideal fluid. However, with incipient ionization of hydrogen (or of hydrogen-helium mixture) the *plasma* viscosity shoots up by several orders of magnitude (cf., e.g., Chapman, 1954; or Oster, 1957); and by further orders of magnitude if the plasma in question becomes partly or wholly *degenerate* (cf. Nishimura and Mori, 1961). In addition, phenomena simulating viscosity are bound to arise in the components of close binary systems if in any parts of their interiors the energy transfer occurs by *turbulent convection*; for the latter may again give rise to viscosity exceeding that of hydrogen plasma per unit mass.

The effects produced by different kinds of viscosity (separately or in concert) are not yet thereby exhausted. For the stars consist not only of mass (consisting of essentially hydrogen-helium plasma), but also of photon gas produced at elevated temperatures, an interaction between matter and radiation will likewise produce effects simulating those of viscosity; and the joint outcome of such effects produces (cf. Kopal, 1978):

(1) a rectification of the axes of rotation of the components of close binary systems to their orbital plane; and

(2) a circularization of the shape of their relative orbit in the course of time.

The rate at which these processes are operative depends on the magnitude of dissipative processes due to viscosity. As long as the angular velocities of axial rotation and orbital revolution have not yet been synchronized, dynamical viscous tides *lag behind* the inviscid ones by amounts depending on the viscosity; and their gradual dissipation generates *heat* which tends to elevate the temperature of the stellar interior. On the observational side, the former can produce *asymmetry of the light curves* of close binary systems (regardless of whether or not the binary in question eclipses); and the heat produced by viscous friction is added to that due to nuclear or other sources). In general, however, the dynamical tides which may be produced by the initial conditions of a new-born binary can be expected to reduce to equilibrium tides, at which viscosity will no longer produce the dynamical effects after it has arrived at the Main Sequence - at least until the degenerate core begins to form near the star's centre where tides are low, but the viscosity of matter enormous.

With all this in the back of our minds, let us return to the more specific problems of observational evidence which (as I gather from the programme of our meetings) should be the subject of considerable discussion: namely, the asymmetry of the light curves exhibited by close eclipsing systems, and (mostly complicated) fluctuations of their orbital periods. As is well known, the principal feature of the light changes exhibited by binary systems which are in equilibrium state should be symmetrical with respect to the conjunctions - between minima as well as within eclipse. Asymmetries arising from orbital eccentricity (if any) are well understood (cf. sec. IV-4 of Kopal, 1979) and not relevant to practical cases; and the sense of asymmetry turns out to fluctuate in time. Moreover, they occur too frequently to be attributed to a rare or unusual case.

In more recent years (as is evident from these Proceedings), the asymmetry of the light curves of close eclipsing systems have frequently been attributed to "star-spots" distributed non-uniformly over the surfaces of their rotating components. Such an explanation has a long history behind it - in fact, it goes back to the pre-telescopic days of our science, when David Fabricius attempted in 1596 to explain in this way the observed asymmetry of newly-detected long-periodic variable Mira Ceti. The first scientific treatment of such a hypothesis goes back to Zöllner (1865), and was extended to three-axial rotation by Gyldén (1880), and carried on to the beginning of the 20th century by Turner (1907, 1908).

The reason why such an approach to our problem was eventually abandoned has been the fact - established first by Bruns (1882) and Russell (1906) that such a process is basically indeterminate; for it cannot provide any *unique* information about the distribution of the requisite hypothetical spots over stellar surfaces.

The principal reason why the early star-spot theories of stellar variability failed was, indeed, the fact that they sought to account for the observations only *descriptively* - in terms of postulated *ad hoc* phenomena - and failed to pay enough attention to the *physical* basis of such phenomena: for instance, to questions concerning the origin of the unequal distribution of brightness over the surface of the respective stars, and to the maintenance of the corresponding lateral temperature gradients.

In more recent times, only one attempt has been made to account for possible existence of such gradients in physical terms - when the present writer (Kopal, 1965) raised a possibility that the "bright spots" - such as invoked heuristically at that time by Binnendijk (1965) to explain the observed asymmetries of light changes of systems of the W UMa-type - may be caused by gas streams emanating from one component and impinging on another. If motion of such streams were supersonic, their stoppage by the target star would produce a stationary shock-wave in front of it. Behind such a shock, a region of enhanced temperature would thus be created - appearing as a "bright spot" - and maintained by a conversion of mechanical energy (i.e., kinetic energy of the gas stream) into heat.

If the requisite gas streams are indeed operative in close binary systems they could, in this manner, indeed account for the formation of bright plages located asymmetrically with respect to the line joining the centres of the two components, and thus give rise (or contribute) to the asymmetry of the light curves observed for many such systems. However, no independent proof that gas streams of the requisite properties actually exist in such systems has so far been forthcoming; nor was a quantitative hydrodynamical theory of such phenomena as yet elaborated. Therefore, any physical theory of the asymmetry of the light curves on these lines - while possible in principle - remains still wholly hypothetical.

It may be added that, as recently as the second decade of this century, Hellerich (1912) or Hagen (1919) discussed a possibility that rotating stars with spotted surfaces could be tentatively identified also with cepheids and other types of physical variables. On the other hand, the efforts to gain a deeper understanding of the variability of stars above their Main Sequence since the 1920's led to a realization that their principal causes are to be sought in the *oscillations* of optical properties of effective stellar photospheres, caused by their instability at certain stages of their evolution.

As is well known today, so long as the stars (including the components of binary systems - be these close or wide) derive their energy from a transmutation of hydrogen into helium, the very high thermal stability of such a process is sufficient to cause them to shine with constant light, and change their external characteristics only on the "nuclear" time-scale (too slow to disclose any such changes within centuries or millenia); the minor periodic variability of the Main-Sequence stars of the type of β CMa or β Cep being due to special causes.

However, for most stars which have evolvedd away from the Main Sequence, and whose evolution unrolls alternatively on the fast nuclear (helium) and dynamical (Kelvin) time-scales, this will, in general, no longer be the case. In particular, certain domains are known to exist in the HR-diagram above the Main Sequence ("instability strips") where any star entering them in the course of its evolution is bound to exhibit physical variability of a particular type (cepheids, long-period variables, etc.); and the question arises as to the relevance of such phenomena also in double-star astronomy. Indeed, no reason is known which would restrict such behaviour to single stars only; the latter are certainly not immune to it! In binary stars the components are, to be sure, gravitationally coupled; and this coupling may affect also their internal structure. In the deep interiors of the stars - where their nuclear energy is almost exclusively produced - the effects of gravitational coupling on internal structure are likely to be negligible; and any fluctuations which may occur there should be entirely unrelated with the orbital period of the system. The photospheric "gating effects", responsible for the variability of the cepheids (see, e.g., Zhevakin, 1953, 1954) may be affected by distortion to a much greater extent; and this should be especially true of "contact" components of the semi-detached binary systems. In such systems, a synchronization (or near-synchronization) of physical variability with orbital motion is more likely to occur; though a more specific investigation of this problem is stil conspicuous by its absence. But whatever may turn out to be the case, a new class of observable phenomena may arise in this way which invites attention; and to these we wish to draw the reader's attention in the present section.

If the variable (oscillating?) component of an eclipsing pair happens to be the more luminous of the two - i.e., the "primary" - its variability may remain discernible throughout the entire orbital cycle - between minima as well as within eclipses; the eclipsing systems AB Cas or Y Cam (exhibiting also variability of δ Sct-type) may be regarded as typical examples of such a situation (others being AI Hya, EN Lac and no doubt others). On the other hand, the oscillating component may happen to be a secondary which is too faint to influence the combined light of the system to a detectable extent.

However, even low luminosity cannot prevent a disclosure of the secondary (eclipsing) component's physical variability if the latter is accompanied with a change of size or shape of its occulting disc within eclipses. Such a situation may well be expected to occur among the typical semi-detached Algol systems, in which a Main-Sequence primary is attended by an evolved mate, occupying a position in the HR-diagram where a star should be expected to oscillate if it were single. Algol itself shows no photometric evidence of instability of either component; but recent observations of fluctuating asymmetry of the minima of U Cep by Olson (1976a,b) strongly suggests that the eclipsing disc of the secondary component may indeed behave in such a manner; and the same is true of U Sge (cf. Olson, 1982), whose light minima due to total eclipses indicate that the secondary's shadow may be changing in size and (or) shape in the course of the minima.

If so, there are certain predictions about the behaviour of such systems which can be made at once. First, the period of free oscillations of the components depends essentially on their internal structure, while that of the orbit (playing the role of forced oscillations) is but very weakly coupled with them. The beat phenomena between the two will, therefore, be in general incongruous; with their relative phase shifting in time.

A superposition of the light changes arising from these sources has recently been investigated by the present speaker (cf. Kopal, 1982); and since their description was incorporated in Section IV.4 of his monograph on *Mathematical Theory of Stellar Eclipses* (Kopal, 1990) which is being brought out by the same Publisher to appear at about the same time as the Proceedings of the present NATO Advanced Study Institute, it seems superfluous to repeat a discussion of the same topic in this place[1]. It should, however, be stressed that everything accomplished in this field represents only the very beginning of this subject; and I venture to hope that our discussions which follow these introductory remarks may contribute towards a further efflorescence of this exciting subject.

References

Binnendijk, L.: 1965, in *Kleine Veröff. Remeis-Sternwarte Bamberg*, 4, (No.40), p.36ff.
Bruns, H.: 1882, *Monatsber. d. Preuss. Akad. d. Wiss. für 1881*, p.48ff.
Chapman, S.: 1954, *Astrophys. J.*, 120, 151.
Gyldén, H.: 1880, *Acta Soc. Sci. Fennicae* (Helsingfors), 9, 345.
Hagen, J. G.: 1919, *Astron. Nachr.*, 209, 33.
Hellerich, J.: 1913, Diss. Berlin.
Kopal, Z.: 1965, in *Kleine Veröff. Remeis-Sternwarte Bamberg*, 4 (No.40); p.52ff.
Kopal, Z.: 1978, *Dynamics of Close Binary Systems*, D. Reidel Publ. Co., Dordrecht and Boston; (pp. XIV+510), Chapters IV and V.
Kopal, Z.: 1979, *Language of the Stars*, D. Reidel Publ. Co., Dordrecht and Boston; (pp. VII+280).
Kopal, Z.: 1982, *Astrophys. Space Sci.*, 87, 149.
Kopal, Z.: 1990, *Mathematical Theory of Stellar Eclipses*, Kluwer Acad. Publ., Dordrecht and Boston; (pp. VI+162).
Nishimura, H. and Mori, H.: 1961, in *Progress Theor. Phys. Japan*, 26, 967.
Olson, E. C.: 1976a, *Astrophys. J.*, 204, 141.
Olson, E. C.: 1976b, *Astrophys. J. Suppl.*, 31, 1-11.
Olson, E. C.: 1982, *Publ. Astron. Soc. Pacific*, 94, 70.
Oster, L.: 1957, *Z. Astrophys.*, 42, 220.
Russell, H. N.: 1906, *Astrophys. J.*, 24, 1.
Turner, H. H.: 1907, *Monthly Not. Roy. Astron. Soc.*, 67, 332.
Turner, H. H.: 1908, *Monthly Not. Roy. Astron. Soc.*, 68, 482.
Zhevakin, S. A.: 1953, *Astron. Zhurnal*, 30, 161.
Zhevakin, S. A.: 1954, *Astron. Zhurnal*, 31, 141, 335.
Zöllner, J. K. F.: 1865, *Photometrische Studien mit besonderer Rücksicht auf die Physische Beschaffenheit der Himmelskörper*, Leipzig; p.256.

[1] A part of its subject was presented orally at these meetings.

ECLIPSING BINARIES AS ASTROPHYSICAL LABORATORIES AND THE STRANGE CASE OF EPSILON AURIGAE

Edward F. Guinan
Dept of Astronomy & Astrophysics
Villanova University
Villanova PA 19085 USA

Sean M. Carroll
Harvard-Smithsonian Center for Astrophysics
60 Garden St.
Cambridge MA 02138 USA

ABSTRACT. **Eclipsing binaries are well known for providing the bulk of accurate data on stellar masses and radii. However, important as these quantities are to stellar physics and evolution, there are several classes of eclipsing binaries whose orbital and physical properties are well suited for them to serve as *astrophysical laboratories* for the study of a wide range of problems. These binaries provide fundamental information about stellar interiors, atmospheres and evolution that can be obtained in no other way. Also, accurate determinations of helium abundances can be obtained from some well suited eclipsing systems. Moreover, in some cases, eclipsing binaries with eccentric orbits and well defined physical properties can even provide tests of current theories of gravity.**

The major part of this paper deals with the enigmatic eclipsing binary ϵ Aurigae. The results from modeling of the light variations during the last eclipse in 1982-1984 are discussed. This analysis indicates that the dark companion to the F supergiant is a large, thin disk of varying opacity that consists chiefly of cool gas and dust. The outer dimension of the disk is about 9 AU, nearly the size of Saturn's orbit. In addition, the central region of the disk, within $\sim$ 1 AU, is essentially free of absorbing material. The properties of the disk are similar to those inferred for protoplanetary disks found around young stars. On the other hand Eggleton and Pringle suggest that ϵ Aurigae is a post-asymptotic giant branch (AGB) system. In this case, the F0 supergiant is not a normal, massive star but a low mass ($\sim$ 1-2 $M_{\odot}$) object that has evolved through the AGB phase of stellar evolution. In this scenario, the disk was produced by gas accreted from the last pulse episode of the AGB star a few thousand years ago. If this is correct, the supergiant is now caught in an extremely rare stage of stellar evolution in which it is a proto-planetary nebula(PPN) object, soon to become a white dwarf. The advantages and disadvantages of the protoplanetary disk model and the PPN model for explaining the properties of ϵ Aurigae are discussed.

C. İbanoğlu (ed.), Active Close Binaries, 7–36.

1. Introduction

Eclipsing binary stars are rich sources of important astrophysical information about properties of stars and also the physical laws which govern them. They are well known for providing the bulk of accurate data on masses, radii and densities of stars. About 4000 eclipsing binaries have been discovered so far and nearly every type of star is represented as a member. These include main-sequence stars, giants and supergiants (with O to M spectral types), subdwarfs, white dwarfs, neutron stars and possibly even black holes. As important as these data are to stellar physics and evolution, some selected eclipsing binaries can provide additional vital information that transcends the determination of stellar masses and radii. These eclipsing binaries have properties that permit them to serve as *astrophysical laboratories* for studying problems that at present can be solved in no other way. Some examples are briefly discussed in the next section. The remainder and major part of the paper is devoted to the unusual, bright 27 year eclipsing binary ϵ Aurigae. The possible importance of ϵ Aurigae to star and planet formation as well as to a rare stage of stellar evolution is discussed.

2. Eclipsing Binaries as Astrophysical Laboratories

2.1 Limb Darkening and Stellar Photospheres

Eclipsing binaries are important laboratories for studying stellar photospheres through the determinations of the limb darkening of the component stars. To obtain accurate values of limb darkening, it is important that the binary have deep, preferably total eclipses and well defined, uncomplicated light curves. Limb darkening coefficients derived from the analysis of the light variations during the eclipses give the distribution of brightness over the apparent disks of the stars. Limb darkening is dependent on the physical properties of the star's photospheres such as opacity (κ_λ), electron pressure (P_e), temperature (T_e) and the temperature gradient (dT/dr).

With the exception of the Sun (and perhaps a handful of nearby stars which have interferometric measures), eclipsing binaries provide the only direct, accurate limb darkening determinations for stars with a wide range of spectral types. It is surprising how few limb darkening studies there are from modern, high precision photometry of well-suited systems. These studies of limb darkening indicate satisfactory agreement with the predictions of modern photospheric models. (see *e.g.* Al-Naimy 1977). For comprehensive discussions of limb darkening see *e.g.* Kopal (1959), Grygar (1965), and Shul'berg (1973).

2.2 Structure of Stellar Atmospheres: Atmospheric Eclipses

Eclipsing binaries which consist of a cool star and a smaller, hotter companion offer the possibility of observing an *atmospheric eclipse.* This occurs when the hot companion moves behind the cool star's atmosphere prior to and just after primary eclipse. During these times the light of the hot star shines through the atmosphere of the eclipsing star along our line of sight. Scattering and absorption of the hot star's light during its passage through the cool star's atmosphere produces a spectroscopic signature of the cool star's atmosphere superimposed on the spectrum of the hotter star. This provides a direct means of studying the structure of the cool star's atmosphere and observing a cross-section of the physical conditions of its atmosphere from its outermost layers down to its photosphere.

The most famous and well studied examples of *atmospheric eclipses* are the ζ Aurigae stars which consist of K to M supergiants and O or B main-sequence or giant companions. A thorough discussion of the optical work on ζ Aurigae stars is given by Wright (1970). More recent studies of these stars in the UV with *IUE* reveal the extent and structure of the envelopes around the cool supergiant components as well as the rates of mass-loss.

The ultraviolet observations of these systems show them to be more complex than indicated from the ground-based photometric and spectroscopic studies. The UV data indicate that winds from the supergiant primaries are accreted by the hotter star in the form of an accretion wake (*e.g.*, Chapman 1981, Ahmad, Chapman and Kondo 1983, Che-Bohnenstengle and Reimers 1986, Ahmad 1989 and references therein). Analyses of the spectral features of these systems yield good estimates of their accretion rates and accretion processes but more importantly yield independent measures of mass-loss rates ($\dot{M} \approx 10^{-6} - 10^{-9} M_{\odot}/yr$) for massive supergiants. Good reviews of recent results are given by Hack and Stickland (1987) and Dupree and Reimers (1987).

Although the ζ Aurigae stars are important laboratories for studying the atmospheric structure and mass-loss of evolved massive supergiants, at present there is only **one** star that can be used in a similar way to study the atmosphere of a main sequence star. That star is the Hyades eclipsing binary V471 Tauri which consists of a K2 V star and hot DA white dwarf with an $\sim$ 12.5 hr orbital period.

V471 Tauri is an eclipsing binary whose physical properties and its membership in the Hyades cluster make it a very important star for the study of many problems of stellar astrophysics and evolution. There are many papers devoted to this star (*e.g.*, see Guinan and Sion 1984; Jensen *et al.* 1986; Skillman and Patterson 1988; İbanoğlu 1989; and references therein.) V471 Tau offers an unique opportunity to study directly the vertical structure of the outer atmosphere of an active K dwarf by using the white dwarf companion as an almost point-like probe. Ultraviolet observations of V 471 Tau made with the *IUE* satellite near primary eclipse reveal the presence of absorption features of C III (λ 1175), Si III + OI (λ 1300), CII (λ 1335), Si IV (λ 1400) and C IV (λ 1550) superimposed on the nearly featureless continuum of the white dwarf (see Guinan *et al.* 1986). These features appear only prior to or after primary eclipse, when the white dwarf is near the limb of the cool star. Also, sometimes the absorption lines are weak or absent even though the white dwarf is close to the cool star's limb. Figure 1 shows representative *IUE* spectra obtained near orbital quadrature and prior to the start of primary eclipse.

Ground-based photometry of the star reveals the presence of a variable photometric wave superimposed on the light variations expected from the eclipse and binary proximity effects (see İbanoğlu 1989). Analysis of the light curves reveals the presence of large starspot regions on the K2 V star which vary in extent and distribution with time. From light curves obtained contemporaneously with the *IUE* observations, it appears that the strengths of the UV absorption features are strongly correlated with the longitudinal distribution of the spots in the sense that when the spots were located near the limb of the K star at primary eclipse, the lines were detected, and vice versa. This indicates that the atmosphere of the K star is inhomogeneous and that the FUV spectral features arise when the line of sight to the hot star intercepts gaseous material lying above (or flowing from) the active regions of the cool star. Analysis of the line strengths indicates a temperature of T $\simeq 10^5$ K, an electron density of Ne $\simeq 10^{10} - 10^{11}$ cm^{-3} and a limiting vertical height above the cool star's surface

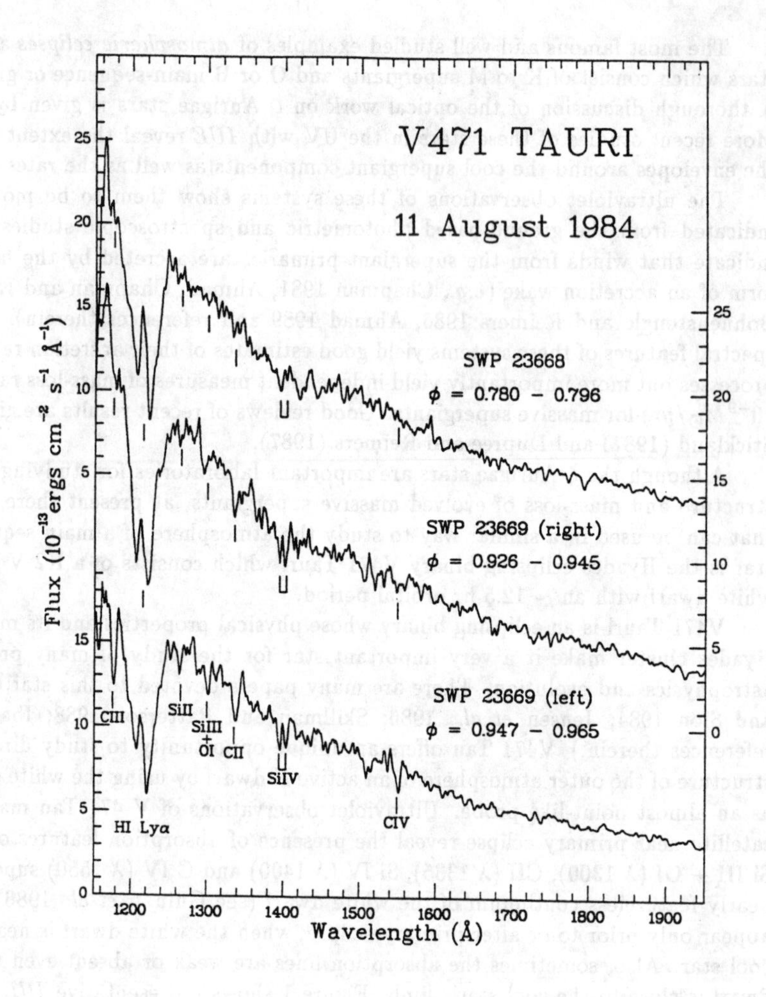

Figure 1. Representative UV spectra of V471 Tauri obtained with the SWP Camera of the *IUE* satellite. These spectra are essentially those of the hot white dwarf component because it overwhelms the light of the cool star at ultraviolet wavelengths. The top spectrum, obtained near orbital quadrature, shows a broad, Lyα feature from the white dwarf's photosphere. It is the only significant spectral feature seen away from primary eclipse. The lower two spectra were obtained prior to the start of primary eclipse. (First contact is at 0.967 phase.) Absorption features of CII, CIII, CIV, OI, SiIII, SiII, and SiIV are clearly seen superimposed on the white dwarf spectrum. These absorption features originate from large-scale structures in the atmosphere of the K star as the light from the white dwarf passes through them.

for the gas of 300,000 - 400,000 km (Guinan *et al.* 1986). The inferred properties of the plasma, along with its apparent location above active sites, indicate that these atmospheric structures may be related to the *cool* loops or active region plumes observed on the Sun (*e.g.* Foukal 1976) but are much more extensive in size. This appears to be the first time that *direct* measures of the physical and geometric properties of atmospheric structures of this kind have been obtained form a star other than the Sun. A model of the system is given in Figure 2.

2.3 Irradiation of Stellar Atmospheres: Learning about Chromospheric Structure and Physics

In binary systems consisting of a cool star and a smaller, hot companion, it is possible to determine the structure of the cool star's atmosphere by studying how the hot star's radiation is absorbed and re-radiated in its atmosphere. Once again V471 Tauri provides an excellent opportunity to probe a cool star's atmosphere by this means. The nearness of the synchronously rotating K2 V component to a hot white dwarf with well-determined physical properties (T_e = 35,000 ± 3,000 K; log g = 8.3 ± 0.3) provides an opportunity for learning about the structure and physical properties of the transition region (TR) and chromosphere of a main-sequence K2 star.

IUE observations of the V471 Tauri obtained by Guinan and Sion over its entire orbit reveal that the net Mg II h+k emission flux is strongly phase dependent. As shown in Fig. 3, the Mg II emission is strongest near 0.50 P when the irradiated hemisphere of the cool star is facing the Earth and smallest when the non-irradiated hemisphere is in view, near primary eclipse (*i.e.* 0.9 P - 0.1 P). The enhancement of the Mg II h+k emission between the irradiated and non-irradiated hemispheres is ~ 2.5 times. Ground based spectroscopic studies by Bois *et al.* (1988) also reveal that the Hα emission is phase locked in the same way as the Mg II emission. However, the study of the continuum at ultraviolet to visible wavelengths indicates **no** significant heating of the K-star's photosphere – *i.e.* there is no appreciable *reflection effect* in the star's continuum. From the *IUE* data, the Mg II h+k surface flux due to the irradiation by the white dwarf is ~ 5.6×10^6 ergs s^{-1} cm^{-2}. This value was obtained by subtracting out the Mg II emission of the non-irradiated side of the star, assuming *that* to be the background chromospheric contribution from the K-star.

Preliminary calculations show that the white dwarf's FUV radiation is sufficient to produce the observed enhancements in the Mg II h+k, and Hα line emissions. However, detailed modeling of the data with the *PANDORA* code (Vernazza, Avrett and Loeser 1981) has not been completed. With the existing *IUE* data, the modeling should yield good estimates of the physical conditions in an active cool star's atmosphere. Figures 4a and 4b show schematic diagrams of the effect of the irradiation of the cool star's atmosphere by the hot white dwarf. This work is being carried out with Eugene Avrett, Sallie Baliunas and John Raymond of the Center for Astrophysics.

2.4a Apsidal Motion: Probing Stellar Interiors

If a star is a member of an eclipsing binary having an eccentric orbit, it is possible to *see* inside the star and determine its internal mass distribution. This is accomplished by determining the rate of apsidal motion of its orbit. The apsidal motion (also known as the rate of change of the longitude of periastron) is determined from observations of the times of

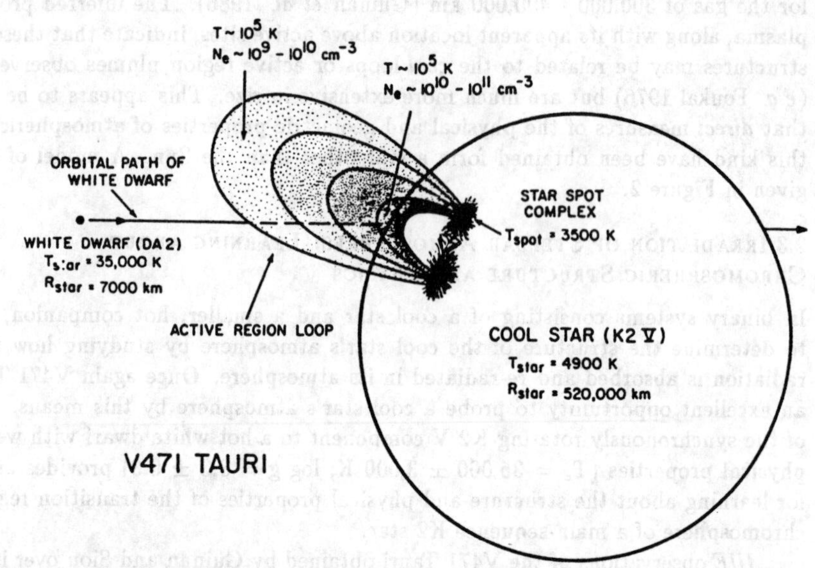

Figure 2. The Surface and Atmospheric Features of V471 Tauri. A schematic diagram of V471 Tauri is given which shows the white dwarf and K2V stars of the binary system. The surface and atmospheric features of the K2 star, inferred from optical photometry and from ultraviolet spectropscopy are shown along with their inferred geometric and physical properties.

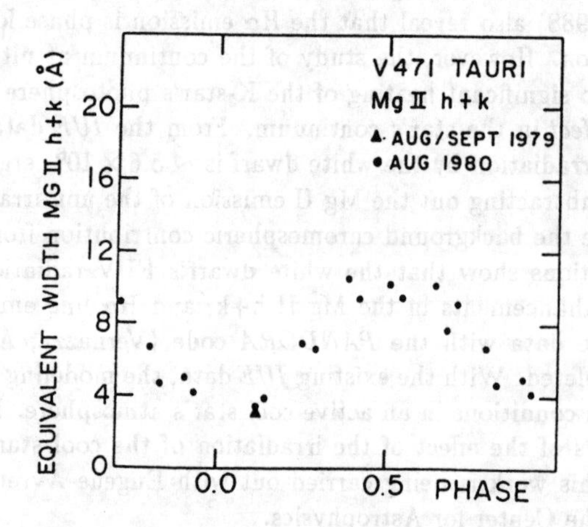

Figure 3. The Phase Dependent Variations of MgII h+k Emissions of V471 Tauri. The equivalent widths of the MgII h+k emission line features for V471 Tauri are plotted against orbital phase. The observations were obtained with the *IUE* satellite during 1979/80 by Guinan and Sion. As shown, the MgII line emission varies nearly sinusoidally over the binary's orbit with minimum and maximum MgII emission occurring at 0.0 phase and 0.5 phase, respectively. No variations in the UV continuum are observed outside primary eclipse.

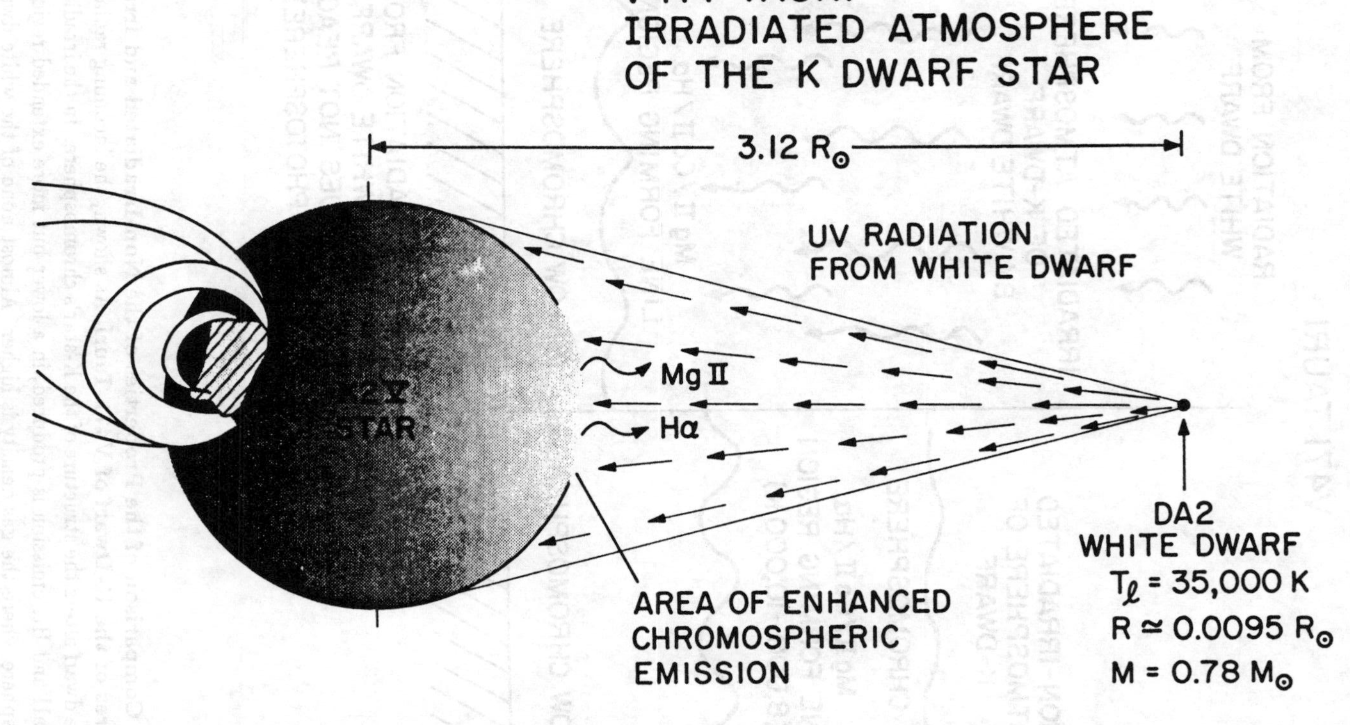

Figure 4a. A Schematic Diagram of V471 Tauri Showing the Origin of the Phase-Dependent MgII Emission Variations. A schematic diagram of V471 Tauri (drawn to approximate scale) is presented which shows the irradiation of the cool star's facing hemisphere by the hot white dwarf. The FUV radiation field from the white dwarf penetrates the cool star's corona and transition-region and produces additional Hα and MgII h+k line emissions. The additional chromospheric emission, produced by the irradiation from the white dwarf, is most visible near 0.5 phase when the irradiated hemisphere of the cool star is in full view.

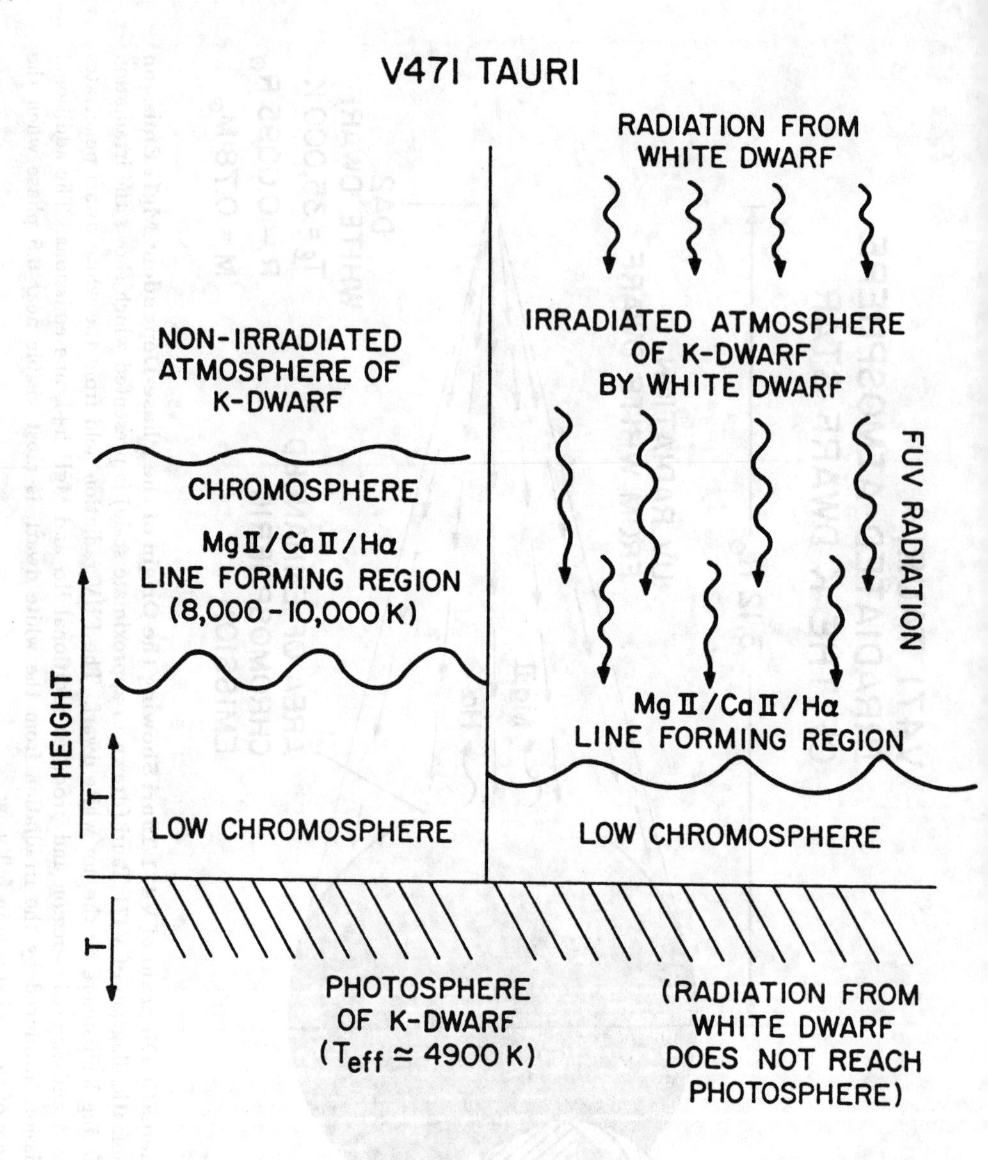

Figure 4b. The Comparison of the Properties of the Non-Irradiated and Irradiated Atmospheres of the K-Dwarf of V471 Tauri. As shown, the incoming radiation from the hot white dwarf alters the structure of the K-star's chromosphere. In the irradiated case, the MgII, CaII and Hα emission is produced, in a lower but more extended region of the star's chromosphere where the gas density is higher. Almost none of the white dwarf's radiation penetrates to the K-star's photosphere.

primary and secondary eclipses over a few decades of time. In an isolated binary system the apsidal motion arises chiefly from the classical quadrupole moment produced by the tidal and rotational distortions in the shapes of the stars. Also, there is a contribution to apsidal motion due to general relativity which for all but a few stars is usually much smaller than that arising from the distorted stars. When the orbital and stellar properties are well known from the analysis of the light and radial velocity curves, the apsidal motion rate yields a determination of the mass distribution inside the stars. Discussion of apsidal motion and its importance to the internal structures of stars can be found in Kopal (1959, 1978), Jeffery 1984; and Giménez and Garcia-Pelayo (1982) and references contained in these books and papers. Except for a handful of binaries with massive components, the overall agreement is good between the internal mass distributions determined from apsidal motion studies and those calculated from modern stellar structure and evolution theory (see Jeffery 1984).

2.4b Apsidal Motion: Eclipsing Binaries as Tests of General Relativity

There are about a dozen eccentric eclipsing binaries where the apsidal motion expected from general relativity is comparable to or greater than that expected from the classical tidal distortion effect. As discussed by Rudjøbing (1959), Koch (1977), and Guinan and Maloney (1985, 1987), these binaries can provide *tests* of general relativity in stronger gravity fields than are available within the solar system. In many of these systems, the theoretical general relativistic apsidal motion rate is hundreds of times larger than the relativistic = 43"/century apsidal motion rate for Mercury-Sun. Furthermore, in a few of these binaries with more massive components, the observed apsidal motion is significantly smaller than expected from general relativistic and classical effects. (see Guinan and Maloney 1985; 1987; and Maloney *et al.* 1989.) The cause of these discrepancies is being studied vigorously but the problem is not yet solved.

Figure 5 illustrates this problem for the eccentric eclipsing binary DI Herculis. Guinan and Maloney (1985) have shown that the observed apsidal motion rate of DI Herculis is about 0.65 deg/100yr. This is about 15% of the theoretically expected apsidal motion rate of 4.27 deg/100yr from general relativity and classical effects.

2.5 Eclipsing Binaries and the Determination of Helium Abundances

Although helium is the second most abundant element in the interstellar medium and in stars that are not highly evolved, its precise fractional abundance is not well determined. Direct spectroscopic determination of the helium content is difficult for most stars, including the Sun, because the helium lines are very weak except for very hot stars. Even for the Sun, the range in the fractional helium abundance (Y= He/ H+He+Metals) is still large using different methods: $0.13 \leq Y_{\odot} \leq 0.35$ (see Novotny 1972 and Bahcall and Ulrich 1988). The helium abundance of the Sun and stars is an important quantity in stellar structure and evolution theory, in fitting theoretical stellar evolution tracks to cluster and determine ages, and it even plays a role in the determining conditions of the early universe.

The most accurate determination of Y yet made for an older star is that obtained by Anderson *et al.* (1988) using the eclipsing binary AI Phe. AI Phe has extremely well determined orbital and physical properties. As discussed by Anderson *et al.* (1988), but first pointed out by Hrivnak and Malone (1984), AI Phe consists of an F7V (1.20 $M_{\odot}$) star and a more evolved K IV (1.24 $M_{\odot}$) component and has an orbital period of P = 24.6 days.

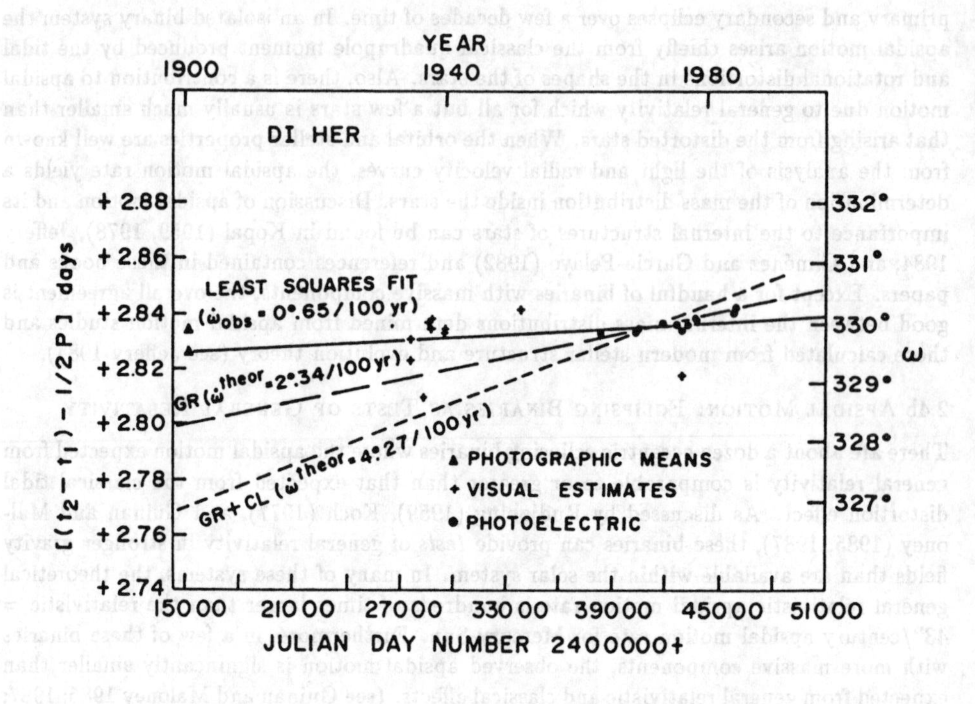

Figure 5. Apsidal Motions of DI Herculis. The plot from Guinan and Maloney (1985) shows the observed and theoretically expected classical and general relativistic apsidal motions of the eccentric eclipsing binary DI Herculis. Least-squares solutions of the timings of primary and secondary eclipses, extending over an 84 year interval yield a smaller than expected apsidal motion rate of $\simeq$ 0.65$\pm$ 0.18 deg/100 yrs. The observed apsidal motion is about 15% of the theoretical apsidal motion of 4.27 deg/100 yr expected from the combined relativistic (2.34 deg/100 yrs) and classical (1.93 deg/100yrs) apsidal motions. This results in a discrepancy of -3.6 deg/100 yr between observation and theory. At present this discrepancy remains unresolved.

AI Phe is caught at a time ($\tau = 4.1 \pm 0.4$ Gyr) when the more massive component has just evolved off the main-sequence while the slightly lower mass companion is temporarily left behind on the main-sequence. With one star near the main-sequence and its slightly more massive partner already on the early giant branch, AI Phe is an ideal binary system for providing an empirical check of stellar evolution models.

AI Phe has well-determined radial velocity and light curves as well as a good metal abundance determination; and the masses, radii and color indices of both stars are known with high precision ($\leq 1\%$). Because of this, the age and helium abundance of AI Phe can be derived with some confidence; Anderson, *et al.*, find a value of Y = 0.27 ± 0.02.

Further, the helium abundance found for AI Phe is in excellent agreement with determinations from two other binary systems. Popper and Ulrich (1986) derive a value of Y = 0.27 ± 0.03 for the Hyades eclipsing binary HD 27130 and more recently Guinan and Burns (1990) find Y = 0.28 ± 0.03 for the eclipsing binary V1143 Cyg. Also, recent theoretical solar interior/evolution models (see *e.g.*, vandenBerg and Laskarides 1987) indicate a value of $Y_{\odot} = 0.27$ for the Sun. Table I gives a summary of the properties of the stars discussed, along with their helium abundances. A more complete discussion is given by Noyes, Guinan, and Baliunas (1990).

Table I. Solar-Type Stars with Well-Determined Helium Abundances

Star	Sp/Mass $M_{\odot}$	Sp/Mass $M_{\odot}$	Age(Gyrs)	[Fe/H]	Y	ref
V1143 Cyg	F5 V 1.39	F5 V 1.35	0.6	+0.08	.28±0.03	1
HD 27130	G8 V 1.06	K2 V 0.765	0.6	+0.12	.27±0.03	2
AI Phe	F7 V 1.196	K0 IV 1.236	4.1	-0.14	.27±.02	3
Sun	G2 V		4.6	0.00	.27±.02	4

1. Guinan and Burns (1990)
2. Popper and Ulrich (1986)
3. Anderson *et al.* (1988)
4. Bahcall and Ulrich (1988)

3. The Strange Case of ϵ Aurigae

3.1 The Mystery

The long period eclipsing binary ϵ Aurigae (F0 Iap + ?) is one of the best-studied and most puzzling stars in the galaxy. We know that it consists of an F0 Ia supergiant and a dark companion which eclipses the primary every 27.1 years – the longest of any known eclipsing binary – and no secondary eclipse has been observed. Beyond these simple facts, the extraordinary amount of data collected has led to few firm conclusions.

The mystery begins with the nature of the eclipsing object. The eclipses last about 1 1/2 years and feature a long, nearly flat bottom; this is typically taken to indicate that the eclipse is total and no light from the eclipsed object is visible. However, the spectrum of the system remains essentially unaltered during the eclipse, and no signatures of the eclipsing object are seen. During the deepest part of the eclipse, the system is approximately half as bright as outside eclipse. Moreover, there is no significant dependence of the depth of the eclipse except at far ultraviolet wavelengths($\lambda < 1400$ Å) and the far infrared where the eclipse becomes shallower. Therefore, the companion must be an unusual object which blocks half of the supergiant's light yet gives a nearly flat-bottomed eclipse, while remaining invisible itself.

Many different interpretations of the unseen companion have been advanced. Kuiper, Struve and Strömgren (1937), for example, proposed that the secondary is a huge, tenuous star which was partially transparent. (For much of this century, astronomy textbooks listed ϵ Aurigae as the largest star in the Universe.) However, it was impossible to physically motivate this model or similar ones which followed. Therefore Huang (1965) proposed that the unseen object was an opaque disk which is perpendicular to our line of sight. Huang's disk is thick, so that it appears as a rectangle in projection - the so-called "brick" model. As it passes in front of the supergiant, it covers half of the star's area, and the eclipse is flat between second and third contact. A modified version of this was given by Wilson (1971) who proposed a thin, semi-opaque disk which is slightly tilted with respect to the orbit of the system.

The recent eclipse in 1982-84 has been very well covered from ultraviolet to infrared wavelengths (see Stencel 1985). A study of the photometric and spectroscopic observations has been conducted recently by Carroll *et al.* (1990) that yields more reliable determinations of the physical properties of the system. In particular, the light changes observed during the eclipse have been successfully modelled by an eclipse of the supergiant star by a large, thin semi-opaque disk viewed nearly edge-on.

Figure 6 shows the photoelectric observations of the 1982-84 eclipse obtained at Villanova University Observatory, supplemented by photometry obtained at Tjornisland Astronomical Observatory (Ingvarsson 1984) during mid-eclipse. As shown in the figure, the brightness of the system increases at all wavelengths, near the middle of the eclipse. This mid-eclipse brightening might at first be thought to result from an increase in luminosity of the star, which is known to have small light variations due to pulsations. However, the pulsations of the supergiant are known to lead to both luminosity and color variations with the star becoming hotter and bluer as it becomes brighter (Gyldenherne 1970; Carroll *et al.* 1990). During the mid-eclipse brightening, no significant change in the color index is observed. Therefore, the brightness increase is most likely a result of less light being blocked by the central region of the disk. Carroll *et al.* (1990) fit the eclipse, including the central brightening, with a thin disk that has a central hole and transitional semi-opaque central region. The fits to the data with different models are shown in Figure 7. The best fit is with the tilted, thin disk model. This analysis indicates that the disk is quite large with an estimated outer radius of about 10 $\times$ the radius of the supergiant, and the radius of the transparent inner hole is nearly 10% of the radius of the disk. Also, the disk is tilted

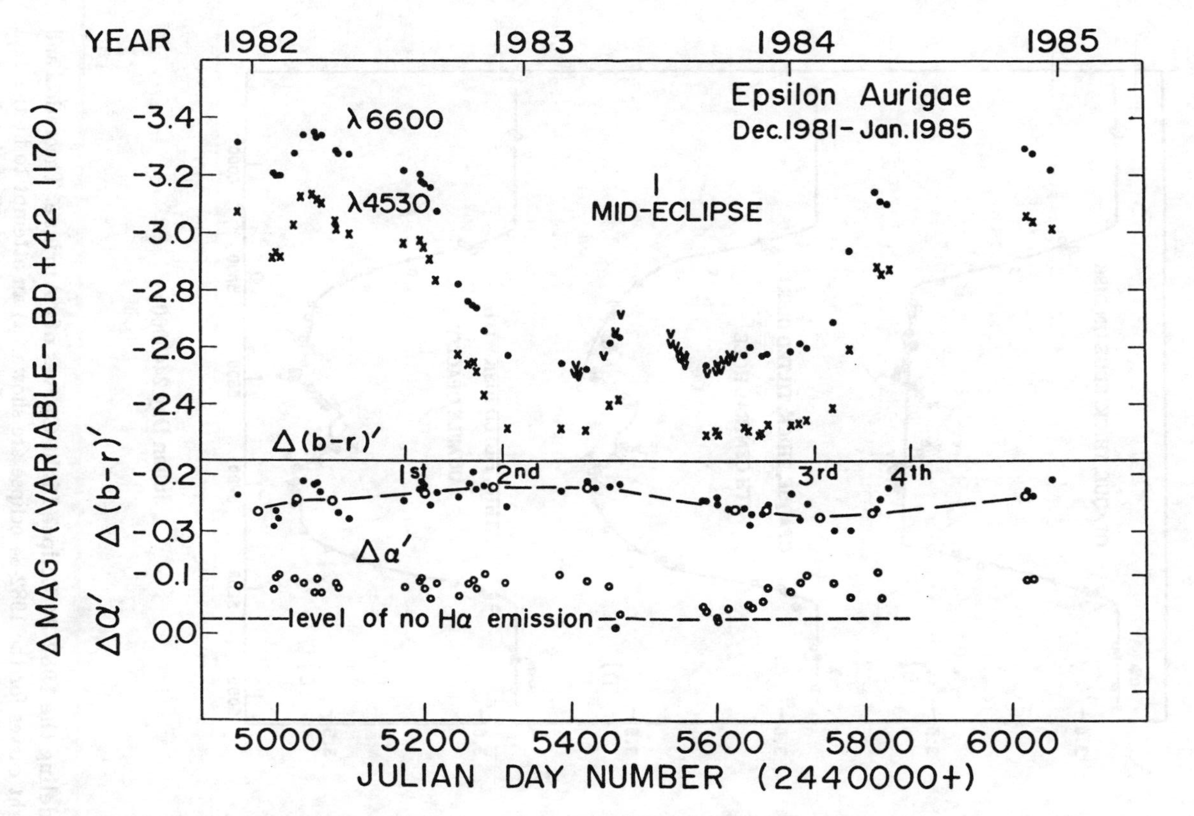

Figure 6. Photoelectric Observations of ϵ Aurigae. The nightly mean differential red (λ 6600) and blue (λ 4530) photoelectric observations of ϵ Aurigae obtained at Villanova University Observatory are plotted against time in the upper panel. The differential Δ (b-r) colors and $\Delta\alpha$' indices are plotted in the bottom panel. Estimates of first - fourth contact times as well as time of mid-eclipse are shown. From the definition of the α-index, more negative values of $\Delta\alpha$' indicate a net increase in Hα emission or a decrease in absorption. The level of no significant Hα emission is indicated. Note that near the center of the eclipse, a significant increase in Hα absorption occurs. Also plotted are V-band observations made at Tjornisland Observatory in Sweden (Invarsson 1984).

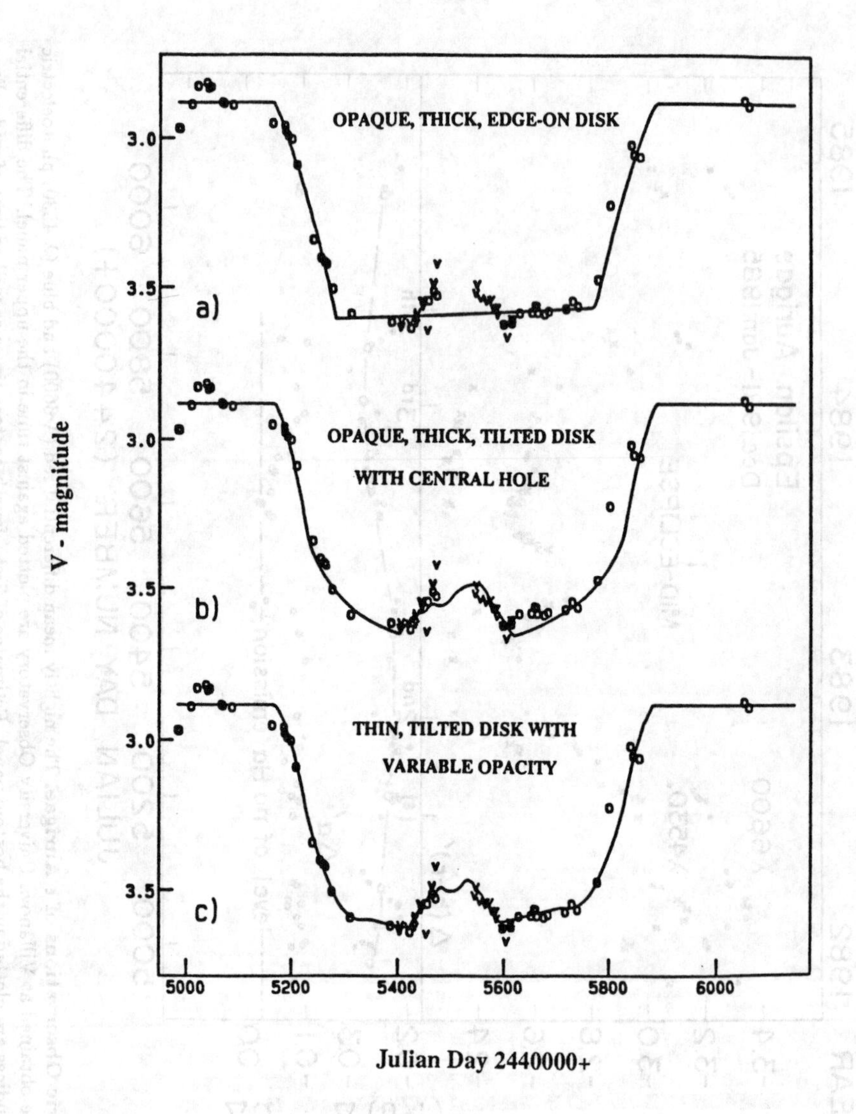

Figure 7. Modeling the 1982-84 Eclipse Light Curves of ϵ Aurigae. Theoretical and photoelectric light curves for the 1982-84 eclipse are shown. a) an attempt to fit the data with Huang's (1965) thick-disk model; this cannot account for the mid-eclipse brightening. b) A thick, tilted opaque disk with small central opening. Although this gives a good fit to the data, we felt that it did not reproduce with sufficient accuracy the flat bottom characteristic of this and previous eclipses. c) an excellent fit is achieved by using a tilted thin disk with variable opacity. The models are given by Carroll *et al.* (1990).

about 2° relative to its orbital plane. As shown in Figure 8, this disk would appear as a thin ellipse with a central hole, a semi-opaque transitional region, and an extensive opaque outer zone. Infrared observations of the system inside and outside eclipse indicate that the disk is dusty and cool with a mean temperature $\sim 475° \pm 50$ K (Backman *et al.* 1984).

While our basic understanding of the morphology of ϵ Aurigae now seems more secure, there still exist many questions about the nature of the system and how this unusual configuration came to be. If the canonical interpretation of the dark companion as a disk is correct, the disk must be extremely large to cause such long eclipses; how did such a disk come to be? Is it a remnant of the formation of the system and perhaps a protoplanetary disk, or is it an artifact of recent mass transfer from the bright star? We know that the supergiant is an F0 Ia star; however, we do not know its evolutionary history. Is it a young massive star that has recently left the main-sequence and is now travelling to the right on the HR diagram, or, as suggested by Eggleton and Pringle(1985), an evolved 1-2 $M_\odot$ star that is a post-asymptotic giant branch (AGB)/proto-planetary nebula (PPN) star that is evolving toward a planetary nebula phase and eventually to become a white dwarf? Supergiants are typically quite massive, and observations of the orbit of ϵ Aurigae indicate that the secondary should be comparable in mass to the primary. Since the disk itself is probably not very massive (compared to the star), the mass must be in the invisible object at the center of the disk. If there is a massive object at the center of the disk, why don't we see it – why is it so under-luminous?

While none of these questions may be answered definitively, the answers which have been proposed have involved many fascinating aspects of modern astrophysics. We will examine these questions in both the context of star and planet formation and stellar evolution to determine how stellar evolution theory can help us unravel the mystery of ϵ Aurigae and how observations of this unique system can help us test and refine our theories. The remainder of the paper is organized as follows. To constrain the possible scenarios of ϵ Aurigae's evolution, we briefly review the relevant observational data and derived facts. Considering these facts leads to two general possibilities, one in which the star is very massive and the supergiant has just evolved off the main sequence (the "high-mass" model) and a more recent suggestion that the star is less massive and more evolved. We then consider these two scenarios in detail, discussing the different ways in which they could be viable, and what makes them important for stellar evolution. Finally, we discuss a problem which plagues both models: the nature of the unseen object(s) at the center of the disk. We conclude by drawing some conclusions and suggesting several provocative questions which remain to be answered as well as new observations that may help solve the strange case of ϵ Aurigae.

3.2. Observations and Scenarios

To speculate on the evolutionary state of the system we need to specify parameters such as the mass and luminosity of the components. To do this, we must examine closely the determinations of the basic data such as the distance to the star.

Controversy begins with the distance and luminosity of the system. Although the star is much too far away to determine its trigonometric parallax, van de Kamp (1978) has determined an astrometric value for the angular distance from the bright star to the barycenter of the system

$$a_1 = 0.''0227 \pm 0.''0010. \tag{1}$$

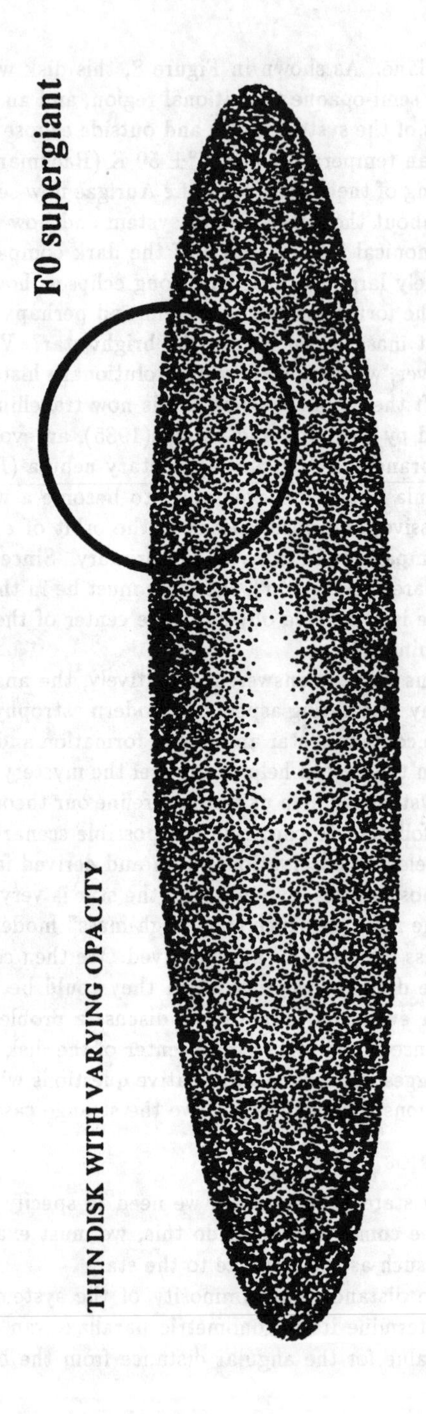

Figure 8. **Tilted, Thin Disk Model for ε Aurigae. The tilted, thin disk model with a central hole and varying opacity is shown. This is the model for ε Aurigae adopted by Carroll *et al.* (1990). The disk is completely opaque near the rim, has a semi-transparent transitional region, and a completely transparent center.**

Meanwhile, Wright (1970) used radial-velocity data to determine a value of $a_1 = 13.2$ A.U. Equating these two values, van de Kamp derives a distance of 580 ± 30 parsecs. With this distance and assuming an interstellar absorption of $A_v = 0.84$ (Morris, 1962), the absolute magnitude of the F supergiant star becomes $M_v = -6.7$ mag.

While the error bars on these crucial numbers seem small, there is reason to doubt their accuracy due to a preponderance of evidence favoring a larger distance. Both the astrometric and spectroscopic values for the semi-major axis are uncertain; for instance, Strand (1959) obtained a value of

$$a_1 = 0.''014 \pm 0.''004, \tag{2}$$

which corresponds to a distance of approximately 1000 pc and an absolute magnitude of $M_v = -7.9$ mag. The amount of interstellar absorption is also uncertain, since ϵ Aurigae is at a galactic latitude of $b = 1.^\circ 18$, very close to the galactic plane. There are, in addition, more indirect measures of luminosity which seem to favor the more luminous value. Stothers (1971) has suggested that ϵ Aurigae is a member of the association Aurigae OB1. This association lies at a distance of 1340 pc with approximate boundaries $l = 168^\circ$ to 178° and $b = -7^\circ$ to $+4^\circ$. The space motions and color excess of Aurigae OB1 are similar to that of ϵ Aurigae. If it were a member, the absolute magnitude of the star would be -8.5 mag. This value is also suggested by the semi-period-color-luminosity relation for supergiants (Burki, 1978). This empirical relation can be expressed as:

$$M_{bol} = 28.76 - 2.63\log(P) - 1.32\log(M) - 7.9\log(T_e) \tag{3}$$

where P is in days and M is in solar masses. ϵ Aurigae is known to pulsate semi-regularly with a characteristic period of approximately ~ 120 days.

If the mass of the supergiant is $\sim 15M_\odot$ (thought to be typical for F0 supergiants) this formula then yields an absolute visual magnitude $M_v = -8.8$ mag. A lower mass yields correspondingly lower luminosities; for example, a mass of $2M_\odot$ yields $M_v = -7.7$ mag. These numbers are corroborated by a relationship discovered by Osmer (1972) between absolute magnitude and the equivalent width of OI at 7774 Å:

$$M_v = -2.62W - 2.55. \tag{4}$$

Osmer lists an equivalent width $W = 2.51$ Å for ϵ Aurigae, which indicates $M_v = -9.1$ mag, with a claimed error of ± 0.5 mag. It should be noted that ϵ Aurigae had the highest value for W of any of the 60 stars in Osmer's study.

Finally, we can consider typical values for M_v for F0Ia stars. However, the values found in the literature are too uncertain to be of much assistance; Blaauw (1963), for example, gives $M_v = -8.5$ as typical for such stars, while Allen (1963) gives $M_v = -6.8$. While the preponderance of evidence indicates a distance of approximately 1000 pc, and an absolute magnitude of at least $M_v = -8.0$, the question is apparently still open and distances as small as 500 and as large as 1500 pc cannot be ruled out.

Besides the luminosity of the primary, the other important stellar parameter which is uncertain at present is the mass of each component. Morris (1962) has determined the mass function of the system to be $f(m) = m_2^3/(m_1 + m_2)^2 = 3.12M_\odot$. Therefore knowing the mass of the primary would determine the mass of the secondary as well. Usually the mass of the primary is taken to be approximately $15M_\odot$, a value consistent with a spectral classification F0 Ia and an absolute magnitude of $M_v = -8.0$. In this case, the mass of the secondary becomes approximately $13M_\odot$, comparable to that of the primary. More recently, however, Eggleton and Pringle (1985) have suggested that the bright star might be a post AGB star with a mass as low as $1.5M_\odot$. The secondary would then be about $5M_\odot$.

More importantly, however, the choice between the two models influences the evolutionary history of the system as well as its dimensions. Reconstructing the evolution of ϵ Aurigae involves interesting and important concepts in stellar theory, including very different physics for the two models. Therefore we should scrutinize these proposals with care, determining what the system is like, how it got to be that way, and what we can hope to learn if one or the other model is correct.

3.3 High-Mass Model – The Case for a Protoplanetary Disk in ϵ Aurigae

Most F0 supergiants are thought to be stars which have recently left the main sequence, and are now in a helium shell burning phase of evolution. Such stars typically have masses near $\sim 15M_\odot$ and luminosities $4.5 - 5.0L_\odot$, although considerable variation is possible. These values are consistent with the observational constraints on the ϵ Aurigae primary, so this is the evolutionary state which has usually been assigned to the star. The mass of the cool object, as has already been mentioned, is $\sim 13 - 14M_\odot$. Let us assume that the distance to the system is 1000 pc; then, with an interstellar absorption of $A_v = 0.84$, the absolute visual magnitude is $M_v = -8.0$ mag. We know that the temperature of the star is approximately 7800° K (Castelli, 1978). The bolometric correction for an F0 supergiant is 0.14 mag, so we obtain $M_{bol} = -8.1$ mag. With $L_\odot = 3.84 \times 10^{33}$ erg/s and the relations

$$M_{bol} - 4.72 = -2.5\log(L/L_\odot) \tag{5}$$

and

$$L = 4\pi R^2 \sigma T_e{}^4 \tag{6}$$

we find $L = 5.2\times 10^{38}$ erg/s and $R = 1.40\times 10^{13}$cm $\simeq 200R_\odot = 0.93$ AU. From the duration of the eclipse, Carroll *et al.* (1990) find that the radius of the disk is approximately ten times that of the star, which means $R_d \approx 2000R_\odot \approx 9.3$ AU, and the radius of the central hole is ~ 0.7 AU. Kepler's law yields a semi-major axis of $a = 27.6$AU $= 5930R_\odot$. It is instructive to consider what the absolute dimensions of the system are under this assumption; a picture of how the system would appear is given in Fig. 9. As shown, the outer dimensions of the disk are nearly the size of Saturn's orbit.

As shown in the figure, these calculations immediately yield an interesting result: while the disk is close to its Roche lobe, the F-star is well inside its Roche lobe. If the assumed evolutionary history of the primary is correct, moreover, the star is as large now as it ever has been. This means that the disk is *not* a result of the primary overflowing its Roche lobe and transferring mass to the secondary. Furthermore, while the spectral features of the primary (Castelli 1978) do show P Cygni profiles which indicate mass loss ($\dot{M} \simeq 10^{-7}$ $M_\odot/yr^{-1}$), it seems to be much too small to result in the presence of the huge disk we observe. This may be somewhat distressing, since disks in binary systems are almost always thought to result from mass transfer. In the high-mass model of ϵ Aurigae, however, this possibility seems to be ruled out; the disk must be a remnant of the formation of the system. In this case, the disk could represent a protoplanetary disk, as first suggested by Kopal (1971).

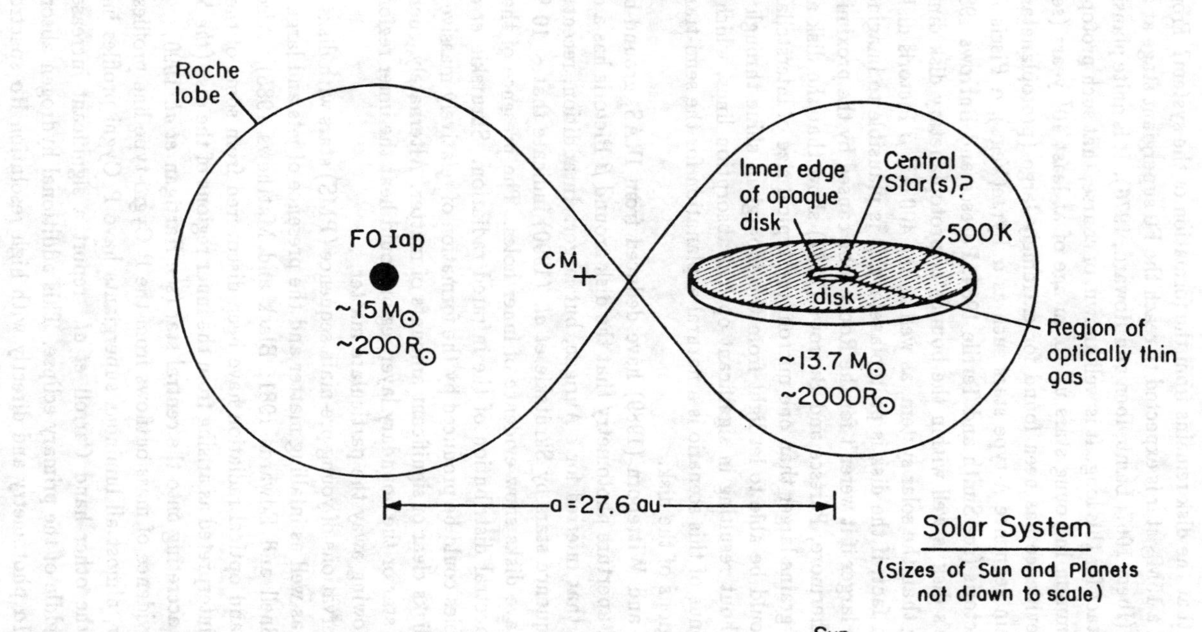

Figure 9. A Scale Model of the ϵ Aurigae System. A scale model of the ϵ Aurigae system showing the relative sizes of the star and disk. The parameters were obtained from the thin disk model with varying opacity from Carroll *et al.* (1990), and an assumed mass for the supergiant of 15 $M_\odot$. For purposes of comparison, the solar system is drawn on the same scale.

How feasible is it that the disk remains from the formation of the system? From stellar evolution calculations, a $15M_\odot$ star is expected to reach the F0 supergiant stage at an age of approximately 10^7 yrs (Iben, 1967; Lamb, Iben and Howard, 1976). It is quite plausible for a large disk to remain intact for this long. It is well known, of course, that such protoplanetary disks appear to be common in young stars up to an age of at least 10^7 years (see Strom *et al.* 1989a,b); and evidence has been found for extremely large (protoplanetary) disks around the nearby main sequence A-type stars such as α Lyrae(Vega), α Piscis Austrinis (Fomalhaut) and β Pictoris (see Smith and Terrile 1984; Paresce and Burrows 1987). The large size of ϵ Aurigae's disk is well within the bounds of protoplanetary disk dimensions; it is somewhat smaller than the solar system, as well as the 400 AU β Pictoris disk (Smith and Terrile, 1984). (In fact, if the disk is protoplanetary, it is plausible to imagine that it would have been much larger if it weren't for the Roche lobe caused by the proximity of the supergiant star.) Furthermore, Paresce and Burrows (1987) show that the disk around β Pictoris is made up of grains larger than one micron, the typical size of interstellar grains. Such a composition would be able to let light from the supergiant shine through the disk during the eclipse without resulting in significant optical absorption lines, which are not observed. An extra bonus of this scenario is a natural explanation for the semi-transparent nature of the inner regions of the disk.

Backman, Gillett and Witteborn (1989) have deduced from IRAS, ground-based IR (5μm), and from multi-aperture photometry that the disk around β Pictoris has a dust-free central hole much like that inferred for ϵ Aurigae, but larger. In addition, recent infrared studies of pre-main sequence stars by Skrutskie *et al.* (1990) indicate that $\sim$ 10 % of the stars sampled which have disks show evidence of inner holes. The presence of these holes is inferred from the spectral distribution of the infrared radiation. Skrutskie *et al.* have suggested that these holes could be produced by the formation of relatively massive planets, which keep the inner disks clear of significant amounts of matter. Alternately, energy from the pre-main-sequence star or the boundary layer region could heat the inner regions of the disk, evaporating or blowing away the particulate matter.

Spectroscopic observations of young pre-main sequence (PMS) stars with disks indicate large outflows (winds) as well as infalling matter and the presence of jets and large bi-polar molecular flows (see Snell and Edwards 1981; Black and Matthews 1985). Also excess continuum ultraviolet and optical radiation have been discovered from some of these PMS objects and have been interpreted as arising from the inner regions of the disk (*the boundary layer*) where matter is accreting onto the central star (see Hartigan *et al.* 1990).

ϵ Aurigae shows evidence of mass outflows from the P Cygni-type line profiles seen in its spectrum. However, almost all luminous supergiants have P Cygni profiles indicating large mass loss. On the other hand Carroll *et al.* report a significant increase in Hα absorption near the middle of the primary eclipse. This additional hydrogen absorption is observed in both the Hα photometry and directly with high resolution Hα spectrospcopy. This additional Hα absorption appears to arise from gas in the central regions of disk when it is seen projected against the supergiant. It is not implausible that gas is flowing out of the central region of the disk analogous to the outflows observed in many pre-main sequence stars with disks. Furthermore, the infrared and ultraviolet properties of ϵ Aurigae are also consistent with the protoplanetary disk hypothesis. The infrared studies of ϵ Aurigae are consistent with the presence of large dust grains, while in the ultraviolet there appears to be an unexplained ultraviolet excess (Boehm *et al.* 1984).

Thus, a natural explanation for the disk in ϵ Aurigae is that it is a protoplanetary/protostellar disk, similar to those inferred for pre-main sequence stars or the disk-like structures discovered around β Pictoris and a couple of other A-type stars. The huge size, low temperature, morphology, and energy distribution of ϵ Aurigae's disk are compatible with it being a protoplanetary object. Moreover, ϵ Aurigae has the spatial and kinematic characteristics of a young (Pop I) star. Carroll *et al.* find that the star is withing 20 pc of the galactic plane and has UVW space motions (relative to the Local Standard of Rest in units of km/s) of $U' = -4$, $V' = -9$, and $W = 0$ when a distance of 900 pc is assumed. A comparison of the properties of ϵ Aurigae's disk with those of protoplanetary disks is given in Table II. The protoplanetary disk scenario for ϵ Aurigae appears promising but needs much more study to be proven.

It is easy to see why the protoplanetary scenario is of interest to star and planetary formation studies: not only do we have a protoplanetary system such as that observed around other stars, but it is in an eclipsing binary, which means that every 27 years an eclipse occurs in which the light of the supergiant shines through the disk, allowing us to analyze its nature closely. Furthermore, the mid-eclipse brightening which was so prominent in the most recent eclipse was much less so in previous eclipses(see Gyldenkerne 1970); this may be an indication of evolution in the disk itself, offering us a unique opportunity to view the changing structure of a protoplanetary system.

Nevertheless, there are difficulties with this interpretation. Most significantly, at the center of the disk lurks a $\sim 13\mathrm{M}_\odot$ object whose spectrum is not observed at any wavelength, even during eclipse. If the center of the disk contained a $13\mathrm{M}_\odot$ main-sequence star, it would certainly be visible, even against the bright contribution of the supergiant; why is it so underluminous? As a possible way out of this and other problems, is the suggestion by Lissauer and Backman (1984) that a close binary system is embedded within the disk. They note that distributing the $\sim 13M_\odot$ of the secondary equally between the two stars results in an object only ~ 10 % as luminous as a massive single star. Such a configuration would be stable if the close binary had a separation of < 5 AU (Pendleton and Black, 1983). If ϵ Aurigae is young, then the proposed close binary inside the disk could consist of protostars or be a protobinary system itself.

One intriguing possibility suggested by Webbink (1985) is that the binary is extremely young($\tau < 10^5$ yrs), and still approaching the main sequence. However, the surface gravity expected for a pre-main-sequence F0 star would place it in luminosity class Iab or Ib, not Ia as is observed. Other possibilities include less massive stars in stages of shell hydrogen burning or core helium burning. However, such stars also are expected to be in a slightly lower luminosity class than is observed for ϵ Aurigae. The one possibility which is not ruled out by such considerations is that the F supergiant is a low mass (1-2 $\mathrm{M}_\odot$) post-asymptotic giant branch (AGB) star which we now consider.

TABLE II. COMPARISON OF THE DISK COMPONENT OF ϵ AURIGAE WITH PROTOSTAR/PROTOPLANETARY DISKS

Property	Protostar/ Protoplanetary Disks	Disk of ϵ Aur
Disk Dimensions		
a. Central Core (=protostar)	$\sim$ 1-10$R_\odot$	?
b. Inner Dust-Free Region (=central holes)	< 1AU $\sim$ 0.7AU	$\sim$ 0.7 AU
c. Opaque Dusty Disk (outer radius)	7 AU - 400 AU typical 30-100 AU	$\sim$ 9 AU (size may be limited by presence of luminous companion and Roche lobe radius.)
d. Dusty Envelope/ Molecular Envelope or Gas Disk	4000-7000 AU	?
Shape of Disk	flattened disk 10% of PMS stars show evidence of central holes	thin disk with central hole; titled $\sim 2^\circ$ relative to orbit
Composition	dust grains $r \geq 1$ micron	dust grains + gas $r \simeq 1$ micron
Temperature	$\simeq$ 50-200 K	450-500 K
Mass(Disk)	$\simeq$ 0.01-0.1$M_\odot$	$\sim$ 0.1-1.0 $M_\odot$
Radio Emission(mm)	yes	?
Infrared Excess	yes	yes
UV Excess (from boundary layer or central star)	yes	yes
Evidence of Outflows	yes	yes(but may be due the Supergiant Companion)
Bipolar Flows	yes	maybe ?
Age	young $\sim$ 1-100 Myr	young $\sim$ 10 Myr

3.4 Low-Mass Model – The Case for ϵ Aurigae Being a Post AGB/PPN Object

It was the problem of the high mass and low luminosity of the cool component which led Eggleton and Pringle to propose that ϵ Aurigae is less massive than previously assumed. They propose that the F0 supergiant component is *not* a normal, massive young star, but a 1-2 $M_\odot$ star that has evolved beyond the asymptotic giant branch (AGB) and is now a post-AGB star/protoplanetary nebula (PPN) object. If the mass of the supergiant is 1-2 $M_\odot$, then the cooler component would have a mass of $\sim$ 5 $M_\odot$ instead of the 13-15 $M_\odot$ assumed in the high mass model. Also, all of the absolute dimensions of the system then correspondingly decrease. Protoplanetary nebular supergiants are typically less luminous than the high-mass stars, perhaps $M_v = -7.0$ mag (consistent with van de Kamp's (1978) distance determination but not with the others). Then the supergiant and the disk would have radii of approximately $150R_\odot$ = 0.7 AU and $1500R_\odot$ = 7.0 AU, respectively. The semi-major axis of the orbit is 17 AU. The supergiant is still well within its Roche lobe, but little closer to it than in the high-mass model.

The post AGB/PPN stage of stellar evolution is very brief for a star as luminous as ϵ Aurigae($\sim 10^3 - 10^4$ yrs) according to Iben (1982) and Schönbernger (1983). However, several mostly high-galactic latitude F and G supergiants, the so-called 89 Her stars, have been suggested as being PPN objects (Bond, Carney and Grauer 1984; Hrirvnak, Kwok and Volk 1989). The majority of these PPN candidates have high space velocities and low metal abundances indicating that they are old disk or even halo objects (see Hrivnak, Kwok, and Volk 1989). Eggleton and Pringle (1985) suggest that the disk of ϵ Aurigae formed recently when the supergiant filled its Roche lobe and transferred mass to its companion. Like Lissauer and Backman (1984), Eggleton and Pringle believe that the secondary star is actually a binary system, making ϵ Aurigae a triple system.

Although this reconstruction of ϵ Aurigae's history differs remarkably from the previous model, it is equally exciting from a stellar evolution point of view. Little is known about post-AGB, protoplanetary nebula (PPN) evolution of moderate-mass stars, and this system may represent a unique opportunity to observe one in action. However, the expected short life times and rarity of such stars causes us to ask how likely it is that this system, already remarkable for other reasons, should contain a PPN object. Is there any observational evidence for such an interpretation? Saito and Kitamura (1986) claim that such evidence can be found in the decreasing size of the primary star. They have performed radial velocity studies which indicate that the star has undergone episodes of catastrophic contraction, and argue from changes in contact times of the eclipse that the supergiant has decreased by 16% of its radius in the last 27 years. If this were true, it would dramatically confirm the hypothesis that the bright star is rapidly contracting. However, it seems unlikely that such a well-observed star could shrink so dramatically so quickly without producing noticeable luminosity and color changes. A further caveat is suggested by the well-known fact that the primary star pulsates (see Carroll *et al.* 1990). This could cause spurious radial velocity determinations and perhaps mimic a collapse of the atmosphere. While this evidence is interesting, it does not represent a *compelling* confirmation of the low-mass model.

Because the lifetime of a post-AGB/PPN star of high luminosity is very brief, the disk of ϵ Aurigae should have formed quite recently when the AGB star underwent a large mass-loss episode. During the mass ejection episode and shortly afterwards, the total luminosity of the binary might be expected to be higher than today. To investigate this possibility, we examined star catalogues from antiquity that contain visual estimates of ϵ Aurigae. These historic magnitude estimates were obtained from the compilations made by Bailey 1843, Flammarion 1882, Peters and Knobel 1915, and Knobel 1917 and are given in Table III. As shown in the table, the historical records indicate that ϵ Aurigae has been near its present luminosity since at least the time of Hipparchos (~ 130 B.C.). Although the early magnitude estimates are not very precise (with errors of ± 0.5 mag), they do indicate that a major change in the star's brightness probably did *not* occur, or went undetected. These observations do not lend support to the low mass model. However, they also do *not* rule it out because a significant brightness change of the star could have gone unnoted or occured prior to recorded history. It should be mentioned, that no significant brightness changes are expected for the high mass model over the last $10^4 - 10^5$ yrs.

TABLE III
VISUAL MAGNITUDE ESTIMATES OF EPSILON AURIGAE FROM HISTORIC CATALOGS

Epoch	Magnitude Estimate	Source(1,2,3,4)
-130 (BC)	4 - 3	Hipparchos Star Catalog: as given in the *Almagest* by Ptolemy (1,2)
+960 (AD)	4	Al Sufi (1,4)
1437	4	Ulugh Beg (3)
1590	4	Tycho Brahe (1,4)
1603	4	Helvelius (1,4)
~ 1700	4	Flamsteed (1,4)
~ 1800	4	Piazzi (1,4)
1840-1880	~ 3.3-4.0	Schmidt/Argelander (4)
1880	3.8	Flammarion (4)
Modern *V* magnitude	2.98-3.80	Carroll *et al.*

1. Bailey (1843)
2. Peters and Knobel (1915)
3. Knobel (1917)
4. Flammarion (1882)

The chief reason for doubting the low mass AGB/PPN model of ϵ Aurigae is that the supergiant component appears to be a normal, luminous extreme Pop I star with no apparent abundance anomalies. For example, no unusually high abundances of CNO processed matter, expected from the dredge-up pulses for a star that has passed through AGB evolution (see Iben 1982), are seen in the spectra of ϵ Aurigae. In addition, its location close to the galactic plane (b $\simeq$ 1°; z $\simeq$ 10-20 pc) and its low space velocity (S$\leq$ 10 km/s) indicate that it is most likely a very young object.

3.5 WHAT'S INSIDE THE DISK?

Almost simultaneously with the proposal of the low-mass model by Eggleton and Pringle, Lissauer and Backman (1984) also proposed that the secondary is a binary embedded within a disk, but in the context of the usual high-mass model. Dynamically, there are advantages and disadvantages to the proposal. An advantage is that orbital resonances with the period of the primary, similar to those which confine Saturn's rings, may serve to define the apparent sharp outer edge of the disk as well as the presence of the inner hole. An additional benefit is that the close binary could transfer angular momentum to the inner disk, keeping the disk itself stable for a longer timescale than would otherwise be expected.

A possible disadvantage, on the other hand, lies in the effect of the central binary on the thickness of the disk. While it is difficult to compute exactly the effect of tidal distortion of the disk due to the close binary, it seems possible that the disk might be warped or puffed up to a thickness incompatible with the tilted thin-disk model. Kumar (1987) has shown that it is possible to imagine a configuration in which the inner binary twists the disk to a thickness that effectively mimics Huang's (1965) thick-disk model. Kumar takes the inclination of the close stars to be $\approx 20°$ with respect to the orbital plane of the wide system. However, as was already noted, the thick-disk model is apparently ruled out by the observation of the mid-eclipse brightening in the recent eclipse. Van Hamme and Wilson (1986), on the other hand, have argued that it is also possible to imagine a configuration with a close binary in which the tidal interactions are sufficiently negligible that a thin disk is still viable. There are actually two effects which would cause the disk to deviate from a simple planar geometry: self-interaction of the disk material and gravitational interaction with the central binary. A simple estimate of the magnitude of the first effect on the ratio of disk thickness to radius, z/r, can be made from the ratio of sound speed to circular velocity (Pringle 1981):

$$\frac{z}{r} = \left(\frac{RTr}{GM\mu}\right)^{1/2} \tag{7}$$

where R is the gas constant per mole, G is Newton's constant, M is the mass of the central binary, and μ is the mean molecular weight. The temperature T of the disk has been measured from IR observations to be approximately 500°K (Backman *et al.*, 1984). Therefore, this ratio for ϵ Aurigae becomes approximately

$$\frac{z}{r} = \frac{0.05}{\mu^{1/2}}. \tag{8}$$

If much of the material in the disk is dust and grains, which we expect for a protoplanetary model, then μ represents the *molecular* weight of a dust particle and would be very high. This forces the disk to be extremely thin, and is thus perfectly consistent with the tilted thin disk model. The gravitational interaction between the disk and the three stars in the system is harder to calculate. Van Hamme and Wilson explored the question by looking at the motion of a test particle in the potential of a hypothetical three-body system: a restricted four-body problem. By performing numerical simulations they found that a disk particle would wander only a very small distance (less than 10^{-3} of the radial distance) if the inner binary were closely aligned with the wide orbit – within approximately 1°. While this is encouraging, it is not clear that such a small inclination can lead to a disk geometry which will explain the light curve, both by eclipsing enough of the primary and letting enough light through during mid-eclipse. However, these simulations lead us to believe that the thin disk with central binary is a very viable alternative.

It is worth noting that, unlike the high-mass model, the low-mass proposal requires the secondary to be a binary, as explained by Webbink (1985). In Eggleton and Pringle's scenario, the primary star reaches its Roche lobe after reaching the AGB. At this stage, the star has a deep convective envelope. At the base of this envelope, hydrostatic equilibrium requires that the pressure scale as P of order M^2/R^4. At the same time, however, adiabatic convective equilibrium requires that, for an ideal gas, the pressure go as P of order $\rho^{5/3}$ of order $M^{5/3}/R^3$. Equating these, we see that the adiabatic response of the star to mass loss is to expand, with its radius going as R of order of $M^{-1/3}$. Meanwhile, the Roche lobe radius also expands with decreasing mass, going as

$$\frac{d(\ln R_L)}{d(\ln M)} \approx 2q - \frac{5}{3}. \tag{9}$$

Here, q is the ratio of the mass of the lobe-filling star to the mass of its companion. If $q > 2/3$, then the Roche lobe will not expand as fast as the star itself. Mass loss then proceeds on a dynamical timescale, resulting in catastrophic mass and angular momentum loss. On the other hand, if $q < 1$, then the star under consideration is less massive than its companion and therefore should have evolved later. The only way to escape this dilemma is to have the secondary consist of a binary: then the primary can be less massive than the secondary considered as a whole, while still being the most massive star in the system.

Thus we may say that the binary secondary is an attractive model, even though it is plagued with the unresolved problem of the gravitational stability of the thin disk. Once again this proposal, if true, makes ϵ Aurigae a fascinating system with which to study stellar evolution, as there are only about 26 triple systems known in which all three stars are close enough to interact within their lifetimes (Fekel, 1981).

Nevertheless, it is far from clear that the attractiveness of this model makes alternatives unworthy of consideration. The only other plausible suggestion of which we know was the proposal that the center of the disk contains a black hole (Cameron, 1971). The motivation for this hypothesis, of course, was the search for an object which could be as massive as $13M_\odot$ without being visible. In fact, ϵ Aurigae was included in the survey of black hole candidates by Trimble and Thorne (1969), who did not accord it serious consideration since there were no observed X-rays or gamma-rays (and there have been none since). However, while such observations would be compelling evidence that a black hole was present, they should not be thought of as necessary to its existence. High-energy radiation would only be prominent in an *accreting* system, and it is not implausible that this isn't happening in ϵ Aurigae. (For example, if the black hole had a less massive close companion, accretion would be suppressed by angular momentum transfer to the disk.) Furthermore, it is still possible that X-rays from a disk are beamed away from us or attenuated, as is believed to occur in the system A0620-00 (Blandford, 1987). Therefore, although the black hole scenario is not absolutely required, it is still a viable alternative to other possibilities.

3.6 Some Concluding Remarks on ϵ Aurigae

ϵ Aurigae presents observational and theoretical stellar astrophysicists with unique opportunities and unique challenges. At present, our knowledge of the mass and luminosity of the system is too uncertain to pinpoint evolutionary state. The two viable options are the high-mass model in which the primary is in a helium shell burning phase, recently off the main sequence, and the low-mass model in which the primary is a post-AGB star and PPN object contracting to a white dwarf state. In the first model, the large disk around the secondary is likely to represent a protoplanetary system at an age of around 10^7 years. The low-mass model explains the disk as the result of recent mass transfer from the primary; in this case, the primary star is of significant interest to stellar evolution studies since it is in such a rare stage of rapid evolution.

The question of the object at the center of the disk remains unresolved. While it is difficult to imagine a single star lurking there unobserved, there are uncomfortable aspects of a binary model or a black hole. The necessity of further study is evident, including search for a secondary eclipse in the infrared and continual monitoring of the star in all wavelengths. Hopefully the future will see us unravel the mystery of this perplexing system.

4. Conclusions

In this paper we have shown the important role that eclipsing binaries play in the modern drama of astrophysics. The selection of topics was of necessity highly personal and strongly biased to subjects and stars near and dear to our hearts and research interests. This account was illustrative rather than comprehensive so that a number of important topics were only briefly covered, or worse yet, not covered at all. Fortunately, this book contains many good examples of binary systems as astrophysical laboratories. These include systems that can be used to study stellar structure and evolution, accretion processes, gas dynamics, and magnetic activity in close binaries.

As we approach the end of this century, the future looks promising for work on close binaries. With automatic photoelectric telescopes (APTs), it is now possible to obtain accurate light curves and lose no sleep over it. There are several APTs already operating, with most of them located in the Southwest USA. Plans are being made to establish a global network of 75 cm APTs (the so-called GNATS: Global Network of Automatic Telescopes) so that continuous coverage of eclipsing binaries (as well as other types of variable stars) could be effectively and efficiently conducted at low cost. The observations would be conducted with the same telescopes, filters and detectors, etc., thus eliminating problems with combining photometric data obtained with different telescopes and instrumentation. The GNAT system would also make it possible to obtain complete light curves of some difficult systems. Binaries with integer day periods or with rapidly changing light curves such as RS CVn systems, cataclysmic binaries, and W Serpentis stars would benefit from a GNAT system.

In addition to increased ground-based coverage, the study of close binaries should benefit from the planned launches of a new generation of orbiting telescopes by NASA, ESA and Japan. By the late 1990's, it should be possible to make X-ray, UV, and infrared observations of faint binaries in our Galaxy as well as in nearby galaxies. This may lead to the possibility of using eclipsing binaries as astrophysical laboratories for exploring differences in the structure, composition, and evolution of stars in different galaxies. It is also possible that eclipsing binaries could play a role in the refinement of the extragalactic distance scale.

5. Acknowledgements

This work was supported in part by grants from NASA: NAG 5-382 and NAG 5-982 to Villanova University. One of us, EFG wishes to thank Drs. Cafer İbanoğlu and Janet Mattei for inviting me to participate in the NATO-ASI meeting on *Active Close Binaries* at Kuşadası, Turkey. We also wish to thank Dr. Elizabeth Jewell for assisting in the preparation of the manuscript and Dr. David Bradstreet for reading the text.

6. References

Ahmad, I. A. 1989, *Ap. J.*, **338**, 1011.

Ahmad, I. A., Chapman, R. D., and Kondo, Y. 1983, *Astron. Astrophys.*, **126**, L5.

Allen, C.W. 1963, *Astrophysical Quantities*, 2d ed. (London, The Athlone Press).

Al-Naimy, H. M. K. (1977), *Ap. Sp. Sci.*, **53**, 181.

Anderson, J., Clausen, J. V., Gustafsson, B., Nordström, B. and VandenBerg, D. A., 1988, *Astron. Astrophys.*, **196**, 128.

Backman, D.E., Becklin, E.E., Cruikshank, D.P., Joyce, R.R., Simon, T., and Tokunaga, A. 1984, *Ap. J.*, **284**, 799.

Backman, D. E., Gillett, F. C. and Witteborn, F. C. 1990, *Ap. J.*, in press.

Bahcall, J. N. and Ulrich, R. K. 1988, *Rev. Mod. Phys.*, **60**, 297.

Bailey, F. 1843, *The Catalogues of Ptolemy, Ulugh Beigh, Tycho Brahe, Halley, and Hevelius*, in *Mem. Roy. Astron. Soc.*, **XIII**.

Blaauw, A. 1963, in *Basic Astronomical Data*, ed. K. A. Strand, (Chicago: Univ. Chicago Press), p. 383.

Blandford, R.D. 1987, in *300 Years of Gravitation*, ed. S.W. Hawking and W. Israel (Cambridge: C.U.P.), p. 277.

Boehm, C., Ferluga, S. and Hack, M. 1984, *Astron. Astrophys.*, **130**, 419.

Bois, B., Lanning, H. H. and Mochnacki, S. W. 1988, *A. J.*, **96**, 157.

Bond, H. E., Carney, B. W., Grauer, A. D. 1984, *Publ. A.S.P.*, **96**, 176.

Burki, G. 1978, *Astron. Astrophys.*, **65**, 357.

Cameron, A.G.W. 1971, *Nature*, **229**, 178.

Carroll, S. M., Guinan, E. F., McCook, G. P. and Donahue, R. 1990, submitted to *Ap. J.*

Castelli, F. 1978, *Astron. Astrophys.*, **69**, 23.

Chapman, R. D. 1981, *Ap. J.*, **248**, 1043.

Che-Bohnenstengle, A. and Reimers, D. 1986, *Astron. Astrophys.*, **156**, 172.

Dupree, A. K. and Reimers, D. 1987, in *Scientific Accomplishments of IUE*, ed. Y. Kondo (D. Reidel Publ. Co., Dordrecht), p. 321.

Eggleton, P.P and Pringle, J.E. 1985, *Ap. J.*, **288**, 275.

Y. Kondo (D. Reidel Publ. Co., Dordrecht), p. 321.
Eggleton, P.P and Pringle, J.E. 1985, *Ap. J.*, **288**, 275.
Flammarion, C. 1882, *Les Étoiles et les Curiositiés du Ciel.*, **Suppl. L'Astronomie Populaire**(Paris).
Fekel, F.C., Jr. 1981, *Ap. J.*, **246**, 879.
Foukal, P. V. 1976, *Ap. J.*, **210**, 575.
Giménez, A. and Garcia-Pelayo, J. M. 1982, in *Binary Stars and Multiple Stars as Tracers of Stellar Evolution* , IAU Colloq. No. **69**, eds. Z. Kopal and J. Rahe (Reidel,
Dordrecht), p. 37.
Grygar, J. 1965, *Bull. Astr. Inst. Czech.*, **16**, 195.
Guinan, E. F. and Burns, J., 1990. in preparation.
Guinan, E. F. and Maloney, F. P. 1985, *A. J.*, **90**, 1519.
Guinan, E. F. and Maloney, F. P. 1987, in *New Generation Small Telescopes*, eds. D. S. Hayes, D. R. Genet, and R. M. Genet (Fairborn, Mesa), p. 383.
Guinan, E. F. and Sion, E. M. 1984, *A. J.*, **89**, 1252.
Guinan, E. F., Wacker, S. W., Baliunas, S. L., Loeser, J. G., and Raymond, J. C. 1986, in *New Insights in Astrophysics: Eight Years of UV Astronomy with IUE*, ESA SP-**263**, 197.
Gyldenkerne, K. 1970, *Vistas Astron.*, **12**, 199.
Hack, M. and Strickland, D. 1987, in *Scientific Accomplishments of IUE*, ed. Y. Kondo (D. Riedel Publ. Co., Dordrecht, Holland), p. 445.
Hartigan, P. Hartmann, L., Kenyon, S. Skrutskie, M. and Strom, S. 1990, *Ap. J. Lett.*, submitted.
Hrivnak, B. J. and Milone, E. F. 1984, *Ap. J.*, **282**, 748.
Hrivnak, B. J., Kwok, S. and Volk, K.M. 1990, *Ap. J.*, in press.
Huang, S.-S. 1965, *Ap. J.*, **141**, 976.
İbanoğlu, C. 1989, *Ap. Sp. Sci.*, **161**, 221.
Iben, I. 1967, *Ann. Rev. Astr. Astrophys.*, **5**, 571.
Iben, I. 1982, *Ap. J.*, **260**, 821.
Ingvarsson, S. I. 1984, *Epsilon Aurigae Compaign Newletter.*
Jeffery, C. S. 1984, *M.N.R.A.S.*, **207**, 323.
Jensen, K. A., Swank, J. H., Petre, R., Guinan, E. F., Sion, E. M., and Shipman, H. L. 1986, *Ap. J.*, **309**, L27.
Knobel, E. B. 1917, *Ulugh Beg's Catalogue of Stars* (The Carnegie Institute, Washington, D.C.).
Koch, R. H. 1977, *A. J.*, **82**, 653.
Kopal, Z. 1971, *Ap. Sp. Sci.*, **10**, 332.
Kopal, Z. 1978, *Dynamics of Close Binary Systems* (Reidel, Dordrecht, Holland).
Kuiper, G.P., Struve, O., and Strömgren, B. 1937, *Ap. J.*, **86**, 570.
Kumar, S. 1987, *M.N.R.A.S.*, **225**, 823.
Lamb, S.A., Iben, I.I., and Howard, W.M. 1976, *Ap. J.*, **207**, 209.
Larson, R.B. 1978, in *Protostars and Planets*, ed. T. Gehrels (Univ. Ariz. Press, Tuscon), 43.
Lissauer, J.J. and Backman, D.E. 1984, *Ap. J. Lett.*, **286**, L39.

Maloney, F. P., Guinan, E. F., and Boyd, P. T. 1989, *A. J.*, **98**, 1800.

Morris, S.C. 1962, *R.A.S.C. Jour.*, **56**, 210.

Novotny, E. 1972, *Introduction to Stellar Atmospheres and Interiors* (Oxford University Press, N. Y.).

Noyes, R. W., Guinan, E. F., and Baliunas, S. L. 1990, in *The Solar Interior and the Atmosphere*, (Univ. Arizona Press), in press.

Osmer, P.S. 1972, *Ap. J. Supp.*, **24**, 247.

Paresce, F. and Burrows, Ch. 1987, *Ap. J. Lett.*, **319**, L23.

Pendleton, Y.J., and Black, D.C. 1983, *A.J.*, **88**, 1415.

Peters, C. H. F. and Knobel, E. B. 1915, *Ptolemy's Catalogue of Stars: A Revision of the Almagest* (The Carnegie Institute, Wash., D.C.).

Popper, D. and Ulrich, R. 1986, *Ap. J.*, **307**, 161.

Pringle, J.E. 1981, *Ann. Rev. Aston. Astrophys.*, **19**, 137.

Rudjøbing, M. 1959, *Ann. Astrophys.*, **21**, 111.

Saito, M. and Kitamura, M. 1986, *Ap. and Sp. Sci.*, **122**, 387.

Schönberner, D. 1983, *Ap. J.*, **272**, 708.

Shul'berg, A. M. 1973, in *Eclipsing Binaries*, ed. V. P. Tsesevich (Halsted Press, NY), p. 72.

Skillman, D. R. and Patterson, J. 1988, *A. J.*, **96**, 976.

Skrutskie, M. F., Dutkevich, D., Strom, S. E., Edwards, S., Strom, K. M. and Shure, M. A. 1990, *A. J.*, **99**, 1187.

Smith, B.A. and Terrile, R.J. 1984, *Science* **226**, 1421.

Snell, R. L. and Edwards, S. 1981, *Ap. J. Suppl.* **45**, 121.

Stencel, R. E. 1985, *The Recent Eclipse of Epsilon Aurigae*, NASA Confer. Publ. **2384**.

Stencel, R. E., Kondo, Y., Bernat, A. P. and McClusky, G. E. 1979, *Ap. J.*, **233**, 621.

Stothers, R. 1971, *Nature*, **229**, 180.

Strand, K.A. 1959, *A. J.*, **64**, 346.

Strom, S. E., Edwards, S., Strom, K. M. 1989a, in *The Formation and Evolution of Planetary Systems*, eds. H. A. Weaver and L. Danly (Cambridge University Press, Cambridge).

Strom, K. M., Strom, S. E., Edwards, S., Cabrit, S. and Skrutskie, M. 1989b, *A. J.*, **97**, 1451.

Trimble, V.L. and Thorne, K.S. 1969, *Ap. J.*, **156**, 1013.

van de Kamp, P. 1978, *A. J.*, **83**, 975.

VandenBerg, D. A. and Laskarides, P. G. 1987, *Ap. J. Suppl.*, **64**, 103.

van Hamme, W. and Wilson, R.E. 1986, *Ap. J. Lett*, **306**, L33.

Vernazza, J. E., Avrett, E. H. and Loeser, R. 1981, *Ap. J. Suppl.*, **45**, 635.

Webbink, R.F. 1985, in *1982-84 Eclipse of Epsilon Aurigae*, ed. R. Stencel (Washington: NASA), 49.

Wilson, R.E. 1971, *Ap. J.*, **170**, 529.

Woodward, P.R. 1978, *Ann. Rev. Astron. Astrophys.*, **16**, 555.

Wright, K.O. 1970, *Vistas Astron.*, **12**, 147.

ACCRETION DISKS IN NON–DEGENERATE BINARIES

Mirek J. PLAVEC
Department of Astronomy
University of California
Los Angeles, CA 90024-1562
U.S.A.

ABSTRACT. Accretion disks play a very important, often dominant, role in cataclysmic variables, that is, in binary systems where the gainer is a degenerate star. In interacting binaries with non–degenerate components, such as the semidetached binaries of the Algol type, the disks play in general a much more subdued role. In ordinary Algols, they are most likely to be detected by the "shell" line absorption they cause. However, in systems with unusually high rates of mass transfer, such as probably are the *W Serpentis stars,* larger and more important disks can be anticipated. A case in point is *Beta Lyrae*, which will be discussed in more detail.

1. Accretion Disks in the Cataclysmics and in Algols

When discussing the accretion disks in non–degenerate close binaries, it is useful to make a brief comparison with cataclysmic variables. The phenomena of sudden and unpredictable outbursts of classical novae and dwarf novae have been successfully explained by assuming that these objects are close binary stars, in which mass is being transferred from a red dwarf onto a degenerate white dwarf. The spectra of a majority of the cataclysmic variables are dominated by the accretion disks. The main–sequence red dwarfs, confined to tiny critical Roche lobes and weakened by mass outflow contribute little luminosity, and, if seen at all, their light shows up only in the red part of the spectrum. The gainers of the material are tiny white dwarfs with high effective temperatures but very small radii, hence their radiation is usually drowned by that of the accretion disk. For the total radiation emitted from one face of an accretion disk that trasfers $\dot{m}$ solar masses per year onto a white dwarf of mass m and radius R, we can write

$$L_d = 0.25\ G\ m\ \dot{m}\ /\ R = 7.853\ m\ \dot{m}\ /\ R,$$

where in the first form everything is expressed in cgs units, while in the second form the mass, radius, and luminosity are in solar units but the rate of mass transfer, $\dot{m}$, is expressed in "microsuny", where 1 μsy= 10^{-6} solar masses per year. The

C. İbanoğlu (ed.), Active Close Binaries, 37–48.

latter unit is convenient since it is an order-of-magnitude estimate of the typical mass transfer rate in the non-degenerate semidetached systems (called here simply *Algols*). For a cataclysmic variable, we can take $m = 1\ m_\odot$, $R = 0.01\ R_\odot$, and $10^{-5} \leq \dot{m} \leq 10^{-2}$ μsy. This gives $0.008 \leq L_d \leq 8\ L_\odot$, while the intrinsic luminosity of the accreting white dwarf is about $0.001 \leq L_{wd} \leq 0.1\ L_\odot$. If the binary system is viewed nearly pole-on, we get the full luminosity of the disk; if the system is seen edge-on and is eclipsing, we will miss much of the disk radiation, but we can usually see a prominent *hot spot* on the edge of the disk. For comparison with the Algols, it is good to realize that the fairly high luminosity of the disk and the low luminosity of the central star are both due to the small radius of the gainer.

In non-degenerate binaries, the situation is quite different, again mainly because of the radius of the gainer. The accreting star is now a main-sequence star (if not larger), which makes it much more luminous, while the luminosity of the disk is smaller because the potential well is much less deep. In absolute values of luminosity, this can be more than compensated for by the mass transfer rate which can be considerably higher than in the dwarf systems. Take as a realistic example the well-studied system of TT Hydrae (Etzel 1988; Plavec 1988): Here we have $m = 2.25\ m_\odot$, $R = 1.9\ R_\odot$, so if $\dot{m} = 1$ μsy, we have $L_d = 9.3\ L_\odot$, a fairly high luminosity; however, the accreting star, being a B9.5 V main-sequence star, emits about 30 $L_\odot$. And the rate of mass transfer is probably smaller than 1 μsy rather than larger, which contributes to the discrepancy.

In addition to the total flux radiated, its wavelength distribution is also important. In the theory of the α-disks (see, e.g., Pringle 1981), the temperature in the disk at a distance r (in solar units) from the center of the gainer is given by the formula

$$T = 4.78 \times 10^{+5}\ \dot{m}^{1/4}\ m^{1/4}\ r^{-3/4}\ [1 - (R/r)^{1/2}]^{1/4}$$

and the maximum effective temperature is reached at $r = 1.36$ stellar radii. Inserting, we get for the above cataclysmic variable T_{max} of at least 30,000 K and possibly significantly higher for higher rates of $\dot{m}$. On the other hand, for a main-sequence gainer such as TT Hydrae, the effective temperature in the disk does not exceed 5,600 K. At the outer edge of the disk, which can stretch out to a distance of about 80% of the Roche radius of the gainer, i.e. to about 8 $R_\odot$, the canonical α-disk will have an effective temperature of only 3,300 K, if we again assume $\dot{m} = 1$ μsy. The secondary component in TT Hydrae has an average effective temperature of some 4,700 K and luminosity about 13 $L_\odot$ (Etzel 1988). We see that in its luminosity contribution, the disk competes at best with the secondary star – actually not even that, since we observe the system nearly edge-on and thereby miss most of the disk radiation.

Figure 1 shows another characteristic of the accretion disks in Algols which tends to diminish their importance even further: We notice that the disk is geometrically quite thin compared to the component stars; at the edge, the disk half-thickness is about 0.7 $R_\odot$, while the stellar radii are 1.9 and about 5 $R_\odot$, respectively.

Thus, in about one third of the time allocated to me, I have only managed to convince you that accretion disks in non-degenerate binaries are of negligible importance! Now I face the hard task to show that it is not quite so.

2. Disks in Algols and Serpentids

Even in ordinary Algols, where the rate of mass transfer is low, the presence of the disks does play a recognizable role. Since we are looking at the systems nearly edge–on, we are unlikely to see much of the disk light, but we will see its absorption effects. Please look again at Figure 1. From very reliable photometric observations of the totally eclipsing system TT Hydrae, Etzel (1988) found that the orbital inclination is 84.4°. As the Figure shows, during the primary eclipse, the secondary star totally eclipses not only the gainer, but its entire accretion disk as well. However, at phases outside primary eclipse, the front portion of the disk is seen projected on part of the globe of the primary (normally I would say "on the disk of the primary", but a double use of the term "disk" would introduce confusion). Since the disk is considerably cooler than the photosphere of the B9.5 V star, we expect to see absorption lines. They are indeed observed, as Etzel's optical and my *IUE* spectra show. In fact, the Fe II absorptions observed in particular in the broad wavelength range between 2000 and 2700 Å, are so deep that most part of the globe of the gainer must be viewed through the veil of circumstellar material which is probably optically thin in the continuum but optically thick in the lines of Fe II and other singly–ionized metals. Fig. 2 shows a segment of the UV spectrum of TT Hydrae compared to a normal B9.5 III star which has no circumstellar disk or shell.

The classical model of an α–disk is too simplified to be useful for studying disks about whose vertical optical thickness in the continuum we are not sure in advance. Dr. Hubený, now at the HAO in Boulder, developed a more general computer code in which each ring in the disk is calculated self–consistently with no *a priori* assumption about optical thickness (cf. a previous paper by Kříž and Hubený 1986). The program uses the powerful method of complete linearization as developed for stellar atmospheres, but still needs additional work for cases where convection plays a decisive role in the disk structure, as is the case for cooler disks.

It is possible to imagine cases of non–degenerate systems where the accretion disk plays a more important role. When other parameters are kept equal but we increase the rate of mass transfer, the disk becomes warmer, thicker, and more luminous. The mass transfer rate can in principle be much higher than the 1 μsy assumed here. Paczynski (1971) has shown that if a star with initial mass m solar units is to lose mass through the Roche lobe overflow, the mass loss rate can, during the so–called rapid phase, become as high as

$$\dot{m} = 0.033 \quad R_l \; L_l \; / \; m_l$$

where again the radius, luminosity, and mass are all expressed in solar units and it is indicated that they refer to the loser; and the resulting rate of mass transfer is again expressed in μsy. Suppose that a 5 solar masses star fills its critical Roche lobe at some time after it has left the main–sequence band and is now traversing the Hertzsprung gap. Its radius increases rapidly from some 9 $R_\odot$ to 40 $R_\odot$ and more, while its luminosity, after reaching a peak of nearly 1,000 $L_\odot$ (in the thick shell burning phase) gradually declines to 300 $L_\odot$. The above equation then shows that the rate of mass transfer can be as high as 50 – 80 μsy.

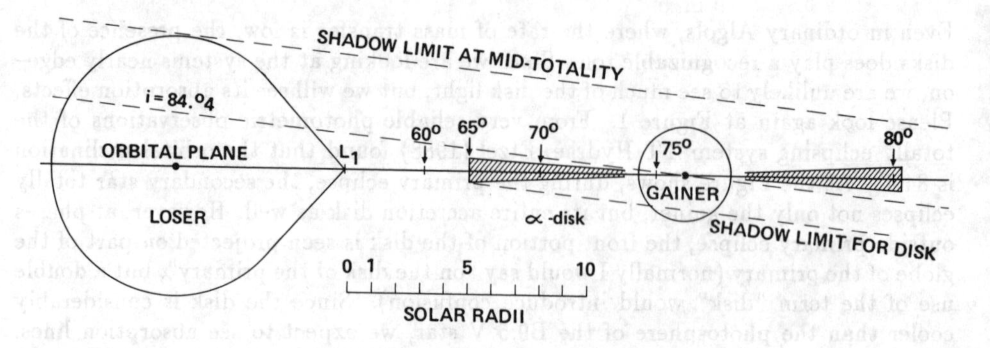

Fig. 1. An edge-on view of TT Hydrae showing the α-disk.

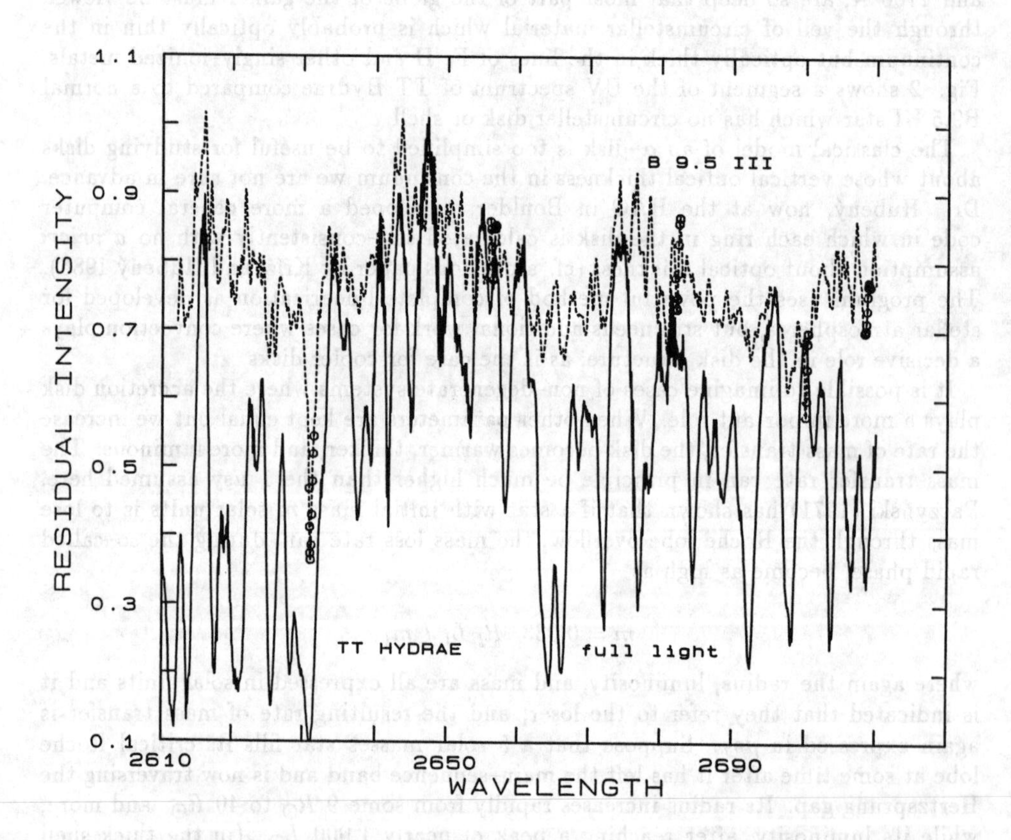

Fig. 2. A segment of the high-dispersion UV spectrum of TT Hydrae, compared to the normal star of the same spectral type (δ Cygni,B9.5 III), clearly shows the deep circumstellar shell absorptions.

If the gainer is a main-sequence star of 2 $m_\odot$, $R = 1.6\ R_\odot$, then the maximum effective temperature in the disk will lie between 10,000 and 20,000 K, and one side of the disk will radiate several hundred solar luminosities. It is tempting to look for this type of interacting binaries that would be non-eclipsing, that is, viewed more nearly pole-on. So far, we have to deal with eclipsing systems, where virtually all this accretion-generated luminosity is from our point of view wasted – emitted into the universe in directions from which we cannot see it. Another aspect of the disks comes into play, namely their geometrical and optical thickness, increased on account of the higher mass transfer rate. In systems with longer periods – tens of days, say – the disk thickness at its edge may well exceed the radius of the accreting star. This disk edge, being rather far from the gainer, remains rather cool, having an effective temperature between 4,000 and 9,000 K, depending on circumstances. Viewing such a system nearly edge-on, we will see an object resembling by its spectrum and luminosity a giant star of spectral type from K to A, as the case may be. *It is my belief that this is fundamentally the explanation for the peculiar interacting systems of the W Serpentis type.* I will discuss them in my second talk. Here I wish to present the case of the famous system of β Lyrae as a very illuminating example.

3. The eternal (?) puzzle of β Lyrae

Since its discovery by Goodricke in 1794, β Lyrae often played an important role in the progress of astrophysics, as its various puzzles stimulated the introduction of new concepts and ideas, as well as rejection of old ones. This role seems not to be over. In keeping with the modern times, I can claim that β Lyrae poses the problem of *missing mass*, although I do not expect that in this particular case the solution will be parallel to the exotic ideas offered for the other cases. The missing mass is the mass of the more massive component. In fact, the mass is most likely exactly there where it should be, but then there is *missing luminosity*, which, of course, is the case with all other cases of *missing mass*; actually, they all should be termed *missing luminosity !*

Today I wish to discuss the possibility that some of the puzzles of β Lyrae may be removed if we assume that a large accretion disk plays an important role in the structure of the system. The work I am reporting is still in progress, and I will give no definite answer to the question if the proposed model with an accretion disk works. However, I feel that the discussion will at least show the problems and possibilities of such a model, perhaps applicable to other systems.

An excellent review of our knowledge about β Lyrae has been published by Sahade (1980). Very few facts about β Lyrae are known with reasonable certainty. One of them is, of course, the length of its orbital period, $P = 12.937$ days. Another safely known parameter is – rather surprisingly – the radial velocity range of the one component that is spectroscopically clearly visible. This is a luminous B star, traditionally classified B8 II, but more recently reclassified as a hydrogen-poor B6 II star (Balachandran *et al.* 1986). Its radial velocity curve is surprisingly smooth, simple, and well defined (Batten and Fletcher 1975), giving $K(B) = 184$ km/s. Combined with the period, $K(B)$ furnishes a reliable mass function, $f(m) = 8.356$

$m_\odot$. And, thirdly, since the system displays fairly deep eclipses, the orbital inclination cannot be very far from 90°. We cannot commit any large error by adopting, for the sake of calculating masses, for example $i = 85°$.

No matter what assumption we make about the mass ratio, the mass of the unknown star will always be large. Solutions with the B star more massive lead to huge masses for both components. Assuming both masses equal gives a mass of 34 $m_\odot$ for each star; taking the B star as twice as massive as the U star leads to masses 150 and 75 $m_\odot$, respectively. Only by assuming the unknown star to be the more massive component can we push both masses into a more plausible range, but even with an extreme mass ratio, the unknown star will have at least 10 solar masses. The actual value of the mass ratio remains largely unknown. Wilson (1974) surveyed all then available indirect estimates, and found a range in q from 1.7 to more than 6, with more weight on the larger side.

My choice of $q = 5.6$ is essentially a personal bias. It gives the following parameters: $m(U)$ =11.7 $m_\odot$, $m(B)$=2.1 $m_\odot$, A=55.6 $R_\odot$, $R(B,z)$= 12.5 $R_\odot$, $R^*(U)$=29.5 $R_\odot$, $K(U)$=33 km s^{-1}. Here A is the separation between the centers of gravity of the two stars, $R(B,z)$ is the polar radius of the B star (its equatorial radius is $R(B,y)$=13.05 $R_\odot$), $R^*(U)$ is the mean radius of the critical Roche lobe of the star U, and $K(U)$ is its predicted radial velocity range. All masses are expressed in solar units, linear distances are in units of solar radius.

Another important qualitative fact about the system of β Lyrae is that its light curve is "of the β Lyrae type", that is, with smooth and continuous variation of light between eclipses. This indicates that the two components are considerably distorted, and not far from contact. The above dimensions are based on a Roche model geometry and assume that the B star fills its critical Roche lobe. This is currently a generally accepted view, although here we are already stepping into the realm of plausible models. The presence of circumstellar material, and the substantial lengthening of the period, make it plausible to assume that β Lyrae is a strongly interacting binary probably of the semidetached Algol-like type. Accepting this idea, we can call the B star the *loser* and the unknown star the *gainer*. Farther on, when we introduce an accretion disk into the system, we will need a name for the entire object, star U and its disk; so we will call it the *secondary (object)*, hence the B star will be the *primary*, in accord with the usual terminology used by spectroscopists.

Experience shows that in typical Algol-like semidetached systems, the gainer looks like a normal main-sequence star, with its radius and luminosity adequate for its mass according to the standard relations. This is definitely not the case in β Lyrae. A good match for our chosen mass $m(U)$ =11.7 $m_\odot$ is the primary component of CW Cephei, listed in the review by Popper (1980) with the following reliable parameters: m=11.8 $m_\odot$, R=5.4 $R_\odot$, T_{eff}=29,500 K, L=19,000 $L_\odot$, M_v=-3.15 mag. Accepting this set of parameters, I will represent the gainer in β Lyrae, assuming for a while that it is a normal star, with the Kurucz (1979) model with T_{eff}= 30,000 K and log g= 4.0. I will also accept the distance to the system to be D = 370 pc, according to Dobias and Plavec (1985). The resulting flux distribution is plotted in Fig. 3 together with the observed flux distributions of β Lyrae, as derived by Plavec (1987). These distributions are based on Wilson's (1974) model in the optical, and

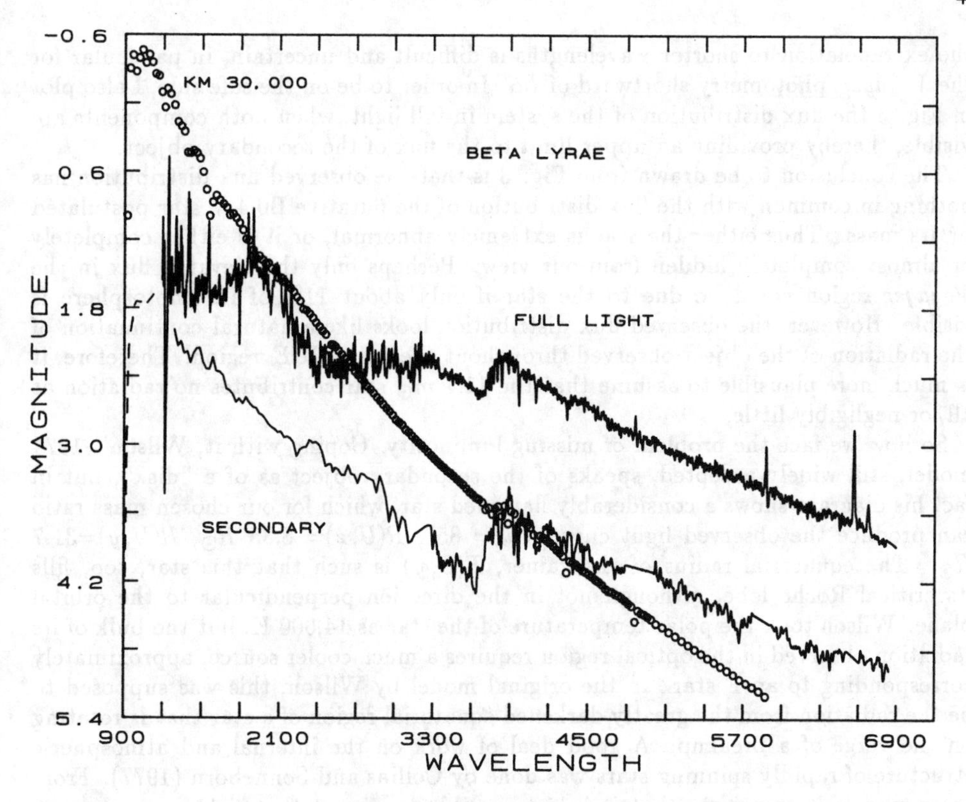

Fig. 3. The observed flux distributions of the components of β Lyrae, compared to the expected flux from a normal B0.4 V star placed at the same distance.

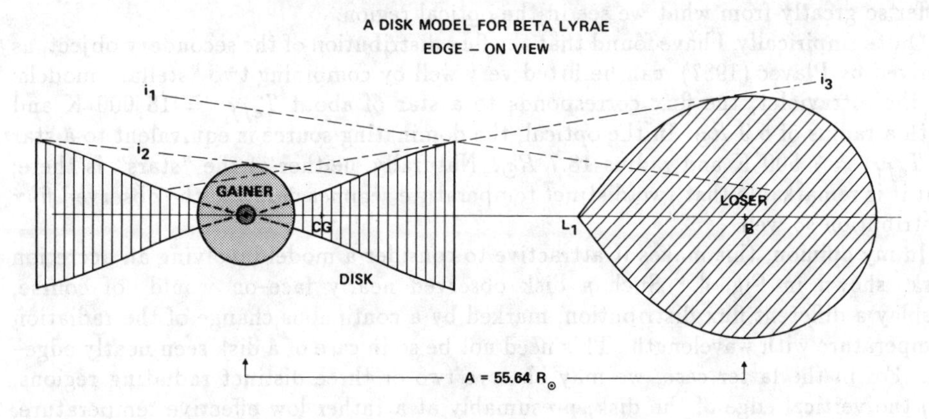

Fig. 4. A model of β Lyrae with the hypothetical accretion disk.

the extrapolation to shorter wavelengths is difficult and uncertain, in particular for the *Voyager* photometry shortward of $L\alpha$. In order to be on the safe side, I also plot in Fig. 3 the flux distribution of the system in full light, when both components are visible, thereby providing an upper limit to the flux of the secondary object.

The conclusion to be drawn from Fig. 3 is that the observed flux distribution has nothing in common with the flux distribution of the putative B0.4 V star postulated by its mass. Thus either the star is extremely abnormal, or it is either completely or almost completely hidden from our view. Perhaps only the elevated flux in the *Voyager* region could be due to the star if only about 11% of its photosphere is visible. However, the observed flux distribution looks like a natural continuation of the radiation of the object observed throughout the entire *IUE* region. Therefore, it is much more plausible to assume that the 11.7 $m_{\odot}$ star contributes no radiation at all, or negligibly little.

So now we face the problem of missing luminosity. Coping with it, Wilson's 1974 model, still widely accepted, speaks of the secondary object as of a "disk", but in fact his diagram shows a considerably flattened star, which for our chosen mass ratio can produce the observed light curve if $i = 85°$, $R(U,z)= 8.35\ R_{\odot}$, $R(U,y)=31.7\ R_{\odot}$. The equatorial radius of the gainer, $R(U,y)$ is such that this star, too, fills its critical Roche lobe, although not in the direction perpendicular to the orbital plane. Wilson took the polar temperature of the star as 14,000 K, but the bulk of its radiation observed in the optical region requires a much cooler source, approximately corresponding to an F star. In the original model by Wilson, this was supposed to be the radiation from the greatly darkened equatorial region of a star that is rotating on the verge of a breakup. A good deal of work on the internal and atmospheric structure of rapidly spinning stars was done by Collins and Sonneborn (1977). From their models it transpires that it is not possible to expect that a 11 $m_{\odot}$ star, no matter how fast it rotates, could have equatorial regions as cool as to imitate an F star. Moreover, Wilson's model was developed well before the advent of ultraviolet satellites, and he could not anticipate that the ultraviolet flux distribution would differ so greatly from what we see in the optical region.

Quite empirically, I have found that the flux distribution of the secondary object, as derived by Plavec (1987), can be fitted very well by combining two "stellar" models: In the ultraviolet, the flux corresponds to a star of about $T_{eff} = 18\ 000$ K and with a radius of 5.9 $R_{\odot}$. In the optical, the dominating source is equivalent to a star of $T_{eff} = 7\ 500$ K and radius 18.7 $R_{\odot}$. Naturally, neither of the "stars" is there; but it is remarkable that two distinct temperature regimes describe the observed flux distribution so well.

In my opinion, this makes it attractive to consider a model involving an accretion disk, shown in Fig. 4. Such a disk observed nearly face–on would, of course, display a different flux distribution, marked by a continuous change of the radiation temperature with wavelength. This need not be so in case of a disk seen nearly edge–on. For in the latter case, we may observe two or three distinct radiating regions: (1) the vertical edge of the disk, presumably at a rather low effective temperature, but possibly with a large radiating surface; (2) the face of the disk, partly eclipsed by the disk edge; (3) and possibly that part of the central star's globe that is not permanently eclipsed by the disk edge. The idea of an accretion disk in β Lyrae is strongly supported by the anticipated high rate of mass transfer. This idea is not new; already Huang (1963) stressed the need for a flattened "disk" at a time when the concept of an accretion disk had not yet emerged. Referring directly to the concept of an accretion disk, Wilson (1981) abandoned his original model of a flattened star in favor of such a disk, but retained certain aspects of the stellar case by assuming that the disk itself has substantial mass and therefore its self-gravitation cannot be neglected. The "massive disk model" was calculated by Wilson for a total disk mass of 0.5 $m_\odot$, but he concludes that a lower mass, perhaps about 0.3 $m_\odot$, would lead to a better agreement with observations. In either case, the disk mass does not substantially reduce the mass of the central star, which with our adopted parameters would be still more than 11 $m_\odot$ and and even in the limit of a very large q it would not fall below 10 $m_\odot$. So where is the radiation from this star? The easiest route of escape is to assume that the star is completely hidden from our sight by the thick disk.

We have found that in order to have a geometrically and optically thick and radially extended disk, it is not necessary to assume any large mass of the disk. Program TLUSDISK developed by Hubený, 1989 May version, calculates self-consistently the vertical structure of a set of concentric rings inside the disk. We have assumed that the regular disk will have its edge at about 0.8 $R^*(U) = 24\ R_\odot$, and certainly not beyond 0.85 $R^*(U)$, since beyond that distance, the perturbing effect of the other star becomes too strong to permit stable periodic orbits. The assumption of a low-mass disk is justified, since even for a rate of mass transfer as high as 100 μsy the total mass of the disk comes out to be only $2.7 \times 10^{-4}\ m_\odot$. Several disk models have been calculated in an attempt to cover a wide range of possible values of the fundamental parameters defining the structure and radiation of the disk, namely the rate of mass transfer $\dot{m}$ and the radius of the central star, $R(U)$. Both are extremely poorly known. Some idea about the mass transfer rate can be obtained from the rate of period increase, which after Herczeg (1973) is 2.59E-4 days (22.35 s) per year. The simplest possible formula assumes a conservative transfer of mass and orbital angular momentum between two mass points, and reads

$$\dot{m} = m(U)\,\dot{P}\;/\;3\,P\,(q-1).$$

With our chosen parameters, we get a mass transfer rate of only 17 μsy. With a lower mass ratio, the same period change gives a higher mass transfer rate; thus for q=3 we get $\dot{m}$= 50 μsy. However, an even greater uncertainty stems from the fact that the mass transfer is definitely not conservative, witness the P Cygni profiles of the numerous strong emissions, nor can be the *orbital* angular momentum conserved as the material is partly accreted on the gainer and spins it up, and partly escapes into space. We have therefore calculated models with mass transfer rates between 10 and 100 μsy.

The problem of the radius of the central star in the disk is perhaps even more difficult. As mentioned above, a regular Algol-type semidetached gainer with a mass of 11.7 $m_\odot$ should have a radius near 5.4 $m_\odot$. A star with a smaller radius is easier to hide; we may perhaps conjecture that the star is greatly flattened by rapid rotation, and consider a 30% smaller radius, i.e. near 4 $R_\odot$, which is still rather large. It is harder to imagine how the gainer could have a radius of only 1 $R_\odot$, as postulated by Polidan (1989). His claim is based on the interpretration of fairly high flux levels near the Lyman limit as observed by Voyager. The only object that could have a mass of 11.7 $m_\odot$ and a radius substantially smaller than the main–sequence radius would be akin to a Wolf–Rayet star, but its radius would not be as low as 1 $R_\odot$, and we would have to anticipate even stronger radiation output from it, to be hidden in the system. Nevertheless, we did calculate disks reaching down to 1 $R_\odot$ in order to test Polidan's claim that the observed flux distribution requires it. The radius of the gainer plays an important role in determining the presence, temperature, and radiation of the innermost parts of the disk, but farther out, it becomes a less important factor.

The investigation is not yet completed, and I am unable to tell you if our fundamental approach promises to move us closer to understanding β Lyrae. I wish to share some thoughts that illuminate the character of the problem. The introduction of an accretion disk into the system of β Lyrae alters some modeling concepts rather drastically. The disks we have calculated are significantly *flared*, that is, they become geometrically rather thick at larger distances from the gainer (see Fig. 4). One consequence is that the largest vertical dimension of the secondary object, which in the two–star model always was at the center of gravity of the gainer, shifts, for sufficiently thick disks, to the disk edge, in our case 24 to 27 $R_\odot$ from the gainer, and thus comes now much closer to the loser. This must affect the eclipse geometry.

If we adopt Wilson's (1974) light curve solutions, we have the following fundamental picture: The disk–like secondary object is assumed to generate constant flux for us, 0.3625 (arbitrary) units in the B band. The flux from the primary varies for two reasons. One is its distorted shape, which leads to a flux variation from the peak value of 0.862 units (at phases 0.25 and 0.75) to 0.6373 in mid–secondary eclipse (phase 0.50). At phases around 0.0, this variation is augmented by a bodily eclipse by the secondary object, and the flux from the primary drops to 0.200. The data available to us do not permit us to separate the two effects accurately. We can estimate that distortion may cause the star to lose about 0.250 light units in primary mid–eclipse, more than in the secondary eclipse, since now we are facing the low–gravity, presumably more darkened region surrounding the L_1 point. Then the full light of the primary at that time would be 0.612, but is only 0.200; therefore approximately 66% of the disk of the primary is eclipsed by the secondary object.

Assume now that the eclipse is caused by the disk edge at a distance R_d from the center of gravity of the secondary. This condition clearly defines a definite orbital inclination i_1 under which we are observing the system; and for given dimensions of the disk, this condition is quite restrictive.

Our other condition is that the central star of the disk should be at all phases hidden by the disk edge. As explained above, we can only guess what the radius of the central star could be; and the minimum inclination i_2 imposed by this condition will greatly depend on our assumption about the size and nature of the gainer. A permanent eclipse of the central star by the disk edge also implies a permanent invisibility of the central parts of the disk; again, the extent largely depends on the size of the star we wish to hide.

And, finally, there is a third condition. Observations show that the eclipse of the secondary object is quite perceptible in the *IUE* ultraviolet, although in view of Polidan's (1988) discussion of the Voyager data, it is arguable if the secondary eclipses exist shortward of 100 nm. Polidan claims that there are no eclipses shortward of about 100 nm, and hence that the hottest part of the disk is permanently visible. This goes contrary to the basic philosophy of my attempted model, namely that a fairly large and luminous star must be hidden by the disk. In Polidan's view the star is unimportant, being small (1 $R_{\odot}$) and contributing rather insignificantly to the hot radiation coming from the inner part of the disk. The implication is that we must look for a different evolutionary story for β Lyrae; it cannot then be an Algol–system, but rather a very transient advanced evolutionary stage of an originally very massive star, which is now similar to a Wolf–Rayet star. However, such a star, if it can exist with such a small radius, will still have an extremely high luminosity. I think that the rather modest, possibly permanently uneclipsed flux observed by the *Voyager*, may be coming from a cloud of ejected material fairly high above the orbital plane, scattering the radiation of the hidden hot star and/or of the inner disk.

There is a third geometrical condition imposed by the disk model. The eclipses of the secondary object may perhaps be imperceptible shortward of 100 nm, but they are definitely observed in the *IUE* ultraviolet, and they are deeper than the primary eclipses. Thus the relative sizes of the primary component and of the accretion disk must be such as to allow eclipses of something fairly warm (representable approximately by an effective temperature near 18,000 K). Presumably the bulk of the *optical* flux from the secondary, resembling an F star, comes from the vertical edge of the disk. However, the radiation of an F star is negligible in the ultraviolet; therefore, in order to obtain perceptible eclipses of the secondary in the ultraviolet, we must postulate that the B star eclipse more than the disk edge; and this imposes another limiting inclination, i_3.

The properties of the disk edge depend very strongly on the mass transfer rate. For $\dot{m} = 100$ μsy, the vertical half–thickness of the disk edge at $24R_{\odot}$ is $8R_{\odot}$ and the effective temperature of the ring is 7,900 K; for $\dot{m} = 30$ μsy these values drop to 5 $R_{\odot}$ and 5,800 K, respectively. The effective thickness of the disk edge plays a very important role.

We are currently studying an additional effect, which we believe may increase the observed flux and be in general quite important for determining the disk structure, and this is the illumination of the disk by the central star. The presumably high radiation flux from the central star heats the disk, increasing its effective temperature and thickness; and a substantial fraction of the flux may be transmitted in our direction.

ACKNOWLEDGEMENTS. It is my pleasure to thank Dr. Ivan Hubený for sharing his computer programs with me, and for numerous valuable discussions. I wish also to thank my wife, Dr. Zdenka Plavec, and Dr. Charles D. ("Tony") Keyes for advice and assistance in computing. The project reported here has been sponsored by NASA–ADP grant U003-88, shared with Dr. Hubený.

References

Balachandran,S.*et al.* (1986). *Mon. Not. R. Astron. Soc.*, **219**, 479.
Batten, A.H. and Fletcher, J.M. (1975). *Publ. Astron. Soc. Pac.*, **87**, 237.
Collins, G.W. II and Sonneborn, G.H. (1977). *Astrophys. J. Suppl.*, **34**, 41.
Dobias, J.J. and Plavec, M.J. (1985). *Astron. J.*, **90**, 773.
Etzel, P.B. (1988). *Astron. J.*, **95**, 1204.
Herczeg, T. (1973). *Inf. Bull. Var. Stars*, , 820.
Kříž, S. and Hubený, I. (1986). *Bull. Astron. Inst. Czech.*, **37**, 129.
Paczynski, B. (1971). *Annu. Rev. Astron. Astrophys.*, **9**, 183.
Plavec, M.J. (1987). *Publ. Astr. Inst. Czech.* 70, **5**, 301.
Plavec, M.J. (1988). *Astron. J.*, **96**, 755.
Pringle, J.E. (1981). *Annu. Rev. Astron. Astrophys.*, **19**, 137.
Polidan, R.S. (1988). *Space Sci. Rev.*, **50**, 85.
Popper, D.M. (1980). *Annu. Rev. Astron. Astrophys.*, **18**, 115.
Sahade, J. (1980). *Space Sci. Rev.*, **26**, 349.
Wilson, R.E. (1974). *Astrophys. J.*, **189**, 319.
Wilson, R.E. (1981). *Astrophys. J.*, **251**, 246.

RESOLVING THE PUZZLE OF THE W SERPENTIS STARS

Mirek J. PLAVEC
Department of Astronomy
University of California
Los Angeles, CA 90024-1562
U.S.A.

ABSTRACT. The W Serpentis subclass of interacting binaries is characterized by strong effects of circumstellar matter on their spectra, light curves, and radial velocity curves. In this work, they are perceived as extreme cases of the semidetached binaries of the Algol type. An attempt is made to get an insight into their structure by studying emission lines in their spectra. Ultraviolet emission lines have been detected by the author in 11 Algols and 6 Serpentids. Evidence is indicated that the lines are formed predominantly by scattering in an induced stellar wind. Relative intensity of Fe II and Fe III emissions appears to be correlated with the spectral type of the gainer. It is suggested that in the Serpentids we see a thick disk nearly edge-on, but that the emission lines indicate the nature of the gainer inside.

1. Why Observe Circumstellar Matter – and How

This report is based on my studies of circumstellar matter in interacting binary systems with non-degenerate components. Since at least the time of Otto Struve, we have known that many close binaries contain substantial amount of circumstellar material, which plays a triple role rather similar to the role played by interstellar matter in stellar and galactic studies. Perhaps the role of the greatest ultimate importance is the fact that interstellar matter is the source from which new stars are born. Then, secondly, for those interested in interstellar matter itself, there are endless sources of excitement and joy in the studies of the structure, physics, and chemistry of the interstellar matter. And, thirdly, perhaps the most general role of interstellar matter is that of an enormous nuisance, interfering at is does with the efforts to establish the intrinsic properties of stars and stellar systems.

Using the analogy between the interstellar and the circumstellar matter, I can now summarize the goals of my project, referring to the above three properties.

C. İbanoğlu (ed.), Active Close Binaries, 49–60.

1.1 EVOLUTIONARY SIGNIFICANCE

The mass–transfering streams in interacting binaries play a Robin Hood role, taking away from stars rich in mass and feeding the poor ones — but one must say that they grossly overextend their zeal in the end. Moreover, when the observed interacting systems are compared with theoretical models, the conclusion seems inescapable that a non–negligible portion of the circumstellar matter we observe is actually smuggled out from the binary system. It is important to identify this component, find its location and extent, and identify the mechanism that drives it out of the system. Particularly promising are the ultraviolet emission lines of fairly highly ionized species, since they bear similarity to stellar wind in luminous stars, and indeed, when observed at high dispersion in β Lyrae, they show similar P Cygni profiles.

1.2 STRUCTURE OF CIRCUMSTELLAR MATTER

It has become customary to use the terms *circumstellar shell* and *circumstellar envelope* when speaking about circumstellar matter. These concepts are extremely vague; when I wanted to learn about the difference between them, I found a statement (I think in a paper by Struve) to the effect that *"The term* shell *does not necessarily imply that the structure is detached from the star"*. We should attempt to interpret them in terms of the modern concepts of *accretion disk, stellar wind, corona* and the like. For this purpose, it is necessary to study the spatial structure and physical state of the circumstellar matter.

Particularly intriguing is the presence (already mentioned above) of fairly strong ultraviolet emission lines emitted by rather highly ionized species, such as N V, Si IV, C IV, Fe III, and so on. The component stars in the Algol systems cannot photoionize these atoms. Actually, we must explain three aspects resulting from the detection of these emission lines: (a) What is the source of the "raw" material? (b) What is the source of the energy that ionizes and excites this material? (c) What is the process by which the ions radiate?

1.3 THE CASE OF W SERPENTIS

Each new observing technique (and each new theoretical discovery) in astrophysics brings along renewed hope that objects deemed intractable in the past can be attacked anew and their riddle resolved. Thus, the ultraviolet spectroscopy with the *IUE* satellite offers a new chance to resolve the "bizarre" eclipsing systems the like of β Lyrae, W Serpentis, W Crucis, or V367 Cygni, which had before defied the optical scrutiny even by the renowned experts at demystifying strange objects. Ultraviolet observations appear to be especially promising, since peculiar features in the optical spectrum hint at the presence of a source in the system hotter than the component stars.

Thus, for example, the only visible component in W Serpentis was classified F5 II by Hack (1963), yet we clearly saw, in optical scans, superposed emission lines not only of hydrogen, but also of neutral helium (Plavec 1980). The ultraviolet spectrum, shown in Fig. 1, does indeed reveal radiation that clearly cannot be emitted by an F star, with emission lines coming from ions that cannot be excited or ionized by the F star either.

However, this observation by itself does not resolve the puzzle. The character of the ultraviolet continuum is not obvious, and one serious obstacle is the uncertainty about the amount of interstellar reddening in front of the object, which lies at galactic latitude of only 1.9° in a region known for chaotic dark clouds. The interstellar bump at 2200 Å is known to be an unreliable indicator, but even if it is reliable, the spectrum around it is so complex that the shape of the interstellar feature cannot be reconstructed with any confidence. Normally, the optical spectral classification would help in estimating both the color excess and the distance to the system; however, there is no guarantee that what we observe is a star: it could well be an optically thick disk, and, in fact, I believe that it is an accretion disk. In Fig. 2, the entire spectrum of W Serpentis from 1200 Å to 7200 Å is compared with the spectrum of an F5 II star; we see that the stellar spectrum matches just only the optical region that was observable in the 1950's. Fig. 3 shows the optical spectrum outside eclipse and deep in primary eclipse. We note the curious thing that the depth of the eclipse decreases rapidly as we go to ultraviolet wavelengths. We must conclude that in the UV, the radiation is coming from a region that is not eclipsed. As to the optical spectrum, the match by an F5 star is satisfactory, but the "shell" line character of the spectrum and its erratic variability speak much more for an accretion disk. No trace of the secondary star has so far been found, and the radial velocity curve of the F object is too distorted by shell lines to be of much use.

Considering all this, I concluded that such a complex system probably cannot be understood if studied in isolation, and that a "collective treatment" of several similar systems may be more promising. Observations of the other binaries of the W Serpentis group, namely SX Cassiopeiae, RX Cassiopeiae, W Crucis, V367 Cygni, and β Lyrae revealed as many similarities as differences, and none of them appears easier to model, except, perhaps, SX Cassiopeiae (Plavec, Weiland and Koch 1982; Andersen *et al.* 1988). Then I expanded the scope of the survey to include the classical Algol systems, in which the properties of the component stars are much easier to establish. However, there is a price to pay: most of the manifestations of the circumstellar environment in the Algols can be observed only during the total eclipse of the primary (hotter) component. Thus there exists a sharp distinction between the W Serpentis stars and the Algols: in the former, the ultraviolet emission lines can be observed at all orbital phases. In the Algols, even the absolutely and relatively brightest UV emissions, in RY Persei, are still about 2.5 mag fainter than the stellar continuum at the corresponding wavelengths.

1.4 PER ASPERA AD ASTRA — 1989 VERSION

Over the eleven years between 1978 and 1989, I have been able to observe 27 interacting eclipsing binaries. The project has advanced slowly, and if Tycho Brahe had been accumulating the positions of Mars at the same pace, we would perhaps still adhere to the Ptolemaic geocentric system. However, Tycho Brahe did not have to compete for observing time with quasar observers. He also was a rich nobleman

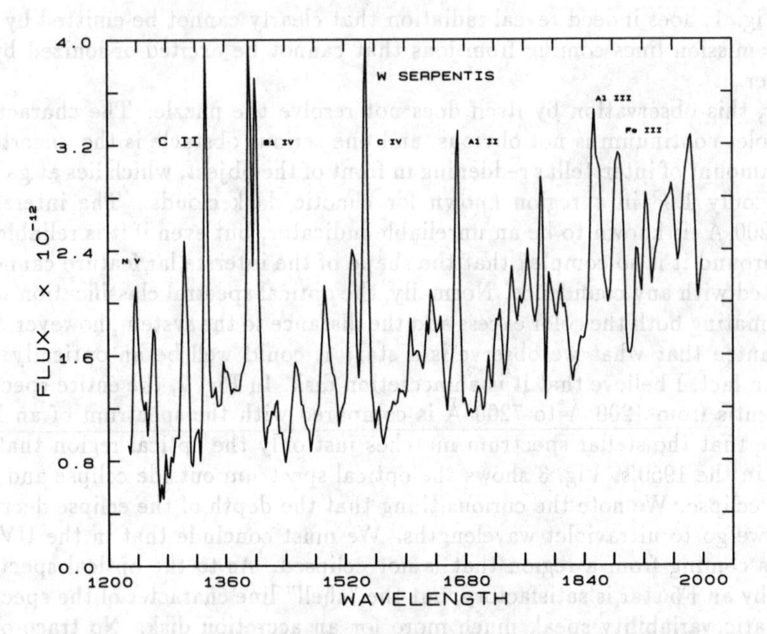

Fig. 1. The far ultraviolet spectrum of W Serpentis, obtained in 1978 with the *IUE SWP* camera. Later spectra showed that the difference in fluxes may be due less to the primary eclipse and more to phase–independent fluctuations.

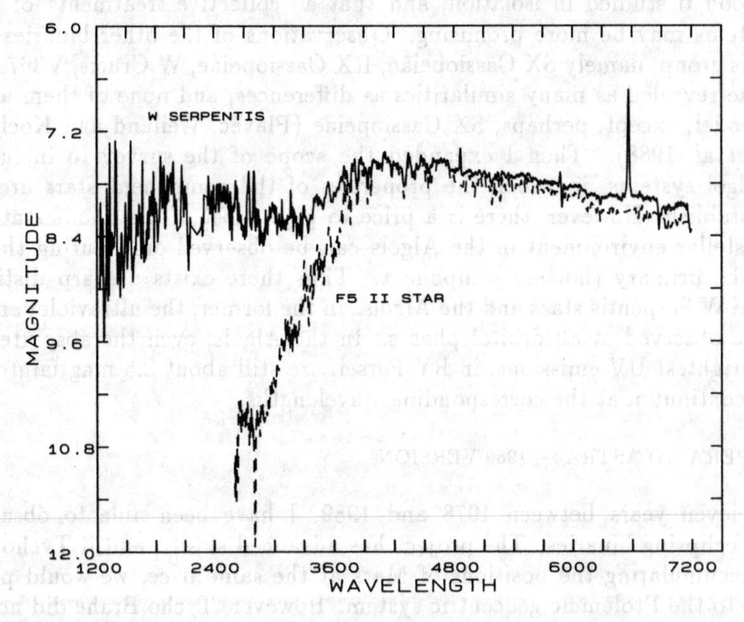

Fig. 2. The spectrum of W Serpentis is compared with an F5 II star; we see that the star provides a satisfactory match just only in the optical region.

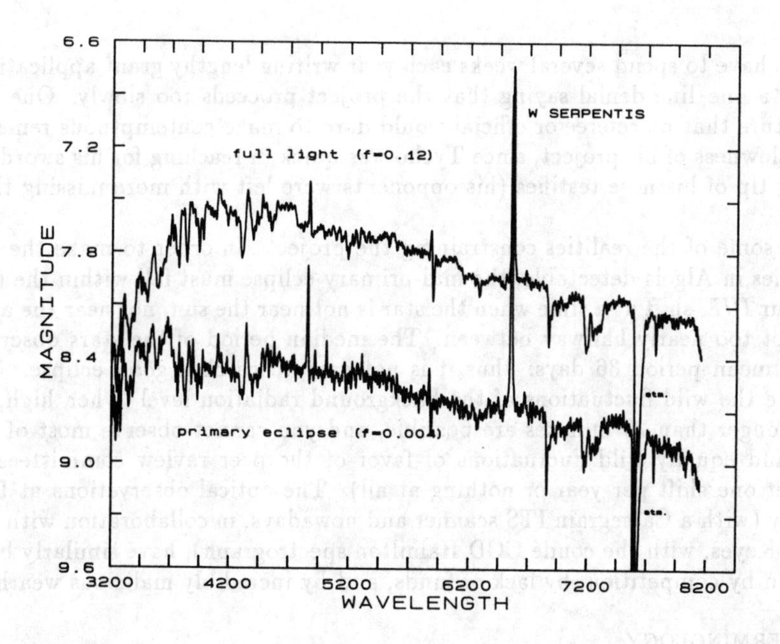

Fig. 3. Optical scans of W Serpentis in full light and in primary eclipse.

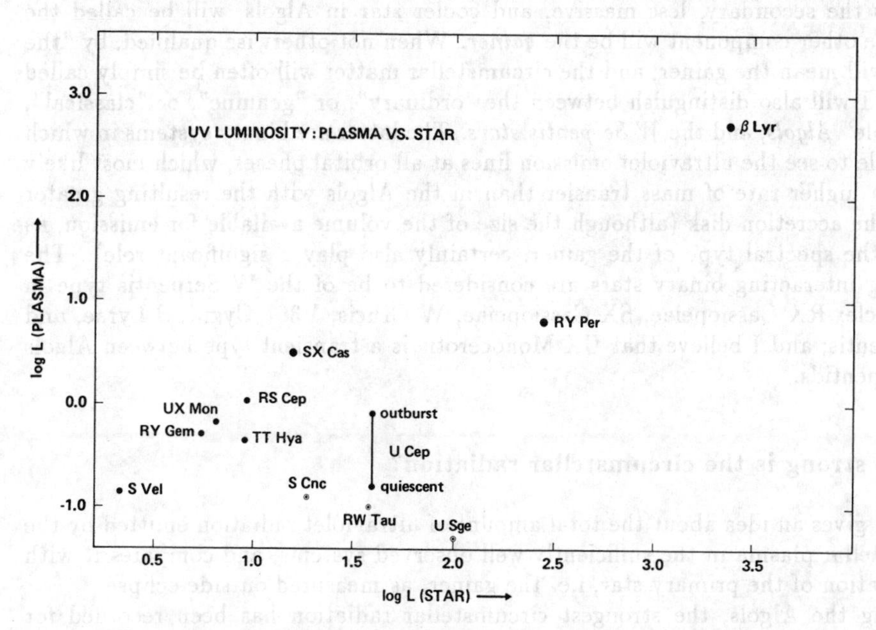

Fig. 4. The UV luminosity in totality, due to circumstellar plasma, is compared with the UV luminosity of the accreting star.

and did not have to spend several weeks each year writing lengthy grant applications only to get a one–line denial saying that the project proceeds too slowly. One can also conjecture that no referee or official would dare to make contemptuous remarks about the slowness of his project, since Tycho was quick in reaching for his sword, as the missing tip of his nose testifies (his opponents were left with more missing than a nose tip).

Here are some of the realities constraining the project. In order to make the UV emission lines in Algols detectable, the mid-primary eclipse must fall within the (US 2) eight-hour *IUE* shift at a time when the star is not near the sun, not near the anti-sun, and not too nearly halfway between. The median period of the stars observed is 10 days, mean period 36 days: thus it is not easy to catch a good eclipse. You must endure the wild fluctuations of the background radiation level (when high, no exposures longer than 20 minutes are possible, and you cannot observe most of the binaries), and equally wild fluctuations of favor of the peer review committees (if low, you get one shift per year or nothing at all). The optical observations at Lick Observatory (with a Cassegrain ITS scanner and nowadays, in collaboration with Dr. Charles D. Keyes, with the coudé CCD Hamilton spectrograph), have similarly been slowed down by competition, by lack of funds, and by incredibly malicious weather.

1.5 SOME TERMINOLOGY

In the following text, I will use the following terminology: The mass–losing star, which is the secondary, less massive, and cooler star in Algols, will be called the *loser*; the other component will be the *gainer*. When not otherwise qualified, by "the star" I will mean the gainer; and the circumstellar matter will often be simply called *plasma*. I will also distinguish between the "ordinary", or "genuine", or "classical", or "simple" *Algols*, and the *W Serpentis stars*. The latter are binary systems in which I was able to see the ultraviolet emission lines at all orbital phases, which most likely signals a higher rate of mass transfer than in the Algols with the resulting greater role of the accretion disk (although the size of the volume available for emission, as well as the spectral type of the gainer, certainly also play a significant role). The following interacting binary stars are considered to be of the W Serpentis type in this article: RX Cassiopeiae, SX Cassiopeiae, W Crucis, V367 Cygni, β Lyrae, and W Serpentis; and I believe that UX Monocerotis is a transient type between Algols and Serpentids.

2. How strong is the circumstellar radiation?

Figure 4 gives an idea about the total amount of ultraviolet radiation emitted by the circumstellar plasma in the sufficiently well observed systems, and compares it with the radiation of the primary star, i.e. the gainer, as measured outside eclipse.

Among the Algols, the strongest circumstellar radiation has been recorded for RY Per, with RS Cep being a distant second. Both stars are very distant objects, near 0.9 kpc from us, and consequently the distance may be affected by a fairly large error; but this cannot affect the result qualitatively. The faintest plasma radiation was found in U Sge and S Cnc. This is not surprising, since both systems are known to be nearly dormant; In S Cnc this is quite understandable, as the loser has only 0.18 solar masses and can hardly lose much more. U Cephei falls close to them if we take the mid–totality spectrum; but I have found (Plavec 1983) that by mid–totality, a large portion of the radiating plasma is eclipsed by the loser. The amount of radiation seen at the beginning of totality is five times larger, and places U Cephei on nearly equal footing with RY Gem or RS Cep. In the outburst, an additional increase was observed (McCluskey, Kondo, and Olson 1988), although the actually observed enhacement of the strongest far UV emission lines increased the total *SWP* flux by only 18% compared to the quiescent flux observed at $\phi = 0.981$. How much the fluxes from other systems fluctuate for the same two causes (eclipses and outbursts) is unknown, although in other systems the eclipse effects seem to be smaller than in U Cep.

Perhaps more informative are the ratios of UV fluxes plasma vs. star shown in Fig. 4, where they are plotted logarithmically. The advantage of these ratios is that they eliminate the uncertainties of interstellar reddening and of distance determination. I find it obvious that the amount of radiating material plays an important role, making S Cnc and U Sge very faint in eclipse and SX Cas relatively very bright. In U Sge, RW Tau, and S Cnc, we get, in total eclipse, less than 1% of radiation observed outside eclipse, while in RS Cep, it is 12%, and in the Serpentids SX Cas and β Lyrae, it is 20% and 34%, respectively. Both numbers are actually lower limits, since the emission lines are visible at all phases and their contribution is almost impossible to subtract completely from the "continuum". There is little doubt that the rate of mass transfer, and no doubt also the volume available to the circumstellar material, plays an important role in these differences between the systems.

However, Fig. 4 tells me that in general, the amount of radiation emitted by the plasma is probably also proportional to the UV radiation of the star; and in that I see my suspicion confirmed that the main process behind the plasma radiation output is scattering of the light of the gainer.

3. Systems with a strong Fe II emitting region

In five spectra, we notice elevated flux levels in the region roughly between 2300 and 2630 Å, and an abrupt drop near 2640 Å. I believe that this enhanced flux is due to superposition of numerous emission lines of Fe II and several other singly-ionized metals like Cr II, Ni II, and Mn II. This explanation is justified as follows. In full light, the star TT Hydrae has been observed with high dispersion, and the spectrum shows very deep and very broad blends in the region mentioned above; then near 2645 Å, there is a region of higher flux, that is lower extinction, extending between 2634 and 2661 Å, after which deeper absorptions gradually set in again. Degraded in the low dispersion, this brighter region produces a spurious and rather striking "emission line" at 2647 Å (see Plavec 1988), while on either side the flux is much depressed by the absorption blends. A comparison with a star of the same spectral type shows that the deep absorptions are shell lines (see my other paper in this volume). In eclipse, a reversal occurs similar to that in the solar chromospheric spectrum at total solar eclipse, and the circumstellar material is seen in emission.

The blends are so broad that it is impossible to separate any individual line that would represent the strength of the Fe II emission. My attempts to isolate the strong Fe II (1) line at 2599 Å failed. Fortunately, the depth of the abrupt drop near 2640 Å seems to be a useful measure of the contribution of Fe II to the emission, since its bottom seems to lie at about the same level as the "continuum" near 1600 Å. Another good indicator is the strength of the broad emission blend at 2750 Å, consisting of Fe II multiplets 62 and 63.

The systems with a conspicous Fe II emission are S Vel ($\Delta m = 2.60$ mag), RY Gem (1.37 mag), UX Mon (1.23 mag), RS Cep (0.96 mag), and TT Hya (0.93 mag). In the parantheses, I give the magnitude difference of fluxes at 2630Å and 2648Å, respectively, that is, I am measuring the height of the "λ2640 discontinuity". The revised spectral types of these gainers are: S Vel, A2.5 V; RY Gem, A0 V; RS Cep, B9.7 V; TT Hya, B9.5 V (see Dobias and Plavec 1988).

The correlation is very nice. Now UX Monocerotis is a rather puzzling system, the hotter component of which has been classified as as late as A7, but is clearly variable. My UV spectra indicate a spectral type variable between B9 and A0. The λ 2640 Å jump in the totality spectrum classifies it as very close to A0 V; the spectrum is shown in Fig. 5.

4. Systems with a strong Fe III emitting region

Four other spectra, namely those of β Lyrae, RY Persei (Fig. 6), RX Cassiopeiae, and W Crucis have a dominating region centered at about λ1900 Å. Three prominent emission lines of Fe III, multiplet 34, are located there (at 1985, 1914, and 1927 Å). These lines by themselves can hardly explain the enormous enhacement of that region seen in β Lyrae, and for a long time I have suspected the presence of an underlying elevated continuum. The source of such a continuum is hard to find; thus it seems that Viotti (1976) was right in attributing this bulge to numerous weaker Fe III lines overlapping in emission; I can only add that Ni II and Ni III also seem to contribute. The Fe III emission reaches up to about 2200Å in β Lyrae and RY Per, but beyond that, where the emission should be dominated by Fe II, the flux level is significantly lower and flat. Also Mg II, whose emissivity (in the purely collisional case) peaks at an electron temperature $T_e = 16{,}000$ K, is weak in β Lyrae and RY Per. All this suggests that the plasma is generally hotter when it surrounds hotter gainers. In RY Per, the gainer is no later than B5 from optical observations (Popper 1980 and private communication), although, rather mysteriously, its ultraviolet spectrum is no earlier than B9. The problem of the representative temperature for the gainer of β Lyrae is well-known; either the embedded star or the inner parts of the disk are certainly hotter than B6.

Now the ultraviolet spectrum of RX Cassiopeiae is rather similar to that of these two stars; thus it is tempting to classify the gainer of RX Cas as an approximately B5–B6 star. However, no such star is seen, and the same ultraviolet spectrum with very strong emission lines and very low (if any) continuum is seen at all phases. In the optical region, we observe two spectra, neither of which is earlier than F (Plavec

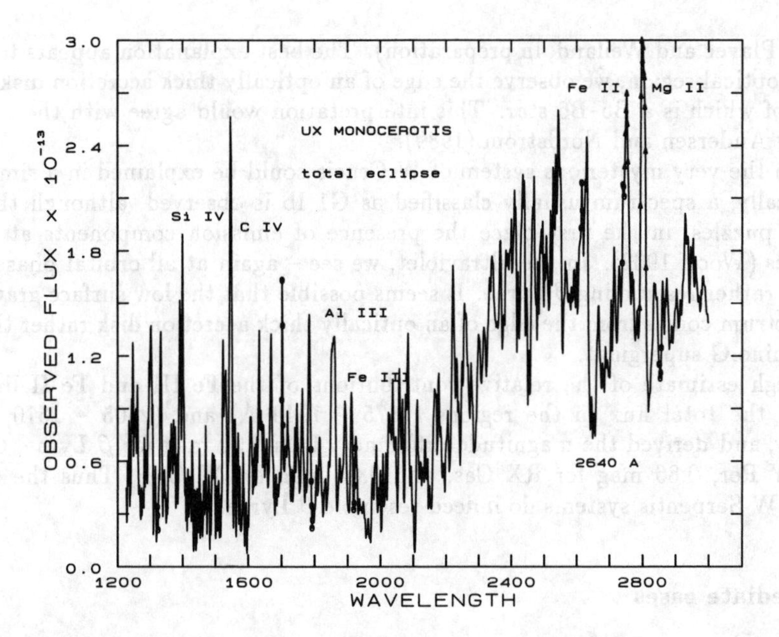

Fig. 5. A representative UV spectrum of a system displaying strong Fe II lines and weak Fe III lines: UX Monocerotis.

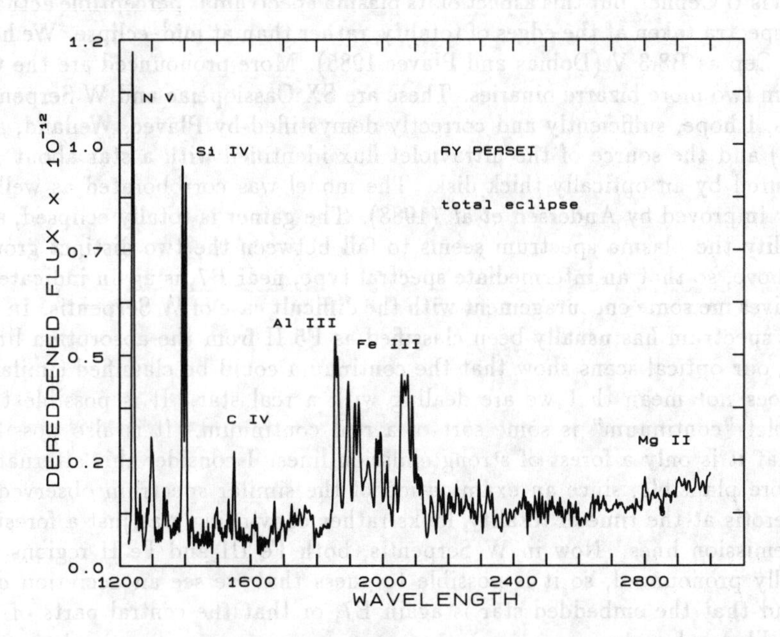

Fig. 6. A representative UV spectrum showing strong Fe III lines and no Fe II lines: RY Persei.

et al. 1981; Plavec and Weiland, in preparation). The best explanation appears to be that in the optical region, we observe the edge of an optically thick accretion disk, at the center of which is a B5–B6 star. This interpretation would agree with the work reported by Andersen and Nordström (1989).

And then the very mysterious system of W Crucis could be explained in a similar way. Optically, a spectrum usually classified as G1 Ib is observed, although there are several puzzles, in the first place the presence of emission components at the Balmer lines (Woolf 1962). In the ultraviolet, we see – again at all orbital phases – a spectrum rather resembling β Lyrae. It seems possible that the low surface gravity optical spectrum comes from the edge of an optically thick accretion disk rather than from a genuine G supergiant.

As a rough estimate of the relative contributions of the Fe III and Fe II lines, I measured the total flux in the regions (1875 - 1940 Å) and (2505 - 2570 Å), respectively, and derived the magnitude difference. It is 0.90 mag for β Lyrae, 0.72 mag for RY Per, 0.86 mag for RX Cas, and 0.84 mag for W Cru. Thus the two mysterious W Serpentis systems do indeed resemble β Lyrae.

5. Intermediate cases

Finally, there are three binary systems which seem to lie halfway between the two types described above, that is, both the Fe II and Fe III regions are mildly prominent. One of them is U Cephei, but this aspect of its plasma spectrum is perceptible actually only in the spectra taken at the edges of totality, rather than at mid–eclipse. We have classified U Cep as B8.3 V (Dobias and Plavec 1985). More pronounced are the two flux bulges in two more bizarre binaries. These are SX Cassiopeiae and W Serpentis. SX Cas was, I hope, sufficiently and correctly demystified by Plavec, Weiland, and Koch (1982) and the source of the ultraviolet flux identified with a star about B7, partly obscured by an optically thick disk. The model was corroborated as well as significantly improved by Andersen *et al.* (1988). The gainer is totally eclipsed, and during totality the plasma spectrum seems to fall between the two distinct groups discussed above, so that an intermediate spectral type, near B7, is again indicated.

All this gives me some encouragement with the difficult case of W Serpentis. In the optical, the spectrum has usually been classified as F5 II from the absorption lines; and indeed, our optical scans show that the continuum could be classified similarly. This still does not mean that we are dealing with a real star. It is possible that the ultraviolet "continuum" is some sort of a real continuum. It is also possible, however, that it is only a forest of strong emission lines. I consider this alternative as much more plausible, since an examination of the similar spectrum observed in UX Monocerotis at the time of totality, looks rather convincingly as just a forest of individual emission lines. Now in W Serpentis, both Fe III and Fe II regions are about equally pronounced, so it is possible to guess that we see an accretion disk edge–on, and that the embedded star is again B7, or that the central parts of the disk radiate like such a star.

6. The Nature of the W Serpentis Stars

The "ordinary" or "classical" Algols display light curves that are fairly smooth and reasonably stable; and the spectra of the component stars are as a rule well observable. The systems W Serpentis, RX Cassiopeiae, W Crucis, and V367 Cygni, on the contrary, display very distorted and variable light curves and their spectra look more like "shell" spectra rather than spectra of ordinary stars. Two systems appear to form a bridge between the two types: UX Monocerotis looks more like and Algol, but with a strangely distorted light curve; and SX Cassiopeiae is a Serpentid, but the character of the stars still seems to be recognizable. Finally, β Lyrae, although its light curve is not strikingly distorted and variable in the optical, becomes much more erratic in the ultraviolet, and by all other properties certainly is a Serpentid.

I believe that the parameter that distinguishes between Algols and Serpentids is the thickness and density of the accretion disk, which again ultimately is due to a significantly higher rate of mass transfer between the components. Ordinary Algols do contain certain amounts of circumstellar matter, as is inevitable, since gas continues to be moved from the Roche–lobe filling loser to the gainer. However, the circumstellar material is not optically thick enough to prevent us from recognizing the nature of the component stars. In some systems, there is very little circumstellar material – for example, in U Sagittae (Dobias and Plavec 1985) and in S Cancri (Etzel and Olson 1985). In other Algols, there is more circumstellar material, as in TT Hydrae, and it can then affect the profiles of the photospheric absorption lines (Etzel 1988) and produce additional, "shell" lines (Plavec 1988).

Although no satisfactory models exist yet, it seems reasonable to assume that in each of the interacting systems, the circumstellar gas is organized in an accretion disk. Since the potential well determining the release of gravitational potential energy is not deep and the rate of mass transfer is not high in ordinary Algols – a rate of $\dot{m}$= 0.1 μsy may be typical – the disk is rather cool and relatively thin. In the Serpentids, I believe that the rates are several orders of magnitude higher, perhaps between 10 and 1,000 μsy. As a consequence, the accretion disk becomes geometrically and optically much thicker, and since we observe these systems very nearly edge–on, the disk may partly (as in SX Cassiopeiae) or completely (as in the other Serpentids) eclipse its central star.

However, part of the circumstellar matter is located well outside of the orbital plane. It is probably being driven out of the system, perpendicularly to the disk, by a wind generated in the interaction between the accreting star and the disk. This component of circumstellar matter manifests itself through emission lines of fairly highly ionized species in the ultraviolet. Being located away from the orbital plane, it is only partly (sometimes perhaps not at all) affected by eclipses. The most important characteristic is, however, that this material sees the central star and the surrounding inner portions of the accretion disk, and transmits this information to us. In this way, we can hope to get a clearer picture about the structure of each of the W Serpentis systems, and use it for modeling its accretion disk.

ACKNOWLEDGEMENTS. I wish to thank Dr. Ivan Hubený for sharing with me his computer codes, and for many helpful discussions. In various aspects of the project, I enjoyed the cooperation with Dr. Charles D. Keyes, my wife Dr. Zdenka Plavec, and Dr. Jan Dobias. It was possible to pursue this project over the years thanks to the continuous (albeit fluctuating) support from the NASA–IUE programs; the disk modeling is currently sponsored by NASA–ADP grant U003-88, shared with Dr. I. Hubený.

References

Andersen,J.,Nordström,B.,Mayor,M.,Polidan,R.S.
(1988). *Astron. Astrophys.*, **207**, 37.
Andersen, J. and Nordström, B. (1989). *Space Sci. Rev.*, **50**, 179.
Dobias, J.J. and Plavec, M.J. (1985). *Publ. Astron. Soc. Pac.*, **97**, 138.
Dobias, J.J. and Plavec, M.J. (1989). *Space Sci. Rev.*, **50**, 340.
Etzel, P.B. (1988). *Astron. J.*, **95**, 1204.
Etzel, P.B.and Olson, E.C. (1985). *Astron. J.*, **90**, 504.
Hack, M. (1963). *Mem. Soc. Astr. Ital.*, **34**, 3.
McCluskey,G.E., Kondo,Y., and Olson,E.C. (1988). *Astrophys. J.*, **332**, 1019.
Plavec, M.J. (1980), in *Close Binary Stars: Observations and Interpretation*
ed. M.J. Plavec, D.M. Popper, and R.K. Ulrich (Dordrecht: Reidel), p. 512.
Plavec, M.J. (1983). *Astrophys. J.*, **275**, 251.
Plavec, M.J. (1988). *Astron. J.*, **96**, 755.
Plavec, M.J. and Dobias, J.J. (1983). *Astrophys. J.*, **272**, 206.
Plavec, M.J., Weiland, J.L., and Koch, R.H. (1982). *Astrophys. J.*, **256**, 206.
Plavec, M.J.*et al.* (1981). *Bull. Am. Astron. Soc.*, **13**, 523.
Viotti, R. (1976). *Mon. Not. R. Astron. Soc.*, **177**, 617.
Woolf, N.J. (1962). *Mon. Not. R. Astron. Soc.*, **123**, 399.

EXTRAGALACTIC INTERACTING BINARIES

Robert H. Koch
Department of Astronomy and Astrophysics
University of Pennsylvania
209 S. 33rd Street
Philadelphia, PA 19104-6394, U.S.A.

ABSTRACT. The effort to discover extragalactic (EG) analogues of galactic interacting binaries is summarized. At present, not all types of close binaries are known outside the Milky Way. The older photographic surveys emphasized discovery of MS-or-brighter binaries and modern ones have concentrated on supernova searches. EG novae continue to be discovered accidentally. Very few light and velocity curves of non-degenerate binaries are presently available. With modern detecting capability most recognized limitations can be overcome technologically but telescope time will probably continue to limit knowledge for the foreseeable future. There is suggested a number of reasons for increasing concentration on EG binaries.

1. INTRODUCTION - THE DISCOVERY ERA

Not all possible types of interacting binaries have yet been discovered in galaxies other than the Milky Way. The pattern of type discoveries might be said to be predictable but, nonetheless, it still contains surprises as the following description will illustrate.

The deflagration/detonation of a white dwarf due to mass transfer from a companion star (Iben and Tutukov 1984, Webbink 1984) or to coalescence of a binary white dwarf (Mochkovitch and Livio 1989), continues to have difficulties satisfying several requirements. However, each mechanism still appears to be a possible interpretation of a Type Ia supernova. If the process does actually occur, it may be said that the subject of extragalactic (EG) binaries first emerged accidentally and unconvincingly on August 17, 1885. Gully (1885) writes: "...je poussai une exclamation à la vue d'une étoile qui occupait la place du noyau [of the Andromeda Nebula]". This object we now call S And. Muller and Hartwig (1920) later considered all the confusing photometric estimates and measures of the following weeks but more accessible and authoritative discussions appear in Jones (1976) and de Vaucouleurs (1985a). (Claims of stellar variability in or near the nucleus of M31 earlier than mid-August 1885 are presumably to be dismissed as meteorological effects or to confusion with the nucleus of

C. İbanoğlu (ed.), Active Close Binaries, 61–68.

the galaxy.) Within a month of its discovery the first spectrum of any supernova was seen by several observers of S And, and their descriptions are also interpreted by de Vaucouleurs (1985b). A few weeks after maximum light, the spectrum appeared as a continuum with yellow and green (and possibly red and blue-green) emissions superposed upon it.

Two additional discoveries, each now catalogued as Type SN I, followed rather quickly in 1895: VW Vir was picked up in NGC 4424 and Z Cen appeared later in NGC 5253.

It is assumed that all novae result from the familiar model of a white dwarf accreting mass from a cool, Roche lobe-filling, non-degenerate donor. Discovery of the first EG nova is usually credited to Ritchey (1917) for the outburst in NGC 6946 although Curtis (1917) pointed out the retrospective confirmation of a nova in NGC 4527 on plates as old as 1900. Ritchey obtained a spectrum of the 1917 nova, substantially after light maximum and remarked on the emission bands overlying the continuum.

As a result of the analysis of Uhuru data by Schreier, et al. (1972), SMC X-1 is the first high-energy binary known outside the Milky Way.

The first known EG close binary that does not contain a collapsed object could be HV 2543 announced as an Algol-type binary by Leavitt (1908). At the distance of the LMC, however, the absolute magnitude of its bright component must be of the order of -6 if the original magnitude estimates are not grossly in error. This, in turn, implies large stellar radii for both stars - so large that these should be inconsistent with a period of 2.414 days and the system would have to be a foreground one. In this case, the first known EG non-degenerate system would be HV 2425. Detection of variability of this object is also attributed to Leavitt but discovery of the nature of the variability is due to McKibben, as Campbell (1938) reports. Of course, many other eclipsing pairs appear on the same plates and a few years thereafter Shapley and McKibben (1942) noted the likely SMC membership of HV 1346. In fact, many of the eclipsing stars represented as foreground objects by Shapley and McKibben are probably Cloud members since the error in the EG distance scale had not yet been discovered when they wrote.

If SN 1987A is or was a close binary, it served to usher in EG "particle" astrophysics as a result of the detection of its neutrino flux.

2. THE SURVEY ERA

2.1. Photographic Surveys for Non-Degenerate Binaries

It is only to be expected that photographic estimation and observation account for practically all work on non-degenerate EG variables but a few visually-estimated light curves probably exist. Information is dominated by the surveys of Gaposchkin (1962) and Baade and Swope (1963, 1965) for M31 and Gaposchkin (1970) for the LMC. In addition to summarizing these surveys, Herczeg (1981) has remarked on the essentially null results from M33 and Draco. From the M31 survey the inventory was 50 eclipsing objects and 17 suspected or possible eclipsing binaries. In the LMC, Gaposchkin's study turned up 76 eclipsing variables. In the process of a Cepheid survey, Tammann and Sandage (1968) uncovered their V55 in NGC2403. Later Sandage (1971) reported Baade's discovery of V31 in IC1613. Among all these discoveries, there is an obvious selection in favor of intrinsically bright systems. The sample is too small to determine whether there exists a galactocentric gradient in M31 for any light curve or binary parameter.

The distribution of the periods of the known non-degenerate systems shows a bias for short-period pairs but the preference is not so severe as from a random selection of an equal number of galactic binaries in Cygnus. Most likely, this difference is due to the relatively long interval between consecutive plates and the inevitable limitation to intrinsically bright systems containing large radius stars orbiting in relatively long periods.

Of these systems, 6 - a representative percentage - move in eccentric orbits as would be expected for nearly ZAMS systems. In fact, essentially every bright stage of binary evolution, excepting the supergiant atmospherically-eclipsing configuration, can be recognized in the samples. It is even possible that the LMC sample of LPV's by Wood, Bessell, and Fox (1983) does actually contain some systems like VV Cep.

2.2. Surveys of Eruptive, Very-Evolved Binaries

The number of EG novae now known is more than 125. Most of these have emerged from the dedicated searches of Hubble (1929), Arp (1956), Borngen (1968), and Rosino (1973) in M31. A considerably smaller number is now known in the Clouds, with a major contribution having come from Graham and Araya (1971). Rosino and Bianchini (1973) found 5 in M33.

This total number is but a fraction of those which could be discovered. Arp's statistic of 26 novae yr^{-1} for M31 is essentially the same as Hubble's from 25 year's earlier. Consider, however, the amazing result by Ciardullo, Ford, and Jacoby (1983): they discovered 4 M31 novae in 2 nights by exploiting H Alp with CCD technology. Even more astonishing is Pritchet and van den Bergh's (1985) record of finding 2 novae in M87 after 300 minutes.

It is true, of course, that the greatest fraction of these stars is known only from fragmentary light curve coverage and, to the author's present knowledge, only 1 EG nova is known to recur.

Barbon, et al. (1984) list 115 EG SN I's. At the time of writing, the present author had counted 43 SN I assignments among those discovered after publication of the Barbon, et al. list. The updated distribution by galaxy type is very similar to that given by Barbon, et al., except in the increased number of S-type galaxies. This results from improved detection efficiency picking up supernovae at increasingly great distances where galaxy classifications are limited. It is obvious that CCD instrumenting of the supernova search program will result in a much greater discovery rate.

Through 1988, there appear to be 6 identified EG binary sources of

X- or X-plus-gamma radiation, 5 in the LMC and 1 in the SMC. As suggested by the (probably galactic) case of H0534-581, it is possible that a certain additional number of EG sources have not yet been discovered in the archival catalogues.

The Einstein survey of M31 by van Speybroeck, et al. (1979) turned up 36 "Population I" sources; this number is now larger but apparently has never been published fully. These authors remark that many more such sources can be found at faint limits and it may be noted additionally that the entire galaxy was not searched. There were also discovered 15 sources certainly or possibly associated with globular clusters. The very low dispersion attempt by Crampton et al. (1984) to find optical counterparts to most of these sources led only to a certain small number of OB stars.

As Herczeg suggests, even the bulge sources may be interacting binaries.

3. BEGINNINGS OF DETAILED ANALYSIS

3.1. Non-Degenerate Binaries

From the survey light curves the only photometric analyses known to the author are those of Russell (1956) and Gaspani (1986a,b). Drawn from photographic estimates, Russell's show limiting values of the light curve parameters and are of historical value only. The data treated by Gaspani are either measures (V55 in NGC2403) and B-calibrated Argelander estimates (V31 in IC1613) and the analytical results cannot be so precise as he indicates. From this era, no velocity curve analyses are known.

The difficulties of obtaining telescope time for dense coverage are evident in the 5 hot eclipsing pairs observed by Davidge (1987, 1988) in the Clouds. In addition, this program seems to show the foreseeable problem of freeing the program star measures of light bias in crowded star fields. It must also be recognized that long histories of comparison stars do not yet exist. At the present stage of understanding these systems, they seem to be already in the post-ZAMS stage with mass transfer already having begun.

More acceptable phase coverage is exemplified in the light curves of the 6 Cloud systems - 1 in common with Herczeg but all different from Davidge's - observed by Jensen, Clausen, and Gimenez (1988). These light curves, based on CCD detection, are most useful in establishing benchmark precision for this type of variable star observing in difficult field conditions. They strongly suggest that CCD correction for crowded field conditions as a supplement to diaphragm isolation with a photocell is not superior to CCD detection for all of the measures of program and comparison stars and background. In terms of binary evolution, these 3 objects resemble those of Davidge.

Batten, Fletcher, and MacCarthy (1989) list radial velocity analyses for 5 Clouds binaries which are not X-ray sources. Among them are AA Dor, which has already had diverse interpretations and the WR system HD5980, for which Massey, Parker, and Garmany (1989) offer a possible resolution of known confusions. In fact, there appear to be 17 WR binaries known in the Clouds at present and they will certainly repay detailed onservation.

3.2. Eruptive and Very-Evolved Binaries

It cannot be said that EG novae have yet returned results that are attainable from galactic novae such as V1500 Cyg and V1668 Cyg.

The light and velocity curves and spectrophotometry of EG supernovae are indispensable to understanding these terminal stages of evolution but for only a minority of them are these data sets satisfactorily in hand. Of course, a sample of only 1, such as SN1987A, resets the standard by which all other results may be judged.

Despite the facts that relatively few high-energy objects are known and that the data sets are not rich, there is a reasonable understanding of the natures of the recognized systems. This information has been developed by the now-familiar methods of analyzing the high-energy eclipses after at least a preliminary period is known and by supplementing these results with visible-band data. The presently-known sample may be summarized: 2 Black Hole candidates (LMC X-1 and LMC X-3); 3 binaries in which a hot, massive, non-degenerate object is accompanied by a neutron star (A0538-66, LMC X-4, and SMC X-1); and 1 system (GBS0526-66) in which the nature of the companion to the neutron star is not known. Obviously, not all the repertoire of their behavior is likely to be known so quickly and these will undoubtedly continue to attract attention.

The magnetic white dwarf systems, BL Hyi and H0534-581, are presumably not outlying members of the Clouds.

4. THE TECHNOLOGICAL PROSPECT: WHAT WE CAN AND CANNOT DO

4.1. Non-Degenerate Binaries and Eruptive Variables

Without doubt, the most satisfying development has been the increasing availability of CCD arrays for radiometry and spectrophotometry. The pace of this development shows no signs of abating.

Coupled with this detector improvement there is the possibility that EG binary studies will command a reasonable portion of the time available to NTT's. Wilson's (1989) description of the initial focal plane performance of the 3.58-m ESO NTT speaks very powerfully of what may be done in faint, crowded fields. For instance, a 10 sec. unfiltered CCD frame on the ESO 3.58-m NTT brings a capability of working all systems brighter than A2 V ones in the LMC, if no account is taken of absorption in the Cloud. With similar assumptions, all O-type and WR binaries in M31 and the WR systems in M81 become available. A factor of 2x increase in exposure time brings access to many Algols in the Clouds.

For the present, Ginga is the only resource available although archival investigations of older data bases may have some measure of success. In the very near future, ROSAT offers an opportunity for some sustained programs.

4.2. Particle Astrophysics from EG Binaries

The neutrino detection from SN 1987A may be summarized as follows: with the Kamiokande II (1987) detector mass of 2.1kT, 12 stellar neutrinos were detected above a threshold of 6 MeV through an interval of about 12 s without statistically-significant directional information. Except for the higher threshold of 20 MeV, the IMB (1987) result is much the same. Inverse-square dilution over the increased distances to the SMC and the Draco, UMi, and Sculptor systems should not forbid detections of comparable intrinsic neutrino fluxes. At the distance of M31, flux is attenuated by a factor of 170x compared to that from the LMC. This may be overcome by increasing detector mass by a comparable factor, but Zatsepin's (1968) criterion for dispersion of neutrinos with finite mass may become important. At the distance of M31 and for a neutrino of m (in units of 10 eV) and energy E (in units of 10 MeV), the time delay for the particle is about $34(E)^{-2}(m)^{2}$ sec. Both the inverse-square dilution and the possible time dispersion have to be overcome against a noise background, which for the present Kamiokande detector, is about 0.1 muon-induced event per sec. It is true that other criteria permit the unambiguous identification of the latter events.

5. THE SCIENTIFIC PROSPECT: WHAT WE CAN LEARN

5.1. EG Binaries just as Close Binaries

In the Milky Way Galaxy we know the single instance of Bet Lyr (with the possibility of R Ara as a second example) exemplifying a critical binary evolutionary stage. Without doubt, we can increase our knowledge of such objects by finding their analogues in other galaxies.

5.2. EG Binaries Supporting General Stellar Studies

There are several contributions which binaries can eventually make to the generalized understanding of stellar evolution within a galaxian environment. These contributions are in train now in the Milky Way Galaxy but with difficulty because of the position of Sun and, in any case, it is difficult to argue acceptably from results of a single sample. (a) With deeper detection and greater focal plane resolution there opens the possibility that young EG binaries will speak to the matter of whether bi-modal star formation is their common or uncommon mode of formation. (b) For a decision on whether there are galactocentric or other gradients in binary formation EG binaries are the natural markers. (c) With increasing confidence in XYZ interpretation of the bulk parameters of close binary components, these stars are also good markers for information regarding chemical gradients within a galaxy. (d) Obviously SN I's will continue to be studied for their own information. (e) Because of their accelerated paces of stellar evolution compared to single stars, EG binaries have much to say about the gas exchange rate back to the interstellar medium of the parent galaxies. (f) The X-ray sources in Milky Way globular clusters have made it very likely that close binaries are not an intrinsic feature of Population I only. The high-energy sources associated with globular clusters in M31 and the novae detected in M87 make this point even more forcefully. The concept of stellar Populations must be generalized to take account of these results.

5.3. EG Binaries Cosmologically

As one understands binary evolution more exactly, the components of these systems offer more standard candles than have been available in the past.

5.4. EG Binaries for Particle Astrophysics

With improving detectability from scaled-up and quieter detectors, it is possible that the time trains of neutrinos from SN I's in the Clouds will be more convincingly interpretable.

ACKNOWLEDGEMENTS

It is a pleasure to thank Mrs. S. Csigi, R. J. Mitchell, and Mme. S. Laloë for technical assistance and W. Forman, M. Friedjung, T. Herczeg, K. Olah, M. F. Struble and R. F. Webbink for scientific advice. I am indebted to K. Lande for the argument regarding detecting neutrino fluxes from sources within the Local Group.

REFERENCES

Arp, H. C., 1956. A.J. **61**, 15.
Baade, W., and Swope, H. H., 1963. A.J. **68**, 435.
_____., 1965. ibid. **70**, 212.
Barbon, R., Cappellaro, E., Ciatti, F., Turatto, M., and Kowal, C. T., 1984. Astr.Ap.Suppl. **58**, 735.
Batten A. H., Flecther, J. M., and MacCarthy, D. G., 1989. Publ. D.A.O.Victoria **17**.
Borngen, F., 1968. A.N. **291**, 19.
Ciardullo, R., Ford, H., and Jacoby, G., 1983. Ap.J. **272**, 92.
Campbell, L. 1938. Pop.Astr. **46**, 343.
Crampton, D., Cowley, A. P., Hutchings, J. B., Schade, D. J., and van Speybroeck, L. P., 1984. Ap.J. **284**, 663.
Curtis, H. D., 1917. Publ.A.S.P. **29**, 180.
Davidge, T. J., 1987. A.J. **94**, 1169.
_____., 1988. ibid. **95**, 731.

De Vaucouleurs, G., 1985a. l'Astr. **99**, 387.
_____., 1985b. ibid. **99**, 447.
Gaposchkin, S., 1962. A.J. **67**, 334.
_____., 1970. SAO Special Rep. No. 310.
Gaspani, A., 1986a, GEOS Cir.,EB No. 14.
_____., 1986b, ibid, No. 15.
Graham, J. A., and Araya, G., 1971. A.J. **76**, 768.
Gully, L., 1885, Ciel et Terre 1, 355.
Herczeg, T. J., 1981, in Binary and Multiple Stars as Tracers of Stellar Evolution. Eds. Z. Kopal and J. Rahe. (D. Reidel: Dordrecht). Ap.Space Sci.Lib. **98**, 145.
Hubble, E. P., 1929. Ap.J. **69**, 103.
Iben, I., and Tutukov, A. V., 1984. Ap.J.Suppl. **55**, 335.
IMB Collaboration, 1987. Phys.Rev.Lett. **58**, 1494.
Jensen, K. S., Clausen, J. V., and Gimenez, A., 1988. Astr.Ap.Suppl. **74**, 331.
Jones, K. G., 1976. J.Hist.A. **7**, 27.
Kamiokande Collaboration, 1987. Phys.Rev.Lett. **58**.
Leavitt, H. S., 1908. Ann.HCO **60**, 87.
Massey, P., Parker, J. W., and Garmany, C. D., 1989. A.J. **98**, 1305.
Mochkovitch, R., and Livio, M., 1989. Astr.Ap. **209**, 111.
Muller, G., and Hartwig, E., 1920. Gesch.Lit.Lichtwechs. **2**, 417.
Pritchet, C., and van den Bergh, S. 1985. Ap.J. **288**, L41.
Ritchey, G.W., 1917. Publ.A.S.P. **29**, 210.
Rosino, L. 1973. Astr.Ap.Suppl. **9**, 347.
Rosino, L., and Bianchini, A., 1973. Astr.Ap. **22**, 461.
Russell, H. N., 1956. Vistas Astr. **2**, 1177.
Sandage, A., 1971. Ap.J. **166**, 13.
Schreier, E., Giacconi, R., Gursky, H., Kellogg, E., and Tananbaum, H., 1972. Ap.J.Lett. **178**, L71.
Shapley, H., and McKibben, V., 1942. Bull.HCO **916**, 19.
Tammann, G. A., and Sandage, A., 1968. Ap.J. **151**, 825.
Wood, P. R., Bessell, M. S., and Fox, M. W., 1983. Ap.J. **272**, 99.
Zatsepin, G. I., 1968. JETP Lett. **8**, 205.

EFFECTIVE GRAVITY-DARKENING OF CLOSE BINARY SYSTEMS

Masatoshi Kitamura

National Astronomical Observatory,
Mitaka, Tokyo, Japan.

ABSTRACT

From a quantitative analysis of the observed ellipticity effect, empirical values of the exponent of effective gravity-darkening have been deduced for the components of close binary systems whose physical parameters are accurately known. The result of analysis indicate: (1) for the MS components of detached systems the empirical exponents are consistent with the existing theories of radiative and convective stellar atmospheres, (2) for the Roche-lobe filling secondary components of semi-detached systems, the empirically deduced effective gravity-darkening is seriously greater than the unity which suggests an excessive darkening due to the mass-loss, and (3) for contact systems, an excess of gravity-darkening, though lesser degree, can be also seen and the relationship between the exponent of effective gravity-darkening and the mass ratio is clearly different between W UMa systems and early-type contact ones, indicating different atmospheric conditions of common envelopes.

1. INTRODUCTION AND BASIC PRINCIPLE

The gravity-darkening exhibited by the distorted stars in close binary systems is one of the important observable phenomena which enable us to probe the structure of stellar atmospheres. This phenomenon is well-known as the ellipticity effect on light curves of close binary systems.

For the general law of gravity-darkening, we can write

$$H \propto g^{\alpha}, \qquad (1)$$

where H denotes the bolometric surface brightness at any arbitrary

C. İbanoğlu (ed.), Active Close Binaries, 69–79.

point on the stellar surface, g the local gravity, and α the exponent of gravity-darkening. As is well-known, we have $\alpha_{rad}=1$ for the surface of a rotationally or tidally distorted star in radiative equilibrium according to von Zeipel's law (1924). However, for the convective stellar atmospheres Lucy (1967) calculated values of α from data of existing stellar models as $\alpha_{conv}=0.28 \sim 0.35$ with the representative value of 0.32.

Observationally the principal effect of gravity-darkening on the light variations of close binary systems is to augment the photometric effects of the ellipticity of the configuration outside eclipse. The most dominant for this ellipticity effect is the cos 2 -term with the coefficient A_2 if the outside-eclipse light variation is expressed by

$$\ell = A_o (1.0 + A_1 \cos\theta + A_2 \cos 2\theta + B_1 \sin\theta + B_2 \sin 2\theta), \qquad (2)$$

where θ denotes the phase angle.

Taking the black-body assumption, the observed quantity A_2 may be expressed in a form as

$$A_2 = A_2(q, i, r_1, r_2, T_1, T_2, u_1, u_2, \lambda ; \alpha_1, \alpha_2), \qquad (3)$$

where q is the mass ratio, i the orbital inclination, r_1 and r_2 the fractional radii, T_1 and T_2 the effective temperatures, u_1 and u_2 the limb-darkening coefficients, λ the observed effective wavelength, α_1 and α_2 the exponents of gravity-darkening and the subscripts 1 and 2 correspond to the primary and secondary components, respectively. The parameters excepting α_1 and α_2 in the parenthesis of the right-hand side of Equation (3) can be known from observations. Based upon models up to the second-order distortion, a rigorous mathematical form of Equation (3) was first deduced by Kopal and Kitamura (1968) and later slightly revised by Kitamura and Nakamura (1983) for a practically more convenient form.

In principle, only one unknown α can be solved empirically from Equation (3). Thus, the soluble cases for α would be as follows.

Case (1). $\alpha_1 = \alpha_2$. If both components of a close binary system have similar spectral types or similar atmospheric structures, we can apply this case.

Case (2). If α_1 for the primary component can be known in advance, an α_2-value for the secondary can be deduced from Equation (3); or vice versa.

Case (3). If the luminosity of the secondary component is very small as compared with that of the primary and may be practically neglected, we can deduce α_1-value for the primary.

Before proceeding to derivation of α-values, the smaller contamination due to the reflection in the observed A_2-value must be eliminated as

$$A_2 = A_{2,obs} - A_{2,ref}, \tag{4}$$

where $A_{2,ref}$ may be evaluated approximately with Rucinski's (1969) values of the bolometric albedo or with the procedure of Budding and Ardabilli (1978). In most actual cases, the $A_{2,ref}$-values calculated with these two methods are found not to be very different.

2. EMPIRICAL GRAVITY-DARKENING OF MS COMPONENTS IN DETACHED SYSTEMS

Table 1 shows detached close binaries used for Case-1 deduction of their α-values. The last column shows empirical α-values deduced for the respective systems. The errors attached to α-values are mainly due to probable errors in the used $A_{2,obs}$-values.

Table 2 shows two detached systems for which Case 2 -deduction may be applied. Here empirical α_2-values of the secondary components in these two systems have been deduced by assuming α_1=1 for the primary with early-type MS spectra. Table 3 shows a result for α_1-value of the primary component of TU Cam deduced by applying Case 3, because the luminosity of the secondary is very small as compared with that of the primary such that L_2/L_1 = 0.97.

All the empirical α-values of MS components shown in Tables 1, 2 and 3 are all plotted in Figure 1. Figure 1 reveals that the radiative equilibrium condition generally holds for the photospheric atmospheres of early-type MS stars down to spectral type A0, because their α-values distribute well around the unity. However, from A-type onward, stars diminish their α-values gradually. From the middle F-type onward, the α-value seems to reach some minimum value between 0.2 and 0.3.

Such a result for empirical gravity-darkening of MS stars deduced from detached systems is generally consistent with not only von Zeipel's law for radiative stellar atmospheres but also Lucy's (1967) theoretical result for convective ones.

Table 1. Empirical α -values of MS detached systems deduced with Case 1.

C.B.	Sp	T	$\alpha_1 = \alpha_2$
V478 Cyg	O9.5V + O9.5V	31000°K	0.99 ± 0.10
V453 Cyg	B0.5V + B0.5V	28500	0.85 ± 0.08
VV Ori	B1V + B4V	26000 + 17300	0.88 ± 0.04*
χ^2 Hya	B8IV + B8.5V	11750 + 11100	0.78 ± 0.01*
CO Lac	B8.5IV + B9V	11350 + 10950	0.87 ± 0.12*
V906 Sco	B9V + B9V	10500	1.01 ± 0.05
MN Cas	A0V + A0V	9600 + 9420	1.04 ± 0.02
WY Hya	A6V + A6V	8000	0.90 ± 0.10
HS Hya	F3V + F3V	6600	0.37 ± 0.05
RZ Cha	F5IV-V + F5IV-V	6500	0.32 ± 0.05
ST Cen	F9V + F9V	6200	0.20 ± 0.03
UV Leo	G0V + G1V	5920 + 5890	0.22 ± 0.02*

*) These have been plotted in Figure 1 for their mean spectral types.

Table 2. Empirical α_2-values of the secondary components of WX Cep and CM Lac in Case 2.

C.B.	Sp	T	α_2
WX Cep	A2V* + A5V	8870 + 8150°K	0.85 ± 0.07
CM Lac	A3V* + A8V	8600 + 7500	0.65 ± 0.10

* With $\alpha_1 = 1$ assumed.

Table 3. Empirical α_1-value of the primary component of TU Cam deduced as Case 3.

C.B.	Sp	T	α_1
TU Cam	A0V + A9	9600 + 7200°K	0.99 ± 0.04
	$L_2/L_1 = 0.97$		

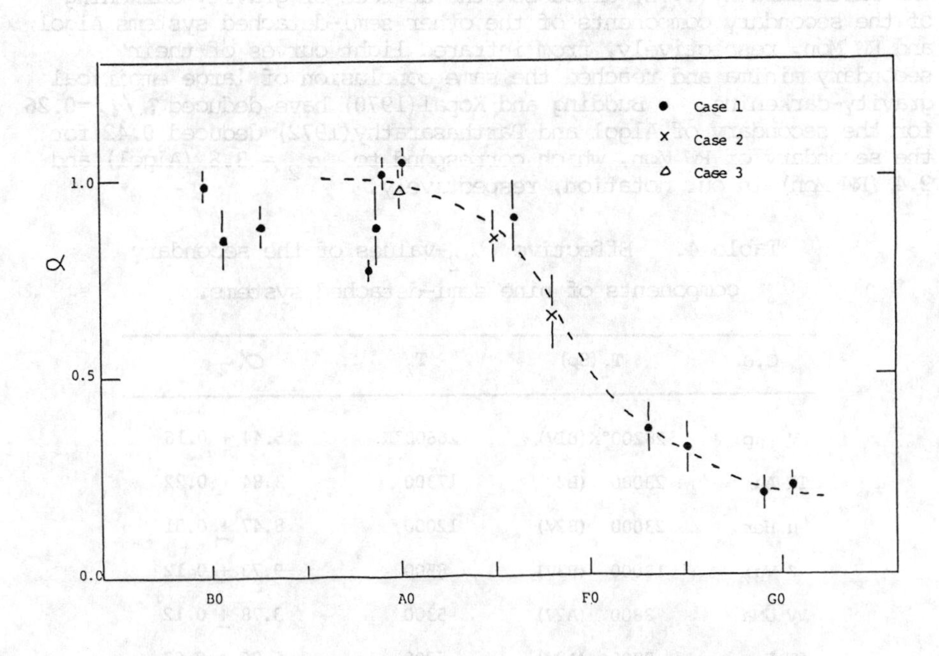

Fig. 1. Empirical gravity-darkening deduced for MS stars in detached close binaries.

3. EFFECTIVE GRAVITY-DARKENING OF THE ROCHE-LOBE-FILLING SECONDARY COMPONENTS OF SEMI-DETACHED SYSTEMS

Inasmuch as we have obtained a clear relationship between the exponent of gravity-darkening and spectral type for MS stars as shown in Figure 1, we may assign the same α-values, corresponding to their spectral types, to the MS primary components in semi-detached systems and, by applying Case 2, we should be able to deduce the effective gravity-darkening for their mates, namely the Roche-lobe-filling secondary components.

Table 4 shows nine well-understood semi-detached systems used for the analysis and the last column gives the α_2-values deduced for the secondary components. The most conspicuous result would be that all the α_2-values thus deduced are seriously greater than the unity.

In this connection, it may be noted that Budding and Kopal(1970) and Parthasarathy(1972) discussed the degrees of gravity-darkening of the secondary components of the other semi-detached systems Algol and RW Mon, respectively, from infrared light curves of their secondary minima and reached the same conclusion of large empirical gravity-darkening. Budding and Kopal(1970) have deduced τ_λ/τ_0=0.26 for the secondary of Algol and Parthasarathy(1972) deduced 0.42 for the secondary of RW Mon, which correspond to α_2 = 3.8 (Algol) and 2.4 (RW Mon) in our notation, respectively.

Table 4. Effective α_2-values of the secondary components of nine semi-detached systems.

C.B.	T_1(Sp)	T_2	α_2
V Pup	28200°K(B1V)	26600°K	5.44 ± 0.16
TT Aur	23000 (B2V)	17300	3.84 ± 0.22
u Her	23000 (B2V)	12000	8.47 ± 0.51
Z Vul	18000 (B3V)	8500	9.73 ± 0.12
VV UMa	8800 (A2V)	5300	3.78 ± 0.12
LT Her	8800 (A2V)	5200	5.86 ± 0.08
RZ Cas	8800 (A2V)	5000	2.25 ± 0.15
V356 Sgr	17900 (B3V)	10340	3.83 ± 0.19
GT Cep	19000 (B3V)	10000	3.00 ± 0.43

No particular correlation could not be found between α_2 and other physical parameters for these secondary components. Even so, such a considerable excess of effective gravity-darkening of the Roche-lobe-filling secondary components over the unity could not be reconsiled by any adjustment of the input physical parameters and the amount of the reflection correction.

Such an excessive gravity-darkening for the Roche-lobe-filling secondaries could be interpretated in terms of energy transport by the mass outflow through the inner Lagrangian point towards the primary, namely as the mass-loss darkening (Unno, Kiguchi and Kitamura 1988).

4. EFFECTIVE GRAVITY-DARKENING OF THE COMPONENTS OF CONTACT SYSTEMS

For a contact system having a common photospheric envelope, we may assume $\alpha_1 = \alpha_2$ and apply Case 1 for the α-derivation. We have analysed six early-type contact systems (whose spectral types are earlier than the middle A) and sixteen W UMa-type systems. For most of these samples, their cross-correlation spectroscopic mass ratios are available.

Table 5 shows the result of analysis for α-values of contact systems. In the table, the temperature indicator θ and the fill-out factor f are defined at the footnote. It is evident from Table 5 that all the α-values of contact systems are considerably larger than Lucy's conventional value 0.32. Kopal (1968a,b) has previously discussed the gravity-darkening based on the first-order theory of distortion for W UMa-type systems and reached a similar conclusion of large gravity-darkening for those systems.

Figure 2 shows a plot of the α-values of the contact systems in Table 5 against their mass ratios. It is clear in Figure 2 that the α-values as a whole decrease with increasing mass ratio for W UMa-type systems while gradually increasing for early-type contact ones. This would indicate that the structure of the common envelope is significantly different between the contact systems of early-type and W UMa-type.

Figure 3 shows the relationship between α-values and effective temperatures for all the relevant contact systems. It is evident that the effective gravity-darkening increases with increasing effective temperature, irrespective of the types of contact systems.

Then, what is the origin of such an excessive gravity-darkening of contact systems (though lesser degree as compared with that of

Table 5. Effective α-values of the components of contact systems.

C.B.	Sp	θ *	q	f **	α
V701 Sco	B2 + B2	0.215	0.99	0.63	2.77
AW Lac	B2 + B4	0.271	1.00	0.71	2.22
BH Cen	B3 + B3	0.285	0.84	0.84	2.54
AU Pup	A0 + A1	0.539	0.64	0.72	1.90
V1073 Cyg	A3 + A5	0.596	0.34	0.07	1.40
V535 Ara	A5 + A5	0.588	0.36	0.03	1.81
XY Boo	F8 + F8	0.840	0.16	0.05	1.74
W UMa	F8 + G0	0.847	0.49	0.05	0.85
TX Cnc	G0 + G0	0.831	0.53	0.21	0.98
RZ Com	G2 + G2	0.869	0.43	0.03	1.20
V502 Oph	G2 + F9	0.850	0.37	0.24	0.90
U Peg	G2 + G2	0.892	0.33	0.15	0.96
YY Eri	G5 + G5	0.916	0.40	0.19	0.82
RW Com	G6 + G6	0.892	0.34	0.17	1.07
TZ Boo	G6 + G6	0.900	0.13	0.1	1.25
XY Leo	K0 + K0	1.075	0.50	0.06	0.65
CC Com	K5 + K5	1.200	0.47	0.20	0.79
V523 Com	K5 + K5	1.172	0.42	0.15	0.78
AW UMa	F1 + F1	0.698	0.074	0.65	1.72
V566 Oph	F4 + F4	0.755	0.24	0.41	1.04
RR Cen	F5 + F5	0.775	0.21	0.35	1.14
Y Sex	F8 + F8	0.730	0.18	0.00	1.85

*) $\theta = 5040/0.5(T_1 + T_2)$.

**) $f = \dfrac{\Omega_{in} - \Omega}{\Omega_{in} - \Omega_{out}}$: if different f-values have been published for a system, the most recent value has been taken.

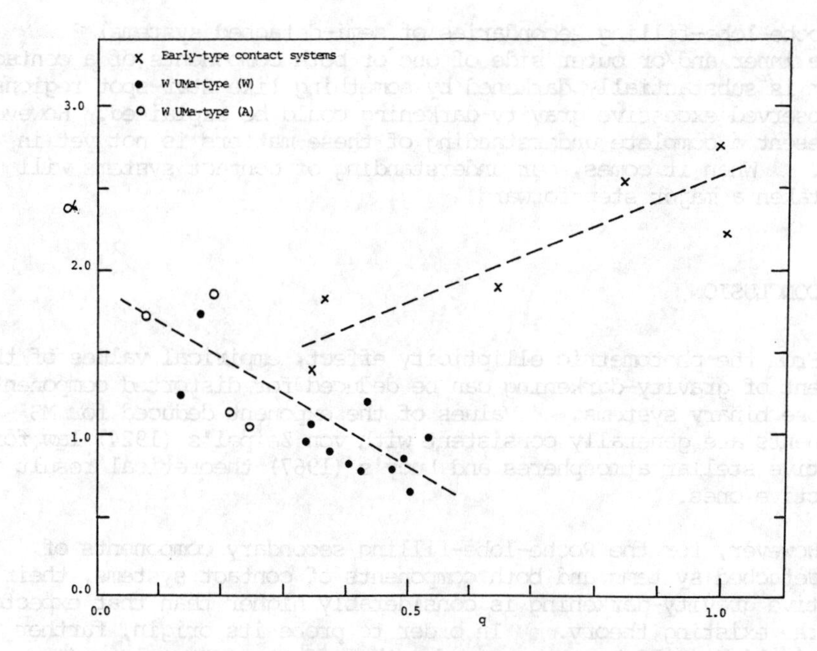

Fig. 2. Relation between empirical α -values and mass ratios for contact systems.

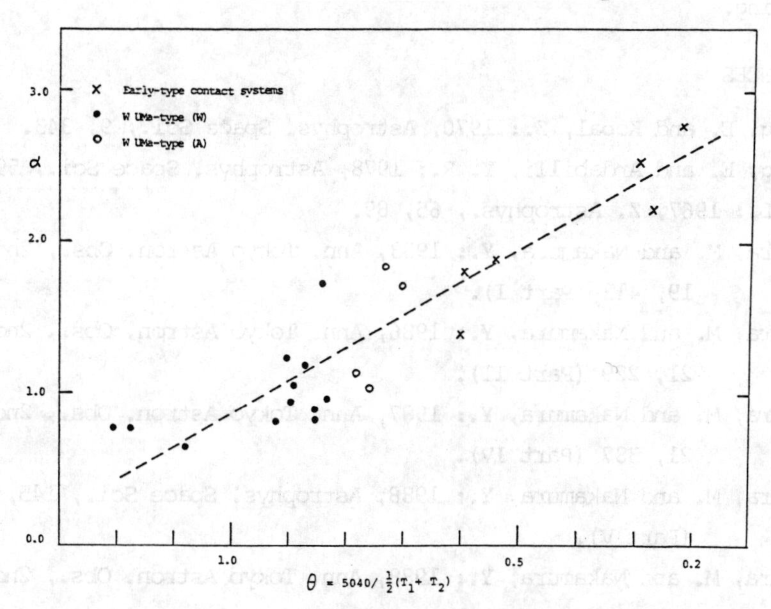

Fig. 3. Relation between empirical α -values and effective temperatures for contact systems.

the Roche-lobe-filling secondaries of semi-detached systems) ?
If the inner and/or outer side of one or both components of a contact system is substantially darkened by something like star-spot regions, the observed excessive gravity-darkening could be explained. However, at present a complete understanding of these matters is not yet in sight. When it comes, our understanding of contact systems will have taken a major step forward.

5. CONCLUSION

From the photometric ellipticity effect, empirical values of the exponent of gravity-darkening can be deduced for distorted components of close binary systems. Values of the exponent deduced for MS components are generally consistent with von Zeipel's (1924) law for radiative stellar atmospheres and Lucy's (1967) theoretical result for convective ones.

However, for the Roche-lobe-filling secondary components of semi-detached systems and both components of contact systems, their effective gravity-darkening is considerably higher than that expected from the existing theory. In order to probe its origin, further investigation would be necessary for the effect of mass loss (or mass transfer) and the effect of star spots on the observed gravity-darkening.

REFERENCES

Budding, E. and Kopal, Z.: 1970, Astrophys. Space Sci., 9, 343.

Budding, E. and Ardabilli, Y. R.: 1978, Astrophys. Space Sci., 59, 19.

Lucy, L.: 1967, Z. Astrophys., 65, 89.

Kitamura, M. and Nakamura, Y.: 1983, Ann. Tokyo Astron. Obs., 2nd ser. 19, 413 (Part I).

Kitamura, M. and Nakamura, Y.: 1986, Ann. Tokyo Astron. Obs., 2nd ser. 21, 229 (Part II).

Kitamura, M. and Nakamura, Y.: 1987, Ann. Tokyo Astron. Obs., 2nd ser. 21, 387 (Part IV).

Kitamura, M. and Nakamura, Y.: 1988, Astrophys. Space Sci., 145, 117 (Part V).

Kitamura, M. and Nakamura, Y.: 1988, Ann. Tokyo Astron. Obs., 2nd ser. 22, 31 (Part VI).

Kitamura, M. and Nakamura, Y.: 1989, Publ. Natl. Astron. Obs. Japan, 1, 43 (Part VII).

Kopal, Z. : 1968a, Astrophys. Space Sci., 2, 23.

Kopal, Z. : 1968b, Astrophys. Space Sci., 2, 166.

Kopal, Z. and Kitamura, M.: 1968, Adv. in Astron. Astrophys. (Academic Press, New York), 6, 125.

Parthasarathy, M. : 1972, Astrophys. Space Sci., 18, 190.

Rucinski, S. M.: 1969, Acta Astron., 19, 245.

Unno, W., Kiguchi, M. and Kitamura, M.: 1988, Proc. IAU Coll. No.108 (Atmospheric Diagnostics of Stellar Evolution), 215.

Zeipel, H. von.: 1924, Monthly Notices Roy. astron. Soc., 84, 702.

BINARY ACTIVITY AT THE TOP OF THE MAIN SEQUENCE: Y CYG AND U OPH

Robert H. Koch
Department of Astronomy and Astrophysics
University of Pennsylvania
Philadelphia, PA 19104-6394, U.S.A.
and
Raymond J. Pfeiffer
Department of Physics
Trenton State College
Trenton, NJ 08650-4700, U.S.A.

ABSTRACT. The binary stars at the top of the main sequence are, in principle, well-known with regard to interior and envelope structure. By mobilizing all available observational data for two of the best-known binaries - Y Cyg and U Oph - it is possible to place limits on the intrinsic activity of these pairs and, by implication, on others similar to them. Each of these binaries shows intrinsic polarization at the level of about 0.3%, traceable to a systemic envelope which may undergo variations as it is fed by stellar winds. The weak winds from each component star intersect between the stars setting up a standing shock of scatterers. Small-scale intrinsic variability in broad- and medium-bandpass light curves has also been seen. In the case of U Oph, residuals from the best theoretical light curve correlate loosely in phase with the distribution of the scatterers modelled from the polarimetry. Three medium-band light curves of U Oph, biased by HI absorption, show effects of variable circumstellar absorption and possibly scattering. Hot, massive main sequence binaries are indeed <u>active</u> ones.

1. INTRODUCTION

It used to be supposed that main sequence close binaries are completely stable in all characteristics, thus validating them as the tools of choice for calibrating the paradigms vested in the mass-luminosity and mass-radius relations. This attitude was eventually supplanted by more selective choosing of only very detached pairs for these purposes. This paper is concerned with quantifying evidence for low-level intrinsic stellar activity near the hot end of the main sequence as exemplified by Y Cyg and U Oph. It is possible that the origin of all the variability is really the interactions of the binary stellar winds. It is not claimed that single, comparably hot and massive stars would show the

C. İbanoğlu (ed.), Active Close Binaries, 81–94.

same effects since these stars would lack the binary interactions.

2. VISIBLE-BAND AND IUE SPECTRA

2.1. Velocities and Spectrophotometry of Y Cyg

From IUE images Morrison and Garmany (1984) measured a significant wind from one of the stars. Profile modelling of the C IV feature led these authors to a terminal velocity which is comparable to the stellar escape velocity. The visible-band spectra of this binary show no evidence of such a wind. One may conclude that judiciously-chosen blue lines yield strengths and velocity patterns that reveal nothing of the circumbinary environment. No further UV work appears to have been done.

2.2. Velocities and Spectrophotometry of U Oph

In neither the study by Olson (1975) nor in any previous work has there been decisive evidence from visible-band spectra of non-photospheric features. In addition, the line profiles are so broad that velocities cannot be measured close to the stellar conjunctions and so the orbit has been treated as circular. Finally, photographic spectrophotometric accuracy is unlikely to discover the 5% to 10% effects described below from medium-band light curves.

The IUE spectra, now in the public domain, have been scrutinized but not analyzed. Since the stars are not very hot, they show neither of the C IV and N V features useful for wind detection. The Si IV doublet is weak and appears entirely photospheric.

3. VISIBLE-BAND POLARIMETRY

3.1. Polarization of Y Cyg

The monitoring of Y Cyg by Koch and Pfeiffer (1989) led to a representation of the mild intrinsic polarimetric variability by invoking a scattering volume between the component stars. This is most conveniently interpreted as a shock from intersecting stellar winds, one of which has been discovered by Morrison and Garmany from the IUE SW images. The wind from the fainter star has not yet been detected in IUE images, in part because the phases of the spectra which have been studied are not appropriate for detecting it. The same data have been treated also by Koch, et al. (1989) according to the (Glasgow) model of Brown, McLean, and Emslie (1978). Their results indicate concentrations of scatterers both within and outside the orbital plane but suffer from failing to treat eclipses and shadowings of the scatterers by the fractionally-large stars.

The polarization of this system has been measured since 1949. From the entire collection of data it is clear that Y Cyg has a long-term polarization variation which is not phase-locked to the apsidal-rotation cycle. This effect is presumably due to the system's variable scattering geometry and/or wind activity, which are now very different than they were formerly.

3.2. Polarization of U Oph

Sporadic polarization measures of U Oph have been taken since the early 1950's. These data, when collected, show the object to be intrinsically variable but without an evident long-term period.

When Pfeiffer and Koch (1977) summarized intrinsic polarization variability as a function of binary evolution, U Oph was the only main sequence system known to show such variability. This conclusion was based almost entirely on Coyne's (1970) visible and near-IR data. These measures may be interpreted to show that the intrinsic visible-band polarization of U Oph cannot be less than 0.2%. Coyne's data show, in addition, that the polarization spectrum of the binary is modified by varying amounts of H self-absorption. Subsequently the object has been monitored at the Flower and Cook Observatory until a semblance of reasonable phase coverage was attained. The evolution of instrumentation and technique over the 8 years of data taking are indicated in Koch and Pfeiffer's concurrent work on Y Cyg.

After an interstellar polarization vector was removed, these data were subjected to the Glasgow model and the results have been tabulated by Koch, et al. (1989). In brief, this study showed an intrinsic systemic polarization of about 0.3%, thus indicating a systemic envelope embedding the pair of stars. Additionally, displaced from the binary line-of-centers by about 30° there appear weak concentrations of scatterers both symmetrically and asymmetrically arrayed with respect to the orbital plane. Since all these data are only broad-band blue ones, nothing may be inferred directly regarding the identity of the scatterers.

In principle, this ignorance may be overcome by appeal to Coyne's multi-bandpass data. Unfortunately, the interstellar polarization is comparable in magnitude to the observed mean value and the orientations of the two electric vectors are very nearly in the same direction. Invariance of the electric vector orientation with respect to wavelength yields a non-flat, non-monotonic complex intrinsic polarization spectrum whereas permitting small (e.g., 2° from one bandpass to another) rotations of the interstellar vector with respect to wavelength generates a variety of intrinsic spectra from flat to classical Be ones. Thus, electron scatterers are permitted but not required.

Fig. 1 shows the intrinsic polarization parameters plotted against Keplerian phase. The measures fill an envelope which is large with respect to the errors of measurement. In order to investigate the data for significant periodicities, discrete Fourier transforms of the intrinsic polarization parameters versus phase were applied to the parameters. The two results demonstrate broadly-peaked maximum relative power at the frequency corresponding to half the Keplerian period. This agrees with the concept of an embedding systemic envelope of scatterers which should produce polarization maxima at the quadrature phases. The power spectra also indicated the presence of a poorly-defined 0.25 period harmonic. This is also expected since light removal during the eclipses of the stars should result in two other peaks in the polarization at these phases. Since about half the systemic light is lost during each eclipse, there should be a two-fold increase in the mean polarizations at these phases, but these changes are not observed. This puts a first constraint on the geometry and density of the envelope. Since the sinusoidal variations of the intrinsic parameters that result from the systemic envelope are only weakly discernible by visual inspection of the data, there must be much high frequency activity in the system, as is also indicated in the power spectra. Such activity may exist in the form of clouds moving through the system or turbulence in the envelope. This high-frequency activity places a second constraint on the geometric distribution and density of the scatterers: the amplitude of the sinusoidal variation in the polarization must be not greater than the scatter-band of the observations - about 0.2%. The observations also impose another third constraint on any model of the system, viz., the mean value of the sinusoidal variations must be about 0.3%. There is yet a fourth observational constraint in that the mean level of the polarization at primary eclipse should be slightly lower than that at secondary eclipse.

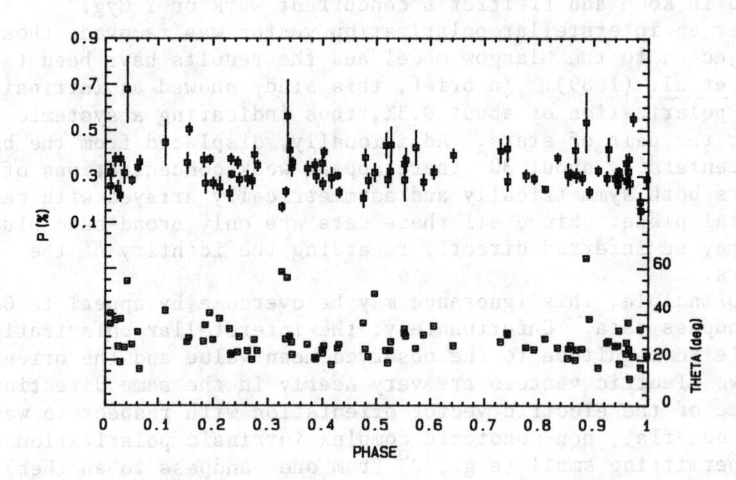

Figure 1. The intrinsic (i.e., freed of the interstellar component) polarization parameters for U Oph. The error bars for p represent probable errors; those for the angular parameter are typically smaller than the symbols.

The Glasgow model assumes the component stars to be point sources. This is clearly not the case for U Oph and the stars eclipse portions of the envelope thereby creating asymmetries in the scattering geometry, which asymmetries change in the course of the revolution of the system. This is a major concern when the scattering envelope is less than a few stellar radii distant from the stars. In addition, the Glasgow model does not account for systemic light variation with phase. Therefore, during the last few years, RJP has been developing a computer code that eliminates these shortcomings in part. This code is a brute-force approach to calculating the polarization that arises from any distribution of circumbinary scatterers. The envelope is represented by a grid of points from which incident stellar light is scattered toward the observer. The linear Stokes parameters are then calculated by summing vectorially over all such points. In order to represent a typical envelope for U Oph 6,000 points have been used. The code takes into consideration eclipsing of points by the stars, variation of systemic light with phase, and orbital inclination. In its present stage of development, the code treats the stars as extended single sources of light for each scatterer.

By trial and error, a model has been found that satisfies all the above constraints except the fourth. It is hoped that additional refinements in the code will result in a model that will satisfy all criteria. The best-fit model at present is a relatively thin shell that roughly conforms to the geometry of the Roche potential surface at a mean distance of 1.5 stellar radii from the photospheres in the orbital plane and extends above and below the orbital plane by about one stellar radius above each photosphere. This shell may be assumed to be an optically thin layer of sufficient density to produce the net observed polarization, but the volume between the shell and the photospheres may well be of lower density than in the shell. The electron number density in the region of the shell situated on the far side of the cooler star from the hotter one is about 6.0×10^{10} cm^{-3}. This is not atypical for circumbinary material. In addition, the model contains a wind shock front that begins near the L_1 point but closer to the cooler star and asymmetrically wraps around and away from the cooler star for about 55°. This detail is in general agreement with Coriolis dynamics. The density in this shock is about twice that found in the more distant shell. The model says nothing about the identity of the scatterers, but it is assumed that they are electrons in a partially ionized gas. The observations indicate that the shell and shock front are not to be considered static distributions of material. Instead, they are the loci of the mean positions of condensations in the wind material and circulating gas. That is, the circumbinary material is highly turbulent with density varying in both time and position. Variable column densities in such an envelope are appropriate for explaining the variable depths of the eclipses that are described below.

3.3. The Polarization of Y Cyg Re-considered

The measure of success achieved for U Oph permits re-examination of the polarization of Y Cyg. While a detailed model has yet to be developed, it may be said that the envelopes for the two pairs must be very similar since the observational constraints are very similar for the two binaries. It may also be expected that the activity of Y Cyg will be greater than that of U Oph since the wind velocity must be more vigorous in the hotter system.

4. ANALYSES OF CONTINUUM LIGHT CURVES

4.1. Visible-band Light Curves of Y Cyg

The most recent published light curve of Y Cyg is that of Magalashvili and Kumsishvili (1959). These observations has been solved three times with indifferent mutual agreement, perhaps due to the indication of mild intrinsic variability outside eclipse. In particular, the results of Giuricin, Mardirossian, and Mezzetti (1980) lead to a slightly evolved characterization of the system.

Currently, the APT facility at Mt. Hopkins is collecting UBV light curves but the observational effort is tedious because of the nearly 3-day period. The data so far in hand, kindly made available in advance of publication, suggest low-level intrinsic variability but it is necessary to await completion of the light curves before a convincing interpretation can be made.

4.2. Visible-band and OAO Light Curves of U Oph

The next-to-last lengthy analysis of the light curve of U Oph is by Koch and Koegler (1977) from their own slotting scanner data and from the OAO fluxes of Eaton and Ward (1973). This study resulted in the surprising description of the hot star as the small one and called into question the main-sequence status of the pair. The light curve appeared complicated by a conspicuous asymmetry of the secondary eclipse, which also contributed to a disposition to give up the accepted understanding of the system. Such a conclusion was challenged by Kämper's (1986) result showing that the eclipsing pair moves in an eccentric orbit, which itself rotates in a period of about 39 years. Without re-solving the light curve, Kämper gives extensive arguments showing the unevolved condition to be the likely one for the system. This conclusion also follows from the circular-orbit analysis by Cester, et al. (1978) of 2 light curves from OAO.

It was decided to re-analyze the ground-based and UV light curves for the eccentric-orbit configuration. Simply because of convenience and zero computer costs, the model chosen was the simple 2-axis ellipsoidal one vested in the code EBOP described, e.g., by Popper and Etzel (1981) and Popper (1984).

The beginning tactical procedure was that recommended by these authors, i.e., isolating a few parameters at a time in order to see the response of the observations to parameter variation. After a phase shift was evaluated, it was found very quickly that the phase-of-minimum displacement term, ecos(OMG), was consistent with Kämper's result from the collected light ephemeris residuals and that the asymmetry term, esin(OMG), could not be evaluated because the asymmetries of the eclipse branches are small. Accordingly, this latter term was calculated and held fixed thereafter. It also emerged quickly that conventional von Zeipel darkening was appropriate within the errors of the gravity coefficient and that the "reflection" terms were best handled by the default option in the code.

These parameters being fixed and the mass ratio established at the accepted value of 0.90, each data set was subjected to variations of 4 parameters: (1) the surface brightness of the star eclipsed at secondary minimum, (2) the radius of the other component, (3) the ratio of the stellar radii, and (4) the orbital inclination. There was some tendency for the representation to move in the direction of the result given by Koch and Koegler but eventually this trend reversed and did not reappear through further trials. Up to 17 iterations of a light curve were pursued in order to seek convergence and mutual agreement within the parameter errors. This condition was eventually satisfied and the light of the third star of the system (already shown to exist by Kämper) was then sought, but its contribution - never greater than 0.01 in normalized units - was always of the same order of magnitude as its error. A final set of iterations fixed the geometrical parameters to be constant for all light curves and solved for improved values of (1) above and a common limb darkening for the two stars at a given bandpass. No attempt was made to discover a lead/lag angular displacement of the long stellar axes against the systemic line of centers.

The results of the analysis appear in Table I. The parenthesized numbers represent 1-sigma values for the last digit in the adjoining value of the parameter. For the geometrical parameters the values are means calculated from those values converged upon independently at each bandpass and their errors are internal ones calculated from the individual values at each bandpass. The errors for $\underline{x}$ emerge from the last set of iterations. For the normalized stellar light values, L_g and L_s 1-sigma errors vary from 0.018 to 0.023.

TABLE I. Light Curve Solutions for U Ophiuchi

a_g = 0.274(2) a_s = 0.239(4) $i = 88.1^o(3)$
b_g = 0.268 b_s = 0.233

WL	J_s	$x_g = x_s$	L_3	L_g	1-sigma
4750	0.911(7)	0.18(4)	+0.012(11)	0.595	0.0053
4476	0.903(6)	0.04(8)	-0.003(3)	0.597	0.0055
4230	0.893(9)	0.11(9)	+0.002(13)	0.600	0.0075
4026	0.913(12)	0.18(12)	+0.009(13)	0.594	0.0106
4250	0.889(10)	0.01(4)	+0.012(13)	0.602	0.0128
3330	0.862(9)	0.32(3)	+0.000(10)	0.608	0.0065
2980	0.843(12)	0.31(6)	-0.002(6)	0.614	0.0097
2460	0.865(11)	0.62(5)	-0.003(5)	0.608	0.0081
1920	0.828(12)	0.82(4)	-0.004(6)	0.618	0.0080
1550	0.826(14)	0.86(4)	+0.011(7)	0.621	0.0109
1380	0.791(22)	0.90(7)	-0.002(10)	0.629	0.0164

It may be noted that these results are completely consistent with a main sequence binary configuration, the large star being the hot and massive one, thus denying the results of the Koch and Koegler analysis. That fallacious result can be traced to mistakes of judgment in interpreting the asymmetry of secondary eclipse as a light complication rather than conclusive evidence of orbital eccentricity and in discarding the descending branch of secondary in the analysis. Even though the true errors of x are not so small as those tabulated, the march of darkening with wavelength is essentially that given originally by Eaton and Ward and subsequently by Koch and Koegler. Because a significant value of "third light" did not reveal itself with decreasing wavelength, it may be concluded that the third star is not a hot white dwarf but a cooler main sequence or still-contracting object. This is consistent also with the young interpretation of the U Oph eclipsing system as indicated by the small orbital eccentricity.

Each of the scanner light curves is 50A wide. From the evidence of Kurucz's (1979) models the light curves at 4750A and 4230A contain about 0.002 per mil line blocking fraction. For the 4476A and 4026A light curves the blocking fractions are 0.012 and about 0.02, respectively. From the photometric parameters of the light curve analysis themselves there is no evidence of non-photospheric radiation - such as an anomalous trend with wavelength of the light balance between the stars - at the 1% level.

This measure of satisfaction is tempered, however, by the actual distribution of the residuals from the visible-band light curves - the nearly pure continuum ones. The averaged residuals are displayed in Fig. 2 and 3 as a function of phase. It is evident that there are non-

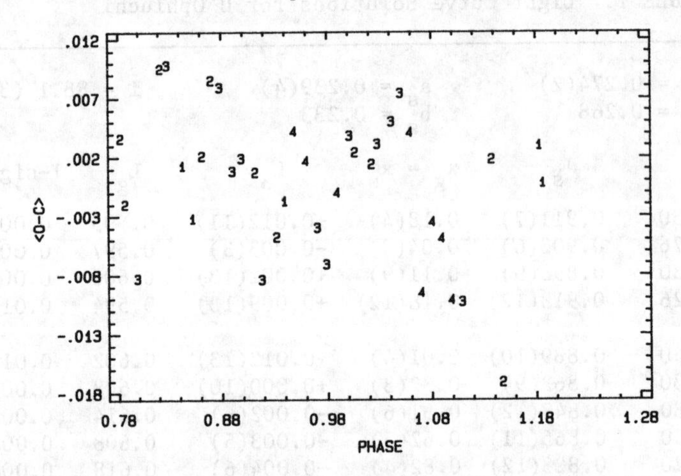

Figure 2. The EBOP residuals, averaged over the 4 light curves, in the half-cycle around primary eclipse. The symbols range from a weight of about 1 through a weight of about 6 in the sequence 1, 2, 3, and 4, respectively.

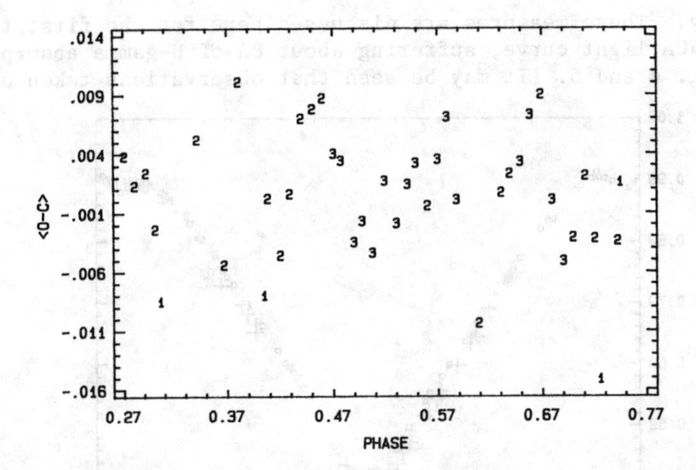

Figure 3. As for Figure 2 over the half-cycle around secondary eclipse.

monotonic, systematic runs of the residuals. These may be described, first of all, as being structured on the rising branch of primary eclipse attaining values as large as -0.01 just before last contact. The residuals also appear non-monotonic and systematic within the secondary eclipse.

Despite the spectral purity of the scanner light curves, it must be admitted that they are based on a relatively small number of data with little night-to-night overlap at most phases outside eclipse. The measures have been studied individually and it was concluded that small-scale night-to-night intrinsic variability does exist but that it is not consistently evident from one light curve to another. Thus, two effects appear enmeshed: sporadic activity at about the level of 0.01 mag and stationary, non-photospherically-distributed flux at a comparable level.

The OAO measures are of varying spectral purity, none so great as the scanner data. These UV observations, taken 6 and 7 years later than the two seasons of scanner data, do not show patterns of residuals consistent with those from the scanner. If there are trends within the UV eclipse data, they are very weak and there are no asymmetries between the levels of maximum light for any UV bandpass.

The light curves of Huffer and Kopal (1951) were sorted into their 1939/1940/1941 and 1947/1948 portions to which were assigned the appropriate values of phase displacement and eclipse asymmetry. These were treated in the manner described above but the errors are too large to add new information to that already gained.

5. ABSORPTION-BIASED LIGHT CURVES

Simultaneously with the observing of the scanner continuum data, the same instrument was used to obtain light curves at 4340A, 3785A, and 3500A, all with a resolution of 50A. These bandpasses contain the H-gamma line, H10 plus H11 lines, and Balmer continuum absorptions, respectively. These measures are discussed here for the first time.

The 4340A light curve, suffering about 8A of H-gamma absorption, is shown in Fig. 4 and 5. It may be seen that observations taken on 1

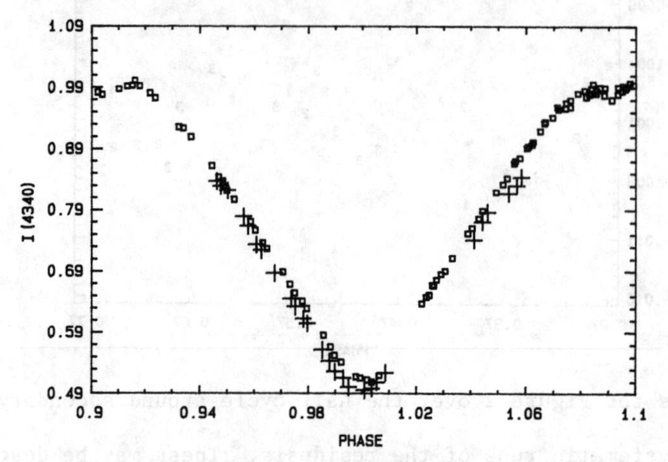

Figure 4. The primary eclipse observations at 4340A. The night of JD2,438,201 is indicated by the plus symbols.

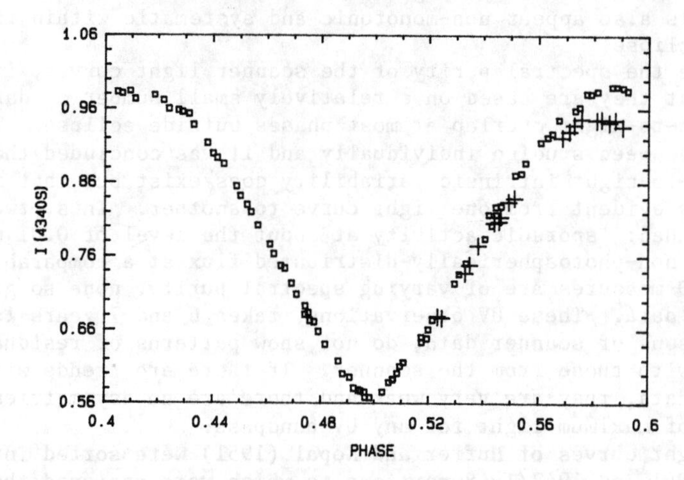

Figure 5. The secondary eclipse observations at 4340A. The night of JD2,438,202 is indicated by the plus symbols.

night fall conspicuously faint with respect to the rest of the primary eclipse. On a second, consecutive night the same effect is sufficiently large to overcome the asymmetry of the secondary eclipse and expunge the evidence for orbital eccentricity. The effect of these data sets is to make each eclipse wider and deeper than ordinarily it is. The observations of the comparison star, when compared to 109 Vir on these nights, are indistinguishable from those of other nights and instrumental calibration was identical for all nights. It is, in fact, impossible for the effect to be due to erroneous location of the scanner window in wavelength for U Oph itself since the effect would be understood to occur in the sense of increased absorption from U Oph and an error in wavelength zero point would bring more continuum within the window and displace line absorption outside it. The effect could, however, be instrumental in the sense that the scanner window might have been erroneously located for the comparison star alone, thus observing more continuum and less line absorption from that star and making magnitude differences too faint. Since the same grating setting was used for both U Oph and the comparison star, this hypothesis cannot be sustained without attributing the same error to the observations of 109 Vir.

The 3785A light curve experiences about 17A of H-line absorption and is even more complicated than the 4340A one. On the one night referred to directly above, the observations are about 0.1 mag fainter than earlier and later measures within the primary eclipse as is shown in Fig. 6. On the second night when data were taken within and following secondary minimum, the observations are faint with respect to the remaining data by more than 0.1 mag.

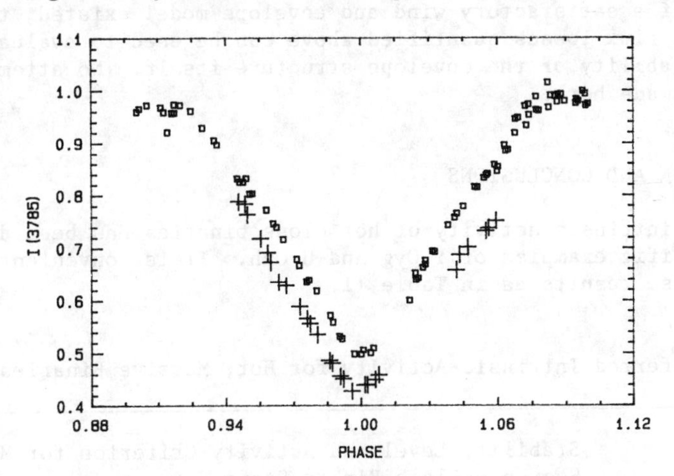

Figure 6. The primary eclipse observations at 3785A. The symbol code is the same as for Figure 4.

Finally, on the same nights the 3500A light curve shows departures from other measures, departures which are intermediate between those of the 4340A and 3785A light curves. Although the noisiest of all the light curves because of the small flux and associated shot noise from the comparison star, this light curve appears to show significant night-to-night transients on other nights as well.

The two faint nights were not included in an average-point 4340A light curve, which was analyzed with the EBOP code. Only a few parameters were treated as variables and their values do not interpolate well within the values developed from the other light curves: the J_s-value is too small and, within its error, the mean darkening is 0.0. The light curve is also noisier than the ones adjoining it in wavelength. The light curves at 3785A and 3500A were not analyzed.

Even though these 3 light curves cannot be analyzed satisfactorily as were those described in Section III, they offer useful information. For 16,000K and log (g) = 4.0, the Kurucz models show that the log of the stellar flux is about 9.64, 9.76, and 9.74 ergs cm^{-2} s^{-1} within the bandpasses of the 4340A, 3785A, and 3500A light curves, respectively. With the same flux units and in the same sense, the non-theoretical light losses on the 2 sensitive nights are about 8.25, 8.78, and 8.35, respectively.

Qualitatively, the light curve features just discussed may be understood to indicate that circumstellar neutral H absorbs and/or scatters more radiation from one star than the other. In fact, the envelope geometry developed from the polarimetric observations can be invoked to explain the photometric anomalies provided it is understood that neutral H (as well as free electrons) exists in the envelope structure. If a satisfactory wind and envelope model existed, the not-insignificant flux losses quantified above can be used to evaluate the sporadic variability of the envelope structure itself. No attempt of this kind is made here.

6. DISCUSSION AND CONCLUSIONS

Evidence for intrinsic activity of hot close binaries has been developed from the specific examples of Y Cyg and U Oph. It is convenient to summarize these results as in Table II.

Table II. Inferred Intrinsic Activity for Hot, Massive Binaries

Observing Technique	Stability Level and Activity Criterion for Main Sequence Close Binary Stars
Visible-band radial velocities	Less than +/- 5 km s^{-1} from measures not at stellar conjunctions
UV spectrophotometry	Average wind flow about 500 km s^{-1}, not known to be variable, for stars of $M > 10_{\odot}$ but not detected for stars of $M < 10_{\odot}$
Visible-band polarimetry	Average level of polarized light at about 0.3% with high-frequency variability at about 0.2% in a peak-to-peak sense; scattered light level > 0.3%

Table II(contd.). Inferred Intrinsic Activity for Hot, Massive Binaries

Observing Technique	Stability Level and Activity Criterion for Main Sequence Close Binary Stars
Visible-band continuum light curves	From simple stellar models systematic residuals of about 0.01 mag phase-locked at selected phases in narrow-band radiometry
UV light curves	Random residuals of about 0.01 mag from simple modelling of intermediate-band radiometry
Visible-band H absorption-biased light curves	Sporadic variability of about 0.1 mag after modelling of narrow-band radiometry

Obviously, it is dangerous to generalize on the basis of a sample size of 2 and particularly so when information for each member of the sample is not the same or, in some cases, not even known. However, the two binaries studied here represent the very best-known unevolved, hot, massive ones in the sky. The prudent course is to search for guidance in effective ways to study others of the same kind.

It is taken as a given that only multiple techniques will yield a complete picture of a binary's activity. There is, however, a qualification on such a statement: if radial velocities are restricted to quadrature phase intervals in order to maximize precision in determining stellar masses, information concerning line asymmetries imposed from the circumstellar environment may remain unknown and a system may be inferred to be simple when it is not. The evidence from the cases studied here and summarized in Table 2 points to the likelihood that well-populated light and polarization curves and UV spectrophotometry do reveal intrinsic activity. This is particularly true if narrow-band observations are available.

The case of U Oph has been particularly informative. The difficult interpretation of the polarimetry led to the concept of a concentration of electron scatterers displaced from the line-of-centers of the binary, which concentration is appropriately located so as to explain the anomalous run of residuals in the continuum-dominated light curves. The scatterers are conceived to work on the light curve so as to scatter from the observer's line of sight about 1% of the light of the hot star as seen when primary eclipse is waning. The same polarimetric data lead to a more distant shell of scatterers embedding the system as a whole. It is presumably in this envelope that one should localize the neutral H which sporadically complicates the other 3 light curves. It may be remembered that neutral H absorption is indeed permitted by the polarization spectrum itself. It should also be remembered that these results emerge in the absence of direct UV evidence for wind effects. These really must exist in order to feed the circumbinary gas but the IUE images give no hint of their existence.

The fundamental weakness in all these conclusions rests in the circumstance that the separate techniques are mostly disjoint in time. The counsel of perfection of collaborative, multiple observational efforts cannot be over-emphasized. Despite this weakness, it has been demonstrated that hot, main sequence binaries are indeed active ones and that it is an unnecessary restriction to apply the term only to cool systems which are well known for their spot and other types of activity.

ACKNOWLEDGEMENTS

It is a pleasure to acknowledge the assistance of R. J. Mitchell, J. M. Gunther, J. C. Maffei, D. Wolinski, and Mrs. S. Csigi. We are grateful to M. F. Corcoran and E. C. Olson who prepared spectra of Y Cyg and U Oph. R.J.P.'s participation in has been supported by a grant from the Distinguished Research Awards Committee of Trenton State College.

REFERENCES

Brown, J. C., McLean, I. S., and Emslie, A. G., 1978. Astr.Ap., **68**, 415.
Cester, B., Fedel, B., Giuricin, G., Mardirossian, F., and Mezzetti, M., 1978. Astr.Ap.Suppl., **33**, 91.
Coyne, G. B., 1970. Spec.Vat.Rich.Astr., **8**, 105.
Eaton, J. A., and Ward, D. H., 1973. Ap.J., **185**, 921.
Giuricin, G., Mardirossian, F., and Mezzetti, M., 1980. Astr.Ap.Suppl., **39**, 255.
Huffer, C. M., and Kopal, Z., 1951. Ap.J., **114**, 297.
Kämper, B.-C., 1986. Ap.Space Sci., **120**, 167.
Koch, R. H., Elias, N. M., Corcoran, M. F., and Holenstein, B. D., 1989. Space Sci.Rev., **50**, 63.
Koch, R. H., and Koegler, C. A., 1977. Ap.J., **214**, 423.
Koch, R. H., and Pfeiffer, R. J., 1989. Publ.A.S.P., **101**, 279.
Kurucz, R. L., 1979. Ap.J.Suppl., **40**, 1.
Magalashvili, N. L., and Kumsishvili, J. J., 1959. Publ.Abastumani Obs., **24**, 13.
Morrison, N. D., and Garmany, C. D., 1984. In Future of Ultraviolet Astronomy Based on Six Years of IUE Research, ed. J. M. Mead, R. D. Chapman, and Y. Kondo (NASA: CP-2349), p. 373.
Olson, E. C., 1975. Ap.J.Suppl., **29**, 43.
Pfeiffer, R. J., and Koch, R. H., 1977. Publ.A.S.P., **89**, 147.
Popper, D. M., 1984. A.J., **89**, 132.
Popper, D. M., and Etzel, P. B., 1981. A.J., **86**, 102.

PERIOD CHANGES AND MAGNETIC CYCLES

Douglas S. Hall
Center of Excellence in Information Systems
Tennessee State University
Nashville, Tennessee 37203
U.S.A.

ABSTRACT. There are periodicities, manifested in a variety of different phenomena in a variety of different types of stars, which fall in the range 10 - 100 years and might all be consequences of solar-type magnetic cycles: the 11-year solar cycle itself; variable Ca II K-line emission in solar-type dwarfs; variable mean brightness in chromospherically active stars, old novas at quiescence, and dwarf novas between outburts; alternating orbital period variations in Algol binaries, W UMa binaries, cataclysmic binaries, and chromospherically active binaries; occurence of flares in flare stars; outburst interval in recurrent novas; changes in outburst interval in dwarf novas; the high-low states in low-mass x-ray binaries like HZ Her; and alternating changes in pulsation period in RR Lyrae and in Cepheid variables. The similarity in time scales suggests a common mechanism. Magnetic cycles have been suggested previously for many of these but not all. For several of them, there never has been an explanation. Encouragingly, cycle lengths predicted by dynamo theory are very similar for most of these cases. Cepheid and RR Lyr variables present a problem: they have no convective envelopes and would not pulsate if they did. At this meeting Mauder suggested turbulence in the outer layers can concentrate flux tubes, other ingredients for a dynamo are present (rotation, differential rotation, and magnetic fields), and organized motion in the radial direction (normally provided by large-scale convection) can be provided by the pulsation itself.

1. Introduction

An interesting phenomenon in astronomy is the occurrence of small but definite changes in rotation periods, pulsation periods, and orbital periods which are otherwise remarkably constant. Some of these changes can be considered understood, at least in broad outline. Many, however, remain completely mysterious.

This paper provides a brief but up-to-date review of these period changes, emphasizing which are understood and which are not. In general it has been difficult to understand alternating period changes (increases

C. İbanoğlu (ed.), Active Close Binaries, 95–119.

followed by decreases) which occur in the same single star or binary system. Most attention will be devoted to presenting the hypothesis that magnetic cycles, which occur on the sun and which have been suspected recently of producing some of the above-mentioned period changes, can explain virtually all of the period changes which have remained a puzzle up to now. This hypothesis is extended to explain a pervasive $\sim$30-year time scale characteristic of many other equally puzzling quasi-periodic phenomena in astronomy.

2. Types of Period Changes

2.1. MONOTONIC PERIOD CHANGES

2.1.1. Mass Transfer. It is now established that the Roche-lobe-filling star in semi-detached binaries is transferring mass to its companion. In the idealized mode of conservative mass transfer, in which the binary's total mass and orbital angular momentum remain constant, the period will increase monotonically if the loser is less massive than the gainer but will decrease monotonically if the loser is more massive than the gainer. A beautiful example of the former is beta Lyrae (Klimek and Kreiner 1973, 1975). A beautiful example of the latter is SV Cen (Rucinski 1976).

The transfer of mass rich in angular momentum onto a mass-gaining star is thought to spin-up the gainer's rotation rate. Stars rotating much more rapidly than synchronously are observed in some Algol-type binaries (Wilson 1989) and in the millisecond pulsars, but we do not observe the rotation period to be variable.

2.1.2. Mass Loss in a Wind or Shell. If one component of a binary system is losing mass approximately isotropically, then the orbital period will change monotonically. Whether the period increases or decreases depends on the mode of mass loss. For example, if the outflowing mass is constrained by magnetic fields to co-rotate with the mass-losing star out to an Alven radius which is sufficiently large, then the orbital period will decrease because a large amount of angular momentum per unit mass is removed from the system (Hall and Kreiner 1980).

Magnetic braking probably is responsible for an increasing rotation period in some single stars. This has been observed clearly in the spin-down of several pulsars but not, to my knowledge, in other types of single stars.

2.1.3. Gravitational Radiation causes a rotating object or orbiting binary system to lose angular momentum. The period will, necessarily, decrease monotonically.

2.1.4. Galactic Acceleration. Because eclipsing binaries are in curved orbits around the center of our galaxy, eclipse timings in principle should show apparent monotonic period changes. These changes will be increases or decreases depending on the location of the binary in our galaxy relative to the sun. Kreiner (1971) searched for evidence of this effect in 137 eclipsing binaries but found it smaller than the uncertainties inherent in his observational material.

2.1.5. Ephemeris Time. If an eclipsing binary has a physically constant orbital period or an RR Lyr variable has a constant pulsation period, and if times of minimum or maximum were recorded in Universal Time rather than Ephemeris Time, then the period should appear to change monotonically as a consequence of the monotonic decrease in the earth's rotation rate. This effect has not, to my knowledge, ever been demonstrated with stellar photometry.

2.2. PERIOD CHANGES IN BOTH DIRECTIONS, STRICTLY PERIODIC

2.2.1. Apsidal Motion. This effect, sometimes called rotation of the line of apsides or precession of the longitude of periastron, causes times of primry minumum to trace out a sinusoidal O-C curve, with an amplitude proportional to the orbital eccentricity. Times of secondary minimum trace out a sinusoidal O-C curve of equal amplitude but exactly 180° out of phase.

The mathematical formulation which determines the period of these sinusoidal O-C curves is often regarded as composed of two sets of terms: classical and relativistic. A recent reference is Guinan (1989). No matter which effect is the dominant one, the resultant apsidal motion period is strictly constant in a given binary system.

2.2.2. Third Body. If an object emitting a periodic light signal is in orbit around another star, then that object will show an O-C curve whose period, amplitude, and shape depend on the period, semi-major axis, and eccentricty of its orbit around the center of mass. The O-C curve will be sinusoidal only if the orbit is circular but the period must remain strictly constant. Objects which emit a periodic light signal would include pulsars, Cepheids, RR Lyr variables, and eclipsing binaries. In the last case, times of both primary and secondary eclipses should trace out identical O-C curves. Complications will, however, arise in a triple system in which the orbit of the close pair is eccentric (Mazeh and Shaham 1979).

2.3. PERIOD CHANGES IN BOTH DIRECTIONS, NOT STRICTLY PERIODIC

For the better part of a century we have been aware of alternating period changes in many eclipsing binaries and also in many pulsating variables of the Cepheid and RR Lyr type. This puzzling phenomenon was first noticed in the first eclipsing binary discovered, namely, Algol itself.

2.3.1. Incorrect Explanations. Throughout most of this century literally hundreds of papers have included proposed explanations for these alternating period changes. Most of them have involved apsidal motion, motion around another star, mass transfer, or mass loss. These explanations have been proven unsatisfactory for several reasons.

In the case of apsidal motion (1) the period changes were not strictly periodic or were not sinusoidal, (2) the eclipsing binary did not have an eccentric orbit, and/or (3) times of primary and secondary minumum did not vary 180° out of phase as required. The previously believed 32-year apsidal motion period in Algol was laid to rest by Soderhjelm (1980) using these argu ents.

In the case of orbit around another body (1) the period changes were not strictly periodic or showed an approximate periodicity different from that of the long-period orbit established by spectroscopic, visual, or astrometric measurements, (2) the inferred mass of the other body was too large to explain why it was invisible in the composite spectrum, (3) the inferred orbit required velocity changes which were not observed, and/or (4) an unreasonably large number of companion stars was required (Abhyankar and Panchatsaram 1984) to explain the observed period changes.

In the case of mass transfer the basic problem is that conservative transfer in a given binary system leads to a monotonic period change, unless the mass flow is supposed to reverse direction, which is unlikely. In the Biermann-Hall model (Biermann-Hall 1973, Hall 1975) the less massive, convective, lobe-filling star was supposed to undergo episodes of sudden mass transfer, on a dynamical time scale, in accordance with something like the Bath instability. The transferred mass, coming over too fast to be accreted immediately by the gainer, would orbit the gainer in a disk, thereby temporarily removing angular momentum from the orbit and causing the orbital period to decrease. Subsequently, on some sort of viscous time scale, that mass in the disk would be accreted and the orbital period would increase. The long-term-average period change would be the same as that in conservative mass transfer theory. This model proved to be wrong because nearly continuous photometric, spectroscopic, and far-ultraviolet measurements simultaneous with observed period decreases showed no evidence of the predicted brief episodes of dramatic mass transfer (Olson 1985).

In the case of mass loss in a wind, the basic problem is that loss from one star in a given binary also leads to a monotonic period change, unless the mode itself (not just the rate) is supposed to undergo substantial cyclical changes, which is difficult to arrange. To explain alternating period changes in the chromospherically active binaries Hall and Kreiner (1980) supposed that mass was lost in an enhanced stellar wind and that it flowed preferentially through the large coronal hole which should overlie the hemisphere opposite the hemisphere covered by large areas of magnetic starspots. Depending on whether the coronal hole was overlying the leading hemisphere or the trailing hemisphere, the orbital period would decrease or increase. As the spotted region migrated in stellar longitude, and the coronal hole with it, the period changes would cycle between increases and decreases. The problem with this so-called rocket model was that the mass loss required to produce the observed period changes was unrealistically large by one or two orders of magnitude, and the predicted periodicity and phasing of the observed period changes were not consistent with the changing starspot longitudes deduced from photometry.

2.3.2. Correct Explanation. The beginning of the idea (which now seems to have been the first one pointed in the right general direction) can be found in a note by Oliver and Rucinski (1978), a paper by Hall and Kreiner (1980), and a comment by Shu (1980). In this picture a change in the radius of one star (or in its radius of gyration) causes a change in rotation period. If tidal torque can reestablish synchronism, then the orbital angular momentum will change and the orbital period with it. Matese and Whitmire (1983) showed, however, that synchronization times computed with the tidal theory of Zahn (1977) are much too long compared to the time scale of the observed orbital period changes. Oliver and Rucinski made their suggestion to explain V471 Tau, Hall and Kreiner considered the idea in connection with the RS CVn binaries, and Shu made his comment in connection with the W UMa binaries.

Soderhjelm (1980) suggested a 32-year magnetic cycle in Algol and speculated that this cyclical activity might modulate the mass transfer is such a way as to produce the observed alternating period changes. His approach, however, veers away from the one discussed above.

Matese and Whitmire (1983) pursued the radius-change idea but argued that the orbital period changes in response to the changing tidal quadrapole force, thereby avoiding the difficulty of requiring a short synchronization time. They considered their model as an explanation for period changes in Algol binaries, RS CVn binaries, and cataclysmic binaries. They were not sure about the nature of the radius changes (homologous? core only? envelope only?) but did mention magnetic activity in the RS CVn binaries and the dynamical instability in the convective envelope (Bath and Pringle 1985) for the cataclysmic binaries.

The next constructive step was taken by VanBuren and Young (1985), who suggested that cyclical changes in one star's magnetic field would, through the action of magnetic pressure, cause cyclical changes in the radius. They proposed this to explain the cyclical period changes in chromospherically active binaries, such as the RS CVn binaries, which are known to display many signatures of extreme magnetic activity (Hall 1976). For this reason they tended to believe the radius changes were restricted to the star's convective envelope. A strong point of their paper was their outline of a method for comparing predictions of their model with quantities observable on a time scale of decades. A flaw in their paper was a factor 10^6 error in their calculation of spin-orbit coupling times using the tidal synchronization theory of Zahn (1977). The relevant coupling times are not around 1 year, as they thought, but rather around 10^6 years, much too long for the 10- to 100-year cyclical period changes which are observed. Apparently they were not aware of the paper by Matese and Whitmire, because they did not cite it.

One year later Van't Veer (1986) suggested a somewhat similar mechanism to explain the alternating period changes in W UMa binaries and RS CVn binaries. He suggested that mass redistribution in the radiative core of one or both stars produced the changes in moment of inertia. Apparently this idea was developed independently, because the paper by Matese and Whitmire was not referenced.

Hall (1987) pointed out the factor 10^6 error of VanBuren and Young but followed with the suggestion that magnetic coupling between the two stars, perhaps with some mass flow between the two stars resulting from enhanced stellar wind, might shorten the coupling time sufficiently. He was not, however, convinced that it could. In his discussion the basic VanBuren-Young mechanism was considered not only for the RS CVn binaries but for the Algol-type binaries as well, because it was realized then that the late-type secondaries in Algol binaries are chromosherically active and in many ways indistinguishable from the late-type components in RS CVn binaries.

Applegate and Patterson (1987) considered in more theoretical detail the problem of time-variable magnetic field, angular momentum, rotation period, and orbital period. They argued that a change in magnetic field would alter the gravitational quadrapole moment of a star and consequently change the orbital period of the binary virtually instantaneously. A scenario like this surely would solve the problem of spin-orbit coupling timescale. In this respect they have duplicated the earlier finding of Matese and Whitmire (1983) but did so apparently independently, without citing the 1983 paper. Applegate and Patterson proposed their results to explain cyclical period changes in V471 Tau; they mentioned the other chromospherically active binaries, the cataclysmic binaries, and the W UMa binaries but not the Algol-type binaries.

Hall (1989) demonstrated that cyclical alternating period changes occur only in semi-detached binaries in which the lobe-filling star is so late in spectral type that it must have a convective envelope. This finding provided support for the basic idea of VanBuren and Young that a magnetic field residing in a star's convective envelope is the fundamental cause.

Bolton (1989) provides a critical, up-to-date review of the magnetic cycle model for explaining orbital period changes. He shows that the newer synchronization theory of Tassoul (1987, 1988) leads to spin-orbit coupling times which are much shorter than those computed with the Zahn tidal synchronization theory. Parameters representative of typical semi-detached Algol-type binaries yielded coupling times as short as a few months, if the lobe-filling star has a convective envelope.

3. QUASI-PERIODICITIES IN OTHER PHENOMENA

In the discussion above it has been shown that cyclical changes in the magnetic field of one star provide the most promising explanation for the heretofore puzzling cyclical alternating orbital period changes observed in chromospherically active binaries, W UMa binaries, Algol-type binaries with convective secondaries, and cataclysmic binaries. Let us extend this idea and propose that similar magnetic cycles can explain quasi-periodicities in other parameters and in other types of stars, single as well as binary. As stated in the introduction, many of these are not otherwise understood. In some, it should be admitted, a connection with magnetic cycles has been suggested already. These suggestions, however, have appeared in a variety of isolated references dealing with quite different areas of astronomy.

3.1. THE SUNSPOT CYCLE

The prototype magnetic cycle is, of course, the 11-year quasi-periodicity with which the number of sunspots waxes and wanes. In phase with this is the phenomenon of latitude drift and the resultant butterfly diagram. And intimately related is the 22-year magnetic cycle.

3.2. CA II EMISSION VARIABILITY

The most directly related stellar counterpart to the solar cycle seems to be the quasi-periodic variability in the strength of emission in the Ca II H and K reversal found in a sample of 99 solar-type dwarfs (Wilson 1978) monitored for almost two decades. Results from this survey have appeared mostly in graphical form and are not easy to analyze quantitatively. We gather from Baliunas (1986) that, in the 99 stars monitored for 18 years, 15% show no variability, 13% show variability with no preferred period, 60% show variability with apparent periods ranging from 2.6 years to about 20 years, and 12% show secular trends in emission strength which suggest possible periods decades in length. Using these percentages along with graphical display (Baliunas 1986, fig. 5) one can estimate that the median cycle length is around 10 years.

Long-term changes in Ca II H and K emission strength, presumably similar to those in the Wilson survey, have been observed in a few chromospherically active binaries and the obvious connection with a solar-type magnetic cycle has been made (Catalano 1989).

3.3. LONG-TERM CHANGES IN MEAN LIGHT LEVEL

The outstanding photometric signature of chromospherically active binaries like those of the RS CVn type is variability with an amplitude of about a tenth of a magnitude in phase with the rotation of the active star. This is produced as a large dark region, covered with starspots, rotates into and out of view. On the other hand, the photometric signature of a magnetic cycle akin to the solar cycle should be long-term (on the order of a decade or decades) changes in the mean brightness level or perhaps the brightness level of the unspotted hemisphere. Reviews of evidence for this sort of cycle can be found in Catalano (1983, table 2), Vogt (1983, table 2), Baliunas and Vaughan (1985), and Hall (1987). Other sorts of long-term photometric cycles may, however, be confused with solar-type magnetic cycles (Hall 1987). One example is the lifetime of a single large starspot region. Such large regions can have lifetimes approaching or sometimes maybe exceeding a decade and may not have a counterpart on the sun. Another example is the apparent long-term cycle which results when two large dark regions, rotating with slightly different periods because they presumably lie at different latitudes, produce a beat period which is typically also around a decade in length.

3.3.1. Chromospherically Active Stars. A list of believable long-term cycles in chromospherically active stars would include the 30-year cycle in CG Cyg (Sowell, Wilson, Hall, Peymann 1987), the 40-year cycle in II Peg (Hartmann et al. 1979), the 50-year cycle in BD +26°730 (Hartmann 1981), the 55-year cycle in both BY Dra and CC Eri (Phillips and Hartmann 1978), the 26-year cycle in AY Peg (Poretti et al. 1986), the 10-year cycle in UX Ari (Wacker and Guinan 1987), the 11.5-year cycle in SS Boo (Wilson et al. 1983), the 10-year cycle in WY Cnc (Sarma 1976), and the 25-year cycle in XY UMa (Geyer 1980).

3.3.2. Dwarfs in Cataclysmic Variables. Virtually all cataclysmic variables are binaries containing a compact object as one component and a convective, lobe-filling star as the other. In most systems, the orbital period is quite short, in which case the convective star is a dwarf. By almost any consideration, such dwarfs should be chromospherically active stars. This was pointed out by Bath and Pringle (1985) but to my knowlegde is not generally appreciated. With the v sini > 5 km/sec criterion first established by Bopp and Fekel (1977), they would definitely qualify. With the small Rossby number criterion, they would qualify as well. The usual signatures of chromospheric activity (photometric variability due to starspots, Ca II H and K emission, strong radio emission, strong coronal x-ray emission, etc.) outlined by Hall (1989) are not apparent because the dwarf star's light is swamped by light from the accretion disk and/or hot spot associated with the compact object. Moreover, the situation is greatly complicated by strong variability and activity in many wavelength regions produced by the accretion disk, the hot spot, the gas stream, and the occasional outbursts themselves. Quite recently, LaDous (1989) described strong far-ultraviolet emission lines in several cataclysmic variables and considered that they may be the familiar signatures of chromospheric activity arising from the dwarf star.

In a very valuable short paper, Bianchini (1988) showed that several old novas and nova-like variables varied in brightness at quiescence with amplitudes around a half magnitude and with periods around a decade. He interpreted this as a consequence of cyclical magnetic activity on the dwarf star, akin to the sun's 11-year cycle. He found cycle lengths of 6.9 years for GK Per, 6.6 years for Q Cyg, 12.6 years for TT Ari, and 3.3 years for V841 Cyg. More recently Mattei (1989) found a cycle of approximately 7 years in the dwarf nova SS Cyg, by analyzing visually estimated magnitudes and looking for changes in the brightness level between outbursts.

3.3.3. The Sun. The sun's bolometric luminosity varies as large sunspots or sunspot groups rotate into and out of view, becoming fainter when spots are seen. Recent results have shown that the sun also varied over the last 11-year cycle, but in the opposite sense. It was bolometrically fainter at the cycle minimum (when fewest spots were present) and brighter at the cycle maximum (when most spots were present). Puzzling though this result is, it suggests long-term changes in mean brightness are to be expected in association with a solar-type magnetic cycle.

3.4. ALTERNATING CHANGES IN ORBITAL PERIOD

This effect has been discussed already, in section 2.3, but let us examine it quantitatively, as it is manifested in the several different types of binaries. The hypothesized magnetic cycle can be estimated from an O-C diagram as the time between one period decrease and the next or twice the time between a period decrease and the next period increase. If only one decrease or increase was observed, then the cycle length is approximately twice the interval of time covered by the observational material.

3.4.1. Algols. Hall (1987) examined the 101 semi-detached binaries listed by Giuricin et al. (1983). As mentioned above, alternating period variations were restricted to those in which the lobe-filling star was convective. For them the statistics were as follow: for 11 (15%) the observational material was too scant to decide, 15 (21%) showed apparently constant periods, 8 (11%) showed monotonically increasing periods, 7 (10%) showed monotonically decreasing periods, and 31 (43%) showed period changes in both directions. The range of cycle lengths found was 7 years to 109 years and the median cycle length was 31 years. The size of a typical period change was $\Delta P/P = 3 \times 10^{-5}$.

3.4.2. Chromospherically Active Binaries. Hall and Kreiner (1980) collected and analyzed all times of minimum available for the eclipsing RS CVn binaries known at that time. Most showed variable periods. The monotonically decreasing periods were interpreted as a result of mass loss in an enhanced stellar wind.

Period changes in both direction were found in six and cycle lengths were estimated: 56 years for RS CVn, 50 years for AR Lac, 60 years for SS Cam, 73 years for SV Cam, 28 years for RT Lac, and 28.5 years for CG Cyg. To this list we can add the 10-year cycle length for V471 Tau (Skillman and Patterson 1988). Note the agreement between CG Cyg's cycle length estimated here and estimated from the mean brightness changes in section 3.3.1. Only in the case of SV Cam can the possibility of a third body not be excluded as an alternative explanation for the period variations (Van Buren 1986).

3.4.3. W UMa Binaries. The most comprehensive treatment of O-C curves for the W UMa binaries is that of Kreiner (1977), although the results were presented only in graphical form. In addition to the 15 listed there, one can add 3 more (RR Cen, RZ Com, CV Cyg) which appeared in Kreiner (1971), again in graphical form. In a few cases, additional references were considered as well: Linnell (1980) for 44 i Boo, Aslan and Herczeg (1984) and Rovithis and Rovithis-Livaniou (1984) for RZ Com, and Rigterink (1972) for U Peg. Note that 44 i Boo is a triple system, as established by astrometric measurements, but additional alternating period changes remain after the effect of the third body is computed and removed from the O-C curve (Linnell 1980).

Period changes, namely, increases or decreases or both, were found in all 18 systems. Estimated cycle lengths were 38.6 years for AB And, 46.8 years for OO Aql, 19.3 years for TY Boo, 86.2 years for 44 i Boo, 53.0 years for RR Cen, 17.8 years for VW Cep, 18.4 years for RZ Com, 67.0 years for CV Cyg, 23.6 years for UX Eri, 26.3 years for AK Her, 28.5 years for SW Lac, 10.7 years for XY Leo, 14.0 years for V502 Oph, 33.6 years for V556 Oph, 22.8 years for ER Ori, 12.5 years for U Peg, 25.3 years for W UMa, and 25.4 years for AH Vir. The median cycle length was 50 years and the size of a typical period change was $\Delta P/P = 3 \times 10^{-5}$.

3.4.4. Cataclysmic Binaries. Alternating orbital period changes have been observed in cataclysmic binaries as well, some more firmly established than others. A list of derived cycle lengths would include 29 years for UX UMa (Nather and Robinson 1974), 5.2 years for RW Tri (Mandel 1965), 30 years or more for both T Aur and EM Cyg (Pringle 1975), and ~10 years for U Gem (Smak 1972). The magnitude of the period change, the average of the first two (best-determined) values, is around $\Delta P/P = 1 \times 10^{-6}$.

3.5. FREQUENCY OF FLARING IN FLARE STARS

Pettersen and Panov (1986) reported an 8.2-year periodicity in the frequency of flaring by the flare star AD Leo. They did suggest that this may be a signature of a solar-type magnetic cycle. According to Van't Veer (1989) at least one other flare star shows a similar cycle.

3.6. FREQUENCY OF OUTBURSTS IN DWARF NOVAS

Studies of long-term photometry of dwarf novas has shown that the time interval between outbursts is variable, with a tendency to cycle between more frequent and less frequent outbursts on a time scale around a decade. Bianchini (1988) analyzed the historical record of outburst times for two dwarf novas, SS Cyg and U Gem, and found periods for this cycle of 7.3 years and 6.9 years, respectively. The record of outburst times for another dwarf nova, SS Aur, presented graphically by Cook (1987), suggests a 16- or 17-year cycle. In the case of SS Cyg, note the close agreement between the 7.3-year cycle given here and the ~7-year cycle found in section 3.3.2 using an entirely different technique. Also, the 6.9-year period given here for U Gem is consistent with the 10-year time scale found in section 3.4.4.

3.7. FREQUENCY OF RECURRENT NOVA OUTBURSTS

It has been argued that the recurrent novas and the classical novas belong to separate groups (Bath and Shaviv 1978). In other words, the recurrent novas are not merely at the high-frequency end of a continuous outburst-frequency spectrum pertaining to all novas in general. Classical novas explode every 10,000 years or so whereas recurrent novas explode every 30 years or so. This 30-year time scale is a complete mystery, according to experts in the field of cataclysmic variables. Wade and Ward (1985) asked "What is the significance of the time interval between eruptions?" They wonder why it is more frequent than in normal novas [10^4 years] and less frequent than in the dwarf novas [months]. And Bath and Pringle (1985) ask "What is the clock...? If we can answer that question we shall at last be making progress."

A list of known recurrent novas would include T CrB, RS Oph, T Pyx, V1017 Oph, U Sco, WZ Sge, and V616 Mon. Years of outburst for these can be found in Hoffmeister, Richter, and Wenzel (1985, Table 30) and in Wade and Ward (1985, table 4.4.2). A histogram of the interval between outbursts has a median of 30 years.

Similarity of orbital period is definitely not a common characteristic of these recurrent novas. Although the orbital period is not known for some, the range is from 82 minutes to 227 days. The compact object's companion star has not been directly observed in every case, but is probably a late-type star (dwarf, subgiant, or giant) filling its Roche lobe in every case.

3.8. THE HIGH-LOW STATES OF HER X-1 = HZ HER

The low-mass x-ray binary Her X-1 is known to cycle between states of high and low x-ray luminosity. This is altogether different from the 35-day on-off cycle, thought to be related to precession of the accretion disk around the neutron star. As the x-ray luminosity emanating from the neutron star or accretion disk varies, the radiant energy impinging on the facing hemisphere of the companion star varies, and hence the optical luminosity of the optical counterpart HZ Her varies as well. Analysis of the archival photograph collection at Harvard University by Jones, Forman, and Liller (1973) yielded information about the time scale of this high-low phenomenon. They said, "Between 1914 and 1957 there does appear to be a tendency for active states to occur every 10 to 12 years." Concerning a possible explanation, they said "An attractive and physically realistic mechanism that can produce such a behavior is an irregular or semi-regular pulsation of the larger, optical star. When the star is in an expanded state, it completely or overfills its Roche lobe and feeds matter into the lobe surrounding the neutron star companion. At other times when the star contracts to a size smaller than its Roche lobe, no mass exchange occurs, and the system enters the inactive state." They did not suggest why the optical companion should vary in radius on that time scale.

Priedhorsky (1986) has reported quasi-periodicities around 1 year in the x-ray luminosity from other low-mass x-ray binaries and suspected a connection with solar-type cycles on the secondary star. These are much shorter than the "10 to 12 years" in HZ Her and may not be the same phenomenon. In any event, the x-ray history of the sources in Priedhorsky's sample was less than 10 years and they are too faint to have been studied with the century-long Harvard photographic archive.

3.9. VARIABLE PULSATION PERIODS IN RR LYRAE VARIABLES

For a long time we have known that these stars have variable pulsation periods. Typically they show alternating changes, for that reason almost surely not related to evolution. Studies of variables in globular clusters show (Goranskij, Kukarkin, Samus 1973) that a large percentage, as large as 93% in one cluster, have variable periods. They also say that period variations occur in all of the RR Lyr variables showing the so-called Blazhko effect, which has been discussed by Szeidl (1976). The percentage of RR Lyrae stars known to show the Blazhko effect is as large as 50% in certain ranges of pulsation period (Szeidl 1976).

Detre and Szeidl (1973) extended the original phenomenological model of Balazs-Detre (1959) to explain the large number of multi-periodic photometric phenomena now known to be associated with the Blazhko effect. The model now involves spots on a rotating star with a magnetic field whose axis is inclined with respect to the rotation axis and which varies in strength with time due to a solar-type cycle. Stothers (1980) provided some theoretical underpinning for the model, the essence of which was a magnetic field, generated by a dynamo, which varied with a solar-type cycle. Such a cycle could account for the variable pulsation period and also for the magnetic field observed in RR Lyrae by Babcock (1958) but not observed by Preston (1967) in 1963 and 1964.

A list of RR Lyrae variables showing variable periods was compiled from papers by Tsesevich (1972), Szeidl (1975, 1976), Olah and Szeidl (1978), and Szeidl, Olah, and Mizser (1986). The 21 variables included AT And, SU Dra, RR Leo, TT Lyn, AR Per, VZ Her, AV Peg, TU UMa, XZ Cyg, RV UMa, RW Dra, RR Lyr, RR Leo, RR Gem, AG Her, SZ Hya, UY Boo, RV CrB, AQ Lyr, V759 Cyg, and SS Tau. In some cases the cycle length, describing the oscillation between shorter-than-average pulsation periods and longer-than-average, was taken from the reference. In other cases cycle length was derived from the O-C curve of times of maximum light with the same technique described in section 3.4. The median cycle length was 55 years and the median period change was $\Delta P/P = 2 \times 10^{-5}$.

3.10. VARIABLE PULSATION PERIODS IN CEPHEID VARIABLES

Also for a long time it has been known that these stars do not have constant pulsation periods. The three-part series of papers on "Photoelectric UBV Photometry of Northern Cepheids" by Szabados (1977, 1980, 1981) is very useful here because period variability or constancy was not a selection criterion for the 106 Cepheids in those three papers. Virtually all types of period changes are observed: monotonic and alternating, gradual and abrupt, increasing and decreasing.

In the case of Cepheid variables one would expect pulsation period to be varying monotonically in response to evolution across the Hertzsprung gap, increasing for those moving redward, decreasing for those moving blueward. Szabados (1983) has shown that, if O-C curves are fit with only a quadratic term to describe a monotonic period variation, observation agrees with theoretical expectation quite well. Those with longer pulsation periods show more rapid (monotonic) period variations, as would be predicted, since the period-luminosity variation tells us the slow pulsators are more luminous and should be evolving more rapidly.

A number of papers have suggested that Cepheids belong to binary systems, in an attempt to use the light-travel-time effect to explain those showing alternating changes in pulsation period. Even though binary Cepheids surely exist, this is not likely the correct approach. (1) Most of the O-C curves which show more than one complete cycle of period variations do not have equal amplitudes, shapes, or periods for successive cycles, as would be required by Keplerian motion. (2) Several Cepheids with demonstrable binary companions show variable O-C curves even after the light-travel-time effect is computed and removed. (3) The observed alternating period variations affect about half of the above-mentioned sample and occur with a relatively narrow range of periods; in the binary idea, about half of a sample might be affected but the range of orbital periods should be quite broad.

From the O-C curves of times of maximum given in the above-mentioned three papers, cycle lengths were determined, again with the same technique discussed in section 3.4. In cases where the O-C curve showed a monotonic period increase or decrease during the interval of observation, it had to be decided whether we were seeing (A) the evolutionary effect or (B) simply an incomplete cycle of the alternating period changes. An O-C curve showing an abrupt change, with the period constant before and after, was taken as possibility B. Sometimes it was difficult to decide whether a relatively small monotonic period change was continuous or abrupt. To help decide, the relation between magnitude of period change versus pulsation period, expected from evolutionary effects, was examined. Those consistent with that relation were taken as possibility A, those not consistent were taken as possibility B. Whenever the existence of a binary companion was certain and the consequent light-travel-time effect could be computed and removed, that was done first.

The result, from consideration of 106 Cepheids, was that 6 had insufficient data on which to make any decision, 42 showed periods constant within the uncertainties of the observational material, 20 showed only continuous period increases or decreases consistent with evolutionary effects, and 38 showed alternating period changes, in some cases on top of the evolutionary change and in other cases exclusively. The median cycle length for the alternating period changes was 50 years. The median period change for those 38 was $\Delta P/P = 9 \times 10^{-5}$.

4. SUMMARY OF OBSERVED PERIODICITIES

Figures 1 through 6 are histograms of cycle lengths for the various phenomena in the various types of stars discussed above. Figure 1 is for

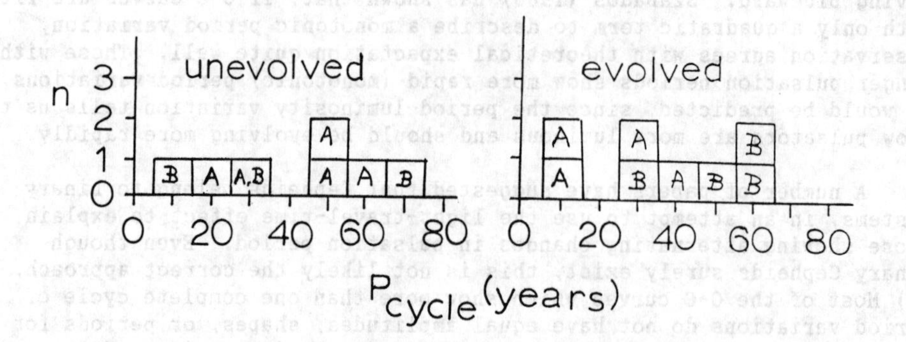

Figure 1. Histogram of cycle lengths in chromospherically active binaries derived from (A) changes in mean luminosity, and (B) changes in orbital period.

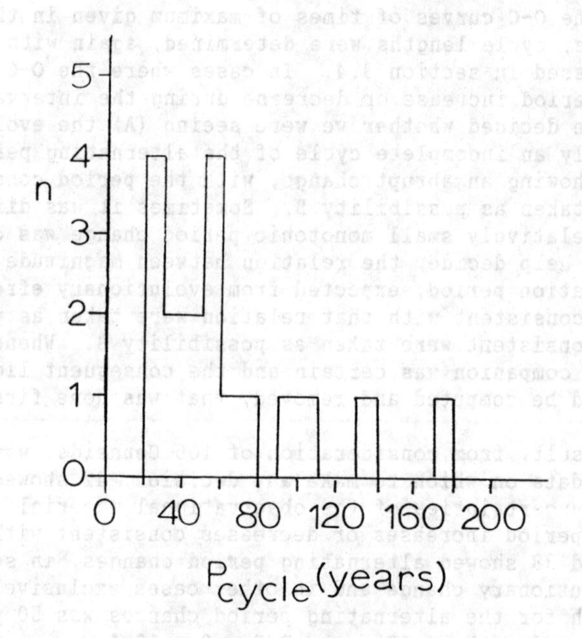

Figure 2. Histogram of cycle lengths in W UMa binaries, derived from changes in orbital period.

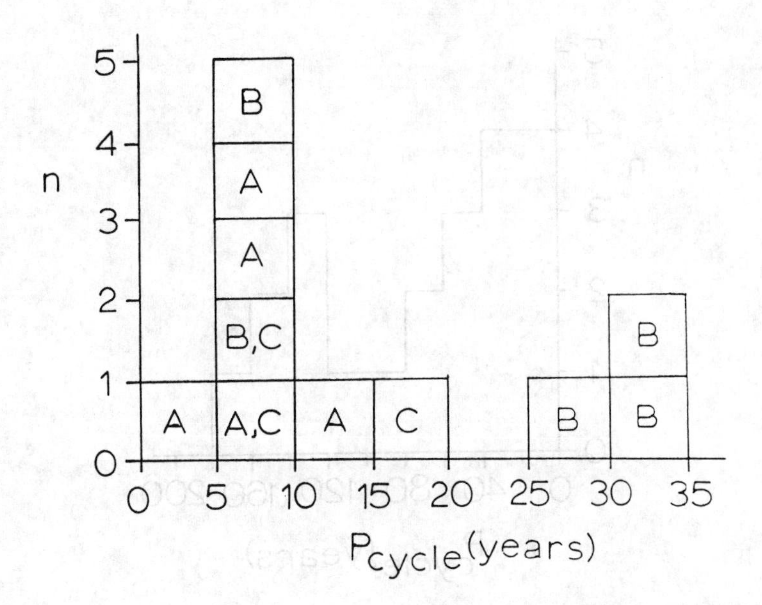

Figure 3. Histogram of cycle lengths in cataclysmic variables, derived from (A) changes in luminosity at quiescence, (B) changes in orbital period, and (C) changes in outburst frequency.

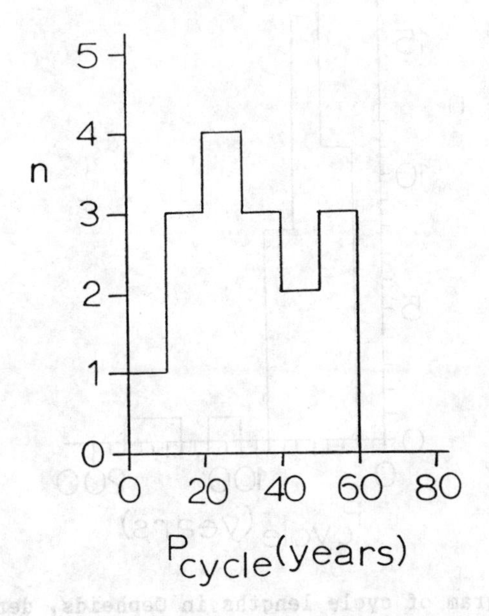

Figure 4. Histogram of interval between outbursts of recurrent novas.

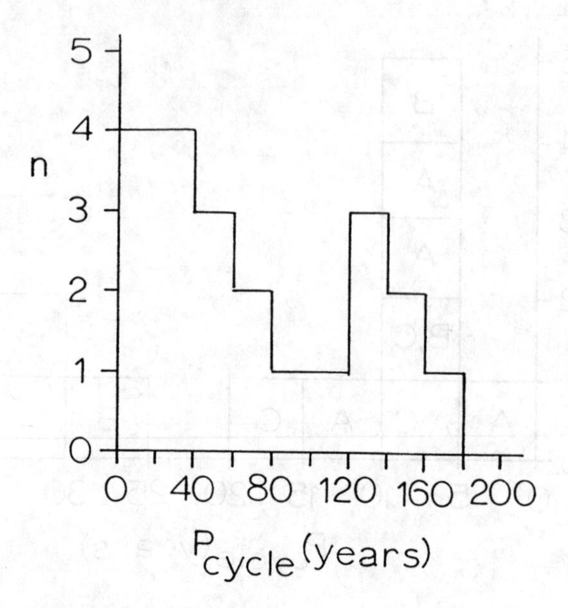

Figure 5. Histogram of cycle lengths in RR Lyrae variables, derived from changes in pulsation period.

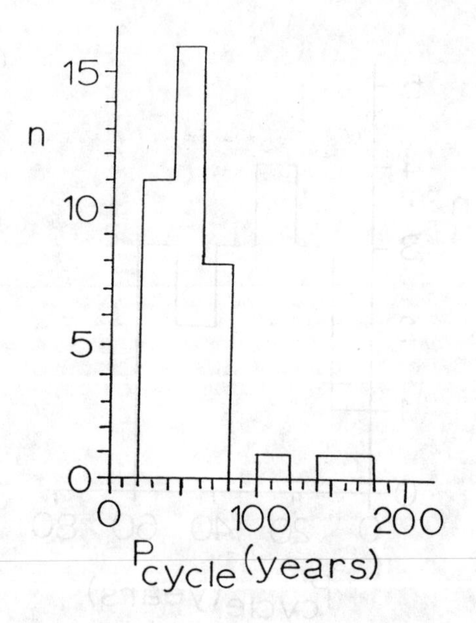

Figure 6. Histogram of cycle lengths in Cepheids, derived from changes in pulsation period.

chromospherically active binaries of the RS CVn and BY Dra type with V471 Tau included as well. Cycle lengths based on mean luminosity variations (section 3.3) and orbital period changes (section 3.4) are included together but identified accordingly. Separate histograms are provided for the unevolved stars (dwarfs) and for the evolved stars (subgiants and giants). Figure 2 is for the W UMa binaries (section 3.4.3). Figure 3 is for dwarf novas, old novas, and nova-like variables. Cycle lengths based on mean luminosity variations (section 3.3.2), orbital period variations (section 3.4.4.), and changes in outburst frequency (section 3.6), are included together but identified accordingly. Figure 4 is for the recurrent novas (section 3.7). Altogether 23 outbursts from the 7 different recurrent novas are represented. Figure 5 is for the 21 RR Lyrae variables discussed in section 3.9. And Figure 6 is for the 38 Cepheids discussed in section 3.10. A histogram for the Algol binaries appeared in Hall (1989, figure 5) and is not repeated. For the solar-type dwarfs discussed in section 3.2, it was not possible to construct a histogram. There was only one object in each of the remaining groups : the sun, the flare star AD Her, and the low-mass x-ray binary HZ Her.

For each group Table 1 summarizes the median cycle length and, in the case of orbital or pulsational period variations, the characteristic size of the change. Note the remarkably narrow range covered by the cycle lengths in Table 1; the shortest and the longest differ by less than one order of magnitude. The narrowness of this range, that fact alone, would suggest a common mechanism is at work.

5. PLAUSIBILITY OF MAGNETIC CYCLES

Is it physically reasonable that a solar-type magnetic cycle alone can be the basic agent responsible for cyclic variability in such a diverse group of phenemona (eclipse times, flares, x-rays, pulsations, cataclysmic explosions, etc.) in such a variety of objects (dwarfs to giants, single stars and binaries)?

5.1. WHICH STARS COULD HAVE MAGNETIC CYCLES ?

Most attempts to understand the sun's magnetic cycle and its various manifestations involve some sort of dynamo. To try and simplify a complex situation, one might say the essential ingredients for a dynamo are convection and rotation. The wide variety of (mostly surface) phenomena is usually termed "solar activity". Other stars similar to the sun show similar activity. Some stars show solar-type activity to a much stronger degree. These are usuaslly termed "chromospherically active", though that term is a bit misleading because the activity is not restricted to the chromophere and because one must understand that "active" implies "very active" compared to the sun. The essential ingredients for extreme solar-type activity seem to be a deep convection zone (deeper than the sun's) and rapid rotation (more rapid than the sun's). Bopp and Fekel (1977) pointed out that equatorial rotational velocity is useful as a parameter to quantify this activity, with 5 km/sec being the threshhold.

Table 1

Median Cycle Lengths

Star	P_{cycle}	$\Delta P/P$
Algols	31 years	3×10^{-5}
Chromospherically cctive binaries		3×10^{-5}
main sequence	50	
evolved	35	-
W UMa binaries	50	3×10^{-5}
Solar-type dwarfs	10	-
Flare stars	‹8›	-
Cataclysmic variables	7	1×10^{-6}
Recurrent novas	30	-
Low-mass x-ray binaries	‹12›	-
Cepheids	50	9×10^{-5}
RR Lyrae variables	55	2×10^{-5}
Sun	‹11›	-

Table 2

Calculated Cycle Lengths

Star	τ_c	d'	l'	P	P'
dwarf	116 days	0.5	0.088	30.8 yrs	44 yrs
sun	25	0.25	0.041	7.6	11
subgiant	92	0.5	0.050	75.6	109
giant	7.3	0.9	0.059	14.0	20

More recently others have suggested using Rossby number as a physically relevant parameter, with a Rossby number around unity corresponding to the sun and smaller Rossby numbers corresponding to stronger activity. Another choice, related to Rossby number but used less commonly, has been dynamo number. See the review by Hartmann and Noyes (1987).

Of the stars mentioned in this paper, there is little doubt that the solar-type dwarfs can reasonably be expected to display solar-type activity, including solar-type magnetic cycles.

The same is true for those which belong to one of the ten groups of stars considered chromospherically active (Hall 1987). That takes in the RS CVn binaries, the BY Dra variables, the flare stars, the W UMa binaries, V471 Tau, and the Algol binaries with convective secondaries.

It has already been argued (in section 3.3.2.) that the secondaries in dwarf novas and old novas must be chromospherically active as well, even though they were not among the ten groups of Hall (1987).

In those recurrent novas for which the spectral type of the secondary has been observed spectroscopically or deduced from multi-color photometry, the secondary is convective. Probably all are. Consideration of their equatorial rotational velocity or Rossby number would make most, probably all, qualify as chromospherically active.

It is not obvious that the lobe-filling star in HZ Her should have a solar-type magnetic cycle. Its mass is 2.0 $M_{\odot}$ (Gursky and Schreier 1975) and its radius, if it exactly fills its Roche lobe, is 3.8 $R_{\odot}$. According to Gilliland (1985), a 2.0 $M_{\odot}$ main-sequence star has a thin, inefficient convective outer layer. The 3.8 $R_{\odot}$ radius is, however, larger than that of a 2.0 $M_{\odot}$ main-sequence star, so it is significantly evolved and would have deeper convection. Straight-forward inferences are difficult to make, because it fills its Roche lobe and has undergone appreciable mass loss during its lifetime.

At first glance one would <u>not</u> expect solar-type magnetic cycles in Cepheids, because they are too early in spectral type to have a convective outer layer. Absence of convection seems quite certain because the presence of convection in the outer envelope can quench the pulsation (Cox 1985). In fact, it is thought that the onset of convection in the outer envelope is what determines the red edge of the pulsational instability strip (Cox 1985). Thus it is a serious question whether the period changes can be a consequence of a magnetic cycle, as has been hypothesized.

In the discussion following the presentation of this paper, Mauder (1989) suggested how this might be possible afterall. Turbulence in the outer envelope, as discussed by Kähler (1989), could provide the initial small-scale twisting of magnetic flux tubes. Though Kähler was concerned primarily with the problem of common envelopes in contact binaries, his results on turbulence are applicable to the radiative envelopes in single stars as well. Rotation and differential rotation can reasonably be presumed present, although neither is readily observed in Cepheids. And the radial pulsation in the outer layers of the star can provide the vertical motion which is provided by the turn-over of large-scale convective cells in convective stars. Consistent with all of this is the direct observation of magnetic fields in Cepheids (Stothers 1979). This may, upon closer examination, prove to be a valid explanation. If not, then we are left with no explanation for the period changes in Cepheids, except for the earlier idea of semi-convection (convective overshoot from the convective core) proposed by Sweigert and Renzini (1979).

Although Stothers (1980) felt that solar-type magnetic cycles were possible in the RR Lyrae variables, he relied on them to have convection in their outer envelopes. As best I can understand the situation, the same argument against convection in Cepheids would rule out convection in RR Lyrae variables as well. But the Mauder mechanism, suggested for the Cepheids, would work just as well for the RR Lyrae stars.

5.2. HOW MAGNETIC CYCLES COULD PRODUCE THE PHENOMENA

It has already been explained in section 2.3.2 how magnetic cycles can act to produce alternating orbital period changes. Bolton (1989) cautions that one should not expect to find clear or simple correlations between these period changes and various other indicators of magnetic activity.

Variability in Ca II H and K emission strength and changes in flaring frequency can reasonably be explained by analogy with the solar behavior. Changes in mean light level probably would be explained in the same way that changes in the sun's mean luminosity are explained, even though that itself has not yet been explained.

The high-low states of HZ Her and also the time interval between recurrent nova outbursts depend on a lobe-filling star to vary slightly in radius throughout the course of a magnetic cycle, in response to changing magnetic pressure, thereby causing overfilling and underfilling of the Roche lobe.

It is not obvious how a magnetic cycle modulates the dwarf nova outburst frequencies. The interval between outbursts (months) is much shorter than the time scale which modulates that interval (decades) and must involve a different mechanism. Probably the outbursts themselves are to be understood in terms of a dynamical instability in the outer convective envelope, with the time interval between outbursts being the thermal relaxation time in the convective envelope, which is on the order of months (Bath and Pringle 1985). No suggestion has been made, however, how a magnetic cycle would act to modulate the thermal relaxation time.

A cyclical change in the pulsation period of a Cepheid or RR Lyrae variable results naturally from the cyclical radius change. A simple way to understand this is to consider the period vs. mean density relation for pulsating variables.

6. PREDICTION FROM DYNAMO THEORY

For several years dynamo theory has been used to account for the sun's 11-year spot cycle or 22-year magnetic cycle. More recently it has been used to account for similar magnetic cycles in similar stars (Hartmann and Noyes 1987). One way to check this for plausibility is to ask if the narrow range of cycle lengths found in section 4 is predicted by dynamo theory for so many different types of stars.

An explicit formula for the frequency of a magnetic cycle is given by Stix (1976). Converting his frequency Ω to cycle length P and dropping a factor 2 so we can compare with the sun's 11-year spot cycle rather than the 22-year magnetic cycle, we get

$$P = d^2/\eta_t , \tag{1}$$

where P is the time the magnetic field needs to diffuse over a characteristic distance d in latitude or longitude and η_t is the turbulent magnetic diffusivity. In his example Stix used the depth of the sun's convective zone for d, so we do the same for our other stars. For η_t he used v l / 3, where v is the velocity of the convective cells and l is the size of the convective cells. For simplicity we will take l to be the pressure scale height H at the base of the convective zone, thus assuming the mixing length equals the pressure scale height. This corresponds to assuming $\alpha = 1$, but that is not important because we will be comparing relative cycle lengths. Because the convective turnover time is defined as

$$\tau_c = H/v , \tag{2}$$

it follows that

$$\eta_t = l^2/3\tau_c . \tag{3}$$

Let us also introduce

$$d' = d/R \tag{4}$$

and

$$l' = l/R , \tag{5}$$

where R is the star's total radius. Then, entering equations 3, 4, and 5 into equation 1 gives

$$P = 3\,\tau_c\,(d'/l')^2 \tag{6}$$

for the length of a magnetic cycle in any convective star.

Gilliland (1985, figure 10) gives convective turnover times τ_c for stars of various mass and evolutionary state. We estimated these from his figure for four types of stars: a 0.5 $M_\odot$ main-sequence star, the sun, a subgiant at the base of the giant branch, and a giant. In doing so we did not rely on his effective temperature scale, which Basri (1986) has noted is in error, but used the fact that trajectories for stars of various masses converge to the same point ($\log \tau_c = 6.9$) at the base of the giant branch, and the fact that the last point calculated for a star of 1.6 $M_\odot$ ($\log \tau_c = 5.8$) is representative of a typical $\sim 10\ R_\odot$ giant. For d' we used realistic estimates for stars at those stages of evolution. For l' we looked at Schwarzschild (1958, tables 28.4, 28.6, and 28.7) and read directly the fractional radius over which the pressure drops by the factor e, starting at the base of the convective zone. More up to date stellar models were not easy to locate in the literature, but would probably give similar values for l'.

These estimates of τ_c, d', and l' are gathered in Table 2, along with the value of P computed with equation 6. Note that the magnetic cycle length given for the sun is shorter than 11 years, a fact well known to solar dynamo theorists (Stix 1976). Following their practice of applying a scale factor to remove the discrepancy, one gets the adjusted cycle lengths P' in the last column of Table 2.

One cannot rely on the absolute values given for P', because $\alpha = 1$ was assumed and because of the arbitrary factor which was applied , but one can look at their relative values. Note that they cover a range of only one order of magnitude, essentially equal to the range observed in Table 1. This argues that magnetic cycles produced by dynamo theory can plausibly be suggested to account for the variety of cycle lengths which were summarized in Table 1.

The four types of stars considered in Table 2 reasonably represent the sun, the solar-type dwarfs, the flare stars, and the convective components in all chrompspherically active binaries, which are all either dwarfs or subgiants or giants. They probably also represent the dwarfs in W UMa binaries, the subgiants and giants in Algol binaries, and the secondaries in the recurrent novas, if filling of the Roche lobe is not a serious complication. It was not possible to apply this plausibility test to HZ Her, the supergiant Cepheids, or the highly evolved RR Lyr variables, because the quantities needed (τ_c or d' or l') were difficult to estimate. Note, however, that Cepheid pulsation periods (10 to 100 days) are included in the range covered by the convective turnover times in Table 2 (7 to 116 days).

7. CONCLUSION

There is a pervasive, remarkably narrow (10- to 100-year) time scale occurring in many different types of phenomena in many different types of stars, both single and binary, evolved and unevolved. A solar-type magnetic cycle is proposed to explain them. For most of them, there is no other plausible explanation. The magnetic cycle explanation is very likely for most of them, and not implausible for any. For a few systems (SS Cyg, CG Cyg, and U Gem) the same cycle length is manifested as two different phenomena. The magnetic cycle explanation represents the most promising direction and should be pursued, both observationally and theoretically. If the explanation proves to be correct, then we can immediately draw on the century-long or longer historical record of mean luminosity changes (the Harvard photographic archive), outburst times (recurrent novas and dwarf novas), and period changes (times minimum for eclipsing binaries and times of maximum for Cepheid and RR Lyr variables) to provide good statistics for magnetic cycle activity as it occurs in a variety of stars other than the sun.

ACKNOWLEDGEMENTS

I am very grateful to Katalin Olah for sending me many useful references on pulsating variables (a new area for me), to Franz Van't Veer for drawing my attenion to a couple of important references I had overlooked, and to Horst Mauder for generously sharing his valuable ideas during the discussion period, This work was supported in part by N.A.S.A. research grant NAG 8-111, while I am on leave from Dyer Observatory, Vanderbilt University, Nashville, Tennessee 37235, U.S.A.

REFERENCES

Abhyankar, K.D. and Panchatsaram, T. 1984, M.N. **211**, 75.
Applegate, J.H. and Patterson, J. 1987, Ap.J. **322**, L99.
Aslan, Z. and Herczeg, T.J. 1984, I.B.V.S. No. 2478.
Babcock, H.W. 1958, Ap.J. Suppl. **3**, 141.
Balazs-Detre, J. 1959, Kleine Veröff. Remeis-Sternw. Bamberg No. 27, 26.
Baliunas, S.L. 1986, in Cool Stars, Stellar Systems, and the Sun, edited by M. Zeilik and D.M. Gibson (Berlin: Springer-Verlag), p. 3.
Baliunas, S.L. and Vaughan, A.H. 1985, Ann. Rev. Astr. Astrophys. **23**,379.
Basri, G. 1986, in Cool Stars, Stellar Systems, and the Sun, edited by M. Zeilik and D.M. Gibson (Berlin: Springer-Verlag), p. 184.
Bath, G.T. and Pringle, J.E. 1985, in Interacting Binary Stars, edited by J.E. Pringle and R.A. Wade (Cambridge: Cambridge Univ. Press), p. 177.
Bath, G.T. and Shaviv, G. 1978, M.N. **183**, 515.
Bianchini, A. 1988, I.B.V.S. No. 3136.
Biermann, P. and Hall, D.S. 1973, Astr. Astrophys. **27**, 249.
Bolton, C.T. 1989, I.A.U. Colloq. No. 107, 311.
Bopp, B.W. and Fekel, F.C. 1977, A.J. **82**, 490.
Catalano, S. 1983, I.A.U. Colloq. No. 71, 343.
Catalano, S. 1989, this volume.
Cook, L.M. 1987, J.A.A.V.S.O. **16**, 11.
Cox, J.P. 1985, I.A.U. Colloq. No. 82, 126.
Detre, L. and Szeidl, B. 1973, I.A.U. No. 21, 31.
Geyer, E.H. 1980, I.A.U. Symp. **88**, 423.

Gilliland, R. 1985, Ap.J. **299**, 286.
Giuricin, G., Mardirossian, F., Mezzetti, M. 1983, Ap.J. Suppl. **52**, 35.
Goranskij, V.P., Kukarkin, B.V., and Samus, N.N. 1973, I.A.U. Colloq. No. 21, 100.
Guinan, E.F. 1989, this volume.
Gursky, H. and Schreier, E. 1975, in Neutron Stars, Black Holes, and Binary X-Ray Systems, edited by H. Gursky and R. Ruffini (Dordrecht: Reidel), p. 175.
Hall, D.S. 1975, Acta Astr. **25**, 1.
Hall, D.S. 1976, I.A.U. Colloq. No. 29, 287.
Hall, D.S. 1987, Publ. Astr. Inst. Czechoslovakia **70**, 77.
Hall, D.S. 1989, I.A.U. Colloq. No. 107, 219.
Hall, D.S. and Kreiner, J.M. 1980, Acta Astr. **30**, 387.
Hartmann, L. 1981, in Solar Phenomena in Stars and Stellar Systems, edited by R.M. Bonnet and A.K. Dupree (Dordrecht: Reidel), p. 487.
Hartmann, L., Londono, C., and Phillips, M.J. 1979, Ap.J. **229**, 183.
Hartmann, L.W. and Noyes, R.W. 1987, Ann. Rev. Astr. Astrophys. **25**, 271.
Hoffmeister, C., Richter, G., and Wenzel, W. 1985, Variable Stars (Berlin: Springer-Verlag).
Jones, C., Forman, W., and Liller, W. 1973, Ap.J. **182**, L109.
Kähler, H. 1989, Astr. Astrophys. **209**, 67.
Klimek, Z. and Kreiner, J.M. 1973, Acta Astr. **23**, 331.
Klimek, Z. and Kreiner, J.M. 1975, Acta Astr. **25**, 29.
Kreiner, J.M. 1971, Acta Astr. **21**, 365.
Kreiner, J.M. 1977, I.A.U. Colloq. No. 42, 373.
LaDous, C. 1989, this volume.
Linnell, A.P. 1980, P.A.S.P. **92**, 202.
Mandel, O.E. 1965, Peremmenye Zvezdy **15**, 474.
Matese, J.J. and Whitmire, D.P. 1983, Astr. Astrophys. **117**, L7.
Mattei, J.A. 1989, this volume.
Mauder, H. 1989, private communication.
Mazeh, T. and Shaham, J. 1979, Astr. Astrophys. **77**, 145.
Nather, R.E. and Robinson, E.L. 1974, Ap.J. **190**, 637.
Olah, K. and Szeidl, B. 1978. Konkoly Obsv. Comm. No. 71.
Oliver, J.P. and Rucinski, S.M. 1978, I.B.V.S. No. 1444.
Olson, E.C. 1985, in Interacting Binaries, edited by P.P. Eggleton and J.E. Pringle (Dordrecht: Reidel), p. 127.
Pettersen, B.R. and Panov, K.P. 1986, in Cool Stars, Stellar Systems, and the Sun, edited by M. Zeilik and D.M. Gibson (Berlin: Springer-Verlag), p. 91.
Phillips, M.J. and Hartmann, L. 1978, Ap.J. **224**, 182.
Poretti, E. et al. [18 authors] 1986, Astr. Astrophys. **157**, 1.
Preston, G.W. 1967, in The Magnetic and Related Stars, edited by R.C. Cameron (Baltimore: Mono Book), p. 26.
Priedhorsky, W. 1986, Astrophys. Space Sci. **126**, 89.
Pringle, J.E. 1975, M.N. **170**, 633.
Rigterink, P.V. 1972, A.J. **77**, 319.
Rovithis, P. and Rovithis-Livaniou, E. 1984, Astr. Astrophys. Suppl. **58**, 679.
Rucinski, S.M. 1976, P.A.S.P. **88**, 244.
Sarma, M.B.K. 1976, Bull. Astr. Inst. Czechoslovakia **27**, 335.

Schwarzschild, M. 1958, Structure and Evolution of The Stars (Princeton: Princeton Univ. Press).
Shu, F.H. 1980, I.A.U. Symp. **88**, 526.
Skillman, D.R. and Patterson, J. 1988, A.J. **96**, 976.
Smak, J. 1972, Acta Astr. **22**, 1.
Soderhjelm, S. 1980, Astr. Astrophys. **89**, 100.
Sowell, J.R., Wilson, J.W., Hall, D.S., and Peymann, P.E. 1987, P.A.S.P. **99**, 407.
Stix, M. 1976, I.A.U. Symp. **71**, 367.
Stothers, R. 1979, Ap.J. **234**, 257.
Stothers, R. 1980, P.A.S.P. **92**, 475.
Sweigert, A.V. and Renzini, A. 1979, Astr. Astrophys. **71**, 66.
Szabados, L. 1977, Konkoly Obsv. Comm. No. 70.
Szabados, L. 1980, Konkoly Obsv. Comm. No. 76.
Szabados, L. 1981, Konkoly Obsv. Comm. No. 77.
Szabados, L. 1983, Astrophys. Space Sci. **96**, 185.
Szeidl, B. 1975, I.A.U. Symp. **67**, 545.
Szeidl, B. 1976, I.A.U. Colloq. No. 29, 133.
Szeidl, B., Olah, K., and Mizser, A. 1986, Konkoly Obsv. Comm. No. 89.
Tassoul, J.L. 1987, Ap.J. **322**, 856.
Tassoul, J.L. 1988, Ap.J. **324**, L71.
Tsesevich, V. 1972, Vistas in Astronomy **13**, 241.
VanBuren, D. 1986, A.J. **92**, 136.
VanBuren, D. and Young, A. 1985, Ap.J. **295**, L39.
Van't Veer, F. 1986, Astr, Astrophys. **156**, 181.
Van't Veer, F. 1989, private communication.
Vogt, S.S. 1983, I.A.U. Colloq. No. 71, 137.
Wacker, S. and Guinan, E.F. 1987, I.B.V.S. No. 3018.
Wade, R.A. and Ward, M.J. 1985 in Interacting Binary Stars, edited by J.E. Pringle and M.J. Wade (Dordrecht: Reidel), p. 129.
Wilson, J.W., Hall, D.S., Henry, G.W., Vaucher, C.A., and Africano, J.L. 1983, A.J. **88**, 1257.
Wilson, O. C. 1978, Ap.J. **226**, 379.
Wilson, R.E. 1989, I.A.U. Colloq. No. 107, 235.
Zahn, J.P. 1977, Astr. Astrophys. **57**, 383.

ABSOLUTE DIMENSIONS OF CLASSICAL ALGOL-TYPE BINARIES

ALVARO GIMENEZ [1] and J. MARIO GARCIA [2]
[1] Instituto de Astrofísica de Andalucía.
Apartado 2144, 18080 Granada. Spain.
[2] Depto. de Física. Universidad Politécnica de Madrid.
Ronda de Valencia 3, 28012 Madrid. Spain

ABSTRACT. Current methods used to obtain absolute dimensions for Algol-type eclipsing binaries are discussed and some alternatives are proposed. The reliability of each procedure is stated through its application to a sample of double-lined systems. We also analize the basic characteristics of the primary and secondary components of the sample against standard evolutionary models for isolated stars. Our results indicate that primaries behave essentially as normal main sequence stars while the well-known "overluminosity" of the secondaries could have its origin in the irradiation of the stars by their primary mates.

1. INTRODUCTION

Algol-type binaries form the largest group of stars among eclipsing binary systems. However only a few of them have well-established basic parameters (masses, radii and temperatures).

The origin of this lack of information is an intrinsic characteristic of Algol binaries: the large difference in luminosity between the component stars implies single-lined spectra. For most of the systems, only new modern detectors are able to measure the radial velocity variations of the fainter component. Light curves show deep primary eclipses, often total, and very shallow secondary eclipses. Emission lines are also frequently found in Balmer lines revealing the presence of high temperature plasma in Algol binary systems.

Recently, Batten (1989) proposed a general definition for these binaries as systems in which only the less massive component fills its Roche lobe, being the other component not degenerate. This definition describes Algols as they are at present, without going into the process undergone to reach their current status. The solution to the underlying paradox - the less massive star is the more evolved one - which have to deal with the problem of drawing the evolutionary history of such systems is thus avoided.

It is generally accepted nevertheless that Algols originate in

C. İbanoğlu (ed.), Active Close Binaries, 121–135.

detached binary systems after the initially more massive star has evolved and reached, at some stage of its expansion, the corresponding Roche lobe. Mass transfer to the still unevolved companion then starts and continues after mass ratio reversal.

The approach to the problem of mass transfer still includes many qualitative aspects. A quantitative description of the physical mechanisms involved is needed in order to derive the total amount of mass transferred and lost. Evolutionary models for this type of stars will lead to the initial parameters of the formely detached binary, but, first of all, we have to know the current status of the system. In this sense, the determination of the present absolute parameters for a wide sample of systems is essential to validate the evolutionary models proposed.

On the other side, Algols, showing relatively unperturbed light curves, are the simplest case of close interacting binaries. They could be considered as a first stage towards genuine "exotic" binaries, where larger mass transfer processes have taken (or are taking) place. A good knowledge of this phase, represented by Algols, will be of great help in understanding the whole evolutionary path followed by close binary systems.

2. STANDARD PROCEDURES TO OBTAIN ABSOLUTE PARAMETERS

Eclipsing binaries for which radial velocity variations of both components are available, make possible a determination of sizes and masses of the component stars with high accuracy.

However, most of the classical Algols are single-lined and, in order to obtain absolute dimensions, some complementary hypothesis must be added to the normal set of available observational parameters. Different assumptions give rise to different procedures (Hall, 1974; Cester et al., 1979; Budding, 1985; García and Giménez, 1989) that we summarize in the following paragraphs.

What we consider as basic absolute parameters are: the masses, the absolute radii and the temperatures of both components. From this fundamental parameters, others such as the absolute magnitudes, the surface gravities, the distance to the system, etc., can be easily calculated.

What we usually have for single-lined eclipsing variables is:

a) the mass function, f(m), derived from the radial velocity curve of the primary.
b) the relative radii of the components, r_i; the inclination, i; the ratio of surface brightness J_2/J_1, and the orbital period, P, obtained from the analysis of the light curve.
c) the effective temperatures, T_i of both components estimated from the colour indices and the light curve.

Typical errors for these parameters are : 15% in f(m), 5% in r_i and 5-10% in T_i .

Table I. List of systems and data adopted in the discussion of the different methods.

Star	P	q	q_{MLR}	q_{SD}	q_G	q_R	q_{ED}	$\log T_1$	$\log T_{1ED}$
TW And *	4.123	0.22	0.22	0.20			0.23	3.866	3.87
RY Aqr *	1.967	0.20	0.19	0.19			0.19	3.881	3.82
KO Aql	2.864		0.21	0.05		0.20	0.21	3.986	
QY Aql	7.230		0.28	0.18			0.30	3.841	
SX Aur ■	1.210	0.54	0.55	0.62	0.44	0.60	0.55	4.342	4.40
TT Aur ■	1.333	0.65	0.66	0.74			0.64	4.373	4.36
IM Aur	1.247		0.30	0.21	0.22	0.25	0.31	3.999	
IU Aur ■	1.811	0.69	0.69	0.79		0.61	0.66	4.505	4.49
Y Cam	3.306		0.21	0.17			0.23	3.869	
S Cnc *	9.485	0.08	0.07	0.09			0.07	3.998	3.98
RZ Cnc *	21.64	0.17	0.19	0.20				3.630	3.79
R CMa	1.136		0.15	0.22			0.16	3.786	
RZ Cas	1.195		0.34	0.37	0.87	0.28	0.33	3.949	
SX Cas *	36.56	0.30	0.30	0.13		0.01	0.30	4.212	4.23
TV Cas	1.183		0.44	0.36	0.19	0.56	0.47	4.016	
RS Cep	12.42		0.37	0.13			0.38	3.918	
XX Cep	2.337		0.16	0.14		0.13	0.16	3.904	
XY Cep	2.775		0.30	0.20			0.31	4.029	
RW CrA	1.684		0.34	0.26			0.34	3.991	
U CrB *	3.452	0.29	0.30	0.32		0.16	0.30	4.198	4.24
RW CrB	0.726		0.27	0.19			0.27	3.845	
AI Cru ■	1.418	0.61	0.64	0.64			0.63	4.384	4.39
SW Cyg	4.573		0.28	0.25		0.04	0.29	3.960	
VW Cyg	8.430		0.28	0.15			0.29	3.960	
WW Cyg	3.318		0.41	0.23			0.42	4.020	
W Del	4.806		0.19	0.26		0.16	0.20	3.968	
Z Dra	1.357		0.23	0.27			0.24	3.941	
TW Dra *	2.807	0.47	0.41	0.45		0.38	0.44	3.921	3.87
AI Dra	1.199		0.43	0.51	0.39	0.57	0.44	4.003	
S Equ	3.436		0.13	0.13	0.17	0.10	0.14	3.920	
AS Eri *	2.664	0.11	0.11	0.13		0.10	0.11	3.934	3.93
RW Gem	2.866		0.36	0.37			0.38	4.093	
RX Gem	12.21		0.24	0.09			0.24	3.929	
RY Gem *	9.301	0.22	0.21	0.18			0.22	3.973	3.99
SV Gem	4.006		0.27	0.34			0.29	4.088	
u Her ■	2.051	0.38	0.36	0.43		0.46	0.37	4.301	4.29
UX Her	1.549		0.25	0.17			0.25	3.957	
AD Her	9.767		0.37	0.17			0.39	3.900	
RX Hya	2.282		0.24	0.30			0.24	3.894	
SX Hya	2.896		0.34	0.37			0.33	3.953	
TT Hya *	6.953	0.27	0.28	0.20		0.02	0.28	3.991	3.99
TW Lac	3.037		0.19	0.36			0.20	3.942	
Y Leo	1.686		0.31	0.45			0.31	3.966	
T LMi	3.020		0.13	0.25			0.13	4.016	
RW Mon	1.906		0.39	0.46			0.40	4.024	
TU Mon *	5.049	0.25	0.28	0.18			0.29	4.292	4.41

Table I. Continued

Star	P	q	q_{MLR}	q_{SD}	q_G	q_R	q_{ED}	$\log T_1$	$\log T_{1ED}$
AU Mon	11.11		0.22	0.27		0.03	0.24	4.122	
FW Mon	3.874		0.31	0.45			0.30	4.124	
UU Oph	4.397		0.18	0.30			0.19	3.992	
AQ Peg	5.548		0.24	0.20			0.25	3.958	
AT Peg	1.146		0.45	0.42			0.46	3.897	
BET Per*	2.867	0.22	0.22	0.17	0.22	0.18	0.23	4.064	4.11
RT Per	0.849		0.25	0.30			0.24	3.828	
RW Per	13.20		0.16	0.06			0.18	3.924	
RY Per *	6.864	0.16	0.15	0.25		0.01	0.15	4.246	4.22
ST Per	2.648		0.18	0.18			0.19	3.956	
AB Per	7.160		0.20	0.29			0.21	3.941	
DM Per *	2.728	0.31	0.32	0.37			0.33	4.201	4.24
Y Psc	3.766		0.23	0.28		0.19	0.24	3.955	
V Pup ■	1.455	0.52	0.56	0.59		0.49	0.55	4.450	4.47
XZ Pup	2.192		0.42	0.44			0.44	3.987	
U Sge *	3.381	0.33	0.36	0.46	0.47	0.37	0.38	4.146	4.24
RS Sgr	2.416		0.39	0.23		0.41	0.41	4.151	
V505 Sgr	1.183		0.51	0.56		0.47	0.52	3.978	
RZ Sct	15.19		0.26	0.23		0.03	0.28	4.097	
LAM Tau*	3.953	0.27	0.25	0.23	0.46	0.27	0.26	4.228	4.26
RW Tau	2.769		0.27	0.25	0.44	0.12	0.28	4.047	
X Tri	0.972		0.55	0.53			0.56	3.945	
TX UMa *	3.063	0.30	0.30	0.33	0.42	0.30	0.33	4.146	4.14
VV UMa	0.687		0.21	0.39			0.21	4.006	
W UMi	1.701		0.47	0.40		0.53	0.51	3.929	
Z Vul *	2.455	0.43	0.38	0.44	0.40	0.33	0.40	4.255	4.20
RS Vul*	4.478	0.31	0.29	0.31	0.41	0.19	0.30	4.180	4.15

From the definition of the mass function we have,

$$m_2 = \frac{f(m)}{\sin^3 i}\left(\frac{1+q}{q}\right)^2 \tag{1}$$

while, the mass ratio gives,

$$m_1 = \frac{m_2}{q} \tag{2}$$

being m_1 and m_2 the masses of the hotter and the cooler components respectively.

Once the mass ratio, q, is known, the other basic parameters can be immediately calculated, since Kepler's law can then be used to calculate absolute radii.

The following methods are used to determine q:

A) It is assumed that the primary component verifies the mass-luminosity relation (MLR) for main sequence stars ($q = q_{MLR}$).

We adopt the MLR in the form

$$\log L = \underset{\pm 2}{0.04} + \underset{\pm 4}{3.93} \log m, \qquad (3)$$

valid for $0.5\ m_{\odot} \leq m \leq 12\ m_{\odot}$, as given by Giménez and Zamorano (1986) who derived it on the basis of Popper's (1980) data.

In this case we can write,

$$\log q_{MLR} = -0.2(\log r_1 + \frac{2}{3}\log P + 2\log T_1 - 6.92) + 0.6\log(1+q_{MLR}) + \\ + \frac{1}{3}\log f(m) - \log(\sin i) \qquad (4)$$

which can be solved through iteration to get q_{MLR}.

Adopting the standard errors given above for the involved observational parameters, the estimated accuracy of this mass ratio is

$$\delta q_{MLR} \approx 0.26\ q_{MLR} \left(\frac{1+q_{MLR}}{3+1.2q_{MLR}} \right)$$

which gives a typical error of about 10% for a typical value of $q \approx 0.3$. The main contribution to this uncertainty comes from the error in the semiamplitude, K_1, of the radial velocity curve, which is often poorly known in Algols.

A weak point of this method lies in our knowledge of the effect of mass accretion on the behaviour of the primary component which could deviate from the adopted MLR

B) It is considered that the secondary component fills exactly its critical Roche lobe, i.e., a semi-detached configuration is assumed ($q = q_{SD}$). In this case, the analysis of the light curve provides the relative radius of the secondary star.

Two different, but essentially equivalent, procedures to obtain q_{SD} are:

B.1) Use tables for Roche surfaces which relate the mass ratio to the shape and size of the critical lobes (Plavec and Kratochvil, 1964; Mochnacki, 1984).

B.2) Optimise the O-C residuals obtained with synthetic ligt curves using Wilson and Devinney code (1971) in mode 5, for different q values.

Expected errors are of the order of 15-20%, mainly due to the high sensitivity of q_{SD} to small variations in the observed value of r_2, particularly if $r_2 \geq 0.2$. Contrary to the case of closer binaries with light curves of the β Lyrae or W UMa type, light variations of Algol systems are dominated by the primary, undistorted, component and, therefore, the information contained in the photometric curve is not generally sufficient to derive q_{SD} with the required degree of confidence. Another uncertainty associated with this method is that the cool component may not actually fill strictly its Roche lobe.

When no radial velocity curves are available or they are of bad quality, a combination of methods A and B above was suggested by Hall (1974). The mass function is not used and, instead, the primary star is assumed to obey the MLR while the semi-detached configuration is forced. Involved errors in this case are probably too large to obtain reliable absolute parameters.

C) It is assumed that the primary star rotates synchronously and that the orbit is circular ($q = q_R$).

Though the first assumption is difficult to accept due to the expected transfer of angular momentum together with mass transfer, it allows us to write

$$v_1 = 50.61\, R_1/P \tag{5}$$

where R_1 is expresed in units of the solar radius and P in days. Combining this equation with Kepler's third law, the definition of the mass function and the relative radius of the primary, we get

$$\log(\frac{1}{q_R}+1) = -2.328 + \frac{1}{3}\log P - \frac{1}{3}\log f(m) + \log (v_1 \sin i) - \log r_1 \tag{6}$$

Considering typical errors in vsini around 10%, permits us to estimate the uncertainty of q_R to be $\sim$30% for a typical value of $q \approx 0.3$. It is well-known, on the other hand, that primaries of Algol binaries do not generally rotate synchronously. Observational evidences are commented in Section 3.

D) It is adopted that the primary star is well reproduced by standard theoretical models of stellar evolution within the main sequence ($q = q_{ED}$).

The effective temperature of the hotter star is known from colour indices and T_1 is used as a fixed input parameter to start an iterative procedure in the log g-log T_e diagram. An initial value of the mass ratio, q_0, combined with the observed mass function gives initial masses for the component stars ($m_{0,1}$, $m_{0,2}$). Then, log g_1 can be calculated and the representation of the primary component on the log g-log T_e plane will provide the theoretical mass for the primary. This value of m_1, combined with f(m) permits now the calculation of a new q. Convergence is reached after few iterations (tipically, 3 or 4).

This method is equivalent to method A, but diminish systematic errors due to evolution from the ZAMS to the TAMS. Estimated accuracies in q are similar to method A ($\sim$5-10%), exception made of stars placed near the TAMS where the estimation of m_1 may not be unique.

E) The equation

$$\log q = \log g_2 - \log g_1 + 2 \log k \qquad (7)$$

where $k=r_2/r_1$, allows the calculation of q ($q=q_G$) if log g_1 is determined independently through line profiles, since log g_2 is directly known from observed parameters (García and Giménez, 1989).

Typical errors in deriving log g_1 from line profiles ($\delta(\log g_1) \approx 0.15$) lead to uncertainties of about 50% in q_G. Therefore, this procedure is not useful unless a determination of surface gravities with $\delta(\log g_1) \approx 0.02$ were posible.

3. RELIABILITY OF THE DIFFERENT METHODS

The validity of the methods given above to estimate q in single-lined Algols can be tested applying them to a group of double-lined systems for which q is directly measured. We have selected a sample of Algol-type eclipsing binaries with double-lined spectra (see Table I) which includes classical Algols, with a late-type secondary subgiant and hot semidetached binaries. We therefore include systems with both cases, A and B, of mass transfer.

The adopted values for masses, temperatures and other observed quantities, have been taken mainly from Popper (1980) and Hilditch and Bell (1987), but have been updated with some more recent analyses found in the literature.

In Table I, q_{MLR} is calculated according to (4) and q_{SD} is obtained using Mochnacki (1984). We assume that the relative radius of the secondary component is identical to the mean radius of the Roche lobe, which is related to the normalized potential, C_1, at the inner lagrangian point.

Figure 1 shows the comparison of q versus q_{MLR} . The agreement between the observed mass ratio, q, and its estimation, q_{MLR}, is generally better than the above quoted uncertainty ($\approx$ 10%). No systematic deviation from the average is found as it would be expected in case of problems with the adopted scale of temperatures . As a consequence, the hypothesis that the MLR actually holds for the primary components, appears to be good enough despite mass accretion. It is true that the mass ratio, q, is less dependent on departures from the MLR than m_1, but the relation appears to hold well in average. We will come back later to this point through the direct comparison of masses.

Figure 2 shows the comparison between q and q_{SD}. A larger scatter is evidenced, particularly for $q_{SD} > 0.13$ ($r_2 > 0.21$). The interagreement is not as good as in the case of q_{MLR}. A more complete sample of Algols has been used to compare q_{MLR} and q_{SD} in Fig. 3. We have added 48 single-lined stars from Budding (1985). In that work, a similar comparison is made, but we have recalculated q_{MLR} and q_{SD} for all the stars, following the criteria explained above. Differences with respect to Budding's q_{SD} values are important for some individual systems.

The estimation of the mass ratio, q_{SD}, is certainly responsible for the large dispersion in Figure 3. Making use of the volume radius instead of the mean radius of the Roche lobe does not change substantially the results.

A significant number of "undersize" subgiants would imply a systematic tendency in Figure 3. However, there is no indication of such trends and suggestions by Budding (1985) are thus not confirmed. On the contrary, Fig. 3 indicates that there is no real class of significantly undersized secondaries as noted by Hall and Neff (1979).

Unfortunately, the number of systems with observed mass ratio for which log g_1 or $v_1$sini is also available, is too low. Therefore, on the basis of our previous results, we decided to compare the estimation of q_G and q_R against q_{MLR} instead of q.

To test method C -synchronous rotation of primary component assumed-, rotational velocities where taken from Van den Heuvel (1970) and Olson (1984). In Figure 4 a comparison of q_R versus q_{MLR} is made. A systematic tendency seems to be present in the sense $q_{MLR} > q_R$ for $q_{MLR} \leq 0.4$. This indicates that roughly synchronous rotation is present for $q_{MLR} > 0.4$, while is faster than synchronous for $q_{MLR} \leq 0.4$. This could be explained in terms of the amount of mass transfer which is expected to be larger for smaller values of q. A larger accretion of angular momentum would, consequently, explain an accelerated rotation.

Nevertheless, small mass ratios also imply large periods and it is known that deviations from synchronism increase with the orbital period since the effectiveness of tidal mechanisms decrease rapidly. In order to show this, we have represented the relative differences of rotational velocities with respect to synchronization, $(v - v_{syn})/v$, versus P and found that binaries with P >3 days are those which show faster than synchronized rotation.

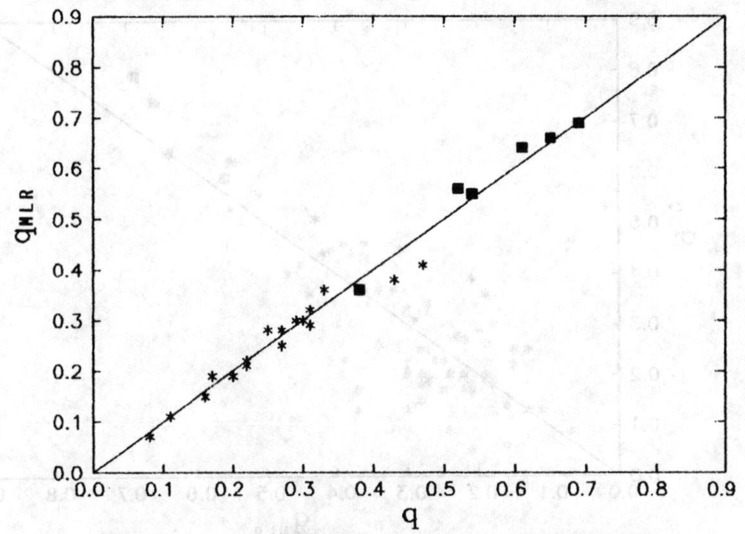

Figure 1.– Comparison of q_{MLR} versus q for classical Algols (*) and hot semidetached binaries (■).

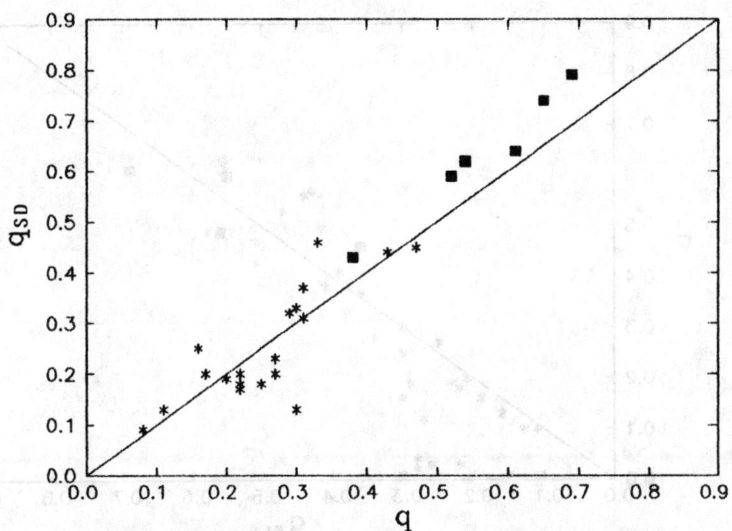

Figure 2.– Comparison of q_{SD} versus q for classical Algols (*) and hot semidetached binaries (■).

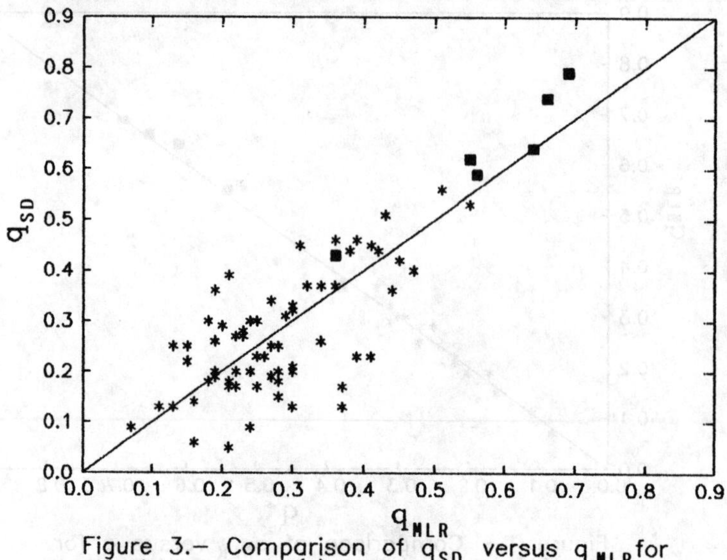

Figure 3.– Comparison of q_{SD} versus q_{MLR} for classical Algols (*) and hot semidetached binaries (■).

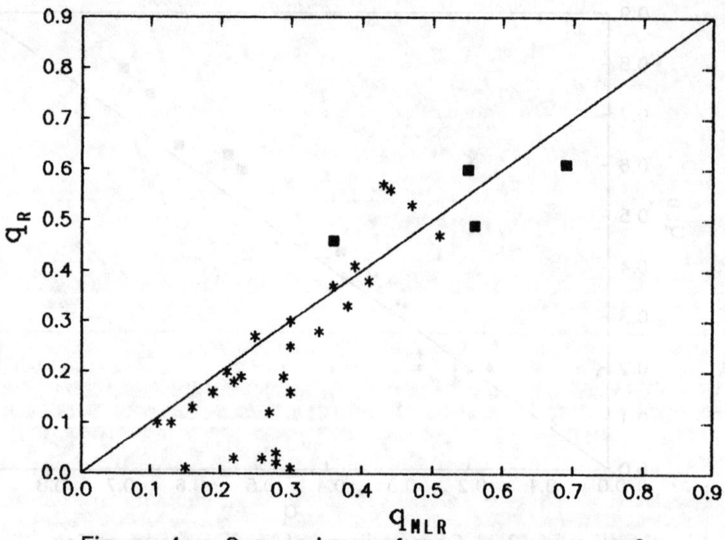

Figure 4.– Comparison of q_R versus q_{MLR} for classical Algols (*) and hot semidetached binaries (■).

In Figure 5 we show the very large scatter found when q_G and q_{MLR} are compared. Errors in log g_1, derived from line profiles are obviously responsible for such dispersion. This method has to be discarded unless a higher accuracy could be reached in measuring log g_1.

For the calculation of q_{ED},we adopt the grid of evolutionary models given by Claret and Giménez (1989) for an initial chemical composition (X,Z)=(0.70,0.02). Figure 6 shows a perfect correlation between q_{MLR} and q_{ED}, as expected. The existence of a small effect either due to evolution from ZAMS to TAMS or probable errors in the MLR can be noticed. $q_{ED} > q_{MLR}$ is expected if primary components are close to the TAMS as it appears to be the case in Figure 6. It should be stressed that real primary stars obtained a significant part of their actual masses through accretion but models for single stars and constant mass have been used.

From the discussion above it is established that q_{MLR} and q_{ED} are the most reliable methods to estimate the actual mass ratio. We prefer q_{ED} because it diminishes systematic errors due to evolution within the main-sequence. Both methods have to be applied with reservations when available radial velocity curves are of poor quality since errors in K_1 have an important effect on the final results.

4. THE COMPONENT STARS OF ALGOL SYSTEMS AGAINST EVOLUTIONARY MODELS

In order to contribute to a better understanding of the evolution and structure of Algol binaries, let us continue considering the sample of binaries with well-established absolute dimensions. The group of semidetached stars used have masses determined from double-lined spectra, radii from photoelectric light curves and temperatures from colour indices. These stars have been marked in Table I with asterisks -classical Algols- or squares -hot semidetached binaries-.

To answer the question of how normal are the primary components of these interanting systems, after they have accreted large amounts of mass, we have analized their behaviour with respect to theoretical models. Evolutionary tracks are taken from Claret and Giménez (1989) for a chemical composition (X,Z)=(0.70,0.02) and α =2.0 .

Individual tracks for observed masses are computed and theoretical effective temperatures (T_{ED}) corresponding to the observed values of log g_1 are estimated.

In Figure 7 we have compared for classical Algols (+) and hot semidetached binaries (⊞), the temperatures of the primary components derived from evolutionary models, T_{1ED}, with the observed values T_1. A good agreement within the involved uncertainties is found. The dispersion around the line log T_1=log T_{1ED} is comparable with that in a similar diagram drawn for well-detached binaries. Ages corresponding to

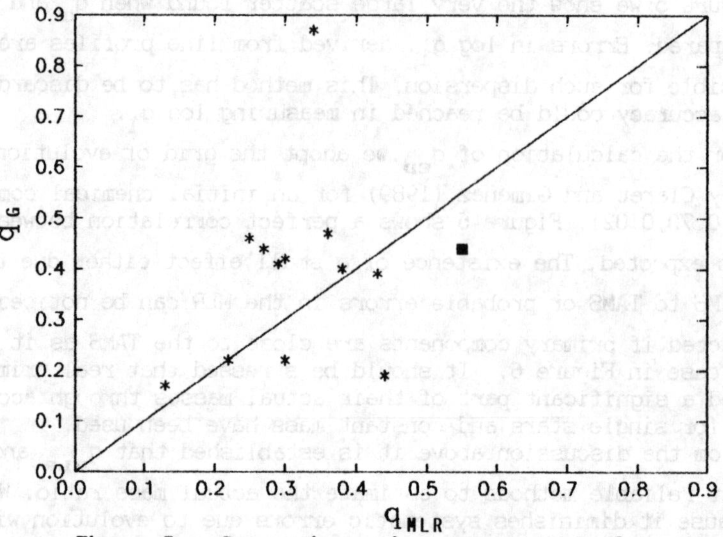

Figure 5.– Comparison of q_G versus q_{MLR} for classical Algols (*) and hot semidetached binaries (■).

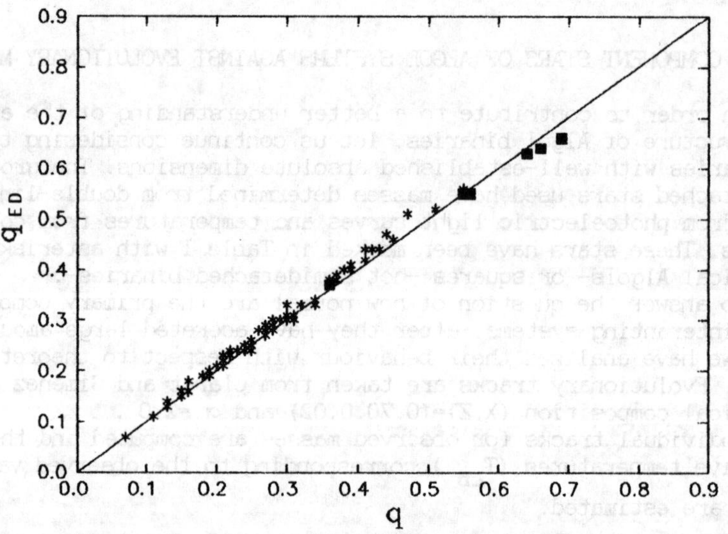

Figure 6.– Comparison of q_{ED} versus q_{MLR} for classical Algols (*) and hot semidetached binaries (■).

isochrones in the $\log g_1 - \log T_1$ plane do not represent, of course, actual values.

We can also estimate the theoretical masses corresponding to the observed values of $\log T_1$ and $\log g_1$ from computed model grids. A good agreement is found, corresponding to the result of the temperature comparison. As a conclusion, we can say that primary components of Algol systems do actually behave like normal main-sequence stars.

With respect to the secondaries it is much more difficult to expect that theoretical evolutionary models can provide much information for a star which have lost large amounts of mass.

When the estimated theoretical masses corresponding to the observed values of $\log T_2$ and $\log g_2$ are plotted against the observed masses, a systematic tendency in the sense $m_{2ED} > m_2$ is clearly noticed. This is described as "overluminosity" in terms of the MLR.

If we calculate instead the theoretical effective temperatures corresponding to the observed masses and $\log g_2$, a systematic trend for observed temperatures larger than predicted is detected (Figure 7), which is equivalent to the effect in masses.

The difference in temperatures is generally smaller than 25% but is larger in some few cases. We have also plotted the relative differences in temperatures, $\delta T_2 = (T_2 - T_{2ED})/T_{2ED}$ against several orbital and photometric parameters of the systems. Only a slight dependence on the mass ratio and on the luminosity of the primary star is observed, which indicates that the effect is larger when the components are more different.

Part of the observed effect in the temperatures of the secondaries could be explained as due to irradiation by the hotter star of the upper layers of the cool component in the way suggested by Claret (1989). Briefly, it is considered that there are two contributions to the observed flux from the secondary: the intrinsic flux of the star and the the reflected energy from the primary. In certain conditions, the absorbed (non reflected) energy produce an increment of its original (unheated) temperature (cf.,e.g., Vaz and Norlund, 1985). This leads to a larger (heated) observed temperature.

Very simple approximations and a typical value of ≈ 0.5 for the albedo of the cooler components allow us to estimate the temperature difference as

$$\delta T_2 \approx \left[1 + 0.25 \frac{F^*}{F} \right]^{1/4} - 1 \tag{8}$$

were F^*/F denotes de ratio between the external and internal flux at the star surface.

In Figure 8 we have plotted the observed δT_2 versus the calculated F^*/F for those stars for which we were able to estimate T_{2ED}. There seems

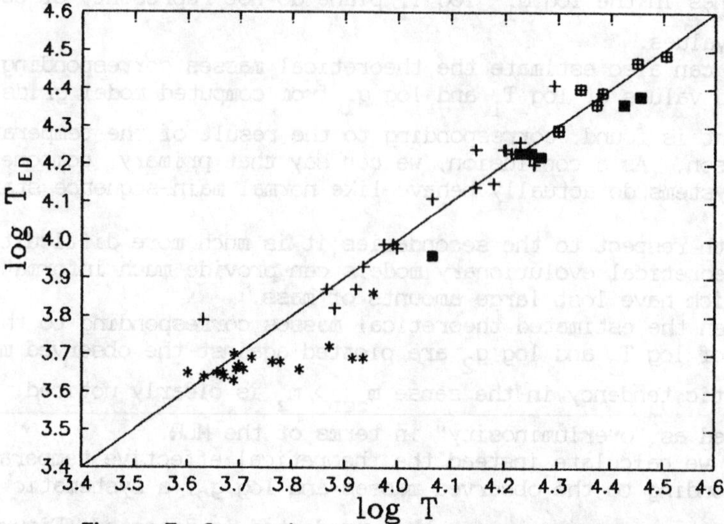

Figure 7. Comparison of the observed log Te versus theoretically predicted values for primaries (+) and secondaries (*) of classical Algols as well as primaries (⊞) and secondaries (■) of hot semidetached binaries.

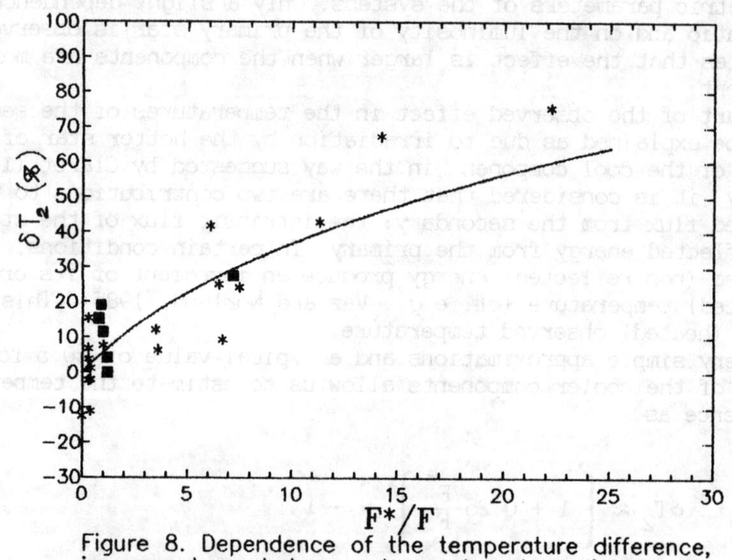

Figure 8. Dependence of the temperature difference, in percentage, between observed and predicted values for the secondaries of classical Algols (*) and hot semidetached binaries (■) versus the ratio of fluxes at their surface due to irradiation.

to be some correlation between δT_2 and F^*/F indicating that the effect of irradiation may explain, at least partly, the observed excess of temperature. The solid line corresponds to equation (8). Large uncertainties are involved, mainly in the computation of F^*/F which can reach up to 60% for standard errors in temperature and relative radii. On the other hand, it is still not excluded that the excess of temperature could be explained on the basis of an unrealistic treatment of the reflection effect through the synthetic light-curve codes used in the photometric analysis.

5. ACKNOWLEDGEMENTS

This work has been supported by Spanish CICYT under project PB87-035.

REFERENCES

Batten,A.H.: 1989, Space Science Reviews,**50**,1.
Budding, E.: 1985, Publ. Astron. Soc. Pacific,**97**,584
Cester,B.,Giuricin,G.,Mardirossian,F.,Mezzetti,M.,Maceroni,C. and Mancuso,S.: 1979, Astron. Astrophys.,**73**,31
Claret, A.: 1989, Private communication.
Claret,A. and Giménez,A.: 1989, Astron. Astrophys. Supl.,**81**,1
García,J.M. and Giménez,A.:1989, Space Science Reviews,**50**,341
Giménez,A. and Zamorano,J.:1986, Anales de Física,**82**,121
Hall,D.S.: 1974, Acta Astron.,**24**,215
Hall,D.S. and Neff,S.G.: 1979, Acta Astron.,**29**,641
Hilditch,R.W. and Bell,S.A.: 1987, Mont. Not. Roy. Astron. Soc.,**229**,529
Mochnacki,S.W.: 1984, Astrophys. J. Suppl.,**55**,551
Olson,E.C.: 1984, Publ. Astron. Soc. Pacific,**96**,376
Plavec,M. and Kratochvil,P.: 1964, Bull. Astron. Czech.,**15**,165
Popper,D.M.: 1980, Ann. Rev. Astron. Astrophys.,**18**,115
Van den Heuvel,E.P.J.: 1970 in Stellar Rotation, A. Sletteback (ed.), Reidel,Dordretch, p.178
Vaz,L.P.R. and Norlund,A.: 1985, Astron. Astrophys.,**147**,281
Wilson,R.E.and Devinney,E.J.: 1971, Astrophys. J.,**66**,605

to be some correlation between $c\ell_i$ and F_1/F_2 indicating that the effect of irradiation may explain, at least partly, the observed excess of temperature. The solid line corresponds to equation (8). Large uncertainties are involved, mainly in the computation of F_1/F_2 which can reach up to 50% for standard errors in temperature and relative radii. On the other hand, it is still not excluded that the excess of temperature could be explained on the basis of an unrealistic treatment of the reflection effect through the synthetic light-curve codes used in the photometric analysis.

5. ACKNOWLEDGEMENTS

This work has been supported by Spanish CICYT under project PB87-035.

REFERENCES

Batten, A.H.: 1989, Space Science Reviews 50, 1.
Bolding, E.: 1985, Publ. Astron. Soc. Pacific, 97, 501.
Cester, B., Giuricin, G., Mardirossian, F., Mezzetti, M., Maceroni, C. and Mancuso, S.: 1979, Astron. Astrophys., 73, 31.
Claret, A.: 1989, Private communication.
Claret, A. and Gimenez, A.: 1989, Astron. Astrophys. Supl. 81, 1.
Garcia, J.M. and Gimenez, A.: 1989, Space Science Reviews 50, 341.
Gimenez, A. and Zamorano, J.: 1985, Anales de Fisica, 82, 121.
Hall, D.S.: 1974, Acta Astron., 24, 215.
Hall, D.S. and Neff, S.G.: 1979, Acta Astron., 29, 641.
Hilditch, R.W. and Bell, S.A.: 1987, Mont. Not. Roy. Astron. Soc., 229, 529.
Mochnacki, S.W.: 1984, Astrophys. J. Suppl., 55, 551.
Olson, E.C.: 1984, Publ. Astron. Soc. Pacific, 90, 376.
Plavec, M. and Kratochvil, P.: 1964, Bull. Astron. Czech., 15, 165.
Popper, D.M.: 1980, Ann. Rev. Astron. Astrophys., 18, 115.
Van den Heuvel, E.P.J.: 1970 in Stellar Rotation, A. Slettebak (ed.), Reidel, Dordrecht, p.178.
Vaz, L.P.R. and Nordlund, A.: 1985, Astron. Astrophys., 147, 281.
Wilson, R.E. and Devinney, E.J.: 1971, Astrophys. J., 166, 605.

RECENT RESULTS OF APSIDAL MOTION STUDIES IN ECLIPSING BINARIES

A. Giménez
Instituto de Astrofísica de Andalucía
Apartado 2144. E-18080 Granada (Spain)

ABSTRACT. A review of possibilities and results in the field of apsidal motion studies is presented. A detailed agreement between observations and theoretical models is not yet available, even taking into account evolutionary effects. Some alternatives are commented requiring future attention by theoreticians as well as when selecting candidate binary systems for detailed analysis.

1. INTRODUCTION

During a previous NATO Advanced Study Institute, organized in 1980 by Prof. Z. Kopal at Maratea (Italy), we discussed the importance and actual possibilities of renewing our efforts in the study of apsidal motions in eclipsing binaries (Giménez, 1981). In this ocassion, I comment the main results obtained during the last years.

A detailed study of stellar structure during hydrogen burning phases is still badly needed. The existence of problems in our understanding of these phases is shown by the detected flux of solar neutrinos, the interpretation of solar oscillations and stellar pulsation, existence of discrepancies in stellar opacity calculations, the parametrization of stellar convection and overshooting, the estimation of mass loss during the main sequence or the location of the termination-age main sequence (TAMS) indicated by open clusters. On the other hand, a detailed knowledge of stellar structure in the main sequence is of primordial importance if we want to perform calculations at more advanced evolutionary phases with certain degree of confidence.

The best observational approach to the problem is, no doubt, the analysis of eclipsing binaries with double-lined spectra (Andersen et al., 1980). This is a well-known old technique which has been renewed with much improved accuracies during the last decade. Furthermore, well-detached main-sequence stars usually present eccentric orbits and are thus expected to show apsidal motion orbital variations. In this

C. İbanoğlu (ed.), Active Close Binaries, 137–143.

way, an important additional source of information to study the internal structure of the stars is provided and we should be able to explain this data in the light of existing theoretical models.

2. INTERNAL STRUCTURE CONSTANTS

The main purpose of the study of apsidal motions is to have an empirical insight into the interior of the stars. Well-known dynamical perturbations in eccentric binaries produce a secular variation of the periastron which can be accurately measured through observations of eclipse timings. The response of a given star to the external potential of its companion depends, of course, on its own internal density concentration and, thus, important information can be derived if the basic gravitational parameters of the problem are known. The current procedure to extract such information is the comparison of observed apsidal motion rates with those derived from calculations based on the assumption of internal structure coefficients, log k_2, as given by theoretical models.

The apsidal motion constants log k_j ($j = 2,3,4$) are defined, for any given distribution of mass along the interior of a star, by the logarithmic derivative of the corresponding order (j) spherical harmonic which describes the perturbed equipotential surfaces. Integration of Radau's differential equation (Kopal, 1978) provides the method to calculate the apsidal motion constants for any given density distribution. Theoretical models (Hejlesen, 1987) show that higher than $j=2$ orders have a negligible contribution to the observed apsidal motion rates and there is an important change in the behaviour of log k_2 depending on the dominant process of energy transport, convective or radiative, within the stellar interior. On the other hand, internal structure constants also depend on the adopted chemical composition and opacities, convection theory, etc.

Evolutionary models show that log k_2 values decrease as the star evolves within the main sequence, i.e. the star becomes more centrally condensed as the envelope expands. The average width of the main sequence is found to be around 0.2 in log k_2, which implies that, if we want to discriminate the location of individual stars within the main sequence, errors in the observational determination of log k_2 should be smaller than around 0.05 and this is only possible by keeping errors in the apsidal motion rate and the relative radii within 2%.

3. OBSERVATION OF APSIDAL MOTIONS

Observational constraints for this type of studies were discussed in some detail by Giménez (1981) and are basically reduced to the condition that accurate absolute dimensions should be available in order that: a) an accurate computation of the external potential and b) an adequate selection of theoretical models for comparison be made. A practical consequence is that the astrophysical usefulness of a long-term project to observe apsidal motion variations is very limited if it is not carried out within the framework of a simultaneous study of absolute dimensions. In particular, stellar radii should be known with high precision since the rate of periastron rotation depends on the fifth power of their relative values. Our analysis should then be restricted to eclipsing binaries with well defined eclipses. A list of binary systems used in our more recent analyses is given in Section 5.

In order to reach the required accuracies, the method of analysis adopted to obtain apsidal motion rates from the observations is also very important. Times of minimum light should be measured, in general, with an accuracy better than 0.0001 days and a realistic function relating the actual position of the eclipses to the orbital parameters is needed. This is done by converting times of minimum light to times of conjunction through the projected apparent separation between stellar centers. Moreover, mean anomalies are related to true anomalies through the orbital eccentricity and the equation of center and we can obtain an explicit equation for the times of minima in terms of the actual position of the periastron and other orbital parameters (Giménez and García-Pelayo, 1983). It should be stressed that attention has to be paid to the truncation of the involved expansion series since, in many cases, terms up to fifth order in the orbital eccentricity are needed.

4. CONTRIBUTIONS TO THE APSIDAL MOTION RATE

The secular motion of the periastron has two main causes: a classical term due to quadrupole effects and a relativistic contribution. In fact, apsidal motion or, for that matter, any secular variation of the orbital parameters is caused by the deviation from Newton's square law of the gravitational potential. Mass points or spherical bodies are at the base of truly keplerian motion but distorted stars, as it is the case in close binaries, can not be reduced to mass points and produce the quadrupole correction to the dynamical behaviour of the orbit. Variational studies of the involved equations show that the orbital elements suffer secular changes but, it was demonstrated by Kopal (1978) that all time derivatives vanish, except that for the position of the periastron, if and only if there is no tidal lag and spin axes of the component stars are perpendicular to the orbital plane. These two later assumptions are quite reasonable according to existing empirical tests as well as available theories of tidal evolution in close binaries.

On the other hand, the so-called classical or quadrupole contribution to the total apsidal motion rate has itself two terms corresponding to the origin of the distortions in the stellar configuration: one term due to tidal elongation and another one due to rotational flattening. Expressions for the apsidal motion in binary systems are given by several authors in terms of the orbital eccentricity, the mass ratio, the relative radii of the component stars, their internal structure constants and their rotational velocities (see e.g., Kopal, 1978).

Nevertheless, even for spherical stars, post-newtonian terms of the gravitational potential produce a secular motion of the periastron in a similar way as the well-known perihelion advance of the planet Mercury. In fact, for very well detached binaries, with relative radii smaller than around 0.1, the apsidal motion rate is actually dominated by the relativistic term due to the dependence of the classical contribution on the inverse of the fifth power of the relative radii. These eclipsing binaries are therefore well suited to test different formulations given by alternative theories of gravitation and general relativity (Giménez, 1985). On the other hand, closer binaries with relative radii larger than around 0.15, show apsidal motions dominated by the classical terms and the derived internal structure constants for these systems are quite independent of the adopted correction for the relativistic contribution.

Let us now concentrate in the study of internal parameters, which are relevant to the astrophysical problems of internal structure. We therefore discard those binaries with important relativistic terms and correct observed apsidal motion rates for small contributions using equations given by General Relativity. We will nevertheless come back later to this point.

5. COMPARISON WITH MODELS

5.1. Using the zero-age main sequence (ZAMS)

In order to interpret the observed values of apsidal motion in a sample of eclipsing binaries compiled during the last ten years, we first calculate empirical values of the second harmonic of the internal structure, averaged for both components, $\log k_2$ (obs). For this purpose, we use the observed apsidal motion period, orbital parameters and absolute dimensions. At the same time, the same absolute dimensions together with available theoretical evolutionary models allow us to predict the corresponding, theoretical, $\log k_2$ (th). A necessary, though not sufficient, condition for the correctness of the adopted theoretical models is a detailed agreement between $\log k_2$ (obs) and $\log k_2$ (th) for the whole range of masses and ages.

A first attempt to perform this kind of comparison for a wide sample of binaries with well-studied parameters was presented during one of the Bamberg colloquia by Giménez and García-Pelayo (1982). Unfortunately, evolutionary models, including internal structure constants, were not available at that time but, the previously known systematic difference between observed values of $\log k_2$ and theoretical predictions for the zero-age main sequence (ZAMS), could be explained in terms of a variation of $\log k_2$ with age. It was shown that the difference between observed $\log k_2$ and theoretical values at the ZAMS presented a high correlation with the corresponding change in log g in such a way that the parameter $\log (k_2R^2)$ is almost independent of evolution within the main sequence for a given mass. Errors in $\log k_2$ for individual systems were found in the range from 0.03 to 0.10 thus showing that the observed systematic effects were real. In this comparison, 55 binary systems were used but, at that time, only seven out of them actually had accurate determinations of their absolute dimensions.

5.2. Using evolutionary models

New models have been computed later which include the internal density concentration coefficients. Hejlesen (1987) published a grid of models during hydrogen burning phases using his computations made during the late seventies. These models were much improved concerning the adopted physics, in particular the opacity library, by Claret and Giménez (1989). Now a days, a direct comparison of observations with models computed for specific stellar masses is possible and we have proceeded in the following way: a) We adopt a grid of evolutionary tracks for initial chemical composition (X,Z) = (0.70,0.02) and standard, but up-dated, physics (e.g., as given by Claret and Giménez, 1989). b) We compute, for the observed stellar masses of each of the components, individual evolutionary tracks and derive from the observed absolute radii, the theoretical predictions for the surface temperature, internal structure constant and evolutionary age. c) Absolute dimensions are checked by comparison, for each binary system, of the evolutionary ages derived for each of the component stars as well as comparing the observed effective temperatures (as deduced from colour indices and light curves) with the theoretically expected values. d) Individual log k_2 theoretical values are then averaged using the same weights as in log k_2 (obs) and then directly compared to each other.

In our comparison with theoretical models we have decided to use only those systems with better determinations of both absolute dimensions and apsidal motion parameters. Contrary to the case of the paper by Giménez and García-Pelayo (1982), differences with respect to evolved models computed specifically for the observed masses and radii require a higher degree of confidence. After rejection of binaries with an important relativistic contribution (some 9 cases), 20 systems were adopted, namely: EM Car, QX Car, OX Cas, PV Cas, V346 Cen, CW Cep, Y Cyg, V380 Cyg, V453 Cyg, V477 Cyg, CO Lac, GG Lup, RU Mon, U Oph, V451 Oph, Tseta Phe, AG Per, IQ Per, V1647 Sgr and V760 Sco. All of them with reliable absolute dimensions. Additional systems with independent determinations of radii were also added for the sake of completeness like AR Cas, GL Car, NO Pup, Alpha Vir and HR 7551. Still under study are several other cases with good apsial motion elements but unreliable absolute dimensions, e.g., DR Vul, FT Ori, etc.

Results of the comparison of absolute dimensions has shown a good interagrement between observed surface temperatures and those corresponding to theoretical models calculated on the base of the observed masses and radii. For a few cases, systematic differences in the effective temperatures were removed by means of small changes in the adopted initial chemical composition. Equally satisfactory was found to be the agreement between the evolutionary ages derived for each of the component stars in all case systems. This can only be interpreted as adequate selection of theoretical models and adopted initial chemical composition. Nevertheless, a systematic difference between observed and predicted values of log k_2 still exists.

Though the agreement between observations and theory with respect to the internal density concentration is certainly better than a few years ago, a detailed agreement has not been found, even taking into account evolutionary effects. Observations indicate that stellar configurations should be more centrally condensed than theoretically predicted. In our previous study (Giménez and García-Pelayo, 1982) this could not be found since the actual rate of change in log k_2 with time was not known and the assumption that the variation in log k_2 was identical to that in log g could explain all available information. Modern evolutionary models nevertheless indicate, as it was also shown by Jeffery (1984), that the mentioned rate is not so large and depends on stellar mass.

6. POSSIBLE SOLUTIONS

The interpretation of the observed systematic difference can only be based on problems with the procedure to obtain log k_2 (obs) and/or log k_2 (th). The observational value of the internal density concentration is derived from the observed apsidal motion rate, corrected for a small relativistic term, and the so-called apsidal motion coefficients, c_i, which reflect the effect of the rotational and tidal terms in the apsidalmotion rate. A systematic error in the estimation of (c_1+c_2) is not supported by either theory or observations. On one hand, all tidal evolution theories indicate that both spin axes should be perpendicular to the orbital plane and tidal lag almost negligible. On the other hand, no correlations has been found between (c_1+c_2) and the difference in log k_2 (observed - theoretical).

Some alternative theories of gravitation (e.g., the non-symmetric theory described by Moffat (1984) have been proposed to explain the case of binaries like DI Her (see paper by Ed Guinan in this volume). Nevertheless, we have found that applying Moffat's equations to our sample produce an even worst interagreement in general though some individual cases could be better reproduced. In particular for AG Per, while general relativity agrees well with observed parameters, the non-symmetric theory of gravitation predicts an apsidal motion rate around 35% slower than observed.

Some other involved parameters, like rotational velocities, have been carefully checked and show good agreement with expected values (Giménez and Andersen, 1983). No simple explanation seems thus to be present. A second alternative is, of course, that models may be systematically wrong and stellar cores may actually be more massive than predicted with standard theories. We are now working on three different possibilities: the effects of rotation in the interior, overshooting in the core and mass loss.

One very suggestive observational result has nevertheless been found. A significant correlation is present between internal structure differences (observed - theoretical values of log k_2) and log g, in the sense that log k_2 differences are largest for the more evolved systems and negligible for those close to the ZAMS. This effect is similar to that found by Giménez and García-Pelayo (1982) but with respect to evolved models instead of just ZAMS configurations. Again it seems that the rate of change of log k_2 seems to be larger according to empirical results than from theoretical models.

Preliminary computations indicate that the existence of overshooting in the core improves the interagreement but, if it is kept within reasonable boundaries, it is certainly not sufficient to explain in a quantitative form the whole range of log k_2 differences. The best observational approach to the problem is probably a detailed study of the more evolved systems, which of course are also the more difficult to analyse, as it was already indicated by Giménez (1984). Overshooting certainly helps by delaying the TAMS in the log k_2 - log g plane since models predict a reversal in the variation of log k_2 toward less centrally condensed configurations after the complete depletion of hydrogen in the core.

REFERENCES:

Andersen, J., Clausen, J.V., Nordström, B.: 1980. IAU Symp. No. 88, eds. M.J. Plavec, D.M. Popper and R.K. Ulrich, Reidel, Dordrecht, p. 81.

Claret, A., Giménez, A.: 1989. Astron. Astrophys. Suppl., 81, 1.

Giménez, A.: 1981. In *Photometric and Spectroscopic Binary Systems*, NATO Adv. Study Inst., Reidel, Dordrecht, p. 511.

Giménez, A.: 1984. In *Observational Tests of the Stellar Evolution Theory*, Reidel, Dordrecht, p. 419.

Giménez, A.: 1985. Astrophys. J., 297, 405.

Giménez, A., Andersen, J.: 1983. *Journées de Strasbourg, 5émé Reunion*. Obs. de Strasbourg, University of Strasbourg, p. 155.

Giménez, A., García-Pelayo, J.M.: 1982. In *Binary and Multiple Stars as Tracers of Stellar Evolution*. Reidel, Dordrecht, p. 37.

Giménez, A., García-Pelayo, J.M.: 1983. Astrophys. Space Sci., 92, 203.

Hejlesen, P.M.: 1987. Astron. Astrophys. Suppl., 69, 251.

Jeffery, S.: Monthly Not. R. Astron. Soc., 207, 323.

Kopal, Z.: 1978. *Dynamics of Close Binary Systems*. Reidel, Dordrecht, Holland.

Moffat, J.W.: 1984. Astrophys. J., 287, L77.

ON THE ORBITAL CIRCULARIZATION OF CLOSE BINARIES

TSEVI MAZEH
The Wise Observatory, Tel Aviv University
and
Harvard-Smithsonian Center for Astrophysics

DAVID W. LATHAM
Harvard-Smithsonian Center for Astrophysics

ROBERT D. MATHIEU
Department of Astronomy, University of Wisconsin - Madison

BRUCE W. CARNEY
Department of Physics and Astronomy, University of North Carolina

ABSTRACT. Recent theoretical and observational work on the tidal circularization of late-type close binaries is reviewed. In a coeval sample of binary systems, short-period binaries are found to have circular orbits, while most long-period systems display eccentric orbits. Two new competing theories of tidal circularization have been suggested recently to account for this phenomenon, both of which are very different from the old theoretical approach. On the observational side, four samples with different ages exhibit transitions between circular and eccentric binaries at distinct periods. The different transition periods, which increase monotonically with the sample age, suggest that the circularization processes are effective during the main-sequence phase of stellar evolution.

C. İbanoğlu (ed.), Active Close Binaries, 145–154.

1.Introduction

For a coeval sample of binary systems with similar components, tidal interaction will tend to circularize the orbits of the close binaries, while the wide binaries will retain their orbital eccentricities. Such a difference between the eccentricity distribution of close and wide binaries has been noticed by Koch and Hrivnak (1981), who considered a sample of all late-type binaries of the Batten *et al.* (1978) catalog. Despite the fact that the sample was not a coeval one, they found that most binaries with periods shorter than 8 days display circular orbits, while most systems with periods longer than this period have orbits with significant eccentricities. The first to study a *coeval* sample of binaries were Mayor and Mermilliod (1984), who considered 33 solar-mass main-sequence spectroscopic binaries in the Hyades, Praesepe and Coma Berenices open clusters. In their seminal work, Mayor and Mermilliod found a sharp transition between circular and eccentric binaries at a period of 5.7 days. Mayor and Mermilliod suggested that the short-period binaries have been circularized either during the short pre-main-sequence contraction phase or throughout the main-sequence lifetime of the sample.

Mathieu and Mazeh (1988) recently pointed out that if the circularization of the short-period binaries takes place when the stars are on the main sequence, a different transition period, hereafter called a cutoff period, is expected for samples of different ages. An observed cutoff period can be used, therefore, as an age indicator; a sample of binaries in an open cluster, for example, can be used to estimate the cluster age. Mathieu and Mazeh pointed out that this method of age estimation is, in principle, independent of the theory of stellar evolution. The use of the new proposed "clock" depends, however, on stellar interior models, on our understanding of tidal circularization, and on the availability of large samples of spectroscopic binaries. Such large samples are becoming available as a result of the stellar radial-velocity spectrometers in full operation at several observatories (see Phillip and Latham 1985). Mathieu and Mazeh even suggested that the new clock can be used to estimate the age of the Galaxy, provided a large sample of spectroscopic halo binaries will be found.

Mathieu and Mazeh (1988) pointed out that the old open cluster M67 also displays a cutoff period, between 10.3 and 11.0 days. The two cutoff periods - 5.7 and about 10.5 days, appear in two open clusters of very different ages: 0.8 and 5 billion years, respectively (e.g. Janes 1988). The two observed cutoff periods are consistent with the notion that the older the sample, the longer its cutoff period.

Given the preliminary state of the circularization theory and its dependence on stellar structure, use of the proposed new chronometer must wait for a substantial improvement of the theory. As pointed out by Mathieu and Mazeh (1988), several observed cutoff periods for binary samples of known evolutionary ages may first be used to confront the timescales of tidal circularization with the better established timescales of stellar evolution theory. After establishing the correct theory of circularization, the age of other samples of binaries could be derived with the new clock, supplying further consistency tests of the theory of stellar evolution.

Since the work of Mathieu and Mazeh (1988), two parallel lines of efforts have been followed. On the theoretical side, two competing treatments of the circularization processes have been published (Tassoul 1988; Zahn and Bouchet 1989), both of which are very different from the previous tidal approach (e.g. Zahn 1966, 1977; Alexander 1973; Lecar Wheeler and McKee 1976; Hut 1981). The work of Zahn and Bouchet is of particular interest, as they claim that the circularization of the late-type binaries takes place mostly during the pre-main-sequence phase of the system. On the observational side, the eccentricity distribution of two other coeval binary samples have been studied (Latham *et al.* 1988a; 1988b; Jasniewicz and Mayor 1988; Mathieu, Walter and Myers 1989). The new data suggest a cutoff period of about four days for pre-main-sequence stars, and a cutoff period of about thirteen days for the halo stars. In this paper we summarize very briefly the new theories of circularization, review the status of the observed cutoff periods, and comment on the possible implications of the new observations for these theories.

2. The Theoretical Controversy

The clock suggested by Mathieu and Mazeh (1988) was based on the theoretical approach of Lecar, Wheeler and McKee (1976), who followed the dissipation theory of Alexander (1973). In this approach one approximates the tidal interaction between two stars in a binary system by assuming that each star retains its equilibrium configuration except for a localized tidal bulge. Due to stellar viscosity, the bulge of each star lags after (or precedes) the line connecting the centers of the two stars with an angle proportional to the apparent angular velocity of the other star. Mathieu and Mazeh (1988) have shown that within this theory, the circularization timescale for late-type stars with convective envelopes is proportional to the orbital period of the binary to the power 16/3. Such a strong dependence on the binary period can account for the sharp transition between the circular and eccentric binaries found in the Hyades/Praesepe and M67 samples.

Mathieu and Mazeh (1988) used the cutoff period of the Hyades to check the numerical values of the constants given by the theory. The cluster age derived from the Hyades/Praesepe cutoff period of 5.7 days was in reasonable agreement with that given by stellar evolution theory, perhaps remarkably so, given the many approximations in the theoretical development. Application of the circularization theory to the M67 eccentricity distribution, however, indicated that the tidal circularization process was more effective than this theory would suggest.

Recently, Tassoul (1988) suggested a completely different mechanism for the circularization of short-period binaries. It involves large-scale, transient meridional currents induced by the tidal distortion of the stellar axial symmetry. The proposed mechanism (Tassoul 1987) was originally suggested to account for the synchronization or pseudo-synchronization observed in early-type stars with radiative envelopes. Tassoul (1988) further claimed that when applied to late-type binaries, this novel mechanism is more efficient than the previously suggested ones. Within this theory, the circularization timescale is proportional to the period of the binary system to the power of 49/12. As this power law is different than the one predicted by the theory of Lecar, Wheeler and McKee (1976), several observed cutoff periods in samples of different ages might be able to decide between the two theories. Tassoul in fact argues that his theory might better explain the M67 cutoff period.

A different approach to the orbital circularization was suggested very recently by Zahn and Bouchet (1989). They argued that for binary systems with masses ranging from 0.5 to 1.25 $M_\odot$, most of the orbital circularization occurs during the pre-main-sequence phase. The subsequent decrease in eccentricity on the main sequence is negligible. Using a new version of the secular equations governing the dynamical evolution of close binaries (Zahn 1989), Zahn and Bouchet predicted that all samples would have a very similar cutoff period, independent of the sample age. The actual value of the cutoff period depends on the initial conditions of the sample at the beginning of the Hayashi track, and should be found in the range between 7.2 and 8.5 days. They claimed that this theory is consistent with the 8 day cutoff period found by Koch and Hrivnak (1981) in the sample of all late-type binaries of the Batten *et al.* (1978) catalog.

Zahn and Bouchet (1989) argued that the cutoff periods derived from the M67 and the Hyades/Praesepe samples are both erroneous. They claimed that the upper limit for the 5.7 day cutoff period of the Hyades found by Mayor and Mermilliod is based only on two single-line binary systems - VB 121 (with orbital period, P, of 5.75 days and eccentricity, e, of 0.35) and KW 181 (P=5.87 days; e=0.35). They suggested that these binary systems have been circularized during the pre-main-sequence phase, and the present eccentricities were acquired only in the late stages of their evolution. In both cases, they pointed out, the unseen secondary is estimated to have a mass smaller than 1 $M_\odot$, and therefore could be a white dwarf. The history of both systems could, thus, include episodes of mass transfer and mass ejection, during which the eccentricity could have been increased. Without VB 121 and KW 181, the cutoff period of the Hyades/Praesepe would be about 8.5 days. Zahn and Bouchet also emphasized the fact, pointed out by Mathieu and Mazeh (1988), that since the work of Mayor and Mermilliod another critical Hyades binary system was discovered. They argued that this system - J331 (P=8.5 days; e=0), stands in an apparent contradiction to the claimed cutoff period of 5.7 days and fits well their theory. Zahn and Bouchet used a similar argument with regard to M67, where the *lower* limit of the cutoff period was set by S986 (P=10.34; e=0). They claimed that this system was circularized very recently, while S986 was expanding rapidly towards the giant branch. Except for S986 the sample of M67 binaries could have a cutoff period of about 8.5 days.

To decide between the different theories of tidal circularization we clearly need several similar homogeneous samples of binaries which cover a substantial baseline in stellar age. New observational input is therefore of particular importance at this stage. In the next section we summarize the accumulated data which address the theoretical controversy.

3. The Observational Status of the Cutoff Periods

In the last two years new observational information about three samples of late-type binaries have been accumulated, all of which suggest a sharp transition between circular and eccentric binaries.

3.1. THE HALO STARS

Latham *et al.* (1988a,b) and Jasniewicz and Mayor (1988) found independently a cutoff period in the metal-deficient stars in the halo of the Galaxy. This sample was presumably formed in the first stages of the Galaxy, and its age is of the order of 10 billion years. The two studies yielded similar cutoff periods: Jasniewicz and Mayor found the cutoff period to be between 12.4 and 18.8 days; Latham *et al.* (1988b) found a cutoff period between 11.7 and 13.7 days. The findings of Latham *et al.* are based on ongoing systematic monitoring of the radial-velocities of about 1500 high proper-motion stars. In this study, which has been conducted at the CfA for several years, the metallicity for most of the stars is also derived (Carney *et al.* 1987; Laird *et al.* 1988). Table I presents the updated results of the CfA study for the metal-deficient systems, as of August 1989. The table lists all binaries for which orbital elements have been already derived, with metallicity $[m/H] \leq -1.0$ and orbital period less than 40 days. Insignificant eccentricities are listed as 0.0. The orbital solutions for most of these binaries have been already published (see Latham *et al.* 1988b). Unpublished orbits are marked with asterisks.

A cutoff period between 12.4 and 13.7 days is suggested by the data of Table I. The lower limit of the cutoff period relies on two circular binaries with similar periods - G176-27 and HD111980 (P=11.7; P=12.4 days, respectively). The upper limit is set by the eccentric system G87-47 (P=13.7 days; e=0.46). Note, however, that the table also includes an almost circular binary - G65-22, at a period of 18.7 days.

Table I. The Halo binaries

Name	Period *(days)*	Eccentricity
CM Dra	1.27	0.0
BD +13°13	1.84	0.0
HD 85091	3.39	0.0
G183-9 *	6.20	0.0
CoD-48°1741	7.56	0.0
BD +5°3080	9.94	0.07
G176-46B *	10.44	0.04
G66-59 *	10.74	0.01
G176-27	11.73	0.02
HD111980	12.43	0.0
G87-47	13.73	0.46
G65-22	18.74	0.04
G253-44 *	19.39	0.52
G88-10	20.63	0.30
G236-38	26.70	0.26
G190-10	30.15	0.23
G206-34	32.39	0.50

3.2. THE PRE-MAIN-SEQUENCE STARS

A compilation of all known low-mass pre-main-sequence (PMS) binaries with orbital determinations was published very recently by Mathieu, Walter and Myers (1989). Their results are listed in Table II. The data suggest a transition between the circular and eccentric orbits at about 4 days. The upper limit for the transition period relies on EK Cep, an eclipsing binary with a main-sequence primary of a mass of 2 $M_\odot$ and a PMS secondary of mass of about 1.2 $M_\odot$ (Popper 1987). Unfortunately, no known PMS system has a period between 4 and 10.4 days. Therefore, the confinement of the transition period to the value of 4 days depends on one binary only.

Table II. The PMS binaries

Name	Period *(days)*	Eccentricity
155913-2233	2.42	0.0
V4046 Sgr	2.43	0.0
V826 Tau	3.91	0.0
EK Cep	4.43	0.11
160905-1859	10.40	0.18
AK Sco	13.61	0.47
P1540	33.73	0.12
162814-2427	35.95	0.49
162819-2423S	89.0	0.41
160814-1857	144.5	0.24

3.3. THE OPEN CLUSTER M67

Extensive radial-velocity observations of the stars in the open cluster M67 (Mathieu *et al.* 1986) yielded eight main-sequence spectroscopic binaries (Mathieu, Latham and Griffin 1989). The binaries, their periods and the orbital eccentricities are listed in Table III. As already noted by Mathieu and Mazeh (1988), a cutoff period between 10.3 and 11.0 days is suggested by the table. The lower limit of the cutoff period still relies on one system - S986. The sample represents a homogeneous set of evolved main-sequence stars, with a cutoff period substantially larger than that of the Hyades/Praesepe binaries.

Table III: The M67 main-sequence binaries

Name	Period *(days)*	Eccentricity
S1234	4.36	0.05
S1024	7.16	0.01
S1045	7.65	0.02
S986	10.3	0.0
S1272	11.0	0.26
S1508	25.9	0.44
S1216	60.4	0.45
S251	951.	0.53

Table IV: The Suggested cutoff periods

Sample	Evolution Age (10^9 *yrs)*	Cutoff Period *(days)*
PMS Stars	0	4
Hyades/Praesepe	0.8	5.7
M67	5	10.3 - 11.0
Halo	10	12.4 - 13.7

4. Discussion

Four samples of coeval spectroscopic binaries exhibit transitions between circular and eccentric orbits at distinct (albeit different) periods. In the four samples, all binaries with substantial eccentricities have periods longer than a period which we have termed a cutoff period. The cutoff periods, summarized in Table IV, range between 4 and 14 days.

A word of caution should be emphasized here. The four samples include two or three systems - BD +5°3080, G176-46B and possibly S1234, which display small eccentricities despite the fact that their orbital periods are shorter than their sample cutoff periods. The samples also include two systems with circular or nearly circular orbits - G65-22 and J331, with periods longer than the corresponding cutoff periods. The existence of these systems is not in conflict with the claimed cutoff periods of Table IV, as we expect some exceptions to the simple picture of a distinct transition between circular and eccentric orbits. Within the transition picture, the only feature which is strictly true is:

ALL binaries with substantial eccentricity, larger than, say, 0.1, have periods longer than the cutoff period.
None of the systems discussed above are in conflict with this statement.

Different interpretations can be suggested for the existence of short-period binaries with small eccentricities and long-period binaries with circular orbits. The short-period binaries with small eccentricities could have a third star in the system (Mazeh 1990), which can induce small eccentricity into a circular orbit (Mazeh and Shaham 1979). The long-period circular binaries might reflect the original distribution of eccentricities of the samples. Presumably, any large enough sample of binaries is formed with some nearly circular systems. We are not aware of any theoretical work which proves that binaries cannot be formed with nearly zero eccentricity. The sample of main-sequence stars considered by Koch and Hrvinak (1981), for example, includes several circular binaries with periods longer than 1000 days, as a close study of their Figure 1 would reveal. Alternatively, some of the long-period circular orbits could have been circularized when the unseen secondary star, presumably now a white dwarf, went through the giant phase.

The monotonic increase of cutoff period with sample age shown in Table IV is most simply interpretable as the product of main-sequence tidal circularization. The data do not seem to be as naturally interpreted within the theory suggested by Zahn and Bouchet (1989), which implies that the cutoff period of any sample should be between 7.2 and 8.5 days, independent of the sample age. Clearly, the halo binaries display a different cutoff period. The M67 sample also seems to display a cutoff period longer than 8.5 days. S986, on which the suggested 10.5 day cutoff period relies, could not have been circularized throughout its evolution off the main sequence (see Mathieu and Mazeh 1988 for detailed discussion). Even without the halo and the M67 samples, we still are left with two or three systems - VB 121 and KW 181 (Hyades/Praesepe binaries), and possibly EK Cep (PMS binary), having eccentric orbits and periods less than 7.2 days.

Based on all these cases, we would like to suggest that the overall evidence is somewhat in favour of an age dependence of the cutoff period, suggesting effective tidal circularization throughout the main-sequence phase of stellar evolution.

In conclusion, the accumulated data supply strong evidence for tidal circularization of late-type short-period binaries. However, the mechanism behind this phenomenon is still not clear. An observational clue for understanding this mechanism is the cutoff period between the circularized and the eccentric orbits in coeval samples of binaries. The present observations suggest that the cutoff period varies with the sample age. More data is needed to find the exact dependence of the cutoff period on the sample age and to reveal the nature of tidal circularization.

ACKNOWLEDGEMENT. We express our deepest thanks to J. Laird for his participation in deriving the metallicities of the halo stars. We are grateful to I. Goldman for critical reading of the manuscript and to Dr. J.-P. Zahn for sending a copy of his fascinating papers in advance of publication.

References

Alexander, M.E. (1973), *Astrophys. Space Sci.*, **23**, 459.

Batten, A.H., Fletcher, J.M. and Mann, P.J. (1978), *Publ. Dom. Astrophys. Obs. Victoria B.C.*, **15**, 121.

Carney, B..W., Laird, J.B., Latham, D.W., and Kurucz, R.L. (1987), *Astron. J.*, **94**, 1066.

Hut, P. (1981), *Astron. Astrophys.*, **99**, 126.

Janes, K.A. (1988), in *Calibration of Stellar Ages*, ed. A.G. Davis Phillip (Schenectady: L. Davis Press) p. 59.

Jasniewicz, G., and Mayor, M. (1988), *Astron. Astrophys.*, **203**, 329.

Koch, R.H. and Hrivnak, B.J. (1981), *Astron. J.*, **86**, 438.

Laird, J.B., Carney, B.W., Latham, D.W., and Rupen, M. (1988), *Astron. J.*, **95**, 1843.

Latham, D.W., Mazeh, T., Carney, B.W., McCrosky, R.E., Stefanik, R.P., and Davis, R.J. (1988a), in *Calibration of Stellar Ages*, ed. A.G. Davis Phillip (Schenectady: L. Davis Press) p. 185.

Latham, D.W., Mazeh, T., Carney, B.W., McCrosky, R.E., Stefanik, R.P., and Davis, R.J. (1988b), *Astron. J.*, **96**, 567.

Lecar, M., Wheeler, J.C., and McKee, C.F. (1976), *Astrophys. J.*, **205**, 556.

Mathieu, R.D., Latham, D.W., Griffin, R.F. and Gunn, J.E. (1989), *Astron. J.*, **92**, 1100.

Mathieu, R.D., Latham, D.W., and Griffin, R.F. (1989), a preprint.

Mathieu, R.D., and Mazeh, T. (1988), *Astrophys. J.*, **326**, 256.

Mathieu, R.D., Walter, F.M., and Myers, P.C. (1989), *Astron. J.*, **98**, 987.

Mayor, M., and Mermilliod, J.-C. (1984), in *Observational Tests of Stellar Evolution Theory*, IAU Symposium 105, eds. A. Maeder and A. Renzini (Dordrecht: Reidel) p. 411.

Mazeh, T., and Shaham, J. (1979), *Astron. Astropys.*, **77**, 145.

Mazeh, T. (1990), *Astron. J*, in press.

Phillip, A.G.D., and Latham, D.W. (1985), *Stellar Radial Velocities*, IAU Coll. 88 ed. A.G. Davis Phillip and David W. Latham (Schenectady: L. Davis Press)

Popper, D.M. (1987), *Astrophys. J. Lett.*, **313**, L81.

Tassoul, J.L. (1987), *Astrophys. J.*, **322**, 856.

Tassoul, J.L. (1988), *Astrophys. J. Lett.*, **324**, L71.

Zahn, J.-P. (1966), *Astron. Astrophys.*, **29**, 489.

Zahn, J.-P. (1977), *Astron. Astrophys.*, **57**, 383.

Zahn, J.-P. (1989), *Astron. Astrophys.*, in press.

Zahn, J.-P. and Bouchet L. (1989), *Astron. Astrophys.*, in press.

STELLAR STRUCTURE AND EVOLUTION WITH RESPECT TO BINARITY

H.Mauder
Astronomical Institute
Waldhäuser Str.64
D-7400 Tübingen
Germany

ABSTRACT. The importance of close binary stars with respect to calculations of stellar structure and evolution is demonstrated for the system CW Eri. The problem of a consistent description of contact binaries is discussed. The TYCHO photoelectric survey should substantially increase the number of known EW stars with small amplitude.

1. INTRODUCTION

With the publication of the pioneering book of M.Schwarzschild "Structure and Evolution of the Stars" in 1958, a large number of theoretical investigations was initiated. Thanks to these efforts, the internal constitution of stars and their evolution up to and including helium burning phases are understood today in great detail. Several problems, however, are not yet solved satisfactorily due to the incomplete knowledge of the physics involved. For instance, the complex network of nuclear reactions in the later stages of stellar evolution with the uncertainties in the cross sections and reaction rates allows for a statistical treatment only. Even in the earliest phases of hydrogen burning there may be some open questions, as can be seen from the solar neutrino problem.
Another difficulty arises from the correct treatment of convective energy transport, especially in the outer layers of stars, where the adiabatic approximation becomes inadequate. In these layers, a complete mixing length theory must be applied with the ratio of mixing length over pressure scale heigth as a free parameter. However, since the radius of a star in hydrostatic equilibrium

C. İbanoğlu (ed.), Active Close Binaries, 155–160.

depends sensitively on this ratio, a precise determination of radii in close binary systems allows for an empirical determination of this parameter, as will be seen below. Another difficulty arises if stellar evolution in very close binary systems is investigated. In several cases the two components may come into physical contact which in turn leads to severe problems with respect to hydrostatic and thermal equilibrium of the binary. On the other hand, a large number of contact binaries is found observationally with several rather unusual properties. Only recently, a possible explanation for the contact binary paradoxon was found, which is in satisfying agreement with the observational results.

2. EVOLUTION IN CLOSE BINARIES

Stellar evolution is characterized by a large increase of the radius as soon as the star evolves away from the main sequence after central hydrogen depletion. Since possible radii in close binary systems are confined by the respective limiting Roche lobes, it is evident, that the evolution of a binary component will differ from the evolution of a single star as soon as the critical lobe is reached. Several authors have given the general scenario of evolution in close binaries, see e.g. Weigert (1968). Subsequent investigations led to an understanding of different types of celestial objects like WR binaries, X-ray binaries or cataclysmic variables as a consequence of mass transfer between the two components. Vicarious for many others see e.g. Hellings and de Loore (1986) or Ritter (1986). The consideration of accretion discs, mass loss and angular momentum loss from the system became increasingly important. The most recent reports on evolution in close binaries will appear in "Highlights in Astronomy 8", IAU General Assembly 1988.
Even in well detached systems, still close to the main sequence, evolutionary effects may be of great interest. In such a case, when double lined radial velocity curves and accurate light curves are available, remarkably precise values for the masses and radii of the two stars can be derived, with relative errors of a few percent. Besides of that, the temperatur difference of the two components can be obtained with high accuracy, say ± 20 K. Therefore, theoretical calculations of stellar evolution can be critically tested against observations, as is demonstrated e.g. by Mauder (1982) for the case of CW Eri. Hejlesen (1980) has published an extensive set of model calculations for stars of different mass, chemical composition and different values for the ratio of mixing length over pressure scale heigth. In his graphs, isochronic lines connect points of equal age on the different evolu-

tionary tracks. This is especially important in the case of close binary systems, since the two components are of equal age. The comparison of the observations with the theoretical graphs, therefore, gives an overdetermination of the problem. If, nevertheless, a unique solution can be found, this will be a strong indication for the correctness of the calculations. In the case of CW Eri, masses and radii were found with an accuracy of about 1% for the masses and 3% for the radii, giving also the surface gravities. The effective temperature of the primary component was derived from the spectrum and the temperature of the secondary was obtained from the light curve solution. In Hejlesens graphs, the evolutionary tracks are plotted as surface gravity over effective temperature. The observed points (g,T) for both components must lie on the respective tracks given for their mass, within the observational uncertainty. This is the first overdetermination which can be removed only by a change of the chemical composition, within reasonable limits. This fixes the relative values for CW Eri to $Y = 0.28 \pm 0.02$ and $Z = 0.020 \pm 0.002$. On the other hand, both components must lie on an isochronic line, which again is an overdetermination. Indeed, an age of 1.5 billion years is found for both components in CW Eri, see the figure in the paper of Mauder (1982), which is an indication, that the calculations of stellar evolution are very precise for the phases of stellar central hydrogen burning. It should be noted, that a unique solution is obtained for a ratio of mixing length over pressure scale heigth, $l/H = 2.0$ only.
In a series of papers from Copenhagen Observatory, similar investigations were performed. As an example, the paper on V 760 Sco by Andersen et al. (1985) is mentioned. It would be interesting, to determine the metal content of these stars from the spectra: in this case, the only remaining adjustable parameter would be the effective temperature as a function of spectral type.

3. CONTACT BINARIES

For the calculation of stellar evolution, spherical symmetry is a sufficient approximation. Even in the phase of Roche lobe overflow and therefore mass transfer from one component to the other, the two components can be approximated by spheres and for the Roche lobes, spheres of equal volume are sufficient. This is no longer true if contact configurations are taken into account, since deviations from spherical symmetry obviously play an important role. Three-dimensional structure equations, however, are too complex to be solved for with present computers. Therefore, approximate solutions must be sought for.

Any theory of contact binaries should explain the basic observational facts:

a) Light curves with almost equally deep primary and secondary minima
b) Empirical period-colour relation
c) Large range in spectral types and total masses
d) Absence of symmetrical systems, i.e. mass ratio $q \neq 1$

Especially difficult to explain is the "light curve paradox", the equally deep minima, which in turn requires approximately equal surface temperature for both components, regardless of the remarkably different masses and radii. Therefore, substantial energy transfer from one component to the other is necessary. In this case, however, extreme difficulties with respect to hydrostatic and thermal equilibrium arise. For late type systems, the W UMa stars, Lucy (1968) developed the common convective envelope model, which solved the problem of hydrostatic equilibrium, but neither the thermal equilibrium problem nor the period-colour relation. In addition, no solution at all was possible for the early type systems. Several attempts were made in the following years to improve the model, including evolutionary cycles on a thermal timescale, but in all cases, if one of the problems was solved, another one increased. Only recently, a promising approach was found by Kähler (1989), which eventually might explain the stability and properties of contact binaries. Kähler starts with a discussion of the general assumptions in the pioneering Lucy model:

(H) Hydrostatic equilibrium
(T) Thermal equilibrium
(E) Equal entropy in the adiabatic part of the common convective envelope
(L) Temperature gradient nabla determined by local quantities

As a consequence of the model, energy flow from the primary to the secondary takes place, which produces satisfying W UMa light curves. However, the period-colour relation is not satisfied, solutions with unequal components are possible only for a total mass of about 2.5 solar masses, and thermal instabilities occur, leading to the theory of thermal relaxation oscillations, which in turn cannot be confirmed by observations. Modifications of the basic assumptions led to the introduction of new difficulties. A relaxation of assumption (E) by e.g. Biermann and Thomas (1972) explained the period-colour relation, but resulted in remarkably different depths of minima in the light curves. Relaxation of assumption (T) by e.g. Lucy (1976), Flannery (1976), Robertson and Eggleton (1977) led to the thermal relaxation oscillations theory, again with unequal depths of minima for a considerable part of the time. Similar difficulties arose from a relaxation of (E) and (T), see e.g. Hazlehurst and Refsdal (1980) and Rahunen (1982). In all cases, energy transfer in the adiabatic part of the common convective envelope is necessary. However, no consistent solution is possible under assumption (H), there is not enough freedom to solve for the light curve paradoxon. Kähler therefore investigated the consequences of a relaxation of (H) and/or (L). In all cases, the light curves are good, if the equal entropy condition is fulfilled, and bad otherwise. This is already a consequence of (H), assumption (E) is therefore redundant! It is for this reason, why all models with hydrostatic equilibrium and relaxation of (E) are inconsistent. Till now, in all models it was assumed, that each component by itself is in hydrodynamic equilibrium and that the masses and radii of each component satisfy euqipotential conditions of a Roche configuration. This led to models with different entropies which cannot be in hydrostatic equilibrium. Kähler suggests, that the mechanism, which tends to equalize the surface brightness of the components might be the process of hydrostatic adjustment. Therefore, (H) must be kept, which automatically solves for the light curve paradoxon, and hydrodynamic processes connected with departures from (L) might be important. It turns out, that (L) must be abandoned and the occurence of turbulence plays an important role. This must not necessarily be the turbulence of convection. Especially turbulent motion components parallel to equipotential layers should occur. As a result, late type as well as early type contact binaries can be explained.

4. SEARCH FOR ECLIPSING BINARIES WITH TYCHO

One of the problems with respect to the origin and evolution of very close binaries, especially contact binaries, is the incompleteness of data. This is due to the fact, that probably most of the variables with amplitudes of 0.3 mag or less are not yet detected. W UMa stars with very small mass ratios like AW UMa, could therefore be rather numerous, as suggested for instance by van t'Veer (1975). This situation could improve substantially within the next few years. The - relatively - successful launch of the HIPPARCOS satellite with the TYCHO project offers the opportunity of a systematical search for new EW binaries of small amplitude. During the lifetime of the satellite, each star of the whole sky up to limiting magnitude B = 11 will be observed photoelectrically for about 100 times, with typically 6 independent observations within 5 hours. With orbital periods of less than a day, some hundred or even a thousand new EW stars with amplitudes between 0.1 and 0.3 mag can be expected, see Mauder and Høg (1987). Till now, less than 100 EW stars of all amplitudes brighter than 11th mag are known. A great progress in the understanding of W UMa stars can be expected from TYCHO results.

5. REFERENCES

Andersen,J.,Clausen,J.V.,Nordström,B.,Popper,D.M.,1985, Astron.Astrophys.151,329
Biermann,P.,Thomas,H.C.,1972,Astron.Astrophys.16,60
Flannery,B.P.,1976,Astrophys.J.205,217
Hazlehurst,J.,Refsdal,S.,1980,Astron.Astrophys.84,200
Hejlesen,P.M.,1980,Astron.Astrophys.Suppl.39,347
Hellings,P.,de Loore,C.,1986,Astron.Astrophys.161,75
Highlights in Astronomy 8, in press
Kähler,H.,1989,Astron.Astrophys.209,67
Lucy,L.B.,1968,Astrophys.J.151,1123 and 153,877
Lucy,L.B.,1976,Astrophys.J.205,208
Mauder,H.,1982, in "Binary and multiple stars as tracers of stellar evolution",D.Reidel Publ.Comp.,p.217
Mauder,H.,Høg,E.,1987,Astron.Astrophys.185,349
Rahunen,T.,1982,Astron.Astrophys.109,66
Ritter,H.,1986,Astron.Astrophys.169,139
Robertson,J.A.,Eggleton,P.P.,1977,Mon.Not.Roy.A.S.179,359
Schwarzschild,M.,1958,Structure and evolution of the stars, Princeton Univ.Press
van t'Veer,F.,1975,Astron.Astrophys.40,167
Weigert,A.,1968,Mitt.Astr.Ges.25,19

An Optimization Method for solutions of Close Eclipsing Binaries

F. Barone[1,*], L. Milano[1,*], G. Russo[1,*]

[1] Dipartimento di Scienze Fisiche dell'Università, I–80125 Napoli, Italy

⋆ Associated to the *Istituto Nazionale Fisica Nucleare*, Italy

Abstract. In this paper we show the performances of the Controlled Random Search of Price (1976) in light curve solution of close eclipsing binaries, in connection with Wilson-Devinney model for light curve synthesis (1971), comparing it with other existing procedures. Examples of solved systems demonstrate the power and reliability of this technique, in particular when applied to simultaneuos solution of light and radial velocity curves for eccentric binary systems.

1 Introduction

Photometric observations at different wavelenghts are one of the most important sources of information about close binary systems. Their analysis, in connection with that of spectroscopic observations, gives, in principle, the possibility of getting all the principal geometrical and physical parameters characterizing the systems. This task is obtained by means of synthesis methods, that, if based on reliable physical model (either analytical or numerical), may allow to obtain very good results.

Two different approaches to this problem exist. The first approach (FD) is based on the analysis of the observed light changes in the frequency domain, while the second one (TD) is based, more conventionally, on the analysis in the time domain, being light curves directly used in the solution of the problem.

The FD procedure, introduced by Kopal (see, for a review, Kopal, 1986), works according to the following scheme. Light curves are transformed by means of Fourier Transform (DFT), obtaining coefficients (harmonics) describing them in the frequency domain. These coefficients are used to evaluate the "*moments of the light curves*", quantitites that are connected to the physical and geometrical parameters of the systems by means of algebraic equations obtained from a previous theoretical analysis. The solution of these equations gives the parameters required. It is evident that the method has very interesting characteristics. Infact, while in the time domain, the relation among the parameters of an eclipsing binary system and its observed characteristics can't be analitycally inverted, in the frequency domain this relation becomes algebraic, allowing algebraic inversion. Moreover it is possible to limit the noise (filtering the data), simply stopping the evaluation of the

C. İbanoğlu (ed.), Active Close Binaries, 161–188.

Fourier Transform after a certain number of harmonics (low pass filtering) or filtering them with suitable numerical filters.

This method, perhaps misunderstood by many scientists, is not too diffused, although it is in principle very powerful and can be used in connection with methods of synthesis in the time domain, giving very useful information for a correct solution.

TD procedures, which we discuss about in this paper, can be described very simply by means of the block diagram of Fig.1. A set of input data (observed light curves) and input parameters (wavelenghts, period, etc.) are given. An algorithm of minimization generates a set of variables (the physical and geometrical parameters of the binary system to solve) for the light curve synthesis algorithm. The light curves generated are compared with the observed ones and the value of a quality coefficient (generally it is the sum of squared residuals) tells us about the goodness of the fitting. If the result is not satisfactory a new set of parameters is generated and the procedure continues (optimization procedure).

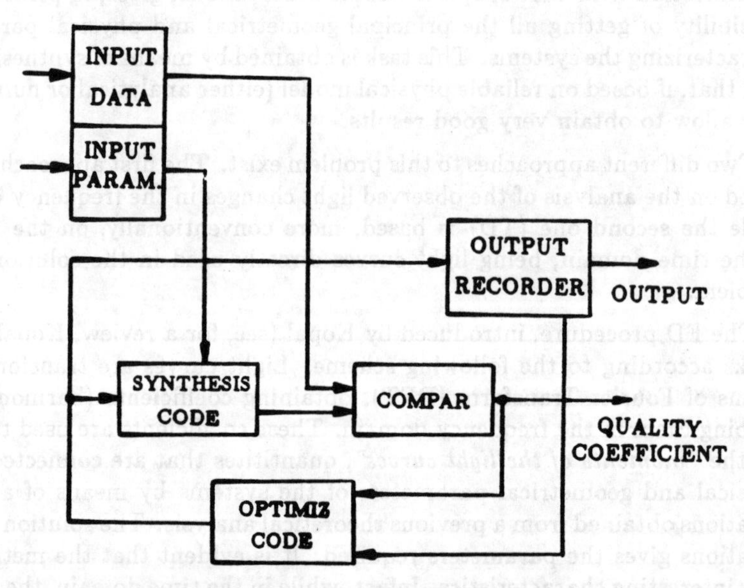

Fig.1 Block diagram of TD procedure

The two very critical parts of this procedure are the light curve synthesis model and the algorithm of minimization.

Many algorithms exist for light curve synthesis. Among them we recall the WINK code (Wood, 1971,1972), that uses the geometry of triaxial ellipsoids; the Wilson and Devinney code and successive improvements (Wilson and Devinney 1971, 1972, 1973; Wilson 1979), whose geometry is based on Roche equipotential surfaces; the LIGHT code of Hill (1979); the Linnell algorithm (1984, 1986). Among them, the model proposed by Wilson and Devinney (hereafter referred to as WD), besides being a good model because of its physical basis, has also been extensively used. Infact it has been the most used by light curves solvers over the last ten years, according to the Report of Commission 42 at the 1988 IAU General Assembly. Diffusion is important as well as goodness of a model. Infact, in this way, it is possibile to obtain solutions not only accurate, but homogeneous with the majority of other author's ones, so that general conclusions can be drawn from the comparison and study of the parameters of different objects. WD is the model used by our group (who experienced more than ten years of usage), and that, we think, overcomes the problem of a good choice of a light curve synthesis model.

The second problem, the minimization procedure, is a very critical part and will be the object of the next section.

2 Optimization Procedure

To solve the problem Wilson and Devinney proposed to use a linear least square algorithm to obtain the parameters of the system in the WD code (WDP). Linearization of the model is imposed through the least square equations of the conditions, where derivatives are computed by finite differences. The extensive use of this differential correction procedure (DCP) showed the existence of sometimes serious problems of convergence, stability and uniqueness of the solutions, that, frequently, lead to completely wrong results.

Two types of difficulties are characteristic of this kind of problem. One is introduced by the differential correction procedure, the other is intrinsic of the problem.

DCP was adequate when proposed in 1971, owing to the lack of high speed computers, but showed additional problems due to the need of a starting point and the evaluation of derivatives. Moreover DCP doesn't allow to escape from a local minimum, i.e. doesn't guarantee that the minimum found is the global one (the real solution), or lead to endless iterations (both situations have been experienced by our group in many years of light curve solutions). Preliminary solution with simpler methods (as the classical nomograph of Russell and Merrill (1952) or inferences from known physical constraints are generally of little avail, because they can seriously bias the final solution. By guess one may estimate the temperature of the primary, from an indipendent set of observations, say spectra or color indices, but, for the above mentioned reasons, the geometrical parameters should be completely unconstrained, even if preliminary values could be obtained with other methods, and that is particularly true for the mass ratio, as shown by Maceroni (1986). The same considerations apply to the "*common*" preliminary classification of the system as detached, semidetached or contact (that must be a priori chosen in WD algorithm). Moreover, the stop criterion adopted in WDP (adjustments smaller than probable errors) has no connection with the stability of the solution; that could also lead to wrong results.

Intrinsic difficulties in solving the problem are often strong correlations among the parameters, or sometimes, the same shape of the function to minimize, which may prevent from an unique determination of some of them.

A study of the properties of the function to minimize in the space of parameters can be done using the WD algorithm and computing grid of solutions (Maceroni et al , 1985; Rafert and Markworth, 1986) but requires prohibitive efforts when the number of critical parameters increases (a grid with a 10 % step over the whole range of m parameters should have 10^m starting points!).

The problem of correlation among parameters (which, at the first order, imply linear correlations among the corresponding correction) is perhaps even more difficult to deal with in the least square approach. Then, infact, one will find that the normal equations are satisfied for a considerable range of values of the correlated unknowns (i.e. the corrections), provided that these values have a certain linear relation to each other. What would be advisable to do in this case is to solve the equations just for a linear combination of the correlated unknowns as other formal solutions would not contain more real information. It has besides to be stressed that correlations cause an increase of probable errors on the corrections, so that the WDP user will be tempted to stop the program very soon. To this regard the method of subsets proposed by Wilson and Biermann (1976), and strongly criticised by De Landtsheer (1983), yelds a way to get the fulfilment of the stop criterion but, again, this has nothing to do with the stability of the solution.

Recently Plewa (1988) proposed to use, as first step, DCP and, to obtain stable solutions, two different local minimization procedures in sequence (Simplex Method and a Variable Metric Method from the CERN library) using the positive definiteness of the covariance matrix, as the stop criterion of calculation. Even if this procedure could give the possibility of obtaining stable solutions, it doesn't solve the more important problem of finding the global minimum (the real solution), if more than one minimum is present.

The right solution of the problem needs a completely different approach. It has been proved that many solution exist, if not for all, at least for many light curves. These solutions can be located very close each other, but can be also scattered: *nothing can be told a priori about them.* Therefore the problem is to choose a suitable algorithm that is not only able to find the global minimum but also to evaluate the local minima that are comparable with the global one.

A brief review of minimization algorithms can be useful to understand what can be the best strategy to use. According to the way of working, minimization algorithms can be divided in two categories:

1) Algorithms for seeking a global minimum;

2) Algorithms for seeking a local minimum;

The purpose of global minimization is to find the point in the search volume V (the volume of the possible solutions) where the objective function (the function to minimize) has its lowest value (global minimum). And while for local minimization, where the reaching of a local minimum is evaluated (the gradient is equal to zero, the second derivative is positive), for global minimization doesn't exist a criterion that ensures the global minimum has been found (Gill et al., 1981; Fletcher, 1987)

Two different strategies govern the distribution of points in the search volume V. The first one (global reliability) takes care of the fact that the global minimum can be everywhere in the volume V. This means no part of the search volume V can be neglected by the minimization procedure: this strategy would suggest a distribution of points in V the most uniform possible. The second one (local refinement) takes care of the fact that the probability of finding a better point in V is bigger around a point with a relatively low value of the objective function: this strategy would lead to create a non-uniform concentration of points around the best ones.

So, for an objective function with a concentrated distribution of minima, local strategy would be more efficient in finding new minima (thus allowing purely local minimization algorithms), but for a scattered distribution, it wouldn't increase the probability of finding a new minimum and this would automatically exclude the use of local minimization technique at the beginning of a procedure of minimization. Owing to the fact that both global reliability and local refinement are important features, the great part of global optimization methods use them together.

The choice of an optimization algorithm suitable for the problem to solve can't be done using an analytical procedure. Infact the mathematical treatment of a global optimization problem means that we have to do assumptions too restrictive that wouldn't allow to reach right results, unless we integrate it with a heuristic approach.

Three classes of global optimization algorithms exist (Törn and Zilinskas, 1989): methods with guaranteed accuracy (Covering Methods), Direct Methods, that use only local information (function evaluations) and Indirect Methods in which the local information is used for building a global model of level sets or of the objective function (see Table I).

Table I

Global Optimization Algorithms	
Method with Guaranteed Accuracy	Covering Methods
Direct Methods	Random Search Methods Clustering Methods Generalized Descent Methods
Indirect Methods	Methods approximating the level sets Methods approximating the objective functions

The possibility of application of Covering Methods is given by the fact that, in general, the objective functions have limited derivatives. Hence, a search on a grid dense enough would guarantee, in principle, to find the global minimum with a fixed accuracy. On the other side the number of points of this grid might be too large on dealing with problems with large dimensionality. Taking into account the time of CPU required for one evaluation of the objective function and the high dimensionality of our problem, it is clear that these methods can't be used in solving it.

On the other side Indirect Methods are not suitable for this task because in our case the techniques of approximating the level sets or the objective function can lead to big errors, owing to the complexity of the problem to solve and the correlations among the parameters.

More suitable are the Direct Methods. The idea of using generalization of local minimization algorithm is included in Generalized Descent Methods. Infact, in past years, many resources were spent in developing local minimization methods. Many of these algorithms have been tested by our group, like Gradient Methods (Fletcher and Reeves, 1964), that evaluate the function and its derivatives at every iteration, and the Variable Metric Methods, like the Davidon Method (1959) modified by Fletcher and Powell (1963), that, in the same way as the methods of differential geometry applied to General Relativity, considers the properties of the objective function as properties of the space of variables. If, for a moment, we don't take care of the fact that these methods are local minimization methods, they are very powerful, but their use can be difficult owing to the unknown correlations among the variables, the constrained nature of the problem and the high dimensionality of the problem. The same considerations are true for methods like the ones of Hooke and Jeeves (1961), Rosenbrock (1960), Powell (1965), Nelder and Mead (1965), etc.

The methods of Generalized Descent, introduced for the first time by Griewank (Griewank, 1981), are a generalization of these algorithms. Among them, some methods are based on the modification of the local descent equation (trajectory methods), while other ones are based on the modifications of the objective function, in order to avoid that the descent would stop in local minima found in previous steps of minimizations (penalty methods). The implementation of the algorithm can be reduced to a choice of a local minimization algorithm and of auxiliary functions whose minimization gives conditions under which the best local minima of the objective function can be found. The auxiliary functions are built using the objective functions and the penalty functions. In general the auxiliary functions become enough flat around the minimum and, for this reason, they are difficult to minimize. This part constitutes an important problem that must be still solved to ensure numerical stability to local minimization algorithms when these are applied to flat functions. A problem of this kind was found when we dealt with the solution of the light curves of the system CW Cas. The problem was very difficult to solve, owing to the fact that the function to minimize was, actually, a very large plateau, around the global minimum. We tested a lot of many different local minimization algorithms and we were not able to find the right minimum, confirming the intrinsic problem of this kind of methods, whilst using a different algorithm (see after), there were no problems for the solution.

Different considerations apply for the random search methods.. While they are quite easy to be implemented on a computer, their efficiency in searching is too low to successfully solve our problem. Infact the convergence to the global minimum is proved by the fact that the probability of reaching the neighbours of the global minimum at the current step is not less than $p > 0$. But p is normally very little, so a great number of function evaluations at random points in V is necessary to obtain a probability that is near 1. But, even if they are not frequently used, on the other side great part of global optimization methods use random search as integrant part of their strategies.

More useful are Clustering Methods which have the following interesting characteristics:

a) it is possible to obtain a sample of points in V that consists of clusters of points around the local minima of the objective function;

b) the points in the sample can be clustered giving clusters that identify the neighbours of local minimizers allowing apply local optimization methods.

c) the procedure a) and b) can be implemented in a way enough efficient on a computer with respect to other methods proposed for the global optimization.

If the global optimization procedure used, is successful, then to perform a local minimization procedure for every cluster, would determine the local minima and, then, the global minimum, too.

The procedure consists of two steps: the sampling step (initialization) and the clustering step (solution). The sampling can be deterministic, using a grid, or random. The principal idea is that of covering the whole volume V in a uniform way. To cluster the points around the minima two different strategies are used. The first one is based on the idea that keeping only the points with relatively low values of the function would lead to form groups around some local minimum (Becker and Lago, 1970; Törn, 1973). The second strategy is that of pushing every point towards a local minimum (Törn, 1973). This last technique with its double effect of removing points with high values creating points with low values would produce clusters of points around all the local minima found in the sample phase.

The clustering methods are very successful in problems of global optimization. One of the reasons is that they make it possible to combine in a very efficient way local and global search, giving enough information on encountered problems in the solution process. Infact, the output of the process, showing the evolutions of the clusters, contains a lot of supplementary information that is very difficult to formalize, but of great importance in the solution of a real problem.

Such methods, generally speaking, have acquired increasing popularity in sciences like analytical chemistry and theoretical physics, but not yet in astronomy. For example, minimization of the energy is needed to find equilibrium configurations of a system. Application of a Direct Search Method has shown (Wille, 1986) that solutions considered as such in the literature were instead only local minima, not the global one.

For our procedure we used a Direct Method: the optimization procedure named Controlled Random Search (CRS) of Price (1976).

3 The Controlled Random Search

We give, now, a brief outline of CRS (for further details, see Price, 1976). Given a function f of m variables whose minima have to be found, one has to assign limits to each of the m variables thus defining an initial search volume V. Hence the minimization procedure can be mathematically expressed by

$$S_{sol} = MIN(f(x_1, x_2, ..., x_m)) \tag{1}$$

where $(x_1, x_2, ..., x_m) \in V$, i.e.

$$\begin{gathered} x_{1_{inf}} \leq x_1 \leq x_{1_{sup}} \\ x_{2_{inf}} \leq x_2 \leq x_{2_{sup}} \\ \dots\dots\dots \\ x_{m_{inf}} \leq x_m \leq x_{m_{sup}} \end{gathered}$$

A number N of trial points is randomly chosen within this domain, consistent with the constraints. The coordinates of every point and the function values are stored in a $N \times (m+1)$ array A whose structure is

$$\begin{pmatrix} x_{11} & x_{12} & .. & x_{1m} & f_1 \\ x_{21} & x_{22} & .. & x_{2m} & f_2 \\ .. & .. & .. & .. & .. \\ x_{m1} & x_{m2} & .. & x_{mm} & f_m \end{pmatrix} \tag{2}$$

After this initialization step, the real minimization procedure begins. A trial point P is selected in a way that depends on the points stored in the array A. Infact $m+1$ points $(R_1, R_2, ..., R_{m+1})$ are randomly extracted from the array and the centroid G of the first m points $(R_1, R_2, .., R_m)$ is evaluated. Then the trial point P is determined following the algebraic operation

$$P = 2 \cdot G - R_{m+1} \tag{3}$$

If the value of the function in P is less than the maximum value stored in A, then P replaces this point, otherwise a new trial point is selected. The algorithm iterates within this scheme. The set of N points of A tends to cluster around the minima, as shown by Price (1976), but can (randomly) reach also zones far from detected minima (a useful feature to get out of the local ones).

The higher the number of lines of the array A, the higher is the probability of finding a global minimum. The higher the number of iterations, the better the definition of minima. The algorithm does not specify a particular stop criterion, leaving the choice to the user. The flow diagram is shown in Fig.2. It is worth spending a few words on the choice of the value N. Although it would be better a high value of N, in order to increase the probability of obtaining a global minimum, on the other hand a high value of N leads to occupy a lot of computer memory and to reduce the speed of convergence of the whole procedure. However there is an optimum value of N for every situation, depending on the complexity of the function to be optimized, on the type of constraints on the variables and on the research volume, so that reduction of the rate of convergence, due to an increasing of N, isn't compensated by a better performance of the algorithm. For these reasons the right choice of N is due to the experience of the user.

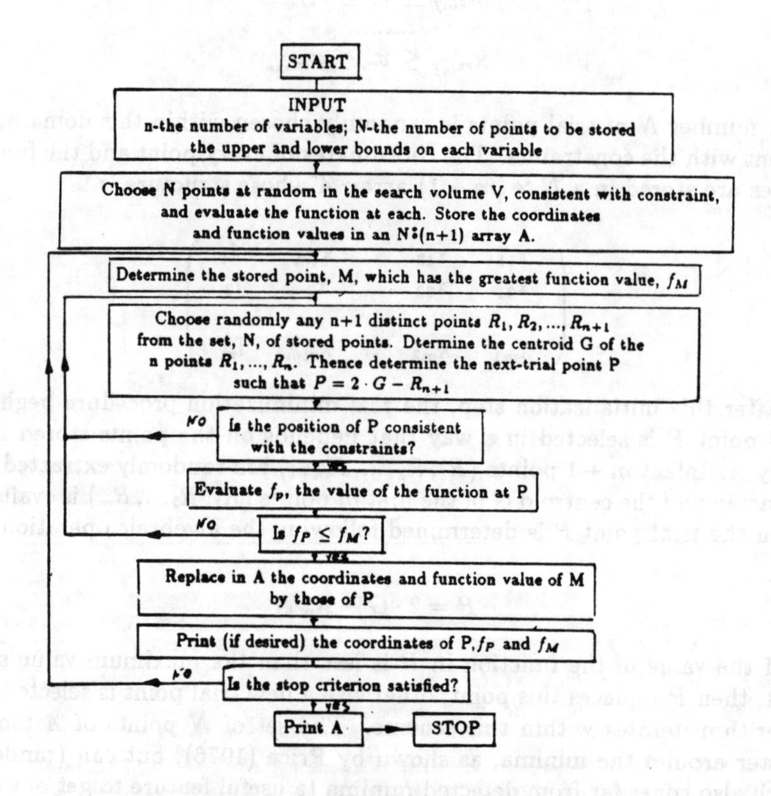

Fig.2 Flow diagram of the Algorithm of Price

We reported here two classical tests for optimization algorithms from which it is possible to understand in which way CRS works. These are bidimensional tests just for sake of explanation, but CRS works very well expecially with a very high number of variables (Barone, Milano, Russo, 1987).

The first test regards the minimization of the function of two variables

$$f(x_1, x_2) = (\mid x_1 \mid -5)^2 + (\mid x_2 \mid -5)^2$$

which has four minima with the same value of $f (f = 0)$.

The analysis of the behaviour of CRS in the evaluation of the minima of this function has the purpose of testing the method in finding many minima at the same time. Clearly this important property is characteristic of this method because local minimization algorithms are able to find just a minimum at a time.

For this test we put $N = 50$ and

$$-1000 \leq x_1 \leq 1000$$

$$-1000 \leq x_2 \leq 1000$$

as search volume.

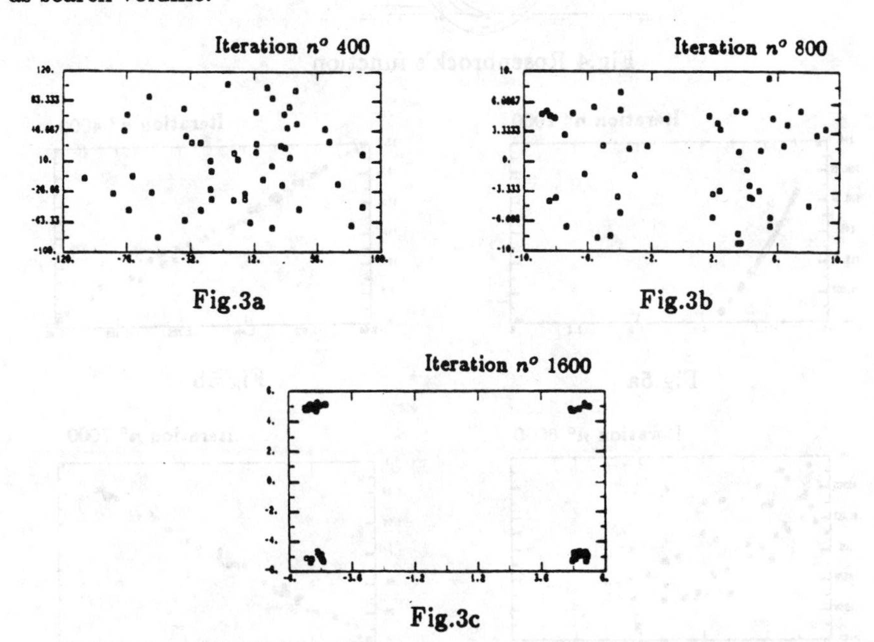

Fig.3a

Fig.3b

Fig.3c

Fig.3 Course of miniminization of $f(x_1, x_2) = (|x_1| - 5)^2 + (|x_2| + 5)^2$

In Fig. 3 it is possible to see the course of minimization. In Fig. 3a note that after 400 iterations the search points are still randomly distributed, while after 800 iterations it is possible to see that the search points are already disposed around the minima (Fig. 3b). Fig. 3c shows the search points after 1600 iterations. It is clear that the four minima have been found.

The next function we reported is the well known " Rosenbrock curved valley " function defined by

$$f(x_1, x_2) = 100 \cdot (x_2 - x_1^2)^2 + (1 - x_1)^2$$

This narrow parabolic valley, shown in Fig. 4, is probably the best known test for minimization algorithms.

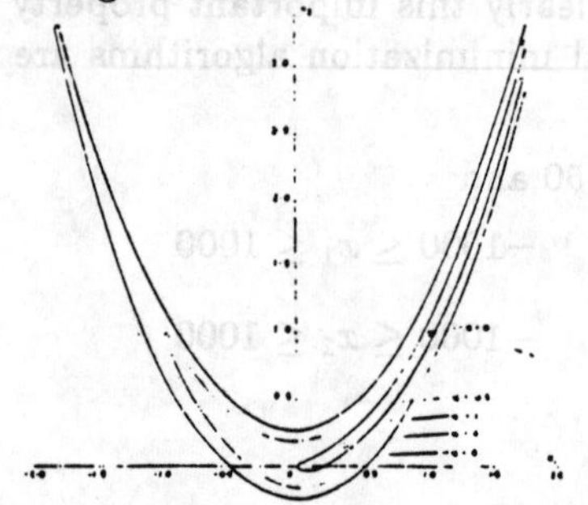

Fig.4 Rosenbrock's function

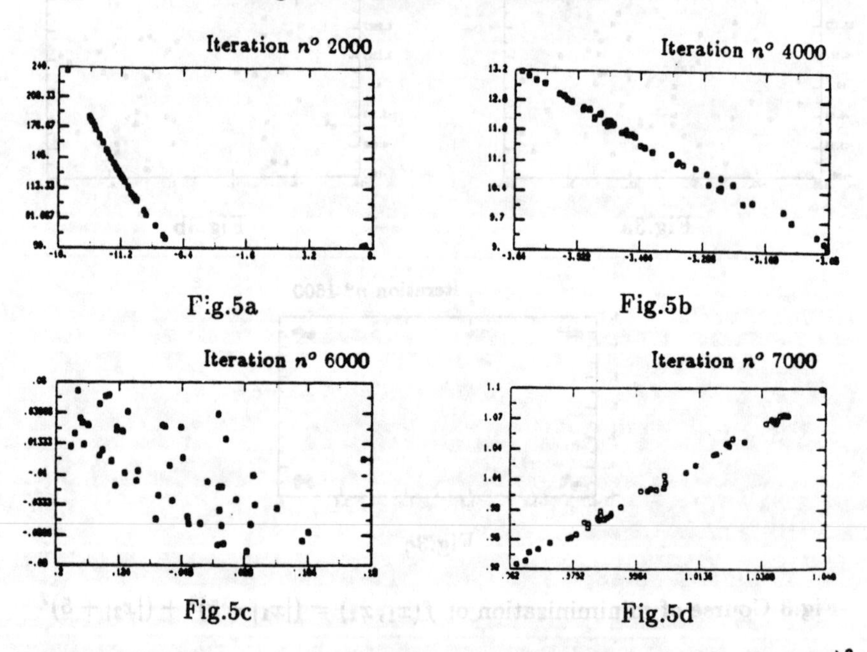

Fig.5a Fig.5b Fig.5c Fig.5d

Fig.5 Course of minimization of $f(x_1, x_2) = 100 \cdot (x_2 - x_1^2)^2 + (1 - x_1)^2$

The bottom of the valley approximately follows the parabola

$$y = x^2 + \frac{1}{200}$$

This function has one minimum in

$$x_1 = 1$$

$$x_2 = 1$$

with $f(x_1, x_2) = 0$.
The search volume chosen is

$$-1000 \leq x_1 \leq 1000$$

$$-1000 \leq x_2 \leq 1000$$

with N=50. In Fig. 5 the course of the minimization is shown.

4 Testing the Method

Before discussing about results of the implementation of the algorithm for light curve solutions, we want to spend a few words about the quality coefficient S of the goodness of fitting, that is, in practice, the objective function. An obvious choice is to fit the data in the least square sense, i.e. minimizing the value of the function (in the general case in which light curves and radial velocity curves are solved together)

$$S = \sum_{j=1}^{l} \frac{1}{n_j} \sum_{i=1}^{n_j} wl_{i,j}(l_{i,obs} - l_{i,comp})_j^2 + \sum_{k=1}^{r} \frac{1}{n_k} \sum_{l=1}^{n_k} wr_{l,k}(r_{l,obs} - r_{l,comp})_k^2 \quad (4)$$

where l is the number of light curves and r is the number of radial velocity curves, $wl_{i,j}$ and $wr_{l,k}$ are the weights of the points of light curves and radial velocity curves, n_j, n_k are the number of observations in each curve, respectively.

An important problem is to choose the values to assign to the weights. From a statistical point of view it seems reasonable to give a value that is 1 when the standard deviation is 0, or 0 when the standard deviation is infinite, with intermediate values for different errors in the measure. In this way the more precise is the datum, the higher is the weight in the context of the minimization.

Actually, the data used are rephased and a mean over groups of points is done, just to avoid having a too great number of points and spending too much time of computer, so it is very difficult to assign a weight following the previous considerations.

Hence, it is preferible to follow an heuristic way. Several years of usage of synthetic light curves generators have demonstrated that in the lightcurves there are points more important than others in the determination of the parameters of the system. For example, the points close the minima of the light curves are very important in the determination of the difference of temperature of the two components ($\Delta T = T_h - T_c$). Hence, a criterium to get a quicker convergence ensuring the reaching of the right parameters is to give a higher weight to these points. But, from another point of view, we have to take care of the instrument used for the measure. Infact, if photoelectric photometry is used, it is important to note that the higher is the value of luminosity, the better is the precision of the measure. So the points of minima are certainly characterized by greater errors than the others. For this reason the weights of the points around the minima should have a lower weight. These two different and opposite effects must be carefully analyzed in such a way to find the best function S to minimize.

The whole procedure was tested in very critical conditions of operations (Barone, Milano, Maceroni, Russo, 1988). To perform this task, we asked F. van't Veer if he could extract from his atlas of about 1500 synthetic light-curves (van't Veer, 1986) some test curves very similar to each other, but coming from different sets of parameters. It is worth noticing that the curves were generated by a different program (see Berthier, 1975) but that the underlying physical model was exactly the same; the only difference is the coordinate system chosen to describe the surfaces: van't Veer code uses in fact cylindrical coordinates, the WDM polar coordinates, but, as we will see in the following, the deviations are negligible. Clearly, as for real systems, we didn't know in advance the results.

The three curves we received, called hereafter VT1, VT2, VT3 were indeed an 'extreme' test: the maximum difference in magnitude of VT2 and VT3 with respect to VT1, chosen as reference curve, is, in fact, of only seven thousandths of magnitudes (Table 2). Fig. 6 shows the phase dependence of VT1, the other two curves are indistinguishable on that scale. Note that, at variance with the usual convention, phase 0 is assigned to the shallower minimum in van't Veer data. We retained his choice, since the WD code can easily handle this situation, for it just requires the star 1 to be that eclipsed at phase 0.

For our test we used as little information as possible on the systems, namely the G5 V spectral type of the hot component, the null albedos ($A_1 = A_2 = 0$) and the constraint of a contact configuration ($\Omega_1 = \Omega_2$). We assumed also from the theory the value of the limb darkening coefficients x_1 and x_2 (Al Namij, 1978) and of gravity darkening exponents g_1, g_2 (Lucy,1967). The initial search domain was defined by the limits on the five parameters shown in Table 3; the range fixed for q allows for both transit and occultation solutions.

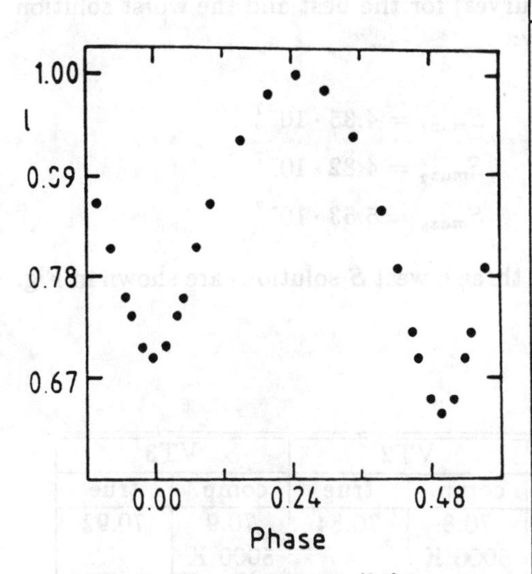

Fig.6 The first test light-curve; ordinates in light units. The curve is symmetrical around phase 0.5.

phase	ℓ_{VT1}	ℓ_{VT2}	ℓ_{VT3}
0.000	0.6918	0.6918	0.6918
0.020	0.7027	0.7040	0.7040
0.040	0.7345	0.7365	0.7359
0.050	0.7558	0.7565	0.7565
0.075	0.8136	0.8113	0.8106
0.100	0.8654	0.8598	0.8606
0.150	0.9324	0.9290	0.9333
0.200	0.9772	0.9790	0.9827
0.250	1.0000	1.0000	1.0000
0.300	0.9863	0.9827	0.9799
0.350	0.9376	0.9315	0.9315
0.400	0.8543	0.8511	0.8527
0.425	0.7907	0.7878	0.7900
0.450	0.7191	0.7178	0.7171
0.460	0.6918	0.6906	0.6893
0.480	0.6480	0.6474	0.6468
0.500	0.6310	0.6310	0.6310

Table 2 Data of the three test lightcurves.

Parameter	Mimimum Value	Maximum Value
i	60.0°	90.0°
T_2	3500 K	7000 K
Ω	1.900	9.164
q	0.10	5.00
L_1	1.00	20.00

Table 3 Search domain for the three test lightcurves.

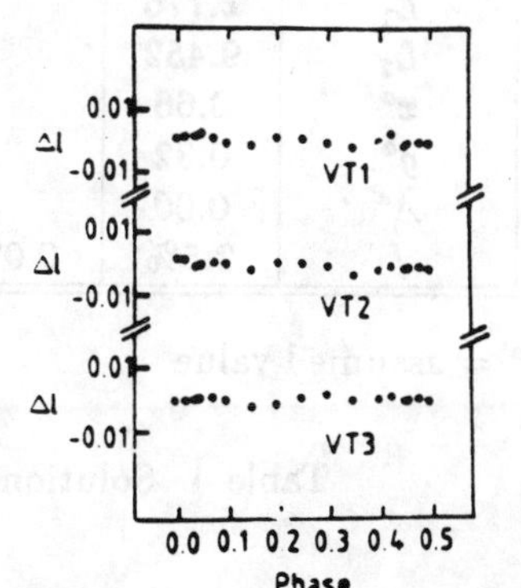

Fig.7 Residuals, in light units, for the three test light curves VT1, VT2, VT3

The best three solutions, obtained after about 30,000 iterations are shown in Table 4 , together with the true values of the parameters. The parameter f of Table 4 is the filling factor, defined as $(\Omega - \Omega_{in})/(\Omega_{out} - \Omega_{in})$; the suffixes are referred respectively to the inner and outer Roche critical surfaces.

The values of S obtained (all the weight are set to 1 owing to the fact that we are solving a theoretical light curves) for the best and the worst solution of the three curves are respectively:

$$S_{min_1} = 4.23 \cdot 10^{-7} \qquad S_{max_1} = 4.35 \cdot 10^{-7}$$

$$S_{min_2} = 4.71 \cdot 10^{-7} \qquad S_{max_2} = 4.82 \cdot 10^{-7}$$

$$S_{min_3} = 5.41 \cdot 10^{-7} \qquad S_{max_3} = 5.53 \cdot 10^{-7}$$

The residuals corresponding to the three lowest S solutions are shown in Fig. 7.

	VT1		VT2		VT3	
parameter	comp.	true	comp.	true	comp.	true
i	75.7	75.38	70.8	70.84	70.9	70.92
T_1^a	5000 K		5000 K		5000 K	
T_2	5226 K	5267 K	5323 K	5296 K	5362 K	5362 K
Ω_1	6.524		3.637		2.990	
Ω_2	6.524		3.637		2.990	
q	2.940	2.857	0.934	0.990	0.560	0.500
L_1	2.776		5.191		6.498	
L_2	9.452		6.719		5.446	
x^a	0.66		0.66		0.66	
g^a	0.32		0.32		0.32	
A^a	0.00		0.00		0.00	
f	2.5%	0.0%	0.9%	0.0%	1.2%	0.0%

^{a} = assumed value

Table 4 Solutions of the three test lightcurves

It is noteworthy that the results obtained after a first run of about 5000 iterations were rather different from the final ones. While, in fact, the temperature difference, the degree of contact and (less well) the inclination reached stability quite early, the mass ratios did not. For instance in the cases of VT1 and VT3 we reached, after this first run a first 'minimum'; then ΔT and i had almost the correct value, but q was still far from it, namely we had for VT1 $q = 2.28$ instead of 2.86 and for VT3 $q = 0.8$ instead of 0.5. The search was in fact temporarily confined in a wide local minimum (see later). Only at the end of a second run (when S was $\approx 5 \cdot 10^{-7}$ and $\Delta S/S \leq 0.05$ all over A) we obtained almost exactly the right parameters. The small differences still found may be easily explained as compensations due to the slight differences between the two light curve synthesis programs. As we mentioned in the previous section, the clustering of the solutions can be visually inspected from plots like those of Fig. 8a, which plot two by two the parameters corresponding to the points in the last search array.[1] These graphs are also useful to study the correlations and to define the indeterminacy intervals of parameters.

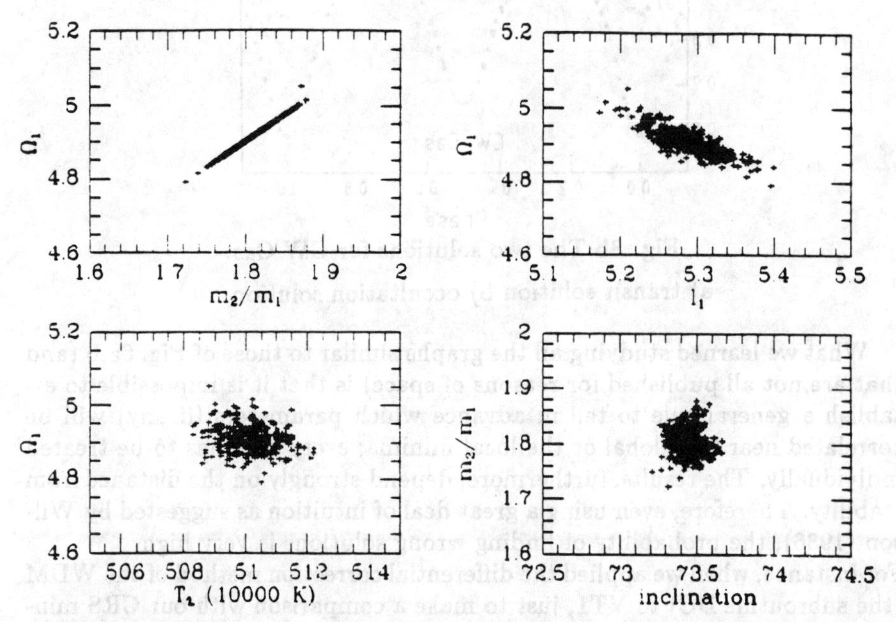

Fig.8a Graphical representation of the search array, for CW Cas occultation solution

[1] **Fig. 8a,b actually refers to the solution of the observed system CW Cas (Fig.8b shows the observed and computed light curves) but, as all these plots have almost the same appearance, we prefer to show, as an example, those related to observed data. Note that in the case of CW Cas, we had only one light curve and the two solutions are characterized by comparable sum of squared residuals, so it would be difficult to prefer one to the other.**

Here too we went through two different stages of definition of the results: after the first 5,000 trials correlations still created problems and we obtained plots with elongated clouds of points describing more or less some linear law (that was almost the rule for the geometrical parameters); these graphs were also remarkably similar to the analogous ones obtained for some 'indeterminate' systems treated by the WDM program (see, for instance, Maceroni 1986). At the convergence, instead, the clouds shrank till reaching the sizes corresponding to the errors of Table 4.

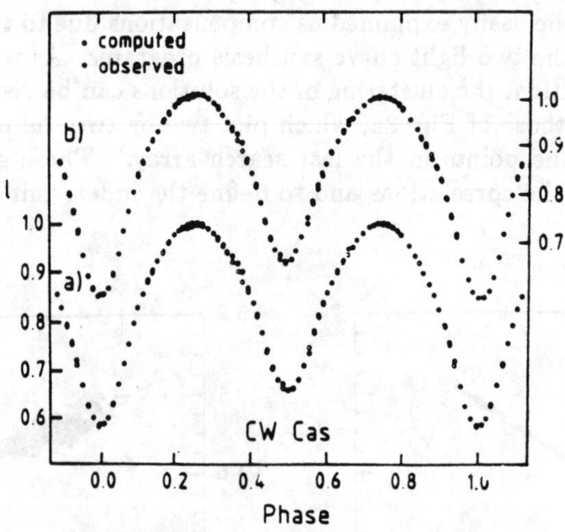

Fig. 8b The two solutions for CW Cas:
a) transit solution b) occultation solution

What we learned studying all the graphs similar to those of Fig. 8a,b (and that are not all published for reasons of space) is that it is impossible to establish a general rule to tell in advance which parameters (if any) will be correlated near the global or the local minima; every case has to be treated individually. The results, furthermore, depend strongly on the distance from stability. Therefore, even using a great deal of intuition as suggested by Wilson (1986), the probability of finding wrong solutions is very high.
For instance, when we applied the differential correction method of the WDM (the subroutine DC) to VT1, just to make a comparison with our CRS minimization, and computed a 'grid' of solutions in the domain $0.3 \leq q \leq 3.333$ we obtained the results of fig. 9. The step used in fig. 9 is small and, moreover, decreases near the minima (which were known in advance); therefore both minima at $q > 1.0$ could be evidenced. A standard grid search, instead, would have had a larger and fixed step: as a consequence the convergence on the wider local minimum would have been probably unavoidable.

The shape of $S(q)$ also explains our initial difficulties with the system; nevertheless the CRS minimization could always find the global minimum; it overcame successfully the difficulties commonly met in contact binary light-curve solutions, therefore we recommend its use in at least all the critical cases.

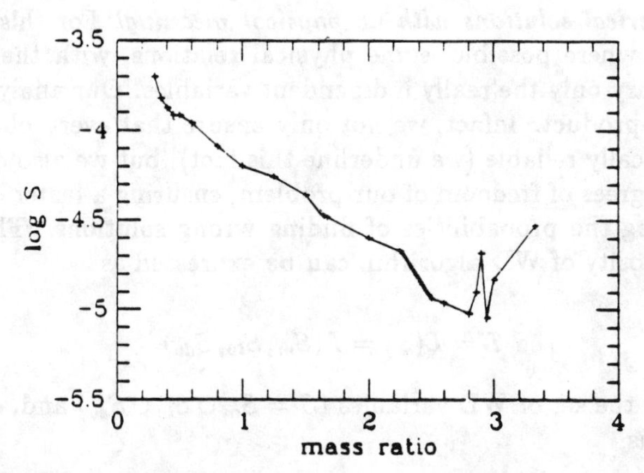

Fig.9 *logn* $\cdot$ *S* versus mass ratio: each point corresponds to a WDP photometric solution with fixed *q*

5 CRS solutions of close binary systems

We firstly applied this procedure to the analysis of close eclipsing binaries with zero eccentricity, using only photometric observations (and, if available, information on the spectral type of the primary). The procedure, after many tests on real systems, using the acquired experience, was improved to obtain faster and very reliable solutions. Infact, synthetic light curve generation by WD algorithm can be expressed, from a mathematical point of view, as a function of 18 variables

$$L = L(i, T_h, T_c, \Omega_h, \Omega_c, q, L_h, L_c, L_3, x_h, x_c, A_h, A_c, g_h, g_c, P, \phi, \lambda) \qquad (5)$$

where i is the inclination of the orbital plane of the system, $T_{h,c}$ are the effective temperature of the two components, $\Omega_{h,c}$ are the Roche lobe potentials, q is the mass ratio ($q = \frac{m_2}{m_1}$), $L_{h,c}$ are the luminosity of the two

components, L_3 is the usual third light, $x_{h,c}$ are the limb darkening coefficients, $A_{h,c}$ are the albedos, $g_{h,c}$ are the gravity darkening coeffcients, P is the orbital period, ϕ is the phase and λ is the effective wavelenght of the photometric system used.

The solution of the problem should determine all these quantities within a certain error. But what happens to the solution if these variables are not really indipendent and we allow them to indipendently vary? Probably *one obtains numerical solutions with no physical meaning!* For this reason, we tried to use, where possible, some physical relations, with the purpose of allowing to vary only the really indipendent variables. Our analysis has a no negligible by-product. Infact, we not only ensure that every obtained solutions is physically reliable (we underline this fact), but we also decrease the number of degrees of fredoom of our problem, ensuring a faster convergence and decreasing the probabilities of finding wrong solutions. Therefore the output luminosity of WD algorithm can be expressed as

$$L = L(S) = L(S_m, S_{iv}, S_{dv}) \tag{6}$$

where S is the set of WD variables ($S = S_m \cup S_{iv} \cup S_{dv}$) and, on the basis of our analysis,

$$S_m = \{\lambda, \phi, P\} \tag{7}$$

is the subset of computing variables,

$$S_{iv} = \{i, T_h, T_c, \Omega_h, \Omega_c, q, L_h, L_3\} \tag{8}$$

is the subset of physical and geometrical variables really indipendent that describe the system, and finally

$$S_{dv} = \{L_c, x_h, x_c, A_h, A_c, g_h, g_c\} \tag{9}$$

is the subset of physical variables that can be considered dependent from variables of the subset S_{iv} in all the modes of computation but modes 0 and 1. About L_c, Wilson considered it as function of the temperature of the secondary T_c by means of the Planck law of black body radiation. Limb darkening coefficients $x_{h,c}$ are computed using Al-Naimy tables (Al-Naimy, 1978), according to the temperature and wavelenght. Gravity darkening coefficients $g_{h,c}$ are set to 1.00 or 0.32 according to the respective temperatures $T_{h,c}$, with the switch at $7200^o K$, the higher value corresponding to Von Zeipel law for a fully radiative envelope (Lucy, 1967). Reflection coefficients $A_{h,c}$ are set to 1.00 or 0.50, corresponding to full or partial reradiation, with the same switch criterion. Clearly all these dependent variables are automatically evaluated at each iteration by our procedure, on the basis of the coordinates of the trial point P. Following this procedure a 15-dimensional problem becomes a 8-dimensional problem.

Another critical point is the right choice of the search volume V. To fasten the solution it should be chosen as little as possible, according to the quantity of extra information we have on the system, not forgetting that some variables are more critical than others. For a right choice of V some rules should be followed.

Nothing we can say a priori about inclination, so that we, generally, put $60^o \leq i \leq 90^o$ taking into account that the system is an eclipsing one and this fact puts an inferior limit to this variable (60°).

T_h needs a more deep analysis. It is obvious that the knowledge of the spectral type of the primary component is very important because T_h is in a range given only by the error in its determination. Actually many authors give to T_h a fixed value and allow only T_c to vary, believing that, on the basis of differential correction procedure, only the difference $\Delta T = T_h - T_c$ can be determined by photometric observations. This is wrong. Infact we have this linear dependence only near the minimum when the Taylor series expansion can be stopped to first order terms, while, far from it, it is certainly not true. Moreover, to give a value to T_h, even if there is an uncertainty of a spectral type, forces the solution and lead to a final ΔT probably different from the reai one, determining T_h and T_c with high errors. We demonstrated that what we think is correct when we analyzed the binary system TT Her (Milano, Barone, Mancuso, Russo, Vittone, 1989). For this system the temperature of the primary was very uncertain. Infact Kaluzny (1985) (JK) assumed $T_h = 7100^oK$ from the (b-y) color index published by Hilditch and Hill (1975) while Kwee and van Genderen (KvG) (1983) assumed $T_h = 9100^oK$ from the spectral type suggested by Sanford (1937). As a consequence the configuration resulting from the two previous analyses were very different: KvG found a contact configuration, while JK determined a set of probable solutions switching the gravity darkening coefficient between 0.32 and 1.00.

We solved simultaneously all the light curves published, assigning to T_h and T_c the ranges

$$5000^oK \leq T_h \leq 10000^oK \tag{10}$$

$$4000^oK \leq T_c \leq 7000^oK \tag{11}$$

The result we obtained and that is worthy of mention is that $T_h \approx 7240^oK$, which is just above the limit generally assumed as a switch from a radiative to convective envelope. This value of T_h implies that the A0 classification given by Sanford was incorrect and that the photometric classification based on the (b-y) color index was F2V. The secondary component is about 2550^oK

cooler, which together with the value of the radii, explains why TT Her is a single lined spectroscopic binary.

In Fig.10 it is shown the graphical representation of the search array for some selected pairs of parameters for the system TT Her, while in Table 5 its derived parameters are shown.

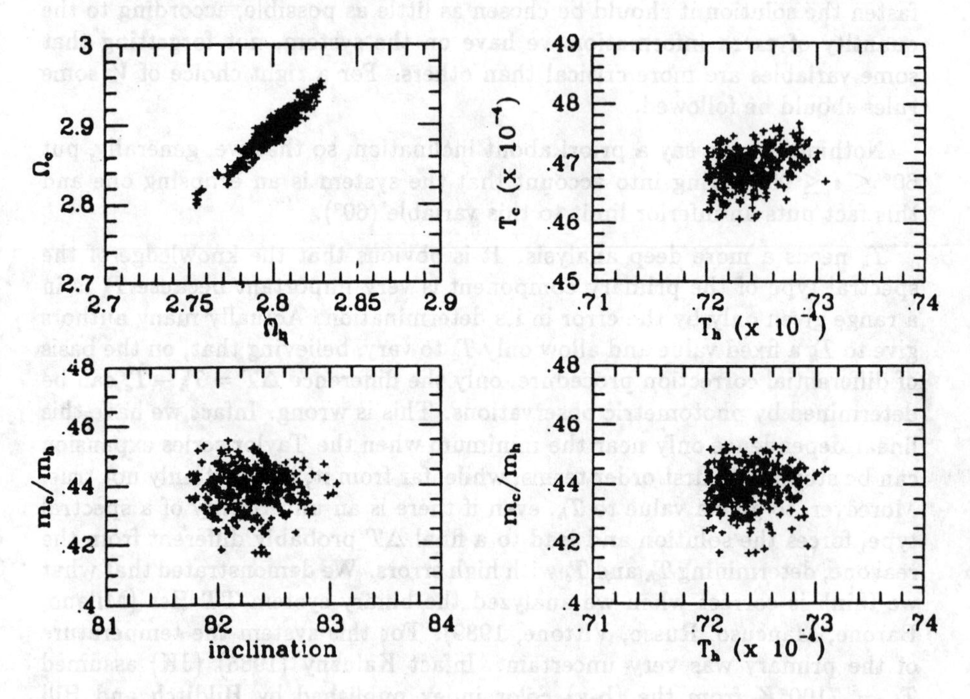

Fig.10 Graphical representation of the search array for TT Her solution

i (degrees)	82.3[illegible] $\pm$ 0.27	r_h(pole)	0.417
Ω_h	2.801 $\pm$ 0.024	r_h(point)	0.519
Ω_c	2.896 $\pm$ 0.027	r_h(side)	0.442
g_h	1.0	r_h(back)	0.467
g_c	0.32	r_c(pole)	0.266
A_h	1.0	r_c(point)	0.312
A_c	0.5	r_c(side)	0.275
T_h(°K)	7239 $\pm$ 24	r_c(back)	0.296
T_c(°K)	4690 $\pm$ 31	r_h(equiv)	0.444
m_c/m_h	0.439 $\pm$ 0.007	r_c(equiv)	0.278
S	0.0003066	l_3	0.000

Table 5. Derived parameters for TT Her

About the range to be chosen for the temperature of the secondary, T_c it is clear that must be $T_{inf} \leq T_c \leq T_h$, so that a program that allows T_h to vary must have a control on T_c of this type.

The ranges for $\Omega_{h,c}$ can be very large and have practically limitations given by the range of the mass ratio. The Wilson-Devinney code offers seven modes of computation, which have been fully described by Leung and Wilson (1977). Several years of usage indicate that modes 2,3,4, and 5 are the most widely used. Mode 2 poses no geometrical constraints, but couples T_c to L_c through Planck's law, while mode 5 poses the additional constraint that Ω_c must be equal to Ω_{in}, the inner critical surface potential, i.e. the secondary is forced to exactly fit its critical lobe, while in mode 4 the primary fills the lobe. In mode 3, the two Ω's are equal, which corresponds to a contact system. A common problem present with the Wilson-Devinney method lies in the fact that a solution with mode 2 may result in a configuration with $\Omega_h \leq \Omega_{in} \leq \Omega_c$, and viceversa, that is one star is inside the critical lobe and the other outside, and, of course, this does not have any physical meaning, So in our procedure we automatically switch from mode 2 to mode 3 according to the evolution of the solution in order to have always meaningful solutions. We choosed not to use mode 4 (and 5) because these are particular cases of mode 2.

The interval of variation of q can be $0 < q \leq 5.0$ and can be divided in two subintervals taking into account that every system has always two minima (transit and occultation), one for $q < 1.0$ and the other for $q > 1.0$. The value of q could be better limited if it were possible to use radial velocity curves to solve together with light curves as we did, for example, for EE Aqr (Covino, Barone, Milano, Russo, Sarna, 1989), where they were available.

For L_h and L_3 there are no particular conditions to fulfil.

It is worth spending a few words on the stop criterion, because, as we said before, CRS algorithm doesn't specify anyone in particular. The most obvious stop criterion might seem that of ending iterations when a stability condition is satisfied, namely when the relative difference between the upper and lower limit of S in A (ΔS %) become smaller than some percent (Milano, Barone, Russo, Sarna, 1989a). But this could not be true for all the systems. Infact, while for some systems with a very deep global minimum it is certainly true, for other systems that have a very shallow one, stopping iterations at a previous fixed ΔS % could prevent from obtaining the right solution. Moreover, when we solved the system VZ Psc (Barone, Milano, Russo, Sarna, 1989b) we found that 3 minima exist, very close each other, needing supplementary iterations before discriminating the right one, or like in systems like V752 Cen (see after) where a stable solution is obtained for a very low ΔS % . For this reason a lot of attention should be paid in the analysis of evolution of clusters. Infact, only when clusters positions in the hyperspace of solutions we can think to have reached the right solution.

At the moment, a new procedure is being developed to analyze also eccentric binary systems by means of simultaneous light and radial velocity curves solutions. The problem becomes more difficult to solve, owing to the increasing in the number of degrees of freedom to determine. The problem becomes, infact, a 15-dimensional one. The program for light and radial velocity curves sysnthesis we use is Wilson algorithm (1979), already used by many authors in systems solutions (Wilson, 1984; Covino, Barone, Milano, Russo, Sarna, 1989). This new procedure has been tested till now in the solution of the binary system V752 Cen (zero eccentricity), solving simultaneously 3 light curves and 2 radial velocity curves. Preliminary results are shown in Fig. 11, 12 and Table 6.

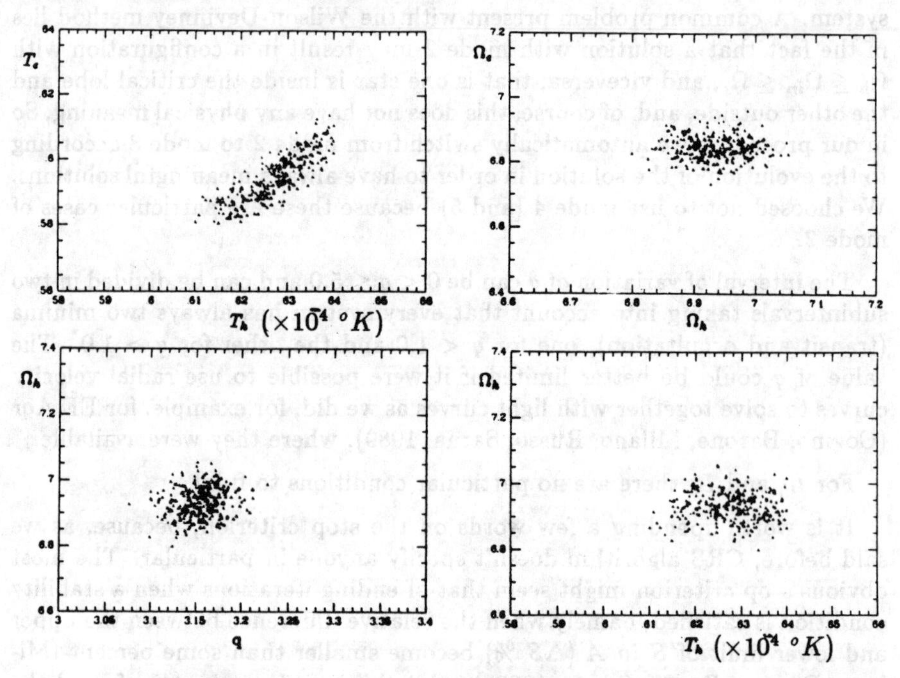

Fig.11 Graphical representation of the search array for V752 Cen solution

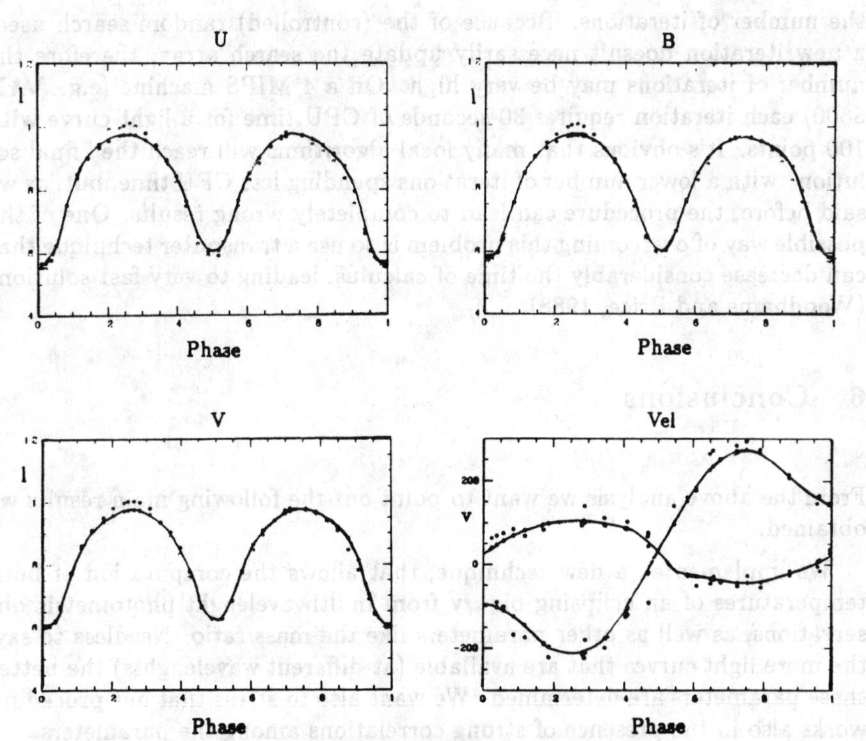

Fig.12 Computed and observed UBV light and radial velocity curves for V752 Cen

i(degrees)	85.7 ± 0.8
Ω_h	6.932 ± 0.047
Ω_c	6.847 ± 0.037
$T_h(^oK)$	6274 ± 67
$T_c(^oK)$	5946 ± 73
m_c/m_h	3.15 ± 0.02
g_h	0.32
g_c	0.32
A_h	0.50
A_c	0.50
$V_{Bar}(km/sec)$	29.2 ± 0.1

Table 6. Derived parameters for V752 Cen

A few words on the time of CPU required for each solution. Infact, the CPU time spent for a solution depends, of course, on the time required for the number of iterations. Because of the (controlled) random search used, a new iteration doesn't necessarily update the search array; therefore the number of iterations may be very high. On a 4 MIPS machine (e.g. VAX 8600) each iteration requires 30 seconds of CPU time for a light curve with 100 points. It's obvious that many local algorithms will reach the "final solution" with a lower number of iterations spending less CPU time, but, as we said before, the procedure can lead to completely wrong results. One of the possible way of overcoming this problem is to use a transputer technique that can decrease considerably the time of calculus, leading to very fast solutions (Woodhams and Price, 1988).

6 Conclusions

From the above analysis we want to point out the following main results we obtained.

We implemented a new technique, that allows the computation of both temperatures of an eclipsing binary from multiwavelenght photometric observations, as well as other parameters like the mass ratio. Needless to say, the more light curves that are available (at different wavelenghts) the better these parameters are determined. We want also to stress that our procedure works also in the presence of strong correlations among the parameters.

About the CPU time required to obtain a solution, it's physiological of every procedure of global optimization. Infact completeness of search and speed of convergence are always two conflicting requirements, and anyone should choose his own compromise: there are many cases in which the method of subsets and the usual WDP can give good results, and there are valid arguments for both ways. We believe that the results obtained by our technique are convincing reasons for its extensive use.

We shall conclude this paper by rewording a sentence by Z. Kopal in his book on "Language of the Stars" and strictly connected with the problem we analyzed: *We should, above all, emphasize that, in the absence of any indipendent check of the correcteness of our result, any confidence in the significance of the results must be based on the justification and internal consistency of the reduction procedures employed to obtain the solution* . In this respect it is also nice to quote another sentence of Kopal in the same book, that applies very well to some criticisms raised against the solution we gave to the problem we analyzed: *who would wonder that, under the actual circumstances, the users of time-honoured precepts were reluctant to abandon them for alternative approaches, requiring at least a good knowledge of "new" topics in the field of data analysis in Astronomy ?.*

References

Al Naimij, H.M.: 1977, *Astrophys. Sp. Sci.* **53**, 181

Barone, F., Mancuso, S., Milano, L.: 1987, *"UBV light curve solutions of TT Her"* in Proceedings of the Tenth European Regional Astronomy Meeting of the IAU, Praga, 24-29 August;

Barone, F., Milano, L., Russo, G.: 1987, *"Optimization of functions with Controlled Random Search Techniques"*, in Mem. S.A.It., **59**, $N^o 3-4$, 3-4 December;

Barone, F., Maceroni, C., Milano, L., Russo, G.: 1988, *Astron. Astrophys.* **197**, 347

Barone, F., Milano, L., Russo, G., Sarna, M.J.: 1989, *Astrophys. Space Sci.* (in press), ESO preprint N^o606, August 1988;

Becker, R.W., Lago, G.V.: 1970, *A global optimization algorithm*, in Proceedings of the 8th Allerton Conference on Circuits and Systems Theory (Monticello, Illinois)

Berthier, E.: 1975, *Astron. Astrophys.* **40**, 237

Covino, E., Barone, F., Milano, L., Russo, G., Sarna, M.J.: 1989, *Monthly Notices Roy. Astron. Soc.* (submitted), ESO preprint N° 634, February 1989

Davidon, W.C.: 1959, 'Variable Method for Minimization', *Argonne National Laboratory Report* ANL-5990 (Rev.)

De Landtsheer,A.C.: 1983, *Astrophys. Sp. Sci.* **92**, 231

Fletcher, R. and Powell, M.J.D.: 1963, *Computer J.* **6**, 163

Fletcher, R. and Reeves, C.M.: 1964, *Computer J.* **7**, 149

Fletcher, R.: 1987, *Practical Methods of Optimization*, John Wiley and Sons, Chichister, 2nd edition

Gill, P.E., Murray, W., Wright, M.H.: 1981, *Practical Optimization*, Academic Press, London

Griewank, A.O.: 1981, *JOTA* **34**, 11

Hilditch, R.W. and Hill, G.: 1975, *Mem. Roy. Astron. Soc.* **79**, 101

Hooke, R. and Jeeves, T.A.: 1961, *J. Ass. Comput. Mach.* **8**, 212

Kaluzny, J.: 1985, *Acta Astronomica* **35**, 327

Kwee, K.K., van Genderen, A.M.: 1983, *Astron. Astrophys.* **126**, 94

Kopal, Z.: 1986, *Vista in Astronomy*, **29**, 295

Leung, K.C. and Wilson, R.E.: 1977, *Astrophys. J.* **211**, 853

Lucy, L.B.: 1967, *Zeits. f. Astrophys.* **65**, 89

Maceroni, C.: 1986, *Astron. Astrophys.* **170**, 43

Maceroni, C., Milano, L. and Russo, G.: 1985, *Monthly Not. Roy. Astron. Soc.* **217**, 843

Milano, L., Barone, F., Mancuso, S., Russo, G., Vittone, A.A.: 1989, *Astron. Astrophys.* **210**, 181

Milano, L., Barone, F., Mancuso, S., Russo, G.: 1988, *Astrophys. Space Sci.* **153**, 273

Nelder, J.A. and Mead, R.: 1965, *Computer J.* **7**, 308

Plewa, T.: 1988 *Acta Astronomica* **38**, 415

Popper, D.M.: 1985, in 'Calibration of fundamental stellar quantities' *I.A.U.Symp. n. 111 ed. by D.S.Hayes, L.E. Pasinetti, A.G. Davis Philips (D. Reidel Publ. Co., Dordrecht Holland)*

Powell, M.J.D.: 1965, *Computer J.* **7**, 303

Price, W.L.: 1976, *Computer J.* **20**, 367

Rafert,J.B. and Markworth, N.L.: 1986, *Astron. Journ.* **92**, 678

Rosenbrok, H.H.: 1960, *Computer J.* **3**, 175

Russell, H.N. and Merrill, J.E.: 1952, *Contrib. Princeton Univ. Observ.* no. **26**

Sanford, R.F.: 1937, *Mont Wilson Contr.*, **574**, 195

Törn, A.: 1973 *Global Optimization as a combination of global and local search*, Gothenburg Business Adm. Studies, **17**, 191

Törn, A., Zilinskas, A.: 1988, *Global Optimization*, Springer-Verlag

van Hamme, W.: 1982, *Astron. Astrophys.* **116**, 27

van't Veer, F.: 1986, *Private Communication*

Wille, L.T.: 1986, *Nature* **324**, 46

Wilson, R.E., Devinney, E.J.: 1971, *Astrophys. J.* **166**, 605

Wilson, R.E., Devinney, E.J.: 1972, *Astrophys. J.* **171**, 413

Wilson, R.E., Devinney, E.J.: 1973, *Astrophys. J.* **182**, 539

Wilson, R.E., Biermann, P.: 1976, *Astron. Astrophys.* **48**, 349

Wilson, R.E.: 1979, *Astrophys. J.* **234**, 1054

Wilson, R.E.: 1983, *Astrophys. Sp. Sci.* **92**, 229

Wilson, R.E.: 1986, in 'Physical Model for Close Binaries and Logical Constraints' *U.S.-China Seminar on Close Binaries ed. by K.C. Leung (Gordon and Breach, New York)*

Wood, D.B.: 1971, *Astron. J.* **76**, 701

Wood, D.B.: 1973, *WINK: a Computer Program for Modeling Non-Spherical Eclipsing Binary Systems*, GFSC Report, Greenbelt, Maryland

Woodhams, F.W.D., Price, W.L.: 1988, *IEE Proceedings* **135**, 214

SPECTROPHOTOMETRY OF ALGOL-TYPE BINARIES

Paul B. Etzel
Department of Astronomy
San Diego State University
San Diego, California 92182-0334, USA

ABSTRACT: High precision spectrophotometric observations of the Algol-type eclipsing binaries S Cnc and TT Hya are presented. A brief discussion of observational techniques and analysis procedures is included. The spectral energy distributions of both components of each system are determined with a refined scan-subtraction process, which uses the result of light-curve synthesis. The stellar components of S Cnc and TT Hya are found to be quite similar; however, in the case of TT Hya there are complications brought on by substantial mass-transfer activity. An ultraviolet excess in the spectrum of TT Hya is demonstrated, which is compared to those found in similar systems.

1. INTRODUCTION

Previously, I discussed the general topic of spectrophotometry of eclipsing binary stars (Etzel 1988a). Because of eclipses, it is possible to determine the spectral energy distributions (SEDs) of the individual components. Here, I will concentrate on Algol-type binaries, which frequently show evidence of mass-transfer activity in the form of circumstellar material. Two totally eclipsing Algol-type systems, S Cancri and TT Hydrae, which have very similar stellar components, will be compared.

S Cancri (P=9.48 days; Sp.=B9.5 V and G8-K0 III) is a rather uncomplicated Algol-type binary. Popper and Tomkin (1984) found the spectroscopic orbits of both components to be circular. In addition, there is no Hα emission, even at totality, although there is a weak and sharp circumstellar component to the Ca II K line. The photometric data of Crawford and Olson (1980) were analyzed by Etzel and Olson (1985), who found the photometric solutions to be relatively uncomplicated. Etzel and Olson also presented additional spectroscopic and spectrophotometric data; they also found photometric and spectroscopic evidence of an extended atmosphere (above the critical Roche lobe) about the cooler component.

C. İbanoğlu (ed.), Active Close Binaries, 189–201.

TT Hydrae (P=6.95 days; Sp.=B9.5 V and G9-K1 III) is more typical of intermediate period Algol-type binaries in that it exhibits many of the classical signatures of mass-transfer activity. Popper (1980, p. 155; 1989) has demonstrated that the velocities for the hotter component are seriously affected by the motions of circumstellar material, while the velocities for the cooler secondary component provides a circular orbit. There is strong Hα emission, even outside eclipse, arising from this circumstellar material (accretion disk) about the primary component (Peters 1980, Etzel 1988b, 1989). The photometry of Kulkarni and Abhyankar (1980, 1981) also shows complications in the U-filter in the forms of an asymmetrical primary eclipse and an ultraviolet excess, which are caused by circumstellar material. In addition to performing independent light-curve solutions on the data of Kulkarni and Abhyankar, Etzel (1988b) also presented and discussed extensive spectroscopic and spectrophotometric data in the context of the true nature of the hotter star, which is at odds with the apparent A2 spectral type. Plavec (1988) has extended the discussion to the satellite ultraviolet spectral region where emission lines arising from a superionized plasma surrounding the mass-gaining star can be observed.

There have been relatively few investigations of Algol-type eclipsing binaries with movable grating scanners. Rhombs and Fix (1976, 1977) searched for ultraviolet light excesses (UVXs) from the systems U Cep, U Sge, and SX Cas. Only for the latter was the search conclusive. Clements and Neff (1979) extended the investigation to λ Tau and V356 Sgr.

The group headed by Mirek Plavec at UCLA has employed the Lick Image Tube Scanner (ITS) and the IUE satellite to extend considerably the range of wavelength coverage for many binary systems. Plavec, Weiland, and Koch (1982) made an extensive study of SX Cas and found that the primary star is more like a B7 V rather than the A5 III classification that is frequently quoted in the literature. Similar investigations have been made on other Algol- and W Serpentis-type binaries including: U Cep (Plavec 1983); U Sge, RW Per, and RX Gem (Dobias and Plavec 1985, 1987a,b); and, RS Cep and RY Gem (Plavec and Dobias 1987a,b). (For RY Gem, it is notable that grating scanner observations were used to extend the earth-based ultraviolet and red ends of the ITS scan obtained during totality). In these cases, direct subtractions of the totality scans from those outside eclipse were employed to determine the SED for the hotter component in the optical region. The large light ratios of the stellar components in the satellite ultraviolet, and the extended baseline of wavelength coverage, minimize systematic scan-subtraction errors.

The ground-based spectrophotometric observations presented here were obtained with the desire to do scan arithmetic. Accurate fluxes are the primary requirement, not spectral resolution. For this reason, movable grating scanners with photon-counting capability were employed. Although grating scanners are relatively inefficient compared to more modern wavelength-multiplexing spectrometers (e.g., ITS), the former are still difficult to surpass in terms of wavelength coverage and precision. Observational procedures were instituted such that the resultant fluxes reproduce the standard system to within 2%.

2. INSTRUMENTATION AND OBSERVATION

Between 1980 and 1982 I obtained photoelectric spectrum scans, in the wavelength range of 3200 to 8090 Å, of Algol-type and related eclipsing binaries. Most of the observations were made at Lick Observatory with the Wampler (1966) scanner on the 0.9-m Crossley telescope. Other data were obtained with similar scanners at Kitt Peak National Observatory (KPNO) on the 0.9-m telescope and at Cerro Tololo Inter-American Observatory (CTIO) on the 1.5-m telescope. All scanners employed dry-ice cooled ITT FW130 photomultipliers and pulse-counting electronics. The typical spectral resolution, as defined by an exit slot at the spectrum focus, was 40 Å.

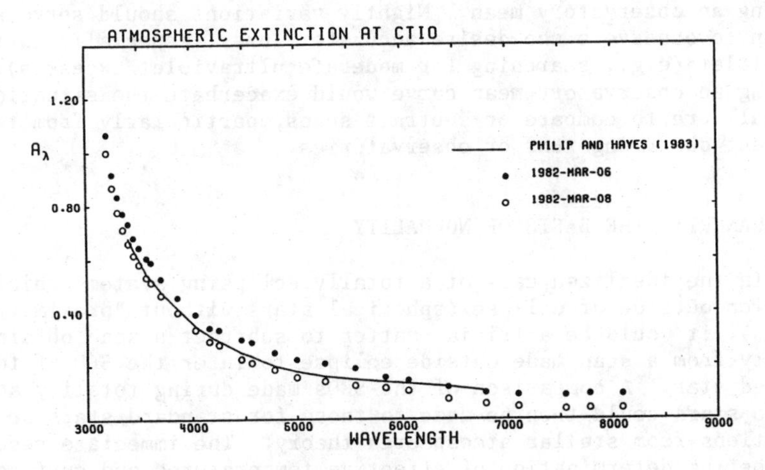

Fig. 1. Atmospheric extinction variations at CTIO.

Extraordinary efforts were taken to retain absolute fluxes on the system of Hayes and Latham (1975) by choosing realistic values of $m_\nu(\lambda5556)$ for all standards relative to a value of $m_\nu(\lambda5556) = 0.00$ for Vega. The main goal is, for a given binary system, to compare inside-eclipse scans directly to those outside eclipse to yield the absolute SEDs for both components. The observational procedures employed were rooted in my experiences with differential photometry for variable stars. The commonly adopted procedure of using mean atmospheric-extinction coefficients, employed by many observers, was avoided. Instead, multiple observations of two standard stars were used to measure extinction nightly. For S Cnc and TT Hya, the stars η Hya and 109 Vir were employed, with θ Crt and θ Vir used as a checks. The procedure of using two or more standards also provides a direct mechanism for evaluating relative values of $m_{\nu}(\lambda5556)$ for the standards. Because of the narrow bandpasses employed, the problems of extinction correction and transformation to the standard system are separable.

The need for the determination of extinction nightly is demonstrated in Fig. 1, which shows the results for two nights during an observing run at CTIO. Also plotted for comparison is a mean extinction curve from Philip and Hayes (1983). The extinction on March 6 was relatively high (e.g., the value of A_{λ} was 0.734 magnitudes per airmass unit at $\lambda3400$). On March 7, there was a change in the weather as a storm system over the Andes, far removed from the observatory, dissipated. The extinction curves for March 8 through 11 were fairly repetitive, but variations in the ultraviolet were significant (e.g., the values for A_{λ} ranged from 0.657 to 0.692 at $\lambda3400$). It is reassuring that during a stable run, the nightly variations are small. The variations in extinction illustrated here demonstrate the necessity of determining the extinction curve nightly, versus the procedure of adopting an observatory mean. Nightly variations should serve as a caution to observers who desire accurate fluxes, especially in the ultraviolet (e.g., searching for moderate ultraviolet excesses). Adopting an observatory mean curve would exacerbate the situation if the goal were to compare or subtract scans, particularly from two different observing runs or observatories.

3. S CANCRI: THE BASIS OF NORMALITY

In the idealized case of a totally eclipsing system exhibiting no variation outside of eclipse (spherical stars without "proximity" effects), it would be a trivial matter to subtract a scan obtained at totality from a scan made outside eclipse to infer the SED of the eclipsed star. A comparison of the SEDs made during totality and for the two stars could then be made to those for standard stars or to predictions from stellar atmospheres theory. The immediate result would be the determination of effective temperatures and surface gravities, which would be determined more reliably than those from standard photometric indices, or from values "determined" from light-curve synthesis. In addition, anomalies in the individual SEDs have the potential for providing information about interstellar reddening, chemical abundance effects, or mass transfer activity.

S Cnc is typical of most genuine Algol-type systems in that its light curve shows variations with phase outside of eclipse that increase in amplitude with longer wavelengths (Crawford and Olson 1980, Etzel and Olson 1985). Usually, the most dominant proximity effect is that of stellar "ellipticity," which is the combination of the variation in projected surface area of a distorted stellar component in concert with limb darkening and gravity darkening. The other major proximity effect is that of "reflection," which is a popular but non-physical term used to describe the atmospheric heating and reradiation of energy by the inner hemisphere of one component from the incident energy received from the other star. For most Algol-type systems, such variations are naturally dominated by the secondary component, which is usually the less-massive, larger, and cooler star of the pair. The first two properties conspire to make the secondary the most distorted, while the last tends to cause the inner and outer hemispheres of the secondary to have the largest observable temperature difference. These variations in light can be modelled with a suitable light-curve synthesis model, such as WINK (Wood 1971 - 1978, Etzel and Wood 1982), or the Wilson-Devinney code (Wilson and Devinney 1971, Wilson 1979).

There are also analogous variations outside eclipse in the observed SEDs with phase for both S Cnc and TT Hya. These variations are consistent with the changes expected in the apparent luminosity of the secondary component as a consequence of proximity effects; namely, they are greatest in the red, and smallest in the ultraviolet. The small variations in the ultraviolet accent the necessity for nightly extinction determinations because nightly variations in extinction can increase dramatically from red to ultraviolet (Fig. 1).

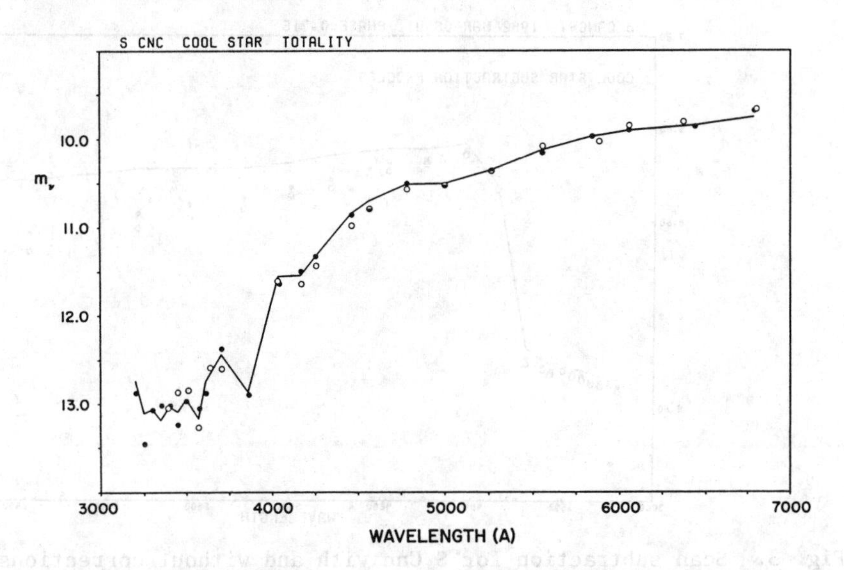

Fig. 2. Spectral energy distribution of S Cnc during totality.

Shown in Fig. 2 are the SEDs of S Cnc as see at totality (filled circles), the K0 III standard star τ CrB (solid line), and theoretical fluxes interpolated from Bell and Gustafsson (1978) for T_{eff} = 4500 K, log g = 2.25 (open circles). The SED for S Cnc is representative of the cooler component's outer hemisphere. The depth of the λ3862 point is very sensitive to spectral type and luminosity class.

Initial efforts to determine the SED for the hot star alone, by direct subtraction of the totality scan from scans taken near quadratures, did not produce consistent results. The average scan difference approximating the hotter component lies roughly between those for the B9 V star θ Crt and the A0 V star 109 Vir. The Paschen continuum is well represented by Kurucz (1979) models with T_{eff} = 10,000 K, but the Balmer jump requires an unrealistically large value of log g = 4.5, which is at odds with the well determined mass and radius. Fixing log g = 4.0 from the mass and radius implies T_{eff} = 10,500 K. A more consistent scan subtraction method is required.

The phase-variable luminosity of the cooler component reaches its minimum value during totality, which is when there is no flux contribution by the reflection effect, and the ellipticity factor is minimal since the cool star projects its smallest area. Using the totality scan to represent the cooler star at all phases for subtraction results in additional flux from the cool star being incorporated into the direct scan differences, which flattens the slope of the Paschen continuum.

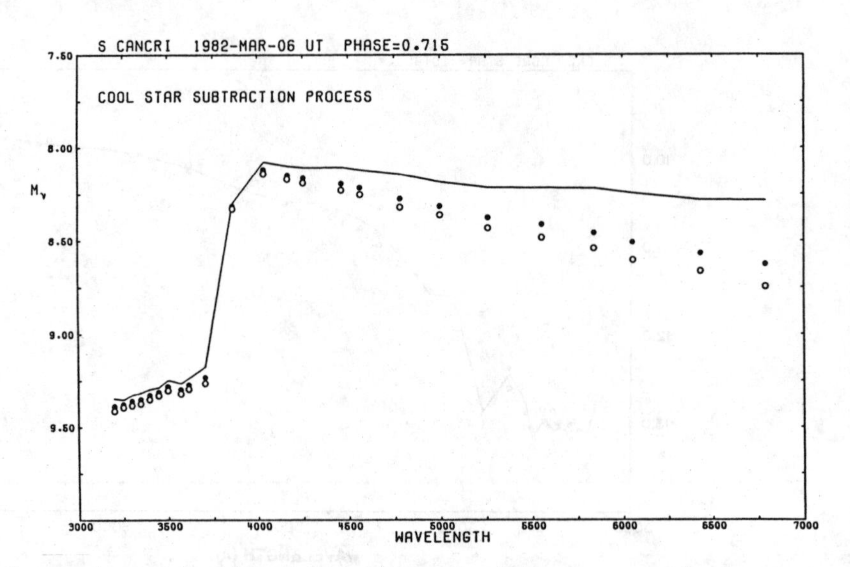

Fig. 3. Scan subtraction for S Cnc with and without corrections.

What we need is the SED for the cool star as it would be seen at any phase, θ, which could be subtracted from the corresponding outside-eclipse scan. To this end, we can compute, by light-curve synthesis, the ratio $L_B(\theta)/L_B(\mathrm{tot})$, for the wavelengths of the scanner observations, from the adopted light-curve solution (Etzel and Olson 1985). Limb-darkening and gravity-darkening coefficients can be assumed from theory. These ratios form a set of corrections to be applied to the observed totality scan to predict the SED of the cool star for each θ for which outside-eclipse scans exist.

Fortunately, these corrections have only a second-order effect on the inferred SEDs for the hot star. They are essentially scaled by the luminosity of the cooler component and are hence more important at longer wavelengths. Figure 3 illustrates the importance of these corrections to the scan-subtraction process, and their affect on the inferred SED of the hot star. Plotted are an observed outside-eclipse scan for the combined light of both stars (solid line), a simple subtraction of the totality scan (filled circles), and the more correct result of subtracting the predicted SED for the cool star at the phase of the outside-eclipse scan (open circles). The steeper slope of the Paschen continuum for the refined scan-subtraction process is assumed to represent best the hot star.

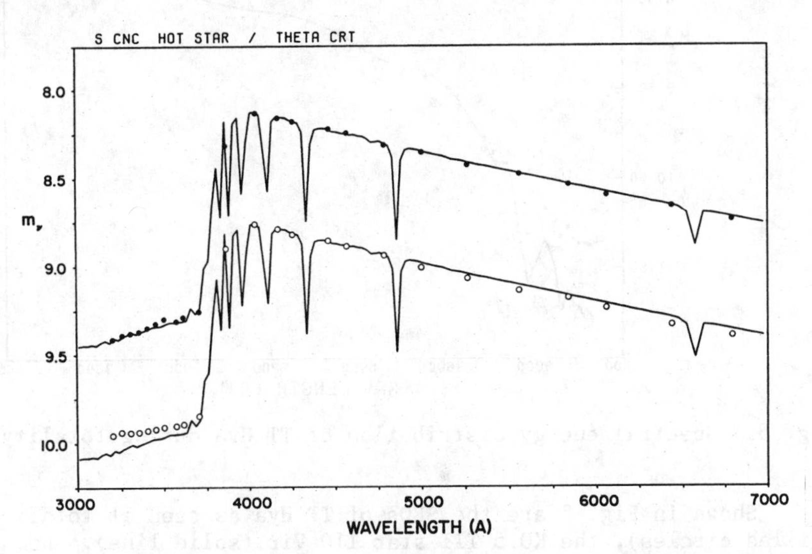

Fig. 4. Spectral energy distribution of the hot star in S Cnc.

This refined scan-subtraction process was applied to all outside-eclipse scans to establish the adopted SED for the hotter star in S Cnc. The standard deviation of an individual flux point is 0.009 mag averaged over the entire wavelength range. This adopted SED is plotted in the upper portion of Fig. 4 (filled circles) along with the B9 V standard θ Crt (open circles) and the Kurucz theoretical fluxes for T_{eff} = 10,500 K and log g = 4.0 (solid line). The hotter component of S Cnc does seem to be slightly cooler than θ Crt, so a B9.5 V spectral type is appropriate for the former star and is also consistent with the measured mass and radius. There appear to be no anomalies in the adopted SED for the hotter star of S Cnc, which is consistent with its lack of significant mass-transfer activity.

4. TT HYDRAE: THE COMPLICATIONS OF REALITY

The approach to TT Hya is essentially the same as for S Cnc since both are totally eclipsing systems. At first glance, the systems are very similar. However, in the case of TT Hya a high level of mass transfer-activity is evident from other data. The goal here will be to note complications in the SEDs, which may result from this activity.

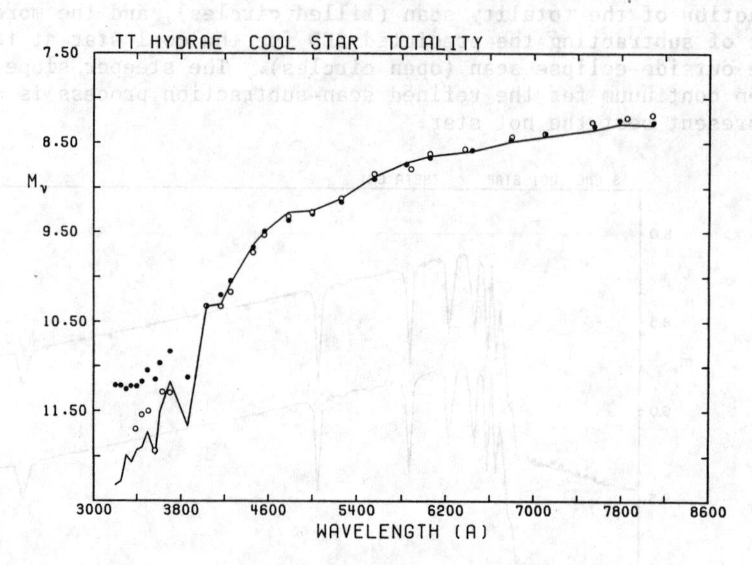

Fig. 5. Spectral energy distribution of TT Hya during totality.

Shown in Fig. 5 are the SEDs of TT Hya as seen at totality (filled circles), the K0.5 III star 110 Vir (solid line), and theoretical fluxes interpolated from Bell and Gustafsson for T_{eff} = 4500 K and log g = 2.63 (open circles). The TT Hya scans were obtained just after mid totality to avoid the complications noted at second contact in the U-filter light curve (Etzel 1988b). The SED for TT Hya at totality represents the outer hemisphere of the cooler component, which appears to be slightly cooler than 110 Vir for wavelengths longward of λ4200. This conclusion is reinforced by the examination of a high resolution spectrum obtained in totality that covers the wavelength region λλ4000 to 4500, which indicates a spectral type of K1 III. The adopted mass and radius were used to fix the value of log g for the theoretical fluxes. All attempts to represent this totality scan with a normal SED that included the fluxes in the range of λλ3200 to 3862 met with failure.

There is a substantial ultraviolet excess (UVX) for $\lambda \leq 3862$ Å, which averages about 0.7 mag. This UVX is emitted by unocculted circumstellar material that surrounds the hotter (mass-gaining) component. It is natural to assume that the UVX and Balmer emission lines have a common origin, although their distributions of flux may be different because of optical depth effects. The UVX plotted in Fig. 6 was found by subtracting the normalized (in the red) SED of 110 Vir from the adopted (mean) scan of TT Hya as observed during the relatively undisturbed portion of totality. The UVX extends from shortward of the Balmer limit to at least 4000 Å and probably to 4200 Å! The immediate result is that the UVX is not caused by Hydrogen recombinations alone.

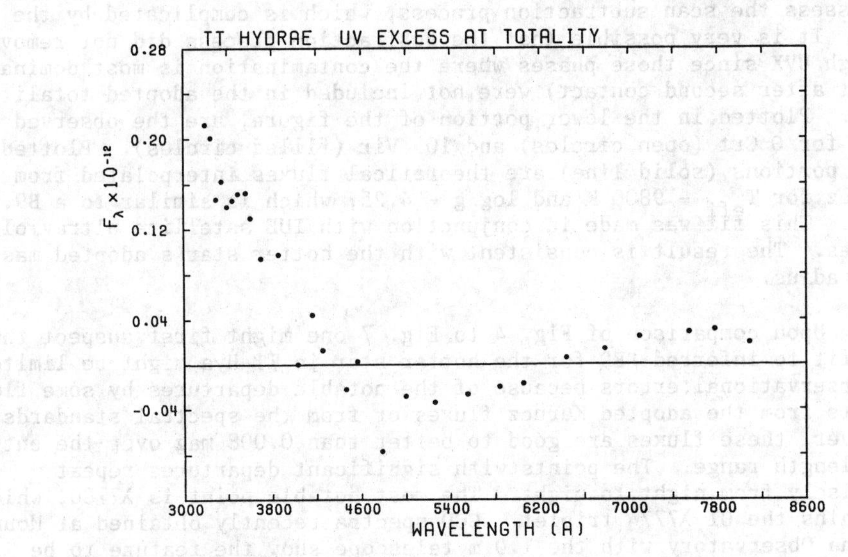

Fig. 6. The ultraviolet excess of TT Hya during totality.

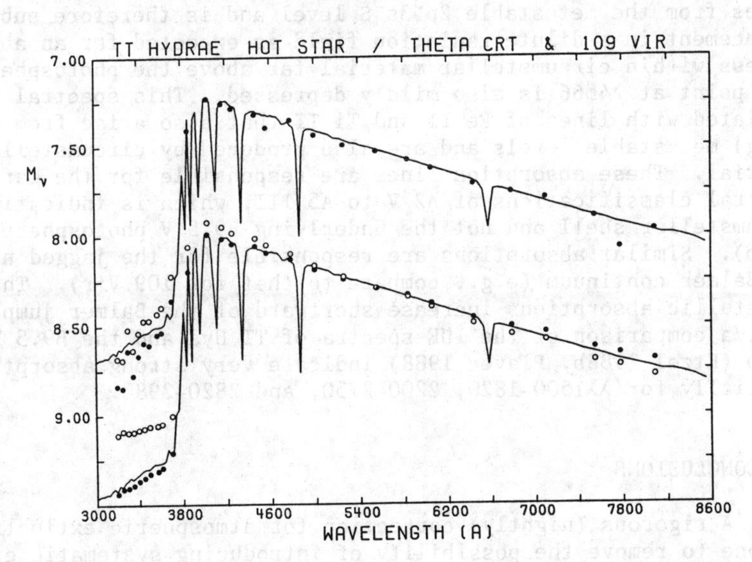

Fig. 7. Spectral energy distribution of the hot star in TT Hya.

We can proceed to determine the SED for the hotter star in TT Hya alone based upon the refined scan-subtraction process developed earlier for S Cnc. The results are shown in Fig. 7. Plotted in the upper portion of the figure, are the adopted SED for the hotter star (filled circles), which is based upon a mean of four nights of outside-eclipse scans, and also the mean outside-eclipse scan for λλ3200-3704 (open circles). The latter group of points are plotted to allow the reader to assess the scan subtraction process, which is complicated by the UVX. It is very possible that the subtraction process did not remove enough UVX since those phases where the contamination is most dominant (just after second contact) were not included in the adopted totality scan. Plotted in the lower portion of the figure, are the observed SEDs for θ Crt (open circles) and 109 Vir (filled circles). Plotted in both portions (solid line) are theoretical fluxes interpolated from Kurucz for T_{eff} = 9800 K and log g = 4.25, which is similar to a B9.5 V star. This fit was made in conjunction with IUE satellite ultraviolet fluxes. The result is consistent with the hotter star's adopted mass and radius.

Upon comparison of Fig. 4 to Fig. 7 one might first suspect that the fit to inferred SED for the hotter star in TT Hya might be limited by observational errors because of the notable departures by some flux points from the adopted Kurucz fluxes or from the spectral standards. However, these fluxes are good to better than 0.008 mag over the entire wavelength range. The points with significant departures repeat precisely from night to night. The most notable point is λ7780, which contains the OI λ7774 triplet. CCD spectra recently obtained at Mount Laguna Observatory with the 1.0 m telescope show the feature to be abnormally strong (for a B9.5 star) and variable in strength, with an average equivalent width of 2.4 Å for the 1989 season. The OI triplet arises from the metastable $2p^3 3s\,^5S$ level and is therefore subject to enhancement in a dilute radiation field as expected for an absorption process within circumstellar material far above the photosphere. The flux point at λ4566 is also mildly depressed. This spectral region is populated with lines of Fe II and Ti II that also arise from (low lying) metastable levels and are also produced by circumstellar material. These absorption lines are responsible for the earlier spectral classifications of A2 V to A5 III, which is indicative of a circumstellar shell and not the underlying B9.5 V photosphere (Etzel 1988b). Similar absorptions are responsible for the jagged nature of the Balmer continuum (e.g., compare to that for 109 Vir). The number of metallic absorptions increase shortward of the Balmer jump. In fact, a comparison of the IUE spectra of TT Hya and the B9.5 V star ν Cap (Etzel 1988b, Plavec 1988) indicate very strong absorptions, especially for λλ1500-1820, 2200-2750, and 2820-2980.

5. CONCLUSIONS

A rigorous (nightly) correction for atmospheric extinction must be done to remove the possibility of introducing systematic effects in the observations, particularly in the ground-based ultraviolet. The analysis of spectrophotometry for a totally eclipsing system requires a second-order correction (determined from a light-curve synthesis code) to extract the SED of the hotter (occulted) component. Failure to apply these corrections results in flattened Paschen continua, which are easily interpreted as cooler photospheric temperatures; although, one could also mistakenly interpret an increasing departure in the near infrared as being caused by free-free emissions. We have found that the stellar components of TT Hya are actually quite similar to those of S Cnc; however, the former system is complicated by presence of circumstellar material. One cannot assume that the SED observed at totality is uniquely stellar. As demonstrated by TT Hya, complications can exist because of mass-transfer activity in the forms of enhanced (shell) absorptions and an ultraviolet excess (UVX); both have their origins in circumstellar material surrounding the hotter component. In the case of TT Hya, the shell absorptions are known to be phase and secularly variable (Etzel 1984). The limited photometry of TT Hya shows the UVX to be secularly variable (Etzel 1988b).

We have seen in the case of TT Hya, that a UVX exists for which hydrogen recombinations alone cannot be responsible because of the presence of flux longward of the Balmer jump. Possible mechanisms to account for this UVX have been suggested before (Etzel 1988b) and they include electron scattering of light from the hotter star and perhaps two-photon emission from a relatively cold (5,000 - 10,000 K) plasma. Such UVX spectra have been observed before in other related systems. Plavec, Weiland, and Koch (1982) clearly confirmed it in the totality spectrum of SX Cas (P=36.6 days). Walker and Popper (Popper 1964) found it in the totality spectrum of KU Cyg (P=38.4 days).

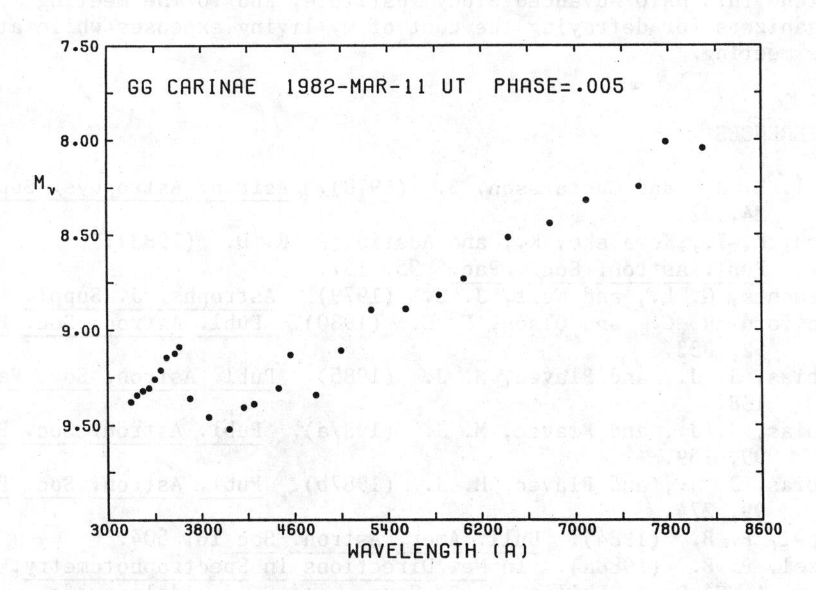

Fig. 8. Spectral energy distribution of GG Car in primary eclipse.

Shown in Fig. 8 is my spectrophotometry of the peculiar eclipsing binary GG Car (P=31.0 days). It has a very obvious UVX as the observed Balmer continuum is in emission! The UVX also extends longward of the Balmer jump to at least λ3800. The system defies a consistent description. Hernandez et al. (1981) list extensive spectral signatures attributed to circumstellar material such as Balmer emissions (to very high order lines) with P Cygni type profiles and emissions of Fe II. Chen et al. (1983) proposed a model with two stellar components with temperatures of 15,000 and 4800 K surrounded by a circumstellar cloud with a temperature of 2800 K to produce free-free emissions in the infrared.

In the future, we can expect more efficient and precise spectrophotometry that will approach the precision of filter photometry. We would then be in a position to do differential spectrophotometry as is routinely done for light-curve studies today. At such a juncture, differential spectrophotometry could eliminate the need for filter photometry if the observations are binned over adjacent wavelength regions for light-curve synthesis. Spectrophotometry has the potential of providing much insight into mass-transfer activity in close binary stars. One area where work is needed in this regard is to accurately measure the UVXs in such systems in the ground-based ultraviolet to match the satellite ultraviolet fluxes. Only the combination of the two instrumental regimes will provide the data needed to specify the origin of UVX.

This work was partially supported by NSF grants AST8500189 and AST8822790. I gratefully acknowledge the NSF for a travel grant to attend this NATO Advanced Study Institute, and to the meeting organizers for defraying the cost of my living expenses while attending the meeting.

REFERENCES

Bell, R. A., and Gustafsson, B. (1978). Astron. Astrophys. Suppl. **34**, 229.
Chen, K.-Y., Kowalske, K., and Austin, R. R. D. (1983). Publ. Astron. Soc. Pac. **95**, 157.
Clements, G. L., and Neff, J. S. (1979). Astrophs. J. Suppl. **41**, 1.
Crawford, R. C., and Olson, E. C. (1980). Publ. Astron. Soc. Pac. **92**, 833.
Dobias, J. J., and Plavec, M. J. (1985). Publ. Astron. Soc. Pac. **97**, 138.
Dobias, J. J., and Plavec, M. J. (1987a). Publ. Astron. Soc. Pac. **99**, 159.
Dobias, J. J., and Plavec, M. J. (1987b). Publ. Astron. Soc. Pac. **99**, 274.
Etzel, P. B. (1984). Bull. Amer. Astron. Soc **16**, 504.
Etzel, P. B. (1988a). In New Directions in Spectrophotometry, A. G. Davis Philip, D. S. Hayes, and S. J. Adelman eds., (L Davis Press, Schenectady, N.Y.), p. 78.

Etzel, P. B. (1988b). Astron. J. **95**, 1204.
Etzel, P. B. (1989). Publ. Astron. Soc. Pac. (in preparation).
Etzel, P. B., and Olson, E. C. (1985). Astron. J. **90**, 504.
Etzel, P. B., and Wood, D. B. (1982). WINK Status Report no. 10. (private circulation).
Hayes, D. S., and Latham, D. W. (1975). Astrophys. J. **197**, 593.
Hernandez, C. A., Lopez, L., Sahade, J., and Thackeray, A. D. (1981). Publ. Astron. Soc. Pac. **93**, 747.
Kulkarni, A. G., and Abhyankar, K. D. (1980). Astrophys. Space. Sci. **67**, 205.
Kulkarni, A. G., and Abhyankar, K. D. (1981). J. Astrophys. Astron. **2**, 119.
Kurucz, R. L. (1979). Astrophys. J. Suppl. Ser. **40**, 1.
Peters, G. J. (1980). In Close Binary Stars: Observations and Interpretations, edited by M. J. Plavec, D. M. Popper, and R. K. Ulrich, (Reidel Publ., Dordrecht), p. 287.
Philip, A. G. D., and Hayes, D. S. (1983). Astrophys J. Suppl. **53**, 751.
Plavec, M. J. (1983). Astrophys. J. **275**, 251.
Plavec, M. J. (1988). Astron. J. **96**, 755.
Plavec, M. J., and Dobias, J. J. (1987a). Astron. J. **93**, 171.
Plavec, M. J., and Dobias, J. J. (1987b). Astron. J. **93**, 440.
Plavec, M. J., Weiland, J. L., and Koch, R. H. (1982). Astrophys. J. **256**, 206
Popper, D. M. (1964). Astrophys. J. **139**, 143.
Popper, D. M. (1980). Annu. Rev. Astron. Astrophys. **18**, 115.
Popper, D. M. (1989). Astrophys. J. Suppl. (in press).
Popper, D. M., and Tomkin, J. (1984). Astrophys. J. **285**, 208.
Rhombs, C. G., and Fix, J. D. (1976). Astrophys. J. **209**, 821.
Rhombs, C. G., and Fix, J. D. (1977). Astrophys. J. **212**, 446.
Wampler, E. J. (1966). Astrophys. J. **144**, 921.
Wilson, R. E. (1979). Astrophys. J. **234**, 1054.
Wilson, R. E., and Devinney, E. J. (1971). Astrophys. J. **166**, 605.
Wood, D. B. (1971). Astron. J. **76**, 701.
Wood, D. B. (1972). Goddard Space Flight Center Report X-110-72-473.
Wood, D. B. (1973 to 1978). WINK Status Reports no. 1 through no. 9. (private circulation).

COMPLICATIONS IN THE BALMER-LINE SPECTRUM OF TT HYDRAE

Paul B. Etzel
Department of Astronomy
San Diego State University
San Diego, California 92182-0334, USA

ABSTRACT: The Balmer lines of interacting binaries cannot be interpreted in the usual sense as diagnostics of stellar photospheres. In an Algol-type binary, the presence of circumstellar material can lead to inconsistent interpretations about the nature of the hotter stellar component. The system TT Hydrae has fairly well established stellar properties. Its Balmer-line spectrum illustrates complications from circumstellar material. Complications at Hα because of the cooler secondary are also evident. Failure to recognize this complication could result in misinterpretations about the circumstellar material.

1. INTRODUCTION

Wyse (1934) first described the spectra of the two components of TT Hydrae. From his low dispersion spectra, he classified the hotter star as A3 knowing that it dominates the light, in the photographic (blue) spectral region, of the system when observed outside of eclipse. At that time, it was assumed that the light observed at totality arose solely from the cooler component. Therefore, the emissions he observed at Hβ, Hγ, and Hδ during totality prompted the "peculiarity" in his classification of dG6p. TT Hya was not alone in his list of eclipsing binaries to show such emissions. In his footnote number 9, he detailed a series of observations of RW Tauri, taken during primary eclipses, which indicated a secular variation in the emission line intensities.

Joy (1942) proposed the fundamental concept behind the currently adopted model to explain the variations of the Balmer emission lines observed around primary eclipse in Algol-type binaries. His paper, which was recently reprinted (Joy 1988), served as the cornerstone of a review by Batten (1988) on the last century of spectroscopic binary star research. Joy's model proposed, for RW Tauri specifically, that the observed (double peaked) emission lines arise from a gaseous ring in motion about the hotter stellar component. During the ingress portions of primary eclipse, the violet (V) lobe is increasingly occulted. During the egress portion, the red (R) lobe is increasingly

C. İbanoğlu (ed.), Active Close Binaries, 203–211.

occulted, in a symmetrical fashion. In the case of RW Tau, the ring (or "accretion disk" in today's vernacular) is completely occulted during totality. This V/R variation has been observed to various degrees in similar interacting binaries.

Models of circumstellar gas dynamics (especially the so-called stream and disk components) have improved greatly since the time of Joy. One of the most successful models is that of Lubow and Shu (1975), which is based upon particle trajectory calculations (i.e., the three-body problem). The bases of their model are the assumptions that the effects of gas pressure are small compared to force of gravity and that there are relatively few interactions for a particle once it is part of the disk. They addressed the flow of material through the L_1 point of the mass-losing component and its subsequent trajectory in a mass-transfer stream towards the mass-gaining component. Most importantly, they addressed the issue of whether the stream would directly impact upon the surface of the gaining component or form a stable accretion disk about it. Their predictions of a minimum disk radius have been fairly successful in describing circumstellar disks in a variety of interacting binaries (Peters and Polidan 1984, and Kaitchuck, Honeycutt, and Schlegel 1985, Peters 1989).

Refined observational data and better understanding of the mass-transfer process have furthered our understanding of accretion disks in Algol-type binaries. Kaitchuck and Honeycutt (1982) have since reobserved RW Tau with high-time-resolution spectroscopy of the emission lines and find: 1) the disk is actually transient, 2) the approximate disk radius is 1.5 times the radius of the hotter star, but variable in size from cycle to cycle, 3) the emission line widths are at least twice that expected from disk rotation alone, and 4) the widths of the profiles suggest supersonic velocities if the broadening mechanism is only turbulence. They also determined an electron density in the disk of $\log N_e \simeq 11$.

2. THE BALMER SPECTRUM OF TT HYDRAE

Sahade and Cesco (1946) first described in detail some of the peculiarities in the outside eclipse spectrum of TT Hydrae. They noted that the very strong hydrogen lines had sharp and deep cores with not very pronounced wings. In addition, the lines of Mg II λ4481 and Si II appeared weak in comparison to the metallic line spectrum. Naftilan (1975) explicitly noted that the hotter component had a shell spectrum and proposed an A5 III spectral type (the giant designation was probably based mostly upon the appearance of the hydrogen-line cores). Bidelman (1976) placed the star on a list of stars with early-type shell spectra. (This list contains many Algol-type and W Serpentis binary stars). Popper (1980, p. 155; 1989) has demonstrated that the velocities determined for the hotter component are seriously affected by the motions of circumstellar material. In fact, a false eccentric orbit results for the hotter component while the velocities for the cooler secondary component demonstrate a circular orbit.

Etzel (1988) discussed the nature of the hotter component of TT Hya from the analyses of photoelectric light curves, continuum spectrophotometry, and coudé spectroscopy in conjunction with Popper's (1982, 1989) radial velocity measures. By assuming that the cooler component (the star with the better determined orbit) fills its Roche lobe, then one can deduce a mass of 2.25 $M_{\odot}$ and a radius of 1.90 $R_{\odot}$ for the hotter component, which leads to a value of log g = 4.23. The ground-based and space-ultraviolet spectrophotometry are both consistent with T_{eff} = 9800 K and log g = 4.25; however, there are departures because of circumstellar material (see also Etzel 1989a). These values of mass, radius, and temperature, along with the derived absolute luminosity, are characteristic of a B9.5 V star, rather than the early to mid A-type that is inferred from the metallic-line spectrum. In addition, the measured strengths of the Mg II λ4481 and Si II lines (and limits on He I) are also consistent with the B9.5 V spectral type. The observed metallic lines are enhanced; they originate in circumstellar material in a dilute radiation field (extended shell) and arise predominantly from transitions in which the lower levels are low-lying metastable states.

TT Hydrae (P=6.95 days) is rather typical of intermediate period Algol-type binaries that exhibit mass-transfer activity. The relative radius of the mass-gaining star is sufficiently small that the mass-transfer stream does not impact. Because of the large volume of the critical Roche surface around this mass-gaining star, a large accretion disk can form. The resulting disk produces strong Hα emission, which is observable even outside eclipse.

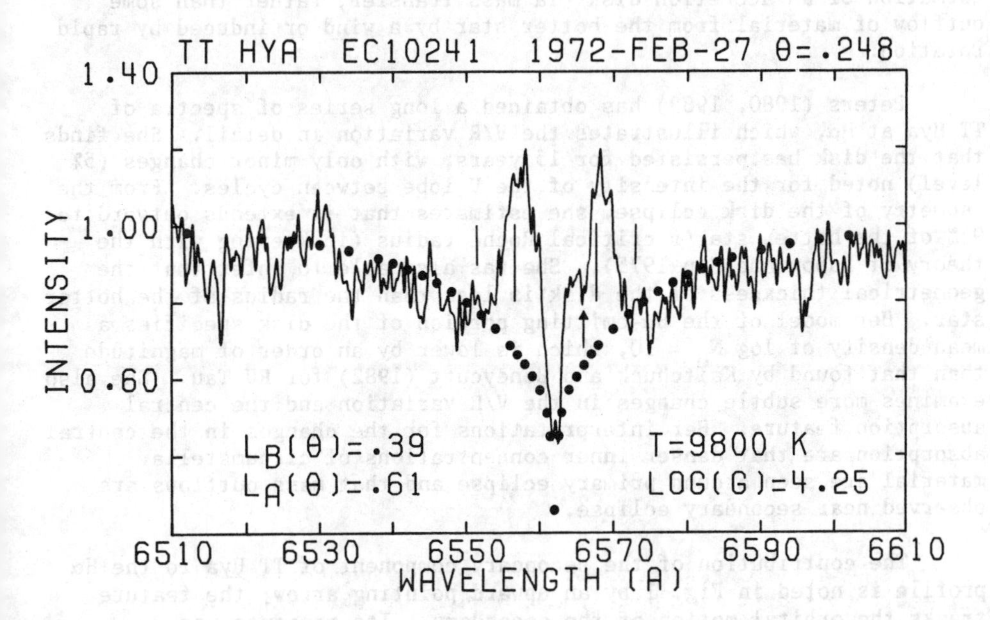

Fig. 1. Hα profile of TT Hya at quadrature compared to theory.

Figure 1 illustrates a typical Hα profile as observed outside eclipse. In the surrounding continuum, the light of the cooler secondary accounts for 39% of the light. A first-order subtraction (continuum only) of the secondary component has been done on this spectrum following the procedure outlined by Etzel (1988) from the results of his photometric and spectrophotometric investigations. It should be noted that this procedure, and the derivative employed by Peters (1989), will not remove strong absorption features of the secondary component. The long vertical fiducials represent the average peak-to-peak velocity separation (510 ± 10 km/sec) for the opposite sides of the accretion disk. The shorter fiducials represent the average velocity separation (990 ± 60 km/sec) of the absorption-emission reversals in both wings. The Keplerian velocity at the surface of the hotter star would be 475 km/sec for the adopted mass and radius. Plotted with the observed profile is a theoretical profile (filled circles) interpolated from Kurucz (1979) for the well-established physical parameters T_{eff} = 9800 K and log g = 4.25. Aside from the obvious complication from the emission components, the fit to the wings is not bad, which leads one to hope that the higher order Balmer lines could be used as diagnostics for the hotter star.

Plavec (1976) compared the Hα emission profile of TT Hya to that found in classical Be stars and noted the V/R variation through the orbital cycle. Indeed, there is a high percentage of interacting binaries that show Be characteristics (Bidelman 1976), although binarity is not a requirement for the phenomenon to exist (Harmanec 1982). The cause of the Be phenomenon in interacting binaries is the formation of an accretion disk via mass transfer, rather than some outflow of material from the hotter star by a wind or induced by rapid rotation.

Peters (1980, 1989) has obtained a long series of spectra of TT Hya at Hα, which illustrates the V/R variation in detail. She finds that the disk has persisted for 15 years, with only minor changes (5% level) noted for the intensity of the V lobe between cycles. From the geometry of the disk eclipse, she estimates that it extends outward to 95% of the hotter star's critical Roche radius (in keeping with the theory of Lubow and Shu 1975). She was also able to infer that the geometrical thickness of the disk is less than the radius of the hotter star. Her model of the Hα-emitting portion of the disk specifies a mean density of log $N_e \simeq 10$, which is lower by an order of magnitude than that found by Kaitchuck and Honeycutt (1982) for RW Tau. She also examines more subtle changes in the V/R variation and the central absorption feature. Her interpretations for the changes in the central absorption are that denser inner concentrations of circumstellar material are seen around primary eclipse and that mass outflows are observed near secondary eclipse.

The contribution of the secondary component of TT Hya to the Hα profile is noted in Fig. 1 by an upward pointing arrow; the feature tracks the orbital motion of the secondary. Its presence was previously noted by Etzel (1988) and will be briefly discussed in the context of its implications. Etzel (1989b) more fully discusses the importance of the contribution of the cooler secondary component at Hα, outlines a more rigorous subtraction process, and lists possible implications in failing to recognize its presence.

For stars with normal stellar photospheres, near spectral type A0, the combined analyses of spectral-energy distributions (or colors) and profiles of lower-order Balmer lines can uniquely determine T_{eff} and log g. Can this method be applied to the hotter components of Algol-type binaries if the wings are relatively uncontaminated? Unfortunately, the inferred (wing) profile for the hotter component in TT Hya is only partially indicative of the underlying stellar photosphere as demonstrated in Fig. 1 where a comparison is made to the theoretically expected profile. Peters (1989) has compared the inferred profiles for the hotter components of TT Hya, SW Cyg, and AD Her directly to spectral standards. What about the usefulness of higher-order members of the Balmer series as diagnostics?

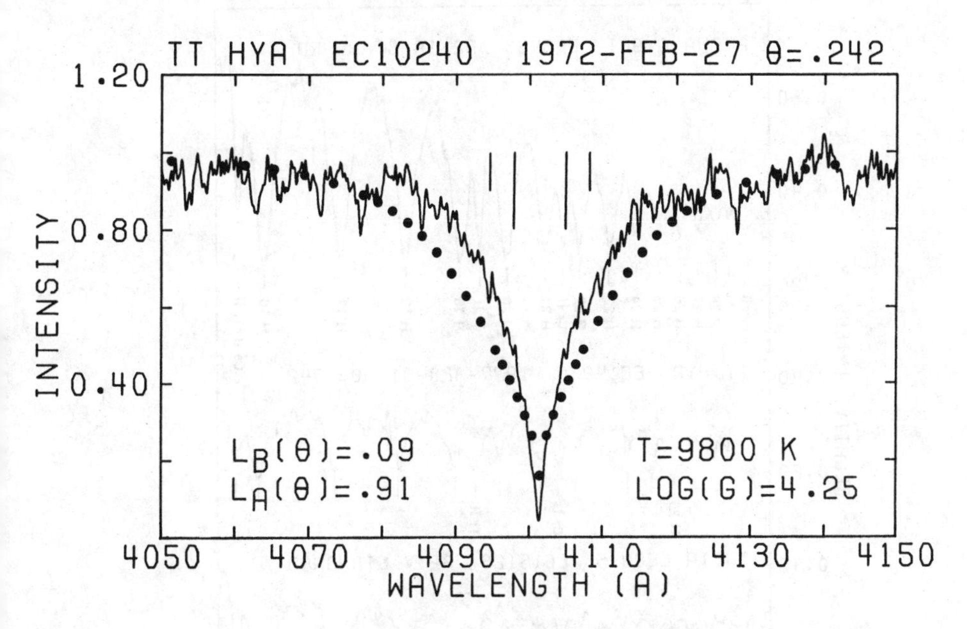

Fig. 2. Hδ profile of TT Hya at quadrature compared to theory.

Outside eclipse in TT Hya, there is weak circumstellar emission partially filling in the Hβ profile (Etzel 1988). There are two emission peaks within the absorption feature, well below the continuum level, that occur at nearly the same velocity spread from the line center as for Hα. The diagnostic value of the Hβ profile is therefore limited for the much the same reasons as for Hα. There is no obvious emission at Hγ outside eclipse in TT Hya, but there are complications by strong metallic-shell absorptions that make profile fitting difficult. The most uncomplicated lower-order Balmer line is Hδ, as shown in Fig. 2. At this wavelength, the cooler secondary only accounts for 9% of the light; its continuum contribution has been removed here. Plotted with the observed profile is the expected theoretical profile (filled circles) interpolated from Kurucz (1979) for T_{eff} = 9800 K and log g = 4.25. Note that there are some metallic-shell lines between 4050 and 4085 Å. The most serious departures are that the wings are not well developed and that the core is sharper and weaker than expected (as noted originally by Sahade and Wood 1946). If one were to take the observed Hγ and Hδ profiles at face value, very inconsistent results would be obtained. For instance, fixing T_{eff} = 9800 K (from the spectral energy distribution) and matching a theoretical profile would require log g = 3.50! Alternatively, fixing log g = 4.25 from the inferred mass and radius (valid for a large range of assumptions, Etzel 1988) would require values of T_{eff} = 12,000 and 14,000 K for Hγ and Hδ, respectively.

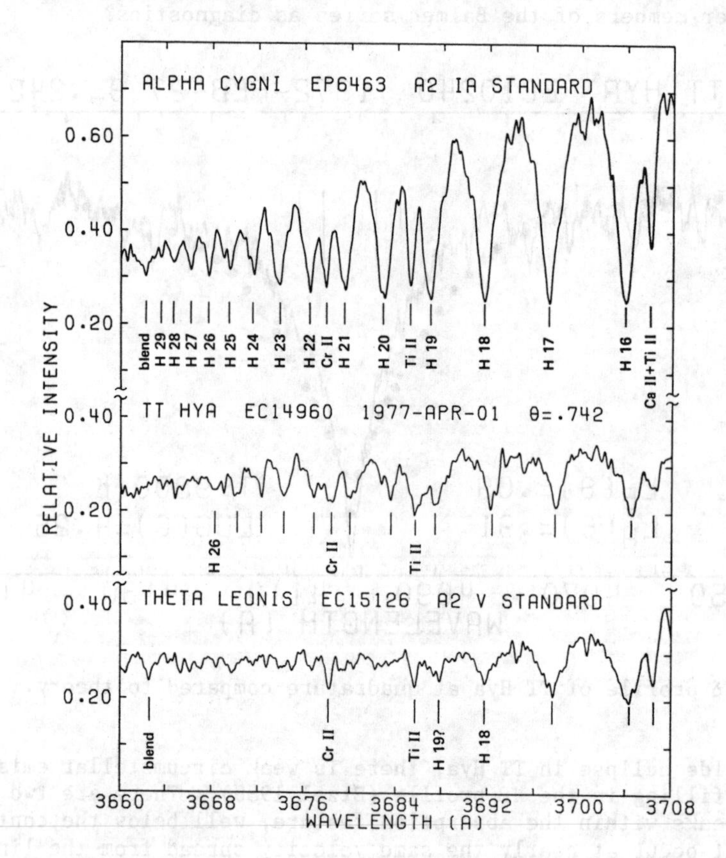

Fig. 3. Balmer-series limit of TT Hya at quadrature compared to spectral standards.

In stars with normal stellar photospheres, the confluence of the Balmer series provides a useful indicator of luminosity class (surface gravity) through the principal quantum number, n, of the last resolvable line. For stars near A0 spectral type, the value of n is strongly correlated with luminosity class. For θ Leo (A2 V), the value is 18; for π Cet (B7 V), the value is 17. For α Cyg (A2 Ia), Groth (1961) determined a value 31; he also determined a value of log g = 1.7 from his curve of growth analysis. I determined a value of n = 29 for α Cyg, with much lower resolution spectra than Groth's. Plotted in Fig. 3 are the Balmer series limits for TT Hya and the spectral standards α Cyg and θ Leo. At this wavelength, the cooler secondary star of TT Hya accounts for only 7% of the observed flux. No correction has been made here for its continuum contribution.

The last resolvable Balmer line in this (Fig. 3) spectrum of TT Hya provides n ≃ 26, which is completely at odds with the inferred main-sequence nature (B9.5 V) of the hotter star. In addition, the value of n is variable from cycle to cycle. An examination of dozens of spectra show the range to be $18 \leq n \leq 28$! The strengths and profiles of the features at the Balmer series limit change with the amount and distribution of circumstellar material surrounding the hotter star. Blending with metallic-shell lines and the available spectrograph resolution do not seem to be limiting factors in the determination of n.

3. DISCUSSION

It seems that in the case of TT Hya, every line in the Balmer series is seriously affected by the presence of circumstellar material. No single model atmosphere can successfully describe the Balmer series. The circumstellar material is not optically thin in the Balmer lines, even far from the line center. The lowest-order lines have obvious central emission components, while the highest-order lines are completely formed by absorptions within the circumstellar material.

The usefulness of the Hα profile as a diagnostic for the hotter component's photosphere is limited to cases where the wings are relatively uncontaminated. In interacting binaries, the greatest usefulness of the Hα profiles is to exploit the central emission as a diagnostic (an integrated measure) of the circumstellar material. Care must be taken to account for the secondary component of intermediate and long period Algol-type binaries. In these cases, the luminosity of the cooler (usually larger) secondary components can become substantial. The photospheric Hα absorption line from the secondary component can blend with the central emission and absorption components to produce effects that might be misinterpreted as being caused by complications from additional sources of circumstellar material (Etzel 1989b).

The confluence of the Balmer series provides us with a direct measure of the density of the circumstellar material. At those times for TT Hya when the value of n ≃ 18, the contamination of the spectrum from circumstellar material must be minimal because that value is what would be expected for a normal B9.5 V photosphere. At other times, the value of n is much larger. Perhaps the vertical structure of the disk changes drastically (Etzel 1984). When n ≃ 28, the contamination must be extensive. If we were looking at a normal stellar photosphere, this value would imply log g ≃ 2! An application of the Inglis-Teller formula (Allen 1973) indicates that $\log N_e \simeq 12.1$, which confirms the non-photospheric nature of the material producing these high-order Balmer absorptions. Note that this implied density is approximately two orders of magnitude higher than that derived for the Hα emitting region as determined by Peters (1980, 1989). There is no reason to assume that these two phenomena arise from the same regions within the circumstellar material. The implication here is that the high-order absorptions are formed much closer to the stellar photosphere than the Hα emission.

The entire Balmer series provides useful diagnostic information for the study of circumstellar material in Algol-type and related binary stars where mass-transfer activity is present.

This work was partially supported by NSF grants AST8500189 and AST8822790. I thank Drs. Daniel M. Popper and Mirek J. Plavec for the loan of their Lick Observatory spectra, which are presented here. I gratefully acknowledge the NSF for a travel grant to attend this NATO Advanced Study Institute, and to the meeting organizers for defraying my living expenses.

REFERENCES

Allen, C. W. (1973). Astrophysical Quantities, third edition, (Athlone, London).
Batten, A. H. (1988). Publ. Astron. Soc. Pac. 100, 160.
Bidelman, W. P. (1976). In Be and Shell Stars, edited by A. Slettebak, (Reidel Publ., Dordrecht), p. 457.
Etzel, P. B. (1984). Bull. Amer. Astron. Soc. 16, 504.
Etzel, P. B. (1988). Astron. J. 95, 1204.
Etzel, P. B. (1989a). "Spectrophotometry of Algol-Type Binaries" In Active Close Binaries, edited by C. Ibanoglu and I. Yavuz, (these proceedings).
Etzel, P. B. (1989b). Publ. Astron. Soc. Pac. (in preparation).
Groth, H.-G. (1961). Z. Astrophys. 51, 231.
Harmanec, P. (1982). In Be Stars, edited by M. Jaschek and H.-G. Groth, (Reidel Publ., Dordrecht), p. 279.
Joy, A. H. (1942). Publ. Astron. Soc. Pac. 54, 35.
Joy, A. H. (1988). Publ. Astron. Soc. Pac. 100, 157.
Kaitchuck, R. H., and Honeycutt, R. K. (1982). Astrophys. J. 258, 224.
Kaitchuck, R. H., Honeycutt, R. K., and Schlegel, E. M. (1985). Publ. Astron. Soc. Pac. 97, 1178.
Kurucz, R. L. (1979). Astrophys. J. Suppl. 40, 1.

Lubow, S. H., and Shu, F. H. (1975). Astrophys. J. **198**, 383.
Naftilan, S. A. (1975). Publ. Astron. Soc. Pac. **87**, 321.
Peters, G. J. (1980). In Close Binary Stars: Observations and Interpretations, edited by M. J. Plavec, D. M. Popper, and R. K. Ulrich, (Reidel Publ., Dordrecht), p. 287.
Peters, G. J. (1989). Space Science Reviews **50**, 9.
Peters, G. J., and Polidan, R. S. (1984). Astrophys. J. **283**, 745.
Plavec, M. J. (1976). In Be and Shell Stars, edited by A. Slettebak, (Reidel Publ., Dordrecht), p. 1.
Plavec, M. J. (1983). Astrophys. J. **275**, 251.
Plavec, M. J. (1988). Astron. J. **96**, 755.
Popper, D. M. (1980). Annu. Rev. Astron. Astrophys. **18**, 115.
Popper, D. M. (1982). Publ. Astron. Soc. Pac. **94**, 945.
Popper, D. M. (1989). Astrophys. J. Suppl. (in press).
Sahade, J., and Cesco, C. U. (1946). Astrophys. J. **139**, 793.
Wyse, A. B. (1934). Lick Obs. Bull. **17**, 37.

AN OPTIMIZATION TECHNIQUE FOR THE FOURIER ANALYSIS OF THE LIGHT CURVES OF TOTALLY ECLIPSING SYSTEMS OF ALGOL-TYPE

P.G. NIARCHOS and G. PANTAZIS
Department of Astronomy
University of Athens
Panepistimiopolis, GR 157 83-Zografos
Athens, Greece

SUMMARY. A fully automatized technique for the evaluation of geometric and photometric elements of Algol type systems exhibiting total eclipses is presented. The method requires observations in at least two wavelengths and enables us to choose the most probable solution from a grid of solutions performed with various possible phase angles of the first contact of the eclipse.

1. INTRODUCTION

The analysis in the Frequency Domain of the light curves of eclipsing variables of any type of eclipse has been developed in a series of papers by Kopal and his collaborators (Kopal, 1975, 1976, 1977) and in a book (Kopal, 1979). Especially, the analysis of light curves of totally eclipsing systems of Algol type can be performed easily and the solution is given in a close algebraic form.

The crucial point in the analysis is the determination of the phase angle θ_1 at which the eclipse sets in. The aim of the present study is to present a simple method for choosing the phase angle for which the most consistent solution is obtained in more than one wavelengths of observations.

2. EQUATIONS OF THE PROBLEM

It is well known from previous literature on the subject, that the first step for the analysis of the observed light changes of eclipsing binary systems in the Frequency Domain is the evaluation of the so-called "moments of the light curves" A_{2m} defined by the equation

$$A_{2m} = \int_0^{\theta_1} (1-l)d(\sin^{2m} \theta), \quad m=0,1,2,... \tag{1}$$

C. İbanoğlu (ed.), Active Close Binaries, 213–218.

where θ_1 stands for the phase angle of the first contact of the eclipse. The A_{2m}'s are obtained from observations by planimetry or otherwise. Then the following well known equation can be used for the evaluation of the elements of the system:

$$A_2 = L_1\bar{C}_3 \tag{2}$$

$$A_4 = L_1(\bar{C}_3 + \bar{C}_2^2) \tag{3}$$

$$A_6 = L_1(\bar{C}_3^3 + 3\bar{C}_2^2\, C_3 + \bar{C}_1\bar{C}_2^2)$$

$$\bar{C}_3 = C_3 \tag{4}$$

$$\bar{C}_2 = [\frac{15-7u_1}{5(3-u_1)}]^{\frac{1}{2}} \quad C_2 \tag{5}$$

$$\bar{C}_1 = \frac{3(35-19u_1)}{7(15-7u_1)}\, C_1 \tag{6}$$

where L_1 is the fractional luminosity of the eclipsed star and u_1 is the linear coefficient of limb darkening. Moreover,

$$C_1 = r_1^2 \csc^2 i \tag{7}$$

$$C_2 = r_1 r_2 \csc^2 i \tag{8}$$

$$C_3 = r_2^2 \csc^2 i - \cot^2 i \tag{9}$$

and

$$r_1^2 = \frac{C_1^2}{(1-C_3)\,C_1+C_2^2} \tag{10}$$

$$r_2^2 = \frac{C_2^2}{(1-C_3)C_1+C_2^2} \tag{11}$$

$$\sin^2 i = \frac{C_1}{(1-C_3)C_1+C_2^2} \tag{12}$$

where $r_{1,2}$ are the fractional radii of the eclipsed and eclipsing component respectively, and

i is the orbital inclination.

Before we apply Equation (1), we must define the phase angle θ_1. An approximate value of it can, of course, be estimated from a plot of the light changes versus time. A more accurate determination of θ_1 is based on the quality of fit to the observations, attainable by trigonometric polynomials with a given number of terms as a function of θ_1 (Kopal, 1986, Elefteriades, 1979).

In the present study we propose a new method for the determination of the phase angle θ_1. The basic idea of that method is the fact that the geometry of the eclipses as well as the geometrical elements (r_1, r_2, i) of the system should be the same in all the wavelengths of observations.

The method consists of the following steps:

i. We plot the available observations in the plane l_λ (light intensity) versus phase, where $\lambda=\lambda_1, \lambda_2, \ldots$ and $\lambda_1, \lambda_2, \ldots$ are the wavelengths of observations.

ii. From a visual inspection of the light curves we specify the phase interval $[p_a, p_b]$ inside of which the phase of the first contact $p = \frac{\theta}{360°}$ lies.

iii. We evaluate the moments A_{2m}'s for each light curve by using as units of light those l_i's which correspond to phases $p_i = p_a + (i-1)\Delta p, i=1,2,\ldots k,(k+1)$, where $\Delta p = \frac{p_b - p_a}{k}$ and k arbitrary positive integer

iv. For each set of moments A_{2m}'s in all the wavelengths, corresponding to a certain unit of light (or phase), we perform the solution in the usual way by using the Equations (1)-(12).

v. For simplicity let us suppose, as usually happens, that we have three light curves corresponding to wavelengths $\lambda_1, \lambda_2, \lambda_3$ (U,B,V). Then we evaluate the quantity

$$q = (r_{1,V}-r_{1,B})^2+(r_{1,V}-r_{1,U})^2+(r_{1,B}-r_{1,U})^2$$

$$+(r_{2,V}-r_{2,B})^2+(r_{2,V}-r_{2,U})^2+(r_{2,B}-r_{2,U})^2 \tag{13}$$

$$+(\sin i_V-\sin i_B)^2+(\sin i_V-\sin i_U)^2+(\sin i_B-\sin i_U)^2$$

where $r_{1,V}$, $r_{1,B}$, $r_{1,U}$: is the fractional radius of the eclipsed component in the three wavelengths,

$r_{2,V}$, $r_{2,B}$, $r_{2,U}$: is the fractional radius of the eclipsing component in the three wavelengths, and

i_V, i_B, i_U : is the orbital inclination in the tree wavelenghts.

vi. It is obvious that the most consistent solution in the three wavelengths is obtained when q takes its lowest value. This value of q corresponds to a phase angle θ_1 which should be considered as the most probable value of the phase angle of the first contact.

3. APPLICATION

The above method has been applied to the Algol-type system U Sge, which exhibits total eclipses. The photoelectric observations of the system made by McNamara and Feltz (1976) have been used in our analysis. The whole procedure has been fully automatized, so that the input data are only the observations (light intensities and phases) in three colours (y,b,u). The phases range from 0 till p_b, where p_b is the right end of the phase interval $[p_a, p_b]$ inside of which the phase of the first contact lies. The light intensities and phases are read from a free hand curve drawn through the observational points. The results from the analysis of the light curves of U Sge are given in Table I.

TABLE I

Solutions of U Sge for various phase angles of the first contact

Phase angle	Filter	r_1	r_2	i	L_1	q
	y	0.251	0.285	90°.18	0.910	
35°.36	b	0.251	0.284	90°.17	0.942	0.00154
	u	0.278	0.279	90°.36	0.971	
	y	0.251	0.286	90°.18	0.910	
35°.06	b	0.251	0.284	90°.17	0.942	0.00127
	u	0.275	0.279	90°.34	0.971	
	v	0.247	0.288	90°.14	0.910	
34°.75	b	0.251	0.285	90°.17	0.949	0.00099
	u	0.270	0.281	90°.30	0.971	
	y	0.243	0.290	90°.09	0.910	
34°.45	b	0.247	0.287	90°.13	0.942	0.00086
	u	0.264	0.283	90°.24	0.971	
	y	0.243	0.290	90°.09	0.910	
34°.14	b	0.238	0.292	90°.02	0.942	0.00072
	u	0.258	0.286	90°.18	0.971	
	y	0.236	0.292	90°.01	0.910	
33°.8	b	0.233	0.294	88°.70	0.942	0.00056
	u	0.251	0.290	90°.03	0.971	
	y	0.239	0.292	90°.01	0.910	
33°.5	b	0.223	0.302	86°.16	0.942	0.00118
	u	0.248	0.291	90°.02	0.971	

cont. TABLE I

Phase angle	Filter	r_1	r_2	i	L_1	q
33°.2	y	0.234	0.295	88°.24	0.909	0.00110
	b	0.217	0.306	84°.14	0.942	
	u	0.240	0.296	87°.82	0.971	
32°.89	y	0.234	0.295	88°.24	0.909	0.00141
	b	0.211	0.312	83°.14	0.942	
	u	0.231	0.302	85°.59	0.971	
32°.58	y	0.229	0.298	86°.68	0.909	0.00161
	b	0.204	0.318	82°.14	0.942	
	u	0.214	0.314	82°.95	0.971	
32°.26	y	0.219	0.307	84°.50	0.909	0.00157
	b	0.197	0.325	81°.11	0.942	
	u	0.202	0.326	81°.16	0.971	
31°.94	y	0.207	0.317	82°.53	0.909	0.00169
	b	0.190	0.333	80°.00	0.942	
	u	0.188	0.341	79°.16	0.971	
31°.63	y	0.192	0.331	80°.46	0.909	0.00547
	b	0.181	0.343	78°.61	0.942	
	u	0.161	0.377	75°.33	0.971	

An inspection of Table I reveals that the quantity q shows a variation for various values of the phase inside the interval $[p_a, p_b]$, where $p_a = 0.0878$ and $p_b = 0.0982$. The lowest value of q was obtained at phase angle $\theta' = 33°.8$, for which the most consistent solution in three wavelengths was obtained. In table II the elements of U Sge evaluated by different methods are given for comparison. The symbols listed have the usual meaning.

TABLE II

Geometric and photometric elements of U Sge*

	Present work	McNamara & Feltz(1976)	Kitamura(1965)	Olson(1987)
r_1	0.236	0.224	0.220	0.223
r_2	0.292	0.295	0.290	0.292
i	90°.0	90°.0	90°.0	89°.04
u	0.6[a]	0.04	-	-
L_1	0.910	0.870	0.903	0.896
L_2	0.090	0.130	0.097	0.104

*: based on yellow observations
a: assumed

It is obvious from Table II that a very good agreement exists between our results and those found by other investigators using different methods and observations.

The advantage of our proposed technique is that it allows a simultaneous treatment of observations in different wavelengths as well as it enables us to specify the phase angle of the first contact by choosing that angle which corresponds to the most consistent solution.

REFERENCES:

Elefteriades, S.: 1979, M.Sc. Thesis, Univ. Manchester.
Kitamura, M.: 1965, in Advances in Astron. Astrophys. 3, 27ff.
Kopal, Z.: 1976, Astrophys. Space Sci. 45, 269.
Kopal, Z.: 1977, Astrophys. Space Sci. 51, 439.
Kopal, Z.: 1979, Language of the Stars, D. Reidel Publ. Co.,Dordrecht, Holland.
Kopal, Z.: 1986, Vistas in Astronomy, **Vol. 29**, pp. 295.
MacNamara, D.H. and Feltz, K.A.: 1976, Publ. Astron. Soc. Pac. 88, 688.
Olson, E.C.: 1987, Astron. J. 94, 1043.

THE ALGOL-TYPE BINARY U CORONAE BOREALIS, A TRIPLE?

J.R.W. Heintze
Sterrekundig Instituut
State University Utrecht
P.O. Box 80.000
3508 TA Utrecht, The Netherlands

ABSTRACT. As yet there is insufficient evidence to disprove the presence of a third component in he U CrB system.

1 Introduction

1.1

U CrB is a well known and well studied Algol-type binary. In the modern literature, estimates of the spectral type of the primary are B5V (Cester et al., 1977) and B7V (Batten et al., 1978) and of the secondary: F8III-IV (Batten & Tomkin, 1980), G0III-IV (Plavec & Polidan, 1976) and G2III-IV (Cester et al., 1977).
For the effective temperature of the primary Cugier (1989) finds 14 800 ± 300 K and Dobias and Plavec (1989): 14 000 K.

1.2

The period is about $3^d\!.45$ and is not constant. Van Gent (1982) has updated the list of moments of primary minimum (O) and determined (O-C) values from Dugan & Wright's (1939) linear emphemeris

$$C = JD_\odot = 2416747.964 + 3.45220416 \times E \tag{1}$$

He has tried to explain the general behaviour of the (O-C) points from 1870 to 1978 by the light-time effect due to an invisivle companion (3rd body in the U CrB system). With future (O-C) data it should be possible to decide whether this concept can be acceped or not. The orbits caused by the 3rd body are highly excentric ($e \approx 0.7$) with a period $P_3 \approx 78$ years and a mass function $f(M_3) \approx 0.01$. Although Van Gent's "light-time" (O-C) curve describes the observations rather well, there

C. İbanoğlu (ed.), Active Close Binaries, 219–235.

are deviations. For example the photographic determinations at E = 1590; 2270; 2840 observed by Dugan & Wright (1939) and at E = 3841 by Gaposchkin (1953). These deviations are about $-0^d.016$; $-0^d.010$; $+0^d.008$ and $-0^d.048$ respectively. The last value has been skipped, however the first 3 ones not. The absolute value of these values is about equal to values recently found (Van Gent, 1989d). For example, in the night 29/30 July 1987 (E = 8756) the observed (O-C) value differs by $+ 0^d.012$ with respect to the "light-time" (O-C) value.
Already in 1983, Frieboes-Conde & Herczeg (1983) have tried to find evidence for a 3rd component in the U CrB system. Tentatively they found $P_3 \approx 26$ years and $f(M_3) \approx 0.008$. Their conclusion at that time was that a periodic representation of the observed (O-C) points is perhaps not impossible; an alternative description should be a set of abrupt period changes.

1.3

If a 3rd body in the U CrB system has to be accepted then the deviations of observed (O-C) values from the "light-time" (O-C) curve indicate the real period variations or period jumps. These variations or jumps $\frac{\Delta P}{P}$ are then smaller than if the 3rd body model is not true. From 1983 to the midst of 1987 the period change $\frac{\Delta P}{P} \approx$ constant $\approx 0.85 \times 10^{-5}$ with respect to the "light-time" (O-C) curve and about 1.22×10^{-5} with respect to the period in formula (1).

1.4

In 1989 Van Gent abandoned the idea of a 3rd component in the U CrB system because the observed (O-C) points from 1983 to 1987 [not only those published by Van Gent but also those by Diethelm (1985 and 1986) and by Hübscher et al. (1985)] did deviate systematically from the "light-time" (O-C) curve (Van Gent, 1989d).

1.5

In this paper evidence will be given that a 3rd body in the U CrB can be present. Deviations between observed and "light-time" (O-C) values can be explained by a superposition of period jumps and/or temporary period changes on the "light-time" (O-C) curve.

2 The lightcurves as observed in the Utrecht Photometric System (≡ UPS)

2.1

UPS lightcurves of U CrB (Van Gent, 1989c) have been obtained in 10 nm wide wavelength bands at the UPS wavelengths/colors 474, 672, 781, 871 nm (Heintze, 1989 and Provoost, 1980) during the years 1981 to 1988. In the following these wavelengths/colors will be abbreviated by 47, 67, 78 and 87.

2.2

From this material, 8 moments of minimum could be determined. These moments can be presented by a linear formula:

$$JD\odot = 2445680.8859(13) + 3.4522482(45) \times (E - 8381), \qquad (2)$$

where E is the same as in formula (1).
The E's at the observed moments of minimum are 8076.5, 8078.5, 8160, 8199, 8327, 8524, 8617 and 8756. Figure 1 of Van Gent (1989c) shows these data together with the graphical expression of (O-C) as calculated with (1)

2.3

The first UPS point (E = 8076.5) causes a shift in (O-C) equal to $\approx -0\overset{d}{.}014$ with respect to the previous photoelectric (O-C) determinations [among others at E = 8 005 (Olson, 1980)]. Between the first and second UPS point an (O-C) shift occurs of $\approx +0\overset{d}{.}\,028$, wheras between the second UPS point and the following ones such a shift is $\approx -0\overset{d}{.}0.20$. The net shift of $\approx -0\overset{d}{.}006$ occurred within 380 days! (see Figure 10d in Van Gent 1989d.)

2.4

According to the results mentioned above, most of the UPS lightcurve measurements could be combined to one lightcurve for each of the wavelengths 47, 67, 78, 87. These lightcurves are presented in Figure 1 in Van Gent (1989a) and in the Figures 2 and 3 in Van Gent (1989c). Three types of primary minima could be distinguished, depending on their depths, denoted here by "high", "medium" and "low", abbreviated in the following by "H", "M" and "L" state. [Van Gent (1989c) called these states: "light excess", "average" and "light deficient" or "undisturbed". Olson (1982) also reports on 3 types of primary minimum: type 1, "undisturbed" and type 2 respectively. Note that Olson's "undisturbed" has another meaning than Van Gent's.]

2.5

In Figure 3 in Van Gent (1989c) all observed (parts of) UPS primary minnima are gathered: 6 of them belong to the "high" state (unfortunately no points around mideclipse have been observed - only branches, which are less steep than in the other states); 11 belong to the "medium" state and 10 to the "low" state. **These latter two states clearly show a flat segment at mideclipse with a width of 0.014 in phase.** An approximate depth at mideclipse of the "high" state can be found by extrapolation.
From Figure 3 in Van Gent (1989c) a phase shift $\Delta\phi = 0.0007$ could be determined at mideclipse. The first term of (2) has to be corrected by $-0\overset{d}{.}0023$, which is a litte bit more than the error given for this term.

3 The interpretation of the UPS lightcurves

3.1 the first approximation

In the following sections the effective temperature will be denoted by T.

As a first aproximation the UPS lightcurves have been interpreted by Van Gent (1989d) assuming the 3rd light to be zero (see section 1.4). He used the Wilson-Devinney code (Wilson & Devinney, 1971, 1973; Wilson, 1979) with which n lightcurves (*i.c.* $n = 4$) and the radial-velocity curves of both components (Batten & Tomkin, 1980) can be solved simultaneously. In this program he introduced some modifications (Van Gent, 1989b, 1989d): For $T_{eff} \geq 5\,500$ K the models of Carbon et al. (1969) were replaced by the Kurucz (1979) models. To describe the fit between observed and calculated lightcurves a quantity χ^2_{red} is introduced:

$$\chi^2_{red} = \frac{1}{N-p} \sum_{i=1}^{N} \frac{(\ell_{obs} - \ell_{calc})_i^2}{\sigma_i^2}, \quad (3)$$

where N = number of normal points (observations)
p = number of parameters to be determined
$N-p$ = degree of freedom
σ_i = standard deviation of each observation.

Van Gent has reduced a) mean UPS lightcurves, where the observations of the three types of primary minimum were lumped to gether and b) the low state UPS lightcurves. Some results are given in Table 1 and Table 2. In the latter table the luminosities of the primary (ℓ_1), the secondary (ℓ_2) and that of the 3rd light (ℓ_3) are given in the same scale such that

$$\ell_1 + \ell_2 + \ell_3 = 1 \quad (4)$$

for each wavelength.
Van Gent has demonstrated that the secondary undoubtedly fills its Roche lobe.

3.2 the second approximation

3.2.1

Van Gent (1989d) assumed $\ell_3 = 0$. He solved the four UPS lightcurves simultaneously with the radial-velocity curves of Batten & Tomkin (1980).

3.2.2

To find out whether it is allowed to assume $\ell_3 = 0$, for each state of the primary minima the lightcurves together with the radial-velocity curves were solved separately assuming $\ell_3 = 0$. For each adopted T_1, it turned out that T_2 is a function of λ. This effect disappears as soon as ℓ_3 is allowed to be $\neq 0$. Therefore a new set of simultaneous solutions has been calculated with $\ell_3 \neq 0$.

3.2.3

It was not clear what T_1 should be. Van Gent (1989d) finds the primary to be a main sequence star with a mass of $\approx$ 5.2 solar masses. In this case $T_1 \approx 16\ 200K$ according to the relations of Strayžis & Kuriliene (1981) as well as to those of Habets & Heintze (1981). As already noted in 1.1, Cugier (1989) finds $T_1 = 14\ 800$ K and Dobias & Plavec (1989) $T_1 \approx 14\ 000$ K. Some results are given in Table 1 and will be discussed in the next section.

3.3 the 3rd approximation

3.3.1

The solutions of the first as well as the second approximation show a sharp primary minimum in the light-curve graphs. This is in contradiction with the observations, which show a flat segment at mideclipse.

3.3.2

It could be, that the 3rd light is not due to a 3rd body, but to radiation coming from a disk around the primary component. To investigate this, solutions wer made of the 4 UPS lightcurves (simultaneously together with the two radial velocity curves) with $\ell_3 \neq 0$ but excluding the observed points of the primary minimum.
For if the 3rd light is associated with the primary, at least part of it will be obscured by the secondary at mid eclipse of the primary minimum. However, the Wilson & Devinney program assumes that a 3rd light is present during all phases. Therefore by leaving away the points of the observed primary minima this problem can be circumvented and the program should be able to find a 3rd light which may partly or completely disappear at primary minimum.

Iterations were carried out for fixed pairs of T_1 and q, T_1 varying from 14 000-20 500 K with steps of 500 K and q varying from 0.278-0.292 with steps of 0.002. The results show, that over the q interval used T_2; Ω_1 and i can be described as linear functions of T_1. In each case a 3rd light has been found, the values being a function of λ. These values increase with T_1. **An interesting result is, that in the light-curve graphs the primary minima show a flat segment at mideclipse.** However the calculated widths are 0.020 in phase, which is larger than the observed one: 0.014 in phase.

3.3.3

Another interesting result is that Ω_1 is much larger (i.e. radius much smaller) than in the previous solutions. As can be seen in Table 1 the radius of the primary component is now smaller than the outer radius of a possible accretion disk,making transient disks possible. Because the primary eclipses are total, possible light of a disk will disappear at mideclipse also.

3.3.4

The light-curve graphs show, that a nice fit can be obtained between observed "L"state - and calculated light-curves (except for the width of the flat segment at mideclipse) for $T_1 \approx 15\ 000$ K at the q values used. For $T_1 < 15\ 000$ K the calculated primary minima are deeper and for $T_1 > 15\ 000$ K they are less deep than the observed ones. It has to be noted, that $T_1 \approx 15\ 000$ is close to the effective temperature of the primary of U CrB as found by Cugier (1989) [section 1.1].

3.3.5

The question that arises now is whether the total 3rd light has to be attributed to a disk or not. If this is the case the luminosity at mideclipse should be ℓ_2 and not $\ell_2 + \ell_3$ [see formula (4)]. So at mideclipse a luminosity equal to $\ell_1 + \ell_3$ should disappear. Light-curve graphs calculated with this condition are always too deep certainly for $T_1 < 15\ 000$ K but also for $T_1 > 15\ 000$ K up to 20 500 K, a temperature which is very likely too hot for the primary. **This means that at any case part of ℓ_3 is coming from a 3rd body that contributes to the observed luminosity in all phases.** This 3rd body does not have to be necessarily part of the U CrB system and may be a background star. It was within the diaphragm of 60″ diameter during the observations.

3.4 the fourth approximation

3.4.1

In this approximation, T_2 as well as Ω_1 are assumed to be the same for "H", "M" and "L" states.

3.4.2

Any solution to be found has to fulfill the condition, that at mideclipse of the primary minimum a flat segment has to be present with a width of 0.014 in phase. However the spacing of the normal- and the calculated points is 0.005 in phase. Therefore quantitatively the condition is that the flat segment should be 0.01 in phase in the calculated light curves.

3.4.3

The procedure followed was, that the light curves were solved as a function of Ω_1; q; i and T_2. The iterations for $T_1, a, \ell_1, \ell_2, \ell_3$[$\Delta\phi$ being taken 0.0007 (section 2.5) and the system velocity -5.4 km/s (Van Gent, 1989d)] were repeated until a stable best fit was obtained between the observed "H", "M" and "L" light curves and the calculated ones. The fit between observations and calculations [see formula (3)] was now defined by

$$\chi^2_{red}(3) = \chi^2_{red}(\text{"}H\text{"}) + \chi^2_{red}(\text{"}M\text{"}) + \chi^2_{red}(\text{"}L\text{"}) \ . \tag{5}$$

It turned out that $\chi^2_{red}(3)$ has quite some local minima in the $\Omega_1; q; i$ and T_2 space. At the moment there is still no complete certainty whether the absolute minimum of $\chi^2_{red}(3)$ has already been reached or not. The input data for the iterations up till now used are for Ω_1 :7.75, 8.0 and 8.25, for q: 0.270-0.320 with steps of 0.005, for i: 82.0, 82.3, 82.7 and 83.0 and for T_2 : 5 100, 5 200, 5 300, 5 400 K.
Some local minima in χ^2_{red} do not fulfill the condition mentioned at the end of 3.4.2. In these cases the flat segment in the primary minimum is either 0 or ≥ 0.02. From the remaining local minima in $\chi^2_{red}(3)$, that with the lowest $\chi^2_{red}(3)$ is reached for $\mathbf{\Omega_1 = 8.0; q = 0.29; i = 82.7}$ **and** $\mathbf{T_2 = 5\ 350K}$ as far as can be concluded now. In Table 1 some results of this semi-final solution are given in the 5th column.

4 Discussion

4.1

Table 1 shows that allowing for a 3rd light in the Wilson-Devinney interpretation program (second approximation) causes the potential of the primary (Ω_1) to increase implying the radius of the primary to decrease by nearly 15%. In the second approximation the radius of the primary becomes already nearly equal to the outer radius of a possible accretion disk. The solution of the 4th approximaton provides a radius of the primary which is $\approx 10\%$ smaller than the outer radius of a possible accretion disk. Now, matter of the secondary can be around the primary without causing a HTAR ($\equiv$ High temperature Accretion Region) (see Figure 1).

4.2

The Wilson & Devinney solution of the 4th approximation dictates $T_2 \approx 5350$ K corresponding to a spectral type of $\approx$ G 3.5 III according to the relations of Straižys and Kuriliene (1981) as well as of Habets & Heintze (1981). However Batten & Tomkin (1980) estimate the spectal type of the secondary to be F8 III-IV [section 1.1]. The above mentioned relations assign to this spectral type a $T_2 \approx 6$ 100 K, 750 K higher than the solution of the 4th approximation provides. Now the Wilson & and Devinney program prints normally for each component an effective temperature not influenced by the other one. Van Gent (1989d) has shown how the local effective temperatures of the secondary change over its surface due to the irradiance from the primary on it. The mean of the local effective temperatures is about 400 K higher than the printed T_2 in the case of U CrB, causing $T_2 \approx 5$ 750 K as the result of the 4th approximation. This corresponds to a spectral type of about G 0.5 III, close to the spectral type G0 III-IV as published by Plavec & Polidan (1976). The nice agreement between these observational- and interpretative results gives confidence in the solution of the 4th approximation. This confidence is necessary for the tentative interpretation in the following sections.

4.3

Some more results of the semi-final solution than given in Table 1 and Table 2 can be found in Table 3. The luminosities of the primary, the secondary and that of the 3rd light were obtained by multiplying the luminosities as given in Table 2 [according to formula (4)] by ℓ_2^{-1} for each wavelength, causing the luminosity of the secondary always to be 1. These luminosities will be denoted by
$\ell'_1; \ell'_2$ and ℓ'_3 respectively.
Note that the luminosity of the secondary is assumed to be constant.
Note that $T_{1,L} > T_{1,M} > T_1, H$.

4.4

As a first step to the interpretation of these data, it will be assumed that the luminosities of the 3rd light in the "L" state, ℓ'_{3L}, are entirely due to a 3rd body. The differences of the $\ell'_{3,M}$ values with those of $\ell'_{3,L}$, denoted by $\Delta\ell'_{3,M}$, are more or less constant: 0.14, 0.14, 018 and 0.23 for the wavelengths 47, 67, 78, 87 respectively

Table 1

Some results of the Wilson-Devinney solutions of the UPS lightcurves in the 4 approximations

	$\ell_3 = 0$		$\ell_3 \neq 0$	
approx.	1 Van Gent (1989d)	2 same procedure as in column 1	3 obs. points of prim. min. excluded in prim. min. a flat segment of 0.02 in phase shows up (too broad).	4 semi-final solution with the condition that the flat segment in the prim. min = 0.01 in phase.
separation of the components: $a(R_\odot)$	18.1	17.6	$25.33 - 25.94 \cdot q$ independent of T_1	17.54
inclination: i	79.3	81.8	$81.623 + 0.575 \cdot \frac{T_1}{10000}$	82.7
mass ratio : q	0.28	0.30	-	0.29
rel. radius secondary : r_2	0.275	0.275		0.274
potential primary : Ω_1	6.16	7.16	$7.94 + 0.465 \cdot \frac{T_1}{10000}$	8.0
rel. radius primary : r_1	0.171	0.146		0.130
outer radius accretion disk	0.146	0.143		0.143
T_1 & T_2	results depend only a little bit on the adopted T_1		$T_2 = 950 + 0.28 \cdot T_1$	see Table 2
$R_2/R_\odot$	4.96	4.82		4.8
$M_2/M_\odot$	1.49	1.41		1.4
$R_1/R_\odot$	3.1	2.57		2.3
$M_1/M_\odot$	5.2	4.71		4.7

Table 2

Best-fit results of the 4th approximation assuming T_2 to be the same for the "H", "M", and "L" state:

$$\Omega_1 = 8.0, \qquad q = 0.29, \qquad i = 82^\circ\!.7, \qquad T_2 = 5350 \text{ K},$$

below: T_1, ℓ_1, ℓ_2 and ℓ_3 as defined according to formula (4):

		λ 47			λ 67			λ 78			λ 87		
state	T_1	ℓ_1	ℓ_2	ℓ_3	ℓ_1	ℓ_2	ℓ_3	ℓ_1	ℓ_2	ℓ_3	ℓ_1	ℓ_2	ℓ_3
"H"	15 300	0.59	0.08	0.33	0.51	0.21	0.29	0.46	0.26	0.28	0.44	0.30	0.26
"M"	15 450	0.63	0.08	0.28	0.54	0.22	0.24	0.50	0.28	0.22	0.48	0.32	0.21
"L"	15 600	0.64	0.09	0.27	0.56	0.22	0.22	0.53	0.29	0.18	0.52	0.34	0.14

Table 3

Best-fit results of the 4th approximation. T_2, Ω_1, q and i the same as in Table 2. The luminosities ℓ'_1, ℓ'_2 and ℓ'_3 are defined by $\ell_1/\ell_2, \ell_2/\ell_2$ and ℓ_3/ℓ_2 respectively.

$\Delta\ell'_{3,\text{"}H\text{"}} = \ell'_{3,\text{"}H\text{"}} - \ell_{3,\text{"}L\text{"}}$ $\qquad$ $\Delta\ell'_{3,\text{"}M\text{"}} = \ell_{3,\text{"}M\text{"}} - \ell'_{3,\text{"}L\text{"}}$

		λ 47				λ 67				λ78				λ87			
state	T_1	ℓ'_1	ℓ'_2	ℓ'_3	$\Delta\ell'_3$	ℓ'_1	ℓ'_2	ℓ'_3	$\Delta\ell'_3$	ℓ'_1	ℓ'_2	ℓ'_3	$\Delta\ell'_3$	ℓ'_1	ℓ'_2	ℓ'_3	$\Delta\ell'_3$
"H"	15 300	7.01	1.	3.85	0.81	2.46	1.	1.41	0.43	1.79	1.	1.07	0.44	1.49	1.	0.87	0.45
"M"	15 450	7.12	1.	3.19	0.14	2.49	1.	1.12	0.14	1.81	1.	0.81	1.18	1.51	1.	0.65	0.23
"L"	15 600	7.21	1.	3.05	0.	2.52	1.	0.98	0.	1.83	1.	0.63	0.	1.52	1.	0.42	0.

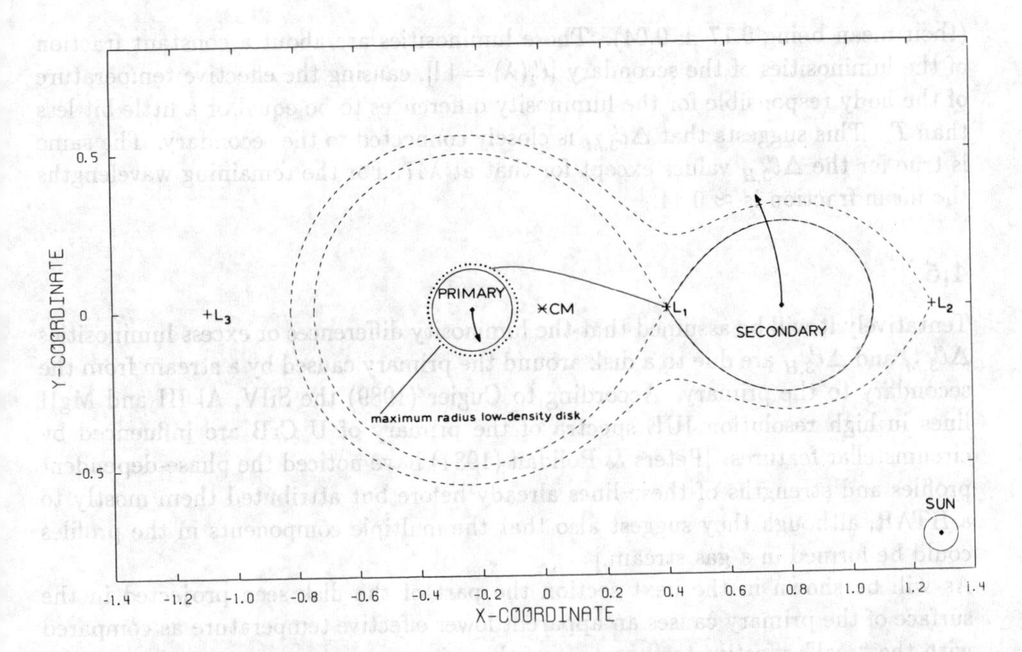

Figure 1: Absolute geometry of U CrB. Also depicted are the inner- and outer Roche lobes with the canonical Lubow-Shu stream trajectory. The dotted circle just outside the primary denotes the outer radius of a possible low-density disk

(their mean being 0.17 ± 0.04). These luminosities are about a constant fraction of the luminosities of the secondary [$\ell'_2(\lambda) = 1$!], causing the effective temperature of the body responsible for the luminosity differences to be equal or a little bit less than T_2. This suggests that $\Delta\ell'_{3,M}$ is closely connected to the secondary. The same is true for the $\Delta\ell'_{3,H}$ values except for that at λ47. For the remaining wavelengths the mean fraction is ≈ 0.44.

4.5

Tentatively it will be assumed that the luminosity differences or excess luminosities $\Delta\ell'_{3,M}$ and $\Delta\ell'_{3,H}$ are due to a disk around the primary caused by a stream from the secondary to the primary. According to Cugier (1989) the SiIV, Al III and MgII lines in high resolution IUE spectra of the primary of U CrB are influenced by circumstellar features. [Peters & Polidan (1984) have noticed the phase-dependent profiles and strengths of these lines already before but attributed them mostly to a HTAR, although they suggest also that the multiple components in the profiles could be formed in a gas stream.]
As will be shown in the next section the part of the disk seen projected in the surface of the primary causes an apparent lower effective temperature as compared with the "real" effective temperature of the primary.

4.6

Whenever the gas stream encircles the primary [this is now possible because $r_1 < r_{disk}$, the effective temperatures of the primary in the "M" and "H" state should be lower than $T_{1,L}$. Table 2 shows that indeed $T_{1,H} < T_{1,M} < T_{1,L}$ in the 4th approximation. If the model still is correct, the fractions (α) of the surface of the primary (S_1), that are obscured by the disk, can be estimated. For if $\ell_{1,L} = S_1 \cdot f_{T_{1,L}}$, where $f_{T_{1,L}}$ is the flux of the primary in state "L", then

$$f_{T_{1,H}} = (1 - \alpha_H) \cdot f_{T_{1,L}} + \alpha_H \cdot f_{T_d} \tag{6}$$

and

$$f_{T_{1,M}} = (1 - \alpha_M) \cdot f_{T_{1,L}} + \alpha_M \cdot f_{T_d} \tag{7}$$

where T_d is the effective temperature of the disk, assumed to be equal to T_2.
Each of the formulas (6) and (7) have to hold for the four UPS wavelengths. So mean α_H and α_M can be calculated.

$$\alpha_H \approx 3.8\% \tag{8}$$

$$\alpha_M \approx 1.9\% \tag{9}$$

This could mean, that $T_{1,M}$ and $T_{1,H}$ have to be regarded as apparent effective temperatures.

4.7

In Figure 2a the mass and the radius of the primary as found in the 4th approximation are plotted in a (mass, radius) diagram. The position is close to the ZAMS relation of Straižys and Kuriliene (1981) indicating a spectral type of about B 4.3. Figure 2b suggests $T_1 \approx 16\ 500$ K .
However the 4th approximation gives $T_{1,L} \approx 15\ 600$ K. Perhaps the observed $T_{1,L}$ has to be regarded as an apparent effective temperature also. Now a similar formula as (6) or (7) can be introduced:

$$f_{T_{1,L}} = (1 - \alpha_L^*) \cdot f_{T_1} + \alpha_L^* \cdot f_{T_d} \tag{10}$$

and for (6) and (7) can be introduced:

$$f_{T_{1,H}} = (1 - \alpha_H^*) \cdot f_{T_1} + \alpha_H^* \cdot f_{T_d} \tag{11}$$

$$f_{T_{1,M}} = (1 - \alpha_L^*) \cdot f_{T_1} + \alpha_L^* \cdot f_{T_d} \tag{12}$$

In this case the mean α values are:

$$\begin{array}{lcl} \alpha_H^* & = & \left.\begin{array}{l} 13.7\% \\ \left.\begin{array}{l} 12.0\% \\ 10.3\% \end{array}\right] 1.7\% \end{array}\right] 3.4\% \\ \alpha_M^* & = & \\ \alpha_L^* & = & \end{array}$$

Note that $\alpha_H^* - \alpha_L^*$ and $\alpha_M^* - \alpha_L^*$ are nearly equal to α_H and α_M [see (8) and (9)].
Observed T_1 have been given by Cugier (1989) and Dobias & Plavec (1989):

$$(14\ 800 \pm 300)\ \text{K and } 14\ 000 \quad \text{respectively.}$$

The latter one from an observed energy distribution (IUE + visible). Assuming $T_{1,real} = 16\ 500$ K, the corresponding α are:

$$\begin{array}{lcl} \alpha_C & \approx & 17\% \\ \alpha_{DP} & \approx & 28\% \end{array}$$

If the model is correct, then the above mentioned low apparent effective temperatures could have been caused just after the arrival of quite some matter from the secondary. Sudden period jumps still take place once in a while.

4.8

The values of $\ell'_{3,L}$ in Table 3 indicate a very high temperature of about 100 000 K of the 3rd body. If the 3rd body is part of the U CrB system, then according to the mass function $f(M_3)$ (see section 1.2) the mass of this 3rd body should be about 0.8 solar masses, assuming the orbit of the binary U CrB to be coplanar with the orbit of the 3rd body. In that case the 3rd body could be a white dwarf.

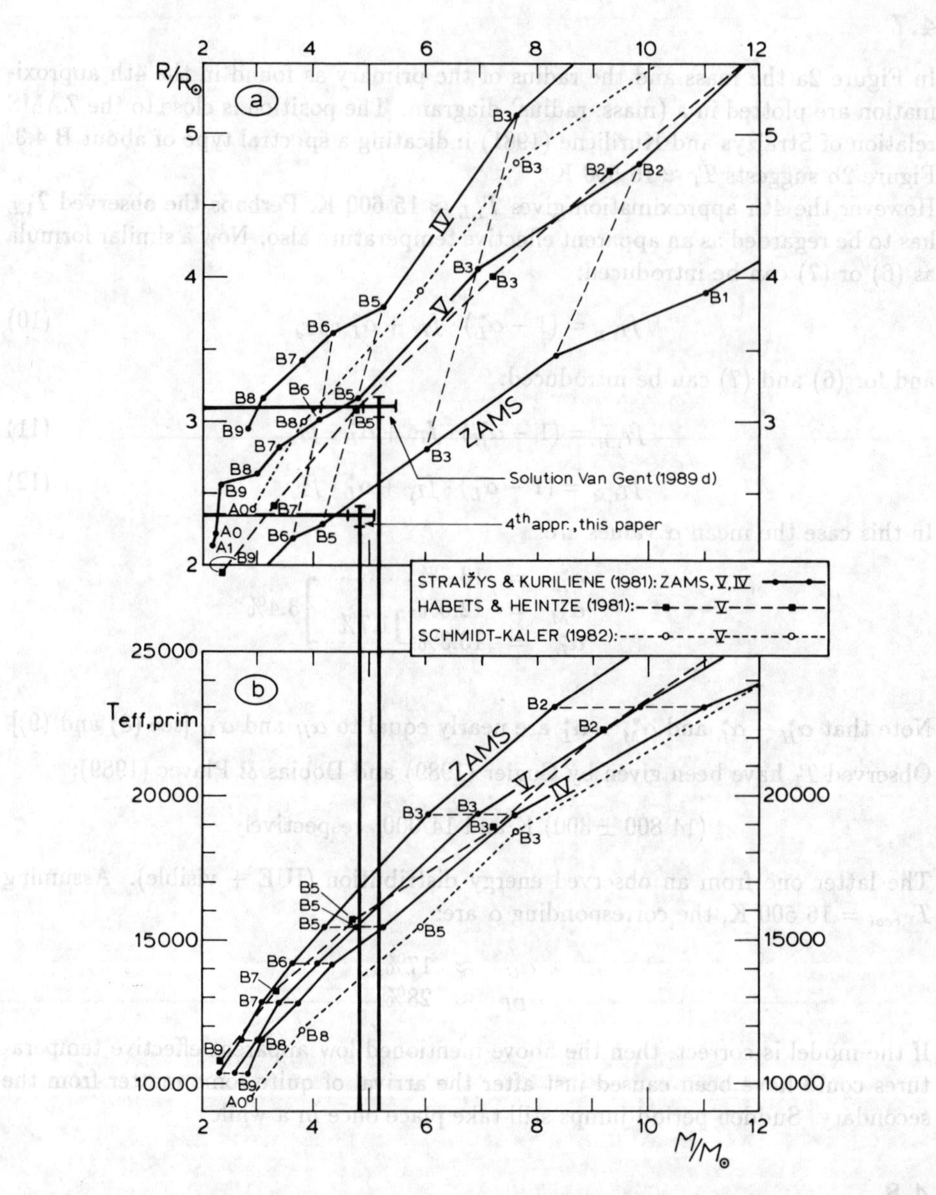

Figure 2: The position of the primary of U CrB in a. the (mass, radius) diagram and b. the (mass, effective temperature) diagram.

4.9

It is assumed that $\ell'_{3,L}(\lambda)$ [lower row in Table 3] respresents the energy distribution of a 3rd body in the U CrB system. However part of these $\ell'_{3,L}$ values can have to be added to the $\Delta\ell'_3(\lambda)$ values [even a $\Delta\ell'_{3,L}$ can exist] lowering as a matter of fact the contribution of the 3rd body light to the total light of the U CrB system. If the $\ell'_{3,L}$ values, as given in the last row of Table 3, represent radiation of a 3rd body in the U CrB system, then at $\lambda = 100nm$ the 3rd body luminosity is estimated to be $\approx 50\%$ of the luminosity of the primary. Observationally this should be noticeable.

5 Conclusion

5.1

It is very hard to draw firm conclusions at the moment. The UPS photometry does not contradict a 3rd component in the U CrB system. If the concept of a 3rd component in the U CrB system is correct, the (O-C) values of the moments of primary minimum should reach maximum values of about $0^d.015$ in the years 2010-2015. However such (O-C) values are already observed in 1987! It is possible that this (O-C) behaviour can be explained by temporary period changes and/or jumps.

5.2

It is not excluded that part of the period changes are caused by mass transfer from the secondary to the primary. The small radius of the primary suggests that this star has been rejuvenated by fast mass transfer. The relatively low effective temperature of the primary as observed suggests that quite some gas of the secondary is still hanging around the primary. The differences in depth of the primary minimum suggest, that still matter is streaming from the secondary to the primary in blobs.

5.3

In any case it is clear that as many as possible primary minima of U CrB should be observed in the next 2-3 decades to follow the (O-C) values and to follow the changes of depth of the primary minima!

Acknowledgements

It is a pleasure to thank dr. J.-P. de Grève for valuable discussions, G.W. Geijtenbeek, J.H.G. Rosenbaum and E.B.J. Van der Zalm for computational assistance,

E. Landré for drawing the figures; E.I. Tokamp for typing the manuscript, and dr. R.H. van Gent for a critical reading of the manuscript.

References

Batten, A.H., Flechter, J.M. & Mann, P.J.: 1978, Publ. Dom. Astrophys. Obs., **15**, no 5

Batten, A.H. & Tomkin, J.: 1980, Publ. Dom. Astrophys. Obs., **15**, 419.

Carbon, D.F., Gingerich, O. & Kurucz, R.L.: 1969, in O. Gingerich (ed.), Theory and Observation of Normal Stellar Atmospheres (M.I.T. Press Cambridge) part V.

Cester, B., Fedel, B. Giuricin, G. Mardirossian, F. & Pucillo, M.: 1977, Astron. Astrophys., **61**, 469.

Cugier, H.: 1989, Astron. Astrophys., **214**, 168.

Diethelm, R.: 1985, BBSAG Bull. no 77.

Diethelm, R.: 1986, BBSAG Bull. no 80.

Dobias, J.J. & Plavec, M.J.: 1989, Space Science Reviews, **50**, 340.

Dugan, R.S., & Wight, F.W.: 1939, Cont. Princeton Obs., **19**, 26.

Frieboes-Conde, H. & Herczeg, T.: 1973, Astron. Astrophys. Suppl. Ser., **12**, 1.

Gaposchkin, S.: 1953, Ann. Harvard Coll. Obs., **113**, 103.

Habets, G.M.H.J. & Heintze, J.R.W.: 1981, Astron. Astrophys. Suppl. Ser., **46**, 193.

Heintze, J.R.W.: 1989, Space Science Reviews, **50**, 257.

Hübscher, J., Lichtenknecker, D. & Mundy, E.: 1985, BAV Mitt. no. 39.

Kurucz, R.L.: 1979, Astrophys. Journ. Suppl. Ser., **40**, 1.

Olson, E.C.: 1980, Comm. 27 IAU Inf. Bull. Var. Stars no 1840.

Olson, E.C.: 1982, Astrophys. J., **259**, 702.

Peters, G.J. & Polidan, R.S.: 1984, Astrophys. J., **283**, 745.

Petrie, R.M.: 1950, Publ. Dom. Astrophys. Obs., **8**, 319.

Plavec, M. & Polidan, R.S.: 1976, in P. Eggleton, S. Mitton & J.A.J. Whelan (eds.), in Structure and Evolution of Close Binary Systems, Reidel,

Dordrecht, p. 289.

Provoost, P.: 1980, Astron. Astrophys. Suppl. Ser. **40**, 129.

Schmidt-Kaler, Th.: 1982, in K. Schaifers and H.H. Vogt (eds.), Landvet-Bömstein Tables, new ser., Springer-Verlag, Berlin-Heidelberg-New York, vol. VI/2b, sect. 4.1

Straižys, V. & Kuriliene, G.: 1981, Astrophys. Space Sc., **80**, 353

Van Gent, R.H.: 1982, Astron. Astrophys. Suppl. Ser., **48**, 457.

Van Gent, R.H.: 1989a, Space Science Reviews, **50**, 265.

Van Gent, R.H.: 1989b, Space Science Reviews, **50**, 371.

Van Gent, R.H.: 1989c, Astron. Astrophys., **77** 471.

Van Gent, R.H.: 1989d, chapter 5, thesis Utrecht.

Wilson, R.E.: 1979, Astrophys. Journ., **234**, 1054.

Wilson, R.E. & Devinney, E.J.: 1971, Astrophys. J., **166**, 605.

Wilson, R.E. & Devinney, E.J.: 1973, Astrophys. J., **182**, 539.

Dordrecht, p. 259.

Prevoost, P., 1980, Astron. Astrophys. Suppl. Ser. 40, 129.

Schmidt-Kaler, Th.: 1982, in K. Schaifers and H.H. Voigt (eds.), Landolt-Börnstein Tables, new ser., Springer-Verlag, Berlin-Heidelberg-New York, vol. VI/2b, sect. 4.1.

Straižys, V. & Kuriliene, G.: 1981, Astrophys. Space Sci., **80**, 353.

Van Gent, R.H.: 1982, Astron. Astrophys. Suppl. Ser. **48**, 487.

Van Gent, R.H.: 1989a, Space Science Reviews, **50**, 265.

Van Gent, R.H.: 1989b, Space Science Reviews, **50**, 371.

Van Gent, R.H.: 1989c, Astron. Astrophys., **77**, 411.

Van Gent, R.H.: 1989d, chapter 5, thesis Utrecht.

Wilson, R.E.: 1979, Astrophys. Journ., **234**, 1054.

Wilson, R.E. & Devinney, E.J.: 1971, Astrophys. J., **166**, 605.

Wilson, R.E. & Devinney, E.J.: 1973, Astrophys. J., **182**, 539.

INTERACTING BINARY CX DRACONIS

P. Koubský, J. Horn, and P. Harmanec
Ondřejov Observatory, 25165 Ondřejov, Czechoslovakia
G.J. Peters
Space Science Center, University of Southern California, Los Angeles, CA 90089-1341, U.S.A.
R.S. Polidan
Lunar and Planetary Laboratory - West, University of Arizona, Tucson, AZ 85721, U.S.A.

CX Dra is one of the best observed Algol-like non-eclipsing binary. Spectroscopic orbit: P=6.69603 days, e=0, K_1=35km.s^{-1}, K_2=150km.s^{-1}, time of the primary minimum - J.D. 42551.293. Primary comp. is B2.5 Ve, secondary comp. is estimated to be mid F star. The emission components to hydrogen and helium lines in the optical region and additional absorption features in the UV resonance lines indicate the presence of circumstellar matter in the system. The behaviour of the UV spectrum of CX Dra was described by Koubský et al. (1987). The most striking features in the UV spectrum are the blue shifted components to Si IV and Al III lines indicating directional mass outflow. They were first reported by Peters and Polidan (1984) and since that time they have been observed on three other occasions around phase 0.3 - 0.4 (450.380, 462.422, 613.350, 623.320) - cycle and phase 0.0 correspond to the above quoted primary minimum. Wonnacott (1988, 1989) tried to interpret this phase-dependent features as a consequence of a stream flowing in the system.

The circumstellar plasma that produces Hαemission in CX Dra has been investigated using data acquired from CCD (KPNO), IDS (UWO), and photographic plates (Ondřejov). The results are summarized in Fig. 1 where the emission strength was binned by 0.05 of the orbital phase in order to remove the long-term variations. The shape of the curve leads to the conclusion that most of the matter emitting Hα emission must be

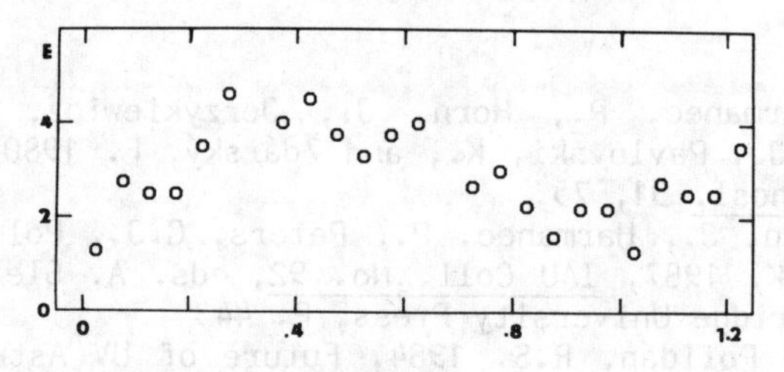

Figure 1. The strength of the emission in Hα in CX Dra versus phase. (0 - abs., 3 - emission continuum level, 6 - strong emission, peak - 1.2 I_c). 113 data since 1975 were binned by 0.05 in phase.

C. İbanoğlu (ed.), Active Close Binaries, 237–238.

inside the system. Note that CX Dra is not an eclipsing binary (Koubský et al. 1980). In Fig. 2 we record H_α and He I 6678 profiles obtained during several cycles when the accretion structure was stable. It

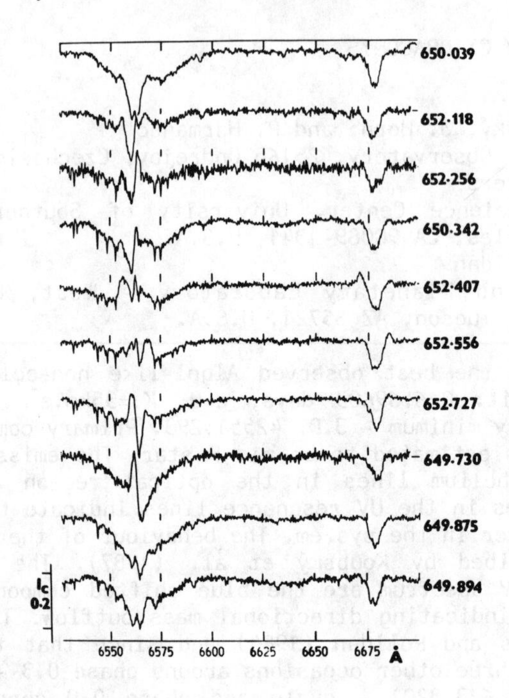

Figure 2. Selected CCD observations throughout several orbital cycles obtained in April 1987.

suggests that there may be two components of the circumstellar matter , one visible only in Hα near phase 0.25 (red lobe) and 0.75 (blue lobe), and a second one visible both in Hα and He 6678. Better phasing (observations of the secondary component) and observations during several orbital cycles are necessary to understand the accretion structure in CX Dra. UWO data were obtained by P.K. Barker.

References

Koubský, P., Harmanec, P., Horn, J., Jerzykiewicz, M., Kříž, S., Papoušek, J., Pavlovski, K., and Žďárský, F. 1980, Bull. Astron. Inst. Czechosl. 31, 75.

Koubský, P., Horn, J., Harmanec, P., Peters, G.J., Polidan, R.S., and Barker, P.K. 1987, IAU Coll. No. 92, eds. A. Slettebak and T.P. Snow, Cambridge University Press, P. 443

Peters, G.J. and Polidan, R.S. 1984, Future of UV Astronomy Based on Six Years of IUE Research, NASA CP2349, p.387

Wonnacott, D. 1988, Decade of UV Astronomy with IUE Satellite. ESA SP 281, I, p.209

Wonnacott, D. 1989, Space Science Reviews 50, 375

DISK-DRIVEN PRECESSION IN ACTIVE CLOSE BINARIES AND IN ACTIVE GALACTIC NUCLEI

J. F. LU
Center for Astrophysics
University of Science and Technology of China
Hefei, Anhui, P. R. China

Two active close binaries, SS433 and Her X-1 are known to contain precessing beams and disks. The disk-driven precession model (Sarazin, Begelman & Hatchett 1980) has been proposed for SS433 which is simultaneously compatible with a 13-day orbital period and a 164-day beam precession period. In this model, a stationary massive outer disk around the compact object causes the compact object's spin axis (plus the inner portion of the disk) to precess due to the Lense-Thirring effect. Accreting matter leaving the inner disk is assumed to be collimated and ejected out along the compact object's axis.

The S-shaped (inversion) symmetry observed in many extended radio sources and in more than 30 percent of all quasars at $z < 1$ (Hutchings, Price & Gower 1981) is suggestive of precession in the beam's orientation. Jet precession might also account for the bending and misalignment of milli-arcsec scale jets observed in some compact radio sources (Linfield 1981). Rotating compact objects (presumably massive black holes) with accretion disks are widely believed to be the power sources associated with active galactic nuclei and quasars. Disk-driven precession may occur in these objects as well.

In the case of SS433, the disk-driven precession model requires a radial inflow rate of $v_r/v_\phi \sim 10^{-10} - 10^{-9}$, corresponding to a disk viscosity parameter of $\alpha \sim 10^{-9} - 10^{-7}$ at r_p (a critical radius in the disk such that the portion of the disk with $r > r_p$ will maintain its orientation and will cause the central object and the inner disk to precess), if the central object's mass is about 1 $M_\odot$, and the accretion rate is about 500 times the Eddington limit. However, the required inflow rate is less extreme for disks around massive black holes. For an extreme Kerr black hole with mass $M \sim 10^8$ $M_\odot$, and the precession period $P \sim 10^6$ yr, it yields that $v_r/v_\phi \sim 10^{-5}$ if $\dot{M}/\dot{M}_{Edd} \sim 500$ is appropriate for SS433 but may bee an overestimate for the extra-galactic case).

What is suggestive of observational test for the model when applied to the extragalactic case is that there might be a (weak) statistical relationship between the precession period and accretion rate, $P \propto \dot{M}^{-3/2}$. If the accretion rate is reflected by the optical luminosity of the source, then it yields that

$$\text{Log } P = \frac{3}{5} M_{abs} + \text{constant}$$

where M_{abs} is the optical absolute magnitude.

C. İbanoğlu (ed.), Active Close Binaries, 239–240.

I collect in Figure 1 all extragalactic radio sources with both the precession period estimated and optical data available. The number of such sources is small for statistical purposes, and uncertainties certainly exist in both the estimation of the precession period and the optical magnitudes taken from the literature. However, Figure 1 still does show an inverse correlation between the precession period and optical luminosity of the central engine in active galactic nuclei. The least squares solution is

$$\text{Log } P = 0.49 M_{abs} + 17.11$$

The agreement with the theoretical prediction is quite good.

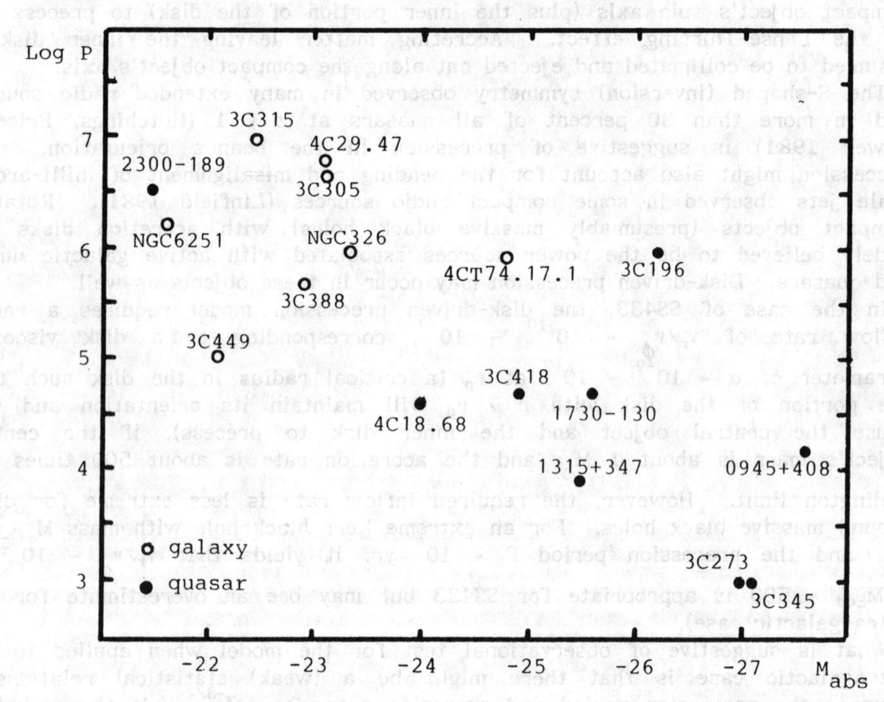

Figure 1

References

Hutchings, J. B., Price, R. & Gower, A. C.: 1988, Astrophys. J., 329, 122.

Linfield, R.: 1981, Astrophys. J., 250, 464.

Sarazin, C. L., Begelman, M. C. & Hatchett, S. P.: 1980, Astrophys. J. (Lett) 238, L129.

NEAR–CONTACT BINARY STARS

J. SCOTT SHAW
University of Georgia
Athens, Georgia 30602
USA

ABSTRACT. A group of close binary stars named **near–contact binaries** is defined. They are divided according to their physical characteristics into to two subclasses: V1010 Ophiuchi binaries and FO Virginis binaries. A list of members of the group, their current observational status, and their physical and evolutionary relationship to W Ursa–Majoris binaries is discussed. The photometric asymmetries in their light curves (the O'Connell effect) are explained by mass transfer from the larger, hot star to the smaller, cool star.

1. Introduction

A perennial problem in the understanding of binary stars has been the W UMa systems. While there now is general agreement on their physical properties – they are short period (< 1 day) systems whose components are in contact – their evolutionary history is not yet well understood. Compounding the problem is the existence of two distinct subclasses of W UMa binaries: the W–type W UMa systems in which the larger, more massive star is the cooler and both stars are usually of later spectral type than F5; and the A–type W UMa systems in which the larger, more massive star is the hotter and is of spectral type A or F. Both of these subclasses have been shown to be old (Guinan and Bradstreet, 1988). If, in fact, their creation involves orbital shrinking through angular momentum loss (AML) caused by magnetic braking (Vilhu, 1982, Rucinski, 1986, Guinan and Bradstreet, 1988), then W UMa systems may be the last stage just prior to coalescence into a single, rapidly rotating star.

If W UMa binaries evolve from detached binary systems whose components are normal main sequence stars by means of AML via magnetic braking or alternately by AML via mass transfer/mass loss, then evidence for this might be found in discovering binary systems in these intermediate stages of evolution. For the last decade a multitude of astronomers have attacked the problem with new and better observations and new and improved theories. The complete list is much too long to enumerate but reviews are given by Mochnacki (1981), van Hamme (1982a,b), Maceroni, Milano and Russo (1985), Kaluzny (1985), Rucinski (1986), Hilditch, King and McFarlane (1988) and Hilditch (1989). Much of the effort has centered on binary systems in which the stars are very close, but one or both of the stars are not filling their Roche surfaces. These systems share some

C. İbanoğlu (ed.), Active Close Binaries, 241–251.

characteristics with both types of W UMa systems: 1) They have periods less than one day and 2) Their strongly interacting components result in proximity effects which produce highly distorted light curves. Near–contact binaries also resemble only the A–type W UMa systems in that the large, more massive star (primary) is the hotter component, while the smaller, less massive star (secondary) is cooler, with spectral type one to two classes later than the primary. Because of such similarities, these near–contact systems have been suspected to be the missing link between detached systems and W UMa binaries. An alternate position in the evolutionary scheme for near–contact systems contends that some of them have already come into contact, but have temporarily detached themselves from their Roche lobes and are "taking a break" from the contact phase. Later they will return to their contact condition.

However, the ways near–contact systems differ from the W UMa systems are just as significant. The most important difference is that they are not in contact. This allows a large temperature difference between the two components. In contact binaries energy flows across the neck of material at the contact allowing both components to come to nearly the same temperature. There have been reports of contact systems with large temperature differences between the components (Lucy and Wilson, 1979, and Kaluzny 1983,1986a–c). However, the physical parameters of the reported systems have been obtained from light curve solutions which 1) have not taken light curve asymmetries into account and/or 2) do not have reliable spectroscopically determined mass ratios. In either or both of these instances the relative sizes of the components are not precisely determined. A small error in radii can result in near–contact binaries mistakenly looking as if they are in contact.

Other differences between the contact and near–contact systems are their ultraviolet spectra and their stages of evolution. Guinan and Shaw (1990) have found that the near contact systems do not have the enhanced ultraviolet emission lines of CIV, NV, etc. found in many W UMa systems (Dupree and Preston, 1981 and Eaton, 1983). The differences in evolution will be mentioned later in this paper.

2. The Near–Contact Systems

Table I contains a list of near–contact binary systems. Information on the systems comes from a large variety of sources including, but not limited to, the review articles by Hilditch, et. al. (1988), Kaluzny (1986c), Mochnacki (1981) and recent work being done by groups at Ankara University, Ege University, University of Athens, and University of Georgia. For the most part there has been a only modest attempt at assessing the reliability of the solutions. However, an effort was made to exclude contact systems, while at the same time including systems which are probably near–contact even though there is a solution in the literature giving them as contact. Such judgments are dangerous under the best circumstances (excellent data, well–behaved systems and accurate solutions) let alone for near–contact systems, and so the list in Table I should be characterized as provisional and open to revision. Borderline judgments on whether the system is contact or near–contact are noted by "C?" under Class in Table I. As will be discussed later many of these systems need more observation and analysis before their near–contact nature can be insured. Finally, also excluded are over 20 systems which exhibit light curves very similar to those in Table I, but do not yet have well–covered, reliable photoelectric light curves or modern analyses. Any photoelectric observers who wish to of expand Table I may contact the author for a list of these "probable near–contact" systems.

TABLE I
Near–contact Eclipsing Binaries

Name	Sp. Type	Period days	Class	Asymmetry Δmag	dP/dt sec/1000yr
AA And	A5	.935		.05	
BL And	A8	.722	V	.012	
BX And	F2	.610	C?		
CN And	F5	.463		.02	3.?
WZ And	F5 + G2	.696		.03	
CX Aqr	F5 + G9	.556	F	.0	
EE Aqr	F1	.509		.0	
ST Aqu	A7	.781		.0	
HL Aur	F4	.623		.0	0.0
ZZ Aur	A0	.601		.0	0.0
DO Cas	A5	.684		.004	
V747 Cen	F2	.537	V?	.005	
WZ Cep	F2	.417	C?	.04	
YY Cet	A8 + G	.790	F	.0	0.0
RV Crv	F2 + K	.747	F	.0	
VZ CVn	F3	.843	V?	.004	
GO Cyg	A0	.717		.0	
KR Cyg	A0 + F5	.845		.0	0.0
DM Del	A3	.845	C?		
BL Eri	G0 + G3	.416	C?	.02	
BW Eri	A8	.638	V	.002	33.?
RU Eri	F0	.632			
TT Her	A0	.912	V	.0	9.3
RS Ind	F0 + K	.624			
ES Lib	A3 + G	.883	V	.015	
SW Lyn	F2	.644	F	.0	0.0
UU Lyn	F3	.468		.0	
TZ Lyr	F3	.529	F?	.015	?
FT Lup	F2 + K3	.470	V?		7.6
V1010 Oph	A7 + F6	.661	V	.025	30.
V392 Ori	A5	.659	F?	.0?	
RT Scl	F2 + K	.511	V	.019	5.6
V525 Sgr	A	.705	V?	.05	
V Tri	A	.585	V?	.015	
RU Umi	F0 + K5	.525		.0	?
AG Vir	A8	.642	C?	.06	?
BF Vir	A2 + K1	.640	F?	.0	0.0
FO Vir	A7	.776	F	.0	
BS Vul	F2	.476		.0?	?

Listed in Table I are: 1) the name of the binary; 2) the spectral type of the hotter star, occasionally accompanied by the spectral type of the cooler star, if available; 3) the period of the system in days; 4) the subclass of the system in the sense C = contact; V = V1010 Oph–type, F = FO Vir–type, with ? designating some uncertainty in the assignment, blanks indicate there is not enough information to even guess at a subclass; 5) the asymmetry measured to be the magnitude difference between Maximum I (phase .25) and Maximum II (phase .75), always in the sense Maximum I brighter than Maximum II; and 6) any measured change in the period, in units of seconds per 1000 years. If the period seemed constant over at least the last 40 years, this was noted by 0.0, while systems with insufficient coverage of their light elements were left blank.

3. The Observations

Definitive determination of the physical parameters of binary stars requires analysis of light curves and radial velocity curves. Unfortunately, the binaries in Table I all have one component much brighter than the other. This results in light curves whose shapes are almost entirely due to the changing aspect of the brighter star and its heating of the facing hemisphere of its cooler companion (the classical reflection effect). The cooler component's own contribution to the light lost in secondary minimum is often less then a few percent of the total light. Thus the determination of the temperature difference of the two components is usually not well determined.

Moreover, the faintness of the cooler component makes it very difficult observe the spectra of both stars in order to obtain the masses from the radial velocities. Nonetheless, groups lead by Hilditch and Hill have been able to use cross–correlation techniques to measure the secondary component's radial velocity. (There are many references. See Hilditch et. al., 1988 for a review and Hill, et. al., 1989 for a recent example). In one other instance Shaw and Guinan (1990) have used the Lyman α emission line of V1010 Oph's cool component to determine the masses of both components. Well determined masses are not only essential in understanding these systems, but they are also needed in the form of the mass ratio (q), to allow accurate solution of the light curves. For this reason, later discussion of the evolutionary status of this class of binaries will be based only on systems with good spectroscopically determined mass ratios.

Photometric problems also abound in the near–contact binaries. Fortunately, high quality, multicolor, photoelectric light curves are becoming more numerous. Most of the systems in Table I have been observed in UBV, although a few such as RT Scl have been observed in uvby (Clausen and Grønbeck, 1977) and those observed at the University of Georgia (AA And, BL And, WZ And, HL Aur, ZZ Aur, SW Lyn and V Tri) have data in the Washington CMVT1T2 system. Many have been observed extensively over more than one season, giving information on short and long term intrinsic variability. A few systems, such as BS Vul (de Bernardi and Scaltriti, 1978) and FT Lup (Mauder, 1982) show intrinsic variation of several hundredths of a magnitude from cycle to cycle. The majority, however, show the same shaped light curve over decades of observation. This is especially surprising since several of the systems have photometric complications in their light curves such as having Maximum I brighter than Maximum II.

The asymmetry of a light curve so that one maximum is higher than the other is sometimes called the O'Connell effect. It has been studied extensively by Davidge and Milone (1984). Although the O'Connell effect is not the general rule for near–contact binaries, it seems to occur regularly. When it does occur, it is in the sense that Maximum I is brighter then Maximum II. Column 5 in Table I lists the brightness difference in the maxima. Interestingly, one of the most extreme cases, V525 Sgr, is one that Fr. O'Connell himself observed (O'Connell, 1950). The brightest system listed in Table I, V1010 Oph, coincidently has one of the largest O'Connell effects of the near–contact binaries. Recent studies of this system and one other (RT Scl) have led to an indication of what is producing the photometric asymmetry for these systems.

Figure 1 shows a representative model of a near–contact binary system. The primary is the hotter, larger and most massive component. The center of mass is marked with a "+" and the direction of motion is shown with arrows. The direction from which the system is viewed with increasing phase is shown increasing clockwise from 0.0 through 0.25, 0.5, and 0.75 back to 0.0. Again, the photometric asymmetries in these systems are such that phase 0.25 is brighter than 0.75. This can be accomplished in one of four ways: brightness can be enhanced on hemispheres "b" or "c" or diminished on hemispheres "a" or "d". Of course various combinations would work as well, but modifying the brightness of only one hemisphere is the simplest model possible.

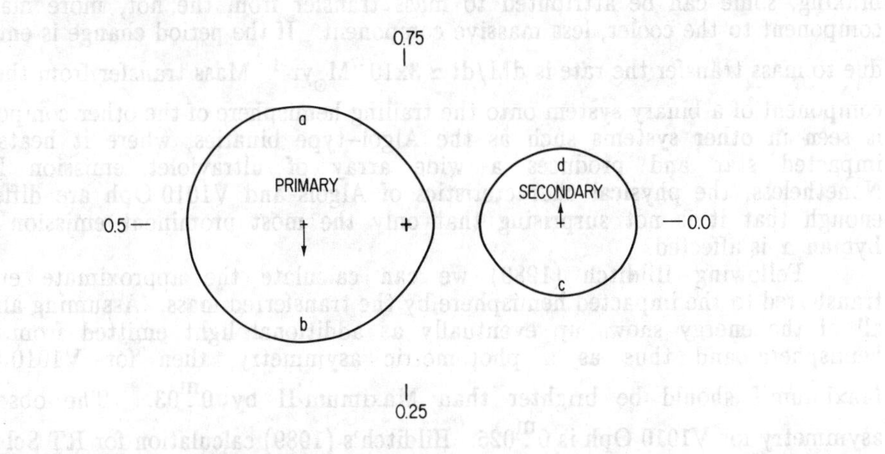

Figure 1. Model of a near–contact binary. The center of mass is marked with a "+" and the direction of motion is shown with arrows. The direction from which the system is viewed with increasing phase is shown increasing clockwise.

The easiest alternative to rule out is the diminution of light at "d". The secondary component of many of these systems contributes less than 1% of the system's light. Yet the asymmetries can be as much as 5% of the system's total light. Even complete obscuration of the secondary would not be adequate to account for the asymmetry. Diminution of the primary's light at "a" is more promising but it would require either cooling of the "a" hemisphere or obscuring material in front of "a". There seems no phenomenon to account for cooling.

However, there is some evidence for absorption. Guinan's (1970) observations of V1010 Oph show a stronger Hβ index at Maximum II than at Maximum I. This could be explained by absorbing material above the photosphere at "a". Alternately it can be explained by enhanced hydrogen emission from hemispheres "b" or "c".

In order to resolve this dilemma Shaw and Guinan (1990) observed V1010 Oph with the IUE satellite. High dispersion, short wavelength spectra were taken at phases 0.25, 0.0, and 0.75. Lyman α emission was seen at 0.25 and 0.75 with radial velocities commensurate with the motion of the cooler component. The total flux of the Lyman α emission line at 0.25 phase was over twice that at 0.75 phase. The measures were taken within three orbital revolutions of V1010 Oph, so it is unlikely the variation is a long term systematic change. The conclusion is that while V1010 Oph's secondary component has the Lyman α emission characteristic of the rapidly rotating, late F star that it is, there is an additional source of Lyman α emission on hemisphere "c".

If there were a common cause for the enhancement of Lyman α emission and the visible photometric asymmetry which could place both of them on hemisphere "c", then we might be close to explaining the O'Connell effect for this subclass of binaries. Such a cause is suggested by V1010 Oph's period change; found by Williams and Guinan (1990) to be 30 sec/kyear (1 kyear = 1000 years). While some of this period decrease could be produced by AML due to magnetic braking, some can be attributed to mass transfer from the hot, more massive component to the cooler, less massive component. If the period change is entirely due to mass transfer the rate is $dM/dt \simeq 3 \times 10^{-7} M_{\odot} yr^{-1}$. Mass transfer from the one component of a binary system onto the trailing hemisphere of the other component is seen in other systems such as the Algol–type binaries, where it heats the impacted star and produces a wide array of ultraviolet emission lines. Nonetheless, the physical characteristics of Algols and V1010 Oph are different enough that it is not surprising that only the most prominent emission line, Lyman α, is affected.

Following Hilditch (1989) we can calculate the approximate energy transferred to the impacted hemisphere by the transferred mass. Assuming almost all of the energy shows up eventually as additional light emitted from that hemisphere and thus as a photometric asymmetry, then for V1010 Oph, Maximum I should be brighter than Maximum II by $0^{m}_{.}03$. The observed asymmetry for V1010 Oph is $0^{m}_{.}025$. Hilditch's (1989) calculation for RT Scl with a mass transfer of $5.5 \times 10^{-7} M_{\odot} yr^{-1}$ is $0^{m}_{.}04$ with the observed asymmetry being $0^{m}_{.}019$. It would be valuable to do the same calculation for the other near–contact systems in Table I, but many do not have period changes or data for such change is lacking. Moreover, most do not have the well determined values for the masses necessary for the calculation. Nonetheless, V1010 Oph and RT Scl suggest that photometric asymmetries in the light curves of near–contact binaries may be explained by mass transfer from the hot component onto the trailing hemisphere of the cool component.

4. Future Needs

To test the above explanation of O'Connell effect in near–contact binaries and to understand their nature and evolution, we most undoubtedly need more information. Table II is a list of near–contact systems indicating the type of new information that would be valuable. As one can see from the scarcity of dP/dt values in Table I, many systems do not have coverage over a long enough time span to ascertain whether they have period changes or not. An x in the second column of Table II indicates binaries in which new times of minimum might allow determination of period changes for that system. An x in the third column indicates the need for a new analytical solution of the the light curves. In several of these cases the photometric asymmetry of the light curve is not incorporated into solution. Unlike the analysis of light curves of some well separated binaries, the analysis of near–contact binary light curves gives very different solutions depending on how photometric asymmetries are taken into account. Reanalysis of the indicated systems with a model that allows non–symmetrical light variation over the surface of the components should result in more reliable physical elements. Binaries with an x in the fourth column need modern, multicolor, photoelectric light curves. In the case of FT Lup a new light curve would monitor its intrinsic variability. In most of the rest of the cases, the only existing curves are visual or photographic.

TABLE II
Near–contact Binaries: What is needed.

Name	Time of Min	Soln	Light Curve	Spectra	Comments
BX And				x	
ST Aqr	x	x			
DO Cas	x				
V747 Cen	x	x		x	Total eclipses
WZ Cep	x	x		x	Total eclipses
V680 Cyg			x		May be contact system
DM Del				x	May be contact system
BL Eri	x				Period change to be confirmed
BW Eri	x			x	May be total eclipses
TT Her	x			x	Suspected period change
RS Ind	x				Has good solution
ES Lib	x				
GR Tau		x	x	x	May be contact but is A3
FT Lup	x		x		Intrinsic variability
V525 Sgr	x		x	x	May be total eclipses
V Tri	x			x	
RU Umi	x			x	Total eclipses
BS Vul	x	x	x	x	

Finally, and most importantly, we need reliable radial velocity curves of these systems which determine the motion of both components. As mentioned before, the major stumbling block to understanding near–contact binary systems is reliable mass determinations. The brightness of the hot component makes it extremely difficult to get the radial velocity of the cooler component. Therefore few systems have both of the masses known. What is more, observers, anticipating only a single line spectroscopic binary with its limited information, have shied away from looking at these systems. Cross–correlation techniques and discovery of the cool component's Lyman α emission in the ultraviolet, recently have yielded some mass determinations and should provide incentive to investigate many more of these systems. While almost all near–contact binaries need spectrographic investigations, those noted in the fifth column of Table II should yield the most new understanding of the near–contact group of binaries.

5. Evolution

Table II's presentation of the great amount of work that still needs to be done provides a proper cautionary tone to the discussion of the evolutionary position of the near–contact binaries. Nonetheless, a few systems are well enough known to suggest some reasonable speculations. As previously mentioned, the evolution of the near–contact binaries has usually been considered in conjunction with that of the W UMa binaries. Whether the near–contact systems are precursors or intermediate stages of W UMa systems is not clear, but the similarity of their physical characteristics leads to the conclusion that their histories are intertwined.

Before describing a near–contact/contact binary evolutionary scenario, we should avoid a major confusion of previous investigators, namely that all of the near–contact binaries are essentially alike. The near–contact binaries are no more a single homogeneous group any more than their cousins the contact binaries are. Unfortunately the important distinctions within the near–contact binaries group don't show up until one has accurate physical properties of the systems. Most crucial is a reliable determination of the mass of the fainter, cooler component. As mentioned before this is very difficult to determine. Nonetheless, Table III contains a few near–contact systems which have spectroscopic mass determinations and reliable light curve solutions. They divide themselves into two types with respect to two criteria: 1) which component is at or near its Roche lobe; and 2) how disparate is the mass and size of the secondary component compared to a main sequence star. Specifically, the near–contact group is divided into two subclasses named after the brightest member of the subclass.

V1010 Ophiuchi Binaries

Primary	Secondary
at or near Roche Lobe	inside Roche lobe
size is normal or a bit evolved	size is 1 – 1.2x oversized for mass

The light curve sometimes shows Maximum I brighter than Maximum II.

FO Virginis Binaries

Primary	Secondary
inside Roche lobe	at or near Roche lobe
size is normal or a bit evolved	size is 2 –3x oversized for mass

The light curve never shows Maximum I brighter than Maximum II.

Table III lists nine systems divided into these two subclasses, with their spectral types, period, asymmetry (O'Connell effect), and any known change of period. In brief, both components of the V1010 Oph systems show little evolution. Perhaps, the hot component is near the so–called Terminal Main Sequence, is at or near its Roche lobe, and is transferring some material onto the cool component. This may result in a slightly larger size for the cool component than is the main sequence norm for its mass. Basically, however, the V1010 Oph systems are only slightly evolved. In contrast, the secondary component of the FO Vir systems is at or near its Roche lobe and is very oversized for its mass. Conversely, this can also be referred to as being undermassive for its size. In this respect the secondary component of the FO Vir binaries resemble the Algol systems which are undermassive because they have undergone extensive mass transfer/loss. If this is the case for the FO Vir binaries, it would indicate that, like Algols, they are very evolved.

TABLE III

V1010 Oph Binaries

Name	Sp T	Period	Asym	dP/dT
TT Her	A0	.912	.0	9.3
ES Lib	A3 + G	.883	.015	
V1010 Oph	A7 + F6	.661	.025	30.
RT Scl	F2 + K	.511	.019	5.6

FO Vir Binaries

Name	Sp T	Period	Asym	dP/dT
CX Aqr	F5 + G9	.556	.0	
YY Cet	A8 + G	.790	.0	0.0
RV Crv	F2 + K	.747	.0	
SW Lyn	F2	.644	.0	0.0
FO Vir	A7	.776	.0	

It is striking that the FO Vir systems have no photometric asymmetry or period change while the V1010 Oph systems show one or both of these. Above, we attributed the photometric asymmetry of the V1010 Oph systems to mass transfer flowing from the hot component which is at or near its Roche lobe. In the FO Vir systems, however, it is the cool component which is at or near its Roche lobe. Thus, the lack of photometric asymmetry in FO Vir systems could be explained by either 1) a much smaller mass flow, consistent with the lack of period changes, or 2) the heating effects of the mass flow making much less difference to the brightness of the hot component.

Finally, what do we know about how the near–contact binaries and the contact binaries fit together in a evolutionary scheme? First, a few important observations. 1) Both W–type and A–type W UMa contact binaries are very old, 1 to 7 Gyr (Guinan and Bradstreet, 1988). 2) V1010 Oph binaries appear to be relatively young, as their A to F hot components are just near the Terminal Main Sequence. 3) Except for not being in contact, the FO Vir binaries are very similar to the A–type W UMa binaries. Both have A to F hot components and the FO Vir systems' cool components are about 2 times overmassive while the A–type W UMa systems' cool components are about 3 times overmassive.

The similarity of the FO Vir near–contact binaries to the A–type W UMa binaries is perhaps the simplest of the above facts to explain. The FO Vir binaries are most likely immediate precursors or temporarily detached stages of the A–type W UMa systems. What is more, the dilemma posed by the substantial age of the A–type W UMa systems versus the early spectral type of their hot components can be resolved by allowing slow mass transfer from the currently cooler, undermassive secondary to increase the mass of the currently more massive, hotter primary. Again, the scenario is a familiar one for Algol systems which have more widely separated components.

All this still leaves unresolved the role of the V1010 Oph systems. Are they precursors of the FO Vir systems? That is, will enough mass be exchanged in V1010 Oph binaries to cause a complete reversal of the primary–secondary masses, sizes and spectral types? It seems unlikely that this could occur without the stars coming into full contact. But if they come into contact, where are all of the late A and early F contact systems? The W–type W UMa systems are in contact but they are much older, much less massive and have both components of much later spectral type than the primary components of the V1010 Oph systems. Further, the W–type W UMa systems also have nearly normal main sequence components like the V1010 Oph systems. However, any evolution of the V1010 Oph systems directly into W–type W UMa systems would require substantial mass loss – over half of the system's entire mass. No evidence of such a large systemic mass loss is seen in either ultraviolet or visible spectra. We conclude that the V1010 Oph binaries do not evolve directly to W–type W UMa binaries. Either the W–type W UMa binaries have still unknown progenitors or there are lengthy intermediate evolutionary stages leading from the V1010 Oph binaries to the W–type W UMa binaries. Possible candidates for intermediate evolutionary stages could be the FO Vir binaries and/or the A–type W UMa systems, but that is very unclear at this point in our understanding.

6. **Summary**

There exists a group of binary stars whose components are near enough to each other to have strong proximity effects like W UMa binaries but are not in contact. The members of this group of near–contact binaries, unlike contact binaries, do not exchange energy through a neck of material connecting the components and therefore they can maintain large temperature differences between the components. The group can be divided into two subclasses: 1) V1010 Oph binaries whose components are nearly normal or slightly evolved main sequence stars with the hot component near or at its Roche lobe and 2) FO Vir binaries who are very evolved and who have their cooler component at or near its Roche lobe and undermassive for its size. The V1010 Oph binaries are relatively early in their evolution while the FO Vir binaries are more evolved. The latter are very similar to A–type

W UMa systems and may be immediate precursors or temporarily detached versions of A–type W UMa systems.

All of the above speculations on the evolution of the near–contact binaries could be put on firmer ground if there were more well observed and well analyzed systems. Critically needed is spectroscopy which yields the masses of both components, most importantly the elusive cooler star. We are just at the beginning of our investigation of the group of near–contact binaries, starting with the recognition of their important differences from W UMa binaries. Much more has to be done, photometrically, spectroscopically and analytically before we can understand this enigmatic group of binary stars.

REFERENCES

Clausen, J. V. and Grønbeck, B. 1977, Astr. Astrophys. Suppl. **28**, 389.
Davidge, T. J. and Milone, E. F. 1984, Astrophys. J. Suppl. **55**, 571.
De Bernardi, C. and Scaltriti, F. 1978, Astr. Astrophys. Suppl. **35**, 63.
Dupree, A. K. and Preston, S. 1981, In The Universe at Ultraviolet Wavelengths, ed. R. D. Chapman, NASA CP–2171, (NASA, Washington, DC), p.333.
Eaton, J. A. 1983, Astrophys. J. **268**, 800.
Guinan, E. F. 1970, Ph.D Dissertation, Univ. of Pennsylvania.
Guinan, E. F. and Bradstreet, D. H. 1988, in "Formation and Evolution of Low Mass Stars", eds. A. K. Dupree and M. T. Lago (D. Reidel, Dordrecht), p. 300.
Guinan, E. F. and Shaw, J. S. 1990, submitted to Astr. J.
Hilditch, R. W., King, D. J. and Mcfarlane, T. M. 1988, Mon. Not. R. Astr Soc. **231**, 341.
Hilditch, R. W. 1989, Sp. Sci. Rev. **50**, 289.
Hill, G., Fisher, W. and A. Holmgren, D. 1989, Astr. Astrophys. **218**, 152.
Kaluzny, J. 1983, Acta Astr. **33**, 345.
Kaluzny, J. 1985, Acta Astr. **35**, 313.
Kaluzny, J. 1986a, Acta Astr. **36**, 113.
Kaluzny, J. 1986b, Acta Astr. **36**, 121.
Kaluzny, J. 1986c, Publ. Astr. Soc. Pacific **98**, 662.
Lucy , L.B. and Wilson, R. E. 1979, Astrophys. J. **231** 510.
Maceroni, C., Milano, L. and Russo, G. 1985, Mon. Not. R. Astr. Soc. **217**, 843.
Mauder, H. 1982, in "Binary and Multiple Stars as Tracers of Stellar Evolution," IAU Coll. No. 69, eds. Kopal, Z. and Rahe, J., (Reidel, Dordrecht), p. 275.
Mochnacki, S. M. 1981, Astrophys. J. **245**, 650.
Mochnacki, S. W., Fernie, J. D., Lyons, R., Schmidt, F. H. and Gray, R. O. 1986, Astr. J. **91**, 1221.
O'Connell, D. 1950, Riverview Publ. **2**,78.
Rucinski, S. M. 1986, in "Instrumentation and Research Programmes for Small Telescopes", IAU Symp. 118, eds. Hearnshaw, J. B. and Cottrell, P. L., Reidel, Dordrecht), p. 421,
Shaw, J. S. and Guinan, E. F. 1983,. Bull. Am. Astron. Soc. **15**, 926.
Shaw, J. S. and Guinan, E. F. 1990, private communication.
Van Hamme, W. 1982a, Astr. Astrophys. **105**, 389.
__________. 1982b, Astr. Astrophys. **116**, 27.
Vilhu, O. 1982, Astron. Astrophys. **109**,17.
Williams, D. B. and Guinan, E. F. 1990, private communication.

OBSERVATIONS AND LIGHT CURVE ANALYSIS OF ECLIPSING BINARIES OF W UMa - TYPE

Rovithis-Livaniou, H., Niarchos, P.G. and Rovithis,P.
Section of Astrophysics-Astronomy and Mechanics
University of Athens
Panepistimiopolis, GR 157 83-Zografos
Athens, Greece

SUMMARY. New light curves in two wavelengths (B and V) of eleven W UMa-type systems, obtained at Kryonerion Observatory in Greece, are presented. The light curves of these systems are analysed by using Frequency Domain techniques and new geometric and photometric elements are given. These elements are combined with the available spectroscopic data of the systems and new absolute elements are derived. Using the latter elements the evolutionary status of these systems is considered on the basis of mass-radius and mass-luminosity diagrams, and the results are compared with previous studies, where the physical parameters of these systems are based on photometric elements derived by other methods.

1. INTRODUCTION

The problem of the origin, structure and evolution of W Ursae Majoris (W UMa) eclipsing binaries has received considerable attention in recent years from many investigators. To face this problem it is necessary to know very accurately the physical parameters of these systems. The determinations of the absolute dimensions (masses, radii, luminosities) of contact systems is a very difficult task. Such a determination would be reliable, if a proper analysis of their light curves is performed in order to get a set of photometric elements which will be used for the derivation of reasonable absolute parameters of these systems.

It is well known that many authors collect, publish and analyse data without paying much attention in the reliability and homogeneity of their data. Of course, the reliability of the absolute dimensions depend strongly, apart from the available spectroscopic data, on the method used to derive the geometric and photometric elements of the systems.

In our analysis we have used the light curves of eleven W UMa systems, which have been obtained by the same instruments (telescope, photometer, etc) during the last decade at Kryonerion Observatory, Greece. These light curves have been analysed by the same method (Kopal's method) to derive the geometric and photometric elements of these systems. Thus, the presupposition of homogeneity of our data has been obtained. Some of our systems have been studied by using different methods of analysis (Van

C. İbanoğlu (ed.), Active Close Binaries, 253–266.

Hamme, 1982a,b; Maceroni et al., 1985) and conclusions have been drawn as far as the evolutionary status of these systems is concerned. Therefore, a comparison can be made, at least for the common systems, between our results and those obtained by a different method of approach.

2. OBSERVATIONS AND DATA OF THE SELECTED W UMa SYSTEMS

All the systems in the present analysis have been observed photoelectrically using the 48-inch Cassegrain reflector at the new Kryonerion Station of the National Observatory of Athens. The telescope was used together with a two-beam, multi-mode photometer. The two intermediate passband filters used were selected to be in close accordance with the standard system U,B,V. Only one of the photomultiplier tubes was used, in front of which the two filters (B,V) were arranged.

REMARKS TO INDIVIDUAL STARS

AB And: It was discovered by Guthnick and Prager in 1927 and its period is variable. Early photoelectric observations of the system were made by Binnendijk (1959); then, it was observed by some other investigators and recently by Rovithis-Livaniou and Rovithis (1981). Struve et al. (1950) describes its spectrum as G5 while according to Landolt (1969) the eclipsing star is probably an unreddened K2 main sequence star.

DO Cas: It was recognized as a variable by Hoffmeister in 1942. Its early photoelectric light curves were obtained by Schneller and Daene (1952) while ours were made during 1979 and 1980. Its period is variable. According to Fehrenbach et al. (1966) the primary should be of spectral type A2II, while according to Hill et al. (1975) A4-5 V; the secondary was estimated as F2 by Koch (1973).

VW Cep: It has both a variable period and a variable light curve. It is the brighter member of the visual binary hz 7 . The system has gained special interest in the recent years. According to Linnel (1980) high excitation far ultraviolet lines have been found in IUE spectra and a low level soft X-ray source has a position error box which includes VW Cep. Many photoelectric observations of the system have been published since Hersey's (1975) study. Spectrosopic data of the system have been reported by Popper (1948), Binnendijk (1966), Kwee (1966) and Hill et al. (1975). Our photoelectric observations have been obtained in the years 1979, 1980.

BV Dra: and BW Dra BV and BW Dra are the two members of the visual binary ADS 9537, which was detected by Batten and Hardie (1965). After its discovery, many light curves have been obtained for both systems. Our photoelectric observations for ADS 9537 were made during 1980, 1981 and 1982; while spectroscopic orbits for both BV and BW Dra were obtained by Batten and Lu (1986).

SW Lac: The system is well known from the variable light curve and its period changes. Numerous visual, photographic and photoelectric observations of this system have been made so far. Infrared observations of the system were made by Jameson and Akinci (1979). Recent spectroscopic observations have been reported by Hill et al. (1975). Our photoelectric observations of the system were made in October 1983.

V502 Oph: After its discovery in 1935 it has been observed several times. Its period is variable. Spectroscopic analyses for the system have been made by Struve and Gratton (1948), Struve and Zebergs (1958) and more recently by King and Hilditch (1984). Our photoelectric observations for V502 Oph were obtained during 1986.

V839 Oph: The system was first observed photoelectrically by Binnendijk (1960) and very recently by Lafta and Grainger (1985). Our photoelectric observations were made in June 1985. No spectroscopic observations of V839 Oph have been published so far.

ER Ori: After its discovery in 1929 by Hoffmeister it has not been observed many times. Kwee (1958) and Binnendijk (1962) have reported sudden changes in its period. According to Struve its spectrum is G1, while in the S.A.O. Catalogue it is referred to as F8. Our photoelectric observations for ER Ori were obtained during 1982 and 1983.

AG Vir: After its discovery by Guthnick and Prager in 1929, it has been the subject of several photometric investigations (Binnendijk 1969; Blanco and Catalano, 1970). Spectroscopic observations of AG Vir have been made by Sanford (1934) and by Hill and Barnes (1972). Our photoelectric observations of the system were made in 1983.

AH Vir: Since its discovery by Guthnick and Prager in 1929, numerous visual, photographic and photoelectric observations of this system have been made. It has been found that the light curve and the period of AH Vir are variable. The system was observed by us in the years 1977, 1978. Spectroscopic observations of AH Vir were made by Chang (1948).

Some data for our sample of W UMa systems, based on spectroscopic observations are given in Table I. Their light curves are presented in Figures 1,2.

3. LIGHT CURVE ANALYSIS

The light curves of all systems have been analysed using Frequencey Domain Techniques, (FDT), which has been developed in a series of papers appeared in Astrophysics Space Science between 1975-79 and subsequently in a book (Kopal, 1979). Thus, we don't like to explain here all the details concerning this kind of Fourier method of light curves analysis, but to outline very briefly the basic steps of it.

According to FDT, the light curve analysis is based on the equations of the moments A_{2m} which are given by:

$$A_{2m} = \bar{A}_{2m} + A_{2m}^{prox} + R_{2m} \qquad m = 1,2,\ldots \tag{3.1}$$

where A_{2m} are the areas substend by the light curves in $1\text{-}\sin^{2m}\theta$ coordi-

TABLE I
(Spectroscopic Data of our sample of W UMa Stars)

Star's Name		BD or HD Number	Period days	Spectral Type	Ks Km/sec	Kg Km/sec	Type*	q_{sp}	Ref
AB	And	BD+36°5017	0.331	G5,G5	265	165	W	0.62	(1)
DO	Cas	BD+59°529	0.684	A4-5, F2	261	73	-	0.28	(2,3,4)
VW	Cep	BD+75°752	0.278	G8, G5	220	90	W	0.41	(5)
BV	Dra	HD 135421	0.350	F9, F8	233.6	93.9	W	0.402	(6)
BW	Dra	HD 135421	0.292	G3, G0	251.5	70.5	W	0.280	(6)
SW	Lac	BD+37°4717	0.320	G3, G1	210	185	W	0.88	(5)
ER	Ori	BD-8°1050	0.423	F5-8	176	95	W	0.54	(2.7)
V502	Oph	HD 150484	0.453	G2, F9	235	95	W	0.40	(8)
V839	Oph	BD+9°3584	0.408	G0	-	-		-	(5)
AG	Vir	BD+13°2481	0.642	A2	272	80	A	0.29	(5,9)
AH	Vir	BD+12°2437	0.407	K0	250	105	W	0.41	(5)

* Based on spectroscopic data

(1): Struve et al. (1950);
(2): Hill et al. (1975);
(3): Koch (1973);
(4): Kaluzny (1985b);
(5): Binnendijk (1970);
(6): Batten and Lu (1986);
(7): Struve (1944);
(8): King and Hilditch (1984);
(9): Hill and Barnes (1972);

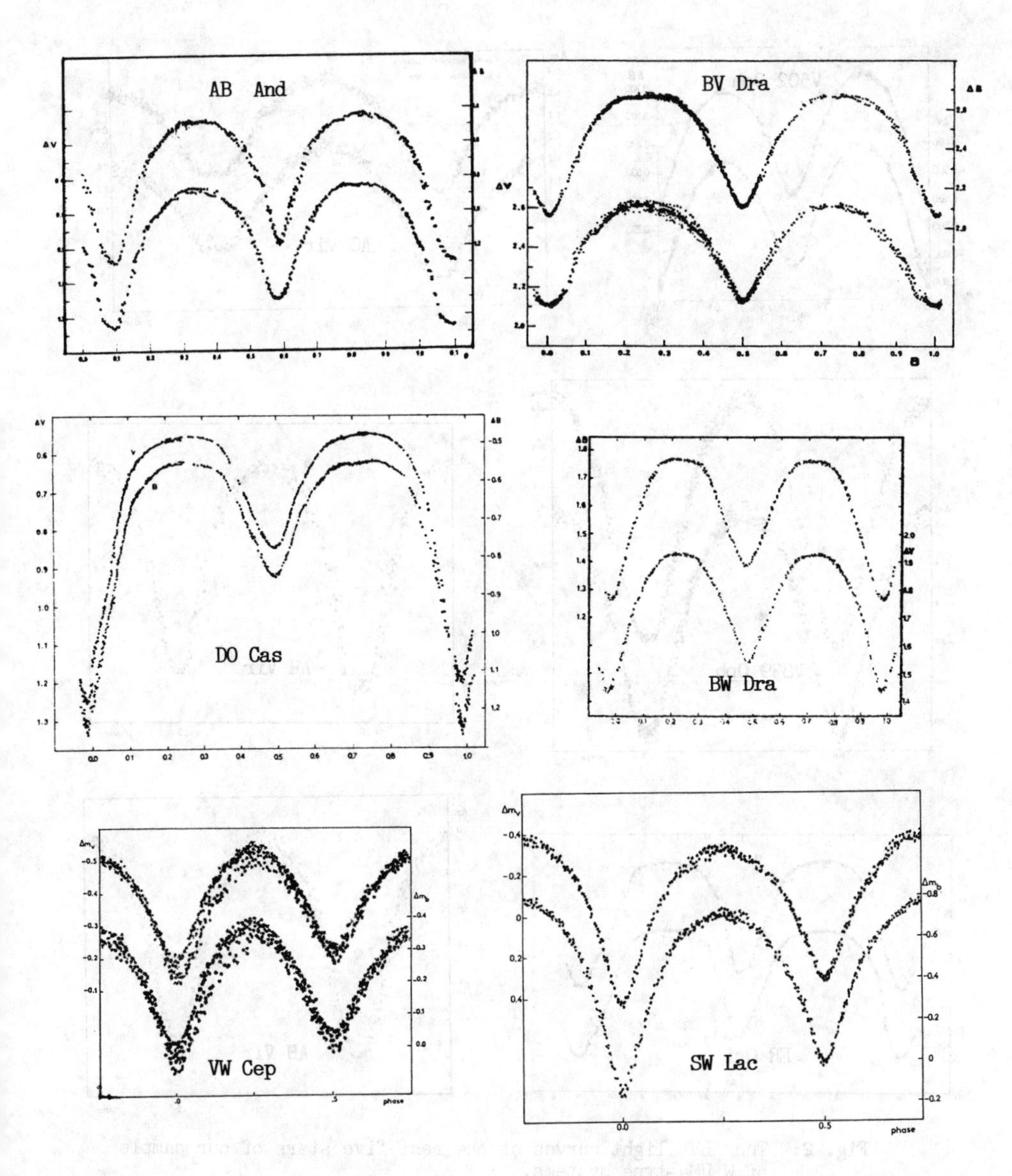

Fig.1: The BV light curves of six stars of our sample of W UMa-type systems.

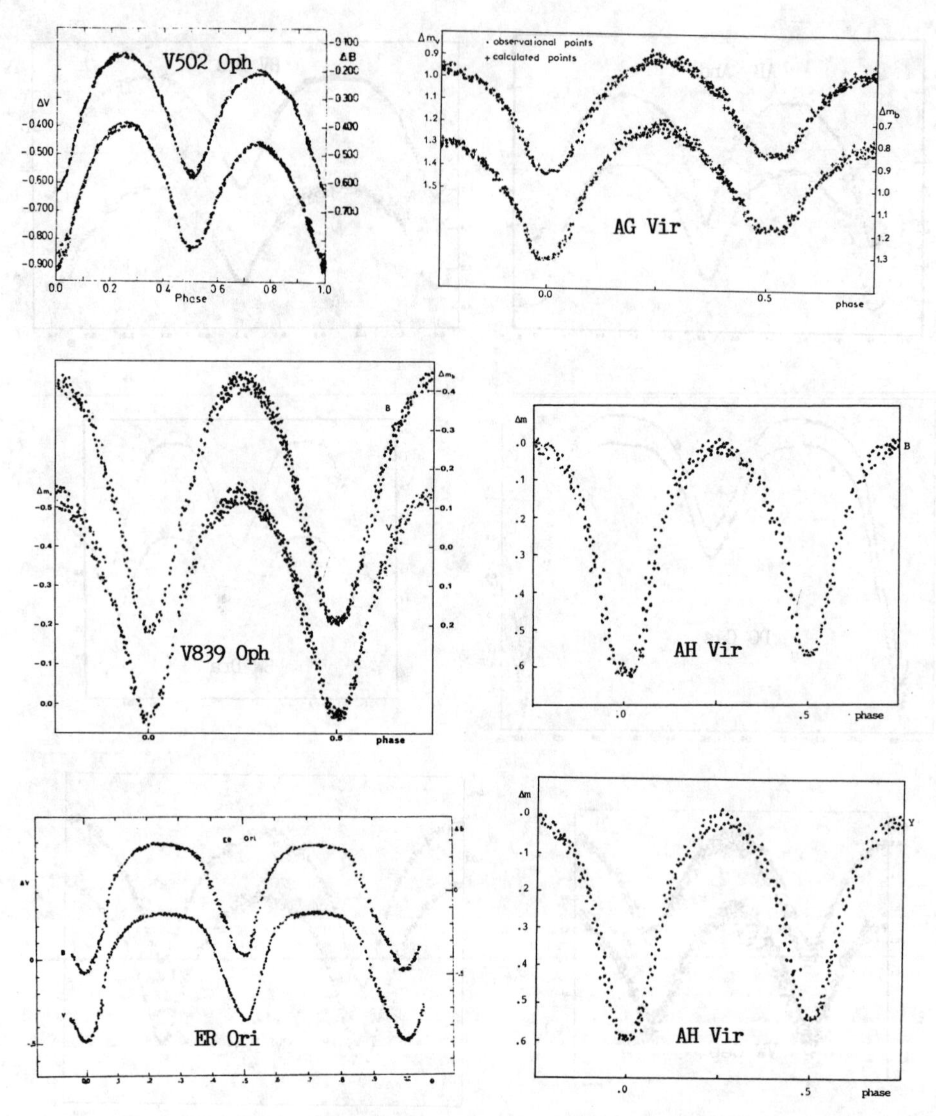

Fig. 2: The BV light curves of the rest five stars of our sample of W UMa-type systems.

nates; l and θ being, as usually, the brightness and the phase angle, respectively. The $\bar{A}_{2m}$ stand for the theoretical expressions of the eclipse moments in terms of the elements of the system and can be regarded as known (Kopal, 1977). The A_{2m}^{prox} represents the "proximity effects" (e.g. Kopal, 1976; Niarchos, 1981) and the R_{2m} denote the photometric perturbations arising from tidal and rotational distortion of both components during eclipses (e.g. Kopal, 1975; Livaniou, 1977; Rovithis-Livaniou, 1983).

The last term R_{2m} in the right hand side of equation (3.1) is generally very small in comparison with the others, thus in the first stage of the analysis it can be ignored. So, the $\bar{A}_{2m}$'s can be found from the known values of A_{2m} and A_{2m}^{prox} and a set of preliminary values of the geometrical elements r_1, r_2 and i can be obtained by the aid of the ratios $g_{2m}(a,c_o)$ (Kopal and Demircan, 1978). Where the a and c_o are constants related with the geometrical elements of the system and from their values one can easily find the type of the respective minimum. Then, the values of the photometric perturbations R_{2m} can be evaluated. This is not so easy in the case of partial eclipses, but at least some indicative values for the R_{2m}'s can be obtained even in that case (e.g. Rovithis-Livaniou, 1978). Thus, a final set of geometric as well as photometric elements can be found.

In Table II the geometrical elements obtained for the eleven systems, which we have analysed using FDT, are presented. In the same Table, the elements found by other investigators, using different methods of analysis and data, are given for comparison. Moreover, in Table II, the type of primary minima -as they came out from our analysis- are given.

Table III represents the absolute elements of the eleven systems - based on their spectroscopic and photometric data given in Tables I and II, respectively - together with those proposed by other investigators for comparison. Especially for V839 Oph there are not spectroscopic observations, thus a photometric mass ratio $q = 0.36$ was evaluated using the well known relation $q = m_s/m_g = (r_s/r_g)^2$.

4. DISCUSSION

Our sample of W UMa systems is characterized by homogeneity as far as the observations and the method of analysis are concerned. By an inspection of Tables II and III it is obvious that the elements found in the present study are in good agreement with those determined by other investigators who used different data and methods of analysis. Moreover, the type of primary minima found from our photometric analysis are the same with those resulted from spectroscopic studies. For AB And and V839 Oph only the results of our work are given in Table III, since, for the latter system, no absolute elements exist in the literature; while for AB And, although there are many, they show great differences which are probably due to the fact that there are some anomalies present in the light curve and the spectroscopic data of the system are ambiguous.

The absolute elements listed in Table III can be used to get an idea as far as the evolutionary status of these systems is concerned. The evolution of W UMa-type systems has been discussed by various investiga-

TABLE II
Geometric Elements of Selected W UMa Systems

Star	r_s	r_g	k	i	Type	Ref.
AB And	0.329	0.396	0.83	84.°4	W	(1)
	0.324	0.409	0.79	80.°8		(2)
DO Cas	0.310	0.460	0.67	82.°8	A	(3)
	0.294	0.502	0.58	81.°8		(4)
VW Cep	0.190	0.483	0.39	65.°4	W	(5)
	0.28	0.48	0.57	60.°0		(6)
BV Dra	0.307	0.467	0.66	77.°3	W	(7)
	0.318	0.472	0.67	76.°3		(8)
BW Dra	0.278	0.474	0.59	76.°5	W	(7)
	0.281	0.499	0.56	74.°4		(8)
SW Lac	0.358	0.388	0.92	80.°2	W	(9)
	0.378	0.402	0.94	78.°9		(10)
ER Ori	0.330	0.410	0.80	81.°9		(11)
	0.328	0.418	0.78	80.°5		(12)
V502 Oph	0.299	0.466	0.46	68.°5	W	(13)
	0.300	0.464	0.65	71.°3		(14)
V839 Oph	0.275	0.464	0.59	74.°9	W	(15)
	0.314	0.427	0.73	76.°9		(16)
AG Vir	0.182	0.505	0.36	72.°5	A	(17)
	0.200	0.495	0.41	77.°2		(18)
AH Vir	0.215	0.503	0.42	74.°8	W	(19)
	0.23	0.46	0.50	80.°8		(20)

(1): Rovithis-Livaniou et al. (1986a) (2): Rigtering (1973)
(3): Rovithis-Livaniou et al. (1986b) (4): Cester et al. (1977)
(5): Niarchos (1984) (6): Binnendijk (1970)
(7): Rovithis et al. (1987) (8): Kaluzny et al. (1986)
(9): Niarchos (1987) (10): Leung et al. (1984)
(11): Rovithis et al. (1986) (12): Liu et al. (1988)
(13): Rovithis et al. (1988) (14): Maceroni et al. (1982)
(15): Niarchos (1989) (16): Lafta et al. (1985)
(17): Niarchos (1985) (18): Blanco et al. (1970)
(19): Niarchos (1983) (20): Binnendijk (1960)

TABLE III
Absolute Elements of Selected W UMa Systems
(Expressed in Solar Units)

System	R_s	R_g	M_s	M_g	$M_{bol,s}$	$M_{bol,g}$	References
AB And	0.92	1.11	1.06	1.71	5.24	4.84	Rovithis-Livaniou et al.(1986a)
	-	-	-	-	-	-	
DO Cas	0.93	2.10	0.59	2.12	3.25	2.18	Present Study
	1.3	2.26	0.57	2.02	4.83	1.83	Kaluzny (1985b)
VW Cep	0.41	1.05	0.33	0.81	6.8	4.9	Niarchos (1984)
	0.61	0.91	0.40	0.93	-	5.09	Kaluzny (1985a)
BV Dra	0.70	1.06	0.39	0.98	5.22	4.31	Rovithis et al. (1987)
	0.76	1.12	0.43	1.04	-	3.82	Kaluzny et al. (1986)
BW Dra	0.52	0.88	0.24	0.86	5.92	4.75	Rovithis et al. (1987)
	0.55	0.98	0.26	0.92	-	4.35	Kaluzny et al. (1986)
SW Lac	0.91	0.98	0.96	1.14	4.88	4.87	Niarchos (1987)
	1.03	1.17	0.97	1.03	-	-	Leung et al. (1984)
ER Ori	0.62	0.77	0.32	0.58	5.47	4.99	Present Study
	0.75	1.06	0.35	0.70	5.31	4.70	Russo et al. (1982)
V502 Oph	0.91	1.41	0.50	1.35	4.60	3.77	Rovithis et al. (1988)
	0.94	1.53	0.56	1.52	-	3.47	Kaluzny (1985a)
V839 Oph	0.71	1.20	0.37	1.02	5.47	4.25	Present Study
	-	-	-	-	-	-	-
AG Vir	0.83	2.28	0.47	1.80	5.50	1.51	Niarchos (1985)
	0.89	2.20	0.59	2.3	5.40	1.70	Hill et al. (1972)
AH Vir	0.63	1.48	0.62	1.47	6.10	4.30	Niarchos (1983)
	0.66	1.34	0.58	1.38	6.39	4.64	Binnendijk (1960)

tors in recent years using different methods of approaching the problem (Webbink, 1977a, 1977b; Mochnacki, 1981; Rahunen and Vilhu, 1982; Vilhu, 1982; Maceroni et al. 1982; Van Hamme 1982a; Kaluzny, 1985a).

The aim of our present work is not to study in detail the evolution of W UMa systems, since our sample is relatively small, but to get an idea of the evolutionary status of our sample on the basis of mass-radius and mass-luminosity relations. For this purpose we used the same diagrams as Kaluzny (1985a) has done: He has adopted the theoretical calibration of ZAMS of VandenBerg and Bridge (1984), who have assumed Y=0.25 for the helium content, Z = 0.0169 (solar) for the heavy element abundances, Fe/H = 0.0 and a value of a = 1.5 for the ratio of the mixing length to the pressure scale height.

Figure 3 shows the diagram log R_g versus log M_g . One can see in this figure that most of the primaries of our sample lie close to the ZAMS. For four systems (AB And, DO Cas, ER Ori and AG Vir) the primaries are far from ZAMS.

Figure 4 represents the diagram $M_{bol,g}$ versus logM_g . Again, most of the primaries lie quite close to ZAMS. Again the systems AB And and ER Ori are far from ZAMS. Two others (DO Cas, AH Vir) lie below the ZAMS in Figure 4, while their position, is above ZAMS in Figure 3.

A possible explanation for the position of AB And and ER Ori is that their masses have been overestimated (for AB And) and underestimated (for ER Ori) due to their poor spectroscopic data. Another possibility is that the chemical content of the stars is different from the solar one. The situation for the systems DO Cas and AH Vir is not quite clear, since their primaries lie among evolved stars in the mass-radius diagram, while they lie below the ZAMS in the mass-luminosity diagram. Especially for the case of DO Cas, which belongs to A sub-class, there is a suggestion from the spectroscopic data that these systems constitute a heterogeneous group with respect to their evolutionary status as well as to the mass of their primary components. For the case of AH Vir a possible explanation is that either its mass has been overestimated or its temperature has been underestimated and this may be due to the poor quality of the spectroscopic data of the system.

Summarising, there are some evidences that primaries of four from our eleven analysed W UMa-type systems deviate from relations fulfilled by typical main sequence stars. It is desirable to obtain accurate spectroscopic observations of these systems to confirm this suggestion. Such observations of high quality are urgently needed, in order to be able to draw conclusions concerning the evolutionary status of contact binaries.

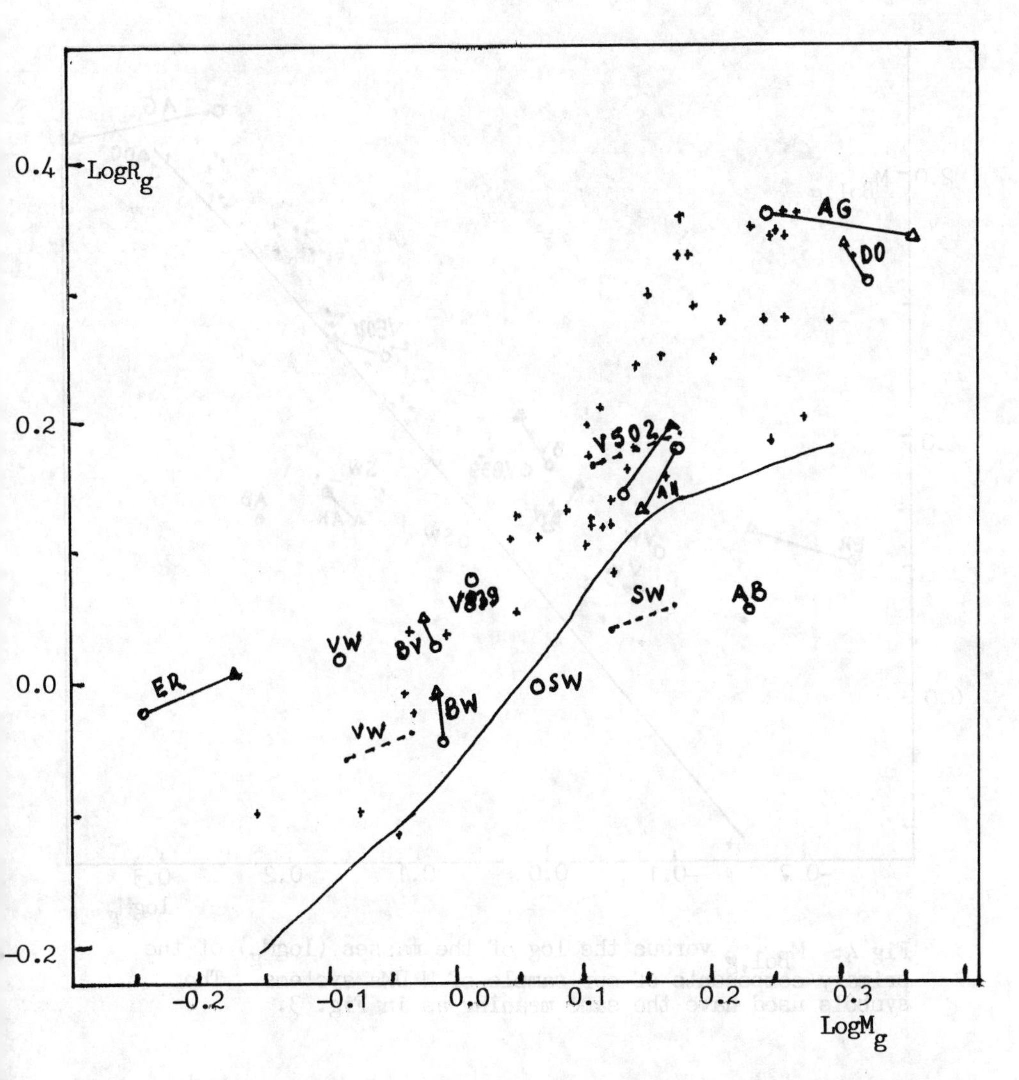

Fig.3: The mass-radius diagram for the primary components of our sample of W UMa systems. Open circles stand for our values and triangles for those from other sources while the broken lines come from Kaluzny's work. The continuous line is the theoretical ZAMS and the crosses denote positions of noncontact Main Sequence Stars.

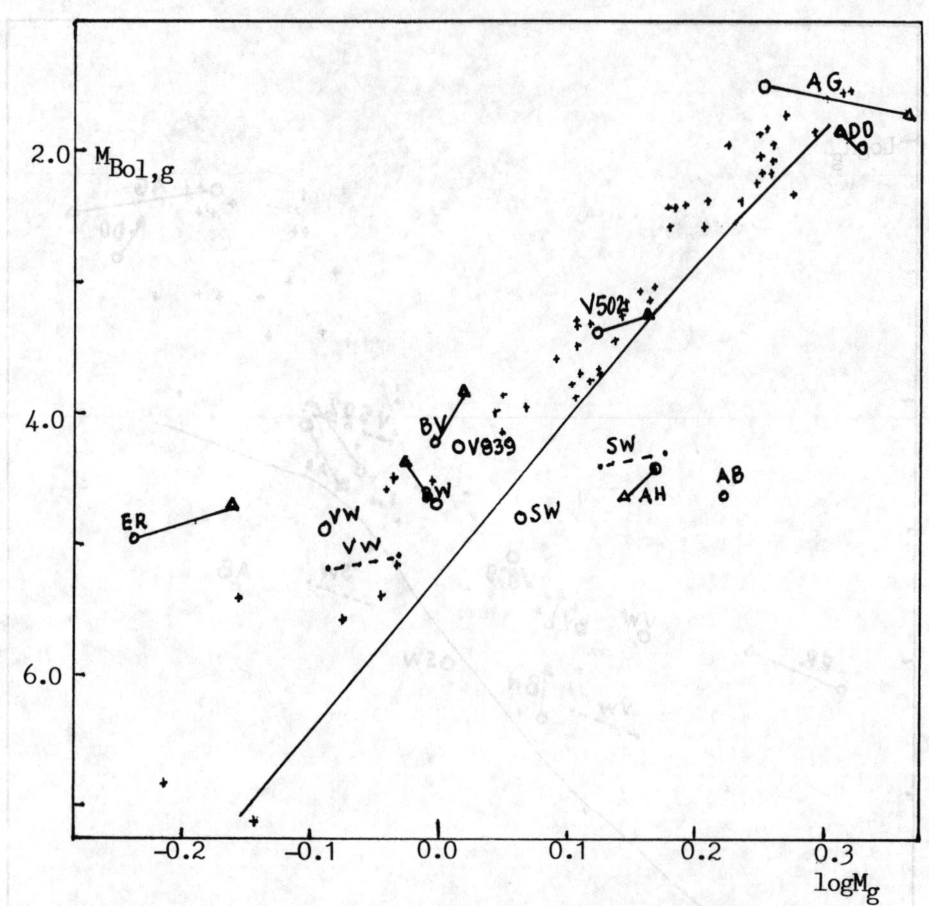

Fig.4: $M_{Bol,g}$ versus the log of the masses ($\log M_g$) of the primary components of our sample of W UMa systems. The symbols used have the same meaning as in Fig. 3.

REFERENCES

Batten, A. and Hardie, R.H.: 1965, Astron. J. 70 , 666.
Batten, A. and Lu, W.: 1986, Publ. Astron. Soc. Pac. 98, 92.
Binnendijk, L.: 1959, Astron. J. 64, 65.
Binnendijk, L.: 1960, Astron. J. 65, 358.
Binnendijk, L.: 1962, Astron. J. 67, 86.
Binnendijk, L.: 1966, Publ. Dom. Astrophys. Obs. 13, 27.
Binnendijk, L.: 1969, Astron. J. 74, 1024.
Binnendijk. L.: 1970, Vistas in Astronomy 12, 217.
Blanco, C. and Catalano, F.: 1970, Ital. Astron. Soc. 41, 343 (No.3)
Cester, B., Giuricin, G., Mardirossian, F. and Pucillo, M.: 1977, Astron. Astrophys. Suppl. Ser. 30, 223.
Chang, Y.C.: 1948, Astrophys. J. 107, 96.
Fehrenbach, Ch., Duot, M., Boulon, J., Rebeirot, E. and Lanoë, C.: 1966, Publ. Obs. Haute Provence 8, 25.
Guthnick, P. and Prager, R.: 1927, Astron. Nachr. 288, 167.
Hershey, J.L.: 1975, Astron. J. 80, 662.
Hill, G. and Barnes, J.: 1972, Publ. Astron. Soc. Pacific 84. 382.
Hill, G., Hilditch, R.W., Younger, F. and Fisher, W.A.: 1975, Mem. Roy. Astron. Soc. 79, 131.
Jameson, R.F. and Akinci, R.: 1973, Mon. Not.R. Astron.Soc. 188, 421.
Kaluzny, J.: 1985a, Acta Astron. 35, 313.
Kaluzny, J.: 1985b, Acta Astron. 35, 327.
Kaluzny, J. and Rucinski, S.M.: 1986, Astron. J. 92, 666.
King, D.J. and Hilditch, R.W.: 1984, Mon.Not.R.Astron.Soc. 209, 645.
Koch, R.H.: 1973, Astron. J. 78, 410.
Kopal, Z.: 1975, Astrophys. Space Sci. 38, 191.
Kopal, Z.: 1976, Astrophys. Space Sci. 45, 269.
Kopal, Z.: 1977, Astrophys. Space Sci. 51, 439.
Kopal, Z.: 1979, "Language of the Stars", D. Reidel Publ. Co., Dordrecht and Boston.
Kopal, Z. and Demircan, O.: 1978, Astrophys. Space Sci. 55, 241.
Kwee, K.K.: 1958, Bull. Astron. Inst. Netherl. 14, 131.
Kwee, K.K.: 1966, Bull. Astron. Inst. Netherl. 18, 448.
Lafta, S.T. and Grainger, J.F.: 1985.: Astrophys. Space Sci. 114, 23.
Landolt, A.U.: 1969, Astron. J. 74, 1078.
Leung, K.C., Zhai, D., Zhang, R.: 1984, Publ. Astron. Soc. Pac. 96,634.
Linnell, A.P.: 1980 , Publ. Astron. Soc. Pac. 12, 202.
Livaniou, H.: 1977, Astrophys. Space Sci. 51, 77.
Liu Qingyao, Yang Yulan, Leung Kam-Ching, Zhai Disheng and Li, Yan-Feng.: 1988, Astron. Astrophys. Suppl. Ser. 74, 443.
Maceroni, C., Milano, L. and Russo, G.: 1982, Astron. Astrophys. Suppl. Ser. 49, 123.
Maceroni, C., Milano, L. and Russo, G.: 1985, Mon. Not. R. Astron. Soc. 217, 843.
Mochnacki, S.W.: 1981, Astrophys. J. 245, 650.
Niarchos, P.: 1981, Astrophys. Space Sci. 76, 503.
Niarchos, P.G.: 1983, Astron. Astrophys. Suppl. Ser. 53, 13.
Niarchos, P.G.: 1984, Astron.Astrophys.Suppl.Ser. 58, 261.
Niarchos, P.G.: 1985, Astron.Astrophys.Suppl.Ser. 61, 313.

Niarchos, P.G.: 1987, Astron. Astrophys. Suppl. Ser. 67, 365.
Niarchos, P.G.: 1989, Astrophys. Space Sci. 153, 143.
Popper, D.M.: 1948, Astrophys. J. 108, 490.
Rahunen, T. and Vilhu, O.: 1982, in Binary and Multiple Stars as tracers of stellar evolution, ed. Z.Kopal and J. Rache (Reidel, Dordrecht) p. 289.
Rigtering, P.V.: 1973, Astron. Astrophys. Suppl. Ser. 12, 313.
Rovithis-Livaniou, H.: 1978, Astrophys. Space Sci. 59, 463
Rovithis-Livaniou, H. and Rovithis, P.: 1981, Astrophys. Space Sci. 76, 465.
Rovithis-Livaniou, H.: 1983, Astrophys. Space Sci. 97, 171.
Rovithis-Livaniou, H. and Rovithis, P.: 1986a, Astron. Nachr. 307, 17.
Rovithis-LIvaniou, H. and Rovithis, P.: 1986b, Astrophys. Space Sci. 119, 381.
Rovithis, P. and Rovithis-Livaniou, H.: 1986, Astron. Astrophys. 155, 46.
Rovithis, P. and Rovithis-Livaniou, H.: 1987, Astron. and Astrophys. Suppl. Ser. 70, 63.
Rovithis, P., Niarachos, P.G. and Rovithis-Livaniou, H.: 1988, Astron. Astrophys. Suppl. Ser. 74, 265.
Russo, G., Sollazzo, C., Maceroni, C., Milano, L.: 1982, Astron. Astrophys. Suppl. Ser. 47, 211.
Sanford, R.F.: 1934, Astrophys. J. 79, 89.
Schneller, H. and Daene, H.: 1952, Astron. Nachr. 281, 25.
Struve, O.: 1944, Publ. Astron. Soc. Pacific 56, 34.
Struve, O. and Gratton, L.: 1948, Astrophys. J. 108, 497.
Struve, O., Horak, H.G., Canavaggia, R., Kourganoff, V. and Colacevich, A.: 1950, Astrophys. J. 111, 658.
Struve, O. and Zebergs, V.: 1958, Astrophys. J. 130, 789.
Van Hamme, W.: 1982a, Astron. Astrophys. 105, 389.
Van Hamme, W.: 1982b, Astron. Astrophys. 116, 27.
Vilhu, O.: 1982, Astron. Astrophys. 109, 17.
Webbink, R.F.: 1977a, Astrophys. J. 211, 486.
Webbink, R.F.: 1977b Astrophys. J. 215, 851.

A RADIAL-VELOCITY SEARCH FOR BROWN DWARFS AND THE LOW-MASS COMPANION OF HD114762

TSEVI MAZEH
The Wise Observatory, Tel Aviv University
and
Harvard-Smithsonian Center for Astrophysics

DAVID W. LATHAM, ROBERT P. STEFANIK
Harvard-Smithsonian Center for Astrophysics

GUILLERMO TORRES
Harvard-Smithsonian Center for Astrophysics
and
Córdoba Observatory, National University of Córdoba

ETI WASSERMAN
The Wise Observatory, Tel Aviv University

ABSTRACT. For the past four years we have monitored the radial velocities of a sample of 24 M dwarfs in order to search for low-amplitude periodic velocity variations that might indicate orbital motion due to low-mass companions. To monitor systematic errors we chose for every star of the sample a comparison star, to be observed next to its project star. We have not yet found a convincing evidence for a *periodic* modulation in the data of the M stars, nor in those of the comparison stars. However, a few candidates for low-amplitude radial-velocity variations can be seen in the data. The best case for a low-mass companion discovered using radial velocities is still HD114762; we briefly report on the status of our study of this IAU standard star. We conclude that substellar companions may be rare compared to stellar companions, but there exist more than one candidate for brown dwarf companions.

C. İbanoğlu (ed.), Active Close Binaries, 267–276.

1. Introduction

Brown dwarfs are substellar objects with mass too small to ignite hydrogen in their core (Tarter 1986). Therefore, they are very dim and their detection is very difficult. Nevertheless, brown dwarfs are being vigorously searched for with different observational techniques in the last few years (e.g. Ianna, Rohde and McCarthy 1988; Becklin and Zuckerman 1988; Forrest, Skrutskie and Shure 1988; Jameson and Skillen 1989; Henry and McCarthy 1989; Stauffer *et al.* 1989). This is because brown dwarfs and the lowest-mass stars might make up the Galactic disc "missing mass" (Bahcall 1986), provided they are numerous enough (Liebert and Probst 1987). In addition, detection of brown dwarfs will enable us to confront the theoretical models (e.g. Nelson, Rappaport and Joss 1986; D'Antona 1986; Burrows, Hubbard and Lunine 1989) with the spectroscopy and photometry of these low-mass objects.

Several teams have been searching for brown dwarfs as unseen low-mass companions to known nearby stars (e.g. Harrington 1986; Marcy and Benitz 1989; Campbell, Walker and Yang 1989). The searches have been utilizing spectroscopic or astrometric observations of bright stars to look for evidence of an unseen companion with a mass lower than 0.08 $M_\odot$ – the theoretical hydrogen-burning limit. Obviously, such a search is also relevant to the quest for extra-solar planets, provided unseen companions could be found with a mass low enough to fall into the planetary regime (e.g. Boss 1986).

In this paper we give a progress report on an ongoing radial-velocity project at the CfA to look for low-mass companions of nearby M stars. We also report on the status of our radial-velocity study of HD114762 which very probably has a brown dwarf as an unseen companion. We conclude by commenting on some of the other searches for brown dwarfs and their results.

2. The CfA Search

For the last four years we have been monitoring a sample of 24 nearby M dwarfs classified by Joy and Abt (1974) as M0 or M0.5 stars. The sample includes all such stars in the Gliese (1969) catalog with declination between $-14°$ and $+64°$, brighter than 9.5 mag in V. (Gls 719 was not included in the sample as it is a double-line spectroscopic binary.) Each of the project stars is being monitored for a low-amplitude radial-velocity periodic modulation. This is done using the digital speedometers (Latham 1985) operated by the Center for Astrophysics (CfA), mainly with the 1.5-m Wyeth Reflector at the Oak Ridge Observatory in Harvard, Massachusetts, and occasionally with the 1.5-m Tillinghast Reflector at the Whipple Observatory on Mt. Hopkins, Arizona. A typical error per measurement is 0.4 to 0.5 km/s. To monitor systematic errors we chose for every star of the sample a comparison star, to be observed immediately before or after its project star. The comparison star is a K or M star within a degree or two of the project star. Hopefully, the velocity difference between the M star and its comparison should be free of systematic errors.

For each of the project stars at least 20 radial-velocity measurements have already been obtained. As the observations accumulated, the data for the project and the comparison stars were searched for a hint of periodic variation, and every suspected star was observed more intensively. No periodic variation with an amplitude of 1 km/s or larger, and with a period of up to a few years was found. Such a periodicity would not have escaped our notice. Variations with smaller amplitude, of about 0.5 km/s, might be present in some of the M stars and the comparison stars.

One of the best examples for a star found in our study with a low-amplitude radial-velocity modulation is HD23712 (= SAO 76206, V = 6.46, B-V = 1.70, U-B = 2.07), a probable K giant which serves as a comparison star to Gls 154. As displayed in Figure 1, where all the measurements as of August 1989 are depicted, the star exhibits a real variation with an amplitude of about 1 km/s and a timescale of about two years. A formal orbital solution of the 38 points yields a period of 615 days and an amplitude of 0.7 km/s. However, to establish undoubtedly the *periodic* nature of the modulation and to find out more about its origin many more observations are needed.

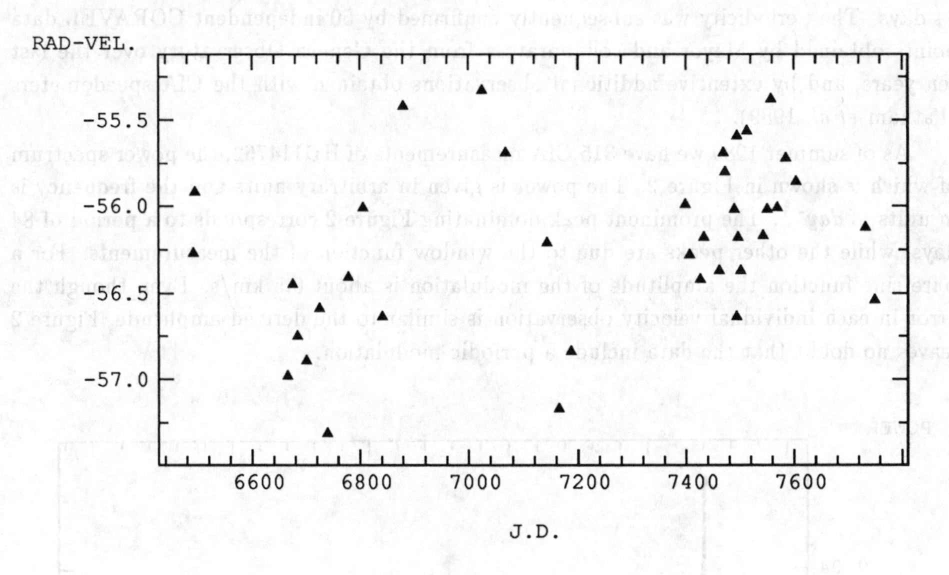

Fig. 1: The radial velocity of HD23712.

3. The Unseen Companion of HD114762

The same CfA instruments have been used for more than eight years to monitor carefully the radial velocities of about a dozen stars with the goal of establishing an improved set of velocity standards for the International Astronomical Union (Mayor and Maurice 1985). One of these standard stars is HD114762 (= BD+18°2700 = SAO 100458; V = 7.3 and B-V = 0.53). The star has a high proper motion and was therefore included in the survey of Carney and Latham (1987), who determine a photometric distance of 28 pc.

We have already reported that in April 1988, as a result of test observations with a new fiber feed for the digital speedometer at the Oak Ridge Observatory, our attention was drawn to a possible variation of HD114762 (Latham *et al.* 1988). An examination of the 63 observations available at that time disclosed a periodic variation, with a period of 84 days. The periodicity was subsequently confirmed by 50 independent CORAVEL data points obtained by Mayor and collaborators from the Geneva Observatory over the last ten years, and by extensive additional observations obtained with the CfA speedometers (Latham *et al.* 1989).

As of summer 1989 we have 315 CfA measurements of HD114762, the power spectrum of which is shown in Figure 2. The power is given in arbitrary units and the frequency is in units of day^{-1}. The prominent peak dominating Figure 2 corresponds to a period of 84 days, while the other peaks are due to the window function of the measurements. For a pure sine function the amplitude of the modulation is about 0.5 km/s. Even though the error in each individual velocity observation is similar to the derived amplitude, Figure 2 leaves no doubt that the data include a periodic modulation.

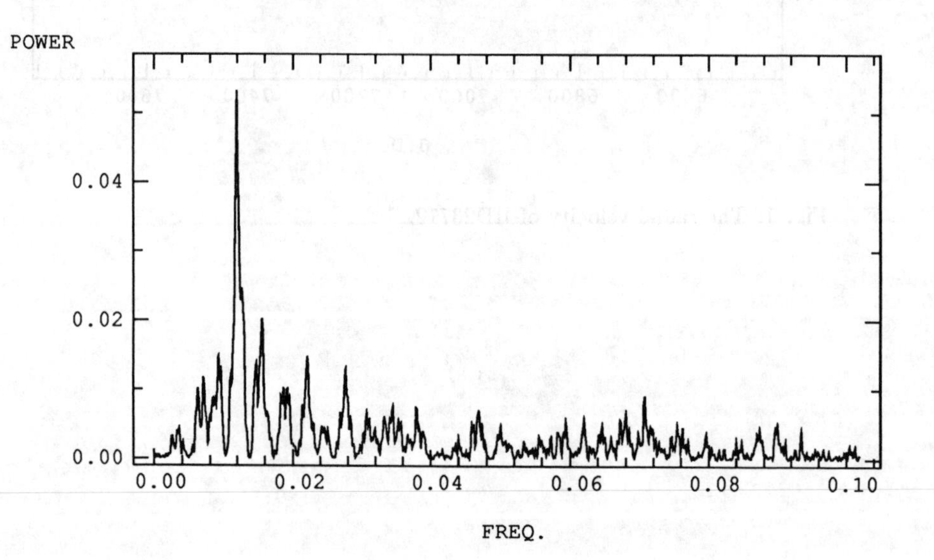

Fig. 2: The power spectrum of the CfA HD114762 data set.

We suggest that the velocity variations seen in HD114762 must be due to orbital motion induced by an unseen companion. The amplitude and the strict periodicity of the modulation can rule out pulsation or surface activity as the mechanism behind the variation of the stellar radial velocity.

Table I lists the parameters of our orbital solutions based on the CfA data alone, and the CORAVEL data alone. The good agreement of the two solutions, without any zero point adjustment, is remarkable.

Table I: The Orbital solutions of HD114762

	CfA	CORAVEL
Period *(days)*	83.78 ± 0.11	83.91 ± 0.09
System velocity *(km/s)*	49.31 ± 0.03	49.39 ± 0.06
Amplitude *(km/s)*	0.55 ± 0.04	0.75 ± 0.12
Eccentricity	0.27 ± 0.06	0.30 ± 0.15
Longitude of periastron *(deg.)*	220 ± 10	280 ± 16
Epoch, T *(JD-2440000)*	5029 ± 3	5033 ± 4
Mass function (10^{-6} $M_{\odot}$)	1.3 ± 0.3	3.1 ± 1.5
Number of observations	315	50
RMS residuals *(km/s)*	0.44	0.39

In Figure 3 we plot the orbital solution for the CfA data set, together with the individual observed velocities.

Assuming the mass of the primary is about 1 $M_{\odot}$, the orbital solution implies that the mass of the unseen companion M_{comp} is

$$M_{comp} \sin i = 0.011 \pm 0.001 \; M_{\odot},$$

where i is the inclination of the orbit. The derived value for M_{comp} is much smaller than 0.08 $M_{\odot}$. Actually, this value is even smaller than 0.02 $M_{\odot}$, the lower limit found theoretically by Boss (1986) for the mass of a companion object that forms by fragmentation in a collapsing cloud. Thus, the unseen companion of HD114762 is a good candidate to be a brown dwarf or even a giant planet. Naturally, we can not be certain of the mass of the companion, since we do not know $\sin i$. This is a basic drawback of the spectroscopic approach to the search for brown dwarfs. However, for the companion of HD114762 to be a main-sequence star, the orbit would have to be within 8 degrees of being viewed face on, which would have less than 1% chance of occurring in a sample of orbits oriented randomly.

One way to exclude a main-sequence companion to HD114762 is to find *another* companion in the system, with a period of the order of 100 days. If HD114762 is a triple

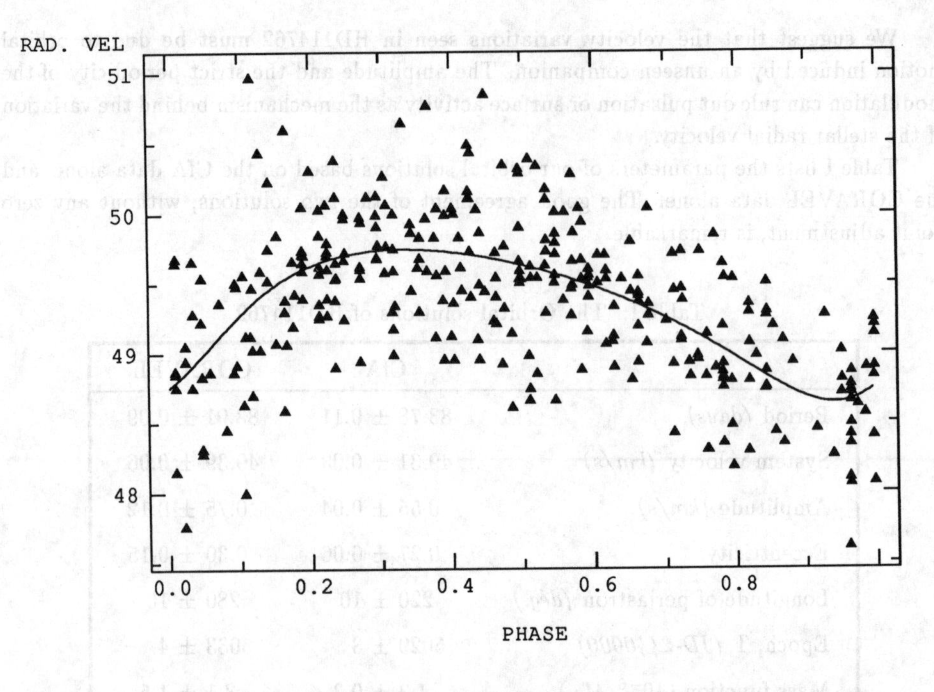

Fig. 3: The radial-velocity curve of HD114762. The continuous line represents the derived solution with the parameters of Table I for the CfA data set.

system with two similar orbital periods, the system is unstable (e.g. Harrington 1977; Bailyn 1987), unless the two unseen companions are substantially less massive than the primary star (Graziani and Black 1981). The solar system, for example, could not have been dynamically stable if the planets had been as massive as the sun. This is the reason why we keep observing HD114762. We would like to find out whether more data might indicate another low-mass object in the system. To look for such an object we search for another periodicity in the radial-velocity data (e.g. Mazeh 1987). The power spectrum of the residuals of the derived orbital solution, based on the 315 CfA data points, is depicted in Figure 4. The units used are the same as in Figure 2. The highest peak, at about 0.02 day^{-1}, corresponds to a modulation with an amplitude of about 0.1 km/s. As the height of this peak is not yet significant, more data are needed to explore this intriguing possibility. At the moment, we can only put an upper limit of about 0.2 km/s amplitude to any additional periodic modulation of HD114762.

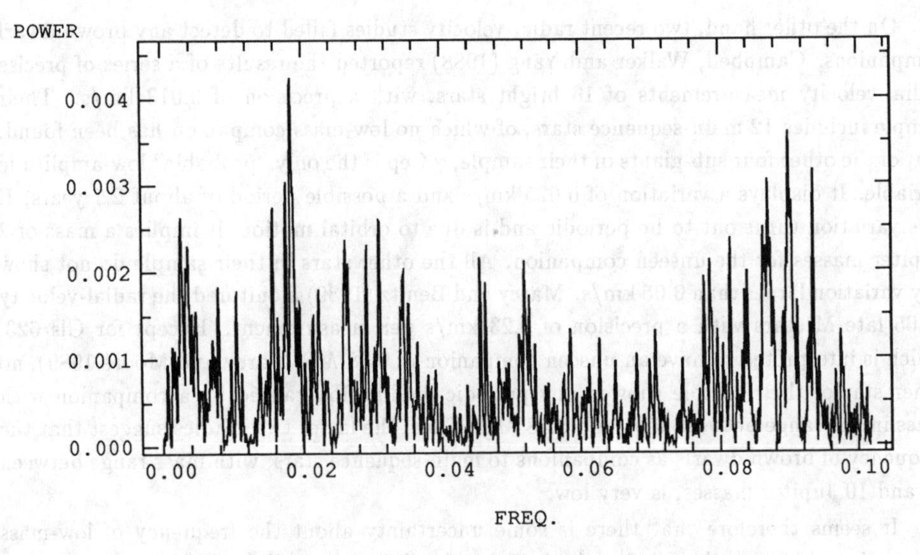

Fig. 4: The power spectrum of the CfA residuals of HD114762.

4. On The Frequency of Brown Dwarfs

How many low-mass companions are there in the solar neighborhood? The best case found by radial-velocity technique seems still to be HD114762. This star has been observed extensively in the last seven years as a new candidate for an IAU radial-velocity standard star (Mayor and Maurice 1985). About 50 candidates have been observed with similar frequency, out of which only HD114762 has so far shown an unquestionable low-amplitude periodic modulation. However, a few other candidates also show small velocity variations which may yield convincing orbital solutions in the future.

We also note that another radial-velocity standard star was found to display a small-amplitude periodic variation. This is HR 152 (McClure *et al.* 1985), one of the four giant stars which were chosen to serve as the fundamental standards for the Cambridge radial-velocity spectrometer (Griffin 1985). If this modulation is of orbital nature, the mass of the companion would be

$$M_{comp} \sin i = 0.026(M_{prim}/M_{\odot})^{2/3} M_{\odot},$$

where i is the inclination of the orbit, and M_{prim} is the mass of the primary. However, as HR 152 is a giant, a different interpretation of the small-amplitude variations is also possible (Irwin *et al.* 1989 and Walker *et al.* 1989).

Our CfA sample, which includes about 50 stars (the M stars and the comparison stars together) may also yield some new candidates for brown dwarfs, in addition to HD23712 mentioned above. Thus, although brown-dwarf companions may be a rare phenomenon, we might have already a few candidates from radial-velocity programs.

On the other hand, two recent radial-velocity studies failed to detect any brown dwarf companions. Campbell, Walker and Yang (1988) reported the results of a series of precise radial-velocity measurements of 16 bright stars, with a precision of 0.013 km/s. Their sample includes 12 main-sequence stars, of which no low-mass companion has been found. Out of the other four sub-giants of their sample, γ Cep is the only "probable" low-amplitude variable. It displays a variation of 0.025 km/s and a possible period of about 2.7 years. If this variation turns out to be periodic and is due to orbital motion, it implies a mass of 2 Jupiter masses for the unseen companion. All the other stars in their sample do not show any variation larger than 0.05 km/s. Marcy and Benitz (1989) monitored the radial-velocity of 65 late M stars with a precision of 0.23 km/s per measurement. Except for Gls 623, which is interpreted to have an unseen companion of 0.08 $M_{\odot}$(Marcy and Moore 1989), no other star of their sample showed any periodic modulation caused by a companion with mass in the range between 0.01 and 0.08 $M_{\odot}$. Thus, the latter two studies suggest that the frequency of brown dwarfs as companions to main-sequence stars, with mass range between 80 and 10 Jupiter masses, is very low.

It seems therefore that there is some uncertainty about the frequency of low-mass companions between the results obtained by the different studies. Clearly, the question will be resolved by results from radial-velocity surveys of large samples of stars with well-defined selection criteria.

ACKNOWLEDGEMENT We thank J. Andersen, J. Caruso, R. Davis, R. McCrosky, O. Rodrigues, G. Schwartz, S. Tokarz, and J. Zajac for help with the observations and data reduction at CfA. This work was supported by the Smithsonian Institution, the U.S. - Israel Binational Science Foundation Grant No. 86-00238, and the Consejo Nacional de Investigaciones Científicas y Técnicas, Buenos Aires, Argentina.

References

Bahcall J.N. (1986), *Ann. Rev. Astron. Astrophys.*, **24**, 577.

Bailyn, C.D. (1987), Ph. D. Thesis, Harvard University.

Becklin, E.E., and Zuckerman, B. (1988), *Nature*, **336**, 656.

Boss, A. P. (1986), in *Astrophysics of Brown Dwarfs*, ed. M.C. Kafatos, R.S. Harrington and S.P. Maran, (Cambridge: University Press) p. 206.

Burrows A., Hubbard, W.B., and Lunine, J.I., (1989), *Astrophys. J.*, in press.

Campbell, B., Walker, G.A. and Yang, S. (1988), *Astrophys. J.*, **331**, 902.

Carney, B.W., and Latham, D.W. (1987), *Astron. J.*, **93**, 116.

D'Antona, F. (1986), in *Astrophysics of Brown Dwarfs*, ed. M.C. Kafatos, R.S. Harrington and S.P. Maran, (Cambridge: University Press) p. 148.

Forrest, W.J., Skrutskie, M.F., and Shure, M. (1988), *Astrophys. J. Lett.*, **330**, L119.

Gliese, W. (1969), *Catalogue of Nearby Stars*, Astronomischen Rechen-Instituts, Heidelberg, No. 22.

Graziani, F., and Black D.C. (1981), *Astrophys. J.*, **337**, 251.

Griffin, R.F. (1985), in *Stellar Radial Velocities*, IAU Coll. 88, ed. A.G. Davis Phillip and David W. Latham (Schenectady: L. Davis Press) p. 121.

Harrington, R.S. (1977), *Astron. J.*, **82**, 753.

Harrington, R.S. (1986), in *Astrophysics of Brown Dwarfs*, ed. M.C. Kafatos, R.S. Harrington and S.P. Maran, (Cambridge: University Press) p. 3.

Henry, T.J., and McCarthy, D.W. Jr., (1989), *Astrophys. J.*, in press.

Ianna, P.A., Rohde, J.R., and McCarthy, D.W. Jr. (1988), *Astron. J.*, **95**, 1226.

Irwin, A.W., Campbell, B., Morbey, C.L., Walker, G.A.H., and Yang, S. (1989), *Publ. Astron. Soc. Pacific*, **101**, 147.

Jameson, R.F., and Skillen, I. (1989), *Mon. Not. R. Astron. Soc.*, **239**, 247.

Joy, A.H. and Abt, H.A. (1974), *Astrophys. J. Supp.*, **28**, 1.

Latham, D.W. (1985), in *Stellar Radial Velocities*, IAU Coll. 88, ed. A.G. Davis Phillip and David W. Latham (Schenectady: L. Davis Press) p. 21.

Latham, D.W., Andersen, J., Geary, J.C., Stefanik, R.P. and Rodrigues, O. (1988), in *Fiber Optics in Astronomy*, Astron. Soc. Pacific Conf., Ser. vol. 3, ed. S.C. Barden, p. 269.

Latham, D.W., Mazeh, T., Stefanik, R.P., Mayor, M., and Burki, G. (1989), *Nature*, **339**, 38.

Liebert, J. and Probst, R.G. (1987), *Ann. Rev. Astron. Astrophys.*, **25**, 473.

Marcy, G.W. and Moore, D. (1989), *Astrophys. J.*, **341**, 961.

Marcy, G.W. and Benitz, K.J. (1989), *Astrophys. J.*, in press.

Mayor, M. and Maurice, E. (1985), in *Stellar Radial Velocities*, IAU Coll. 88, ed. A.G. Davis Phillip and David W. Latham (Schenectady: L. Davis Press) p. 299.

Mazeh, T., Kemp, J.C., Leibowitz, E.M., Meningher, H. and Mendelson, H. (1987), *Astrophys. J.*, **317**, 824.

McClure, R.D., Griffin, R.F., Fletcher, J.M., Harris, H.C. and Mayor, M. (1985), *Publ. Astron. Soc. Pacific*, **97**, 740.

Nelson, L. A., Rappaport, S. A. and Joss, P. C. (1986), in *Astrophysics of Brown Dwarfs*, ed. M.C. Kafatos, R.S. Harrington and S.P. Maran, (Cambridge: University Press) p. 177.

Stauffer, J.R., Hamilton, D., Probst, R.G., Rieke, G.H. and Mateo, M.L. (1989), a preprint.

Tarter, J.C. (1986), in *Astrophysics of Brown Dwarfs*, ed. M.C. Kafatos, R.S. Harrington and S.P. Maran, (Cambridge: University Press) p. 121.

Walker, G.A.H., Yang, S., Campbell, B., and Irwin, W. (1989), *Astrophys. J. Lett.*, **343**, L21.

PHOTOMETRY OF SOME CLOSE BINARIES AT THE ANKARA UNIVERSITY OBSERVATORY

Z. Müyesseroğlu, O. Demircan, E. Derman, S. Selam

Ankara University Observatory
Science Faculty
06100 Beşevler ANKARA/TURKEY

ABSTRACT: Three close binary systems YY Eri, V456 Oph and AW Peg are discussed on the basis of UBV photoelectric observations carried out at the Ankara University Observatory in 1988 and 1989.

1. INTRODUCTION

A well organized team at the Ankara University Observatory carry out UBV photoelectric observations of some bright (brighter than 11th magnitude) close binaries of Algol, RS CVn and W UMa type. The light curves in three colors of more than thirty close binaries were completed in last two years. The observed binaries are Algol, AB And, BX And, OO Aql, V346 Aql, UX Ari, CK Boo, 44i Boo, KR Cyg, V1073 Cyg, LS Del, BV Eri, RZ Eri, YY Eri, AK Her, HS Her, RX Her, FG Hya, SW Lac, AM Leo, AP Leo, DH Leo, UV Leo, β Lyr, V451 Oph, V502 Oph, V456 Oph, V566 Oph, V839 Oph, AW Peg, UV Psc, AW UMa, AH Vir and ER Vul. The light curves of some short period systems (such as AK Her, CK Boo, OO Aql, LS Del, ER Vul) were obtained in every observing season. No clear night (in fact no clear portion of the sky at any night) is missed unless the system is out of order. Some successful fourth year undergraduate students help the observers as night assistant. The telescope used for the observations is a 30 cm Maksutov type with EMI 9798QB photomultiplier attachment. The continuity in observations aims deducing more information on physical mechanisms causing the color and time dependent variations in the light curves.

C. İbanoğlu (ed.), Active Close Binaries, 277–285.

In the present contribution we discussed three close binaries on the basis of preliminary results from the analysis of our photoelectric observations. The binaries are YY Eri, V456 Oph and AW Peg.

2. YY ERIDANI

YY Eridani (HD 26609) is a partially eclipsing W-type contact binary system. The system was observed photometrically by Cillie (1951), Huruhata et al. (1953) and Purgathofer (1960) and Eaton (1986). Huruhata et al. (1953), Binnendijk (1965) published the Russell model solutions of photometric observations. Struve (1947) determined the mass ratio spectroscopically, q=0.59, which is much larger than the newly determined value of q=0.401. The latter value was obtained by a simultaneous solution of new radial velocity and photometric light curves (Nesci et al. 1986). Eaton (1986) could obtained better simultaneous fit to his V and (V-I) light curves, with a starspot on the more massive component. But, his solution is quite different from that of Nesci et al. In Eaton's solution the fill out parameter is too large, the temperatures of both components and mass ratio are larger (see Table 2 for comparison).

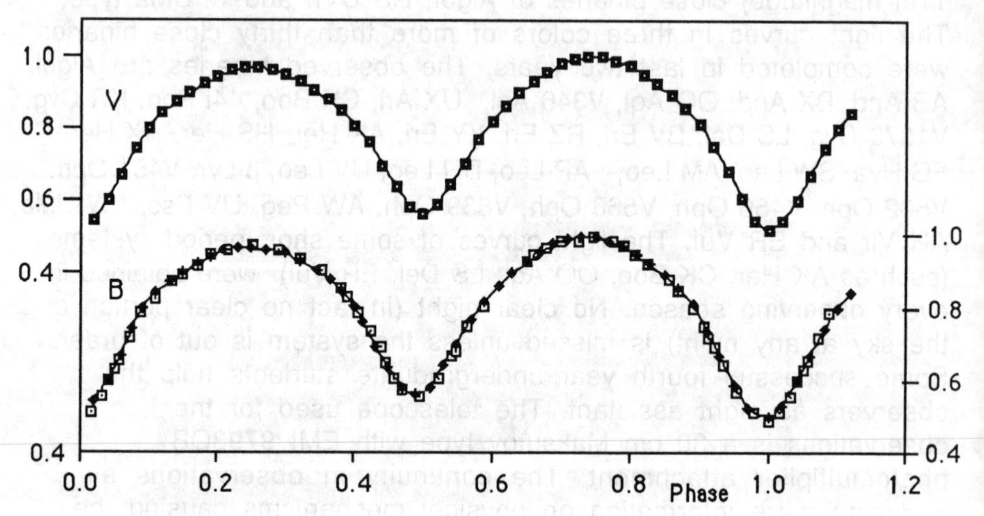

Figure 1. Fit to the YY Eri normal point light curves in V and B.

It is important to note that in our observations the levels of the successive maxima are not equal, and difference in depth between primary and secondary minimum is about 0.068 magnitude in V. By an inspection of all available light curves it was seen that the maximum and minimum levels of the light curves are changing continuously. The depths of the successive minima, and difference between the light levels of successive maxima and minima for different light curves in V are given in Table 1. It is evident from Table 1 that the light distribution over the surfaces of the component stars is not homogeneous which produces first the unequal maxima.

Table I. The light curve variations of YY Eri in V filter.

	Cillie (1951)	Huruhata (1953)	Purgathofer (1960)	Eaton (1976)	Present (1988)
MinI- MinII	0.070	0.076	0.059	0.081	0.068
MaxI-MaxII	0.000	-0.010	-0.010	0.000	-0.030
MaxI- MinI	0.693	0.699	0.693	0.700	0.675
MaxI- MinII	0.621	0.623	0.634	0.619	0.607
MaxII- MinI	0.693	0.688	0.683	0.700	0.706
MaxI- MinII	0.621	0.612	0.624	0.619	0.638

Moreover the variations of the light levels at maxima and minima in time require the variations (in area, temperature and location) of the inhomogeneities.

In our observations the difference at maxima level is greatest. There could be a cool spot region (as it was assumed by Eaton, 1986) on the more massive star seen face on at phase 0.25. For the solution we first normalized our light curves to the secondary maximum. Then taking the solution of Nesci et al. as the initial parameters we applied Mode 3 of the Wilson-Devinney code. A good fit is obtained to our observations (except around phase 0.25). By assuming a circular cool spot on the more massive component we further minimize the χ^2 and obtained better fit to the observations. Our solution with spot model are given in Table 2 together with the solutions of Eaton and Nesci et al. Figure 1 show the fitting of the synthetic light curves to our normal points in V and B filters. It is interesting that we obtained the best fit with $q = 0.58$ although

Table II. Synthetic light curve parameters of YY Eri.

Parameter	Eaton	Nesci et al.	Present work
i(degrees)	80.8	82.5	79.4
$q(=m_c/m_h)$	0.50	0.401	0.58
$\Omega_h=\Omega_c$	- - -	5.839	4.738
T_h(°K)	5850	5600	5600
T_c(°K)	5675	5362	5397
$L_h/(L_h+L_c)$	- - -	0.351	0.423
$L_c/(L_h+L_c)$	- - -	0.649	0.577
r_h	---	0.311	0.345
r_c	- - -	0.469	0.439
f(% overcontact)	84	15	18
T_{spot}(°K)	4610	- - -	5084
χ^2	- - -	- - -	0.0032

we tried the spectroscopic value of 0.40 . The radius of the circular spot in our solution is 26°, and spot center is on the equator. The radii in Table 2. (and thereafter) are the mean radii of the respective Roche figures. Our solutions is by no means final, but shows that unequal temperature requirement for the components is not enough to have a good fit, to the present light curves of this system. A model to account unequal maxima on the light curves should be considered.

3. V456 OPHIUCHI

V456 Ophiuchi (BD+ 08°3814) is a detached binary of A5 spectral type (Kukarkin et al., 1969). It seems the first photoelectric minimum observations of this binary were obtained in 1973 by one us (O.D.) in BV filters (cf. Kızılırmak and Pohl, 1974,1975). The minimum times were delayed with respect to the given photographic ephemeris in the General Catalogue of Variable Stars (GCVS), by about four hours. New light elements, determined by using mostly our observations, was given by Diethelm (1981), as

$$\text{Min I} = \text{Hel.JD. } 2441897.532 + 1^d.0159996E.$$

The GCVS 1985 edition cites Diethelm's light elements and notes a variable period for the system. No spectroscopic observations exists for V456 Oph. We completed first photoelectric light curves of the system in B and V filters in 1988. By applying a simple spherical model approach to the primary eclipse observations we found the first approximations to the eclipse elements which identify the primary minimum as a transit eclipse. Thus the system is not a classical algol since the larger star is hotter and more luminous. This is possible only if the component stars of the system are not evolved off the main sequence in the H-R diagram.Assuming the spherical model

Table III. Synthetic light curve parameters of V456 Oph

Parameter	W-D model	Spherical Model
i(degrees)	89.5	89.9
$q(=m_c/m_h)$	0.914	---
Ω_h	4.9255	---
Ω_c	5.0446	---
T_h(°K)	8500	---
T_c(°K)	6200	---
$L_h/(L_h+L_c)$	0.65	0.63
$L_c/(L_h+L_c)$	0.34	0.37
r_h	0.253	0.250
r_c	0.232	0.236
χ^2	0.0031	0.00012

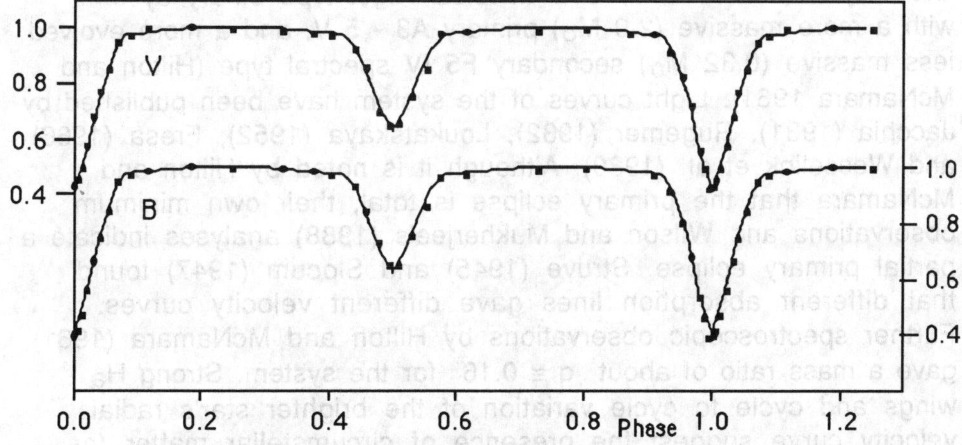

Figure 2. Fit to the V456 Oph normal point light curves in V and B

solution as the initial set of parameters we applied Wilson-Devinney (1971) code to our observations. The mass ratio of the system was estimated to be $q = 0.94$ by using an empirical relation between the ratio of radii k and mass ratio q for the systems with MS components. The relation is given by Demircan and Kahraman (1989) as $k = q^{0.935}$. However, we applied the Wilson-Devinney code for four different initial values of q (=0.2, 0.5, 0.7 and 0.9). The effective temperature of hotter component was assumed 8500°K (according to the given spectral type A5 of the system in the GCVS 1969) as the initial value. We obtained the same geometrical elements from B and V observations. Our final solution for V and B observations are given in Table 3, together with the solution by simple spherical model approach. The normal observational points and best theoretical fit which are obtained by the final solution are shown in Figure 2. for the V light curve. The final value of q in the solutions confirm that the components of the system are both in the MS. It is interesting that the resulting elements from the simple spherical model approach agree well with the final solution obtained by the much sophisticated model of Wilson-Devinney (1971). Although the eclipse observations look partial, the solution tells us that V456 Oph is a totally eclipsing system but total phase lasts only about 9 minutes.

4. AW PEGASI

AW Pegasi (HD 207956) is a classical Algol type binary system with a more massive (2.0 $M_{\odot}$) primary A3 - 5 V, and a more evolved less massive (0.32 $M_{\odot}$) secondary F5 IV spectral type (Hilton and McNamara 1961). Light curves of the system have been published by Jacchia (1931), Rugemer (1932), Loukatskaya (1952), Fresa (1966) and Wesselink et al. (1980). Although it is noted by Hilton and McNamara that the primary eclipse is total, their own minimum observations and Wilson and Mukherjee's (1988) analyses indicate a partial primary eclipse. Struve (1945) and Slocum (1947) found that different absorption lines gave different velocity curves. Further spectroscopic observations by Hilton and McNamara (1961) gave a mass ratio of about $q \cong 0.16$ for the system. Strong H_{a} wings and cycle to cycle variation of the brighter star's radial velocity curve suggest the presence of circumstellar matter (or probably a disc) around hotter component. An analysis of the Fresa's

(1966) observations by Fresa and Cester et al. (1978) resulted with almost same elements which are quite different from the elements obtained by Wilson and Mukherjee (1988). The binary was observed mostly by one of us (E.D.) in UBV filters at Ankara University Observatory. We solved B and V light curves by using the Wilson-Devinney (1971) code. First, we saw that the theoretical

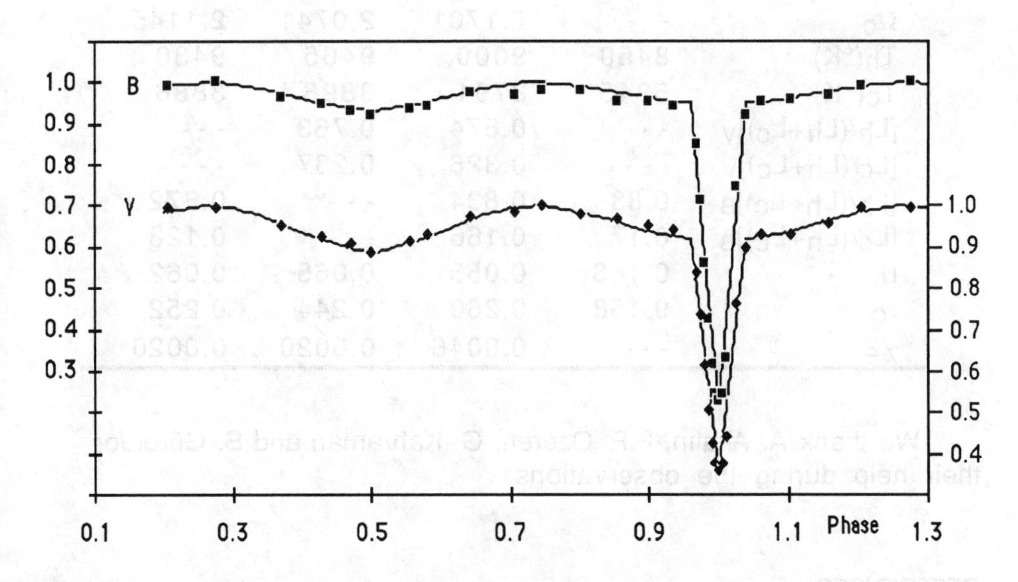

Figure 3. Fit to the V456 Oph normal-point light curves in V and B.

light curve formed by Cester et al.'s solution do not fit at all with our observations. Secondly, we confirmed that the secondary component fills its Roche lobe completely. Thus at least a slow rate mass transfer trough the first Lagrangian point may be present. Our final solutions for V and B light curves are given in Table 4, together with the solutions of Wilson and Mukherjee, and Cester et al. It is seen that our solutions for V and B light curves not much different from each other and from Wilson and Mukherjee's solutions. However, light contributions of the components are different, because we obtained slightly (≈400 °K) hotter primary. Figure 3 show the fitting of the synthetic light curves to our normal points in V and B filters.

Table IV. Synthetic light curve parameters of AW Peg

Parameter	Cester et al.	W&M (B,V)	Present work V	B
i(degrees)	86.9	79.05	79.80	79.73
q ($=m_c/m_h$)	0.16	0.1753	0.1394	0.1542
Ω_h	---	20.23	18.173	18.685
Ω_c	---	2.1701	2.0741	2.1145
T_h(°K)	8450	9000	9465	9430
T_c(°K)	5845	3791	3896	3886
$[L_h/(L_h+L_c)]_V$	---	0.674	0.763	---
$[L_c/(L_h+L_c)]_V$	---	0.326	0.237	---
$[L_h/(L_h+L_c)]_B$	0.83	0.834	---	0.872
$[L_c/(L_h+L_c)]_B$	0.17	0.166	---	0.128
r_h	0.166	0.055	0.065	0.062
r_c	0.158	0.260	0.244	0.252
χ^2	---	0.0046	0.0020	0.0020

We thank A. Akalın, F.F. Özeren, G. Kahraman and B. Gürol for their help during the observations.

REFERENCES:

Binnendijk, J.:1965, Astron. J., **70**, 209.
Cester, B., Fedel, B., Giuricin, G., Mardirossian, F., Mezzetti, M.:1978, Astron. Astrophys., **62**, 291.
Cillie, G. G.:1951, Harvard Coll. Obs. Bull., **920**, 41.
Demircan, O., Kahraman, G.:1989, in preparation.
Diethelm, R.:1981, BBSAG Bull. no:57.
Eaton, J. A.:1986, Acta Astronomica, **36**, 79.
Fresa, A.:1966, Osservatorio Astronomica di Capodimonte, Ser.II, Vol.V, No: 16.
Hilton, W. B., McNamara, D. H.:1961, Astrophys. J., **134**, 839.
Hoffmeister, C.:1935, Astronomische Nachrichten, **255**, 405.
Huruhata, M., Dambara, T., Kitamura, M.:1953, Annual Tokyo Ast. Obs. Sec. Ser. Vol III, 4, 227.

Jacchia, L.:1931, Beobachtungs-Zirkular der Astronomischen Nachrichten, **3**, 55.
Kızılırmak, A., Pohl, E.:1974, IBVS no: 937.
Kızılırmak, A., Pohl, E.:1975, IBVS no: 1053.
Kukarkin, B. V., et al.:1969, GCVS, Vol. II, Moscow
Loukatskaya, F. I.:1952, Variable Stars, **9**, 57.
Nesci, R., Maceroni, C., Milano, L., Russo, G.:1986, Astron. Astrophys., **159**, 142.
Purgathofer, A. and I.:1960, Mitt. Univ. Sternwarte Wien, no:10, 211.
Rugemer, H.:1932, Beobachtungs-Zirkular der Astronomischen Nachrichten, **14**, 74.
Slocum, L. F.:1947, Astrophys. J., **105**, 350.
Struve, O.:1945, Astrophys. J., **102**, 74.
Struve, O.:1947, Astrophys. J., **106**, 92.
Wesselink, T., Van Paradijs, J., Steller, R. F. A., Meurs, E. J. A., Kester, D.: 1980, IBVS no:1800 .
Wilson, R. E., Devinney, E. J.:1971, Astrophys. J., **166**, 606.
Wilson, R .E., Mukherjee, J.:1988, Astron. J., **96**, 747.

CIRCULARIZATION, SYNCHRONIZATION, AND DIFFERENTIAL ROTATION IN CHROMOSPHERICALLY ACTIVE BINARIES

Douglas S. Hall
Gregory W. Henry
Center of Excellence in Information Systems
Tennessee State University
Nashville, Tennessee 37203
U.S.A.

ABSTRACT. Our data base includes chromospherically active binaries for which P_{ROT} is known accurately from photometry of the starspot wave. The SU UMa binaries are hypothesized to have starspots, in which case $P_{SUPEROUTBURST} = P_{MIGRATION}$.

Evolved binaries are shown to be in excellent agreement with the new circularization theory of Tassoul; the evolved star's radius cleanly separates the eccentric from the circular orbits.

Several binaries with eccentric orbits appear to be in pseudosynchronous rotation, in agreement with theory, but two or three systems surprisingly show $P_{ROT} = P_{ORB}$ and $P_{ROT} \neq P_{PSEUDO}$.

About 20 % of our sample rotate asynchronously, with $\Delta P > 10\%$. Tassoul's new synchronization theory predicts that evolved stars (base of the giant branch or above) should rotate synchronously unless the $P_{ORB} > 1$ yr. The several examples of grossly asynchronous rotation in binaries with $P_{ORB} \approx 1$ month (HD 181809, λ And, AY Cet) are therefore a puzzle. We suggest differential rotation with depth as an explanation, in which a co-rotating depth would be analagous to the previously defined co-rotating latitude. This suggests using observed (surface) rotation as a probe of internal rotation profiles.

Differential rotation as a function of latitude is examined for a large number of stars, with the coefficient k deduced from the small differences between observed P_{ROT} and P_{ORB} in synchronously rotating systems; k decreases sharply with decreasing Rossby number Ro. This is contrary to the prediction of linear dynamo theory (k proportional to 1/Ro) but consistent with non-linear dynamo theory, which expects k to be nearly suppressed by extremely rapid rotation. For the SU UMa systems, k is 600 times smaller than the solar value $k_{\odot} = 0.189$.

1. Introduction

We investigate circularization, synchronization, and pseudosynchronization, in chromospherically active (CA) binaries and differential

C. İbanoğlu (ed.), Active Close Binaries, 287–307.

rotation in CA binary and single stars. Each of these four has a major section in this paper. Previous discussions of some of these phenomena, based on fewer data and older theories, include Zahn (1977), Hut (1981), Scharlemann (1982), Giuricin, Mardirossian, and Mezzetti (1984ab, 1985), Hall (1986ab), and Habets and Zwaan (1989). Recently a newer theory has been developed for synchronization (Tassoul 1987) and circularization (Tassoul 1988) but not yet compared extensively with observation, at least not for binaries containing evolved late-type stars.

Although these four phenomena can be discussed for binaries in general, we consider only the CA stars because their rotation periods can be determined very precisely by photometry of the light variation produced by longitudinally uneven starspot coverage (Hall 1987). Vsini determinations, on the other hand, result in rotation periods which are relatively imprecise, especially for slowly rotating stars, and require that the star's radius and inclination be known or (as is usually the case) assumed.

CA binaries are additionally useful in discussing these four phenomena. First, although they do not comprise a group which is homogeneous in evolutionary state or chronological age, most of them do contain one evolved star (Hall 1987). Consequently, that star's position in the HR Diagram (or, equivalently, its radius) can indicate the age of the binary. An individual star evolves so little while on the main sequence that (unless it happens to be a cluster member) its age is practically indeterminate, i.e., it may lie anywhere within its (long) main-sequence lifetime. Second, it turns out that CA binaries, especially those of the RS CVn type, are statistically likely to be eclipsing SB2 systems (Morgan and Eggleton 1979). That means many of them will have directly determined masses and radii, which is useful in quantitative comparison between theory and observation.

In our investigation our data base included the 72 CA stars, binary and single, observed spectroscopically and photometrically from South Africa by Balona (1987), Collier Cameron (1987ab), Lloyd Evans and Koen (1987), and Collier Cameron, Lloyd Evans, and Balona (1985), the 168 CA binaries and 37 candidate CA binaries in the catalogue of Strassmeier et al. (1988), the 5 rapidly rotating (most of them CA) single stars for which Strassmeier and Hall (1988) analyzed four years of photometry, the 49 binaries (most of them CA) for which Strassmeier et al. (1989) analyzed 4 years of photometry, and the 50 suspected variables (binary and single, most of them CA) for which Hooten (1989) and Hooten and Hall (1990) analyzed available photometry spanning up to 11 years in a few cases. These were updated in a few cases with more recent references. It goes without saying that many of the stars mentioned above appear in more than one of the references.

Our investigation of differential rotation includes other types of stars, both binary and single, which are known (Hall 1987) or hypothesized (in this paper) to be CA: Algol binaries, W UMa binaries, SU UMa binaries, BY Dra variables, FK Com variables, T Tau stars, and the sun.

2. Circularization

Some binaries have eccentric orbits and some do not. We want to understand why. Our basic approach is as follows. It is generally presumed that any eccentricity present at the origin of a binary decays with an e-folding circularization time scale τ_{CIRC}. In most circularization theories τ_{CIRC} is most strongly dependent on R/a, where R is the radius of the larger star and a is the semi-major axis of the orbit. Thus the current eccentricity of a binary's orbit should be determined by its original eccentricity (which cannot exceed unity), the binary's age, and a time integral of some function of R/a. For stars less than 3 $M_\odot$ or so, evolution after the Z.A.M.S. involves an expansion in radius which is monotonic with time, except for a minor glitch or two. In the absence of significant mass transfer (i.e., before Roche lobe overflow) or mass loss, there is little change in a. That means the current eccentricity of a binary containing a main-sequence or evolved star less than about 3 $M_\odot$ can be expressed as a function of the original eccentricity, the value of a, and the current value of R.

We want to compare observation with the new circularization theory of Tassoul (1988), in which

$$\tau_{circ} = 94000 \cdot 10^{-N/4}\ (1+q)^{2/3}\ g^{-2}\ L^{-1/4}\ M^{23/12}\ R^{-5}\ P^{49/12}, \quad (1)$$

where τ_{CIRC} is in years. The value of N is 0 if the outer envelope of the larger star is radiative and 10 if convective, q is mass ratio in the sense smaller star to larger star, g is the relative radius of gyration of the larger star, and P is orbital period in days. L, M, and R are the luminosity, mass, and radius of the larger star, all in solar units. In this version of Tassoul's formulation, the parameter a has been removed with the help of Kepler's Third Law.

Evolutionary tracks of Maeder and Meynet (1989) give R and L as a function of time for stars of various M and also tell whether a star is radiative or convective at any given evolutionary stage. For a given M and P (assumed constant), an assumed value of q = 1, and values of N and g appropriate for a given evolutionary stage, we have used numerical integration to determine the time t when a binary's eccentricity should have dropped to 1% of its original value. This would correspond to circularization, to within the limits of observational error. The radius at this time, which we call R_{CRIT}, was also noted. This process was repeated for values of M between 1 $M_\odot$ and 3 $M_\odot$ and for values of P between 0.3 day and 300 days.

The results are summarized in Figure 1, a plot of R_{CRIT} versus P_{ORB}. The smallest value of P_{ORB} shown corresponds to a contact binary on the Z.A.M.S. Although only two values of M are shown, for clarity, we actually had results for 1, 1.15, 1.3, 1.5, 1.7, 2, 2.5, and 3 solar masses. For a given value of M and P_{ORB}, one can read R_{CRIT}. The meaning of all this is as follows. If the larger star in an evolved binary has a radius larger than R_{CRIT}, its eccentricity should have decayed to

1% of its original value or smaller. If its radius is smaller than R_{CRIT}, then it should have still have an appreciable fraction of its original eccentricity. The small-slope portion of each curve results from slow evolution on the main sequence, the steep-slope portion results from rapid evolution across the Hertzsprung gap, and the intermediate-slope portion results from evolution on the giant branch, which is intermediate in speed. The large separation between the two steep-slope portions results from the fact that stars of 1.3 $M_\odot$ and less are convective while on the main sequence (N = 10) whereas stars of 1.5 $M_\odot$ and more are not (N = 0).

To compare theory with observation, we considered only binaries with one or more evolved (subgiant or giant) component. For all those observed to have appreciably eccentric orbits (e > 0.05) we got M and R from available sources, mostly from the Strassmeier et al. (1988) catalogue but also from a few more recent sources, such as Popper (1988). Values of $M\sin^3 i$ and $R\sin i$ were used when an estimate of i was known. If only a luminosity class was known, we assigned 4 $R_\odot$ to class IV, 7 $R_\odot$ to class IV-III, and 10 $R_\odot$ to class III. For the few Algol-type (i.e., semi-detached) CA binaries like RZ Cnc, AR Mon, and RT Lac, we used $M = (M_1+M_2)/2$, effectively assuming the mass transfer had been

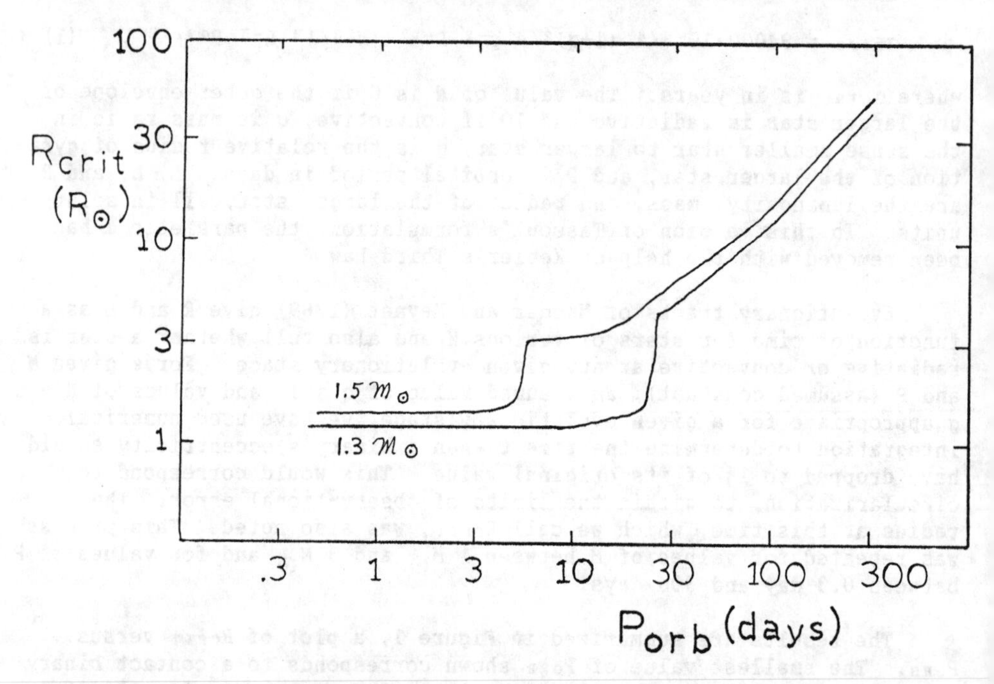

Figure 1. R_{CRIT} versus P_{ORB}, both plotted on a log scale. The beginning of each curve corresponds to a contact binary on the Z.A.M.S.; the end corresponds to $R_{CRIT} = R_{ROCHE}$ for the evolved star. If $R > R_{CRIT}$, the orbit should be circularized; if $R < R_{CRIT}$, it should be eccentric.

conservative. There are so many more binaries with effectively circular ($e < 0.05$) orbits that we could restrict our sample to those with directly determined masses and radii or with $M\sin^3 i$, $R\sin i$, and i known.

The result of the comparison is shown in Figure 2, a plot of observed eccentricity versus the ratio R/R_{CRIT}. Different symbols differ-

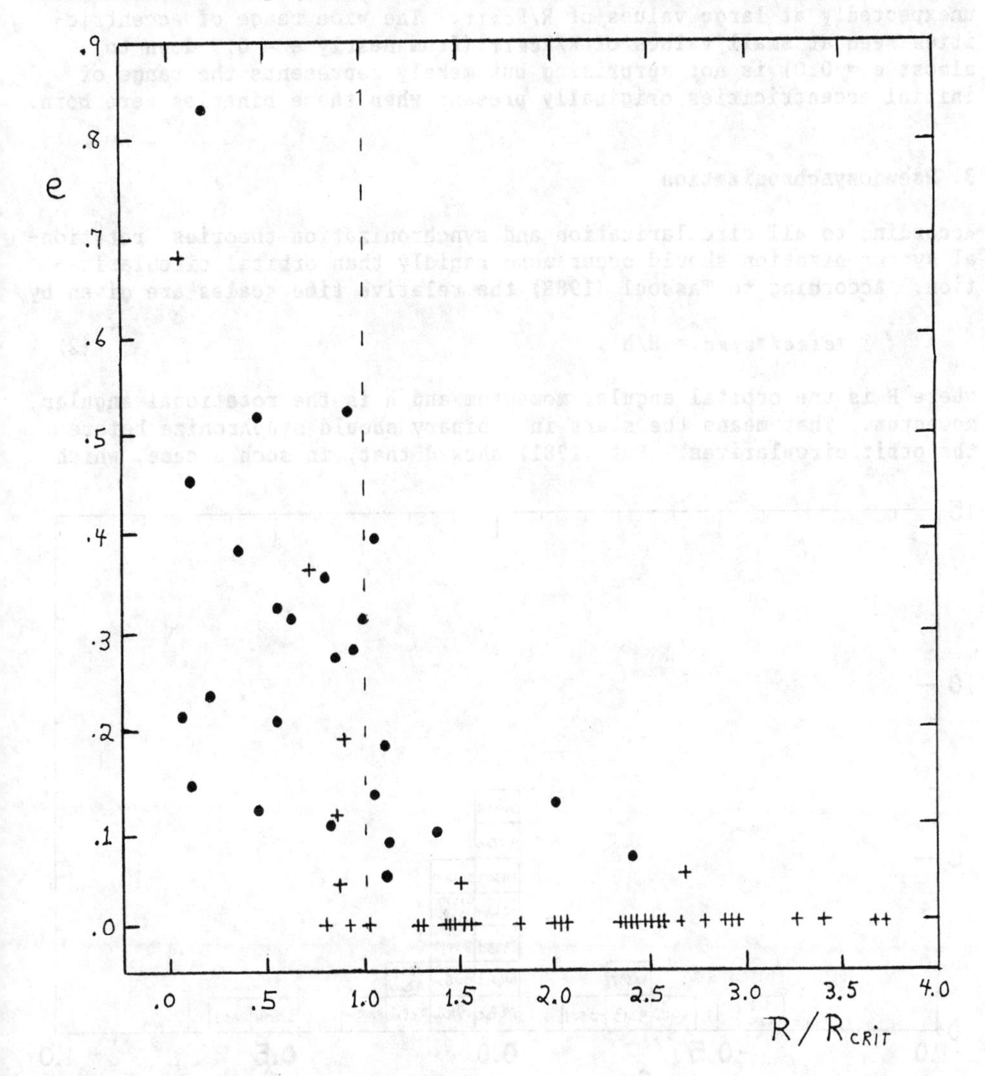

Figure 2. Observed eccentricity versus the ratio R/R_{CRIT}. Crosses are binaries with well-determined masses and radii; filled circles are those with estimated masses and radii. The dashed line at $R = R_{CRIT}$ should separate binaries with circular orbits (to the right) from those with eccentric orbits (to the left). It does.

entiate between those with directly determined values of M and R and those with R estimated from the luminosity class. The dashed line at $R/R_{CRIT} = 1$ should, if the Tassoul circularization theory is correct, rather cleanly separate binaries with eccentric orbits (to the left) from those with circular orbits (to the right). The separation indeed is remarkably clean. Only one or two non-zero eccentricities occur unexpectedly at large values of R/R_{CRIT}. The wide range of eccentricities seen at small values of R/R_{CRIT} (from nearly e = 0.9 down to almost e = 0.0) is not surprising but merely represents the range of initial eccentricities originally present when these binaries were born.

3. Pseudosynchronization

According to all circularization and synchronization theories, rotational synchronization should occur more rapidly than orbital circularization. According to Tassoul (1988) the relative time scales are given by

$$T_{CIRC}/T_{SYNC} = H/h \,, \tag{2}$$

where H is the orbital angular momentum and h is the rotational angular momentum. That means the stars in a binary should synchronize before the orbit circularizes. Hut (1981) showed that, in such a case, which

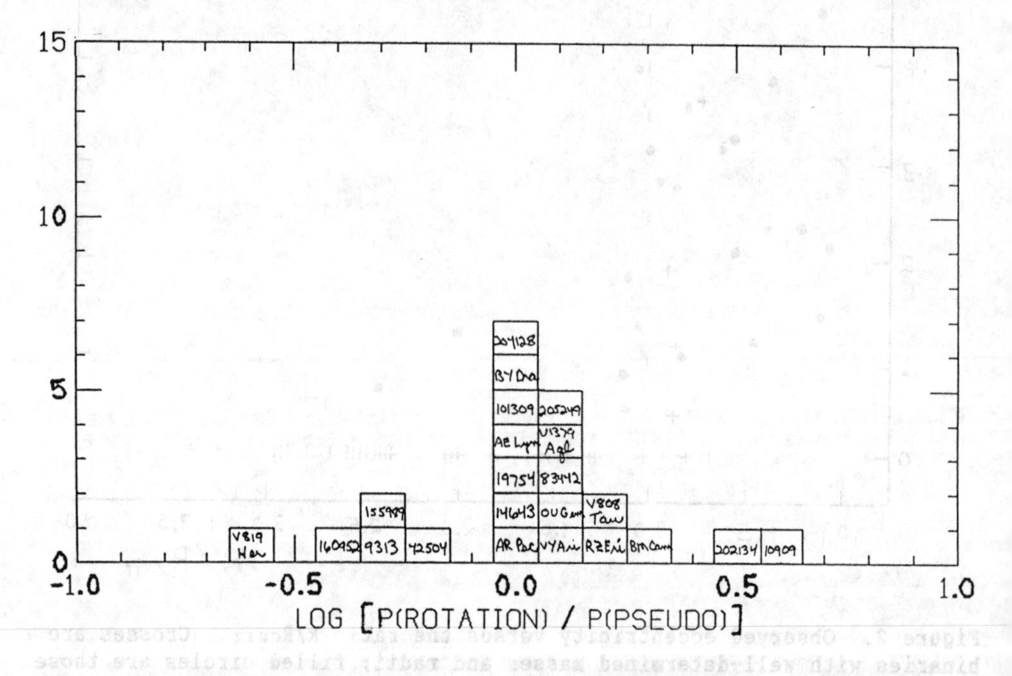

Figure 3. A histogram of P_{ROT}/P_{PSEUDO} for binaries with significantly eccentric (e > 0.05) orbits. The spike around $P_{ROT}/P_{PSEUDO} = 1$ would indicate those systems rotating pseudosynchronously.

has been called pseudosynchronous rotation, the rotational period is not equal to the mean orbital period (as it is in a circular orbit) but is shorter. He gave an expression for P_{PSEUDO}/P_{ORB} as a function of the orbital eccentricity (Hut 1981, eq. 42). It turns out that P_{PSEUDO} is nearly equal to the effective orbital period at periastron.

Pseudosynchronism is observed in many early-type binaries (Giuricin, Mardirossian, and Mezzetti 1985) and in several CA binaries with eccentric orbits (Hall 1986b).

We look for pseudosynchronism in our larger sample of CA binaries by plotting a histogram of P_{ROT}/P_{PSEUDO} for those with significantly eccentric (e > 0.05) orbits and with P_{ROT} known from photometry. This is Figure 3. The spike at $P_{ROT} = P_{PSEUDO}$ should be those exhibiting pseudosynchronous rotation, and in fact many of them are ones mentioned earlier by Hall (1986b).

Something curious happens, however, when we plot a histogram of P_{ROT}/P_{ORB} for the same CA binaries with eccentric orbits, in Figure 4. There is spike here also, at $P_{ROT} = P_{ORB}$. If pseudosynchronous theory is correct, then the orbital period should no longer be special and there should be no spike at that value. To be honest, we should point

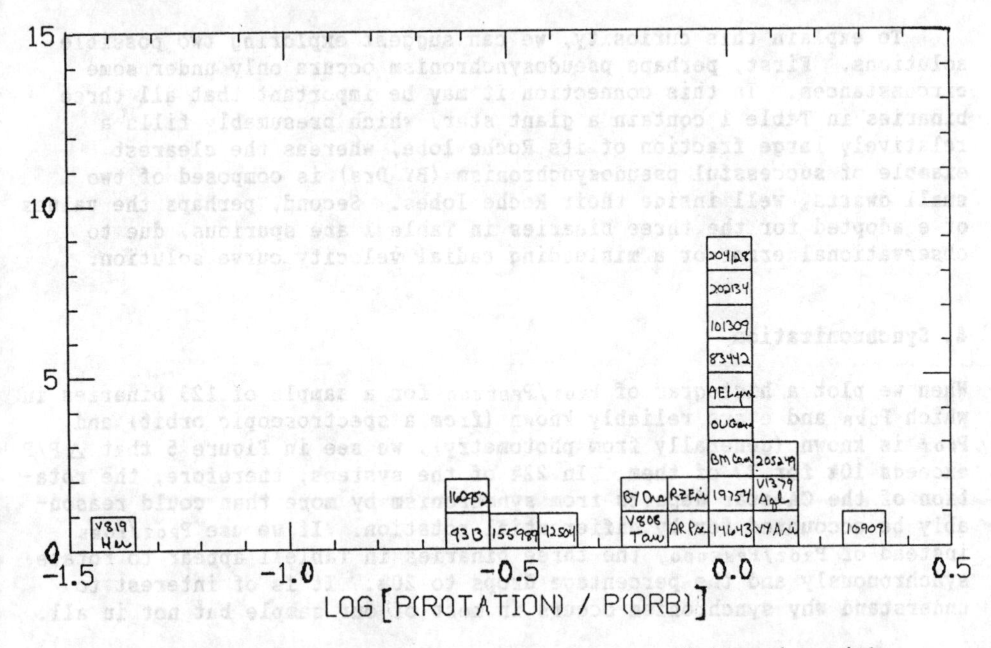

Figure 4. A histogram of P_{ROT}/P_{ORB} for the same 22 binaries with eccentric orbits. The spike around $P_{ROT}/P_{ORB} = 1$ seems to indicate several systems have synchronized with the orbital period, contrary to the prediction of pseudosynchronous rotation theory. See Table 1.

Table 1

Three Examples of Non-Pseudosynchronous Rotation

Binary	sp.tp.	ecc.	P_{ROT}	P_{ORB}	P_{PSEUDO}
BM Cam	K0 III	0.35	81^{d}.75	79^{d}.93	44^{d}.88
HD 83442	K2 IIIp	0.13	54.95	52.27	47.45
HD 202134	K1 IIIp	0.52	61.73	63.09	20.91

out that several stars are in both spikes; these are the ones with relatively small eccentricities, in which case the test is not sensitive. There are, however, two and maybe three clear cases where P_{ROT} is equal to P_{ORB} to high precision and very different from the expected value of P_{PSEUDO}. Details on these three are given in Table 1, taken from Hall and Busby (1989) for BM Cam and from Strassmeier et al. (1988) for the other two.

To explain this curiosity, we can suggest exploring two possible solutions. First, perhaps pseudosynchronism occurs only under some circumstances. In this connection it may be important that all three binaries in Table 1 contain a giant star, which presumably fills a relatively large fraction of its Roche lobe, whereas the clearest example of successful pseudosynchronism (BY Dra) is composed of two small dwarfs, well inside their Roche lobes. Second, perhaps the values of e adopted for the three binaries in Table 1 are spurious, due to observational error or a misleading radial velocity curve solution.

4. Synchronization

When we plot a histogram of P_{ROT}/P_{PSEUDO} for a sample of 123 binaries in which P_{ORB} and e are reliably known (from a spectroscopic orbit) and P_{ROT} is known (generally from photometry), we see in Figure 5 that $\Delta P/P$ exceeds 10% for 27 of them. In 22% of the systems, therefore, the rotation of the CA star departs from synchronism by more than could reasonably be accounted for by differential rotation. If we use P_{ROT}/P_{ORB} instead of P_{ROT}/P_{PSEUDO}, the three binaries in Table 1 appear to rotate synchronously and the percentage drops to 20%. It is of interest to understand why synchronism occurs in most of our sample but not in all.

It was pointed out already (Hall 1986a) that orbital period alone is a poor predictor of asynchronous rotation if used as a single parameter. There is <u>synchronous</u> rotation in many binaries with quite <u>long</u> orbital periods and grossly <u>asynchronous</u> rotation in several binaries with moderately <u>short</u> orbital periods. For example P_{ROT}/P_{ORB} = 1.05 in

HR 7428, with P_{ORB} = 108^{d}57 (Gessner 1989), whereas P_{ROT}/P_{ORB} = 4.67 in HD 181809, with P_{ORB} = 13^{d}05 (Hall and Pazzi (1987).

We want to compare observation with the new synchronization theory of Tassoul (1987), in which

$$\tau_{SYNC} = 535 \cdot 10^{-N/4}\ (1+q)q^{-1}\ L^{-1/4}\ M^{5/4}\ R^{-3}\ P^{11/4}, \quad (3)$$

where τ_{SYNC} is the e-folding synchronization time scale in years (although Tassoul used the symbol t_{SD}). The value of N is 0 if the outer envelope of the larger star is radiative and 10 if convective, q is mass ratio in the sense smaller star to larger star, and P is orbital period in days. L, M, and R are the luminosity, mass, and radius of the larger star, all in solar units.

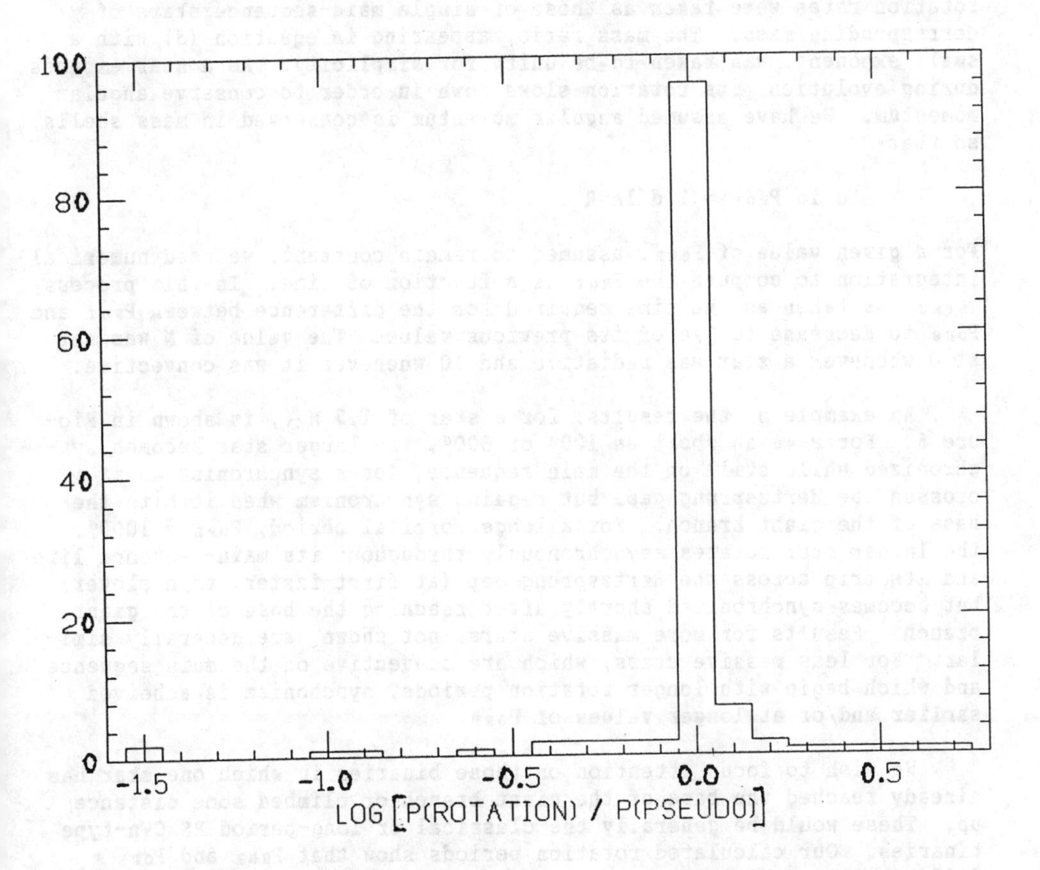

Figure 5. A histogram of P_{ROT}/P_{PSEUDO} for a sample of 123 binaries, with circular and eccentric orbits. The 96 within the central spike are rotating synchronously or pseudosynchronously, within ± 10%, whereas the other 27 are clearly rotating asynchronously.

It can be seen immediately that orbital period is only one of several parameters. Many factors act to determine the present rotation rate of a star in an evolved binary. Even though mass appears in equation (3) with an exponent not much different from unity, mass strongly influences the rotation rate of a star on the main sequence: around one month for convective stars but only one day or less for radiative stars. The times spent in various evolutionary stages (main sequence, Hertzsprung gap, giant branch, etc.) differ by orders of magnitude. And the radius, which appears in equation (3) with a large exponent, changes by at least two orders of magnitude as a star evolves from the main sequence to the tip of the giant branch.

We have used the evolutionary models of Maeder and Meynet (1989) to specify R and L as functions of time for stars of various mass. Initial rotation rates were taken as those of single main-sequence stars of corresponding mass. The mass ratio, appearing in equation (3) with a small exponent, was taken to be unity for simplicity. As a star expands during evolution, its rotation slows down in order to conserve angular momentum. We have assumed angular momentum is conserved in mass shells, so that

$$d \ln P_{ROT} = 2 \, d \ln R \, . \tag{4}$$

For a given value of P_{ORB}, assumed to remain constant, we used numerical integration to compute the P_{ROT} as a function of time. In this process, τ_{SYNC} was taken as the time required for the difference between P_{ROT} and P_{ORB} to decrease to 1/e of its previous value. The value of N was set at 0 whenever a star was radiative and 10 whenever it was convective.

An example of the results, for a star of 1.7 $M_\odot$, is shown in Figure 6. For P_{ORB} as short as 100^d or 300^d, the larger star becomes synchronized while still on the main sequence, loses synchronism as it crosses the Hertzsprung gap, but regains synchronism when it hits the base of the giant branch. For a longer orbital period, $P_{ORB} = 1000^d$, the larger star rotates asynchronously throughout its main-sequence life and its trip across the Hertzsprung gap (at first faster, then slower) but becomes synchronized shortly after reaching the base of the giant branch. Results for more massive stars, not shown, are generally similar. For less massive stars, which are convective on the main sequence and which begin with longer rotation periods, synchonism is acheived earlier and/or at longer values of P_{ORB}.

We wish to focus attention on those binaries in which one star has already reached the base of the giant branch or climbed some distance up. These would be generally the classical or long-period RS CVn-type binaries. Our calculated rotation periods show that P_{ROT} and P_{ORB} differ by less than 1% by the time the evolved star has reached the base of the giant branch, models 14 or 15 of Maeder and Meynet (1989), provided the orbital period is less than about 1 year. The actual limit is somewhere between 300 and 450 days, depending on the assumed mass. Then, because radius increases sharply as a star evolves up the giant branch, synchronism becomes progressively tighter as time goes on. In the limit where $P_{ROT} \approx P_{ORB}$, it can be shown that

$$\Delta P/P_{orb} = 2\ \tau_{SYNC}\ d\ln R/dt \tag{5}$$

is a good approximation, where $\Delta P = P_{ROT} - P_{ORB}$. Figure 7 illustrates this progressive tightening of synchronism. Each curve for a given value of P_{ORB} begins with the larger star at the base of the giant branch and terminates when that star fills up its Roche lobe. If a smaller value q = 0.5 had been used instead of q = 1, all of the curves would be shifted up by only 0.2 in the log of $\Delta P/P$. Although a 1.3 $M_\odot$ star is used in this example, curves for other masses between 1 $M_\odot$ and 4 $M_\odot$ would differ by less than 0.2 in the log of $\Delta P/P$. Note that $\Delta P/P$ must exceed several percent, let us say 5%, in order to indicate asynchronous rotation; a difference less than that becomes indistinguishable from the effects of differential rotation. Thus we see that a star at the base of the giant branch should be expected to rotate asynchronously only if the orbital period is longer than about 750 days, and that the minimum orbital period will be even longer if the star has evolved any distance up the giant branch.

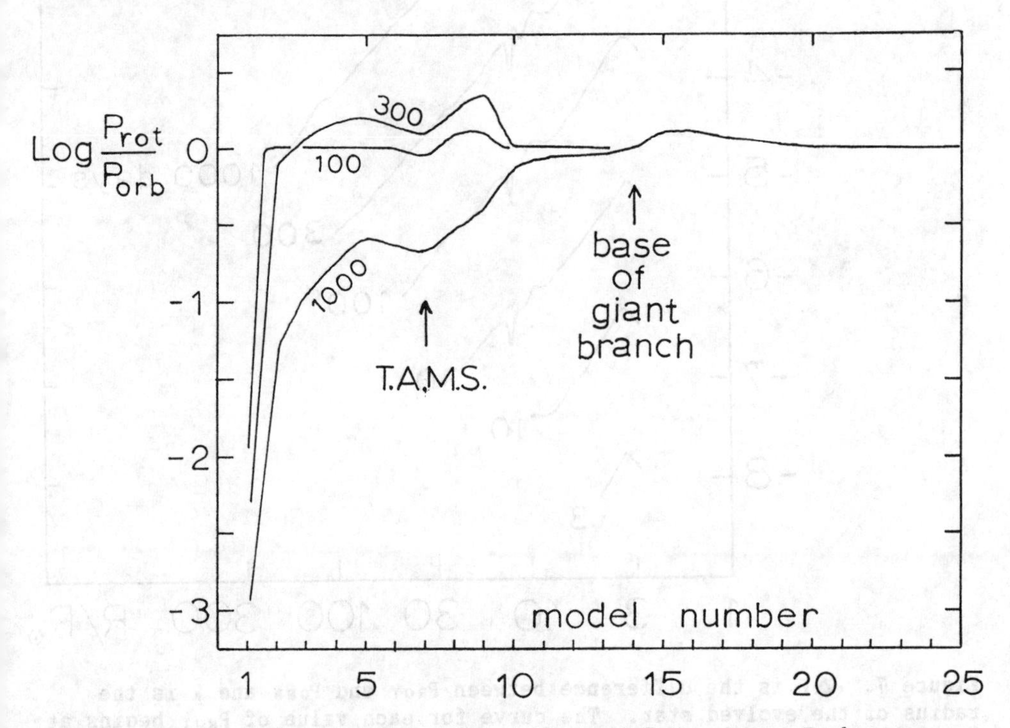

Figure 6. The evolution of P_{ROT}/P_{ORB} for a 1.7 $M_\odot$ star. Each curve begins with P_{ROT} = 1ᵈ1 on the Z.A.M.S. The abscissa is stellar model number from Maeder and Meynet (1989). The T.A.M.S. (model 7) and the base of the giant branch (model 14) are indicated.

Here we come to an unanticipated puzzle. Many of the well known examples of grossly asynchronous rotation, in which one star is a late-type giant, have orbital periods much shorter than 750 days. Three of these are HD 181809 (K1III, P_{ORB} = 13ᵈ.05, $\Delta P/P$ = 3.67), lambda And (G8IV-III, P_{ORB} = 20ᵈ.52, $\Delta P/P$ = 1.78), and AY Cet (G5III, P_{ORB} = 56ᵈ.8, $\Delta P/P$ = 0.36). If we place the evolved star in each case at the base of

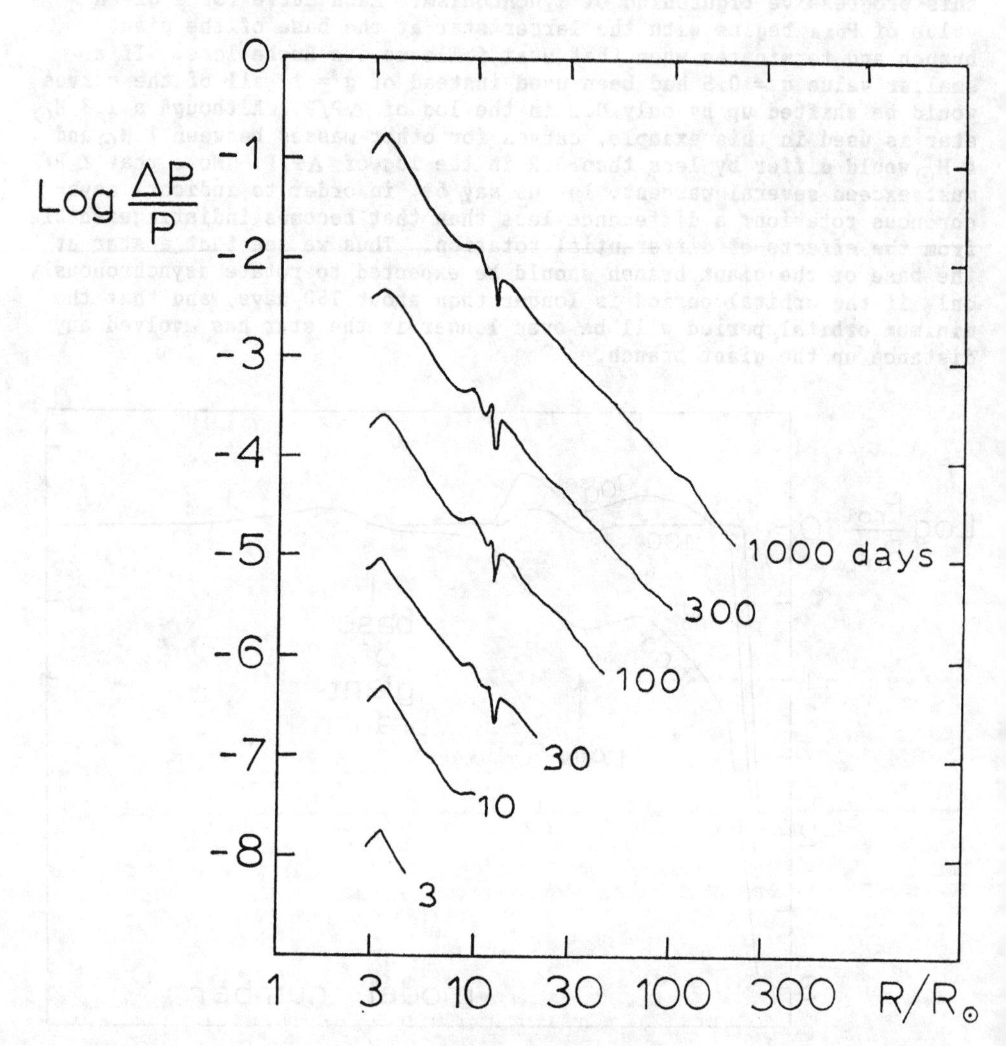

Figure 7. ΔP is the difference between P_{ROT} and P_{ORB} and R is the radius of the evolved star. The curve for each value of P_{ROT} begins at the base of the giant branch and ends with Roche lobe overflow. This example considers a star of 1.3 $M_{\odot}$ in a binary with q = 1, but the results are not sensitive to mass or mass ratio. For measurably asynchronous rotation log $\Delta P/P$ must be -1.3 or greater.

the giant branch, the expected values of $\triangle P/P$ would be only 7×10^{-7}, 3×10^{-6}, and 4×10^{-5}, respectively, many orders of magnitude less than is observed. Any evolution up the giant branch would make the expected values of $\triangle P/P$ even smaller. So, how to understand this puzzle?

Scharlemann (1982) showed that the forces which act to enforce synchronism are not strong enough to suppress differential rotation, i.e., the functional relation between rotation and stellar latitude. (Though he was working with the synchronization theory of Zahn (1977), perhaps the same conclusion would hold if the Tassoul (1987) theory had been used). His calculations supported the qualitative suggestion made years ago by Hall (1972) that a differentially rotating star in a binary system is synchronized on the average, in which case $P_{ROT} = P_{ORB}$ at a "co-rotating latitude" ϕ_{COROT}, $P_{ROT} < P_{ORB}$ at $\phi < \phi_{COROT}$, and $P_{ROT} > P_{ORB}$ at $\phi > \phi_{COROT}$. Scharlemann (1982) calculated possible values of ϕ_{COROT} under various assumptions. We propose here that this picture can be extended to include the functional relation between rotation and radial distance in towards the star's center. See Figure 8. Again the star is synchronized on the average, but now the average includes all layers. Thus there should be a co-rotating depth (r_{COROT}) at which $P_{ROT} = P_{ORB}$, with faster rotation at deeper layers ($r < r_{COROT}$) and slower rotation at shallower layers ($r > r_{COROT}$). We would guess that the "average" is such that the moment of inertia contributed by the deeper layers and the moment of inertia contributed by the shallower layers would be equal but opposite in sense. Work by others (Gray 1988) makes it clear that one should expect some redistribution of angular momentum between mass shells initially rotating at different rates as a star evolves off the main sequence, but not complete redistribution, which would result in solid-body rotation. Thus there should be differential rotation as a function of radial depth in the evolved stars we are considering here.

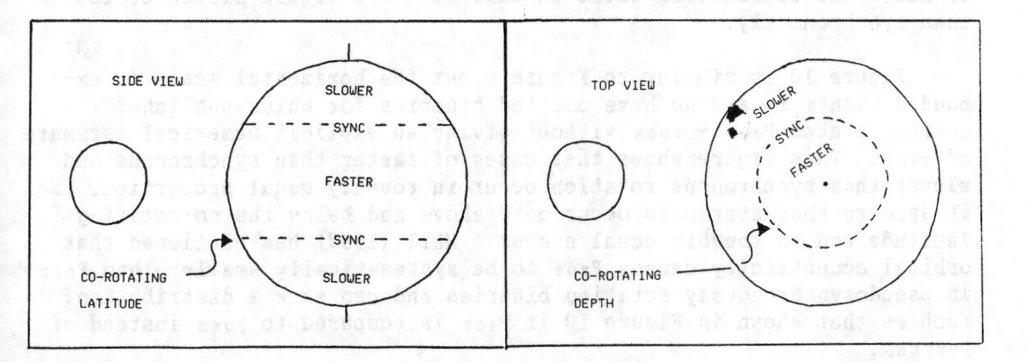

Figure 8. The grossly asynchronous rotation seen in evolved binaries like HD 181809, λ And, and AY Cet might be a result of differential rotation. The evolved star would be synchronized on the average, but deeper layers would rotate faster than synchronously and outer layers slower. The co-rotating depth would be analagous to the co-rotating latitude proposed by Hall (1972).

The important consequence of all this is that an evolved star can be rotating synchronously on the average, and hence be consistent with the tight synchronism predicted by the Tassoul (1987) theory, but be rotating much more slowly than synchronously on its surface, and hence be consistent with the slow rotation rates deduced from photometry of its spotted surface layers.

If our suggestion is basically correct, then we have a new tool for probing the internal rotation profile of a star. Although it will not be possible to go uniquely from an observed value of P/P to a rotation profile, it should be possible to use a theoretically computed rotation profile $\omega(r)$ or P(r) and mass profile M(r) to compute the ratio $P_{SURFACE}/P_{COROT}$, which should be equal to the observed ratio P_{ROT}/P_{ORB}.

5. Differential Rotation

In section 4 we showed that P_{ROT} differed from P_{ORB} (or P_{PSEUDO}) by more than 10% in about 20% of a sample of 123 stars for which P_{ROT} and P_{ORB} were both known. These binaries were judged to be rotating asynchronously. The other 80%, all 96 stars within the large spike straddling $P_{ROT}/P_{PSEUDO} = 1$ in Figure 5, can be considered rotating synchronously.

Those 96 stars are shown in another histogram, Figure 9, with a greatly expanded horizontal scale. This shows that small but non-zero differences between P_{ROT} and P_{PSEUDO} persist. The explanation first proposed by Hall (1972) is generally accepted now, namely, that we are seeing the effects of differential rotation. In this explanation the late-type star rotates differentially as a function of latitude, probably similar to the sun, and a starspot situated at any latitude above or below the co-rotating latitude will rotate a little slower or faster than synchronously.

Figure 10 is similar to Figure 9 but the horizontal scale is expanded even more and we have omitted binaries for which published sources stated $P_{ROT} \approx P_{ORB}$ without giving an explicit numerical estimate of P_{ROT}. This figure shows that cases of faster than synchronous and slower than synchronous rotation occur in roughly equal proportion. So it appears that starspots occur both above and below the co-rotating latitude and in roughly equal number. Hall (1987) has cautioned that orbital eccentricity causes P_{ROT} to be systematically smaller than P_{ORB} in pseudosynchronously rotating binaries and can skew a distribution such as that shown in Figure 10 if P_{ROT} is compared to P_{ORB} instead of P_{PSEUDO}.

It has been known for some time, even when differential rotation in the prototype RS CVn was first suggested (Hall 1972), that the degree of differential rotation is much less than that in the sun. A summary of differential rotation rates in a larger sample (Rodono 1986, figure 6) showed this to be a general phenomenon among spotted stars in binaries. Because Scharlemann (1982) showed that the forces which act to synchronize rotation in close binaries are too weak to suppress or even diminish the degree of differential rotation, we were left with no explanation for the phenomenon.

In order to investigate this phenomenon in greater detail, we have determined values of the differential rotation coefficient k, which appears in the equation

$$P_{\phi} = P_{EQ} / (1 - k \sin^2\phi) , \qquad (6)$$

for a sample of 83 binary and single stars. The method used to convert $\Delta P/P$ into k was identical to that used by Hall and Busby (1990). We used P_{PSEUDO} instead of P_{ORB} in computing $\Delta P/P$, except for the three stars in Table 1. The only exceptions were BY Dra, FF And, and FK Aqr. Starspot modelling of BY Dra has shown that its starspots cover a latitude range of virtually 90°, whereas a range of 45° had been assumed implicitly by Hall and Busby (1990, figure 2). It, FF And, and FK Aqr are

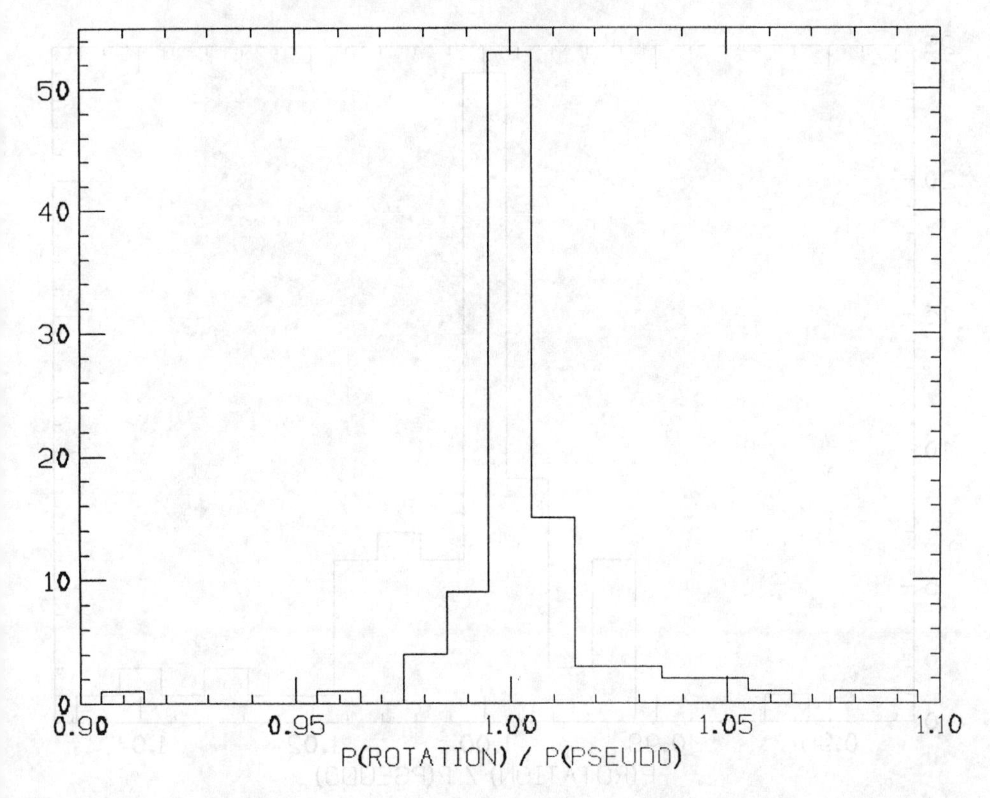

Figure 9. A histogram of P_{ROT}/P_{PSEUDO} expanded to show the small, less than 10%, differences indicative of differential rotation as a function of stellar latitude. All 96 systems shown here were included within the central spike of the histogram in Figure 5.

the only three stars in our sample with spectral types of late K or M. The single stars and some classes of binaries not normally included in discussions of starspots or differential rotation require a little elaboration.

For U Cephei we are interpreting the ≈ 6-year cycle of brightness changes at totality as a migration curve produced by the spotted G8 IV star, as originally suggested by Olson (1984, 1985). Migration periods had earlier been found in two W UMa-type binaries, 718 days in VW Cep (Leung and Jurkevich 1969) and 500 days in W UMa itself (Rigterink 1972). For the sun itself we used the canonical value k = 0.189. For other single stars, like FK Com (Guinan, Robinson, and Wacker 1986) and the T Tau star V410 Tau (Vrba, Herbst, and Booth 1988) we got $\Delta P/P$ by comparing the range of rotation periods with the mean rotation period.

For the SU UMa-type cataclysmic binaries we make the audacious hypothesis that the period of the super-outbursts is equivalent to a

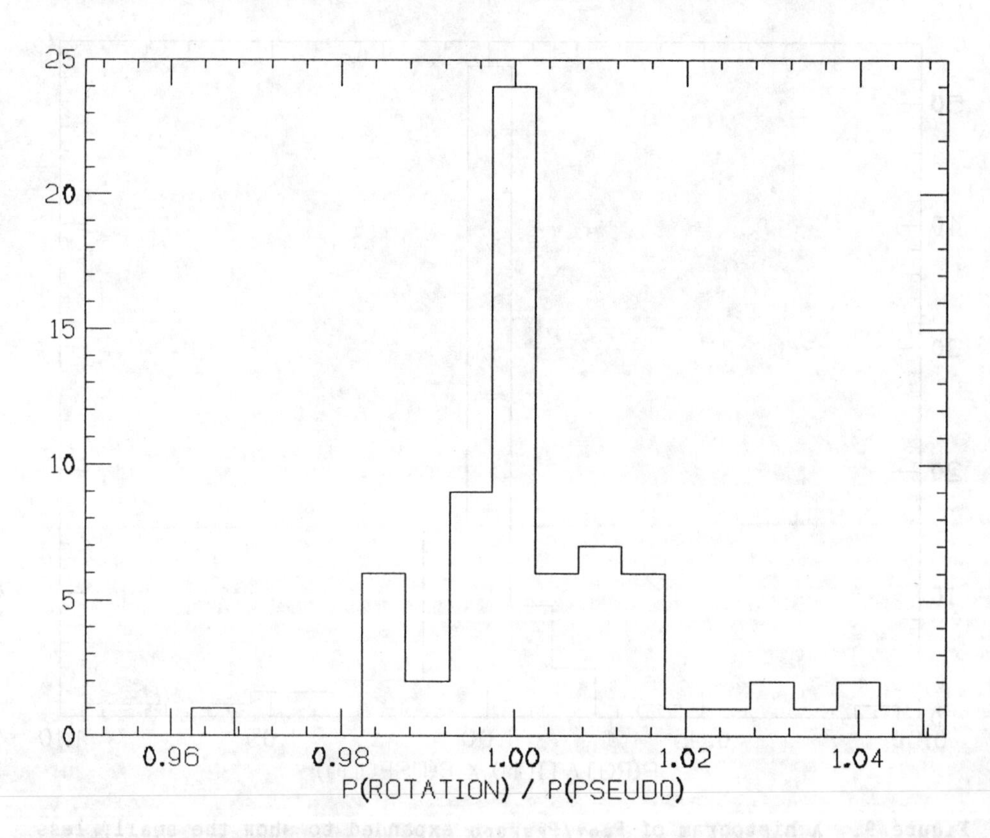

Figure 10. Another histogram of P_{ROT}/P_{PSEUDO} expanded even more to show that starspots occur above and below the co-rotating latitude in roughly equal proportion.

migration period, in which case P_{ROT} can be derived from equation (1) of Hall and Busby (1990); for those in which P_{ORB} is not known, we have used the approximation $P_{ORB} \approx P_{SUPERHUMP}$. The basis for our hypothesis is as follows. Suppose there is a localized starspot region on the cool lobe-filling component which differential rotation causes to migrate. Periodically the starspot region would move around to the inner Lagrangian point and face the companion star. That would trigger an episode of enhanced mass transfer which would produce a super-outburst. The mechanism should be very similar to the one proposed by Olson (1984) to account for the cycle of enhanced mass transfer activity in U Cephei which is correlated with its above-mentioned $\approx$ 6-year migration period. There is support for our hypothesis. First, most SU UMa binaries show several different values of $P_{SUPEROUTBURST}$, each one lasting around 2000 days (Vogt 1980, Mattei 1982). This is consistent with the birth and death of different starspot regions, each with its own P_{ROT}. We have computed k from the range in P_{ROT} exhibited by V346 Cen, VW Hyi, Z Cha, YZ Cnc, AY Lyr, and EK TrA. Second, the 2000-day duration is believably identified as a starspot lifetime. Inserting mean values for the six stars (k = 0.00017 and P_{ROT} = $0^{d}.075$) into equation (5) of Hall and Busby (1990) we would predict starspot lifetimes of 2000 days if the spot radii (not known for these stars) were between 8° and 18° depending on where they lay in latitude. These are quite typical of spot diameters in other CA stars.

Linear dynamo theory predicts that the differential rotation coefficient k should be a function of the Rossby number, defined as

$$Ro = P_{ROT} / \tau_{C} , \quad (7)$$

where τ_C is the turnover time for large convective cells which circulate within the convective zone of a star. For this reason we have computed or estimated Ro for each of our 83 stars, binary and single, using the calculations of Gilliland (1985) after correcting his effective temperature scale as explained by Hall (1990). In many cases the mass and radius of the spotted star were known directly. For unevolved stars on the main sequence, we assumed masses and radii typical of main-sequence stars of corresponding spectral type. For evolved stars in SB1 binaries we assumed a mass typical for similar (i.e., RS CVn-type) binaries of known mass (Popper and Ulrich 1977) and adopted radii of 4 $R_{\odot}$, 7 $R_{\odot}$, and 10 $R_{\odot}$ for luminosity class IV, IV-III, and III, respectively. For the late-type components in the SU UMa binaries we used the Gilliland model for log M = -0.3.

Although, as we have said already, Scharlemann (1982) did not think that synchronization forces could suppress or diminish k, we ask whether this conclusion would hold if one used the Tassoul (1987) synchronization rather than the Zahn (1977) theory, which Scharlemann had used. Since synchronization forces in either theory are proportional to the Roche-lobe-filling factor f = R/R_{ROCHE}, to some power, we computed or estimated f for each of the 83 stars in our sample, using the same masses and radii we had used to compute or estimate Ro. For the Algol-type binary U Cephei, the W UMa binaries, and the SU UMa binaries f = 1; for the sun and the other two single stars, f = 0.

Figure 11 is a plot of log k versus log Ro. We have fit the points with two straight-line segments joined at log Ro = 0, log k = -0.9, i.e. at Ro = 1, k = 0.125. The slopes of the two segments, determined by linear least squares, are -0.8 + 0.8 and +0.84 + 0.06. These fits give us the following empirically derived relations between k and Ro :

$$k = 0.125\ Ro^{-0.8} \quad \text{for Ro} > 1 \qquad (8)$$

and

$$k = 0.125\ Ro^{0.84} \quad \text{for Ro} < 1 . \qquad (9)$$

Note the sense of these correlations: as rotational velocity increases (as Rossby number decreases), differential rotation at first becomes stronger (larger k), reaches a maximum around Ro = 1 (k = 0.125), and then progressivly diminishes (smaller k). The rms deviation from these fits is + 0.4 in log k, which is only a factor 2.5, remarkably good considering the various uncertainties and approximations involved in determining or estimating k and Ro. The sun's observed value k = 0.189 is too large, vis à vis these fits, but only by a factor 1.5.

We mentioned above that the lobe-filling factor f might influence k, perhaps contributing to its observed diminution. Therefore we performed regression analysis with two parameters, log Ro and f. The result showed a marginally significant correlation with f, indicating that more nearly contact stars have slightly diminished differential rotation. The magnitude of the diminution was such that k is smaller by a factor 2 for contact stars (f = 1) compared to widely detached or single stars (f = 0), but the standard error was about half of the correlation coefficient. When the second parameter was included in regression analysis, the exponent in equation (9) decreased slightly, from 0.84 to 0.79. We should stress that rapid rotation (small Rossby number) is far more important in determining the diminution in k. The decrease from Ro = 1 to Ro = 0.0008 (corresponding to the SU UMa systems) is a factor 400, whereas the decrease from f = 0 to f = 1 is only a factor 2.

We worried that inclusion of the SU UMa binaries may have had an unduly strong influence on these correlation coefficients, because their stars have by far the shortest rotation periods (smallest Rossby numbers) and are all filling their Roche lobes as well (f = 1). Therefore we repeated the analysis with them omitted. The resulting coefficients, from both the linear regression and the regression with two parameters, differed from the original coefficients by considerably less than the standard deviations in all cases.

Can we understand this marked diminution in k? There have been a few attempts to study the influence of a stellar dynamo on differential rotation. One example is Belvedere, Paterno, and Stix (1980). In this paper they give an expression for the angular rotational velocity as a function of radial depth and latitude. In this expression their term $\varepsilon\omega_2$ seems to correspond to our differential rotation coefficient k, as defined in equation (6). They further show that $\varepsilon\omega_2$ is proportional to $\omega l^2/\nu$, where ω is a star's mean angular rotational velocity, l is mixing length, and ν is turbulent kinematic viscosity. In that paper they equate ν with $u_c l$, where u_c is convective velocity. Given that $u_c = l/\tau_c$, $1/\omega$ is proportional to P_{ROT}, and $Ro = P_{ROT}/\tau_c$, it follows that k should be inversely proportional to Rossby number or, in other words, directly proportional to inverse Rossby number (k : 1/Ro).

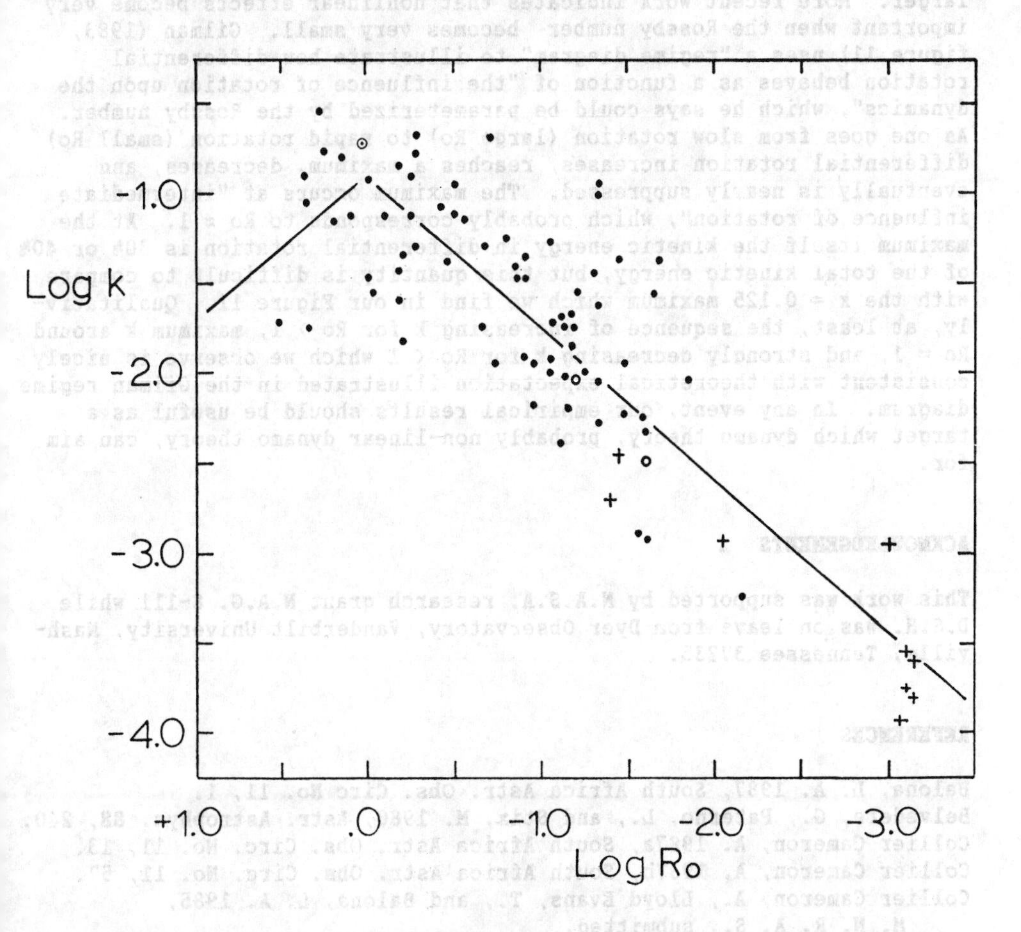

Figure 11. A plot of log k versus log Ro. Differential rotation is strongly suppressed (k gets smaller) as the angular velocity increases (Ro gets smaller). Different symbols indicate detached binaries (●), contact binaries (+), the sun (☉), two other single stars (○). The two straight lines represent equations (8) and (9).

The observed behavior of k as a function of Ro, seen in Figure 11, is in marked conflict with that theoretical expectation. Equation (8) shows that, for Ro > 1, k might increase in proportion to 1/Ro; but equation (9) shows that, for Ro < 1, k clearly <u>decreases</u> in proportion to 1/Ro. If one were to extrapolate k : 1/Ro from Ro = 1 to Ro = 0.001, then the discrepancy between theory and observation would be five and a half orders of magnitude in k.

The resolution of the conflict might be as follows. The formulation used by Belvedere, Paterno, and Stix (1980) implicitly assumed weak dynamo action, which is the case with Rossby numbers around unity or larger. More recent work indicates that nonlinear effects become very important when the Rossby number becomes very small. Gilman (1983, figure 11) uses a "regime diagram" to illustrate how differential rotation behaves as a function of "the influence of rotation upon the dynamics", which he says could be parameterized by the Rossby number. As one goes from slow rotation (large Ro) to rapid rotation (small Ro) differential rotation increases, reaches a maximum, decreases, and eventually is nearly suppressed. The maximum occurs at "intermediate influence of rotation", which probably corresponds to Ro ≈ 1. At the maximum itself the kinetic energy in differential rotation is 30% or 40% of the total kinetic energy, but this quantity is difficult to compare with the k = 0.125 maximum which we find in our Figure 11. Qualitativly, at least, the sequence of increasing k for Ro > 1, maximum k around Ro = 1, and strongly decreasing k for Ro < 1 which we observe is nicely consistent with theoretical expectation illustrated in the Gilman regime diagram. In any event, our empirical results should be useful as a target which dynamo theory, probably non-linear dynamo theory, can aim for.

ACKNOWLEDGEMENTS

This work was supported by N.A.S.A. research grant N.A.G. 8-111 while D.S.H. was on leave from Dyer Observatory, Vanderbilt University, Nashville, Tennessee 37235.

REFERENCES

Balona, L. A. 1987, South Africa Astr. Obs. Circ No. 11, 1.
Belvedere, G., Paterno, L., and Stix, M. 1980, Astr. Astrophys. **88**, 240.
Collier Cameron, A. 1987a, South Africa Astr. Obs. Circ. No. 11, 13.
Collier Cameron, A. 1987b, South Africa Astr. Obs. Circ. No. 11, 57.
Collier Cameron, A., Lloyd Evans, T., and Balona, L. A. 1985, M. N. R. A. S., submitted.
Gessner, S. L. 1989, private communication.
Gilliland, R. L. 1985, Ap. J. **299**, 286.
Gilman, P. A. 1983, I.A.U. Symposium **102**, 247.
Giuricin, G., Mardirossian, F., and Mezzetti, M. 1984a, Astr. Astrophys. **134**, 365.

Giuricin, G., Mardirossian, F., and Mezzetti, M. 1984b, Astr. Astrophys. **141**, 227.
Giuricin, G., Mardirossian, F., and Mezzetti, M. 1985, Astr. Astrophys. Suppl. **59**, 37.
Gray, D. F. 1988, Lectures on Spectral-Line Analysis: F, G, and K Stars (Arva, Ontario: The Publisher), lecture 5.
Guinan, E. F., Robinson, C. R., and Wacker, S. W. 1986, in Cool Stars, Stellar Systems, and the Sun, edited by M. Zeilik and D. M. Gibson (Berlin: Springer-Verlag), p. 304.
Habets, G. M. H. J. and Zwaan, C. 1989, Astr. Astrophys. **211**, 56.
Hall, D. S. 1972, P.A.S.P. **84**, 323.
Hall, D. S. 1986a, in Cool Stars, Stellar Systems, and the Sun, edited by M. Zeilik and D. M. Gibson (Berlin: Springer-Verlag), p. 40.
Hall, D. S. 1986b, Ap. J. **309**, L83.
Hall, D. S. 1987, Pub. Astr. Inst. Czechoslovakia **70**, 77.
Hall, D. S. 1990, this volume.
Hall, D. S. and Busby, M. R. 1989, in Remote Access Automatic Telescopes, edited by D. S. Hayes and R. M. Genet (Mesa: Fairborn Press), in press.
Hall, D. S. and Busby, M. R. 1990, this volume.
Hall, D. S. and Pazzi, L. 1987, B.A.A.S. **19**, 713.
Hooten, J. T. 1989, Master's Thesis, Vanderbilt University.
Hooten, J. T. and Hall, D. S. 1990, Ap. J. Suppl., in press.
Hut, P. 1981, Astr. Astrophys. **99**, 126.
Leung, K. C. and Jurkevich, I. 1969, B.A.A.S. **1**, 251.
Lloyd Evans, T. and Koen, M. J. 1987, South Africa Astr. Obs. Circ. No. 11, 21.
Maeder, A. and Meynet, G. 1989, Astr. Astrophys. **210**, 155.
Mattei, J. A. 1982, B. A. A. S. **14**, 879.
Morgan, J. G. and Eggleton, P. P. 1979, M. N. R. A. S. **187**, 661.
Olson, E. C. 1984, in Advances in Photoelectric Photometry, Volume II, edited by R. C. Wolpert and R. M. Genet (Fairborn: Fairborn Press), p. 15.
Olson, E. C. 1985, I.A.P.P.P. Comm. No. 19, 6.
Popper, D. M. 1988, A. J. **96**, 1040.
Popper, D. M. and Ulrich, R. K. 1977, Ap. J. **212**, L131.
Rigterink, P. V. 1972, A. J. **77**, 230.
Rodono, M. 1986, in Cool Stars, Stellar Systems, and the Sun, edited by M. Zeilik and D. M. Gibson (Berlin: Springer-Verlag), p. 475.
Scharlemann, E. T. 1982, Ap. J. **253**, 298.
Strassmeier, K. G. and Hall, D. S. 1988, Ap. J. Suppl. **67**, 439.
Strassmeier, K. G., Hall, D. S., Zeilik, M., Nelson, E. R., Eker, Z., and Fekel, F. C. 1988, Astr. Astrophys. Suppl. **72**, 291.
Strassmeier, K. G., Hall, D. S., Boyd, L. J., and Genet, R. M. 1989, Ap. J. Suppl. **61**, 141.
Tassoul, J. L. 1987, Ap. J. **322**, 856.
Tassoul, J. L. 1988, Ap. J. **324**, L71.
Vogt, N. 1980, Astr. Astrophys. **88**, 66.
Vrba, F. J., Herbst, W., and Booth, J. F. 1988, A. J. **96**, 1032.
Zahn, J. P. 1977, Astr. Astrophys. **57**, 383.

DO THERMALLY DECOUPLED LATE-TYPE CONTACT BINARIES REALLY EXIST?

CARLA MACERONI
Osservatorio Astronomico di Roma
Viale del Parco Mellini 84, I-00136 Roma, Italy

FRANS VAN 'T VEER
Institut d' Astrophysique - CNRS
98 bis Boulevard Arago, 75014 Paris, France

ABSTRACT: We have explored the possibility that the thermal decoupling found in B-type contact binaries (Wilson and Lucy, 1979) could be a spurious effect from light curve solution performed with an inadequate model, i.e. a model not including starspots. To test this hypothesis we computed a synthetic light curve from parameters of a thermally coupled W UMa binary, with a dark spot on the surface of the primary component affecting the the second half of the light curve. We solved this perturbed section with a spotless model. The outcoming solution, of excellent quality, had indeed all the characteristics of a typical B-type system. We also used a spotted model to fit the light curve of the overcontact, thermally decoupled system VZ Psc, and we were able to get a solution corresponding to a semi-detached configuration.

1 Introduction

The latest developments of the studies on stellar surface activity have shown that surface inhomogeneities (dark spots, plages) are widely present in the atmospheres of late type stars. Dark spots have been observed in single stars and in RS CVn binaries (Hall, 1976, Rodonò et al., 1986). These spots may easily cover 20 % of the stellar surface (Vogt, 1983) but more extreme coverage has also been found.

Dark spots are also a straightforward explanation of the cyclic perturbations of the light curve of some W UMa systems. This idea was already proposed long time ago (Binnendijk, 1960), and has recently been applied, among others, to the well studied system VW Cep (Bradstreet, these proceedings). The presence of dark spots also on W UMa binaries is not surprising, as their components possess the two main characteristics needed for efficient magnetic activity: high rotational velocity

C. İbanoğlu (ed.), Active Close Binaries, 309–319.

and convective envelope. Besides, they show other marks of the same phenomenon: strong Ca II H and K emission lines and enhanced UV and X fluxes.

In RS CVn binaries spots are easily detected, for their effect is superimposed on the light curve of a detached system, with flat long maxima, and they appear as a migrating wave-like perturbation. However, their parameters cannot be uniquely determined only from light curve solutions, for spots of different size, position and temperature may produce the same effect.

Spot detection is more difficult in contact binaries, with their continuously varying light curve, and the problem of uniqueness of spot parameters is still more severe. Besides these stars may well have hot spots on their surface, in particular around the connecting neck, where mass and energy exchange is going on. So, looking at just one perturbed light curve, there will always be the problem to decide which part is unperturbed.

This difficulty can be overcome by monitoring the system during a suitable time interval. However it is not easy to extend observations over more than ten years, as for case of VW Cep. Such long term observations are not avaliable for most contact binaries. On the other hand, the wrong assumption concerning the unperturbed part of the light curve may lead to spurious results, for the light curve solution based on an unspotted model may converge to a wrong set of parameters. The most sensitive parameters will be, as we will see in the following, the temperature difference between the components and the contact parameter. For spot sizes as those found in other late type stars the effect may be conspicuous.

In the next sections we will examine some cases where the problem of misinterpreted spots may have played an important role.

2 Thermally decoupled contact systems: a simulation

The thermally decoupled contact binaries, also called B-type after Lucy and Wilson (1979) are a group of a dozen systems whose properties are in contrast with the current theories of contact binary structure.

Lucy and Wilson, when finding the first members of this group, were looking for contact binaries in the broken contact stage. These objects should exist and be observable, according to the Lucy (1976) thermal relaxation oscillations theory, but had never been observed at that time. They were expected to be both geometrically and thermally in a phase of broken contact, i.e. semi-detached systems (or perhaps in very marginal contact) and with a light-curve with unequal minima. The systems they found, and the others later added to the list, have indeed the second property but not the first, as their degree of contact is in the same range of that of normal W UMa binaries.

Several B-type systems have been analyzed by Kaluzny, in a series of papers (Kaluzny 1986b and references therein). In table 1 we have collected, from his list, the late-type ones. We also added the system FS Lup studied by Milano et. al. (1987). The systems analyzed by Lucy and Wilson and by Kaluzny have some common features. First of all the observational material is rather poor for most

Table 1. Thermally decoupled late type systems.

name	Sp	P	Δm	ΔT	f%	A_2	Ref	Notes
W Crv	G5	$0.^d388$	$0^m.1$	660	35	0.5	1	asymmetries removed
				610	17	2.85		A-type assumed
WZ Cep	F2	$0^d.417$	$0^m.05$	1000	50	0.5	3	points with $0.1 \leq \Phi \leq 0.45$
				1000	21	$\simeq 3$		discarded, A-type assumed
BE Cep	G2	$0^d.424$	≈ 0	–	–	0.5	5	type not determined
				900	23	3.75		
FS Lup	G3	$0^d.381$	$0^m.05$	1120	33	0.5	6	most symmetric l.c. chosen
				1300		2.72		
AU Ser	F8	$0^d.386$	$0^m.05$	300	8	2.72	4	points with $0.08 \leq \Phi \leq 0.46$
								discarded, A-type assumed
CN And	F5	$0^d.463$	$0^m.04$	1000	10	0.5	2	points with $\Phi \leq 0.5$ discarded

1) Lucy and Wilson , 1979, 2) Kaluzny, 1983, 3) Kaluzny, 1986, 4) Kaluzny, 1986a, 5) Kaluzny 1986b, 6) Milano et al., 1987.

Table 2. Simulations of a B-type l.c. solution

	dark sp. model	B-type solution I		B-type solution II	
i	78°	75.9^o	$\pm 0^o.6$	79.6^o	$\pm 0^o.1$
q	1.84	1.84		0.55	
$g_1 = g_2$	0.32	0.32		0.32	
T_1	5087 K	5087 K		5087 K	
T_2	4770	4350	± 30 K	4376	± 7 K
A_1	0.50	0.50		0.50	
A_2	0.50	0.50		3.8	± 0.1
f	20 %	48.0 %	± 0.3	33.0 %	± 0.1
$L_1/(L_1 + L_2)$	0.450	0.584	± 0.004	0.7930	± 0.0001
$x_1 = x_2$	0.62	0.62		0.62	
θ_{sp}	35°				
ϕ_{sp}	153°				
ρ_{sp}	35°				
T_{sp}/T_2	0.75				
A_{sp}/A_2 %	8.6%				

Errors listed are probable errors.

of them. Second the authors, in the light curve analysis assumed, in a rather arbitrary way, that, when the maxima were of unequal height, it was the lower one that corresponded to an "immaculate" stellar surface, while the higher was produced by a hot spot. Third, the "perturbed" points where discarded from the set used for the light curve solution.

The derived solution have also some interesting common characteristics:

a) when the value of the albedo coefficient of the secondary star, A_2, is fixed to the appropriate value for the observed spectral type or color (usually $A_2 = 0.5$), the degree of contact obtained from the solution is rather high, even in comparison to that of normal W UMa stars.

b) the contact parameter decreases by a factor of ≈ 2 when the A_2 coefficient is also adjusted. In this case A_2 reaches an unphysical value ($A_2 > 1$), in most solutions $2 \leq A_2 \leq 4$.The quality of the fit with this unphysical values, usually measured from the sum of weighted squared residuals, S, is remarkably better.

c) the temperature difference, often around thousand degrees, turns out to be only slightly sensitive to the type of solution.

The values obtained in these two cases are collected, when avaliable, in table 1. In this table the quantity Δm indicates the magnitude difference between the maximum following the primary minimum ($\max_I$) and that following the secondary one ($\max_{II}$). The quantity ΔT is defined as the temperature difference between star 1 (the one eclipsed at primary minimum) and star 2.

The system FS Lup has a longer series of observations, having been systematically observed during two years by Terzan and Didelon (1985). The light curves are rather variable in shape, Milano et al. chose for their solution the most symmetric one (same height of maxima). Of course the symmetry could just be produced by a particular distribution of spots whose combined effect was similar at quadratures.

As we were asking ourselves if the results on these B-type solutions could be due to the assumptions made and to the way of treating the data, we decided to perform a quite simple test. We computed, with the latest distributed (1979) version of the Wilson and Devinney program , which synthetizes light curves from spotted stars, a light curve (at $\lambda = 5500$ Å) corresponding to a normal W UMa system of the Binnendijk (1970) W subtype. This model system had a spot on the larger star, and the light curve had maxima of unequal height; the spot parameters were chosen so that $\Delta m \simeq 0^m.05$, as for most observed cases. The parameters of the model are reported in table 2, the light curve and the spot effect on it are shown in fig. 1. In the following we will often call this the "observed" light curve. That is of course an improper denomination, but we will indeed treat its points as real observations.

According to the procedure applied by Kaluzny, we assumed that the first half of the "observed light curve" ($\Phi < 0.5$) was affected by a hot spot and we discarded the corresponding points. Therefore, the solution was performed only on the second half and with a model without spots. We adjusted the inclination, i, the effective temperature of the secondary component T_2, the luminosity of the primary, L_1, and the surface potential Ω [but in the tables we report the more convenient parameter f, defined as $f = (\Omega - \Omega_{in})/(\Omega_{in} - \Omega_{out}]$. A solution was reached, as probable errors were larger than computed corrections and S did no longer improve with iterations. But the quality of the solution, performed with fixed $A_2 = 0.5$, was very poor, as the value of S was rather large, $S = 0.0236$.

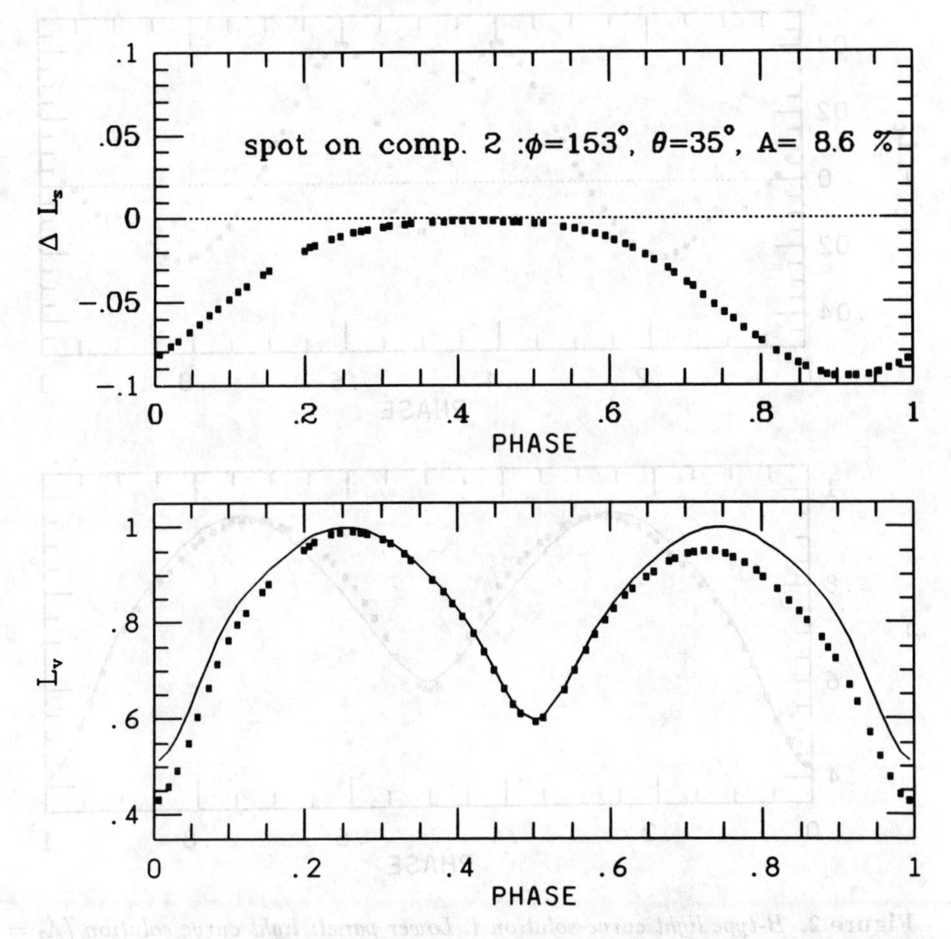

Figure 1. *The synthetic light curve used for the simulation of a B-type solution. Lower panel, continuous line: light curve in absence of spots; points: light curve with dark spot perturbation. Upper panel: the spot effect as a function of phase. The uneven phase spacing of points corresponds to the spacing of normal points of an observed system (CW Cas).*

Figure 2 shows the solution and the residuals. The points and curve in the phase interval $0 \leq \Phi \leq 0.5$ are a mirror image of those in the interval $0.5 \leq \Phi \leq 1.0$, included just to show a complete cycle. It is worth noticing that the residuals have a somewhat sinusoidal pattern which is very similar to that shown by the residuals from solutions of real systems, also performed with $A_2 = 0.5$ (BE Cep and FS Lup, see fig.3 of Kaluzny ,1986b, and fig.7 of Milano et al., 1987. Note that this last figure shows the computed minus observed intensities).

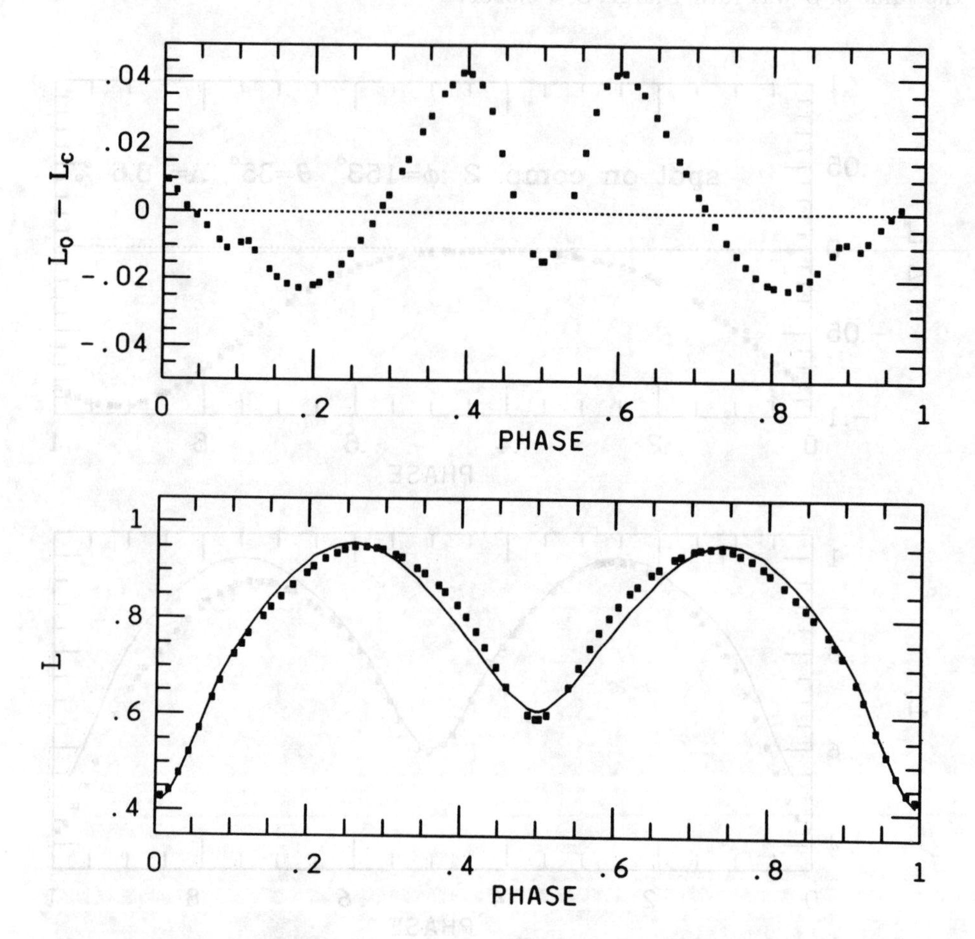

Figure 2. *B-type light curve solution I. Lower panel: light curve solution (A_2 = 0.5), upper panel: residuals. The curve and points in the phase interval $\Phi < 0.5$ are mirror images of $\Phi \geq 0.5$.*

The parameters of this B-type solution I, i.e. with fixed A_2, appear in the second column of table 2. It is indeed a B-type system: large temperature difference (ΔT = 737 K) and large degree of contact (f = 48 %).

The second step of the simulation was to adjust also A_2. We also made the additional hypothesis that the system was of A-type. For this reason the value of the mass ratio q, not adjusted in the solution, is simply the inverse of the initial one.

The B-type solution II, shown in the third column of table 2, is of much better quality, with $S = 0.00169$, i.e. smaller by more than a factor of ten than that of solution I. A_2 is in the expected range of "unphysical" values found for real systems and the contact parameter reduces from $f = 48$ % to $f = 33$ %.

The light curve solution and the residuals are shown in fig. 3. Here the complete set of "observed" points is shown, even if the fit was performed only on the phase interval $\Phi > 0.5$.

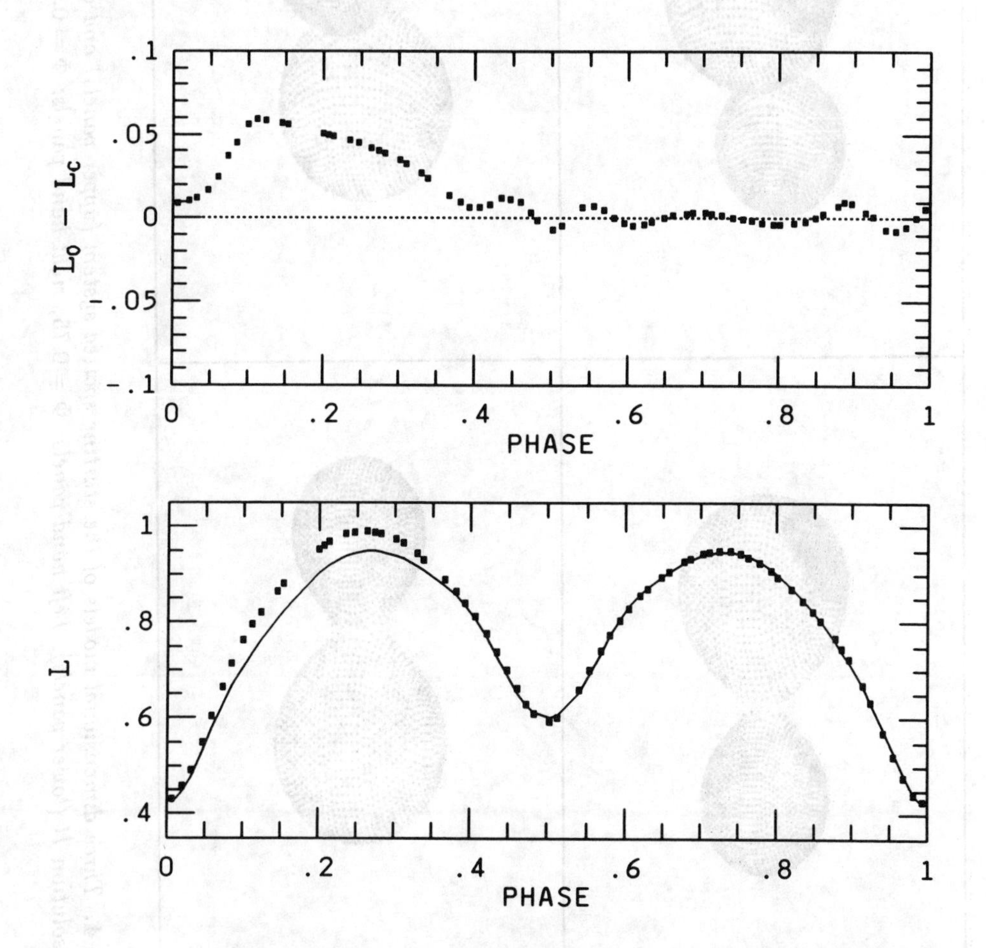

Figure 3. *B-type light curve solution II. Lower panel: light curve solution ($A_2 = 3.8$), upper panel: residuals. Also this solution was performed only on the second half of "observed" light curve.*

A straightforward explanation of the systematic positive values of the residuals

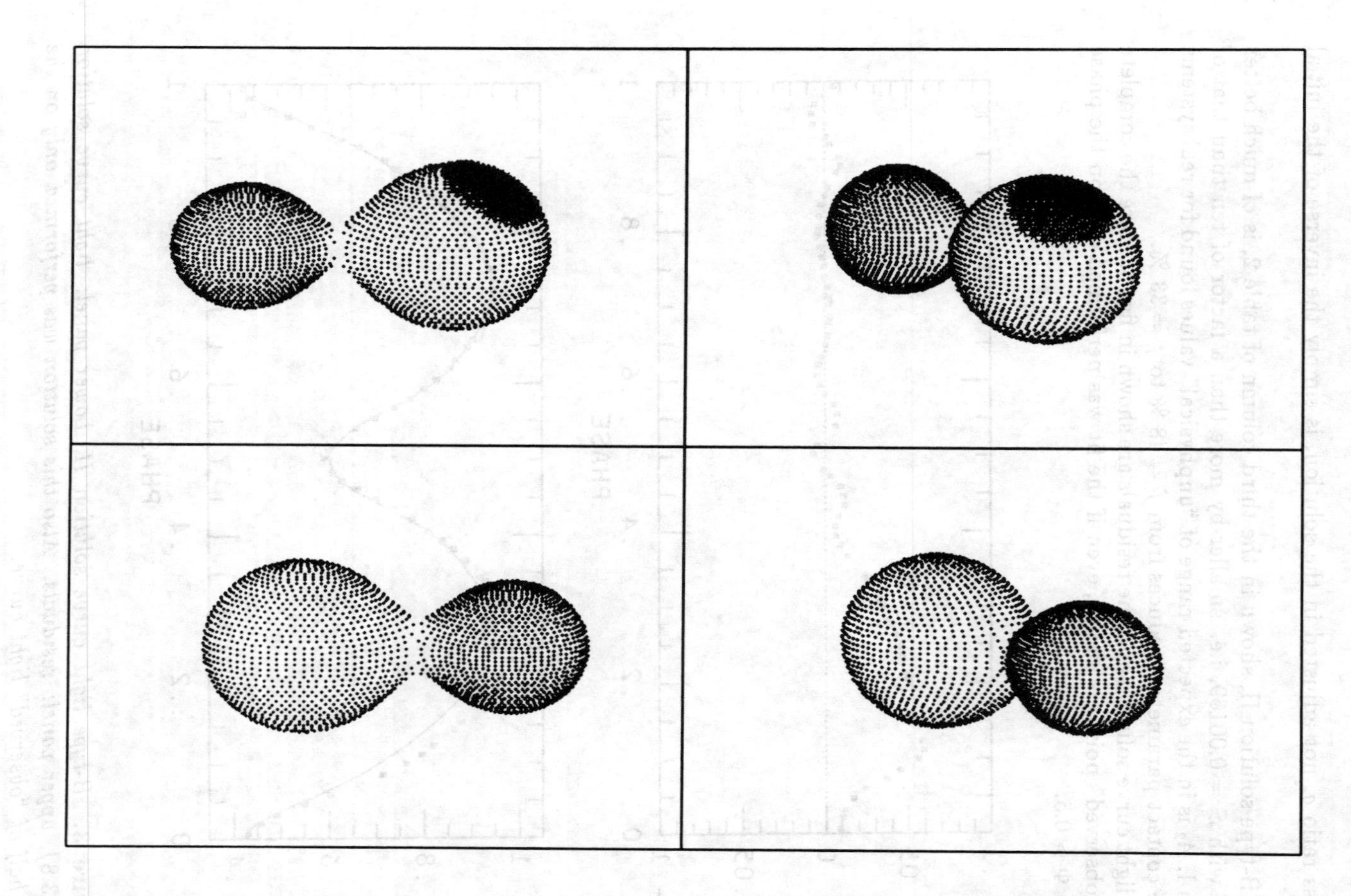

Figure 4. *Three dimensional models of the starting spotted system (upper panels) and of the one corresponding to B-type solution II (lower panels). Left hand panels:* $\Phi = 0.75$, *right hand panels:* $\Phi = 0.875$.

for $0.0 < \Phi < 0.4$, would have been (for a real system), that a hot spot is in view. That is the hypothesis of Kaluzny for all the system with unequal maxima he analyzed.

Finally figure 4 shows how the initial spotted system and that coming from the solution II look like, at two different orbital phases.

3 An observed system: VZ Psc

The very short period ($P = 0^d.261$), late type (K3-K5) system VZ Psc is the most extreme member of the B-type contact systems. The solution recently published by Hrivnak and Milone (1989), based on the light curve secured by Bradstreet (1985) and their own radial velocity curves, is characterized by extreme f and ΔT. The first reaches the value of 74 %, the second about 1200 K. Besides, the mass ratio obtained from spectroscopic data is not far from one, $q = 0.92$. These parameters, coupled to an inclination of only 36°, yield absolute dimensions difficult to understand, for the stars come out too massive and too large for their spectral type.

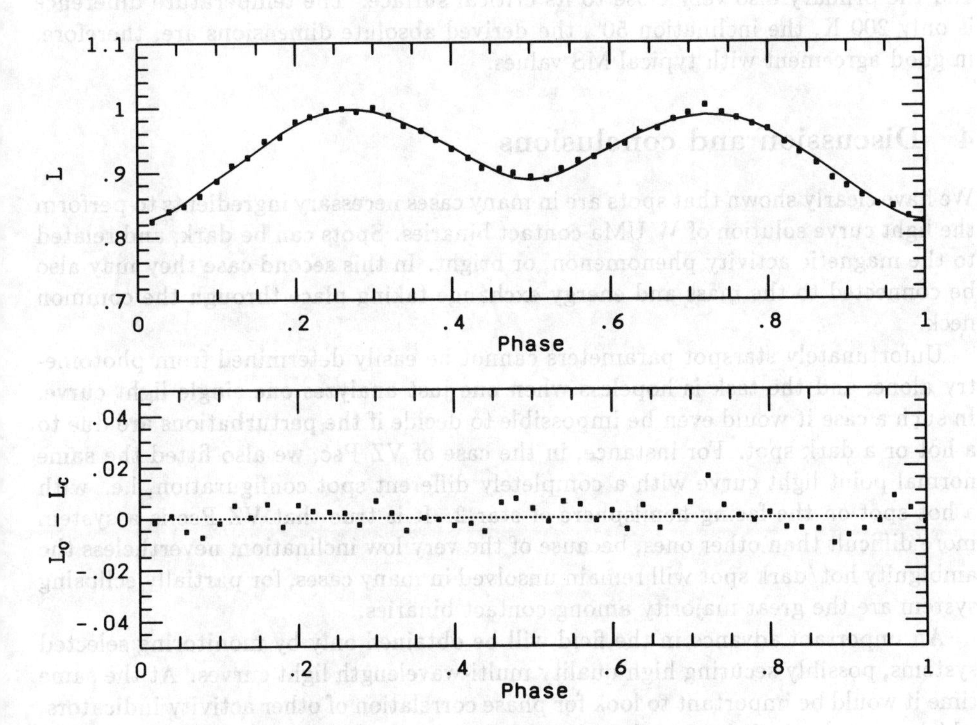

Figure 5. *Light curve solution and residuals for VZ Psc*

We decided to look for an alternative solution assuming that the difference of depth between the two minima was due to dark spots and not to a peculiar temperature distribution. We selected a set of input parameters for the iterative procedure assuming: a) equal temperature of the components, in agreement with spectral type, b) MS radii, c) just detached configuration.

The spots had to be placed on the surface by trial and error, for the differential correction method does not allow the adjustment of spot parameters. A suitable input model, giving some approximate fit of the light curve, showed two spots on the back hemisphere of star 2. The first, at colatitude $\theta = 85^o$ and longitude $\phi = 135^o$, had a temperature ratio, with respect to the local temperature in absence of spots, T_s/T_l, of 0.7. Its area, A, was 3.4 % of that of the star. The second, positioned at $\theta = 85^o$ and $\phi = 210^o$, had $T_s/T_l = 0.7$ and $A = 3.6$ %.

The solution derived from these input values was of better quality than that presented by Hrivnak and Milone, which we recomputed with our surface grid and weighting scheme, the value of S was 20 % smaller.

We will not give here all the details of the solution, which will be published elsewhere (Maceroni et al., 1989). It is sufficient here to say that the system comes out to be semi-detached, with the secondary (here the smaller star) filling its lobe, and the primary also very close to its critical surface. The temperature difference is only 200 K, the inclination 50°, the derived absolute dimensions are, therefore, in good agreement with typical MS values.

4 Discussion and conclusions

We have clearly shown that spots are in many cases necessary ingredients to perform the light curve solution of W UMa contact binaries. Spots can be dark, and related to the magnetic activity phenomenon, or bright. In this second case they may also be connected to the mass and energy exchange taking place through the common neck.

Unfortunately starspot parameters cannot be easily determined from photometry alone, and the task is hopeless when one just analyzes one single light curve. In such a case it would even be impossible to decide if the perturbations are due to a hot or a dark spot. For instance, in the case of VZ Psc, we also fitted the same normal point light curve with a completely different spot configuration, i.e. with a hot spot on the facing hemisphere of star 2. It is true that VZ Psc is a system more difficult than other ones, because of the very low inclination, nevertheless the ambiguity hot/dark spot will remain unsolved in many cases, for partially eclipsing system are the great majority among contact binaries.

An important advance in the field will be obtained only by monitoring selected systems, possibly securing high quality multi-wavelength light curves. At the same time it would be important to look for phase correlation of other activity indicators, which can greatly help to solve the uniqueness problem (Andrews et al., 1988, Guinan, these proceedings).

It is important to stress that spot solutions should be done on light curves covering small time interval, possibly just a cycle, and that normal point light curves from observations covering several years should be avoided. It is not true,in fact, that solutions of mean light curves are relevant to some mean physical model of the binary.

We also feel that the so called good "spotless" solutions from individual symmetric light curve can be questionable, unless the light curve is known to be very stable in time. There exists, in fact, a sort of selection effect produced by choosing a symmetric curve from a set of perturbed ones as corresponding to immaculate star surfaces. It could of course just correspond to an accidental spot distribution producing similar effect at symmetric phases.

Finally, concerning B-type systems, we think that these systems should be much better studied and observed before assuming that the thermal decoupling really exist. It is worth noticing that the hot-contact thermally decoupled systems, also presented by Kaluzny in the quoted series of papers, can also be explained in an alternative way (Shaw, these proceedings). So it would be probably wiser to devote new efforts to observations and interpretations than to theories of thermal decoupling.

References

- Andrews, A.D. et al.: 1988, *Astron. Astrophys.* **204**, 177
- Binnendijk, L.: 1960, *Astron. J.*, **65**, 358
- Binnendijk, L.: 1970, *Vistas in Astronomy,* **12**, 217
- Bradstreet, D.H.: 1985, *Astrophys. J. Suppl.* **58**, 413
- Hall, D.S.: 1976 *I.A.U. coll. 29*, ed. W.S. Fitch, Reidel, Dordrecht, p. 287
- Hrivnak,B.J., Milone, E.F.: 1989, *Astron. J.* **97**, 532
- Kaluzny, J.: 1983 *Acta Astron.* **33**, 345
- Kaluzny, J.: 1986 *Acta Astron.* **36**, 105
- Kaluzny, J.: 1986a *Acta Astron.* **36**, 113
- Kaluzny, J.: 1986b *Astron. J.* **98**, 662
- Lucy, L.B.: 1976, *Astrophys. J.* **205**, 208
- Lucy, L.B., Wilson, R.E.: 1979, *Astrophys. J.* **231**, 502
- Maceroni, C., Van Hamme, W., Van 't Veer, F.: 1989, submitted to *Astron. Astrophys.*
- Milano, L., Russo, G., Terzan, A.: 1987, *Astron. Astrophys.* **183**, 265
- Rodonò, M. et al.: 1986, *Astron. Astrophys.* **165**, 135
- Terzan, A., Didelon, P.: 1985, *Inf. Bull. Var. Stars* No. 2716
- Vogt, S., S.: 1983, *Activity in Red Dwarf Stars*, Byrne, P. B., Rodonò, M. eds., Reidel Publ. Co., p.137
- Wilson, R.E.: 1979, *Astrophys. J. Suppl.* **234**, 1054

THE EVOLUTION OF THE PERIOD AND SYNCHRONIZATION FOR CLOSE G-TYPE MS BINARIES

FRANS VAN 'T VEER
Institut d' Astrophysique - CNRS
98 bis Boulevard Arago, 75014 Paris, France

CARLA MACERONI
Osservatorio Astronomico di Roma
Viale del Parco Mellini 84, I-00136 Roma, Italy

ABSTRACT: We give some preliminary computations of the variation of the synchronization parameter $f = P_K/P$ and the orbital period P_K with time for late type close binaries. Angular momentum loss by stellar wind from the components and angular momentum transfer due to the torque exerted by the tidal bulge are both taken into account. The uncertainties introduced by the treatment of the equilibrium tide and the estimation of the angular momentum loss are discussed. It is argued that the expected non-homogeneous rotation of the components has only a negligible effect on the results of the computations

1 Introduction

Recent researches on late type binaries and single stars have stressed the importance of dark spots related to magnetic activity on the surface of these stars. For the fast rotating stars these dark inhomogeneities may cover up to 20 or 30% of the surface of the star (see different contributions in these proceedings). Along with the discovery of these spots the problem is raised how their presence can be understood from the dynamo theory and in particular from the rotational velocity distribution necessary for development of the dynamo.

Synchronization parameters can be estimated in certain cases from the motion of spot groups with respect to the orbital motion. They can be computed from the theory of tidal interaction and the equations of angular momentum loss for rotating components.

The main purpose of this paper is to give the evolution of the synchronization and the orbital period of a typical old MS binary with two identical components of spectral type G5V. We will take into account the recent results concerning angular

C. İbanoğlu (ed.), Active Close Binaries, 321–329.

momentum loss and use the equations developed for the equilibrium tide by Zahn (1977) and Campbell and Papaloizou (1983) (see also Zahn, 1989).

2 The angular momentum loss (AML)

It is known since some decades that the sun and magnetically active solar type stars are losing AM carried away by stellar winds. Skumanich (1972) was the first to derive from the rotation of G-type stars of galactic clusters of known age a differential equation for the spin-down of these stars.

The exponential equation is of the type

$$\dot{\omega} = c\omega^3 \tag{1}$$

where ω is the angular velocity of the rotation of the outer layers of the stars, and can be related to the AM of the star by the equation

$$\dot{H} = xI\dot{\omega} \tag{2}$$

where I is the inertial moment

$$I = (kr)^2 m \tag{3}$$

of the star with radius r, mass m and relative radius of gyration k. During the evolution from ZAMS to TAMS we consider I as constant. Strictly speaking this is not true, but the uncertainty introduced by the factor x is more important. This factor takes into account the extension of the convective envelope and its coupling with the inner part of the star. When there is no coupling with the underlying radiative core we may write

$$x = (I - I_c)/I \tag{4}$$

where I_c is the inertial moment of the radiative core.

Equation (1) is valid in the velocity interval for which it is determined. This interval extends from about 10 to 2 km/s. The components of close binaries however are rotating with equatorial velocities up to $v_{eq} = 200$ km/s, so we need a different equation for the AML of the more rapidly rotating components. Such an equation has been derived by us (van 't Veer, Maceroni, 1988, 1989; hereafter VM1, VM2) from the statistics of late type binary periods. The bimodal distribution of these binaries (Farinella et al., 1979) has been explained in a satisfactory way by a differential equation of the first order

$$\dot{\omega} = c'\omega \tag{5}$$

along with the hypothesis that strong spin-orbit coupling sets in abruptly at $P_K = 3^d$ corresponding to $v_{eq} = 20$ km/s for a G5V star. The value of c' has been determined from lifetime discussions (VM2) and we thus also possess the AML equation for the interval 20 to 200 km/s. The missing part (10-20 km/s) could not be covered by observational results and we propose a power law relating the inner limits, 10 and 20 km/s, of the two intervals.

This steep interpolated AML function is uncertain, but its use is restricted to a small velocity interval and for this reason its influence on the results of dynamical computations is limited.

So the adopted AML function, given in fig.(1), is composed of 3 parts: a part of high AML, but slowly varying with ω, for rapid rotators, an intermediate part, and the well known cubic law of AML extending towards the slowest observed rotators.

We finally want to mention that the great AML found for high angular rotations is confirmed by studies on the rotational decay of members of very young clusters (Stauffer and Hartmann, 1987).

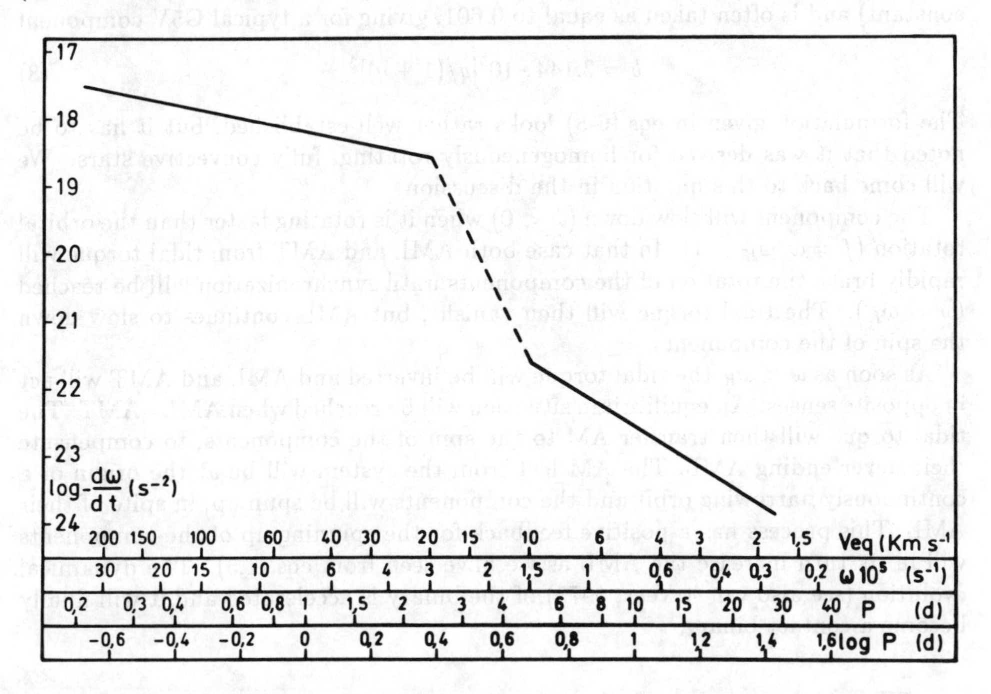

Figure 1. *The variation of rotation of a G5V type star, resulting from stellar winds (AML). The angular velocity ω of the visible layers decreases with time. The AML can be found with the aid of eq.(2).*

3 The angular momentum transfer (AMT) produced by tidal torque

The torque due to AML, as it was discussed in the preceding section, only depends on the angular velocity of the components and the magnetic properties of their convective envelope.

The tidal torque on the components, however, depends on the angular velocity of the components as well as on the orbital velocity. It has been formulated, among others, by Zahn (1979) and Campbell and Papaloizou (1983). They find expressions which may be written as follows,

$$\dot{\omega} = b(\omega_K - \omega)\omega_K^5, \tag{6}$$

where

$$b = q^2 6k_2/k^2 (mr^2/L)^{-1/3} (GM/r^3)^{-2} \tag{7}$$

mainly depends on the spectral type of the components and their mass ratio q. The constant $6k_2/k^2$ depends on the internal structure (k_2 is the second apsidal motion constant) and is often taken as equal to 0.601, giving for a typical G5V component

$$b = 2.144 \cdot 10^5 [q/(1+q)]^2. \tag{8}$$

The formulation given in eqs.(6-8) looks rather well established, but it has to be noted that it was derived for homogeneously rotating, fully convective stars. We will come back to this question in the discussion.

The component will slow down ($\dot{\omega} < 0$) when it is rotating faster than the orbital rotation ($f = \omega/\omega_K > 1$). In that case both AML and AMT from tidal torque will rapidly brake the rotation of the components until synchronization will be reached ($\omega = \omega_K$). The tidal torque will then vanish , but AML continues to slow down the spin of the components.

As soon as $\omega < \omega_K$ the tidal torque will be inverted and AML and AMT will act in opposite senses. An equilibrium situation will be reached when AML=AMT. The tidal torque will then transfer AM to the spin of the components, to compensate their never ending AML. The AM lost from the system will be at the origin of a continuously narrowing orbit and the components will be spun up, in spite of their AML. This process has a positive feedback for the spinning up of the components will in its turn increase the AML as we have seen from eqs.(1,5). The dynamical evolution (see also van 't Veer, 1979) of the binary is accelerated and it will finally become a contact binary.

4 The time scale of the AML-controlled dynamical evolution

The time necessary for the evolution into contact depends on the AML from the system and hence on both the initial spin of the components and the initial orbital period. From fig.(1) we can see how the AML increases ($\dot{\omega}$ decreases) with the rotation of the visible outer layers.

Fig.(2) shows how the AMT from orbit to spin and vice-versa is related to stellar (P) and orbital (P_K) period. The derivatives $\dot{\omega}$ given by the curves of fig.(2) are computed with the eqs.(6- 8) and their absolute value increases with the distance from the f=1 line and the fourth power of $1/P$. The sign changes when the $f = 1$ line is crossed.

In fig.(3) we present the dynamical evolutionary track in the $P_K - P$ diagram for a detached binary with initial of $P_K = 3^d$ and pf $P = 0^d.24$ (corresponding to $v_{eq} = 200$ km/s),as computed by stepwise iteration.

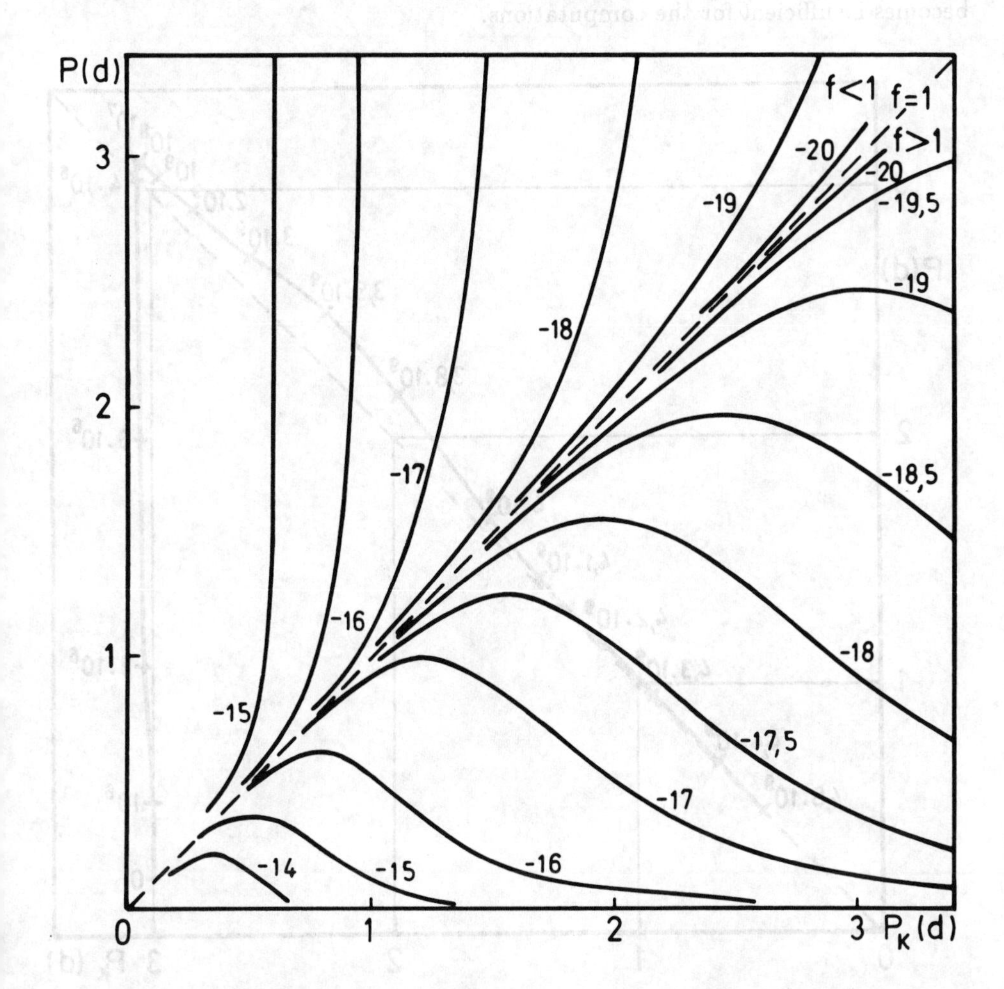

Figure 2. *The variation of rotation of a G5V type component resulting from the tidal torque (AMT). The parameter $f = P_K/P$ defines the degree of synchronization. the curves are labeled by the parameter $\dot{\omega}$. For $f < 1$ we have $\dot{\omega} > 0$ and conversely.*

We remark the rapid evolution during the first 10^7 years with an orbital period slowly increasing from 3 to $3^d.06$ because of the absorption of AM from the spin of the stars. After the overshoot from $f > 1$ to $f < 1$ at about $5 \cdot 10^6$ y, when both AML and AMT become very small, a slow decrease of P and P_K sets in, until a new acceleration towards a smaller orbit will follow after about 10^9 y.

The computation was stopped at $P_K = 0^d.5$, when physical interaction between the atmospheres of the two components begins to set in and the tidal torque alone becomes insufficient for the computations.

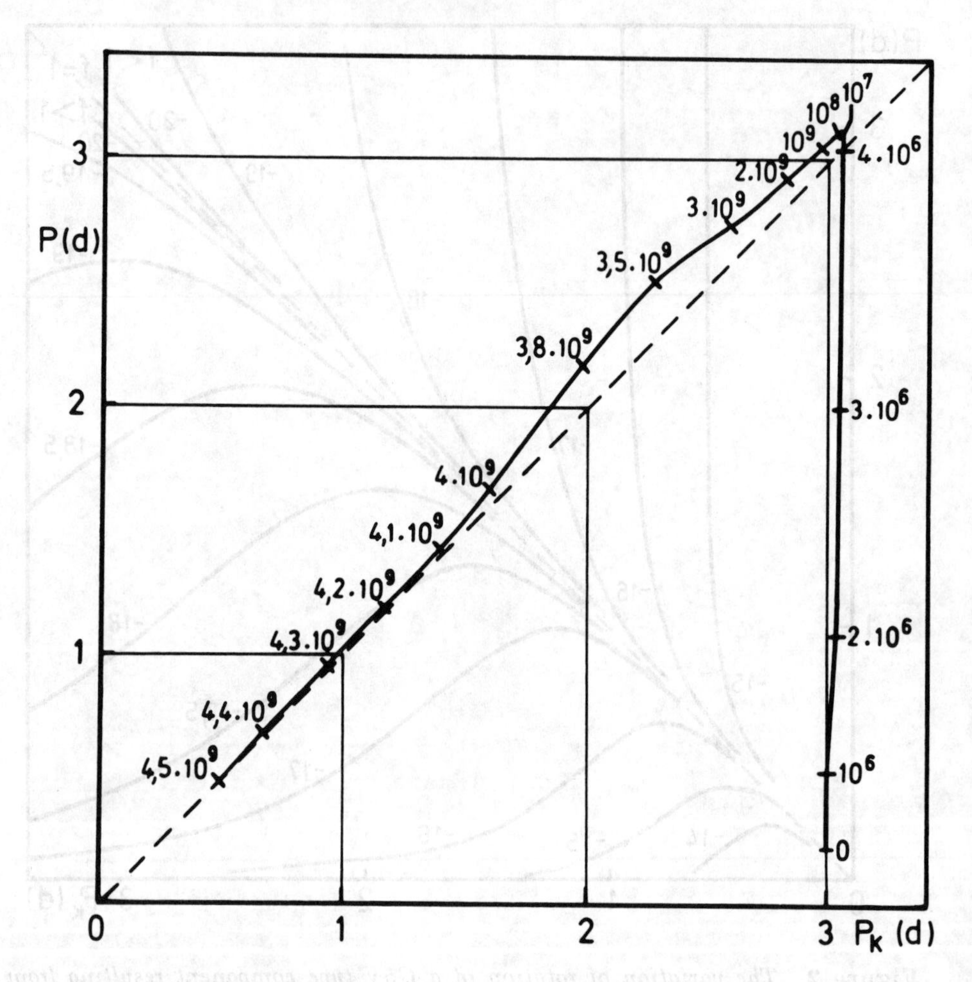

Figure 3. *The dynamical evolution of a detached binary computed with the eqs.(1-8) given in this paper. The time marks are expressed in years after the origin at the ZAMS.*

When the same computation is done for binaries with initial period greater than $3^d.5$ we find that the age of our galaxy is not sufficient to bring the two components together. However the track after the overshoot in the $P_K - P$ diagram will follow the same path.

Indeed this part of the track is defined by the compensating tidal and AML torques. That means that binaries with standard components (say G5V) and equal periods will sit on the same location in the $P_K - P$ diagram. The only difference is their age which depends on the initial parameters of the system.

So it is possible now to compute for these binaries with standard components the synchronization parameter f as a function of P_K alone. The result is given in fig.(4).

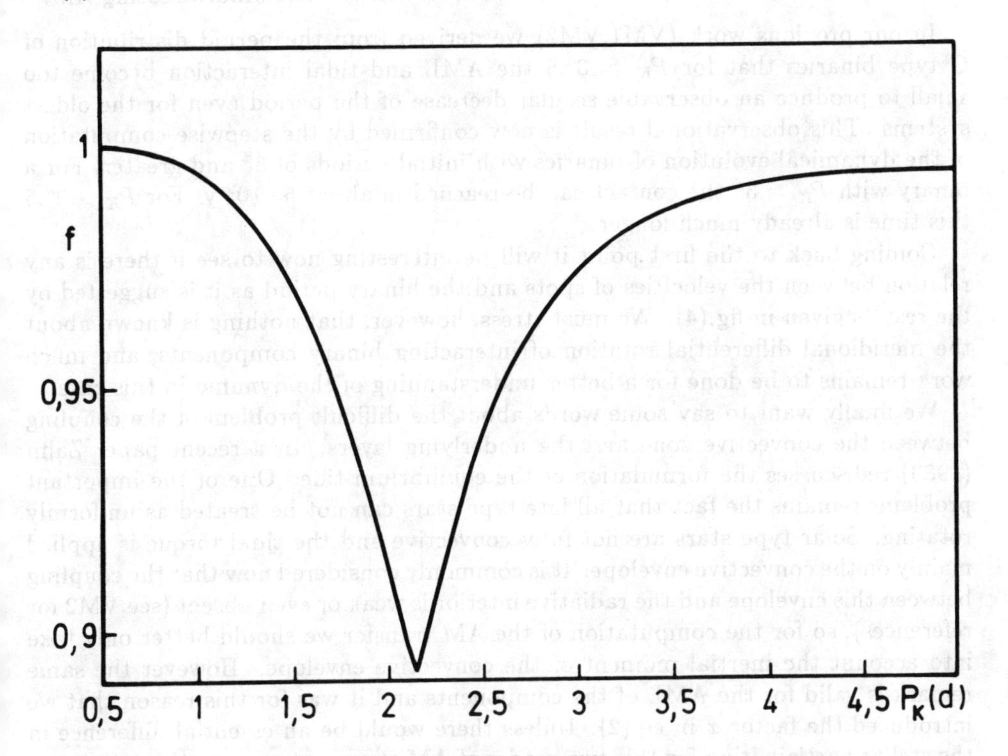

Figure 4. *The synchronization parameter f as a function of the orbital period.*

The synchronization goes through a minimum situated at $P = 2^d.1$. The existence of this minimum can be understood from torques acting on the components which depend on ω, ω_K and $\omega - \omega_K$. The location and the depth of this minimum depends however on several constants and conditions as will be discussed in the next section.

5 Discussion

We considered the dynamical evolution of a detached binary with two identical magnetically active components losing AM carried away by corotating stellar winds, and transferring AM from orbit to spin or vice versa by virtue of the torque exerted by the tidal bulge of the component.

Our principal aims were:

1. The determination of a synchronization parameter that can be used for studies of the migration velocity of spot groups in close binaries.

2. A better estimation of the dynamical evolution of close binaries losing AM.

In our previous work (VM1,VM2) we derived from the period distribution of G type binaries that for $P_K > 3^d.5$ the AML and tidal interaction become too small to produce an observable secular decrease of the period even for the oldest systems. This observational result is now confirmed by the stepwise computation of the dynamical evolution of binaries with initial periods of 3^d and greater. For a binary with $P_K = 3^d$ the contact can be reached in about $5 \cdot 10^9$ y. For $P_K > 3^d.5$ this time is already much longer.

Coming back to the first point it will be interesting now to see if there is any relation between the velocities of spots and the binary period as it is suggested by the results given in fig.(4). We must stress, however, that nothing is known about the meridional differential rotation of interacting binary components, and much work remains to be done for a better understanding of the dynamo in this case.

We finally want to say some words about the difficult problem of the coupling between the convective zone and the underlying layers. In a recent paper Zahn (1989) rediscusses the formulation of the equilibrium tide. One of the important problems remains the fact that all late type stars can not be treated as uniformly rotating. Solar type stars are not fully convective and the tidal torque is applied mainly on the convective envelope. It is commonly considered now that the coupling between this envelope and the radiative interior is weak or even absent (see VM2 for references), so for the computation of the AM transfer we should better only take into account the inertial moment of the convective envelope. However the same remark is valid for the AML of the components and it was for this reason that we introduced the factor x in eq.(2). Unless there would be an essential difference in the stellar participation for this two modes of AM transport we may think that our computations are not strongly affected by the badly known parameter x.

References

- Campbell,C.G., Papaloizou, J.: 1983, *Monthly Notices Roy. Astron. Soc.* **204**, 433

- Farinella,P., Luzny,F., Mantegazza,L.,Paolicchi,P.: 1979, *Astrophys. J.* **234**, 973

- Linnell,A.P.: 1980,2nd *Cambridge workshop on cool stars, stellar systems and the sun*, t.2, eds. M.S.Giampapa, L.Golub p.65

- Skumanich,A.: 1972, *Astrophys. J.* **171**, 565

- Stauffer,J.R., Hartmann,L.W.: 1987, *Publ. Astron. Soc. Pac.* **98**, 1233
- van 't Veer,F.: 1979, *Astron. Astrophys.* **80**, 287
- van 't Veer,F., Maceroni,C.: 1988, *Astron. Astrophys.* **199**, 183 (VM1)
- van 't Veer,F., Maceroni,C.: 1989, *Astron. Astrophys.* **220**, 128 (VM2)
- Zahn,J.-P.: 1977, *Astron. Astrophys.* **57**, 383
- Zahn,J.-P.: 1989, *Astron. Astrophys.* **220**, 112

THE A-TYPE W UMA SYSTEM AW UMA

J.R.W. Heintze, J.J.M. In 't Zand and F. van 't Veer
Sterrekundig Instituut
State University Utrecht
P.O. Box 80 000
3508 TA Utrecht, The Netherlands

Institut d'Astrophysique
98bis, Boulevard Arago
75014 Paris
France

ABSTRACT. For this contact binary a rather reliable mass ratio, metal abundance as well as masses, dimensions and effective temperatures of the components as the distance to Earth could be determined. The common envelope is in convective equilibrium. It is a suitable test case for the theory of energy transport from the more massive – to the less massive component by circulation currents.

1 Introduction

AW UMa is the W UMa system with the smallest known mass ratio up to now: ≈ 0.07. According to Paczynski (1964) its spectral type must be somewhere between F0 and F2. His light curve measurements reveal at maximum light:

$$V = 6.84 \quad B - V = +0.35 \quad U - B = -0.0057,$$

whereas the apparent visual magnitude of the primary is

$$m_{V_1} = 7.08 \text{ mag}, \tag{1}$$

being the V magnitude in the secondary minimum, because then the secondary is completely obscured by the primary.

The galactic latitude is about +71°. Therefore it can be assumed that

$$E(B - V) = 0 \tag{2}$$

(Mochnacki, 1981).

In this paper some new results of IUE measurements and of photometry in the wavelengths 474, 529, 672, 781 and 871 nm of the Utrecht Photometric System (≡ UPS) will be presented.

C. İbanoğlu (ed.), Active Close Binaries, 331–346.

2 Photometry and the period

2.1 The photometric observations

In the winter seasons 1983/1984, 1984/1985 and 1985/1986 AW UMA has been observed in the wavelengths 474, 579, 672, 781 and 871 nm of the Utrecht Photometric System (≡ UPS) [see Heintze (1989) and Provoost (1981)].

2.2 The period

From 10 well-covered minima their moments of minimum (O) were determined with the method of Van Gent (1989), which is a refinement of Kalv's (1979) method. These times were compared with calculated ones (C) according to the ephemeris of Dworak and Kurpinska (1975):

$$C = JD_{\odot} = 2438044.7812 \quad + 0.43873235 \times E \tag{3}$$

The (O-C) values as found by Hrivnak (1982) show clearly a decrease in the period, already noted by Woodward et al. (1980) and Kurpinski Winiarska (1980) with respect to the measurements before E ≈ 12 618. In 't Zand (1990) has shown that the (O-C) values as found from the UPS measurements and formula (3) are a straight continuation of Hrivnak's determinations. A combination of his observations and those in the UPS results in an ephemeris.

$$C = 2438044.8117 + 0.43872934 \times E \quad \text{for } E > 13440 \tag{4}$$

This formula covers a time interval from E = 13441 to E = 19287 (included). In that time interval also (O-C) values have been published by Hart et al. (1979), Istomin et al. (1980), Kurpinska-Winiarska (1980), Mikolajewska & Mikolajewski (1980) and Srivastava & Padalia (1986). However, the latter ones deviate very much (about 0.013) from all the other ones and have been omitted as suggested by In 't Zand et al. (1990).

In order to find at least two linear ephemeri for which as much as possible observed moments of minimum could be taken together, it turned out to be appropriate to combine the points No. 1 to No. 20 (included) and the points No. 20 to No. 52 (included) (omitting the numbers 39, 41, 42 and 45). In this way the following ephemeri were obtained:

$$C = 2438044.7813 + 0.43873231 \times E \quad \text{for } E < 10363 \tag{5}$$

$$C = 24438044.8127 + 0.43872928 \times E \quad \text{for } E > 10363, \tag{6}$$

formula (5) being nearly equal to formula (3) and formula (6) to formula (4). According to the formulae (5) and (6) a period jump of $\frac{\Delta P}{P} = -6.91 \times 10^{-6}$ should have occurred around E ≈ 10363. The phases of the UPS observations as well as those of the IUE spectra (see section 3) were calculated with formula (6). For a numerical as well as graphical presentation of the UPS light curves see In 't Zand et al. (1990), in which also three possible explanations are given for the period jump. In Table 1 (O-C) values of all determinations of moments of primary minimum mentioned here are given.

2.3 The UPS light curves

A remarkable feature of these light curves at the UPS wavelengths 474, 579, 672, 781 and 871 nm is, that they are equal to each other within the observational errors. In section 3.3.3 the effective temperature of the primary is estimated to be 7150 K. This value was used as a constant input parameter in the Wilson & and Devinney program to interpret the UPS light curves. Test iterations on each of the light curves separately showed that the light curves at λ = 579, 672, 782 and 871 could be solved simultaneously. Best fits were obtained if the 3rd light was taken zero. The light curve at λ = 474 nm provided calculated parameters deviating from those as obtained from thhe light curves at the other wavelengths a little bit more than the error given by the program. This light curve was excluded from the final interpretation.

Nevertheless it was difficult to find a final solution because near this solution several local minima were present in the quantity χ^2_{red}, with which the fit between the obser-vaitons and calculations is measured (see my paper on U CrB in these proceedings). There In 't Zand et al. (990) made iteration scans in the (q, i) subspace, q varying from 0.067 to 0.091 with steps of 0.004 and i from 76°.1 to 84°.1 with steps of 2°. For each pair 10 iterations turned out to be necessary. The absolute minimum ocurred to be at q = 0.074 and i = 78°.1. Some results are shown in Table 4. All results will be given in In 't Zand (1990).

3 The IUE observations and the effective temperature of the primary

3.1 The observations

During the first UPS observing season in the night April 3/4 1984, low dispersion IUE spectra in the wavelength region 200-300 nm were obtained at the phases: −0.066, −0.005, +0.116, +0.214, +0.294, +0.351, +0.401 and +0.456 by W.E.C.J. Van der Veen. The integration time was always about 0.002 in phase. The phases were calculated with ephemeris (6).

Table 1

(O-C) values of moments of minima as calculated with several ephemeri

no.	ref.	moment of minimum in JD⊙−2400000	E	(O-$C_{(3)}$)	(O-$C_{(5)}$)
1	[8]	38045.0020	0.5	+0.0014	+0.0013
2	¦	38045.8785	2.5	+0.0005	+0.0004
3	¦	38046.9740	5.0	−0.0009	−0.0010
4	[8]	38089.9707	103.0	+0.0001	−0.0000
5	[5]	38487.6825	1009.5	+0.0010	+0.0009
6	[5]	38501.7195	1041.5	−0.0014	−0.0015
7	[1]	41333.5178	7496.0	−0.0011	−0.0009
8	¦	41336.5898	7503.0	−0.0002	−0.0000
9	¦	42074.5380	9185.0	+0.0002	+0.0004
10	¦	42091.4302	9223.5	+0.0012	+0.0014
11	¦	42096.4745	9235.0	+0.0001	+0.0003
12	¦	42107.4425	9260.0	−0.0003	−0.0000
13	¦	42108.5393	9262.5	−0.0003	−0.0001
14	¦	42134.4249	9321.5	+0.0001	+0.0004
15	¦	42140.3470	9335.0	−0.0007	−0.0004
16	[1]	42148.4648	9353.5	+0.0006	+0.0008
17	[10]	42151.7539	9361.0	−0.0008	−0.0006
18	[10]	42152.8495	9363.5	−0.0021	−0.0018
19	[10]	42153.7287	9365.5	−0.0003	−0.0001
20	[1]	42461.5000	10067.0	+0.0002	+0.0005
period jump		at ≈ 42591.25	≈10363.0		(O-$C_{(6)}$)
21	[10]	43580.70074	12618.0	−0.0053	+0.0020
22	[10]	43621.7220	12711.5	−0.0055	+0.0021
23	[6]	43941.7714	13441.0	−0.0113	−0.0015
24	[3]	43945.7190	13450.0	−0.0123	−0.0025
25	[2]	43945.7220	13450.0	−0.0093	+0.0005
26	[2]	43948.7927	13457.0	−0.0097	+0.0001
27	[3]	43954.7158	13470.5	−0.0095	+0.0004
28	[4]	43966.3420	13497.0	−0.0097	+0.0002
29	[3]	43970.7281	13507.0	−0.0110	−0.0010
30	[3]	44274.7702	14200.0	−0.0104	+0.0017

no.	ref.	moment of minimum in JD⊙−2400000	E	(O-$C_{(3)}$)	(O-$C_{(5)}$)
31	[3]	44277.8396	14207.0	−0.0121	+0.0000
32		44283.7634	14220.5	−0.0112	+0.0010
33	[3]	44292.5358	14240.5	−0.0134	−0.0012
34	[6]	44294.5093	14245.0	−0.0142	−0.0020
35	[7]	44320.39378	14304.0	−0.01495	−0.0025
36	[7]	44343.43098	14356.5	−0.01120	+0.0014
37	[3]	44608.8622	14961.5	−0.0131	+0.0014
38	[3]	44664.7993	15089.0	−0.0143	+0.0015
39	[9]	45768.1950	17604.0	−0.0305	— *)
40	[11]	45773.4679	17616.0	−0.0224	+0.0002
41	[9]	45783.1060	17638.0	−0.0364	— *)
42	[9]	45795.1712	17665.5	−0.0363	— *)
43	[11]	45809.4445	17698.0	−0.0218	+0.0010
44	[11]	45814.4873	17709.5	−0.0245	−0.0016
45	[9]	45821.2742	17725.0	−0.0379	— *)
46	[11]	46040.6519	18225.0	−0.0264	−0.0019
47		46047.6727	18241.0	−0.0253	−0.0008
48		46073.5220	18345.5	−0.0235	+0.0013
49		46100.5407	18361.5	−0.0246	+0.0004
50		46497.5911	19266.5	−0.0269	+0.0007
51		46499.5648	19271.0	−0.0275	+0.0002
52	[11]	46506.5845	19287.0	−0.0275	+0.0002

*) see In't Zand et al., 1990

[1] Dvorak & Kurpinska, 1975
[2] Hart et al., 1979
[3] Hrivnak, 1982
[4] Istomin et al., 1980
[5] Kalish, 1965
[6] Kurpinska-Winiarska, 1980
[7] Mikolajewska & Mikolajewski, 1980
[8] Paczynski, 1964
[9] Srivastava & Padalia, 1986
[10] Woodward et al., 1980
[11] In't Zand et al., 1990

$C_{(3)}$, $C_{(5)}$ and $C_{(6)}$ calculated moments of minimum according to formula (3), (5) and (6) respectively.

3.2 An IUE light curve

In general the appearance of these spectra is changing quite a bit from phase to phase. Only just longward of the Mg II lines in the wavelength region from about 282 to about 284 nm (this is 10 pixels wide) the shape of the spectrum looked repeatable (it is a local maximum in the psuedo continuum). The mean fluxes in the wavelength region 282-284 nm of the available spectra provide part of an IUE light curve covering the phases −0.116 to +0.456. This partial light curve is equal to the same part of the UPS light curves within the observational errors, suggesting that the effective temperature is constant over the common envelope. For in this case at each wavelength the observed luminosity is proportional to the projected area of the common envelope in the sky. A more or less constant effective temperature over the whole common envelope will be achieved if the common envelope is in convective equilibrium. For then the exponent in Von Zeipel's theorem is small. As can be seen from Table 4, the difference in effective temperature between primary and secondary is of the order of (200 ± 140) K acording to the standard output of the Wilson & Devinney progam. In reality this difference is smaller whenever the reflection effect is taken into acount (see section 4.2 of my U CrB paper in these Proceedings). Therefore it may be assumed that the common envelope is in convective equilibrium even at spectral type F0-F2.

3.3 Metal abundance, effective temperature of the primary and the distance of the system to Earth

3.3.1 Information from the spectrum at secondary minimum

According to the interpretation of the UPS light curves, the secondary of AW UMa is already totally eclipsed at phase 0.456, at which phase a low dispersion long wavelength IUE dispersion was obtained. So the IUE spectrum at that phase is a spectrum of the primary alone. This absolute calibrated IUE spectrum has been compared with energy distributions calculated by Kurucz (1979), $F_{Kur}(\lambda_i)$. Because the resolution of the IUE low dispersion spectra is about 45 × larger than the resolution of the Kurucz' energy energy distribution, the IUE fluxes $f_{IUE}(\lambda_i)$ were binned into wavelength intervals used by the models [see section 2.2 of In 't Zand (1990)]. For the best-fit Kurucz model, the following equation should hold for each λ_i:

$$\frac{F_{Kur}(\lambda_i)}{f_{IUE}(\lambda_i)} = \left(\frac{d}{R_{v,1}}\right)^2 = c_i = \text{constant,} \qquad (7)$$

where d = distance of AW UMa to Earth and $R_{v,1}$ = volume radius of the primary component (see Table 4).

Normally c_i is not constant. Now the best-fit model is that model for which c_i is as constant as possible or if

$$\frac{1}{N}\sum_{i=n}^{N} c_i = < c > \pm\varepsilon, \tag{8}$$

for which ε reaches a minimum value.

Because $F_{Kur}(\lambda_i)$ depends on T_{eff}, log g and log Z/$Z_\odot$ [(Z/$Z_\odot$) being the metal abundance of the star considered with respect to that of the Sun], this holds for ε also. So

$$\varepsilon = \varepsilon(T_{eff_1},\ \log g_1,\ \log Z_1/Z_\odot) \tag{9}$$

During the iterative procedure to find the final solution log g_1, converged to 4.1. For a detailed description of the procedure followed see In 't Zand 1990. Here shortly will be indicated that successively for log $Z_1/Z_\odot$ was taken −1.0; −0.5; 0 and +0.5. For each of these values T_{eff_1} was taken successively 6000 + n × 100 K (n = 1,2,...,20) and for each of these combinations ε was calculated. For each log $Z_1/Z_\odot$ value considered a T_{eff} could be found for which ε reached a minimum value: ε_{min}. The four ε_{min} values (one for each log $Z_1/Z_\odot$ considered) turned out to be nearly equal to each other. The relation between T_{eff} and log $Z_1/Z_\odot$ as obtained form the four "constant" ε_{min} values turned out to be linear:

$$T_{eff_1} = 7155(\pm 8) + 1070(\pm 12) \times \log Z_1/Z_\odot. \tag{10}$$

Each of the 4 solutions defines a value for $\frac{d}{R_{v,1}}$ [see formula (7)] and therefore a distance d. In Table 2 the four log $Z_1/Z_\odot$ values for which above mentioned calculations were carried out are given in the first column; the T_{eff_1} values as calculated from (10) in the second column and the distances , d, as obtained from (7) in the third column. In the fourth column the absolute visual magnitude of the primary, $M_{V,1}$, as calculated with formula (11) is given.

$$M_V = m_V + 5 - 5\log\ d, \tag{11}$$

m_V being taken from section 1 assuming the interstellar extinction to be zero.

Table 2

The effective temperature, the distance to Earth and the absolute visual magnitude of the primary of AW UMa as a function of the metal abundance, assuming log $g_1 = 4.1$

(1) log $Z/Z_\odot$	(2) T_{eff} (K)	(3) d (pc)	(4) M_{V_1} (mag)	(5) M_{bol_1} (mag)	(6) BC	(7) $M^*_{V_1}$ (mag)	(7)-(4) ΔM_{V_1}
−1.0	6100	61	3.15	3.276	−0.131	3.407	0.26
−0.5	6600	75	2.70	2.934	−0.068	3.002	0.30
0.0	7150	90	2.31	2.587	+0.002	2.585	0.27_5
+0.5	7700	105	1.98	2.265	+0.043	2.221	0.24

column (5) from column (2) and $R_{v,1}/R_\odot = 1.78$ (see Table 4)
column (6) from Table 2 of Buser & Kurucz (1978) and formula (14)
column (7) = column (5) - column (6)

Now $M^*_{V_1}$ can also be estimated with

$$M_{bol} = M_{bol_\odot} + 10 \log T_{eff\odot} - 5 \log R/R_\odot - 10 \log T_{eff} \tag{12}$$

and

$$M^*_V = M_{bol} - BC. \tag{13}$$

In the fifth column of Table 2 the M_{bol_1} are given according to formula (12), using T_{eff} from column (2) of Table 2 and the volume radius of the primary $R_{v,1}$ (see Table 4). Reliable values for BC are not yet available. They have been taken from Buser & Kurucz' (1978) Table 2 [being equal to those in Table VB of Kurucz (1979)] assuming log g_1 = 4.1 and the four log $Z_1/Z_\odot$ values and using the relation given by Buser & Kurucz:

$$BC = BC_T + 0.1, \tag{14}$$

where BC_T are the tabulated values of Buser & Kurucz (1979).
The BC and the M_{V_1} values as found in this way are given in the sixth and seventh column of Table 2 respectively. The last column of Table 2 gives the differences between the two M_{V_1} values. This difference is more or less constant:
(0.27 ± 0.03) mag and could be compensated for by enlarging the distance as given in the third column of Table 2 by a factor 1.13. However the luminosity of the system can change also (see end of section 5).

As already stated before, the BC's are not yet known accurately. This is illustrated in Table 3. There T_{eff}, $M/M_\odot$, $R/R_\odot$, log g and BC values are given as a function of spectral type (A8V-G8V) according to Straižys & Kuriliene (1981) and to Habets & Heintze (1981). The differences between the two relations with respect to T_{eff}, $M/M_\odot$, $R/R_\odot$ and log g are relatively small, however with respect to the BC

Table 3

T_{eff}, M/M$_\odot$, R/R$_\odot$, log g and BC as a function of spectral type according to [a] Straižys & Kuriliene (1981) and [b] Habets & Heintze (1981) for mainsequence stars A8-G8 assuming log Z/Z$_\odot$ = 0.

	(1) T_{eff} (K)			(2) M/M$_\odot$			(3) R/R$_\odot$			(4) log g			(5) BC		
	[a]	[b]	(c)	[a]	[b]	(d)	[a]	[b]	(e)	[a]	[b]	(f)	[a]	[b]	(g)
G8	5250	5370	5310	0.91	0.83	0.87	0.83	1.00	0.91_5	4.56	4.36_5	4.46	-0.13	-0.57	-0.244
G5	5500	(5575)	5540	0.95_5	0.90_5	0.93	0.91	(1.05)	0.98	4.49	4.35_5	4.42_5	-0.09	-0.46	-0.153_5
SUN		5770		1.00			1.00			4.44					-0.108
G2	5790	5770	5780	1.05	0.98_5	1.02	1.02	1.11	1.065	4.40	4.34_5	4.39_5	-0.07	0.34	-0.106_5
G0	5940	5950	5945	1.05	1.04	1.04_5	1.07	1.15	1.11	4.39	4.33_5	4.37	-0.05	-0.27	-0.087
F8	6150	6150	6150	1.10	1.09	1.09_5	1.15	1.19	1.17	4.35	4.32_5	4.34	-0.03	-0.21	-0.069_5
F5	6500	6150	6150	1.20	(1.19_5)	1.20	1.29	(1.27)	1.28	4.28	4.31	4.31	-0.02	-0.10	-0.043
F2	7000	7040	7020	1.35	1.31	1.33	1.41	1.34	1.37_5	4.26	4.30	4.29	0.01	-0.07	-0.012_5
F0	7290	7420	7355	1.44	1.40	1.42	1.41	1.40	1.40_5	4.28	4.29	4.30	0.02	-0.10	-0.0008
AW UMa	see Section 3.3.3.			1.51			1.78			4.1				see Table 2	
A8	(7780)	7800	7780	(1.59)	1.53	1.56	(1.50)	1.48	1.49	(4.27)	4.28	4.29	(0.02)	-0.16	-0.0003

Values between brackets are interpolated values.
The values in the columns (c), (d) and (e) are the mean of the values at [a] and [b] in the columns (1), (2) and (3) respectively.
The values in column (f) were calculated from those in the columns (d) and (e) with the formula $\log g = \log M/M_\odot - 2 \log R/R_\odot + \log g_\odot$.
The values in column (g) were taken from Table 2 of Buser & Kurucz (1978) and formula (14) assuming $\log Z/Z_\odot = 0$.

they are large (columns (5) [a] and (5) [b]). For comparison, the BC's as obtained from Table 2 of Buser & Kurucz (1978) and from formula (14) are given in column (5) (g).

3.3.2 Information from the $[(B-V), (U-B)]$ diagram

In Figure 1 that part of the $[(B-V), (U-B)]$ diagram is shown for which $0 \leq (B-V) \leq +0.55$ and $+0.25 \geq (U-B) \geq 0.15$. The long dashed lines in this figure define the upper and lower boundary of all luminosity class V and IV stars of spectral type A3 to F8 as published in the Bright Star Catalogue (Hoffleit & Jaschek, 1982) and as far as they lie in this Figure. The stars that mark these boundaries are indicted by open squares, their BS numbers and spectral types being given also.

The position of a star in this diagram is determined mainly by its intrinsic color indices, its rotation (v sini) and the interstellar extinction. The intrinsic color indices depend on the chemical composition (metal abundance).

In the diagram, positions of stars are given for which metal abundances [Me/H] and [Fe/H] have been determined by Nissen (1981) and by Nissen (1981) and Saxner & Hammerbäck (1985) respectively. Fortunately these two (in an observational way differently obtained) abundances agree very satisfactorily with each other. The metal abundances are given in this Figure.

Now it is tried to find in Figure 1 a borderline between negative and positive metal abundances. The threefold line tentatively gives part of such a borderline. This line is more or less arallel to the upper and lower boundaries and lies more or less in the midst between them. The position of AW UMa is close to the threefold line in Figure 1. This position is not affected by interstellar reddening (see section 1).

The rotational velocity of the primary of AW UMa is about 200 km/s. In figure 1 a vector is pointing to AW UMa. This vector starts at a point where – according to the calculations of Maeder (1971) – a nonrotating star of the same mass as the primary of AW UMa should be. This vector is also more or less parallel to the upper and lower boundaries in Figure 1.

3.3.3 The effective temperature of the primary of AW UMa

On account of the two facts mentioned above the metal abundance of AW UMa is adopted to be solar: $\log Z_1/Z_\odot = 0$. Therefore according to formula (10);

$$T_{eff_1} \simeq 7150K$$

This is in agreement with Table 3. For according to this table, a main sequence star of the same mass as the primary of AW UMa (1.51 solar masses) should have a $T_{eff} \simeq 7630$ K. The radius of the primary of AW UMa (1.78 solar radii) is about

1.22 × larger than that of a main sequence of 1.51 solar masses (1.46 solar radii) causing for the primary of AW UMa:

$$T_{eff_1} \simeq 6910K$$

which is close to the value of 7150 K found above in which the error is 100 K. An absolute energy distribution in the visual and near-infrared is hardly needed.

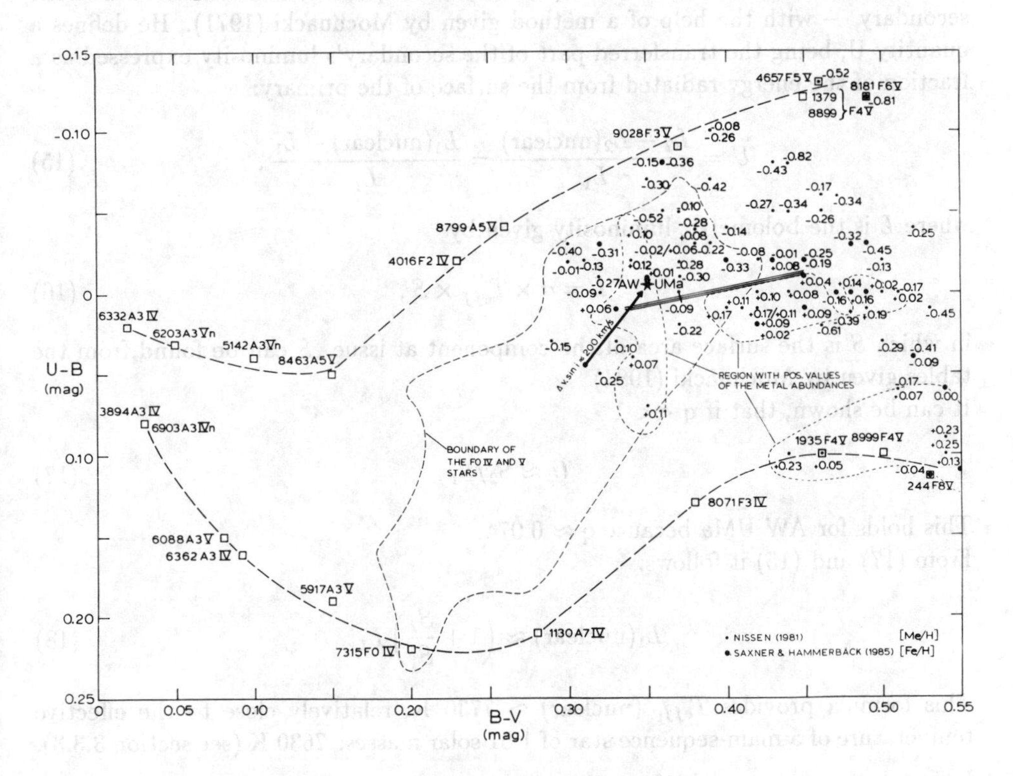

Figure 1: Part of the $[(B-V), (U-B)]$ diagram. For explanation see section 3.3.2

4 Results of the analysis

In the paper of In't Zand et al. (1990) the details of the analysis will be given. The mass of the secondary is only 0.11 solar masses. Therefore it is questionable whether still nuclear reactions are going on in its interior. If any, T_{eff_2} (nuclear) $\leq$ 3000 K, much lower than the observed 7 000 K. In Table 4 some results for AW UMa are given.

5 Energy transport from the primary to the secondary

To explain the high T_{eff_2} with respect to its mass, energy has to flow from the primary to the secondary. According to In 't Zand (1990), it is possible to find T_{eff_1} (nuclear) — this is the effective temperature of the primary if not connected to the secondary — with the help of a method given by Mochnacki (1971). He defines a quantity U, being the transferred part of the secondary's luminosity expressed as a fraction of the energy radiated from the surface of the primary:

$$U = \frac{L_2 - L_2(\text{nuclear})}{L_1} = \frac{L_1(\text{nuclear}) - L_1}{L_1}, \tag{15}$$

where L is the bolometric luminosity given by

$$L = \sigma \times T_{eff}^4 \times S, \tag{16}$$

in which S is the surface area of the component at issue. S can be found from the tables given by Mochnacki (1984).
It can be shown, that if q << 1

$$U \approx S_2/S_1 . \tag{17}$$

This holds for AW UMa because q ≈ 0.074.
From (17) and (15) it follows:

$$L_1(\text{nuclear}) \approx \left(1 + \frac{S_2}{S_1}\right) L_1 . \tag{18}$$

This formula provides T_{eff_1} (nuclear) ≈ 7730 K, relatively close to the effective temperature of a main-sequence star of 1.51 solar masses: 7630 K (see section 3.3.3).

In any case energy transfer from the primary to the secondary is necessary to cause a $T_{eff_2} \approx 7000$ K. It is hoped that the theory of Kähler (1980) can be applied on this system. In this theory circulation currents transport the energy from one to the other component. These currents need space. As figure 2 shows in the case of AW UMa the neck between the two components in relatively large, hopefully large enough for the circulation currents.

6 Final remarks

At the beginning of this study on AW UMa it was hoped to find changes in the lightcurves. If any there must have been only small ones: all lightcurve raw data lie within a band of about 0.06 mag for each UPS wavelength, indicating small seasonal

Table 4

Some results for AW UMa

	Paczynski (1976)		McLean (1981)	Rensing et al. (1985)		
rad.-vel. curves						
disp. (Å/mm)	75		20	12		
γ0	−1		−17	−0.8		
K1 (km/s)	28		29	22.2		
K2	–		423	–		
q	–		≈ 0.07	–		
lightcurves	"					UPS
λ	U, B, V					474; 579; 672; 781; 871 nm
analysis by	Mauder (1972)	Wilson & Devinney (1973)				in 't Zand et al. (1990)
i						78.1
a/R⊙						2.86
q				if 0.070	if 0.079	0.074
M1/M⊙	2.38	3.1	3.9 (lower limit)	1.7	1.2	1.5
M2/M⊙				0.12	0.09	0.11
Rv,1/R⊙						1.8
Rv,2/R⊙						0.61
log g1/log g2						4.1 / 3.9
Teff1 (K)	7000*)					7150
L1						7.2
nuclear Teff1 (K)						7330
nuclear L1						8.1
gravity darkening	as fixed parameter: Lucy's (1967) value					0.08
	as free parameter: a better fit could be obtained with					0.11

*) Mochnacki & Doughty (1972) assumed $T_1 = 7300$ from spectral type
Woodward et al. (1980) assumed $T_1 = 7175$ from u; v; b; y; r photometry

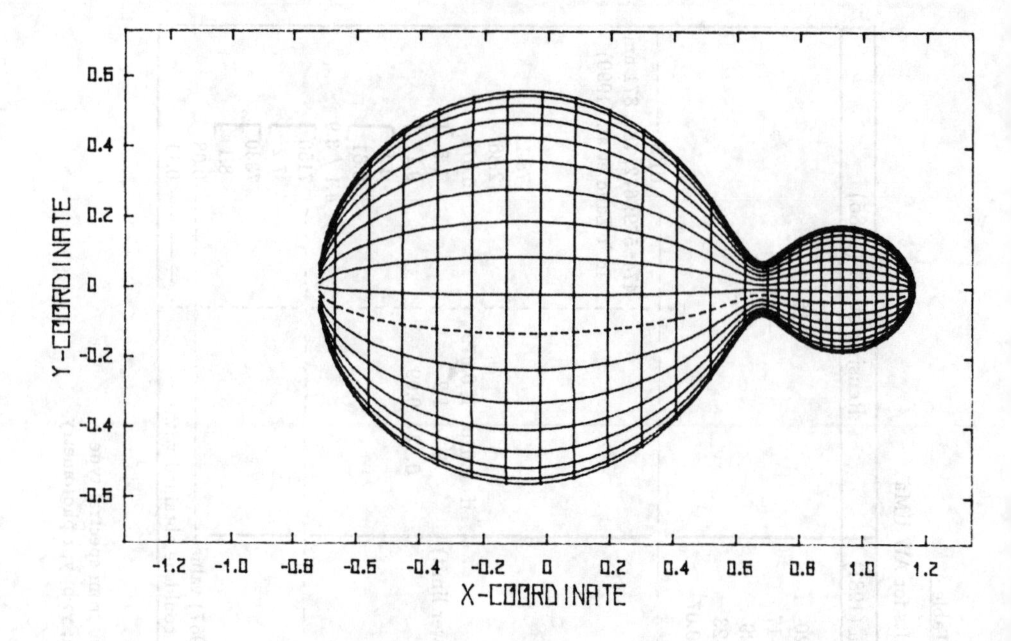

Figure 2: The model for AW UMA.

variations.

During the conference Derman & Demircan (1989) showed me U, B, V lightcurves of AW UMa which they obtained in the nights 1989 February 20/21 and 21/22 and March 17/18, 18/19 and 20/21. These lightcurves were not repeatable at all, showing luminosity variations of at least 0.1 mag within a time interval corresponding to 0.2 in phase and and lasting sometimes at least 0.6 in phase.
This means that the apparent visual magnitude of the primary of AW UMa (m_{V_1} = 7.08 mag) as measured by Paczynski (1961) [see formula (1)] does not have to hold for the time interval in which the UPS lightcurves were obtained. This causes uncertainties in M_{V_1} in column (4) of Table 2.

Acknowledgement

It is a pleasure to thank G.W. Geijtenbeek, J.H.G. Rosenbaum and E.B.J. van der Zalm for computational assistance, E. Landré for drawing the figures, E.I. Tokamp for typing the manuscript.

References

Buser, R. & Kurucz, R.L.: 1978, Astron. Astrophys. **70**, 555

Derman, E. & Demircan, D.: 1989, private communication

Dworak, T.Z. & Kurpinska, M.: 1975, Acta Astronomica,**35**, 417

Habets, G.M.H.J. & Heintze, J.R.W.: 1981, Astron. Astrophys. Suppl. Ser. **46**, 193

Hart, M.K., King, K., McNamara, B.R., Seaman, R.L., Stoke, J.: 1979, Cie. 27 IAU, Inf. Bull., Var. Stars, No. 1701

Heintze, J.R.W.: 1989, Space Sci. Rev., **50**, 257

Hoffleit, D. & Jaschek, C.: 1982, The Bright Star Catalogue, 4th ed., Yale Univ. Obs., New Haven, (Conn.)

Hrivnak, B.J.: 1982, Astrophys. J., **260**, 742

In 't Zand, J.J.M., Heintze, J.R.W. & Van 't Veer, F.: 1990, submitted to Astron. Astrophys.

Istomin, L.F., Orlov, L.M., Kulagin, V.V.: 1980, cie. 27 IAU, Inf. Bull. Var. Stars, No. 1802

Kähler, H.: 1986, Mitt. Astron. Gesellschaft, **67**, 85

Kalish, M.S.: 1965, Publ. Astron. Soc. Pacific, **77**, 36

Kolv, P.: 1979, Tartu Astrofüüsika Obs., Teated, No. 58, 64

Kurpinska-Winiarski, M.: 1980, Cie 27 IAU, Inf. Bull. Var. Stars, No. 1843

Kurucz, R.L.: 1979a, Astrophys. J. Suppl. Ser., **40**, 1

Kurucz, R.L.: 1979b, in A.G. Davis Philip (ed.), Problems of Calibration of Multicolor Photometric, Dudley Obs. Rep. **14**, 363

Kurucz, R.L.: 1981, in A.G. Davis Philip & E.D. Hayes (eds.), Astrophysical Parameters for Globular Clusters, Dudley Obs. Re.p. **15**, 289

Lucy, L.B.: 1967, Zs. Ap. **65**, 89

Maeder , A: 1971, chapter 3, thesis Genève

Mauder, H.: 1972, Astron. Astrophys.˙**17**, 1

McLean, B.: Monthly Notices Roy. Aston. Soc., **195**, 931

Mikolajewska, J. & Mikolajewski, M.: Cie 27 IAU, Inf. Bull. Var. Stars. No. 1812

Mochnacky, S.W.: 1981, Astrophys. J. **245**, 650

Mochnacky, S.W. & Doughty, N.A.: 1972, Monthly Notices Roy. Astron. Soc. **156**, 51

Nissen, P.E.: 1981, Astron. Astrophys. **97**, 145

Poezynski, B.: 1964, Astron. J. **69**, 124

Provoost, P.: 1980, Astron. Astrophys. Suppl. Ser. **40**, 129

Rensing, M.J., Mochnacki, S.W., Bolton, C.T.: 1985, Astron. J., **90**, 767

Saxner, M. & Hammarbäck, G.: 1985, Astron. Astrophys. **151**, 372

Srivastava, R.K. & Padalia, T.D.: 1986, Astrophys. Space Sci., **120**, 121

Straižys, V. & Kuriliene, G.: 1981, Astrophys. Space Sci. **80**, 353

Van Gent, R.H.: 1989, chapter 3, thesis Utrecht

Von Zeipel, H.: 1924, Monthly Notices Roy. Astron. Soc. **84**, 665, 684, 702

Wilson, R.E. & Devinney, E.J.: 1975, Astrophys. J., **166**, 605

Wilson, R.E. & Devinney, E.J.: 1973, Astrophys. J. **182**, 539

Woodward, E.J., Koch, R.H. & Eisenhardt, P.R.: 1980, Astron. J. **85**, 50

ROTATION AND ACTIVITY

Santo Catalano
Institute of Astronomy
University of Catania
Viale A. Doria I-95125 Catania Italy

ABSTRCT: The review summarizes the present knowledge of the rotation activity connection from different magnetic activity indicators and in different class of stars. The important role of active binaries like RS CVn and Algols in studying activity is emphasized. Implication and constraints of the observed activity-rotation relations to dynamo models are also discussed.

1 Introduction

Due to the lack of spatially resolved informations, the presence of active magnetic regions on the surface of late type stars can be inferred from a number of indirect observations. The solar analogy is a useful reference point to devise the diagnostic tools for stellar activity studies. Taking advantage of detailed observations of solar activity manifestations we may presume that hypothetical observations of the Sun at stellar distances would show long and short term variability of chromospheric, transition region and coronal spectral diagnostics. However a straight- forward extension of the solar-stellar analogy has to take into account instrumental sensitivity limitations, the effect of integrated light observations, the limits on time and spectral resolution and the different environment where the phenomena develop. As a matter of fact, the Sun, which is considered a good example of active star, when it is compared to other stars of the same or similar spectral type appears to be a marginally active star.

While the term activity is generally associated with variability, intrinsic variables like pulsating variables, irregular and semi regular variables are not considered active stars. From solar observations, we know that regions of strong, closed magnetic fields (flux tubes) are the basic structural elements of solar activity, where enhanced non-radiative heating take place. Stellar activity, include those phenomena that develop when local strong magnetic field modify the local momentum and energy balance to produce observable effect in a stellar atmosphere. Therefore, magnetic fields and intrinsic variability are the characteristic and identifyng aspects of *activity*. Stars showing variability interpreted as pure hydrodynamics process that develop on thermal time scale or magnetic A-type stars which show intrinsically stable magnetic fields and luminosity, are not considered as active stars in the sense specified above.

Stellar activity manifestations like the solar ones include dark photospheric spots, seat of strong magnetic field, which manifest as light changes in a time scale of the stellar rotational period or in a long term amplitude variations; bright chromospheric or coronal plages observable as enhanced emission in chromospheric lines or X-ray flux modulated by the rotation and long term cycles; flare eruption characterized by millisecond to hour light and/or lines brightenings.

The association of solar bright chromospheric and transition region plages with strong local magnetic field, and the brighter coronal X-ray emission largely confined to closed magnetic loops does suggest

C. İbanoğlu (ed.), Active Close Binaries, 347–362.

that these enhancements in emission are due to large non thermal heating resulting from conversion of magnetic energy into heat. The presence of chromospheric, transition region and coronal emission and their average strength, by analogy with the Sun, are commonly considered as a measure of the average *magnetic activity* level of late type stars. Since the early studies of Ca II emission in late type stars by O.C. Wilson (1963, 1966) it was realized that faster rotators show brighter H and K emission cores, i.e. they are more active. The key role of rotation in determining stellar activity has been demonstrated in a number of subsequent works (e.g. Skumanich 1972, Bopp and Fekel 1977, Pallavicini et al. 1981, Marilli and Catalano 1984, Noyes *et al.* 1984, Marilli *et al.* 1986). The dynamo mechanism (Parker 1955, 1979) for the generation of surface magnetic fields through field amplification in the interaction of rotation and convection provides the theoretical ground on the role played by rotation in determining the stellar magnetic activity. However, other parameters that determine the dynamical structure of stellar interior, like the depth of convection zone and the characteristic time of convection appear to play a critical role in the magnetic dynamo process.

Since rotation, convection depth, and convection velocity depend on the mass and age of stars it is not easy to separate the effect of rotation from the influence of stellar parameters in the magnetic field generation. Based on the largely increased data base of rotation and magnetic activity tracer measurements (from the photosphere to the coronae) several approaches for separating the effect of rotation from internal stellar properties on observed magnetic activity have been pursued.In this respect an important contribution came from the observation of active binaries like RS CVn. For systems with $P_{orb} \leq 20$ days the tidal force produce synchronous rotation (i.e. $P_{rot} = P_{orb}$), so that the rotation rate does not depend on stellar mass and age contrary to what happens on the main sequence. This allows to study the chromospheric and coronal activity independently of the age parameter (Basri 1987). The stellar activity - rotation connection has recently been discussed and main advances summarized in several papers (see e.g. Catalano 1984, Baliunas and Vaughan 1985, Rosner *et al.* 1985, Hartmann and Noyes 1987, Rodonó 1987, Linsky 1988).

2 Activity indicators and rotation

The various manifestations of the stellar activity in some case appear to be more characteristic of certain groups of type of stars so it is not easy to disclose their dependence on rotation or whether they are mainly driven by other stellar parameter. Let me summarize here the present status for the more conspicuous of them in order to try to sketch a global view of stellar magnetic activity.

- Star spots

Among the phenomena characterizing *Stellar activity* large - scale dark spots are *detected* from light variability in late type binaries and single stars. Nearly sinusoidal light variations, modulated by stellar rotation, indicative of an asymmetric distribution of spots have been observed in several type of stars, e.g. RS CVn, BY Dra, M dwarfs in young clusters and solar type stars. Stellar spots, like solar spots, are dark and cool relative to the stellar photosphere, so their detection against the bright photospheric background requires that the brightness contrast relative to the photosphere be large enough. Conclusive evidence that spots are dark and cool come from the reddening of color indices and strengthening of spectral indicators such as the TiO band (Ramsey and Nations 1980). Further evidence is given by the Doppler imaging techniques (e.g. Vogt and Penrod 1983). Spotted areas, covering less than 1% to 30 – 40% of the projected stellar disk, and spot temperatures cooler than the average photosphere by about 300-1500 K have been derived modeling multicolor light curve. Two or more spotted areas, including even polar spots are often needed to reproduce the observed light variations, (figure 1) (e.g. see Rodonó *et al.* 1986).

In analogy with sunspot, there are evidence that stellar spots are regions of strong magnetic field. However no direct measurements are possible of the magnetic field in these spots, because they are dark and thus contribute very weakly to the integrated stellar light. So that it is not available any statistical relation between the possible magnetic field and the parameter of stellar spots.

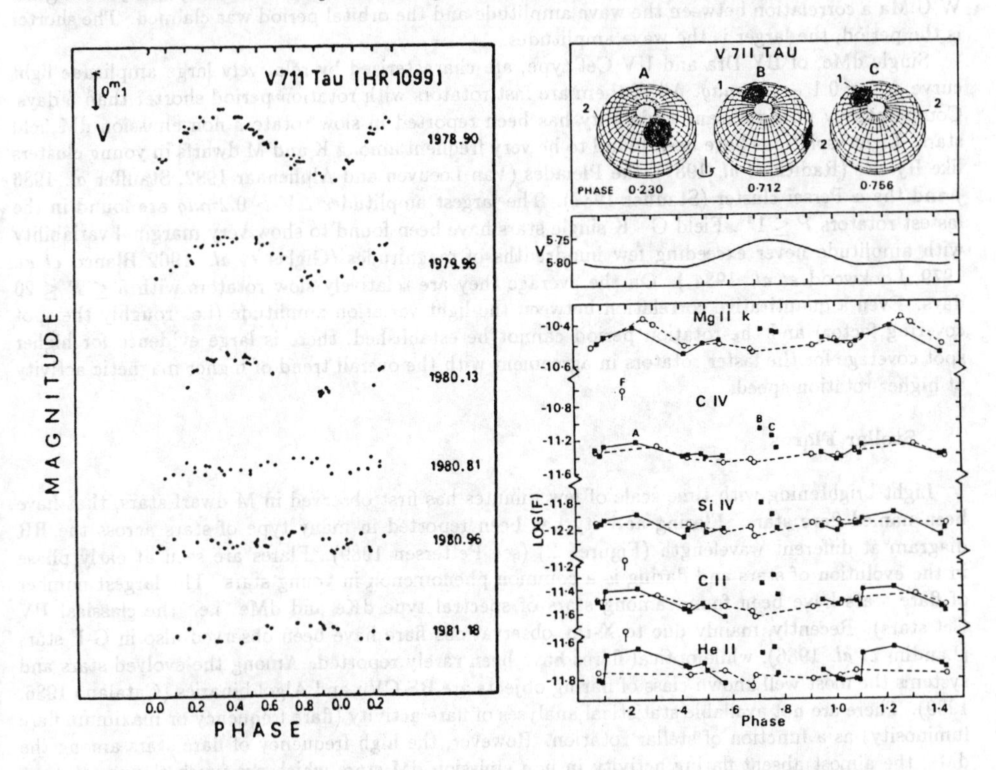

Figure 1. V light-curve of the RS CVn binary HR 1099 (left panel), Showing the Characteristic Photometric wave and its remarkable seasonal variations (from Rodonò 1983). Right panel example of spott modeling (top) and, chromospheric and TR line variation in anti-phase with the optical light-curve (from Andrews et al 1988).

Light curve amplitudes are highly variables both on time scales of the order of 10 days, due to the evolution and decay of localized magnetic structures, and of the order of 1-10 years, as a result of long term activity cycles. Taking the maximum amplitude of light variation as parameter for the spot activity we can investigate if there is any dependence from the rotation period.

Selected as a group the RS CVn stars after correction for the light contribution of the unspotted companion (Catalano Frisina Rodonò 1980), show the largest amplitude light curves $\Delta V = 0.5 - 0.6mag$. However there is no evidence for dependence of the amplitude from the orbital period (i.e. the rotation period, because the systems are synchronized.) (Catalano 1983). Only among the W U.Ma a correlation between the wave amplitude and the orbital period was claimed. The shorter is the period, the larger is the wave amplitudes.

Single dMe, of BY Dra and UV Cet type, are characterized by relatively large amplitude light curve $\Delta V = 0.1 - 0.3mag$. All of them are fast rotators with rotation period shorter than 5 days. Comparatively no significant variability has been reported in slow rotators non emission dM field stars. Light variations have been found to be very frequent among K and M dwarfs in young clusters like Hyades (Radick *et al.* 1987), the Pleiades (Van Leeuven and Alphenaar 1982, Stauffer *al.* 1986) and the α Persei cluster (Stauffer 1987). The largest amplitudes $\Delta V > 0.2mag$ are found in the fastest rotators $P \leq 1^d$. Field G - K single stars have been found to show very marginal variability with amplitude never exceeding few hundredths of magnitudes (Giglas *et al.* 1962 Blanco *et al.* 1979, Lookwood *et al.* 1984). On the average they are relatively slow rotators with $5 \leq P \leq 20$ days. Even a quantitative correlation between the light variation amplitude (i.e. roughly the spot covering factor) and the rotation period cannot be established, there is large evidence for higher spot coverage for the faster rotators in agreement with the overall trend of higher magnetic activity at higher rotation speed.

- Stellar Flares

Light brightening with time scale of few minutes has first observed in M dwarf stars, that have been named flare stars. Flaring activity has been reported in many type of stars across the HR diagram at different wavelength (Figure. 2) (see Pettersen 1989). Flares are seen at early phase in the evolution of stars and flaring is a common phenomenon in young stars. The largest number of flare stars have been found among stars of spectral type dKe and dMe (i.e. the classical UV Cet stars). Recently, mainly due to X-ray observations flare have been observed also in G-F stars (Landini *et al.* 1986), while optical flares have been rarely reported. Among the evolved stars and systems the most well known class of flaring objects are RS CVn and Algol binaries (Catalano 1986, 1989). There are not available statistical analyses of flare activity (flare frequency or maximum flare luminosity) as a function of stellar rotation. However, the high frequency of flare stars among the dMe, the almost absent flaring activity in non emission dM stars which are much slower rotators, and the aging effect clearly indicate the rotation as an important parameter for flare activity.
Convection parameters appear also to affect the flare activity. Pettersen (1989) found a well defined empirical relationship between flare luminosity and the volume of convection zone for a number of solar neighbourhood stars. A clear parallel behavior separate dMe and dM, the latter showing a smaller luminosity that can be accounted for by the slower rotation (Figure 3). A plot as a function of the Rossby number may probably put all the data on the same sequence, confirming that also flaring activity depend on the same parameter as the other activity indicators.
Comparison with evolved stars is not very easy because of the lack of comparable data. However giant stars in synchronized binary systems like RS CVn and Algol shows sizable flare activity in chromospheric and coronal indicators, while single stars of the same spectral type do not. The latter, due to the braking during the main sequence life time and the volume expansion off the main sequence are much slower rotators with respect to the similar counter part in the binary systems.

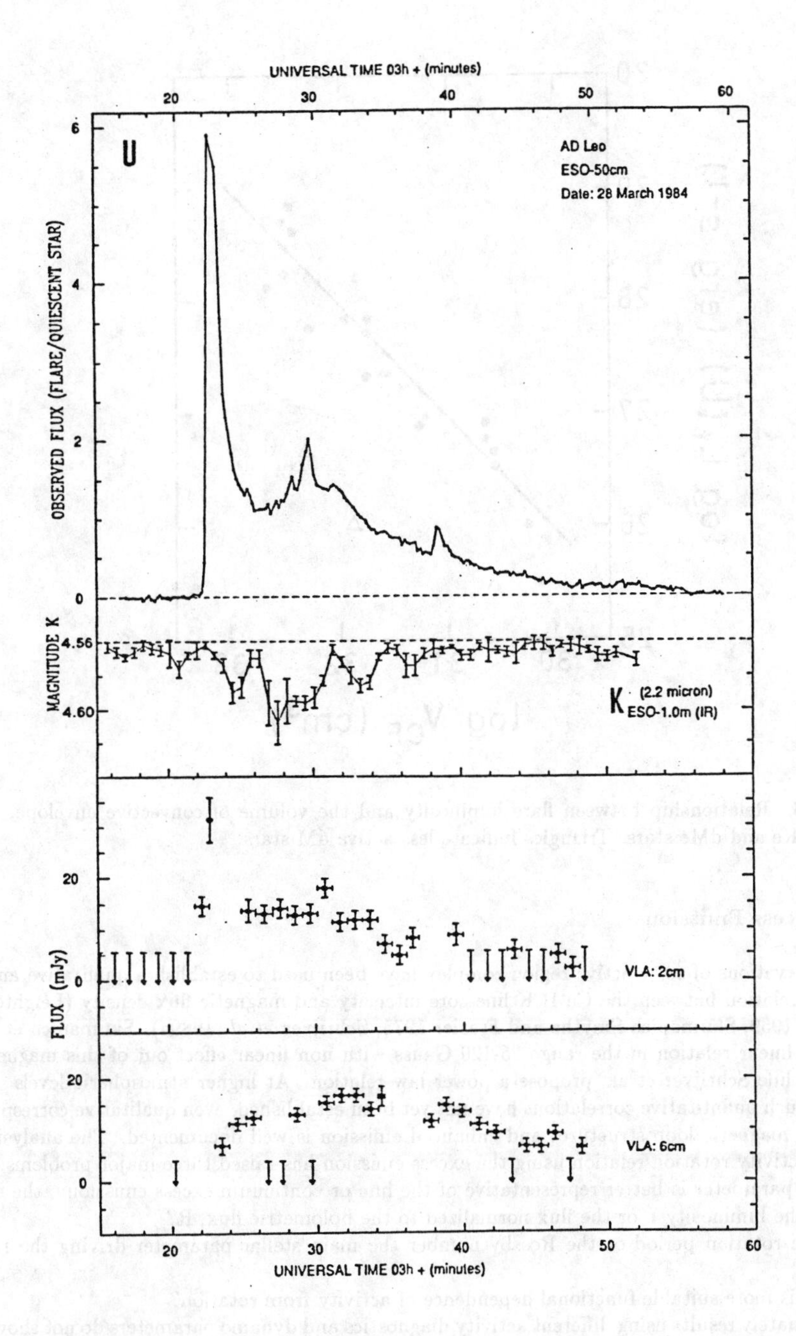

Figure 2. Simultaneous U, K, 2 cm and 6 cm observations of an intense flare in AD Leo (dM3.5e) (from Rodonó et al. 1989)

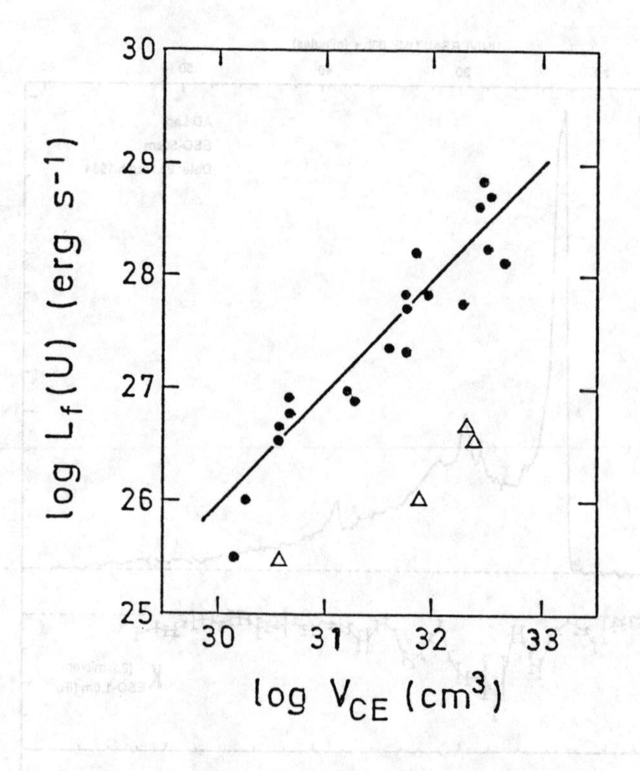

Figure 3. Relationship between flare luminosity and the volume of convective envelope, for very active dKe and dMe stars. Triangles indicate less active dM stars.

- **Excess Emission**

Observations of solar active region complex have been used to establish a qualitative and quantitative relation between the Ca II K line-core intensity and magnetic flux density (Leighton 1959, Howard 1959, Skumanich Smythe and Frazier 1975, Schrijver *et al.* 1989a). Skumanich et al. proposed a linear relation in the range 25-120 Gauss with non linear effect out of this magnetic field range, while Schrijver et al. propose a power law relation. At higher atmospheric levels, TR and corona such quantitative correlations have not yet been established, even qualitative correspondence between magnetic loop structures and enhanced emission is well documented. The analysis of the stellar activity rotation relation using the excess emission has raised three major problems:
a) what parameter is better representative of the line or continuum excess emission: the emission flux F, the luminosity L or the flux normalized to the bolometric flux, R?
b) is the rotation period or the Rossby number the main stellar parameter driving the magnetic activity?
c) what is more suitable functional dependence of activity from rotation?
Unfortunately results using different activity diagnostics and dynamo parameters do not show always the expected agreement. On the other hand there is not a unique criterion suggested by dynamo theory or observation constraints to parameterize stellar magnetic activity from excess emission.

However even still there are a large number of uncertainties and discrepancies the correlation between excess emission from chromosphere, TR, corona and rotation appear to give the more useful information for possible quantitative or functional dependence of magnetic activity from rotation.

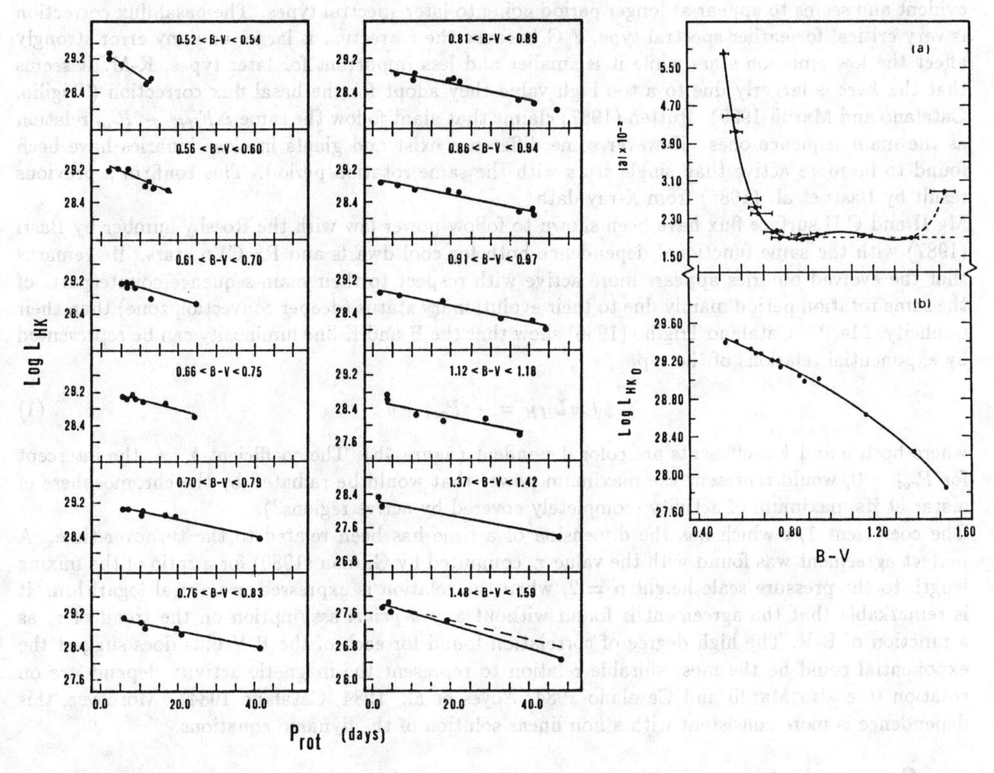

Figure 4. Log L_{HK} versus rotation period for narrow B-V intervals, linear fits log L_{HK} = -aP_{rot}+b are shown as solid lines(left). Coefficients *a* and *b* of the fit plotted as a function of B-V are also shown (right) (from Marilli et al. 1986)

- *Chromospheric indicators*

The simple proportionality between the Ca II emission and the angular rotation Ω suggested by Skumanich (1972) has further been studied using more accurate data. Noyes et al. (1984) showed that the normalized Ca II H K line flux normalized to the surface flux $R'_{HK} = F'_{HK} / \sigma T^4_{eff}$ (where F'_{HK} is the surface flux in H and K lines measured with Mt. Wilson photometer and corrected for the photospheric contribution outside the H_1 and K_1) is a smooth function of the Rossby number. $R_o = P_{rot}/\tau_c$. The convective turn-over time τ_c was adopted from Gilman (1980) calculated one scale height above the bottom of convection zone. A similar relation was found by Hartmann et al. (1984) using Mg II h and k line fluxes normalized to the bolometric flux. The correlation appears to be independent from the stellar mass, and it apply to main sequence stars. A power law relation represents equally well the corrected Ca II chromospheric flux F_{HK} as a function of the rotation period P_{rot} (Noyes *et al* 1984), but with a slight larger scatter.

Zwaan and co-workers (Zwaan 1986, Rutten 1987) using almost the same data but a different correction for the *basal flux* found color depended relations of the excess flux ΔF and the rotation period. Their relations show a *knee* that is very pronounced at early spectral types, becomes less evident and seems to appear at longer period going to later spectral types. The basal flux correction is very critical for earlier spectral type, F G because the correction is larger and any error strongly affect the low emission stars while it is smaller and less important for later types, K-M. It seems that the *knee* is largerly due to a too high value they adopt for the basal flux correction (Trigilio, Catalano and Marilli 1989). Rutten (1987) claims that giant follow the same $\Delta F_{HK} - P_{rot}$ relation as the main sequence ones. However some difference exist and giants in close binaries have been found to be more active than single stars with the same rotation period. This confirm a previous result by Basri et al. (1985) from X-ray data.
Mg II and C II surface flux have been shown to follow power law with the Rossby number by Basri (1987) with the same functional dependence both for cool dwarfs and RS CVn stars. He remarks that the evolved binaries appears more active with respect to their main-sequence counterparts of the same rotation period mainly due to their evolutionary status (deeper convection zone) than their duplicity. Marilli, Catalano Trigilio (1986) show that the H and K line luminosity can be represented by exponential relations of the type

$$logL_{HK} = -aP_{rot} + b \tag{1}$$

where both a and b coefficients are color dependent (figure 4b). The coefficient b, i.e. the intercept for $P_{rot} = 0$, would represent the maximum power that would be radiated by the chromosphere of a star at its maximum of activity (completely covered by active regions?).
The coefficient 1/a which has the dimension of a time has been related to the turnover-time. A perfect agreement was found with the value τ_c computed by Gilman (1980) for a ratio of the mixing length to the pressure scale height $\alpha = 2$, when the relation is expressed in natural logarithm. It is remarkable that the agreement is found without any *a-priori* assumption on the trend of τ_c as a function of B-V. The high degree of correlation found for each of the B-V bins does suggest the exponential could be the most suitable relation to represent the magnetic activity dependence on rotation (see also Marilli and Catalano 1984, Noyes et al. 1984, Catalano 1984). Moreover, this dependence is more consistent with a non linear solution of the dynamo equations.

- *Coronal indicators*

Since the work of Ayres Marstad and Linsky (1981) it has been confirmed by many authors (see Oranje, Zwaan and Middelkoop 1982, Marilli and Catalano 1984, Schrijver 1983, 1986) that line flux from the chromosphere (Ca II, Mg II, Si II) transition region (C II, CIV, Si IV) and the corona (X-ray) follow very tight power-law correlations. The slopes of these power-law relations are generally found to steepen when hotter diagnostics are compared with cooler diagnostics. Soft X-rays relative to the low temperature chromospheric diagnostic present the greatest slope ($\approx$ 2.5). This does suggest that it is not necessary to study *activity* in all the various diagnostics available to gain fundamental knowledge in its dependence on relevant parameters. However, since some diagnostic can be difficult to be observed in some type of stars due to contrast, effect instrument sensitivity limits and, etc. , it could be of interest to mention some of the results obtained using the coronal diagnostics like X-ray and radio continua. This two diagnostics are of great importance for discussing rotation activity relation in binaries since they are free of photospheric contamination of the companion as it happens with many of chromospheric diagnostics for systems containing early type stars.

Controversial correlations have been presented for coronal diagnostics. Ayres and Linsky (1980) from a limited sample of 15 stars found that $R_x = L_x/L_{Bol}$ was a steep function of Vsin i increasing roughly as $(Vsin\ i)^3$ Pallavicini et al. (1981) show that the X-ray emission luminosity L_x for single stars follows a quadratic dependence from the rotational velocity and that the RS CVn systems

roughly fall in the relation as a group but do not show specific correlation. Walter and Bowyer (1981) show that RS CVn stars exhibit a nice linear relation between R_x and the angular velocity. Walter (1982) found that to different power law for the slow and rapid rotators is a better description of the data, with a change in the slope at rotation period about 12 day. The slope derived for F, G and K rapidly rotating stars ($P \leq 12$ days) are identical within the errors, and are identical to that derived for the RS CVn binaries. An alternative description of the data given by Walter (1982) involves an exponential decay of R_x with Ω. X-ray emission luminosity from young star in the Pleiades is either consistent with a power law or an exponential dependence (Caillout and Helfand 1985) in agreement with previous report on Hyades stars (Stern *et al.* 1981). Marilli and Catalano (1984) show that when exponential relation between emission luminosity and rotation period are adopted an unique functional dependence hold for all chromospheric TR and coronal diagnostics with slope consistent with the flux to flux correlations. Mangeney and Praderie (1984),making an interesting attempt for a global description of the activity, show that the X-ray flux could be related by a single power-law to an effective Rossby number $R_* = \frac{1}{2V_m/\Omega}L$ ($V-m$ is the maximum convective velocity and L the total depth of convection zone). The remarkable result is that the relation seems to hold for main sequence stars of spectral type from O to M.
Binaries have been left aside in many of these discussions mainly because of the complication of the dependence of stellar parameters themselves on rotation in the RS CVn as pointed out by Rengarajan and Verma (1983) and the possible effect of tidal interaction and emission. The lack of correlation between L_X and rotation shown by Majer et al. (1986) for RS CVn binaries according to Basri (1987) results from the use of the non appropriate parameter L_x to describe activity. Considerations on surface X-ray flux lead Basri *et al.* (1985) to show that a coherent interpretation for the coronal emission of these stars is possible.
The ***radio emission*** flux is an additional important diagnostic of the coronal magnetic activity of stars. The interpretation of stellar radio data is not straightforward as a coronal diagnostic. As a comparison, radio radiation from solar active regions depends on both atmospheric level location and observing frequency. For example, the emission from an active region a 2 cm, 6 cm and 20 cm originates respectively near the feet, legs and top of magnetic loops (Lang, Willson and Gaizauskas 1983). Moreover the emission mechanisms appear to be different at the different frequencies.The 20 cm emission is apparently the bremsstrahlung emission of coronal loops,the 2 cm seems to be the thermal bremsstrahlung emission from the transition region and the 6 cm emission is probably due to gyroresonant emission of thermal electron in the magnetic field . A number of recent survey have increased greatly the number of sources and have established classes of radio stars. The surveys related to the active stars we are interested on have shown high frequency of detection on RS CVn (Morris and Mutel 1988, Drake *et al.* 1989, Slee *et al.* 1987) M dwarfs (Bookbinder 1988, Jackson *et al.* 1987), Algols (Umana *et al.* 1989), PMS stars (Feigelson 1987).

Surprising results in the context of general ideas of stellar magnetic activity are given by survey of F-G and K dwarfs, which like the M dwarfs show both direct and indirect evidence for strong dynamo-generated magnetic field, and therefore should be sources of radio emission. Careful surveys on K dwarfs (Bookbinder 1987, Linsky and Gary 1983) as well as of F dwarfs (Bookbinder and Walter 1987) have resulted in a lack of any detections. Evidence exists for detection on G star (Seaquist and Taylor 1985). These results prevent any study of radio emission activity and rotation for main sequence stars. The lack or marginal detection of single G-K giants (Drake and Linsky 1986 Catalano et al. 1989) do not allow any speculation on the effect of gravity and convection parameters.

Some interesting results, however come from the studies of RS CVn and Algol binaries. Using the average radio luminosity Drake et al. (1989) were not able to find any correlation with the orbital period of RS CVn. Also Morris and Mutel (1988) did not found any correlation except perhaps for systems with periods shorter than 10 days(see also Mutel, this volume). Stewart et al. (1988) and Slee and Stewart (1989) found some evidence for correlation between the surface radio flux density and the rotation period. I think it is important to mention that instead of the average quiescent radio flux, they adopted as parameter the maximum emission flux ever observed for each system.

This is a very uncertain parameter and is more representative of the flare activity rather than the average magnetic activity indicator we are discussing in this paragraph.Radio emission from Algol system is apparently not dependent from orbital period (Umana et al. 1989).

3 Magnetic field

Detailed confirmation and refinement of stellar magnetic activity observation and theory require information on the actual strength and extent of magnetic fields on stellar surface. Unfortunately, such data have been very difficult to obtain, partly because of inadequate observational techniques. Measurement analyzing line polarized light (Babcock 1958, Vogt 1980, Brown and Landstreet 1981) have been inconclusive, most likely owing to complex magnetic topologies. A major advance was made by Robinson (1980), who propose to derive the magnitude of stellar vector magnetic field from careful study of the profile broadening of magnetically sensitive line (large Landé factor).
Contradictory results were obtained on the early studies based on Robinson method. Some stars with only moderate level of chromospheric activity showed filling factor of nearly 90% (e.g. ϵ Eri), others, quite inactive stars showed similar amounts of magnetic flux. Anyway, Marcy (1984) found a correlation between the Ca II chromospheric emission and the magnetic field intensity for late type stars. Indeed, Gray (1985) in an examination of all magnetic measurements on cool stars available at that date, noted that the product, field strength to filling factor, Bf was a constant independent of spectral type and rotation velocity.
A more refined method for the measurement of magnetic fields in cool stars that take into account radiative transfer effect and the exact Zeeman patterns has been developed by Saar (1987). From the new measurements several trends in the magnetic parameters have emerged. There is a significant anti-correlation between B and T_{eff} , which is interpreted to be the result of the equipartition magnetic field strength (i.e.,B=B_{eq} = (8 π P_{gas})) (Saar and Linsky 1986). Apparently, the external photospheric gas pressure confines flux tubes and limits B. Thus G-K stars appear to have magnetic field of 1500-3000 Gauss while cool M stars are found to have magnetic field up to 5000 gauss. The $B = B_{eq}$ relation is consistent with many theoretical models (e.g. Parker 1979, Belvedere et al. 1980, Belvedere 1985) Saar and Linsky do not found any clear dependence of B from the stellar angular velocity Ω, or with the inverse Rossby parameter, $\tau_c\Omega$. On the other hand, they found that the filling factor f, is well correlated with Ω (Figure 5) and shows no obvious dependence on T_{eff}. The scatter is not reduced when f is plotted against $\tau_c\Omega$. A simple power law fit yield $f \propto \Omega$ for $\Omega \leq 0.25 days^{-1}$. A two power law fit represent equally well the data, but the magnetic filling factor seems to saturate at f = 0.80 (Linsky and Saar 1987). This saturation effect seems to be consistent with the saturation of magnetic activity indicators at high angular velocity noted by some investigators (Vilhu 1984, 1987). An important result from magnetic field measurements come from time variation. The active dwarfs ϵ Eri and ξ Boo A show time-variable f values, but B appears roughly constant in time. Coordinated observation of magnetic field and chromospheric activity tracers (Ca II H and K, Mg II, CIV, CII) show rotational modulation of emission line and magnetic fluxes in phase one another (Saar et al. 1988). This result represent the first direct evidence of a connection between photospheric magnetic fields and outer atmospheric emission other than in the Sun. In close agreement with the solar result the stellar Ca II emission flux has been found to scale with the magnetic flux as $< Bf >^{0.6}$ (Saar and Schrijver 1987).

4 The rotation-activity dependence and dynamo models

On the theoretical ground mean-field dynamo theories are characterized by dimensionless parameter, like the dynamo number N_D (Parker 1979) defined as the squared ratio of diffusive damping

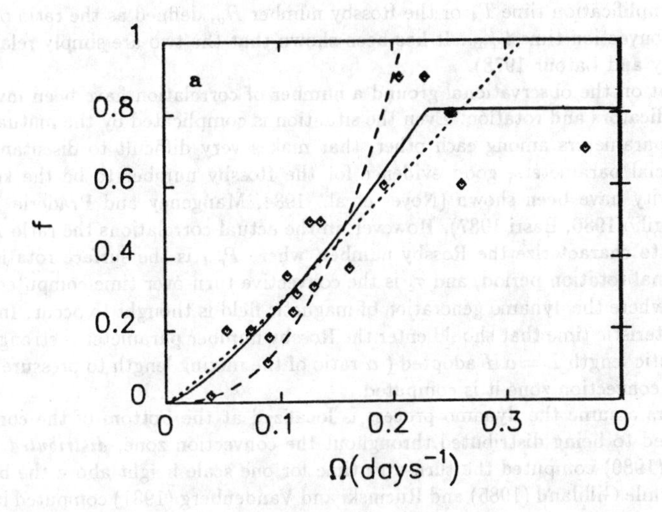

Figure 5. Filling factor versus Ω (diamonds). Single (f ∝ $\Omega^{1.1}$; short dashed line) and double (f ∝ $\Omega^{1.3}$ for $\Omega \leq 0.25$ days^{-1}, plus saturation at f = 0.8 for f > o.25 days$_{-1}$; solid line) power law fits. Durney and Robinson (1982) model is also shown (f ∝ $\Omega^{2.5}$ long dashed line) (from Linsky and Saar 1987).

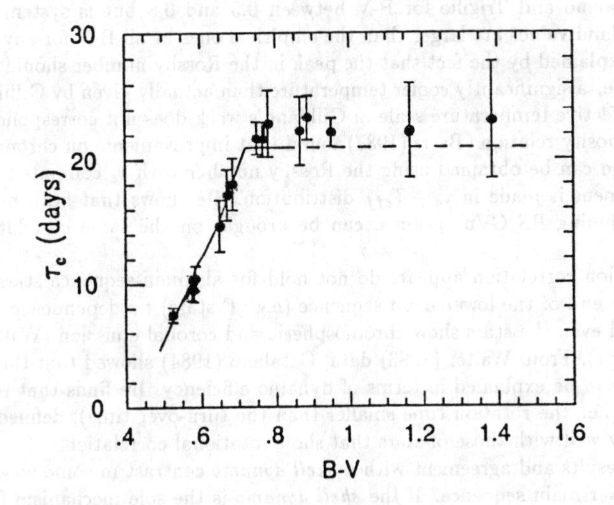

Figure 6. The convective turn-over time versus B-V. Thesolid line is from Gilman(1980) for α = 2; the dashed line is an extrapolation of Gilman's value for B-V>0.85. Points are the inverse of the *observed a* values in Figure 4. (from Catalano 1988)

time T_D to the amplification time T_A or the Rossby number R_o, defined as the ratio of the rotation time t_{rot} to the convection time t_{conv}. It has been shown that the two are simply related such that $N_D \approx R_o^2$ (Durney and Latour 1978).

We have seen that on the observational ground a number of correlation have been investigated between activity indicators and rotation. Even the situation is complicated by the mutual dependence of global stellar parameters among each other, that makes very difficult to disentangle the effect of individual crucial parameters, good evidence for the Rossby number to be the key parameter of magnetic activity have been shown (Noyes et al. 1984, Mangeney and Praderie 1984, Marilli, Catalano and Trigilio 1986, Basri 1987). However, in the actual correlations the ratio $R_o = P_{rot}/\tau_c$, is generally used to characterize the Rossby number, where P_{rot} is the surface rotation period instead of the internal rotation period, and τ_c is the convective turn-over time computed deep on the convection zone, where the dynamo generation of magnetic field is thought to occur. In addition the convective characteristic time that should enter the Rossby number parameter is strongly dependent on the characteristic length $L = \alpha H$ adopted (α ratio of the mixing length to pressure scale length) and where in the convection zone it is computed.

Several authors assume the dynamo process is localized at the bottom of the convection ***shell dynamo*** as opposed to being distributed throughout the convection zone, *distributed dynamo*. For example Gilman (1980) computed the turn-over time for one scale height above the bottom of the convection zone while Gilliland (1985) and Rucinski and Vandenberg (1987) computed it at half scale height. The present data on activity rotation correlation, even with some uncertainties seem to give useful indication on the dynamo model and parameters. Noyes et al. (1984) found that a choice of $\alpha = L/H = 1.9$ minimize the scatter of $R\prime_{HK}$ versus R_o. Marilli, Catalano and Trigilio (1986) show that the empirical turnover time deduced from exponential representation of Ca II luminosity versus the rotation period for main sequence stars agree very well with Gilman (1980) turn-over time τ_c for $\alpha = 2$. Catalano (1988) argues that τ_c values calculated by Rucinski and Vandenberg (1987) for α=1.6 at half scale height above the bottom of convection zone agree with the empirical τ_c by Marilli Catalano and Trigilio for B-V between 0.5 and 0.8 but is systematically larger for $B - V \geq 0.8$. Gilliland values are larger than the empirical ones at all B-V for any values of α. This partially can be explained by the fact that the peak in the Rossby number should occur at the base of giant branch; i .e. a significantly cooler temperature than actually given by Gilliland (Basri 1987). Apparently the effective temperature scale in Gilliland's work does not correspond to the empirical temperature-luminosity relation. Basri (1987) found that improvements on chromospheric flux and rotation correlation can be obtained using the Rossby number with τ_c computed by Gilliland after an ad hoc adjustment is made in $\tau_c - T_{eff}$ distribution. He shows that both main sequence and evolved stars, including RS CVn systems, can be brought on the same correlation with a fairly reduced scatter.

The activity-rotation correlation appears do not hold for all main-sequence stars with convection zone. At the early end of the lower main sequence (e.g. F stars) no dependence on rotation or on R_o has been found even the stars show chromospheric and coronal emission (Walter 1983, Wolff et al. 1986, Wolff 1987). From Walter (1983) data, Catalano (1984) showed that the limits where the correlation starts can be explained in terms of dynamo efficiency. He finds that regions of efficient dynamos $R_o \leq 1$ (i.e. the rotation time smaller than the turn-over time), defined by τ_c computed for α=2 agree very well with those of stars that show rotational correlation.

These promising results and agreement with a ***shell dynamo*** contrast in some way with result from the late end of lower main sequence. If the ***shell dynamo*** is the sole mechanism for magnetic field generation in late type main sequence stars then, this mechanism would cease to operate with the onset of full convection in stellar interior as would the magnetic activity-rotation relation.However, a number of results on late M dwarfs and T Tau stars show that magnetic activity is operating in full convective stars (Giampapa and Liebert 1986, Doyle 1987, Bouvier 1986) as summarized by Giampapa (1987). Similar, to the result shown for the early end of main sequence are those of Bopp et al. (1981). They found that those M dwarf stars that have equatorial rotational velocity (v_e

sin i) greater than 5 Km/sec tend to show H_α emission and sometime exhibit evidence for cool spot modulation. By contrast, Those M dwarfs rotating more slowly than 5 Km/sec are non-dMe stars and do not show evidence for photometric variations, due to cool spots.
In summary, there is substantial evidence for a *rotation-activity* connection even for full convective stars. This implies that regenerative dynamo action does not require the existence of a convectively stable boundary of the base of the convection zone and that a *distributive* dynamo can operate in full convective late type stars though rotation is fast enough to switch on an efficient dynamo.

5 Concluding remarks

Even if we are still far away from having a consistent observational and theoretical understanding of the activity-rotation connection, the present overview clear demonstrate that it is very promising to investigate the dependence of activity on rotation in different class of active stars in order to get constraint on:

- the key activity parameter (rotation rate Ω, Rossby number?)

- The convective turn-over time

- The functional dependence of emission on rotation (power law, exponential law?)

- The ratio between the mixing-length and the pressure scale height

- In a more general sense to the models of stellar convection zones

- The evolution of stellar activity and the problems connected to the evolution of rotation.

Acknowledgements: this work has been supported by the Ministero dell'Universitá della Ricerca Scientifica e Tecnologica through the University of Catania, The Catania Astrophysical Observatory and the CNR (Gruppo Nazionale di Astronomia) under contract No. 88.00349.02 The work has been produced using the Catania Astronet computing facilities. I would like to thank Grazia Umana for discussion and her help and miss Cinzia Spampinato for kindly typing the manuscript. It is a pleasure to thank prof. C. Ibanoglu for inviting me to review this stimulating subject and for his warm hospitality in this beautifull country.

6 References

- Andrews,A.D., et al.: 1988, *Astron. Astrophys*, **204**, 177.
- Ayres,T.R., Linsky,J.L.: 1980, *Astrophys. J.*, **241**, 279.
- Ayres,T.R., Marstad,N.C., Linsky,J.L.: 1981, *Astrophys. J.*, **247**, 545.
- Babcock,H.W.: 1958, *Astrophys J. Suppl.*, **3**, 141.
- Baliunas S.L., and Vaughan A.H.: 1985 *Ann. Rev. Astron. and Astrophys.*, **23**, 379.
- Basri,G.S.: 1987, *Astrophys. J.*, **316**, 377.
- Basri,G.S., Laurent,R., Walter,F.M.: 1985, *Astrophys. J.*, **298**, 761.
- Belvedere,G., Chiuderi,C., Paternò,L.: 1981, *Astron. Astrophys.*, **96**, 369.

- Belvedere, G.: 1985, *Solar Physics*, **100**, 363.
- Blanco,C., Catalano,S., Marilli,E. Rodonò,M.: 1974, *Astorn. Astrophys.*, **33**, 257.
- Blanco,C., Catalano,S., Marilli,E.: 1979, *Astron. Astrophys. Suppl. Series*, **36**, 297.
- Bookbinder,J.A.: 1988, in *Activity in Cool Stars Envelopes*, O.Havnes, B.R. Pettersen, J.H.M.M. Schmitt, J.E.Solheim eds., Kluwer Academic Publishers, p. 257.
- Bookbinder,J.A., Walter,F.M.: 1987, in *Cool Stars, Stellar Systems and the Sun*, J.L. Linsky, R.E. Stencel eds., Springer-Verlag Berlin, p. 260.
- Bopp B.W., Fekel F.: 1977, *Astron. J.*, **82**, 490.
- Bopp,B.W., Noah,P.V., Klimke,A., Africano,J.: 1981, *Astrophys. J.*, **249**,210.
- Bouvier,J.: 1986 *Adv. Space Res.*, **6**, No. 8, 175.
- Brown,D.N., Landstreet,J.D.: 1981, *Astrophys. J.*, **246**, 899.
- Budding,E., Kadauri,T.H.: 1982, ?
- Caillault,J.P., and Helfand,D.J.: 1985, *Astrophys. J.*, **289**, 279.
- Catalano,S.:1983, in *Activity in Red-Dwarf Stars*, P.B. Byrne and M. Rodonò eds., Reidel Dordrecht, Holland, p.343.
- Catalano S.: 1984, in *Space Research Prospects in Stellar Activity and Variability*; A. Mangeney and F. Praderie eds., Observatoire de Paris-Meudon **p. 243**.
- Catalano,S.: 1986, in Flares: Solar and Stellar P.M. Gondhalekar ed., RAL-86-085, p. 105
- Catalano,S.: 1988, *Irish Astron. J.*, **18**, 265.
- Catalano,S.: 1989, This meeting
- Catalano,S., Frisina,A., Rodonò,M.: 1980, *I.A.U. Symposium N. 88*, 405.
- Catalano,S., Gibson,D.M., Rodonò,M.: 1989, *Astron. Astrophys.*, submitted.
- Doyle,J.G.: 1987, *Mont. Not. R. Astr. Soc.*, **224**, 1p.
- Drake,S., Linsky,J.L.: 1986, *Astron. J.*, **91**, 602.
- Drake,S., et al.: 1989,*Astrophys. J.* in press
- Durney,B.R., Latour,J.: 1978, *Geophys. Astrophys. Fluid Dyn.*, **9**, 241.
- Durney,R.B., Robinson,R.D.: 1982, *Astrophys. J.*, **253**, 290. - Feigelson,E.D.: 1987, in *Cool Stars, Stellar Systems and the Sun*, J.L. Linsky, R.E. Stencel eds., Springer-Verlag Berlin, p. 455.
- Giampapa,M.S.: 1987, in *Cool Stars, Stellar Systems and the Sun*, J.L. Linsky, R.E. Stencel eds., Springer-Verlag Berlin, p. 236.
- Giampapa,M.S.,Liebert, J.: 1986,*Astrophys. J.*, **305**, 784.
- Giclas,H., Burnham,R., and Thomas,N.: 1962, *Lowell Obs. Bull.*, **5**, 257.
- Gilliland,R.: 1985, *Astrophys. J.*, **299**, 286.
- Gilman,P.: 1980, in *Stellar Turbulence, IAU Coll. No 51*, D.F. Gray, J.L. Linsky eds., Springer-Verlag New York, p. 19.
- Glebocki,R., Stawikowski,A.: 1988, *Astron. Astrophys*, **189**, 199.
- Gray D.F.: 1985, *Publ. Astron. Soc. Pacific*, **97**, 719..
- Hartmann,L., Baliunas,S.L., Dunca,D.K., Noyes, R.W.: 1984, *Astrophys. J.*, **279**, 778.
Hartmann L.W., Noyes R.W.: 1987, *Ann. Rev. Astron. and Astrophys.*, **25**, p. 271.
- Howard,R.:1959,*Astrophys. J.*, **130**, 193.
- Jackson,P., White,S., Kundu,M.: 1987 in *Cool Stars, Stellar Systems and the Sun*, J.L. Linsky, R.E. Stencel eds., Springer-Verlag Berlin, p. 103.
- Landini M., Monsignori-Fossi B.C., Pallavicini R., Piro L.: 1986 *Astron. Astrophys.*, **157**, 217.
- Lang,K., Willson,R., Gaizauskas,V.: 1983, *Astrophys. j.*, **267**, 455.
- Leighton,R.B.: 1959, *Astrophys. J.*, **130**, 366.
- Linsky,J.L., Ayres,T.R.: 1978, *Astrophys. J.*, **220**, 619.
- Linsky,J.L. Gary,D.E.: 1983, *Astrophys. J.*, **274**, 776.
- Linsky,J.L., Saar,S.H.: 1987, in it Cool Stars, Stellar Systems and the Sun, J.L. Linsky, R.E. Stencel eds., Springer-Verlag Berlin, p. 44.
- Linsky J.L.: 1988 in *Multiwavelength Astrophysics*, F. Cordova ed., Cambridge University Press, p..

- Lockwood, G.W., Thompson,D.T., Radick,R.R., Osborn,W.H., Baggett,W.E., et al.: 1984, *Publ. Astron. Soc Pacific*, **96**, 714.
- Majer, P.,Schmitt,J.H.M.M., Golub,L., Harden,F.R., Rosner,R.: 1986, *Astrophys. J.*, **300**,360.
- Mangeney,A., Praderie,F.: 1984, *Astron. Astrophys.*, **130**, 143.
- Marcy,G.W.: 1984, *Astrophys. J.*, **276**, 286.
- Marilli E., Catalano S.: 1984, *Astron. Astrophys.*, **133**, 57.
- Marilli,E., Catalano,S., Trigilio,C.: 1986, *Astron. Astrophys.*, **167**, 297.
- Morris,D.H., Mutel,R.L.: 1988, *Astron. J.*, **95**,204.
- Noyes R.W., Hartmann L.W., Baliunas S.L., Dunkan D.K., Vaughan A.H.: 1984 *Astrophys. J.*, **279**, 763.
- Oranje,B.J., Zwaan,C., Middelkoop,F.: 1982, *Astron. Astrophys.*, **110**,30.
- Pallavicini,R., Golub,L., Rosner,R., Vaiana,G.S., Ayres,T.R., Linsky,J.L.: 1981, *Astrophys. J.*, **248**,279.
- Parker E.N.: 1955 *Astrophys. J.*,**122**, 293.
- Parker,E.N.: 1979, *Cosmical Magnetic Fields: Their Origin and Their Activity.*, Clarendon Press Oxford.
- Pettersen,B.R.: 1989, *IAU Coll. N. 104, Solar Physics*, in press.
- Radick,R.R., Thompson,D.T., Lockwood, G.W., Duncan,D.,K., and Baggett,W.E.: 1987, *Astrophys. J.*, **321**,459.
- Ramsey L.W., and Notions H.L.: 1980 *Astrophys. J.*, **239**, L121.
- Rengarajan,T.N., Verma,R.P.: 1983, *Mon. Not. R. Astr. Soc.*, **203**, 1035.
- Robinson,R.D.: 1980, *Astorphys. J.*, **239**, 961.
- Rodonó M.: 1983, *Adv. Space Res.*, 1, **N. 9**, 225.
- Rodonó, M., Cutispoto,G., Pazzani,V., Catalano,S., Birne, P.B., Doyle, J.G., Butler,C.J., Andrews,A.D., Blanco,C., Marilli,M., Linsky,J.L., Scaltriti,F., Busso,M., Cellino,A., Hopkins,J.L., Okazak Hayashi, S.S., Zeilik,M., Henson,G., Smith,P., Simon,T.: 1986, *Astron. Astrophys.* **165**, 135.
- Rodonó M.: 1986a, in *Cool Stars, Stellar Systems and the Sun*, eds. M. Zeilik and D.H. Gibson, Springer-Verlag Berlin, **p. 475**.
- Rodonó M.: 1987, in *Solar and Stellar Physics proceedings of the 5th European Solar Meeting*, E.H. Schroter and M. Schussler eds., Springer Verlag Berlin p..
- Rodonó,M., et al.: 1989, *IAU Coll. No. 104, Poster Papers*, B.M. Haish and M. Rodonø' eds., Catania Astrophysical Observatory Special Publication, P. 53.
- Rucinski,S.M., Vandenberg, Don A.: 1986, *Publ. Astron. Soc. Pacific*, **98**, 669
- Rosner R., Golub L., and Vaiana G.S.: 1985 *Ann. Rev. Astron. and Astrophys.*, 23, 413.
- Rutten,R.G.M: 1987 *Astron. Astrophys.*, **177**, 131.
- Rutten, R.G.M., Schrijver.C.J.: 1987, *Astron. Astrophys.*, **177**, 155.
- Saar,S.H.: 1987, in *Cool Stars, Stellar Systems and the Sun*, J.L. Linsky, R.E. Stencel eds., Springer-Verlag Berlin, p. 10.
- Saar,S.H. ,Houvelin,J., Giampapa, M.S., Linsky,J.L., Jordan,C.:1988, in *Activity in Cool Stars Envelopes*, O.Havnes, B.R. Pettersen, J.H.M.M. Schmitt, J.E.Solheim eds., Kluwer Academic Publishers, p. 45.
- Saar,S.H., Linsky,J.L.: 1986, *Adv. Space Res.*, **6**, No. 8, 235.
- Saar,S.H., and Schrijver,C.J.: 1987, in Cool Stars, Stellar Systems and the Sun, J.L. Linsky, R.E. Stencel eds., Springer-Verlag Berlin, p. 38.
- Schrijver,C.J.: 1983, *Astron. Astrophys.*, **127**, 289.
- Schrijver.C.J.: 1986. *PhD Thesis*, Univ. Utrecht, Neth.
- Schrijver,C.J.:1987,in *Cool Stars, Stellar Systems and the Sun*, J.L. Linsky, R.E. Stencel eds., Springer-Verlag Berlin, p. 135.
- Schrijver,C.J., Cotè,J., Zwaan,S.H., Saar,S.H.: 1989a, *Astrophys. J.*, **337**, 964.
- Schrijver C.J., Dobson A.K., Radick R.R.: 1989b, *Astrophys. J.*, **341**, 1044.
- Seaquist,E.R., Taylor, A.R.: 1985, *Astron. J.* , **90**, 2049.

- Skumanich,A.: 1972, *Astrophys. J.*, **171**, 565.
- Skumanich,A., Smythe,C., Frazier,E.N.: 1975, *Astrophys. J.*, **200**, 747.
- Slee,O.B., Nelson,G.J., Stewart,R.T., Wright,A.E., Innis,J.L., Ryan,S.G., Vaughan,A.E.: 1987, *Mon. Not. R. Astr. Soc.*, **229**, 659.
- Slee,O.B., Stewart,R.T.: 1989, *Mon. Not. R. Astr. Soc.*, **236**, 129.
- Stauffer,J., Dorren,J., and Africano,J.: 1986, *Astron. J.*, **91**, 1443.
- Stauffer,J.: 1987, *Publ. Astron. Soc Pacific*, **99**, 471.
- Stern,R.A., Zolcinski,M.C., Antiochos,S.K., Underwood,J.H.: 1981, *Astrophys. J.*, **249**, 647.
- Stewart,R.T., Innis,J.L., Slee,O.B., Nelson,G.J., Wright,A.E.: 1988, *Astron. J.*, **96**, 371.
- Trigilio,C., Catalano,S., Marilli, E.: 1989, preprint.
- Ulmschneider,P.: 1989, *XI ERAM* Tenerife July 1989 (in the press).
- Ulmschneider,P., Stein,R.F.: 1982, *Astron. Astrophys.*, **106**, 9.
- Umana,G., Catalano,S., Rodonò,M., Gibson,D.,M.: 1989, in *Algol's IAU Coll. No. 107*, in press.
- Van Leeuven,F., and Alphenaar,P.: 1982, *ESO Messenger*, No. 28, p.15.
- Vilhu,O.: 1984, *Astron. Astrophys.*,**133**, 117.
- Vilhu,O.: 1987 in *Cool Stars, Stellar Systems and the Sun*, J.L. Linsky, R.E. Stencel eds., Springer-Verlag Berlin, p. 110.
- Vogt,S.S.: 1980, *Astrophys. J.*, **240**, 567.
- Vogt S.S., and Penrod G.D.: 1983, *Pub. Astron. Soc. Pacific*, **95**, 565.
- Walter,F.M.: 1982, *Astrophys. J.*, **253**,745.
- Walter,F.M.: 1983, *Astrophys. J.*, **274**,794.
- Walter,F.M., Bowyer,S.: 1981, *Astrophys. J.*, **245**, 671.
- Walter,F.M., Schrijver,C.J.:1987, in *Cool Stars, Stellar Systems and the Sun*, J.L. Linsky, R.E. Stencel eds., Springer-Verlag Berlin, p. 262.
- Wilson O.C.: 1963, *Astrophys. J.*, **138**, 382.
- Wilson O.C.: 1966, *Astrophys. J.*, **171**, 695.
- Wolff,S.C.: 1987, in *Cool Stars, Stellar Systems and the Sun*, J.L. Linsky, R.E. Stencel eds., Springer-Verlag Berlin, p. 223.
- Wolff,S.C., Boesgaard,A.M., Simon,T.: 1986, *Astrophys. J.*, **310**, 360.
- Zwaan,C.: 1986, in *Cool Stars, Stellar Systems and the Sun*, M. Zeilik, D.M. Gibson eds., Springer-Verlag Berlin, p. 19.

SOLAR-LIKE MAGNETIC STRUCTURES IN ACTIVE CLOSE BINARIES

Bernard H. Foing

ESA/ESTEC Space Science Dept, postbus 299, 2200 AG Noordwijk, Netherlands
and Institut d'Astrophysique Spatiale (IAS), BP 10, 91371 Verrières-le-Buisson, France

SUMMARY. We review some aspects about solar observations of small-scale magnetic structures such as flux tubes and coronal loops and how active complexes of similar structures can be observed in solar-like stars and in active close binaries. We describe also different diagnostics for the study of these structures: magnetic field measurements, photometric and spectroscopic rotational modulation techniques and indirect imaging methods. We stress the need for multiwavelength multisite observations of these fast varying active objects, and the usefulness of modeling/interpretative techniques previously applied to the solar context.

Partly based on observations obtained at ESO, CFH, OHP observatories, with the IUE and EXOSAT satellites, and with the Transition Region Camera.

1. Solar Magnetic Structures

1.1 Small scale magnetic fluxtubes

Solar physicists have learned the critical role played by magnetic fields in heating the chromosphere and corona, in confining the geometry of observed fine structures, and also in underlying solar activity phenomena on very different spatial and time scales. The correlation between the magnetic field and its atmospheric manifestations can be quantified through spatially resolved observations on specific regions. Magnetic fields in the solar photosphere are currently measured from differences in the line profiles of magnetically sensitive lines (with high Landé g factor) in opposite circular polarisation. High spatially resolved observations of the magnetic field in the photosphere have shown the existence of

C. İbanoğlu (ed.), Active Close Binaries, 363–375.

elementary field concentration in "flux tubes" of 1.5 kG in size less than the arcsec. These tubes, which originate from the coupling between convection and the magnetic field in the subphotosphere, are swept by the supergranular motions in the boundaries, and appear cospatial with the 0.2" filigree emission in white light (Dunn 1973) or the enhanced emission network observed in chromospheric spectroheliograms. The correlation between the chromospheric emission and the magnetic flux has been shown (Skumanich et al 1975). The flux tubes appear thus as the elementary blocks which collective distribution determine the large scale properties of active regions. It is however necessary to know the small scale structure of "individual flux tubes" in order to understand their physical properties and to predict observables at larger scale. For different components of activity, semi-empirical models have been constrained by ultraviolet continua or chromospheric line spectra to represent the height variation of temperature, densities, radiation quantities , energy balance and other physical parameters (Vernazza et al 1981, Lemaire et al 1981, Avrett 1984).
The vertical structure of these flux tubes can be diagnosed with high spatial resolution observations of emission excesses at different wavelengths corresponding to different temperatures. For instance 1" resolution pictures obtained with the Transition Region Camera (Bonnet et al 1980, Foing & Bonnet 1984a&b, Foing, Bonnet & Bruner 1986) in the 160 nm Ultraviolet continuum formed near the temperature minimum region, in the Ly α line in the base of the transition region at 20000 K, or in the CIV transition region line at 10^5 K allow to span different altitudes for the diagnostic of these flux tubes and loops.

1.2 Models of thin flux tube and faculae

The fine structure of the chromospheric network shows up as bright elementary grains on the 160 nm continuum pictures. We developed a thin flux tube model in order to interpret quantitatively the observed contrast of the UV emission as the effect of the presence of the magnetic field. The excess emission in the network element can be understood as the effect of the difference of brightness temperature at optical depth $\tau=1$ inside and outside the network element. We have shown (Foing & Bonnet 1984a) that due to the different depth of formation (Wilson effect) the enhancement is proportional to the outer temperature gradient, the scale height and to $\log(1-1/\beta_e)$ where $\beta_e=8\pi P_e/B_i^2$ is the ratio between the external gas pressure and internal magnetic pressure. The thin flux tube model rely on a series of approximations that are valid between the photosphere and the temperature minimum :

i) thin tube approximation allowing to neglect the magnetic tension forces

ii) horizontal and vertical hydrostatic equilibrium that takes into account the magnetic pressure $B^2/8\pi$ at all heights:

$P_i + P_i$ turb+ $B_i^2/8\pi = P_e + P_e$ turb and $dP/dz=-\rho g$

iii) thermalisation of the fluxtubes with the surroundings (Foing 1983):

assuming that the radius R of the radius be much smaller than the thermalisation length, in this case the mean free path of optical photons, and that the radiative time scale τ rad be smaller than a typical dynamical time scale τ dyn =H/V (for bringing matter from a scale height H with velocity flow V along the internal tube).

As a consequence, $T_e=T_i$ and $dP_i/P_i dz=-\rho_i g=-m_i g/kT_i = dP_e/P_e dz$.

On the base of these assumptions, one shows that the stratification inside and outside the tube is the same and that the β_e ratio $8\pi P_e/B^2$ remains constant with height. If the average scale height for the gas pressure is H, then $P_i=(\beta_e-1)B^2/8\pi = P_i°\exp(-h/H)$, then the variation of field and radius of the tube with height are $B=B°\exp(-h/2H)$ and $R=R°\exp(h/4H)$. The comparison of these predicted quantities at different heights for a flux tube of B°=1.2kG and R°=150km at photospheric level has been made consistently to the observations obtained by the TRC and on white light high resolution ground based images (Foing, Bonnet & Bruner 1986).

The observed facular contrast agree with a value $ß_e=8\pi P_e/B_i^2$ of about 2, suggesting equipartition between the internal gas presuure and the magnetic pressure in the tube.

For stellar faculae that would be made of accumulation of similar fluxtubes, this analysis can be applied. Thus the observed brightness is $fS(T + \Delta T\phi) + (1-f)S(T)$ where f is the filling factor which may vary with height due to the tube divergence, and $\Delta T\phi$ the excess temperature brightness in the flux-tube $\Delta T\phi=\Delta h\ .dT/dz$ is proportional to the gradient of the temperature (of the source function S) in the quiet region and to the Wilson height depression $\Delta h=H\log(1-1/ß_e)$. This model predicts an enhanced emission for the stellar faculae as well as a bluer colour in the photometry. Such effects have just been reported in this conference by Guinan (on HR 1099) and by Olah (on HK Lac).

1.3 Observations of coronal loops

Rocket photographs have shown that the material at Te> 2 10^6K is in the form of loop structures connecting regions of opposite polarity observed with ground based magnetographs. Loops have been observed also at the limb and on the disk by the Transition Region Camera in Ly alpha or C IV at ranges of temperatures from 2 10^4 to 10^5K (Bonnet et al 1980), showing the coexistence of multitemperature plasmas. Also dark absorbing loops are visible at apparent coronal heights. On the Ly α filtergrams obtained by the TRC, the loop structure of active regions is well shown, and we can observe several dark or white threadlike structures of length smaller than 10 " extending out of active regions or crossing the network boundaries. This shows again that the upper chromosphere and corona cannot anymore be described by an homogeneous plane parallel model and that the basic structures are loops.The fact that they are observed both in absorption or in emission on the disk suggests that a very heterogeneous distribution of the plasma temperature and electronic density delineates the loop magnetic field. Scaling laws relating the length, pressure, maximum temperature of coronal loops (e.g. Rosner et al 1978) have been proposed that can be of relevance for stellar coronal studies. A review of coronal loops models can be found in Golub (1983).

2. Magnetic activity of solar like stars

The magnetic activity of several classes of late-type stars has been particularly studied, both observationally and theoretically. In solar-like late-type dwarfs, the magnetic field seems to be amplified by the dynamo effect resulting from the coupling between convective motions and the differential rotation. Also, the modulation with rotation phase of the stellar flux and of some chromospheric indicators of activity indicates the presence of photospheric spots and chromospheric active regions. In analogy with the Sun, the appearance of spots could be attributed to intense sub-photospheric magnetic fields which inhibit the convective energy transport until the surface. Also long term periodical variations indicate solar-like activity cycles (Wilson,1978) on late-type stars. Some empirical relations were established (Mangeney, Praderie 1984, Noyes et al 1984) between the coronal or chromospheric activity and the rotational velocity through the Rossby number suggesting that the coronal/chromosphere heating is related to the magnetic field produced through a dynamo mechanism. However, little is known on the actual distribution of those magnetic fields and activity phenomena on these stellar surfaces. Some recent global magnetic field measurements (Saar et al 1987) have allowed to start deriving mean magnetic intensities and surface filling factor for active dwarfs.

2.1 Measurements of stellar magnetic fields

Sunspots, plages, flares, and the non-radiatively heated chromospheres/coronae all owe their existence to the presence of magnetic fields. Similar phenomena in late-type stars, and in close binaries in particular, indicate that a substantial amount of magnetic flux exists in their photospheres as well (see Linsky 1985). This knowledge is however only qualitative, the detailed physics behind the fundamental role of magnetic fields in these phenomena is still unknown. The lack of understanding is partly due to the small number of magnetic field measurements available for late-type stars. In particular the very active RSCVn stars and close binaries are underrepresented. Only five dMe stars (Saar and Linsky 1985, Saar Linsky and Giampapa 1987) and only one RSCVn (Giampapa, Golub and Worden 1983) have been detected to date. Yet due to their high activity, they hold the keys to understanding the extremes of stellar magnetic phenomena. Preliminary indications are that magnetic field pressure ($B_i^2/8\pi$) scale with photospheric pressure, a finding which is consistent with our modeling of solar-like flux-tubes and observations indicating equipartition. Saar (1987) finds also that surface filling factors are proportional to stellar angular velocity, saturating at small periods. These results are quite uncertain, however, due to the small number of stars and the difficulty of the measurements. Measurements of B and f on rapidly rotating BY Dra and RSCVn stars will help verify these relations and can be used to test flux tube models (e.g. Galloway and Weiss 1981, Foing et al 1986), coronal loops models (e.g. Rosner et al 1978, Golub 1983), dynamo models (e.g. Durney and Robinson 1982), empirical activity theories (e.g. Vilhu 1984), MHD wave heating theories for stellar outer atmospheres (e.g. Ulmschneider and Stein 1982), and rotational evolutional theories (e.g. Bohigas et al 1986) all of which require detailed information on f and B values.

Optical measurements of magnetic parameters on close active binaries, RSCVn, dMe and BY dra star are hindered by considerable line blending (dMe and BY Dra), rapid rotation (RSCVn and close binaries) and small magnitude of the magnetic broadening. Since there is less blending in the infrared and the Zeeman splitting scale as λ^2, it is favourable to measure magnetic field on RSCVn and dMe stars using infrared spectra. With a line profile modelling in the 2.2 micron region (in particular the TiI multiplet), Saar and Linsky (1985), Saar (1988) compared the differential broadening between low and high magnetically sensitive lines. The improved version compared to the Robinson (1980) technique includes both radiative transfer effects and compensation for line blending through the use of low-activity comparison stars of similar spectral type. High S/N (>50), high resolution spectra (>40000) are required. Infrared measurements have the additional advantage of allowing detection of separate plage and spot magnetic components (if they exist), since the cooler spot is thus easier to detect against the weaker photospheric continuum at 2 microns. Such a separation of magnetic contribution from plage and spot components has never been made for a star other than the sun. Currently the only instrument capable of such observations is an infrared Fourier transform spectrometer.

2.2 Rotational modulation by active structures

Periodic or quasi-periodic low-amplitude continuum or emission line fluxes are observed in late-type active stars. Many are members of close binary systems, and it is likely that the synchronisation of short period binaries enforces dynamo effect and thus the activity. The flux variation appears as a distortion wave which can vary in shape and amplitude over a few months. These variations are attributed to unevenly-distributed surface inhomogeneities, whose visibility is modulated by the star rotation.

From analytical modeling of these lightcurves with surface distribution of starspots, typical results give 10-40% spotted area coverage , and spots temperatures 400-1500 K cooler than the immaculate photosphere. These parameters could arise from observational selection, as smaller or hotter spots would escape detection, and because rotational modulation methods on integrated fluxes are biased towards large structures. This lack of geometrical resolution can be improved, by using Doppler imaging techniques that provide a better resolution and detection capabilities for smaller spots.

The observation of solar-like activity phenomena has suggested the existence of solar-like activity cycles in stars. From Harvard archival plates, Hartmann et al (1982) have found possible photometric variations on timescales of decades in late-type active stars. The observed variability can be attributed to long term variations of spots filling factor, like during the solar cycle. Also, systematic monitoring can allow to observe the birth, decay of spots or activity complexes, and the long-term arrangement of their geometry. Surface structures are present also at chromospheric and coronal heights. Active regions can thus be studied through rotational modulation in chromospheric, transition region or coronal fluxes.In addition eclipsing systems give the possibility to deduce the distribution of surface or vertically extended structures.

3. Magnetic structures in RSCVn systems

RSCVn are binary detached systems with periods typically between 1 and 14 days, being synchronised by tidal effects, and generally composed with a subgiant primary (of type around K0 IV) and a dwarf secondary (around G5V). The review of properties for these systems can be found in (Hall 1980); the most important photometric characteristics of these systems is the quasisinusoidal distortion of the lightcurve. Its slow migration towards decreasing orbital phases was discovered on RSCVn itself by Catalano and Rodonò (1967). Among the proposed explanations, the spot model appear as the best established: the variations (except an eventual eclipse) are attributed to the rotational modulation of a nonuniform distribution of photospheric spots. The rotational modulation of chromospheric emission in the Ca II H and K lines was shown for several RSCVn systems in the Mt Wilson H and K variability survey (Vaughan et al 1981). Results from a joint IUE and ground based observations (Rodonò et al 1987) have shown for the RSCVn system II Peg that the chromospheric and transition region emission reaches a maximum when the visible photometry is at minimum, thus indicating that chromospheric plages cover, in first approximation, an area associated to the photospheric spots.

A further step in the study of the activity of these systems is to obtain the large scale distribution of the magnetic field on these active stars. Previous attempts to measure the magnetic field on these stars, using deconvolution of line profiles with different sensitivity to the Zeeman effect gave only marginal results until now. These measurements are made very difficult by the large rotational broadening (in general > 30 km/s) for these systems, and by the possibility that the magnetic field on the surface of the giant or subgiant (following the equipartition argument) may be lower than in the solar case. Thus, a crucial information relies on the knowledge of the spatial distribution of phenomena associated with these magnetic fields. The Doppler imaging method allows the localisation and reconstruction of the spots, active structures distribution, with constraints from the velocity profile information, in addition to the photometry. The observation of the rotational modulation of spectroscopic profiles and the application of reconstruction techniques must permit: i) to obtain the configuration of the spots on these stars (polar, equatorial?), ii) to compare them with the signature of the chromospheric, transition region or coronal emission, iii) to calculate extrapolated magnetic fields from these constraints and iv) to derive the vertical stratification and energy balance of active structures. Also large scale magnetic structures can be followed over months and years to track differential motions, and constrain theories of internal rotation and dynamo.

Also, circumstellar material can be observed in absorption due to ejected clouds of cool material transiting in front of a stellar disk, as described by Collier-Cameron, Robinson (1989). The distribution of corotating clouds can also be derived from an inverse mapping technique of "skewing" technique. They deduced the structure of the clouds from continuous repeated observations at high resolution, in H alpha, and a more recent campaign showed also absorption transients in lines of Ca II, Mg II and Na I and even through H alpha photometry (Collier-Cameron et al 1990, Foing et al 1990).

4 INDIRECT IMAGING OF FAST-ROTATING STARS

4.1 Stellar active structures

The evidence for stellar photospheric spots can be obtained in the case of RSCVn or BY Dra systems from the photometric light curve periodic modulation. Migration of these photometric waves can also be observed as an indication of the change in the spot distribution over the surface. Different modelling methods of spotted surfaces have been implemented for describing the photometric observations (e.g. Rodonò et al 1987).
In particular the presence of plages and spots has been inferred from periodic low-amplitude photometric and spectral feature variations due to rotational modulation of spot-plage visibility. Systematic long-term observations for RSCVn and other active stars, have shown almost sinusoidal light curves to become multipeaked or even flat, suggesting variations in the spot number and distribution over the stellar surface. Spot and plage modeling (Byrne et al 1987) indicates that their physical characteristics are close to the solar ones, but they can cover up to 50% of the stellar surface. From simultaneous optical and IUE observations (Marstadt et al 1982, Rodonò et al 1986), a close spatial correlation between spots and chromospheric / transition region plages is apparent. High resolution spectroscopic observations of lines at different phases of the rotational period of active stars obtained with IUE or from the ground has allowed to develop "spectral imaging" on IUE Mg II data (Walter et al, Neff Ph.D. 1987) , or "Doppler imaging" techniques for recovering the spatial distribution of surface activity over the star (Vogt and Penrod 1983, Jankov, Foing 1987).
The objectives of such studies are to obtain the geometric distribution of activity phenomena, to understand the differences with their solar equivalent, to model the active and quiescent atmospheric regions, to study the correlation between the structures observed at different heights, and monitor the changes associated with active region behaviour, cyclic activity, dynamo phenomena and differential rotation. Basic considerations about the imaging of spotted stars from high resolution high signal to noise spectroscopy of photospheric or chromospheric lines are described afterwards.

4.2 Indirect imaging methods

The past decade has seen a very strong effort to understand and spatially resolve the surface structures and environment of late-type stars. As the techniques of interferometric imaging have still a limited angular resolution with the available baselines, indirect imaging techniques are necessary. An access to the information of quasistationary surface (or vertically extended) external structures is possible through the observation and interpretation of temporal photometric or spectroscopic variability along the rotational phase of the star, as well as during possible eclipses in binary systems. The intensity rotational modulation gives a one dimensional projection in longitude of the surface structures. In a method developed by Deutsch (1970) for chemically peculiar stars, the variation of line equivalent widths can be adjusted by parameters describing the development in spherical harmonics of the stellar surface inhomogeneities. This method did not make full use of the profile, mainly due to the low spectral resolution and low signal to noise ratio (S/N) available then on photographic plates spectra. The advent of high S/N observations with CCD and reticon detectors has stimulated the development of quantitative mathematical methods for studying the spectroscopic indirect imaging. With the Doppler imaging technique it is possible from the line profile disturbances observed at high spectral resolution and with adequate phase coverage, to obtain an information not only in longitude but also in latitude about surface stellar structures. Different formulation of the Doppler imaging method have been proposed or applied to various observations. Khokhlova (1985) calculates the line profile with the Doppler shifted contributions due to surface inhomogeneities; a Lagrange multiplier method is used to minimise an error functional between calculated and observed profiles, with a stabilisation Tikhonov functional. Jankov (1987), Jankov and Foing (1987) gave a mathematical formulation for the indirect imaging, and compared the reconstruction efficiency for different regularisation factors. Vogt et al (1987) express the relation between local surface intensities and the observed spectral profile using a matricial relation.

4.3 Comparison with observations

Several results have been obtained on Ap stars and RSCVn type stars by Vogt et al (1987, 1988), Khokhlova and collaborators (1984,1986). However, some aspects deserve further work, such as: i) the importance of the noise on the solution, ii) the possible biases in the reconstruction, iii) the position dependent resolution and accuracy for a characterised set of observations, and iv) the role of uncertainties in the stellar matrix of transformation parameters. A simple variant of the Doppler imaging method has been applied by Gondoin (1986), on observations of HR1099, by identifying bumps components in the profile and following their velocity changes with rotational phase. Recently, Neff (1987), Walter (1987) and collaborators have developed a spectral imaging method adjusting IUE MgII emission spectra of the system AR LAC with a minimal number of components. A description of Doppler and spectral imaging methods and results is given by J.Neff in the Kusadasi conference.

5 Multiband observations of stellar flares

The flares are the most complex and violent phenomena occurring on these stars. Flare events have typical timescales of 10s for rise and 10^3s for decay. Simultaneous photometry and spectroscopy of flares have shown emission line enhancements of different species to take place as the response of the stellar atmosphere to the flare energy release. We refer here to the proceedings of the IAU Colloquium 104 on Solar and Stellar Flares (Haisch, Rodonò eds, 1989) and to Catalano in his review at this ASI, for an overview of the flare observations and corresponding theories. Flare observations give diagnostics of : 1) what is the energy budget for a typical sample of flares; 2) what are the respective roles of radiation from the corona (as shown by the X rays), conductive losses through the transition region (EUV) and expansion (as indicated from velocity fields measurements); 3)what are the temperatures, densities and volumes of the hot flaring plasma. Fundamental issues for flare research are adressed in Foing (1989), with relevant stellar flare spectroscopic diagnostics, concerning the energy transport mechanism (particle beam vs heat conduction); the atmospheric response to flares; mass motions, ejected components and momentum balance; microflaring, flaring and the heating of coronae; statistics and recurrence of flares.

6 Need for Multiwavelength Multi Site Observations

The scientific need for multiwavelength multisite observations of these intrinsically variable active stars, has promoted the organization of coordinated campaigns of observations, employing the various areas of expertise of different collaborating groups (in our case Armagh and Catania observatories, IAS/LPSP, JILA and Lockheed), and their access to large telescopes and satellite observatories. Also, a project for a MUlti SIte COntinuous Spectroscopy network (MUSICOS) was set up to provide round-the-clock observations of such objects (Catala, Foing 1988). This work requires the analysis of different sets of data, the development and exchange of new methods and software for data handling, reduction, processing and archiving. Finally, we need to discuss the theoretical analysis of data corresponding to very different wavelengths and emission mechanisms, in relation with solar physics and astrophysics (MHD, stellar atmospheres, etc).

This collaborative research was set in the context of programmes accepted for ESA/NASA satellites (IUE, EXOSAT) and for large ground based telescopes at ESO, CFHT, OHP and US observatories.The unexpected longevity of the IUE satellite together with the quantity of the accessible archives for IUE and EXOSAT has, along with the access to new ground-based telescopes, given a further perspective to the program. The future launch of the space telescope (especially the HRS spectrograph) and of ROSAT will give us soon instruments for these multiwavelength observations. In view of the competition for observing time on ST, it is necessary to prepare these scientific programmes well and to insure that the overall necessary expertise exists (coordination of space/ground based observations, data reduction, archival, analysis, scientific interpretation) from both solar/stellar aspects. Other sophisticated instruments on Very Large Telescopes, and with future space missions (cf Foing review on Perspectives for ground-based and space research on Close Binaries in this conference) will enable us to push the research on stellar activity and close binaries beyond the present limits, and to adress related problems in broader astrophysical contexts.

7 CONCLUSION

Active regions and flares reflect the underlying stellar magnetic field, and contribute a significant heating towards the upper stellar atmospheres of active close binaries. Their study is of particular importance for the theory of stellar dynamos, chromospheres and coronae. Magnetic field measurements, involving detailed modelling of high signal-to-noise spectra, should become available on close binaries. The use of solar experience in the field of semi-empirical modelling of the chromosphere/corona, and of flux tubes and coronal loops is very useful for the study of magnetic structures in active close binaries. In the framework of solar evidence of spatial heterogeneity (as chromospheric plages and network), the rotational modulation of fluxes, asymetries and spectral profiles allows to diagnose active regions on moderately rotating stars. For active fast rotating stars, the line profile changes can map the information on the geometrical surface structure distribution. We stressed the need to coordinate future synoptic observations at all accessible wavelengths, and with multi-site networks of spectrophotometers, for these objects which are highly variable on all timescales from seconds to years.

References

Avrett, E.H.: 1984. *"Small Scale Processes in Quiet Stellar Atmospheres"*, Ed. S. Keil.
Bohigas, J. et al: 1986, *Astron. Astrophys.*, **157**, 278.
Bonnet , R.M. et al.: 1980. *Astrophys. J. (Letters)*, **237**, L47
Byrne, P.B. et al.: 1987. *Astron. Astrophys.*, **180**, 172.
Catala, C., Foing, B.H. Eds: 1988, *1st Workshop on Multi-Site Continuous Spectroscopy*
Catalano, S., Rodonò, M.: 1967, *Mem. Soc.Astron.Ital.*, **38**, 395
Char, S., Foing, B.H.: 1989, in *"Modeling the Stellar Environment"* , Edit. Frontieres, Gif
Collier-Cameron, A., Robinson, R.D.: 1989, *MNRAS* **236**, 57
Collier-Cameron, A. et al: 1990, submitted
Crivellari, L. et al. : 1987. *Astron. Astrophys.* **174**, 127.
Deutsch, A.J.: 1970, *Astrophys.J.* **159**, 985.
Dunn R. : 1973. *Solar Physics* **33**, 281.
Durney, B.R., Robinson, R.D.: 1982, *Astrophys. J.* **253**, 290.
Foing, B.H. : 1983. Thesis Univ. Paris VII-Meudon-LPSP.
Foing, B.H. and Bonnet, R.M.: 1984a, *Astrophys. J.*, **279**, 848.
Foing, B.H. and Bonnet, R.M.: 1984b, *Astron. Astrophys.*, **136**, 133.
Foing, B.H.et al: 1986a, in *The Lower Atmosphere of Solar Flares*, Ed D.Neidig, p.319
Foing, B.H., Bonnet, R.M. and Bruner, M.E. : 1986b. *Astron. Astrophys.* **162**, 292.
Foing, B.H.: 1989, *Solar Phys.* **121**, 117-133
Foing, B.H. et al: 1989, *Astron. Astrophys. Suppl.*, **80**, 189
Foing, B.H. et al: 1990 , in preparation
Galloway, M.S. and Weiss, N.O.: 1981, *Astrophys. J.* **243**, 945.
Giampapa, M.S., Golub, L., and Worden, S.P.: 1983, *Astrophys. J.(Letters)* **268**, L121
Golub, L.: 1983, in IAU Colloquium 71, Activity in Red Dwarf Stars, Eds. B. Byrne and M.Rodonò (Dordrecht: Reidel), p. 83
Hall, D.S.: 1980, in *Solar Phenomena in Stars and Stellar Systems*, eds. R.M. Bonnet and A.K. Dupree, Reidel, p.431
Gondoin P. : 1986. *Astron. Astrophys.* , **160**, 73
Gouttebroze, P. , Leibacher, J.: 1980. *Astrophys. J.* **238**, 1134.
Gouttebroze, P. et al. : 1986. *Astron. Astrophys.* **154**, 154.
Guinan, E.: 1990, NATO/ASI proceedings, Kusadasi conference on *Active Close Binaries*
Haisch, B.M. and Rodonò, M. Eds: 1989, *"Solar and Stellar Flares" in Solar Phys.***121**

Hartmann, L.et al : 1980, in Bonnet and Dupree, p.487

Jankov S. : 1987. DEA Univ Paris VII/LPSP.

Jankov S., Foing B.:1987 in *"Cool Stars, Stellar Systems and the Sun"*, J.L.Linsky, R. Stencel eds

Khokhlova, V.L., Rice, J.B., Wehlau, W.M.: 1986, *Astrophys. J.* **307**, 768

Khokhlova, V.L.:1985, Astrophys. Space Phys. Rev. **4**, 99

Khokhlova, V.L., Pavlova, V.M.: 1984 *Sov. Astron. Lett.* **10**, 158

Lemaire, P. et al.: 1981. *Astrophys. J. Suppl.* **45**, 350.

Linsky, J.L.: 1985, *Solar Phys.* **100**, 333

Machado, M.E., Avrett, E.H., Vernazza,J.E., Noyes, R.W.: 1980, *Astroph.J.* **242**, 336

Mangeney, A., Praderie, F.:1984, *Astron. Astrophys.* **130**, 143.

Marstadt, et al.: 1982, NASA Conf. Publ. **2238**, 554.

Neff, J. : 1987. PhD Thesis, JILA , Boulder.

Neff, J.:1990, NATO/ASI proceedings, Kusadasi

Noyes, R.W. et al: 1984, *Astrophys. J.* **279**,763.

Olah, K. : 1990, NATO/ASI proceedings, Kusadasi conference on *Active Close Binaries*

Robinson, R.D.: 1980, *Astrophys. J.* **321**, 496.

Rodonò, M. et al.: 1986. *Astron. Astrophys.* **165**, 135.

Rodonò, M. et al.: 1987. *Astron. Astrophys.* **176**, 267.

Saar, S.: 1987, PhD thesis, University of Colorado

Saar, S., Linsky, J.L.: 1985, *Astrophys. J.(Letters)* **299**, L47..

Saar, S., Linsky, J.L., Beckers, J.M.: 1987, *Astrophys. J.* **302**, 777.

Skumanich, A., Smythe C. and Frazier, E.: 1975. *Astrophys. J.* , **200**,747.

Ulmschneider, P. and Stein, R.:1982, *Astron. Astrophys.* **106**, 9.

Vaughan , A.H. and Preston, G.W.: 1980. *P.A.S.P.* , **92**, 385.

Vaughan, A.H. et al : 1981, *Astrophys. J.* , **250**, 276.

Vernazza, J.E., Avrett, E.H., Loeser, R.: 1981. *Astrophys. J. Suppl.* **45**, 635.

Vilhu, O.: 1984, *Astron. Astrophys.* **133**, 117.

Vogt S. and Penrod H.: 1983. *P.A.S.P.* **95**, 565.

Vogt, S.S., Penrod, G.D., Hatzes, A.P.: 1987, *Astrophys. J.* **321**,496

Vogt, S.S.: 1988, IAU Symp. 132, eds. G. Cayrel & M. Spite, 253

Walter, F. et al.: 1987. *Astron. Astrophys.* **186**, 241.

Wilson, O.C.: 1978, *Astrophys.J.* **226**, 379

STARSPOT LIFETIMES

Douglas S. Hall
Michael R. Busby
Center of Excellence in Information Systems
Tennessee State University
Nashville, Tennessee 37203
U.S.A.

ABSTRACT. Starspot lifetimes (t_c) are computed with the assumption that they are disrupted by the shearing effect of differential rotation and compared with observed lifetimes (t_o) for 40 spots on 17 different stars. The differential rotation coefficient (k) for each star is estimated from the range of rotation periods exhibited by its spots. The radius of each spot (r_s) is estimated from its maximum wave amplitude (Δm). Small spots ($r_s < 20°$) die before they are disrupted ($t_o < t_c$) but larger spots are in fact disrupted by differential rotation ($t_o = t_c$).

1. Introduction

Perhaps the most striking observational feature of the RS CVn binaries, and other chromospherically active stars as well, is the migrating wave. The phenomenon first became clear in the 1963 and 1964 light curves of RS CVn (Chisari and Lacona 1965). Hall (1972) explained the wave in terms of a dark starspot region concentrated on one hemisphere of one of the two stars, presumed to be rotating synchronously with the orbital motion, and explained the wave migration in terms of solar-type differential rotation. For a review of waves and wave migration on other chromospherically stars, single and binary, see Hall (1987).

There are still several important questions which need answering before we can say we understand these large starspot regions (Hall 1987). First, why is the area covered with spots so large, up to 50% of a hemisphere in II Peg (Doyle et al. 1988), compared to less than a few tenths of a percent the sun? Second, why is the starspot coverage restricted in area rather than distributed uniformly in stellar longitude? Third, why do stars have only one or two, but not more than two, large starspot regions? Does this imply a pervasive longitudinal sector structure with only two active sectors? Fourth, does a starspot region consist of one or a few very large spots or does it consist of many smaller spots, each analagous to a typical sunspot or sunspot pair?

C. İbanoğlu (ed.), Active Close Binaries, 377–392.

Fifth, why does a large starspot region live so long (up to a decade or more) compared to sunspots (days) or solar active regions (one or two months)?

The last of these questions is the topic of this paper. Throughout most of this paper, the term "starspot" or "spot" will refer to the large dark region which produces the wave. Representative values of t_o, observed starspot lifetimes, will be extracted from the literature and compared to t_c, lifetimes calculated with the tentative assumption that the shearing effect of differential rotation disrupts a spot. We will understand more about the physics of a large starspot once we know whether $t_o < t_c$ or $t_o = t_c$ or $t_o > t_c$.

2. Observed Spot Lifetimes

The evolution of a dark region or spot proceeds as follows. A spot forms at a particular latitude and longitude. The resulting wave has an amplitude m proportional to the area of the spot and a period P(rot) determined by the spot's mean latitude and the star's differential rotation law. In general, P(rot) will differ slightly from the orbital period P(orb) in the case of a binary or from the mean rotation period ⟨P(rot)⟩ in the case of a single star. The wave amplitude may rise smoothly to a maximum and then decrease, as a spot blossoms and fades. A new spot forms, in general, at a different latitude and longitude. A new spot can appear before the previous spot has vanished.

The photometric history of a spotted variable star is often displayed on a "migration curve" and accompanying "wave amplitude curve". The former is simply a plot of θ(min) vs time, where θ(min) is phase of wave minimum and phase is computed with an ephemeris using P(orb) or ⟨P(rot)⟩ as the period. The so-called migration period is given by

$$1/P(\mathrm{migr}) = 1/P(\mathrm{rot}) - 1/P(\mathrm{orb})\ . \qquad (1)$$

The latter is simply Δm versus time. The life of one spot appears in a migration curve as a straight-line segment, the slope of which indicates its rotation period and the length of which indicates its lifetime. The rise and fall of the wave amplitude should occur within this lifetime.

Several spotted stars have been analyzed in this way, resulting in a determination of lifetimes, rotation periods, and maximum wave amplitudes for one or more different spots. Examples are BY Dra (Oskanyan et al. 1977), V478 Lyr (Hall, Henry, Sowell 1990), and DM UMa (Mohin et al. 1985). A particularly nice example of a multi-spot migration curve, shown in Figure 1, is that of RT Lac, which is discussed in more detail in section 4.11. Note the 8 straight-line segments corresponding to 8 different spots during the 80-year interval of observation. Their lifetimes have ranged from 4 years (the 1932 spot) to 16 years (the 1965 spot). Their periods have ranged from 0.17 % faster than P(orb.) in 1903 to 0.25 % slower than P(orb.) in 1938.

Values of t_o for several other spotted variables were extracted from the literature. If the original analysis did not yield an explicit determination of spot lifetime, the migration curve and wave amplitude curve were reexamined in the manner described above.

3. Calculated Spot Lifetimes

Four quantities are needed to compute t_c, the time required for the shearing effect of differential rotation to disrupt a spot: (1) the differential rotation coefficient k as it appears in the equation

$$P\phi = P_{eq} / (1 - k \sin^2 \phi), \quad (2)$$

where ϕ is stellar latitude, (2) the rotation period of the star at its equator, P_{eq}, (3) the radius r_s of the spot, assumed to be circular, and (4) the latitude of spot center, ϕ_s.

3.1. Equatorial Rotation Period P_{eq}

For our purposes here we need only an approximate value of P_{eq}. Therefore, in the case of a binary, the known P(orb) will suffice and, in the case of a single spotted star, the mean P(rot) will suffice.

3.2. Differential Rotation Coefficient k

On the sun this quantity, which has the value k = 0.189 (Allen 1963, section 86), can be measured directly by tracking sunspot motions. On other stars, which are not spatially resolved, it must be inferred. In principle starspot latitudes (gleaned from spot modelling) together with starspot rotation periods (gleaned from migration curves) could evaluate k in equation (2). The practical difficulty is that spot latitudes are

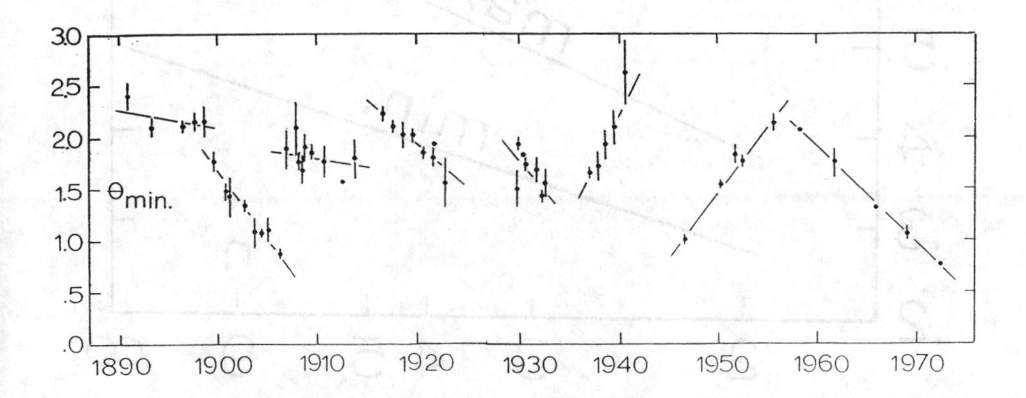

Figure 1. Migration curve for RT Lac. Each straight-line segment corresponds to one of the eight starspots.

poorly determined in all spot modelling techniques (Vogt 1983) and few spotted variables have been modelled anyway.

One can, however, make use of the observed range of P(rot). From equation (2) we get

$$(P_{max} - P_{min}) / P_{eq} = k \ . \ f \ , \tag{3}$$

where

$$f = \sin^2\phi_{max} - \sin^2\phi_{min} \ . \tag{4}$$

With a total range of 45° assumed for ϕ, f lies between 0.5 and 0.7 depending on whether the dark region covers the equatorial, intermediate, or polar latitudes of the star. In the above example of RT Lac's eight spots, $(P_{max} - P_{min}) / P_{eq} = 0.0042$. If $0.5 < f < 0.7$, then we have $0.0060 < k < 0.0084$. Note that this is differential rotation 25 times less extreme than the solar case.

Whereas eight different spots have been observed on RT Lac, fewer will have been observed on stars with shorter photometric histories. For them P_{max} and P_{min} will not represent the full 45° latitude range

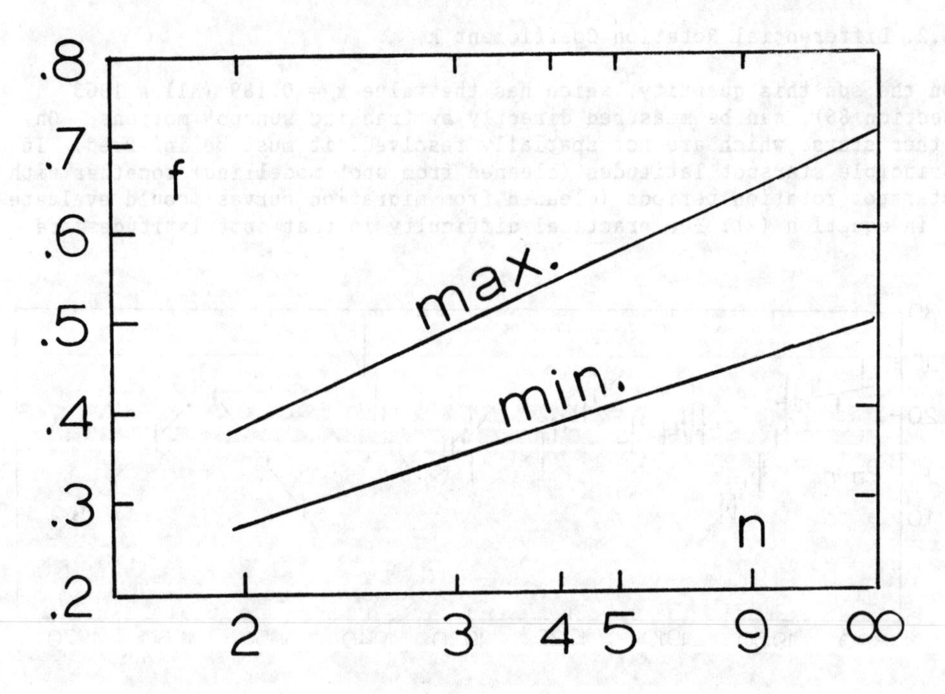

Figure 2. The factor f used in equation (3) to evaluate the differential rotation coefficient k, where n is the number of starspots observed.

and the appropriate values of f will be correspondingly smaller. This is illustrated in Figure 2. When only one value of P(rot) has been measured, one can estimate $\Delta P/P$ by comparing P(rot) with P(orb). This amounts to comparing the latitude of that one spot with the corotation latitude, defined by Hall (1972) and evaluated by Scharlemann (1982). In such a case, it is probably sufficient to enter n = 2 into Figure 2.

3.3. Radius of the Spot r_s

Wave amplitude Δm and spot radius r_s can be related with a simple spot model which takes into account limb darkening, foreshortening effects, inclination of the rotation axis, and temperature difference between spotted and unspotted photospheric regions. Figure 3 shows the relation in the case of $x = 0.6$, $i = 90°$, and $T_{spot} = 0°$ K.

If a spotted star is single or occurs in an SB1, then the observed wave amplitude can be used directly in Figure 3. If a companion star contributes appreciable luminosity to the system, however, then the observed wave amplitude must be increased by an appropriate factor, inversely proportional to the relative luminosity of the spotted star, L.

Temperatures measured for spots in several stars (Vogt 1983) show starspots are not black. The flux ratio in the V-band wavelength region

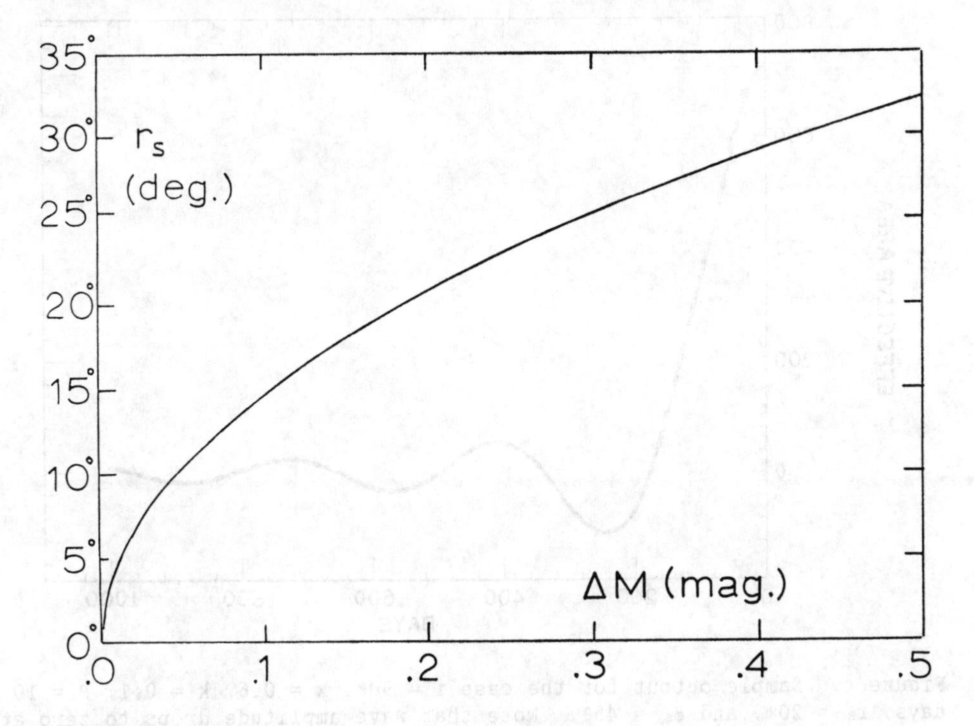

Figure 3. Starspot radius r_s as a function of wave amplitude.

is typically around 0ᵐ.2. Therefore, spot radii derived from wave amplitudes in the V bandpass using Figure 3 should be increased by the factor

$$[\ 1/(1-0.2)\ ]^{1/2} = 1.12\ .$$

3.4. Latitude of Spot Center ϕ_s

As explained above, starspot latitudes in general are not known. Therefore, we are forced to compute t_c for a range of possible starspot latitudes, from a spot touching the equator to a spot touching the pole.

3.5. Calculating t_c

The simple starspot model described in section 3.3 was used to follow the disruption of a spot by differential rotation. The original area of the spot remains constant but the longitude of a strip at latitude ϕ_i varies as the star rotates according to equation (2). Wave amplitude at a given time is the difference between flux received when the longitude of spot center faces earth and flux received from the opposite hemisphere. When wave amplitude drops to zero, after time t', the spot is considered disrupted and its life terminated.

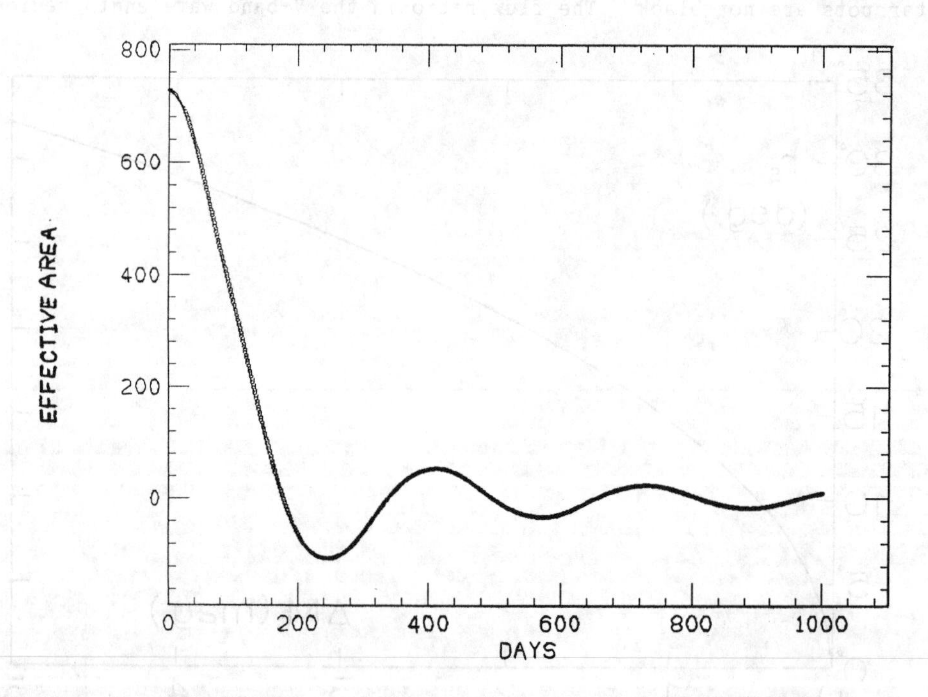

Figure 4. Sample output for the case $i = 90°$, $x = 0.6$, $k = 0.1$, $P = 10$ days, $r_s = 20°$, and $\phi_s = 45°$. Note that wave amplitude drops to zero at $t' = 180$ days.

Fixed input parameters were $x = 0.6$, $i = 90°$, $k = 0.1$, and $P_{eq} = 10$ days. Variable parameters were spot radius r_s and latitude of spot center ϕ_s. Figure 4 shows sample output for the case of a spot 20° in radius centered at latitude 45°. The general results are summarized in Figure 5, where t' has units of days and r_s has units of degrees. Note that t' has an upper and lower limit for any given r_s. For an arbitrary star, with different values of P_{eq} and k,

$$t_c = 0.01\ t'\ P_{eq}\ /\ k\ . \tag{5}$$

4. Parameters for 17 Spotted Stars

The 17 stars listed in Table 1 were considered. The first column is the star's name. The second column is the star's P_{eq}, approximated as P(orb) or mean P(rot). The third column is the relative luminosity of the spotted star, used to correct the observed wave amplitude. The fourth and fifth columns contain the range $P_{max} - P_{min}$ covered by the n spots which were observed. And the last column is the value of k derived from $\Delta P/P$ with the appropriate value of f read from Figure 2.

Values of t_c calculated for spots on those 17 stars are given in Table 2. The first column is the star's name. The second column identifies the spot with a median epoch. The third column is that spot's

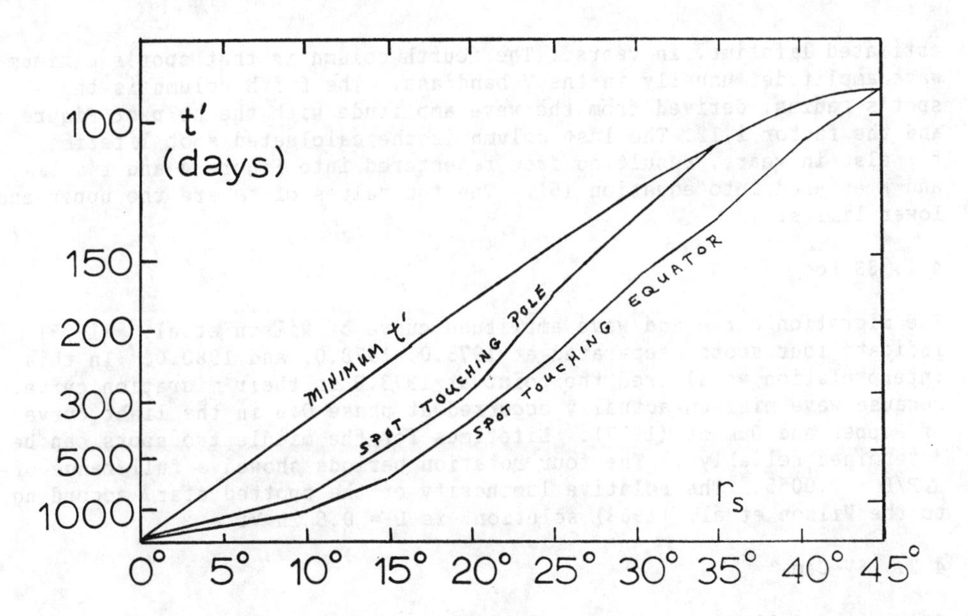

Figure 5. Calculated spot disruption times t' (plotted proportional to 1/t') versus spot radius r_s. The maximum and minimum values of t' depend on the assumed ϕ_s.

Table 1

Parameters for 17 Stars

Name	P_{eq}	L	$\Delta P/P$	n	k
SS Boo	$7^{d}.606$	0.6	0.0055	4	0.012
SV Cam	0.5931	0.947	.0022	4	.0047
BM Cam	79.93	SB1	.019	2	.058
FK Com	2.3981	single	.005	4	.011
CG Cyg	0.631	0.5	.00025	3	.00059
V1764 Cyg	40.13	SB1	.0063	1	.021
BY Dra	3.827	0.75	.105	14	.175
EK Eri	335.	single	--	1	.051
σ Gem	19.423	SB1	.0115	4	.0244
MM Her	7.95	0.415	.0038	2	.0116
RT Lac	5.074	0.5	.0042	8	.0078
HK Lac	24.4284	SB1	.0115	4	.0244
V478 Lyr	2.13	1.00	.0095	9	.0173
VV Mon	6.05	0.64	.0048	1	.016
V1149 Ori	53.58	SB1	.006	1	.02
DM UMa	7.492	SB1	.002	2	.0067
HD 181943	385.3	single	--	1	0.044

estimated lifetime, in years. The fourth column is that spot's maximum wave amplitude, usually in the V bandpass. The fifth column is that spot's radius, derived from the wave amplitude with the help of Figure 3 and the factor 1.12. The last column is the calcluated spot lifetime t_c, also in years, resulting from r_s entered into Figure 4 and t', P_{eq}, and k entered into equation (5). The two values of t_c are the upper and lower limits.

4.1. SS Boo

The migration curve and wave amplitude curve of Wilson et al. (1983) indicate four spots, separated at 1975.0, 1978.0, and 1980.0. In this interpretation we altered the point at 1973.0 in their migration curve, because wave minimum actually occurred at phase 0.6 in the light curve of Popper and Dumont (1977). Lifetimes for the middle two spots can be determined reliably . The four rotation periods showed a full range of $\Delta P/P$ = 0.0055. The relative luminosity of the spotted star, according to the Wilson et al. (1983) solution, is L = 0.6 in V.

4.2. SV Cam

Starspot modelling by Zeilik, De Blasi, and Rhodes (1988) is the source here. In four cases a sequence of light curves appeared to give a short migration curve segment which could be attributed to one spot: 7-8-9, 12-13-14-15, 16-17-18, and 20-21-22. Their rotation periods, determined

Table 2

Parameters for Starspots

Star	Spot	t_o	Δm	r_s	t_c
SS Boo	1976	3	0^{m}14	25°	2.55 - 4.07
	1979	2	.20	30	2.14 - 2.85
SV Cam	1983	0.33	--	32	0.40 - 0.52
BM Cam	1981	9	.15	20	6.62 - 13.57
	1985	9	.15	20	6.62 - 13.57
FK Com	--	1	.2	23.5	0.94 - 1.63
CG Cyg	1967	4.5	.10	23.5	4.46 - 7.71
	1973	3	.07	19.7	5.25 - 11.04
	1979	4	.08	20.9	4.95 - 9.70
V1764 Cyg	1983	8	.09	15.7	11.89 - 30.92
BY Dra	1974.6	0.12	.026	9.5	0.23 - 0.99
	1974.8	0.20	.037	11.8	0.18 - 0.61
	1975.6	0.044	.019	8.4	0.25 - 1.17
	1975.7	0.046	.028	10.1	0.21 - 0.87
EK Eri	1984	20	.16	21	30.22 - 58.94
σ Gem	1978	3.5	--	32.3	1.38 - 1.78
	1981	3.2	--	25.9	1.69 - 2.61
	1983	4.0	--	28.7	1.52 - 2.12
	1984	1.7	--	25.6	1.75 - 2.72
MM Her	1977	3	.13	29	2.32 - 3.17
	1980	2	.11	22	2.50 - 3.75
RT Lac	1895	9	.08	20.9	3.07 - 6.02
	1903	8	.12	25.6	2.56 - 4.01
	1910	7	.07	19.7	3.18 - 6.65
	1918	9	.17	30.4	2.12 - 2.80
	1932	4	.12	25.6	2.56 - 4.01
	1938	7	.07	19.7	3.18 - 6.65
	1952	13	.16	29.5	2.17 - 2.93
	1965	16	.11	24.5	2.58 - 4.26
HK Lac	1978	4.3	--	30.7	3.24 - 4.25
	1978	4.0	--	26.8	3.70 - 5.56
	1983	3.3	--	27.0	3.65 - 5.41
	1983	3.4	--	28.0	3.51 - 5.06
V478 Lyr	1985	0.47	.04	10.6	1.14 - 4.40
	1985	0.29	.08	14.7	0.81 - 2.23
VV Mon	1979	6	.06	16	2.30 - 5.94
V1149 Ori	1983	10	.40	32.6	8.20 - 10.65
DM UMa	1980	5	.32	29.5	3.73 - 5.04
	1983	5	.23	25.1	4.37 - 7.00
HD 181943	1984	30	0.145	20	42.06 - 86.23

by least squares, showed a full range of $\Delta P/P = 0.0022$. In the fourth case, the light curve immediately following (no. 23) showed that the original spot had disappeared; this made it possible to estimate a spot lifetime. Spot radii were taken directly from Zeilik et al. (1988, table 2) but adjusted to compensate for our adopted flux ratio. The relative luminosity of the spotted star, L = 0.947, was taken from Zeilik et al. (1988, table 3).

4.3. BM Cam

Hall and Busby (1989) analyzed all 1979 through 1988 photometry. Their revised orbital period, derived from the ellipticity effect, was P(orb.) = $79^{\rm d}.93$. Two spots were found, one present in 1979 but vanishing in 1984, the other beginning in 1984 and still present in 1988. Their maximum wave amplitudes were both around $0^{\rm m}.15$. Their rotation periods were $82^{\rm d}.5$ and $81^{\rm d}.0$, yielding a difference of $\Delta P/P = 0.019$.

4.4. FK Com

This single star, the prototype of its class, shows a mean rotation period of $2^{\rm d}.3981$. Changes in the migration period and wave amplitude during four years of photometry (Guinan, Robinson, and Wacker 1986) suggest four different spots were observed. Each one reached a maximum wave amplitude of about $0^{\rm m}.2$ and lived an average of only 1 year. The four rotation periods covered a range of $\Delta P/P = 0.005$.

4.5. CG Cyg

The migration curve and wave amplitude curve of Sowell, Wilson, Hall, and Peyman (1987) show the presence of three spots, separated at 1969.5 and 1976.5. Their maximum wave amplitudes were $0^{\rm m}.10$, $0^{\rm m}.07$, and $0^{\rm m}.08$, respectively. The shortest and longest rotation period differed by $\Delta P/P = 0.00025$. Extrapolation of the three wave amplitude curves yielded relatively good spot lifetimes. Various published light curve solutions are not in close agreement, so we do not know reliably the relative luminosity of the two stars. Moreover, because they have somewhat similar dimensions, either one might be the spotted star. Therefore, we have estimated L = 0.5 .

4.6. V1764 Cyg = HD 185151

Photometry and analysis by Lines, Lines, Kirkpatrick, and Hall (1987) show a spot wave in addition to a slightly larger ellipticity effect. During the 6 years only one spot seems to have been present, reaching a maximum wave amplitude of $0^{\rm m}.09$. Its rotation period of $39^{\rm d}.878$, compared to the orbital period of $40^{\rm d}.13$, yields $\Delta P/P = 0.0063$.

4.7. BY Dra

Photometry by Oskanyan et al. (1977) showed small-amplitude variations which continued for finite intervals of time before suffering a phase jump. We take these intervals to be spot lifetimes. Six different spots were apparent but two of them gave only lower limits for lifetimes, because of gaps in the photometric coverage. The rotation periods of these six and eight others, summarized by Oskanyan et al. (1977, figure 3), cover a range of $\Delta P/P = 0.105$. Assuming the K4 V star has the spots, we adopt L = 0.75.

4.8. EK Eri = HR 1362

Strassmeier, Hall, Barksdale, Jusick, Henry (1990) show that this G8 IV single star had one spot during the last 10 years. The rotation period was 335 days while the amplitude dropped monotonically from $0^m.16$ to $0^m.02$. We cannot determine k directly but can estimate it from the star's Rossby number R. A G8 IV star should have a convective turnover time of $\tau_c = 90$ days according to Gilliland (1985). This, along with P(rot) = 335 days, gives R = 3.7. The empirical relation

$$k = k_{\odot} R^{-1} \tag{6}$$

of Hall and Henry (1990) for stars with Rossby numbers greater than unity would suggest k = 0.051 for HR 1362.

4.9. Sigma Gem

Spot modelling by Strassmeier et al. (1988) of 10 years of photometry was reinterpreted to show that seven spots were present during that time. Their maximum radii were derived from the longitude and latitude boundaries of the rectangular spots given by Strassmeier et al. (1988, table 3). The seven rotation periods had a range of $\Delta P/P = 0.023$. For four of the spots, good estimates of their lifetimes could be made; for the other three spots, only lower limits.

4.10. MM Her

The migration curve and wave amplitude curve of Sowell et al. (1983) show two spots were present, one before 1979.0 and one after that. Their maximum wave amplitudes were similar: $0^m.13$ and $0^m.11$. Their rotation periods differ by $\Delta P/P = 0.0038$. Extrapolation of the two wave amplitude curves yields relatively good spot lifetimes. L = 0.415 for the spotted star comes from the Sowell et al. (1983) light curve solution.

4.11. RT Lac

This spotted eclipsing binary has an especially long photometric history, but its migration curve as originally presented by Haslag (1976) and Hall and Haslag (1976) was not interpreted correctly. The migration curve shown in Figure 1 is more nearly correct. The eight straight-line segments correspond to eight different spots during the interval 1891 to 1972. Their lifetimes have ranged from 4 to 16 years. Their periods have ranged from $\Delta P/P = -0.0017$ to $+0.0025$, where Δ is in the sense observed minus orbital. The light curve solution (Eaton and Hall 1979) indicates the two stars are comparable in luminosity, so we take L = 0.5 for the spotted star. Maximum wave amplitudes come from Haslag (1976, table 6) although his semi-amplitudes had to be doubled. Ten years of valuable photoelectric photometry since 1978 has been presented by Evren (1989) but his analysis did not include values of Θ(min) which could be included in Figure 1.

4.12. HK Lac

Olah et al. (1985, 1986) have provided spot modelling of all photometry obtained since 1967. A plotting of their spot longitudes and radii versus time indicates four spots. Their individual rotation periods show a full range of $\Delta P/P = 0.0115$. The maximum spot radii in Table 1 come directly from Olah et al. (1985, table III) and Olah et al. (1986, table III).

4.13. V478 Lyr

Spot lifetimes, rotation periods, and maximum wave amplitudes are taken directly from the analysis of Hall, Henry, and Sowell (1990). Excluding those with only lower limits on the lifetime, we are left with spots C and D'. The nine different rotation periods cover a range of $\Delta P/P = 0.0095$. Because the G8 V primary dominates the system's light and is the spotted star, we take L = 1.00.

4.14. VV Mon

The first 5 points in the migration curve and wave amplitude curve of Busso et al. (1985, figure 4) belong to one spot, which faded away after 1982. The maximum wave amplitude was $0^m.06$ and the rotation period differed from the orbital period by $\Delta P/P = 0.0048$. A light curve solution by Popper (1988) gave L = 0.64 for the spotted star.

4.15. V1149 Ori = HD 37824

Strassmeier, Hall, Boyd, and Genet (1989) present photometry from 1983-84, 1984-85, and 1985-86 and describe earlier photometry dating back to 1979. One spot grew monotonically in wave amplitude from $0^m.04$ to $0^m.40$ in 6 years. Its times of wave minimum yielded a rotation period of $53^d.9 \pm 0^d.3$. The orbital period of P(orb) = $53^d.58$ comes from Balona (1987). Comparing the two gives $\Delta P/P = 0.006$.

4.16. DM UMa

Six years of photometry analyzed by Mohin et al. (1985) show two spots present during that time, one waning and the other waxing. They produced maximum wave amplitudes of $0^m.32$ and $0^m.23$, respectively, but had similar rotation periods, both shorter than the $7^d.492$ orbital period by $\Delta P/P = 0.002$.

4.17. HD 181943

Hooten et al. (1989) showed that this single G8 IV star had one spot present during the last 10 years of photometry. The rotation period was 385 days but, unlike HR 1362, the amplitude remained constant at 0ᵐ145. This suggests a spot lifetime much longer than the baseline of the observations, perhaps 30 years. With τ_c = 90 days for a G8 IV star (Gilliland 1985) and P(rot) = 385 days, we get R = 4.3. This, with the Hall-Henry relation in equation (6), suggests k = 0.044 for HD 181943.

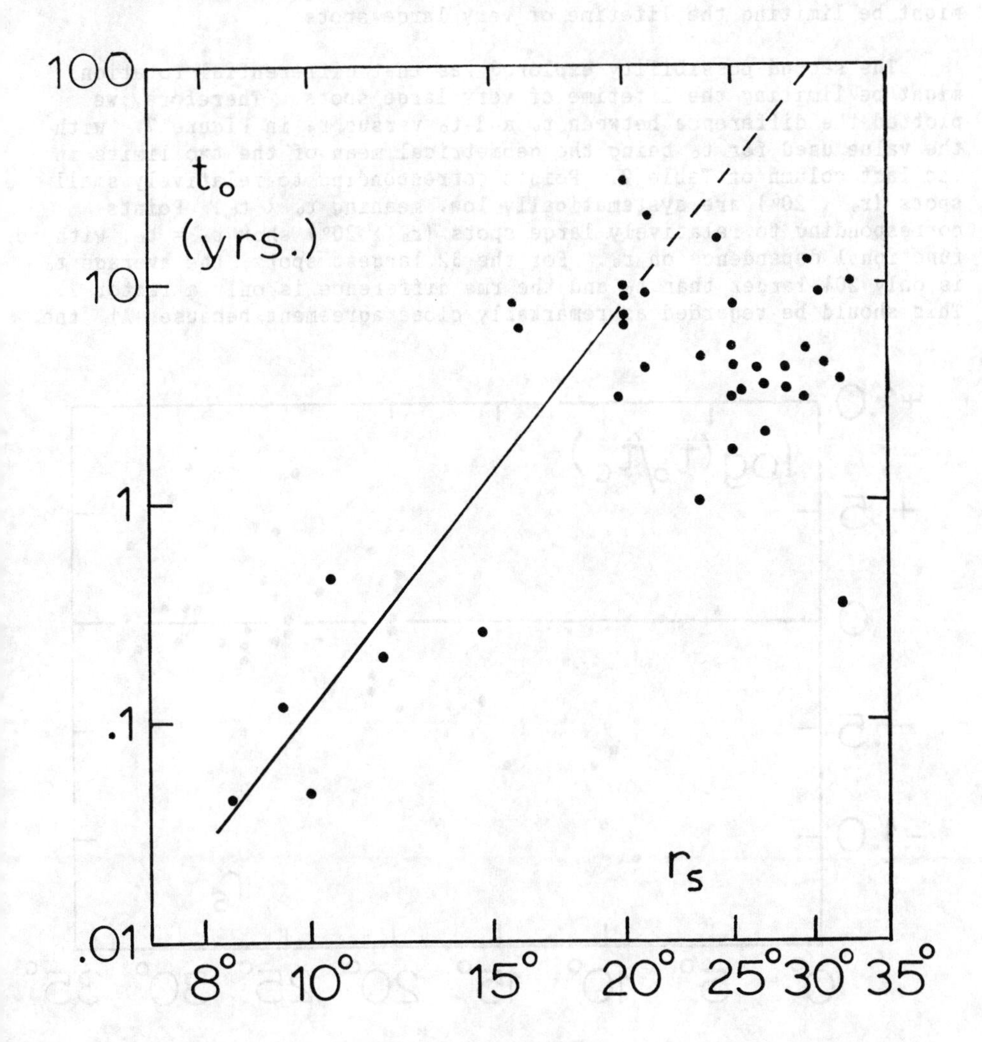

Figure 6. Observed spot lifetimes t_o (plotted on a log scale) versus spot radius r_s (plotted proportional to r^2). Note that larger spots live longer for r_s < 20° but not for r_s > 20°.

5. Comparing t_o and t_c

The first possibility explored was that spot lifetime was proportional to maximum spot area. Therefore, observed spot lifetimes t_o were plotted versus r_s^2, in Figure 6. There is some tendency for small spots to have short lifetimes, somewhat in line with the similar relation known to hold for large active regions on the sun (Allen 1973, section 88). But the "turn over" around $r_s = 20°$ indicates that very large spots have shorter lifetimes. This suggests that some other mechanism might be limiting the lifetime of very large spots.

The second possibility explored was that differential rotation might be limiting the lifetime of very large spots. Therefore, we plotted the difference between t_o and t_c versus r_s in Figure 7, with the value used for t_c being the geometrical mean of the two limits in the last column of Table 2. Points corresponding to relatively small spots ($r_s < 20°$) are systematically low, meaning $t_o < t_c$. Points corresponding to relatively large spots ($r_s > 20°$) show $t_o = t_c$, with no functional dependence on r_s. For the 32 largest spots, the average t_o is only 20% larger than t_c and the rms difference is only a factor 2. This should be regarded as remarkably close agreement because (A) the

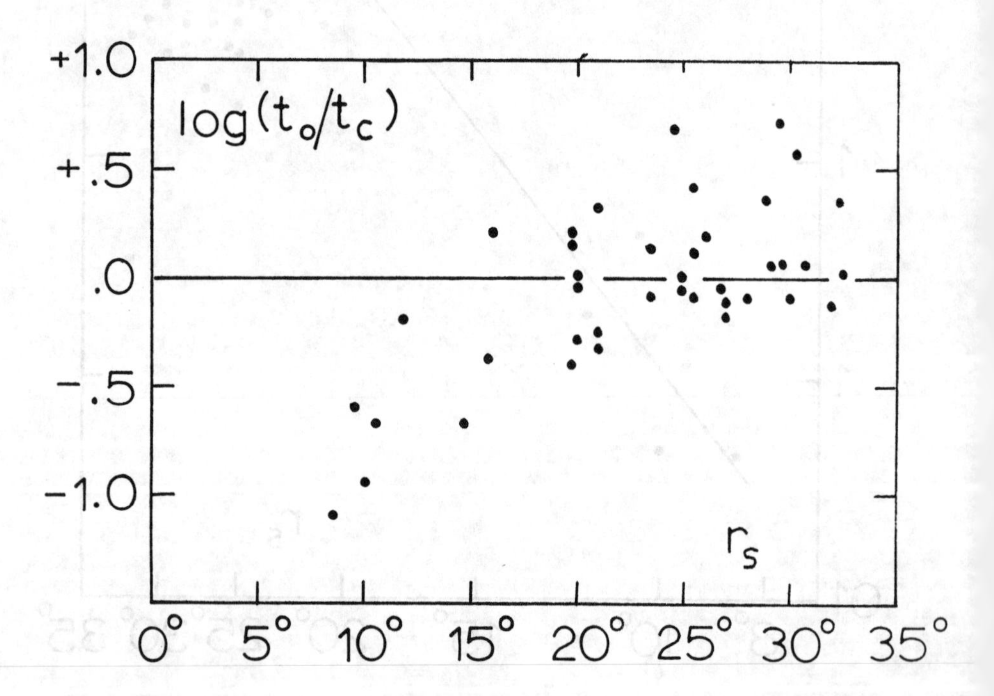

Figure 7. The logarithmic difference between t_o and t_c versus r_s. Note that large spots ($r_s > 20°$) have shear-limited lifetimes ($t_o = t_c$) but smaller spots ($r_s < 20°$) do not ($t_o < t_c$).

range in t_c (from Table 2) is around 2X and the factor f (used to get k) is probably uncertain by that much also, and because (B) each value of t_o also must be uncertain by a similar amount, largely due to inadequacies in the observational material.

This agreement ($t_o = t_c$ for the 32 larger spots) cannot be a result of a triviality, because two of the three parameters which determine t_c cover enormous ranges and so does the observed lifetime itself. From Table 1 we see $0.631 < P_{eq} < 385$ days (600X), $0.00059 < k < 0.058$ (100X) and $0.3 < t_o < 30$ years (100X).

6. Conclusions

Figure 6 shows that relatively smaller spots die before they can be disrupted by differential rotation. No firm statement about that time scale or a possible physical mechanism for it is proposed, but there seems to be a proportionality between spot area and spot lifetime, perhaps analagous to one pertaining to solar active regions.

Figure 7 shows no tendency for a spot to live longer than its shear-limited lifetime. Thus there are no excessively long lifetimes requiring the hypothesis of permanent anchor lines between surface spots and a rigidly rotating deeper layer.

The remarkably close agreement between t_o and t_c (for larger spots) indicates that lifetimes of large spots are in fact limited by shear. This observation is consistent with either of two scenarios: (1) a large spot originates in a deep layer which may be rigidly rotating, is magnetically disconnected after awhile, and is disrupted by differential rotation characteristic of the surface, or (2) a large spot is not disconnected magnetically but the deeper layers from which it originated approximately mirror the same differential rotation law which applies to the surface layers.

These conclusions could be bolstered if we had longer series of nearly continuous photometry for more spotted stars, coupled with the proper analysis to yield reliable differential rotation coefficients, spot lifetimes, and maximum spot radii.

ACKNOWLEDGEMENTS

This work was supported by N.A.S.A. research grant N.A.G. 8-111 while D.S.H. was on leave from Dyer Observatory, Vanderbilt University, Nashville, Tennessee 37235.

REFERENCES

Allen, C.W. 1963, Astrophysical Quantities (London: Athlone).
Balona, L. 1987, S.A.A.O. Circ. No. 11.

Busso, M., Scaltriti, F., and Cellino, A. 1985, Astr. Astrophys. **148**,29.
Chisari, D. and Lacona, G. 1965, Memorie Soc. Astr. Italiana **36**, 463.
Doyle, J.G., Butler, C.J., Morrison, L.V., and Gibbs, P. 1988, Astr. Astrophys. **192**, 275.
Eaton, J.A. and Hall, D.S. 1979, Ap.J. **227**, 907.
Evren, S. 1989, Astrophys. Space Sci., in press.
Gilliland, R. 1985, Ap.J. **299**, 286.
Guinan, E.F., Robinson, C.R., and Wacker, S.W. 1986, in Cool Stars, Stellar Systems, and the Sun, edited by M. Zeilik and D.M. Gibson (Berlin:Springer-Verlag), p. 304.
Hall, D.S. 1972, P.A.S.P. **84**, 323.
Hall, D.S. 1987, Publ. Astr. Inst. Czechoslovakia **70**, 77.
Hall, D.S. and Busby, M.R. 1989, in Remote-Access Automatic Telescopes, edited by D.S. Hayes and R.M. Genet (Mesa: Fairborn Press), in press.
Hall, D.S. and Haslag, K.P. 1976, in Multiple Periodic Variable Stars, Part II, edited by W.S. Fitch (Dordrecht: Reidel), p. 331.
Hall, D.S. and Henry, G.W. 1990, this volume.
Hall, D.S., Henry, G.W., and Sowell, J.R. 1990, A.J., in press.
Haslag, K.P. 1976, Master's Thesis, Vanderbilt University.
Hooten, J.T. et al. [29 authors] 1989, I.A.P.P.P. Comm. No. 38, in press.
Lines, H.C., Lines, R.C., Kirkpatrick, J.D., and Hall, D.S. 1987, A.J. **93**, 430.
Mohin, S. et al. [9 authors] 1985, Astrophys. Space Sci. **115**, 353.
Olah, K. et al. [12 authors] 1985, Astrophys. Space Sci. **108**, 137.
Olah, K. et al. [13 authors] 1986, Astrophys. Lett. **25**, 133.
Oskanyan, V.S., Evans, D.S., Lacy, C., and McMillan, R.S. 1977, Ap.J. **214**, 430.
Popper, D.M. 1988, A.J. **96**, 1040.
Popper, D.M. and Dumont, P.J. 1977, A.J. **82**, 216.
Scharlemann, E.T. 1982, Ap.J. **253**, 298.
Sowell, J.R., Hall, D.S., Henry, G.W., Burke, E.W., and Milone, E.F. 1983, Astrophys. Space Sci. **90**, 421.
Sowell, J.R., Wilson, J.W., Hall, D.S., and Peyman, P.E. 1987, P.A.S.P. **99**, 407.
Strassmeier, K.G. et al. [24 authors] 1988, Astr. Astrophys. **192**, 135.
Strassmeier, K.G., Hall, D.S., Boyd, L.J., and Genet, R.M. 1989, Ap.J. Suppl. **69**, 141.
Strassmeier, K.G., Hall, D.S., Barksdale, W.S., Jusick, A.T., and Henry, G.W. 1990, Ap.J., in press.
Vogt, S.S. 1983, in Activity in Red Dwarf Stars, edited by P.B. Byrne and M. Rodono (Dordrecht: Reidel), p. 137.
Wilson, J.W., Hall, D.S., Henry, G.W., Vaucher, C.A., and Africano, J.L. 1983, A.J. **88**, 1257.
Zeilik, M., De Blasi, C., and Rhodes, M. 1988, Ap.J. **332**, 293.

TIDAL TORQUES AND MAGNETIC STELLAR WINDS IN RS CVn BINARIES

Ronald F. Webbink
Department of Astronomy, University of Illinois
1011 W. Springfield Ave., Urbana, Illinois 61801, USA

Michael S. Hjellming
Department of Physics and Astronomy, Northwestern University
2145 Sheridan Rd., Evanston, Illinois 60208, USA

ABSTRACT. The theory of tidal dissipation by convective turbulence is applied to a sample of 39 RS Canum Venaticorum-type binaries showing small departures from pseudosynchronism. In an equilibrium model, tidal torques replenish the spin angular momentum lost by the active stars in these binaries through a magnetically-coupled stellar wind. However, contrary to expectation, in most systems tides appear to be extracting angular momentum *from* the active components, rather than imparting it to them. The presence among these binaries of previously undetected small orbital eccentricities ($e \lesssim 0.05$) would suffice to resolve this discrepancy, since the pseudosynchronous rotation frequency then slightly exceeds the orbital frequency.

1. INTRODUCTION

One of the most vexing problems in understanding the evolution of binary stars is the issue of angular momentum loss. Many types of systems show clear evidence of having lost a significant fraction of their initial angular momentum in arriving at their present state. In some cases, such as the cataclysmic binaries, only a small fraction of the initial angular momentum can have been preserved. Theorists appeal to the process of common envelope evolution — the dissipation of the angular momentum of an embedded binary star within a surrounding, more slowly-rotating common envelope (Paczyński 1976; Meyer and Meyer-Hofmeister 1979) — to produce the dramatic orbital contraction needed to produce such systems. In others, such as the Algol-type binaries, losses have evidently been much more modest, but even so typically drain one-third to one-half the initial angular momentum over the lifetime of the system (*e.g.*, Plavec 1973; Refsdal, Roth, and Weigert 1974). Efficient angular momentum loss could also explain the dearth of detached low-mass binaries at short orbital period ($P \leq 3^d$; Popova, Tutukov, and Yungelson 1982; van't Veer and Maceroni 1988) and suppress the thermally detached phase in the thermal relaxation oscillator model

C. İbanoğlu (ed.), Active Close Binaries, 393–409.

of W Ursae Majoris binaries.

In some cases, the level of activity and mass transfer observed among certain types of binaries points clearly to the existence of an angular momentum loss mechanism driving the whole process. Notable examples of this phenomenon are the cataclysmic binaries and low-mass X-ray sources. The donor stars in these interacting binaries are typically but a fraction of the mass of the accreting white dwarfs or neutron stars, and so low in mass that their own evolutionary time scales generally far exceed a Hubble time. Mass transfer rates should scarcely exceed 10^{-10} $(L_2/L_\odot)$ $M_\odot$ yr^{-1}, where L_2 is the intrinsic luminosity of the donor star, unless driven by systemic angular momentum loss. The mass transfer rates needed to produce the observed accretion disk luminosities or X-ray luminosities of these systems are typically more than an order of magnitude greater (see, *e.g.*, Patterson 1984). Of known angular momentum loss mechanisms, only general relativistic gravitational radiation is calculable from first principles, and it fails by one to two orders of magnitude to explain the inferred angular momentum loss rates in most of these systems. Theorists are left to appeal here as well to phenomenological loss mechanisms which can only be calibrated empirically.

The problem, of course, is that it is not even remotely possible to observe directly angular momentum loss from a star or binary system. To measure the mass loss rate is extremely difficult or impossible, except among very luminous stars and the Sun itself. No means exist to determine the transverse as well as radial velocity of outflowing material. Instead, we must resort to methods which are more or less indirect in order to estimate angular momentum losses or loss rates. Such methods include orbital period changes, which can be measured very accurately, and direct estimates of mass transfer rates in interacting binaries, which are subject typically to large errors. Inferences may also be drawn from evolutionary calculations. However, these methods are all afflicted by severe uncertainties, since the dynamical effects of angular momentum loss upon the binary orbit are difficult to separate from those of mass loss, and systemic angular momentum losses can be masked by temporary angular momentum redistribution within a binary.

Another widely adopted approach is to appeal to the same angular momentum loss mechanisms among binary stars as are inferred to operate among single stars. It is observed that late-type main sequence stars typically rotate much more slowly than do early-type main sequence stars, and that the mean rotation velocities of late-type dwarfs in clusters decrease with increasing cluster age. An early study of solar-type dwarfs (Skumanich 1972) produced an empirical relation between rotation velocity, v_{rot}, and stellar age, t, of the form,

$$v_{rot} = f\ 10^{14}\ t^{-0.5}\ \mathrm{cm\ s^{-1}}, \tag{1}$$

where f = 0.73 according to Skumanich (1972), or f = 1.78 according to Smith (1979). This spindown is generally attributed to the braking action of stellar magnetic fields, which enforce corotation of an outflowing stellar wind out to many stellar radii. This large moment arm imparts a relatively high specific angular momentum to mass which is lost, and so spins down the star quite efficiently.

The so-called "Skumanich law" (Eq. 1) was applied to low-mass X-ray binaries (and, by extension, cataclysmic binaries) by Verbunt and Zwaan (1981). Mochnacki (1981) applied a similar model, based upon the magnetic braking theory of Belcher and MacGregor (1976), to contact binaries, again with an extension to cataclysmic binaries (and Algol binaries as well). The physical argument is that, in a short-period close binary system, tidal action imparts a high rotation rate on a lobe-filling, or nearly lobe-filling star. The Skumanich law implies that such rapidly-rotating stars suffer high angular momentum loss rates, i.e., that the magnetic stellar wind brakes the rotation of such stars very efficiently. However, this loss is continuously replenished by tidal action, which draws angular momentum from the binary orbit and deposits it in the rotation of the lobe-filling star. The rotation of the star remains nearly synchronously locked to the binary orbit, which gradually contracts as angular momentum is drawn from it.

Verbunt and Zwaan derived from Eq. 1 an angular momentum loss rate, $\dot{J}_{VZ}$, from the wind source (presumed to be the donor star, component 2),

$$\dot{J}_{VZ} = -0.5\ 10^{-28}\ f^{-2}\ k^2\ M_2\ R_2{}^4\ \omega^3 \quad g\ cm^2\ s^{-2} , \tag{2}$$

where M_2 and R_2 are the mass and radius, respectively, of that star, k its dimensionless radius of gyration, and ω its rotational angular frequency. They showed that such a loss rate is of the requisite order of magnitude to account for observed mass transfer rates among low-mass X-ray binaries and cataclysmic binaries.

A serious concern is that Eq. 1 is based on a fairly homogeneous sample of stars. Even if it adequately describes the dependence of angular momentum loss rate of stellar rotation rate for that sample (*cf*. Hartmann and Noyes 1987; Stauffer 1987), one does not really know how the factor f depends on stellar parameters. Physical arguments can be mustered to this issue, for example the braking model of Belcher and MacGregor (1976) which Mochnacki (1981) used, and more recently that of Mestel and Spruit (1987) employed by McDermott and Taam (1989). However, for lack of a complete, self-consistent dynamo model, which inevitably requires a comparably rigorous treatment of convection, key elements of all existing magnetic braking prescriptions must be supplied empirically — the braking models are "tuned" to fit observational constraints.

An example of the tuning of magnetic braking models may be found among a number of theoretical models which attempt to reproduce the "period gap" among cataclysmic binaries. This term refers to the nearly total absence of cataclysmic binaries with orbital periods between 2 and 3 hours. In simple evolutionary models, the donor low-mass main sequence star is gradually eroded, sliding down the main sequence while continuously filling its Roche lobe. The orbital period of the binary at any point in this sequence is, in hours, roughly equal to one-tenth the mass of the donor star, in solar masses, and there is thus, naively, no reason not to expect cataclysmic binaries to populate the 2 - 3 hour orbital period interval in numbers comparable to those of greater or shorter orbital periods. The magnetic stellar wind model proposes to explain this by driving mass transfer above the gap so rapidly that the donor star departs from the main sequence: its deep outer convection zone cannot radiate it energy away rapidly enough to keep the star in thermal equilibrium, and so it becomes over-sized for its mass. As the star reaches the upper edge of the period gap, it becomes fully convective; its magnetic field is largely expelled, or radically reconfigured, as it loses its anchor in the radiative core. The torque from the magnetic stellar wind suddenly subsides, greatly reducing the rate of orbital contraction and allowing the donor star to contract into thermal equilibrium within its Roche lobe. Subsequent angular momentum losses, due to gravitational radiation and perhaps to weak residual magnetic braking, eventually draw the Roche lobe of the donor in upon it as the orbital period shrinks to 2 hours (the lower edge of the gap), and mass transfer resumes at a much smaller rate. The location and especially the width of the period gap are therefore sensitive functions of the magnitude of magnetic wind torques in systems just above the period gap (Rappaport, Verbunt, and Joss 1983; Spruit and Ritter 1983; Hameury, *et al*. 1988, McDermott and Taam 1989).

A common feature of the various attempts described above to constrain magnetic braking laws is their statistical nature: rather than to attempt to determine torques in individual systems, researchers use their collective properties in a statistical sense. However, there are another class of objects in which one might hope to determine empirically what net torques exist within individual binaries, and deduce from that their angular momentum loss rates. These are the active close binaries (RS CVn and BY Dra systems), in which one can measure independently, with great accuracy, the orbital angular frequencies of the binaries and the rotational angular frequencies of one or perhaps even both of their components. Since the active components typically have deep surface convection zones, models of turbulent dissipation, consistent with the mixing-length theory of convection, provide a well-defined basis on which to relate departures from synchronism to the net rate of tidal angular momentum transport within the system.

2. METHOD

The intent of the present study is to explore the dependence of magnetic stellar wind torques in RS CVn systems on the evolutionary state of the active component, which is presumably the source of the wind. Magnetic activity on these stars is manifested by starspots, which produce migrating photometric disturbances in their lightcurves. The difference between the rotation period of the active component and the orbital period of the binary can therefore be measured with considerable accuracy from the migration rate of the photometric disturbances through the lightcurve. In systems in a near-synchronous state, according to the magnetic braking model, a steady state exists in which the departure from synchronism is just sufficient to provide a tidal torque, $\dot{J}_{tide}$, on the active star equal to its angular momentum loss rate in its magnetically-controlled wind. This torque extracts angular momentum from the binary orbit, and we thus have

$$\dot{J}_{orb} = -\dot{J}_{tide} = \dot{J}_2 \quad . \tag{3}$$

We calculate the tidal torque directly from the departure of the rotational angular velocity of the active component, Ω, from the orbital angular frequency, ω, according to the formalism developed by Zahn (1977):

$$\dot{J}_{tide} = 6\,\frac{k_2}{\tau_F}\,q^2 M_2 R_2{}^2 \left(\frac{R_2}{A}\right)^6 \left[(\omega - \Omega) + e^2\left(\frac{27}{2}\,\omega - \frac{15}{2}\,\Omega\right) + O(e^4)\right] \quad , \tag{4}$$

where k_2 is the apsidal motion constant of the active star (component 2: our designation of components 1 and 2 is the reverse of Zahn's), M_2 its mass, R_2 its radius, q the mass ratio (M_1/M_2), A the orbital separation, e the orbital eccentricity, and τ_F the frictional time scale of the convective envelope of the active star, which we approximate as

$$\tau_F{}^{-1} \approx \frac{4\pi}{3m_{ce}} \int_{r_{bce}}^{r_{tce}} \rho\, v_{conv}\, \ell_{mix}\, dr \quad , \tag{5}$$

where m_{ce} is the mass of the convective envelope, r_{tce} is the radius of the top of the convective envelope, r_{bce} the radius of the bottom of the convective envelope, ρ the local gas density, v_{conv} the local convective velocity derived from mixing length theory, and ℓ_{mix} the local mixing length. [N.B.: The frictional time scale, τ_F, defined by Eq. 5 is a simple approximation, consistent to order of magnitude with mixing length theory. See, *e.g.*, Zahn 1989.]

A grid of stellar evolution models (Hjellming 1989) was used to evaluate the structure constants which appear in Eqs. 4 and 5. The model sequences used take no account of stellar mass loss, but this is unlikely to be a serious defect, as the systems used in this study are all detached binaries. A ratio of mixing length to pressure scale height $\ell_{mix}/H_P = 1.5$ was assumed. The values of k_2 and τ_F are sensitive functions of effective temperature, but much weaker functions of luminosity, as illustrated in Figures 1 and 2.

The basic observational data for this study were drawn from the catalog of chromospherically active stars of Strassmeier, *et al.* (1988a), supplemented by the data of Strassmeier, *et al.* (1989). We selected systems for which the dimensionless radius of the active star, R_2/A, was known or could be estimated with some confidence, and for which independent determinations of the orbital period and rotational period of the active star were available. Since we are interested specifically in determining magnetic wind torques, and to do so we must presume that the rotation rate of the active star is in an equilibrium state in which magnetic braking is balanced by tidal acceleration (*cf.* Eq. 3), only systems approximating a pseudosynchronous state were retained in our sample. In this state, the bracketed term on the right-hand side of Eq. 4 vanishes (see Hut 1981 for higher-order terms in

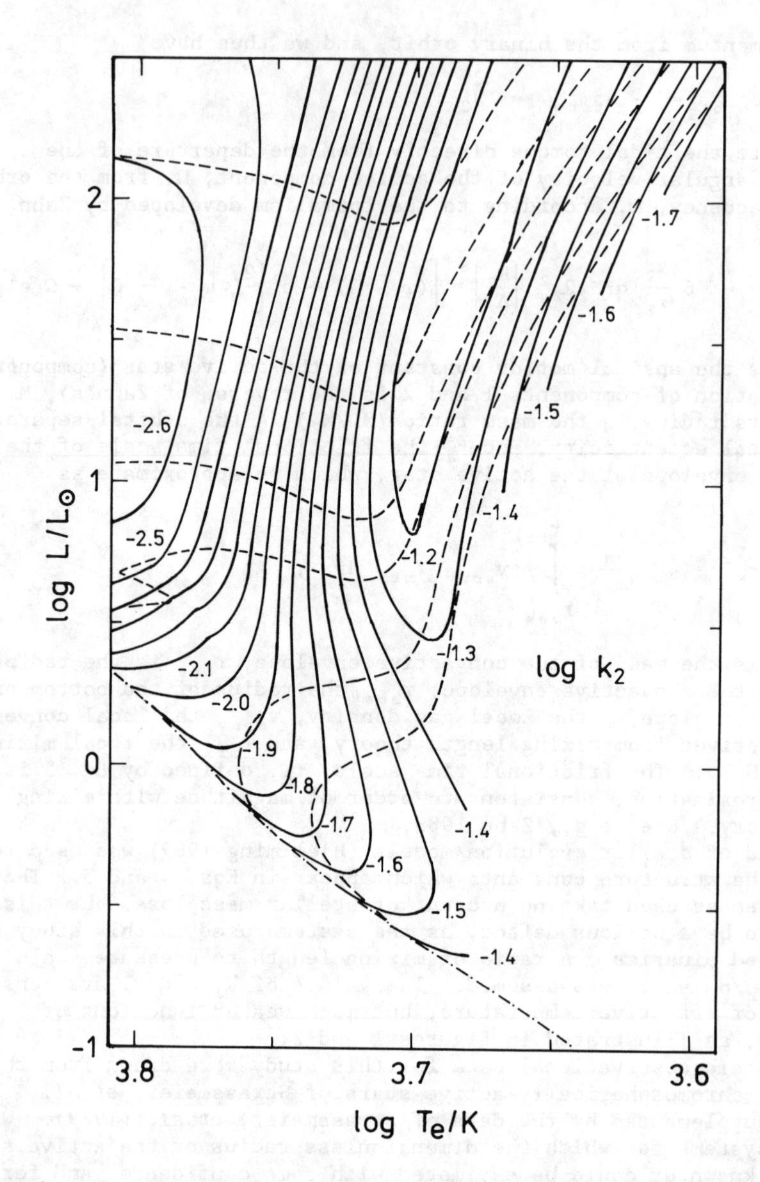

Figure 1. Apsidal motion constants, k_2, derived from evolutionary calculations, as a function of position in the HR diagram. The solid lines are contours of constant k_2, labeled by the value of log k_2. Dashed lines mark, in order of increasing luminosity, the evolutionary tracks of stars of masses 0.75, 0.90, 1.25, 1.50, 2.0, and 3.0 $M_\odot$, respectively. The dash-dotted line marks the zero-age main sequence.

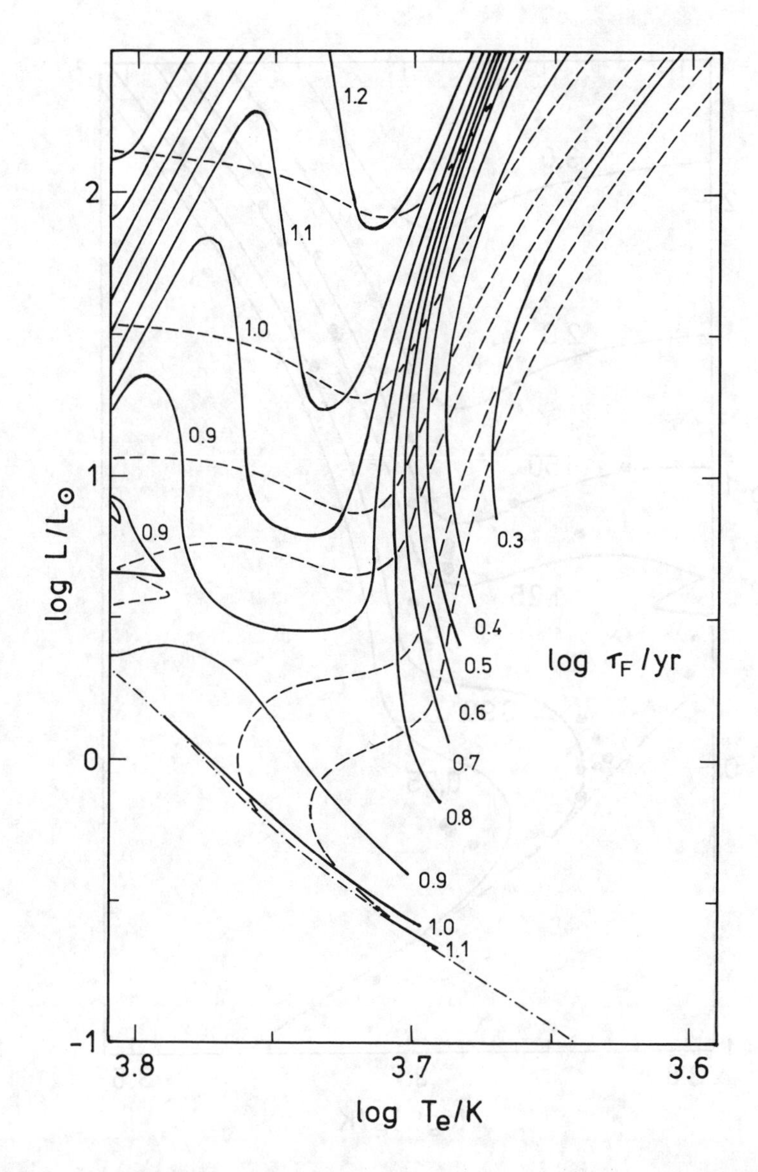

Figure 2. Frictional time scales, τ_F, in the convective envelopes of stellar evolution models. The solid lines are contours of constant τ_F, labeled by the value of log τ_F in years. Dashed lines mark evolutionary tracks, and the dash-dotted line the zero-age main sequence, as in Figure 1.

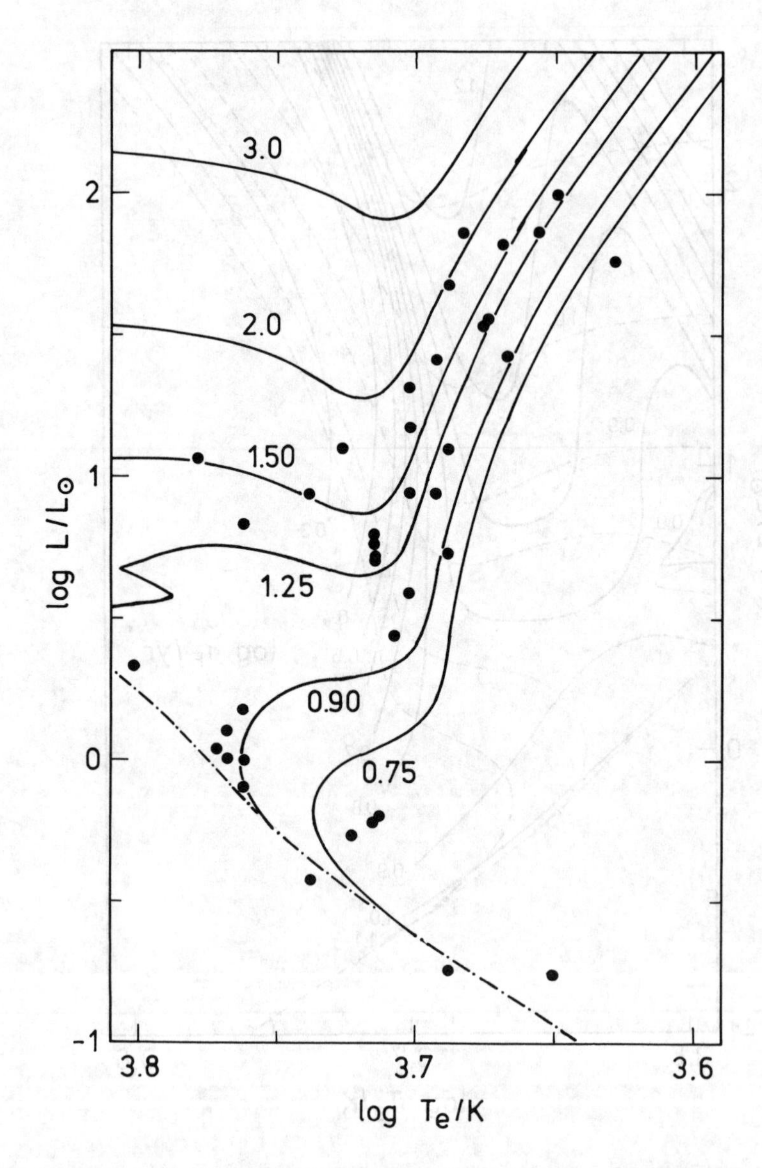

Figure 3. The 39 systems in Table I (filled circles) compared with evolutionary tracks (solid lines, labeled by stellar mass in $M_{\odot}$) in the HR diagram. The dash-dotted line marks the zero-age main sequence.

orbital eccentricity). In practice, pseudosynchronism is easily distinguishable from non-equilibrium rotation, because in this state the absolute value of the bracketed term is typically of order $\lesssim 10^{-3}\ \Omega$, but is otherwise of the same order as Ω.

The 39 systems which met these criteria are listed in Table I, with the adopted mass, radius, effective temperature, and bolometric luminosity of the active components listed in the second through fifth columns, respectively, the mass ratio in the sixth column, orbital period, eccentricity, and inclination in the seventh through ninth columns, and computed tidal torque on the active component (positive in the same sense as the orbital motion) in the tenth column. The location of these systems in the HR diagram is compared with model evolutionary tracks in Figure 3. Masses listed in italics were deduced from the position of the star in the HR diagram, and the mass ratios (also in italics) from the assumed mass of the active component, the spectroscopic mass function, and the orbital inclination. Orbital eccentricities and inclinations in italics are also assumed values.

3. RESULTS

In most RS CVn systems, the lightcurve disturbances produced by starspots migrate toward earlier orbital phase: the active components in these systems rotate more rapidly than synchronously. Tides in the majority of systems listed in Table I would thus appear to be *removing* angular momentum from the active components, rather than restoring the putative losses in magnetic stellar winds, as illustrated in Figure 4 where the computed tidal torques are compared with the simple braking model of Verbunt and Zwaan (1981).

It is difficult to understand how this conclusion can represent the true state of affairs in these binaries. Any sink of orbital angular momentum in the system (including losses to the rotation of the binary components as they expand during evolution) produces an *increase* in the orbital frequency, and will thus tend to produce lagging tides on both components in an equilibrium tidal state.

One possible explanation for the discrepancy is that starspots on an active star in fact drift with respect to the true rotation of the stellar envelope. It is virtually impossible to refute this hypothesis directly. Even on the Sun, precise measurements of its rotation period have used sunspots as markers. However, even if spots drift significantly with respect to the stellar envelope, we should expect that they track the rotation of the stellar magnetic field, which in the presence of any magnetic braking should exert a retarding torque on the stellar envelope, not an accelerating one. While caution is in order in equating precisely spot rotation with that of the surrounding stellar envelope, we can therefore reject this explanation.

On the Sun, we see clear evidence of differential rotation, with the solar equator rotating more rapidly than its poles. A similar equatorial acceleration could in principle explain the apparent super-synchronous rotation of many of the active components in this study. There is a natural selection in favor of highly-inclined binaries for

TABLE I. COMPUTED TIDAL TORQUES IN RS CVn BINARIES

System	M_2 ($M_\odot$)	R_2 ($R_\odot$)	log T_e (K)	log L_2 ($L_\odot$)	q	P_{orb} (d)	e	i (°)	$\dot{J}_{tide}$ ($M_\odot R_\odot^2 yr^{-2}$)
UX Ari	1.09	3.	3.715	0.77	0.89	6.4379	*0.0*	60.	-1.9(-6)
CQ Aur	2.0	8.7	3.676	1.53	0.80	10.6215	*0.0*	*90.*	-9.3(-3)
SS Boo	1.00	3.28	3.702	0.79	1.00	7.6061	0.05	88.8	+6.2(-5)
SS Cam	3.4	7.0	3.676	1.34	1.12	4.8242	*0.0*	78.6	-1.9(-3)
SV Cam	0.7	0.7	3.651	-0.75	1.43	0.5931	0.0	80.	-1.0(-7)
RU Cnc	1.5	5.	3.702	1.16	0.99	10.1730	0.0	89.5	-1.3(-4)
WY Cnc	0.93	1.	3.762	0.00	0.57	0.8294	0.0	*90.*	-6.4(-7)
RS CVn	1.40	4.10	3.715	1.04	0.96	4.7979	0.0	87.	-3.9(-4)
BI Cet	≥0.8	~0.9	3.762	-0.09	1.09	0.5158	0.0	~30.	+1.5(-4)
σ^2 CrB	1.12	1.22	3.802	0.33	1.03	1.1398	0.022	28.	+3.9(-5)
CG Cyg	0.52	0.88	3.723	-0.27	1.00	0.6311	*0.0*	81.8	+3.0(-5)
V1762 Cyg	*1.3*	9.	3.695	1.64	*0.95*	28.59	0.04	60.	-2.6(-4)
EI Eri	3.8	2.6	3.762	0.83	0.37	1.9472	0.0	46.	-5.9(-6)
σ Gem	*1.3*	*16.7*	3.649	2.00	*0.63*	19.6046	0.0	60.	-1.1(-1)
Z Her	1.10	2.60	3.715	0.64	1.12	3.9928	0.0	83.13	-8.7(-5)
MM Her	1.27	2.83	3.738	0.81	0.98	7.9603	0.04	86.35	+2.2(-6)
V772 Her	1.04	1.0:	3.772	0.04	0.57	0.8795	0.045	82.7	+6.9(-6)
V792 Her	1.46	12.28	3.683	1.86	0.96	27.5384	0.0	80.	-6.0(-3)
DK Hya	0.79	0.9	3.663	-0.49	0.99	6.8657	0.0	~65.	-1.3(-4)
IL Hya	*1.6*	12.	3.679	1.83	*1.21*	12.908	0.05	>30.	+3.5(-4)
BQ Hyi	*0.98*	*2.1*	3.708	0.44	*0.73*	18.379	0.06	*90.*	+1.4(-8)
RT Lac	0.78	4.2	3.726	1.10	1.86	5.074	0.0	89.	-3.7(-4)
AR Lac	1.31	3.1	3.715	0.80	1.00	1.9832	0.0	87.	-3.1(-3)
HK Lac	*1.6*	*9.7*	3.688	1.68	*0.58*	24.4284	0.01	70.	-2.6(-5)
DH Leo	0.83	0.97	3.715	-0.21	0.67	1.0704	0.0	78:	-3.4(-6)
GX Lib	*0.95*	8.	3.667	1.43	*0.68*	11.1345	0.0	58:	-1.6(-2)
AE Lyn	1.64	3.14	3.778	1.06	0.98	11.0764	0.107	60.	-1.9(-7)
V478 Lyr	*0.77*	1.0	3.712	-0.20	*0.32*	2.1305	0.0	~67.	+1.4(-7)
VV Mon	1.5	6.0	3.715	1.37	0.89	6.0506	0.033	86.	+5.9(-3)
EZ Peg	0.93	*0.6*	3.762	-0.44	1.01	11.6598	0.0	~25.	-1.6(-13)
LX Per	1.32	2.8	3.715	0.71	0.99	8.0382	0.0	87.4	-2.5(-6)
SZ Psc	1.8	5.08	3.702	1.17	0.74	3.9659	0.04	74.	+1.3(-2)
AR Psc	1.53	3.3	3.738	0.94	0.77	14.300	0.19	~27.	+9.1(-8)
V965 Sco	>2.0	14.	3.629	1.76	0.99	30.969	0.0	<72.	-6.3(-5)
V471 Tau	0.8	0.6	3.688	-0.74	0.75	0.5212	0.0	*90.*	-6.7(-5)
V711 Tau	1.4	3.9	3.702	0.94	0.80	2.8377	0.0	33.	+7.6(-4)
V837 Tau	1.0	1.1	3.768	0.11	0.62	1.9299	0.035	65.	-2.3(-5)
XY UMa	0.95	0.98	3.768	0.01	0.74	0.4790	*0.0*	85.	+8.1(-6)
ER Vul	0.98	1.23	3.762	0.18	1.08	0.6981	0.02	71.	-9.7(-5)

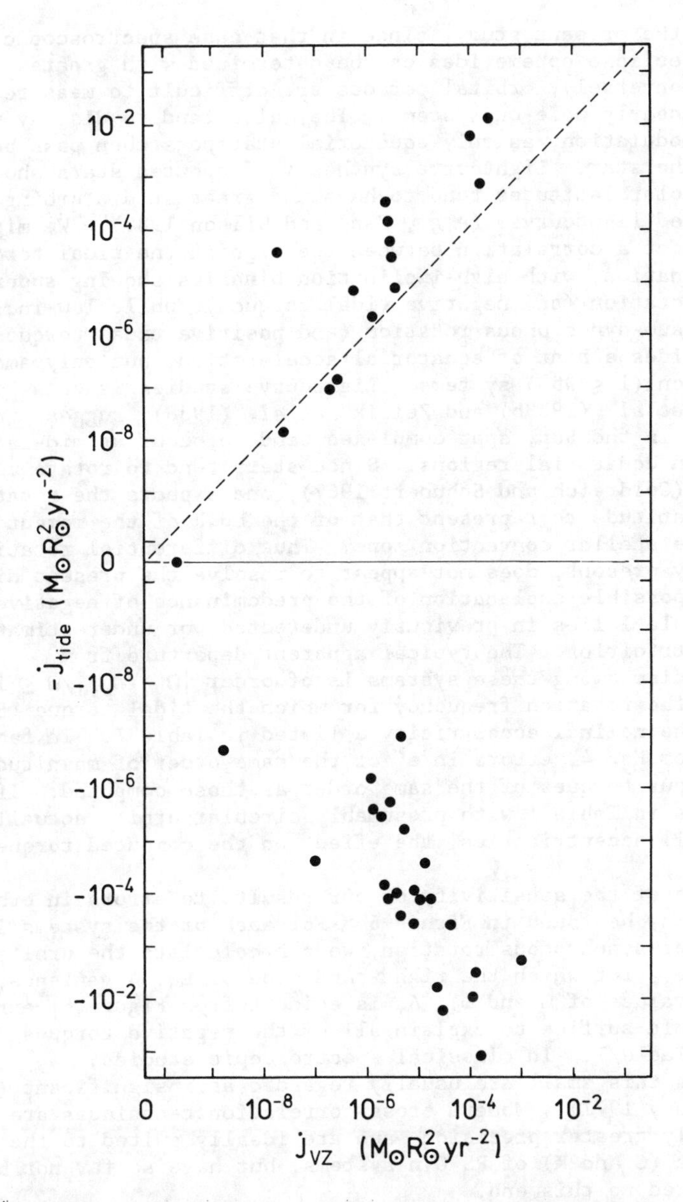

Figure 4. Computed tidal torques for the systems in Table I compared with the empirical relation, $\dot{J}_{VZ}$, derived by Verbunt and Zwaan (Eq. 2) for solar-type dwarfs. The negative of $\dot{J}_{tide}$ is plotted, since this quantity should equal the magnetic wind torque (after a small correction for evolutionary expansion) on an active component in rotational equilibrium. Systems would fall along the dashed diagonal line according to Eqs. 2 and 3.

inclusion in the present study, since in that case spectroscopic or (especially) eclipse ephemerides can be determined with greatest precision. Conversely, orbital periods are difficult to measure in low-inclination, nearly pole-on systems. They also tend to display weaker photometric modulation, as only equatorial starspots then pass behind the limb of the star. Lightcurve syntheses of spotted stars show that spots at subsolar latitudes tend to have the greatest disturbing effect on the computed lightcurves (*e.g.*, Kang and Wilson 1989). We might therefore expect a correlation between the sign of the tidal torque and orbital inclination, with high-inclination binaries showing super-synchronous rotation (and negative tidal torques), while low-inclination systems show sub-synchronous rotation (and positive tidal torques). Figure 5 provides a hint of equatorial acceleration, but only among very low-inclination ($i \lesssim 35°$) systems. Lightcurve studies, such as those of Strassmeier, *et al.* (1988b) and Zeilik, *et al.* (1988), suggest in any case that, as in the Sun, spot complexes tend to occur at mid-latitudes, rather than in equatorial regions. Since stars tend to rotate uniformly on cylinders (Goldreich and Schubert 1967), one expects the rotation rate at mid-latitude to represent that of the bulk of the moment of inertia of the stellar convection zone. Thus differential rotation, while possibly present, does not appear to resolve the present dilemma.

A third possible explanation of the predominance of negative torques in Table I lies in previously undetected, or underestimated orbital eccentricities. The typical apparent departure from pseudosynchronism among these systems is of order $|\Omega - \Omega_{ps}|/\Omega \lesssim 10^{-3}$, where Ω_{ps} is the rotation frequency for which the tidal torque (Eq. 4) vanishes at the nominal eccentricity e listed in Table I. In fact, as is evident from Eq. 4, errors in e^2 of the same order of magnitude can produce spurious torques of the same order as those computed. If many of the systems in Table I with presumably circular orbits actually had even very small eccentricities, the effect on the computed torques could be profound.

A measure of the sensitivity of our results to errors in orbital eccentricity may be found in Figure 6. For each of the systems in Table I showing super-synchronous rotation, we can calculate the orbital eccentricity e_{ps} for which the right-hand side of Eq. 4 vanishes, given the observed values of Ω and ω. As is evident from Figure 6, errors of $\lesssim 0.05$ in e would suffice to explain all of the negative torques appearing in Table I. In classical spectroscopic studies, eccentricities this small are usually regarded as insignificant (*e.g.*, Lucy and Sweeney 1971). Modern cross-correlation techniques are capable of considerably greater precision, and are ideally suited to the spectral types (G and K) of RS CVn systems, but have so far not been widely exploited to this end.

Small but finite orbital eccentricities might also be manifested in sinusoidal, periodic variations in eclipse timings, with semi-amplitude eP_{orb}, due to apsidal motion. A comprehensive survey of orbital period variations in RS CVn binaries by Hall and Kreiner (1980) shows clear evidence of coherent period variations in a number of systems, though the cause of those variations remains controversial. However, the search for apsidal motion is complicated by the fact that the expected

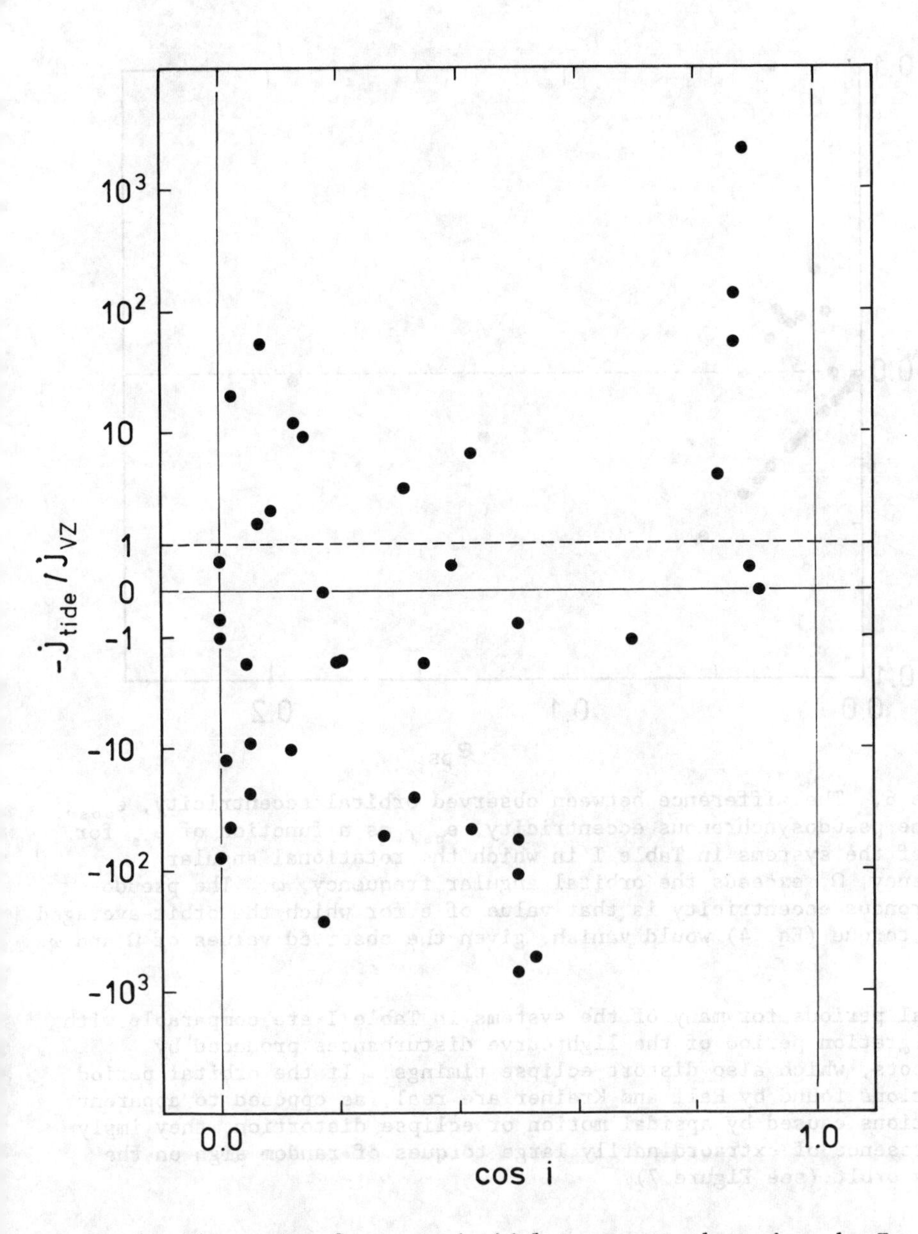

Figure 5. The ratio of computed tidal torque to that given by Eq. 2, for each of the systems in Table I, as a function of the cosine of the orbital inclination. Differential rotation at the surfaces of the active components should be manifested in a correlation of $-\dot{J}_{tide}/\dot{J}_{VZ}$ with cos i.

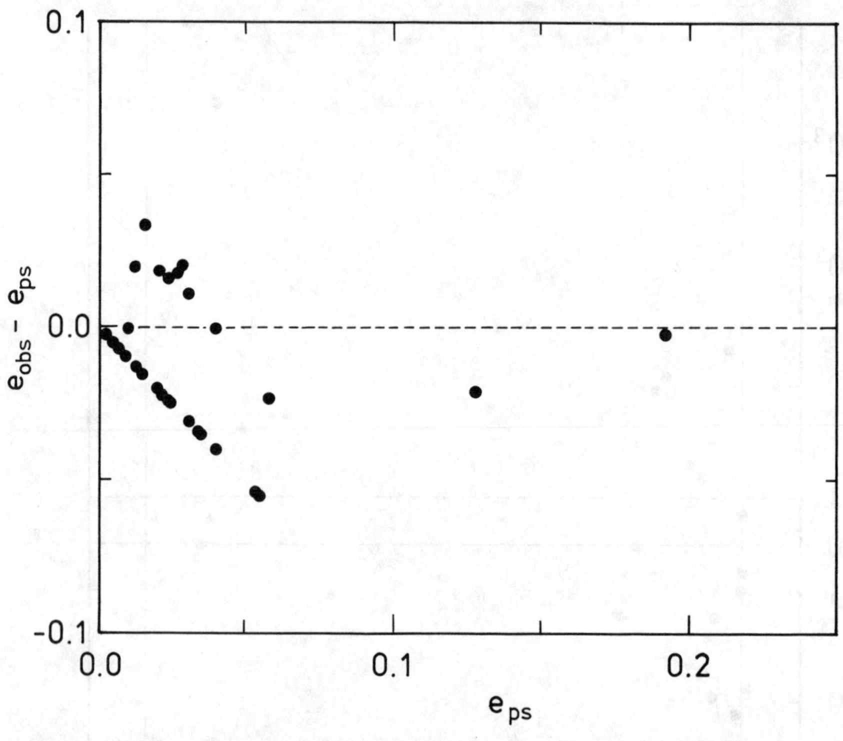

Figure 6. The difference between observed orbital eccentricity, e_{obs}, and the pseudosynchronous eccentricity, e_{ps}, as a function of e_{ps} for each of the systems in Table I in which the rotational angular frequency, Ω, exceeds the orbital angular frequency, ω. The pseudosynchronous eccentricity is that value of e for which the orbit-averaged tidal torque (Eq. 4) would vanish, given the observed values of Ω and ω.

apsidal periods for many of the systems in Table I are comparable with the migration period of the lightcurve disturbances produced by starspots, which also distort eclipse timings. If the orbital period variations found by Hall and Kreiner are real, as opposed to apparent variations caused by apsidal motion or eclipse distortion, they imply the presence of extraordinarily large torques of random sign on the binary orbit (see Figure 7).

4. CONCLUSIONS

The possibility that small, hitherto undetected orbital eccentricities persist in most RS CVn binaries could explain the problematic results presented here. Their confirmation will require

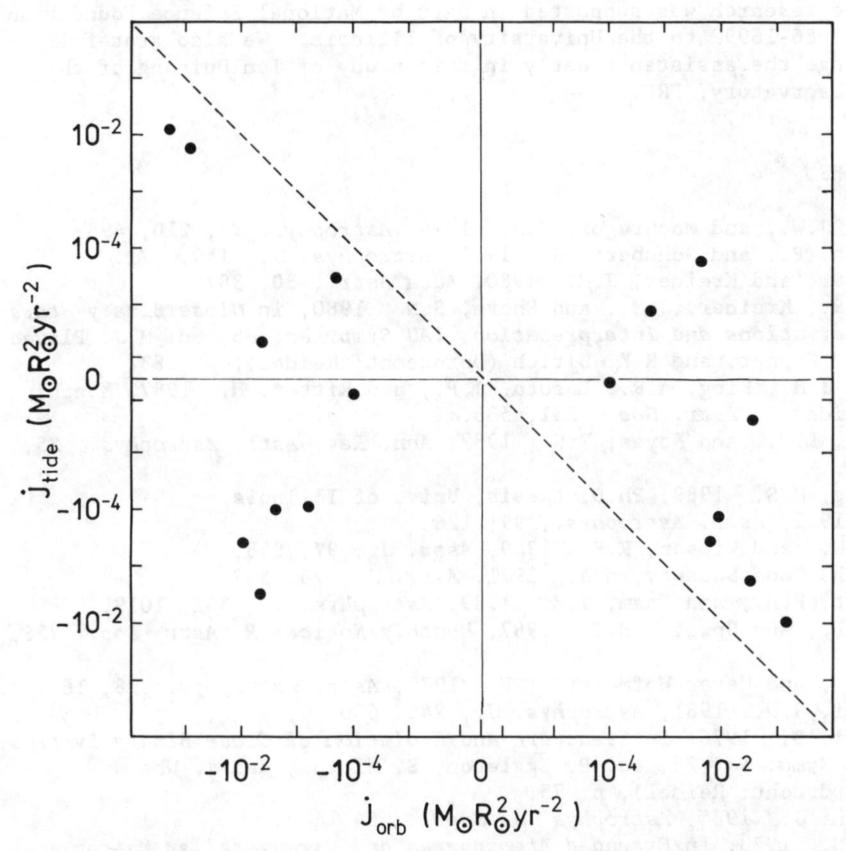

Figure 7. Tidal torques from Table I compared with orbital torques calculated from the observed period derivatives (Hall, Kreiner, and Shore 1980). In the magnetic braking model described in the text, these torques should be equal in magnitude and opposite in sign (as marked by the dashed line). No such correlation is present, and in fact the inferred orbital torques typically exceed the tidal torques by one to three orders of magnitude.

spectroscopic orbits of great precision, with errors in $e^2 < 10^{-3}$ (and preferably an order of magnitude smaller), but orbital determinations of this precision are now feasible.

Finally, it is worth noting that, apart from problematic differences of sign, the tidal torques computed here are typically comparable in magnitude, or perhaps only slightly greater than those predicted by the simple braking law of Verbunt and Zwaan. Alternative braking laws which predict angular momentum loss rates more than an order of magnitude higher under the range of conditions spanned by our sample can be excluded.

This research was supported in part by National Science Foundation grant AST 86-16992 to the University of Illinois. We also gratefully acknowledge the assistance early in this study of Tan Huisong of the Yunnan Observatory, PRC.

REFERENCES

Belcher, J.W., and MacGregor, K.B. 1976, *Astrophys. J.*, **210**, 498.
Goldreich, P., and Schubert, G. 1967, *Astrophys. J.*, **150**, 571.
Hall, D.S., and Kreiner, J.M. 1980, *Acta Astr.*, **30**, 387.
Hall, D.S., Kreiner, J.M., and Shore, S.N. 1980, in *Close Binary Stars: Observations and Interpretation, IAU Symp. No. 88*, ed. M.J. Plavec, D.M. Popper, and R.K. Ulrich (Dordrecht: Reidel), p. 383.
Hameury, J.M., King, A.R., Lasota, J.P., and Ritter, H. 1987, *Monthly Notices R. Astr. Soc.*, **231**, 535.
Hartmann, L.W., and Noyes, R.W. 1987, *Ann. Rev. Astr. Astrophys.*, **25**, 271.
Hjellming, M.S. 1989, Ph.D. thesis, Univ. of Illinois.
Hut, P. 1981, *Astr. Astrophys.*, **99**, 126.
Kang, Y.W., and Wilson, R.E. 1989, *Astr. J.*, **97**, 848.
Lucy, L.B., and Sweeney, M.A. 1971, *Astr. J.*, **76**, 547.
McDermott, P.N., and Taam, R.E. 1989, *Astrophys. J.*, **342**, 1019.
Mestel, L., and Spruit, H.C. 1987, *Monthly Notices R. Astr. Soc.*, 226, 57.
Meyer, F., and Meyer-Hofmeister, E. 1979, *Astr. Astrophys.*, **78**, 167.
Mochnacki, S.W. 1981, *Astrophys. J.*, **245**, 650.
Paczyński, B. 1976, in *Structure and Evolution of Close Binary Systems, IAU Symp. No. 73*, ed. P. Eggleton, S. Mitton, and J. Whelan (Dordrecht: Reidel), p. 75.
Patterson, J. 1984, *Astrophys. J. Suppl.*, **54**, 443.
Plavec, M. 1973, in *Extended Atmospheres and Circumstellar Matter in Spectroscopic Binary Systems, IAU Symp. No. 51*, ed. A.H. Batten (Dordrecht: Reidel), p. 216.
Popova, E.I., Tutukov, A.V., and Yungelson, L.R. 1982, *Astrophys. Space Sci.*, **88**, 55.
Rappaport, S., Verbunt, F., and Joss, P.C. 1983, *Astrophys. J.*, **275**, 713.
Refsdal, S., Roth, M.L., and Weigert, A. 1974, *Astr. Astrophys.*, 36, 113.
Skumanich, A. 1972, *Astrophys. J.*, **171**, 565.
Smith, M.A. 1979, *Publ. Astr. Soc. Pacific*, **91**, 737.
Spruit, H.C., and Ritter, H. 1983, *Astr. Astrophys.*, **124**, 267.
Stauffer, J.R. 1987, in *Cool Stars, Stellar Systems, and the Sun*, ed. J.L. Linsky and R.E. Stencel (Berlin: Springer), p. 182.
Strassmeier, K.G., Hall, D.S., Zeilik, M., Nelson, E., Eker, Z., and Fekel, F.C. 1988a, *Astr. Astrophys. Suppl.*, **72**, 291.
Strassmeier, K.G., Hall, D.S., Eaton, J.A., Landis, H.J., Barksdale, W.S., Reisenweber, R.C., Powell, B.E., Fernandes, M., Zeigler, K.W., Renner, T.R., Wasson, R., Nielsen, P., Louth, H., Chambliss, C.R., Fortier, G., Grim, B.S., Turner, L.C., Stelzer, H.J.,

Slauson, D.M., Fried, R.E., Brettman, O.H., Engelbrektson, S., Krisciunas, K., and Miles, R. 1988b, *Astr. Astrophys.*, **192**, 135.

Strassmeier, K.G., Boyd, L.J., Hall, D.S., and Genet, R.M. 1989, *Astrophys. J. Suppl.*, **69**, 141.

van't Veer, F., and Maceroni, C. 1988, *Astr. Astrophys.*, **199**, 183.

Verbunt, F., and Zwaan, C. 1981, *Astr. Astrophys.*, **100**, L7.

Zahn, J.-P. 1977, *Astr. Astrophys.*, **57**, 383; 1978, *Astr. Astrophys.*, **67**, 162.

Zahn, J.-P. 1989, *Astr. Astrophys.*, **220**, 112.

Zeilik, M., De Blasi, C., Rhodes, M., and Budding, E. 1988, *Astrophys. J.*, **332**, 293.

FLARES ON RS CVn AND RELATED BINARIES

Santo Catalano
Institute of Astronomy
University of Catania
Viale A. Doria, Città Universitaria
I-95125 Catania, Italy

None can wash its feets twice in the same water
because
everything is changing, everything is turning

Herakleitos
EPHESUS 2nd Century A.D.

ABSTRACT. The observational aspects of the flare events in RS CVn binaries are reviewed. The main results from observations in the optical continuum, optical and UV lines, X-Ray and radio wavelength suggest that ourtburst are time linked at all spectral band. Observed physical parameters are consistent with, but more energetic two-ribbon solar flares. VLBI measurements indicate that radio flares take place in regions of the stellar size compatible with giant loop structures. Occurence of flares in Algol and W UMa systems are discussed in the context of the RS CVn phenomenology they present.

1 Introduction

For decades the only known flare stars were either the Sun or dMe stars (UV Cet type). In recent years an observational breakthrough has occurred thanks to new technologies and facilities that give access to other wavelength (Radio, UV and X-ray) other than optical. Flaring appears to be a very common phenomenon across the H-R diagram (see Pettersen 1989 for a summary) evėn with different temporal frequency. Several hundreds flaring objects have been cataloged in Orion and the Pleiades, few have been detected among T Tau stars and in star formation complexes, from optical observations. Flares have also been seen on stars close to the main sequence and in post T Tau star phase.
The idea that flaring is associated with youth has largely changed by the results from the X-ray satellites EINSTEIN and EXOSAT. The first flare on a normal solar type star (π^1 UMa; GO V) was just observed by EXOSAT (Landini et al. 1986). Other main sequence stars, σ^2 CrB, SV Cam, XY UMa, have been found flaring. Actually they are F and G type binaries generally considered as RS CVn systems. Among the evolved stars RS CVn binaries, where typically one component is a sub-giant or a giant, are the most well known class of flaring object . Giantic flares at X-ray and radio wavelength have been reported from many of these systems and Algol systems (see Catalano 1986 for a summary). These flare events involved energies much larger than similar phenomena on

C. İbanoğlu (ed.), Active Close Binaries, 411–429.

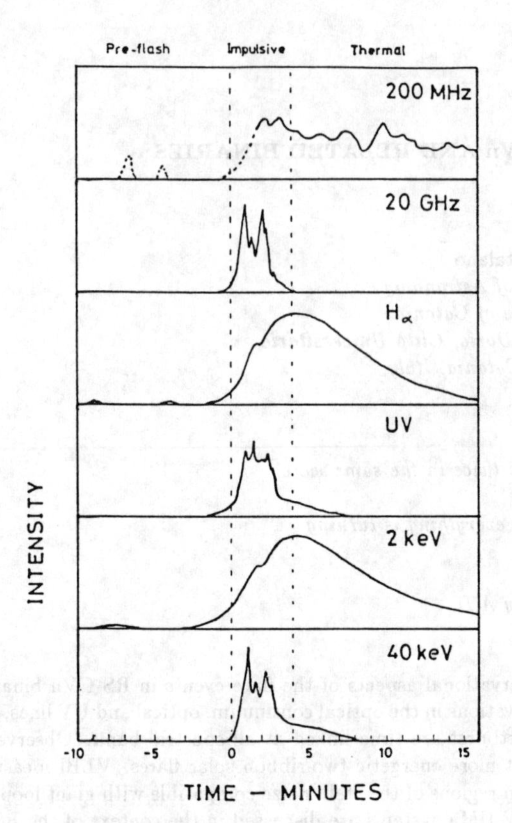

Figure 1. A schematic representation of the intensity-time variations ofdifferent wavelength emissions during a typical solar flare.

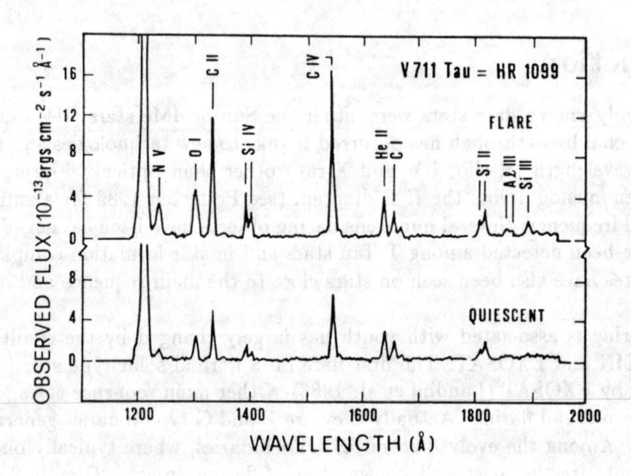

Figure 2. IUE short wavelength spectra of HR 1099 during and out of flare (adapted from Rodonò et al 1987)

the Sun, even the temporal profile and the derived temperatures and densities were quite similar to those observed in solar flares. Therefore, not only are such new observations important because they extend the range of the flare phenomenon to ever more different stellar environments and range of energies, but more important because flaring appears to be an universal MHD process, not yet well understood.
Stellar as solar flare develop at the different atmospheric levels, so they are seen and monitored through the different diagnostics. However while for the Sun observations both at high spatial and high temporal resolution have been collected and the temporal evolution at the different wavelength fairly well established (Figure 1), not so good co-temporal observations exist for RS CVn systems. In order to get some idea of similarity or difference with the solar case let me summarize the present state of flare observations in binary systems, together with their interpretation and modeling.

2 Photospheric flares

Continuum optical or white light flares are not very frequent on the Sun. Only 60 white light flares have ever been observed in the past one hundred years (Neidig and Cliver 1983). Thousands of dMe stellar white light flares have been observed. Although the total optical (U-B-V) photometry coverage of the RS CVn systems is very high, many systems have been systematically observed for several years, no clean evidence has been gathered for optical flare detection in regular and long-period systems, even during radio outburst. The only unquestionable flare events have been observed in the short-period binaries SV Cam (Patkos 1981) and XY UMa (Zeilik et al. 1983). Flare-like deeps seem to be frequently observed in the short period system ER Vul(G0 V + G5 V) (Keskin et al. this volume).
Several claims for detection have not been confirmed by contemporaneous observations of different authors (e.g. Guinan et al. 1979, Srivastava (1983), Bartolini et al. 1983) or can be ascribed to near permanent features in the light curve (Rodono' 1986). No changes to 0.01 - 0.02 mag have been reported concurrently with UV flare in λ And (Baliunas et al 1984) or giant flare in HR 1099. (Chambliss et al 1978). A brightening of about 0.1 was seen in the photometry made with the IUE Fine Error Sensor (FES) at the time of the 1981, October 3 radio outburst, (Linsky et al 1989) but no changes were detected in contemporaneous ground-based observations (Rodonò et al 1986). Contribution from UV emission line can explain the apparent detection of flare with the FES.
The detection of optical flares is strongly biased by contrast effect, as has been demonstrated for UV Cet star, for which later spectral-type stars appear to flare more frequently than earlier type ones. The active components of RS CVn binaries are K Type stars so that one might argue that only very powerful flares can be seen but these are very rare. Moreover in some cases the active K component contributes only a small fraction to the light of the whole system (for example,only 10% in the case of HR 5110). However in XY UMa and SV Cam the active components are of spectral type G2 V an KOV respectively. This suggests that the contrast effect is not the determinant factor, and the non detection of optical flare in regular and long-period systems may be an intrinsic effect.
In the extreme solar flares, there is a close correspondence and white light emission consistent with heating of upper chromosphere by non-thermal electron or proton beaming (Kane et al 1985). Solar white light emission can reach L_{WL} 10^{28} erg/sec and the primary radiation in X-rays is typically only $L_X \simeq 10^{26} - 10^{28}$ erg/sec. X-ray flares in RS CVn have average peak luminosity $L_X \simeq 10^{31} - 10^{32}$ erg/sec (see below Table II). If white light emission scales as for the Sun, it may be as large as $L_{WL} \simeq 10^{32} - 10^{33}$, erg/sec i.e. it would be detectable at least in the U band. The lack of white light flare detection in RS CVn can be used to give constraint on the electron or proton beam penetration, and therefore give information on which mechanism is operating.

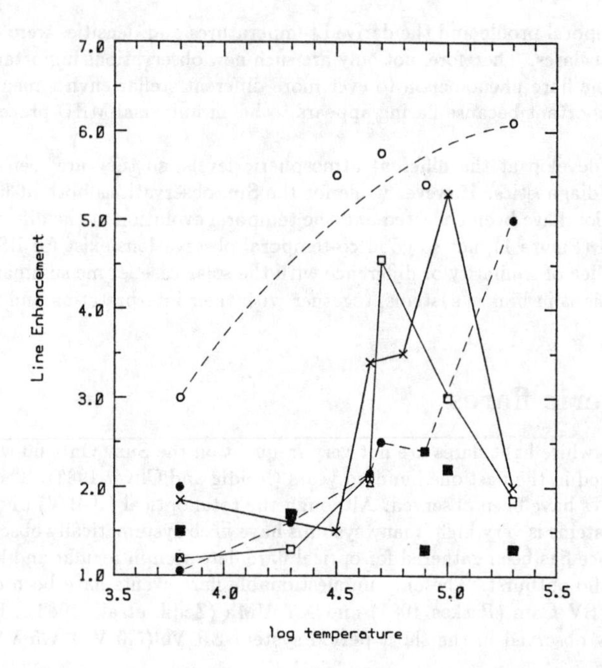

Figure 3. Line enhancement as a function of temperature for flares on selected objects. Solid lines IM Peg (*open squares*) and λ And (*crosses*); dashed lines UX Ari (*open circles*) and HR 1099 (*closed circles*). A typical solar flare (triangles) and a flare on Gliese 867A (closed squares) are plotted for comparison.

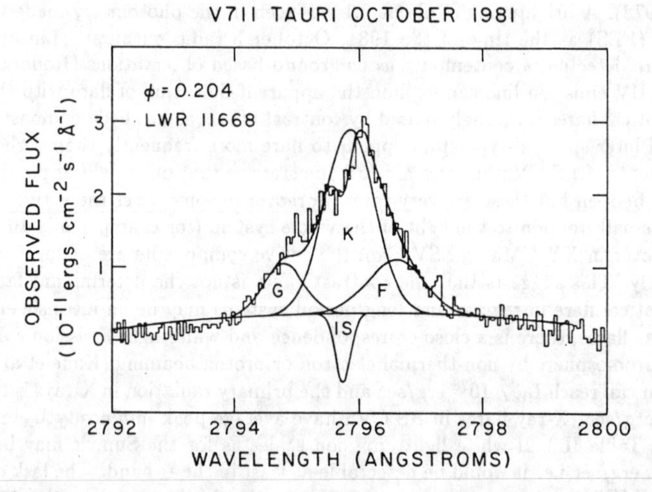

Figure 4. The observed flare peak Mg II k line profile an a four gaussian fit. The G star, K star, and Interstellar Medium gaussian are constrained to be the quiescent ones, while the radial velocities are those predicted for this phase. The fourth gaussian (F) is unconstrained and can be ascribed to the flare. It is centered at 90 ± 30Km/sec relative the K star. (Linsky et al 1989)

3 Chromospheric and TR Flares

Chromospheric and Transition Region lines appear as strong emission features in the spectrum of RS CVn stars except Hydrogen H_α that can be either in absorption or in emission. Quiescent emission levels are much stronger than in similar spectral type single stars. Chromospheric and TR flares have been detected as enhanced ultraviolet emission lines (Figure 2) by means of the Copernicus and IUE satellites in several systems (see Table 1). Simon et al (1980) noted that the high-temperature TR lines (e.g. C IV and Si IV) strengthen by larger factors than the low temperature chromospheric lines (e.g. Mg II, O I, Si II). An enhancement about twice larger for TR lines was also found for flare in λ And (Baliunas et al. 1984), HR 1099 (Linsky et al. 1989), but larger factors are also common (Figure 3). This larger increase in TR lines is consistent with the flux to flux correlations observed for late type stars (see Ayres Marstad and Linsky 1981).

Peak luminosity for RS CVn chromospheric and TR flares, cumulating the contribution of all the UV emission lines is in the range 5 to 60x10^{30} erg/sec (Linsky et al 1989) that has to be compared with the luminosity of 28x10^{24} erg/sec of the strong solar flare of 1973 Sep. 5 (Doyle and Raymond 1984). However the majority of the radiative output at ultraviolet wavelengths is concentrated in the Ly-α and Mg II h and k line (Baliunas et al. 1984, Weiler et al. 1978, Linsky et al. 1989). The ultraviolet continuum, especially the 1580-1630 Å region has also been found to brighten during flares (Baliunas et al. 1984, Buzasi et al. 1987).

Time scale of chromospheric flare, as well as rising and decay time are not well established because of low resolution time and fragmentarity of observations. Baliunas et al. (1984) estimate for the UV lines a rising time of 2.5 hour and a decay time of 3 hour. Similar time-scales have been found for the flare in AR Lac (Neff et al. 1989) and a longer duration, ~ 20 hour, for the 3 October 1981 flare on HR 1099 (Linsky et al. 1989), with a *1/e* decay time scale of roughly 12 hour for the Mg II and 5 hours for the TR lines. These time scales are much longer than the 10-20 minute time scales typical of solar compact flares, but are comparable to solar two-ribbon flares (Canfield et al. 1980).

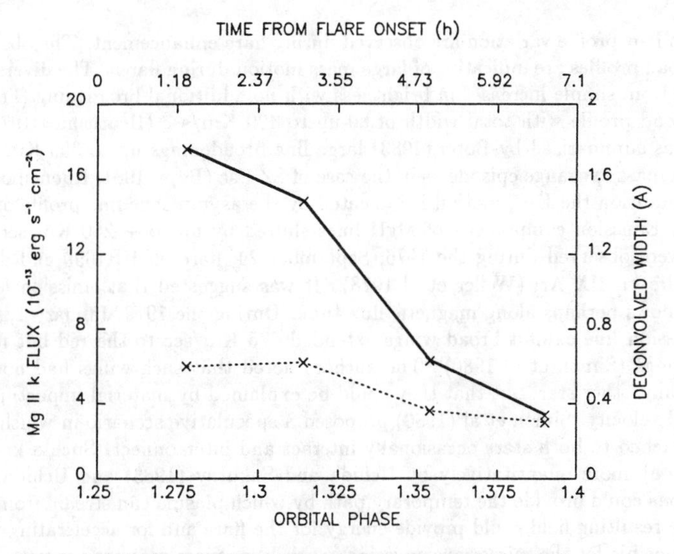

Figure 5. The flux (solid line) and the width (dashed line) of the flaring component on Mg II AR Lac spectra determined by multigaussian fit. (Neff et al. 1989).

The high peak flux intensification and the long duration of chromospheric flares imply total energy output over the chromosphere transition region layer exceeding 10^{35} erg (e.g. $1x10^{35}\lambda$ And (Baliunas et al 1984), $2.4x10^{36}$ HR 1099 Linsky et al. 1989). The H-α, like other chromospheric lines is enhanced during radio-flares, and it appears as an emission feature also in systems that normally shows it in absorption (Bopp 1983). Enhancement of H-α have been observed concurrently with radio and other chromospheric lines fluxes, even on time scale of several days (Weiler et al 1978). The emission equivalent width (EW) shows typical enhancement by a factor of 2, in agreement with enhancement for other chromospheric lines. Fraquelli (1984), using H-α data obtained with concurrent radio monitoring over an interval of several years, demonstrate that there is a good correlation (r = 0.94) between the H-α flux and the log of radio flux from HR 1099.

Table I. RS CVn UV Flares: Peak Luminosity

System	Date	L_{TR} (10^{30}	L_{MgII} erg	$L_{Ly-\alpha}$ sec^{-1})	Reference
HR 1099	24 Sept 1976		3.8	1^a	Weiler et al. (1978)
HR 1099	3 Oct 1981	2.3	2.4	8.1	Linsky et al. (1989)
UX Ari	1 Jan 1979	6.6	13.	43	Simon et al. (1980)
λ And	6 Nov1982	0.83	1.3	2.1	Baliunas et al. (1984)
AR Lac	5 Oct 1983	~0.1	0.25		Walter et al. (1987)
AY Cet	6 Dec 1983	4.5	3.0	3.0	Simon et al. (1987)
IM Peg	5 Jul 1985	70			Buzasi et al. (1987)
Sun	5 Sept 1973	12	(2.3)	7.5 (10^{24})	Canfield et al (1980)

^{a} not corrected for interstellar absorption

- *Mass motion*

Significant H-α profile variation are observed during flare enhancement. The observed asymmetric or vary broad profiles are indicative of large mass motion during flares. The diverse shapes of H-α profiles range from simple increases in brightness with no additional broadening (Fraquelli 1984) to symmetric broad profile with total width of 80-up to 400 Km/sec (Hearnshaw 1978, Furelind and Young 1978) as summarized by Bopp (1983) large line broadenings up to 300 Km/sec can be also the result of a mass exchange episode as in the case of SZ Psc (Bopp 1981, Huenemoerder and Ramsey 1984) In addition the H-α, motion is indicated by the asymmetric line profile of MgII observed during flares. Emission components of MgII lines shifted by up to + 250 Km/sec relative to the KO IV star were observed during the 1976, September 24, flare on HR1099 and during the 1976, October 2, flare on UX Ari (Weiler et al 1978). It was suggested that emission features indicate high velocity flows perhaps along magnetic flux tube. During the 1979, January 1, flare on UX Ari the MgII emission line exhibit broad wings extended 475 Km/sec to the red but no enhancement of the blue wings (Simon et al 1980). The authors noted that such wings had not been detected heretofore in any other star, but that they could be explained by material impacting on the G star at the free-fall velocity. Simon et al (1980) proposed a speculative scenario in which large magnetic flux tubes attached to both stars occasionally interact and interconnect. Such a kind of model has been worked out more quantitatively by Uchida and Sakuray (1983) and Uchida (1986). These interconnections could provide the temporary path by which plasma can stream from one star to the other, and the resulting field could provide energy for the flare and for accelerating the high energy particles responsible for the microwave emission. Such large mass motions are not comparable with chromospheric down flow seen in the solar flares (Ichimoto and Kurokawa 1984) where maximum velocity of 30-100 Km/sec are detected just before the microwave peak, and then decreases rapidly. Velocities of 30 Km/sec more consistent with the solar one, have been determined from a multigaussian fit analysis of the Mg II K line profile during flares on HR 1099 (Linsky et al 1989) and AR Lac (Neff et al 1989). The flux of kinetic energy at flare peak due to the down flow and turbulence during the 1981 October 3 flare (Linsky et al 1989) was estimated to be about 2×10^{32} erg/sec , far exceeding the flare radiative luminosity. Considerations on the line width of the flare component near the peak flux, allow to estimate minimum size of the flaring area. From the deduced filling factor of 0.01 Neff et al (1989) estimate that the Mg II K line surface flux in the flare of AR Lac is 2.7×10^{8} erg cm^{-2} sec^{-1} , nearly 100 times higher than the mean non-flaring level from the G star. In several occasions it has been reported the occurrence of flare near the light minimum, i.e. when the magnetic spotted emisphere is facing the observer.

- *Physical properties of chromospheric flaring Plasma*

Physical properties of plasma during flares have been investigated in few occasions through high resolution IUE spectra. The electron density in the flaring plasma is estimated using density sensitive line flux ratio, like the Si III] 1892 Å/ C III] 1909 Åand Si IV (1396 + 1403 Å) / C III 1909 Å. The derived flare electron density for the 1981 October 3 flare on HR 1099 (Linsky et al. 1989), $N_e = 1 \times 10^{11} cm^{-3}$ is an order of magnitude larger than the pre-flare value. The emission measure analysis based on the above flare electron density leads to a flare volume between 8×10^{29} and $36 \times 10^{29} cm^3$. Assuming a loop structure for the emitting plasma, separation of the loop foot-points of 15-25% of the K star radius have been derived. This size, given the lower surface gravity of the K sub giant, is comparable with the analogous largest solar active region loops. If the flare comes from several flaring loops, as the radio observations suggest smaller loop sizes can be derived.
Whether an observable flare will occur depends on two factors. First, enough energy must be stored in the pre-flare configuration and second, there must be a possibility for an instability. The total energy available for a flare must be stored in the pre-flare magnetic field configuration at the moment the instability occurs. The largest solar flares have been found to develop in the so called two-ribbon configuration. The hypothesis that energetic stellar flares are closely similar to the solar two-ribbon flares has been called upon several times and its formalism has been applied to single stars (Poletto et al 1988).
Adopting a two-ribbon flare model it has been shown that the maximal storage of energy is obtained when the filament is located between both binary components (Van den Oord 1988). Application to II Peg flare of 1983 February 2 by Doyle et al (1989), for which they estimate a total radiative loss of 2.4×10^{36} erg from the chromosphere and TR and corona (excluding hydrogen line radiation), leads to the requirement that the pre-flare energy has to be stored in a filament longer than the stellar radius possibly located between the two components.

4 Coronal flares

Flares from coronae of RS CVn stars have been easily detected at radio and X-ray wavelengths. Several events have been reported in the literature and summary is given in Catalano (1986) Pallavicini and Tagliaferri (1989).

- Radio flares

Flare activity at radio wavelengths is one of the most outstanding characteristics of RS, CVn binaries. It has been widely discussed, together with emission mechanism, in various recent reviews (Hjellming and Gibson 1980, Gibson 1980, Feldman 1983, Kuijpers 1985, Mullan 1985, Mutel, this volume).

- *Time-scale*

Radio-flares in RS CVn systems exhibit a large variety of time scale, flares as short as 4-5 minutes (Brown and Crane 1978), together with a very long-lasting single flare such as the 1976, November 9 - November 15, flare on HR 1099 (Hjellming and Brown 1979) or the 1974 August 11-14 on UX Ari (Gibson, Hjellming and Owen 1975) Typical flares have a duration of 24 hour with a decay time 3 times greater than the rise time. High states of continuous flaring are common in several systems (Feldman et al 1978). These time scales are much longer than solar flares. So there is evidence that stars with weak gravitational gradients, e.g. stars which (nearly) fill their Rocke lobes in close binary systems, or supergiants, appear to exhibit flares which are both stronger and of longer duration than those in objects with strong gravitational gradients, e.g. solar-type dwarfs. A scaling relation as g^{-2} has actually been proposed for the duration of optical flares in M dwarfs (Pettersen et al 1984).

- *Peak luminosity*

An Emission luminosity at ν = 5 GHz of about 1000 LU (1 LU =10^{15} erg sec^{-1} Hz^{-1}) is frequently observed in strong radio-flares. Evidence is given for a cut-off in the peak luminosity, at $\simeq$ 2000 LU (Feldman 1983). This means, that with a median quiescent level of 40 LU, enhancement factors of about 25-30 are very common, while factors around 50 are reached during superflares.

- *Spectra*

The quiescent flat spectrum (generally measured between 2.96 GHz and 5 GHz) steepens during the onset of flare with spectral index values greater for large than for small flares. During the decay the spectrum frequently becomes inverted ($\alpha < 0$) indicating the nonthermal nature of the radio emission. Multifrequencies observations of the HR1099 event of February 1978 show that the flare peaks around 5 GHz (Feldman et al. 1978). Dynamic spectra have been determined from 5 frequency observations of a double event in UX Ari (Hjellming and Gibson 1980) and 3 frequency observations of AR Lac (Rodonò et al 1984). In both cases, one important feature is that the low frequency portion retains the typical slope of a self-absorbed radio source, and that the variable events therefore show up mainly at the highest frequencies. This is the reason why flares are generally seen at higher frequencies, with a lower frequency limit around 2 GHz.

- *Polarization*

Circular polarization is frequently very high (100%) during outburst and is more variable at higher frequencies (Spangler 1977, Owen et al 1976). A major portion of an event of H1099 showed up only in left circular polarization at 2695 MHz (Brown and Crane 1978). Variations with a time scale of minutes in the left circular polarization were observed during different outbursts of HR 1099 by Brown and Crane (1978) and by Fix et al (1980). The former authors showed that the polarized flux exhibits an oscillatory behaviour with an approximated 4-5 minute cycle.
At the higher frequency the flux and polarization are consistent with a gyrosynchroton source of mildly relativistic electrons while at lower frequencies 2.7 GHz the strong enhancement seen at one polarization requires a more coherent emission process such as the cyclotron maser.

- *Source size and location*

From the spectral peak frequency (5 GHz) and synchroton loss time, physical parameters of the source during outburst have been derived. Rodonò et al (1984) estimate a minimal radius of the sources R (sources)> 0.76 R_* for the outburst of AR Lac. However recent high spatial resolution observations using multistation VLBI arrays have allowed the source size of 8 RS CVn binaries to

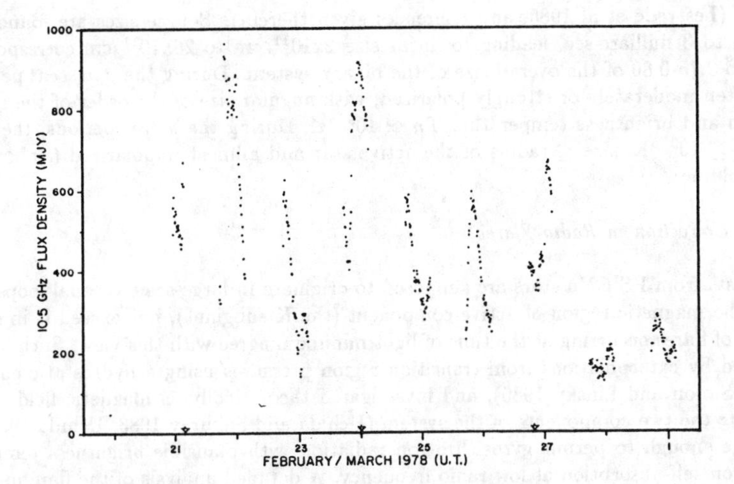

Figure 6. Behavior of HR 1099 at 2.8 cm during a period of unusual strong flaring activity (from Feldmann et al 1978).

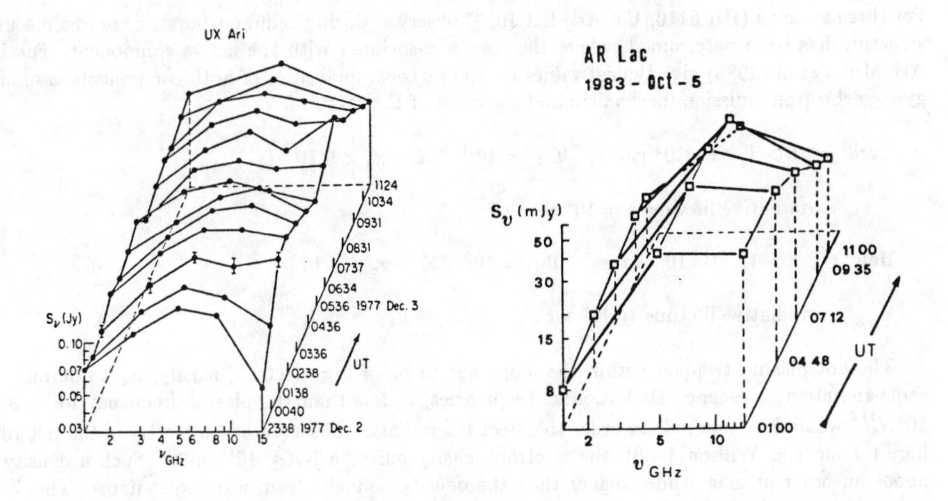

Figure 7. Dynamic spectra of radio outburst on UX Ari at five frequencies (Hjellming and Gibson 1980), and AR Lac at three frequencies (Rodono' et al. 1984)

be determine (Lestrade et al 1985, and references given therein). Source sizes are found to be in the range 0.5 to 3 milliarc-sec, leading to linear size $2x10^{11}$ cm to $26x10^{11}$ cm corresponding on the average to 0.75-0.80 of the overall size of the binary system. During the *quiescent* periods, the emission is often moderately or strongly polarized, with angular size of the order of the size of the binary system and brightness temperature $T_B \simeq 10^9$ °K. During the *active* periods, the source is compact, $T_B \geq 10^9$ °K, size $\leq$ radius of the active star and allmost unpolarized ($\pi < 10\%$)(see Mutel, this volume)

- *Physical Condition in Radio-Flares*

Radio bursts from RS CVn stars are supposed to originate in large-scale coronal loops that are anchored in the magnetic region of active component (the K subgiant), but exceed it in size. The many reports of flares occurring at the time of light minimum agree with this view. Such loops have been suggested by extarpolations from transition region pressures using a hydrostatic equilibrium scaling law (Simon and Linsky 1980), and investigated theoretically as magnetic field structures interconnecting the two components of the system (Uchida and Sakuray 1983, Uchida 1986). They would be large enough to permit gyrosichrotron radiation with plausible brightness temperatures and synchrotron self-absorbtion at low radio frequency. A detailed analysis of the flare in HR 1099 exclude any model with a uniform total density of electrons in the emitting volume while an electron density scaling as $N_R \propto R^{-\alpha}$ in a uniform field B = 100 gauss perfectly fits the observations (Borghi and Chiuderi-Drago 1985). Solutions with a spatial varying magnetic field have also been investigated (Klein and Chiuderi-Drago 1987).
For three systems (HR 5110, UX Ari, HR 1099) observed during radio outburst, a core halo source structure has been determined, where the core is associated with the active component. For UX Ari, Mutel et al (1985) give typical values for the physical quantities of both components assuming gyrosynchrotron emission mechanism and electrons of $E \simeq$ 1-10 MeV.

Core: $\tau > 1$, $B=10^2$ gauss, $T_B = 10^{10}$ °K $\pi_c < 0.10$

radiative lifetime $t = 10^4$ sec

Halo: $\tau > 1$, B=10 gauss, $T_B = 10^9$ °K $\pi_c > 0.10$

radiative lifetime $t=10^6$ sec

The hot plasma trapped within the loops has to be of low electron density, N_e to permit the radio radiation to escape. Radiation at frequencies, ν, less than the plasma frequency $\nu_P = 8.9x\ 10^3 N_e^{1/2}$ would be absorbed. In order to detect the radiation observed at 1465 MHz in the HR 1099 flare (Lang and Willson 1988), the electron density must be $N_e \leq 10^{10} cm^{-3}$. Such a density is about an order of magnitude smaller than the density deduced from ultraviolet flares. This is in agreement with the suggestion that the latter originate close to the stellar chromosphere and TR in sources that are much smaller than the star in size.
Even several flares have concourrently been observed at radio and UV wavelength, simultaneous VLA-IUE obsevations of UX Ari and HR 1099 indicate that sometime intense UV flare occur when no radio flares are detected and that rado f;ares are observed when no counterpart is detected at ultraviolet wavelength (Lang and Willson 1988, Willson and Lang 1986).

- **X-Ray Flares**

The systematic survey with the HEAO-1 and Einstein observatories showed that RS CVn systems have hot coronae (10^6 - 10^7 °K) with soft X-ray luminosities in the range 10^{30} - 3.10^{31} erg/sec

(Charles, Walter Boyer 1979,Walter et al. 1980, Charles 1983).
X-ray flares were early observed in the history of X-ray emission from stellar sources, six of which were detected flaring by *Ariel V* (Pye and McHardy 1983). A simultaneous radio, H-α, Ly-α and X-Ray outburst from HR 1099 was first observed with *Copernicus* satellite in 1976 by White, Sanford and Weiler (1978). During this flare, the peak X-Ray luminosity was 2.2x10^{31} erg/sec, whereas the observed Lyα luminosity was 1x10^{30} erg/sec and the peak Mg II luminosity was 3.8x10^{30} erg/sec, (i.e. the peak X-Ray luminosity is several time that of the total ultraviolet emission). In addition to that one, several flares have been reported for HR 1099. The transient *Uhuru* X-Ray source 4U0336+017 and the *Ariel V* source 3A0335+001 have been identified with HR 1099 (McHardy et al 1981). These identifications, due to the huge flux, imply that giant X-Ray bursts occur in this system, during which the X-Ray intensity rises by factor of $\geq$ 10. Pye and McHardy (1983) have identified also the *Ariel V* fast X-Ray transient source AT0339+063 with HR 1099 and realize that a huge X-Ray outburst with a peak X-Ray luminosity of $\simeq$ 6x10^{32} erg/sec and lasting for $\sim$ 7h was detected. Occurrence of the X-Ray flaring activity in HR 1099 had been recently reported by Barstow (1985), and Agrawal and Vaidya (1988). To date tenths of X-Ray flares have been observed in several RS CVn Systems, (Table II), many of them have been found flaring more than one time.

The flare detected by *Ariel V* on HR 1099 (Pye and McHardy 1983) as well as those in σ Gem (Pye and McHardy 1980) and BD +61 1211 (Charles et al 1979) were extremely energetic events that rleased 9x10^{36}, 3.5x10^{38}, 4x10^{37} erg respectilevy. The total energy released in these flares is one or two order of magnitude larger than that found in any other detected flare. This even may therefore represent a class of very energetic but rare stellar flares, which are distinct from the less energetic ($\leq$) and more frequent flares, where the enhancement is many times the quiet level, up to L_{flare}/L_{quiet} = 35-40

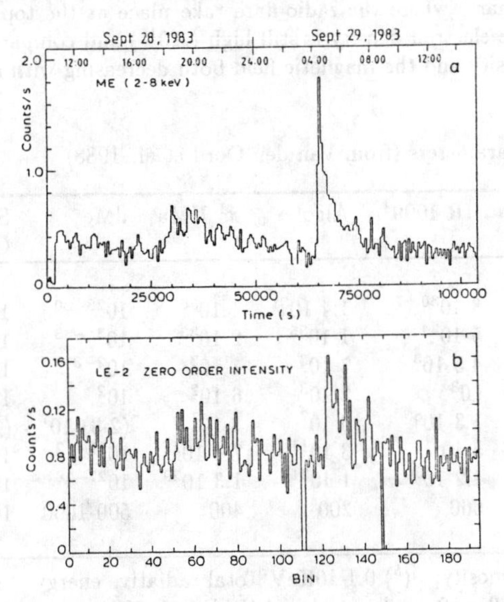

Figure 8. Light-curve of σ CrB obtained with ME(a) and LE(b) experiment on board of EXOSAT (Van den Oord et al. 1988)

A preflare is observed about 10 minutes before the main flare on σ CrB (Brinkman et al. 1985, Van den Oord 1988) (Figure 8) as well as in the case of π^1 UMa (Landini et al. 1986). A small flare follows, 2-3 hours later, the main flare of HR 1099 in 1976, September 24, (White et al 1978). These few examples are indicative of low level flaring associated with strong 10^{32} peak flares.
The estimated frequency of X-Ray flares is largerly uncertain. A frequency less than one per year up to 2/year has been suggested for the major flares $E_{tot} \simeq 10^{37}$ erg (Schwartz et al 1978, Pye and McHardy 1983). By contrast low energy flares $E_{tot} \leq 10^{35}$ seem to occur quite frequently (e.g. a rate of more than one flare per day has been estimated for HR 1099 (Agrawal and Vaidya 1988)).

The time evolution is longer than in solar X-ray flares but generally shows the same behaviour; a fast rise, a fast decaying phase with e-folding time of 50 minutes - 1 hour and a slow decay lasting up to 4-5 hours. The total energy release during the flare is of 4 to 5 order of magnitude larger than in the solar case i.e about 10^{36} erg. The volume emission measure at the flare peak is between 10^{52} and 10^{54} cm^{-3}.
In the hypothesis of optically thin plasma cooling by radiation, and assuming that the cooling time is equal to the emission decay time (Moore et al. 1980) the flare temperature, density and volume are estimated. The inferred linear dimension is some tenth of the stellar radius i.e. the flare develops in magnetic confined loops. Equipartition consideration allows lower limits to the magnetic field strength of the order of a few hundred gauss to be estimated (Table 2).
Although many of the X-Ray and radio flares in RS CVn systems have been classified as two-ribbon flares, the flare observed in σ CrB appears more compatible with a compact flare due to its short rising time, only 3 min(Van den Oord et al 1988).
Comparison of the average plasma parameters during X-Ray flares and those of radio flares clearly show that the two events take place in different part of the corona. However the data are consistent with a loop or loops scenario where the radio-flare take place at the top of the loop, the X-Ray flare in the legs where the electron density is still high (10^{11}) and comparable with the TR region density. The electron density and the magnetic field both decreasing with height.

Table II. X-Ray Flare Parameters (from Van den Oord et al. 1988)

Parameter	σ CrB	HR 1099$^+$	Algol	π^1 UMa	dMe	Sun Compact	Sun Two-ribbon
L_X(erg/sec)^a	9.4 10^{30}	8 10^{30}	1.4 10^{31}	1 10^{30}	10^{28-30}	10^{26-27}	10^{27-28}
E_X(erg)^b	2.4 10^{34}	5 10^{34}	1 10^{35}	2 10^{33}	10^{31-33}	10^{29-31}	10^{32}
t_D(sec)	1.7 10^3	4.5 10^3	7 10^3	1 10^3	10^{2-3}	10^3	10^4
t_{rise}(sec)	180	10^3	2 10^3	6 10^2	10^2	10^2	10^3
T_{max}(K)	9.5 10^7	1.3 10^8	6 10^7	3 10^7	(2-3) 10^7	(1-3) 10^7	
N_e(cm^{-3})	6 10^{11}	5 10^{11}	3 10^{11}	7.4 10^{11}	10^{11-12}	10^{11-12}	10^{10-11}
V(cm^3)	6.9 10^{29}	1.2 10^{30}	1 10^{31}	1.3 10^{29}	10^{27-28}	10^{26-27}	10^{28-29}
B(Gauss)	630	660	200	400	500-1000	100	100

(^{a}) 0.1-10 KeV peak luminosity; (^{b}) 0.1-10KeV Total radiative energy
($^+$) data for 17 Feb. 1980 flare from Agrawal and Vaidya (1988)

5 Algol Systems

Although Algol system are different from RS CVn mainly from the evolutionary point of view, they exhibit some of the RS CVn phenomena. So it seems to be of interest to give here a short comparative analysis. Changes in the optical light curve are seen in many Algol systems. U Cep is the best documented case. Brightening of the G star with a time scale of days is followed by light losses outside eclipses and in primary eclipse egress (Olson 1985). However the *spikes* which are associated with mass flows, determining sometimes long periods of *activity*, cannot be considered as true flare events.

Strong H-alpha emission has been seen during periods of activity (Batten et al 1975, Plavec and Polidan 1975) with a profile denouncing *turbolent* velocity components of 100 Km/sec (Crawford 1981).

At radio wavelengths few Algol systems have been detected on early surveys (Hjellming and Gibson 1980). A recent VLA search gives a high detection percentage ($\simeq$ 75%) at a level of few mJy (Umana et al. 1989a). The *quiescent* radio properties of Algol and RS CVn binary systems are very similar, as average luminosity and spectra (Umana et al. 1989a, Umana et al. 1989b), though the radio flux density and the degree of circular polarization of Algol's may be more time variable (Gibson 1975).

Characteristics of Algol radio-flares have been reviewed by Hjellming and Gibson (1980): a) The events exhibit nearly all the same structure with a decay time twice the rise time; b) the typical spectral index between $\alpha = 0$ steepens as the event rises, peaks at about $\alpha = 1$ and flattens out again as the event decays. The source size measured from VLBI observation during two events on Algol, reaching 0.7 and 1.02 Jy, was found to be $8\text{x}10^{11}$ cm and $1.5\text{x}10^{12}$ cm (Clark, Kellermann and Shaffer, 1975, Clark et al 1976). This size is somewhat larger than the stellar disk, $4.5\text{x}10^{10}$ cm, but close to the separation of the stars, the semimajor-axis beeing 10^{12} cm.

Two exceptional events occurred on Algol during VLBI observations have recently been reported by Lestrade et al. (1989). The first was a high brightness, broadband outburst observed simultaneously at 2.3 and 8.4 GHz. The souce size was strongly frequency dependent: 3.0 $\pm$ 0.3 and 0.7 $\pm$ 0.3 time the radius of the K-subgiant at 2.8 and 8.4 GHz respectively. The corresponding brightness temperatures were $T_B \sim 1.5\text{x}10^{10}$ ° K and $T_B \sim 3\text{x}10^{10}$ ° K. A model of the emission region assumed to be spheric symmetric, inhomogeneous synchrotron source with a power-law electron energy spectrum leads to magnetic field strengths of $\sim$ 10 gauss and electron effective energy of 3 Mev. The radial dependence of the magnetic field and the relativistic electron number density are found to be B(r) $\propto r^{-1.5}$ and N(r) $\propto r^n$ with n $\simeq$ 0.$\pm$0.5 during moderate activity and n $\simeq$ 2.5(+0.5 -3.) during the outburst. The index found for B(r) suggests that the magnetic field has a complex structure that is not consistent with a dipole or a tangled field.

The second exceptional event was a short duration ($\sim$ 15 min), highly left circularly polarized ($\pi_c \simeq$ 50% and high brightness temperature ($T_B > 5\text{x}10^{9}$°K) outburst at 1.66 GHz. The combination of large fractional circular polarization and high brightness temperature implies a coherent emission process, which was modelled using an electron-cyclotron maser model. The inferred magnetic field was B $\sim$ 300 gauss.

When radio fluxes from Algol are plotted as a function of orbital phase it appears that the radio-flares preferably occur outside eclipses, Figure 9. This effect might indicate that the sources of flares are located in between the two stars, i.e. flares take place along a gaseous stream flowing from the cooler to the hotter component.

Detection of a X-Ray flare in Algol in 1984, August 19, was reported (Parmar et al. 1985, White et al. 1986) (Figure 10). The e-folding decay time is about twice the rise time, 97 and 40 minutes respectively. The peak luminosity $1.4\text{x}10^{31}$ erg/sec is nearly five times smaller than in RS CVn flares and the total energy release in the 0.1-10 KeV band, $2.5\text{x}10^{34}$ erg, is about two orders of magnitude smaller. The hardness ratio reaches a peak about 18 minutes before the peak of the 2-6 KeV light curve,

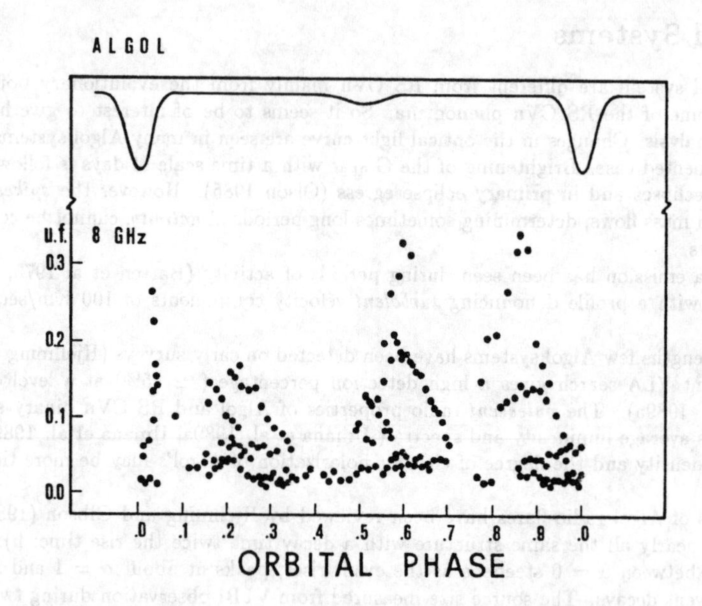

Figure 9. Density flux at 8.085 GHz of ALGOL plotted versus orbital phase.

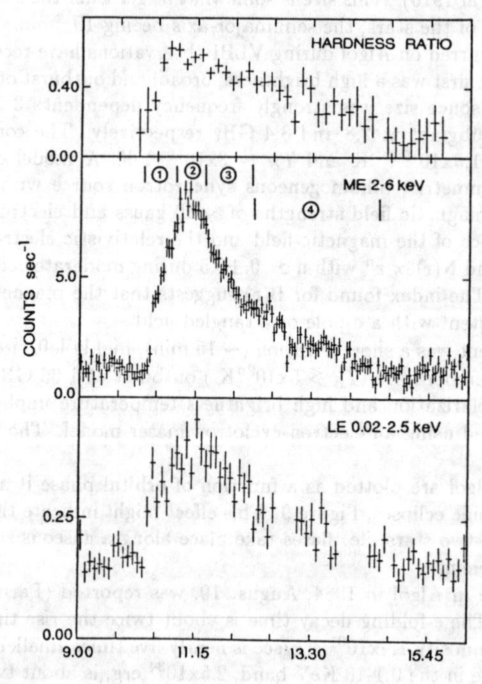

Figure 10. X-Ray flare event of ALGOL obtained with ME(b) and LE(c) experiment on board of EXOSAT and the spectral hardness ratio (White et al 1986).

giving a maximum temperature of $6x10^7$ °K. Flaring plasma parameters from White et al (1986) (Table 2), show a high similarity with RS CVn flare parameters.

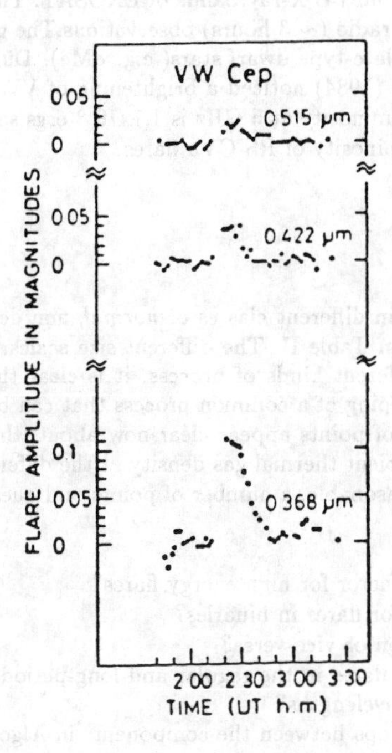

Figure 11. The flare like event of VW Cep observed by Egge and Pettersen (1983).

6 W UMa Systems

W UMa are rather different systems from RS CVn binaries due to the common envelope connecting both components, however spot characteristics have been detected (Binnendijk 1970, Eaton, Wu and Rucinski 1980, Bradstreet this volume). Hall (1976) discussed them in the context of stellar spots and Russo et al (1982) considered their evolutionary stage in relation to the short-period RS CVn binaries. Optical flares detections in W UMa were reported more than 30 years ago in 44 i Boo B (Eggen 1948) and U Peg (Huruhata 1952). The suspicion that they result from imperfections in the photometers has been cleared up by the observation of a well defined flare, in the UV band (center 3300 A; width 50 A) lasting 7 minutes, in W UMa (Kuhi 1964). More recently, Egge and Pettersen (1983) observed a flare in VW Cep in the U B V bands (Figure 11). The rise time was shorter than five minutes, but the decay to half of the peak was 13 minutes, giving a total duration of about 35 minutes. The authors estimated a peak luminosity in the U band of $7x10^{30}$ erg/sec and a total energy during the flare of $4x10^{33}$ erg. A disturbance in the light curve of 44 i Boo with a flare-like shape was recently reported by Hopp and Witzgmann (1982). The detection of X-ray flares has recently well established. The slow X-ray variability seen in VW Cep (Dupree 1983) might have some flare origin. The variation which has a time scale of the orbital period and does not follow the optical light curve exhibits a maximum enhancement of a factor of two. A X-ray and microwave flare was detected in this active system by Vilhu et al. during contemporaneous EXOSAT and VLA observations. The peak flare emission first appared in the microwave region, followed a few minutes later by its appearance in the ME and LE X-ray bands of EXOSAT. The duration is shorter in the X-ray (~1 hour) and longer in the radio (~ 3 hours) observations.The general shape of this flare is similar to that seen on most other late-type dwarf stars(e.g. dMe). During a VLA survey of 12 W UMa systems Hughes and McLean (1984) noticed a brightening of VW Cep from 0.50 mJy to 4.7 mJy in about 2 hours. The peak luminosity at 5 GHz is 1.7×10^{15} ergs sec^{-1} Hz^{-1}, i.e. 2-3 orders of magnitude lower than the peak luminosity of RS CVn flares.

7 Conclusions

Differences and analogies of flares in different classes of *normal*, non degenerate stars and systems including the Sun can be verified in Table II. The different size scales, time scales and energy involved are obvious. Instead of different kinds of process, it is clear that they represent different *laboratory* conditions for the developing of a common process that can be excited by different kind, of instability. Although a number of points appear clear now about, the condition and location of flares, the typical temperature, ambient thermal gas density at the different atmospheric levels, and the loop model appear the most reasonable, a number of points and questions can still be rised and many others need to be clarified:

a) is the binarity the determinant factor for high energy flares?
b) is the mass motion a condition for flares in binaries?
c) does the flare trigger mass motion or vice-versa?
d) Why, we do not observe optical flares in the regular and long-period RS CVn system, although they appear very active at other wavelengths?
e) are there interacting magnetic loops between the components in Algol's?

In any case, flares appear a very powerful tool for exploring the stellar magnetosphere and the system environment of RS CVn and other binaries. Detailed studies of many more flares in RS CVn binaries are however necessary to improve our understanding of the heating mechanism of the flare plasma and the radiation processes involved. It is worth to notice that within all this amount of observation there is not a flare well observed at all spectral bands so as to allow an accurate timing of the onset and evolution of the event. Therefore,I think it is still very profitable to spend time in looking for flares in *Active Binaries*, mainly in coordinated programmes at many wavelengths.

Acknowledgements: this work has been supported by the Ministero dell'Università della Ricerca Scientifica e Tecnologica through the University of Catania, The Catania Astrophysical Observatory and the CNR (Gruppo Nazionale di Astronomia) under contract No.88.00349.02 The work has been produced using the Astronet computing facilities. I would like to thank Grazia Umana for her unvaluable help and miss Cinzia Spampinato for kindly typing the manuscript . It is a pleasure to thank prof. C. Ibanoglu for inviting me at this stimulating meeting and for his umparallel warm hospitality.

8 References

- Agrawal,P.C., Vaidya,J.:1988, *Mon. Not. R. Astr. Soc.*, **235**,239.
- Ayres, T., Marstad, N., Linsky, J.L.:1981, *Astrphys. J.*, .
- Ambruster, C., Snyder, W.A., Wood, K.S.: 1984, *Astrophys.J.* ,**284**, 270.
- Baliunas, S.L., Guinan, E.F., Dupree, A.K.: 1984, *Astrophys,J.*, **282**, 733.

- Bartolini, C., Blanco C., Catalano S., et al. 1983 *Astron. and Astrophys,.*
- Batten, A.H., Fisher, W.A., Baldwin, B.W., Scarfe, G.D.: 1975 *Nature*, **253**, 174.
- Binnendijk, K.L.: 1970, *Vistas in Astronomy* , **12**, 217.
- Blanco, C., Catalano, S., Marilli, E., Rodonò, M.: 1983 in *Activity in Red-Dwarf Stars*, IAU Colloquium No.71, P.B., Byrne, M.Rodonò eds, Reidel, Dordrecht,bf p.387.
- Bopp, B.W.: 1983, in *Activity in Red-Dwarf Stars*,IAU Colloquium No.71, P.B., Byrne, M.Rodonò eds, Reidel, Dordrecht, **p.363**.
- Bopp, B.W.: 1981, *Astron.J.*,**86**, 771.
- Borghi,S., Chiuder Drago,F.: 1985, *Astron. Astrophys.*, **143**' 226.
- Brinkman A.C., Gronenschild, E.H.B.M., Mewe R., McHardy, I., Pye, J.P.: 1985, *Adv.Space Res.*, **Vol. 5**, p.65.
- Brown, F.N., Crane, P.C.: 1978,*Astron.J.*,**83**, 1504.
- Buzasi, D.L., Ramsey L.W., Huenemoerder, D.P.:1987 *Ap. J.*, **322**, 353.
- Burke, E.W., et al.: 1980,*Astron.J.*, **85**, 744.
- Byrne, P.B., et. al.: 1982, *Proc. Third European IUE Conference*, ESA SP-176, **p.125**.
- Canfield,R.C., et al.:1980, in *Solar Flares*, P.A. Sturrock ed., Colorado Associated University Press, **p.451**.
- Catalano, S.: 1986, in *Flares: Solar and Stellar*, P.M., Gondhalekar ed., RAL-86-085, **p. 105**.
- Catalano, S.: 1983,*Activity in Red-Dwarf Stars*, IAU Colloquium No.71 P.B., Byrne, M.Rodonò eds, Reidel, Dordrecht, **p.343**.
- Catalano S., Rodonò, M.: 1967,*Mem.Soc.Astron.Ital.*, **38**, 345.
- Chambliss, C.R., Hall, D.S., Landini, H.J., Louth, H., Olson E.C., Renner, T.R., Skillman, D.R.: 1978,*Astron.J.*, **83**, 1514.
- Charles, P.A., Walter, F., and Bowyer, S.: 1979, *Nature*,bf 282, 691.
- Charles, P.A.: 1983, *Activity in Red-Dwarf Stars*, IAU Colloquium No. 71, P.B., Byrne, M. Rodonò eds, Reidel, Dordrecht, **p.415**.
- Cheng, C.C.: 1984,*Mem.Soc.Astron.Ital.*, **55**, 663.
- Clark, B.G., Kellerman, K.I., Shaffer, D.: 1975 *Astrophys.J.*, **198**, L123.
- Clark, B.G., et al.: 1976, *Astrophys.J.*, **206**, L107.
- Crawford, R.C.: 1981 *Ph.D. Thesis*, UCLA.
- Doyle, J.G., Raymond :1984, *Solar Physics*, **90**, 97.
- Doyle, J.G., Byrne, P.B., Van den Oord, G.H.J.:1989 *Astron. Astrophys.*, in press.
- Dupree, A.K.: 1983 in *Activity in Red-Dwarf Stars*, IAU Colloquium No.71, **p.447**.
- Eaton, J.A., Wu, C.G., Rucinski, S.M.: 1980 *Astrophys.J.*, **239**, 919.
- Eggen, O.J.: 1948,*Astrophys.J.*, **108**, 15.
- Egge, K.E., Pettersen, B.R.: 1983, in *Activity in Red-Dwarf Stars*, IAU Colloquium No. 71, P.B., Byrne, M. Rodonò eds, Reidel, Dordrecht, **p.481**.
- Feldman, P.A., Taylor, A.R., Gregory, P.C., Seaquist, E.R., Balonek, J.J., Cohen, N.L.: 1978, *Astron.J.*, **83**, 1471.
- Feldman, P.A.: 1983, *Activity in Red-Dwarf Stars*, IAU Colloquium No. 71, P.B., Byrne, M. Rodonò eds, Reidel, Dordrecht, **p.429**.
- Fix J.D., Claussen, M.J., Doiron, D.J.: 1980 *Astron.J.*, **85**, 1238.
- Forman, W., et al.: 1978, *Astrophys.J. Suppl.*, **38**, 357.
- Fraquelli, D.: 1984, *Astrophys.J.*, **276**, 243.
- Furelind, I., Young, A.: 1978,*Astron.J.*, **83**, 1527.
- Garcia, M., Baliunas, S.L., Conroy, M., Johnston, M.D., Ralph, E., Roberts, W., Schwartz, D.A., Tonry, J.: 1980, *Astrophys.J.*, **240**, L107.
- Gibson, D.M.: 1975, *Ph. D. Thesis*, University of Virginia.
- Gibson, D.M., Hjellming, R.M., Owen, F.N.: 1975, *Astrophys.J.*, **200**, L99.
- Gibson, D.M.: 1980, in *Solar Phenomena in Stars and Stellar Systems*, Bonnet R.M., Dupree, A.K., eds., Reidel Dordrecht, **p.145**.

- Guinan, E.F., Dorren, J.D., Siah, M.J., Koch, R.H.:1979, *Astrophys.J.*, **229**, 296.
- Hall, D.S.: 1976, *IAU Colloquium No.29*, W.S.Fitch ed., Reidel, Dordrecht, **p.287**.
- Hearnshaw, J.B.: 1978, *Astron.J.*, **83**, 1531.
- Hopp, U., Witzgmann S.: 1982, *Astrophys. Space Sci.*, **83**, 171.
- Huenemoerder, D.P., Ramsey, L.W: 1984, *Astron.J.*, **89**, 549.
- Hughes, V.A., McLean, B.J.: 1984, *Astrophys.J.*, **278**, 716.
- Huruhata, M.: 1952, *Publ.Astron., Soc. Pacific*, **64**, 200.
- Hjellming, R.M., and Gibson, D.M.: 1980, in *Radio Physics of the Sun*, Kundu, M.R., and Gergely T.E., eds., Reidel Dordrecht, **p.209**.
- Hjellming, R.M., Webster, E., Balick, B.: 1972 *Astrophys.J.*, **178**, L139.
- Kane, S.R., Lowe, J.J., Neidig, d.f., Cliver, E.W.:1985, *Astrophys. J. (letter)*, **290**, L45.
- Klein, K.L., Chiuderi Drago,F. : 1987, *Astron. Astrophys*, **175**, 179.
- Kuhi, L.: 1964, *Publ.Astron.Soc.Pacific*, **76**, 430.
- Kuijpers, J.: 1985, in *Radio Stars*, Hjellming, R.M., Gibson, D.M. eds, Reidel, Dordrecht, **p.3**.
- Ichimoto, K., Kurokawa, H.:1984, *Solar Physics*, **93**, 105.
- Landini, M., Monsignori Fossi, B.C., Pallavicini, R., Piro, L.: 1986, *Astron.Astrophys.*, **157**, 217.
- Lang, K.R., Willson, R.F.:1988, *Astrophys. J.*, **328**, 610.
- Lestrade, J.F., Mutel, R.L., Phillips, R.B., Webber, J.G., Niell, A.E., Preston, R.A.: 1984, *Astrophys.J.*, **282**, L23.
- Lestrade, J.F., Mutel, R.L., Preston, R.A., Phillips, R.B.: 1985, in *Radio Stars*, Hjellming, R.M., Gibson, D.M., eds., Reidel Dordrecht, **p.275**.
- Lestrade, J.F., Mutel, R.L., Preston, R.A., Phillips, R.B.: 1989 preprint.
- Linsky, J.L.: 1983, in *Cool Stars, Stellar Systems and The Sun*, S.L. Baliunas and L. Hartman eds., Springer-Verlag Berlin, **p.244**.
- Linsky, J.L., Neff, J.E., Brown, A., Gross, B.D., Simon, T., Andrews, A.D., Rodonò, M., Feldman, P.A.:1989, *Astron. Astrophys.*, **211**, 173.
- Marstad, N., et al.: 1982, in *Advances in Ultraviolet Astronomy: Four years of IUE Research*, Kondo, Y., Mead, J.M., Chapman, R.D., eds, NASA Conf. Pub. 2238, **p.554**.
- McHardy,I.M., Pye,J.P.: 1981, *Space Sci. Rev.*, **30**, 457.
- Milano, L.: 1981, in *Photometric and Spectroscopic Bynary Systems*, Carling E.B., Kopal, Z.eds., Reidel, Dordrecht, **p.331**.
- Moore, R. et al.: 1980, in *Solar Flares*, P.A. Sturrock ed., (Boulder, Colorado Associated University Press) **p.341**.
- Mullan, D.J.: 1985, in *Radio Stars*, Hjellming, R.M., Gibson, D.M., eds, Reidel Dordrecht, **p.173**.
- Mutel, R.L., Lestrade, J.F., Preston, R.A., Phillips, R.B.: 1985, *Astrophys.J.*, **289**, 262.
- Neff, J.E., Walter, F.M., Rodonò, M., Linsky, J.L.:1989, *Astrophys. J.*, **215**, 79.
- Neidig, D.F., Cliver, E.W.:1983, *AFGL-TR 83-0257*.
- Olson, E.C.: 1985, in *Interacting Binaries*, Eggleton, P.P., Pringle, J.E., eds, Reidel Dordrechts **p.127**.
- Owen, F.N., Jones, T.W., Gibson, D.M.: 1976, *Astrophys.J.*, **210**, L27.
- Pallavicini, R., Tagliaferri, G.:1989 in *IAU Coll. 104, Poster Volume*, - B.M. Haisch, M. Rodonò eds., Catania Astrophysical Observatory Special Publication, **p. 17**.
- Parmar, A.N., Culhane, J.L., White, N.E., Vanaden Oord, G.H.J.: 1985, Adv. Space Res., Vol.5, p.69.
- Patkos, L.: 1981, *Astrophys.Lett.*, **22**, 1.
- Plavec, M.J., Polidan, R.S.: *Nature*, **253**, 175.
- Pettersen,B.R., Calerman, L.A., Evans, D.S.:1984, *Astrophys. J. Suppl.*,**54**, 375.
- Poletto, G., Pallavicini, R., Kopp, R.A.:1988, *Astron. Astrophys.*, **201**, 93.
- Pye, J.P., McHardy, I.M.: 1980, *BAAS*, **12**, 855.
- Pye, J.P., McHardy, I,M.:1983,*Mon. Not. R. Astron. Soc.*, **205**, 875.
- Rodonò, M.: 1986, in *Cool Stars, Stellar System, and the Sun*, M. Zeilik, D.M. Gibson eds.,

Springr-Verlag Berlin, **p. 475**.
- Rodonò M., et al.: 1984, *Proc.4th European IUE Conference ESA*, SP-218, **p.247**.
- Rodonò, M., et al. : 1986 *Astron. and Astrophy,* **165**, 135.
- Rodonò, M., et al. : 1987, *Astron. Astrophys.*, **176**, 267.
- Russo, G., Milano, L., Mancuso, S.: 1983, in *Activity in Red Dwarf Stars*, IAU Colloquium No. 71, P.B., Byrne, M. Rodonò eds, Reidel, Dordrecht, No. 71 **p.463**.
- Rust, D.M., et al.: 1980, In *Solar Flares*, Sturrock, P.A., ed., Colorado Associate University Press, Boulder, **p.273**.
- Schwartz, D.A., Garcia, M., Ralph, E., Doxsey, R.E., Johnston, M.D., Mc Hardy, I.M., Pay, J.P.: 1981, *Mon.Not.R. Astron.Soc.*, **196,** 95.
- Simon, T., Linsky, J.L., Schiffer, F.H. III: 1980, *Astrophys.J.*, **239**, 911.
- Spangler, S.R.: 1977, *Astron.J.*, **82**, 169.
- Srivastava, R.K.: 1983 *IAU Inf.Bull. Variable Stars*, No. 2450.
- Stern, R.A., Underwood, J.H., Antrochos, S.K.: 1983, *Astrophys.J.*, **264**, L55.
- Uchida, Y.: 1986, *Astrophys. Space Sci.*, **118**, 127.
- Uchida, Y., and Sakurai, T.: 1983, in *Activity in Red-Dwarf Stars*,IAU Colloquium No. 71, P.B., Byrne, M. Rodonò eds, Reidel, Dordrecht, **p.629**.
- Umana, G., Catalano, S., Rodonò, M., Gibson, D.M.:1989a in *Algols*, IAU Coll. 107, in press.
- Umana, G., Catalano, S., Rodonò, M., Hjellming, R.M.:1989b (preprint).
- Van der Oord, G.H.J.:1988, *Astron. Astrophys.*, **205**, 167.
- Van der Oord, G.H.J., Mewe, R., Brinkman, A.C.:1988, *Astron Astrophys.*, **205**, 181.
- Vilhu, O., Caillault, J-P., Heise, H.: 1988, *Astrophys. J.*, **330**, 922.
- Walter, F., Cash, W., Charles, P.A., Bowyer, S.: 1980, *Astrophys.J.*, **236**, 212.
- Weiler, E.J. et al.: 1978, *Astrophys.J.*, **225**, 919.
- White, N.E., Culhane, J.L., Parmar, A.N., Kellet, B.J., Kahn, S., Van der Oord, G.H.J., Kuijpers, J.:1986, *Astrophys. J.*, **301**,262.
- White, N.E., Sanford, P.W., Weiler, E.J.: 1978, *Nature*, **274**, 569.
- Willson, R.F., Lang, K.R.:1986, in *New Insightis in Astrophysics*, ESA SP-263, **p.57**.
- Zeilik M., Elston, R., Henson, G., Smith, P.: 1983 in *Activity in Red-Dwarf Stars*, IAU Colloquium No. 71, P.B., Byrne, M. Rodonò eds, Reidel, Dordrecht, **p.411**.
- Zirin H., Neidig, D.: 1981 *Astrophys.J.*, **248**, L45.

[illegible]ringer-Verlag Berlin, p. 477.
Rodonò, M. et al. [illegible] Proc. 4th European IUE Conference ESA SP-218, p. 247.
Rodonò, M. et al. [illegible] Astron. Astrophys. [illegible]
Rodonò, M., et al. [illegible] Astron. Astrophys. [illegible]
[illegible] 1983, in Activity in Red Dwarf Stars, IAU Colloquium No. [illegible] Reidel, Dordrecht, p. 361.
Ross, D.M. [illegible] 1986, in Cool Stars, Stellar Systems, and the Sun, [illegible] Cambridge University Press, [illegible] p. 274.
[illegible] D.A., [illegible] M., [illegible] M.R., [illegible] F.C., 1987, Mon. Not. R. astr. Soc. [illegible]
Simon, T., [illegible] 1990, Astrophys. J. 330 [illegible]
[illegible] R. [illegible] 109.
[illegible] R. [illegible] 1985, [illegible] Stars [illegible] 159.
Stern, R.A., Underwood, J., Antiochos, S.K., 1983, Astrophys. J. 264, L55.
[illegible] 1980, Astrophys. Space Sci. 71, 163.
[illegible] Y. and [illegible] 1983, in Activity in Red Dwarf Stars, IAU Colloquium No. 71, [illegible] Reidel, Dordrecht, p. 320.
[illegible] 1983, in [illegible] IAU Coll. 102, in press.
[illegible] 1987 (in print).
Van den [illegible] 1991, Astron. Astrophys. [illegible]
Van den Oord, G.H.J., [illegible] Astron. Astrophys. 220, 158.
[illegible] 1988, Astrophys. J. 324, 924.
[illegible] 1990, Astrophys. J. [illegible]
[illegible] 1978, Astrophys. J. 225, 519.
[illegible] van den Oord, G.H.J., [illegible]
[illegible] 1988, Astrophys. J. [illegible]
White, N.E., Sanford, P.W., Weiler, E.J., 1978, Nature 274, 569.
[illegible] 1986, in [illegible] ESA SP-263, [illegible]
[illegible] 1983, in Activity in Red Dwarf Stars, IAU Colloquium No. 71, [illegible] Reidel, Dordrecht, p. 415.
[illegible] 1991, Astron. Astrophys. [illegible]

ACTIVITY AND EVOLUTION IN RS CVn SYSTEMS

Osman DEMİRCAN
Ankara University Observatory
Science Faculty, Astronomy Department
06100 Tandoğan, Ankara

ABSTRACT: Activity and evolutionary status of the component stars in RS CVn type binaries have been reexamined by using the absolute dimensions of 31 systems. Most of the components (mostly the cooler ones) were found much cooler than expected from mass-luminosity relation. Many supporting evidences were noted on this reddening problem of RS CVn systems. Relying on the masses and radii, the MK classes and thus effective temperatures and luminosities of the component stars in the sample have been revised.

It was realized that RS CVn type binaries are divided in two distinct groups according to their x-ray luminosities L_x. For the higher level emission group ($\log L_x > 30.85$), the orbital period P, mass ratio q and distance a between the component stars are relatively smaller and the equatorial velocity v and Roche lobe filling percentage RL% are relatively larger. The X-ray surface flux F_x first increases with increasing RL%, until RL% $\sim$ 0.60 and then tend to decrease with increasing RL%. The fast rotating G dwarfs with RL% $\sim$ 0.60 have maximum F_x values.

Non linear correlations between $\log F_x$, and log g and log RL%, and linear correlations between $\log F_x$ and Log M, log v, log R, log P and log RL were obtained. It seems the mass is most responsible parameter in determining the activity level, since the mean x-ray surface flux $\bar{F}_x = L_x / [R^2_h + R^2_c]$ is also strongly correlated with the total mass of the system. However, as a result of interdependences between all the activity related parameters and large errors in the activity data and absolute parameters, we could not separate the individual effects on the activity. This important step toward better understanding dynamo generated magnetic activity can be accomplished by the application of some new techniques.

C. İbanoğlu (ed.), Active Close Binaries, 431–465.

1. INTRODUCTION

RS CVn systems are known to be emission line detached binaries with late type components. The cooler components which are slightly more massive and almost fill their Roche lobes are evolved to the base of the giant branch and thus mostly have luminosity classes of IV or III, while the other slightly less massive and hotter components are evolved to near central hydrogen exhaustion (Morgan and Eggleton 1979), and thus mostly have the luminosity classes of V or IV.

As an operational definition, these systems have been found to show interesting forms of activity in every wavelength region in which they have been observed. They exhibit (i) variable radio emission, (ii) quasisinusoidal distortion waves in their visible and near infrared light curves, (iii) strong emission in Ca II H and K and Balmer lines of hydrogen, (iv) strong emission in the satellite UV observations, (v) soft x-ray emission, and (vi) occasional flares in every wavelength region (Hall 1976, 1981, 1987).

The most promising model for the types of activity associated with the RS CVn stars is that which postulates the presence of extreme solar-type surface activity, assumed to occur mostly on the cooler components in these systems (e.g. Eaton and Hall 1979, Shore and Hall 1980, Bopp and Talcott 1980, Vogt 1983).

However, since solar-type surface activity is believed today to be exhibited at different levels by all late type (later than about F0) stars with convective envelopes, the classification of some binaries as RS CVn type according to mostly activity criteria is found to be artificial. Thus, the defining characteristics of this class of binaries have been extended to include practically all late type Ca II emission binaries (Bopp and Talcott 1980, see also Morgan and Eggleton, 1979).

The properties of classical RS CVn binaries have recently been reviewed by Bopp (1983), Catalano (1983), Charles(1983), Vogt (1983), Linsky (1984), Rodono (1986), Hall (1987), and Montesinos et al. (1988).

In the present work, we questioned, ones more the evolutionary status and activity parameters of Hall's (1976) group of RS CVn type binaries.

2. DATA

In order to investigate further the general properties and aspects of the activity and evolutionary status of the component stars in RS CVn type binaries we have extracted first the most accurate data for 31 systems from "A Catalog of Chromosphericaly Active Binary Stars" (hereafter CABS) prepared by Strassmeier et al. (1988). The orbital periods within the sample range from 0.5 to 104 days, with spectral types of the components from mid F to mid K dwarfs, subgiants and giants. Twenty eight of the systems are present in Hall's (1981) list and two of the remaining (namely AY Cet and V471 Tau) have white dwarf components. Twenty six of the systems are eclipsing binaries with mostly total eclipses, thus having relatively more reliable absolute parameters. A list of binaries considered is given in Table 1, together with the spectral types of the components, orbital periods and eclipse type of the systems. For each system considered, the mass and radius of the component stars, distance, orbital inclination and soft x-ray luminosity are available together with some other (such as M_V, B-V, a sin i) incomplete data in CABS. The luminosity classes of some of the components are not given in CABS. They were assumed initially to be dwarfs. To derive the effective temperatures T_{eff} and luminosities L of the components, a calibration by de Jager and Nieuwenhuijzen (1987) according to spectral type have been used. The absolute parameters for the components of Z Her, RS CVn, SZ Psc, LX Per, WW Dra, VV Mon, MM Her and RZ Eri have been replaced by the revised values given by Popper (1988). The absolute bolometric magnitudes M_{bol}, surface fluxes F and surface gravities g of the component stars were then calculated by well known formula:

$$M_{bol}=4.72-2.5 \operatorname{Log} L, \qquad (1)$$
$$\operatorname{Log} F=4^{*}\operatorname{Log} T_{eff}-4.25, \text{ and} \qquad (2)$$
$$\operatorname{Log} g=4.44+\operatorname{Log} M-2^{*}\operatorname{Log} R, \qquad (3)$$

where F and g have cgs units, while L, M , and R are in solar units. The orbital radius a, angular rotation velocity w, equatorial rotation velocity v, Roche lobe radius RL and Roche lobe filling percentage RL%=R/RL of the component stars were also calculated. In the calculation of rotation velocities we have used the the photometric rotation periods for the asynchronous rotators (namely components

Table 1. Active binaries considered in this work

NAME	Sp. Ty. (h)	Sp. Ty. (c)	P (orb.)	Eclipse
CF TUC	G0 V	K4 IV	2.80	Partial
AY CET	WD	G5 III	56.82	None
UV PSC	G4-6 V	K0-2 V	0.86	Partial
LX PER	G0 IV	K0 IV	8.04	Total
V711 TAU	G5 IV	K1 IV	2.84	None
V471 TAU	WD	K2 V	0.52	Total
RZ ERI	F5 IV	K2 III	39.28	Total
ALFA AUR	F9 III	G6 III	104.02	None
CQ AUR	G2	K0	10.62	Total
SV CAM	G2-3 V	K4 V	0.59	Partial
VV MON	G2 IV	K0 IV	6.05	Total
SS CAM	F5 V-IV	K0 IV-III	4.82	Total
AR MON	G8 III	K2-3 III	21.21	Total
GK HYA	F8	G8 IV	3.59	Total
RU CNC	F5 IV	K1 IV	10.17	Total
RZ CNC	K1 III	K3-4 III	21.64	Total
TY PYX	G5 IV	G5 IV	3.20	Partial
RW UMA	F8 IV	K1 IV	7.33	Total
DQ LEO	A6 V	G5 IV-III	71.69	None
RS CVN	F5 IV	K0 IV	4.80	Total
SS BOO	G0 V	K1 IV	7.61	Total
RT CRB	G0	K0-2	5.12	Partial
TZ CRB	F6 V	G0 V	1.14	None
WW DRA	G2 IV	K0 IV	4.63	Partial
Z HER	F4 V-IV	K0 IV	3.99	Partial
MM HER	G2 IV	G8 IV	7.96	Partial
AW HER	G0	K1 IV	8.80	Total
ER VUL	G0 V	G5 V	0.70	Partial
RT LAC	G9 IV	K1 IV	5.07	Total
AR LAC	G2 IV	K0 IV	1.98	Total
SZ PSC	F8 IV	K1 IV	3.97	Partial

of AY Cet, RZ Eri, Alpha Aur, DQ Leo) from Fekel and Eitter (1989) and for the remaining ones, the orbital periods from CABS.

3. EVOLUTIONARY STATUS

Observational data for the components of 31 systems (except the two WD's) have been plotted in the mass-radius (M-R) and mass-luminosity (M-L) diagrams in logarithmic scale. An important feature is seen in the latter diagram (Fig. 1) where in the theoretical ZAMS and TAMS relations were drawn by using the accurate theoretical models of Maeder and Meynet (1988): Some of the components are found much below the ZAMS line. For example, four giants (both components of α Aur and the hotter components of AR Mon and RZ Cnc) are in the early MS position in the diagram. In fact it was realized in the M-L diagram that, in general, larger the mass lower the luminosity (and thus temperature) than expected according to theoretical M-L relation. It is interesting that none of the components of Z Her, RS CVn, SZ Psc, LX Per,WW Dra, VV Mon, MM Her and RZ Eri whose absolute parameters have been critically revised by Popper (1988) are located below the MS in this diagram.

In the M-R diagram no such strange feature is seen: the giants according to MK class are all in the giant location, the subgiants are in subgiant location and dwarfs are between the ZAMS and TAMS lines (Fig. 2). Only two hotter dwarf components (of AW Her and RW UMa) are slightly lower than the ZAMS locus. Most probably radii of these components are not correct. The bolometric fluxes obtained through the Boltzmann law ($F = \sigma T^4$) and $F = L/(4\pi R^2)$ are also quite different for most of the component stars (Fig. 3).

Thus it is clear that the radiative quantities are not consistent with the masses and radii of these stars. If, for example, both radii and luminosities are accurate enough for our sample, then it is normally not expected to have at the same time proper positions in M-R and conflicting positions (according to MK classes) in M-L diagrams. We believe the radii for our sample are much more accurately determined (since they are all eclipsing binaries) than the luminosities. Remember that the luminosities (and thus surface fluxes and effective temperatures) were evaluated according to MK classes, and we know that determination of MK classes only from optical spectra is full of difficulties owing to the blending and broadening of the lines and variable nature of the continuum. In fact,the photospheric and chromospheric activity in these stars could well perturb the true MK type. Thus any MK classification

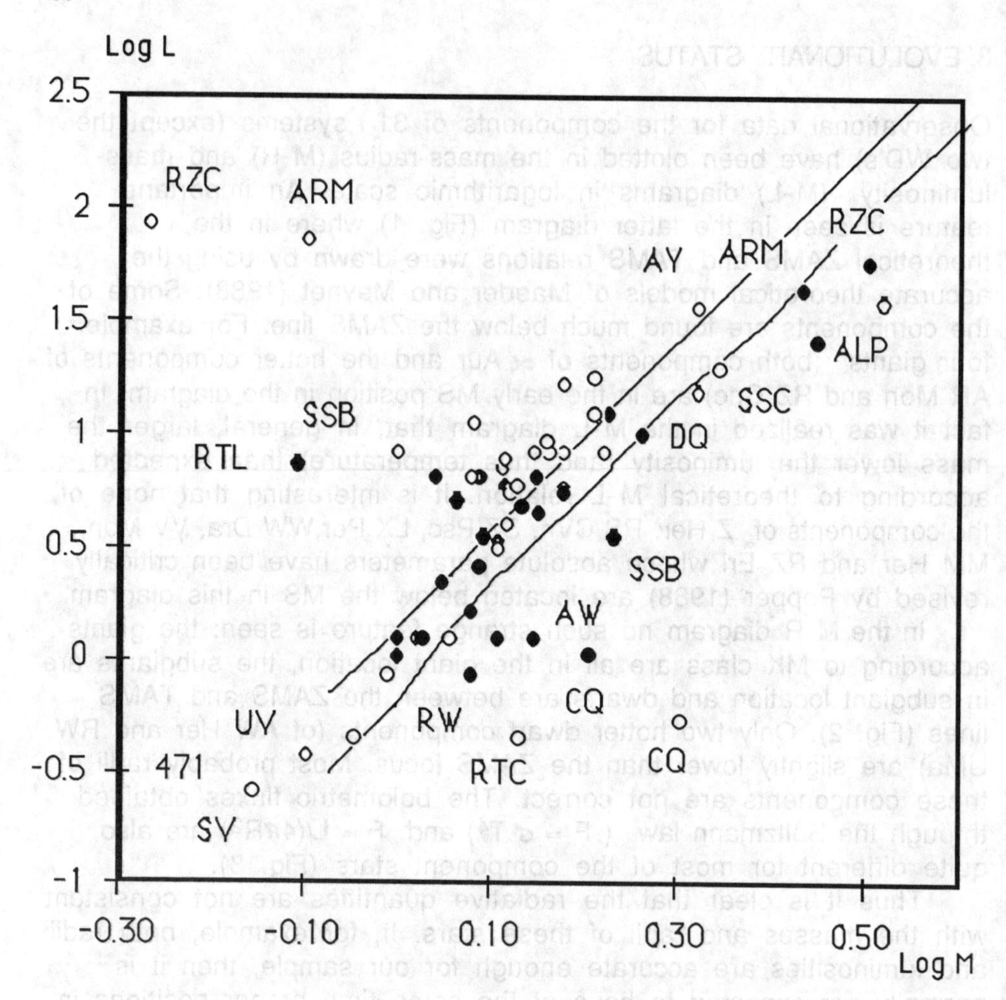

Figure 1. The M-L diagram for the components of RS CVn systems, before the revision of radiational parameters.The theoretical ZAMS and TAMS lines from the accurate models by Maeder and Meynet(1988) were also shown. Some of the components locate unexpectedly below the ZAMS line. The mass loosing components of algol type systems AR Mon, RZ Cnc and RT Lac are on the upper left corner of the diagram. Here and all in the following figures full and open diamonds are for the hotter and cooler components, respectively.

without considering the UV, red and infrared portions of the continuum may not be reliable (see e.g. Shore and Adelman,1984).

Assuming the masses and radii of our stars are accurate enough, we decided to revise the MK classes and thus effective temperatures and luminosities of the stars in our sample by using the calibrations of Straizys and Kuriliene (1981) in graphical form. We found this calibration of MK classes is more complete for mass, radius and luminosity. We first plot our stars on the Sp-Log M, Sp-Log R, Sp-Log L and Sp-Log T_{eff} planes where the calibration lines for dwarf, subgiant and giant stars are also shown. Relying on the mass and radius, the luminosity classes of our stars were fixed on the first two diagrams. To find a unique position relative to the MK lines on the first two diagrams, sometimes the radius of the star should also be revised. The fixed relative position of the star in Sp-Log L diagram then gives us the most probable value of Log L, which is used to find Teff of the star with known L and R values in solar units, by

$$\text{Log } T_{eff} = [\text{Log } L - 2 \text{ Log } R]/4 + 3.762 \qquad (4)$$

where 3.762 = Log T_{sun} . The final Sp - Log T_{eff} diagram can be used then to find the revised consistent spectral type Sp of the star.

Let us consider an extreme example, RT Lac whose hotter component is slightly below the dwarf line in Sp - Log M plane and above the subgiant line in Sp - Log R plane . Thus the mean position gives its luminosity class as V-IV. The cooler component is above the subgiant but not close to the giant line in both Sp - Log M and Sp - Log R diagrams. Thus it has luminosity class IV. The mean relative positions obtained from these two diagrams give us Log R = 0.60 and Log L = 1.10 in solar units for the cooler star. Equation (4) then yields Log T_{eff} = 3.74 for this star which is converted to an MK class of G2 IV with the help of Sp - Log T_{eff} diagram. We realize that mass losing components in Algol-type systems (like AR Mon and RZ Cnc) are much affected by the mass transfer process and do not obey the M-R and M-L relations obtained from well detached eclipsing binaries. We did not attempt any revision on the absolute dimensions of the hotter component of RT Lac, since it fills more than 80% of its first critical Roche lobe, (Milone 1977, Eaton and Hall 1979, Huenemoerder and Barden 1986). Thus, by this revision, the supposed cooler component with MK

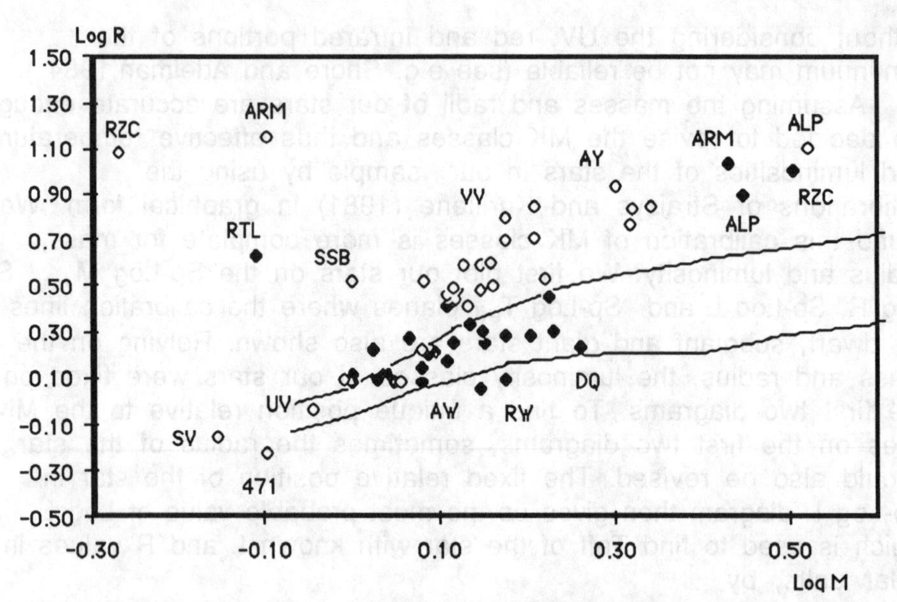

Figure 2. The M-R diagram for the components of RS CVn systems Other explanations are just as in Figure1

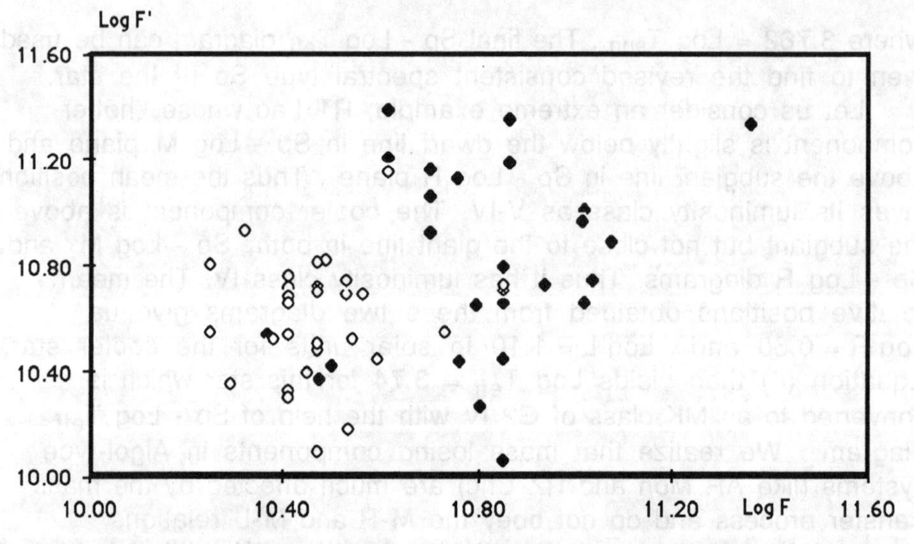

Figure 3. The bolometric surface flux F obtained through the Boltzmann Law ($F = \sigma T_{eff}^4$) versus $F' = L/4\pi R^2$, before the revision of radiational parameters of RS CVn systems. It is seen that F and F' are quite different for most of the component stars.

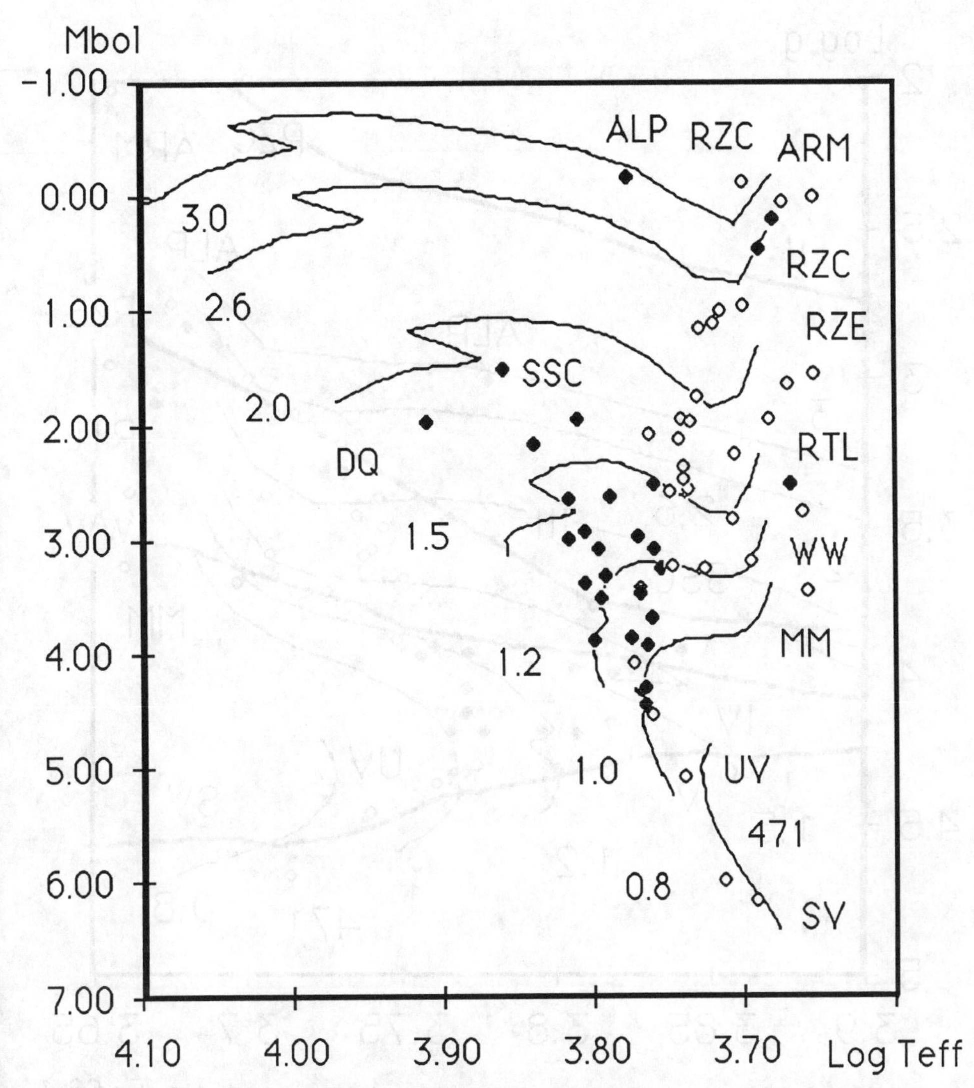

Figure 4. The H-R diagram for the components of RS CVn systems, after the revision of radiational parameters. The evolutionary tracks are from Vanden Berg (1985). The cooler components of MM Her, WW Dra, RZ Eri and AR Mon on the right ofthe diagram are seen much cooler than expected through the evolutionary tracks. The radiational parameters of these components were not revised (see text).

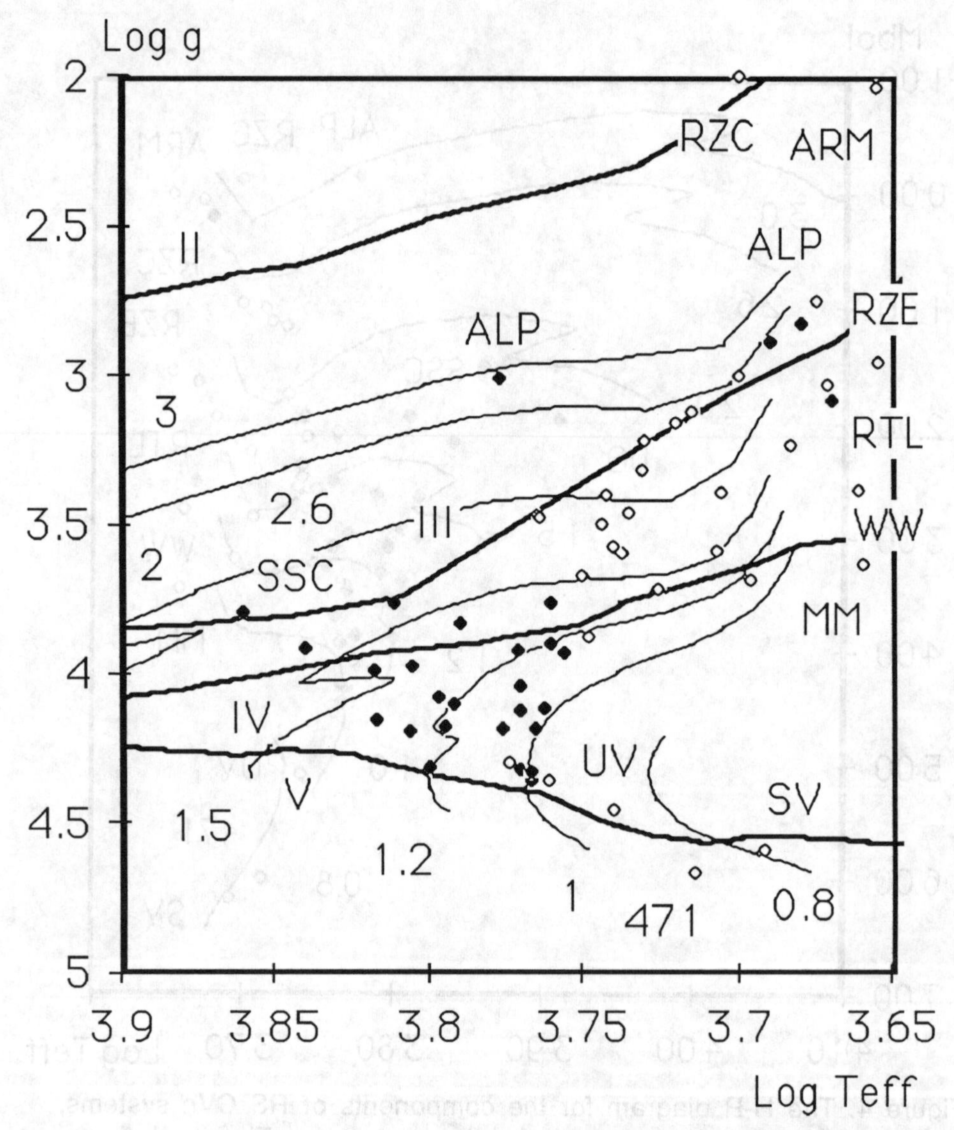

Figure 5. The Log T_{eff} - Log g diagram for the components of RS CVn systems, after the revision of radiational parameters. Other explanations are just as in Figure 4. The lines for the mean luminosity classes (from Straizys and Kuriliene, 1981) were also shown. The mass loosing cooler components of RZ Cnc and AR Mon are seen luminosity class II giants in this diagram.

class K1 IV turned out to be the hotter component of the system. The radius of this component increased from 3.4 to 4.0 R_{sun} , and T_{eff} from 4600 to 5400 °K giving rise to a 48% increase in the luminosity and thus the star became G2 subgiant. The colour inconsistency for the components of this system has been noted before in above cited references. Huenemoerder and Barden noted that "in a (3100 - V)-(B - V) colour - colour plane, the K star falls among G8 stars." Probably a scattering shell or disk about this mass accreting G2 IV star causes the star to be observed as K1 subgiant.

We applied the process to all the stars in our sample and thus revised the radiational parameters, and radii. Then, all the related parameters (such as F, M_{bol}, g, w, v, k and RL%) have been re-calculated accordingly. The key results related to the evolutionary status of the components of the systems are as follows:

1. As known before (see e.g. Morgan and Eggleton, 1979), the hotter components are mostly close to TAMS or slightly above it, started burning hydrogen in thick envelope, while the cooler components are evolved mostly close to the base of the giant branch, burning hydrogen in the thin outer envelopes (Fig. 4 and 5).

2. The absolute dimensions are mostly in the following intervals:

$0.90 < M < 2.25\ M_{sun}$; $1.00 < R < 6.30\ R_{sun}$

$0.80 < L < 32\ L_{sun}$; $4500 < T_{eff} < 7100$ °K

$3.0 < \text{Log } g < 4.4$

3. The hotter and cooler components are quite distinct with a few exceptions in R , Log g, and T_{eff} (or F_{bol}):

$(R)_{hotter} < 2.5\ R_{sun}$; $(R)_{cooler} > 2.5\ R_{sun}$

$(T_{eff})_{hotter} < 5650$ °K ; $(T_{eff})_{cooler} < 5650$ °K

$(\text{Log } g)_{hotter} > 3.7$; $(\text{Log } g)_{cooler} < 3.7$

$(F_{bol})_{hotter} > 5.6\ 1010$ ergs cm-2 sn-1

$(F_{bol})_{cooler} < 5.6\ 1010$ ergs cm-2 sn-1

4. Most of the components (mostly the cooler ones) are observed much cooler than expected values from the M-L relation (Fig. 1). We have more to say on this point, later.

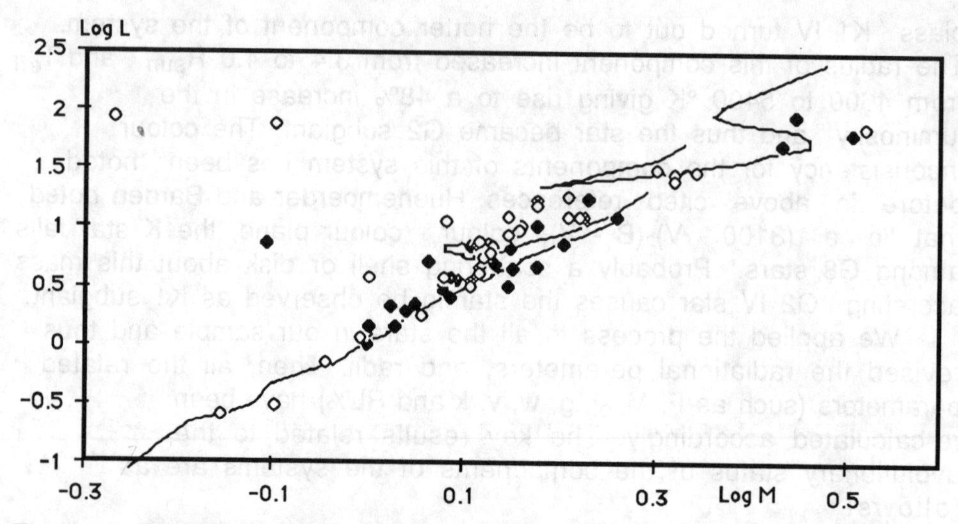

Figure 6. The M-L diagram for the components of RS CVn systems, after the revision of radiational parameters. The lines for the mean luminosity classes (from Straizys and Kuriliene, 1981) were also shown.

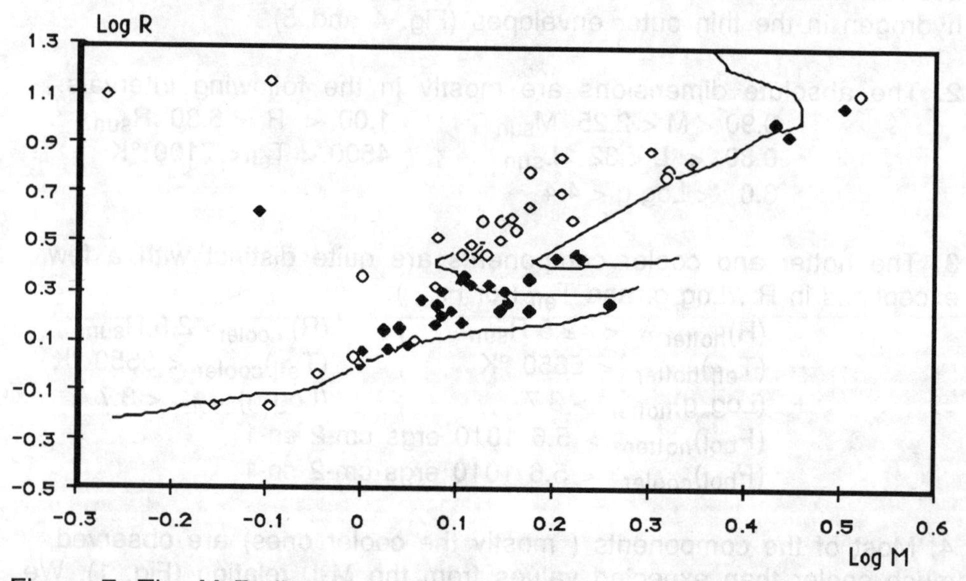

Figure 7. The M-R diagram of RS CVn systems. The lines for the luminosity classes (from Straizys and Kuriliene, 1981) were also shown.

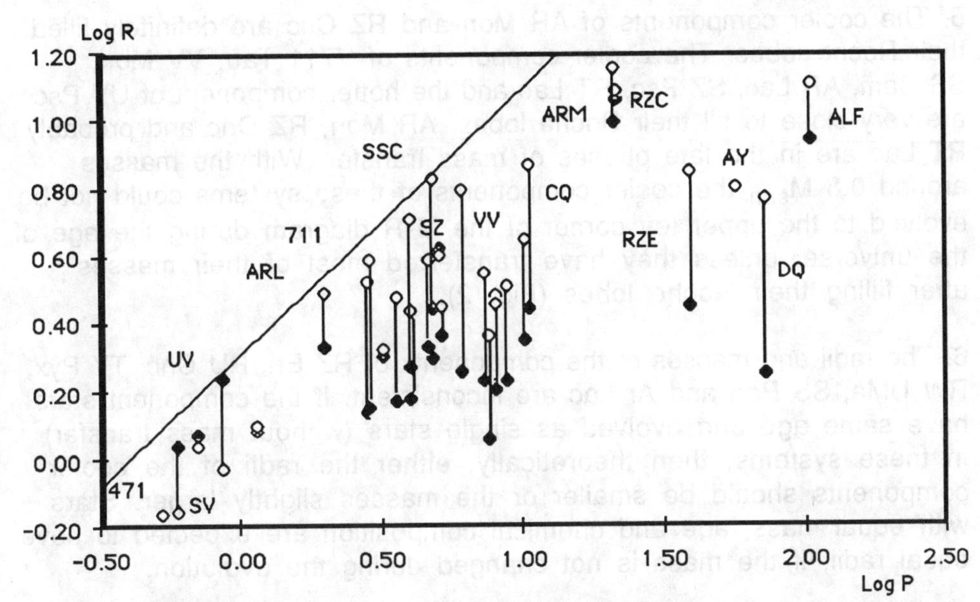

Figure 8. The P-R diagram of RS CVn system. The upper bound for the radius is drawn by the Roche Lobe radius. The larger size stars preferentially accommodate in longer period systems.

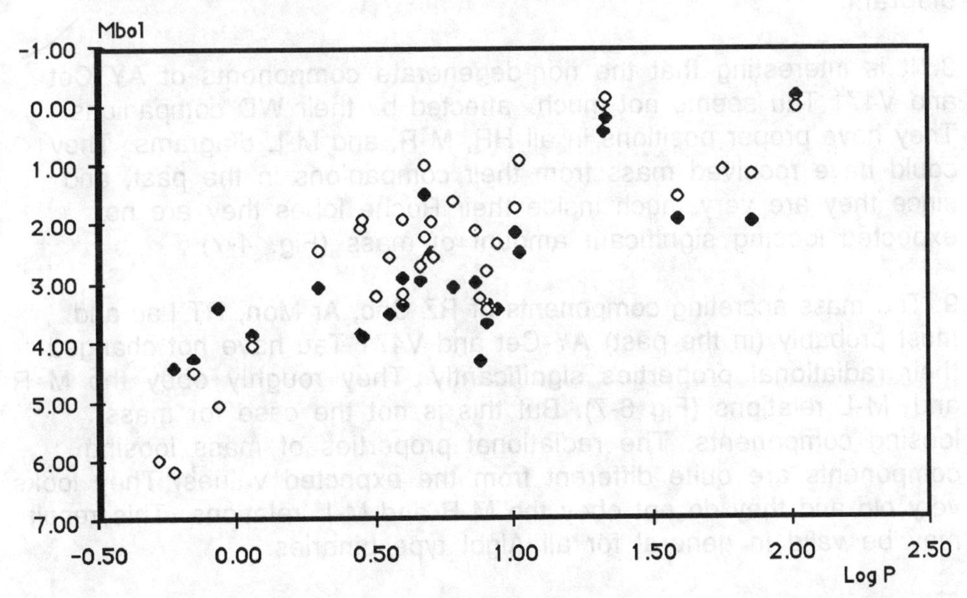

Figure 9. The P-M_{bol} diagram of RS CVn systems. The more luminous stars preferentially accommodate in longer period systems.

5. The cooler components of AR Mon and RZ Cnc are definitely filled their Roche lobes. The cooler components of V711 Tau, VV Mon, SS Cam, AR Lac, SZ Psc, RT Lac and the hotter component of UV Psc are very close to fill their Roche lobes. AR Mon, RZ Cnc and probably RT Lac are in the late phases of mass transfer. With the masses around 0.8 M_{sun} the cooler components of these systems could not be evolved to the upper left corner of the M-R diagram during the age of the universe, unless they have transferred most of their masses after filling their Roche lobes (Fig. 2).

6. The radii and masses of the components of RZ Eri, RU Cnc, TY Pyx, RW UMa, SS Boo and Ar Lac are inconsistent. If the component stars have same age and evolved as single stars (without mass transfer) in these systems, then theoretically, either the radii of the cooler components should be smaller or the masses slightly larger. Stars with equal mass, age and chemical composition are expected to have equal radii, if the mass is not changed during the evolution.

7. The mass loosing cooler components of RZ Cnc and AR Mon are luminosity class II giants according to their surface gravities (Fig. 5), but they are among the luminosity class III giants in the H-R diagram.

8. It is interesting that the non-degenerate components of AY Cet and V471 Tau seems not much affected by their WD companions. They have proper positions in all HR, M-R, and M-L diagrams. They could have received mass from their companions in the past, and since they are very much inside their Roche lobes they are not expected loosing significant amount of mass (Fig. 4-7).

9. The mass accreting components of RZ Cnc, Ar Mon, RT Lac and most probably (in the past) AY Cet and V471 Tau have not changed their radiational properties significantly. They roughly obey the M-R and M-L relations (Fig 6-7). But this is not the case for mass loosing components. The radiational properties of mass loosing components are quite different from the expected values. They looks very old and they do not obey the M-R and M-L relations. This result may be valid in general for all Algol type binaries.

10.The larger, more massive, more evolved and more luminous stars accommodate prefentially in larger period systems with larger orbital radius (Fig. 8-9), and as a selection effect these systems are observed in larger distances.

11.With the exception of RZ Cnc, AR Mon, RT Lac, AY Cet and V471 Tau, the mean mass ratio $q = M_h/M_c$ of our systems is $q = 0.92\pm0.10$. Fig. 10 shows the evolution of the systems in k-q diagram. The empirical MS is shown by the dashed line (cf. Demircan and Kahraman, 1989). $k = R_h/R_c > 1.00$ only for UV Psc, SV Cam and ER Vul. It is interesting that all these three systems have their components in the MS. Having larger and more massive components hotter in these systems, they have to be located before the hottest points of their evolutionary tracks in the HR diagram. After passing this point, the larger and more massive component will be the cooler component of the system, and only then the system would satisfy Hall's criteria to become an RS CVn system. If we do not consider therefore these three systems then essentially all the others are located in one side of the MS line, around q = 0.92 in Fig. 10. This, also shows the evolutionary status of the classical RS CVn systems. The normal evolution of these detached binaries off the MS involve a decrease of k with constant q because the more massive component evolves faster, and its radius increases more quickly. Most of the systems are located in the lower part of the k-q diagram shown in Fig. 10. This is another strong evidence that the classical RS CVn systems are evolved binaries. The value of k decreases through the evolution of the system, from about 1.0 in the MS up to 0.3 when the larger mass star fills the Roche lobe in subgiant stage. After then the value of q starts increasing because of the Roche lobe overflow and the system shifts rightward in the k-q diagram. It seems the distance from the q = 0.92 line in the k-q diagram is a measure of mass transfer though the Roche lobe overflow.

4. ACTIVITY

As noted before the solar type atmospheric activity is fairly general properties of not only RS CVn type but all systems containing a late type (later than about FO) star (Morgan and Eggleton, 1979), and the enhanced activity in classical RS CVn's in comparison to single stars is attributed to the enforced faster rotation of the component stars through the binary synchronization (Hall, 1976, 1981, 1987). Recent reviews on the atmospheric activity of these binaries are already cited in the first section. We shall concentrate in this section on the defining parameters of the activity which were deduced through the correlation analysis between various measures of activity and relevant stellar parameters.

It is known mostly from the single MS stars that (i) the level of activity is not determined only by the position of the star in the HR diagram (Rosner, Golub and Vaiana, 1985). (ii) The primary determining parameter for activity level is found to be the stellar rotation and thus age (Kraft 1967, Skumenich and Eddy 1981, Noyes et al. 1984, Pallavicini et al. 1981). (iii) The activity-rotation relations flatten out (activity saturates) at high rotational velocities (Walter 1983, Caillaut and Helfand 1985, Vilhu and Walter 1987). (iv) The best measure of activity is the surface activity flux (Zwaan 1986, Basri 1987, Rutten and Schrijver 1987). (v) Activity diagnostics at various levels in the atmospheres are tightly correlated (Schrijver 1985). (vi) The structure of the convection zone plays an important role in determining the activity levels (Noyes et al. 1984, Hartmann et al. 1984, Manganey and Praderie 1984). (vii) The rotation, rate of mass loss in stellar wind, metal abundance and Rossby number $R_0 = P_{rot}/t$, (where t is the convective turnover time scale) and atmospheric activity are all found age dependent quantities for the late type dwarfs within 25 parsecs of the Sun(Barry 1988).

The relation between atmospheric activity level and Rossby number R_0 links the action of a magnetic dynamo to stellar magnetic fields, rotation and convection (Durney and Latour, 1978). However in many studies, inclusion of the theoretical convective turnover time (t) resulted in no significant improvement in the activity correlations (Basri 1987, Rutten and Schrijver, 1987), although it places the correlation on a more theoretical bases.

Analysis on the larger sample of single MS stars yielded that coronal x-ray luminosity L_x is strongly correlated with the photospheric radius (Fleming et al. 1989) which implies the bolometric luminosity L_{bol}, effective temperature T_{eff} and thus stellar mass M correlations with the activity luminosity L_a (Micela et al. 1985, Bookbinder, Galub and Rosner, 1986). The role of area in surface activity of RS CVn systems have been noted independently by

Majer et al. 1986, and Demircan, 1986. Demircan points out that the surface area (in fact R) probably represents the most important dynamo parameter t for the late type evolved stars in RS CVn binaries (e.i. R $\propto$ t). Surprisingly, no clear L_a-rotation period correlation, in RS CVn binaries were found (Majer et al. 1986; Demircan 1986, 1987a, 1987b, 1988; Basri 1987, Morris and Muttel 1988) unless the activity flux or the normalized activity luminosity (L_a/L_{bol}) is used as the activity indicator. In fact, L_a tends to increase with increasing rotational period only because of the period-radius dependence of the evolved components of RS CVn systems (Demircan 1987a). Similarly, Stauffer and Hartmann (1987) note that emission activity do not show strong dependence on rotation in older, more slowly rotating single stars. However the activity-rotation correlation exists if the activity flux is used in RS CVn systems (Demircan 1987a, Basri 1987, Slee and Stewart 1989).

Theories for the generation of acoustic and magnetic waves, which may heat the chromosphere and corona, suggest that the output flux should be dependent upon stellar T_{eff} (and thus L_{bol} and M) (cf. Ulmschneider and Stein 1982). Evidence in RS CVn systems was found that L_x is not independent of L_{bol} (and thus M) (Demircan 1987b).

It is expected in general that at least a part of the scatters in activity-rotation correlation could be attributed to the mass and composition dependence of the activity, because these parameters determine the structure and dynamical behavior of the stellar interiors. Separating these effects on the activity will be an important step toward understanding dynamo generation of magnetic fields. Unfortunately, this is hard to do because stellar parameters (particularly in binaries) are generally correlated with each other, and the theory is not sufficiently advanced to tell us what to look for. On the other hand, in addition to the large errors in the most easily obtained activity indicators (such as chromospheric or coronal emission luminosities or fluxes) these indicators may bear a complex relation to the underlying magnetic activity.

A systematic search for various activity parameters was carried out in this work by correlation analysis between the coronal emission as activity indicator and the revised absolute parameters for 31 binaries discussed in the previous section.

The coronal x-ray luminosities L_x are extracted from CABS, and L_x, and F_x have been applied in the analysis as measures of coronal activity. We assumed in the calculation of surface activity flux F_x that the x-ray emitting regions around a star are homogeneously distributed and their mean radius (coronal radius) is proportional to the photospheric radius R of the star.

To find out which stellar parameters are responsible for determining the level of activity in RS CVn systems, we started with plotting Log L_x from the system against possible activity related parameters. Table 2, summarizes appeared correlations. Whenever the possible activity related parameter is not the systemic parameter like P and d, but for the individual component stars like M, v, L, etc. then Log L_x was plotted against that parameter for hotter or cooler components separately, assuming all the x-ray emission comes from hotter or cooler components of the systems, respectively. Log L_x was also plotted, whenever it is meaningful, against the summation, (or average) of that parameter for individual components like $[R^2_h + R^2_c]$ or $[(T_h + T_c)/2]$.

Several interesting points are apparent in these correlations. First, not only the cooler more evolved but also the hotter components are both responsible for the coronal x-ray emission from RS CVn systems, because the correlation is much better when Log L_x is plotted against the summation (e.g. $RL_h + RL_c$) or average [e.g. $T_{eff} = (T_h + T_c)/2$] of a stellar parameter for individual components of the system. The correlation between L_x and d in the sense that larger the distance higher the L_x value is definitely induced by the d - R correlation noted before as the selection effect: the larger more evolved giants accommodate preferentially in longer period systems (Fig. 8, 9) with larger a, and in addition such systems are observed in larger distances. The both Log L_x - Log v and Log L_x-Log L correlations are largely due to the v-R and L-R relations, because there is no strong Log L_x - Log P and Log L_x - Log T_{eff} correlations. x-ray emission of the systems containing giant, mass transferring (RL% ≈ 1), and or relatively early MS stars is very much lower in comparison to that of other systems.

An important feature is seen particularly in Log L_x - RL^2, Log L_x - Log P and Log L_x - Log a correlations that the systems are divided in two distinct groups according to x-ray emission level (Figure 11). The systems with giant components do not obey such

Table 2. Correlations between Log L_x of the system and possible activity related parameters

Correlation coeff.	Correlated parameter
r <0.20	Log P, Log w, $M_h + M_c$
0.20 < r <0.40	DM, log g, Log T_{eff} ,Log a,Log v, Log L,Log d
r >0.40	$RL\%_h^2+RL\%_c^2$, $R_h^2+R_c^2$, $RL_h^2+RL_c^2$, Log $\overline{T}_{eff}$

distinction. The higher level emission group (Log L_x > 30.85) is formed by the systems UV Psc, AR Lac, CF Tuc, V711 Tau, GK Hya, RS CVn, SZ Psc and RT Lac. The other systems in Table 1, with the exception of systems with giant components are in the lower level emission group. It is interesting that there is a significant gap in x-ray emission level between these two groups. There is no system with 30.80 < Log L_x < 30.94 (or 6.3 10^{30} < L_x < 8.7 10^{30} ergs s^{-1}) in our sample. There is only one binary (out of 63 detections) in CABS with L_x = 6.75 10^{30} ergs s^{-1}: TZ Tri A = 6 Tri but whose both components are giants. It is not clear with the present data whether this gap is real, related with the origin and evolution of the binaries or just due to a selection effect. Mean values of the stellar parameters for two distinct groups of RS CVn systems are given in Table 3.

Table 3. The mean values of the stellar parameters for the high and low level x-ray emission RS CVn groups.The group 1 is the high level x-ray emission group.

	Group 1		Group 2	
Parameter	Hot Comp.	Cool Comp.	Hot Comp.	Cool Comp.
$M(M_{sun})$	1.17	1.36	1.25	1.29
T_{eff}(°K)	6000	5300	6150	5200
RL%	0.46	0.75	0.38	0.54
v(km/s)	39	57	32	41
Log g	4.0	3.5	4.1	3.7
$a(R_{sun})$		12		16
q		0.86		0.97
P (days)		3.2		4.7

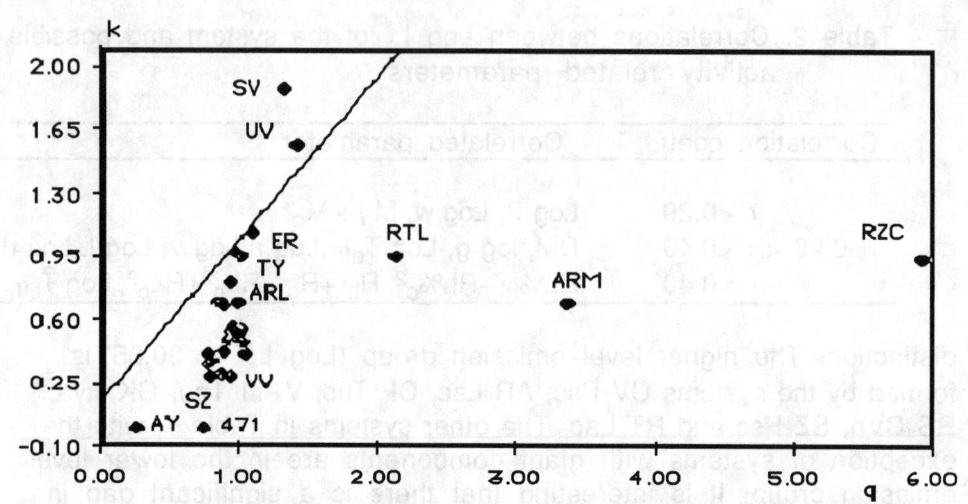

Figure 10. The k-q diagram of RS CVn systems. Empirical MS line (from Demircan and Kahraman, 1989) is also shown. See text for explanation.

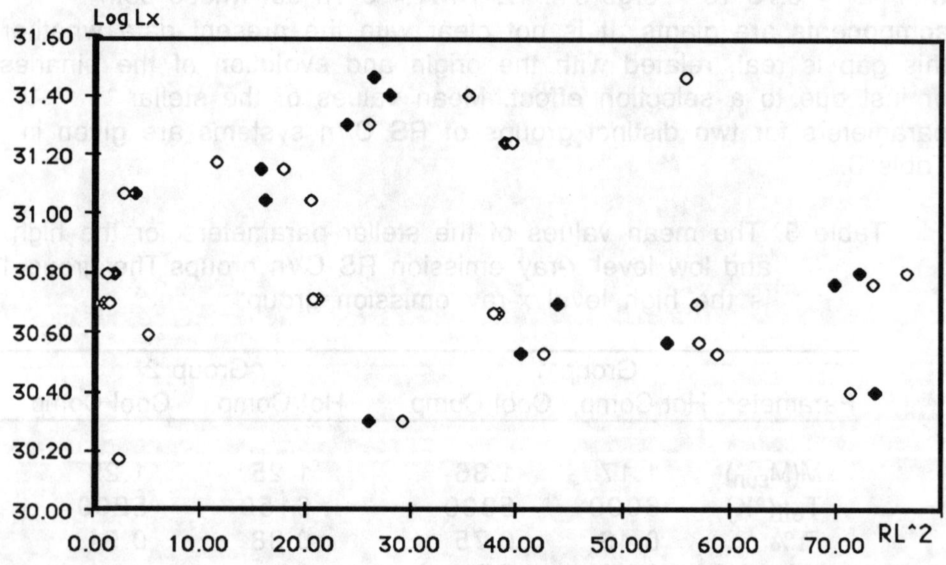

Figure 11. The x-ray luminosity (of the system) against the Roche Lobe radius square for the components of RS CVn systems. The components with $RL^2 > 80$ are not included in the diagram. Two distinct groups of RS CVn systems are formed according to their x-ray luminosity. See text and Table 3 for further explanation.

It is seen in Table 3 that the most responsible parameters for determining two distinct groups of RS CVn systems in x-ray emission level are the Roche lobe filling percentage RL%, rotation velocity v, gravity acceleration g, and the separation a between the components. The parameter RL% contains information on the radius R, mass ratio $q = M_h/M_c$, and the separation a between the components. It is known that larger the parameter a, smaller the chance to have larger RL%. It is interesting to note in Table 3 that the component stars of the systems in the second group are closer each other in physical characteristics. But since the separation a between the components is larger in this group, the synchronized rotation period of the components (through the Kepler's Law) is larger and thus the components of these systems are relatively slower rotators. In addition, the RL% parameter of the stars in the second group is relatively smaller, mostly because of the larger separation a between the component stars; since RL%=R /af (M_h/M_c). Having almost identical components the mean mass ratio q for the second group is almost unity (≈0.97) while it is 0.86 for the first group. It is thus seen in Table 2 and Table 3 that the activity determining parameters in binaries are found to be R, P, v, q, g and a. The g dependence of the activity represents the evolutionary effect on the atmospheric activity. The parameters R, P, and v are interdependent quantities ($v = 2\pi R/P$). Therefore it is not clear at present whether one or all of these parameters are responsible in the production levels of stellar activity. It is interesting to realize that almost all of the activity related quantities cited above are represented by a single parameter in binaries, which is RL%. This single parameter contains the information of size, mass, tidal interaction, age, evolution and rotation altogether. On the basis of this realization and considering the strong correlation (Fig. 12) between Log L_x and $(RL\%^2_h + RL\%^2_c)$, we deduced that each component in binaries should contribute to the atmospheric emission as proportional to their Roche lobe filling percentages (Demircan 1986, 1987a). This important deduction enable us to decouple the x-ray luminosity of the systems into the component stars:

$$(L_x)_{h,c} = x_{h,c} L_x$$

where

$$x_{h,c} = (RL\%)^{\alpha}_{h,c} / [(RL\%)^{\alpha}_{h} + (RL\%)^{\alpha}_{c}]$$

defines the x-ray luminosity contribution of the hotter and cooler components of a binary system. We have calculated x-ray luminosities of the individual components in our sample by using this formulation in terms of the known quantities L_x, $(RL\%)_h$ and $(RL\%)_c$. The power α is not known at present. We assumed $\alpha = 2$, because then the activity contributions of the components agree well with the known cases. For example Mg II h+k emission contributions of the components of AR Lac was estimated to be %70 and %30 (cf. Kızıloğlu et. al. 1983), while our calculation with $\alpha = 2$ yields these quantities as %68 and %32.

For the binaries V471 Tau and AY Cet, the x-ray emission contribution of the non-degenerate components were taken as %20 as estimated value for the system V471 Tau (Swank 1986). We have plotted Log L_x of the individual components against the activity related parameters M, R, P, L, T_{eff}, v, RL, Log g and RL%. The correlations with RL, v, and Log P are highly complex but almost have the same form. As an example Fig. 13 shows Log L_x - v correlation. Up to $v \approx 50$ kms^{-1}, L_x increases with increasing v, but for $v > 50$ kms^{-1}, L_x tends to decrease with increasing v. Similarly L_x increases with increasing P, up to $P \approx 2.0$-2.5 days in Log L_x- Log P correlation. No regression analysis has been attempted on these complex correlations. The correlations with other quantities seem linear and with the exception of Log L_x - Log M, for all the other correlations, the correlation coefficient r is in between 0.45 and 0.65. According to these correlations L_x decreases with increasing M, R, L, RL%, T_{eff} and g. The Log L_x- Log Lbol correlation seems mostly due to Log L_x - Log T_{eff} correlation, but there is also the area (or radius R) effect, because r for the Log L_x - Log T_{eff} correlation is 0.62 while it is 0.46 for the Log L_x - Log R correlation. Log L_x - Log T_{eff} correlation is shown as an example in Figure 14. The correlation coefficient in Log L_x - Log M correlation is only 0.16.

We have also estimated x-ray surface fluxes from the individual components by assuming the x-ray emitting coronal regions are homogeneously distributed and the mean coronal radius is proportional to the photospheric radius R of the component stars. Thus, it can be written that

$$(F_x)_{h,c} \propto (L_x)_{h,c} / R^2_{h,c} .$$

In our estimates of $(F_x)_{h,c}$ we normalized $(L_x)_{h,c}$ by the photospheric surface area of the component stars. Plots of evaluated Log F_x values against the activity related parameters revealed interesting results. First, it was confirmed (after Zwaan 1986, Basri 1987, Rutten and Schrijver 1987) that the surface emission flux is the best measure of activity to produce the lowest scatter in the correlation analysis. Second, the crude assumption we applied as the x-ray emitting regions are homogeneously distributed and the mean coronal radius is proportional to the photospheric radius is sensible. The results of the two dimensional regression analysis are given in Table 4. Surprisingly almost no correlation is found between F_x and T_{eff}, but in average, cooler the star, lower the F_x value is seen in the scatter diagram.

The correlation with RL% is rather complex, although it looks quadratic in form F_x first increases with increasing RL%, until RL% ≈ 0.55-0.60, and then tend to decrease with increasing RL%. Early type more massive stars and the components of mass transferring algols have relatively lower values of F_x. The Log F_x - Log g correlation (Figure 15) is similar (but more pronounced) to Log F_x - Log T_{eff} correlation: smaller the g, lower the F_x value. The fast rotating G dwarfs with RL% ≈ 0.60 have maximum F_x values. The upper bound in Log F_x - Log g correlation decreases rather slowly with decreasing Log g up to Log g ≈ 3.0 which corresponds to the giant branch for the coolest stars in the sample. Then Log F_x drops sharply from about 7.20 to 5.40 at Log g ≈ 2.70. The mass losing components of algols (namely the cooler components of AR Mon and RZ Cnc in our sample) do not obey the Log F_x - Log g and Log F_x - Log M correlations. It is seen in Table 4 and Figure 16-18 that the regression equations for the correlations with Log R, Log P and Log RL are almost same. This is definitely induced by the interrelations between R, P and RL. The correlation with Log v is weaker (Figure 19) in comparison to others. It may well be induced by the P and R dependence of v and deformed by other relations such as M - L. A careful inspection of other correlations in Table 4 and Figure 20-21 shows also that at least part of the correlations with Log M and Log $(M_h + M_c)$ may also be induced by the M - R relation. $\bar{F}_x$ in Table 4 is the mean x-ray surface flux from the system obtained by dividing L_x (from the system) to the sum of the radius squares of the component stars:

$$\bar{F}_x = L_x / [R^2_h + R^2_c] .$$

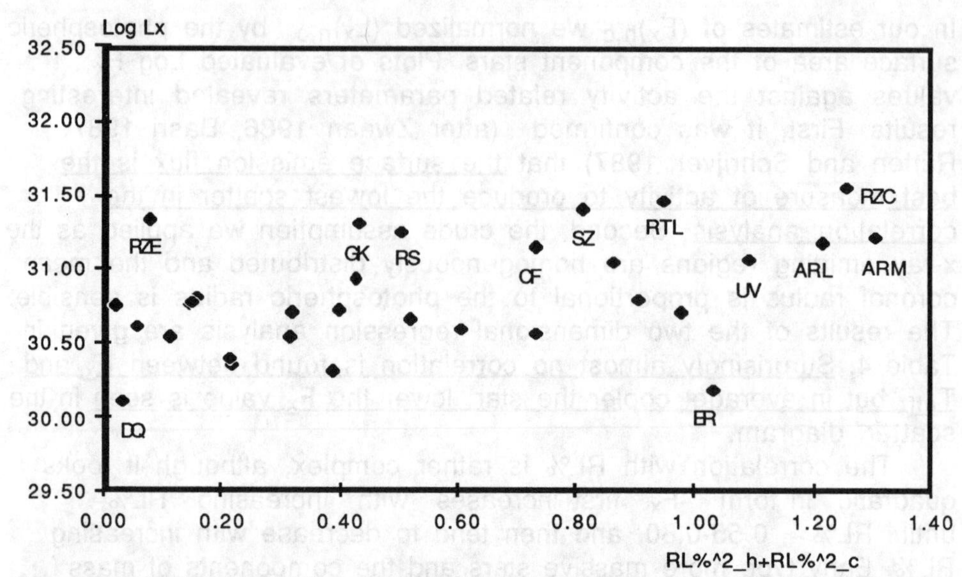

Figure 12. The x-ray luminosity (of the system) against the sum of the squares of Roche Lobe filling percentages for the components of RS CVn systems. ER Vul and RZ Eri are slightly out of correlation probably because of the wrong L_x values due to the large errors in their distances ($L_x = 4pd^2f$).

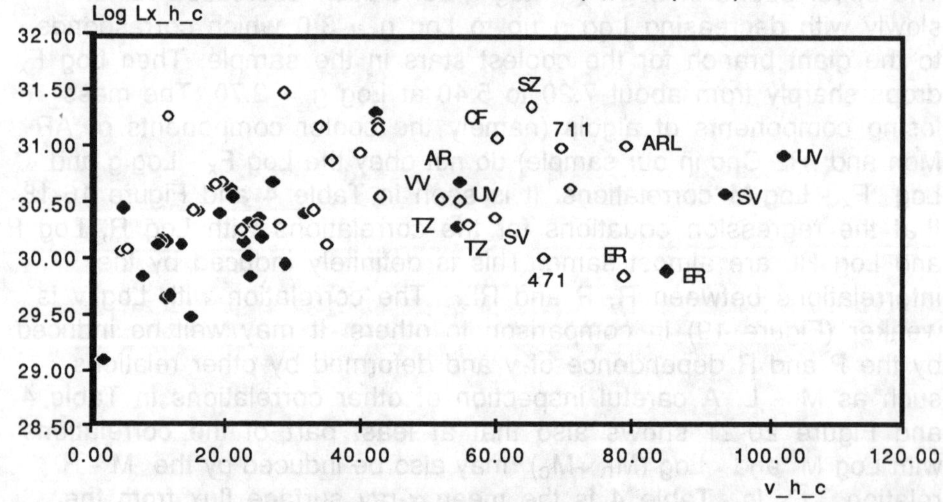

Figure 13. The x-ray luminosity against the equatorial velocity v for the components of RS CVn systems. L_x tends to increase with increasing v, up to about v≈50 kms^{-1}, then L_x tends to decrease with still increasing v.

Table 4. Two-dimensional regression results between F_x and other parameters in logarithmic scale.

Parameter	Regression Equation	Correlation Coeff.
M	$\text{Log } F_x = 7.30 - 3.67 \text{ Log } M$	0.83
$M_h + M_c$	$\text{Log } \bar{\bar{F}}_x = 9.80 - 4.26 \text{ Log}(M_h + M_c)$	0.84
v	$\text{Log } F_x = 5.54 + 0.86 \text{ Log } v$	0.53
R	$\text{Log } F_x = 7.33 - 1.30 \text{ Log } R$	0.69
P	$\text{Log } F_x = 7.44 - 0.93 \text{ Log } P$	0.80
RL	$\text{Log } F_x = 7.82 - 1.25 \text{ Log } RL$	0.83
g	Not linear, complex	-
RL%	Not linear, complex	-
T_{eff}	Not significant	-
F_{bol}	Not significant	-

It is interesting that both F_x- M and $\bar{\bar{F}}_x$ -(M_h +M_c) correlations (Figure 20 - 21) are very tight with the same correlation coefficients. M is involved in all R, P and RL through the following relations:

$$M_1 + M_2 = a^3 / P^2$$
$$RL = af\ (M_h / M_c)$$
$$R = 1.053\ M^{0.935}$$

where the last relation is empirical M-R relation for the MS stars (cf. Demircan and Kahraman, 1989). Hence, the F_x - R, F_x - P and F_x - RL correlations (Fig. 16 - 18) may well be induced inversely by M dependence of R, P and RL. On the other hand, comparison of the correlations of L_x and F_x with other parameters show that the sense of the correlation is inverted by the R (in fact area) effect, in the case of Log g, Log R and Log T_{eff} correlations, when changing the activity indicator from L_x to F_x. This situation is ambiguous in the case of Log M, Log v, Log P, Log RL and Log RL% because Log L_x-Log M, Log L_x-Log v, Log L_x-Log P, Log L_x-Log RL and Log F_x-Log RL% correlations are either weak and/or non linear.

Because the stellar parameters cited above are highly interdependent on each other it is still not clear with the present data whether one or all are responsible at different rates in the production level of stellar activity. Other emission activity indicators (in UV or IR) for different groups of stars with reliable

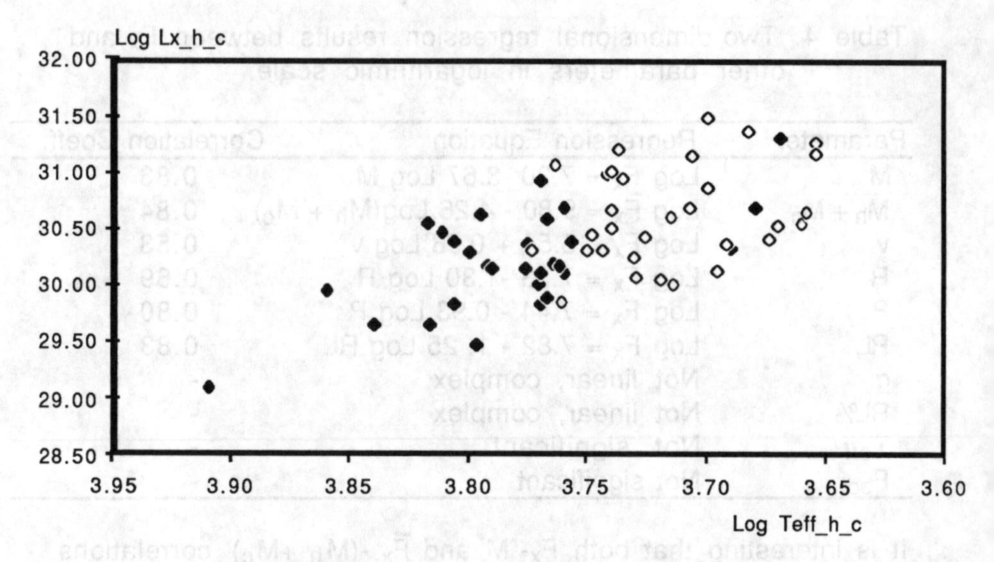

Figure 14. The x-ray luminosity against the effective temperature for the components of RS CVn systems. The cooler the star, larger the L_x values.

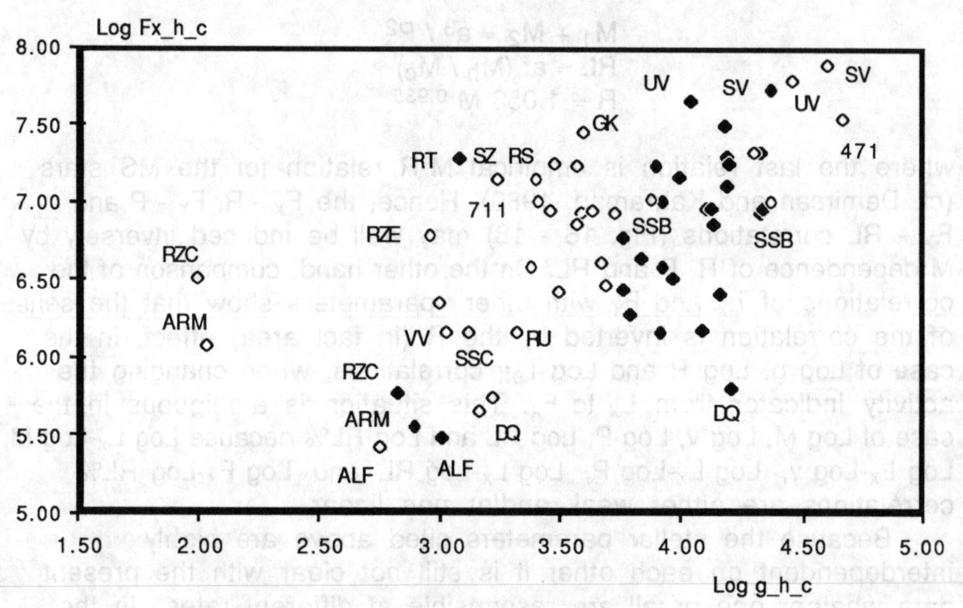

Figure 15. The x-ray surface flux against the gravity g for the components of RS CVn systems. Closer the star to MS, larger the F_x value. See text for further explanation.

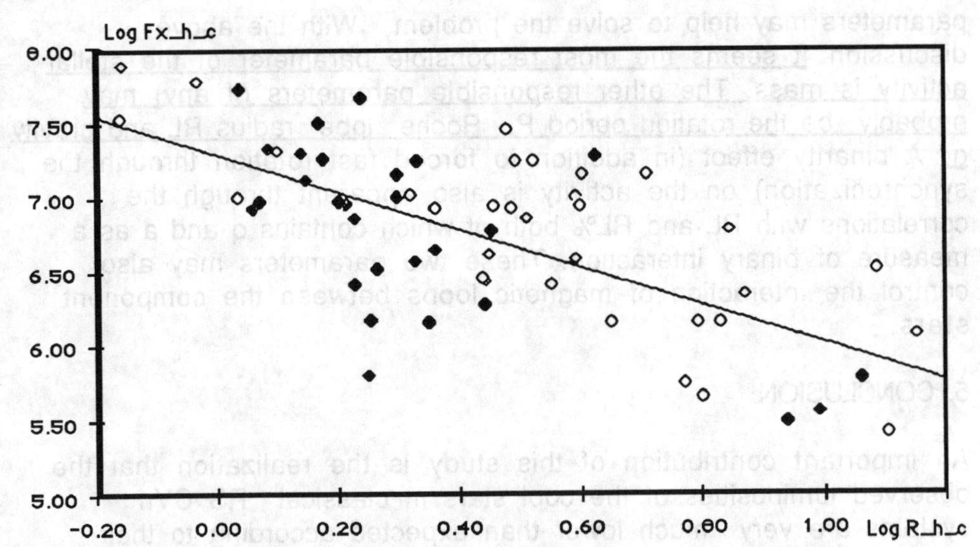

Figure 16. The x-ray surface flux against the radius R for the components of RS CVn systems. The larger radius stars have lower F_x values.

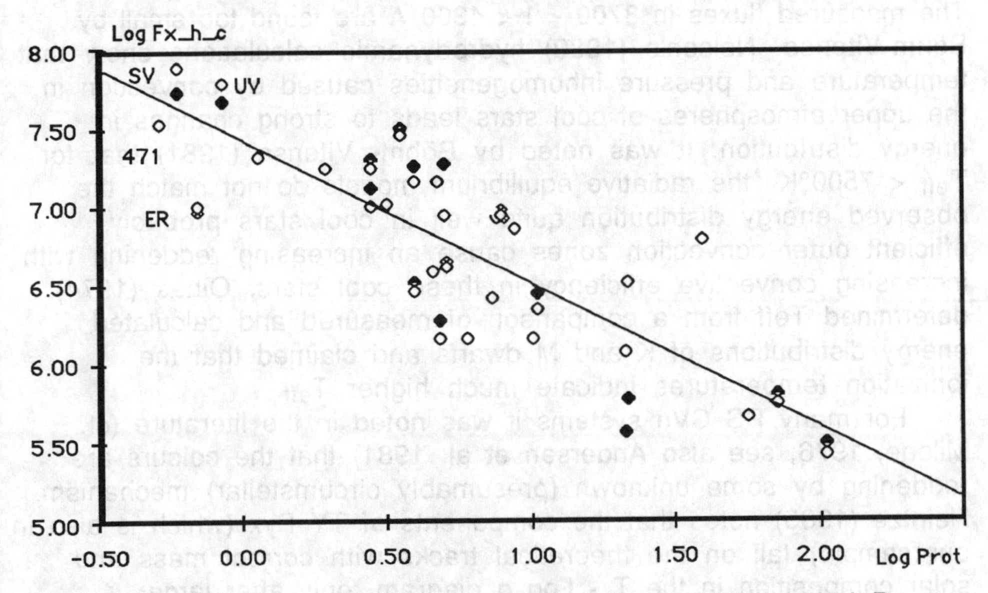

Figure 17. The x-ray surface flux against the rotation period P_{rot} for the components of RS CVn systems. The longer period systems have lower F_x values.

parameters may help to solve the problem. With the above discussion it seems the most responsible parameter of the stellar activity is mass. The other responsible parameters (if any) may probably be the rotation period P, Roche lobe radius RL and gravity g. A binarity effect (in addition to forced fast rotation through the synchronization) on the activity is also apparent through the correlations with RL and RL% both of which contains q and a as a measure of binary interaction. These two parameters may also control the interaction of magnetic loops between the component stars.

5. CONCLUSION

An important contribution of this study is the realization that the observed luminosities of the cool stars in classical RS CVn systems are very much lower than expected according to their masses and radii. Some supporting evidences were noted in Section 3 on this reddening problem of RS CVn systems. It is in fact known (cf. Böhm-Vitense, 1977) in general that there are discrepancies between measured and calculated energy distributions of cool stars. The measured fluxes in 3700 < l < 4300 A are found too small by Böhm-Vitense. Nelson's (1980) hydrodynamic calculations show that temperature and pressure inhomogeneities caused by convection in the upper atmospheres of cool stars leads to strong changes in energy distribution. It was noted by Böhm- Vitense (1981) that for T_{eff} < 7500°K the radiative equilibrium models do not match the observed energy distribution quite well in cool stars probably efficient outer convection zones cause an increasing reddening with increasing convective efficiency in these cool stars. Oinas (1974) determined Teff from a comparison of measured and calculated energy distributions of K and M dwarfs and claimed that the ionization temperatures indicate much higher T_{eff}.

For many RS CVn systems it was noted in the literature (cf. Milone, 1976; see also Andersen et al. 1981) that the colours are reddening by some unknown (presumably circumstellar) mechanism. Heintze (1985) notes that the components of TY Pyx (which is also in our sample) fall on the theoretical tracks with correct mass and solar composition in the T - Log g diagram, only after large corrections in T values.

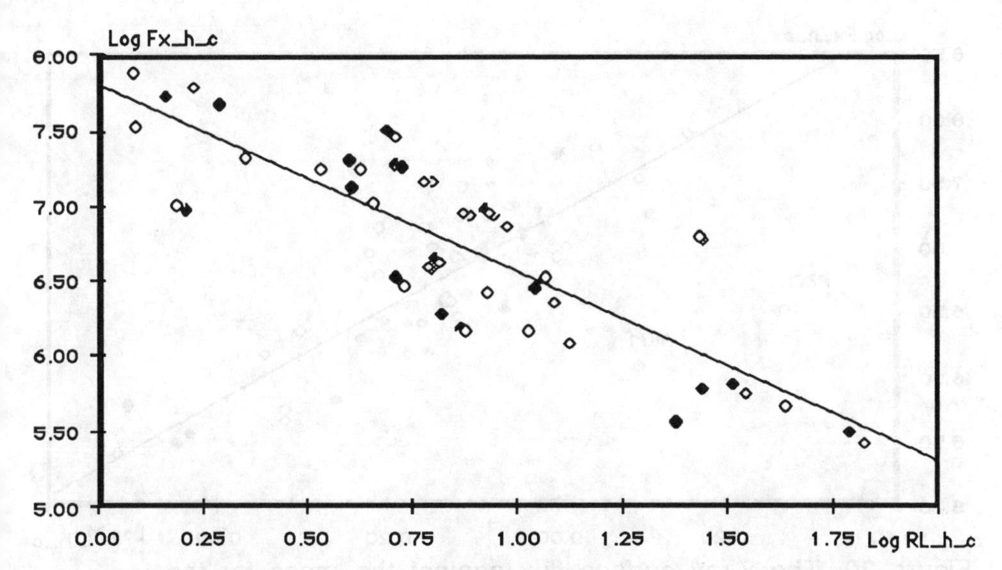

Figure 18. The x-ray surface flux against the Roche Lobe radius RL for the components of RS CVn systems. The stars with larger RL have lower F_X values.

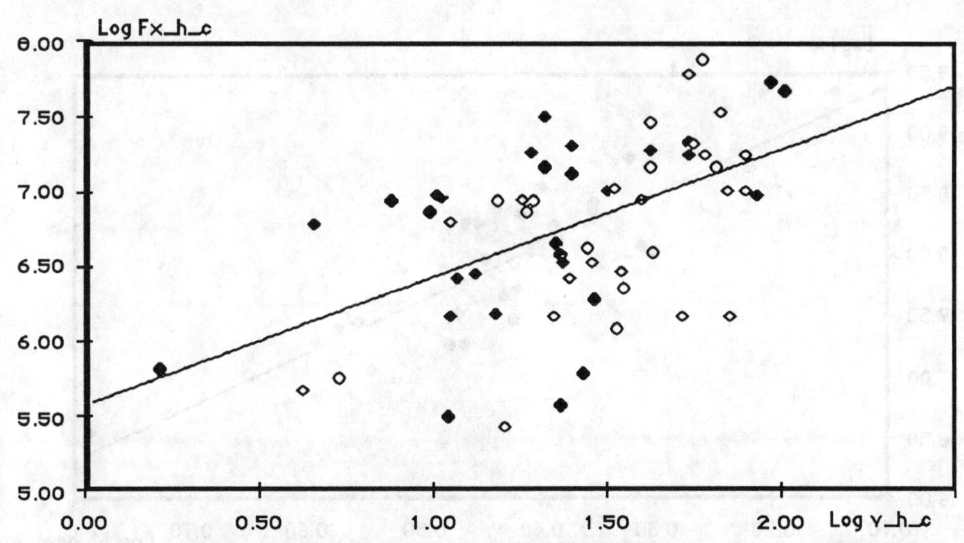

Figure 19. The x-ray surface flux against the equatorial velocity v for the components of RS CVn systems. The stars with higher v have relatively larger F_X values. The correlation is relatively weaker (see text).

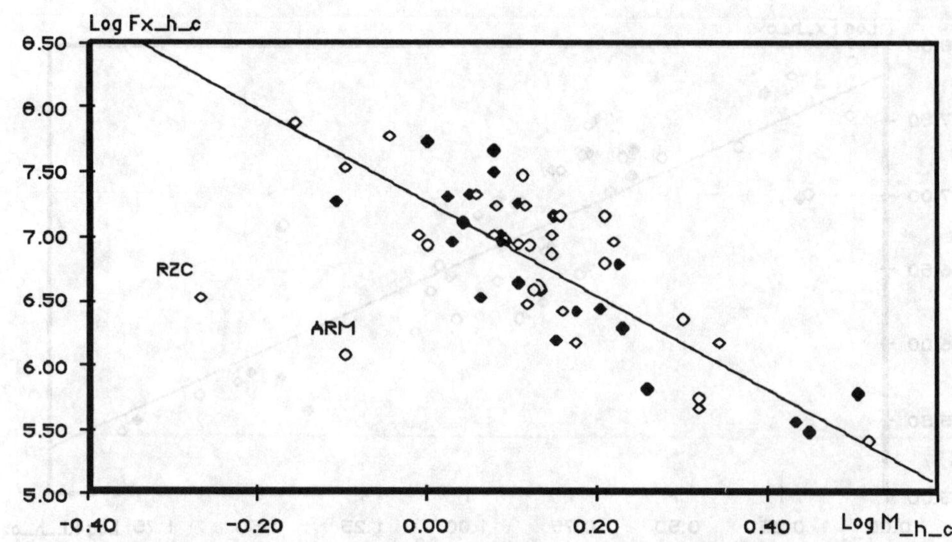

Figure 20. The x-ray surface flux against the mass for the components of RS CVn systems. The larger mass stars have lower F_x values. The mass loosing cooler components of RZ Cnc and AR Mon do not obey the correlation.

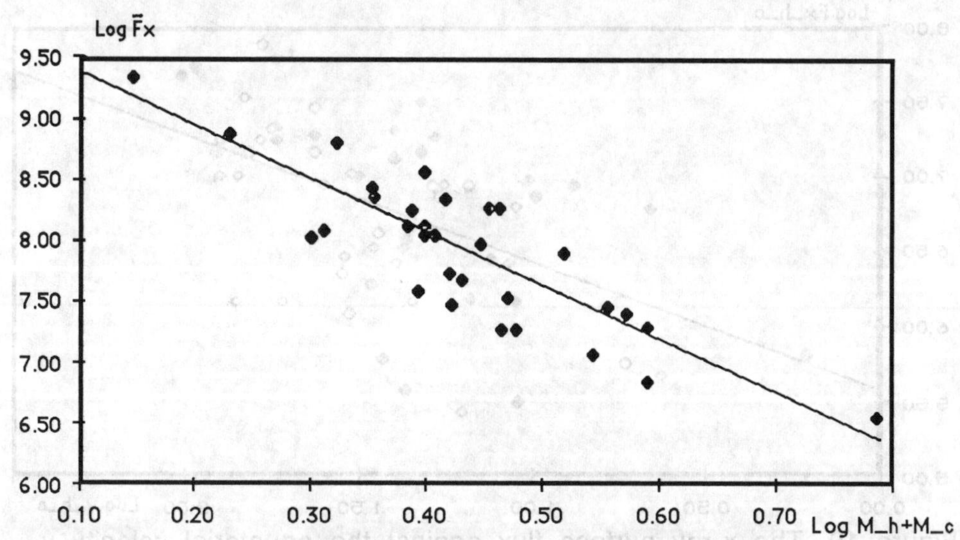

Figure 21. The mean x-ray surface flux $\bar{F}_x = L_x/(R_h^2 + R_c^2)$ against the total mass for the RS CVn systems. Almost the same correlation between F_x and M (as shown in Figure 20) is obtained. Here each full diamond represents a binary system.

Infrared observations of RS CVn systems have so far produced a confused situation, in which some authors (Antonopoulou 1983, Antonopoulou and Williams 1984, Berriman et al. 1983) do not find any significant excess, while others (Hall, Montle and Atkins 1975, Milone 1976, Verma et al. 1983) argue in favour of their existence. Recently, Busso et al. 1987, 1988 found clear evidences that some of the RS CVn's (e.g. VV Mon, RU Cnc, CF Tuc) have rather large phase independent infrared excess whose spectral distribution suggests presence of thin absorbing material around.

It is, in fact, known by observational evidences (Hartmann and Rosner,1979) and theoretical calculations (Spruit and Weiss,1986) that the total luminosities (and thus T_{eff}) of chromosphericaly active stars change as they go through the starspot cycle. It was shown that this is also true for the Sun (e.g. Hudson, 1988), but such periodic change in the luminosity (or T_{eff}) do not account for the large reddening in RS CVn systems, because we found almost all the systems are cooler than expected.

The concept of absorbing circumstellar material can be thought of as density, temperature and pressure inhomogeneities in chromosphere and corona. We think it is true, just as Plavec (1985) notes for the W Ser stars "the circumstellar hot turbulent shell is more extensive and probably also denser, and may be forming a kind of a superchromosphere" particularly around subgiant and giant components of RS CVn systems. The hot turbulent dense plasma on the top of the closed magnetic loops may well be the source of reddening in RS CVn systems. Recently, Collier, Cameron and Robinson (1989) found a strong evidence for the rapidly rotating G8 - K0 dwarf AB Dor that transient absorption features originate in cool dense clouds embedded in and corotating with the hot extended corona. Since the cloud densities may be as much as 10^7 times greater than would be expected at corona, they may well perturb the true MK type (and thus T_{eff}. and L_{bol}) most likely toward cooler values in optical region of the spectrum.

Other, important contribution of the present work is the discovery of direct activity-mass correlation. A strong correlation exist when the x-ray surface activity flux is plotted against the mass for the components of classical RS CVn systems. As far as we are aware, no such direct correlation between activity and mass is known in the literature for any group of active stars.

We found that the classical RS CVn's are divided in two distinct groups according to x-ray luminosities. For the higher level emission group (Log $L_x > 30.85$), the parameters P, q and a are relatively smaller and v and RL% are relatively larger. Thus, the responsible parameters for determining two distinct groups of RS CVn systems in L_x are found to be M, q, a, P, R, v and g. The binarity effect in activity is involved not only in P and v (through the synchronization) as largely believed, but also in q and a which may control the rate of interaction of the magnetic loops between the component stars. Such interaction in turn is thought to cause extra chromospheric and coronal emission. Since the parameter RL% contains the information of size, mass, tidal interaction, age, evolution, and rotation altogether, then each component in binaries contribute to the activity luminosity as proportional to their RL% values. The mass loosing components of algol type systems have very much lower values of activity luminosities, while the mass accreting components of the same systems obey the activity correlations. We found, in general, from observational data that the mass accreting components of algols do not change their radiational properties much.

The correlations with F_x are much tighter, indicating that the surface activity flux is the best measure of activity level in cool stars. However, for the evolved components of RS CVn systems not the equatorial rotation velocity v, but rotation period P, turns out to be the more responsible parameter (after mass) in determining the activity level. We think, the F_x - RL and F_x -RL% correlations contain at least some binarity effect (through q and a) on the activity.

As a result of interdependences between all the activity related parameters and large errors in the activity data and absolute parameters we could not separate the individual effects on the activity. This important step toward better understanding dynamo generated magnetic activity can be accomplished by the application of some new techniques (e.g. multi dimensional regression analysis) to the more accurate data.

ACKNOWLEDGEMENT:I thank Ethem Derman for stimulating discussion on the spectral types of the components of RS CVn systems, and for the considerable help in preparing this paper.

REFERENCES

Andersen, J., Clausen. J. V., Nordström, B., Reipurth, B.:1981, Astron. Astrophys., **101**, 7.

Antonopoulou, E., Williams, P. M.:1984, Astron. Astrophys.,**135**, 161.

Antonopoulou, E.:1983, Astron. Astrophys., **120**, 473.

Barry, D. C.:1988, Astrophys. J., **334**, 436.

Basri, G.:1987, Astrophys. J., **316,** 377.

Berriman, G., De Campli, W. M., Werner, M. W., Hatchett, S. P.:1983, Monthly Notices Roy. Astron. Soc., **205**, 859.

Bookbinder J., Golub, L., Rosner, R.:1986, in Cool Stars, Stellar Systems and the Sun, eds. M. Zeilik, D. Gibson, p 97. Springer-Verlag, Berlin.

Bopp, B.W., Talcott, J.C.:1980, Astron. J., **85**, 55.

Bopp, B.W.:1983, in IAU Colloq. No 71, Activity in Red Dwarf Stars, eds. P.B. Byrne, M. Rodono, p 343. D.Reidel, Holland.

Böhm-Vitense, E.:1981, Ann. Rev. Astron. Astrophys., **19**, 295.

Böhm-Vitense. E.:1977, Astrophys. J., **223**, 509.

Busso, M. et al.:1987, Astron. Astrophys., **183**, 83.

Busso, M. et al.:1988, Monthly Notices Roy. Astron. Soc., **234**, 445.

Caillault, J. P., Helfand, D. J.:1985, Astrophys. J., **289**, 279.

Catalano, S.:1983, in IAU Colloq. No 71, Activity in Red Dwarf Stars, eds. P.B. Byrne, M. Rodono, p 363. D.Reidel, Holland.

Charles, P.A.:1983, in IAU Colloq. No 71, Activity in Red Dwarf Stars, eds. P.B. Byrne, M. Rodono, p 415. D. Reidel, Holland.

de Jager, C., Nieuwenhuijzen, H.:1987, Astron. Astrophys.,**177**, 217.

Demircan, O.:1986, I. B. V. S. No: 2969.

Demircan, O.:1987a, Astrophys. Space Sci., **136**, 201.

Demircan, O.:1987b, Astrophys. Space Sci., **137**, 195.

Demircan, O.:1988, in Hot Thin Plasmas in Astrophysics, ed. R. Pallavicini, p 147. Kluwer Acad. Publ. Dordrecht.

Demircan, O., Kahraman, G.:1989, in preparation.

Durney, B. R., Latour, J.:1978, Geophys. Ap. Fluid Dyn., **9**, 241.

Eaton, J.A., Hall, D.S.:1979, Astrophys. J., **277**, 907.

Fekel, F. C., Eitter, J. J.:1989, Astron. J., **97**, 1139.

Hall, D. S., Montle, R. G., Atkins, H. L.,:1975, Acta Astron., **25**, 125.

Hall, D. S.:1976, in Multiple Periodic Variable Stars, IAU Col. NO 29, ed. W. S. Fitch, p 287. D. Reidel, Holland.

Hall, D. S.:1981, in Solar Phenomena in Stars and Stellar Systems, eds. R. Bonnet, A. K. Dupree, p 431. D. Reidel, Holland.

Hall, D. S.:1987, in Astrophys. (10th the European Regional Astr. Meeting of the IAU), ed. P. Harmanec, p 77. Praha, Czech.

Hartmann, L., Rosner, R.:1979, Astrophys. J., **230**, 802.

Hartmann, L. W., et al.: 1984, Astrophys. J., **279,** 778.

Heintze, J. R. W.:1985, in IAU Symp. No. 111, Calibration of Fundamental Stellar Quantities, eds. D. S. Hayes, L. E. Pasinetti, A. G. D. Philip, p. 433, D. Reidel, Holland.

Hudson, H. S.:1988, Ann. Rev. Astron. Astrophys., **26**, 473.

Huenemoerder, D. P., Barden, S. C.:1986, Astron. J., **91**, 583.

Kızıloğlu,Ü., Derman E., Ögelman H., Tokdemir F.: 1983, Astron. Astrophys., **123**, 17.

Kraft, R. P.:1967, Astrophys. J.,150 , 551.

Linsky, J.L.:1984, in Cool Stars, Stellar Systems and the Sun, eds. S. Baliunas, L. Hartmann, p 244. Springer-Verlag, Heilderberg.

Maeder, A., Meynet, G.:1988, Astron. Astrophys. Supp. Ser., **76**, 411.

Majer, P., et al.:1986, Astrophys. J., **300,** 360.

Manganey, A., Praderie, F.:1984, Astron. Astrophys., **130**, 143.

Micela, G., et al.:1985, Astrophys. J., **292**, 172.

Milone, E. F.:1977, Astron. J., **82**, 998.

Milone, F. F.:1976, Astrophys. J. Supp. Ser., **31**, 93.

Montesinos, B., Gimenez, A., Fernandez-Figueroa, M.J.:1988, Monthly Notices Roy. Astron. Soc., **232**, 261.

Morgan, J. G., Eggleton, P.P.:1979, Monthly Notices Roy. Astron. Soc.,**187**, 661.

Morris, D. H., Muttel, R. L.:198, Astron. J., **95**, 204.

Nelson, G. D.:1980, Astrophys. J., **238**, 659.

Noyes, R. W., et al.:1984, Astrophys. J., **279**, 763.

Oinas , V.:1974, Astrophys. J. Supp. Ser., **27**, 391.

Pallavicini, R., et al.: 1981, Astrophys. J., **248**, 279.

Popper, D. M.:1988, Astron. J., **96**, 1040.

Rodono, M.:1986, in Highlights of Astronomy, ed. J.P. Swings, p 429. D.Reidel, Holland.

Rosner, R., Galub, L.,Vaiana, G. S.:1985, Ann. Rev. Astron. Astrophys., **23**, 413.

Rutten, R. G. M., Schrijver, C. J.:1987, Astron. Astrophys., **177**, 155.

Schrijver, C. J.:1985, Space Sci. Rev., **40**, 3.

Shore, S. N., Adelman, S. J.:1984, Astrophys. J. Suppl. Ser., **54**, 151

Shore, S.N., Hall, D.S.:1980, in IAU Symp. 88, Close Binary Stars: Observations and Interpretation, eds. M. Plavec, R. Ulrich, p 389. D.Reidel, Holland.

Skumenich, A., Eddy, J. A.:1981, in Solar Phenomena in Stars and Stellar systems, eds. R. Bonnet, A. K. Dupree, p. 349. D. Reidel, Holland.
Slee, O. B., Stewart, R. T.:1989, Monthly Notices Roy. Astron. Soc., **236**, 129.
Spruit, H. C., Weiss. A.:1986, Astron. Astrophys., **266**, 167.
Stauffer, J. R., Hartmann, L.:1987, Astrophys. J., **318**, 337.
Straizys, V., Kuriliene, V.:1981, Astrophys. Space Sci., **80**, 353.
Strassmeier, K.G. et al.:1988, Astron. Astrophys. Suppl. Ser.,**72**, 291.
Ulmschneider, P., Stein, R. F.:1982, Astron. Astrophys., **106**, 9.
Verma, R. P. et al.:1983, Astrophys. Space Sci., **97**, 161.
Vilhu, O., Walter, F. M.:1987, Astrophys. J., **321**, 958.
Vogt, S.S.:1983, in IAU Colloq. No 71, Activity in Red Dwarf Stars, eds. P.B. Byrne, M. Rodono, p 137. D.Reidel, Holland.
Walter, F. M.:1983, Astrophys. J., 274, 794.
Zwaan, C.:1986, in Cool Stars, Stellar Systems and the Sun, eds. M. Zeilik, D. Gibson, p 19. Springer-Verlag, Berlin.

DECIPHERING LONG-TERM PHOTOSPHERIC AND CHROMOSPHERIC ACTIVITY ON THE CONTACT BINARY VW CEPHEI

David H. Bradstreet
Department of Physical Science
Eastern College
St. Davids PA 19087
USA

Edward F. Guinan
Department of Astronomy & Astrophysics
Villanova University
Villanova, PA 19085
USA

ABSTRACT. During 1987, 1988, and 1989 we have been conducting a coordinated effort of ground-based photometry and UV spectroscopy of the bright W UMa contact binary VW Cephei. The photometry has indicated the presence of large, cool starspot regions on the larger, cooler star of the system. IUE observations were conducted to search for possible phase dependent variations of the strong chromospheric and transition region emission lines that could correlate with the location and extent of the starspots deduced from the light curves. From these data we found strong but not conclusive evidence to support the physical and geometrical connection between the photospheric starspots and the overlying chromospheric and transition region emissions. In addition, we have completed the measurements of all useable archival spectra of VW Cep back to 1978 to investigate the long-term variations of the UV line emissions. This work indicates systematic seasonal changes in the UV emission level with a time-scale of a few years which could be attributed to an activity cycle. From high dispersion spectra obtained in 1988 and 1989 we detected a sharp apparently stationary emission feature superimposed over the broad Mg II h + k emission line profiles of the two stars. VW Cep represents an extreme case for studying stellar dynamo theory and long-term ground based photometry coupled with IUE observations should play a crucial role in the understanding of magnetic fields and activity cycles in rapidly rotating solar-like stars.

1. Introduction

The W Ursae Majoris-type contact binary systems are the most common close binary known, with a space density of ~ 10^{-4} of all stars (Budding 1982). These are binaries whose orbital periods are generally shorter than 2/3 of a day and whose components are in contact with their Roche limiting surfaces. Light variations arise from the mutual eclipses and from the highly distorted figures of the stars caused by tidal interaction and rotational flattening. The component stars of W UMa systems are generally in the range of F-K spectral types and lie near or just above the main sequence. Their rapid rotation and

C. İbanoğlu (ed.), Active Close Binaries, 467–484.

inferred magnetic dynamos produce ultraviolet and x-ray chromospheric and coronal emission that are among the brightest in surface flux of cool stars (Eaton 1983; Rucinski and Vilhu 1983; Cruddace and Dupree 1984).

Observations of W UMa-type binaries with the IUE have shown that these stars are rich sources of ultraviolet emission. They generally have large surface fluxes of high temperature lines such as C II, C IV, N V, and Si IV as well as moderate temperature (10^4 K) chromospheric emission lines of Mg II, Ly α, C I, O I, and Si II. In analogy with the chromospherically active RS CVn stars, these line emissions are thought to arise from dynamo generated magnetic fields since these stars possess tidally enforced rapid rotation and deep convective zones, at least for the shorter period systems. However, because of their strong tidal interactions, components of W UMa systems should not possess significant differential rotation (both radial as well as latitudinal) which is believed to be an important component of stellar dynamo theory (Parker 1986; Gilman 1980).

The light curves of most of the shorter period W UMa systems, of G to K spectral type, display markedly asymmetric light curves. Modeling of light curves with large, cool spotted regions can account for the observed asymmetries (Bradstreet 1985; Yamasaki 1982). The presence of starspots on these stars is not unexpected because of their short orbital and rotational periods. However, the W UMa-type stars are not as active as simple extrapolations of trends for detached binaries (the RS CVn stars) would indicate (Rucinski and Vilhu 1983). This could indicate a saturation of the dynamo mechanism in these stars.

2. VW Cep - An Intriguing W UMa System

VW Cep (HD 197433) is one of the brightest ($V_{max} \sim +7.8$ mag), best observed short period W UMa binaries. It consists of ~ G5V and ~ K0V components in contact with their Roche limiting surfaces. Photoelectric photometry has been carried out on this star since 1948. Frequently, the light curves exhibit large asymmetries of up to 0.07 mag at visible wavelengths (e.g., Kwee 1966). Since 1978, VW Cep has been monitored photoelectrically at Villanova University Observatory. The observations were made with a photometer mounted on the 38-cm reflector using intermediate band filters. These light curves were combined with others obtained over the same time interval and found to change systematically with time. To illustrate the large changes in the light curve we plot the yellow light curves of October 1986 and November 1989 in Figure 1. In October 1986 the secondary minimum was ***deeper*** than the primary minimum due to the visibility of dark starspots near phase 0.5P. In November 1989 the primary minimum was the deepest ever recorded and indicated a concentration of large starspots on the facing hemisphere of the cooler, eclipsing star. Secondary minimum of this light curve is near its brightest level indicating that the spots are indeed mostly out of view at this phase. Also, comparable changes in the brightness levels occur *outside* the eclipses.

During 1979-1981 asymmetries are evident in the light curves with the brightness of maximum I (at 0.25P) and maximum II (at 0.75P) as well as minimum I (0.00P) and minimum II (0.50P) changing up to 0.07 mag (in yellow) with a characteristic period of ~ 155 ± 5 days. The light curves obtained during 1982-84 appear to be disturbed the least amount. Beginning in 1985, and continuing through 1987, the light curves were asymmetric with max I being ~ 0.06 mag brighter than max II as shown in the top panel of

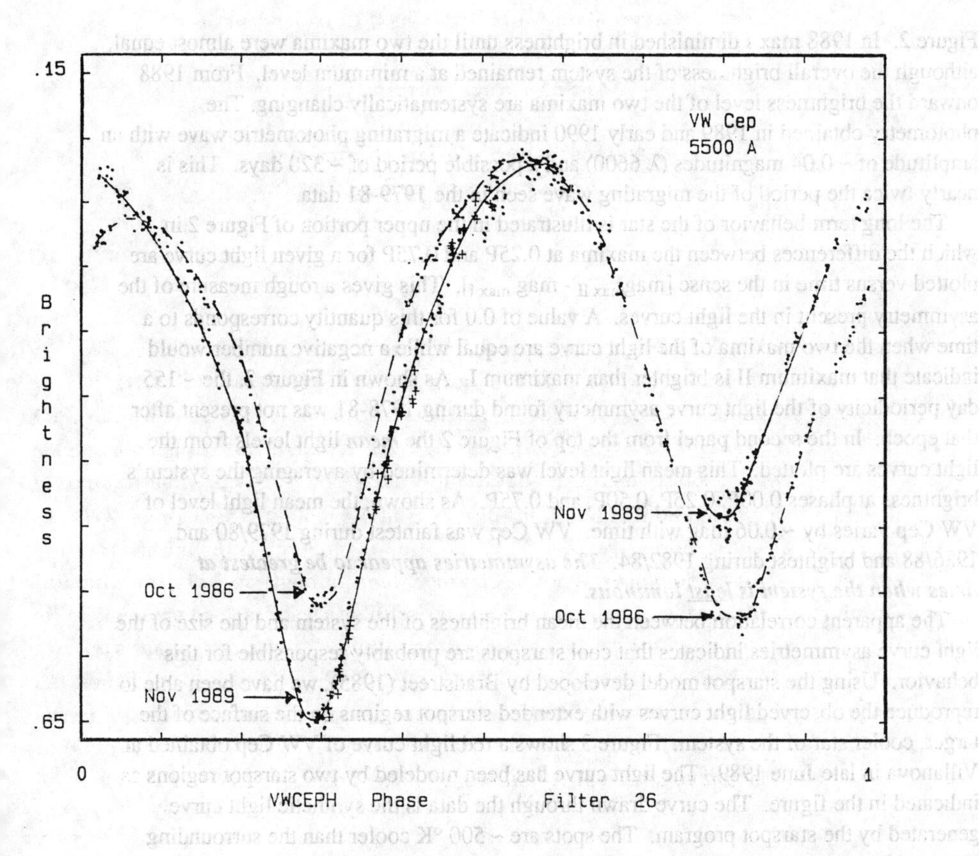

Figure 1. Light curves of VW Cep obtained using a yellow filter (λ5500) at Villanova University in October 1986 and November 1989. The secondary eclipse depth of the October 1986 light curve is ***deeper*** than its corresponding primary eclipse due to the visibility of spots near phase 0.5P at that time. The primary eclipse of the November 1989 light curve is the deepest ever observed at Villanova, again due to the visibility at that phase of active regions on the larger, cooler star.

Figure 2. In 1988 max I diminished in brightness until the two maxima were almost equal, although the overall brightness of the system remained at a minimum level. From 1988 onward the brightness level of the two maxima are systematically changing. The photometry obtained in 1989 and early 1990 indicate a migrating photometric wave with an amplitude of ~ 0.04 magnitudes (λ 6600) and a possible period of ~ 320 days. This is nearly twice the period of the migrating wave seen in the 1979-81 data.

The long term behavior of the star is illustrated in the upper portion of Figure 2 in which the differences between the maxima at 0.25P and 0.75P for a given light curve are plotted versus time in the sense [$mag_{max\ II}$ - $mag_{max\ I}$]. This gives a rough measure of the asymmetry present in the light curves. A value of 0.0 for this quantity corresponds to a time when the two maxima of the light curve are equal while a negative number would indicate that maximum II is brighter than maximum I. As shown in Figure 2, the ~ 155 day periodicity of the light curve asymmetry found during 1978-81 was not present after that epoch. In the second panel from the top of Figure 2 the *mean* light levels from the light curves are plotted. This mean light level was determined by averaging the system's brightness at phases 0.00P, 0.25P, 0.50P, and 0.75P. As shown, the mean light level of VW Cep varies by ~ 0.06 mag with time. VW Cep was faintest during 1979/80 and 1986/88 and brightest during 1982/84. ***The asymmetries appear to be greatest at times when the system is least luminous.***

The apparent correlation between the mean brightness of the system and the size of the light curve asymmetries indicates that cool starspots are probably responsible for this behavior. Using the starspot model developed by Bradstreet (1985), we have been able to reproduce the observed light curves with extended starspot regions on the surface of the larger, cooler star of the system. Figure 3 shows a red light curve of VW Cep obtained at Villanova in late June 1989. The light curve has been modeled by two starspot regions as indicated in the figure. The curve drawn through the data is the synthetic light curve generated by the starspot program. The spots are ~ 500 °K cooler than the surrounding photosphere of the star and cover several percent of the area of the larger component. By careful monitoring of the changing asymmetries in the light curves we have been able track the migrating spot regions over the surface of the larger component. The 155 and 320 day cyclic behavior observed during 1978-81and 1988-89, respectively, appear to arise from a small difference between the rotational period of the starspot forming region and the orbital period of the binary. These wave migration periods are apparently the beat period between the orbital period and the rotational period of the photosphere region where the spots were located at this time. For example, the 320 day migration period indicates that the region on the star where the spots are located has a period 21 seconds longer (0.1%) than the assumed synchronous rotation period of the stars. The migration of the spotted region around the surface of the cooler star produces the varying light levels of the maxima and minima of the light curve. At times, when the spots are small, such as in 1982-84, the asymmetries are smallest and the mean brightness of the system is greatest, and vice versa as in 1979-80 and 1986-88. A spot cycle of ~ 8-9 years may be present in which the spots attained their maximum areal extent during 1979-80 and 1987-88. This possible spot cycle produces the long-term variation in the mean light level of the system as shown in the second panel from the top of Figure 2. Figure 4 shows a plot of all of the red photoelectric observations obtained at Villanova from 1978 - 1989 (~ 4000 points). The error of each observation is ≤ 0.008 magnitude (about the size of the plot symbol). The apparent scatter

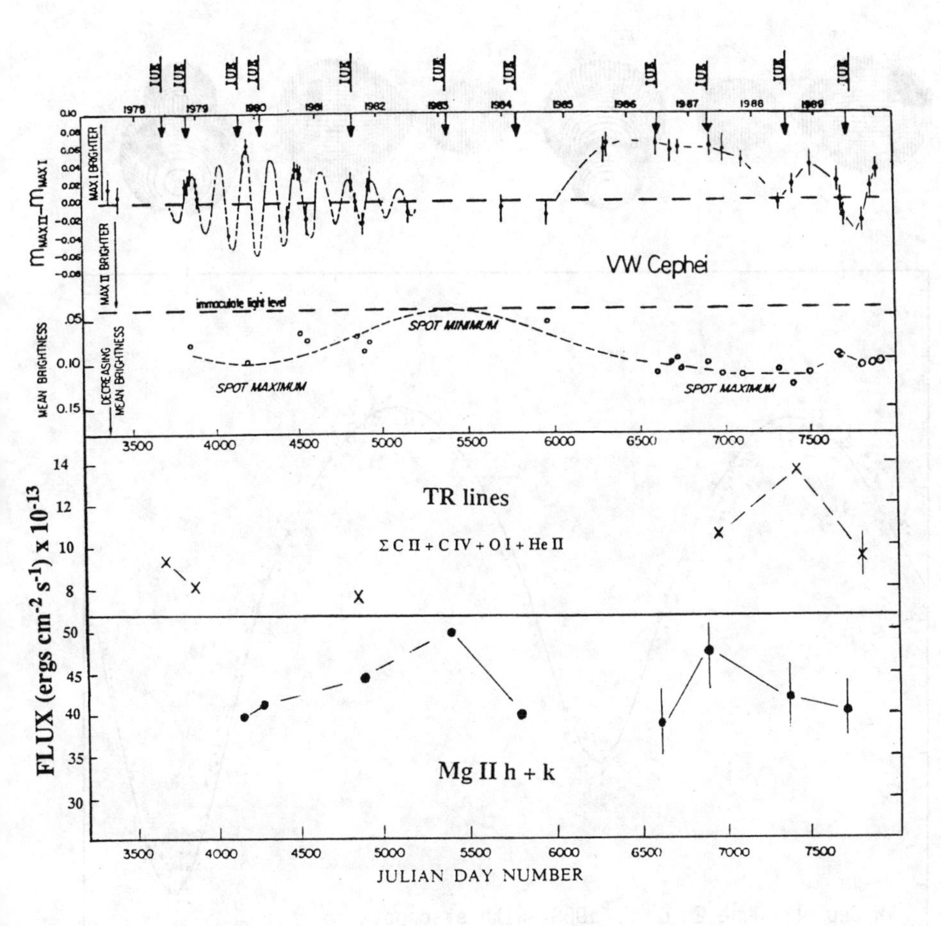

Figure 2. The long term photometric behaviour of VW Cep is shown. The top panel represents the changing of the levels of maximum light in the sense [$mag_{max\ II}$ - $mag_{max\ I}$] where max II and max I are 0.75P and 0.25P, respectively. The second panel from the top shows the mean light of the system. The third panel from the top plots the sum of the integrated emission fluxes of the transition-region (TR) lines C II, C IV, O I, and He II. The bottom panel displays the integrated emission fluxes of the Mg II h + k lines.

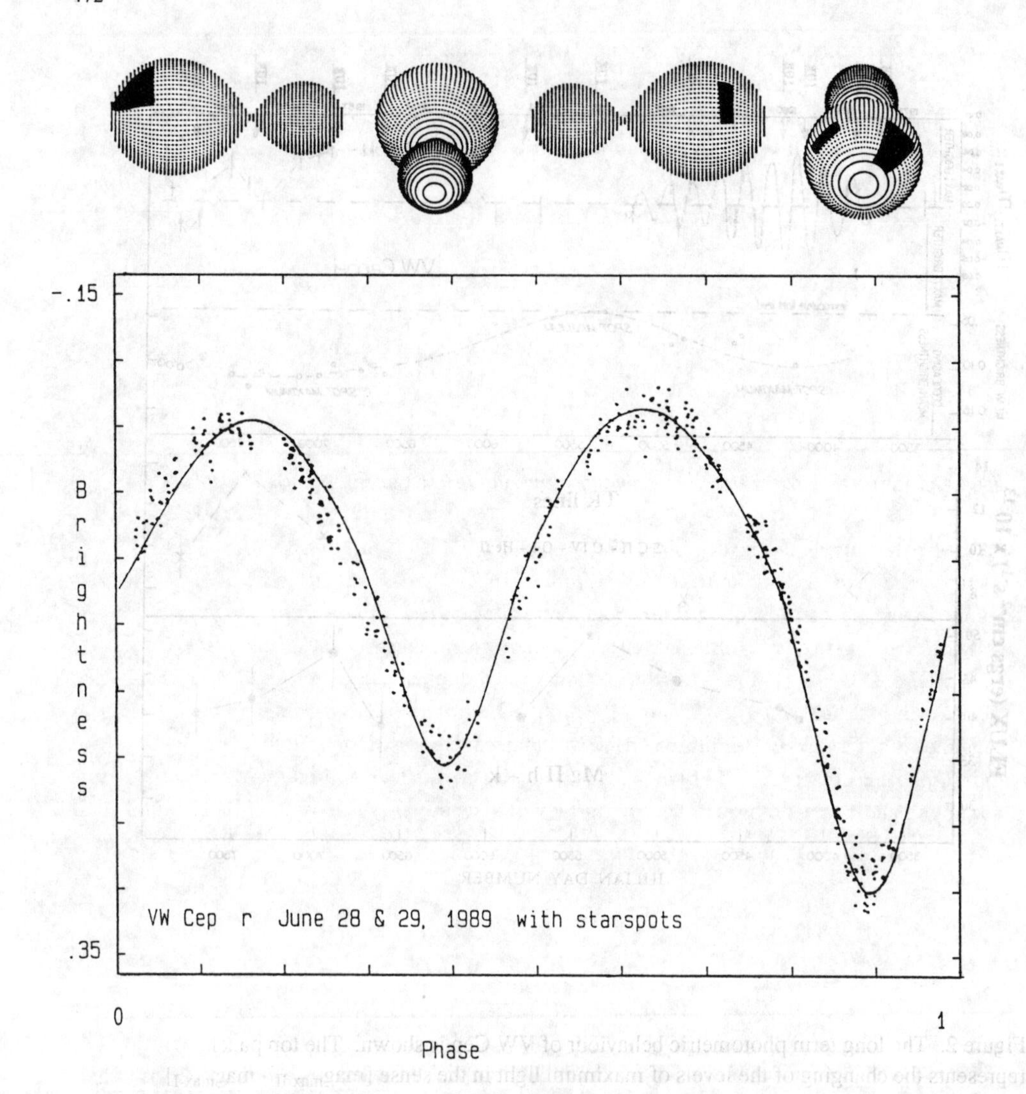

Figure 3. The red light curve of VW Cep obtained in June 1989 at Villanova two days prior to IUE observations. The asymmetries and unusually deep primary eclipse were modeled with cool starspots as shown in the three dimensional representations of VW Cep displayed at different orbital aspects corresponding to the phases in the light curve directly beneath them. The solid curve drawn through the data represents the theoretical light curve generated by the starspot program.

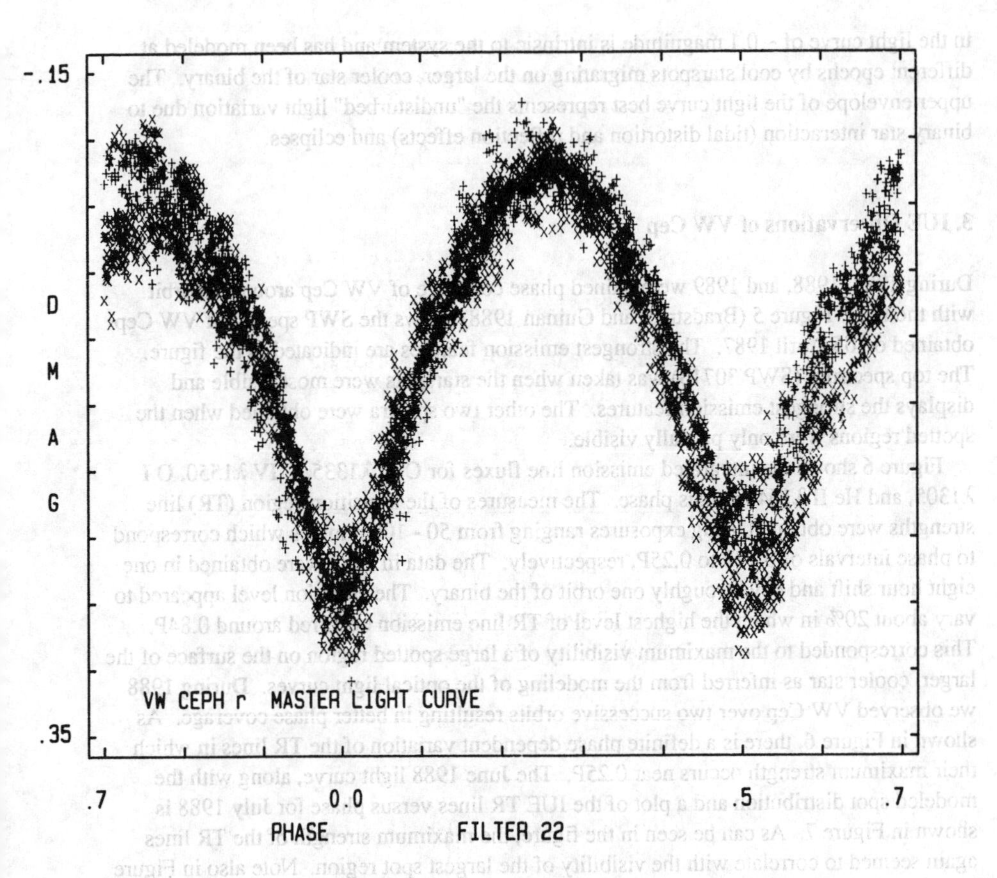

Figure 4. Photoelectric observations of VW Cep using a red filter (λ6600) obtained at Villanova University from 1978-mid 1989. Over 4000 observations are plotted. The apparent 0.1 magnitude scatter in the data is intrinsic to the binary and is attributed to changing and migrating spot regions on the larger, cooler star.

in the light curve of ~ 0.1 magnitude is intrinsic to the system and has been modeled at different epochs by cool starspots migrating on the larger, cooler star of the binary. The upper envelope of the light curve best represents the "undisturbed" light variation due to binary star interaction (tidal distortion and reflection effects) and eclipses.

3. IUE Observations of VW Cep

During 1987, 1988, and 1989 we obtained phase coverage of VW Cep around its orbit with the IUE. Figure 5 (Bradstreet and Guinan 1988) shows the SWP spectra of VW Cep obtained on 05 April 1987. The strongest emission features are indicated in the figure. The top spectrum (SWP 30713) was taken when the starspots were most visible and displays the strongest emission features. The other two spectra were obtained when the spotted regions were only partially visible.

Figure 6 shows the integrated emission line fluxes for C II $\lambda1335$, C IV $\lambda1550$, O I $\lambda1305$, and He II $\lambda1640$ versus phase. The measures of the transition-region (TR) line strengths were obtained using exposures ranging from 50 - 100 minutes which correspond to phase intervals of 0.12P to 0.25P, respectively. The data in 1987 were obtained in one eight hour shift and cover roughly one orbit of the binary. The emission level appeared to vary about 20% in which the highest level of TR line emission occurred around 0.84P. This corresponded to the maximum visibility of a large spotted region on the surface of the larger, cooler star as inferred from the modeling of the optical light curves. During 1988 we observed VW Cep over two successive orbits resulting in better phase coverage. As shown in Figure 6, there is a definite phase dependent variation of the TR lines in which their maximum strength occurs near 0.25P. The June 1988 light curve, along with the modeled spot distribution and a plot of the IUE TR lines versus phase for July 1988 is shown in Figure 7. As can be seen in the figure, the maximum strength of the TR lines again seemed to correlate with the visibility of the largest spot region. Note also in Figure 6 that the overall emission levels in 1988 were ~ 30-40% higher than those observed in 1987. This may be due to an increase in magnetic activity in the TR region in 1988. In 1989 we secured better phase coverage over four orbits and found a definite phase dependent variation of the TR emission line strengths as seen in Figure 6. During 1989 the lowest emission was measured near the bottom of primary minimum and the greatest at the quadratures of the orbit which indicates that the active regions may be partially eclipsed during primary minimum. The mean level of the TR line emission in 1989 is ~ 40% lower than in 1988 and comparable to that found in 1987. Unlike the previous two years where there appeared to be a correspondance between the visibility of spot regions and the TR emission line strengths, in June/July 1989 the largest spot (as inferred from the light curve analysis depicted in Figure 3) appears to be in view near phase 0.0P. There is strong evidence for phase dependence in the strengths of the TR emission lines, but the phase dependence itself varies from year to year both in morphology and mean strength. As shown in the top panel of Figure 2, the morphology and mean light level of the optical light curve also underwent significant changes during 1987-1989 which we have attributed to changing and migrating cool spots.

We have analyzed all of the useable IUE observations of VW Cep in the archives. The bottom two panels in Figure 2 show the mean values of the TR lines and the mean

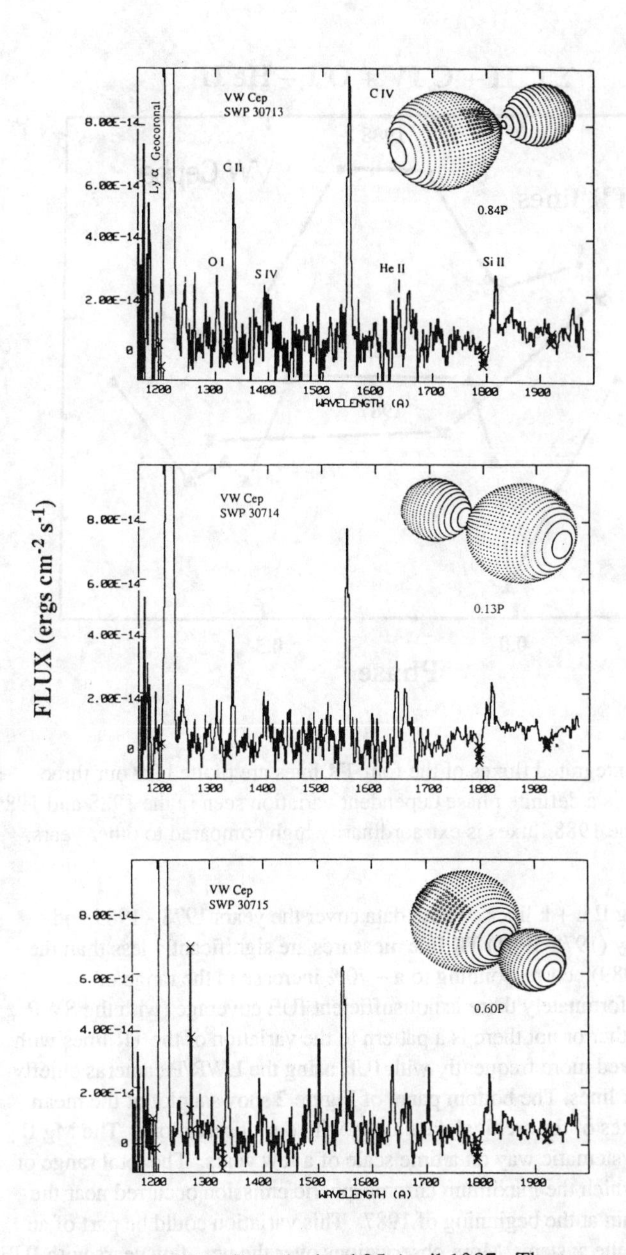

Figure 5. SWP spectra of VW Cep obtained 05 April 1987. The top spectrum (SWP 307213) was centered on 0.84P when the starspots were most visible. The middle spectrum (SWP 30714) was centered at 0.13P, and the bottom spectrum (SWP 30715) was centered at 0.60P. The insets in the upper righthand corners show representations of VW Cep at the appropriate phase of mid-exposure.

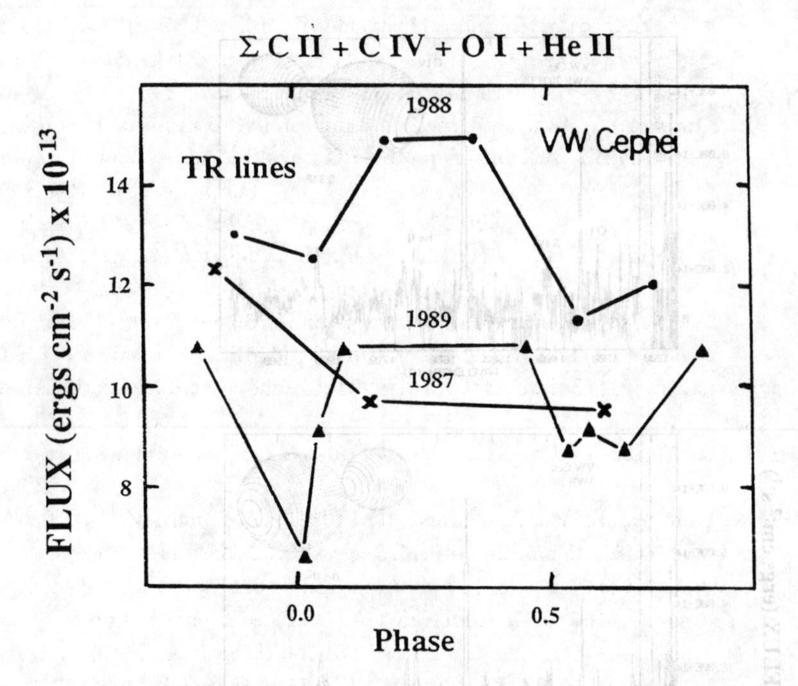

Figure 6. The sum of the integrated fluxes of the four TR lines are plotted for our three previous IUE runs. There is a definite phase dependent variation seen in the 1988 and 1989 data. The mean level of the 1988 fluxes is extraordinarily high compared to other years.

integrated fluxes of the Mg II h + k lines. These data cover the years 1978 - 1989 and include our data. The early (1978-1981) TR line measures are significantly less than the more recent data (1987-1989), corresponding to a ~ 70% increase in the range of integrated line fluxes. Unfortunately there is not sufficient IUE coverage (with the SWP camera) to determine whether or not there is a pattern to the variation of the TR lines with time. VW Cep was observed more frequently with IUE using the LWR/P cameras chiefly to measure the Mg II h + k lines. The bottom panel of Figure 2 shows a plot of the mean integrated Mg II h + k fluxes of the archival data along with our observations. The Mg II lines appear to vary in a systematic way on a time scale of a few years. The total range of variation is ~ 25-30% in which the maximum chromospheric emission occurred near the beginning of 1983 and again at the beginning of 1987. This variation could be part of an activity cycle operating in the system. More observations over the next few years with IUE will be crucial in defining this long term variation with the possibility of confirming an activity cycle in the chromosphere.

An examination of the overall long term behavior of the photospheric, TR, and chromospheric activity in VW Cep (i.e., Figure 2 taken as a whole) does not indicate any obvious correlations among them. This is both perplexing and interesting in light of the expectations of dynamo-related magnetic activity. Perhaps as indicated in the recent studies

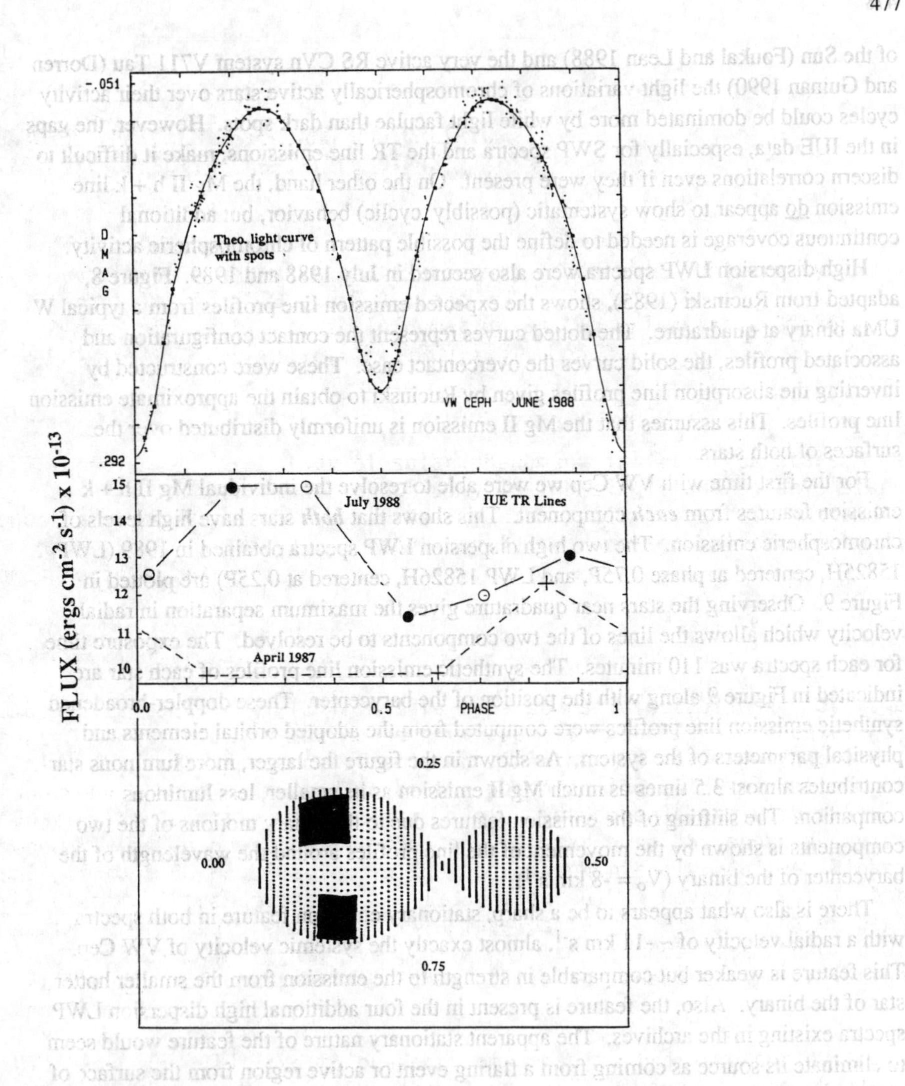

Figure 7. The top panel is the red (λ6600) light curve of VW Cep obtained in June 1988 at Villanova. The solid curve represents the theoretical light curve *with* starspots. The size and locations of the spots are shown in the bottom panel from a pole-on view of the binary. The middle panel shows the combined emission fluxes of C II, C IV, O I, and He II plotted versus phase that were obtained in July 1988. The closed circles represent the SWP images obtained during the first shift, the open circles those secured during the second shift. Note that the largest peak of emission occurs between 0.18P - 0.35P, when the largest starspot is most visible.

of the Sun (Foukal and Lean 1988) and the very active RS CVn system V711 Tau (Dorren and Guinan 1990) the light variations of chromospherically active stars over their activity cycles could be dominated more by white light faculae than dark spots. However, the gaps in the IUE data, especially for SWP spectra and the TR line emissions, make it difficult to discern correlations even if they were present. On the other hand, the Mg II h + k line emission do appear to show systematic (possibly cyclic) behavior, but additional continuous coverage is needed to define the possible pattern of chromospheric activity.

High dispersion LWP spectra were also secured in July 1988 and 1989. Figure 8, adapted from Rucinski (1985), shows the expected emission line profiles from a typical W UMa binary at quadrature. The dotted curves represent the contact configuration and associated profiles, the solid curves the overcontact case. These were constructed by inverting the absorption line profiles given by Rucinski to obtain the approximate emission line profiles. This assumes that the Mg II emission is uniformly distributed over the surfaces of both stars.

For the first time with VW Cep we were able to resolve the individual Mg II h + k emission features from ***each*** component. This shows that ***both*** stars have high levels of chromospheric emission. The two high dispersion LWP spectra obtained in 1989 (LWP 15825H, centered at phase 0.75P, and LWP 15826H, centered at 0.25P) are plotted in Figure 9. Observing the stars near quadrature gives the maximum separation in radial velocity which allows the lines of the two components to be resolved. The exposure time for each spectra was 110 minutes. The synthetic emission line profiles of each star are indicated in Figure 9 along with the position of the barycenter. These doppler-broadened synthetic emission line profiles were computed from the adopted orbital elements and physical parameters of the system. As shown in the figure the larger, more luminous star contributes almost 3.5 times as much Mg II emission as its smaller, less luminous companion. The shifting of the emission features due to the binary motions of the two components is shown by the movement of the line profiles around the wavelength of the barycenter of the binary ($V_o = -8$ km s^{-1}).

There is also what appears to be a sharp, stationary emission feature in both spectra with a radial velocity of ~ -11 km s^{-1}, almost exactly the systemic velocity of VW Cep. This feature is weaker but comparable in strength to the emission from the smaller hotter star of the binary. Also, the feature is present in the four additional high dispersion LWP spectra existing in the archives. The apparent stationary nature of the feature would seem to eliminate its source as coming from a flaring event or active region from the surface of either star. However, we cannot rule out that this feature originates from a region close to the hemisphere of the larger component near the neck connecting the two stars. Since this region overlies the barycenter it would be minimally doppler broadened and radial velocity shifted. The idea of an active region near the neck has been discussed by Vilhu and Heise (1986). They interpreted their x-ray data of VW Cep as enhanced coronal activity on the larger star as well as from additional x-ray activity apparently coming from the neck region.

We have undertaken a preliminary analysis to model the observed high dispersion Mg II emission profiles and the apparent stationary feature. This analysis is summarized in Figure 10 for the k-line of LWP 15826H which was obtained over 0.10P - 0.40P in phase. The top panel of Figure 10 shows the four gaussians used in the modeling. The curves marked "STAR 1" and "STAR 2" correspond to the gaussians which best fit the synthetic profiles previously depicted in Figure 9. At phase 0.25P the more massive, cooler star

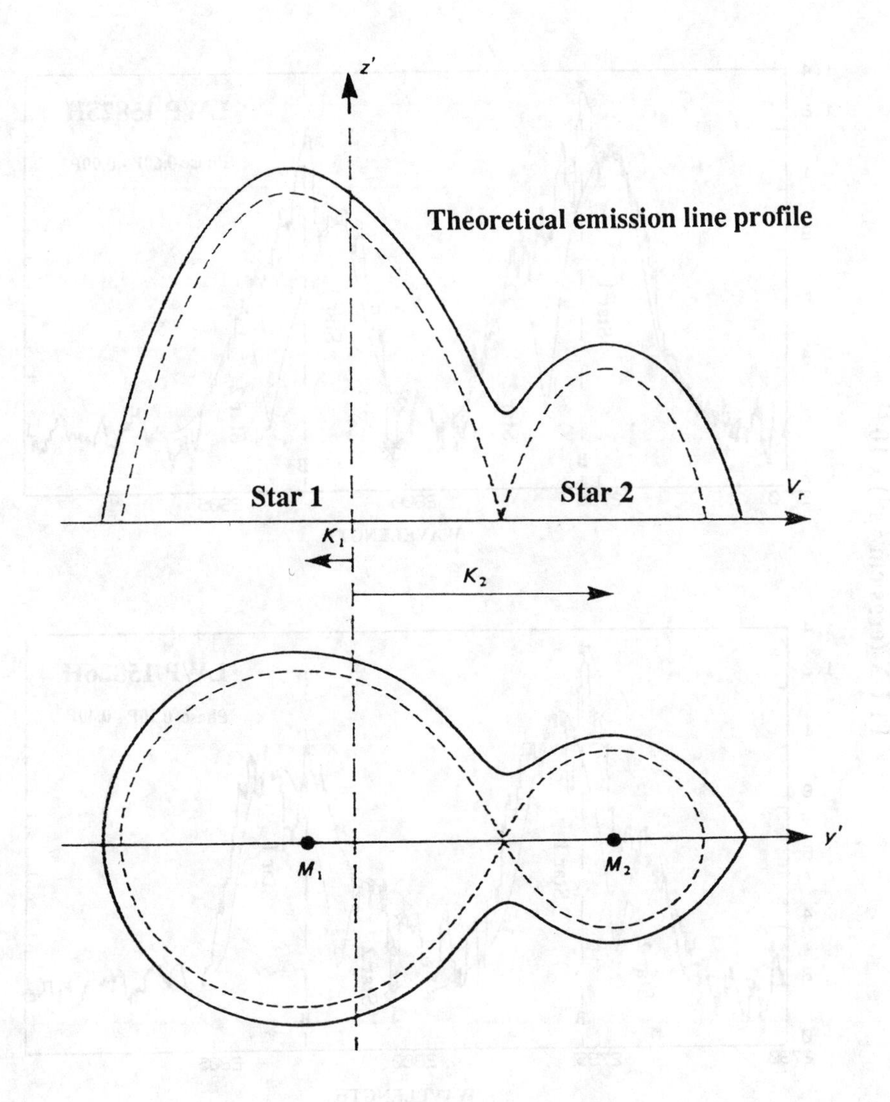

Figure 8. The theoretical emission line profile expected for contact binaries at quadrature. The dotted curves represent the contact configuration and associated profile, the solid curves the overcontact case. The z axis goes through the barycenter. Adapted from Rucinski (1985).

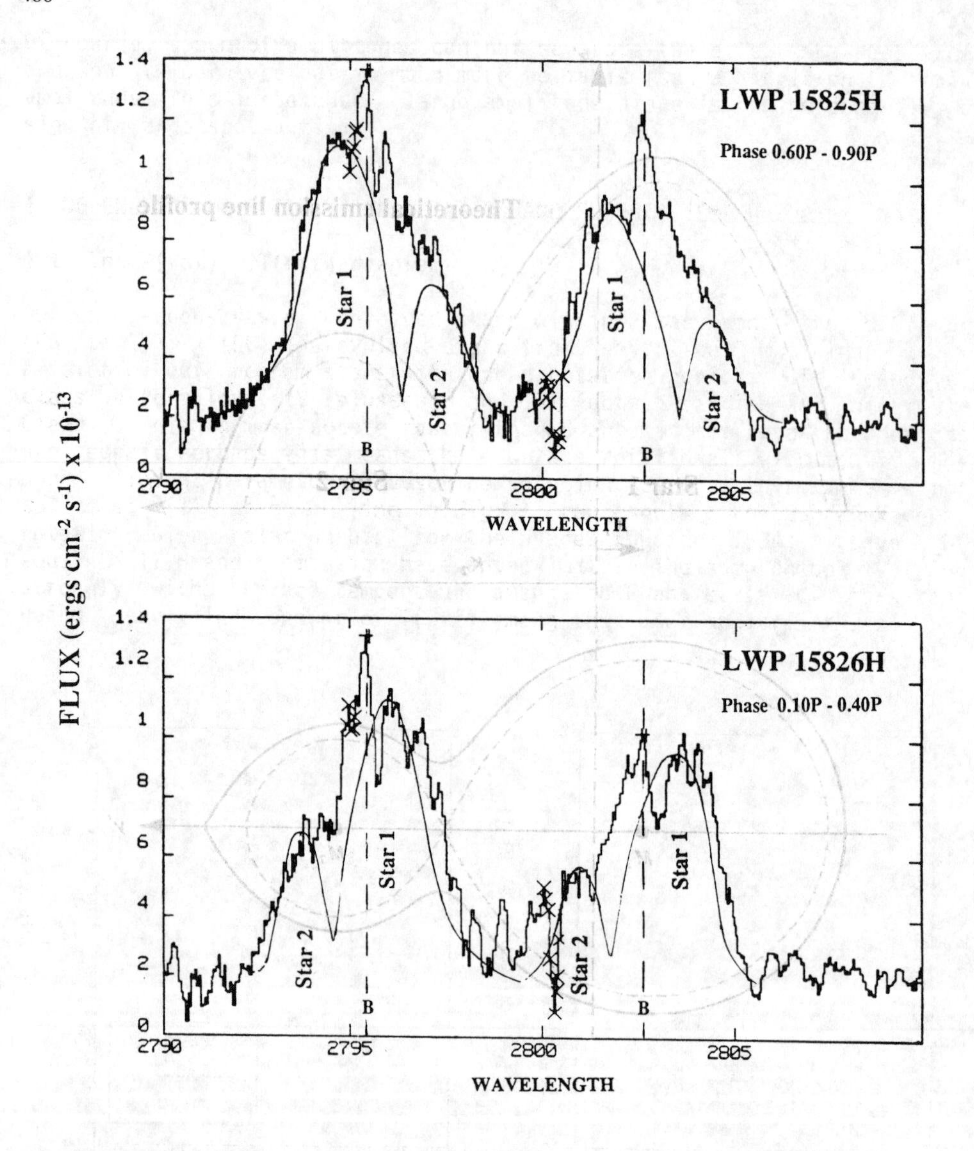

Figure 9. Two high dispersion IUE LWP spectra showing the Mg II h + k features. The top spectra, LWP 15825H, was exposed for 110 minutes covering the phases 0.60P - 0.90P. The bottom spectra, LWP 15826H, had the same exposure time but covered phases 0.10P - 0.40P. The smooth curves designated Star 1 and Star 2 represent the synthetic line profiles calculated from the light curve models integrated over these same phase intervals. The dotted lines marked B are the position of the barycenter of the system. Note the excess emission peaking near the barycenter. This emission could arise from the neck region of the binary or from circumbinary gas as discussed in the text.

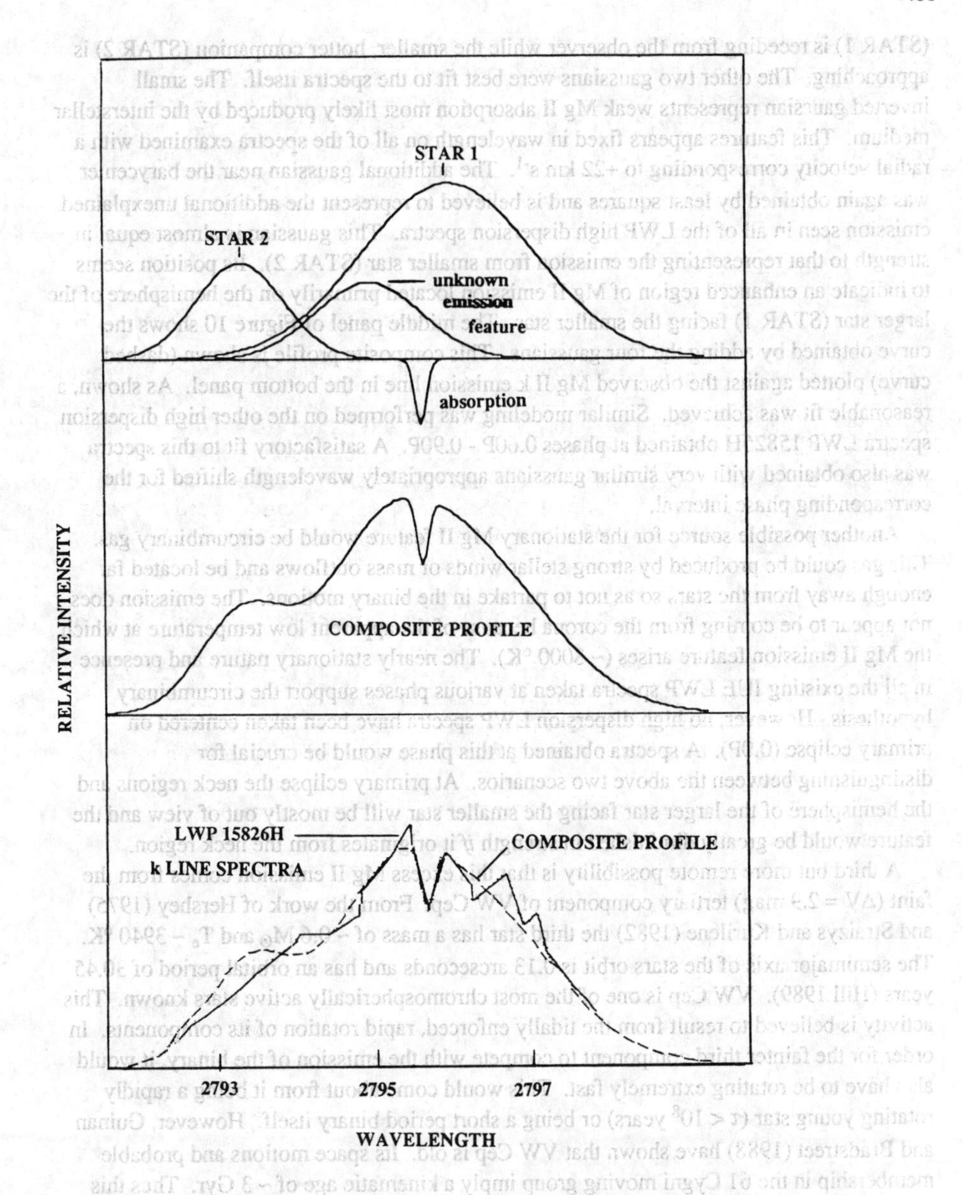

Figure 10. The top panel shows the four gaussians used to model the Mg II k line in LWP 15826H. The curves marked STAR 1 and STAR 2 represent the best fit gaussians to the synthetic line profiles given in Figure 9. The middle panel shows the result of adding the four gaussians. The bottom panel shows the composite profile (dashed curve) compared to the actual Mg II k line spectra (solid curve).

(STAR 1) is receding from the observer while the smaller, hotter companion (STAR 2) is approaching. The other two gaussians were best fit to the spectra itself. The small inverted gaussian represents weak Mg II absorption most likely produced by the interstellar medium. This features appears fixed in wavelength on all of the spectra examined with a radial velocity corresponding to +22 km s^{-1}. The additional gaussian near the barycenter was again obtained by least squares and is believed to represent the additional unexplained emission seen in all of the LWP high dispersion spectra. This gaussian is almost equal in strength to that representing the emission from smaller star (STAR 2). Its position seems to indicate an enhanced region of Mg II emission located primarily on the hemisphere of the larger star (STAR 1) facing the smaller star. The middle panel of Figure 10 shows the curve obtained by adding the four gaussians. This composite profile is shown (dashed curve) plotted against the observed Mg II k emission line in the bottom panel. As shown, a reasonable fit was achieved. Similar modeling was performed on the other high dispersion spectra LWP 15825H obtained at phases 0.60P - 0.90P. A satisfactory fit to this spectra was also obtained with very similar gaussians appropriately wavelength shifted for the corresponding phase interval.

Another possible source for the stationary Mg II feature would be circumbinary gas. This gas could be produced by strong stellar winds or mass outflows and be located far enough away from the stars so as not to partake in the binary motions. The emission does not appear to be coming from the corona because of the apparent low temperature at which the Mg II emission feature arises (~ 8000 °K). The nearly stationary nature and presence in all the existing IUE LWP spectra taken at various phases support the circumbinary hypothesis. However, no high dispersion LWP spectra have been taken centered on primary eclipse (0.0P). A spectra obtained at this phase would be crucial for distinguishing between the above two scenarios. At primary eclipse the neck regions and the hemisphere of the larger star facing the smaller star will be mostly out of view and the feature would be greatly diminished in strength *if* it originates from the neck region.

A third but more remote possibility is that this excess Mg II emission comes from the faint ($\Delta V = 2.9$ mag) tertiary component of VW Cep. From the work of Hershey (1975) and Straizys and Kurilene (1982) the third star has a mass of ~ 0.6 $M_{\odot}$ and $T_e \sim 3940$ °K. The semimajor axis of the stars orbit is 0.13 arcseconds and has an orbital period of 30.45 years (Hill 1989). VW Cep is one of the most chromospherically active stars known. This activity is believed to result from the tidally enforced, rapid rotation of its components. In order for the fainter third component to compete with the emission of the binary, it would also have to be rotating extremely fast. This would come about from it being a rapidly rotating young star ($\tau < 10^8$ years) or being a short period binary itself. However, Guinan and Bradstreet (1988) have shown that VW Cep is old. Its space motions and probable membership in the 61 Cygni moving group imply a kinematic age of ~ 3 Gyr. Thus this seems to rule out that the tertiary companion is a young and active star. We cannot rule out that the third component is itself a short period binary, but there is no evidence supporting this unlikely possibility.

4. Conclusions

Although W UMa systems have been studied extensively in the past, much remains to be learned concerning their origin, evolution, and intrinsic variability. VW Cep, perhaps the most intensively studied contact binary, continues to mystify us with its ever-changing variations. It seems that VW Cep supports the adage that the more we observe a star, the more we need to observe it. Contact binaries have typically been chosen as short-term observing projects because of their short periods. Our work demonstrates the need for sustained, systematic observations of W UMa binaries over many years. Individual multiwavelength light curves should be secured in the shortest time possible. Light curves accumulated from observations spread over several weeks or months will tend to blur and conceal short-term intrinsic variations which may be important in interpreting evidence of stellar activity such as starspots and flares. Hopefully, continued ground-based monitoring coupled with IR, UV and X-ray satellite observations will provide enough clues to finally unravel the dilemmas of these perplexing binaries.

This work was partially supported by grants from NASA, NAG 5-382 and NAG 5-982, which we gratefully acknowledge. We also wish to thank the local organizing committee, especially Dr. Cafer İbanoğlu, for their wonderful hospitality and exceptional planning which produced the memorable NATO-ASI meeting on Active Close Binaries, held in Kuşadasi, Turkey. We also wish to thank Brian Deeney of Villanova University for helping to analyze IUE spectra. DHB thanks Eastern College for support and release time to attend the meeting.

REFERENCES

Bradstreet, D.H. 1985. *Ap.J. Suppl.* **58**, 413.

Bradstreet, D.H. and Guinan, E.F. 1988. *In A Decade of UV Astronomy with the IUE Satellite*, ESA SP-281, p. 303.

Budding, E. 1982. In *Binary Stars as Tracers of Stellar Evolution* eds. Z. Kopal and J. Rahe; (Reidel: Dordrecht) p. 351.

Cruddace R.F. and Dupree, A.K. 1984. *Ap.J.* **277**, 263.

Dorren, J.D. and Guinan, E.F. 1990. *Ap.J.* , **348**, 703.

Eaton, J.A. 1983. *Ap.J.* **268**, 800.

Foukal, P. and Lean, J. 1988. *Ap.J.* **328**, 347.

Gilman, P.A. 1980. In IAU Colloq. 51, *Stellar Turbulence,* eds. D. Gray and J.L. Linsky (Springer: New York), p. 19.

Guinan, E.F. and Bradstreet, D.H. 1988. In *Formation and Evolution of Low Mass Stars* ed. A.K. Dupree and M.T. Lago (Reidel: Dordrecht), NATO ASI, p. 345.

Hershey, J.L. 1975. *Astron. J.*, **80**, 662.

Hill, G. 1989. *Astron. Astrophys.*, **218**, 141.

Kwee, K.K. 1966. *Bull. Astron. Inst. Netherlands Suppl.* **1**, 245; 265.

Parker, E.N. 1986. In *Lecture Notes in Physics*, No. 254, "Cool Stars, Stellar Systems, and the Sun", ed. M. Zeilik and D.M. Gigson (Springer-Verlag: New York), p. 341.

Rucinski, S. 1985. *In Interacting Binary Stars* eds. J.E. Pringle and R.A. Wade (Cambridge), p. 85.

Rucinski, S. and Vilhu, O. 1983, *Mon. Not. R.A.S.*, **202**, 1221.

Straizys, V. and Karilene, G. 1982. *Astrophys. Sp. Sc.*, **80**, 353.

Vilhu, O. and Heise, J. 1986. *Ap. J.* , **311**, 937.

Yamasaki, A. 1982. *Astrophys. Sp. Sci.* **85**, 43.

SYNOPTIC DOPPLER IMAGING AND PHOTOMETRY OF SPOTTED CHROMOSPHERICALLY ACTIVE STARS

K.G.Strassmeier
Department of Physics and Astronomy, Vanderbilt University, Nashville, TN 37235, U.S.A.; and
Center of Excellence in Information Systems, Tennessee State University, Nashville, TN 37209-1561, U.S.A.

ABSTRACT. We present an observational program for studying stellar surface features on chromospherically active (CA) late-type stars. Synoptic observations are currently conducted for the RS CVn binary HD 26337 = EI Eri. Other stars will be added in the near future. Our first photospheric Doppler maps show a polar spot. Our broad-band photometry detected a long-term variability of the mean light level, occasional rapid light curve changes on a time scale of only a few rotation cycles (approximately a week), and some evidence for differential rotation ten times smaller than on the Sun. Phase-dependent variations in the Mg II k line profiles will be used to map the chromospheric structure as well.

1. INTRODUCTION

We have initiated a long-term international campaign of simultaneous multicolor photometry and Doppler imaging to monitor stellar surface features on the G5 IV star HD 26337. Our goal is to track a starspot as it develops, obtaining 2-3 Doppler maps a year accompanied by broad-band photometry and high-resolution spectroscopy of proxy magnetic field indicators. Collaborating in this campaign are G. Cutispoto (U. Catania), A. Hatzes (U. Texas), J. E. Neff (GSFC), J. Rice (Brandon U.), M. Rodono (U. Catania), I. Tuominen (U. Helsinki), S. S. Vogt (Lick), W. H. Wehlau (U. Western Ontario), and myself. A first paper is currently in press (Strassmeier 1990).

2. PHOTOMETRIC SPOT MODELING

V photometry of HD 26337 exists since the discovery of its light variability in 1979 (see Hall et al. 1987). The star is currently one of the many targets of the Vanderbilt University APT (Automatic Photoelectric Telescope) jointly operated with Tennessee State University on Mt. Hopkins, Arizona.

C. İbanoğlu (ed.), Active Close Binaries, 485–491.

2.1. The unspotted magnitude

Any light curve modeling requires knowledge of the unspotted ("immaculate") brightness of the star, a quantity which, together with the spot temperature, controls the total spotted area and ironically is hard to determine and physically not really understood. We just don't know how the energy blocked by a spot or spot group as large as 40% of a stellar hemisphere will be redistributed in the unperturbed photosphere. Very likely this "unspotted" magnitude is not a constant value but changes gradually with time and I would not be surprised if the time scale involved is comparable to a star's "activity" cycle.

So it is common to choose the brightest observed magnitude and assume it to be the unspotted magnitude. Very obviously this is dangerous because (1) there might be still spots on the surface, and (2) one might have missed the maximum brightness due to too short a baseline in time. In the case of HD 26337 we have 10 years of continuous photometry (see section 2.2.) and, due to a lucky coincidence, the brightest magnitude ever observed occured during our simultaneous Doppler imaging observations in Dec./Jan. 1987/88. The light curve solution, however, agreed only qualitatively with the Doppler imaging solution: the line profiles were filled in at all phases while the overall V brightness had a maximum value. Solving first for the line profile variations and then computing a theoretical light curve showed that the observed mean V light level was fainter by 0.057 mag (twice the photometric amplitude at that particular time). Obviously the star was still spotted though it was in its brightest stage ever observed. We then adjusted this "least spotted" magnitude by the amount of 0.057 mag to obtain a better estimate of the "unspotted" magnitude. This is an important result for stellar spot modeling: although we had 10 years of data we still underestimated the unspotted magnitude by an amount which changed the generated spot models considerably.

2.2. Long-term behavior

Figure 1 summarizes our findings. Altogether 25 light curves were used in the analysis:

- o There is a secular trend in the mean V-light level. So far the pattern has not repeated. A likely explanation is the stellar analogon of the solar 11-year spot cycle. A period, if one exists, must be larger than the baseline of our photometry, i.e., 10 years.
- o We observe migration rates of the spots on the surface between -3.0 and +1.7 deg/day yielding dw/w=0.02, a factor ten smaller than on the Sun.
- o There seems to be no conclusive evidence for the existence of prefered longitude positions or sector structure of the spots on HD 26337.
- o The total spotted area might undergo cyclic variations of approximately 3-5 years.

3. DOPPLER IMAGING OF HD 26337

3.1. Independent application of different codes

Twenty-one spectra were obtained at KPNO with the coudé feed telescope and at NSO with the McMath solar telescope within twenty nights in Dec./Jan. 1987/88. Surface maps were derived independently by two groups using different techniques: by Strassmeier (1990) using the "iterative" technique (Strassmeier 1988), and by Rice and Wehlau (1989, private communication) using the "inversion" technique (Rice, Wehlau, and Khokhlova 1989). Both computer codes produced more or less the same map (Figure 2). The spot configuration consists of a spot situated directly at the pole with several attached small bumps descending towards lower latitudes. These preliminary results are very promising. Two additional maps using the same data but different regularization forms in the inversion process are currently being produced independently by Illka Tuominen and Steve Vogt.
Meanwhile we have data for two additional images from the 1988-89 observing season and further observations are scheduled for fall 1989 and early 1990.

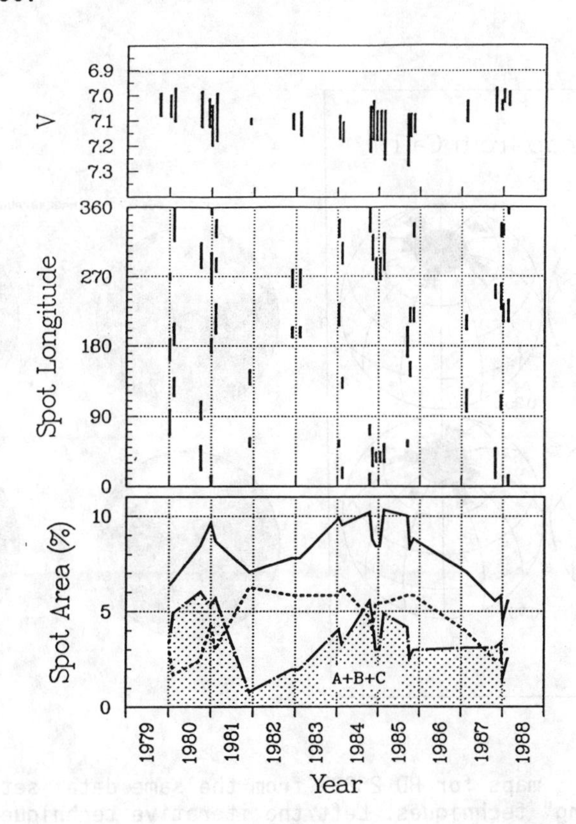

Fig. 1: Long-term behavior of (from top to bottom) mean V brightness, spot longitude, and spotted area for HD 26337 within the past ten years.

3.2. How secure is the identification of a polar spot on HD 26337?

Figure 3 is a simple demonstration of this. Here we compare three theoretical line profiles with an observed Ca I 6439 A profile (plusses). Profile "A" is due to the circular dark polar spot, profile "B" due to a circumpolar dark ring, and profile "C" due to an equatorial dark ring. If the line-profile cores are filled in at <u>all</u> rotation phases, profile "A" is the only possible fit to the data. The dotted profile in Fig. 3 is the undistorted profile, i.e., a profile with no spots. In fact this is an observed μ Her (also G5 IV, <u>vsini</u>=1.2 km/s, macroturbulence broadening = 4.5 km/s, Gray and Nagar 1985) Ca I 6439 A profile spun up to match <u>vsini</u> of HD 26337. Thus the Doppler map obtained by Strassmeier (1990) is based on a "null" star convolution technique rather than on disk integration of local theoretical line flux profiles from model atmospheres. In the case of HD 26337, ignoring the possible difference between the equivalent width in the spot and in the surrounding photosphere is not a particular problem because the spots are very cool (dT=1800 K). The contribution to the intrinsic line flux is therefore neglectable.

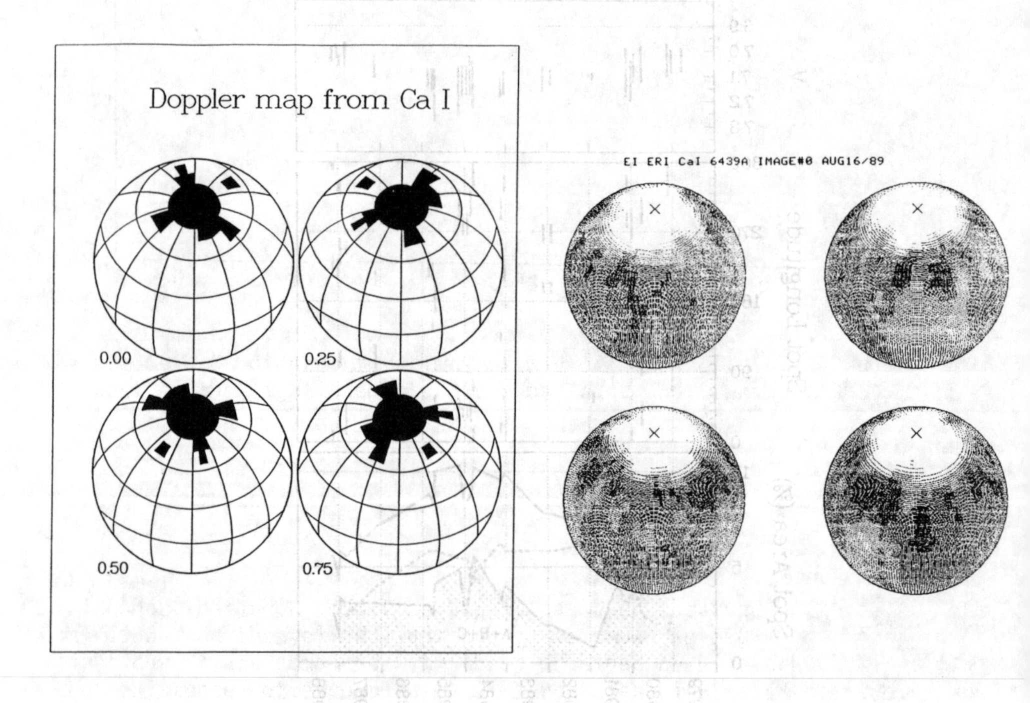

<u>Fig. 2</u>: Doppler maps for HD 26337 from the same data set but using different "imaging" techniques. Left the iterative technique, right the inversion technique. Note that the direction of rotation is different for the two maps and also that the right map is a "negative" plot, i.e., the white area is actually the dark spot, a.s.o..

Secondary evidence for the existence of a polar spot comes from photometry. By assuming the brightest observed magnitude as the unspotted magnitude (see section 2.1.) we would have been forced to introduce either a band of spots (e.g. "C" in Fig. 3), as Bopp and Noah (1980) did for II Peg, or a circumpolar or polar spot in order to keep up with the low light level observed during 1984-85. Clearly a band of spots or any other homogeneous distribution of little spots all over the surface can not explain the line-profile core filling.

4. ROTATIONAL MODULATION OF PROXY INDICATORS

4.1. Mg II k line fluxes

Neff et al. (1989) obtained 24 well-exposed high-resolution spectra of the Mg II k emission line during a four day period in September 1988, i.e., within two consecutive rotation cycles. From the first round of fits, they found that there are no systematic asymmetries in the k-line profiles, in agreement with a contemporaneous ground-based almost "flat" V-light curve. While the position and width of the line is relatively constant, its flux varies by 20% but does not correlate with either the far-ultraviolet fluxes or the visible light (Neff et al. 1990).

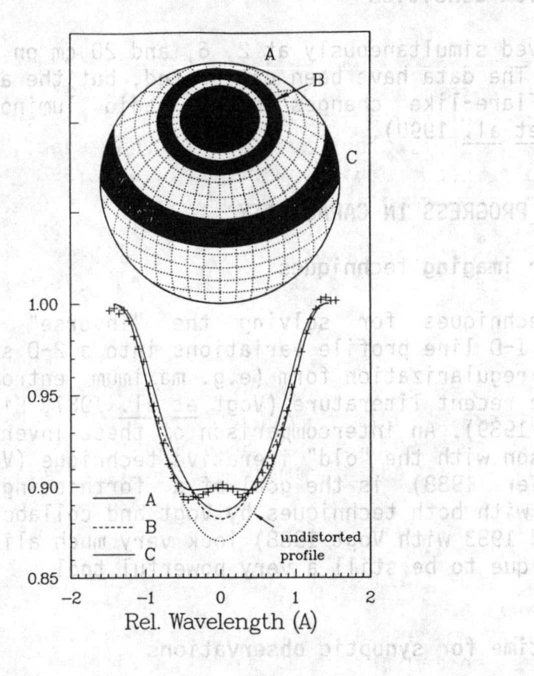

Fig. 3: A comparison of three theoretical line profiles (A, B, and C) with an observed profile (plusses) of HD 26337. The dotted line is an observed profile of μ Her spun up to match the rotational velocity of HD 26337. A spot situated at the pole (profile "A") is apparently the best representation of the observation.

4.2. Ca II H and K line fluxes

These observations were made simultaneous to our Doppler imaging observations in late 1987. Only one half of a rotation cycle has been covered. Thus no definite statement with respect to rotational modulation can be made. However, we note that the flux at phase 0 was about 50% larger than at phase 0.5 measured the night before. Note also that the simultaneously observed light curve had a minimum around phase 0. This is good evidence that plage-like regions overlay spotted regions.

4.3. H-alpha core emission

Four H-alpha spectra obtained in 1983 (Fekel, private communication) on four consecutive nights showed a very weak, almost absent, absorption feature. Core-emission equivalent width changes of up to 100% were observed from night to night. Clearly the H-alpha line of HD 26337 deserves more attention especially in relation to simultaneous Doppler imaging from photospheric indicators.

4.4 VLA radio flux densities

Jim Neff observed simultaneously at 2, 6, and 20 cm on 16, 18, and 19 September 1988. The data have been calibrated, but the analysis is still in progress. Flare-like changes in the radio luminosity have been observed (Neff *et al.* 1990).

5. FUTURE WORK, PROGRESS IN CAPABILITY

5.1. New Doppler imaging techniques

Several new techniques for solving the "inverse" problem, i.e., converting the 1-D line profile variations into a 2-D surface map with some sort of regularization form (e.g. maximum entropy), have been published in the recent literature (Vogt *et al.* 1987, Rice *et al.* 1989, Piskunov *et al.* 1989). An intercomparison of these inversion techniques and a comparison with the "old" iterative technique (Vogt and Penrod 1983, Strassmeier 1988) is the goal of a forthcoming paper. Recent images obtained with both techniques by Vogt and collaborators (compare Vogt and Penrod 1983 with Vogt 1988) look very much alike and proofed the "old" technique to be still a very powerful tool.

5.2. Telescope time for synoptic observations

Understanding of the nature and behavior of stellar active regions requires long-term high-resolution high-S/N spectroscopy along with multicolor photometry. APTs are an ideal design for this task. Spectroscopy, on the other hand, is still nightly work for at least one astronomer. In the recent years, however, spectroscopic observations were made possible by the National Solar Observatory (NSO) McMath-CCD synoptic night-time program at Kitt Peak. I would like to use the possibility here to thank NSO and especially the resident observer, Paul Avellar, for the good work.

5.3. Other stars for Doppler imaging

Several spotted stars are currently Doppler imaged: HR 1099, UX Ari, HD 199178, HD 32198, and of course HD 26337. With bigger telescopes we could observe down to magnitude 10, maybe 11, or by increasing the instrumental resolution, image brighter stars with lower vsini, say 25 km/s. Several test spectra with the 0.9 m coudé feed telescope of low vsini stars at a resolution of only 40,000 show promising line profile distortions due to spots, e.g., for HD 106225 and DM UMa. Others gave negative results, e.g., for HD 108102.

This research was partially supported by NASA grant NAG 8-111 to Tennessee State University.

REFERENCES

Bopp, B. W., and Noah, P. V. 1980, Pub. A.S.P., **92**, 717.
Gray, D. F., and Nagar, P. 1985, Ap. J., **298**, 756.
Hall, D. S., Osborn, S. A. G., Seufert, E. R., Boyd, L. J., Genet, R. M., and Fried, R. E. 1987, A. J., **94**, 723.
Neff, J. E., Strassmeier, K. G., Rodono, M., and Cutispoto, G. 1990, Astr. Ap., to be submitted.
Neff, J. E., Walter, F. M., Skinner, S. L., Brown, A., Strassmeier, K. G., Rodono, M., Cutispoto, G., Jankov, S., and Char, S. 1989, Bull. A.A.S., **21**, 747.
Piskunov, N. E., Tuominen, I., and Vilhu, O. 1989, Astr. Ap., submitted.
Rice, J. B., Wehlau, W. H., and Khokhlova, V. L. 1989, Astr. Ap., **208**, 179.
Strassmeier, K. G. 1988, Ap. Space Sci., **140**, 223.
Strassmeier, K. G. 1990, Ap. J., in press (January issue).
Vogt, S. S. 1988, in IAU Symp. 132, The Impact of Very High S/N Spectroscopy on Stellar Physics, ed. G. Cayrel de Strobel and M. Spite (Dordrecht: Kluwer), p. 253.
Vogt, S. S., and Penrod, G. D. 1983, Pub. A.S.P., **95**, 565.
Vogt, S. S., Penrod, G. D., and Hatzes, A. P. 1987, Ap. J., **321**, 496.

INFRARED STUDIES ON ACTIVE BINARIES AND CIRCUMSTELLAR MATTER

F. SCALTRITI
Osservatorio Astronomico di Torino
Strada Osservatorio 20
I-10025 Pino Torinese (Torino) - Italia

ABSTRACT. The infrared properties of RS CVn-type systems are reviewed. The following points are examined: a) amplitude of the wave-like distortion, b) spectrophotometry, c) infrared excesses. In particular, point c) has been explored in some detail owing to its relevance as far as the evolutionary scenario of RS CVn's is concerned.

1. INTRODUCTION

In the last years a remarkable interest of the astronomical community has been devoted to chromospheric active late-type stars, and among them to RS CVn-type variables. This fact allowed to recognize that solar-like phenomena may be present in stars whose spectral type is similar to the solar one. It is known that the activity manifests itself in a wide range of wavelengths, from UV to radio bands, involving the whole atmosphere of the star; in particular, at photospheric level a simple tracer of the emergence of the buoyant magnetic flux tubes into dark spots is the luminosity changes (caused by the perturbed regions) which are modulated by differential rotation and meridional circulation, giving rise to wave-like fluctuations in the lightcurves and long-term cycles of photospheric variability (see e.g. Rodono' 1981, Catalano 1983, Busso et al. 1986). The features of the distortion wave and its temporal evolution have been followed in time by means of an extensive amount of optical photometry which allowed also to develop models concerning either spot temperatures and location and size of the active regions responsible of the observed fluctuations.

In what follows the known properties of RS CVn-type binaries in the infrared are briefly reviewed, extending the information up to IRAS bands, when possible. Moreover, an outline of the most recent findings concerning the possible presence of cool matter in those systems is given.

C. İbanoğlu (ed.), Active Close Binaries, 493–499.

2. INFRARED PROPERTIES OF RS CVN'S

At present a large amount of data based on infrared (IR) photometry of active binaries exists; full-cycle lightcurves or surveys on several objects have been performed. According to the findings of various authors, we have got IR information about: a) wave amplitude, b) spectrophotometry, c) (possible) excesses. Let us comment on each of the raised points; the analysis is based on the literature I was able to collect. I apologize for possible omissions.

2.1 Wave amplitude

Even from photometry in the optical we know that generally the amplitude of the fluctuation diminishes from U to I bands; for example (but other systems can be found showing the same characteristics) ESO unpublished observations, carried out by the author in April 1985, show that the amplitudes in U and I are 0.12 and 0.05 mag, 0.06 and 0.02 mag, 0.14 and 0.09 mag for the binaries V824 Ara, TY Pyx, PZ Tel, respectively.

This trend is confirmed (but see also below) if we compare contemporary observations in optical and IR. This is the case for CG Cyg (Bedford et al. 1987), UV Psc (Antonopoulou 1987b, and Vivekananda Rao and Sarma 1983), TY Pyx (Antonopoulou 1983, and Vivekananda Rao et al. 1981), V711 Tau (Antonopoulou and Williams 1980, and Bartolini et al. 1983), HR 7275 (Zeilik et al. 1983, and Fried et al. 1982); for II Peg extensive photometry by Lazaro et al. (1987) from U to L bands shows an amplitude variation from 0.15 to about 0.02 mag.

A contrasting result has been found in UX Ari by Hall et al. (1975): the wave amplitude is larger at L wavelength (0.2 mag) than at U-band (0.1 mag). Moreover, uncertain conclusions can be drawn for AD Cap (Antonopoulou 1987a) and CF Tuc (Antonopoulou 1987a, and Scaltriti et al. (unpublished)).

It has to be taken into account that, if we estimate the fractional contribution of the light originating in the spot to the total light received due to (spot+star) (see Vogt 1981), this fraction is larger at longer wavelengths where the contribution of the spots to the total light comes out to be more and more important. On the other hand, following Berriman et al. (1983), a single spot at a temperature of 2000 K and covering 30% of the surface of a K0 subgiant would make IR colours redder than those of an unspotted star by an amount comparable with the observational error.

2.2 Spectrophotometry

There is a little amount of data on spectrophotometric curves in the infrared. This kind of information is given here just to have a more complete scenario at long wavelengths. Figure 1 shows spectrophotometry of two RS CVn's obtained in the range 1.4-2.5 micron with a Circular Variable Filter at La

Silla (ESO) (Busso et al. (1987); for each binary two trends are plotted, corresponding to the phases of maximum and minimum visibility of the spots. Those spectrophotometric data do not show appreciable and systematic variations in the flux distribution with phase. Though in some spectral bands differences slightly higher than observational error are present, their behaviour is not simple also because they are characterized by opposite directions in different wavelengths. It is clear that more data are needed, particularly on binaries showing larger wave amplitudes than RU Cnc and VV Mon; in any case, these results point out that the influence of spots in the IR is certainly lower than in the optical part of the spectrum, where the presence of modulations in the lightcurve is evident (see also the considerations in 2.1).

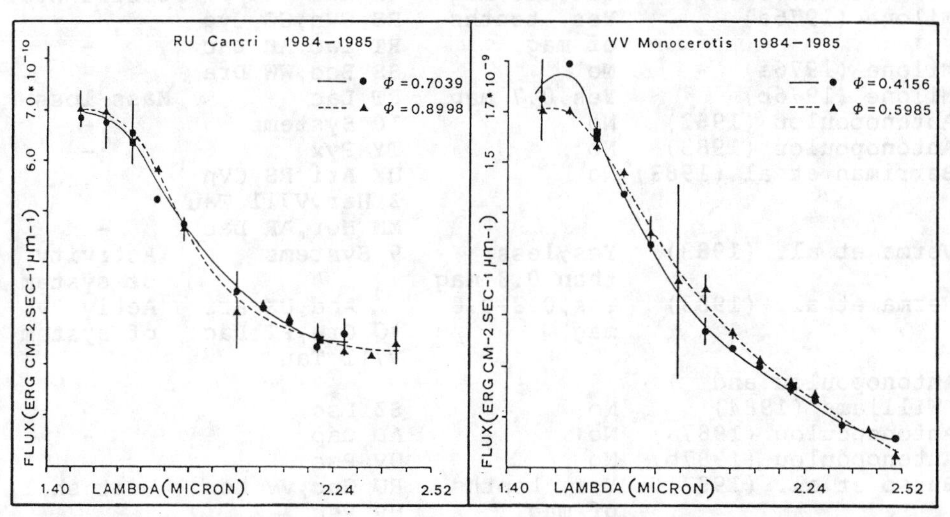

Figure 1 - IR spectrophotometry of RU Cnc and VV Mon at diffe= rent phases. Error bars are shown when appreciable.

2.3 Infrared excesses

In my opinion this is the most important point also for the implications on evolution. In a paper by Berriman et al. (1983) it was shown that the combination of V-band and IR observations is useful to search for possible IR excesses and for classification purposes of the component stars whose spectral types can be uncertain due to blending problems. As far as IR excess is concerned, the situation is confused because some authors do not find significant excess while others argue in favour of its existence. In Table 1 the present situation is summarized, as I could find in the literature. It can be noticed that the excess, when present, extends to the IRAS bands at [12], [25], [60] microns (Verma et al. 1987, Busso et al. 1988). A variety of methods have

Table 1

Summary about Infrared excesses

Reference	IR excess Amount	Binary System	Source of the excess
Atkins and Hall (1972)	Yes,0.5 mag	RS CVn,Z Her AR Lac,SZ Psc	Cooler star
Atkins and Hall (1972)	No	LX Per	-
Hall et al.(1975)	Yes,0.5 mag	UX Ari	Cooler star
Milone (1976a)	Yes, tenths of mag.	RS CVn,CG Cyg RT Lac,AR Lac	-
Milone (1976a)	No	SS Boo,WW Dra	-
Milone (1976b)	Yes,0.7 mag	RT Lac	Mass loss
Antonopoulou (1982)	No	10 Systems	-
Antonopoulou (1983)	No	TY Pyx	-
Berriman et al.(1983)	No	UX Ari,RS CVn Z Her,V711 Tau MM Her,AR Lac	-
Verma et al. (1983)	Yes,less than 0.3 mag	9 Systems	Activity of system
Verma et al. (1983)	Yes,0.3-0.8 mag	λ And,UX Ari σ CrB,RT Lac V711 Tau	Activity of system
Antonopoulou and Williams (1984)	No	SZ Psc	-
Antonopoulou (1987a)	No	AD Cap	-
Antonopoulou (1987b)	No	UV Psc	-
Busso et al. (1987)	Yes, tenths of mag.	RU Cnc,VV Mon UV Psc	Dust shell
Busso et al. (1987)	No	TY Pyx	-
Wang et al. (1987)	Yes	Z Her,V711 Tau UX Ari,SZ Psc	-
Wang et al. (1987)	No	37 Systems	-
Arevalo et al. (1988)	Yes, tenths of mag	ER Vul	Matter in the system
Busso et al. (1988)	Yes,0.2-0.7 mag	CF Tuc	Dust shell
Busso et al. (1988)	No	λ And,UX Ari AR Lac	-
Lazaro (1988)	Yes,0.2 mag in (L-M)	II Peg	Matter in the system
Busso and Scaltriti (1989)	Yes	WY Cnc,GK Hya AR Mon,AR Psc	Dust shell
Busso and Scaltriti (1989)	No	CQ Aur,RZ Cnc DK Dra,RZ Eri DH Leo,GX Lib	-

been employed in order to search for IR excesses; among them we quote: a) the comparison of the expected location in some colour-colour plot according to the known spectral types of the components with that derived from the observations (see, for example, Berriman et al. 1983, Lazaro 1988), b) the comparison between the observed energy distribution (in an interval of wavelengths as wide as possible) with the trend expected from the spectral types (see, for example, Antonopoulou and Williams (1984), Busso et al. (1987,1988)). Owing to the larger number of constraints involved in the method b), we show in Figure 2 the expected and observed flux distributions obtained GK Hya; it is seen that, even allowing a change in the spectral types given in the literature, no agreement can be found. For other systems the energy trend can be simply accounted for by two stellar spectra. According to the analysis performed by Busso et al. (1987,1989), the only possibility in order to reconcile expectation and observations is that of a thin dust shell surrounding the system at a given distance (with physical conditions, T(dust) = 1000-2000 K, optical depth less than 0.01, composition = dirty silicates (Jones and Merril 1976). A similar model has been applied by Lazaro (1988) for II Peg.

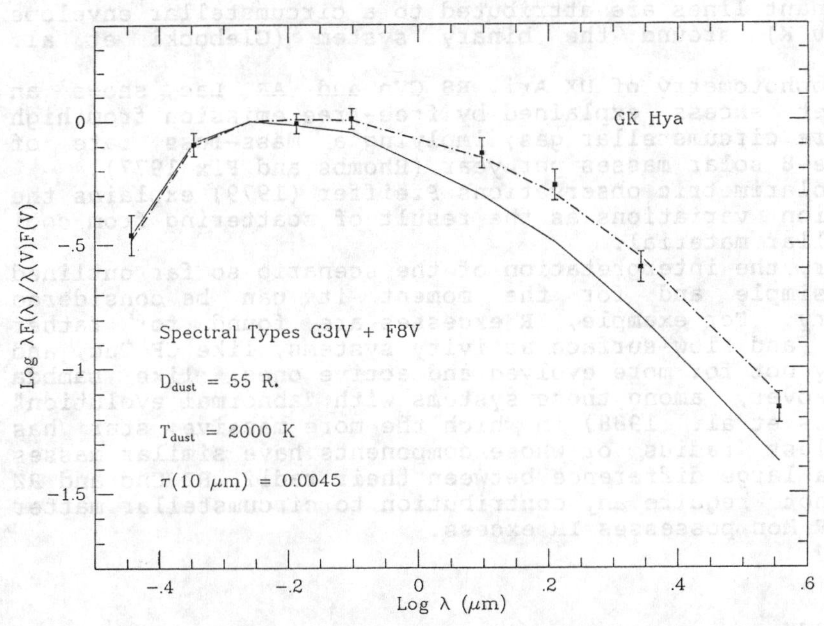

Figure 2 - Observed (squares) and predicted (solid line) energy distribution of GK Hya; the dash-dotted line represents the fit with a thin dust shell.

According to the data listed in Table 1 and some others not yet published (Busso and Scaltriti 1989), among the known RS

CVn-type binaries (about one hundred), we may select roughly 40% of them possessing photometry from ultraviolet to IR (and possibly IRAS bands) so allowing a careful analysis concerning IR excess. We may infer that about 25% of the RS CVn stars show evidence of an excess in the infrared.

3. CONCLUSIONS

Even if the scenario is far from being clarified, the IR excess in RS CVn-type stars is best explained by the presence of some sort of circumstellar matter. This is not surprising because, following the evolutionary scenario outlined by Popper and Ulrich (1977), the activity in RS CVn-type binaries developes when they move through Hertzsprung gap; this phase is accompanied by mild mass exchange (10e-11 solar masses per year) and perhaps mass loss.

On the other hand, also observations on single objects point towards the existence of circumstellar matter. In RT Lac one of the components fills 80-90% of its Roche lobe and H-alpha observations suggest mass loss through the Lagrangian points L1 and L2 and are consistent with intermittent mass transfer (Huenemoerder 1985). For Lambda And asymmetries in emission peaks and relative shifts of h3 and k3 absorption in MgII resonant lines are attributed to a circumstellar envelope (T = 12000 K) around the binary system (Glebocki et al. 1986).

Spectrophotometry of UX Ari, RS CVn and AR Lac shows an ultraviolet excess explained by free-free emission from high temperature circumstellar gas, implying a mass-loss rate of 10e-7, 10e-8 solar masses per year (Rhombs and Fix 1977).

From polarimetric observations Pfeiffer (1979) explains the polarization variations as the result of scattering from cool circumstellar material.

However, the interpretation of the scenario so far outlined is not simple and for the moment it can be considered preliminary. For example, IR excesses are found for rather unevolved and low-surface activity systems, like CF Tuc, and apparently not for more evolved and active ones, like Lambda And; moreover, among those systems with "abnormal evolution" (Montesinos et al. 1988) in which the more massive star has the smallest radius or whose components have similar masses but show a large difference between their radii, RZ Cnc and RZ Eri do not require any contribution to circumstellar matter whereas AR Mon possesses IR excess.

BIBLIOGRAPHY

- Antonopoulou,E.: 1982, Hvar Obs. Bull. 1,55.
- Antonopoulou,E.: 1983, Astron. Astrophys. 120,85.
- Antonopoulou,E.: 1987a, Astron. Astrophys. Suppl. 68,521.
- Antonopoulou,E.: 1987b, Astrophys. Space Sci. 135,335.
- Antonopoulou,E.,and Williams,P.M.: 1980, Astrophys. Space

Sci. 67,469.
- Antonopoulou,E.,and Williams,P.M.: 1984, Astron. Astrophys. 135,61.
- Arevalo,M.J.,Lazaro,C.,and Fuensalida,J.J.: 1988, Astron. J. 96,1061.
- Atkins,H.L.,and Hall,D.S.:1972,Publ.Astr.Soc.Pacific 84,638.
- Bartolini,C. et al.: 1983, Astron. Astrophys. 117,149.
- Bedford,D.K.,Fuensalida,J.J.,and Arevalo,M.J.: 1987, Astron. Astrophys. 182,264.
- Berriman,G.,De Campli,W.M.,Werner,M.W.,and Hatchett,S.P.: 1983, Mon. Not. R. astr. Soc. 205,859.
- Busso,M.,Scaltriti,F.,and Cellino,A.: 1986, Astron. Astrophys. 156,106.
- Busso,M.,Scaltriti,F.,Persi,P.,Robberto,M.,and Silvestro, G.: 1987, Astron. Astrophys. 183,83.
- Busso,M.,Scaltriti,F.,Persi,P.,Ferrari-Toniolo,M.,and Origlia,L.: 1988, Mon. Not. R. astr. Soc. 234,445.
- Busso,M.,and Scaltriti,F.: 1989, in preparation.
- Catalano,S.: 1983, in Activity in Red-dwarf Stars, eds. P.B. Byrne and M. Rodono', Reidel, Dordrecht, p. 343.
- Fried,R.E.,et al.: 1982, Astrophys. Space Sci. 82,181.
- Glebocki,R.,Sikorski,J.,Bielicz,E.,and Krogulec,M.: 1986, Astron. Astrophys. 158,392.
- Hall,D.S.,Montle,R.E.,and Atkins,H.L.:1975,Acta Astr.25,125.
- Huenemoerder,D.P.: 1985, Astron. J. 90,499.
- Jones,T.W.,and Merril,K.M.: 1976, Astrophys. J. 209,508.
- Lazaro,C.,Arevalo,M.J.,and Fuensalida,J.J.: 1987, Astrophys. Space Sci. 134,347.
- Lazaro,C.: 1988, Astron. Astrophys. 193,95.
- Milone,E.F.: 1976a, in Multiple Periodic Variable Stars, IAU Colloquium no. 29, ed. W.F. Fitch, Academic Press.
- Milone,E.F.: 1976b, Astrophys. J. Suppl. 31,93.
- Montesinos,B.,Gimenez,A.,Fernandez-Figueroa,M.J.: 1988, Mon. Not. R. astr. Soc. 232,361.
- Pfeiffer,R.J.: 1979, Astrophys. J. 232,181.
- Popper,D.M.,and Ulrich,R.K.: 1977, Astrophys.J. 212,L131.
- Rhombs,C.G.,and Fix,J.D.: 1977, Astrophys. J. 216,503.
- Rodono',M.: 1981, in Photometric and Spectroscopic Binary Systems, eds. E.B. Carling and Z. Kopal, Reidel, Dordrecht, p. 285.
- Verma,R.P.,Ghosh,S.K.,Iyengar,K.V.K.,Regarajan,T.N., Tandon,S.N.,and Daniel,R.R.:1983,Astrophys.Space Sci.97,161.
- Verma,R.P.,Iyengar,K.V.K.,and Rengarajan,T.N.: 1987, Astron. Astrophys. 177,346.
- Vivekananda Rao,P.,and Sarma,M.B.K.:1981,Acta Astron.31,107.
- Vivekananda Rao,P.,and Sarma,M.B.K.: 1983, J. Astrophys. Astr. 4,161.
- Vogt,S.S.: 1981, Astrophys. J. 20,327.
- Wang Gang, Hu Jing-yao, Qian Zhong-yu,and Zhou Xu: 1987, Chin. Astron. Astrophys. 11,328.
- Zeilik,M.,Elston,R.,Henson,G.,and Smith,P.: 1983, Inf. Bull. Var. Stars, no. 2333.

Infrared photometry of RS CVn short-period systems

M.J. Arévalo and C. Lázaro

Instituto de Astrofísica de Canarias.
38200 La Laguna Tenerife. SPAIN

We are carrying out an infrared monitoring program of RS CVn systems devoting special attention to the short-period group. The observations are made with the 1.5 m C.S. telescope at the Observatorio de Izaña (Tenerife, Canary Islands) using an infrared single-channel photometer with an InSB detector cooled with liquid nitrogen. The system is compatible with Johnson (Arribas and Martinez–Roger, 1986).

We report here completed light curves in the J and K filters of five of the eight systems currently included as components of the RS CVn short-period group. (see Milano 1981, Milano et al 1986 or Budding and Zeilik 1987 for a review of its main properties). For all of them we have studied the infrared colours in order to obtain the possible IR excess. We have made also a new determination of the geometrical elements using the program developed by Budding and Zeilik (1987). If the observed irregularities in its light curves are interpreted as due to stars spots we can expect its contribution at longer wavelengths to be lower. In this way the resulting parameters from the IR light curves could improve previous solutions inferred from visible photometry. In the following sections we summarize the results for the observed systems.

WY Cnc

We observed this system during April 1989 (Arévalo and Lázaro, 1989a). From the analysis of the IR colours we obtained J-K=0.59 for levels out of eclipses J-K=0.53 at secondary eclipse and J-K=0.58 at the primary eclipse. If we adopt the spectroscopic classification G5V+M0V for the systems components (Awadalla and Budding, 1979), as the secondary is a total eclipse, we derive a J-K colour for the primary component redder by 0.14 mag than the expected for its spectral class (Koornneef, 1983). Out of eclipse colours are redder by 0.16 mag than the combination of both component. Scaltriti (this ASI) has also confirmed the existence of IR excess in WY Cnc.

Light curve solutions (Figure 1 and Table I) give geometrical elements close to those obtained from visible photometry by Awadalla and Budding (1979) or Naftilan (1987).

XY UMa

Unfortunately the primary minimum is not covered in the 1988-89 observations of XY UMa (Arévalo and Lázaro, 1989a), therefore we cannot improve previous geometrical

C. İbanoğlu (ed.), Active Close Binaries, 501–508.

TABLE I. Infrared light curves solutions of WY Cnc

J Filter	K Filter
r_1=0.230±0.002	r_1=0.257±0.002
r_2=0.138±0.002	r_2=0.148±0.002
i=86 (Fixed)	i=86 (Fixed)
L_1=0.843±0.003	L_1=0.804±0.003
L_2=0.129±0.003	L_2=0.156±0.003
χ^2= 249	χ^2= 672
N=94	N=102

TABLE II. Infrared light curves solutions of XY UMa

J Filter	K Filter	J Filter
r_1=0.339±0.002	r_1=0.341±0.001	r_1=0.358±0.001
r_2=0.166±0.001	r_2=0.172±0.001	r_2=0.201±0.001
i=88.2 (Fixed)	i=88.2 (Fixed)	i=77.2 (Fixed)
L_1=0.853±0.002	L_1=0.844±0.001	L_1=0.814±0.003
L_2=0.094±0.002	L_2=0.122±0.001	L_2=0.127±0.002
χ^2= 857	χ^2= 465	χ^2= 831
N=156	N=157	N=156

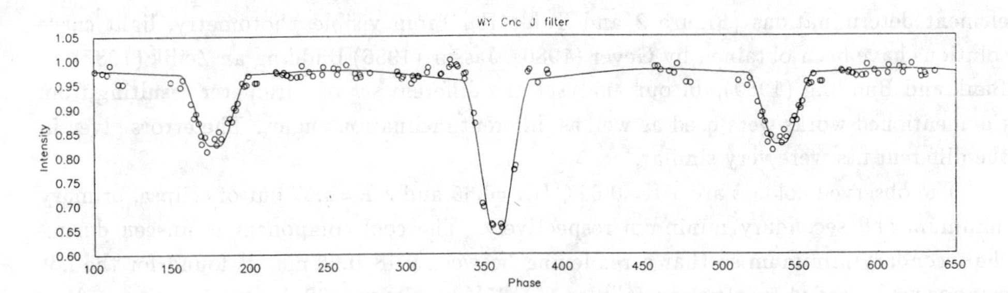
WY Cnc J filter
Intensity
Phase

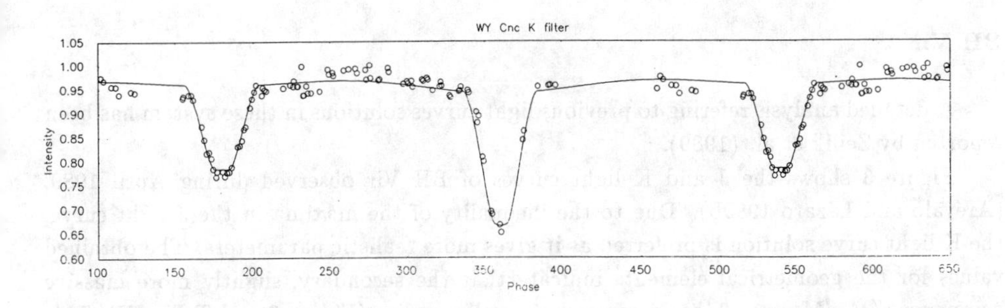
WY Cnc K filter
Intensity
Phase

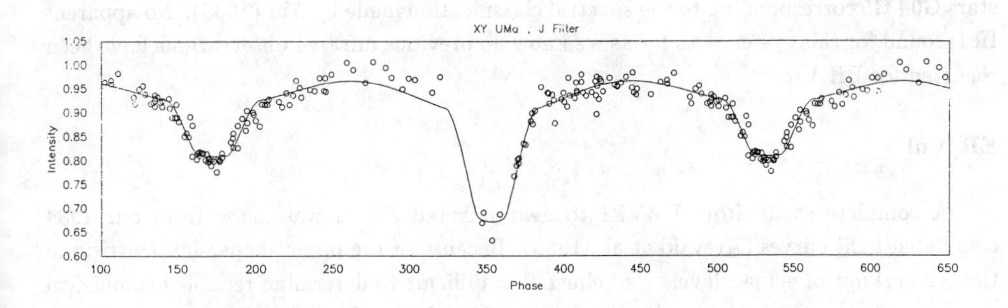
XY UMa , J Filter
Intensity
Phase

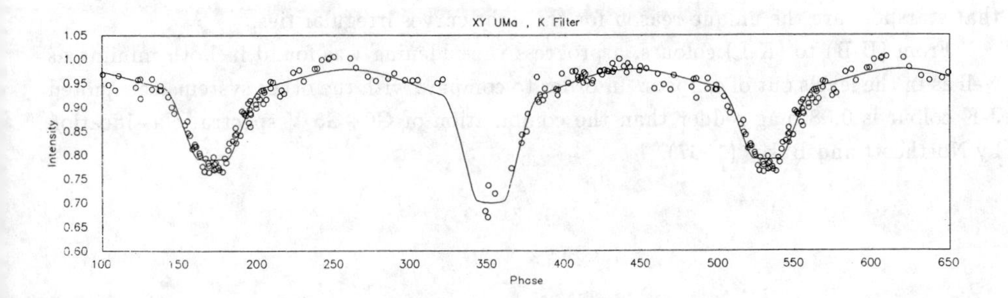
XY UMa , K Filter
Intensity
Phase

element determinations (Figure 2 and Table II). From visible photometry, light curve solutions have been obtained by Geyer (1980), Jassur (1986) Budding an Zeilik (1987), or Bank and Budding (1989). In our analysis, the different set of parameter resulting from the mentioned works were used as well as different inclination angles. The errors given in the different fits were very similar.

The observed colours are J-K=0.62 , J-K=0.65 and J-K=0.57 out of eclipse, primary minimum and secondary minimum respectively. The cool component is unseen during the secondary minimum so that a reddening between 0.18–0.26 mag is found for the hot component classified spectroscopically as G0-G5V by Geyer (1980). Out of eclipse colour is redder by about 0.17 mag than the combination of G0-G5 V + K5 V stars.

BH Vir

A detailed analysis refering to previous light curves solutions in these system has been reported by Zeilik et al. (1989).

Figure 3 shows the J and K light curves of BH Vir observed during April 1989. (Arévalo and Lázaro 1989b). Due to the inequality of the maxima in the J light curve, the K light curve solution is preferred as it gives more realistic parameters. The obtained values for the geometrical elements indicate that the secondary, slightly more massive component (M_2/M_1)=1.02, has a somewhat smaller radius (Figure 3 and Table III). This result was also obtained by Budding and Zeilik (1987) and recently by Zeilik et al. (1889).

The infrared colours are in good agreement with the combination of two main sequence stars G0+G5 corresponding to the spectral classification made by Abt (1965). No apparent IR is found for this system. As far as we know no previous infrared observations have been reported for BH Vir.

ER Vul

A complete study from UBVRI to near infrared JHKL was made from our 1984 completed light curves (Arévalo et al., 1988). Because of the larger dispersion (intrisic to the system) out of eclipse levels we believe it is difficult to determine reliable geometrical elements. An important point is that in the 1984 observations the K light curve shows much more unequal maxima than J or H which could be contradictory with the assumption that starspots are the unique reason for the light curves irregularities.

From (U-B) to (K-L) colours, a progressive reddening was found in both minima as well as in the levels out of eclipses. In order to compare with the other systems the quoted J-K colour is 0.08 mag redder than the combination of G0+G5 V spectral classification by Northcott and Bakos (1967).

TABLE III. Infrared light curves solutions of BH Vir

J Filter	K Filter
r_1=0.241±0.001	r_1=0.249±0.001
r_2=0.227±0.001	r_2=0.228±0.001
i=86.8 (Fixed)	i=86.8 (Fixed)
L_1=0.574±0.002	L_1=0.568±0.002
L_2=0.380±0.002	L_2=0.397±0.002
χ^2= 626	χ^2= 311
N=149	N=149

TABLE IV. Infrared light curves solutions of CG Cyg

J Filter	K Filter
r_1=0.223±0.001	r_1=0.239±0.001
r_2=0.197±0.001	r_2=0.216±0.001
i=82.5 (Fixed)	i=82.5 (Fixed)
L_1=0.652±0.002	L_1=0.559±0.002
L_2=0.307±0.002	L_2=0.393±0.002
χ^2= 524	χ^2= 162
N=77	N=78

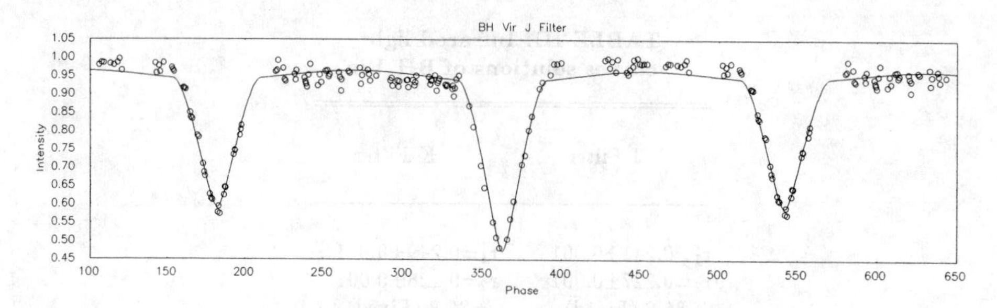
BH Vir J Filter
Intensity
Phase

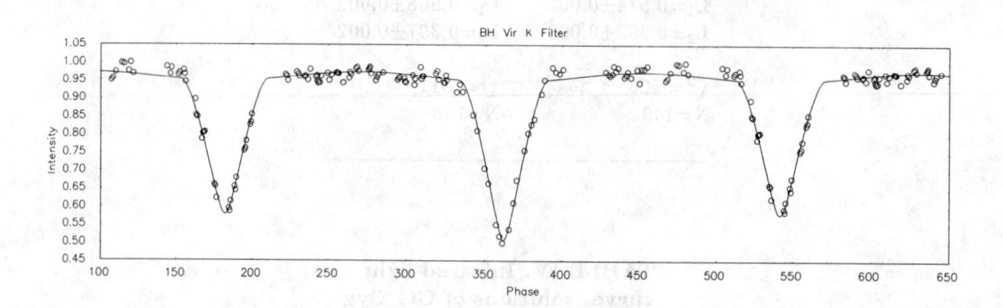
BH Vir K Filter
Intensity
Phase

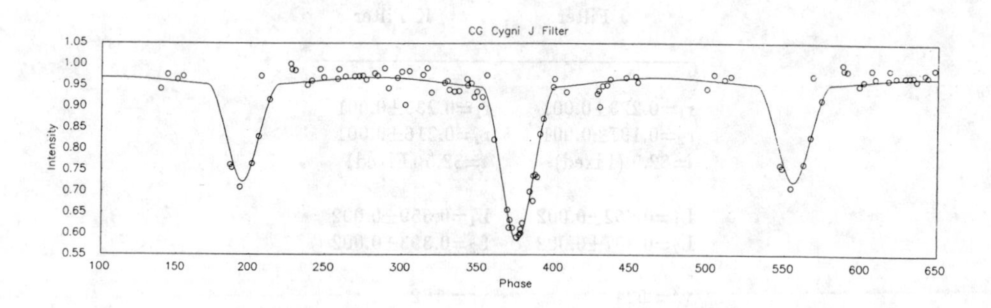
CG Cygni J Filter
Intensity
Phase

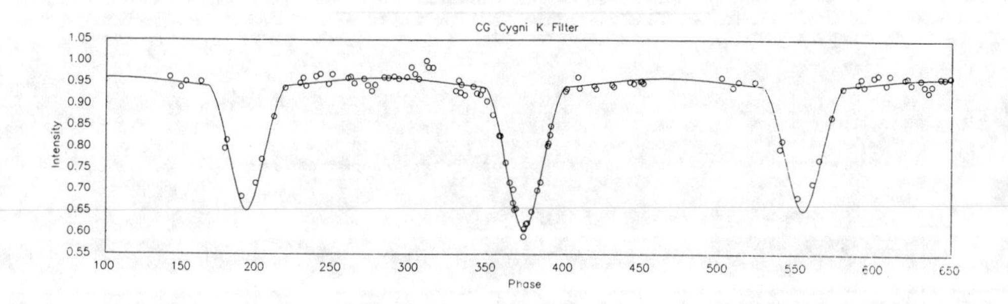
CG Cygni K Filter
Intensity
Phase

CG Cyg

The existence of infrared excess in this system was well stablished by earlier workers as Atkins and Hall (1972), Milone (1975), Milone and Naftilan (1980), or Bedford et al (1987). From our recent photometry carried out in July 1989 we reconfirm an apparent infrared excess of about 0.20 mag for the levels out of eclipse and for both eclipses.

The geometrical elements inferred from the J and K light curves (Figure 4 and Table IV) are slightly different, but both fits give smaller secondary component, in close agreement with the solution reported by Budding and Zeilik (1987). Another previous absolute elements for CG Cyg has been determined and the hipothesis of the transit or occultation solutions discussed (Naftilan and Milone 1985; Jassur, 1980; Sowell et al., 1987)

To conclude, we may report that in four (WY Cnc, XY UMa, ER Vul and CG Cyg) of the five studied systems it is found significative infrared excesses compared to the observational photometric errors. For two of then, XY UMa and WY Cnc, this excess seems to be originated in the primary component, considered to be the most active in both systems.

For the other three systems of the RS CVn short-period group, IR observations have been made for SV Cam and UV Psc while, as far as we know, there is not published IR observations for RT And. In SV Cam (Cellino et al. 1984) found an IR excess while in the case of UV Psc, Antonopoulou (1987), from complete light curves does not derive any apparent excess, but Busso et al. (1987) point out a possible infrared excess for this system.

The solutions of the IR light curves give primary eclipses correspond to transits rather than occultations.

The autours are grateful to Professors M. Zeilik and E. Budding who have made available the light curve fitting programs.

References:

Abt, H.A., (1965) Publi. Astron. Soc. Pacific **77**, 367.

Antonopoulou, E. (1987) Astrophys. Space Sci. **135**, 335.

Arévalo, M.J., Lázaro,C. and Fuensalida, J.J. (1988) Astron. Journal ,**96**, 1061

Arévalo M. J. and Lázaro, C. (1989a) in press.

Arévalo M.J. and Lázaro, C. (1989b) in preparation.

Arribas, S. and Martinez–Roger, C. (1987), Astron. Asptrophys. Suppl.**70**, 303.

Atkins, H.L., Hall, D.S. (1972) Publ. Astron. Soc. Pacific **84**, 638.

Awadalla,N.S. and Budding, E. (1979). Astrophys. Space. Sci. **63** ,479

Banks, T., and Budding E., (1989) IAU Inf. Bull. Var. Star 3304

Bedford, D.K., Fuensalida, J.J., Arévalo, M.J. (1987) Astron. Astrophys. **182**, 264.

Budding, E. and Zeilik, M. (1987), Astrophys. J. **319**, 827

Busso, M., Scaltriti, F., Persi, P., Robberto, M., and Silvestro, G. (1987) Astron. Astrophys. **183**, 83.

Cellino, A., Scaltriti, F., Busso, M. (1984) Astron. Astrophys. **144**, 315.

Geyer, E.H., (1980) in IAU Symposium 88, *Close Binary Star: Observations interpretations.* ed M.J. Plavec, D.M. Popper and R.K. Ulrich (Dordrecht Reidel) p. 423.

Jassur, D.M.Z. (1986) Astrophys. Space. Sci.**128**,369

Jassur, D.M.Z. (1980) Astrophys. Space. Sci. **67**, 19

Koornneef, J. (1983b) Astron. Astrophys **128**, 84

Milano,L. (1981) *in Photometric and spectroscopic Binary System*, ed E.B Carling and Z. Kopal (Dordrech: Reidel), P.331.

Milano, L., Mancuso, S., Vittone, A., D'Orsi, A., Marcozzi, S. (1986) Astrophys. Space Sci. **124**, 83.

Milone, E.F. (1975) *in multiple Periodic Variable Stars*, ed B. Szeidl, Academic, Budapest, p. 321

Milone, E.F. and Naftilan, S.A. (1980) IAU Symp. *88 Close Binary Star: Observations and interpretations.* ed M.J. Plavec, D.M. Popper and R.K. Ulrich (Dordrecht: Reidel) p. 419 Naftilan,S.A. and Milone, E.F. (1979) Astron. J. **84**, 1218

Naftilan, S.A. and Milone, E.F. (1985) Astron. Journal **90**, 761

Naftilan, S.A. (1987) Astron. J. **94**, 1327

Northcott, R.J. and Bakos, G.A. (1967) Astron. J. **72**, 89.

Sowell, J.R., Wilson, J.W., Hall, D.S. and Peyman P.E. (1987) Publ. Astron. Soc. Pacific **99**, 407.

Zeilik,. Ledlow, M., Rhodes, M., Arévalo, M.J., and Budding, E. (1989) in press

ON CHROMOSPHERIC EMISSION AND PHOTOMETRIC VARIABILITY OF ACTIVE LATE-TYPE STARS

K. G. Strassmeier
Department of Physics and Astronomy, Vanderbilt University,
Nashville, TN 37235, U.S.A.; and
Center of Excellence in Information Systems, Tennessee State University, Nashville, TN 37209-1561, U.S.A.

ABSTRACT. To pursue the study of solar phenomena in late-type stars, we have investigated chromospheric activity in 100 single and binary stars of spectral type F6 to M2 and luminosity class III, IV, and V. Attempts are made to "quantify" starspot activity and correlate it to chromospheric activity.

1. INTRODUCTION

Chromospheric activity (CA) in late-type stars is established by the presence of emission in the cores of the Ca II H and K resonance lines but also by the presence of emission in the Balmer H-alpha line. Starspots have been confirmed to be the phenomenological explanation of RS CVn-type photometric variability, i.e., photospheric activity (PA). A frequently asked question is: does any CA star also has spots and vice versa? Of the 168 CA binaries in the "Chromospherically Active Binary Star" catalog (Strassmeier <u>et al.</u> 1988, updated) 21 have no sufficient photometry. Of the remaining 147 binaries, 9 (i.e. 6%) have been reported constant in light, i.e., have wave amplitudes less than 0.01 mag at the time of observation. These statistics support a correlation between PA and CA.

In this paper I would like to present some results from two surveys: one spectroscopic study (100 stars) at Ca II H and K and H-alpha (Strassmeier <u>et al.</u> 1990a) and one photometric study (80 stars) mainly in the <u>V</u>-band pass.

2. ABSOLUTE Ca II H AND K FLUXES; H-ALPHA CORE EMISSIONS

2.1. Accuracy of our fluxes/equivalent widths

Moderately high resolution (0.2-0.5 A) CCD spectra have been obtained at KPNO, DAO, McDonald, and Ritter Observatory.
Altogether we observed 21 SB2/3, 31 SB1, and 48 single stars. The measurements of the Ca II surface fluxes followed the procedure outlined by Linsky <u>et al.</u> (1979). The mean deviations of observations which have

C. İbanoğlu (ed.), Active Close Binaries, 509–514.

been obtained at KPNO and DAO are within the range of 20%. The agreement with fluxes obtained by other authors (partially using different calibration techniques) is mostly better than 40%. We estimate the internal precision to be 10-15%.

The H-alpha core emission was determined by subtraction of the equivalent width of an in-active star of the same spectral type and luminosity class. The internal precision for spectra of in-active stars taken on the same night is 4 mA for the KPNO/coudé feed data, 6 mA for the 2.1 m McDonald data, and 10 mA for the 1m Ritter echelle data. The external accuracy is estimated from spectra taken at different epochs to be 40-60 mA depending on the instrumental resolution.

2.2. Problems with composite spectra

The spectroscopic presence of a companion star dilutes the emission from the other (active) component but also the observed colors and the continuum flux at Ca II H and K. Altogether, errors of up to 50 % for some of the double-lined systems are possible. In a follow-up study we determined individual spectral types and luminosity classes of 12 CA binaries with composite spectra by means of the "spectrum synthesis" technique (Strassmeier and Fekel 1990). In RS CVn-type systems, typically an unevolved F star and an evolved G or K star, we found that the luminosity class for the hotter component is in most cases IV rather than V.

2.3. V-R color excess

The Linsky-Willstrop flux calibration relates the V-R color of a star to the integrated flux per Angstroem in the 3925-3975 A bandpass. What V-R color should be used? The observed color which might be affected by color excess, spots, and binarity? Or the color derived from the spectral type which also might be affected by spectral-type misclassification?
Since most of our program stars had no observed V-R color the decision was easy to make. To be on a consistent scale throughout the sample, we have used Johnson's (1966) V-R - spectral type relation. Just for a little "nightmare": a difference of 0.1 mag in V-R relates to a difference in the absolute emission line flux of 250%!!

3. BROAD-BAND PHOTOMETRY

3.1. Long-term observations

Doug Hall organized long-term photometric observations of many RS CVn systems beginning in late 1983 when Louis Boyd's 25 cm APT (Automatic Photoelectric Telescope) was put into regular service.

The results have been published in a series of papers entiteled "Photometric Variability in CA Stars" (Strassmeier and Hall 1988a,b; Strassmeier et al. 1989). In late 1987 the 40 cm Vanderbilt University APT on Mt. Hopkins, Arizona went into operation and all program stars were transfered to this telescope.

Photometry of 50 additional CA candidates has been analyzed by Hooten (1989). Rotation periods were found for 34 stars. This brings the total number of CA stars with spot activity to 250 (singles and binaries).

3.2. Definition of a spot-activity parameter R

Let $\underline{V}_0$ be the magnitude at light maximum for a given $\underline{V}$ light curve, $d\underline{V}(A)$ the $\underline{V}$ amplitude due to a spot A and, if the light curve is double humped or asymmetrically single humped, $d\underline{V}(B)$ the $\underline{V}$ amplitude due to a second spot B; $\underline{V}_{0,max}$ be the unspotted magnitude. Then

$$R_V = V_{0,max} - V_0 + dV(A) + dV(B) - \text{const.}$$

where the constant is the light contribution of the unspotted component.

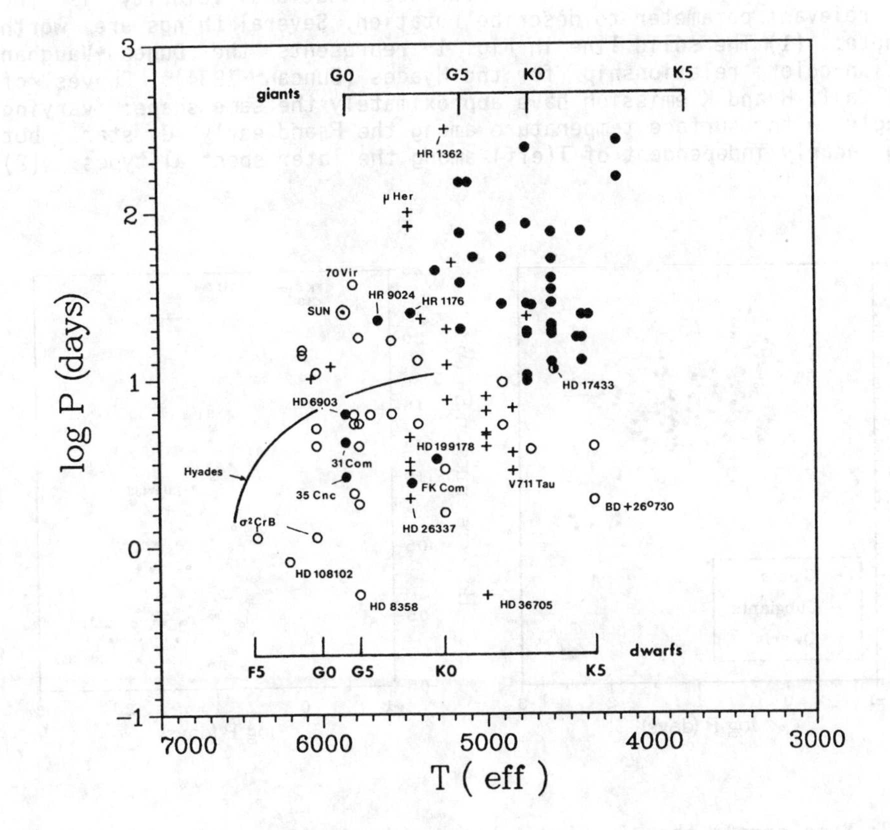

Fig. 1: Part of our sample in the P(rot)-T(eff) plane. Dots are luminosity class III, plusses are class IV, and circles are class V. The solid line is the rotation-color relationship for the Hyades dwarf stars. It is approximately a line of constant activity. The lower the line in the diagram the higher the Ca II emission. See also the text.

Of course, a sum of amplitudes can not separate the effects of spot area and spot temperature but is much more reliable than a single photometric amplitude. This is because a large amplitude alone is not necessarily a sign of large spot activity.

4. DEPENDENCE OF ACTIVITY ON ROTATION AND SURFACE TEMPERATURE

4.1. The P(rot) - T(eff) diagram

Our sample consists of unevolved stars within a mass range from 0.67 M_{O} (K5) to 1.3 M_{O} (F6) and evolved stars from, say, 1 to 3 M_{O}. Figure 1 is a plot of our program stars with known rotation periods of luminosity class V (circles), IV (plusses), and III (dots), singles and binaries. Clearly, giants must rotate faster than dwarfs if the rotation periods are equal. Perhaps this means that surface rotational velocity is the more relevant parameter to describe rotation. Several things are worth to note: (1) The solid line in Fig. 1 represents the Duncan-Vaughan rotation-color relationship for the Hyades (Duncan 1984). Curves of equal Ca II H and K emission have approximately the same shape: varying strongly with surface temperature among the F and early G stars but being nearly independent of T(eff) among the later spectral types. (2)

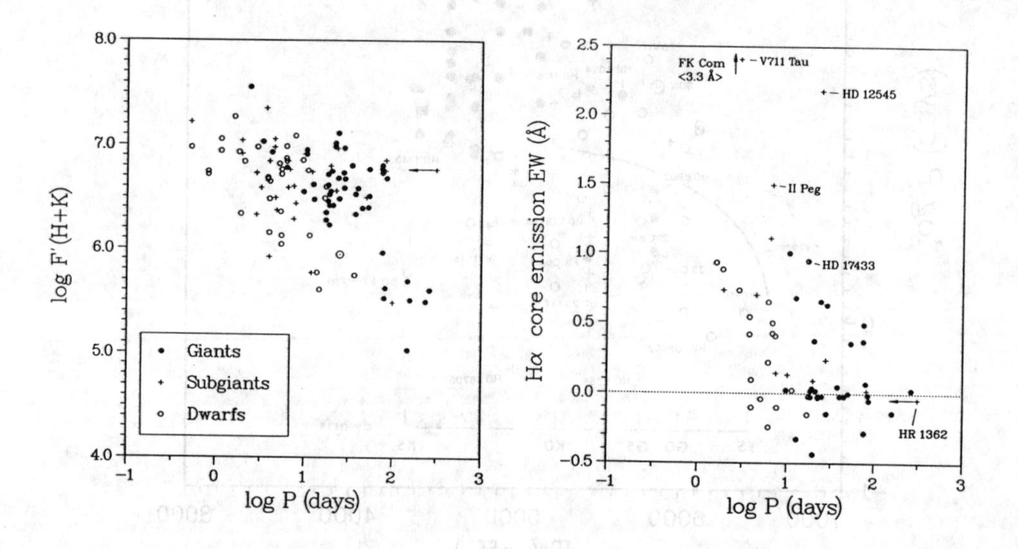

Fig. 2: Same sample than in Fig. 1 but in the Ca II emission line flux - P(rot) plane. The "+" with an arrow is HR 1362 pointing to the position if P(rot) would be one half the observed value. See the text.
Fig. 3: The H-alpha core emission - P(rot) plane. The point with an arrow is again HR 1362. There is the same separation of luminosity classes as seen in the Ca II fluxes in Fig. 2 due to the different radii.

HR 1362 stands out of the general trend in that it is overactive by more than an order of magnitude than implied from its position in the P(rot)-T(eff) plane. Its rotation period from 11 years of photometry is 335 days (Strassmeier et al. 1990b). (3) A group of five giant stars with very short rotation periods and relatively high temperatures stand out from the rest: HD 6903, 31 Com, 35 Cnc, FK Com, and HD 199178. All are single stars of spectral types G1-3 III and are within the most active systems among our sample.

4.2. The F'(H+K) - P(rot) diagram

In Figure 2 we plot the sum of the corrected Ca II H and K indices versus rotation period. Excluded are SB2s and SB3s. There is clear evidence that evolved stars are generally more active than main-sequence stars of the same rotation period. The data also show that a cool evolved star in a close binary system is generally more active than a single star, simply because it can have a higher rotation period due to tidal coupling. This is not necessarily true for the evolved stars.

4.3. The H-alpha core emission - P(rot) diagram

The H-alpha activity - rotation relation in Figure 3 is quite similar to that of the Ca II surface fluxes. Figure 3 also shows the same separation of luminosity classes: the evolved stars are more active than their main-sequence counterparts of the same rotation period.

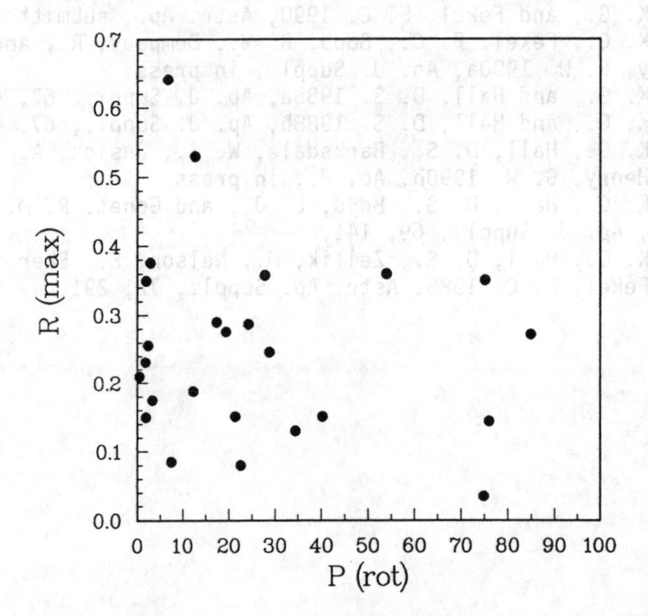

Fig. 4: Spot-activity parameter R_V versus rotation period for a sample of 27 stars with sufficient photometric data. The notation "max" for R means the largest ever observed value has been plotted.

4.4. The spot activity - P(rot) diagram

Spot modeling of multi-color photometric data seems to me to be the most promising technique to quantify starspot activity. Unfortunately, only a handful stars have photometry with a reasonably long baseline in time, say >10 years, in order to determine the "unspotted" magnitude (see Strassmeier 1990, and also this proceedings). What about the other approximately 240 stars? Approximately half of them have only a single light curve. Of the remaining 100 systems only 20 have a longer baseline in time (say 5 years). In Figure 4 we plot 27 stars (singles and single-lined binaries). Obviously there is no correlation with rotation period.

This research was partially supported by NASA grant NAG 8-111 to Tennessee Stare University.

REFERENCES

Bopp, B. W. 1983, in IAU Colloq. 71, Activity in Red Dwarf Stars, ed. P. B. Byrne and M. Rodono (Dordrecht: Reidel), p. 363.
Duncan, D. K. 1984, in Cool Stars, Stellar Systems, and the Sun, ed. S. L. Baliunas and L. Hartmann (Berlin: Springer), p. 128.
Hooten, J. T. 1989, Master Thesis, Vanderbilt University.
Johnson, H. L. 1966, Ann. Rev. Astr. Ap., **4**, 193.
Linsky, J. L., Worden, S. P., McClintock, W., and Robertson, R. M. 1979, Ap. J. Suppl., **41**, 47.
Strassmeier, K. G. 1990, Ap. J., in press.
Strassmeier, K. G., and Fekel, F. C. 1990, Astr. Ap., submitted.
Strassmeier, K. G., Fekel, F. C., Bopp, B. W., Dempsey, R., and Henry, G. W. 1990a, Ap. J. Suppl., in press.
Strassmeier, K. G., and Hall, D. S. 1988a, Ap. J. Suppl., **67**, 439.
Strassmeier, K. G., and Hall, D. S. 1988b, Ap. J. Suppl., **67**, 453.
Strassmeier, K. G., Hall, D. S., Barksdale, W. S., Jusick, A. T., and Henry, G. W. 1990b, Ap. J., in press.
Strassmeier, K. G., Hall, D. S., Boyd, L. J., and Genet, R. M. 1989, Ap. J. Suppl., **69**, 141.
Strassmeier, K. G., Hall, D. S., Zeilik, M., Nelson, E., Eker, Z., and Fekel, F. C. 1988, Astr. Ap. Suppl., 72, 291.

LONG-TERM OBSERVATIONS OF ACTIVE BINARIES

C. İbanoğlu
Ege University Observatory
Bornova - İzmir
Turkey

ABSTRACT. The photoelectric photometry of selected 14 chromospherically active binaries has been made between 1978 and 1989. SV Cam, UV Psc, V471 Tau and ER Vul are short, and RS CVn, RT CrB, WW Dra, Z Her, MM Her, RT Lac, AR Lac, II Peg, LX Per, and SZ Psc are long period RS CVn type binaries. To reveal the photometric peculiarities of RT Lac and V471 Tau, they have been observed extensively. Almost all of the systems (except RT CrB) showed wave-like distortion in their light curves. The light curve and orbital period variations are discussed. A period for the variations of the brightness at mid-primary of RT Lac was estimated to be six years. The mean total brightness of V471 Tau seems to be changed with a period of about 18 years and an amplitude of about $0^{m}.15$. The orbital periods tend to decrease continuously. The light variations at mid-eclipses with an amplitude of wave-like distortion occur in short time intervals and seem to be the common property of all RS CVn type binaries. Starspot hypothesis appears to be inadequate for explaining all the phenomena observed in chromospherically active close binaries.

1. Introduction

The existence of a group of eclipsing binaries has firstly been noticed by Struve (1946). These stars differ from the others with respect of their spectroscopic properties. In the spectra of these stars, the emission lines of ionized calcium were stronger than those which are of same spectral types. The components of these type binaries are generally late G or early K type stars. Later spectral investigations have not only increased their number but also revealed their properties gradually. The investigations made by Popper (1970) indicated that the masses of the components in this group are close to each other and, H and K emissions of ionized calcium are produced by the subgiant K0 companions.

In the middle of 1960s, the existence of irregular variations in the light curves of these type binaries have also been noticed. At first, the observers of Catania Observatory Catalano and Rodono (1967, 1969) announced that there is a wave-like distortion with an amplitude of about $0^{m}.2$ in the light curves (outside eclipses) of RS CVn. Moreover, this wave-like distortion has displaced towards the decreasing orbital phases. The change in the depth of primary eclipse and the displacement of secondary eclipse seemed to be associated with the wave-like distortion. The earliest discoveries were of Oliver's (1974) who

C. İbanoğlu (ed.), Active Close Binaries, 515–544.

proposed that the wave-like distortion and its displacement towards the decreasing orbital phases are the common property of the binaries with Ca II emissions.

As it is known the common properties of so-called RS CVn binaries have been summarized by Hall (1976). Soon after, an international campaign has been started for investigating these stars from very short to radio waves. Therefore, a group consists of C. İbanoğlu, M. Kurutaç, Z. Tunca, A. Y. Ertan, S. Evren and O. Tümer started to observe 14 RS CVn binaries at Ege University Observatory in May 1978. However, due to the intensive program of the observatory it was impossible to observe each star in every observing season. So, an observing list was made for each year. Thus, it was tried to obtain the full light curve of the selected systems. Some parts of the observational results obtained with this research have been published seperately or being prepared for publication. In early 1980 Kurutaç, Ertan and Tümer departed from us and in 1983 C. Akan and V. keskin joined to us.

In this paper I want to present the light curves of some binaries and discuss outside light variations, period of migration and light curve analysis. My talk will concentrate mainly on two stars RT Lac and V471 Tau.

2. Observational Data

The observations of the selected active binaries were made with the 48 cm Cassegrain reflector of the Ege University Observatory. An EMI 9781 A photomultiplier and the B and V filters of the UBV system were used. In Table 1 the systems which were chosen to observe were given with their orbital periods.

TABLE 1: The selected RS CVn-type binaries

Stars	max	min	Sp.	P
SV Cam	$8^{m}.40$	$9^{m}.11$	G5V+G3V	$0^{d}.593$
RS CVn	7.93	9.14	F4IV+K0IV	4.798
RT CrB	10.20	10.82	G0	5.117
WW Dra	8.30	8.95	G2IV+K0IV	4.630
Z Her	7.30	8.18	F4IV-V	3.993
MM Her	9.45	10.43	G2-5	7.960
RT Lac	8.84	9.89	G9IV+K1IV	5.074
AR Lac	6.08	6.77	G2IV+K0IV	1.983
II Peg	7.18	7.78	K2IV-V	6.703
LX Per	8.10	8.93	G5IV+G5IV	8.038
SZ Psc	7.18	7.72	K1IV+F8V	3.966
UV Psc	8.91	10.05	G2	0.861
V471 Tau	9.40	9.71	K0V+DA	0.521
ER Vul	7.27	7.49	G0V+G5V	0.698

Table 2 indicates the years of the observations made for each system. As it is seen from this table the more thoroughly observed stars are RT Lac, AR Lac, LX Per, SZ Psc, UV Psc, V471 Tau, ER Vul and II Peg. Among these stars V471 Tau has been observing each year since 1973; and RT Lac since 1978. These two systems seem to be very important because of their unusual light variabilities.

TABLE 2. The years of the observations of individual systems.

Star	1978	79	80	81	82	83	84	85	86	87	88	89
SV Cam								+				
RS CVn		+										
RT CrB		+	+	+								
WW Dra			+									
Z Her	+	+	+	+								
MM Her		+	+			+	+	+				
RT Lac	+	+	+	+	+	+	+	+	+	+	+	+
AR Lac	+	+	+	+	+	+						
II Peg						+	+			+	+	+
LX Per		+	+	+	+	+						
SZ Psc		+	+	+								
UV Psc				+	+		+	+	+	+	+	+
V471 Tau	+	+	+	+	+	+	+	+	+	+	+	+
ER Vul				+	+		+	+	+	+		

3. Period Changes

Systematic residuals in the timings of minima of eclipsing binaries have been widely interpreted as changes in their orbital period. Most of active close binaries indicate period changes in a plausible time interval. The changes of the orbital periods of RS CVn stars seem to be very important characteristics of their properties. A number of mechanisms have been proposed for explaining these changes. Among these, the angular-momentum transfer between the stars and orbit via mass transfer, magnetic fields, or tidal coupling and very recently three-body theory have mostly been taken into consideration.

Mass transfer/mass loss as a cause of observed period changes appears to be easily excluded since almost all of the binaries are detached and there are no certain evidences of mass loss. Van Buren and Young (1985) propesed a mechanism by which magnetic activity cycles drive structural changes in the convective star which then lead to spin - orbit angular momentum exchange through the action of tidal torques. As it is known from solar analogy at the times of greatest magnetic activity, and hence strongest Ca II line emission, the pressure support from the magnetic field is also at maximum. At these times the radius of the star increases, therefore moment of inertia, and then the rotational period increases. Thus, if the stellar magnetic activity indicators are at maximum the O-C diagram will have the largest positive slope.

Later on, Van Buren (1986) proposed three-body theory for the period changes in RS CVn systems. SV Cam, AR Lac and V471 Tau are given as samples. In two of these systems (SV Cam and V471 Tau) the predicted masses of third bodies are too low, namely 0.35 and 0.05 $M_\odot$, respectively. However in the third case, AR Lac, the mass for the third body is 2.7 $M_\odot$ and is nearly equal to binary mass. We should point out that only 40 % of the RS CVn type binaries indicated definite period changes over the time interval they have been observed, typically sevaral decades.

The systems, which indicate systematic O-C changes during the intervals of observations are RT Lac, AR Lac, SZ Psc and V471 Tau. In all of these systems the O-C variations indicate the orbital periods are decreasing continuously. The rates of decreaments are as follows:

RT Lac : 62 s.century^{-1}
AR Lac : 17 s.century^{-1}
SZ Psc : 179 s.century^{-1}
V471 Tau : 1 s.century^{-1}

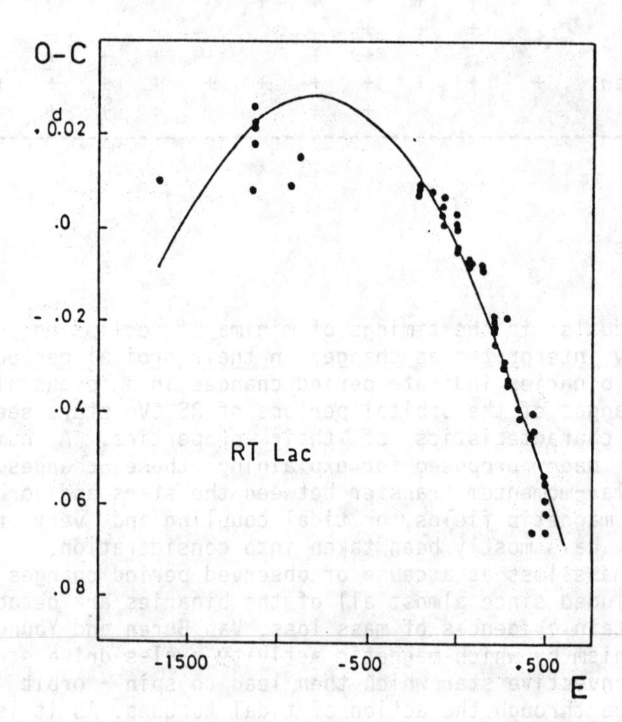

Figure 1. The O-C variations of RT Lac during last years (cf. Evren, 1989).

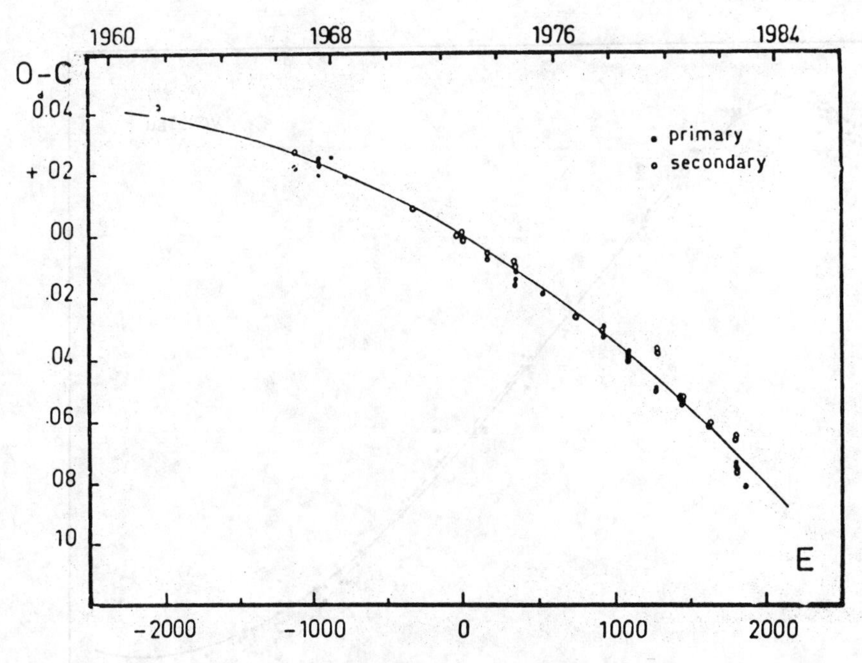

Figure 2. The O-C curve for both primary and secondary minimum of AR Lac. Dots and circles denote primary and secondary, respectively. The continuous line represents the computed curve (cf. Evren et al., 1983).

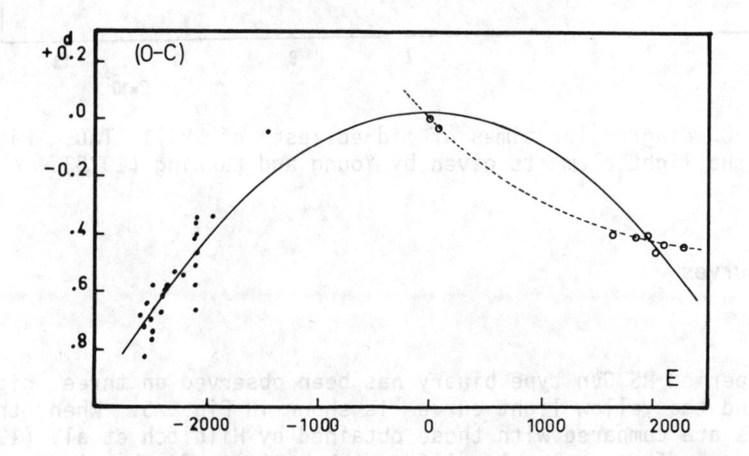

Figure 3. The O-C curve of SZ Psc. Dots, circles denote photographic or visual and photoelectric timings, respectively. The continuous and broken lines represent the O-C curves computed by using only photoelectric minimum times, respectively (cf. Tunca, 1984).

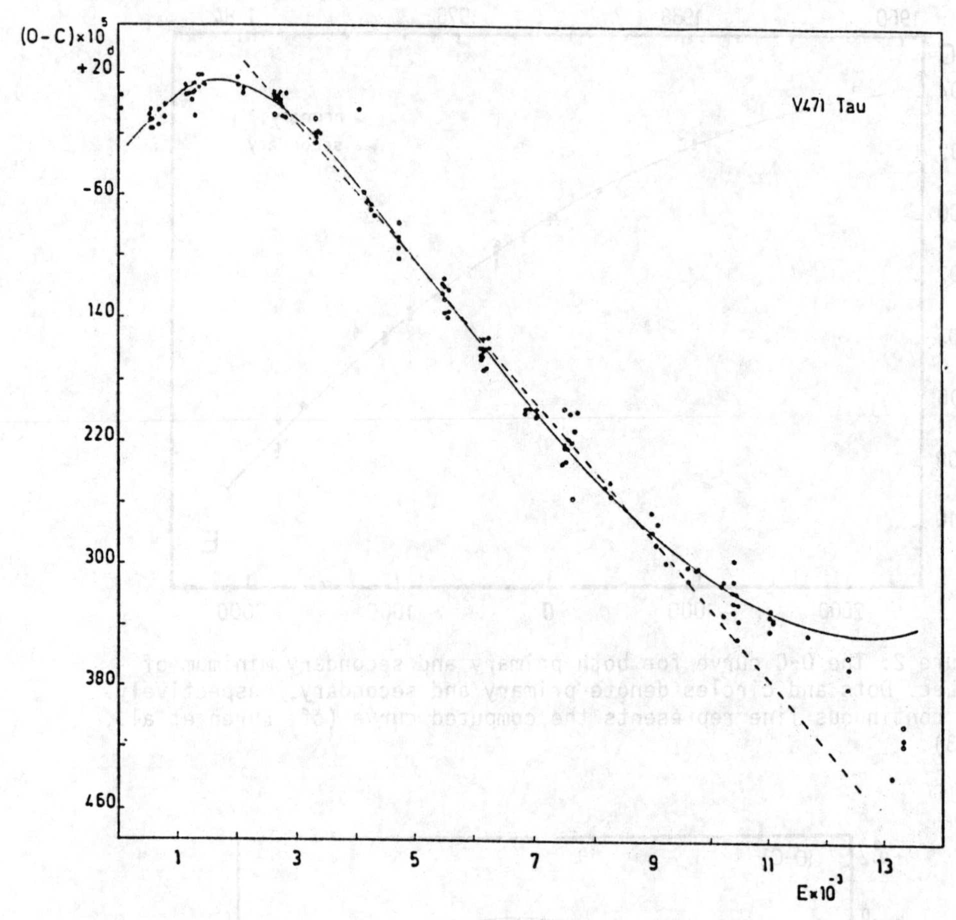

Figure 4. O-C diagram for times of mid-eclipse of V471 Tau, with respect to the light elements given by Young and Lanning (1975).

4. Light Curves

SV Cam :
This short period RS CVn type binary has been observed on three nights in 1985 and the yellow light curve is shown in Fig. 5. When these light curves are compared with those obtained by Hilditch et al. (1979) the existence of a completely different behaviour of the individual light curves obtained in different epochs is clearly noticeable. According to the observations of Hilditch et al. all ligth curves coincide to within 0.01 mag at mid-secondary eclipse, which normally

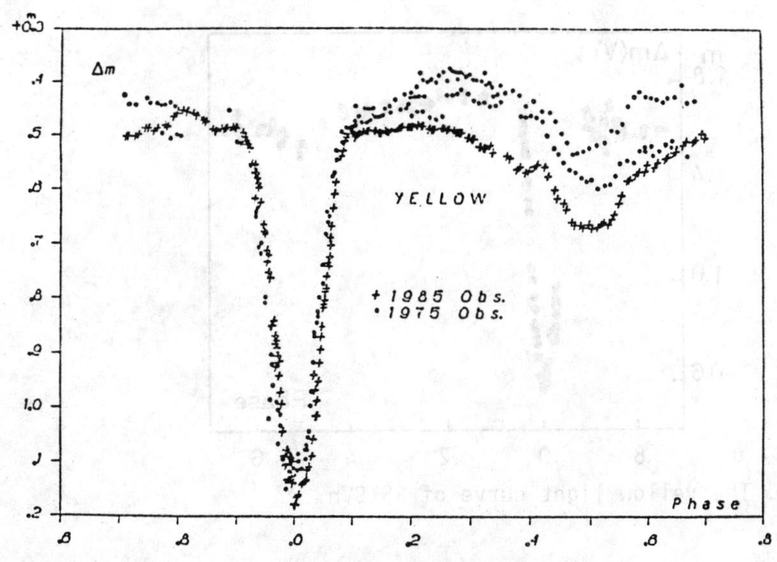

Figure 5. V light curves of SV Cam. Dots are Çelikezer's observations obtained in 1975.Plus signs refer to the observations obtained in 1985.

urged them to conclude that the light curve variations are due to intrinsic variability of the cooler component which is behind the hotter star at phase 0.5. Contrary to this situation, the light curves obtained in our observatory do not have indications of any coincidence at mid-secondary eclipse. The photoelectric observations obtained by us evidently suggest that the primary component should have a significant role on the light curve changes. This is also consistent with conclusions of Cellino et al. (1985) and Patkos (1981). On the other hand, momentary brightenings at phase 0.41 and 0.81 are also detectable.

RS CVn :
The blue and yellow light curves obtained in 1979, and the yellow light curve is shown in Fig. 6. The wave - like distortion at outside eclipse is clearly seen. The amplitude of the wave is about $0^m.09$ in blue, $0^m.14$ in yellow light. The minumum of the wave falls approximately at phase 0.72.

RT CrB :
This faint object was observed in 1979 ,1980 and 1981. The yellow light curve of the system is shown in Fig. 7. The light curve of RT CrB evidently shows that the proximity effects are too small to be taken into account. In addition, there is no indication about the light variation due to the wave-like distortion.

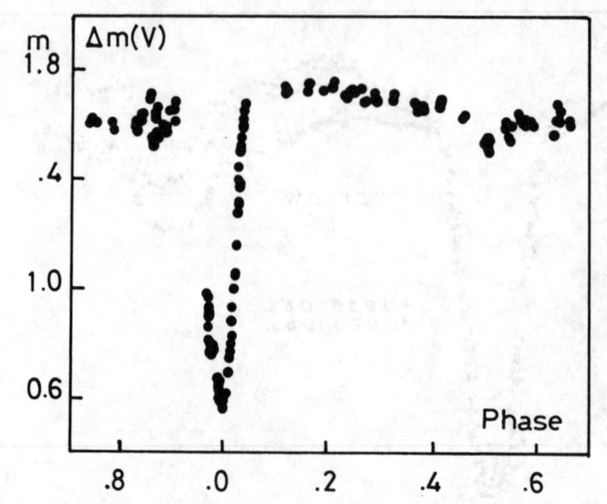

Figure 6. The yellow light curve of RS CVn.

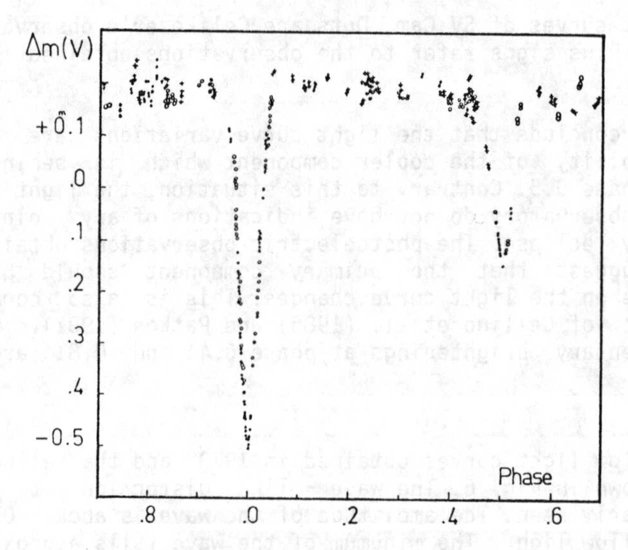

Figure 7. The yellow light curve of RT CrB. Plus, circles and dots denote the observations obtained in 1979,1980 and 1981,respectively.

WW Dra :
The eclipsing binary WW Dra, the brighter component of the visual binary ADS 10052, was observed in 1980 and the yellow light curve is shown in Fig. 8. The wave-like distortion at outside eclipse is noticable.

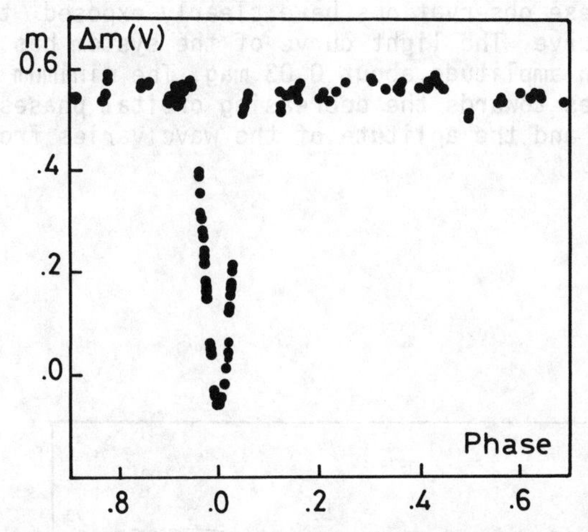

Figure 8. The yellow light curve of WW Dra obtained in 1980.

Z Her :
The observations were made during a period of four years between 1978 and 1981. Two-colour light curves of the system are shown in Fig. 9.

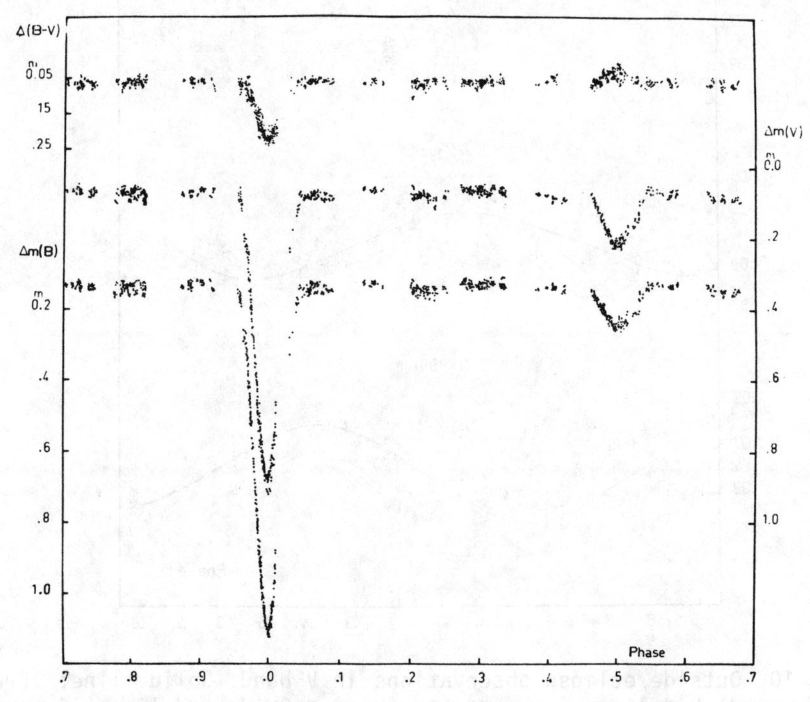

Figure 9. Photoelectric light and color curves of Z Her.

Despite some gaps, these observations have clearly exposed the main feature of the light curve. The light curve of the system has a wave-like distortion with an amplitude about 0.03 mag. The minimum of the distortion wave displaces towards the decreasing orbital phases with a period of about 1.4 yr and the aplitute of the wave varies from $0^m.017$ to $0^m.038$ in yellow band.

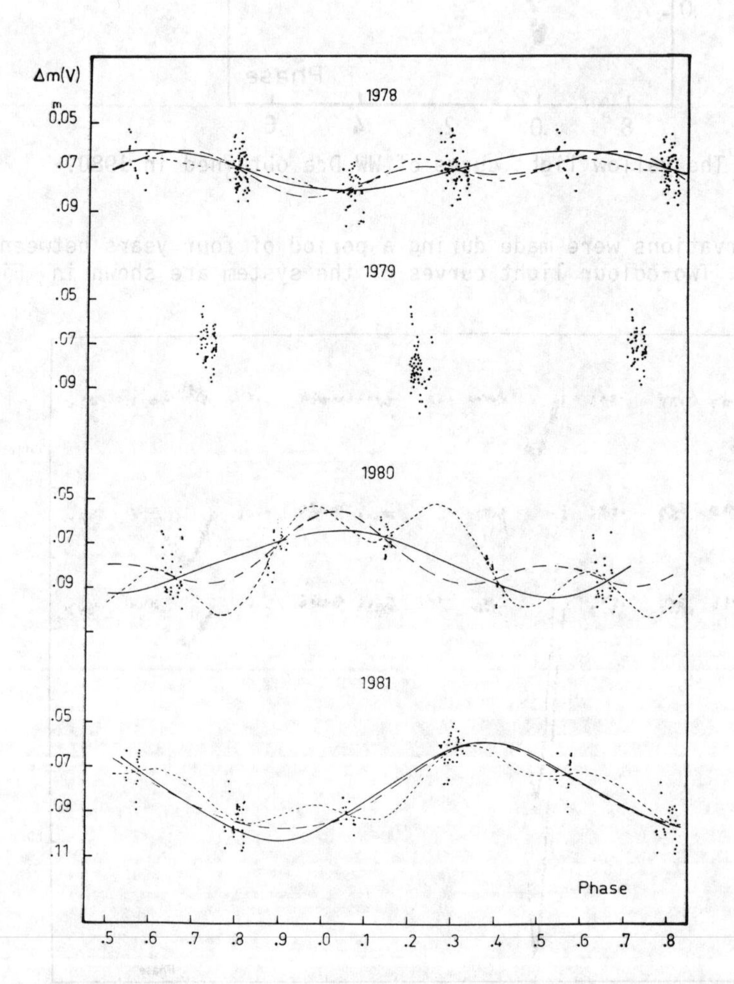

Figure 10. Outside eclipse observations in V band. Solid line: Free-hand curve; dashed line: Computed curve for n=2; dotted line: Computed curve for n=3.

MM Her :
This system was observed in 1979, 1980, 1983, 1984 and 1985. The light curves obtained in 1985 are shown as an example.

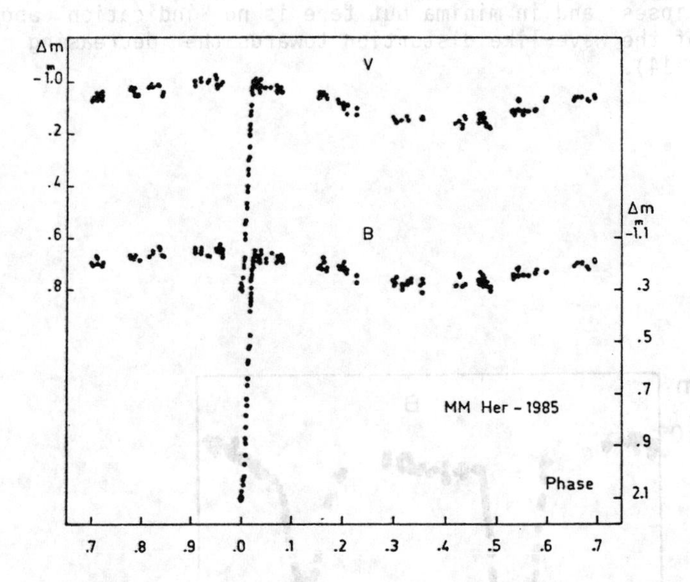

Figure 11. The blue and yellow light curves of MM Her obtained in 1985.

The wave amplitude has been increased about $0^m.08$ from 1976 to 1986 as shown in Fig. 12. On the other hand, the wave sweeps out the light curve with a period of about 3.57 years.

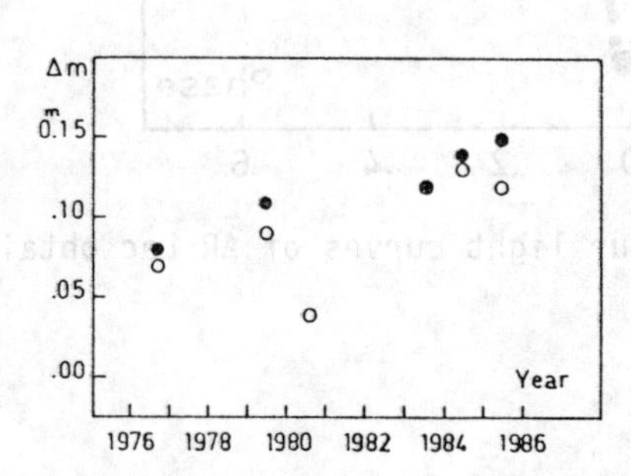

Figure 12. The wave amplitude variations of MM Her in B and V. Dots and circles denote the amplitudes in V and B , respectively.

AR Lac :
The light curves of the system were obtained in 1978, 1979, 1980, 1981 and 1982. As an example the 1980 and 1981 light curves are shown in Fig. 13a, b. The outside eclipse light variations are very similar to each other in successive two years. The distortions were seen at both outside eclipses and in minima but tere is no indication about the migration of the wave-like distortion towards the decreasing orbital phases (Fig. 14).

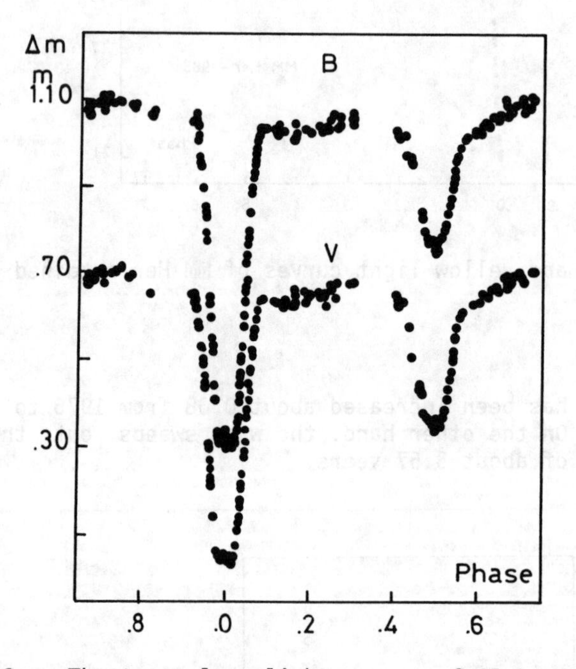

Figure 13 a. The two-colour light curves of AR Lac obtained in 1980.

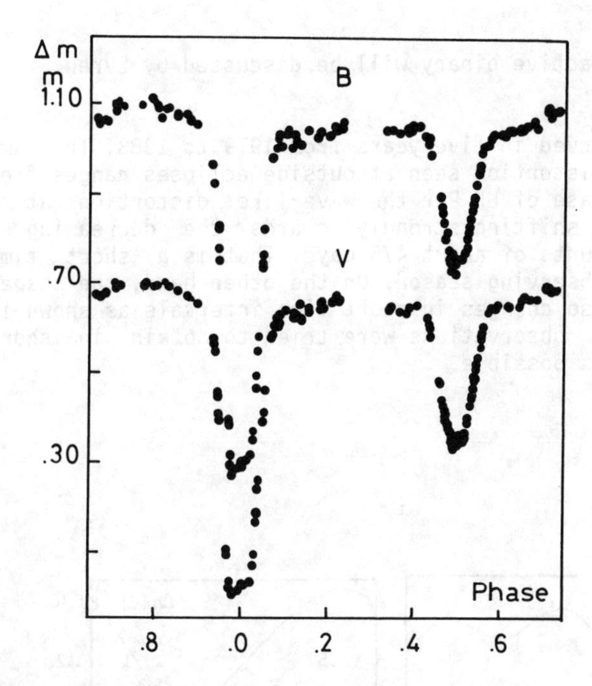

Figure 13 b. The two-colour light curves of AR Lac obtained in 1981.

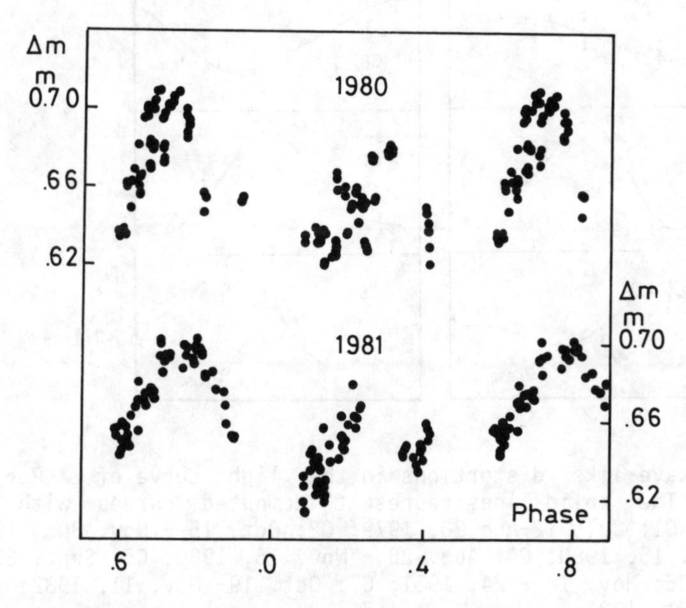

Figure 14. Outside eclipse light variations of AR Lac in yellow light.

II Peg :
This non-eclipsing active binary will be discussed by Evren.

LX Per :
The system was observed in five years from 1979 to 1983. The amplitute of the wave-like distortion seen at outside eclipses ranges from $0^m.03$ to $0^m.08$. In the case of LX Per the wave-like distortion at outside eclipses has been shifting strongly towards the decreasing orbital phases, which amounts of about 475 days. That is a short time when compared with the observing season. On the other hand, the shape of the distortion wave also changes in short time intervals as shown in Fig. 15. Therefore the observations were tried to obtain in short time intervals as soon as possible.

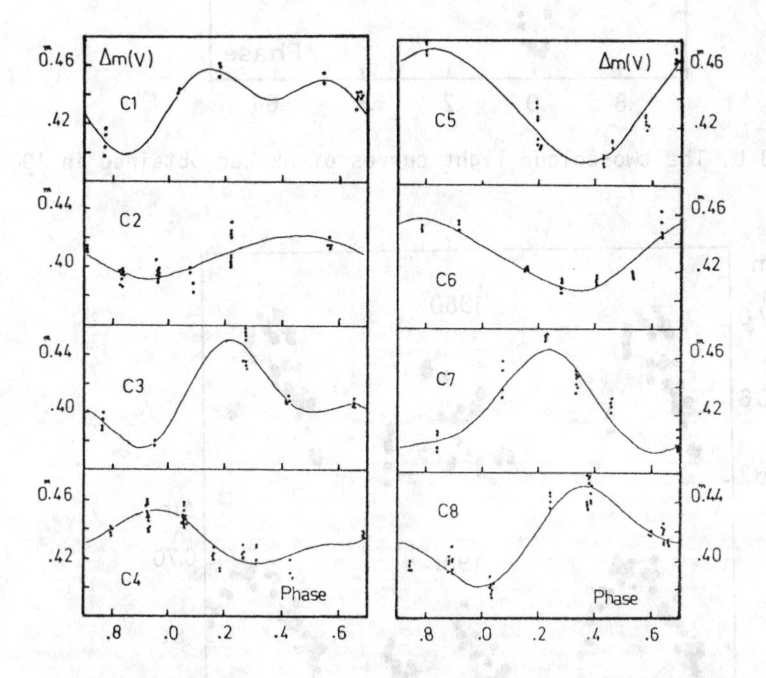

Figure 15. The wave-like distortions in the light curve of LX Per in yellow light. The solid lines represent computed curves with the Fouries series. C1: July 12-Aug.26, 1979; C2: Oct. 15 - Nov. 30, 1979; C3: Jan. 23-Feb. 15, 1980; C4: Aug. 26 - Nov. 16, 1980; C5: Sept. 23 - Oct. 4, 1981; C6: Nov. 18 - 24, 1981; C7: Oct. 19- Nov. 11, 1982; C8: Sept. 7 - Oct. 21, 1983.

SZ Psc :
The system was observed in 1979 and 1981. The light curves obtained in 1981 are shown in Fig. 16. The amplitudes of the light variations at outside eclipses are $0^m.18$ and $0^m.21$ in B and V colours.

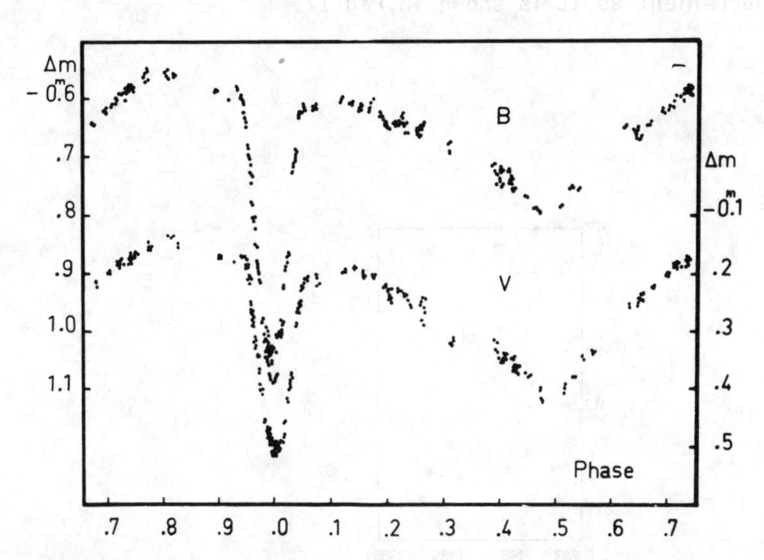

Figure 16. The blue and yellow light curves of SZ Psc.

UV Psc :
Will be discussed by Akan.

ER Vul :
Will be discussed by Keskin.

5. Light Changes During the Eclipses

The wave-like distortion on the light curves of RS CVn type binaries generally affects the primary eclipses, in which the spotted, cooler stars are infront of the hotter ones. However, at the secondary eclipses the cooler stars are behind the hotter ones and therefore we do not expect any change due to the spots on the cooler star. The following systems show variations at the primary eclipses.

RT CrB :
Although, there are no significant variations at outside eclipses during successive two years the primary eclipse is deeper about $0^m.09$ (in V) in 1981 than the one in 1980.

Z Her :
The light variations in primary eclipse, with an amount equals the wave amplitude, namely 0.03 mag, have been taking place in short time intervals (i.e. a month) in the case of Z Her. These short period changes make difficult the light curve analysis accurately.

MM Her :
Although the total brightness of the system was remained constant between 1983 and 1985, the depth of the primary eclipse indicated a regular decrement as it is shown in Fig 17.

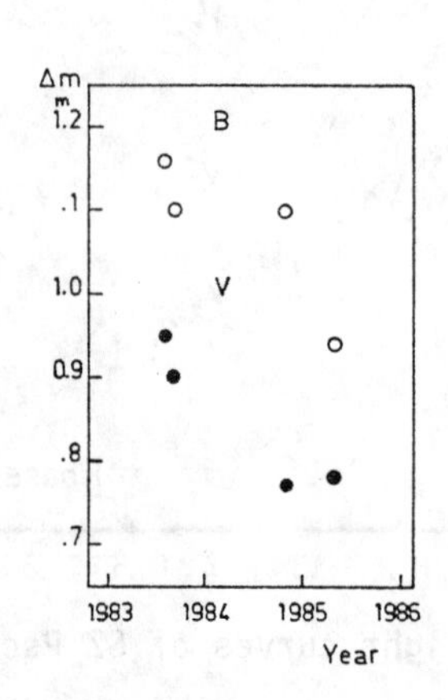

Figure 17. The depth variations of primary minima of MM Her.

AR Lac:

The brightness at total eclipse has been changed about $0^{m}.04$ in B and $0^{m}.02$ in V within a month. In addition the slope between the second and third contact also varies with time. Such variations occur at annular, secondary eclipse but with a very smaller scale. The variations at total and annular eclipses are shown in Fig. 18. and 19.

SZ Psc :

At mid-primary eclipse the brightness varies in short time intervals. These variations exceed 0.06 mag.

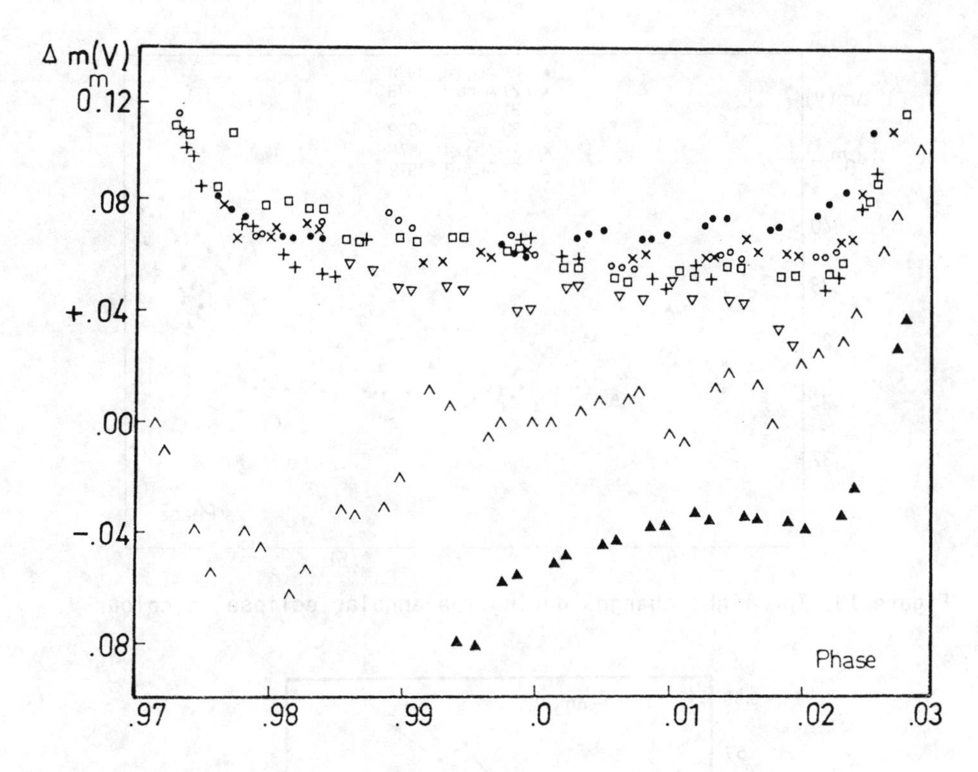

Figure 18. The light changes during the totality, in colour V. The symbols denote the nightly observations:

● August 18, 1978	□ September 3 , 1978
○ August 24, 1978	▽ September 11, 1978
× August 26, 1978	▲ August 6 , 1979
+ August 30, 1978	^ August 28, 1979

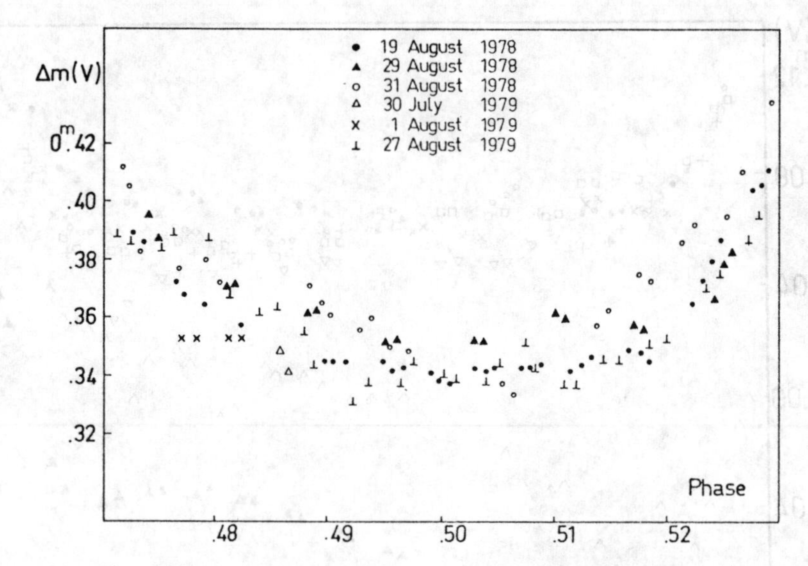

Figure 19. The light changes during the annular eclipse in colour V.

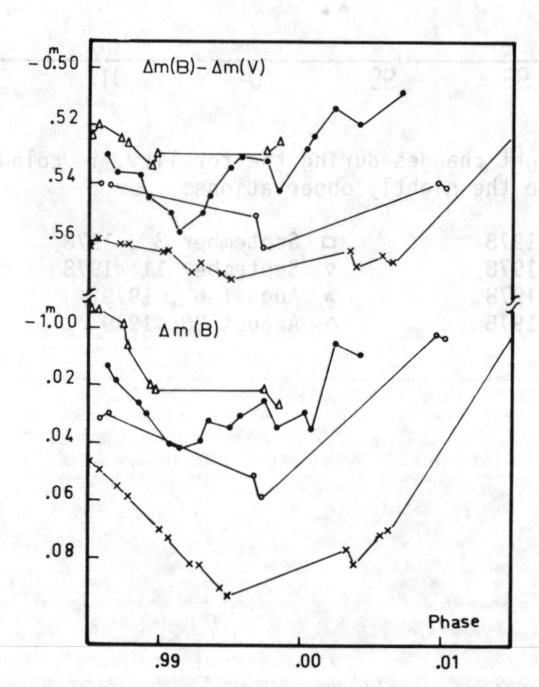

Figure 20. The B light and colour changes at mid-primary minimum of SZ Psc. The symbols denote the nightly observations as: Δ =JD 2444878, • =JD 2444882, o = JD 2444894, x = JD 2444930.

6. Light Variations of RT Lac

The eclipsing binary RT Lac is known as active binary in a wide wavelength interval. It has a distorted light curve like other RS CVn type binaries. Although we have no precise method for solving such distorted light curve both of the approximations made by Milone (1977) and, Eaton and Hall (1979) indicated that the less massive secondary is larger than the more massive, late-type compenent. If these solutions are accepted to be correct, it seems to be very difficult to explain the bluer primary and redder secondary minimum. This apparent paradox has been tried to explain with an envelope, which surrounds the more massive component, by Hall and Haslag (1976), and Milone (1977), while Milone and Naftilan (1980) proposed circumstellar matter.

RT Lac has many arguments unsolved up to date:

1) The large differences in the brightnesses of the maxima,
2) The bluer primary and redder secondary minimum,
3) Precise analysis of the light curves and actual dimensions of the components,
4) Mass-transfer, or envelope around one of the compenents,
5) Period changes,
6) Phase-depended and phase-independent infared excesses,
7) Unusual variations of the brightnesses at maxima and,
8) Unusually luminous radio source when compared with the other RS CVn binaries.

Hall and Taylor (1971), Hall and Haslag (1976), Shore and Hall (1978), Tunca et al. (1983) tried to explain light curve variations with star-spot hypothesis. They suggested a period for the migration from 4 to 30 years. Evren et al. (1985) reported that starspot hypothesis is insufficient alone to explain all the phenomena observed. The extensive phase and time coverage H alpha spectroscopy made by Huenemoerder (1985) indicated evidences for intermittent gas flow in this semi-detached system. Later on, Huenemoerder and Barden (1986) obtained the spectra of the system in the H alpha and H beta regions and concluded that RT Lac is a mass-transfer system.

This peculiar system was observed in successive 12 years from 1978 to 1989. The light curves are shown in Fig. 21. As it is noticed the light curve of the system shows dramatic variations. Glancing at the bottoms of the primary and secondary minima, one easily notices that the brightnesses at both mid-minima vary with time. The brightnesses at mid-primary and secondary were plotted against the years and are shown in Fig. 22 and 23.

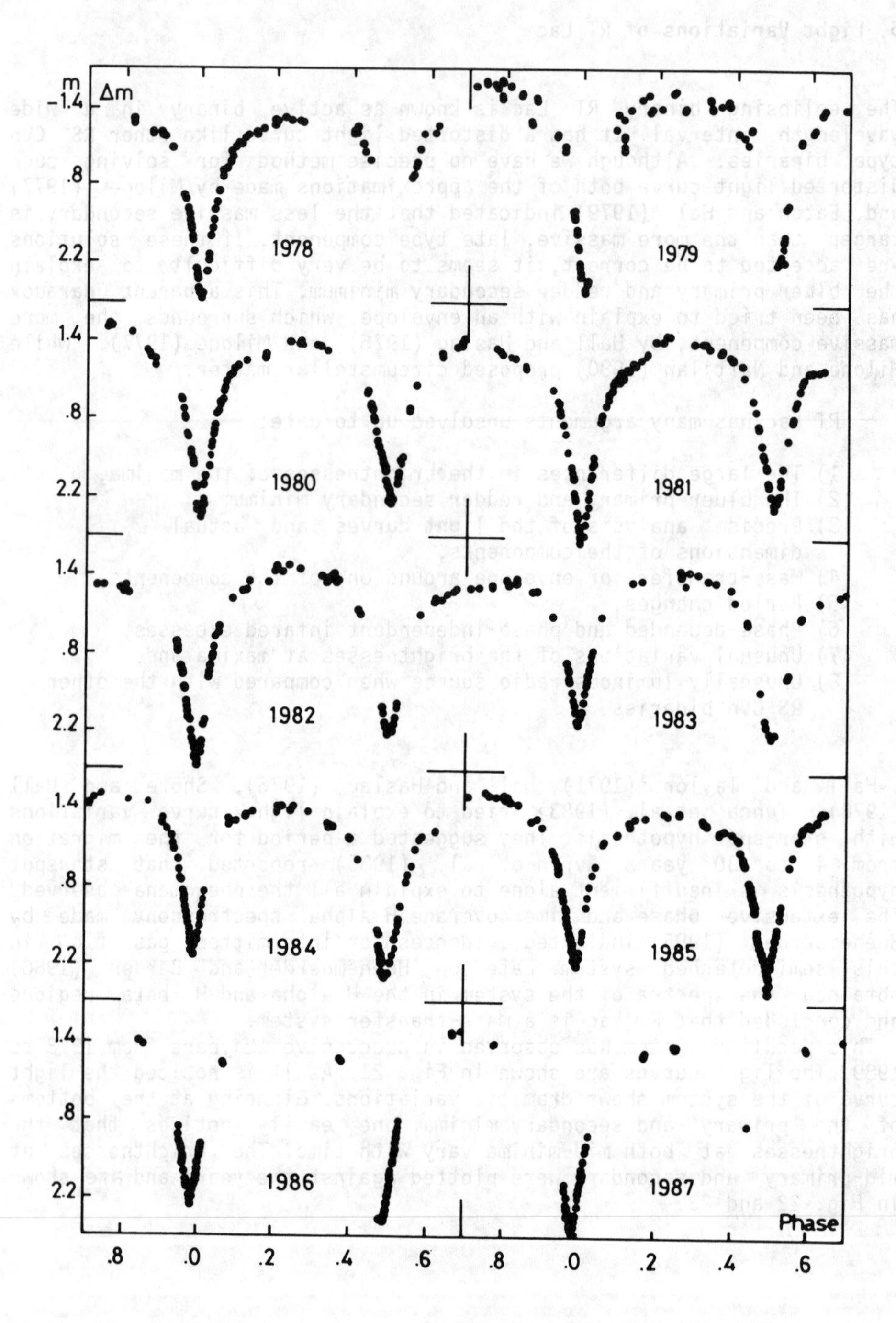

Figure 21. The light curves of RT Lac obtained in last 10 years.

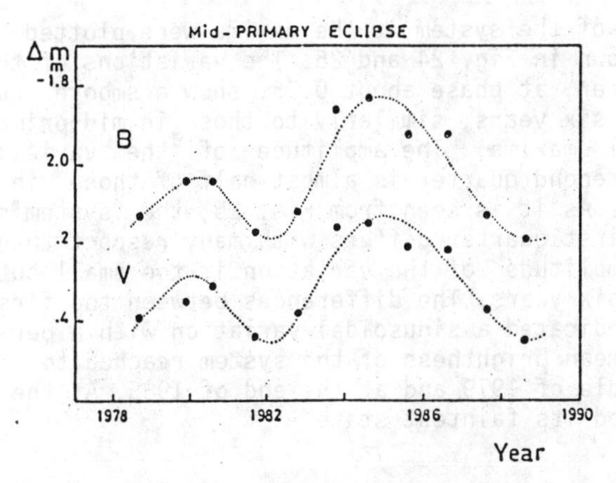

Figure 22. The variations of the brightness of RT Lac at mid-primary.

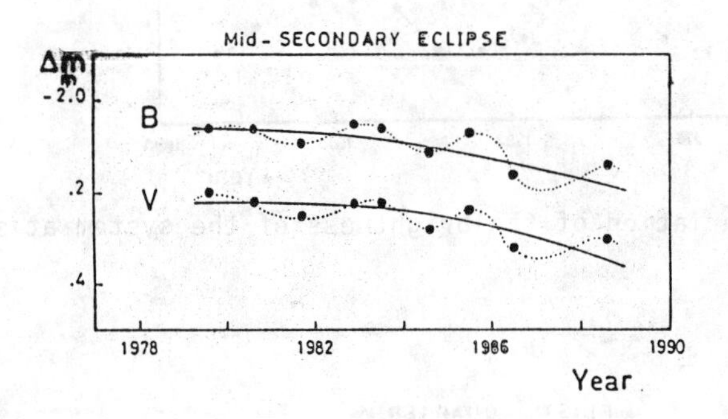

Figure 23. The variation of the brightness of RT Lac at mid-secondary.

The variation of the brightness at mid-primary seems to be periodic, which amounts to about six years, with an amplitude 0.36 mag in B and 0.32 mag in V. On the other hand, Fig. 23 shows that the brightness at mid-secondary minimum seems to be decreasing continuously since 1979. In addition to this variation there is an indication of a quasi-periodic variations with an increasing amplitude superimposed on the continuously decreasing one. The brightness at mid-secondary has been decreased about 0.10 mag during the last 12 years.

The brightnesses of the system at the maxima were plotted against the years and are shown in Fig. 24 and 25. The variations of the brightness at second quarter, at phase about 0.75, show a smooth curve with a period of about six years, similarly to those in mid-primary, but it has two unequal maxima. The amplitude of the variations of the brightness at second quarter is almost half of those in mid-primary (e.i. 0.18 mag). As it is seen from Fig. 25, the system's brightness variation at first quarter differs with many respect than at second quarter. The amplitude of the variation is too small but it has a period of about six years. The differences between the first and second quarters also indicated a sinusoidal variation with a period of about six years. The mean brightness of the system reached to its maximum value in the middle of 1979 and at the end of 1985. At the end of 1982 the system reached its faintest state.

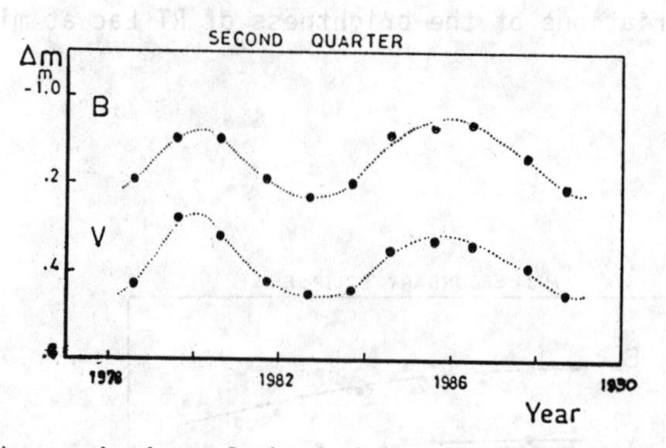

Figure 24. The variation of the brightness of the system at second quarter.

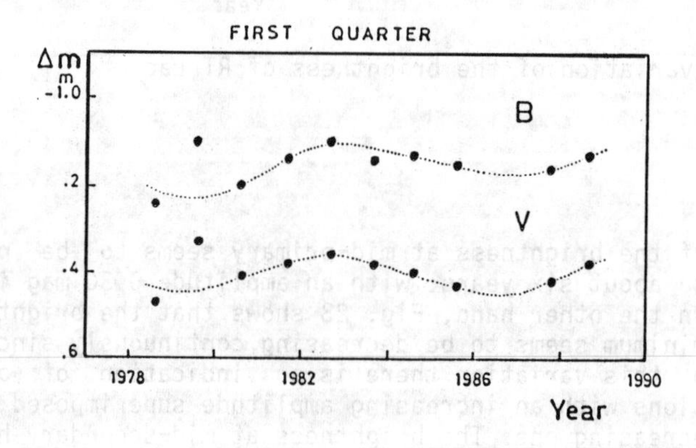

Figure 25. The variation of the brightness of the system at first quarter.

The mean colours of the system at four different phases of the light curve are shown in Fig. 26. The colour variation at mid-primary follows the variations of brightnesses at mid-primary and first quarter. The colour at mid-primary is about 0.07 mag bluer in 1980 and 1985 than in 1982. Whereas there is no indication about variation of the colour at mid-secondary.

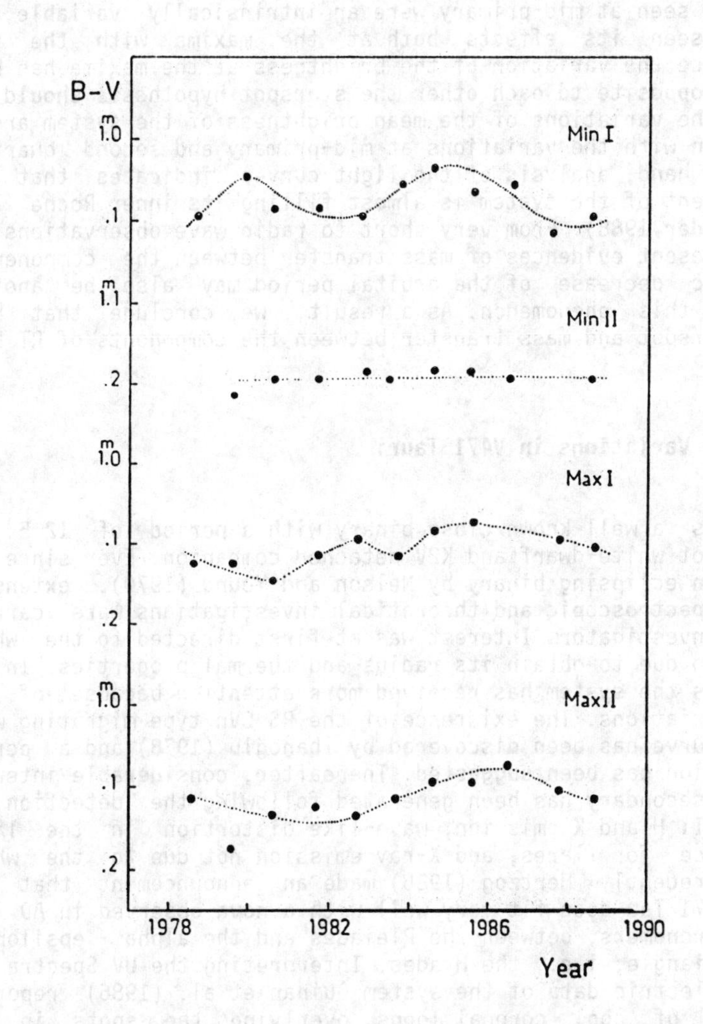

Figure 26. The colour variations of RT Lac at mid-eclipses and maxima.

The wave-like distortions seen in the light curves of RS CVn type binaries can easly be explained with a simple starspot model. The light curves obtained successively during the last 12 years clearly indicate that the system brigtnesses at both maxima and minima are changing with time. The variations occurred at mid-primary minimum are being at least three times larger than those of secondary minimum. Therefore, the variations in the light curve of the system should be arosen from the star seen at primary minimum. In fact, the variations seen at second quarter seem to follow the variations at mid-primary eclipse with an equal period. The colour variations corresponding to these phases also confirmed the brightness variations. However, the variations at first quarter are oppositely changed with respect to the variations seen at second quarter and mid-primary minimum.

If the star seen at mid-primary were an intrinsically variable, we should have seen its effects both at the maxima with the same direction. Since the variation of the brightness at the maxima has been almost in the opposite to each other the starspot hypothesis should not be excluded. The variations of the mean brightness of the system are in same direction with the variations at mid-primary and second quarter. On the other hand, analysis of the light curves indicates that the hotter component of the system is almost filling its inner Roche lobe (cf. Huenemoerder,1988). From very short to radio wave observations of the system present evidences of mass transfer between the components. The systematic decrease of the orbital period may also be another indicator of this phenomenon. As a result, we conclude that both phenomena, starspot and mass transfer between the components of RT Lac, take place.

7. Light Curve Variations in V471 Tauri

V471 Tauri is a well known close binary with a period of 12.5 hr, containing a hot white-dwarf and K2V detached companion. Ever since its discovery as an eclipsing binary by Nelson and Young (1970), extensive photometric, spectroscopic and theoretical investigations were carried out by many investigators. Interest was at first directed to the white dwarf companion due to obtain its radius and thermal properties. In the last ten years the system has received more attention because of the light curve variations. The existence of the RS CVn type migrating wave in the light curve has been discovered by İbanoğlu (1978) and a period for the migration has been suggested. Thereafter, considerable interest in V471 Tau secondary has been generated following the detection of variable Ca II H and K emission, wave-like distortion in the light curve, evidence for flares, and X-ray emission not due to the white dwarf. Very recently Hertzog (1986) made an announcement that the location of V471 Tau does fit very well with a nova observed in AD 396 by chinese astronomers, between the Pleiades and the alpha - epsilon - lambda Tau triangle; i.e., the Hyades. Interpreting the UV Spectra and visible photoelectric data of the system Guinan et al. (1986) reported the existence of cool coronal loops overlying the spots in the atmosphere of the K dwarf. UV observations indicated that chromospherically active K-star has been surrounded with a hot plasma extending $5x10^5$ km above its surface. The EXOSAT X-ray observations of the system have been interpreted by Jensen et al. (1986). They have reported the detection of short X-ray fluxes from both the white dwarf and the K dwarf, the discovery of 9.25 min pulsation from the white dwarf and the discovery of orbital phase related soft X-ray dips. These dips occur on the ingress side at phase 0.83 and on the egress side at phase 0.17, which correspond to the triangular Lagrangian points of the binary orbit. At these phases the L_4 and L_5 points are near the line-of-sight to the white dwarf.

The orbital period of the system has been continuously decreasing since 1973. Analysing the O-C diagram many investigators suggested that abrupt period changes might be occurring in the system. Taking into account the interpretations based on the abrupt period changes are not appropriate and do not fit with the later observations Beavers et al. (1986) suggested a light-time effect.

V471 Tau was observed with the same instruments since 1973. During the observations about 40 light curves of the system were obtained. Using the orbital elements obtained by Ibanoglu (1978) the light variations due to the tidally distorted K dwarf have been subtracted out from all of the light curves, and the actual shape of the wave-like distortion was revealed. Using the resultant light variations the wave minima have been derived. The period of the wave migration determined by us is 182.17 days. However, Guinan and Sion (1981) suggested a period of about 247 days and recently a value of about 372 days has been proposed by Skillman and Pattersen (1988). As it is seen from Fig. 27 the shift of the wave minimum is noticeable within a month. The light curves obtained by us indicate that, with no doubt, the period of migration is about six months.

The mean brightness of the system, excluding the eclipse of the white dwarf, has been plotted versus the years in Fig. 28. The mean brightness seems to indicate a quasi-periodic variation with a period of about 18 years and an amplitude of about 0.15 mag. The first and the last light curves of the system, obtained by us, are shown in Fig. 29. The total brightness of the system has been increased of about 0.19 mag in both B and V during the last 16 years.

The red dwarf-white dwarf eclipsing binary V471 Tau is one of the interesting active binaries and is a ballerina of Hyades cluster as stated by Skillman and Patterson (1988). The Hyades membership of the system suggests that the ancestor of the white dwarf must have had a mass of at least two solar masses which is the mass at present turnoff of the cluster. Whereas, the total mass of the system is smaller at present than that value. If so, extensive mass loss in the course of its evolution should need to be taken place. High-resolution UV spectroscopy obtained by IEU by Bruhweiler and Sion (1986) revealed the existence of a high-velocity expanding gas around the system. On the other hand, the soft X-ray dips observed at phase 0.15, 0.18 and 0.85 were attributed to the material located at the triangular Lagrangian points (Jensen et al. 1986). Both UV and X-ray observations indicate some evidences about the existence of an absorbing material around the

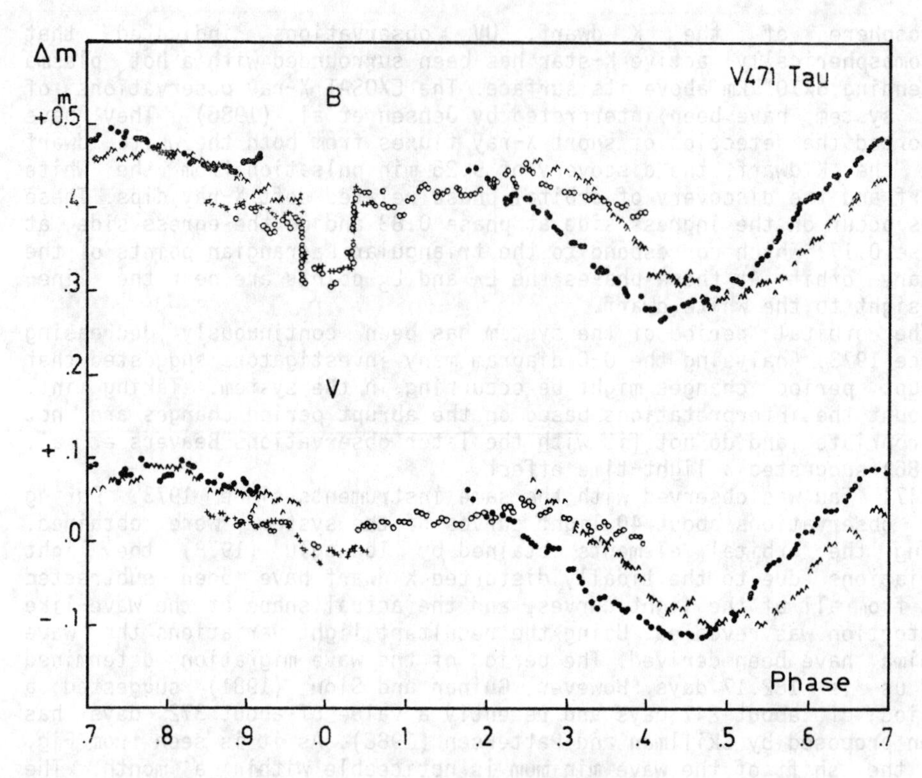

Figure 27. The blue and yellow light curves obtained in 1985.

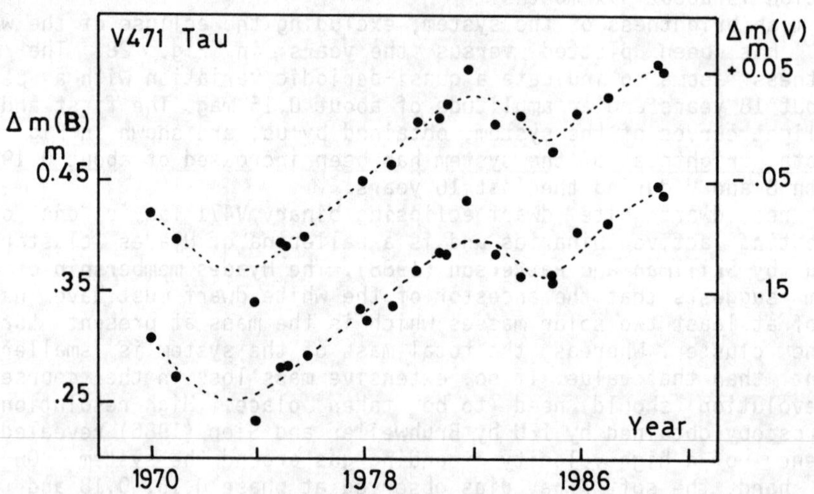

Figure 28. The variations of the mean brightness of V471 Tau versus the years.

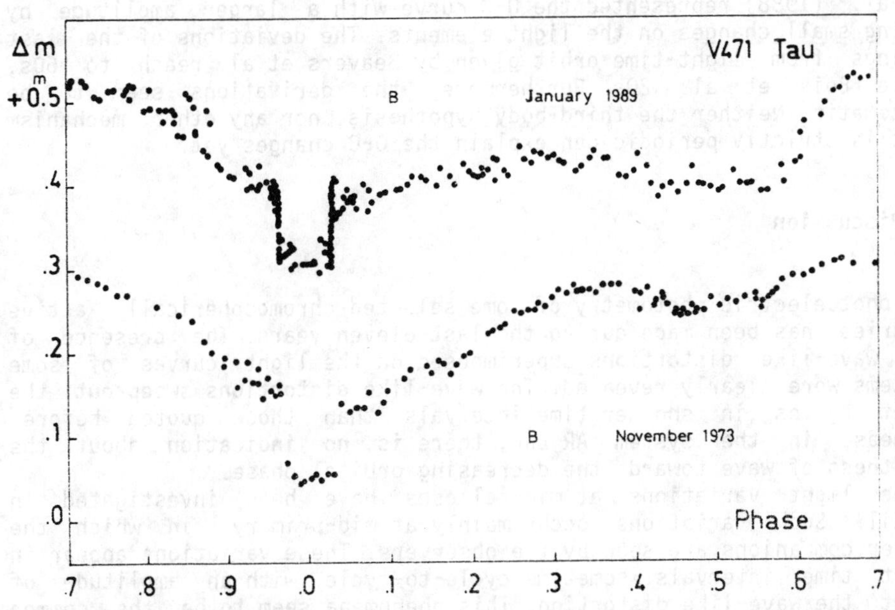

Figure 29. The blue light curves of the system obtained in November 1973 and in January 1989.

system at circumbinary distances. It seems, however, not to be a natural way to attribute the brightness variations to these events. The variations in the total brightness of the system were attributed to the strong magnetic avtivity by Skillmann and Patterson (1988). In this case, by the analogy with the Sun, the variation of the mean brightness would tend to be periodic behaviour as suggested by Evren et al. (1986). The brightness decreases occurred between 1970 and 1973 and also between 1982 and 1985 are very similar to each other and it may be associated with a strong magnetic activity. However, the general shape of the mean brightness variations appears to be uncorrelated with the activity of the red dwarf alone.

The O-C diagram of V471 Tau is shown in Fig. 4. Though some investigators propose that the minimum timing deviations are best represented by straight line segments, others suggest a quadratic or continuous curve. In the first case, sudden mass exchange or mass loss from the system should be happened. In the latter case, the period of the system would decrease slowly and could be related to either a mass exchange or the light-time effect caused by a third body orbiting the system. Beavers et al. (1986) tried to represent the O-C diagram with a light-time effect and derived the parameters of the orbit. Later, Bois et al. (1988) represented the O-C curve with a larger amplitude by making small changes on the light elements. The deviations of the last timings from light-time orbit given by Beavers et al. reach to 60s, while Bois et al. 20s. Furthermore, the derivations seem to be systematic. Neither the third-body hypothesis, nor any other mechanism that is strictly periodic can explain the O-C changes yet.

8. Discussion

The photoelectric photometry of some selected chromospherically active binaries has been made during the last eleven years. The presence of the wave-like distortions superimposed on the light curves of some systems were clearly revealed. The wave-like distortions sweep out the light curves in shorter time intervals than those quoted before. Whereas, in the system AR Lac, there is no indication about the shiftness of wave towards the decreasing orbital phase.

The light variations at mid-eclipses have been investigated in detail. Such variations occur mainly at mid-primary, in which the cooler companions are seen by the observers. These variations appear in short time intervals, sometime cycle-to-cycle, with an amplitude of about the wave-like distortion. This phenomena seem to be the common property of all RS CVn type binaries.

In most of the RS CVn-like systems the orbital periods show changes in a few decades. Mass transfer and mass loss have been accepted as a mechanism for providing torques which cause period changes in binary stars. However, in the case of RS CVn-like binaries mass transfer and mass loss appear to be unlikely, because they are generally detached systems and consist of main sequence and subgiant companions. Therefore, the period changes have been tried to explain by magnetic activity. In this case, an increase in the mean magnetic field throughout the convection zone provides an additional pressure support and increases the radius of the star. For the conservation of angular momentum the convective star must reduce its rotation rate. As a result by transferring the angular momentum from the star's rotation to the orbit the orbital period will be lengthened. When the field decays, the period shortens (Van Buren and Young, 1985). According to this hypothesis the orbital period of RS CVn-like systems will be largest during the active portion of the magnetic cycle and smallest during the quiescent part. The activity cycle can approximately be deduced from the variations of the light curves of RS CVn-type systems. The activity cycle has been derived for some binaries and it appears that there is no correlation between the period changes and activity cycles. On the other hand, the three-body theory should not be considered for explaining the timing residuals of the RS CVn-type binaries.

The most interesting systems in the observing list were RT Lac and V471 Tau. In the system RT Lac, a large amount of light variations occurs mainly in mid-primary eclipse. Similar variations appear at second quarter but in a lesser scale. Whereas, the variations at first quarter are opposite to them. This behaviour of the variation of the light suggests that the cooler -also larger- component of RT Lac is spotted. The difference of the amplitudes of variation at quarters may arise from the mass transfer from G9 companion to K1 companion. Mass falling surface of the K1 companion is seen by the observer during the first quarter.

The white-dwarf red-dwarf eclipsing binary V471 Tau has been observed from 1973 to 1989. The mean brightness of the system has decreased up to 1973, it has increased up to 1982 and decreased up to 1985 and it started to increase again. The system has brightened by about $0^{m}.19$ since 1973. The brightness decrements occurred between 1970 and 1973 and 1982 and 1985 are very similar to each other. These variations may be associated with a strong magnetic activity. However, the general shape of the mean brightness variations appears to be uncorrelated only with the activity of the red dwarf.

Unfortunately starspot hypothesis appears to be inadequate for explaining all the phenomena observed in chromospherically active close binaries.

References

Beavers, W.I., Herczeg, T.J., and Lui,A.: 1986, Astrophys.J. **300**, 785.

Bois, B., Lanning, H.H., and Mochnacki, S.W.: 1988, Astron.J. **96**, 157.

Bruhweiler, F.C., and Sion, E.M.: 1986, Astrophys. J. Lett. **304**, L21.

Catalano, S., and Rodono, M.: 1967, Mem. Soc. Astron. Ital. **38**, 395.

Catalano, S., and Rodono, M.: 1969, Non-Periodic Phenomena in Variable Stars, ed. L. Detre, Academic Press, Budapest, p.435.

Çelikezer, N.: 1976, Master's thesis, Ege University, Bornova, İzmir.

Cellino,A., Scaltriti,F., Busso,M.: 1985, Astron. Astrophys. **144**, 315.

Eaton, J., and Hall, D.S.: 1979, Astrophys. J. **227**, 907.

Evren, S.: 1989, Astrophys. Space Sci. (in press).

Evren, S., İbanoğlu,C., Tunca,Z., and Tümer,O.: 1986, Astrophys. Space Sci. **120**, 97.

Evren, S., İbanoğlu, C., Tümer, O., Tunca, Z., and Ertan, A.Y.: 1983, Astrophys. Space Sci. **95**, 401.

Evren, S., Tunca, Z., İbanoğlu, C., and Tümer, O.: 1985, Astrophys. Space Sci. **108**, 383.

Guinan, E.F., and Sion, E.M.: 1981, Bull. Am. Astron. Soc. **13**, 817.

Guinan, E.F., Wacker, S.W., Baliunas, S.L., Loeser, J.G., and Raymond, J.C.: 1986, in 'New Insights in Astrophysics: Eight Years of UV Astronomy with IUE', ESA SP-263, 197.

Hall, D.S.: 1976, in W. Fitch (ed.) 'Multiple Periodic Variable Stars' IAU Colloq. 29.

Hall, D.S., and Haslag, K.P.: 1976, in W. Fitch (ed.) 'Multiple Periodic Variable Stars', IAU Colloq. 29.

Hall, D.S., and Taylor, M.C.: 1971, Bull. Am. Astron. Soc. **3**, 12.

Hertzog, K.P.: 1986, The Observatory **106**, 38.

Hilditch, R.W., Harland, D.M., and Mc Lean, B.J.: 1979, Monthly Notices Roy. Astron. Soc. **187**, 797.

Huenemoerder, D.P.: 1985, Astron. J. **90**, 499.

Huenemoerder, D.P.: 1988, Publ. Astron. Soc. Pacific 100, 600.

Huenemoerder, D.P., and Barden, S.C.: 1986, Astron. J. **91**, 583.
İbanoğlu, C.: 1978, Astrophys. Space Sci. **57**, 219.
Jensen, K.A., Swank, J.H., Petre, R., Guinan, E.F., Sion, E.M., and Shipman, H.L.: 1986, Astrophys. J. Lett. **309**, L27.
Milone, E.F.: 1977, Astron. J. **82**, 998.
Milone, E.F., and Naftilan, S.A.: 1980, in 'Close Binary Stars: Observations and Interpretations', IAU Symp. No.**88**, ed. M.J. Plavec and R.K. Ulrich (Reidel, Dordrecht).
Nelson, B., and Young, A.: 1970, Publ. Astron. Soc. Pacific **82**, 695.
Oliver, J.P.: 1974, Ph.D. Thesis, Univ. of California, Los Angeles.
Patkos, L.: 1981, Astrophysical Letters **22**, 1.
Popper, D.M.: 1970, in 'Mass Loss and Evolution in Close Binaries', IAU Colloq. **6**, 13.
Shore, S., and Hall, D.S.: 1978, Report at RS CVn Workshop, Secorro, N.M.
Skillman, D.R., and Pattersen, J.: 1988, Astron. J. **96**, 976.
Struve, O.: 1946, Ann. Astrophys. **9**, 1.
Tunca, Z.: 1984, Astrophys. Space Sci. **105**, 23.
Tunca, Z., İbanoğlu, C., Tümer, O., Ertan, A.Y., and Evren, S.: 1983, Astrophys. Space Sci. **93**, 431.
Van Buren, D.: 1986, Astron. J. **92**, 136.
Van Buren, D., and Young, A.: 1985, Astrophys. J. Lett. **295**, L39.
Young, A., and Lanning, H.H.: 1975, Publ. Astron. Soc. Pacific **87**, 461.

SIMILARITIES AND DIFFERENCES BETWEEN THE LIGHT VARIABILITY OF THE K GIANT AND DWARF ACTIVE BINARIES HK LAC AND BY DRA

K. Olah
Konkoly Observatory
1525 Budapest, P.O. Box 67
Hungary

ABSTRACT. The long term light and color (B-V, U-B) variations of BY Dra and HK Lac are discussed. Cyclic behaviour of the visual and U-B color variation of HK Lac has been found with a period of 5.4 years, showing 0.3 phase shift between the V(mean) and U-B cycles. A possible explanation of the feature is given.

1. INTRODUCTION

The light variability of the two binaries HK Lac (K0III+F1IV?) and BY Dra (K4V+K7.5V) due to starspot activity, is well known and investigated in the literature (see e.g. Olah and Hall, 1988, for HK Lac and Poe and Eaton, 1985, for BY Dra). Basic data of the two systems can be found in Strassmeier et al.'s (1988) catalog, under numbers 126 (BY Dra) and 154 (HK Lac). Using some unpublished and new observations of the two systems we try to describe the observed similarities and differences of their variability. The time base we used is the same for the two stars: from J.D. 2443000 (1977) till the present. From this date more or less continuous ultraviolet observations were already available which were necessary for our investigations.

2. OBSERVATIONS

The photometric data were taken from the literature and were supplemented by our data (partly unpublished). Our observations at Konkoly Observatory between 1977-1987 were made by the 60 cm and 50 cm telescopes using integrating photometers in the standard Johnson UBV system, and in 1988 by the 1 m telescope mounted with a thermoelectrically cooled (-20 ^{o}C) UBV$(RI)_{KC}$ photon counting photometer.

C. İbanoğlu (ed.), Active Close Binaries, 545–549.

3. RESULTS

The light and color long term variability of BY Dra and HK Lac are shown in Figures 1 and 2. The upper panels display the ΔV values: the horizontal and vertical extent of the rectangles give the time bases of the observational data and the amplitudes, respectively.

The observed amplitudes of the dwarf binary BY Dra are compressed by the substantial light contribution of the secondary component. If we consider Lp/Ls = 1.93 (Vogt, Fekel, 1979) then the real amplitudes should be 1.5 times higher (supposing, that only the primary has spots). The color variations of BY Dra show only fluctuations. The B-V color curves have small amplitudes while sometimes high amplitude U-B variations were measured.

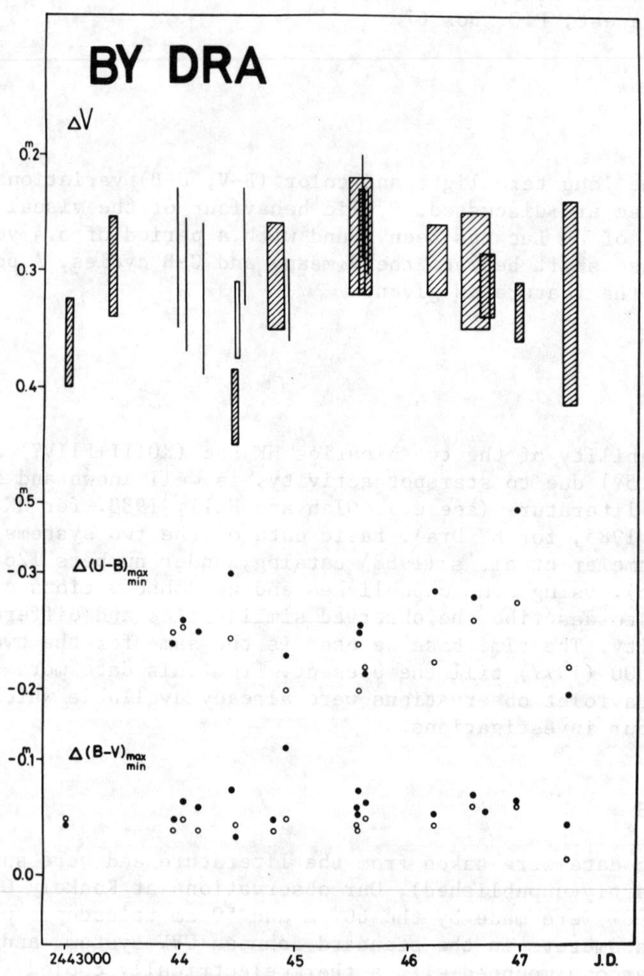

Figure 1: Light and color variations of BY Dra. Symbols of the Δ(B-V), Δ(U-B) colors: dots - maximum, circles - minimum values.

The giant star HK Lac shows cyclic behaviour in the mean V brightness and this variability is even more evident in the U-B color (see Figure 2). A long term trend is seen in the B-V data without any apparent correlation to the visual and U-B variability.

The results of the light and color long-term variability of BY Dra and HK Lac are summarized in Table I.

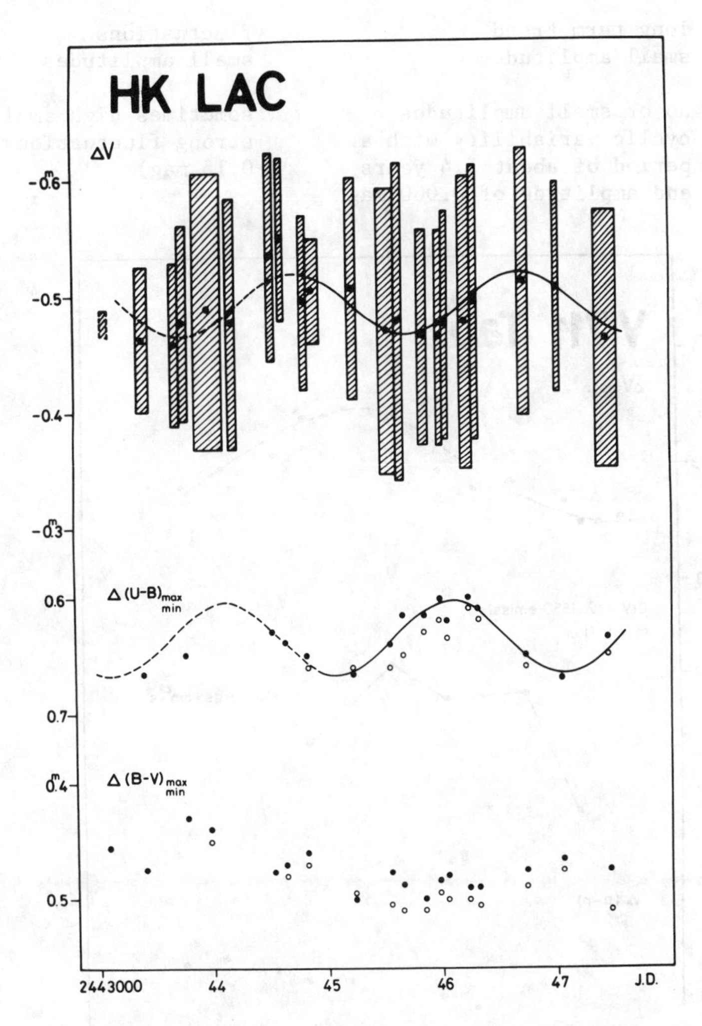

Figure 2: Light and color variations of HK Lac. Symbols as in Figure 1. Dots in the upper panel show the mean ΔV values.

TABLE I

	HK LAC	BY DRA
V	amplitude up to 0.2-0.25 m cyclic variability in the mean V (P = 5.4 years, amplitude is 0.052 mag)	amplitude up to 0.2-0.25 m slow changes in the mean V
B-V	long term trend small amplitudes	fluctuations small amplitudes
U-B	no or small amplitudes cyclic variability with a period of about 5.4 years and amplitude of 0.060 mag	sometimes high amplitudes strong fluctuations (within 0.15 mag)

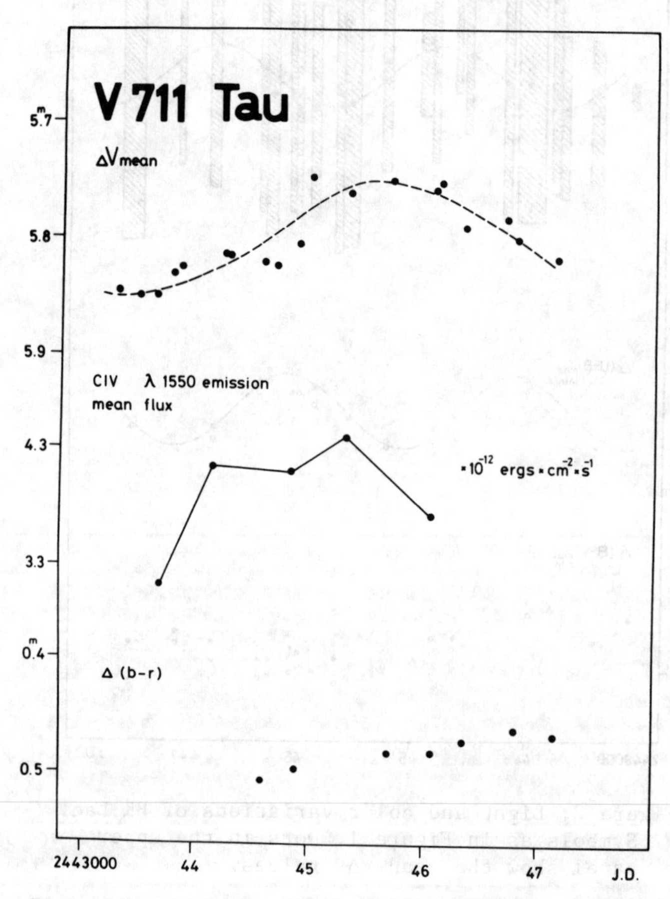

Figure 3: Variations of V711 Tau in mean V light, in CIV (λ1550) mean flux and in Δ(b-r) color.

Investigating the cyclic variations of HK Lac, a simple sine-wave was fitted to the V(mean) and U-B(max) data, which gave an average period of about 5.4 years with 0.052 mag amplitude in V and 0.060 mag amplitude in U-B. The existence of a quasi-periodic variation of 5.5 years cycle length in U-B has been found by Olah (1989). Interesting is the phase shift between the V(mean) and U-B cycles. The V(mean) curve is about 0.3 phase precedes the U-B cycle and not 0.5 phase, as suspected, if we think that the spot maximum (visible light minimum) corresponds to the chromospheric activity maximum (UV light maximum). After the maximum of the spot activity of HK Lac around 2445800-900, the star started to brighten, while the U-B light was still increasing till about 2446200. When the chromospheric activity reached its minimum level around 2447100, the spot activity has already started to increase.

A similar feature, but more pronounced, was seen in V711 Tau by Dorren and Guinan (1989, see Figure 3), using their optical and IUE data. They found that with increasing chromospheric activity, the visual brightness of the star was also increasing. A possible explanation of this result is given by Dorren and Guinan (1989): the visual light variability is a result of a competition between the darkening effect of the spots and the brightening effect of the white-light faculae. It is quite possible that a similar mechanism is operating in HK Lac which may explain the 0.3 phase shift between the V(mean) and U-B variability: the facular white light contribution makes the visual light brighter in the time of the activity maximum.

ACKNOWLEDGEMENT. My sincere thanks are due to Dr. E.F. Guinan for his valuable comments and discussions.

REFERENCES

Dorren, J.D., Guinan, E.F., 1989, Astrophys. J., in press

Olah, K., 1989, Proc. IAU Coll. No. 104 (Solar and Stellar Flares, ed. by B.M. Haisch and M. Rodono), Poster Papers, Catania Special Publ. p. 135

Olah, K., Hall, D.S., 1988, Comm. Konkoly Observatory, No. 91 (Vol. 10, Part 5), p. 121

Poe, C.H., Eaton, J.A., 1985, Astrophys. J. 289.644

Strassmeier, K.G., Hall, D.S., Zeilik, M., Nelson, E., Eker, Z., Fekel, F.C., 1988, Astron. Astrophys. Suppl. 72.291

Vogt, S.S., Fekel, F., 1979, Astrophys. J. 234.958

RECENT MEASURES OF CaII H AND K CHROMOSPHERIC FLUXES IN A SAMPLE OF RS CVN SYSTEMS

J.E. Armentia[*], M.J. Fernández-Figueroa[*], M. Cornide[*], E. de Castro[*] and J. Fabregat[+]
* *Departamento de Astrofísica*
Facultad de Ciencias Físicas
Universidad Complutense de Madrid
E-28040 Madrid (Spain)
+ *Departamento de Matemática Aplicada*
Facultad de Ciencias Matemáticas
Universidad de Valencia
Burjassot, Valencia (Spain)

ABSTRACT. In this contribution we present the current status of a high resolution H and K spectroscopic survey of RS CVn and BY Dra systems which is being performed at our Department. The reduction and calibration procedures used in the extraction of emission fluxes are analyzed. We discuss the preliminary results for three of the RS CVn systems: HR 5110, Z Her and RT Lac.

1. INTRODUCTION

In 1985 we started a project devoted to the study of spectroscopic properties of chromospherically active binary stars,in the region of CaII H and K lines, principally for RS CVn and BY Dra systems. As it is widely known, the emission profiles on these absorption lines are the most noticeable tracers in the optical domain of the presence of active chromospheres in late-type stars.

The aim of such a survey is to collect a wide sample of emission flux measurements with which several studies can be performed:

- existence of variations related with the orbital phase, indicating the presence of active regions in the active star;
- long term variations, informing about the different activity levels, connected with the existence of activity cycles in these objects;
- finally, a large sample of fluxes is needed in order to study the chromospheric activity phenomenon and its correlation with structural parameters of the stars.

C. İbanoğlu (ed.), Active Close Binaries, 551–560.

2. OBSERVATIONS

The observational programme is being carried on two telescopes:
- 2.2m Telescope at the German Spanish Astronomical Observatory in Calar Alto (Almería, Spain), using the Coudé B&C Spectrograph and a CCD as detector. November 1986, January 1988 and July 1989.
- Isaac Newton Telescope (INT) at the Observatorio del Roque de los Muchachos (La Palma, Spain), using the IDS and an IPCS as detector. June 1985, June 1987 and July 1988.

Both configurations allow a resolution of 0.2 Å/pixel in the region of CaII lines.

The list of the systems observed is shown in Table I, with the number and date of the spectra obtained along with the status of their study. There are a total of 52 systems observed, 25 of them in more than one epoch and 30 with more than 2 spectra per run.

Partial results of this survey have already been published (Fernández-Figueroa et al., 1986; Castro et al., 1989)

3. DATA REDUCTION

The wavelength-calibrated spectra are extracted from IPCS and CCD images, using the standard reduction procedures for these kind of images in IHAP and MIDAS packages, available at the Villafranca Satellite Tracking Station Computer Center (ESA, Madrid). For some of the images, an optimized extraction algorithm (following Horne, 1986) has been tested, finding no important improvement of the S/N ratio (the lowest S/N ratios are near 50).

The spectra are then corrected from atmospheric extinction by means of the semiempirical method by Hayes and Latham (1975). We have compared this method with the extinction curve from La Palma for INT spectra and also with the photometric extinction coefficients obtained for Calar Alto Observatory by Fabregat (1989), having found no significative deviations in any case.

The flux calibration is performed using spectra of standard stars from Oke and Gunn (1983) and from Barnes and Hayes (1984). After this calibration the flux units are erg cm^{-2} s^{-1} $Å^{-1}$. Night to night discrepancies found for the standard stars are within a 5%. For some of the systems (those having V-R determinations for each component) we have been able to calculate the continuum flux at 3950.5 using the relation from Pallavicini et al. (1988) and the agreement is also within a 10%.

The emission fluxes are being measured reconstructing the absorption profile below the peak(s) following the method of Blanco et al. (1974). Since the emission usually originates at the cooler star of the system, whose contribution to the continuum in the region is very low, this procedure can be the only useful in most of the cases, because an integration of the line from zero would generally take much of the flux coming from the hotter companion.

Table I.
Resume of the observations

Star	Obs.	Status	Star	Obs.	Status
33 Psc	1 (NOV/86)	A	ε UMi	1 (JUL/89)	A
13 Cet	4 (NOV/86)	A	V792 Her	1 (JUN/87)	A
HD 8357	3 (NOV/86)	A		1 (JUL/89)	A
HR 1099	2 (NOV/86)	A	HR 6469	1 (JUN/87)	A
	(JAN/88)	A		2 (JUL/88)	A
HD 26337	1 (JAN/88)	A		1 (JUL/89)	A
12 Cam	5 (NOV/86)	A	29 Dra	1 (JUN/87)	A
CQ Aur	2 (NOV/86)	A		1 (JUL/89)	A
SV Cam	6 (NOV/86)	A	Z Her	2 (JUN/85)	P (a)
σ Gem	3 (JAN/88)	A		1 (JUL/88)	W
54 Cam	3 (NOV/86)	A		4 (JUL/89)	W
	1 (JAN/88)	A	MM Her	1 (JUL/88)	A
RZ Cnc	1 (JAN/88)	A		1 (JUL/89)	A
DH Leo	2 (JAN/88)	A	V772 Her	10 (JUL/88)	P (b)
ξ UMa	3 (JAN/88)	A		2 (JUL/89)	A
93 Leo	2 (JAN/88)	A	HD 166181	2 (JUN/85)	P (a)
HR 4665	1 (NOV/86)	A		6 (JUL/88)	A
	3 (JAN/88)	A	HR 7125	1 (JUL/89)	A
AS Dra	1 (JUN/85)	P (a)	V775 Her	5 (JUL/88)	A
HD 108102	2 (JUN/85)	P (a)	V1762 Cyg	1 (JUL/89)	A
	3 (NOV/86)	A	HR 7428	2 (JUL/88)	A
UX Com	2 (JUN/85)	P (a)	V1764 Cyg	1 (JUL/89)	A
RS CVn	5 (JUN/85)	P (a)	ER Vul	5 (JUL/88)	A
	2 (JAN/88)	A		3 (JUL/89)	A
HR 5110	4 (JAN/88)	W	HR 8283	1 (JUL/88)	A
	1 (JUL/89)	W	RT Lac	1 (NOV/86)	W
RV Lib	1 (JUN/87)	P (b)		2 (JUL/88)	W
SS Boo	1 (JUN/87)	P (b)		1 (JUL/89)	W
	1 (JUL/88)	P (b)	HK Lac	1 (JUL/89)	A
GX Lib	1 (JUN/87)	A	AR Lac	2 (JUL/89)	A
	1 (JUL/89)	A	V350 Lac	1 (JUL/89)	A
RT CrB	1 (JUN/87)	P (b)	IM Peg	1 (JUL/88)	A
	1 (JUL/88)	P (b)		1 (JUL/89)	A
RS UMi	1 (JUN/87)	A	SZ Psc	3 (NOV/86)	A
	1 (JUL/88)	A		1 (JUL/88)	A
TZ CrB	3 (JUN/85)	P (a)		1 (JUL/89)	A
	2 (JAN/88)	A	λ And	1 (JUL/89)	A
WW Dra	1 (JUN/87)	P (b)			
	1 (JUL/88)	P (b)			
	1 (JUL/89)	A			

Notes:
A reduced, flux calibrated, analysis being performed
P published
W this work
a) Fernández-Figueroa et al., 1986
b) Castro et al., 1989

For some systems (as it is described further for HR 5110) the absorption feature can be "isolated" from the spectra, because the emission line arising from the active cool component is shifted in wavelength during the orbital period, leaving different parts of the absorption profile free of emission.

This criterion, however, seems to overestimate the photospheric contribution (Linsky and Ayres, 1978) and some other errors can arise, as quoted in a previous paper (Fernández-Figueroa et al., 1986)

Finally, surface flux calculations are carried out for those systems with determinations of radii and distance, or in those cases for which we can apply the calibrations of Blanco et al. (1982) or Pallavicini et al. (1988).

4. RESULTS

We present and discuss the results for three of the observed systems (HR 5110, Z Her and RT Lac). The stellar parameters used are shown in Table II, and the flux measures for each observation in table III.

Table II
Stellar parameters

Name	d(pc)	P_{orb}(days)	i(°)	SpT	$M/M_{\odot}$	$R/R_{\odot}$	eclipse
HR 5110 (BH CVn)	53	2.6132	8.9	F2IV	1.5	3.10	none
				K2IV	0.8	2.85	
Z Her	75	3.9928	83.13	F4V-IV	1.23	1.69	partial
				K0IV	1.10	2.60	
RT Lac	205	5.0740	89.0	K1IV	0.78	4.2	total
				G9IV	1.66	3.4	

References:
Strassmeier et al.,1988, Astron.Astrophys.Suppl.Ser.,72,291, and references therein
Batten et al.,1989, Publ.Domin.Astrophys.Observ.,XVII

Table III
Results

Name	JD observation	Phase	λ	f_{obs}	F_{sup}
HR 5110	244 7188.7125	0.188	K 3933.9	38.58	2.62
			H 3968.8	29.63	2.01
	244 7189.7458	0.583	K 3933.2	32.07	2.18
			H 3968.1	20.89	1.42
	244 7191.7167	0.338	K 3934.1	28.24	1.92
			H 3968.9	21.80	1.48
	244 7192.7257	0.724	K 3933.3	33.30	2.26
			H 3968.0	27.45	1.87
	244 7721.36.46	0.021	K 3933.9	34.38	2.34
			H 3968.6	20.77	1.41
Z Her	244 7372.4479	0.979	K 3933.2	9.62	1.57
			H 3967.9	5.87	0.96
	244 7722.4632	0.641	K 3932.2	8.58	1.40
			H 3966.9	6.33	1.03
	244 7723.4285	0.882	K 3932.5	9.01	1.47
			H 3967.3	7.46	1.22
	244 7724.4319	0.134	K 3934.2	8.38	1.37
			H 3969.2	8.35	1.36
	244 7725.4549	0.390	K 3934.3	8.22	1.34
			H 3968.9	7.07	1.15
RT Lac	244 6760.2694	0.87	K_c 3934.7	2.07	0.96
			K_h 3932.9	3.02	2.16
			H_c 3970.4	1.94	0.91
			H_h 3968.8	2.27	1.62
	244 7369.6590	0.970	K_T 3932.7	5.31	*
			H_T 3967.2	4.98	*
	244 7372.6076	0.551	K_T 3932.8	4.29	*
			H_T 3967.4	3.93	*
	244 7725.6069	0.121	K_c 3931.7	4.12	1.93
			K_h 3933.3	4.03	2.88
			H_c 3966.4	2.75	1.29
			H_h 3968.1	2.82	2.01

Notes:
Observed fluxes int units of 10^{-13} erg cm^{-2} s^{-1}.
Surface fluxes in units of 10^{-6} erg cm^{-2} s^{-1}.
* At these phases the emissions are blended; only the total emission fluxes are given

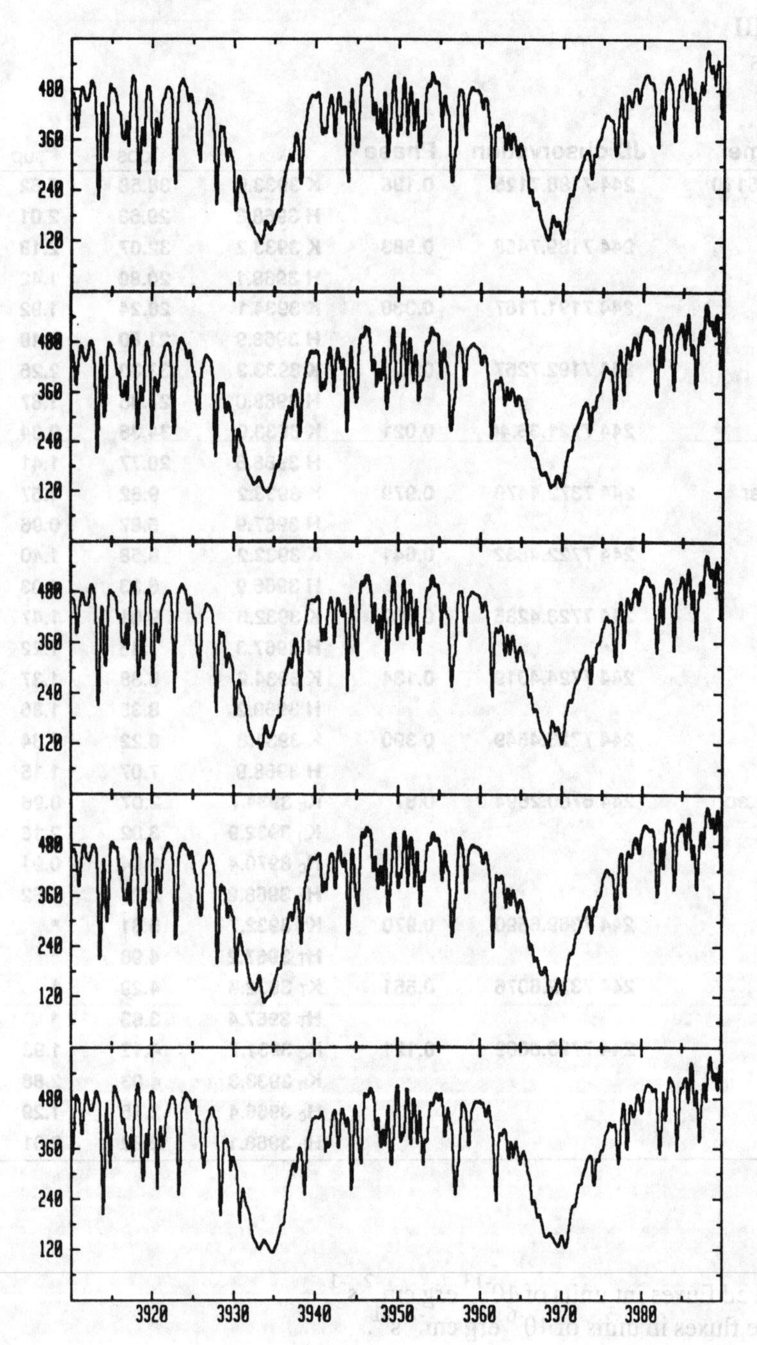

Figure 1.
Spectra of HR 5110. Phases are (from top to bottom): 0.188, 0.583, 0.338, 0.724 and 0.021. Flux in units of 10^{-13} erg cm^{-2} s^{-1}.

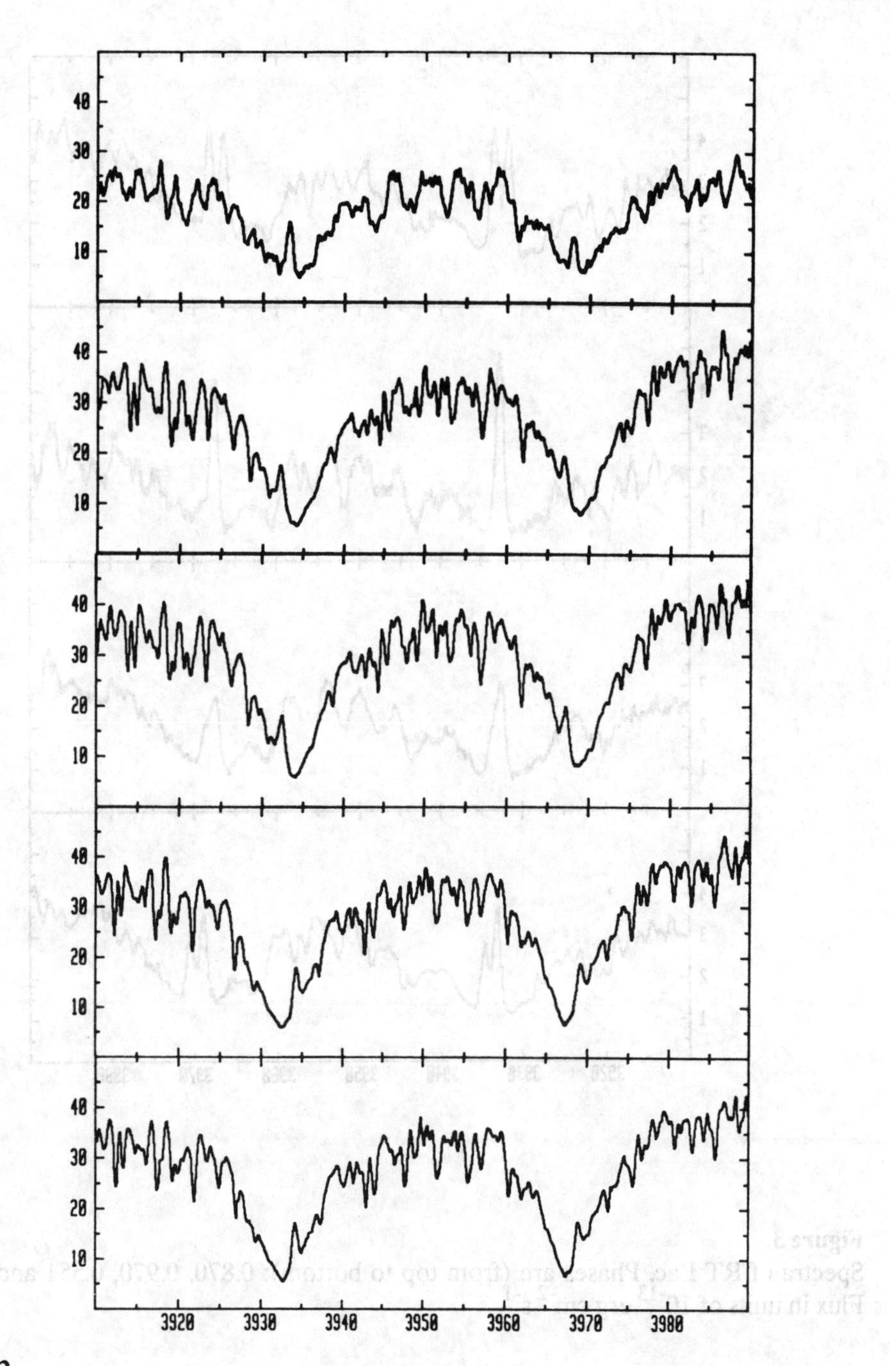

Figure2.
Spectra of Z Her. Phases are (from top to bottom): 0.979, 0.641, 0.882, 0.134 and 0.390. Flux in units of 10^{-13} erg cm^{-2} s^{-1}.

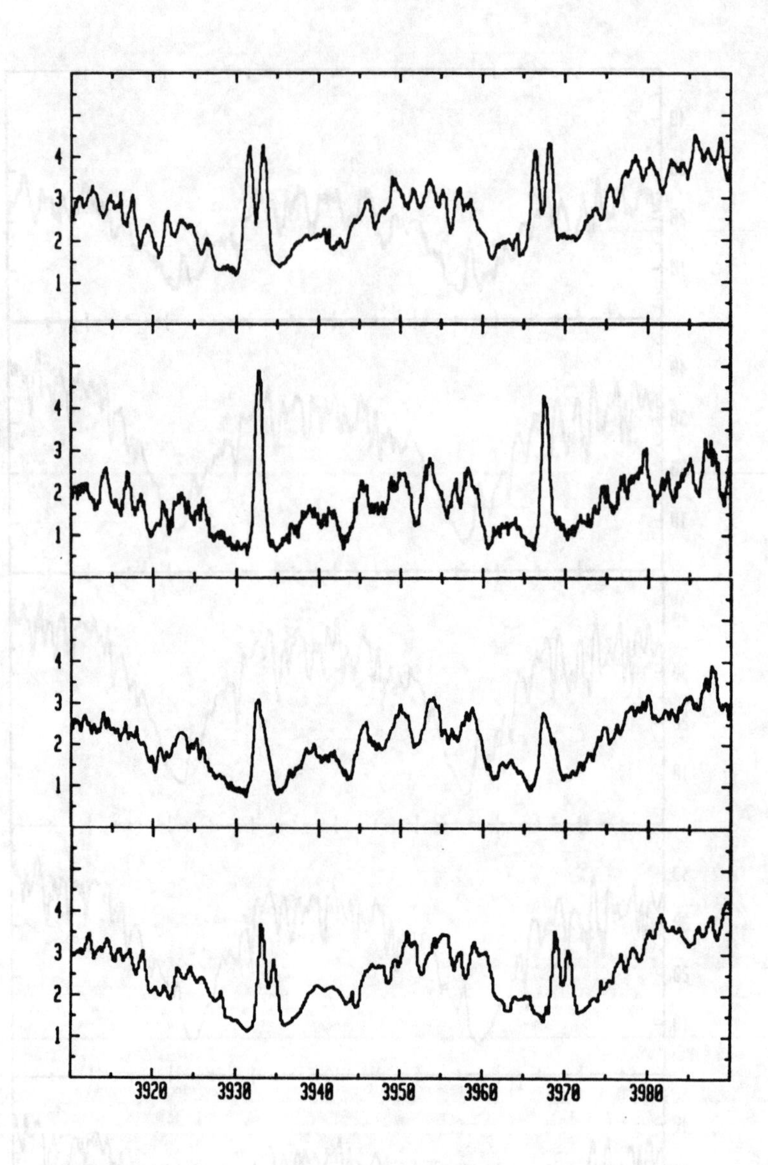

Figure 3.
Spectra of RT Lac. Phases are (from top to bottom): 0.870, 0.970, 0.551 and 0.121. Flux in units of 10^{-13} erg cm^{-2} s^{-1}.

HR 5110 (BH CVn)

This bright single-lined spectroscopic binary is classified as a RS CVn star because, although its light variation is due to a reflection effect rather to a migrating wave, the system presents radio outbursts and UV chromospheric lines common to other RS CVn stars.

The primary (spectral type F2IV) is twice as massive as the secondary, being this one (spectral type K2IV) almost filling its Roche lobe. This semi-detached configuration is characteristic of the Algol systems (Dorren and Guinan, 1980), but as the inclination is very low no eclipses are produced.

We have taken four spectra of this system at Calar Alto in January 1988 and one more in July 1989.

As it can be seen in figure 1, the emission of the active cool star shifts along the absorption during the orbital period, while the photospheric spectrum remains unshifted, because the mass relation is quite low. It is remarkable the presence of $H\varepsilon$ in absorption, which can be confused with a second emission peak.

A rough estimation of the contribution of the K star to the flux in the region gives a 5% of the flux from the F star, value within the calibration error. Actually, the spectrum does not vary along the period, exception made of the emission, so we can use the different spectra to reconstruct the photospheric profile of H and K lines. This allow us to separate the emission, measuring the fluxes. No special differences are found between the spectra at various phases nor at the two epochs of observation.

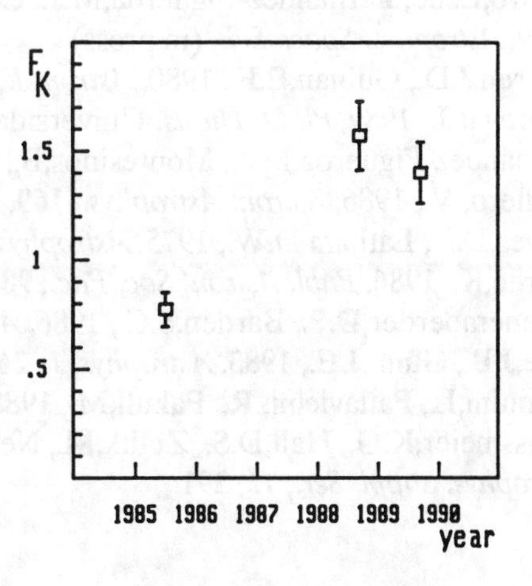

Figure 4.
Evolution of the emission flux in Z Her from 1985 to 1989.
(Units of 10^{-13} erg cm^{-2} s^{-1})
Error bars indicate a 15% of the flux.

Z Her

This system is a partially eclipsing spectroscopic binary, coming the main contribution to its spectrum from the F4IV-V star. The cool companion (K0IV) is responsible of the emission, resembling what occurred with HR 5110.

One spectrum has been obtained at La Palma in July 1988, at phase 0.979 during the primary eclipse (the hot star is behind, a 40% of its surface occulted). The other four spectra were taken in Calar Alto in July 1989 (figure 2)

The flux variations in the period are within the calibration errors, but the activity level seems to be higher in 88-89 than in June 1985 (Fernández-Figueroa et al.,1986), as it is shown in figure 4.

RT Lac

Both components of this system present emission in H and K lines. Their spectral types are K1IV and G9IV and the cool star is almost filling its Roche lobe (Huenemoerder and Barden, 1986), being less massive and larger than the primary.

We have taken four spectra of this system: one in Nov 86, two in Jul 88 during primary and secondary eclipses, and one more in Jul 89 (figure 3).

It is normally accepted (Strassmeier et al.,1989) that the cool star is more active, as can be seen in the spectrum of Nov 86. In Jul 89, however, both components have a similar emission.

References

- Barnes,J.V., Hayes,D.S., 1984, *IRS Standard Manual*
- Batten et al., 1989, *Publ. Domin. Astrophys. Observ.*, **XVII**,1
- Blanco,C., Bruca,L., Catalano,S., Marilli,E., 1982, *Astron. Astrophys.*, **115**, 280
- Blanco,C. Catalano,S., Marilli,E., Rodono,M., 1974, *Astron. Astrophys.*, **33**, 257
- Castro,E. de, Fernández-Figueroa,M.J., Cornide,M., Armentia,J.E., Reglero,V., 1989, *Astrophys. Space Sci.*, (in press)
- Dorren,J.D., Guinan,E.F., 1980, *Astron. J.*, **85**, 1082
- Fabregat,J., 1989, *Ph.D. Thesis*, Universidad de Valencia (Spain)
- Fernández-Figueroa,M.J., Montesinos,B., Castro E. de, Rego, M. Giménez, A., Reglero, V., 1986, *Astron. Astrophys.*, **169**, 219
- Hayes,D.S., Latham,D.W., 1975, *Astrophys. J.*, **197**, 593
- Horne,K., 1986, *Publ. Astron. Soc. Pac.*, **98**, 609
- Huenemoerder,D.P., Barden,S.C., 1986, *Astron. J.*, **91**, 583
- Oke,J.B., Gunn,J.E., 1983, *Astrophys. J.*, **266**, 713
- Pasquini,L., Pallavicini, R., Pakull,M., 1988, **191**, 253
- Strassmeier,K.G., Hall,D.S., Zeilik,M., Nelson,E., Eker,Z., Fekel,C., 1988, *Astron. Astrophys. Suppl. Ser.*, **72**, 291

PHOTOMETRIC VARIABILITY OF II PEGASI

S. EVREN
Ege University Observatory
Bornova, İzmir
Turkey

ABSTRACT. In this work, B and V photometry of the RS CVn- type binary II Peg is presented. The light curves obtained in 1988 and 1989 have different shapes and amplitudes. The amplitude values obtained by us and by several investigators so far were collected, and the period of the amplitude variation was rougly found to be four years.

1. Introduction

II Peg (=HD 224085), spectral type K2 IV-V, is an active, non-eclipsing RS CVn-type single-line spectroscopic binary with a photometric period of approximately 6.7 days. The photometric variability was first noted by Chugainov (1973) and has since been confirmed by several investigators, e.g. Vogt (1981) and Rodono et al.(1983). Based on the photometric variability, this system was first classified as a BY Dra - type variable, however a more detailed study by Rucinski (1977) revealed that it was probably an RS CVn-type star. Rodono et al.(1983) analysed the available photometric data from 1974 to 1981 and a crude interpretation of the light curve indicated two spotted regions migrating at different rates towards decreasing orbital phases. Photometry and spot-model analysis have been presented by Vogt (1981), Rodono et al.(1986), Boyd et al.(1987), Evren (1988), Doyle et al. (1988) and recently by Casas et al.(1989). The smallest spot amplitude of 0.05 mag in V was observed on II Peg in 1983 by Evren (1988). The largest spot amplitudes of $\Delta V = 0.50$ mag in V were observed in late 1986 by Doyle et al. (1989) and in early 1989 by Casas et al.(1989). According to Evren (1988), the amplitude seems to vary with three different periods as 10, 6, and 4 years. Furthermore, Doyle (1988) calculated the spot migration rate from late 1981 to early 1986 as to be only 0.03 rotations per year.

Here, we report photometric observations taken in 1988 - 1989 and suggest a new period for the amplitude variation.

C. İbanoğlu (ed.), Active Close Binaries, 561–572.

2. Observations

The differantial B, V photometry presented in this paper was made with the 48 cm. Cassegrain telescope in Ege University Observatory. The photometer head consists of an uncooled EMI 9781A photomultiplier. BD+28°4666 was taken as the comparison star. All the observations were made between July 8 and October 18 in 1988 and between February 9 and September 3 in 1989. The orbital phases were computed using the light elements given by Hall and Henry (1983) as

$$J.D. = 24\ 43030.24 + 6\overset{d}{.}724183.E.$$

Since the light curves of the system change very rapidly during even one observing season, the observations were made in different time intervals during each year. These time intervals are listed in Table 1, where the first column indicates the observing years and the second column the number of groups made in these years. The third and the fourth column give the time intervals and the number of the observing nights for these groups, respectively. The light curves of the system in B and V are shown in Figures 1 and 2 for 1988, and in Figures 3, 4, and 5 for 1989.

TABLE 1. Distribution of the observing nights over years

Year	Group	Time Interval	Nights number
1988	I	July 8 - 21	10
1988	II	Sept.5 - Oct. 18	14
1989	I	February 9 - 21	6
1989	II	July 3 - 9	7
1989	III	July 21 - Sept. 3	23

3. The Light Curve Variations

In the light curve obtained between July 8 and 21, 1988 (first group for 1988), we have a minimum at about phase 0.36 and a maximum at about phase 0.73. Meanwhile, we have a distinctive hump at about phase 0.10. However, in the second group of observations (from Sept. 5 to Oct. 18 for the same year) the brightness of the system reaches its maximum value at about phase 0.30 and its minimum value at about phase 0.70. But we do not have any hump in this second group.

The more interesting light curves appear in 1989. The maximum brightness occurs at about phase 0.68 in the first group of observations, it then increases continuously to phase 0.75 in the second group of observations, and increases again up to phase 0.78 in

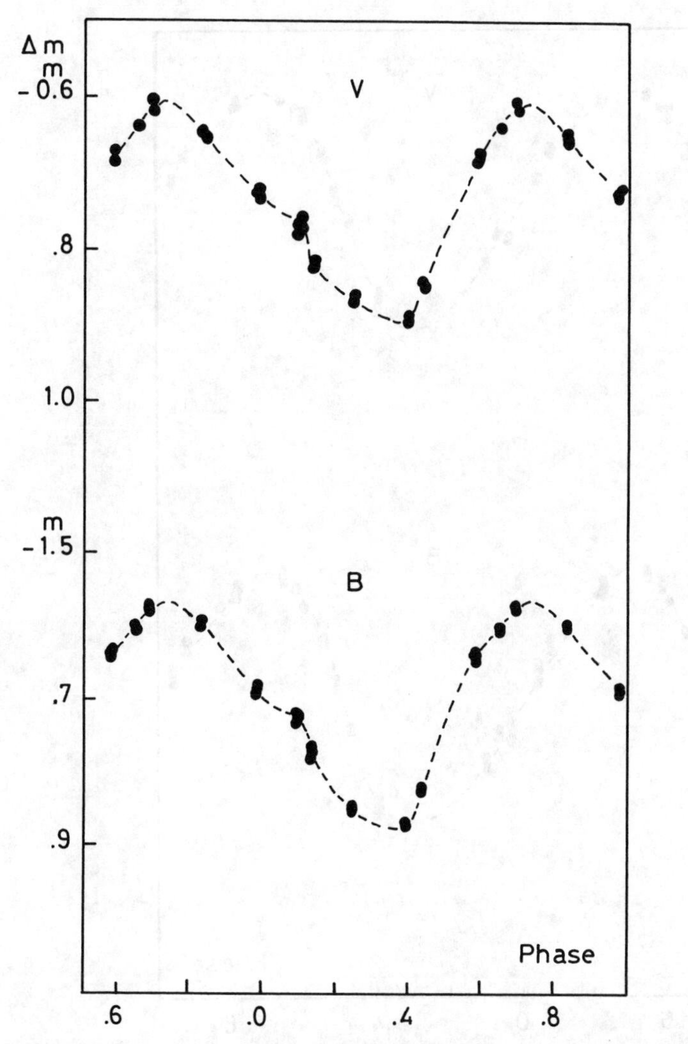

Figure 1. The light curves obtained between July 8 and 21, 1988 (Group I, 1988).

the third group of observations. Meanwhile, the minimum brightness of the system occurs at phase 0.28 in the first group of observations while it occurs at about phase 0.33 in the second and third group of observations. In the last light curve, we again have a hump at about phase 0.05.

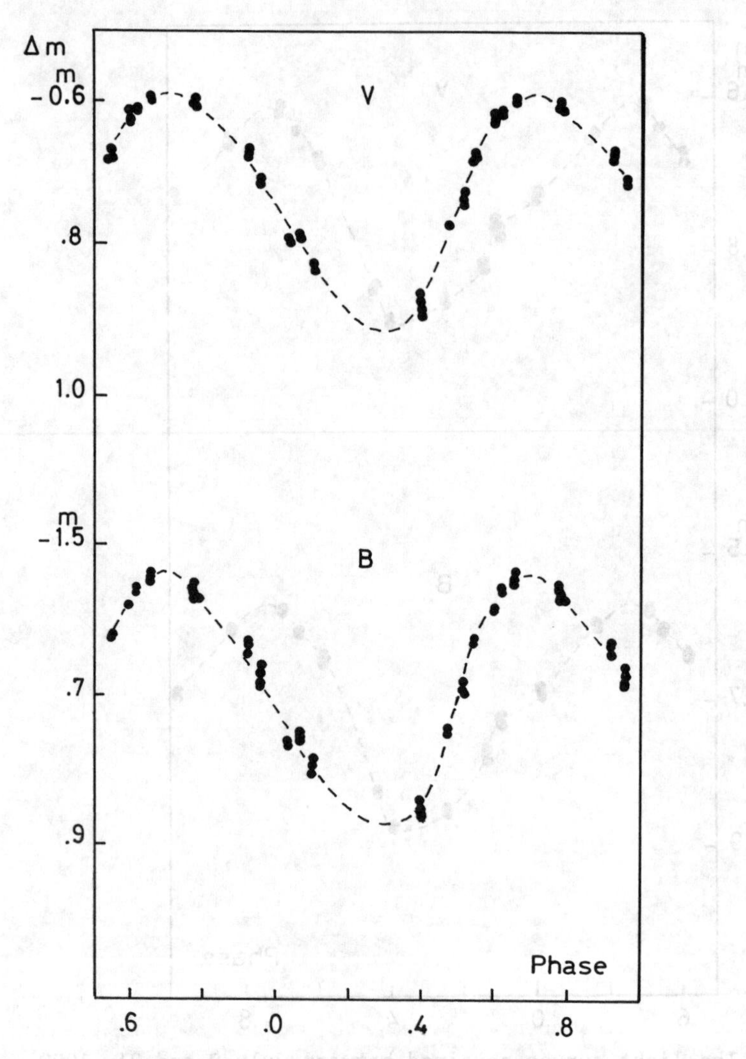

Figure 2. The light curves obtained between September 5 and October 18, 1988 (Group II, 1988).

4. The Amplitude Variation and Conclusions

II Peg is an active RS CVn system which has different-amplitude light variations even in the same observing season. In 1988, the first group light curve has an amplitude of 0.28 mag while the second group has 0.31 mag in yellow light. Next year, the amplitudes of the consecutive

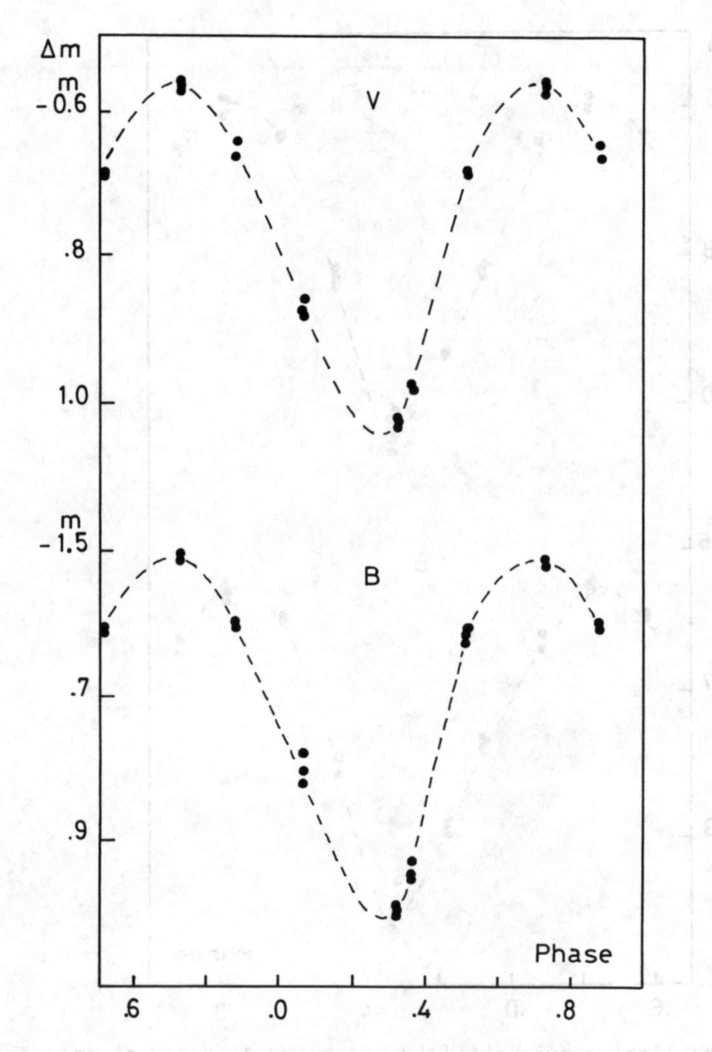

Figure 3. The light curves obtained between February 9 and 21, 1989 (Group I, 1989).

light curves decrease, amounting to six percent of a magnitude (0.48 mag for the first group; 0.44 mag for the second group; 0.42 mag for the third group in V colour).

The phases of the minimum and the maximum light of the light curves obtained in 1988 seem to migrate towards the decreasing orbital phases with different rates. The comparison of the light curves obtained in this year is shown in Figure 6. Meanwhile, the phase of the maximum

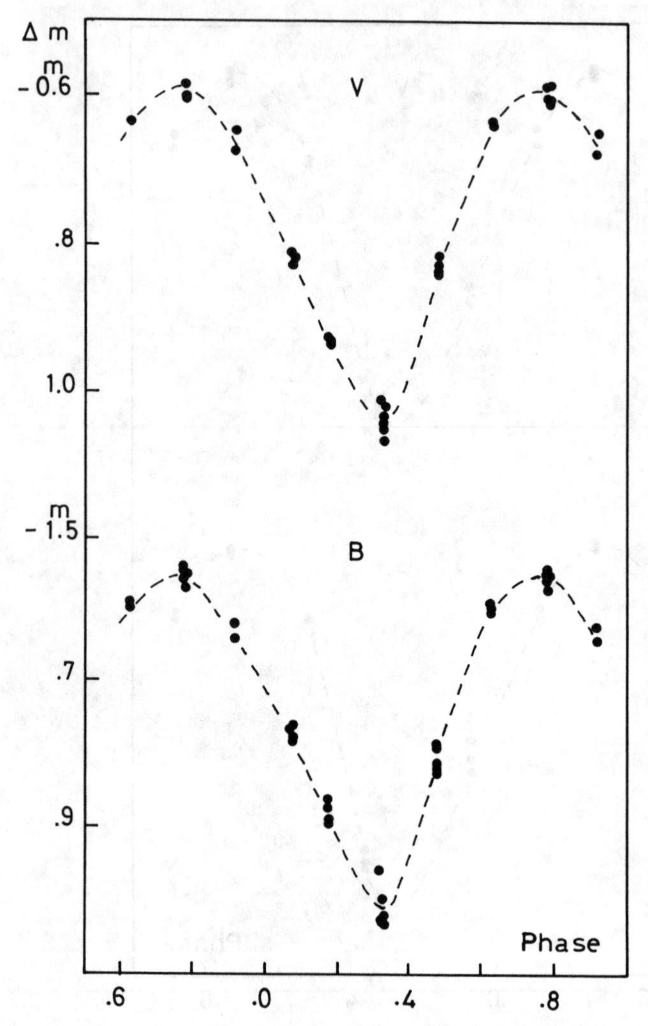

Figure 4. The light curves obtained between July 3 and 9, 1989 (Group II, 1989).

light of the light curves obtained in 1989 shifts towards increasing orbital phases. However, the phase of the minimum light remains roughly constant. The comparison of the light curves obtained in 1989 is shown in Figure 7.

Since the system has been observed from 1974 until now (1989), we have different amplitudes obtained in V colour by several authors. All of the amplitudes together with their observing years are given in

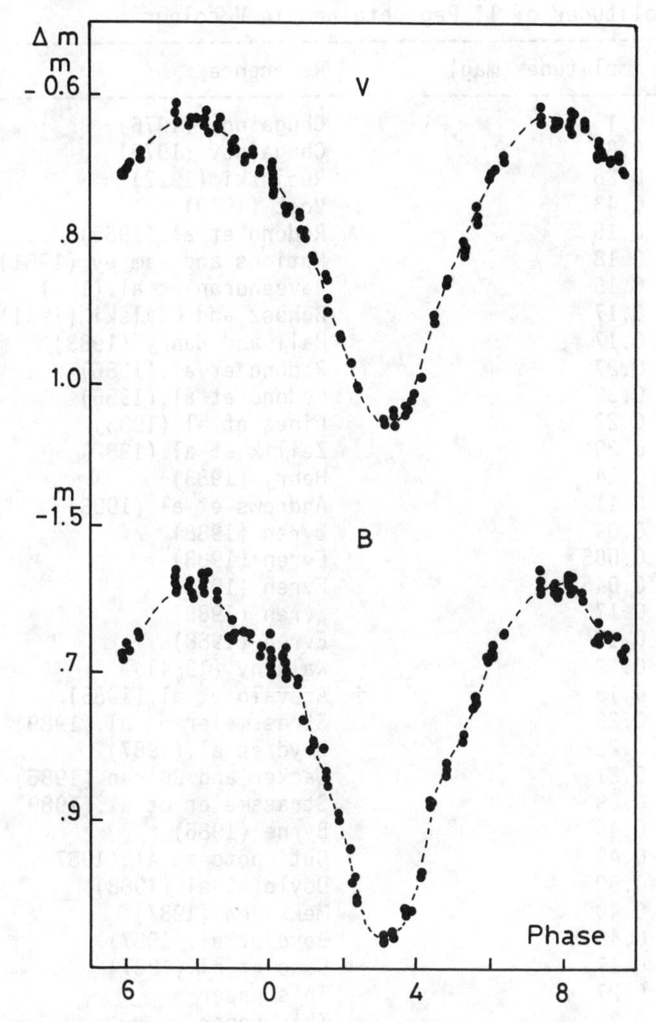

Figure 5. The light curves obtained between July 21 and September 3, 1989 (Group III, 1989).

Table 2 and the amplitude variations are plotted against the years in Figure 8. As it is seen from Figure 8, the system's amplitude dropped its minimum values in 1980, 1984, and 1988. On the other hand, it reached its maximum values in 1978, 1982, and 1986. So, we may expect the next maximum in 1990. Therefore, we may roughly assign a four-year period for this event. However, Bohusz and Udalski (1981) gave a 8 or 10-year period for the amplitude variation. It does not fit to our

TABLE 2. The amplitudes of II Peg obtained in V colour

Year	Amplitude (mag)	Reference
1974.05	0.1	Chugainov (1976)
1974.65	0.32	Chugainov (1976)
1976.80	0.26	Rucinski (1977)
1977.65	0.43	Vogt (1979)
1979.55	0.15	Rodono et al.(1980)
1979.82	0.18	Nations and Ramsey (1981)
1980.00	0.15	Raveendran et al.(1981)
1980.65	0.17	Bohusz and Udalski (1981)
1980.73	0.17	Hall and Henry (1983)
1981.73	0.27	Rodono et al.(1986)
1981.76	0.30	Rodono et al.(1986)
1981.76	0.22	Lines et al.(1983)
1981.83	0.20	Zeilik et al.(1982)
1982.1	0.24	Henry (1983)
1982.6	0.11	Andrews et al.(1988)
1983.58	0.09	Evren (1988)
1983.68	0.085	Evren (1988)
1983.76	0.045	Evren (1988)
1984.56	0.17	Evren (1988)
1984.64	0.15	Evren (1988)
1984.65	0.12	Kaluzny (1984)
1984.68	0.15	Arevalo et al.(1985)
1985.	0.23	Strassmeier et al.(1989)
1985.87	0.23	Boyd et al.(1987)
1985.88	0.21	Wacker and Guinan (1986)
1986.	0.29	Strassmeier et al.(1989)
1986.71	0.45	Byrne (1986)
1986.77	0.49	Cutispoto et al.(1987)
1986.78	0.50	Doyle et al.(1988)
1986.79	0.40	Mekkaden (1987)
1986.96	0.40	Boyd et al.(1987)
1987.75	0.17	Cano et al.(1987)
1987.79	0.27	This paper
1988.56	0.29	This paper
1988.74	0.31	This paper
1988.78	0.41	Pajdosz et al.(1989)
1988.91	0.43	Casas et al.(1989)
1989.07	0.50	Casas et al.(1989)
1989.13	0.48	This paper
1989.50	0.44	This paper
1989.61	0.42	This paper

value. So, this system needs future observations. Light curves must be obtained several times in a season. Also it would be better to observe the system simultaneously spectroscopically and photoelectrically.

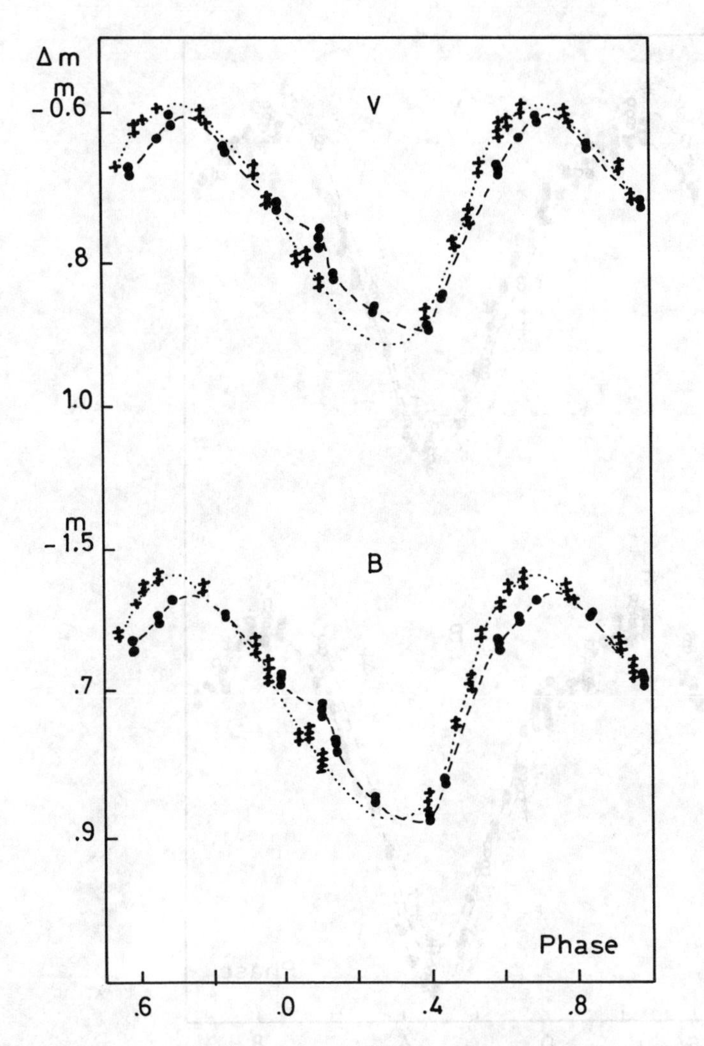

Figure 6. The comparison of the light curves obtained in 1988. The symbols correspond to seperate groups: (•) Group I, (+) Group II.

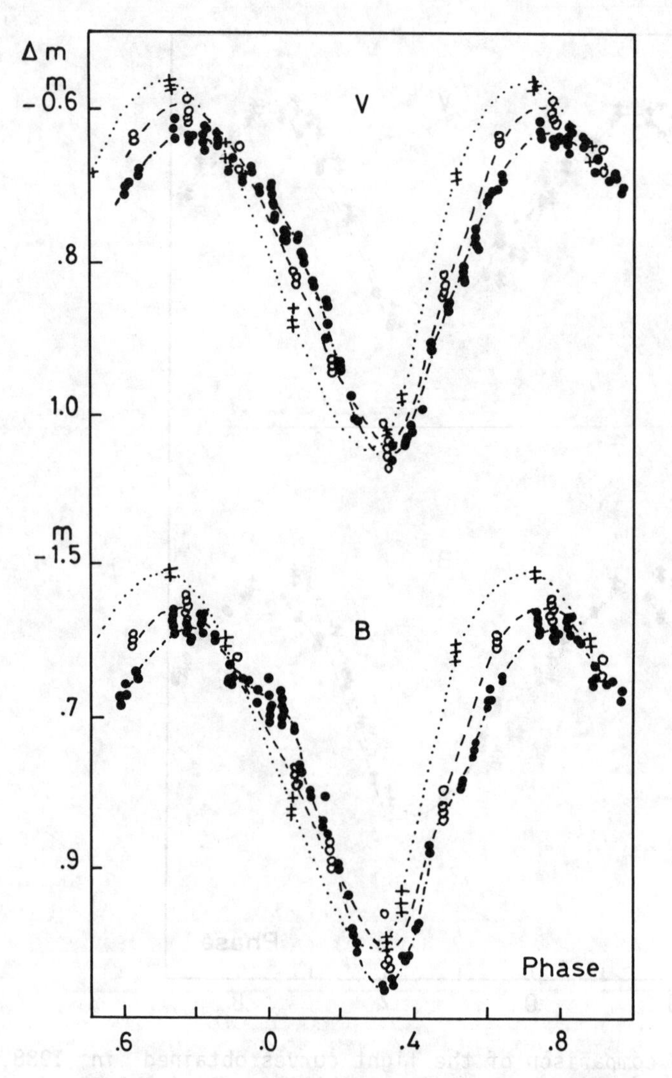

Figure 7. The comparison of the light curves obtained in 1989. The symbols correspond to separete groups: (+) Group I, (o) Group II, (•) Group III.

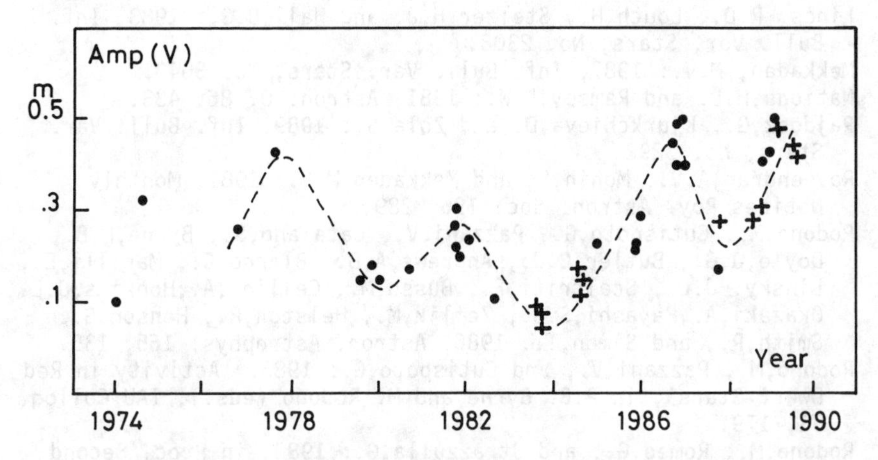

Figure 8. The amplitude variation of II Peg in V colour. The dashed line is the free-hand curve which gives a period of about 4 years. Crosses stand for the amplitudes obtained by us.

References

Andrews,A.D., Rodono,M., Linsky, J.L.,Brown,A., Butler,C.J., Catalano, S., Scaltriti,F., Busso,M., Il-Seong Nha, Oh,J.Y., Henry,M.C.D., Hopkins,J.L., Landis,H.J., and Engelbrektson,S.: 1988, Astron. Astrophys. **204**, 177.

Arevalo,M.J., Lazaro,C., and Fuensalida,J.J.: 1985, Inf. Bull. Var. Stars, No.2840.

Bohusz,E. and Udalski,A.: 1981, Acta Astron. **31**, 185.

Boyd,P.T.,Garlow,K.R.,Guinan,E.F., Mc Cook,G.P., Mc Mullin,J.P., and Wacker,S.W.: 1987, Inf. Bull. Var. Stars, No. 3089.

Byrne,P.B.: 1986, Inf. Bull. Var. Stars, No. 2951.

Cano,J.A., Casas,R., Gallart,C., Gomez,J.M., Jariod,E., and Peracaula,M.:1987, Inf. Bull. Var. Stars, No.3107.

Casas,R., Gomez-Forrellad, J.M. and Tomas,L.: 1989, Inf. Bull. Var. Stars, No. 3330.

Chugainov, P.F.: 1973, Izv. Krymsk. Astrofiz. Obs. **48**, 3.

Chugainov, P.F.: 1976, Izv. Krymsk. Astrofiz. Obs. **54**, 89.

Cutispoto,G., Leto,G., Pagano,I., Santagati,G., and Ventura,R.: 1987, Inf. Bull. Var. Stars, No. 3034.

Doyle,J.G.: 1988, Astron. Astrophys. **192**, 281.

Doyle,J.G., Butler,C.J.,Morrison,L.V. and Gibbs,P.: 1988, Astron. Astrophys. **192**, 275.

Evren, S.: 1988, Astrophys. Space Sci. **143**, 123.

Hall,D.S. and Henry,G.W.: 1983, Inf. Bull. Var. Stars, No. 2307.

Henry, G.W.: 1983, Inf. Bull. Var. Stars, No. 2309.

Kaluzny, J.: 1984, Inf. Bull. Var. Stars, No. 2627.

Lines, R.D., Louth,H., Stelzer,H.J. and Hall,D.S.: 1983, Inf. Bull. Var. Stars, No. 2308.
Mekkaden, M.V.: 1987, Inf. Bull. Var. Stars, No. 3043.
Nations,H.L. and Ramsey,L.W.: 1981, Astron. J. **86**, 433.
Pajdosz,G., Kjurkchieva,D. and Zola,S.: 1989, Inf. Bull. Var. Stars, No. 3292.
Raveendran,A.V., Mohin,S. and Mekkaden,M.V.: 1981, Monthly Notices Roy. Astron. Soc. **196**, 289.
Rodono,M., Cutispoto,G., Pazzani,V., Catalano,S., Byrne,P.B., Doyle,J.G., Butler,C.J., Andrews,A.D., Blanco,C., Marilli,E., Linsky, J.L., Scaltriti,F., Busso,M., Cellino,A.,Hopkins,J.L., Okazaki,A.,Hayashi,S.S., Zeilik,M., Helston,R., Henson,G., Smith,P., and Simon,T.: 1986, Astron. Astrophys. **165**, 135.
Rodono,M., Pazzani,V., and Cutispoto,G.: 1983, 'Activity in Red Dwarf Stars', in P.B. Byrne and M. Rodono (eds.), IAU Colloq. **71**, 179.
Rodono,M., Romeo,G., and Strazzulla,G.: 1980, in Proc. Second Europen IUE Conference, ESA SP-157, p.55.
Rucinski,S.M.: 1977, Publ. Astron. Soc. Pacific **89**, 280.
Strassmeier,K.G., Hall,D.S., Boyd,L.J., and Genet,R.M.: 1989, Ap. J. Suppl. **69**, 141.
Vogt, S.S.: 1979, Publ. Astron. Soc. Pacific **91**, 616.
Vogt, S.S.: 1981, Astrophys. J. **247**, 975.
Wacker,S.W. and Guinan,E.F.: 1986, Inf. Bull. Var. Stars, No. 2970.
Zeilik,M., Elston,R.,Henson,G., Schmolke,P. ve Smith,P.: 1982, Inf. Bull. Var. Stars, No. 2177.

Activity cycle in the RS CVn system II Peg

C. Lázaro.

Instituto de Astrofísica de Canarias.
38200 La Laguna, Tenerife, Spain

Abstract

We have analyzed visible light curves of II Peg (HD 224085), covering in time from 1974 to 1989, with a two spots simulating program.

The 1989 light curve shows an exceptional concentration and size of the spots , with the highest amplitude ever known, and reaching at the maximum a magnitude near the brightest state historically measured.

Based on the photometric light curves, we suggest the existence of differential rotation on the star, a magnetic cycle of about 14 years and the existence of two longitudinal active bands (around 90° and 270°) , where the spots appear preferentially.

Introduction

The star II Peg is a single lined spectroscopic binary, with a period of about 6.72^d, wich has been observed photometrically with regularity in the last decade.

The light curves of the system show clearly one or two minima, with morphology changing typically in a few months. These modulations are usually interp[reted as the darkening effect of colder photospheric regions as the star Arotates.

The absence of photometric binarity of the system reduces some of the ambiguity found in the analysis of spots activity in other RS CVn systems, and facilitates its modelling and interpretation. By this reason, we think this system is specially well suited for the study of the spots distribution on the star's surface, wich gives us one of the most interesting results of this work.

Now, the number of published light curves of II Peg is enough to permit a global analysis of them, in order to derive some general conclusions of statistical significance on the geometrical properties of spots and the long term photometric evolution of the system.

We have collected light curves of II Peg available from the literature, covering from 1974 to 1989 (Chugainov 1976, Rucinski 1977, Nations and Ramsey 1981, Vogt 1981, Henry 1983, Hall and Henry 1983, Lines et al 1983, Kaluzny 1984, Cutispoto et al. 1987, Boyd et al. 1987, Cano et al. 1987, Lázaro et al. 1987, Andrews et al 1988, Doyle et

C. İbanoğlu (ed.), Active Close Binaries, 573–578.

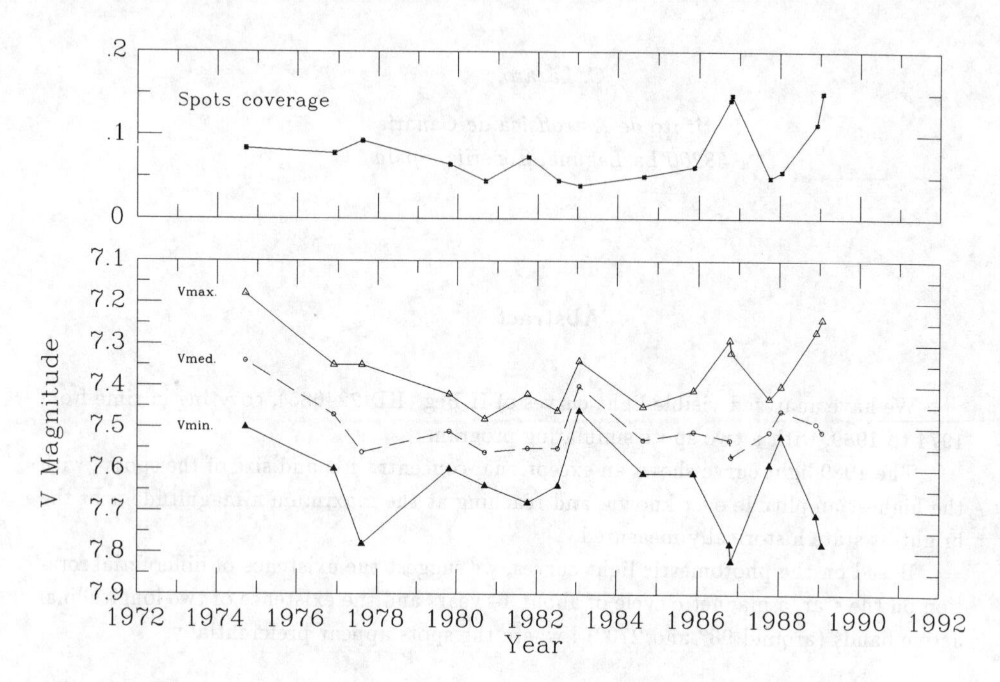

Figure 1. Long term evolution of V magnitudes and spots coverage.

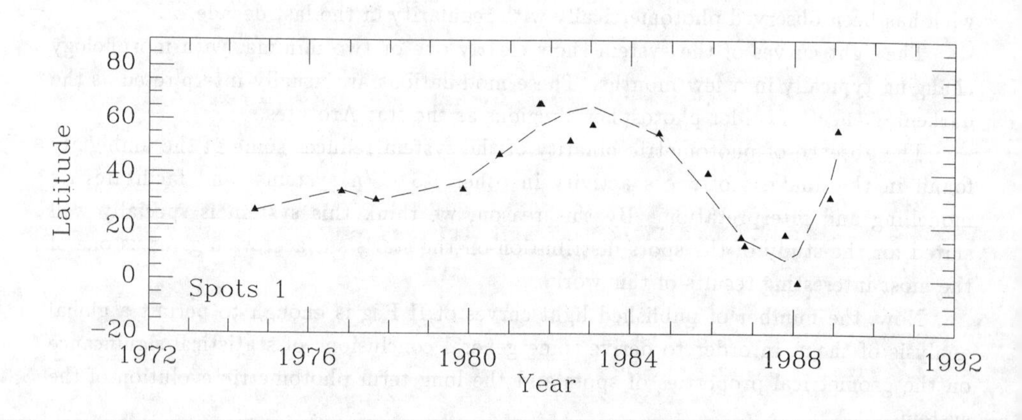

Figure 2. Spots latitude of the biggest groups.

al 1988, Gomez 1989). Unfortunately, no more than one light curve is usually available for each year, and most of the published light curves have a small number of points, but they contain enough information for a general description of the biggest spots groups and their temporal evolution.

For the analysis of the light curves, we have used a two spots program written by E. Budding for analysing spots on eclipsing binary systems, adopted with minor changes. This code has been already described in detail (Budding and Zeilik, 1987). In the fitting procedure we have only left as free parameters those concerning the geometry of the spots: longitude, latitude and radius for the two spots. The longitude and radius of the spots are very well determined, however the latitude has been found in general to be affected by a rather big uncertainty, and its best fit value is sometimes dependent on its initial value.

The inclination angle of the system is unknown, but Vogt (1981) suggests an angle much higher than 34°. The analysis of the photometric light curves is quite unsensitive to the assumed inclination of the system, but for several of the light curves we find slightly better fits with i around 65° than for i=30° . We have made the analysis with i=60°. This value determines the resultant latitudes of the spots, and to lesser extent their radius, but the other resultant parameters such as longitudes are very scarcely affected.

For the limb darkening coefficients we have adopted values adequate to the spectral type of the visible component (K2 V-IV), interpolated from those tabulated by Al-Naimy (1981). In visible the value used has been: u_V=*0.71*.

For the analysis of the light curves we have adopted as reference level the maximum of each light curve, variable from curve to curve. The alternative procedure, considering an assumed unspotted photosphere magnitude as the reference level, leads unavoidably to derive big polar spots. Also, we find impossible to derive sensed solutions for some of the light curves with the last procedure.

Results

The temporal evolution of the maxima and minima V magnitudes are shown in Figure 1, together with the spots coverage along the years. The star shows a secular brightening of the maximum since 1980 wich is reaching close to the highest state reported for 1974 by Chugainov (1976). The new photometry for the beginning of 1989 shows an exceptional state of concentration of spots over one of the hemispheres, and a spots coverage stronger than any previously recorded. It is appreciated that the maximum value reached by the spots coverage is increasing, with stronger short variations in the last years. This runs parallel with higher maxima and fainter minima in the light curves for the last years, indicating a progressive concentration of the spots groups. The secular trend of $V_{max.}$ suggests a possible cycle of about 15 years.

The light curves have been phased with the ephemerides given by Rucinski (1977). In Figure 2, we show the resulting latitudes for the spots groups (being the spots 1 the

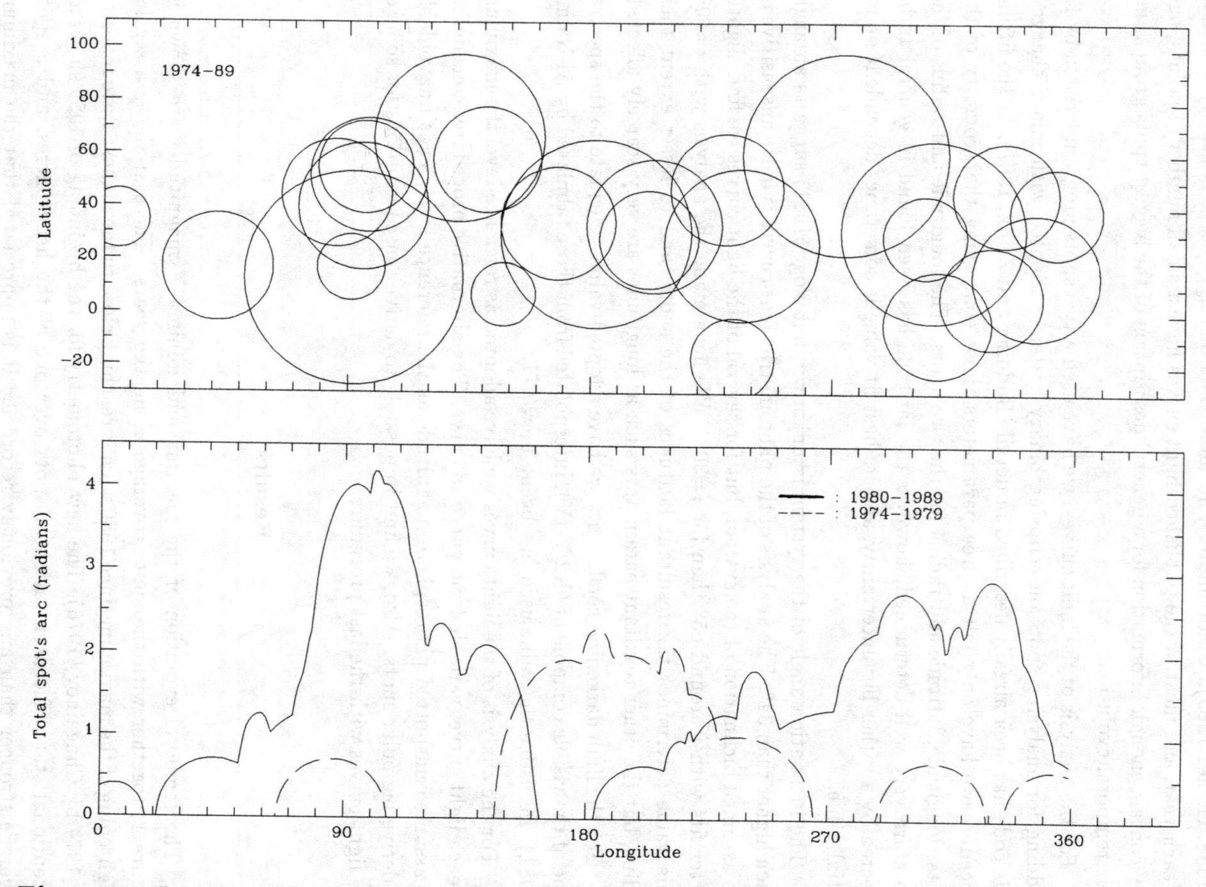

Figure 3. Longitude-Latitude distribution of all the spots during the period 1974–1989 (upper) and total subtended arc (down).

biggest). The bigger spots groups cover a wide range from low to intermediate values. The secondary groups are more uncertain in their parameters as in some cases they do not produce secondary minima but only assymmetries on the primary ones. So, they can partially reflect the inadecuacy of circular spots to represent the real, probably irregular, primary groups. The latitude of spots 1 show a maximum around 1982 and a minimum at about 1988. If it is periodic, it implies a cycle of about 13 years.

By other hand, when we represent the longitude of the photometric maxima against the year, using the ephemerides of Raveendran et al. (1981) for phasing the light curves, we find a cyclic modulation superposed on a linear decreasing trend. The linear trend shows that at least some of the latitudes on the star rotates non syncronized with the orbital period. The modulation, with a period of about 13 years and maxima and minima close to those in the Latitude-Year diagram, suggests an effect analogous to the "butterfly" diagram of the Sun spots.

In Figure 3 (upper) we have represented the derived spots in a Longitude-Latitude diagram for both spots groups, for all the analyzed light curves. It can be seen a preference of the big spots groups to appear in longitudes around 90° and 270°, avoiding the 0° longitudes (facing the unseen secondary). This effect is better seen in the lower figure, where we have added the arcs subtended in latitude by all the spots at each longitude on the star.

A similar effect has been found by Zeilik et al. (1987) for the starspot activity on SV Cam, suggesting that it may be a general property of Active Close Binaries.

References

Andrews A.D. et al.: 1988, Astron. Astrophys., 204, 177.
Al-Naimy H.M.K.: 1981, Astron. Astrophys. Suppl. Ser., 43, 85.
Budding E., Zeilik M.: 1987, Astrophys. J., 319, 827.
Cano J.A., Casas R., Gallart C., Gomez J.M., Jariod E., Peracula M.: 1987, Inf. Bull. Variable Stars, No. 3107.
Chugainov P.F.:1976, Izv. Krymsk. Ap. Obs. 54, 89.
Cutispoto G., Leto G., Pagano I., Santagati G., Ventura R.: 1986, Inf. Bull. Variable Stars, No. 3034.
Doyle J.G., Butler C.J., Morrison L.V., Gibbs P.: 1988, Astron. Astrophys., 192, 275.
Gomez J.M.: 1989, private comunication.
Hall D.S., Henry D.W.: 1983, Inf. Bull. Variable Stars, No. 2307.
Henry D.W.: 1983, Inf. Bull. Variable Stars, No. 2309.
Kaluzny J.: 1984, Inf. Bull. Variable Stars, No. 2627.
Lázaro C., Arévalo M.J., Fuensalida J.J.: 1987, Astrophys. and Space Sci., 134, 347.
Lines R.D., Louth H., Stelzer H.J., Hall D.S.: 1983, Inf. Bull. Variable Stars, No. 2308.
Nations H.L., Ramsey L.W.: 1981, A. J., 86 433.

Raveendran A.V., Mohin S., Mekkaden M.V.: 1981, Mon. Not. Roy. Astron. Soc., 196, 289.

Rucinski S.M.: 1977, Publ. Astron. Soc. Pacific, 89, 280.

Vogt S.S.: 1981, Astrophys. J., 247, 975.

Zeilik M., De Blasi C., Rhodes M., Budding E.: Fifth Cambridge Workshop on "Cool Stars, Stellar Systems and the Sun", Boulder, Colorado, 1987. Lecture Notes in Physics, Springer-Verlag.

Acknowledgements:

The author is thankful to Professors M. Zeilik and E. Budding, who have made available their spots analysis code.

LOW FLUX LINE PROFILES ON THE IUE SPECTRA OF II PEG

Akif Esendemir, Ümit Kızıloğlu
Middle East Technical University
Physics Department
Ankara 06531, TURKEY

ABSTRACT. The He II λ1640 Å emission feature on the IUE spectra of II Peg has been compared with the observed x-ray flux. The atmospheric activity level is determined by comparing the several diagnostic line intensities with the Solar case. Gaussian convolution is applied to the high dispersion IUE spectra and several low flux emission profiles were reported to be exist, including the temperature diagnostic line Mg I λ2852 Å.

1. Atmospheric activity and x-ray dependence of He II flux for II Peg

For the He II λ1640 emission, two formation mechanisms have been proposed. First being the collisional excitation (Athay, 1965; Jordan, 1975) indicating a temperature $\simeq$8 10^4 K, and second being recombination following photoionization by coronal x-rays (Zirin, 1976). In the case of quite Sun, the contribution of recombination to the He II flux is $\simeq$30%. However it's contribution increases up to 80% in more active regions (Rego et al., 1983). Another contributor to 1640 Å feature is the Fe II λ1640.15 emission (43rd multiplet) and could account for 50% of 1640 Å intensity in quiet Sun (Jordan, 1975; Boland et al., 1975; Kohl, 1977). In order to check the amount of contribution by Fe II λ1640.15 emission, Fe II λ1612.8 line can be used since it has the same multiplet number.

Although the intensity fraction caused by the recombination is not known, the activity dependence of recombination fraction can be used as a measure which is 30%, 60%, 80% in the quiet, active and superactive Sun, respectively. This classification is correlated with C IV λ1550 line intensity in the different solar regions (Rego et al., 1983). On the basis of the C IV sensitivity to the activity level, Hartman et al. (1982) have elaborated a criterion to classify Sunlike stars into three activity categories, using the ratio of the C IV stellar surface flux to the quiet Sun. The stars for which the C IV $F/F_{\odot}$ surface ratio lies in the range $1-10$ are classified as active, whereas the lower and higher ratios are classified as quiet and superactive, respectively.

C. İbanoğlu (ed.), Active Close Binaries, 579–589.

Table 1: SWP low resolution images of II Peg

IUE Image No	Year Day	Start Time	Exposure (Mins)	Julian Day 2440000+	Orbital phase
SWP 6362	79/245	19:18	40	4119.3180	0.934
SWP 9531	80/198	14:02	30	4437.0950	0.191
SWP10328	80/284	19:05	100	4523.3296	0.015
SWP14999	81/258	10:44	70	4862.9713	0.524
SWP15147	81/274	15:46	100	4879.1918	0.936
SWP15151	81/275	03:41	80	4879.6811	0.009
SWP15160	81/276	02:41	80	4880.6393	0.151
SWP15166	81/276	18:45	100	4881.3157	0.252
SWP15171	81/277	09:20	80	4881.9165	0.341
SWP15182	81/278	11:14	50	4882.9856	0.500
SWP15196	81/280	09:14	40	4884.8984	0.784
SWP19165	83/032	23:14	80	5367.4958	0.552
SWP19167	83/033	03:30	50	5367.6632	0.577
SWP19168	83/033	05:11	50	5367.7336	0.587
SWP19174	83/033	22:13	50	5368.4431	0.693
SWP19175	83/033	23:55	80	5368.5245	0.705
SWP19180	83/034	12:16	85	5369.0403	0.782
SWP19184	83/035	00:53	80	5369.5646	0.859
SWP19187	83/035	06:33	80	5369.8009	0.895
SWP19192	83/035	16:29	80	5370.2147	0.956
SWP19202	83/036	12:31	80	5371.0491	0.080
SWP19205	83/036	18:07	80	5371.2829	0.115
SWP19208	83/036	22:58	40	5371.4711	0.143
SWP19209	83/037	00:31	40	5371.5351	0.153
SWP19211	83/037	04:53	40	5371.7174	0.180
SWP19215	83/037	22:45	30	5372.4582	0.290
SWP19216	83/038	00:32	90	5372.5535	0.304

1.1. DETERMINATION OF THE DIAGNOSTIC LINE FLUXES

Keeping in mind that, the absolute flux scale for IUE is considered to be accurate up to 10% (Bohlin, 1986) and the maximum scatter for optimally exposed 5 Å wide continuum bands is 5% (Bohlin and Grillmair, 1987), the IUE spectra of II Peg listed in Table 1 are analyzed. Phases of the observations are calculated by using the ephemerides:

$$JD_{\sigma}=2,442,021.7264+6^{d}.724464E. \qquad \text{Raveendran et al. (1981)}$$

The identified lines of SWP low resolution spectra are shown on an average spectrum on Figure 1. Low resolution spectra show strong emission features which

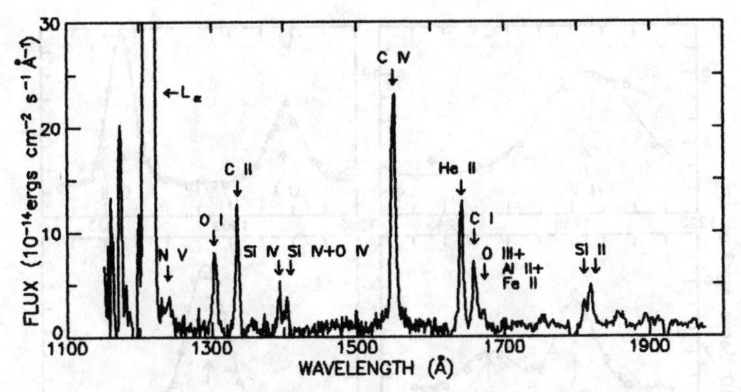

Figure 1: Prominent emission lines of II Peg

seem to vary in magnitude throughout the observations. As seen from the figure the well-exposed transition region lines (e.g. C IV, Si IV, C II) and chromospheric lines (e.g. Si II, O I and C I) exist on the spectrum. Due to the 6 Å resolution, some of the emission lines are unresolved or partially resolved multiplets.

The emission line fluxes for each spectrum are calculated by single or multiple gaussian profile fits depending on whether the emission profile is blended or not. In order to preserve the continuum flux level, smooth wavelength intervals at the wings of the emission profiles are included into the curve fitting process, wherever it is possible. N V fluxes are significantly uncertain due to the strong L_α geocoronal line emission. As an illustration of the process, fitted and the original profiles for image SWP15147 (ϕ=0.936) are shown on Figure 2.

1.2. ACTIVITY MEASURE AND He II FLUX

The surface fluxes are calculated by assuming a 4π geometry for the emitting zone, with a $30pc$ star distance depending on the parallax measured by Jenkins (1963), and $2.2R_\odot$ radius for the visible component of II Peg. This radius calculated using (V–R) values given by Vogt (1981) and the empirical relations given by Barnes et al. (1978). Ratios of the II Peg surface fluxes to the quite Sun values are shown on Figure 3, in temperature and phase domain. N V, C IV, Si IV, C II, Si II and C I lines, for which the formation temperatures and quite Sun surface fluxes known (Linsky et al., 1978) are used to construct the graph. Observed He II flux values which peaks in the phase interval of $0.5 - 0.6$ are shown on Figure 4.

The flux ratios of II Peg are highly temperature and phase dependent as seen from the Figure 3. In the temperature range of Log(T), 3.5–4.0 the emission strength is $\sim$ 10 times more than the quite Sun values. This interval corresponds to chromospheric regions and it is the main emission zone of Si II, C II lines, in the case of Sun (Linsky et al., 1978). The transition region, from where C II, Si IV,

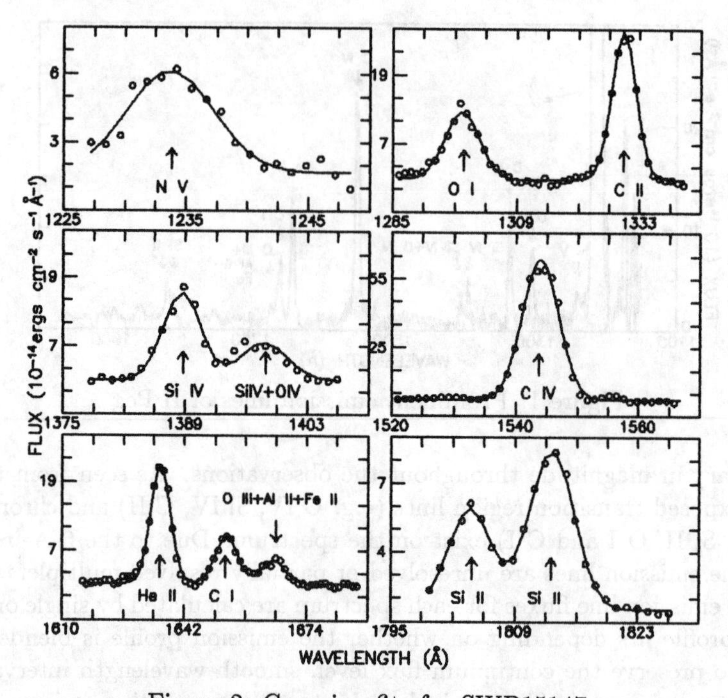

Figure 2: Gaussian fits for SWP15147

Circles represent the original IUE data and continuous line is the result of a single or multiple gaussian fit depending on the emission profile(s) respectively.

C IV, N V lines are expected, which lies in the Log(T) range of 4.0–5.5, seems to be much more active with respect to Sun. Especially N V and C IV emission fluxes are strongly enhanced, about 100 to 200 times as compared to the Sun. For typical K stars analogous enhancement is less, about 5–30 times in the case of ε Eri and only 1–6 times in the case of α Cen B (Udalski and Ruciński, 1982). Such a high C IV ratio , classifies II Peg as a chromosphericaly superactive system. Although the emission strengths differ qualitatively and quantitatively from that of stars having similar spectral types, it should also be noted that observed trend on Figure 3 is analogous to that observed for other RS CVn stars (Simon and Linsky, 1980).

In the phase interval of 0.4–0.7, fluxes are enhanced by a factor of two at least with respect to the quite intervals and this enhancement applies to all emission lines. The flux enhancement of the very active regions of the Sun with respect to the quite Sun are of the order of 12 for N V, 4 for C IV, 5 for Si IV, 8 for C II, 3 for Si II and C I (Rodonò, 1984).The flux enhancements on Figure 3 are at the order of 3 for N V, 5 for C IV, 7 for Si IV, 6 for C II, 3 for Si II and C I. The similarity of these ratios for II Peg and Sun suggest similar mechanisms for the enhancement.

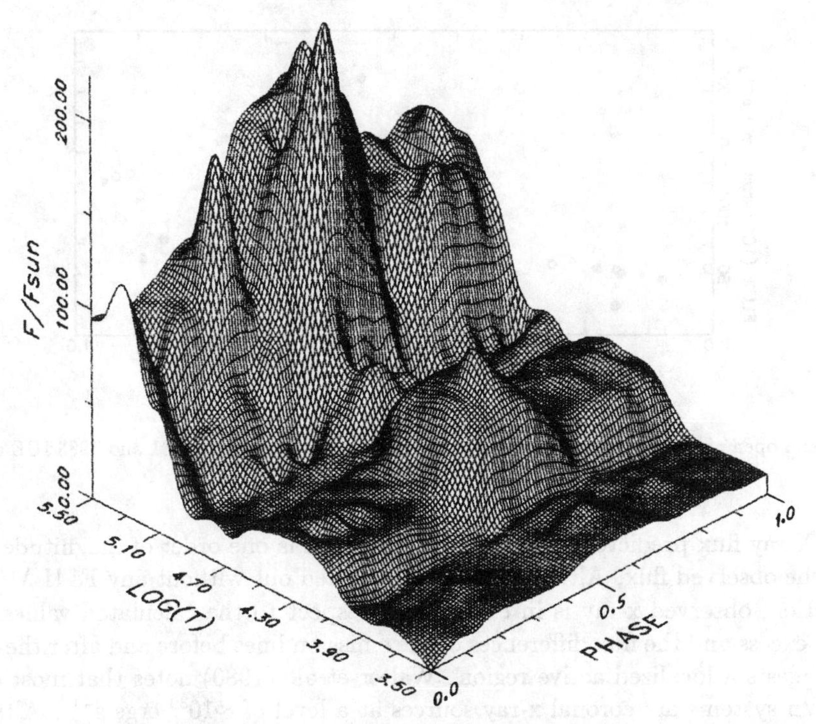

Figure 3: Comparison of II Peg line fluxes with quite Sun

1.3. X-RAY LUMINOSITY PREDICTION

Rego et al. (1983) derived an empirical relationship (Eq. 1) between the x-ray flux (F_x) and the recombination fraction of He II surface flux (F_r) through the compilation of 14 similar stars as :

$$F_x = 90F_r(He\ II)^{0.94} \qquad (erg\ cm^{-2}\ s^{-1}) \tag{1}$$

Recombination fraction of the He II flux is assumed to be 80% since we have classified II Peg as a chromospherically superactive star. The λ1640 Å feature is used without any Fe II correction due to the absence of the Fe II diagnostic line in the spectra of II Peg. Rego's empirical relation for II Peg gives L_x values ($\times 10^{30}$ ergs s^{-1}) of 2.8±0.5 between ϕ=0.0 and ϕ=0.4, 2.7±0.2 between ϕ=0.6 and ϕ=0.9, and $\sim 7.4 \pm 0.7$ at the maximum He II intensity (ϕ =0.552).

The observed x-ray luminosity of II Peg on 1977 December 29 by HEAO 1 satellite was $(3.98 \pm 1.2) \times 10^{31}$ ergs s^{-1} (Walter et al., 1980). Raveendran's ephemerides places the 1977 December 29 observations into the phase interval of 0.166–0.315, where the predicted x-ray luminosity is $(2.8 \pm 0.5) \times 10^{30}$ ergs s^{-1}. It is still not possible to predict such a high x-ray flux even if with an assumption of 100% He II flux is due to the recombination.

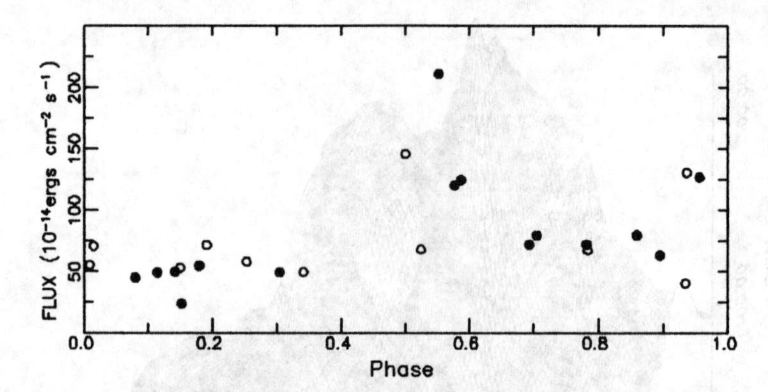

Figure 4: He II fluxes at Earth

Filled and open circles represent the integrated He II fluxes from 1979–1981 and 1983 IUE spectra respectively.

X-ray flux predicted from He II B_α emission is one order of magnitude lower than the observed flux. Although the work carried out without any Fe II λ1640.15 correction, observed x-ray is in excess with respect to the calculated values. The global excess and the flux differences of the emission lines before and after the phase 0.5 suggests a localized active region. Walter et al. (1980) notes that most of the RS CVn systems are coronal x-ray sources at a level of $\sim 10^{30}$ ergs s^{-1} . Although that value coincide with the x-ray values predicted from the He II emission, it is more probable (Linsky, 1977) that the x-ray emission is due to the plasma confined in the dipolar magnetic fields expected to be associated with the star spots.

2. Other Low Flux Emission Profiles of II Peg

The resonance lines of Mg II, commonly called the h (λ2802.7) and k (λ2795.5) lines, have played an important role in the study of the chromospheres of cool stars. Since the ionization potential of Mg^+ is 3.2 eV larger than that of Ca^+ and the solar abundance of magnesium is about 15 times that of calcium (Basri et al., 1979), the Mg II resonance lines are indicative of chromospheric properties at somewhat higher temperatures than the more easily observed Ca II H and K lines, and special attention has been given to Mg II emission as an appropriate measure of chromospheric activity (Ruciński, 1985).

In the case of II Peg, single component Mg II emission is found to be correlated to the appearance of spot concentrated active hemisphere (Rodonò et al., 1986,1987, Andrews et al., 1988, Byrne et al., 1989, Doyle et al., 1989). The spot models used for the solution of light curves, assumes an effective temperature of $\sim$4650 K for the unspotted star surface and $\sim$1000 K cooler spots.

The available IUE spectra of II Peg covering the 1981, 1983 and 1985 obser-

Table 2: LWR high resolution images of II Peg

IUE Image No	Year Day	Start Time	Exposure (Mins)	Julian Day 2440000+	Orbital phase
LWR11555	81/258	09:40	60	4862.8944	0.527
LWR11652	81/274	15:18	25	4879.1456	0.929
LWR11655	81/275	03:19	25	4879.6370	0.002
LWR11667	81/276	02:00	35	4880.5956	0.145
LWR11673	81/276	18:03	35	4881.2646	0.244
LWR11684	81/277	17:44	35	4882.2511	0.391
LWR11690	81/278	10:53	30	4882.9527	0.495
LWR11714	81/280	08:35	35	4884.8695	0.780
LWR15159	83/033	00:40	40	5367.5417	0.559
LWR15161	83/033	04:26	40	5367.6986	0.582
LWR15167	83/033	23:09	40	5368.4785	0.698
LWR15172	83/034	11:28	40	5368.7007	0.731
LWR15175	83/035	02:18	40	5369.6132	0.867
LWR15177	83/035	05:50	40	5369.7563	0.888
LWR15181	83/035	17:54	40	5370.2597	0.963
LWR15183	83/035	22:06	40	5370.4347	0.989
LWR15191	83/036	13:54	40	5371.0931	0.087
LWR15195	83/036	22:23	30	5371.4424	0.139
LWR15196	83/036	23:45	40	5371.5167	0.150
LWR15197	83/037	01:16	30	5371.6049	0.163
LWR15202	83/037	23:20	26	5372.4813	0.293
LWR15203	83/038	01:08	30	5372.5576	0.305

vations are compiled by Doyle (1988). Although the enhancements of the transition region lines was found to be correlated with the active hemisphere, this conclusion was largely biased by the 1981 and 1983 data-sets (Doyle et al. 1989). The IUE spectra of 1986 showed an inverse correlation with the spot concentration for which the mean spot temperature was assigned to be ~3700 K where the unspotted temperature was assumed to be 4650 K. Byrne (1988) pointed out there is an evidence to suggest that this temperature assignment may be incorrect.

The temperature dependence of Mg I resonance line at λ2852 Å is known. That line appears to be in absorption for hotter stars (61 Cyg B, HD 1326 A, and HD 95735) but in emission in the cooler stars (YZ CMi and UV Cet) (Linsky et al., 1982). Effective temperatures of 61 Cyg B, HD 1326 A, and HD 95735 are 4130 K, 3780 K and 3720 K respectively. The cooler stars YZ CMi and UV Cet have the effective temperature of 3380 K and 3200 K (Linsky et al., 1982). Although Mg I line reported to be tentative absorption profile from the low resolution IUE spectra (Udalski and Ruciński, 1982), the high resolution IUE spectra of II Peg (Table 2) do not show a clear profile of Mg I line.

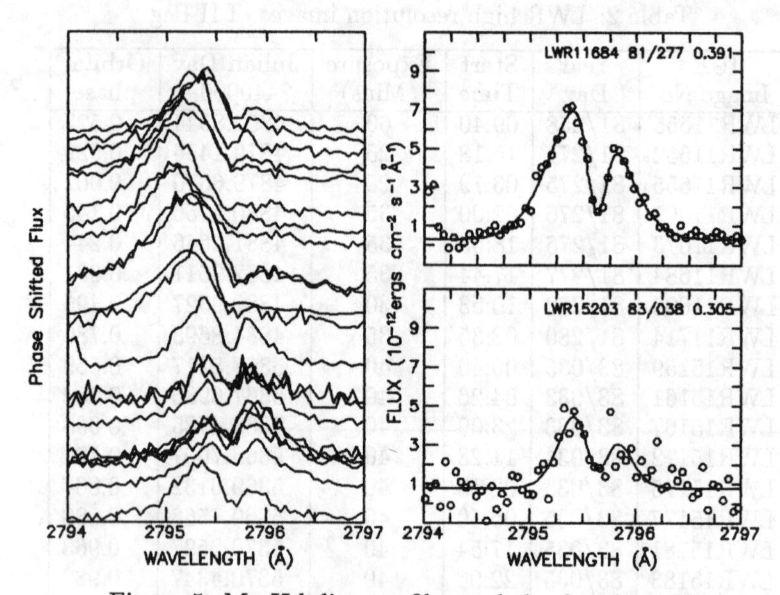

Figure 5: Mg II k line profiles and the fitted profiles.
For the visual interpretation of the Doppler-shifted line profiles, 4 10^{-13} times respective phase added to the flux values and plotted on the left panel. The best and the worst fits to the Mg II k line profiles are plotted on the right panels together with the spectral data.

2.1. DETERMINATION OF THE DOPPLER SHIFTS

One method for searching low flux line profiles due to the specific component of a binary system, is the coaddition of the spectra after a velocity correction due to that component. In order to apply the method to II Peg, Doppler corrections for each spectrum determined from the Mg II k line profiles (Fig. 5). The emission cores of Mg II profiles that move on the wavelength axis with respect to the unresolved interstellar absorption feature, reflect the radial velocity changes of the visible component. To determine the Doppler corrections for each spectrum, k line profiles have been fitted by the difference of two gaussians; first to represent the stellar emission and the other to represent the interstellar absorption. The routine we have employed iterates least square solutions with the amplitudes of the two gaussians, their widths, their central wavelengths and the continuum as free parameters. In some phases, interstellar absorption feature is very near to the redward wing of the emission profile and fitting the model becomes difficult. Regarding that, emission core wavelengths of the spectra obtained between the phase interval of 0.67 and 0.80 are fitted by using the average gaussian widths of the $\phi = 0.582$ and $\phi = 0.867$ spectra. The spectra LWR15202 ($\phi = 0.293$) and LWR15203 ($\phi = 0.305$) have low signal to noise ratio and the determined velocities are unreliable.

The velocities calculated from the emission profiles are plotted on Figure 6 together with the synthetic curve based on the velocity parameters given by Vogt

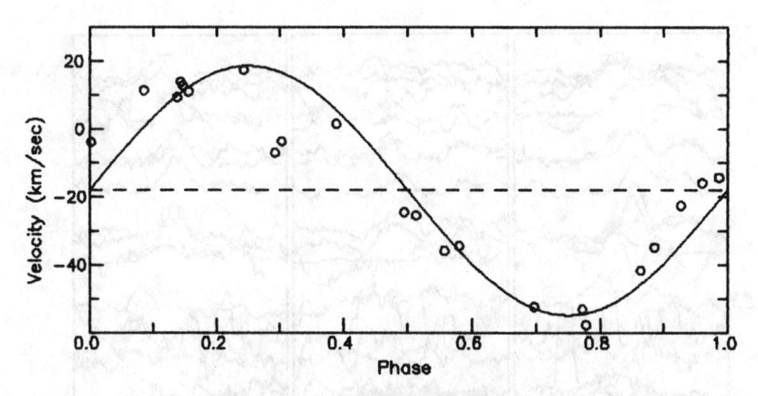

Figure 6: Radial velocity curve

(1981). The measured velocities mostly coincide with the theoretical values.

2.2. A SEARCH FOR THE WEAK LINE PROFILES

In order to determine the other wavelength intervals to be searched, high resolution spectra first corrected with the determined velocities and then averaged. Several wavelength regions on the averaged spectrum appeared as candidates to be searched. In order to obtain a clear visual interpretation, the spectra convolved with a gaussian profile. FWHM of the gaussian profile is selected to be 1.5 times of the average data interval for high resolution spectra. After such a process the velocity and the flux information would be lost partially so only the existence of the lines and the Doppler-shifts can be discussed.

Almost all of the candidate lines show the radial velocity of the visible component (Fig. 7). Although the 2621 Å region bumps over the average spectrum, the structure is not so distinct and it is included into the figure for the sake of completeness. The emission profiles near 2625 Å and 2631 Å are identified as the U19 multiplet of Mn II (λ2625.6, λ2632.3). Also the 2628 Å profile can be attributed to the 1A multiplet of Mn II (λ2632.3). We couldn't identify the 2599 Å emission. However the Mg I (λ2852) resonance line is found to be in emission.

3. CONCLUSION

II Peg is classified as a superactive system by comparing the emission lines with that of the quite Sun. Although the II Peg enhancement ratios are in agreement with the active/quite Sun ratios, the strength of the C IV and N V flux suggest that either II Peg contains more material near the layer T$\simeq$2 10^5 K or has more extended transition region. Predicted x-ray flux based on the He II emission, is one order less than the observed flux indicating another possible contributing mechanism besides the recombination.

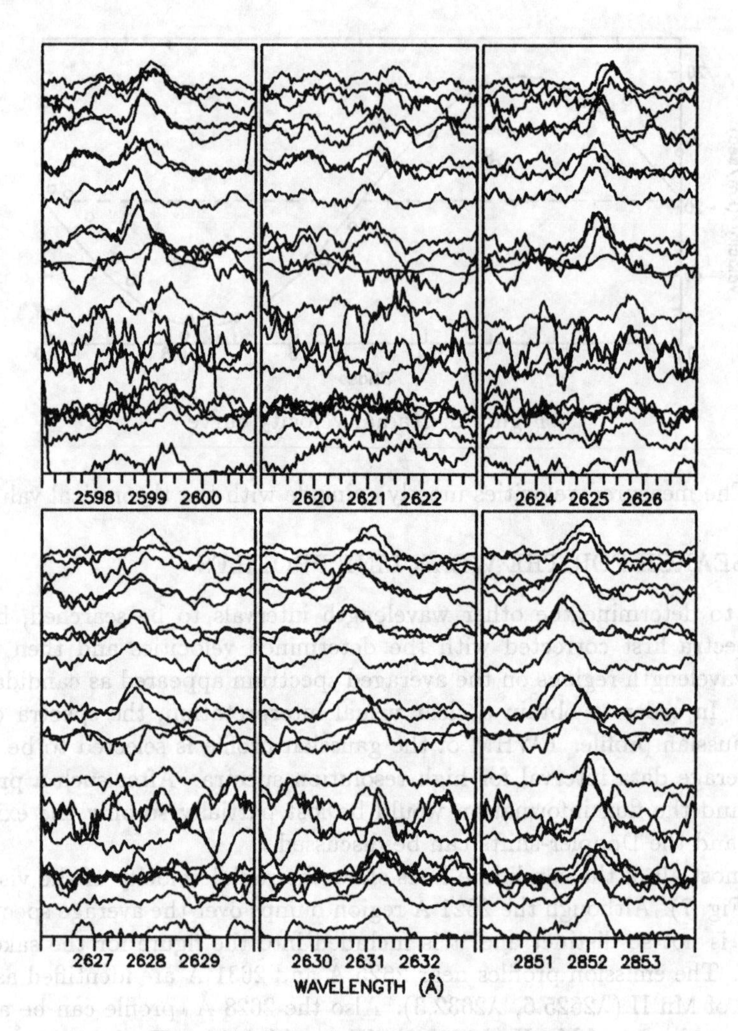

Figure 7: Doppler-shifted line profiles
For the visual interpretation 4 10^{-13} times respective phase added to the flux values as in Figure 5

Several metallic emission lines are found in emission, including Mg I resonance line. The effective unspotted surface temperature of II Peg (4650 K) is high enough to expect Mg I in absorption. Possibly, Mg I emission is due to the cooler spots. The correlation between the Doppler shifts of Mg I profiles and the spot appearance at the limp of the star could not be studied due to the information lost during the applied gaussian convolution. The effective temperature of stars which contains Mg I in emission is ~3300 K. That temperature, which is possibly the spot temperature, is ~1350 K cooler than the assigned unspotted surface temperature of II Peg.

References

Andrews, A.D., et al., 1938. A.A., 204, 177.
Athay, R.G., 1965. Ap. J., 142, 755.
Barnes, J.G., et al., 1978. Mon. Not. Astr. Soc., 183,285.
Basri, G.S., et al., 1979. Ap. J., 230, 924.
Bohlin, R.C., 1986. Ap. J., 308, 1001.
Bohlin, R.C., Grillmair, C.J., 1987. IUE NASA Newsletter, 33, 28.
Boland, B.C., et al., 1975. Mon. Not. Astr. Soc., 171, 697.
Byrne, P.B., 1988. Proc. 5th Cambridge workshop on Coll Stars, Eds. Linsky, J.L., Stencel, R.E., Springer, Berlin, Heidelberg, New York, p. 491
Byrne, P.B., et al., 1989. A.A., 214,227.
Doyle, J.G., et al., 1988. A.A., 192, 281.
Doyle, J.G., et al., A.A., 1989, 223, 219.
Hartman, L., et al., 1982. Ap. J., 252, 214.
Jenkins, L.F., 1963. General Cat. of Trig. Stellar Parallaxes, Yale Univ. Obs.
Jordan, C., 1975. Mon. Not. Roy. Astr. Soc., 170, 429.
Kohl, J.L., 1977. Ap. J., 211, 958.
Linsky, J.L., 1977. in "The Solar Output and Its Variation", Ed. White, O.R., Colorado Press, p. 477.
Linsky, J.L., et al., 1978. Nature, 275, 398.
Linsky, J.L., et al., 1982. Ap. J. 260, 670.
Raveendran, A.B., et al., 1981. Mon. Not. Roy. Astron. Soc., 196, 289.
Rego, M. et al., 1983. A.A., 119, 227.
Rodonò, M., 1984. in "Cool Stars, Stellar Systems and the Sun", Eds. Baliunas, S.L., Berlin, Newyork, p. 475.
Rodonò, M., 1987. A.A., 176, 267.
Rodonò, M., et al., 1986. A.A., 165, 135.
Ruciński, S.M., 1985. Mon. Not. Roy. Astron. Soc., 215, 615.
Simon, T., Linsky, J.L., 1980. Ap. J., 241, 759.
Udalski, A., Ruciński, S.M., 1982. Acta Astronomica, 32, 315.
Vogt, S.S., 1981. Ap. J., 247, 975.
Walter, F.M., et al., 1980. Ap. J., 236, 212.
Zirin, H., 1976. Ap. J., 208, 414.

A PHOTOMETRIC STUDY OF UV PISCIUM

M. C. AKAN
Ege University Observatory, Bornova, İzmir, Turkey

ABSTRACT. The light curves of the eclipsing system UV Psc were obtained in B and V colours between 1981 and 1989 at Ege University Observatory. The amplitude of the wave-like distortion seems to vary with a four or five year period in blue colour whereas this value seemed to be doubled in yellow. The migration period of the wave-like distortion is found to be roughly one year.

1. INTRODUCTION

Based on photographic observations Huth (1959) published the first light curve of UV Psc. Following the discussion of Oliver (1974), the system has been included in the short-period group of RS CVn-type binaries by Hall (1976). The spectroscopic record of UV PSc is due to Popper (1969,1976) and the system is listed as a double-lined binary with emission from both components present in the H and K lines of Ca II.

Sadik (1979) obtained photoelectric light curves of the system and according to his light curve analysis he stated that the irregularities in the light curve were caused by a locally hotter, rather than a cooler, region on one of the components. Further photoelectric light curves of the system have been obtained by Zeilik et al. (1981,1982). In the first of these papers, they stated that they found no evidence for an asymmetrical distortion wave, but emphasised the existence of an activity in the system that causes radical changes in its light curve; in the latter, however, they drew attention on the existence of an asymmetrical distortion wave and also a large asymmetry in the secondary eclipse. Based on their analysis of photometric data, Busso et al. (1986) discussed the presence of cycles of variability in UV Psc similar to what is found in other members of the same class. Vivekananda Rao and Sarma (1984) observed the system in UBV and suggested that the hotter component in the system could be an intrinsic variable. The same was also proposed by Antonopoulou (1987) who observed the system in the infrared.

C. İbanoğlu (ed.), Active Close Binaries, 591–600.

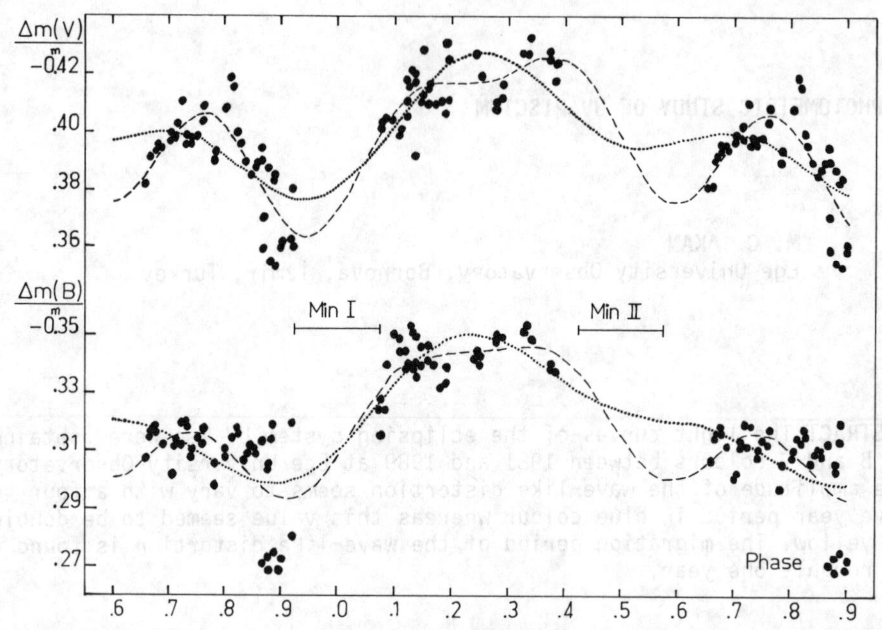

Fig. 1 . The distortion waves at outside eclipses for 1981. Tiny dots and dashed lines represent the computed curves for second-order and third-order approximations, respectively. The larger dots stand for observations.

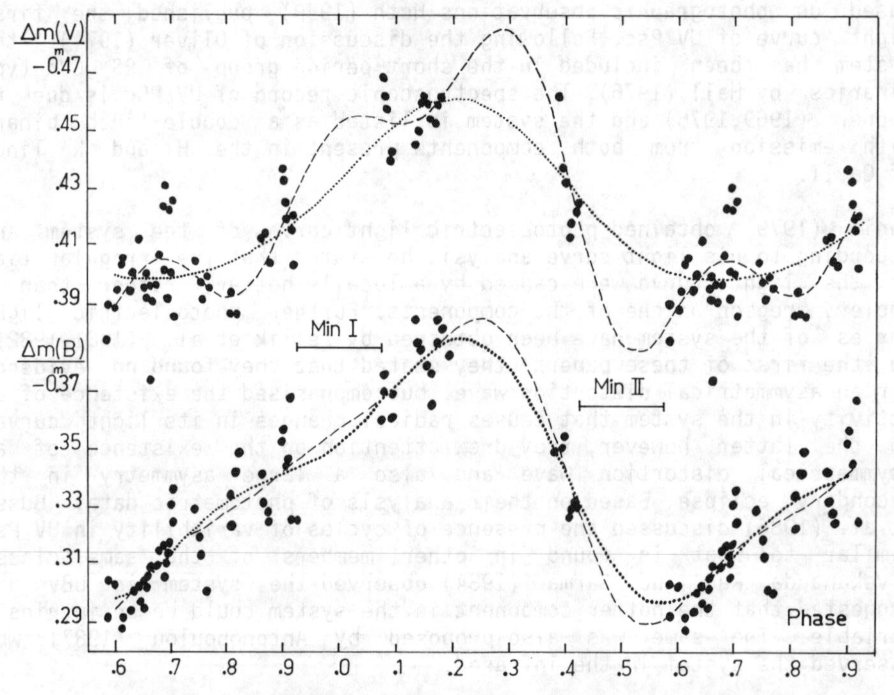

Fig. 2. The distortion waves at outside eclipses for 1982. The symbols are the same as those used in Figure 1.

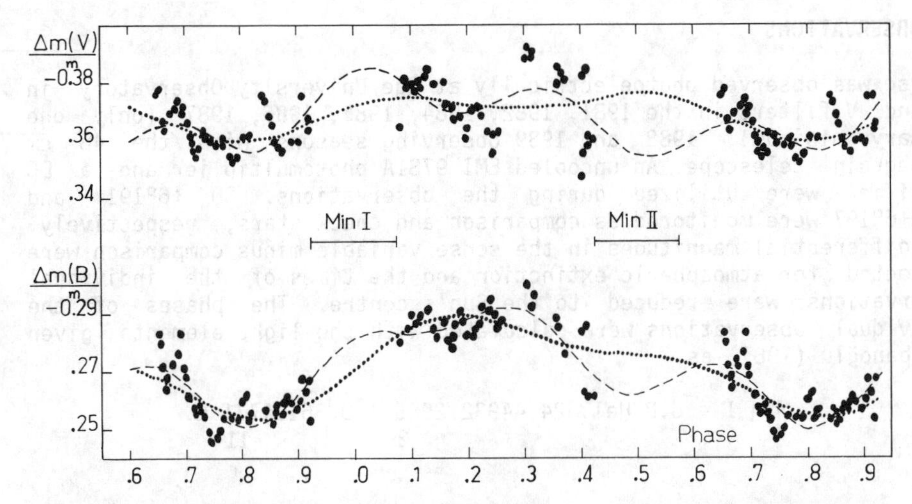

Fig. 3. The distortion waves at outside eclipses for 1984. The symbols are the same as those used in Figure 1.

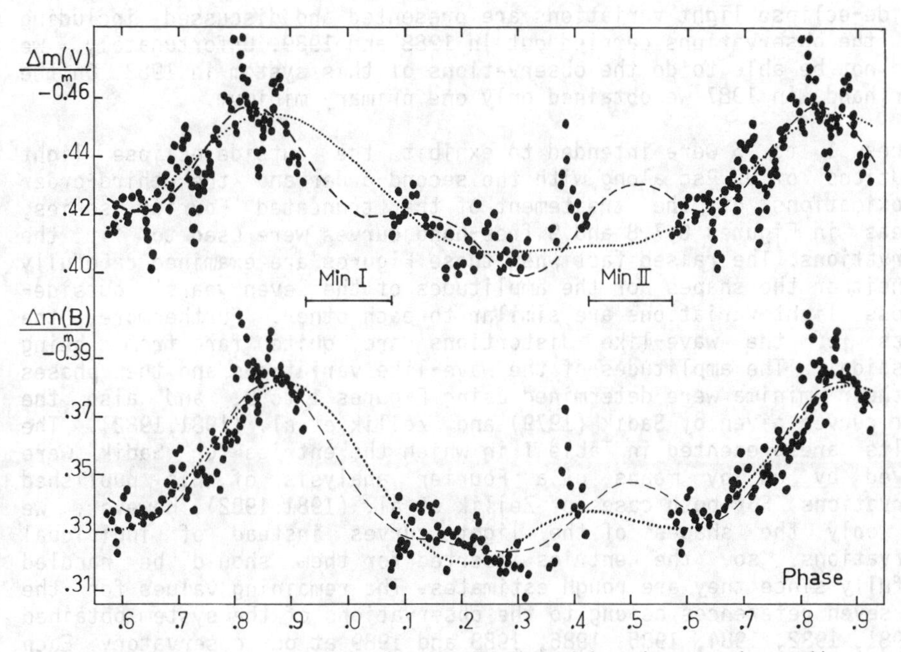

Fig. 4. The distortion waves at outside eclipses for 1985. The symbols are the same as those used in Figure 1.

2. OBSERVATIONS

UV Psc was observed photoelectrically at Ege University Observatory in B and V filters in the 1981, 1982, 1984, 1985, 1986, 1987 (only one primary minimum), 1988 and 1989 observing seasons with the 48 cm Cassegrain telescope. An uncooled EMI 9781A photomultiplier and a DC amplifier were utilized during the observations. BD +6°191 and BD +6°197 were monitored as comparison and check stars, respectively. The differential magnitudes in the sense variable minus comparison were corrected for atmospheric extinction and the times of the individual observations were reduced to the Sun's centre. The phases of the individual observations were calculated with the light elements given by İbanoğlu (1987) as,

$$\text{Min I} = \text{J.D.Hel. } 24\,44932.298\underset{2}{5} + 0^{d}.8610477\underset{11}{1}\ E.$$

3. OUTSIDE-ECLIPSE LIGHT VARIATIONS

The light curves of UV Psc obtained in 1981, 1982 and 1984 have been presented by İbanoğlu (1987) and the light curves obtained in 1985 and 1986 have been presented by Akan (1988). So, in this text only the outside-eclipse light variations are presented and discussed including also the observations carried out in 1988 and 1989. Unfortunately, we could not be able to do the observations of this system in 1983. On the other hand, in 1987 we obtained only one primary minimum.

Figures 1 to 5 were intended to exhibit the outside-eclipse light variations of UV Psc along with the second-order and the third-order approximations of the treatement of the truncated Fourier series, whereas in Figures 6,7,8 and 9 free-hand curves were used to fit the observations. The raised fact when these Figures are examined carefully is neither the shapes nor the amplitudes of the seven years' outside-eclipse light variations are similar to each other. Furthermore, the shapes of the wave-like distortions are quite far from being sinusoidal. The amplitudes of the wave-like variations and the phases of their minima were determined using Figures 1 to 9, and also the light curves given by Sadik (1979) and Zeilik et al. (1981,1982). The results are presented in Table I in which the entries for Sadik were derived by us by means of a Fourier analysis of his published observations. For both cases of Zeilik et al. (1981,1982), however, we have only the shapes of the light curves instead of individual observations, so, the entries inserted for them should be handled carefully since they are rough estimates. The remaining values for the last seven references belong to the observations of the system obtained in 1981, 1982, 1984, 1985, 1986, 1988 and 1989 at our observatory. Each entry in the first coloumn stands for the middle of the time interval in which the observations were obtained.

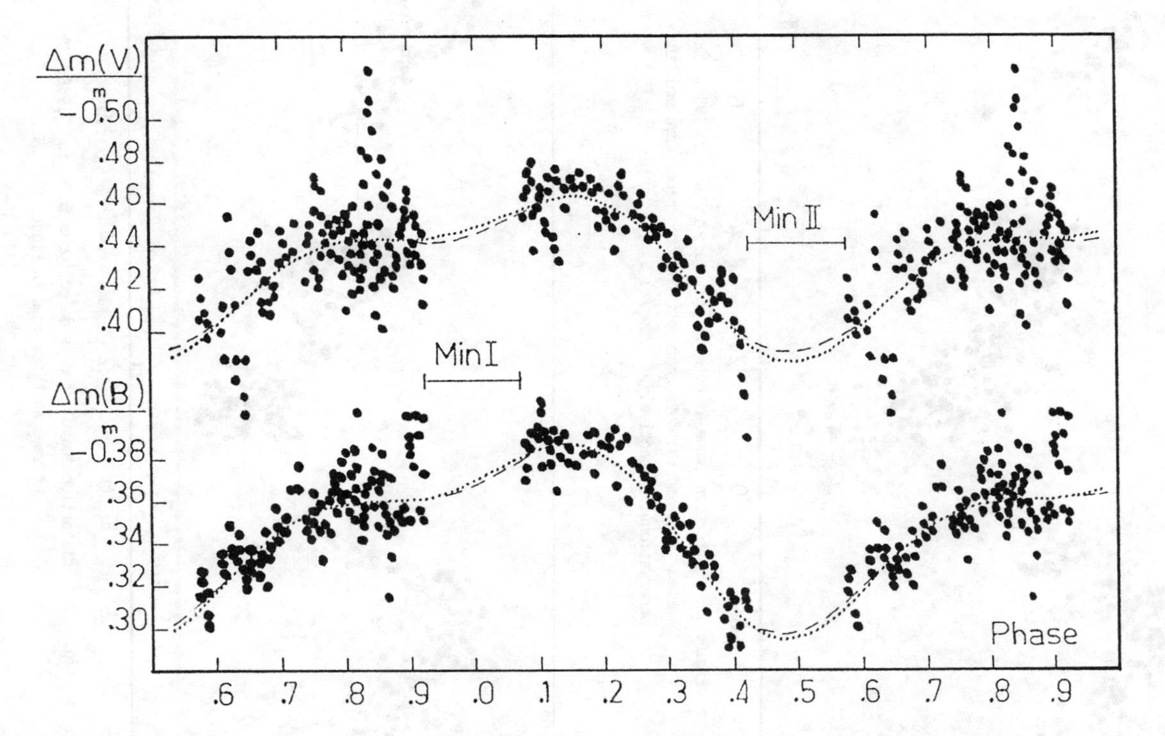

Fig. **5**. The distortion waves at outside eclipses for 1986. The symbols are the same as those used in Figure 1.

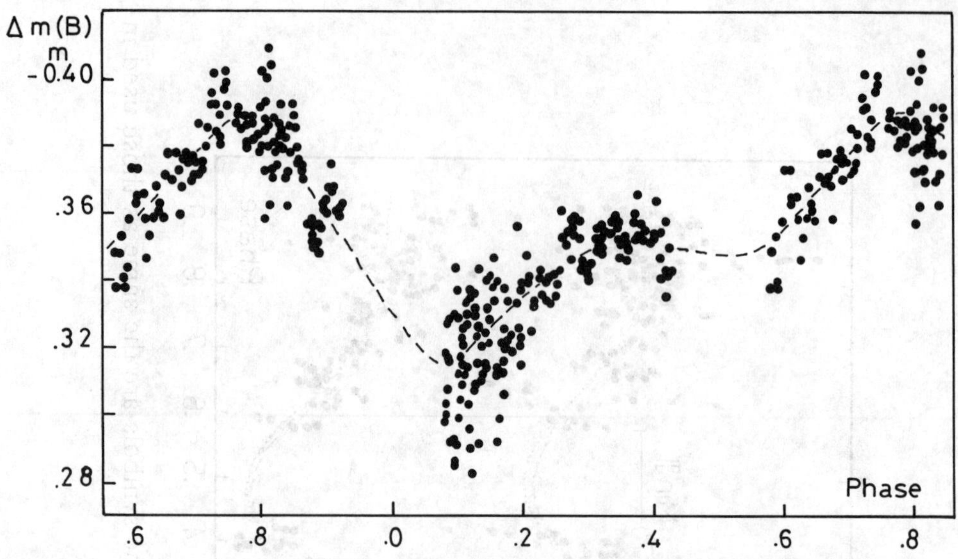

Fig.6. The distortion wave at outside eclipses in B for 1988. The dashed line and the dots represent the free-hand curve and the observations, respectively.

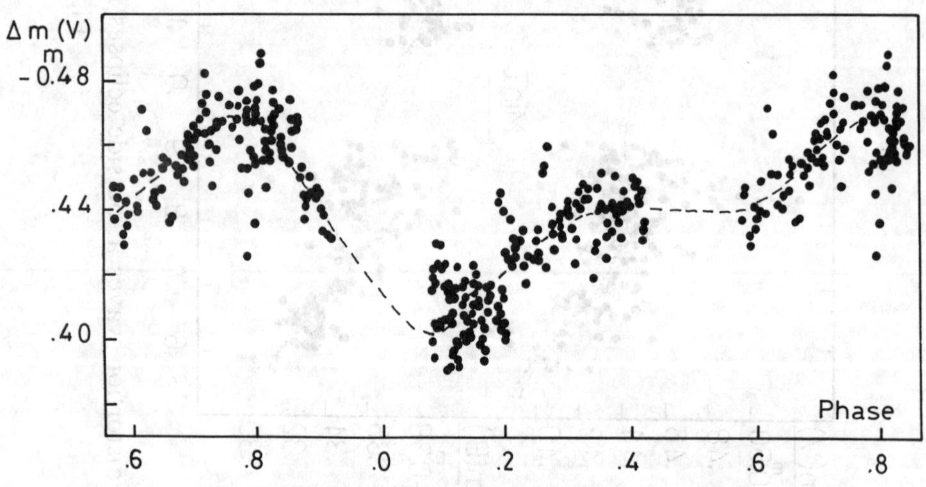

Fig.7. The distortion wave at outside eclipses in V for 1988. The symbols are the same as those used in Figure 6.

TABLE I
The amplitudes of the wave-like variations and the phases of their minima

Year	Amp.(B)	Amp.(V)	$_{min}$(B)	$_{min}$(V)	Ref.
1977.9	$0^m.094$	$0^m.094$	$0^p.575$	$0^p.575$	S
1980.8	0.066	0.058	0.250	0.300	Z1
1981.8	0.071	0.068	0.900	0.900	Z2
1981.9	0.051	0.051	0.900	0.935	A
1982.8	0.091	0.062	0.575	0.675	A
1984.8	0.036	0.016	0.750	0.800	A
1985.7	0.062	0.047	0.325	0.300	A
1986.9	0.093	0.078	0.490	0.490	A
1988.7	0.090	0.075	0.100	0.100	A
1989.6	0.080	0.075	0.500	0.500	A

S:Sadik(1979);Z1:Zeilik et al.(1981);Z2;Zeilik et al.(1982); A:Author.

By use of the values pertaining to the amplitudes in Table I the amplitude variation of the wave-like distortion is presented in Figure 10 for both colours. The amplitude variation range is 0.058 and 0.078 mag in blue and yellow colours, respectively. This distribution seems to have peaks for 1977.9, 1982.8 and 1986.9 in B colour, which may be an indication of a four or five year periodicity for this phenomenen. In yellow light, however, the situation is somewhat complicated; what we have in this case is a slight peak instead of a distinguished one for 1982.8, so, the value for the periodicity of the same event seems to be roughly doubled in V colour.

Using the $_{min}$ values given in Table I we have attempted to determine the migration period of the wave-like distortion, which is assumed moving towards decreasing orbital phases in each successive cycle. As can be easily noticed in Figures 1 to 7, most of the light curves have more than one minimum. Therefore, we always took into consideration only the deeper ones for each light curve. In order to estimate the migration period of the wave-like distortion for UV Psc, the $_{min}$ values have been plotted versus years and shown in Figure 11. From this distribution, a one year period can be attributed for the migration period.

4. DISCUSSION

For UV Psc, three different migration periods for the wave-like distortion had been proposed by Akan (1988) and another one by İbanoğlu (1987), previously. Without comprising the observations made later than 1986, in the first of these papers the most possible figure was emerged to be the one out from 0.4, 0.7, 2.3 years; in the latter, however, a period between 1.5 and 2 years had been proposed. The same value had also been appeared in the paper by Keskin et al. (1987).

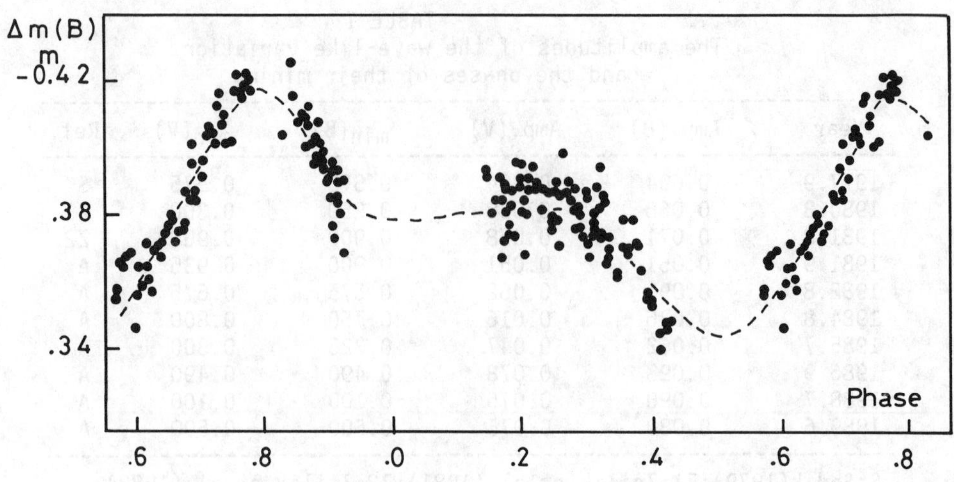

Fig.8. The distortion wave at outside eclipses in B for 1989. The symbols are the same as those used in Figure 6.

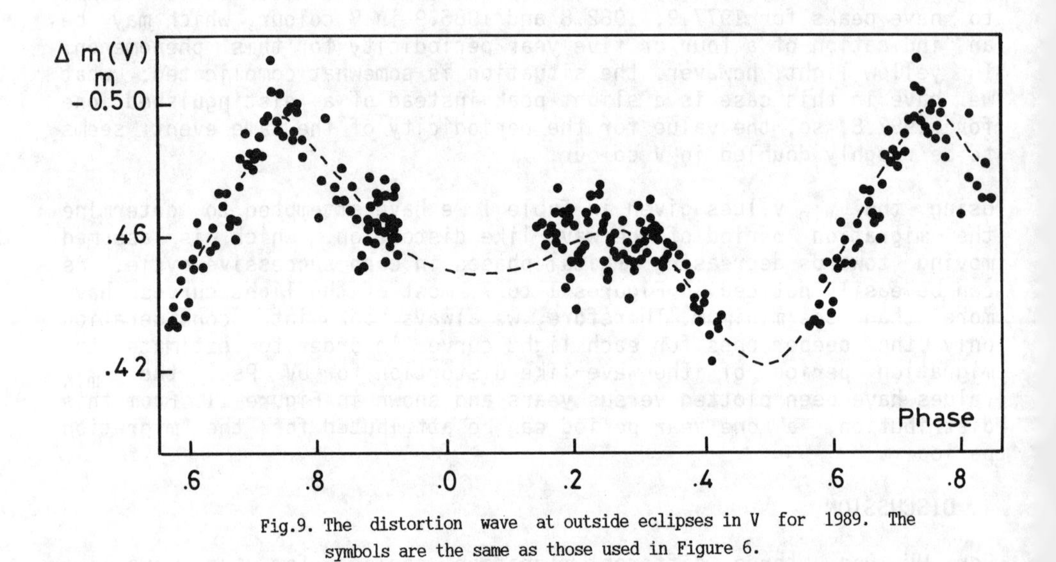

Fig.9. The distortion wave at outside eclipses in V for 1989. The symbols are the same as those used in Figure 6.

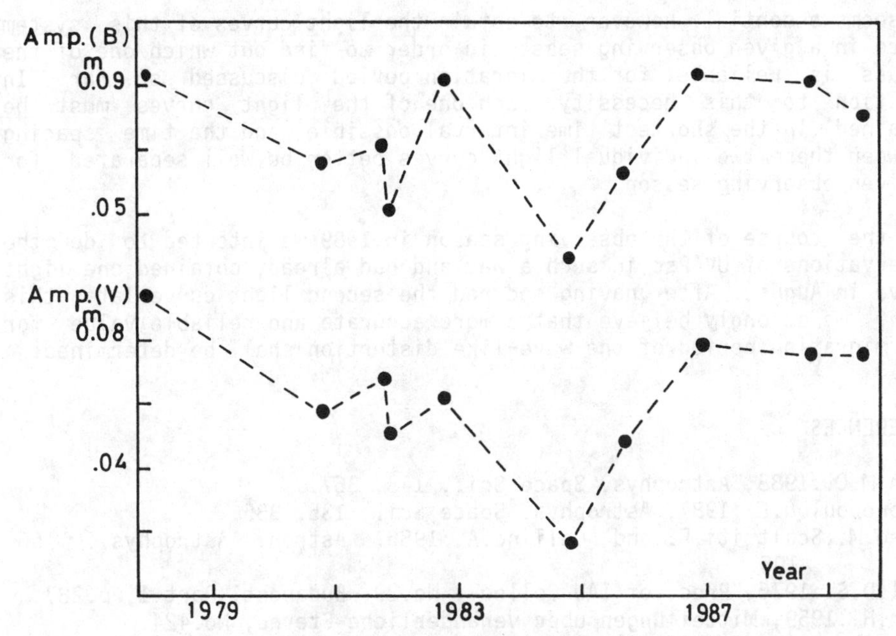

Fig.10. The amplitude variation of the wave-like distortion.

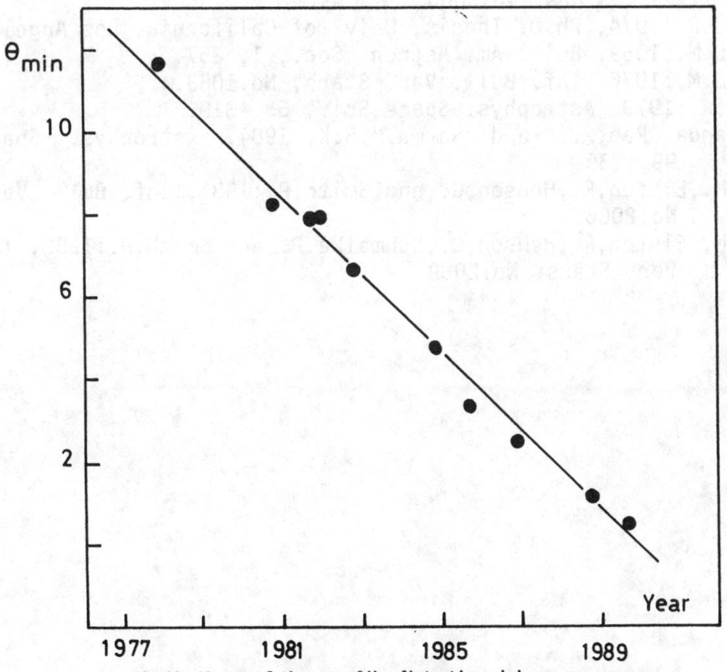

Fig.11. Phases of the wave-like distortion minima versus years.

It seems essential, however, to obtain the light curves of this system twice in a given observing season in order to find out which one of the values is reliable for the migration period discussed so far. In addition to this necessity, each one of the light curves must be obtained in the shortest time interval possible, and the time spacing between these two individual light curves better be well separated for a given observing season.

In the course of the observing season in 1989 we intented to do the observations of UV Psc in such a way and had already obtained one light curve in August. After having secured the second light curve late this year, we strongly believe that a more accurate and reliable value for the migration period of the wave-like distortion shall be determined.

REFERENCES

Akan,M.C.:1988, Astrophys. Space Sci., 143, 367.
Antonopoulou,E.:1987, Astrophys. Space Sci., 135, 335.
Busso,M.,Scaltriti,F.,and Cellino,A.:1986, Astron. Astrophys., 156, 106.
Hall,D.S.:1976, Proc. of IAU Colloq., No.29, Budapest, Part 1, p.287.
Huth,H.:1959, Mitteilungen über veränderliche Sterne, No.424.
İbanoğlu,C.:1987, Astrophys. Space Sci., 139, 139.
Keskin,V.,Akan,M.C.,İbanoğlu,C.,Tunca,Z.,and Evren,S.:1987, Inf. Bull. Var. Stars, No.3060.
Oliver,J.P.:1974, Ph.D. Thesis, Univ. of California, Los Angeles.
Popper,D.M.:1969, Bull. Am. Astron. Soc., 1, 257.
Popper,D.M.:1076, Inf. Bull. Var. Stars, No.1083.
Sadik,A.R.:1979, Astrophys. Space Sci., 63, 319.
Vivekananda Rao,P., and Sarma,M.B.K.:1984, Astrophys. Space Sci., 99, 239.
Zeilik,M.,Elston,R.,Henson,G.,and Smith,P.:1981, Inf. Bull. Var. Stars, No.2006.
Zeilik,M.,Elston,R.,Henson,G.,Schmolke,P.,and Smith,P.:1982, Inf. Bull. Var. Stars, No.2089.

NOVAE AND THEIR NON SPHERICALLY SYMMETRIC EXPLOSIONS

Marina Orio
Max Planck Institut für Astrophysik, Garching, FRG
and
Osservatorio Astronomico, Pino Torinese, Italy

Abstract

After a brief review of the results obtained in modelling the outbursts of novae and of the problems still to be solved, I describe different approaches to the meaning and the treatment of a non spherically symmetric outburst, a situation that is likely to occur very often. A work on *localized* thermonuclear runaways is presented.

1 - The current status of the nova theory

The explosive properties of CNO hydrogen burning in an electron degenerate gas (Schatzman, 1950) had already been known for 14 years when Kraft, in 1964, studied 10 well known nova systems finding out evidences of binarity in 7 of them. The following step was to propose that *all* novae must be close binaries in which a degenerate object, a white dwarf, accretes material. At the bottom of the accreted layer (because of compression and mixing with the WD material) the electron gas becomes highly degenerate. If the conditions of the accretion allow a sufficient degree of degeneracy, in fact, the pressure eventually depends only on the density and not on the temperature. The gas heated by the thermonuclear reactions does not expand and cool and the thermonuclear reactions become uncontrolled - the CNO burning energy generation rate being for a while almost proportional to the 18th power of the temperature. The runaway that follows is the only acceptable explanation of the nova phenomenon, whose characteristics are ununderstandable in terms of any phase of single star evolution. A nova occurs therefore in a system that seems in all analogous to that of *dwarf* novae, of which J. Mattei and C. La Dous have given a clear picture in this conference, but the outburst must be powered by thermonuclear instead of gravitational energy. It took several years to develop detailed full scale models for the outburst of novae, due to the difficulties in coupling stellar evolution with hydrodynamics. The efforts in this direction eventually produced a class of models that successfully explains the general characteristics of the nova outbursts (see in particular Starrfield et al, 1974, Prialnik et al, 1978).

C. İbanoğlu (ed.), Active Close Binaries, 601–610.

It turns out that a white dwarf needs to accrete only some $10^{-5} M_{\odot}$ in a period variable from hundreds (tens?) of years to 10^5 to reach the runaway conditions. This means that the burst can be recurrent in a system till the reservoir of fuel of the secondary finishes. The high frequency of novae (about 30 yr^{-1} in M 31) is thus justified. By now also full cycles of bursts have been calculated (see Prialnik, 1986). It is very important that the simulations do reproduce the energetic budget of a nova. Also the velocities of the material are obtained in extremly good agreement with the estimated values of a few hundreds to a few thousands km/sec. The mass ejected in the simulations is usually around $10^{-5} M_{\odot} yr^{-1}$, but with certain sets of parameters can even be a factor 100 inferior, and the same amount is evaluated from the observations. A powerful outburst is obtained more easily with enhanced values of CNO: in reality the spectra show sensibly enhanced CNO abundances. The Ne and Mg lines in the spectra of many novae can be understood in terms of an outburst on a white dwarf of neon-oxygen-magnesium. These results are remarkably good. Our picture is, however, not completely clear yet. Important questions on which people are currently working are:

- do classical novae have discs, as it was always assumed till some time ago, or can accretion be driven along the lines of force of a magnetic field (see Williams, 1989)?
- is the nuclear time scale too long to justify the rapid return to minimum luminosity of novae, as also the first x-ray observations seem to imply (see Ögelman, 1989) or do we have to admit another mechanism that helps to eject all the envelope?
- what is the role of the secondary while it is embedded in the ejected shell ? (Remember: the nova shell as the dimensions of a supergiant whereas the typical separation between the components is only of the order of a solar radius!)
- why do novae seem to show a cylindrical rather than spherical symmetry, with most of the matter ejected at the equator and in polar caps?
- is mass transferred non continuously or at a not continuous rate (the so called *hibernation* periods)? Does this mean that novae and dwarf novae really occur in the same systems at different phases?
- can we justify with a thermonuclear runaway (TNR) the short recurrence times of recurrent novae?
- Finally, it is necessary to obtain synthetic spectra and at least to know the *expansion* opacities of novae to be able to compare the theoretical with the observed light curves and to understand the spectra.

2 - The lack of spherical symmetry

A serious, and perhaps the greatest drawback in all the nova modelling is that accretion is surely a non spherically symmetric phenomenon. If there are magnetic novae, with fields and orbital periods like those of AM Her stars, the matter is accreted onto two hot polar caps. Whether it can then distribute uniformly all over the surface is rather doubtful, and a problem that has not been adressed yet; in any case we can expect the polar caps to be much hotter then the rest of the envelope. The case of accretion through a disc has been studied since the two works of Durrisen, 1977, and Kippenhahn & Thomas, 1978: these authors suggested that there can only be equatorial belts and a uniform envelope will never be reached. Further work, however, has made this picture more complicated. King & Shaviv, 1982, showed that for CV's with high accretion rates the boundary layer could become unstable and expand to form a corona that eventually falls on the white dwarf. Local stability analysis by Mc Donald, 1983, Hanawa & Fujimoto, 1988, Livio & Truran, 1988, has not reached definitive results, but indicates that spreading of the matter all over the surface is possible and put the accent on the importance of chemical enrichment due to shear mixing (this phenomenon helps to obtain the outburst). So there should be significant chemical gradients along the surface.

I will now review the work that has been done on the possibility that the thermonuclear runaway is localized, and successively describe the models I have developed together with Giora Shaviv. Non spherically symmetic outbursts have been modelled already years ago for neutron stars, but for white dwarfs the time scales, the extended portion of the surface on which the inhomogeneities are likely to be, the presence of an extended atmosphere, the radial extension of the envelope make the problem much more complex to study.

The evidences of a peculiar shape of the nova shells was the first characteristic that suggested that the burst might not occur with spherical symmetry (Tutukov & Yungelson, 1972). The profiles of most lines in novae spectra, splitted in two or four components, are interpreted with the hypothesis that the densities in different parts of the envelope should differ considerably and that the matter is eject in preferred directions, along the equatorial plane and/or at the poles. The photographs of nova shells (it can be seen well in the Duerbeck Atlas of Nova Shells) seem to streghten this hypothesis. More evidence is in the bidimensional CCD spectra that show evidence of different densities and probably even chemical compositions in different portions of the envelope. Infrared studies show that the same nova can form both C-rich and O-rich grains (Gehrz 1989), that need a different C/O ratio to form. This fact could easily be explained by a localized outburst. Bandiera & Foccardi presented the $H\alpha$ evolution of Nova Cen 1986 and concluded that the blobs in the envelope must have formed at the time of the outburst.

Alternative explanations for the peculiar spatial structure have been proposed: collisions with the disc, rotation and magnetic fields acting on the ejected material, but specially the collision-interaction with the secondary could be important (see Livio et al., 1989). However, if the different chemical abundances in different portions of the shell will be confirmed this is going to be an important proof in favour of a *localized* thermonuclear runaway ($LTNR$).

Such a localized outburst could be, of course, of particular importance also for a number of other factors. Studying it can also change our knowledge of some of the relevant range of parameters for the nova phenomenon.

3 - Different approaches to the problem

Tutukov & Yungelson (1972) first suggested that most novae are really due to LTNR's because a homogeneous temperature distribution on a nova shell is very difficult to reach. On the contrary, even a slow rotation of the white dwarf will cause a cylindrical rather than spherically symmetric temperature distribution of the material on the surface (the poles being hotter). This will result in a TNR in the polar regions. Only subsequently the interaction of matter ejected at the poles causes a shock wave in the equatorial plane. Also white dwarfs with non negligible magnetic field will have pressure gradients in the envelope, and the TNR will develop only in a portion of the surface. Tutokov & Yungelson indicated rough analytical estimates of the delay time of the outburst at the poles and at the equator, depending on the parameters of the system.

Mitrofanov, 1979, and Shara, 1982, had a different approach. Mitrofanov proposed that dwarf novae are also due to thermonuclear energy, the flash developing only in a portion of the accreted envelope. The different energy of the burst would thus only be due to the much smaller amount of burned material. Shara developed this idea and estimated under which conditions the thermalization time scales are long enough to fail to smooth the matter distribution. He compared the time scale of the thermonuclear burning of hydrogen in the CNO cycle with the time necessary to reach thermal equilibrium, evaluating the physical quantities at the point at which enough matter has been accreted to reach the critical pressure for the runaway. At this stage the phenomenon that dominates the conductivity is conduction: it becomes more important than radiation. Writing the lateral flux with the diffusion approximation Shara was able to compare the two different time scales and his conclusions were that *massive* and *hot* white dwarfs have a definite tendency to develop LTNR's, because the conductive time scale is longer than the typical CNO burning times, even for a minimal temperature inhomogeneity in the accreted envelope. Shara predicted that these LTNR's would result in sort of vulcanoes. A chimney with hot lava would be formed, convection helping the vertical expansion of the hot front, while pressure waves on the sides would inhibit the lateral expansion. Subsequently, however, for the strong gravitational force the material would fall over the external surface of the envelope and cool, emitting EUV and soft x-rays in the process, while the chimney would contract more rapidly than the burning front expands. Shara's prediction was also that the site would stay hot and in it the next LTNR could develop in few weeks or months.

The physics of the phenomenon is, however, even more complex. Fryxell & Woosley (1982) studied the finite propagation time of multidimensional thermonuclear runaways. The lateral burning velocity of the propagating front is the sum of the conductive and the convective velocity. The velocity of turbulent convection is

$$v_{turb.conv} = \sqrt{\frac{h_p}{\tau_b}},$$

where h_p is the pressure scale height and τ_b the time scale of the burning fuel.

This velocity results negligible in comparison with the radiative and conductive heat transmission during most of the secular evolution of novae, but by the time the CNO burns at a fast rate, at temperatures $T > 15 * 10^6$K and densities of the order of $10^4 g/cm^3$, it has increased of even a factor 10^7! For such densities and temperatures typical of a nova approaching an outburst, the time for the deflagration to propagate is of a few days. This poses serious problems to the interpretation of nuclear, rather than gravitational, energy source powering the outburst of a dwarf nova, but implies a classical nova with a smooth rise to the optical maximum, lasting - like in many observed cases - for much longer than the few hours necessary for the turnover time scale of the β decaying nuclei (what determines the peak of the bolometric light in the spheri-symmetric models). It means also that the outburst of novae is a rather more complicated phenomenon than in the current models and needs 2-dimensional modelling.

Kutter & Sparks, 1987, and 1989, have developed a 1-dimensional model in which some effects of the non spheri-symmetrical accretion can be computed. They devised how to include in their code the centrifugal force and the effect of shear mixing on the abundances; moreover they assumed that accretion occurs only in a belt subtending an angle of 27.5° above and under the equator and therefore mutipled all the energy sources (gravitational, shear, nuclear) by a factor $\theta/4\pi$. Although in their first models they did not obtain mass ejection, a result due either to the too early release of the heat and/or to the lowering of the mechanical confinement by the centrifugal forces, a model developed later with a WD of 1 $M_\odot$ and an accretion rate of 4 10^{-10} $M_\odot yr^{-1}$ ended in an ejection of $10^{-4} M_\odot$.

The importance of this work is specially in having shown how the non spherically symmetric effects can change the range of parameters of the different physical quantities involved in the outburst, in particular strong explosions might be obtained also with intermediate mass white dwarfs and intermediate accretion rates.

4 - A model for the lateral flows

In this section I will describe a project that I am carrying on with Giora Shaviv to study LTNR's. The first question is, of course, whether the conditions are really favourable for a *localized* developement of the runaway or not. With simple analytical estimates Shara obtained the important result that it is possible. His parametrical study, however, cannot take all the long phase of secular evolution into account. For this reason our first step was a simple 1 1/2 -D model to compute the meridional heat flows in the envelope, while this is being accreted, its structure and densities change continuously and the nuclear fuel is slowly burned. These computations were also intended to give the parameters for a starting model in 2-D, a type of computation in which no parametrical study is possible because of lenght and coast. A first formulation of the equations can be found in Orio (1987) and Orio & Shaviv (1988) while a revised and more complete version is in Orio & Shaviv, 1989. Our model is analogous to that of Nozakura & Fujimoto (1986) for the X-ray burst on a neutron star, with the difference that in our case we had to take into account the geometry of the curved surface, the size of the region in which there is a temperature inhomogeneity being a non negligible portion of it.

The first parameter of the model is the mass acccretion rate, $\dot{m}$, and the equation of continuity yields the change of column density Σ:

$$\frac{\delta\Sigma}{\delta t} = \frac{\dot{m}}{4\pi R^2}.$$

The column density determines the pressure at the bottom of the accreted envelope and from it we determined the density ρ_b at the bottom, where the burning occurs:

$$P = f(\rho_b, T_b, X, Z) => \rho(P_b),$$

T_b being the temperature at the base of the envelope, X the hydrogen abundance and Z the abundance of heavy elements. To simplify we considered as heavy elements a mixture of C,N, and O with the average molecular weight of the three elements.

The equation of energy can be written as a function of the temperature, which is our next unknown variable:

$$\rho c_v \frac{\delta T}{\delta t} = -\frac{P}{\rho}\frac{\dot{\Sigma}}{\Sigma} + \epsilon\rho - \frac{1}{R}\frac{\delta F}{\delta\theta} - F_r,$$

where F is the meridional flux and F_r the radial flux. For the first we can consider the diffusion approximation:

$$F = -\frac{4ac}{\kappa\rho}\Delta T^4$$

where the opacity κ is given by $1/\kappa = 1/\kappa_{rad} + 1/\kappa_{cond}$. For the radial term of the flux in a certain range of temperature and density the plane parallel approximation can be considered:

$$F = -\frac{ac(T_c^4 - T_0^4)}{3\kappa\Sigma^2}$$

where the subscript c indicates the core, 0 the atmosphere. During all the p-p cycle and in a first phase of the CNO cycle this approximation is not a suitable one for the material burning on a white dwarf, but the full scale 1-D calculations show that the radial flux is very little. A possibility of parametrization that we have attempted is to consider the energy flowing radially as a constant fraction of the energy generated in the burning (Orio 1987).

Finally, we had to include an equation for the variation of hydrogen abundance due to the burning:

$$\frac{\delta X}{\delta t} = -\frac{\epsilon(\rho, T)}{Q}.$$

Obviously the kinetic terms are missing in the equation of energy and also an equation of motion should be included to have a full model, but our aim was here only to determinate the conditions at the onset of the outburst. In this stage we stopped our calculations when the temperature reached values of the order of 10^8 and the dynamical effects became important. For the same reason we could afford neglectig convection. A full equation of state from stellar evolution codes was included, the radiative and conductive opacities were taken from Kovetz & Shaviv, 1979. For the energy generation rate we used simple analytical approximations, taking into account that at high temperatures the nuclear time scale is the one of the β decaying nuclei. To consider the geometry of the curved surface, we integrated all the physical quantities over the volume of a torus of material in the envelope, $dV = 2\pi R^2 sin\theta dr$.

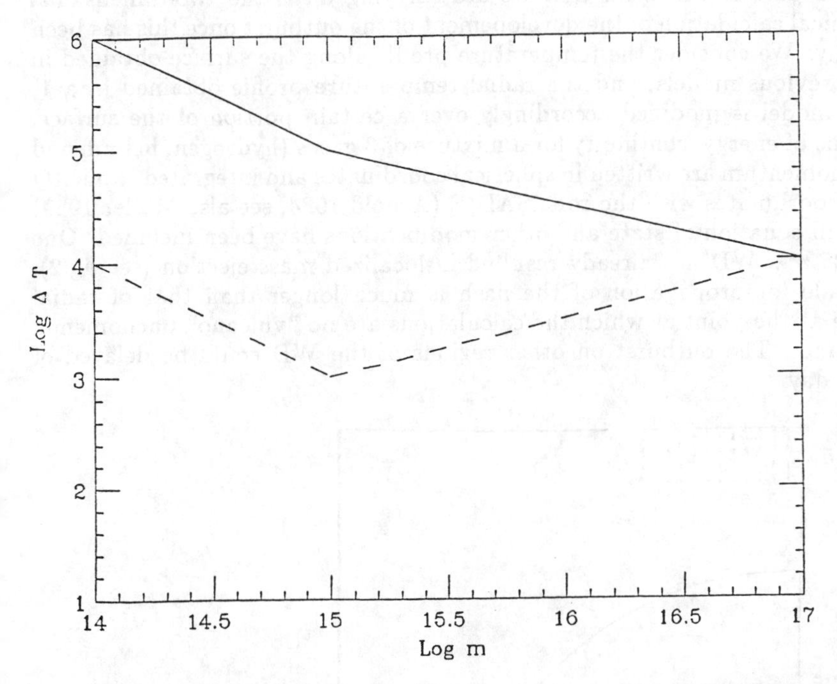

Fig. 1 - The minimal ΔT that does not smooth out and produces a localized outburst on the surface of a WD of 1.3 $m_\odot$ (full line) and 0.8 $m_\odot$, as a function of the mass accretion rate in g/sec.

With perturbations (temperature gradients) of different amplitude and wavelength the parametrical study yielded the following results:

1) there is a minimal ΔT for which, with a given $\dot{m}$ and m_{WD}, the flash developes only in a small portion of the surface. Its propagation would only be due to the turbolent velocity, and this explains the slow rise to maximum of many novae;

2) the flash tends to be localized on hot (*luminous*) white dwarfs;

3) the minimal ΔT that is necesary for a *localized* outburst decreases with increasing mass accretion rate only for massive white dwarfs (see fig.1), whereas for intermediate masses, whose envelope has lower densities, the interplay between the conductivity that tends to thermalize and the nuclear energy generation is very delicate and critically dependent on the prameters. For a white dwarf of 0.8 $M_\odot$

the LTNR is more likely to occur at intermediate mass accretion rates, of the order of 10^{15} g/sec (see fig.1);

4) at accretion rates of the order that is typical for dwarf novae LTNR's do not usually occur, or at least they need large temperature gradients (several million degrees) and very luminous WD's, so the LTNR cannot be a good general explanation for the phenomenon of dwarf novae.

A second part of the work that we are carrying on is the two-dimensional hydrodynamical calculation of the developement of the outburst once this has been ignited locally. We consider the temperature profile along the surface obtained in one of the previous models, and the radial temperature profile obtained in a 1-D full scale model is modified accordingly over a certain portion of the surface. The equations of energy, continuity for a mixture of 3 gases (hydrogen, helium and CNO) and momentum are written in spherical coordinates and integrated explicitly in spherical coordinates with the code SADIE (Arnold 1984, see also Müller 1989) in which a full equation of state and other modifications have been included. One model of 1.25 $m_\odot$ WD has already resulted in localized mass ejection (see fig.2). The time scale for propagation of the flash is much longer than that of radial ejection, and at the point at which the calculations are no "vulcano" phenomenon has been found. The outburst on other regions of the WD could be delayed of more than 1 day.

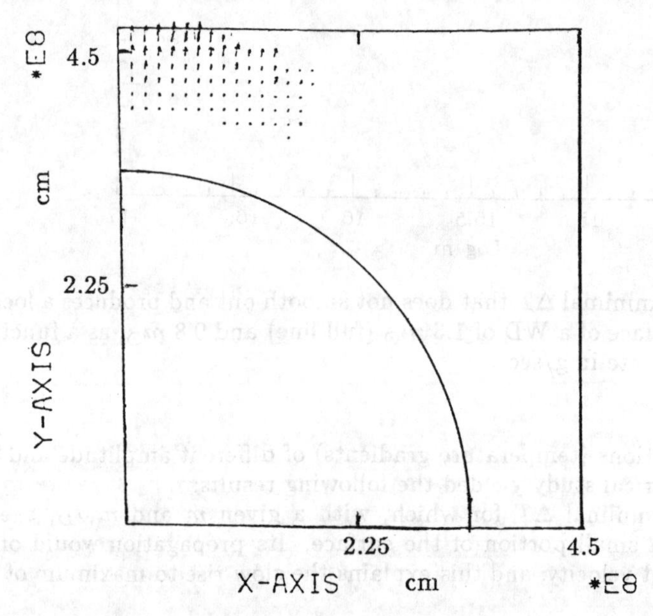

Fig. 2 - The velocity field soon after the beginning of the runaway ($t \simeq$ 10*sec.*): matter is ejected only in the equatorial plane. The maximum velocity is about $10^7 cm/sec$.

A future project will be to calculate a 2-D dynamical model also from a much earlier stage, to check the validity of our assumptions on the amount of energy that flows radially and to include small radial expansions of the envelope during the phase of secular accretion.

5 - Conclusions

To work on the 2-dimensional effects of the thermonuclear runaway on white dwarfs one has to face many physical and numerical problems, the numerical work is moreover very time consuming. The velocities are mostly subsonical and the Courant time step very small if one works with explicit codes, to work in two dimensions with an implicit code poses several numerical problems. The paths explored so far, however, indicate how important it can be. Many conclusions that have been reached on the range of physical parameters that are typical for novae and on the characteristics of the outburst could still be modified by a more thorough exploration of this phenomenon. In the future, after the very good result obtained by 1-dimensional computations in outlining the main features of the mechanism, for further progress substantial effort must be dedicated to the understanding of the non spherically symmetric effects.

References

Arnold, C. N., 1985, *Ph.D. thesis*, University Microfilms, Ann Arbor, Michigan

Bandiera, Foccardi, P., *Messenger no.* 52

Duerbeck, H. W., 1987, *Messanger no.* 50

Durrisen, R.H., *Ap.J.* 1977, **213**, 145

Fryxell, B.A., Woosley, S. E., *Ap. J.*, **261**, 332

Gehrz, R., *proceeding of the I.A.U. Coll. 122*, to be published

King, A., Shaviv, G., 1984, *Nature*, **308**, 519

Kippenhahn, R., Thomas, H.C., 1978, *A. & A.*, **63**, 265

Kutter, G.S., Sparks, W. M., 1987, it Ap. J., **321**, 386

Kutter, G.S., Sparks, W. M., 1989, *Ap. J.*, **340**, 985

Ladous, C., *this conference*

Livio, M., Truran, J. W., 1987, *Ap. J.*, **318**, 316

Mc Donald, J , 1984, *Ap.J.*, **283**, 241

Mitrofanov, I. G., 1980, *I.A.U. Symp.88, "Close Binary Systems", ed. M. Plavec, D. Popper, R.K. Ulrich, (Dordrecht:Reidel)*, p.431

Müller, E., *Jour. of Comp. Phys.*, **79**, 277

Nozakura T., Ikeuchi S., Fujimoto, M. Y., 1984, *Ap. J.* **286**, 221

Ögelman, H.B., *proceedings of the I.A.U. Coll. no. 122*, to be published

Orio,M., Shaviv, G., *"Mass Outflows from Stars and Active Galactic Nuclei"' L. Bianchi and R. Gilmozzi eds.*, Kluwer Academical Publisher

Orio, M., 1987, Ph.D. thesis at the *Technion*, *MPA* preprint no. 335

Orio, M., Shaviv, G., submitted to *A. & A.*

Shara, M. M., 1982, *Ap. J.*, **261**, 649

Shara, M., Livio, M., Moffat, A.F.J., Orio, M., 1986, *Ap. J.*, **314**, 653
Prialnik, D., 1986, *Ap.J.*, **310**, 222
Prialnik, D., Shara, M. M., Shaviv, G., 1978, *Ap. J.*, **62**, 339
Starrfield, S., Sparks, W.M., Truran, J.W., 1974, *Ap. J.*, **311**, 163
Tutokov, A., Yungelson, L., 1974, *Nauchnye Informatsii*, **29**, 134
Williams, R., 1989, *preprint*

OPTICAL PROPERTIES OF CATACLYSMIC VARIABLE STARS

JANET AKYÜZ MATTEI
American Association of Variable Star Observers (AAVSO)
25 Birch Street
Cambridge, Massachusetts 02138
USA

ABSTRACT. An overview is given of the optical behavior of cataclysmic variables, including the various types, the canonical model of these close binary systems, the causes of eruptions, short-term variations, and orbital periods, and correlations of light curve parameters with physical properties.

1. Introduction

Cataclysmic variable stars are close binary systems, most of which undergo brightenings - eruptions or outbursts - at intervals ranging from about 10 days to hundreds of years, and at frequencies ranging from once to many times, depending on the type of the system. The amplitude and the duration of the eruptions vary from 2 to 18 magnitudes and from a few days to a few months, again depending upon the type.

They are spectroscopic binary systems with orbital periods ranging from about 76 minutes to 16 hours in which a late-type star of spectra ranging from G to M (main sequence or slightly evolved) fills its Roche lobe and transfers mass through the inner Lagrangian point to the primary star, which is a white dwarf. Due to its angular momentum the transferred material accumulates in a disk around the white dwarf. A shock front is formed as a bright spot (hot spot) near the outer edge of the disk where the stream of material from the secondary impacts the disk.

In this paper an overview will be given of the optical observational and physical properties of these stars, as well as of the correlations among light curve parameters. Excellent review papers have been published that give thorough information (Robinson 1976; Cordova and Mason 1981; Patterson 1979; Webbink et al. 1987). The most extensive review of dwarf novae and novalike stars has been written by la Dous (1989) as part of the Monograph Series on Nonthermal Phenomena in Stellar Atmospheres. Ritter (1987) has published an excellent compilation of the physical and observational properties of cataclysmic variables. Duerbeck's (1987) book, *A Reference Catalogue and Atlas of Galactic Novae*, gives properties, references, and finder charts from the Palomar Sky Survey Prints.

C. İbanoğlu (ed.), Active Close Binaries, 611–627.

2. Classification

The classification of cataclysmic variables is based on the behavior of light variation. Novae, recurrent novae, dwarf novae, and nova-like variables comprise the class of cataclysmic variables. Although symbiotic stars are sometimes also considered to be part of this class, these stars will not be discussed here. In this overview emphasis will be placed on dwarf novae.

In most cases the classification is quite clear. However, for some members of this class, as more information has become available the classification has been changed over the years, as in the cases of WZ Sge being reclassified from recurrent nova to dwarf nova and DQ Her, originally classified as nova due to its bright eruption in 1936, being classified recently also as intermediate polar prototype.

Most of the information on the long-term optical behavior of these variables has been obtained by observers of the American Association of Variable Star Observers (AAVSO), the Royal Astronomical Society of New Zealand, and other variable star observing groups around the world.

2.1. NOVAE

These stars have had outbursts only once within our recorded history, in which the amplitudes ranged from about 8 to 18 magnitudes and during which between 10^{44} and 10^{45} ergs of energy were emitted. Depending upon the rate of decline to 2 magnitudes from maximum, novae are divided into the categories fast, in which the star fades by 2 magnitudes in 25 days or less, moderate, in which it fades by 2 magnitudes in 26 to 80 days, and slow, in which it fades by 2 magnitudes in 81 days or more. Figure 1 is an AAVSO light curve of the fast nova V1668 Cyg (Nova 1978), and Figure 2 the AAVSO mean light curves of the moderate novae V533 Her (Nova 1963) and V446 Her (Nova 1960). Some novae, about 120 days after maximum, show a sudden fading of several magnitudes in the optical. They recover in about a month, reaching a secondary maximum which is fainter than the brightness level before the fast fading, and then slowly continue to fade. Ney and Hatfield (1978) conclude that in these novae the grains in the ejecta increase in size during the fast fading, thus decreasing the optical light from the system. In time, as ejected material expands, the system appears brighter. Figure 2 also shows the AAVSO mean light curve of DQ Her, the famous nova of 1936, which is the best example of this phenomenon.

Some novae show significant light variation in their declines, as seen in Figure 3, the AAVSO mean light curve of GK Per (Nova 1901). After several decades, this interesting nova started to have minor outbursts, like a dwarf nova, at intervals of several years, as seen in Figure 4.

2.2. RECURRENT NOVAE

These stars have outbursts more than once, the recurrence frequency varying from star to star and ranging from about 10 to more than 100 years. The amplitude of the outbursts ranges from 7 to 9 magnitudes, during which 10^{43} to 10^{44} ergs of energy are emitted. The rise to maximum is extremely fast, usually within 24 hours, and the decline may last several months. The light curves of recurrent outbursts of such a star are almost identical, as may be seen in the superimposed light curves of the 1898, 1933, and 1958 outbursts of RS Oph in Figure 5.

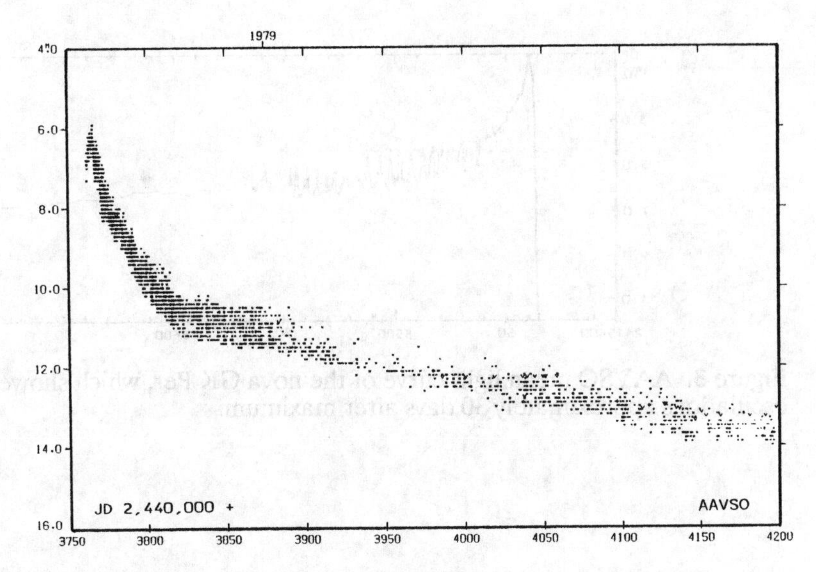

Figure 1. AAVSO light curve of the fast nova V1668 Cyg. Each dot is one observation.

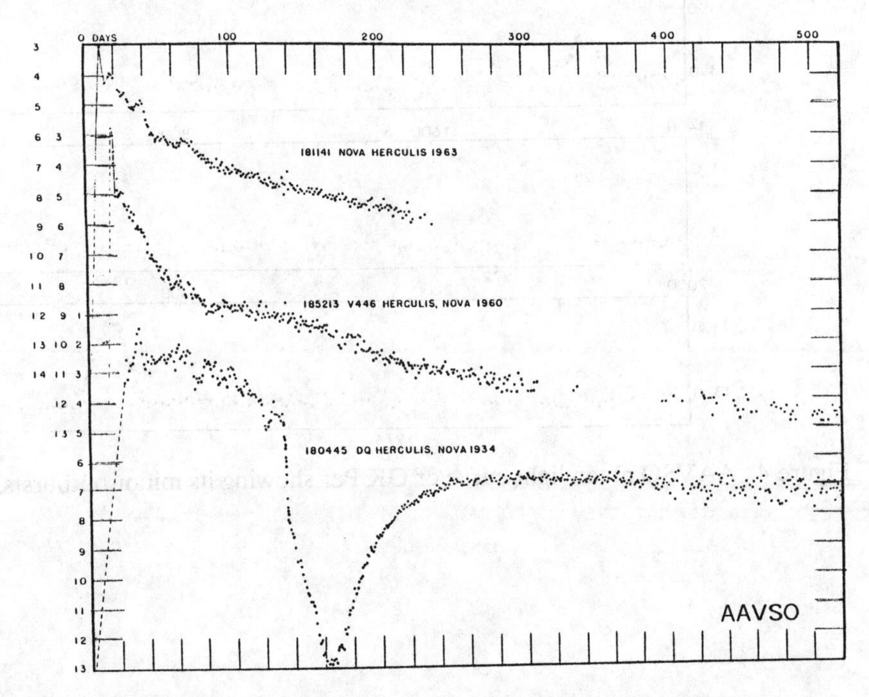

Figure 2. AAVSO mean light curve of the moderately fast novae V533 Her, V446 Her, and DQ Her. DQ Her showed a rapid decline in brightness and a secondary maximum.

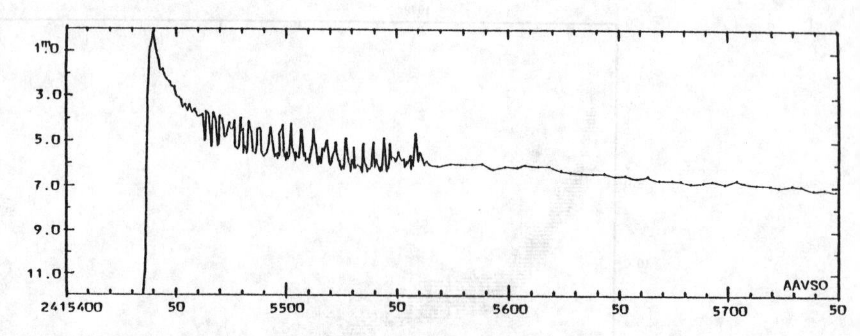

Figure 3. AAVSO mean light curve of the nova GK Per, which showed rapid oscillations approximately 30 days after maximum.

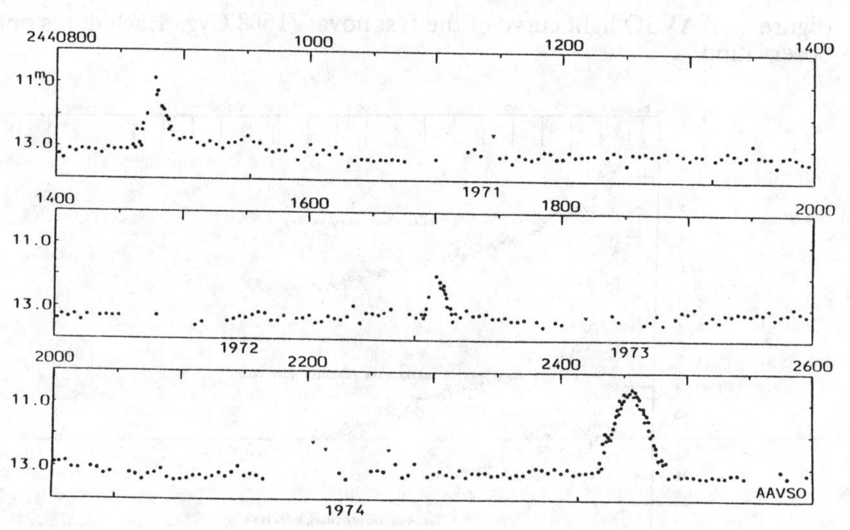

Figure 4. AAVSO mean light curve of GK Per showing its minor outbursts.

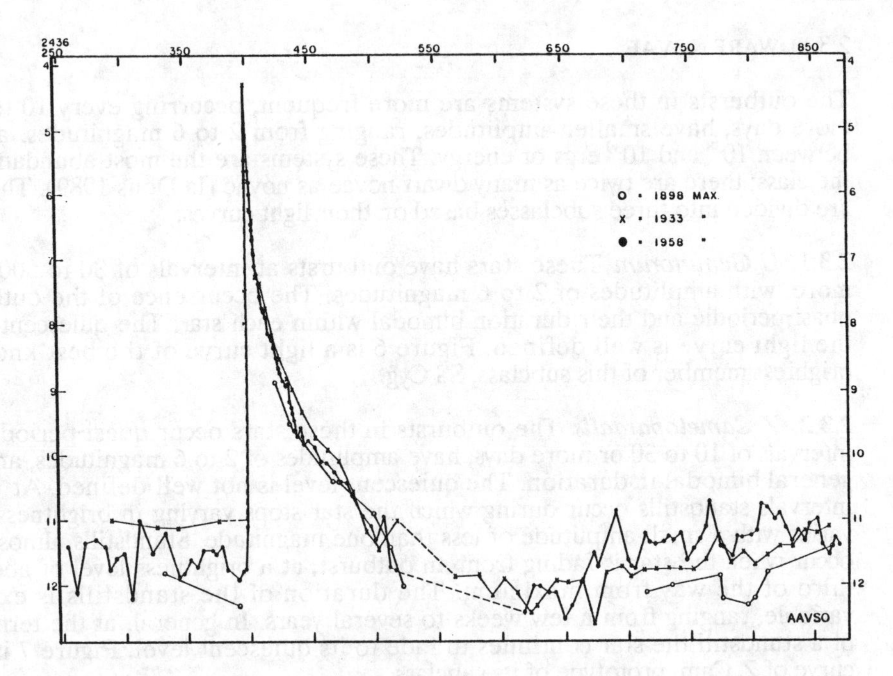

Figure 5. AAVSO mean light curve of the recurrent nova RS Oph with three outbursts (1898, 1933, 1958) superimposed.

2.3. DWARF NOVAE

The outbursts in these systems are more frequent, occurring every 10 to 500 or more days, have smaller amplitudes, ranging from 2 to 6 magnitudes, and emit between 10^{38} and 10^{39} ergs of energy. These systems are the most abundant among the class; there are twice as many dwarf novae as novae (la Dous 1989). These stars are divided into three subclasses based on their light curves.

2.3.1. *U Geminorum.* These stars have outbursts at intervals of 30 to 500 days or more, with amplitudes of 2 to 6 magnitudes. The occurrence of the outbursts is quasi-periodic and their duration bimodal within each star. The quiescent level of the light curve is well defined. Figure 6 is a light curve of the best known and brightest member of this subclass, SS Cyg.

2.3.2. *Z Camelopardalis.* The outbursts in these stars occur quasi-periodically at intervals of 10 to 30 or more days, have amplitudes of 2 to 6 magnitudes, and are in general bimodal in duration. The quiescent level is not well-defined. At random intervals standstills occur during which the star stops varying in brightness and/or varies with a small amplitude of less than one magnitude. Standstills almost always occur when the star is fading from an outburst, at a brightness level of about one-third of the way from maximum. The duration of the standstills is extremely variable, ranging from a few weeks to several years. In general, at the termination of a standstill the star continues to fade to its quiescent level. Figure 7 is a light curve of Z Cam, prototype of its subclass.

2.3.3. *SU Ursae Majoris.* These stars have two distinct types of outbursts - narrow outbursts of 1 to 3 days duration which occur every 10 to 30 days or more, and wide outbursts (superoutbursts) which are 0.5 to 1 magnitude brighter than the narrow outbursts and occur less frequently, after every 10 to 15 narrow outbursts. The superoutbursts are periodic, so that each star has two or more superoutburst periods. Each period lasts from 3 to 30 cycles, with an abrupt change from one period to the other. Figure 8 is an AAVSO light curve of an SU UMa star, AY Lyr.

The signature of this subclass is the occurrence of superhumps, small-amplitude, periodic oscillations which have periods 2 to 3 percent longer than the orbital period of the system and which are observed only during superoutbursts.

Some SU UMa stars have infrequent narrow outbursts that occur several hundred days apart. Superhumps seen during superoutbursts (Bond, Kemper, and Mattei 1982; Robinson et al. 1987), together with the distinct light curve of narrow and wide outbursts, confirm membership in this subclass.

2.4. NOVA-LIKE VARIABLES

These stars do not show distinct eruptive activity, and in fact some even have anti-eruptions during which the star fades in brightness by 2 to 5 magnitudes. In some ways they resemble dwarf novae at certain stages of their activity. Some have strong magnetic activity. Some members of this class may also be classified as another type of cataclysmic variable due to their iight curves. Their observable features are in agreement with the assumption of the Roche model for the explanation of their physical nature. These stars are divided into five subclasses depending upon their characteristics (la Dous 1989).

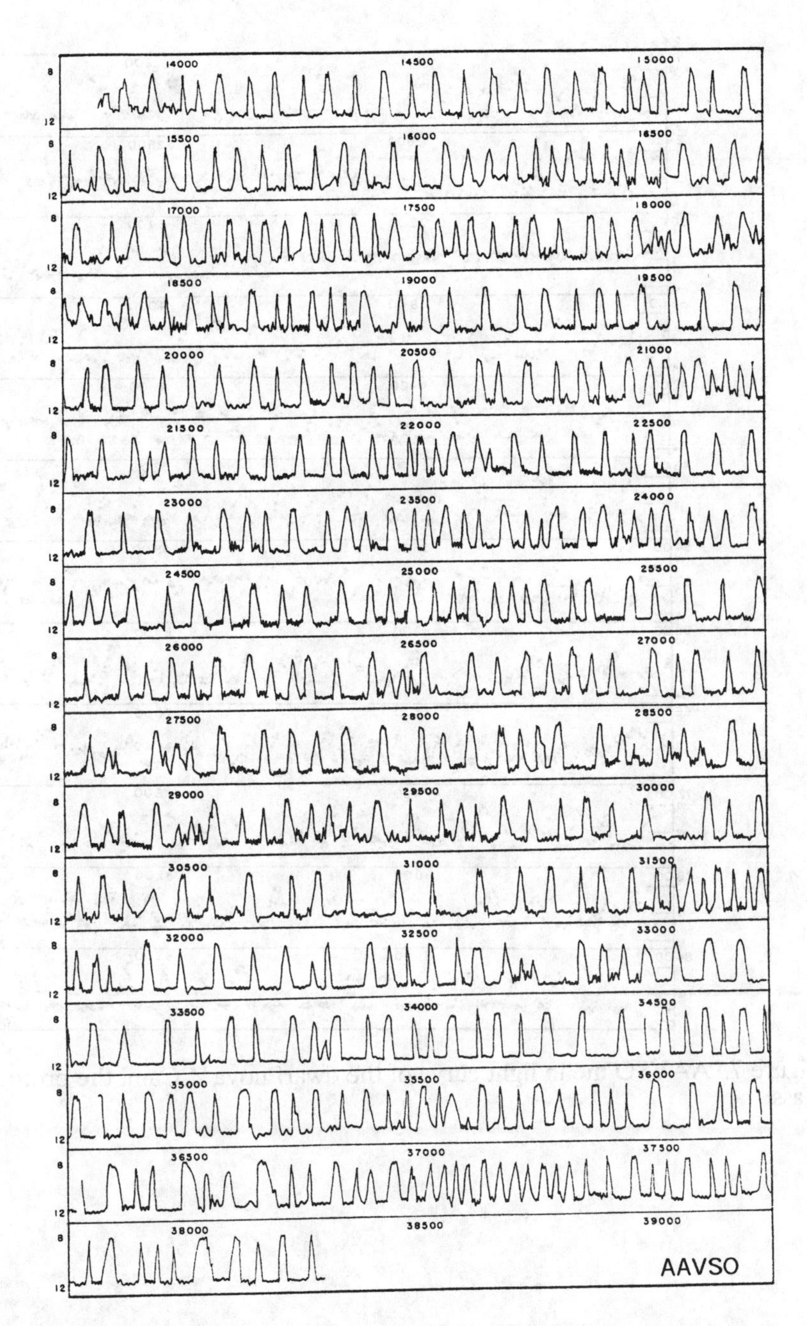

Figure 6. AAVSO mean light curve of the U Gem-type dwarf nova SS Cyg.

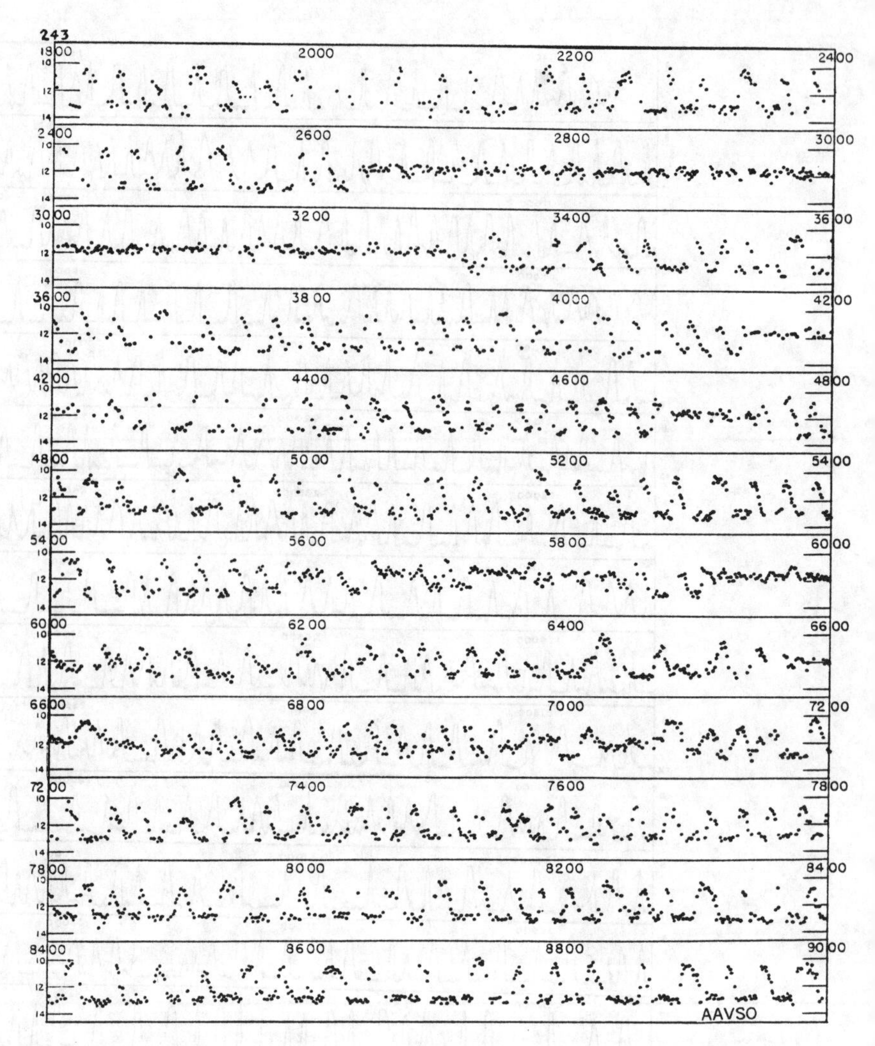

Figure 7. AAVSO mean light curve of the dwarf nova Z Cam, the prototype of its class.

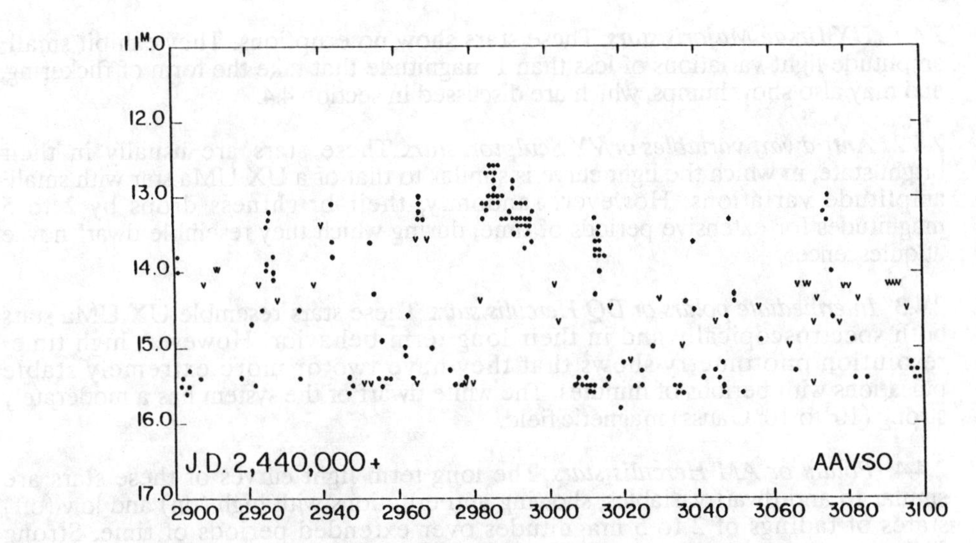

Figure 8. AAVSO light curve of the SU UMa-type dwarf nova AY Lyr.

2.4.1. *UX Ursae Majoris stars.* These stars show no eruptions. They exhibit small-amplitude light variations of less than 1 magnitude that take the form of flickering, and may also show humps, which are discussed in section 4.4.

2.4.2. *Anti-dwarf variables or VY Sculptori stars.* These stars are usually in their bright state, in which the light curve is similar to that of a UX UMa star with small-amplitude variations. However, randomly, their brightness drops by 2 to 5 magnitudes for extensive periods of time, during which they resemble dwarf novae at quiescence.

2.4.3. *Intermediate polars or DQ Herculis stars.* These stars resemble UX UMa stars both spectroscopically and in their long-term behavior. However, high time-resolution photometry shows that they have two or more extremely stable pulsations with periods of minutes. The white dwarf of the system has a moderately strong (10^5 to 10^6 Gauss) magnetic field.

2.4.4. *Polars or AM Herculis stars.* The long-term light curves of these stars are similar to anti-dwarf variables, showing anti-eruptions with high (on) and low (off) states of fadings of 2 to 5 magnitudes over extended periods of time. Strong flickering activity is also seen in their light curves. The characteristic feature of these stars is the very strong optical circular and linear polarization.

2.4.5. *AM Canum Venaticorum stars.* These stars have no outbursts, but show flickering and quasi-periodic oscillations. Their spectra are characterized by the absence of hydrogen lines and the presence of strong helium lines. They have the shortest orbital periods of cataclysmic variables, less than 80 minutes. They are believed to consist of two white dwarf components.

Table I shows the types and outburst characteristics of the cataclysmic variables mentioned above.

3. Causes of Outbursts

The cause of outbursts varies from type to type. In novae the cause of outbursts is believed to be thermonuclear runaway. Hydrogen transferred from the main sequence secondary star slowly accumulates on the surface of the white dwarf, forming an envelope around the white dwarf. In time, with slowly arriving matter at the base of the accreted hydrogen shell, the density increases so that the gas becomes degenerate. When density and temperature become so great that hydrogen-burning nuclear reactions are initiated, the enhanced carbon, nitrogen, and oxygen (CNO) from the interior of the white dwarf act as catalysts (Starrfield et al. 1972). In a typical nova outburst 10^{-5} to 10^{-4} solar masses are ejected (Webbink 1989).

In dwarf novae the brightening of the disk is responsible for the outbursts. One model attributes this brightening to a hydrodynamic instability that occurs in the cool secondary, causing an increase in mass flux into the accretion disk, which in turn causes the material in the disk to accrete onto the white dwarf and so create the outburst. The competing model attributes the accretion of matter onto the white dwarf to an instability in the disk caused by a change in viscosity and/or temperature (Webbink 1989). Observational evidence exists supporting both models, and it even may be that within a star both mechanisms may be responsible

TABLE I

CLASSIFICATION OF CATACLYSMIC VARIABLE STARS
TYPES AND OUTBURST CHARACTERISTICS

TYPE	Δ_M	ENERGY OUTPUT (ERGS)	RECURRENCE
NOVA	8 - 18	10^{44} - 10^{45}	NO RECURRENCE
RECURRENT NOVA	7 - 9	10^{43} - 10^{44}	10 - 100+ YEARS
DWARF NOVA			
U GEM	2 - 6	10^{38} - 10^{39}	30 - 500+ DAYS
SU UMA	2 - 6	10^{38} - 10^{39}	10 - 30+ DAYS
Z CAM	2 - 6	10^{38} - 10^{39}	10 - 50 DAYS
NOVALIKE			
UX UMA	---	---	---
ANTI-DWARF	2 - 5	---	ANTI-ERUPTION
DQ HER	---	---	---
AM HER	2 - 5	---	ANTI-ERUPTION
AM CVN	---	---	---

for outbursts.

4. Short-term Variations

High-speed photometry has shown that along with large-amplitude outbursts, cataclysmic variables also show short-term, small-amplitude variations: flickerings, coherent oscillations, quasi-coherent oscillations, humps, and superhumps.

4.1. FLICKERING

These random, small-amplitude variations, ranging typically from about 0.1 to 0.5 magnitude and having periods between 20 seconds and many minutes, have been seen in dwarf novae as well as in the quiescent stage of novae and nova-like variables.

The source of flickering has been claimed to be a hot spot on the outer edge of the accretion disk produced by a stream of gas transferred from the secondary (Robinson 1976). However, detailed eclipse observations of the dwarf novae HT Cas, V2051 Oph, and RW Tri suggest that the source of flickering may be the hot inner disk close to the white dwarf (Patterson 1981; Warner and Cropper 1983; Horne and Stiening 1985; la Dous 1989).

In the eclipsing dwarf nova U Gem, high-speed photometry shows that while flickering is strong before the eclipse, it is greatly reduced or disappears during eclipse. Warner and Nather (1971) point out that it is the hot spot that is eclipsed in this system, and the source of flickering is this hot spot.

During outburst the flickering activity is continuous, suggesting that the outburst does not affect the hot spot.

Flickering activity is seen in different wavelengths with no significant differences in the pattern except that the amplitude of variation is larger in the shorter wavelengths (la Dous 1989).

4.2. COHERENT OSCILLATIONS

During outbursts of dwarf novae, extremely periodic coherent oscillations with periods ranging from 7 to 39 seconds, very small amplitudes ranging from 0.006 to 0.02 magnitude, and coherence times typically 10^3 cycles, or several hours, have been observed (Robinson 1976; Robinson and Nather 1979; Patterson 1979; Hildebrand et al. 1981). These oscillations have been recorded during normal outbursts and superoutbursts of dwarf novae, except during the standstills of Z Cam stars. They have never been detected during the quiescent states of dwarf novae (la Dous 1989).

Oscillations have been observed during eclipses in several systems (Warner 1976; Patterson 1981; Schoembs 1986). Patterson found that in the SU UMa-type dwarf nova HT Cas the oscillations undergo a phase shift of -360°, and the amplitude is significantly reduced before mid-eclipse. Schoembs found that in the SU UMa system OY Car the oscillations are almost totally eclipsed after mid-eclipse.

The period of the oscillations changes from star to star and from outburst to outburst of the same star. The period varies during an outburst so that it is constant during maximum brightness, and as the star declines in brightness from maximum the period changes and the oscillations disappear. During the October 1979 outburst of SS Cyg, Hildebrand et al. (1981) detected oscillations with a period of 7.29 seconds. During the six days of observations of the outburst the coherence changed and the period lengthened to 8.54 seconds. As the star declined to quiescence these oscillations disappeared. According to Cordova and Mason (1981) the oscillations may not be absent but may be combined with the other small-amplitude variations of the star at this time.

Several scenarios have been suggested to explain these oscillations, including the rotation of the white dwarf (Paczynski 1978), inhomogeneities in the inner part of the accretion disk (Bath 1973), and non-radial pulsations of the white dwarf (Robinson 1976; Cordova and Mason 1981). It is safe to conclude that the source of these oscillations is somewhere in the area of the inner disk, boundary layer, and/or the white dwarf (la Dous 1989).

4.3. QUASI-PERIODIC OSCILLATIONS

During outbursts of dwarf novae, other small-amplitude, short-period variations are seen. These are quasi-periodic oscillations that have longer periods ranging from 23 to 413 seconds and amplitudes ranging from 0.005 to 0.02 magnitude. As the name suggests, these oscillations are not strictly periodic and their period may last only one or two cycles, after which both the period and the amplitude may change abruptly. Quasi-periodic and periodic oscillations may appear simultaneously in some systems, as in the cases of SS Cyg and RU Peg. Within the same object the period may vary significantly from outburst to outburst. In U Gem, Robinson and Nather (1979) found the period in one outburst to be 150 seconds, while Patterson (1981) during another outburst found quasi-periodic oscillations of 24 seconds.

Robinson and Nather (1979) found that during an eclipse of U Gem, quasi-periodic oscillations were present throughout the eclipse. Thus, they suggest that the source of quasi-periodic oscillations may not be the hot spot, as that is eclipsed. Furthermore, the abrupt change in oscillation period precludes the pulsations of the white dwarf being the source. Patterson (1981) summarizes most of the suggested scenarios. Cordova and Mason (1981) tabulate the cataclysmic variables in which periodic and quasi-periodic oscillations have been observed.

4.4. HUMPS

A significant number of cataclysmic variables show humps, or brightenings of the light curve with amplitudes from 0.2 to 0.3 magnitude. Humps are periodic; their period is equal to the orbital period of the system. In non-eclipsing systems humps help to determine the orbital period of the system. Within a system, the amplitude of the humps may vary from cycle to cycle. The amplitude may increase while the system is bright. For example, the amplitude of the humps in the light curves of AH Her and RX And increased by 40 percent during their outbursts (Warner 1976).

Flickering may be superimposed on the humps and increase in amplitude around the maximum of the hump. In some systems, in addition to orbital humps, intermediate humps of smaller amplitude are also present. These intermediate humps last from one end of one main hump to the beginning of the next (Haefner, Schoembs, and Vogt 1979; Schoembs and Hartmann 1983).

Robinson (1976) suggests that because during an eclipse of U Gem both the humps and flickering disappeared, the source that gives rise to these variations is the same and is the hot spot.

4.5. SUPERHUMPS

This feature is seen only in the light curves of SU UMa-type dwarf novae, and only 2 to 3 days after the start of a superoutburst. They are periodic and their period is 1 to 3 percent longer than the orbital period of the system. The amplitude is 25 to 40 percent larger than that of humps, and reaches 0.2 to 0.3 magnitude. Superhumps continue throughout a superoutburst, and as the star starts to fade and the superoutburst to end, the superhumps decrease in their amplitude and then disappear completely (Haefner, Schoembs, and Vogt 1979). In some systems the period of the superhumps decreases towards the end of the superoutburst (Vogt, Krzeminski, and Sterken 1980; Patterson 1981; Bond, Kemper, and Mattei 1982).

The relationship of the superhumps to the orbital period implies that for SU UMa systems for which the orbital periods are not known, if superhumps are observed during superoutbursts then the orbital periods can be determined.

The cause and the origin of both superoutbursts and superhumps in SU UMa systems are not yet known, although various models have been put forth. The most plausible current models combine both disk instability and mass transfer instability which gives rise to temporary deformation of the accretion disk (la Dous 1989).

5. Orbital Periods

The orbital period is an important physical parameter in these systems that enables the determination of other physical parameters, such as the mass and the radius of the system. Mass loss from the secondary component which causes changes in the orbital period is an important agent for secular evolution.

Some of the systems are double-lined spectroscopic binaries, but most of them are single-lined. Thus orbital periods are obtained spectroscopically from radial velocity curves, from the eclipse timings of those systems that have eclipses, from the periods of the humps or superhumps obtained with high-speed photometry, and from correlations with the light curve morphology.

The orbital periods of cataclysmic variables range from 76 minutes to 16 hours, except that of the recurrent nova T CrB at 227 days and the old nova GK Per at 2.00 days. Most of them have periods less than 10 hours. There is a sharp cut-off at around 80 minutes, and a gap between 2.25 hours and 2.83 hours.

There is an interesting distribution to the orbital periods in that a) the shortest orbital period is 76 minutes; b) all U Gem- and Z Cam-type dwarf novae, DQ Her- and VY Scl-type stars, and all novae have periods larger than 3 hours, above the period gap; c) all SU UMa-type dwarf novae have orbital periods of less than 2 hours, except for TU Men with 2.82 hours, which is in the period gap; d) AM Her-type stars tend to have orbital periods of less than 2 hours, below the period gap; and e) AM CVn-type stars have the shortest periods of all and lie below the minimum orbital period. Diagrams of the distribution of the orbital periods in la Dous (1989) show vividly these relationships and highlight the outstanding features of the period gap and the sharp cutoff of the periods.

The period gap may exist because cataclysmic variables are not formed having periods in the gap and do not evolve into it, or because their orbital periods change quickly and thus they move through the gap so fast that detection probability decreases greatly, or because for some time the system may be a detached binary in which mass transfer stops and thus the system does not have the recognizable characteristics of a cataclysmic variable (la Dous 1989).

The upper part of the period gap may be explained by the cessation of magnetic braking, detaching the secondary from its Roche lobe so that mass overflow stops. Gravitational radiation then brings the secondary and its Roche lobe back into contact and the period decreases to about 2 hours. The orbital period keeps decreasing due to gravitational radiation until at the minimum orbital period of around 80 minutes the secondary mass has decreased enough so that the star cannot maintain nuclear energy generation any longer and becomes degenerate.

Thus the tendency for certain subclasses of cataclysmic variables to appear at one end of the gap or the other suggests that the distribution of the orbital periods may have physical significance and evolutionary causes.

6. Correlations Between Light Curves and Physical Parameters

For decades investigators have made attempts to find relationships among various parameters of the light curve such as amplitude, outburst interval, rise and decline times, outburst duration, level of quiescence, and lengths of intervals preceding and following outbursts. Some of the findings are more convincing than others. Thorough investigation using long-term data on different types of cataclysmic variables is needed for more definite results.

Initially, Kukarkin and Parenago (1934) obtained a relationship between the amplitude and the recurrence time of outbursts in dwarf novae and recurrent novae in that the longer the recurrence time the larger the amplitude. Recently, Antipova (1987) carried this investigation further and obtained a relationship between the energy spent during an outburst and the recurrence time.

Bailey (1975) found a relationship between the orbital period and the decline rate from an outburst of dwarf novae. Mattei and Klavetter (1981) confirmed this relationship and further found that there could be two separate relationships, one between orbital period and decline rate in SU UMa-type stars, and a similar one in other types of dwarf novae. Warner (1987) obtained a relationship between the absolute brightness during outburst and the orbital period of dwarf novae in that the longer the orbital period the brighter the absolute brightness of the system.

Possible correlations between outburst interval and the kind of outburst have been made. In the cases of SS Cyg and CN Ori, the width of the outburst, and in FQ Sco, BI Ori, and U Gem the maximum brightness of the outburst was correlated with the length of the interval following the outburst (Bath and van Paradijs 1983; Isles 1976; Gicger 1987; la Dous 1989). For VW Hyi both total duration and the maximum brightness, for UZ Ser and TU Men the width, and for WW Cet the maximum brightness were correlated with the quiescent interval preceding outburst, but not the following one (Gicger 1987; Smak 1985; van der Woerd and van Paradijs 1987; la Dous 1989). For SS Aur the width was correlated with the maximum brightness and the length of the quiescence preceding outburst (Mattei et al. 1986). For AY Lyr and UU Aql no correlations have been found between either the preceding or following quiescence interval and the maximum brightness (Danskin and Mattei 1978; Davis and Mattei 1981). Szkody and Mattei (1984) obtained correlations among rise and decline times, total duration, and orbital period on 21 dwarf novae. Van Paradijs (1983) obtained correlations between the width of the outbursts and the orbital period.

Kiplinger et al. (1988) and, independently, Bianchini (1988), found periodicity in the brightness level of SS Cyg with periods of 7.2 and 7.3 years, respectively.

Light curves of cataclysmic variables hold a wealth of information that has significant bearing on the physics of these systems, and more thorough analyses should be carried out. For results to be significant in these correlative studies, however, it is essential that long-term and homogeneous data are used.

7. Acknowledgements

I acknowledge with gratitude the financial support provided by the NATO Advanced Study Institute that made it possible for me to attend the conference on Active Close Binaries.

8. References

Antipova, L.I. (1987), *Astrophys. Space Sci.*, **131**, 453.
Bailey, J. (1975), *Journ. Brit. Astron. Assoc.*, **86**, 30.
Bath, G.T. (1973), *Nat. Phys. Sci.*, **246**, 84.
Bath, G.T. and van Paradijs, J. (1983), *Nature*, **305**, 33.
Bianchini, A. (1988), *Inf. Bull. Var. Stars*, No. 3136.
Bond, H., Kemper, E.,and Mattei, J.A. (1982), *Astrophys. Journ.*, **260**, L79.
Cordova, F. and Mason, K. (1981), *Accretion Driven Stellar X-Ray Sources*, Cambridge U. Press, Cambridge.
Danskin, K. and Mattei, J.A. (1978), *Journ. Amer. Assoc. Var. Star Obs.*, **7**, 1.
Davis, J.F. and Mattei, J.A. (1981), *Journ. Amer. Assoc. Var. Star Obs.*, **10**, 1.
Duerbeck, H.W. (1987), *A Reference Catalogue and Atlas of Galactic Novae*, D. Reidel Publ. Co.
Gicger, A. (1987), *Acta Astron.*, **37**, 29.
Haefner, R., Schoembs, R., and Vogt, N. (1979), *Astron. Astrophys.*, **77**, 7.
Hildebrand, R., Spillar, E.J., and Stiening, R.F. (1981), *Astrophys. Journ.*, **243**, 223.
Horne, K. and Stiening, R.F. (1985), *Mon. Not. Roy. Astron. Soc.*, **216**, 933.
Isles, J.E. (1976), *Journ. Brit. Astron. Assoc.*, **86**, 327.
Kiplinger, A., Mattei, J.A., Danskin, K., Morgan, J.E. (1988), *Journ. Amer. Assoc. Var. Star Obs.*, **17**, 34.
Kukarkin, B.V. and Parenago, P.P. (1934), *Var. Stars*, **4**, 251.
la Dous, C. (1989), in *Cataclysmic Variables and Related Objects*, ed. M. Hack, NASA/CNRS Monograph Series on Nonthermal Phenomena in Stellar Atmospheres, in press.
Mattei, J.A., Cook, L.M., McMahon, A.P., and Foster, M.R. (1986), *Journ. Amer. Assoc. Var. Star Obs.*, **15**, 1.
Mattei, J.A. and Klavetter, J. (1981), to be published.
Ney, E.P. and Hatfield, B.F. (1978), *Astrophys. Journ.*, **219**, L111.
Paczynski, B. (1978), 'Nonstationary Evolution in Close Binaries', in A.N. Zytkow (ed.), *Second Symposium of Problem Comm. Physics and Evolution of Stars, Warsaw, Poland, June 20-25 1977*, Polish Scientific Publ., Warsaw, 89.
Patterson, J. (1979), *Astron. Journ.*, **84**, 804.
Patterson, J. (1981), *Astrophys. Journ. Suppl.*, **45**, 517.
Ritter, H. (1987), *Astron. Astrophys. Suppl.*, **70**, 335.
Robinson, E.L. (1976), *Ann. Rev.*, **14**, 119.
Robinson, E.L. and Nather R.E. (1979), *Astrophys. Journ. Suppl.*, **39**, 461.
Robinson, E.L., Shafter, A.W., Hill, J.A., Wood, M.A., Mattei, J.A. (1987), *Astrophys. Journ.*, **313**, 772.

Robinson, E.L. and Warner B. (1972), *Month. Not. Roy. Astron. Soc*, **157**, 85.
Schoembs, R. (1986), *Astron. Astrophys.*, **158**, 233.
Schoembs, R. and Hartmann, K. (1983), *Astron. Astrophys.*, **128**, 37.
Smak, J. (1985), *Acta Astron.*, **35**, 357.
Starrfield, S., Truran, J.W., Sparks, W.M., Kutter, G.S. (1972), *Astrophys. Journ.*, **176**, 169.
Szkody, P. and Mattei, J.A. (1984), *Publ. Astron. Soc. Pac*, **96**, 988.
van der Woerd, H. and van Paradijs, J. (1987), *Mon. Not. Roy. Astron. Soc.*, **224**, 271.
van Paradijs, J. (1983), *Astron. Astrophys.*, **125**, L16.
Vogt, N., Krzeminski, W., and Sterken, C. (1980), *Astron. Astrophys.*, **85**, 106.
Warner, B. (1976), "Observations of Dwarf Novae" in **IAU Symp.** 73, *Structure and Evolution of Close Binary Systems*, P. Eggleton, S. Mitton, J. Whelan (eds.), D. Reidel Publ., 85.
Warner, B. (1987), *Mon. Not. Roy. Astron. Soc.*, **227**, 23.
Warner, B. and Brickhill, A.J. (1978), *Mon. Not. Roy. Astron. Soc.*, **182**, 777.
Warner, B. and Cropper, M. (1983), *Mon. Not. Roy. Astron. Soc.*, **203**, 909.
Warner, B. and Nather, R.E. (1971), *Mon. Not. Roy. Astron. Soc.*, **152**, 219.
Webbink, R.F. (1989), *American Scientist*, 77, 248.
Webbink, R.F., Livio, M., Truran, J.W., Orio, M. (1987), *Astrophys. Journ.*, **314**, 653.

IUE SPECTRA OF CATACLYSMIC VARIABLES: OBSERVATIONAL RESULTS AND THEORETICAL IMPLICATIONS

C. A. LA DOUS
Institute of Astronomy
University of Cambridge
Madingley Road
Cambridge, CB3 OHA
England

ABSTRACT. The impact of observations with the IUE Satellite on research in dwarf novae and nova-like stars is reviewed. Various kinds of observational phenomena are presented and their theoretical implications are discussed. Special emphasis is put on so far unsolved theoretical problems they pose, as well as on the still largely unexploited collective statistical information content of the data archive.

1. Introduction

The history of cataclysmic variables dates back for over 300 years, when in 1670 the nova CK Vul was detected during its outburst. In 1855 the first dwarf nova, U Gem, was detected. For many more decades to come, the only property of cataclysmic variables that could be observed, was the outburst activity, and later spectra with low temporal as well as spectral resolution followed. In any event, since no periodic pattern could be found in these stars, all attempts to explain the underlying physics failed. Only when almost 100 years after the detection of the first dwarf nova, in the 1950es the binarity of several cataclysmic variables could be established, and thus finally some periodic pattern was found, the big break-through in the understanding of these stars was possible with the application of the Roche model (Crawford and Kraft, 1955; 1956).

On the theoretical side there seems to be no way to get around accepting the Roche Model as the basic model for all cataclysmic variables. This is, we believe that all novae, dwarf novae, and nova-like stars consist of a white dwarf primary component and a secondary component that loses mass preferentially into the Roche lobe of the primary star. In most cases the secondary is a cool star on, or close to, the main sequence, but some cases are known where it seems to be noticeably evolved. And provided the white dwarf's magnetic field is not too strong, conservation of angular momentum of the transfered material leads to the formation of an accretion disk around the white dwarf.

Anything beyond this basic model, however, i.e., details of the disk structure, extent, temporal stability, and emitted radiation, details of the nature of the secondary star and whether and how it is influenced by being confined to the tight Roche lobe, details on what the interface between the disk and the white dwarf might look like, and what it does to its degenerate material to obtain a shell of non-degenerate accreted material, how the various system components interact by irradiation and with their magnetic fields, all these are still very controversial, and essentially unsolved questions. In addition, as will be seen below, recent observations, rather than clarify, tend to even complicate the picture.

Although novae, dwarf novae, and nova-like stars in their physical nature all seem to be about the same kind of objects, it is not yet clear why the exhibited phenomena are so different. Also, because of this difference, research tends to fractionate in novae on one hand (dealing mostly with the violent large-scale outbursts and the ejected shells of material), and in dwarf novae and nova-like stars on the other (dealing with the binary system, the accretion disk, and their properties).

C. İbanoğlu (ed.), Active Close Binaries, 629–638.

Here I want to follow this distinction too and constrain myself to the discussion of the latter two groups of stars ... keeping in mind, however, that results are likely to apply to novae as well.

In both dwarf novae and nova-like stars several sub-classes of objects are distinguished, mostly on the basis of their photometric appearance. Only in the case of AM Herculis and DQ Herculis stars the reason for their being different from the others seems to be the strength of the magnetic fields of the white dwarfs, and in the case of AM Canum Venaticorum stars, the primary as well as the secondary component seem to be white dwarfs. In all other cases, there are some guesses as to why they might exhibit the observed differences, but a lot of clarifying work remains to be done. For what I am going to discuss, these classifications are largely irrelevant and, unless explicitly stated differently, I am going to treat dwarf novae and nova-like stars as one single group of objects, and the term 'cataclysmic variable' in the following ought to be understood as equivalent to 'dwarf novae and nova-like stars'.

In the following I will briefly present details of the Roche model, discuss the various radiation sources, and what the understanding of their nature was before IUE results became available. Results from the x-rays I will largely leave aside since this radiation originates from very different parts of the system than the UV, the optical, and the IR radiation. In section III the general pattern and appearances of the IUE spectra is presented, to the extent they are known today. Section IV will deal with photometric aspects of the UV radiation, and section V with particular spectroscopic features as the continuous flux distribution and its relation to the optical flux, the line radiation in quiescence and outburst, and theoretical implications. And, finally, in section VI, the probable benefits and first results are presented of an attempt to explore the collective statistical information as contained in the IUE archive of dwarf novae and nova-like stars.

2. Cataclysmic Variables Before IUE

I will not attempt here to review all the various photometric and spectroscopic patterns of behaviour that are exhibited by cataclysmic variables in the optical, but a brief general introduction might be useful for those who are not too familiar with them.

Photometrically, dwarf novae undergo semi-periodic so-called outbursts, i.e., brightness increases by typically 3 to 5 magnitudes, on time scales of some 10 to 100 days. These are accompanied by spectacular changes in the appearance of the spectra, which (in the optical) change form exhibiting essentially strong emission lines of hydrogen and helium (sometimes also weak absorptions of a cool star) during quiescence (the low brightness state), to showing almost exclusively strong, wide Balmer absorptions during outburst maximum, with rise and decline being the respective transition phases. On time scales of hours, in many objects photometric as well as spectroscopic variability due to the orbital revolution becomes apparent. Favourable inclination angles provided, such phenomena as the orbital hump and single or double eclipses can be observed, in particular during the low brightness state. They also are accompanied by changes in the spectral appearance, which, however, is considerably less spectacular than what is seen during the outburst cycle. On even shorter time scales on occasions and in some objects pulsations and oscillations can be observed. For a more detailed summary of these features as well as of what is seen at other wavelengths, see la Dous (1989a).

The entire formulation of the Roche model for cataclysmic variables is based on observations in the optical. As was stated above, one of the primary assumptions is that all these objects are binaries. The model was formulated by deduction from observed features, but for the understanding of the advantages and shortcomings of dealing (exclusively, as it were) with the optical, it is more instructive to look at the contributions to the integral radiation of the system which are emitted by the various components of a cataclysmic system.

When doing so, it immediately becomes apparent that the optical radiation is a convolution of radiation from the accretion disk (which, depending on the distance of the emitting area from the central star, radiates in all wavelengths form the UV to the IR), the secondary star (which is strongest at IR wavelengths), and the hot spot (which has its radiation maximum in the optical and often shows up as the so-called hump in the optical light curves); under special conditions even contributions from the white dwarf might be non-negligible. On the other hand, when considering the UV flux, essentially only the accretion disk is contributing; all other radiation sources to a good degree of approximation can be neglected. Furthermore, far and away most of the entire flux emitted by a cataclysmic system (maybe with the exception of the x-rays, but neither observationally not theoretically it is entirely clear what is happening there) is emitted at UV wavelengths. Also, it are those parts of the disk right in the vicinity of the white dwarf which entirely dominate the UV, while at the same time they are essentially unseen in the optical.

In other words, the optical radiation, the one on which still most research in cataclysmic variables is based, has the advantage that in a way almost all components of the system can be investigated, though only, in general, with some degree of inaccuracy due to contamination by all the other sources. In the UV, however, it is only the disk that is seen. On this very aspect, the disk in the UV, I want to concentrate in the following.

3. Dwarf Novae and Nova-Like Stars in the UV - The General Appearance

Traditionally, and this also means: in the optical, phenomena observed in cataclysmic variables are broadly divided into two classes: spectroscopic and photometric appearance (In principle there also is the third aspect of polarimetric appearance, which, however, has any real relevance only for the strongly magnetic AM Herculis stars, and certainly, since currently unobservable, non at all at UV wavelengths; so I will forgo this here.) Photometric aspects furthermore conveniently are divided according to the time scales involved, into phenomena seen during the outburst cycle (on time scales of days, weeks and longer), those seen on orbital time scales (hours), and variability within minutes or seconds. For spectra no similarly complete information is available, thus the classification, if any is seriously considered at all, rather is into continuum and line radiation. Tentatively I will try to follow this in the UV too, and we will see in what ways the patterns there are similar to, or different from, the optical, and where the gaps are.

Before doing this, it might be useful to say a few words about the IUE satellite, emphasizing the aspects that are important in our context. The *International Ultraviolet Explorer* is a 45 cm orbiting spectroscopic telescope. It was launched on January 26, 1978, and, as I write this, still is enjoying good health. Spectra can be taken in the effectively usable ranges of 1200-2000 Å and 1900-3000 Å. For each of these there is an option for high (some 0.1 - 0.3 Å/pixel) and low (some 6 - 7 Å/pixel wavelength resolution. In addition, the *Fine Error Sensor* can be used to obtain an indication of the optical flux at the time of observations. (More detailed information on the Satellite can be found, for instance in Boggess and Wilson (1987) and Harris and Sonneborn (1987).) For cataclysmic variables, due to their relatively low brightness, almost exclusively the low resolution mode has to be used. To-date over 2000 spectra of more than 100 dwarf novae and nova-like stars of all sub-classes and at all brightness states have been taken, and most of them are of quite good quality. All those low resolution spectra that were taken until the end of 1987 have been published in la Dous (1989b). Thus, they constitute a unique, homogeneous set of observations of sizable dimensions.

4. Photometric Observations in the UV

In the optical, observations of the outburst light curves of, in particular, dwarf novae are the most complete among the data available on cataclysmic variables. This holds concerning the length of the time base as well as the completeness of coverage. It needs to be emphasized and gratefully acknowledged here that this is true almost exclusively due to the immense effort and activity of the amateur astronomers all over the world who have monitored the stars night by night for decades and in regular publications as well as upon special request make them available to us.

It almost follows naturally from the above, that nothing comparable is available at UV wavelengths. Some of the brightest targets have been observed for over a decade ever and ever again, but in no case anything like an outburst light curve can be obtained. It is clear that the UV flux does change conspicuously, roughly in phase with the optical flux but this is about as far as any safe statement can go for the time being.

On orbital time scales, in the optical the information is not nearly as complete as for the outburst state, mostly, of course, due to the fact that observations are restricted to occasional observing runs. The best coverage of any cataclysmic variable in this respect is of the dwarf nova CN Ori which was observed almost uninterruptedly (i.e. 24 hours/day) for one complete quiescent state (10 days) and a good deal of the following outburst (Mantel et al., 1988). The much more normal case, however, is that an object is observed for one or two orbital cycles at a time, maybe repeatedly so on consecutive nights, but not for very long. So, in spite of some decades of high time resolution photometric observations, there are still very many gaps in our knowledge of the phenomenology.

Again, in the UV even far less information is available. This is mostly due to a combination of the low brightness of cataclysmic variables and the fact that IUE is a spectroscopic telescope: the exposure times for spectra in general are on the order of the orbital periods (some hours). Only for some of the brightest systems limited phase resolution can be obtained. By averaging the flux in suitable wavelength bands, light curves can be produced.

In general it appears that there is some low amplitude photometric variability on orbital time scales, but in most cases it does not seem to be related to the orbital phase. In particular, in dwarf novae there is no clear evidence for the orbital hump to be present, even in cases where it is strong in the optical. Incidentally, this detection is quite in agreement with the notion that the temperature of the spot is around 10000 K, which should leave it with little flux to emit at UV wavelengths. In striking contrast to this are observations of the anti-dwarf nova TT Ari, where the hump is seen to decrease between 5000 and 3000 Å, but then strongly reappears at shorter wavelengths (Jameson et al., 1982, and Guinan and Sion, 1981). It is not yet understood what this means for the physics of the hot spot in general, as well as in TT Ari in particular.

Since already there are problems with obtaining orbital resolution in the UV, it is no wonder that absolutely nothing is known about oscillations or pulsations at these wavelenghts. At times and in some objects they are rather prominent both in the optical and at x-rays, thus they are likely to also exist in the UV.

5. Spectroscopic Properties

When looking at the flux emitted by cataclysmic variables it is important to realize that there are not only many different radiation sources in the system, but also, and in the case of the UV radiation in particular, the flux emitted from the accretion disk is a very strong function of the distance from the white dwarf. When approaching the central star, the temperature of the gas increases appreciably and consequently the emitted radiation is much bluer. In general, the UV radiation originates entirely from within a few stellar radii, while the optical is influenced mostly by areas farther away.

5.1. THE CONTINUOUS RADIATION

Turning now to spectra, the situation is much better than in the case of photometry. The mount and quality of data available in the optical and in the UV is comparable. Still, in spite of the over 2000 IUE spectra that exist, deplorably little has been published of, and on, them. They now all have become available in the form of a catalogue (la Dous, 1989b), but besides that the literature still is rather sparse. This holds for observations of individual objects, and to an even much higher degree for the statistical information content of the archive.

Let us first turn to individual investigations, and to begin with to the continuous flux distribution. The continuum flux in cataclysmic variables if falling continuously from about Lyα all the way to the IR. (Voyager observations of some dwarf novae show that between Lyα and the Lyman edge at 912Å the continuum is strongly flattened (Polidan and Holberg, 1984). This was not expected on theoretical grounds, and still is not really understood theoretically. Since this wavelength range is not reliably observable with the IUE it should not concern us here). The observed flux distribution is much unlike that of 'normal' individual stars (i.e. somewhat similar to a black body) but rather well agrees with the theoretical predictions of the flux emitted by an accretion disk (in a very rude approximation a power law). This gives strong support to the notion that the bulk of radiation, and almost all of the UV radiation from (most) cataclysmic variables is emitted by the accretion disk. Still, however, neither the quality of observations is in any way good enough nor is the theoretical understanding of the physics of accretion disks well enough advanced to allow for somewhat reliable determination of any kind of system, or only disk, parameters from comparison of theory and observations.

Qualitatively, the UV flux distribution in dwarf novae during quiescence is noticeably flatter (redder) than during the outburst when the flux increases in intensity the more the shorter the wavelength. No exception to this pattern so far could be found in any dwarf nova, while for instance in the anti-dwarf nova MV Lyr the UV flux distribution appears the steeper the lower the overall flux level is (Szkody, Downes, 1982). Whether or not this is universal for the low-brightness states of nova-like stars cannot yet be clarified. Also it is not clear yet how this behaviour can be understood in terms of the Roche model ... which, in its current form, predicts the sort of changes seen in dwarf novae.

At rise to an outburst in dwarf novae two different kinds of behaviour have been observed. One is that the optical and UV flux start to rise about simultaneously (whether or not this means exactly at the same time or only approximately so, could not yet been determined observationally) and keep on rising together. The other is that there is a rather pronounced delay in that the optical flux starts rising several hours before the UV, and in turn has well reached maximum level when the UV flux still is increasing (Verbunt, 1984; van Amerongen et al, 1987). Verbunt (1987) published a compilation of such contemporaneous observations in the optical and in the UV in which he reports on seven objects. The IUE archive now contains many more rise observations than he refers to which have not yet been properly investigated as to this aspect. First results which I recently obtained in connection with investigations of the Mg II resonance line (to be reported in somewhat more detail below) indicate, however, that the UV delay, as it is commonly referred to, is by far the more common case in dwarf novae, while only in relatively few cases the optical and UV flux rise simultaneously. A particularly interesting case in this context is SS Cyg (mostly probably because it is the brightest dwarf nova and thus has been observed rather often and minutely): this is the only documented object so far in which both types of outburst, i.e. simultaneous as well as delayed rise in the UV, were seen (Verbunt, 1987). It may or may not be significant that the UV delay was seen on the rise to a normal outburst in SS Cyg (i.e. an outburst with a fast flux rise in the optical), while the simultaneous rise at optical and UV wavelengths was seen in the case of an anomalous (i.e. slow-rise) outburst (la Dous, 1987c).

Decline from an outburst always only has been seen to proceed simultaneously at all wavelengths. In the case of VW Hyi and WX Hyi it was found that in the UV the decline continued until onset of the next outburst, while no such trend could be detected in U Gem (Hassall et al., 1985; Verbunt et al., 1987). No similar investigations have been carried out yet about other systems.

All these observations can be explained qualitatively when it is assumed that in the case of an UV delay the outburst starts in the outer disk, far away from the white dwarf, while in the case of simultaneous rise it starts in the inner disk. The basic assumption for this, and in fact one which gained considerable support from these observations, is that the outburst is nothing happening instantaneously to the entire disk but rather something that takes its origin at some distance from the white dwarf (also not necessarily at the outer rim) and then spreads out. The reasons for the different effects on the observed flux is simply due to the outer disk being cool, thus its radiation is primarily seen at optical wavelengths; when the outburst starts there and the flux rises, it takes some time for it to proceed inwards towards the hotter regions nearer to the white dwarf, thus only with some delay would the effect be felt at UV energies. If on the other hand the outburst starts in a region in the disk which is hot enough to contribute to both the optical as well as the UV, both wavelength ranges are affected simultaneously, no matter how fast the outburst spreads through the disk. Again, however, for the time being, neither theory nor observations are well enough advanced, or understood, for any more qualitative statements to be justified.

5.2. THE LINE RADIATION

The line radiation of quiescent dwarf novae in the optical is characterised by in generally rather pronounced Balmer emission lines, and also He I and He II often are seen in emission; in objects with orbital periods in excess of some 5 or 6 hours also the cool absorption spectrum of the secondary component becomes visible. In almost all well investigated cases radial velocity curves could be determined, and often characteristic variations of the line profiles on orbital time scales can be seen. It should be stressed, however, that in spite of the many observations that have been obtained of optical spectra, it still is by no means clear whether these changes, with the only exception of those related to eclipses, are stable features or whether they are merely transitory features in the accretion disk which then vary due to orbital aspect variations, but on the whole are not stable. It would be very simple to establish the answer to this issue by monitoring an object spectroscopically for several consecutive nights ... but it has not yet been done. Comparison of various published spectra of the same (any) object do suggest, though, that the line emission of dwarf nova accretion disks in the optical (and the same holds for the UV, too) is not a stable characteristic feature of an object. There is indication, however, that for each object the general character, i.e. for instance whether the emission lines are strong or rather weak, is maintained.

In the UV the general pattern by and large is very similar: in most dwarf novae the UV quiescent spectrum is dominated by strong emission lines of the UV resonance lines of Si III, Si IV, C II, C III, C IV, Mg II, and sometimes Al III; mostly also the Balmer α line of He II is visible. There are some other objects, however, which rather exhibit weak absorptions, or maybe only C IV is in emission. Again, while the particular details of the spectra change on orbital time scales as well as over long time scales, the general character of the spectrum of an object seems to be preserved.

No relation seems to exist between the appearance of the optical and the UV spectrum of a dwarf nova in that strong lines in one wavelength range may or may not be accompanied by strong ones in the other. This supports the notion that the radiation emitted in these two regimes largely originates from two physically independent loci in the system, i.e. primarily from the outer or inner accretion disk, respectively.

During outburst, in the optical, the Balmer emissions turn into strong, broad absorptions. Around maximum mostly they exhibit pure absorption profiles, while during decline the emissions gradually grow until the quiescent spectrum is restored.

Again, in the UV the pattern is very similar in that during outburst all lines turn into absorptions. Here the pattern seems to be (though clearly further and more detailed investigation is necessary) that the absorptions seen during outburst tend to be the stronger the stronger the emissions are during the quiescent state. During decline, as in the optical, slowly the quiescent spectrum is restored, though in the UV it seems that rather first the absorptions disappear and only then the emission components appear, but again, this is a point that requires further detailed investigation ... as for many other unsolved questions concerning the behaviour in the UV, the observations do exist, but have not yet been looked at from this point of view.

A special case are the line profiles of C IV, Si IV , and sometimes N V during the outbursts. Mostly they exhibit appreciably blue-shifted profiles or even, in particular in the case of C IV, often P Cyg profiles ... while all the other lines, often also Si IV, have symmetric profiles. These profiles are clearly variable on orbital time scales but also clearly not as a function of the orbital phase (see, e.g. Szkody, Mateo, 1986). An exception form the very high-inclination (double-eclipsing) systems which during outburst exhibit only emission lines; these, however, are effected in different ways by the eclipse (Naylor et al, 1987).

This phenomenon of the P Cyg and related lines was investigated by Drew (1986, 1987) and Mauche and Raymond (1987). It turns out that qualitatively the line profiles can be understood when it is assumed that they originate from a, probably bi-polar, wind that is driven out by radiation pressure from the center of an optically thick accretion disk. This works much like the P Cyg profiles seen in hot stars, as a projection effect, but the different geometry of the disk, as compared to a star, as the underlying continuum source, turnes out to be of crucial importance for the reproduction of the observed line profiles. Furthermore, the eclipse observations of the double-eclipsing system OY Car during a superoutburst, which demonstrated that the C IV emission line is less affected by the eclipse that other lines, suggests that there also is some kind of a corona in the system which is likely to not be entirely due to the wind but rather originates from or above the disk.

All these observations seem to support the notion that the accretion disk is largely optically thin during quiescence (thus the prevailing emission lines), although in some objects (those with absorption spectra in the UV) it apparently is optically thick in the center near the white dwarf. During outburst it becomes optically thick throughout and often winds are ejected from areas very close to the white dwarf. In any event the structure of the accretion disk appears to be variable, suggesting some kind of a random and variable patchy appearance which accounts for the variations on orbital time scales that are not stable over longer periods of time.

In the optical as well as in the UV nova-like stars, concerning the appearance of their spectra, can be found anywhere between the extreme of dwarf novae in quiescence and in outburst: they can exhibit emission as well as absorption spectra, or a combination, or may even change in quality from one epoch to another. From the few cases known it appears that the anti-dwarf novae in the high state much resemble dwarf novae during the outburst, while during low state they look rather like quiescent systems. In both wavelengths ranges the AM Herculis systems form an exception in that they invariably emit very strong emission lines.

The vast majority of IUE spectra of dwarf novae and nova-like stars has been taken in the low-resolution mode, but for some of the brightest sources it was possible to also obtain some high-resolution spectra. In them very sharp, narrow absorption components of the UV resonance lines become visible (Cordova, 1986). For the most part they seem to originate from the interstellar gas between the systems and the Sun. In the very well observed system SS Cyg, there also is some evidence that they might originate in an H II region surrounding the system which is irradiated by the x-rays emitted from the dwarf nova (Mauche et al., 1988).

6. A Statistical Investigation of the IUE Spectra

6.1. A PLEADING FOR ARCHIVAL RESEARCH

Research in cataclysmic variables these days is very much split into two aspects: modelling and observations. Unfortunately there is much less interaction between those two than one should think, or would like, to be the case. Admittedly, the theory, whether it is reproducing the outburst light curves or the observed spectra, is fairly intricate, involves a lot of not-well-understood physics, and thus requires the implicit or explicit assumption of values of many free parameters. The result is that in fact it often is possible to theoretically fit observations with suitable choices of the many free parameters and derive some system parameters, but then at the same time it is not at all clear that these parameters are in any way reliable and what can be learned from such a procedure. It is my conviction that one of the chief reasons for this dilemma is that even from the point of view of observations it is not at all clear what really is to be reproduced theoretically. In other words: even though the number of observations of cataclysmic variables that are available is huge, concerning spectra as well as photometric observations, in fact it is so huge that it has become virtually impossible to keep all the individual features in mind that are seen in all the stars at different times. As a result of this, everybody involved in this work cherishes some ideas of what spectra or light curves look like in general, supplemented by information on some specific objects. In effect, this can be considered as a first attempt to extract statistically significant features from all observations one happens to be aware of, and then maybe try to model those in order to reproduce 'the general features'. Alternatively it is attempted to reproduce some specific observation of some object one happens to have observations of, running the risk that a lot of time and energy is spent in reproducing features that well might be very important for this particular object, but of rather secondary significance for the entire class. It is not surprising that both of these approaches in the past couple of years have proven to not yield the universally applicable results that were hoped for.

What is needed is a totally new approach, a statistical approach that is able to reliably cope with large amounts of data and to reliably determine which of the many observed features are significant for all, or at least a large number, of objects ... and thus, before anything else, should be reproduced theoretically. With such statistically significant 'average' observations (that by no means need to have been actually observed, but well can be idealized spectra or light curves) plus some understanding of the range of possible variations, it should be possible to, on the theoretical side, safely rule out those parameters as physically unreasonable that yield significantly different numerical results than this 'average'.

So, what is required is many data, to do statistics with. These are data from many objects as well as observations of individual objects over a long time. There are several sources for this, like the big existing archives (IUE, EXOSAT, ...); amateur observations providing lots of outburst light curves of dwarf novae over, in many cases, several decades; there are lots of unpublished observations of all kinds; and, finally, many data have been published in the literature but possibly can be looked at from some new point of view.

6.2. FIRST RESULTS: THE Mg II RESONANCE LINE

One of the nicest data bases we currently have available certainly is the IUE archive. It contains many observations of many objects over a long time, that have all been obtained in the same way. Unfortunately, so far hardly anything has been done with it in the sense of statistical research. One attempt by Verbunt (1987), where he checked the most obvious aspects for correlations, using a rather limited data base, yielded mostly negative results. But clearly with more data more can be done, and obviously it also will be necessary to not only look for the obvious but follow what the data have to tell.

Several months ago I started a project the main aim of which it is to see what collective information on cataclysmic variables can be obtained from a statistical analysis of the entire archive on these stars. Besides the relatively easy availability of these data, the main incentive to use IUE is that the most flux from cataclysmic variables is emitted in just the wavelength range that is observable with the IUE satellite, and that it is pretty clear that this radiation originates almost exclusively from the inner parts of the accretion disk, rather than beings a convolution of contributions from many system components. Consequently, it should be comparatively easy to find the significant features and to come up with physical explanations for them.

In preparing this work, I collected all the spectra that had been taken until the end of 1987 (to be updated eventually) and reduced, corrected, and dereddened them in as homogeneous a way as possible. Characteristic features like continuum fluxes, line fluxes, equivalent width, etc. were measured in all spectra. System parameters were taken from the literature. And the optical light curves around the times of IUE observations were kindly provided to me by the AAVSO.

The first problem I decided to turn to was the nature and origin of the Mg II resonance line at 2800 Å. The reason for this decision partly was that this supposedly is a rather limited problem which would not involve dealing with all the other resonance lines at the same time (although whether or not this really holds still remains to be seen), and partly because so far it seems to be an entirely open question what the origin is of this line.

In ordinary stars the presence of Mg II emission signifies chromospheric activity. Thus, the two possible candidates for origin in a cataclysmic system are the secondary star or the outer parts of the accretions disk far away from the white dwarf.

As to the strictly observational appearance, without invoking any theoretical concepts, the pattern is that:

Mg II appears strongly during quiescence and is weak or absent during outburst maximum;

during decline and quiescence there are substantial flux and equivalent width variations on orbital time scales, but they are not obviously related with the orbital period (due to the very small equivalent widths no measurements are possible at maximum and no appropriate data are available for the rise);

there is no evidence for systematic flux changes during the outburst cycle;

the disappearance of Mg II at the top of outburst appears to be merely due to little contrast to the rising continuum;

in most objects the equivalent width of Mg II undergoes a very characteristic loop-like development during the course of an outburst; the effect is considerably stronger when W_λ is looked at as a function of the optical continuum flux (m_{FES}) than of the UV flux;

details of the loop are different for different objects as well as for different outbursts of the same object;

there are exceptions to this general behavioural pattern.

Another way of looking at the data is to compare the strength of the Mg II line with what is seen in 'normal' stars. Since it is not known so far how much Mg flux is emitted from the disk, it shall be assumed that it all is emitted by the secondary component. This flux, however, cannot be compared with the emission from single cool main sequence stars since their rotation velocity (on the order of 10 km/s) is far less than the bound rotation of the secondaries in cataclysmic systems (typically some 130 km/s) and this is likely to have rather strong effects on the coronal activity of the stars. There are, however, the fast-rotating cool W UMa stars which posess orbital (and thus rotational) periods of almost the right size. The strength of the Mg II resonance line in them was investigated by Rucinski (1985). When comparing his data with those of quiescent dwarf novae, the latters' coronal activity in most cases appears to be significantly stornger than what would be expected in analogy to W UMa stars. There still are many difficulties, partly profound theoretical problems about the depth of the convection zone, involved and much more careful checking of the data will be necessary before the above tentative result can be considered well established. There is no particular need why it would have to be assumed that the secondaries in cataclysmic variables are physically substantially different from the components of cool W UMa stars, so the conclusion from the above tentative result is that there is good indication that a non-negligible fraction of the Mg II flux is emitted by the accretion disk. This in turn would mean that some kind of 'coronal activity' also is present in the disk. Furthermore, since the line strength does change on, partly, rather short time scales, but since at the same time there is no indication that it varies as a function of the outburst cycle, this corona seems to be variable in its structure as well as essentially independent of the workings in the actual disk that are connected with the outburst activity.

These are first results and not yet entirely secured ones, and so far they only are based on data from dwarf novae, while clearly nova-like stars will have to be considered too before on can arrive at any reliable conclusions. The main reason for presenting them here is to demonstrate the potential benefits of doing archival research of this sort, one way of doing it, and to hopefully convince you that it is very rewarding to start looking at observational data from also this point of view.

References

van Amerongen, S., Damen, E., Groot, M., Kraakman, H., and van Paradijs, J.: 1987, *Mon. Not. R. Astr. Soc.*, **225**, 93

Boggess, A., and Wilson, R.: 1987, in Y. Kondo, *Exploring the Universe with the IUE Satellite*, Reidel Publ. Co., p. 3

Córdova, F.A.: 1986, in *The Physics of Accretion onto Compact Objects*, ed. K.O. Mason, M.G. Watson, and N.E. White, Springer-Verlag, p. 339

Crawford, J.A., and Kraft, R.P.: 1955, *Publ. Astr. Soc. Pac*, **67**, 337

- : 1956, *Astroph. J.*, **123**, 44**123**, 44

la Dous, C.: 1989a, in *Cataclysmic Variables and Related Objects*, ed. M. Hack, NASA/CNRS Monograph Series on Nonthermal Phenomena in Stellar Atmospheres, p. 15, in press

- : 1989b, *Space Science Reviews* **49**, 425 (full version to appear in 1990)

- : 1989c, *Mon. Not. R. Astr. Soc.* **238**, 935

Drew, J.E.: 1986, *Mon. Not. R. Astr. Soc.* **218**, 41 p

- : 1987, *Mon. Not. R. Astr. Soc.* **224**, 595

Guinan, E.F., and Sion, E.M.: 1981, in *The Universe at Ultraviolet Wavelengths. The First Two Years of International Ultraviolet Explorer*, Proc. of a Symposium held at the NASA Goddard Space Flight Center, May 7-9, 1980, NASA Conf. Publ. **2171**, p. 477

Hassall, B.J.M., Pringle, J.E., and Verbunt, F.: 1985, *Mon. Not. R. Astr. Soc.* **216**, 353

Jameson, R.F., Sherrington, M.R., King, A.R., and Frank, J.: 1982, *Mon. Not. R. Astr. Soc.* **200**, 455

Harris, A.W., andSonneborn, G.: 1987, in Y. Kondo (ed.), *Exploring the Universe with the IUE Satellite*, Reidel Publ. Co., p. 729

Mantel, K.-H., Marschhaeuser, H., Schoembs, R., Haefner, R., and la Dous, C.: 1988, *Astr. Astroph.* **193**, 101

Mauche, C.W., and Raymond, J.C.: 1987, *Astroph. J.* **323**, 690

Mauche, C.W., Raymond, J.C., and Córdova, F.A.: 1988, *Astroph. J.* **335**, 829

Naylor, T., Bath, G.T., Charles, P.A., Hassall, B.J.M., Sonneborn, G., van der Woerd, H., and van Paradijs, J.: 1987, *Mon. Not. R. Astr. Soc.* **231**, 237

Polidan, R.S., and Holberg, J.B.: 1984, *Nature* **309**, 528

Ricinski, S.M.: 1985, *Mon. Not. R. Astr. Soc.* **215**, 615

Szkody, P., and Downes, R.A.: 1982, *Publ. Astr. Soc. Pac.* **94**, 321

Szkody, P., and Mateo, M.: 1986, *Astrophys. J.* **301**, 286

Verbunt, F.: 1987, *Astr. Astrophys. Suppl.* **71**, 339

Verbunt, F., Hassall, B.J.M., Pringle, J.E., Warner, B., and Marang, F.: 1987, *Mon. Not. R. Astr. Soc* **225**, 113

Binaries and Pulsars as gravitational Wave Sources: Possibility of Direct Detection By Ground Based Experiments

L. Milano[1,*], F. Barone[1,*], G. Russo[1,*]

[1] Dipartimento di Scienze Fisiche dell'Università, I-80125 Napoli, Italy

* Associated to the *Istituto Nazionale Fisica Nucleare*, Italy

Abstract. We analyzed periodic astrophysical sources as gravitational continuous waves emitters. Particular attention received binaries with highly eccentric orbits and cataclysmic variables either computing their dimensionless strain amplitude spectra or trying to establish if indirect evidence of Gravitational radiation could be detected from observations of their period variations. The trials were unsuccessfully, but this point must be carefully checked by further studies. From this analysis we conclude that interesting astrophysical sources for detection at low frequencies ($\nu > 10Hz$) are pulsars. A scenario of ground based planned experiments of direct detection by large base two wave interferometric gravitational wave antennas is also given. Particular attention was given to the presentation of the italo-french project VIRGO to implement such type of GW antenna at very low frequencies.

1 Introduction

During the past years great efforts have been made in the field of direct detection of Gravitational Waves (hereinafter GW).

In this lecture we shall try to summarize the scientific motivations of this challenging field of research and we hope to give a scenario of the experimental projects that, almost surely, will be implemented in the world during the end of this century, i.e. during the years '90's. We are well aware that, until now (1989) nearly all the knowledge of the distant Universe is coming from observations of Electromagnetic Waves(EW) (from radio, γ rays). Cosmic rays and neutrinos observations are complementary informations coming from underground experiments (neutrinos detection). The opening of a completely new channel will increase considerably or even revolutionize our comprehension of the Universe.

Such is precisely the aim of the present active research about GW. The success will mean the beginning of a new era of research in Astrophysics: the born of "GRAVITATIONAL ASTRONOMY".

Gravitational wave research consists of two equally important aspects:

1) A theoretical one, i.e. gravitational radiation theory

2) An experimental one, i.e. gravitational radiation detectors.

Gravitational wave detection was pioniered by J. Weber (1961). In about thirty years of active research much progress has been accomplished, and at

C. İbanoğlu (ed.), Active Close Binaries, 639–691.

the very beginning of 1990 a collaboration/competition might start among different groups to implement detectors of sufficient sensitivity to get the first direct detection of a gravitational signal from an astrophysical source. Until now no clear evidence of gravitational signals has yet been obtained but this negative result is not a concern for gravity theorists for two reasons:

1) current theoretical estimates about the gravitational waves predict amplitudes on the earth two or three times smaller than the sensitivity of today's best detector.

2) a clear indirect evidence of the existence of GW comes from reaction effects in the motion of binary systems constituted of gravitationally condensed stars. The first credited indirect evidence is constituted by the binary pulsar PSR 1913+16 (Taylor, Fowler, Mc Cullough, 1979, Taylor, Weisberg: 1982, Taylor et al. 1989; Damour, Deruelle: 1981; Damour, 1983)

The present lack of direct detection of GW, if not a concern for the theorists, is a challenge for the experimentalists who are now in the conditions to get the (amplitude) sensitivity of detectors in the level where theorists expect interesting signals to exist.

We shall divide this lecture in three parts:

1) Brief summary of the theory of GW; type of sources and dimensionless strain evaluations.

2) Trials to get out indirect evidence from eclipsing binaries.

3) Direct methods of detection:

a) Weber bar antennas

b) Interferometric antennas

2 Brief summary of the GW theory

2.1 Solutions of the wave equations

The basic physical nature of gravitational radiation (hereinafter GR) can be understand with an approach that is quite similar to the one used for EM radiation. Since the basic field equations of gravity are nonlinear, GR is usually discussed in the context of the weak field approximation, i.e. the total metric is decomposed into a flat space 'background' metric plus a small perturbation due to the waves:

$$g_{\mu,\nu} = g^{(o)}_{\mu,\nu} + h_{\mu,\nu} \tag{1}$$

where $g^{(o)}_{\mu,\nu}$ is the usual galilean metric and $h_{\mu,\nu} \ll 1$ is the perturbation. It can be shown that with this approximation the field equations of the general relativity can be reduced to

$$\Box^2 h_{\mu,\nu} = 0 \tag{2}$$

where

$$\Box^2 = \Delta^2 - \frac{\partial^2}{\partial t^2} \tag{3}$$

This is the standard equation of the waves and its solution gives under reasonable constraints, as solutions two indipendent components constituting the two polarizations states. Usually the two polarizations are represented in terms of the 'plus' and 'cross' basic states which have a simple representation for a wave propagating in the x_3 direction (see fig. 1).

The form of the general plane wave equation can be written:

$$h_{\mu,\nu} = A_+ h_+ e^{\pm i K_\alpha x^\alpha} + A_\times h_\times e^{\pm i K_\alpha x^\alpha} \tag{4}$$

where h_+ and $h_\times$ are :

$$h_+ = \begin{pmatrix} 0 & 0 & 0 & 0 \\ 0 & 1 & 0 & 0 \\ 0 & 0 & -1 & 0 \\ 0 & 0 & 0 & 0 \end{pmatrix} \qquad h_\times = \begin{pmatrix} 0 & 0 & 0 & 0 \\ 0 & 0 & 1 & 0 \\ 0 & 1 & 0 & 0 \\ 0 & 0 & 0 & 0 \end{pmatrix}$$

and A_+ and $A_\times$ are two arbitrary constants. Since the principle of equivalence states that the physics is the same in a sufficiently small region of all reference frames, there is no way to tell whether a given observer is in an accelerated frame or experiencing a gravitational field. Tidal effects are one way to determine the difference. Using the masses to mark points in the space time, and as the passage of a GW distorts the spacetime, we can see the distortion by measuring the relative motions of the masses. The relevant consequence is that two free masses experiment a dimensionless strain:

$$h_{x_1 x_2} = 2\frac{\Delta L_{12}}{L_{12}} \tag{5}$$

by a GW travelling along x_3 axis for the action of the Riemann tidal force : $F_R = M\ddot{h}_{\mu,\nu} L$ (see fig. 1)

In the general case there is a linear combination of h_+ and $h_\times$.

In EM theory, the lowest radiator is a dipole, in GW theory it is possible to show that the lowest radiator is a mass quadrupole. The formula for the total power radiated into by a GW quadrupole source is:

$$L_{GW} = \frac{dE}{dt} = -\left(\frac{G}{5c^5}\right)\left(\frac{d^3Q_{\alpha,\beta}}{dt^3}\right)^2 \tag{6}$$

where

$$Q_{\alpha,\beta} = \int \rho(x_{\alpha,\beta} - \frac{1}{3}\delta_{\alpha,\beta})dV \tag{7}$$

is the quadrupole momentum, ρ is the mass density, and G the gravitational constant.

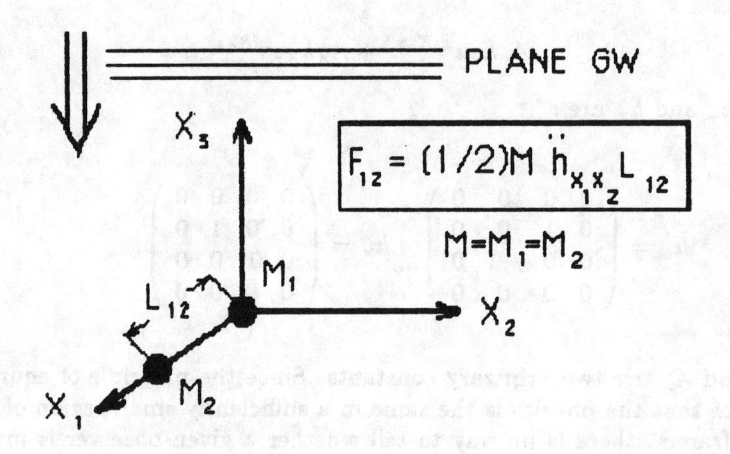

Fig.1

From a quick estimate of the relative strengths of gravitational and electromagnetic radiation

$$L_{GW} = \left(\frac{G}{5c^5}\right)\left(\frac{d^3Q_{\alpha,\beta}}{dt^3}\right)^2 \tag{8}$$

$$L_{ED} = \left(\frac{2}{3c^3}\right)\left(\frac{d^2Q_{\alpha,\beta}}{dt^2}\right)^2 \tag{9}$$

we obtain:

$$\frac{L_{GW}}{L_{ED}} \approx \left(\frac{G}{5c^5}M^2r^4\omega^6\right)\frac{2}{3c^3}q^2r^2\omega^4 =$$
$$= \frac{3}{10}\left(\frac{GM^2}{q^2}\right)\frac{\omega r^2}{c^2} = \frac{3}{10}\frac{GM^2}{q^2}\left(\frac{v}{c}\right)^2$$
$$= 7.2 \cdot 10^{-44}\left(\frac{v}{c}\right)^2$$

The symbols M, q and r are the characteristic mass, charge and size of the source (electron). From that estimate we can understand why GW are not routinely generated in laboratories.

The quadrupole formula is the basis for most astrophysical GW source calculations. An estimate for the metric perturbation $h_{\mu,\nu}$ at the antenna can be made from the flux for a plane wave and, after some algebra we have:

$$L_{GW} = \frac{dE}{dt} \approx \left(\frac{G}{5c^5}\right)M^2r^4\omega^6 \tag{10}$$

so

$$h \approx \left(\frac{GM}{rc^2}\right)\left(\frac{r}{R}\right)\left(\frac{v}{c}\right)^2 \tag{11}$$

For a solar mass neutron star at the center of the Galaxy, $\frac{GM}{rc^2} \approx 0.2$, $\frac{r}{R} \approx 3 \cdot 10^{-17}$, and $\omega r = v \approx c$, the estimated strain amplitude is $h \approx 7 \cdot 10^{-18}$. This sets the size of the effect.

3 Types of sources

The sources can be divided into three categories: periodic, impulsive and stochastic. Each category of sources is examined in terms of the metric perturbation h,the quantity of greatest interest to an experimentalist. The bulk of the discussion will center on periodic sources since the detection of such sources is the subject of this lecture and one of the main the aim of the VIRGO project. (Bradaschia et al. 1989)

3.1 Binary star systems

Binary star systems are known to radiate GW because they have an intrinsic time varying quadrupole moment. About half of all stars in our Galaxy are thought to be in binary systems, so the existence of this type of sources is both guaranteed and plentyful. A binary system must be composed of compact objects for GR to be an important energy loss mechanism, and it is not known what fraction of all binaries fit this description.

According to the classical Peters and Mathews (PM) analysis (1963), the GW luminosity of a bin source at the n-th harmonic ($n \geq 2$) of its orbital frequency is:

$$L_n^{+,\times} = 2.2 \cdot 10^{45} \cdot \left[\frac{m_1 \cdot m_2}{(m_1 + m_2)^{1/3}}\right]^2 \cdot f_{orb}^{10/3} \cdot g^{+,\times}(n,e) \qquad [erg \cdot sec^{-1}] \quad (12)$$

wherein e is the orbital eccentricity, $m_{1,2}$ are the masses (in units of $M_\odot$), and the subscripts $+, \times$ refer to the (fundamental) polarization states vive respectively.

The universal functions $g^{+,\times}(n,e)$ are

$$\begin{aligned} g^{+}(n,e) &= \frac{n^4}{384} \cdot \{\frac{16}{n^2} \cdot J_n^2(ne) + 7 \cdot (1-e^2) \cdot [J_{n-2}(ne) - 2 \cdot J_n(ne) + J_{n+2}(ne)]^2 + \\ &+7\{\frac{2}{n} J_n(ne) + 2e[J_{n+1}(ne) - J_{n-1}(ne)] + J_{n-2}(ne) - J_{n+2}(ne)\}^2\} \end{aligned} \quad (13)$$

$$\begin{aligned} g^{\times}(n,e) &= \frac{5 \cdot n^4}{384} \cdot \{(1-e^2) \cdot [J_{n+2}(ne) - 2 \cdot J_n(ne) + J_{n-2}(ne)]^2 + \\ &+\{J_{n-2}(ne) - J_{n+2}(ne) + 2e[J_{n+1}(ne) - J_{n-1}(ne)] + \frac{2}{n} \cdot J_n(ne)\}^2\} \end{aligned} \quad (14)$$

For a circular orbit only the 2nd harmonic is emitted, whilst for $e \neq 0$ the largest luminosity L_n occurs at $n = N_{max}(e)$ (see fig. 2) for both polarizations.

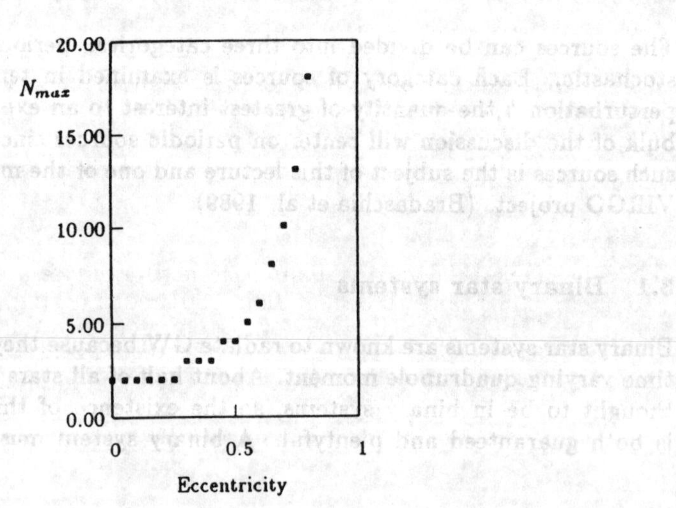

Fig.2

It is convenient to let

$$G_{max}(e) = g^{+}(N_{max}, e) + g^{\times}(N_{max}, e) \tag{15}$$

$$\bar{g}^{+,\times}(n, e) = \frac{g_n^{+,\times}(n, e)}{G_{max}(e)} \leq 1 \tag{16}$$

Note that as e increases from 0 to ≈ 0.25 the luminosity of the 2nd (most intense) harmonic and hence the value of $G_{max}(e)$ decreases, as an effect of spectral widening.

The GW luminosity of a binary of known parameters can be obtained for each and any harmonic of the orbital frequency by combining eq.s 12, 15, 16 The universal functions $\bar{g}^{+,\times}(n, e)$ have been computed for $e = 0, 0.25, 0.50, 0.75$, and are shown in fig.s 3 to 6, respectively.

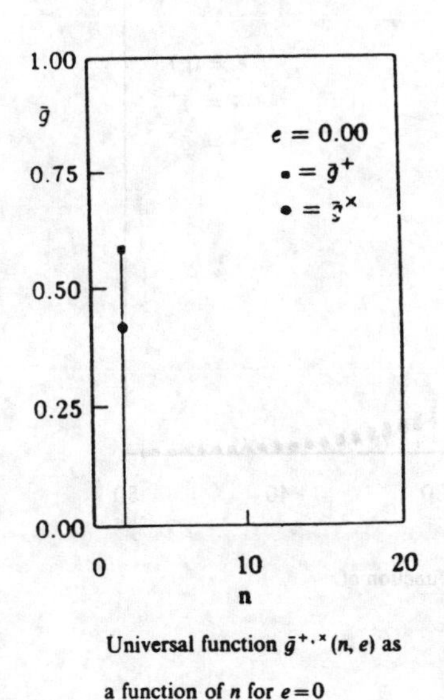

Universal function $\bar{g}^{+,\times}(n, e)$ as a function of n for $e = 0$

Fig.3

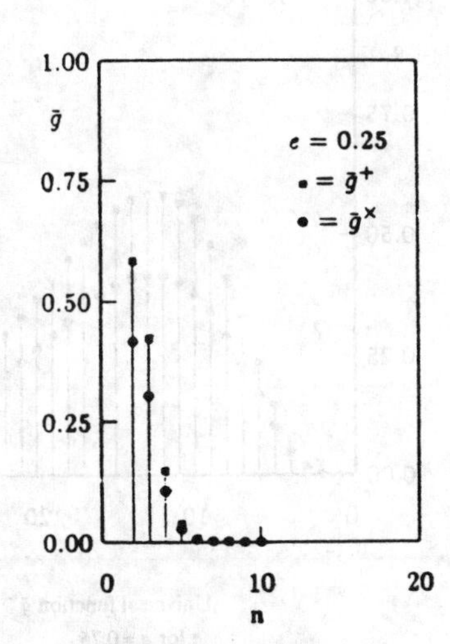

Universal function $\bar{g}^{+,\times}(n, e)$ as a function of n for $e = 0.25$

Fig.4

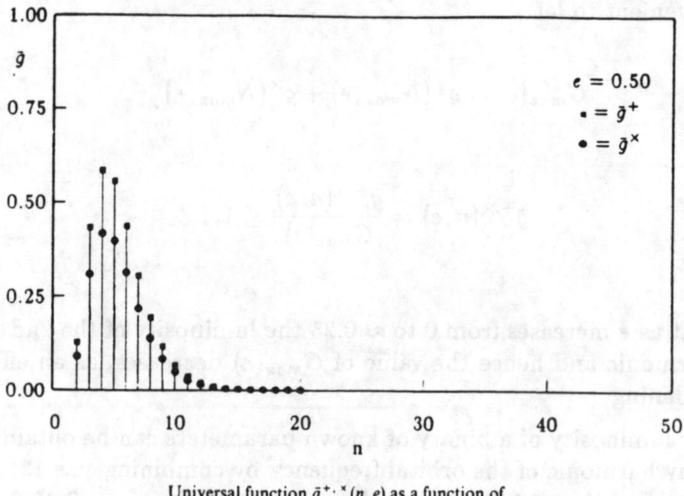

Universal function $\bar{g}^{+,\times}(n, e)$ as a function of n for $e=0.50$

Fig.5

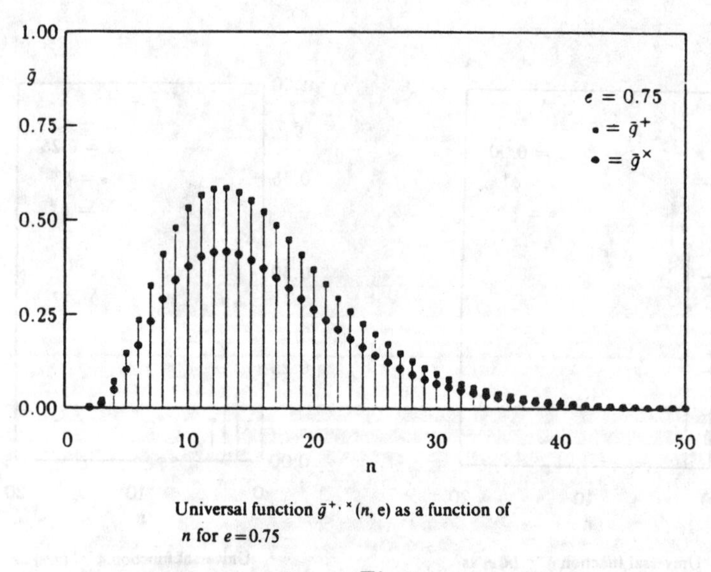

Universal function $\bar{g}^{+,\times}(n, e)$ as a function of n for $e=0.75$

Fig.6

As far as the direction properties of the gravitational radiation are concerned, we introduce the gravitational radiation diagrams

$$D_n^{+,\times}(\theta,\phi) = \frac{\frac{d}{d\Omega}\langle L_n^{+,\times}\rangle\,|_{\theta,\phi}}{\frac{L_n^{+,\times}}{4\cdot\pi}} \tag{17}$$

wherein $D_n^{+,\times}(\theta,\phi)$ is the angular power density at the n-th harmonic ($n \geq 2$) of the orbital frequency in the (θ,ϕ) direction divided by the mean angular power density, and Ω is the solid angle.

The qualitative features of the overall GW directivity diagrams are shown in fig.s 7 and 8. The $\phi = \frac{n\cdot\pi}{2}$ diagrams are all practically coincident; the ×-component power flux vanishes in the $\theta = \frac{\pi}{2}$ plane; the maximum directivity is always ≈ 2.5, irrespective of n and e.

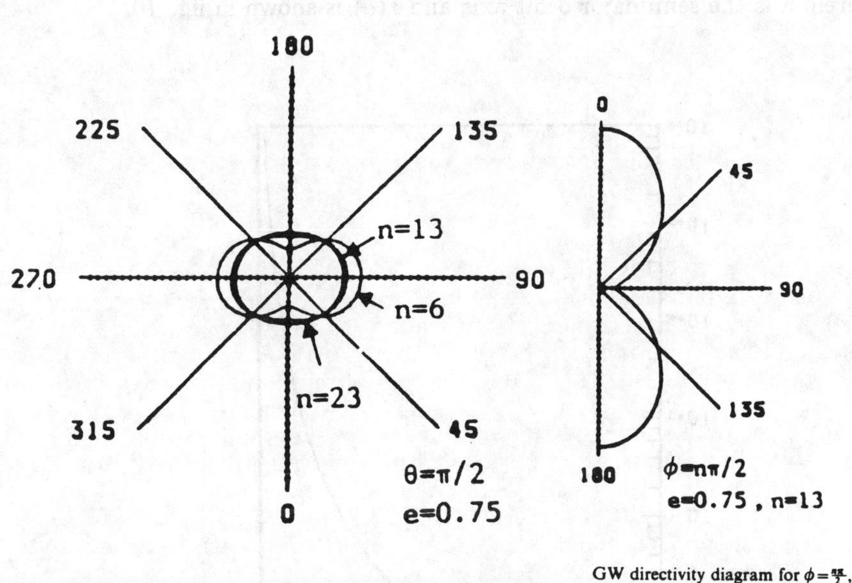

GW directivity diagram for $\theta=\frac{\pi}{2}$, $e=0.75$, $n=6$, $n=13$, $n=23$ (+polarization)

Fig.7

GW directivity diagram for $\phi=\frac{n\pi}{2}$, $e=0.75$, $n=13$ (+polarization)

Fig.8

As an effect of GW emission, secular changes in the orbit are produced: the orbit shrinks and circularizes, and correspondingly the GW spectrum narrows and its peak moves toward higher frequencies.

The shrink of the orbit is usually described in terms of the (time averaged) rate of change of the orbit semimajor axis,viz:

$$\langle \frac{da}{dt} \rangle = \frac{256}{5} \cdot \pi^2 \cdot \frac{G^2 \cdot M_\odot^2}{c^5} \cdot f_{orb}^2 \cdot m_1 \cdot m_2 \cdot \psi(e) \qquad [cm \cdot sec^{-1}] \tag{18}$$

where $\psi(e)$ is shown in fig. 9, and all other symbols have been previously defined.

On the other hand, the eccentricity changes accordingly to a simple law:

$$e(a) = e(a_o) \cdot \frac{\tau(a)}{\tau(a_o)} \tag{19}$$

wherein a is the semimajor orbit axis and $\tau(a)$ is shown in fig. 10.

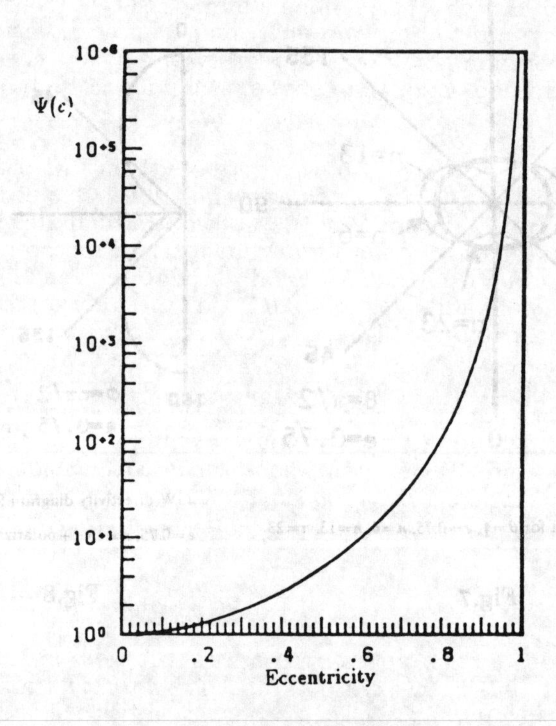

$\psi(e)$ as a function of eccentricity e

Fig.9

On the basis of eq.s 18, 19 the binary mean life turns out to be

$$\tau \approx 10^5 \cdot \frac{(m_1 + m_2)^{1/3}}{m_1 \cdot m_2} \cdot f_{orb}^{-8/3} \cdot \frac{1}{\psi(e)} \qquad [sec] \tag{20}$$

By comparing eq. 20 with a suitable estimate of the binary birth rate, one can obtain an estimate of the binary abundance as a function of orbital frequency and orbital eccentricity.

Evolutionary binary stars periods are thought to be never shorter than $\approx$ 1hour which is obtained equating the Helmoltz time to the GR time-scale (Rappaport et al., 1983). However several binaries with shorter periods are known, which could have been borne after collisional events, including HZ 29, three $\approx$ 40 minutes binaries (Rappaport et al., 1984) and a fresh discovered X-ray source binary with a period of eleven minutes (King and Watson, 1985).

An estimate of the maximum orbital frequency of a binary is obtained using Kepler law relating the orbit semi-major axes to the orbital frequency and assuming that tidal disruption would destroy the binary at $\approx 5\cdot$(radius of largest member). Table 1 is accordingly obtained. Note that extreme compact binaries could hardly been classified as periodic sources, since life-time is less than period.

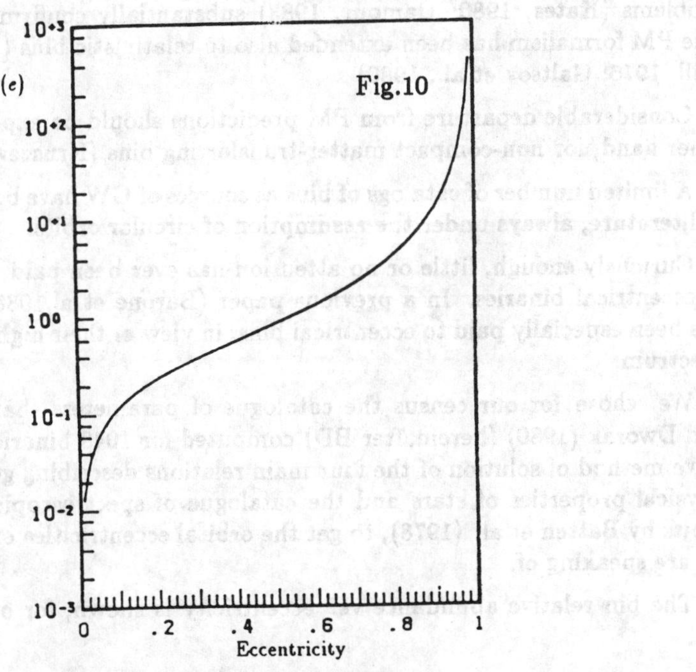

$\tau(e)$ as a function of eccentricity e

Note also that the life of a super-fast contact black-hole binary would not exceed 1 sec, typically, which makes it rather difficult to be observed.

Table 1

Maximum orbital frequency ($m_1 \approx m_2 \approx 1 M_\odot$)

Type	Radius (cm)	f_{ORB} (s^{-1})	τ (s)
Ordinary	10^{11}	$2\,10^{-4}$	10^{16}
White-dwarf	10^{9}	$2\,10^{-1}$	10^{8}
Neutron star	10^{6}	$5\,10^{3}$	10^{-4}
Black hole	$3\,10^{5}$	$3\,10^{4}$	10^{-6}

The PM analysis of GW emission from binaries was developed on the assumption of point masses, slow-motion and weak-field (i.e. implying Keplerian orbit and dominant quadrupole radiation (Peters-Mathews,1963; Peters,1964)). Arguments against the validity of quadrupole formulas and naive weak-field expansion for gravitationally-bound systems have been raised by several Authors (Coperstock and Lim, 1985; Ehlers, 1986). However, more rigorous matched asymptotic expansion solutions of the two-(compact)body problems (Kates, 1980; Damour, 1983) substantially confirm PM results. The PM formalism has been extended also to relativistic bins (Wagoner and Will, 1976; Galtsov et al., 1980)

Considerable departure from PM predictions should be expected, on the other hand, for non-compact matter-transferring bins (Kruszewsky, 1966).

A limited number of catalogs of bins as sources of GW have been published in literature, always under the assumption of circular orbits.

Curiously enough, little or no attention has ever been paid to the subset of eccentrical binaries. In a previous paper (Barone et al.1988a) attention has been especially paid to eccentrical bins, in view of their highly populated spectrum.

We chose for our census the catalogue of parameters that Brancewitz and Dworak (1980) (hereinafter BD) computed for 1048 binaries by an iterative method of solution of the four main relations describing geometric and physical properties of stars and the catalogue of spectroscopic binary star orbits by Batten et al. (1978), to get the orbital eccentricities of the systems we are speaking of.

The bin relative abundance vs. eccentricity is shown, for our sample in

fig. 11. The list of all bins with $e \geq 0.20$ belonging to the sample is given in Table 2. For each bin the GW emission on the lowest 9 harmonics has been computed (see, e.g., Table 3, 4, 5, 6), using PM results.

Table 2

Subset of eccentrical binaries

Name	Eccentricity	Orbital frequency (days^{-1})	GW flux on Earth (erg s^{-1} cm^{-2})
V1143 Cyg	0.54	0.1309	$0.4047\ 10^{-14}$
SX Cas	0.50	0.0272	$0.1875\ 10^{-18}$
SW Cma	0.50	0.0991	$0.5239\ 10^{-16}$
NY Cep	0.49	0.0655	$0.1547\ 10^{-15}$
AZ Cas	0.49	0.0003	$0.6693\ 10^{-23}$
ζ Aur	0.41	0.0010	$0.3242\ 10^{-21}$
μ Sgr	0.40	0.0055	$0.2504\ 10^{-17}$
α Crb	0.40	0.0576	$0.5811\ 10^{-15}$
W Ser	0.37	0.0706	$0.4337\ 10^{-16}$
RZ Eri	0.36	0.0255	$0.2224\ 10^{-18}$
VV Cep	0.35	0.0001	$0.1227\ 10^{-22}$
BL Tel	0.31	0.0013	$0.2907\ 10^{-21}$
UY Vir	0.30	0.5169	$0.7343\ 10^{-14}$
RX Gem	0.30	0.0819	$0.8456\ 10^{-18}$
ω^2 Cyg	0.30	0.0009	$0.5235\ 10^{-20}$
V 477 Cyg	0.30	0.4261	$0.8925\ 10^{-14}$
SW Cyg	0.30	0.2187	$0.2898\ 10^{-16}$
RW Tau	0.29	0.3612	$0.2669\ 10^{-15}$
RU Mon	0.28	0.2790	$0.1608\ 10^{-15}$
KU Cyg	0.25	0.0260	$0.6727\ 10^{-19}$
AR Cas	0.25	0.1648	$0.6255\ 10^{-14}$
AQ Peg	0.24	0.1803	$0.3498\ 10^{-17}$
β Per	0.23	0.3488	$0.1148\ 10^{-12}$
RY Per	0.22	0.1457	$0.5479\ 10^{-16}$
V 380 Cyg	0.22	0.0805	$0.3665\ 10^{-14}$
AB Per	0.21	0.1397	$0.2393\ 10^{-17}$
TU Mon	0.20	0.1981	$0.2325\ 10^{-15}$
W Del	0.20	0.2081	$0.9244\ 10^{-17}$
ZZ Cas	0.20	0.8042	$0.1387\ 10^{-13}$

Table 3

Parameters computed for V1143 Cyg

Name	Period (days)	Masses ($M_\odot$)	Ecc.	Distance (pc)
V1143 Cyg	7.6408	1.08 1.07	0.54	30

No. of harmonic	Flux on Earth (ergs^{-1} cm^{-2})	Dimensionless amplitude
2	$0.9677\ 10^{-16}$	$0.1818\ 10^{-21}$
3	$0.4043\ 10^{15}$	$0.2478\ 10^{-21}$
4	$0.6383\ 10^{-15}$	$0.2335\ 10^{-21}$
5	$0.6996\ 10^{-15}$	$0.1956\ 10^{-21}$
6	$0.6307\ 10^{-15}$	$0.1547\ 10^{-21}$
7	$0.5030\ 10^{-15}$	$0.1184\ 10^{-21}$
8	$0.3688\ 10^{-15}$	$0.8874\ 10^{-22}$
9	$0.2542\ 10^{-15}$	$0.6550\ 10^{-22}$
10	$0.1672\ 10^{-15}$	$0.4781\ 10^{-22}$

Table 4

Parameters computed for α Crb

Name	Period (days)	Masses ($M_\odot$)	Ecc.	Distance (pc)
α Crb	17.3599	2.76 0.99	0.40	26

No. of harmonic	Flux on Earth (ergs^{-1} cm^{-2})	Dimensionless amplitude
2	$0.8550\ 10^{-16}$	$0.3883\ 10^{-21}$
3	$0.1680\ 10^{-15}$	$0.3629\ 10^{-21}$
4	$0.1463\ 10^{-15}$	$0.2540\ 10^{-21}$
5	$0.9154\ 10^{-16}$	$0.1607\ 10^{-21}$
6	$0.4770\ 10^{-16}$	$0.9669\ 10^{-22}$
7	$0.2213\ 10^{-16}$	$0.5644\ 10^{-22}$
8	$0.9468\ 10^{-17}$	$0.3231\ 10^{-22}$
9	$0.3818\ 10^{-17}$	$0.1824\ 10^{-22}$
10	$0.1471\ 10^{-17}$	$0.1019\ 10^{-22}$

Table 5

Parameters computed for ω^2 Cyg

Name	Period (days)	Masses ($M_\odot$)	Ecc.	Distance (pc)
ω^2 Cyg	1147.4	22.6 8.15	0.30	217

No. of harmonic	Flux on Earth (ergs^{-1} cm^{-2})	Dimensionless amplitude
2	$0.1831\ 10^{-20}$	$0.1188\ 10^{-21}$
3	$0.1938\ 10^{-20}$	$0.8147\ 10^{-22}$
4	$0.9606\ 10^{-21}$	$0.4302\ 10^{-22}$
5	$0.3463\ 10^{-21}$	$0.2066\ 10^{-22}$
6	0.1045 10[illegible]	$0.9458\ 10^{-23}$
7	$0.2814\ 10^{-22}$	$0.4207\ 10^{-23}$
8	$0.6998\ 10^{-23}$	$0.1836\ 10^{-23}$
9	$0.1642\ 10^{-23}$	$0.7904\ 10^{-24}$
10	$0.3682\ 10^{-24}$	$0.3369\ 10^{-24}$

Table 6

Parameters computed for V477 Cyg

Name	Period (days)	Masses ($M_\odot$)	Ecc.	Distance (pc)
V477 Cyg	2.3470	2.10 1.49	0.30	175

No. of harmonic	Flux on Earth (erg s^{-1} cm^{-2})	Dimensionless amplitude
2	$0.3125\ 10^{-14}$	$0.3174\ 10^{-21}$
3	$0.3309\ 10^{-14}$	$0.2177\ 10^{-21}$
4	$0.1640\ 10^{-14}$	$0.1150\ 10^{-21}$
5	$0.5911\ 10^{-15}$	$0.5522\ 10^{-22}$
6	$0.1784\ 10^{-15}$	$0.2528\ 10^{-22}$
7	$0.4803\ 10^{-16}$	$0.1124\ 10^{-22}$
8	$0.1195\ 10^{-16}$	$0.4906\ 10^{-23}$
9	$0.2803\ 10^{-17}$	$0.2112\ 10^{-23}$
10	$0.6285\ 10^{-18}$	$0.9003\ 10^{-24}$

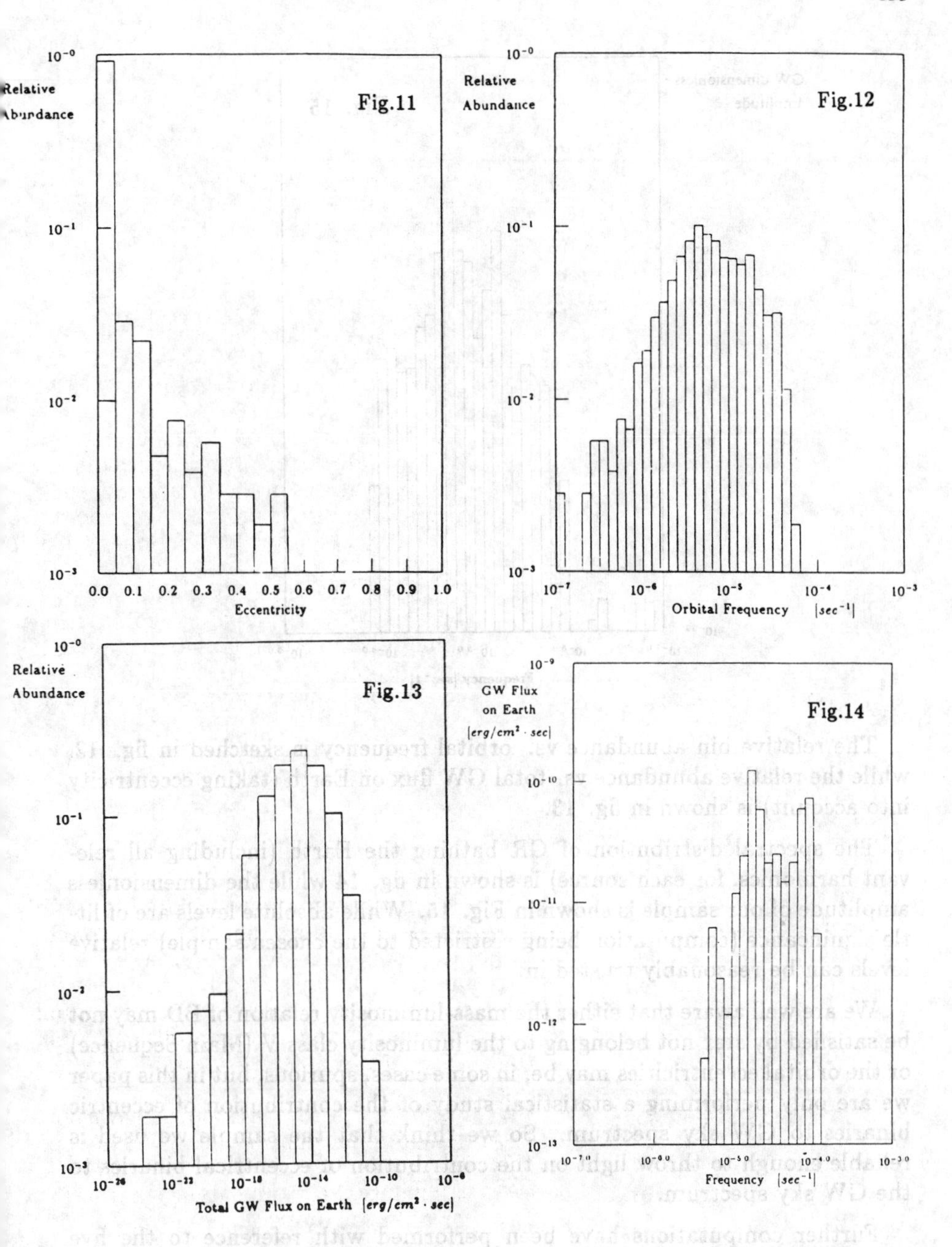
Fig.11
Relative Abundance
10⁻⁰
10⁻¹
10⁻²
10⁻³
0.0 0.1 0.2 0.3 0.4 0.5 0.6 0.7 0.8 0.9 1.0
Eccentricity
Fig.12
Relative Abundance
10⁻⁷ 10⁻⁶ 10⁻⁵ 10⁻⁴ 10⁻³
Orbital Frequency [sec⁻¹]
Fig.13
Total GW Flux on Earth [erg/cm² · sec]
Fig.14
GW Flux on Earth [erg/cm² · sec]
Frequency [sec⁻¹]

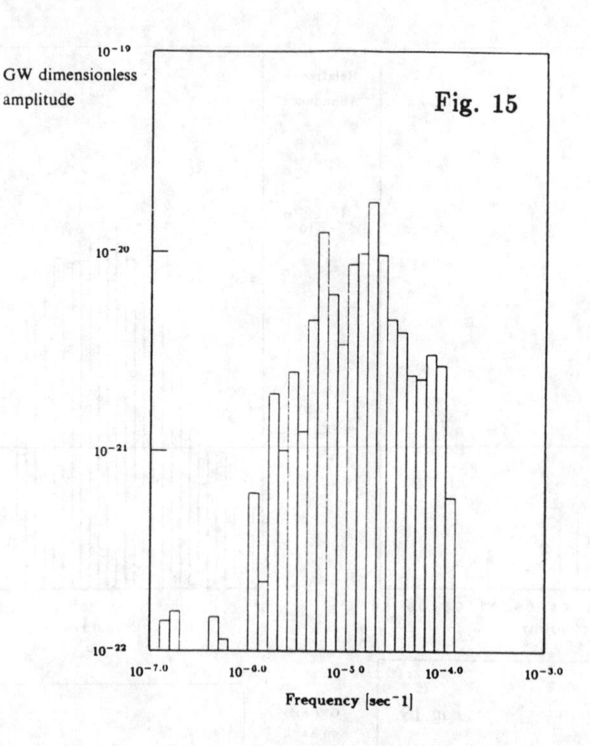

The relative bin abundance vs. orbital frequency is sketched in fig. 12, while the relative abundance vs. total GW flux on Earth (taking eccentricity into account) is shown in fig. 13.

The spectral distribution of GR bathing the Earth (including all relevant harmonics, for each source) is shown in fig. 14 while the dimensionless amplitude of our sample is shown in Fig. 15. While absolute levels are of little significance (computation being restricted to the chosen sample) relative levels can be reasonably trusted in.

We are well aware that either the mass-luminosity relation of BD may not be satisfied by star not belonging to the luminosity class V (Main Sequence) or the orbital eccentricities may be, in some cases, spurious, but in this paper we are only performing a statistical study of the contribution of eccentric binaries to GW sky spectrum. So we think that the sample we used is reliable enough to throw light on the contribution of eccentrical binaries to the GW sky spectrum.

Further computations have been performed with reference to the five known binary pulsars (1986). The corresponding orbital parameters were taken from (M.Damashek et al., 1982; R.N.Manchester et al., 1983; V.Boriakoff et al., 1983; G.H.Stokes et al., 1985; L.A.Rawley et al., 1986) and are reported in Table 7

Table 7. Subset of eccentrical binary pulsars

Name	Ecc.	Period (days)	Flux on Earth ($erg\,s^{-1}\,cm^{-2}$)
PSR 0655+64	$4.0\ 10^{-5}$	1.0287	$0.8066\ 10^{-14}$
PSR 0820+02	0.012	1232.400	$0.1623\ 10^{-25}$
PSR 1913+16	0.617	0.3229	$0.2414\ 10^{-13}$
PSR 1953+29	0.000	117.3490	$0.2836\ 10^{-23}$
PSR 2303+46	0.658	12.3395	$0.9352\ 10^{-18}$

Table 8. Parameters computed for PSR 1913+16

Name	Period (days)	Masses ($M_{\odot}$)	Ecc.	Distance (pc)
PSR 1913+16	0.3229	1.4 1.4	0.617	5200

No. of harmonic	Flux on Earth ($erg\,s^{-1}\,cm^{-2}$)	Dimensionless amplitude
2	$0.1321\ 10^{-15}$	$0.8980\ 10^{-23}$
3	$0.8804\ 10^{-15}$	$0.1545\ 10^{-22}$
4	$0.1846\ 10^{-14}$	$0.1678\ 10^{-22}$
5	$0.2593\ 10^{-14}$	$0.1591\ 10^{-22}$
6	$0.2956\ 10^{-14}$	$0.1416\ 10^{-22}$
7	$0.2964\ 10^{-14}$	$0.1215\ 10^{-22}$
8	$0.2722\ 10^{-14}$	$0.1019\ 10^{-22}$
9	$0.2346\ 10^{-14}$	$0.8409\ 10^{-23}$
10	$0.1927\ 10^{-14}$	$0.6859\ 10^{-23}$

Table 9. Parameters computed for PSR 2303+46

Name	Period (days)	Masses ($M_{\odot}$)	Ecc.	Distance (pc)
PSR2303+46	12.3395	1.4 1.4	0.658	2300

No. of harmonic	Flux on Earth ($erg\,s^{-1}\,cm^{-2}$)	Dimensionless amplitude
2	$0.1860\ 10^{-20}$	$0.1288\ 10^{-23}$
3	$0.1674\ 10^{-19}$	$0.2575\ 10^{-23}$
4	$0.4094\ 10^{-19}$	$0.3020\ 10^{-23}$
5	$0.6509\ 10^{-19}$	$0.3046\ 10^{-23}$
6	$0.8314\ 10^{-19}$	$0.2869\ 10^{-23}$
7	$0.9294\ 10^{-19}$	$0.2600\ 10^{-23}$
8	$0.9493\ 10^{-19}$	$0.2299\ 10^{-23}$
9	$0.9085\ 10^{-19}$	$0.1999\ 10^{-23}$
10	$0.8275\ 10^{-19}$	$0.1717\ 10^{-23}$

Note that the mass values are somewhat uncertain, being constrained only by the knowledge of the mass function, with the exception of PSR 1913+16 and PSR 2303+46.

Two binary pulsars show fairly large eccentricity values. Accordingly, the corresponding lowest GW harmonic luminosities, fluxes on Earth and metric deviations have been computed, and are reported in Table 8 and 9.

Thus ordinary binary system, although guaranteed sources, seem not likely to be very important sources of periodic GR for earth based antennas because the radiation from the orbital motion occurs at frequencies outside of the sensitivity passband. However, there are several closely related sources that require the presence of a binary system. The class of close X-ray binaries may be good candidate for accretion-driven emission of periodic GR at high frequencies through the mechanism suggested by Wagoner (1984).

3.2 Binary cataclysmic variables (BCV) and binary X-ray sources

The BCV (classical novae,recurrent novae, dwarf novae and nova-like) and binary X-ray sources are very interesting GW radiators. The orbital periods of these objects span from few minutes to several hours. To try to evaluate the GR background of these two class of binaries we studied the statistical sample distributions of the relevant parameters (GR frequency fe, L_{GW}, gravitational flux F, dimensionless amplitude h) to try to get informations on the parent distributions, in other words on the total number N_0 and on the GR background from BCV in our Galaxy(Barone,F., D'Ambrosio, F.C., Milano,L., Russo,G., 1989, hereinafter BDA). The basic data on 137 binary systems (97 BCV,20 X-ray binaries and 20 'related objects', i.e. binary pre-cataclysmic systems) were obtained from the catalogues by Ritter (1984) and by Patterson (1984). In our analysis ,taking into account possible selection effects,we made the following hypotheses:

a) indipendency of the classes of BCV's from the sides of the period gap (i.e. with $P \leq 2$h and $P \geq 3$h);

b) masses of secondaries assumed according to table 10 (see for details BDA)

Table 10

Mean values and mean square residuals of masses for the components of BCV's

	$P \leq 2$ h	$P \geq 3$ h
m1	0.86	0.86
Δ m1	0.27	0.29
m2	0.15	0.63
Δ m2	0.06	0.46

c) number of BCV's with $2h \leq P \leq 3h$ assumed to be negligible.

With these hypotheses and taking into account that we have data in our sample of BCV's in a sphere of radius 500 pc around the earth we cannot estimate the distribution density function in our galaxy, but assuming the density $\rho(r) = \rho_0 = const$ we can estimate the maximum and minimum number N_0 of BCV's in our galaxy from the upper and lower limit of the volume. From the analysis we made in BDAMRU we have: $\rho_0 = 10^{-6} pc^{-3}$ in good agreement with the value derived by Patterson(1984) and then :

$$10^6 \leq N_0 \leq 10^8 \tag{21}$$

According to the Peters and Mathews formula and assuming the orbital eccentricity $e = 0$ we have $g_n(0) = \delta_{n2}$ i.e. GR is on the second harmonic of the orbital frequency so

$$f_{eGW} = \frac{2}{P} Hz \tag{22}$$

and

$$L_{GW} = 2.2 \cdot 10^{45} \frac{(m_1 \cdot m_2)^2}{(m_1 + m_2)^{\frac{2}{3}} P^{\frac{10}{3}}} [erg/sec] \tag{23}$$

with the masses in solar unit and the orbital period in seconds;we have also for the flux F and the dimensionless amplitude h :

$$F = 8.4 \cdot 10^{-39} \frac{L_{GW}}{d^2} [erg/sec \cdot cm^2] \tag{24}$$

$$h = 2.4 \cdot 10^{-16} \frac{m_1 \cdot m_2}{d(m_1 + m_2)^{\frac{1}{3}} P^{\frac{2}{3}}} \tag{25}$$

From the computations,we performed on the sample of BCV's(BDAMRU), we can derive the following mean values shown in table 11 of the gravitational parameters according the relations written above.

Table 11

Mean gravitational parameters of BCV's		
frequency	$2.1 \cdot 10^{-4}$	Hz
L_{GW}	10^{30}	erg/sec
Flux	$5.2 \cdot 10^{-12}$	$erg/sec \cdot cm^2$
h	10^{-21}	

An interesting result of our analysis is that L_{GW} varies of four order of magnitude in the sample while the flux and h vary less than in the case of different sources like the pulsars in which we have a much greater variation of those parameters. This fact is a consequence of the small range of variation both of masses and periods of BCV's and of the small dimension of the sample we could have to analyze in only 500 pc all around the earth.This means that we can assume the mean values, we determined, as a characteristic emission for what concern the luminosity and a kind of upper limit for the dimensionless amplitude h of BCV's.We shall return on these points at the end of the discussion.Examining the trend of the luminosity versus the orbital period we note that there is a nearly uniform decrease (increase) of the luminosity with the increase (decrease) of the orbital period (frequency).

Owing to the small range of variation for m_1 and the stronger dependence of L_{GW} from the period we can assume that the gravitational luminosity depends mainly from the periods of BCV's. We can, with this crude hypothesis, estimate in a very simple way, the global GR from BCV's of our Galaxy:

$$E_{tot} = 4.8 \cdot 10^{31} N_o [erq/sec] \tag{26}$$

$$5 \cdot 10^{37} \leq E_{tot} \leq 5 \cdot 10^{39} [erg/sec] \tag{27}$$

where for N_o we assumed the upper and lower limits quoted above for the total number of BCV's in our Galaxy. We can also estimate the maximum flux and the maximum dimensionless amplitude from the mean value of the gravitational luminosity

$$L_{gwm} = 4.8 \cdot 10^{31} [erg/sec] \tag{28}$$

$$F_{max} = 8.4 \cdot 10^{-39} \frac{L_{gwm}}{d_o^2} [erg/sec.cm^2] \tag{29}$$

$$h_{max} = 5.0 \cdot 10^{-39} (\frac{L_{gwm}}{f_e d_o})^{\frac{1}{2}} \tag{30}$$

Where f_e and d_0 are the emission frequency and the distance in parsecs respectively. These results, taking into account the hypotheses we were compelled to do, state with a fairly good confidence the upper limits of the gravitational parameters for BCV's.

Concerning the X-ray binaries of low mass and the precataclysmic binaries we did not perform the same analysis as for the BCV's owing to the very small amount of data we had so we give only the mean values of the parameters shown in table 12.

Table 12

Mean Gravitational Parameters			
	X-ray Bin.	Precatacysm. bin.	
freq.	$2.1 \cdot 10^{-4}$	$2.2 \cdot 10^{-5}$	Hz
L_{GW}	$2.0 \cdot 10^{30}$	$5.1 \cdot 10^{28}$	erg/sec
Flux	$1.2 \cdot 10^{-14}$	$2.4 \cdot 10^{-13}$	erg/sec cm^2
h	10^{-22}	$7.2 \cdot 10^{-22}$	

3.3 Pulsars

To solve the problem of evaluating the upper and lower limit of dimensionless amplitude by pulsars in such a way to have suitable parameters for the project of a long base interferometric GW antenna,we analyzed a sample of 330 pulsars (Barone,F., Milano,L., Pinto ,I. , Russo,G. 1988; hereinafter BM). We are well aware that the main problem in computing the L_{GW}, F_{GW} and h spectra is the determination of the product $J\theta_w\omega\epsilon_o$ of the moment of inertia J, of the wobble angle $\theta\omega$, and of the oblateness ϵ_o.

Spinning neutron stars emit GW via two different mechanisms. Newborn PSRs would radiate away most of their deviation from axisymmetry in strong bursts of gravitational radiation (Thorne, 1969). Furthermore, and more important from the viewpoint of detecting GWs, they would emit *continuous* (CW) gravitational radiation if they are not axisymmetric, with the GW frequency equal to twice the spinning frequency (Ipser, 1970).

The gravitational luminosity (both $\times$ and $+$ polarization contributions included) is given by (Douglass and Braginsky, 1979):

$$L_{GW} = \frac{32G}{5c^5} \cdot \epsilon^2 J^2 \cdot (2\pi f)^6 \tag{31}$$

where G is Newton constant, f the spinning frequency, c the velocity of light *in vacuo*,

$$J = J_{11} + J_{22} \tag{32}$$

J_{11},J_{22} being the xx and yy quadrupole moments (rotation around $0\vec{z}$ assumed) and:

$$\epsilon = \theta_w \cdot \epsilon_o \tag{33}$$

θ_w being the so-called *wobble* angle, which describes the misalignment between the star's rotation and symmetry axes and ϵ_o is the *oblateness* defined by:

$$\epsilon_o = \frac{(J_{11} - J_{22})}{J} \tag{34}$$

The corresponding angular luminosity-distribution (both polarizations included) is:

$$\frac{dL}{d\Omega} = \frac{G}{\pi c^5} \cdot J^2 \epsilon_o^2 \theta_w^2 (2\pi f_g)^6 \cdot \left[4 \cdot (1 - sin^2\theta) + \frac{1}{2} sin^4\theta\right] \tag{35}$$

where $f_g = 2f$, and θ is the polar angle.

Assuming isotropic radiation, the GW flux F_{GW} reaching the Earth is:

$$F_{GW} = \frac{L_{GW}}{4\pi R^2} \tag{36}$$

where R is the Earth distance from the source. For a monochromatic wave with frequency f_g the flux is related to the dimensionless amplitude h via:

$$h \approx \left(\frac{G}{\pi^2 c^3}\right)^{\frac{1}{2}} \cdot \frac{L_{GW}^{\frac{1}{2}}}{f_g R} \tag{37}$$

Substituting (31) into (37) we obtain:

$$h = 4\pi^2 \left(\frac{32}{5}\right)^{\frac{1}{2}} \frac{G}{c^4} \cdot \frac{J\epsilon f^2}{R} \tag{38}$$

or, in more convenient units,

$$h \approx 8.1 \cdot 10^{-28} \cdot \left(\frac{J_{[gcm^2]}}{3 \cdot 10^{+44}}\right) \cdot \left(\frac{\epsilon_{[rad]}}{10^{-6}}\right) \cdot \left(\frac{100}{R_{[pc]}}\right) \cdot \left(\frac{f_{[Hz]}}{10}\right)^2 \tag{39}$$

At present more than 330 pulsars are known in the Galaxy while the estimated total number is about $(70 \pm 17) \cdot 10^3$ (Lyne et al., 1985).

Our statistics have been deduced from the 330 PSRs catalog by Manchester and Taylor (1981, hereinafter MT), plus some sparse data on PSRs with spinning frequency greater than $10\,Hz$: PSR 1937+21 (Backer et al., 1982; Backer and Kulkarni, 1983; Ashworth et al., 1983), PSR 1855+09 (Segelstein et al.,1986), PSR 1953+29 (Boriakoff et al., 1983; Rawley et al., 1986), PSR 1830-08 (Clifton and Lyne, 1986).

Our results are collected under Fig.s 16 to 22.

The relative PSR abundance *vs.* distance from the Earth (in kpc) has been computed and is shown in Fig.16.

From knowledge of PSR galactic latitudes, the PSR galactocentric radial distance and height distributions have been also computed, and are shown in Fig.s 17 and 18, respectively.

It is important to recognize the agreement between our results and those obtained from a much smaller sample (149 PSRs as compared to 330) by Manchester and Taylor (1977) (statistically homogeneous samples, or same selection effects!).

The same agreement is observed as regards Fig. 19, showing the PSR relative abundance *vs.* rotational frequency.

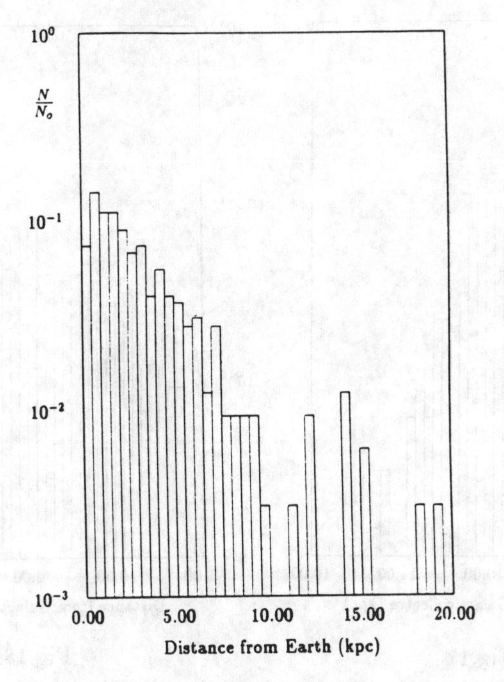

Fig.16

Note, in this connection, that PSRs could be *recycled*, i.e. resulting from the final coalescence of binaries (van den Heuvel, 1984a, 1984b; de Kool and van Paradjis, 1987). An estimate of the *minimum* period P_{min} for such objects (Gosh and Lamb, 1979; Henrichs, 1983) gives the following upper and lower limits:

$$inf(P_{min_{[sec]}}) \approx \left(\frac{B_{[gauss]}}{10^9}\right)^{\frac{6}{7}} \tag{40}$$

$$sup(P_{min_{[sec]}}) \approx \left(\frac{B_{[gauss]}}{2 \cdot 10^{11}}\right)^{\frac{1}{2}} \tag{41}$$

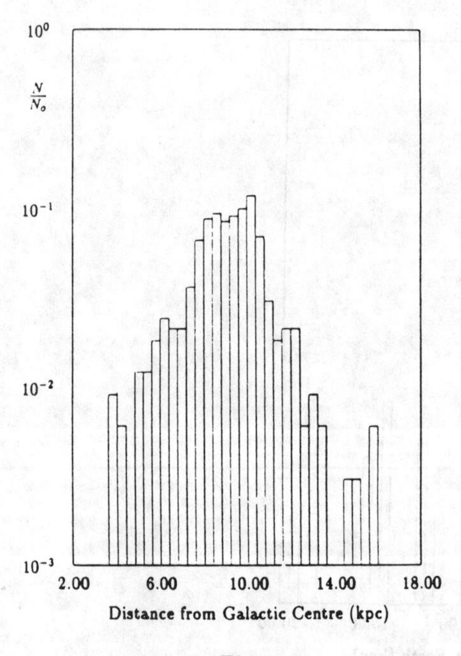

Fig.17

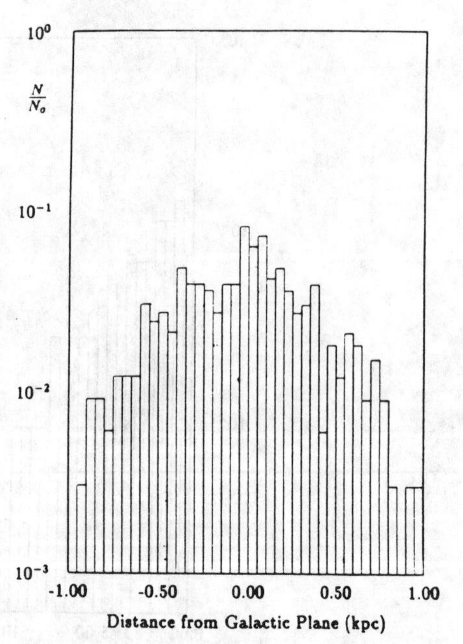

Fig.18

in agreement with experimental evidences (Taylor and Stinebring, 1986).

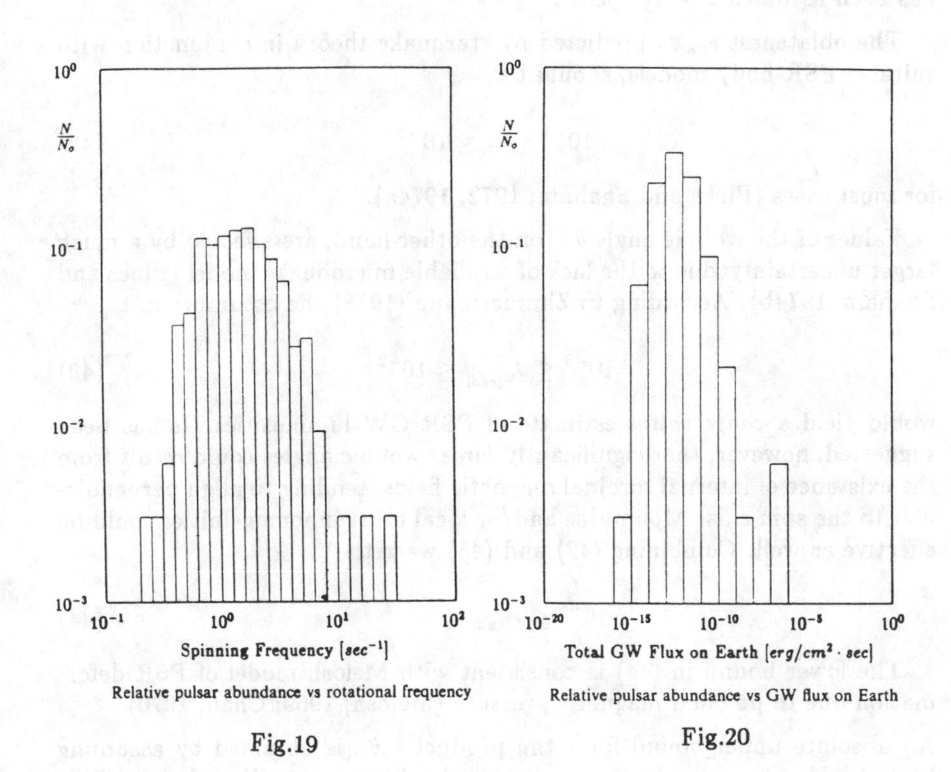

Relative pulsar abundance vs rotational frequency

Fig.19

Relative pulsar abundance vs GW flux on Earth

Fig.20

According to recent work (Stollman, 1987; Chevalier and Emmering, 1986; Proszinsky and Przybycien, 1985; Lyne et al., 1985), the natural logarithm of the surface magnetic field, B, should be gaussian-distributed with average value $3.2 \cdot 10^{12}$ gauss and a standard deviation of 0.69. The observed existence of two distinct maxima in the PSR period distribution, roughly at $0.5s$ and $1.25s$, could be explained by resorting to two different spindown mechanisms, with or without magnetic decay (Peng et al., 1982).

Computing the L_{GW}, F_{GW} and h spectra requires knowledge of the product $J\epsilon_o\theta_w$.

The value of J should be in the range $3 \cdot 10^{44} \leq J_{[gcm^2]} \leq 3 \cdot 10^{45}$, from hadronic equations of state (Pandharipande et al., 1976) corresponding to

PSR masses in the range $0.2M_\odot \div 3.0M_\odot$. This uncertainty is not so relevant because it gives only an order of magnitude difference in the computation of h and two orders in the computation of F_{GW} and L_{GW}. In the following it has been assumed $J = 10^{45} gcm^2$.

The oblateness ϵ_o, as predicted by starquake theory in conjunction with suitable PSR-body models, should be:

$$10^{-3} \leq \epsilon_o \leq 10^{-2} \tag{42}$$

for most cases (Pines and Shaham, 1972, 1974a).

Values of the wobble angle θ_w, on the other hand, are affected by a much larger uncertainty, due to the lack of a reliable microquake model (Pines and Shaham, 1974b). According to Zimmermann (1978) the assumption:

$$10^{-3} \leq \theta_{w[rad]} \leq 10^{-2} \tag{43}$$

would yield a *conservative* estimate of PSR GW luminosities. It has been suggested, however, that significantly *larger* wobble angles could result from the existance of internal toroidal magnetic fields, tending to align perpendicular to the spin axis. Mountains and/or local crust inhomogeinities could be effective as well. Combining (42) and (43) we get:

$$10^{-6} \leq \epsilon_{[rad]} \leq 10^{-4} \tag{44}$$

The lower bound in (44) is consistent with Melosh model of PSR deformation due to poloidal magnetic pressure (Melosh, 1969; Chau, 1970)

An absolute upper bound for ϵ the product $\epsilon_o \theta_w$ is obtained by assuming that the observed spindown rate is completely due to GW emission. This has been done by several authors (Press and Thorne, 1972; Groth, 1975; Zimmermann, 1978). However many concurrent mechanisms have been hypothesized to account for pulsar slowdown, including magnetic dipole radiation at the spinning frequency (Manchester and Taylor, 1977), neutrino (cyclotron) radiation from superfluid vortexes (Peng et al., 1982), longitudinal current flow along the open magnetic field lines (Beskin et al., 1984). The first of these mechanisms should be the dominant one, at least in radio-PSRs, but possibly for the early stages of pulsar life (Ostriker and Gunn, 1969). Neutrino emission, on the other hand, has been claimed as the dominant breaking mechanism for PSRs with periods in excess of 1 sec.

On the basis of the data reported in MT and of the above assumptions: i) the PSR relative abundance *vs.* GW flux on Earth; ii) the GW flux on the Earth *vs.* GW frequency; iii) the GW dimensionless amplitude on Earth *vs.* frequency have been computed and are shown in Fig.s 21, and 22, respectively. Assuming that the chosen sample is statistically significant, the effect of including the whole PSR population, would simply raise the flux levels by a factor of the order ≈ 100, and, correspondingly, the dimensionless amplitudes by a factor ≈ 10, at most.

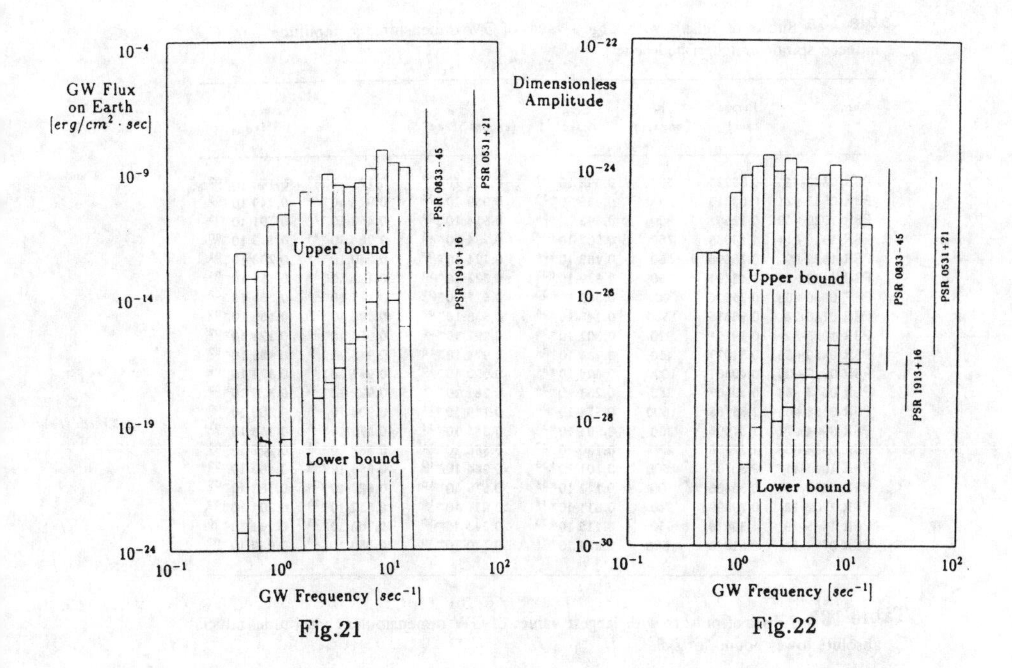

Fig.21

Fig.22

Our computations do not include the effect of Doppler frequency shift due to PSR recession from the galactic centre. The average estimated velocity should be $\approx 100 km/s$, thus producing a fractional Doppler shift of $\approx 3 \cdot 10^{-4}$ in the worst case, which is negligible in the present context.

In Table 13a the twenty pulsars in MT with the highest values of dimensionless amplitude, computed under the assumption of a pure GW induced spindown, are reported. A *different* table is obtained on the basis of the value of $\frac{f^2}{R}$ (Prentice and ter Haar, 1969) as suggested by Eq. (38) and is shown in Table 13b where the absolute lower bound for ϵ has been used. In

Table 14a,b the estimated pulsar gravitational radiation from pulsars with spinning period less than $0.1s$, that are of primary interest in GW detection by seismic-noise-limited Earth-based laser interferometers are collected, and computed under the same assumptions.

In Table 15 data for the Crab and Vela pulsars (Zimmermann, 1978) are reported for comparison.

Table 13a Subset of pulsars with largest vaues of GW dimensionless amplitude: pure GW induced spindown (cherished belief)

Name	Period (sec)	R ($parsec$)	L_{GW} ($ergsec^{-1}$)	F_{GW} ($ergcm^{-2}sec^{-1}$)	h	$\epsilon = \epsilon_o\ \theta_w$ (rad)
PSR 0833-45	0.08925	500	$0.692\ 10^{+37}$	$0.232\ 10^{-06}$	$0.121\ 10^{-23}$	$0.180\ 10^{-02}$
PSR 0531+21	0.03313	2000	$0.459\ 10^{+39}$	$0.959\ 10^{-06}$	$0.911\ 10^{-24}$	$0.749\ 10^{-03}$
PSR 1929+10	0.22652	800	$0.393\ 10^{+34}$	$0.514\ 10^{-08}$	$0.456\ 10^{-24}$	$0.701\ 10^{-03}$
PSR 1916+14	1.18088	760	$0.507\ 10^{+34}$	$0.734\ 10^{-10}$	$0.284\ 10^{-24}$	$0.113\ 10^{-00}$
PSR 1822-09	0.76896	560	$0.463\ 10^{+34}$	$0.124\ 10^{-09}$	$0.240\ 10^{-24}$	$0.298\ 10^{-01}$
PSR 1133+16	1.18791	150	$0.879\ 10^{+32}$	$0.327\ 10^{-10}$	$0.191\ 10^{-24}$	$0.151\ 10^{-01}$
PSR 0950+08	0.25306	90	$0.558\ 10^{+33}$	$0.577\ 10^{-09}$	$0.171\ 10^{-24}$	$0.368\ 10^{-03}$
PSR 0740-28	0.16675	1500	$0.143\ 10^{+36}$	$0.533\ 10^{-09}$	$0.108\ 10^{-24}$	$0.169\ 10^{-02}$
PSR 1055-52	0.19711	920	$0.301\ 10^{+35}$	$0.297\ 10^{-09}$	$0.955\ 10^{-25}$	$0.128\ 10^{-02}$
PSR 0450+55	0.34073	450	$0.236\ 10^{+34}$	$0.974\ 10^{-10}$	$0.945\ 10^{-25}$	$0.185\ 10^{-02}$
PSR 0919+06	0.40361	1000	$0.824\ 10^{+34}$	$0.690\ 10^{-10}$	$0.941\ 10^{-25}$	$0.574\ 10^{-02}$
PSR 0655+64	0.19567	280	$0.263\ 10^{+34}$	$0.281\ 10^{-09}$	$0.922\ 10^{-25}$	$0.370\ 10^{-03}$
PSR 0154+61	2.35165	1600	$0.574\ 10^{+33}$	$0.188\ 10^{-11}$	$0.904\ 10^{-25}$	$0.300\ 10^{-00}$
PSR 0904+77	1.57905	330	$0.501\ 10^{+32}$	$0.385\ 10^{-11}$	$0.870\ 10^{-25}$	$0.268\ 10^{-01}$
PSR 0834+06	1.27376	430	$0.130\ 10^{+33}$	$0.588\ 10^{-11}$	$0.867\ 10^{-25}$	$0.227\ 10^{-01}$
PSR 1001-47	0.30707	1600	$0.301\ 10^{+35}$	$0.984\ 10^{-10}$	$0.855\ 10^{-25}$	$0.483\ 10^{-02}$
PSR 0656+14	0.38486	400	$0.111\ 10^{+34}$	$0.579\ 10^{-10}$	$0.823\ 10^{-25}$	$0.183\ 10^{-02}$
PSR 1702-18	0.29899	740	$0.611\ 10^{+34}$	$0.934\ 10^{-10}$	$0.812\ 10^{-25}$	$0.201\ 10^{-02}$
PSR 1845-19	4.30818	530	$0.115\ 10^{+32}$	$0.343\ 10^{-12}$	$0.708\ 10^{-25}$	$0.261\ 10^{-00}$
PSR 0727-18	0.51015	1500	$0.563\ 10^{+34}$	$0.210\ 10^{-10}$	$0.656\ 10^{-25}$	$0.959\ 10^{-02}$

Table 13b Subset of pulsars with largest values of GW dimensionless amplitude: fiducial absolute lower bound for $\varepsilon_0 \theta_w$

Name	Period (sec)	R ($parsec$)	L_{GW} ($ergsec^{-1}$)	F_{GW} ($ergcm^{-2}sec^{-1}$)	h	$\epsilon = \epsilon_o \cdot \theta_w$ (rad)
PSR 0833-45	0.08925	500	$0.215\ 10^{+31}$	$0.719\ 10^{-13}$	$0.670\ 10^{-27}$	$0.100\ 10^{-06}$
PSR 0531+21	0.03313	2000	$0.821\ 10^{+33}$	$0.172\ 10^{-11}$	$0.122\ 10^{-26}$	$0.100\ 10^{-06}$
PSR 1929+10	0.22652	800	$0.803\ 10^{+28}$	$0.105\ 10^{-13}$	$0.650\ 10^{-27}$	$0.100\ 10^{-06}$
PSR 1916+14	1.18088	760	$0.400\ 10^{+24}$	$0.580\ 10^{-20}$	$0.252\ 10^{-29}$	$0.100\ 10^{-06}$
PSR 1822-09	0.76896	560	$0.525\ 10^{+26}$	$0.140\ 10^{-18}$	$0.806\ 10^{-29}$	$0.100\ 10^{-06}$
PSR 1133+16	1.18791	150	$0.386\ 10^{+24}$	$0.144\ 10^{-18}$	$0.126\ 10^{-28}$	$0.100\ 10^{-06}$
PSR 0950+08	0.25306	90	$0.413\ 10^{+28}$	$0.427\ 10^{-14}$	$0.463\ 10^{-27}$	$0.100\ 10^{-06}$
PSR 0740-28	0.16675	1500	$0.505\ 10^{+29}$	$0.188\ 10^{-15}$	$0.640\ 10^{-28}$	$0.100\ 10^{-06}$
PSR 1055-52	0.19711	920	$0.185\ 10^{+29}$	$0.183\ 10^{-15}$	$0.747\ 10^{-28}$	$0.100\ 10^{-06}$
PSR 0450+55	0.34073	450	$0.693\ 10^{+27}$	$0.287\ 10^{-16}$	$0.511\ 10^{-28}$	$0.100\ 10^{-06}$
PSR 0919+06	0.40361	1000	$0.251\ 10^{+27}$	$0.210\ 10^{-17}$	$0.164\ 10^{-28}$	$0.100\ 10^{-06}$
PSR 0655+64	0.19567	280	$0.193\ 10^{+29}$	$0.206\ 10^{-14}$	$0.249\ 10^{-27}$	$0.100\ 10^{-06}$
PSR 0154+61	2.35165	1600	$0.642\ 10^{+22}$	$0.210\ 10^{-22}$	$0.302\ 10^{-30}$	$0.100\ 10^{-06}$
PSR 0904+77	1.57905	330	$0.700\ 10^{+23}$	$0.538\ 10^{-20}$	$0.324\ 10^{-29}$	$0.100\ 10^{-06}$
PSR 0834+06	1.27376	430	$0.254\ 10^{+24}$	$0.115\ 10^{-19}$	$0.383\ 10^{-29}$	$0.100\ 10^{-06}$
PSR 1001-47	0.30707	1600	$0.129\ 10^{+28}$	$0.423\ 10^{-17}$	$0.177\ 10^{-28}$	$0.100\ 10^{-06}$
PSR 0656+14	0.38486	400	$0.334\ 10^{+27}$	$0.175\ 10^{-16}$	$0.451\ 10^{-28}$	$0.100\ 10^{-06}$
PSR 1702-18	0.29899	740	$0.152\ 10^{+28}$	$0.232\ 10^{-16}$	$0.404\ 10^{-28}$	$0.100\ 10^{-06}$
PSR 1845-19	4.30818	530	$0.170\ 10^{+21}$	$0.506\ 10^{-23}$	$0.271\ 10^{-30}$	$0.100\ 10^{-06}$
PSR 0727-18	0.51015	1500	$0.616\ 10^{+26}$	$0.229\ 10^{-18}$	$0.684\ 10^{-29}$	$0.100\ 10^{-06}$

Table 14a GW radiation from pulsars with period less than 0.1 s: pure GW induced spindown (cherished belief)

Name	Period (sec)	R ($parsec$)	$\epsilon = \epsilon_o \; \theta_w$ (rad)	L_{GW} ($ergsec^{-1}$)	F_{GW} ($ergcm^{-2}sec^{-1}$)	h
PSR 1937+21	0.00156	2000	$0.486\ 10^{-08}$	$0.177\ 10^{+37}$	$0.370\ 10^{-08}$	$0.266\ 10^{-20}$
PSR 1855+09	0.00536	350	$0.106\ 10^{-07}$	$0.513\ 10^{+34}$	$0.350\ 10^{-09}$	$0.282\ 10^{-20}$
PSR 1953+29	0.00613	3500	$0.225\ 10^{-07}$	$0.103\ 10^{+35}$	$0.702\ 10^{-11}$	$0.456\ 10^{-27}$
PSR 0531+21	0.03313	2000	$0.749\ 10^{-03}$	$0.459\ 10^{+39}$	$0.959\ 10^{-06}$	$0.911\ 10^{-24}$
PSR 1913+16	0.05903	5200	$0.806\ 10^{-05}$	$0.166\ 10^{+34}$	$0.513\ 10^{-12}$	$0.119\ 10^{-26}$
PSR 1830-08	0.08528	14000	$0.452\ 10^{-03}$	$0.537\ 10^{+30}$	$0.245\ 10^{-10}$	$0.118\ 10^{-25}$
PSR 0833-45	0.08925	500	$0.180\ 10^{-02}$	$0.692\ 10^{+37}$	$0.232\ 10^{-06}$	$0.121\ 10^{-23}$

Table 14b GW radiation from pulsars with period less than 0.1 s: fiducial absolute lower bound for $\varepsilon_0\theta_w$

Name	Period (sec)	R ($parsec$)	$\epsilon = \epsilon_o \cdot \theta_w$ (rad)	L_{GW} ($ergsec^{-1}$)	F_{GW} ($ergcm^{-2}sec^{-1}$)	h
PSR 1937+21	0.00156	2000	$0.100\ 10^{-05}$	$0.753\ 10^{+41}$	$0.157\ 10^{-03}$	$0.550\ 10^{-24}$
PSR 1855+09	0.00536	350	$0.100\ 10^{-05}$	$0.458\ 10^{+38}$	$0.312\ 10^{-05}$	$0.266\ 10^{-24}$
PSR 1953+29	0.00613	3500	$0.100\ 10^{-05}$	$0.204\ 10^{+38}$	$0.140\ 10^{-07}$	$0.203\ 10^{-25}$
PSR 0531+21	0.03313	2000	$0.100\ 10^{-06}$	$0.821\ 10^{+33}$	$0.172\ 10^{-11}$	$0.122\ 10^{-26}$
PSR 1913+16	0.05903	5200	$0.100\ 10^{-06}$	$0.256\ 10^{+32}$	$0.794\ 10^{-14}$	$0.148\ 10^{-27}$
PSR 1830-08	0.08528	14000	$0.100\ 10^{-06}$	$0.282\ 10^{+31}$	$0.120\ 10^{-15}$	$0.263\ 10^{-28}$
PSR 0833-45	0.08925	500	$0.100\ 10^{-06}$	$0.215\ 10^{+31}$	$0.719\ 10^{-13}$	$0.672\ 10^{-27}$

Table 15

	Standard Crab	PPS Crab[1]	Vela
P(s)	0.033	0.033	0.089
$\omega_{GW}(s^{-1})$	380	380	140
R(pc)	2,000	2,000	500
I(gcm^2)	$4 \cdot 10^{44}$	$2 \cdot 10^{45}$	$2 \cdot 10^{45}$
ϵ_o	$2 \cdot 10^{-4}$	$4 \cdot 10^{-4}$	$3 \cdot 10^{-3}$
θ_w	10^{-2}	10^{-2}	10^{-2}
$L_{GW}(ergs^{-1})$	$5 \cdot 10^{32}$	$4 \cdot 10^{34}$	$1 \cdot 10^{34}$
max	$1 \cdot 10^{35}$	$1 \cdot 10^{37}$	$2 \cdot 10^{37}$
min	$3 \cdot 10^{29}$	$8 \cdot 10^{31}$	$9 \cdot 10^{30}$
$F_{GW}(ergcm^{-2}s^{-1})$	$1 \cdot 10^{-12}$	$9 \cdot 10^{-11}$	$4 \cdot 10^{-10}$
max	$4 \cdot 10^{-10}$	$4 \cdot 10^{-8}$	$1 \cdot 10^{-6}$
min	$4 \cdot 10^{-16}$	$1 \cdot 10^{-13}$	$2 \cdot 10^{-13}$
h	$1 \cdot 10^{-27}$	$9 \cdot 10^{-27}$	$5 \cdot 10^{-26}$
max	$2 \cdot 10^{-26}$	$2 \cdot 10^{-25}$	$3 \cdot 10^{-24}$
min	$2 \cdot 10^{-29}$	$3 \cdot 10^{-28}$	$1 \cdot 10^{-27}$

[1] PPS Crab: see Pandharipande V.R., Pines D., Smith R.A., 1976

(from Zimmermann, 1978)

4 Trials to get indirect evidence of GR from astrophysical sources

4.1 Close binary systems

It is well known that GW emission must lead to observable effects on the orbital elements of revolving close binaries, namely on the orbital periods. According to the analysis of Peters and Mathews for elliptical orbits we have:

$$\langle P' \rangle = -\frac{96}{5}(2\pi)^{\frac{8}{3}}\frac{G^{\frac{5}{3}}}{c^5} m_1 m_2 (m_1 + m_2)^{-\frac{1}{3}} P^{-\frac{5}{3}} f(e) \tag{45}$$

where $f(e) = (1 + \frac{73}{24}e^2 + \frac{37}{96}e^4)/(1 - e^2)^{\frac{7}{2}}$ and $\langle \Gamma' \rangle = \frac{dP}{dt}$ is computed over an orbital cycle.

Taking account of the values of the constants and measuring the period P in days,the masses in solar masses after some algebra (i.e. introducing the mass ratio $q = \frac{m_2}{m_1}$) we have:

$$\langle P' \rangle \approx -1.7 \cdot 10^{-14} m_1^{\frac{5}{3}} q (1 + q)^{-\frac{1}{3}} P^{-\frac{5}{3}} f(e) \tag{46}$$

To compute the spectrum of $\langle P' \rangle$ we can use the harmonic analysis of the Keplerian motion so we have:

$$P'(n, e) = -1.7 \cdot 10^{-14} m_1^{\frac{5}{3}} q (1 + q)^{\frac{1}{3}} P^{-\frac{5}{3}} g(n, e) \tag{47}$$

where P is measured in days, m_1 is in solar unit and the function $g(n, e)$ is:

$$\begin{aligned} g(n,e) \;=\; & \frac{n^4}{32}\{[J_{n-2}(ne) - 2eJ_{n-1}(ne) + \frac{2}{n}J_n(ne) + 2eJ_{n+1}(ne) - J_{n+2}(ne)]^2 \\ & +(1 - e^2)[J_{n-2}(ne) - 2J_n(ne) + J_{n+2}(ne)]^2 + \frac{4}{3n^2}J_n^2(ne)\} \end{aligned} \tag{48}$$

where $J_n(.)$ are Bessel function of first kind and $f(e) = \sum_{n=1}^{\infty} g(n, e)$. Provided that the systems are not contaminated by period changes resulting from other causes gravitational waves could be revealed in an indirect way from observations of times of minimum light of eclipsing binaries. There are three problems, we must analyze, to throw light on this possibility:

a) what is the precision needed on the times of minimum.

b)is it possible to take account both of mass loss or apsidal motion and GW emission on the period changes?

c) what systems are the most suitable candidates?

We can do a quick estimate of the size of the effect according to (46) evaluating the period shortening and the corresponding phase gain over a 20 years of observation for typical eclipsing binaries. The results are shown in table 16.

Table 16

V1143 Cyg	$e \cong 0.54$	$P \cong 7.64$	$\Delta\Phi \cong 5 \cdot 10^{-4} sec$
V447 Cyg	$e \cong 0.33$	$P \cong 2.35$	$\Delta\Phi \cong 2\, sec$
VZ Sge	$e = 0.00$	$P \cong 0.057$	$\Delta\Phi \cong 120\, sec$

Taking of the observational constraints as it is possible to see the results are negative. In other words, mass-loss, mass transfer, tidal forces surely mask the presence of a period shortening by GW emission from binaries.

Our concluding remark is that the only credited indirect evidence of GW emission remains the binary pulsar PSR 1913+16 (Taylor et al. 1982, 1989).

5 Direct methods of detection

This lecture has been conceived to give account to the astrophysical community of the progress made during the last years towards the realization of GW antennas in the range of the sensibilities in which we expect to be GW signals from astrophysical sources. We shall examine the main problems concerning the implementation of a Weber bar antenna and of a long base interferometric laser antenna (hereinafter LBIA),avoiding to do a review of the field ,since excellent review articles exist both on generation and on detection of GW (Braginsky and Rudenko, 1978; Douglass and Braginsky, 1979; Weiss, 1979; Thorne, 1983; Thorne, 1987; Hough et al.,1987; Giazotto, 1988). We would like only to outline the main problems either solved or to be solved especially concerning LBIA.

It is well known that to measure the dimensionless strain amplitude $h_{\mu,\nu}$ taking account of the Riemann tidal force two main methods are possible: using an harmonic oscillator (i.e. a Weber bar antenna) to reveal the resonances induced by a GW impinging on the oscillator or interferometrically detecting the anisotropy of space induced by the action of a GW on two almost free test masses electromagnetically coupled and separated by a distance L.

5.1 Weber bar antennas

J. Weber is the pioneer of the experimental research on the direct detection of GW from astrophysical impulsive sources. He started in 1960 implementing the first harmonic oscillator constituted of an aluminium bar at room temperature with a very high Q. Many groups followed him all over the world implementing bar antennas more and more sophisticated. Among the active groups involved in bar antennas implementation a particular mention merit the group of the university of Roma 'La Sapienza' whose leaders have been E. Amaldi and successively G. Pizzella. Their work produced some of the more sophisticated antennas actually operative in the world with the adequate sensitivity to detect GW impulsive signals.

Just for sake of completeness we want to recall that being $h^{tt} = h_o e^{i\Omega t}$ the time varying dimensionless strain amplitude at the circular frequency $\Omega = 2\pi\nu$ of the impulsive signal of duration $\Delta t = \frac{1}{\nu}$ impinging on the antenna whose circular frequency of resonance is $\omega_o^2 = 2k/M$ being k a factor depending on the structure of the material constituting the resonator and M the resonator mass we have a spectral signal and a thermal spectral noise:

$$\Delta\tilde{x}_s = h_o L \frac{\Omega^2}{-\Omega^2 + i\frac{\Omega}{\tau} + \omega_o^2}$$

$$\Delta\tilde{x}_N = h_o L \frac{\left(\frac{8KT\omega_o}{MQ}\right)^{\frac{1}{2}}}{-\Omega^2 + i\frac{\Omega}{\tau} + \omega_o^2}$$

in which τ is the relaxation time of the material,L is the rest length of the resonator and h_0 is the maximum dimensionless strain amplitude of the GW,while T is the temperature of the bar and Q is its mechanical factor of quality. To have a useful signal to noise ratio we must have:

$$h_o \geq \frac{1}{\Omega^2 L}\left(\frac{8KT\omega_o}{MQ}\right)^{\frac{1}{2}} \tag{49}$$

From the above relations it is clear that to achieve good sensitivities it must lower the temperature and to have mass and Q of the resonator as high as possible. Actually with temperature of the bar $T = 4.2^o K$, L=3m M=2370 Kg and $\nu = \frac{\Omega}{2\pi} = 10^3$ Hz the group of Rome reached $h_o \approx 3 \cdot 10^{-20}$($h_o \approx 3 \cdot 10^{-18}$ for 1 ms pulse). The next step will be a bar at $T = 50mK$ with $h_o \approx 1 \cdot 10^{-21}$ (NAUTILUS bar antenna).

The problems with these types of antennas are connected with the very narrow bandwidth (of the order of some tenth of Hz) and also with the necessity of achieve very small temperatures to lower the thermal noise. Concerning the bandwidth and the work frequency if we would use a bar antenna at very low frequency,say 10 Hz,with the same sensitivity of the one at 10^3 Hz it is required a temperature of $4 \cdot 10^{-8} K$ due to the $\frac{1}{\Omega^2}$ term and in the meantime we always have a very narrow bandwidth. As we shall see the way to escape from these serious limitations is constituted by the implementation of interferometric antennas that provide the means to realize true wide band GW antennas.

5.2 Interferometric antennas

The basic idea for the implementation of an interferometric antenna to detect GW is, in principle, very simple and was clearly stated for the first time by Gertsenshtein and Pustovoit (1963):"...it should be possible to detect GW by the shift of the bands in an optical interferometer" (see also Giazotto, 1988). Weiss (1972) produced the first complete theoretical work on the noises competing with GW signals. Great merits of Weiss are also the ideas either to use the Optical Delay Line (Herriot, 1964) in the arms of a laser based Michelson interferometer (this has the effect of increasing the equivalent length of the arms in such a way that for a length arm of, say, $3km$ it is possible to obtain an equivalent length of nL (where n is the number of the laser beam round trips in the cavity), or to use fast light phase modulation to take account of the laser's amplitude fluctuations. The first experimental attempt to implement this type of detector is merit of Forward (1978) who obtained a spectral strain sensitivity of $h > 2 \cdot 10^{-16}(Hz)^{\frac{-1}{2}}$ for $\nu > 2kHz$ implementing a small size,low power Michelson interferometer actively controlled for locking the interferometer to a fringe. Successively the group of Max Planck Institute in Munich, the group of Glasgow University and the one in Caltech realised prototypes of increasing size and sensitivities (Billings et al. 1979; 1983; Hough et al., 1983; Drever et al. 1983;). The Munich group following the Weiss's idea implemented a Michelson interferometer with ODL in the arms, obtaining with their 30 m interferometer a spectral strain dimensionless amplitude $h > 8 \cdot 10^{-20}(Hz)^{-\frac{1}{2}}$, while the groups of Glasgow and Caltech following an elegant idea of Drever (Drever et al. 1980; 1981;) realised two Michelson interferometers having Fabry-Perot cavities in their arms, obtaining sensitivities of $h \approx 1.2 \cdot 10^{-19} Hz^{-\frac{1}{2}}$ in Glasgow (Ward et al., 1987) and $h \approx 5 \cdot 10^{-19} Hz^{-\frac{1}{2}}$ (Spero, 1986) in Caltech. To increase the interferometer's sensitivity several optical schemes have been conceived to lower the loss of power from the cavities by reusing the light escaping from the input port of the beam splitter (Drever, 1982)i. e. the light power recycling scheme, experimentally tested with success (Rudiger et al., 1987; Man et al., 1987), or schemes to increase the sensitivity to periodic signals like the syncronous recycling (Drever, 1981), or the dual recycling (Meers, 1988) or the detuned recycling (Vinet et al., 1988). Apart the optical schemes, in our opinion, a great progress towards the detection of GW at very low frequencies $\nu \geq 10$ Hz is also constituted by the implementation of efficient seismic noise filters (Del Fabbro et al., 1987). Just for sake of completeness we shall expose, briefly, the basic principles and the limiting sensitivity of an interferometric detector of GW, then, we shall give account of the main progress concerning the different type of interferometers (ODL and FP) and the noise sources i.e.:

a) Delay Lines (ODL) and Fabry Perot(FP) interferometers.

b)Recycling schemes.

c)Laser noises.

d)Refraction index fluctuations.

e)Thermal noise.

f)Seismic noise.

g)Chaos effects induced by radiation pressure fluctuations.

h)Cosmic ray background and electric and magnetic field noises.

Finally we shall give account of the characteristic of the long base antenna constituting the VIRGO project (Bradaschia et al., 1989).

We said above that since the principle of equivalence states that the physics is the same in a sufficiently small region of all reference frames, there is no way to tell whether a given observer is in an accelerated frame or experiencing a gravitational field. Tidal effects are one way to determine the difference. Using the masses to mark points in the space time, and as the passage of a GW distorts the spacetime, we can see the distortion by measuring the relative motions of the masses. The relevant consequence is that two free masses experiment a dimensionless strain:

$$h_{x_1,x_2} = 2\frac{\Delta L_{12}}{L_{12}} \tag{50}$$

by a plane GW travelling along x_3 axis for the action of the Riemann tidal force : $F_R = M\ddot{h}_{\mu,\nu}L$ (see fig.1). So owing to quadrupolar character of GW we have differential strain increasing, say, along x_1 while decreasing along x_2 for a GW travelling along x_3 perpendicularly to the plane x_1x_2. An interferometer in the plane x_1x_2 can be used to detect Δ L. Any GW, causing unequal or differential mode, changes the two light paths and can be detected by either a transient or a periodic phase shift. Writing the perturbation of the metric tensor as:

$$h_{x_1x_1} = -h_{x_2x_2} = h(\omega)e^{i(kx_3-\omega t)} \tag{51}$$

all other components h_{ij} being null we have the transfer function of the ideal interferometer : $\Delta t(\omega)/h(\omega) = tt \cdot sinc(\frac{\omega tt}{2})e^{-i\omega \cdot tt/2}$ being tt the time travel of the light in each interferometer arm and sinc y=sin y/y (Brillet,1985). Assuming n roundtrips of the light in each arm of length L,we have tt=2nL/c and a maximum of the transfer function when $\omega \cdot tt = \pi$ with a phase shift: $\Delta\phi_{max}(\omega) = \frac{2\pi}{\lambda}h(\omega)\frac{\pi}{\omega}$. From the above formulae it is easy to see the main features that an interferometric antenna must have:

1)the length of the interferometer arms ,to increase the sensitivity h,must be the largest practical one,up to $nL \approx$ 200Km as equivalent length, being $\omega \cdot tt = \pi$, $\omega = 2\pi 10^3 Hz$. This means lengths of the antenna from 1Km $\div$ 10 Km and a ligth path folding from 10 to 200 times to fulfil the condition $\omega \cdot tt \approx \pi$.

2)to avoid acoustical noise and refraction index fluctuations the beam paths must be evacuated to a high vacuum (lesser than $10^{-6} \div 10^{-7}$ *torr*).

3)the system constituted by mirrors and beamsplitter to approximate the condition of free falling bodies, required from the theory, must be isolated the one from the other and ,to avoid seismic noise, from the earth.Pendular suspensions of all the masses(beamsplitter and mirrors) constituting the interferometer to a vibration isolated stand can solve the problem and adequate servosystems must be implemented to maintain the alignment of the apparatus and to prevent oscillations of the pendula.

4)the basic GW antenna will be the multipass two waves Michelson interferometer,whose arms are constituted either of ODL or FP cavities.In a kilometric antenna the illumination is a crucial problem because using broadband light,like the one coming from an argon laser,it is hard to realize length differences between the two arms less than half a wavelength of the light.This problem was solved,as we shall see, using stabilized frequency argon or Nd-Yag lasers.

a)Optical Delay Lines (ODL) and Fabry Perot(FP) interferometers.

The necessity of increasing the sensitivity of the interferometer to a phase shift due to a GW signal,comes from the existence of noises which are present even if the system (laser,beamsplitter,mirrors) is perfectly aligned and isolated (phase noises or internal noises) or from external factors other than a GW signal(displacement noises or external noises). First of all let us remember the problem of the ultimate sensitivity in this type of antennae constituted by the quantic nature of the light,i.e. the photon counting noise or Poisson noise or shot noise.Due to the anticorrelated fluctuations Δn of the photon number n in the interferometer arms according to the uncertainty relation we have a phase shift $\Delta\Phi_s \geq \frac{1}{\Delta n}$.

For a photon coherent state $\Delta n = (n)^{\frac{1}{2}}$,hence $\Delta\phi \geq \frac{1}{n^{\frac{1}{2}}} \cong \left(\frac{h\nu_o}{P_e t}\right)^{\frac{1}{2}}$ where h is the Planck's constant, ν_o the laser frequency, P_e the light power in the interferometer arms and t the measurement time.To overcome the effect of this and other noises two different optical schemes have been chosen allowing the beams to bounce back and forth in the optical cavities i.e. ODL and FP cavities.

In the Optical Delay Line scheme (firstly studied by Herriot et.al.,1964) the laser beam is entering the two optical cavities and bounces 2N times between the mirrors with the purpose of increasing the signal to noise ratio S/N of the GW signal to photon counting noise(see fig. 23).In an ODL the number of output photons per unit time will be

$$n = R^{4N}\left(\frac{P}{2h\nu}\right)(1 - cos\phi)$$

where R^2 is the intensity reflectivity of the mirrors,h the Planck's constant,ν

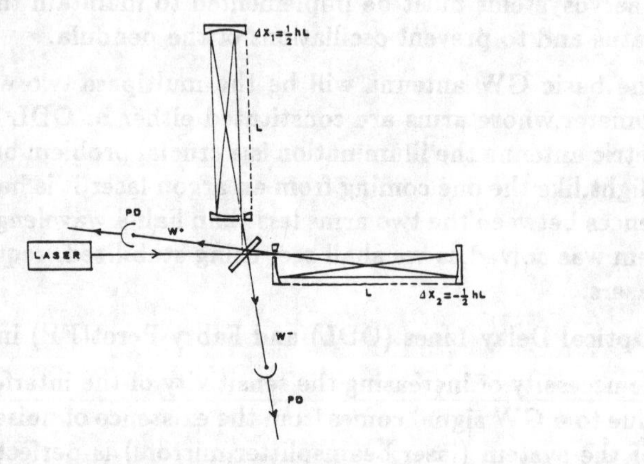

Fig 23 In the ODL scheme ,the laser beam is entering the two optical cavities and bounces 2N times to increase the S/N ratio of GW to the shot noise.

(from Bradaschia et al., 1989)

the light frequency and ϕ the phase difference between the two arms.

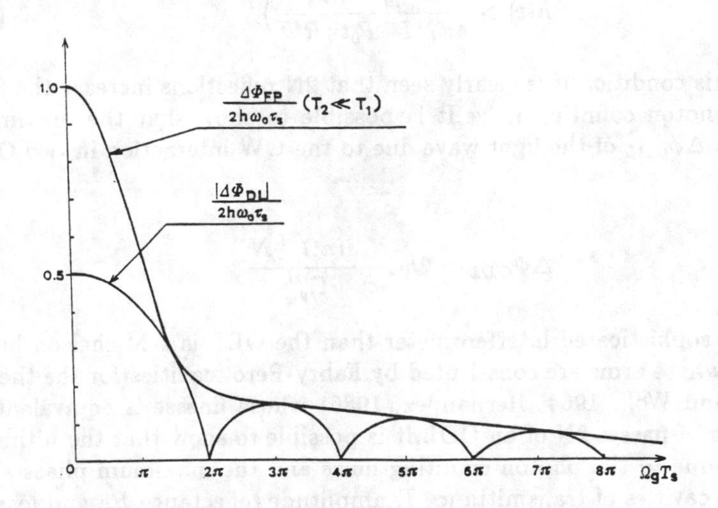

Fig 24 Comparison between the phase shift due to GW amplitude h of a FP ($T_2 \ll T_1$) and an ODL interferometer having the same storage time.
(from Bradaschia et al., 1989)

A phase shift $\Delta\phi$ produces a change in the number of photons

$$\Delta n = R^{4N}\left(\frac{P_o}{2h\nu}\right)\Delta\phi sin\phi$$

The laser source fluctuations of the number of photons obey to Poisson statistics,so the mean square fluctuation is :

$$\delta n^2 = \frac{n}{t}$$

for a measurement time t,and a phase shift $\Delta\phi$ will be observable under the condition:

$$\Delta n^2 \geq \delta n^2 \tag{52}$$

from which we can derive the condition of the ultimate sensitivity h(t) for a mirror displacement Δx,for detection photodiode efficiency η,for $\Omega_g \frac{L}{c} \ll 1$ being Ω_g the circular gravitational frequency of the signal:

$$h(t) > \frac{\lambda}{4\pi NL}\left(\frac{h\nu_o}{P_0 t\eta R^{4N}}\right)^{\frac{1}{2}} \tag{53}$$

From this condition it is clearly seen that 2N reflections increase the S/N ratio for photon counting noise.It is possible to show that the maximum phase shift $\Delta\phi_{ODL}$ of the light wave due to the GW interaction in two ODL is :

$$\Delta\Phi_{ODL} = 2h\omega\frac{L}{c}\frac{sin\Omega_g \frac{L}{c} N}{\Omega_g \frac{L}{c}} \tag{54}$$

A more sophisticated interferometer than the ODL is a Michelson interferometer whose arms are constituted by Fabry-Perot cavities(for the theory see Born and Wolf, 1964; Hernandez, 1986) whose finesse is equivalent to the number of passes 2N of an ODL.It is possible to show that the ultimate sensitivity due to the photon counting noise and the maximum phase shift for two FP cavities of transmittance T_i,amplitude reflectance R_i, and loss B_i of the mirrors are given by :

$$\tilde{h} \approx \frac{\lambda}{2\pi LF}\left(\frac{h\nu}{P}\right)^{\frac{1}{2}} \cong 1.5 \cdot 10^{-23}\,(Hz)^{-\frac{1}{2}}$$

and

$$\Delta\phi \cong \omega h\tau_s \frac{2T_1^2}{T_1^2+T_2^2}\left(\frac{R_2}{R_1}\right)^{\frac{1}{2}}(1+\Omega_g^2\tau_s^2)^{-\frac{1}{2}}$$

respectively. From fig. 24 it is possible to see the behaviour of the phase shift due to the GW amplitude h of a FP ($T_1 \ll T_2$) and an ODL interferometer having the same storage time $\frac{L}{c}$.When $\frac{\Omega_g L}{c} \gg 1$ the phase shifts are comparable.Nevertheless the advantages of FP are: more versatility, a smaller transverse size of the beam with a reduction of the diameter of the vacuum tubing and, perhaps, a lower sensitivity to scattered light.Their drawbacks can be summarized in the necessity of more servoloops,some loss of contrast,and a better laser frequency stabilization and a better vacuum.Actually (1989) most of the project to implement GW antennae are oriented to use FP cavities.

From the above discussion it is also clear that CW laser with very high power are needed(typically the goal is a CW laser power larger than 500 Watt) to lower the shot noise level.To fulfil this need there are nowdays several schemes of the socalled light recycling technique,about which we shall discuss in the next paragraph.

b)Recycling schemes.

The basic idea of recycling,as we said above, was proposed by Drever (1982):in a resonant optical cavity containing the interferometer ,provided that the losses are small and the cavity is kept on resonance with the incoming monochromatic light,there is a power build-up which results in a reduction of the shot noise (see fig.25).The realization of this idea can be done in several ways,according to the geometry of the GW antenna (ODL or FP cavities) and was successfully tested experimentally (Man, Shoemaker, Brillet, 1988),while the theory has been exposed in a very elegant and exhaustive paper by Vinet, Meers, Man, Brillet (1988).From the work of these authors we can derive the important conclusions that:

a) recycling is the adequate solution to the lack of power of the existing CW single frequency lasers;it is possible to obtain with high quality optics power increase up to 100 times in a wideband device using standard power recycling,and up to 1000 times in a narrow band interferometer using syncronous recycling.

b) The recently proposed techniques of detuned recycling and of dual recycling ,offer the possibilty of realizing 'intermediate' configurations,where both the bandwidth and the sensitivity are tunable,being their sensitivities to the optical losses of the interferometer components different.In this way it is the possible the optimization of device performances according to the desired interferometer response without requiring any other change than an offset in two servoloops.That is the case of the FP specific detuned recycling

c)Laser intensity noise.

Representing the laser power as:

$$P(t) = P_0 + \delta P(t) \tag{55}$$

we consider the laser instantaneous power P(t) coming out owing to the instantaneous fluctuations $\delta P(t)$ of the laser mean power P_0.The effects on the interferometric signal could be suppressed by proper design (perfect symmetry + use of a dark fringe).Since this conditions cannot be fulfilled a standard phase modulation technique is used, giving a frequency transposition of the signal at a few megahertz,where the laser does not have excess noise.But coupling into the signal of power fluctuations around the GW frequency can happen also for radiation pressure effect , radiometer effect and also for the thermal expansion of the mirror coatings.The lower the frequencies the larger the noise spectral density associated to these effects. It is possible to show that assuming an asymmetry factor β and being $\delta P(t)$ the laser power fluctuations, with frequency modulation, in a zone where the laser power fluctuations are at the shot noise level, we have a limit on h

$$h = \beta \cdot \frac{\Delta x}{L} \cdot \frac{\delta P}{P_o} \tag{56}$$

In a large kilometric antenna the beam size will be of the order of 10^{-1} m with the purpose of minimizing the size on the far mirror. This implies the use of modulating large aperture Pockels cells, unpractical for being carried by the test masses. This requirement can be circumvented using an external modulation scheme (see fig 26b), in which a small fraction of the incident light is sent, through a Pockels cell, to interfere with the beam containing the GW signal. The Pockels cell is modulated at a frequency where the laser power noise has reached the shot noise and the signal is obtained by making syncronous detection with the modulation signal. In this scheme the noise can be slightly higher (Man, 1988) than in the internal modulation one, but the external modulation has the advantage to bring a net sensitivity improvement enhancing the recycling factor.

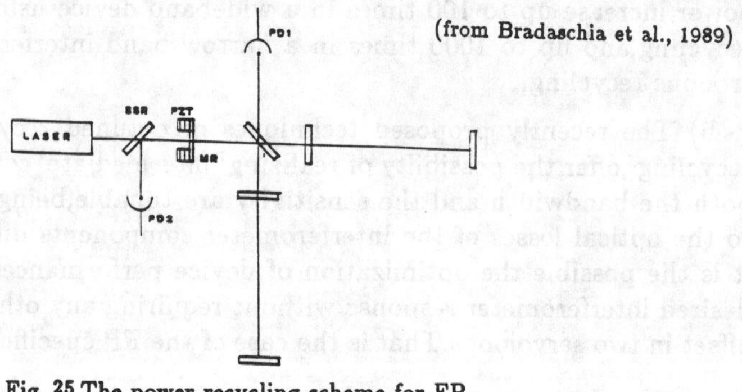

Fig. 25 The power recycling scheme for FP.

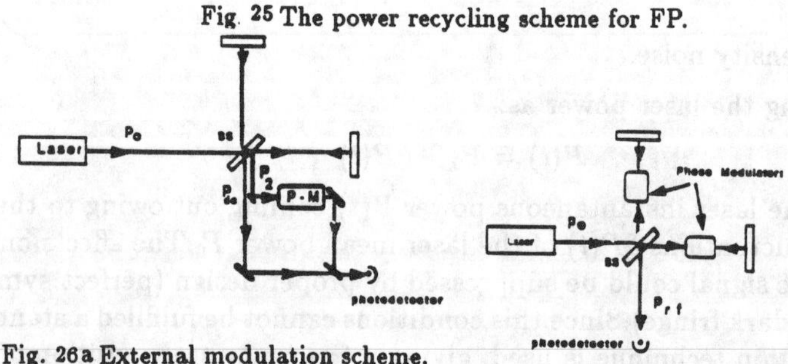

Fig. 26a External modulation scheme.

Fig. 26 b. Internal phase modulation scheme.

d)Laser linewidth noise.

Laser frequency fluctuations produce phase noise in an interferometer with arms having unequal length.If ν_0 and $\Delta\nu$ are the laser frequency and the r.m.s. frequency fluctuation respectively,then the r.m.s. phase fluctuation due to an arm lenght difference ΔL is:

$$\Delta\phi \cong \frac{2\pi}{c}\Delta\nu\Delta L \tag{57}$$

It is very important to avoid that rays having large ΔL interfere.In a multipass ODL interferometer the light hitting the mirrors is scattered by the reflecting coating and enters into the optical path of one of the other ODL beams.This phenomenon ,even if the scattered beam intensity is of the order of $\epsilon \cong 10^{-4} \div 10^{-5}$ of the incident one ,may create a large background because the interference of the scattered beam with the main one has an amplitude proportional to the square root of the scattered light intensity.

Different methods have been adopted to get rid of this phenomenon (Schilling et. al., 1981; Schnupp et.al., 1985; Rudiger et.al., 1981). The one that is widely adopted consists in operating a reference FP cavity (Drever et.al. 1983;Hough et.al.1987;Shoemaker et.al. 1987)fed with a small fraction of the laser light modulated at the frequency ν_m by means of a Pockels cell (see fig.26b).If the laser frequency is tuned to one of the FP resonances the reflected light has the two sidebands at frequency $\pm\nu_m$ having opposite sign amplitudes ,giving zero output in a photodiode .If the laser frequency fluctuates the two sidebands amplitudes will not cancel anymore and give a signal in the photodiode,which can be syncronously detected.The signal is proportional to the laser frequency displacement $\Delta\nu$ with respect to the FP resonance frequency.It can be fed to a laser intracavity Pockels cell and to a piezoelectric trasducer PZT which moves one of the laser mirrors to stabilize the frequency.The limiting noise is the shot noise ,and the linewidth, for a lossless cavity ,is:

$$\Delta\nu \geq \frac{1}{2\pi\tau_s}\left(\frac{h\nu}{P_s t}\right)^{\frac{1}{2}} \tag{58}$$

where τ_s is the reference cavity storage time, P_s is the power used in the stabilization circuit and t is the observation time.Using this technique with some improvements different groups (Billing et al. 1983; Shoemaker et al. 1985; Shoemaker et al. 1989) obtained final integrated linewidth of 10^6 smaller than an unstabilized laser (see fig.27). The same technique applied to FP antennas requires faster servoloops because, even in the case of perfectly equal arms there is incomplete cancellation of the laser width noise, unless the mirror transmittance and losses in the two arms are equal.

e)Lateral beam jitter noise.

If the beam splitter is not symmetrical between the two interferometer's arms ,but deviates by an angle $\delta\alpha$, then a lateral beam jitter δx will produce the phase shift (Billing et. al. 1979):

$$\Delta\phi \cong 2\delta\alpha\,\delta x\frac{4\pi}{\lambda} \tag{59}$$

Two methods have been adopted for reducing this noise:

the first use a mode cleaner (Rudiger et. al., 1981; Meers, 1983), while the second, a simpler one, even if 30-50 % of the laser power gets lost,is the use of a monomode optical fiber coupler as suggested by R. Weiss of MIT.Experimental results obtained from the implementation of these two techniques are shown in fig.28 (Shoemaker et. al., 1985)

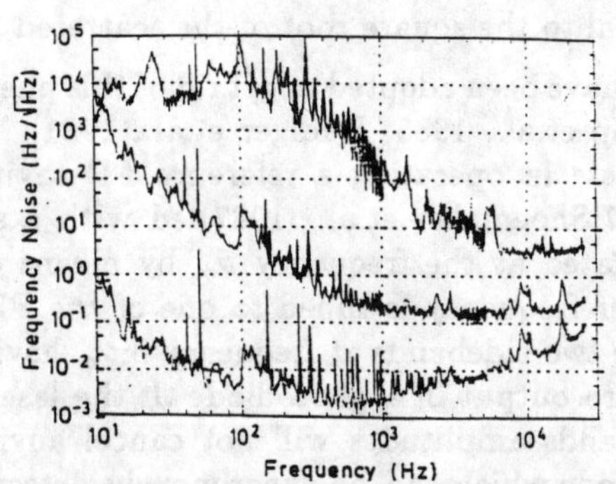

Fig. 27 From the top to the bottom are shown the unstabilized laser spectral line density ,the stabilized laser spectral line density with a reference FP cavity,while in the lower part is shown the stabilized laser spectral line density after the combined effects of the reference FP and the total ODL path (from Shoemaker et.al. ,1985).

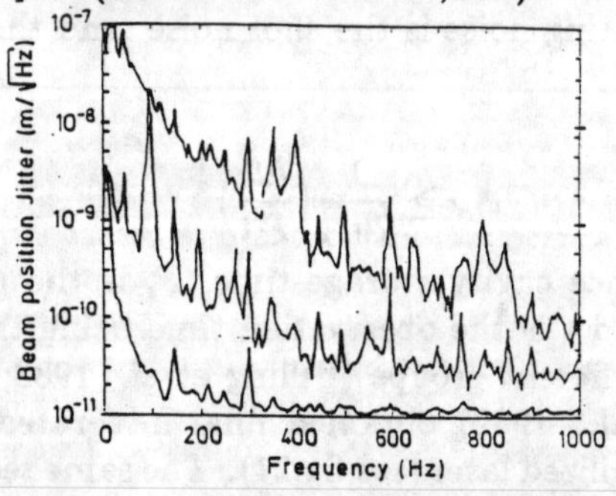

Fig 28 The lateral beam jitter (from Shoemaker et. al. 1985) as measured with a position sensing photodiode;the upper curve is the unfiltered laser beam, the middle one represents the beam jitter after a mode cleaner and the lower represents the jitter after a monomode optical fiber.

f) Refraction index fluctuations.

This phase noise originates from the fluctuations of the refractive index in the interferometer's vacuum pipes. The laser light is bouncing between the mirrors of either the FP or ODL system; the number of residual gas molecules contained in the light beam path is then fluctuating almost in a Poissonian way hence varying the refraction index (Brillet, 1984, 1985; Hough et. al. 1986). It is possible to show that in a multipass ODL there are N beams with uncorrelated fluctuations and the strain sensitivity limited by this noise will be:

$$h = \left(4\alpha^2 \frac{\rho}{LN} \pi D v\right)^{\frac{1}{2}} \tag{60}$$

whereas in a FP system:

$$h = \left(4\alpha^2 \frac{\rho}{L} \pi D v\right)^{\frac{1}{2}} \tag{61}$$

being α the atomic polarizability, v the average velocity of the atoms across the beam, D the beam diameter, ρ the average density of the atoms. This effect can be decreased by increasing L and D and by decreasing the pressure. So a high or ultrahigh vacuum is needed, unless one considers unreasonably large beams.

g) Thermal noise.

The mass of the mirror is driven by the stochastic forces produced by the thermal noise. Considering both the forces acting on the mirror suspensions and those producing an excitation of the mirror normal modes it is possible to show that in the frequency domain we have a condition on the measurable spectral strain $h(\omega)$:

$$\tilde{h} > \frac{1}{L}\left(2K\tilde{T}\sum \frac{1}{M_i Q_i \omega_i^3}\right)^{\frac{1}{2}} \tag{62}$$

where the sum is extended to both the masses number Mi of the mirrors and to the longitudinal modes of circular frequencies ω_i and mechanical quality factors Q_i at the temperature T. From the formula above it is clear that the higher the mirror masses as well as the Q's the lower is the thermal noise.

h) seismic noise.

In the frequency range between 10Hz and 1000 Hz, the ground motions are mostly generated by local disturbances such as wind, breaking ocean waves and human made noise. Their amplitude is then quite variable and can be approximated sufficiently well using the formula :

$$x_T = \frac{a}{\nu^2} \frac{m}{Hz^{\frac{1}{2}}} \tag{63}$$

being x_T the r.m.s. spectral displacement of the suspension point of a mono or multistage pendulum.The factor 'a' is $\cong 10^{-9}$ at a depth of $\cong 10^3$ m up to $\cong 10^{-6}$ at the earth surface in a relatively quiet place.

The use of active seismic isolation schemes is strongly limited by the difficulty of making multiple three-dimensional (3 D) systems,this necessity is dictated by the fact that a non isolated degree of freedom reintroduces the seismic noise even if others degrees of freedom are isolated.

For this reason passive schemes have been adopted ,able to isolate in vertical direction as well (Giazotto,1987;Shoemaker et. al. 1987) .

The basic idea is to use a multiple stage pendulum with the masses supported by springs. It can be shown that the frictionless transfer functions F for both the vertical and horizontal directions can be brougth to the following canonical form :

$$F = \prod_{n=1}^{N} \frac{\omega^2}{\left(\Omega^2 + \omega_n^2\right)} \tag{64}$$

being $\nu_n = \frac{\omega_n}{2\pi}$ the n-th mode frequency and N is the number of masses.

In the interferometric antenna aiming to reach very low frequencies ($\nu \geq 10Hz$) the seismic isolation requires a very careful design with the purpose of avoiding mechanical resonances falling into the interval $10 < \nu < 1000Hz$;these are produced mainly by springs 'rocking and normal modes.

A full scale suspension system conceived for the VIRGO project has been built and tested at the INFN Pisa laboratory.This system is called Super Attenuator (SA).The SA system consists in a cascade of 7 vertical gas springs ,each weighting 100 Kg,connected to vertical wires each 0.7 m long (see figg.29 and 30). The SA can sustain heavy 400 Kg mirrors, which can be useful to lower the thermal noise, and produce 3-dimensional attenuation of $\approx 10^{-9}$ at 10 Hz (Giazotto,1989; Del Fabbro et.al., 1987; 1988).

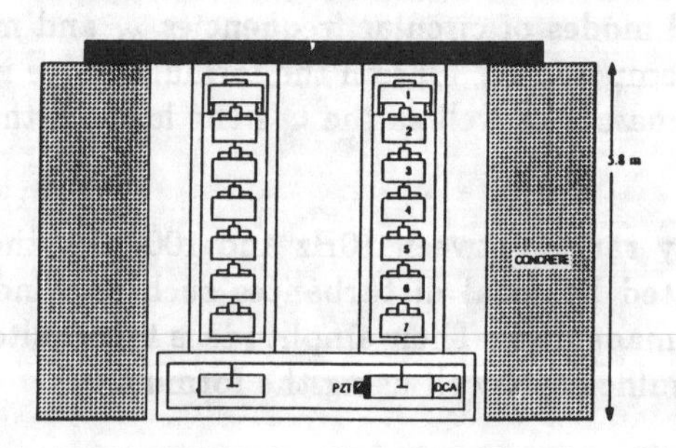

Fig. 29 Schematic diagrams of the experimental apparatus.

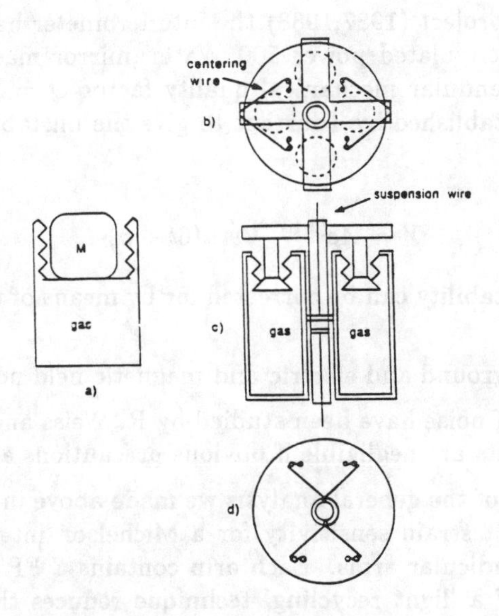

Fig.30 Schematic diagrams of the gas spring and its practical realization.

From the tests made in Pisa on two 400 Kg masses suspended to SA's an absolute upper limit less than $10^{-11}m(Hz)^{\frac{1}{2}}$ for $10 < \nu < 200Hz$ and the transfer functions (TF) were measured.These TF applied to the measured seismic noise spectral displacement should allow in a 3 Km long interferometric antenna a spectral strain sensitivity $h < 1.9 \cdot 10^{-20}(Hz)^{-\frac{1}{2}}$ at $\nu = 10Hz$ and $h < 7.5 \cdot 10^{-19}/\nu^2(Hz)^{-\frac{1}{2}}$ for $\nu > 20Hz$. This gives at the Vela (Crab) pulsar frequency the limit $h < 3.\cdot 10^{-25}(h < 4.\cdot 10^{-26})$ for a 1 year integration time.

i)Chaos effects induced by radiation pressure fluctuations.

In a FP interferometer radiation pressure fluctuations can produce multistability (Deruelle and Tourrenc,1984; Tourrenc and Deruelle 1985; Aguirregabiria and Bel ,1987) and this can be seen from the equation of motion of say the M_2 mirror:

$$\ddot{x} + \frac{\Omega}{Q}\dot{x} = -\Omega^2 x + \frac{2\mid\Psi\mid^2}{M_c} = -\Omega^2 x + \frac{2P}{M_c}\frac{sin^2\theta}{1 + cos^2\theta + 2cos\theta[\frac{4\pi}{\lambda}(D_s + x) - \mu]} \qquad (65)$$

where M is the M_2 mirror mass, Ω the pendulum circular frequency and Q the mechanical quality factor.(see fig.31)

In the VIRGO project (1987,1988) the interferometer having arm lenght 3Km,$\lambda = 1\mu m$,recirculated power 500 Watts ,mirror masses 400 Kg , finesse F=40 and pendular mechanical quality factor $Q = 10^6$ the retarded effects has been established by Tourrenc to give the unstable mirror motion equation:

$$Y = Ae^{(2\cdot 10^{-3}t)} sin(6t + \phi) \tag{66}$$

This type of instability can be corrected for by means of the mirror active feedback.

j)Cosmic ray background and electric and magnetic field noises.

These sources of noise have been studied by R. Weiss and his conclusions are that these effects are negligible if obvious precautions are taken.

Taking account of the general analysis we made above in figg.32, 33 there are shown the limit strain sensitivity for a Michelson interferometer ,with 3 Km long ,perpendicular arms. Each arm contains a FP cavity of finesse F=40. The use of a 'light recycling' technique reduces the required laser power from 500 Watts to 10 Watts .The laser source is frequency and amplitude stabilized using very fast ,ultrahigh gain ,shot-noise limited servo-loops.The critical components of the interferometer are seismically isolated by multidimensional SA.The position and the alignment of the optical com-

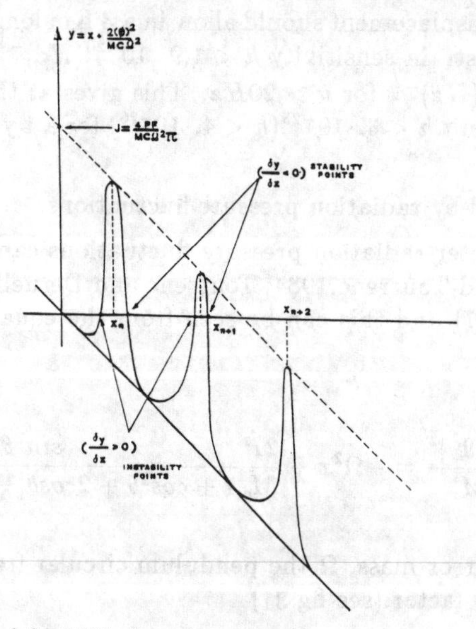

Fig.31 Plot of the RHS of equation (66) ;if the laser power P increases the peak at x_{n+2} may cross the y=0 axis creating a new stability point.

ponents are maintained by local servo-loops under global computer control.All the interferometric system operates under a high vacuum ,in order to avoid acoustic noise and index fluctuations. A suitable site for the construction of VIRGO has been found in Cascina near Pisa.

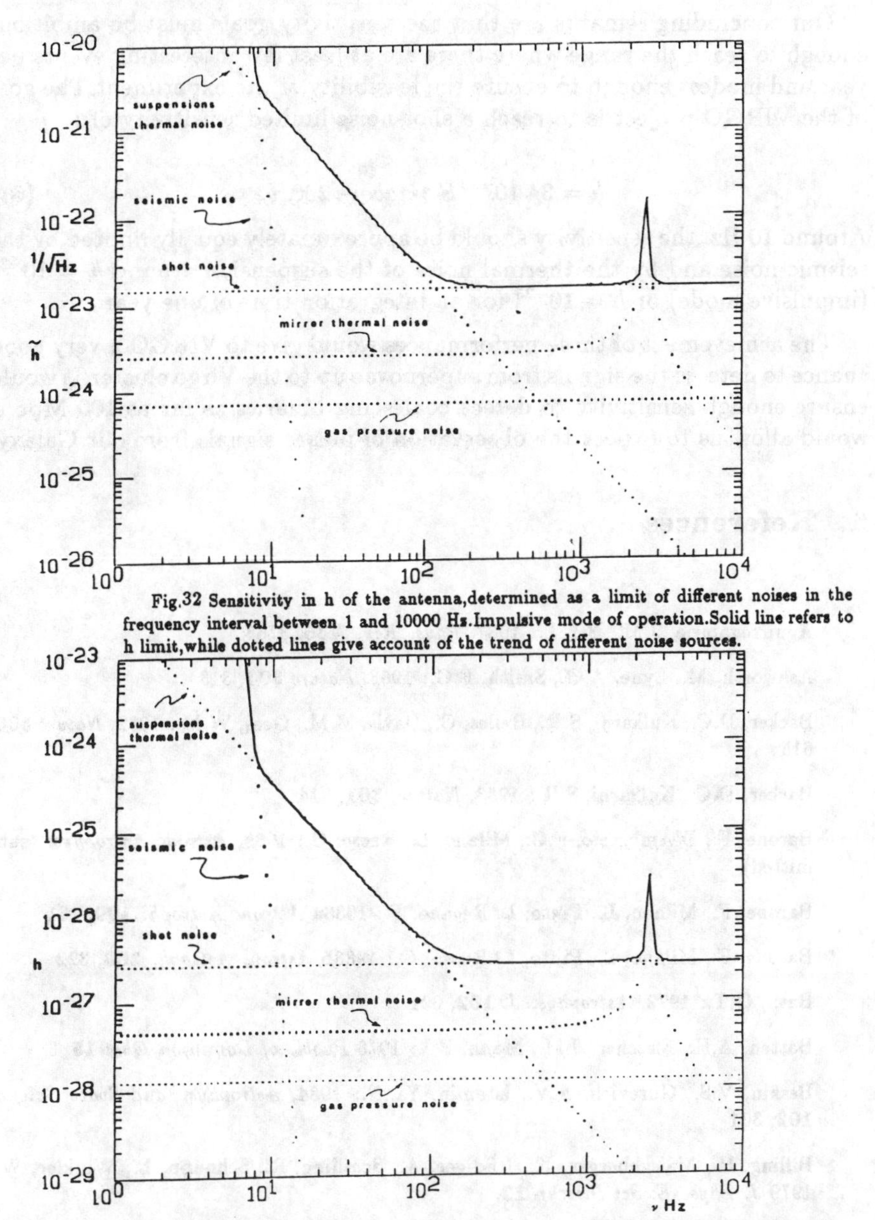

Fig.32 Sensitivity in h of the antenna,determined as a limit of different noises in the frequency interval between 1 and 10000 Hs.Impulsive mode of operation.Solid line refers to h limit,while dotted lines give account of the trend of different noise sources.

Fig. 33Sensitivity in h of the antenna (solid line),determined as a limit of different noise sources (dotted lines) in the frequency interval between 1 and 10000 Hs (integrated mode for periodic sources ;one year integration time)

Data acquisition consists in running the apparatus and registering the fluctuations of the phase difference between the beam reflected by the two 'gravito optic' FP transducers,with a bandpass of 10 Hz to 3 KHz,while monitoring the environment and the apparatus.

Our concluding remarks are that the sensitivity goals must be ambitious enough to reach the range where there are at least few interesting events per year,and modest enough to ensure the feasibility of the experiment.The goal of the VIRGO project is to reach a shot-noise limited sensitivity of :

$$h = 3 \cdot 10^{-23} Hz^{\frac{1}{2}} \; above \, 200 Hz \tag{67}$$

Around 10 Hz ,the sensitivity should be approximately equally limited by the seismic noise and by the thermal noise of the suspension around $h = 10^{-21}$ (impulsive mode) or $h = 10^{-25}$ for an integration time of one year.

The achievement of these performances would give to VIRGO a very good chance to detectc the signals from supernovae up to the Virgo cluster.It would ensure enough sensitivity to detect coalescing binaries as far as 100 Mpc.It would allow us to expect the observation of pulsar signals from our Galaxy.

6 References

Aguirregabiria, J.M., Bel, L.: 1987, *Phys. Rev.* **A36**, 3768

Ashworth, M., Lyne, A.G., Smith, F.G.: 1983, *Nature* **301**, 313

Backer, D.C., Kulkarni, S.R., Heiles, C., Davis, M.M., Goss, W.M.: 1982, *Nature* **300**, 615

Backer, D.C., Kulkarni, S.R.: 1983, *Nature,* **301**, 314

Barone, F., D'Ambrosio, F.C., Milano, L., Russo, G.: 1989, *Astron. Astrophys.* (submitted)

Barone, F., Milano, L., Pinto, I., Recano, F.: 1988a *Astron. Astroph.* **199**, 161

Barone, F., Milano, L., Pinto, I., Russo, G.: 1988b *Astron. Astroph.* **203**, 322

Bath, G.T.: 1972, *Astrophys. J.* **162**, 621

Batten, A.H., Fletcher, J.M., Mann, P.J.: 1978 *Pubbl. of Dominion Obs.* **15**, 5

Beskin, V.S., Gurevich, A.V., Istomin, Ya.,N.: 1984, *Astrophys. and Space Science* **102**, 301

Billing, H., Maischberger, K., Rüdiger, A., Schilling, R., Schnupp, L., Winkler, W.: 1979 *J. Phys. E: Sci Instrun* **12**,

Billing, H., Winkler, W., Rüdiger, A., Schilling, R., Maischberger, K., Schnupp, L.: 1983, *The Munich gravitational wave detector using laser interferometry* from Quantum Optic, Experimental Gravity and Measurement Theory, ed. P.Meyestre and M.O. Schully (Plenum Publ. Cort.)

Boriakoff, V. Buccheri, R., Fauci, F.: 1983, *Nature* **304**, 417

Born, Wolf, E.: 1964, Principles of Optics, Pergamon Press, Oxford, 1964

Bradaschia, C., Del Fabbro, R., Di Virgilio, A., Giazotto, A., Kautsky, H., Montelatici, V., Passuello, D., Brillet, A., Cregut, O., Hello, P., Man, C.N., Manh, P.T., Marraud, A., Shoemaker, D., Vinet, J.Y., Barone, F., Di Fiore, L., Milano, L., Russo, G., Solimeno, S., Aguirregabiria, J.M., Bel, H., Duruisseau, J.P., Le Denmat, G., Tourrenc, Ph., Capozzi, M., Longo, M., Lops, M., Pinto, I., Rotoli, G., Damour, T., Bonazzola, S., Marck, J.A., Gourghoulon, Y., Fuligni, F., Iafolla, V., Natale, G., Holloway, L.E.: *"The Virgo Project"*, Proposal for the italo-french Interferometer for Detection of Gravitational Waves, Italy, June 1989

Braginsky, V.B., Rudenko, V.N.: 1978, *Phys. Reports* **46**, N^o5, 165

Braginsky, V.B., Thorne, K.S.: 1985, *Nature* **316**, 610

Brancewitz, H.K., Dworak, T.Z.: 1980, *Acta Astron.* **30**, 501

Brillet, A.: 1984, *The interferometric detection of gravitational waves*, Lecture Notes in Phys. **212**, Gravitation, Geometry and Relativistic Physics, Proc. Aussois, France

Brillet, A.: 1985, *Ann. Phys. Fr.* **10**, 219

Chau, W.Y.: 1970, *Astrophys. J.* **147**, 664

Chevalier, R.A., Emmering, R.T.: 1986, *Astrophys. J.* **304**, 140

Clifton, T.R., Lyne, A.G.: 1986, *Nature* **320**, 43

Cooperstock, F., Lim, P.H.: 1986, *Can. J. Phys.* **55**, 265

Damashek, M. et al.: 1982, *Astroph. J.* **253**, L57

Damour, T.: 1983, *Phys. Rev. Lett.* **51**, 1019

Damour, T., Deruelle, N.: 1986, *Ann.Inst. H. Poincaré (Phys. Teor.)* **44**, 263

Del Fabbro R., Di Virgilio A., Giazotto A., Kautzky H., Montelatici V., Passuello D., Stefanini A., Barone F., Bruzzese R., Cutolo A., Longo M., Milano L., Pinto I., Solimeno S., Bordoni F., Fuligni F., Iafolla V., Brillet A., Nary Man C., Shoemaker D., Tourrenc P., Vinet J.Y.: 1987, Antenna interferometrica a grande base per la ricerca di Onde Gravitazionali, Proposal, INFN, PI/AE 87/1, Pisa (Italy)

de Kool, M., Van Paradijs, J.: 1987, *Astron. Astrophys.* **173**, 279

Deruelle, N., Tourrenc, P: 1984, in *Gravitation, Geometry and Relativistic Physics* **212**, Proceedings, Aussois, France

Douglass, D.H., Braginsky, D.V.: 1979, in *Einstein Centenary Volume*, S.W. Hawking and W. Israel, ed.s, Cambridge Univ. Press, p 90

Drever, R.W.P., Ford, G.M., Hough, J., Kerr, I.M., Munley, A.J., Pugh, J.R., Robertson, N.A., Ward, H.: *Proc. of the Ninth International Conference on General Relativity and Gravitation*, Jena, ed. E.Schmutzer, Deutscher Verlag der Wissenschften, Berlin

Drever, R.W.P., Hough, J., Munley, A.J., Lee, S.A., Spero, R., Whitcomb, S.E., Ward, H., Ford, G.M., Hereld, M., Robertson, N.A., Kerr, I., Pugh, J.R., Newton, G.P., Meers, B., Brooks III E.D., Gursel, Y.: 1981, *Optical cavity laser interferometers for gravitational wave detection* in Laser Spectroscopy V. ed. A.R.W. McKellar, T.Oka, B.P. Stoicheff (Springer, Berlin, Heidelberg, New York)

Drever, R.W.P.: 1981, in "Gravitational wave detectors using laser interferometers and optical cavities - ideas, principles and prospects" invited lecture given at Nato Advanced Study Institute on Quantum Optics and general relativity, Bad Windsheim, Germany, August

Drever, R.W.P.: 1982 *Interferometric detectors for gravitational radiation* for "Gravitational Radiation" ed. by T.Piran and N.Deruelle, Proceedings of the Les Houches Summer Institute, June

Drever, R.W.P., Hall, J.L., Kowalsky, F.V., Hough, J., Ford, G.M., Munley, A.J., Ward, H.: 1983, *Appl. Phys.* **B31**, 97

Drever, R. et al.:1986, *Abst. cont. papers at GR-11*, paper 18:13

Eastbrook, F.B., Walquist, H.D.: 1975 *GRG* **5**, 439

Eastbrook, F.B., Walquist, H.D.: 1977, *Acta Astron.* **5**, 5

Ehlers, J. et al.:1976, *Ann. N.Y. Acad. Sci.* **208**, L77

Faller, J.A., Bender, P.L.: 1984 in *"Precision Measurements and Fundamental Constants"*, NBC Spec. Publ. **617**,689

Galtsov et al., Phys. Lett. 77a:387 (1980).

Gertsenshtein, M.E., Pustovoit, V.I., 1963, *Sov. Phys. JETP* **16** *N°*2, 433

Giazotto, A.: 1988, *Phys. Lett. A* **128**, 241

Ghosh, P., Lamb, F.K.: 1979, *Astrophys. J.* **234**, 296

Groth, L.: 1985, in *Neutron Star, Black Holes and Binary X-ray Soruces*, H.Gursky, R.Ruffini eds., Reidel

Hellings, R.W. et al.: 1981, *Phys. Rev.* **D23**, 844

Hellstrom, C.W.: 1978, in *"Statistical Theory Of Signal Detection"*, Pergamon Press, Oxford, ch. 7

Henrichs H.F.: 1983, in "Accretion Driven Stellar X-Ray Sources", W.H.G. Lewin and E.P.J. van den Heuvel eds., Cambridge Univ. Press

Hernandez, G.: 1986, Fabry-Perot Interferometers, Cambridge University Press, Cambridge.

Herriot, D, Kogelnik, H., Kompfner, R.: 1964, *Applied Optics* **3**

Hils, D. et al., paper 15:13 in Abstracts of Contributed Papers at GR-11, Stockholm (1986).

Hough, J., Meers, B.J., Newton, G.P., Robertson, N.A., Ward, H., Schutz, B.F., Drever, R.W.P., Mason, R., Pollard, C., Tolcher, R., Bellenger, D.W., Bernnett, J.R.J., Corbett, I.F., Percival, M.D.: "A British long baseline gravitational wave observatory", GWD/RAL/86-001.

Hough, J.: "Report on Round Table Discussion-Gravitational wave Detectors", to be published in Proceedings of NATO Advanced Research Workshop on Gravitational Wave Data Analysis, Cardiff, July 1987.

Hough, J., Meers, B.J., Newton, G.P., Robertson, N.A., Ward, H., Schutz, B.F., Corbett, I.F., Drever, R.W.P.: 1987, *Vistas in Astronomy* **Vol 30**, 109

Ipser R.: 1970, *Astrophys. J.*, **166**, 175

Kates, R.E.: 1980 *Phys. Rev.* **D22**, 1871

King, A., Watson, M.: 1986, *Nature* **323**, 105

Kruszewsky, A.: 1966, *Adv. Astron. Astroph.* **233**, 1966

Livas, J.: 1986 *Poster-Paper 03.09*, GR-11

Lyne A.G., Manchester R.N., Taylor J.H.: 1985, *Mon. Not. Roy. Astron. Soc.*, **213**, 613

Man, C.N.: "G.W. Data Analysis Proceeding, Cardiff, July, 1987

Man. C.N., Shoemaker, D., Brillet, A.: 1988, "External Modulation and recyclings", European Collaboration Meeting on Interferometric Detection of GW, Sorrento, September 1988.

Manchester R.N., Taylor J.H.: 1977, *Astrophys. J.* **215**, 885

Manchester R.N., Taylor J.H.: 1981, *Astron. J.* **86**, 1953

Manchester, R.N., Newton, L.M., Cooke, D.T., Backus, P.R., Damashek, M., Taylor, J.H., Condon, J.J.: 1983, *Astrophys. J.* **268**, 832

Melosh H.J.: 1969, *Nature* **224**, 781

Meers, B.J.: 1983, *Optics Comm.* **47**, 237

Meers, B.J.: 1988, *Phys. Rev. D* **D38**, 2317

Meers, Man Brillet, 1989

Ostriker J.P., Gunn J.E.: 1969, *Astrophys. J.* **157**, 1935

Pandharipande V.R., Pines D., Smith R.A.: 1976, *Astrophys. J.* **208**, 550

Patterson, J.: 1984, *The Astrophys. Journal Suppl.* **54**, 443

Peng Q.-y., Huang K.-l., Huang J.-h.: 1982, *Astron. Astroph.* **157**, 258

Peters, P.C., Mathews, J.: 1963, *Phys. Rev.* **131**, 435

Peters, P.C.: 1964, *Phys. Rev.* **136**, B1224

Pines D., Shaham J.: 1972, *Nature Phys. Sci.* **235**, 43

Pines D., Shaham J.: 1974a, *Nature* **248**, 483

Pines D., Shaham J.: 1974b, *J. Comm. Astrophys. Space Phys.* **2**, 37

Prentice A.F.R., ter Haar D.: 1969, *Mon. Not. Roy. Astron. Soc.* **146**, 423

Press, W.L., Thorne, K.: 1972, *Ann. Rev. Astron. Astrophys.* **10**, 335

Proszynsky, M., Przybycien, D.: 1984, in *Proc. Workshop on Millisecond Pulsars*, NRAO, Greenbank, USA

Rappaport, S., Joss, P.: 1983, *Astrophys. J.* **270**, L73

Rappaport, S., Joss, P.: 1984, *Astrophys. J.* **283**, 232

Rawley, L.A. et al.: 1986, *Nature* **319**, 383

Ritter, H.: 1984, "Catalogue of Cataclysmic Binaries"

Rudiger, A., Schilling, R., Schnupp, L., Winkler, W., Billing, H., Maischberger, K.: 1981, *Optica Acta* **28**, 641

Rudiger, A., Schilling, R., Schnupp, L., Shoemaker, D., Winkler, W., Leuchs, G., Maischberger, K.: "The Garching 30-meter prototype and plans for a large gravitational wave detector", Proc 18th Texas Symposium on Relativistic Astrophysics, Chicago, 1986, ed. Melville P.Ulmer, World Scientific, 1987

Segelstein, D.J., Rawley, L.A., Stinebring, D.R., Fruchter, A.S., Taylor, J.H.: 1986, *Nature* **322**, 714

Shoemaker, D., Winkler, W., Maischberger, K., Rudiger, A., Schilling, R., Schnupp, L.: "Progress with Garching 30-meter prototype for a gravitational wave detector", MPQ 100, 1985

Shoemaker, D., Brillet, A., Man, C.N., Cregut, O., Kerr, G.: 1989, *Optics Letters* (submitted)

Spero, R.: "The Caltech Laser-interferometer gravitational wave detector", Proceedings at the Fourth Marcel Grossmann Meeting on the Recent Developments of General Relativity, R.Ruffini (ed.), Elsevier Svi. Publ. B.V., 1986

Stokes, G.H., et al.: 1985, *Nature*

Stollmann, G.M.: 1987, *Astron. Astrophys.* **178**, 143

Taylor, J.H., Fowler, L.A., Mc Culloch, P.M.: 1979, *Nature* **277**, 437

Taylor, J.H., Weisberg, J.M.: 1982, *Astrophys. J.* **253**, 908

Taylor, J.H., Stinebring, D.R.: 1986, *Ann. Rev. Astron. Astrophys.* **24**, 285

Thorne, K.S.: 1969, *Astrophys. J.* **158**, 1

Thorne, K.S.: 1969, *Astrophys. J.* **158**, 997

Thorne, K.S.: 1983, *Gravitational radiation*, eds N.Deruelle and T. Piran, North-Holland, Amsterdam, p.1

Thorne, K.S.: 1987, *300 years of Gravitation*, eds. S.W. Hawking and W. Israel, Cambridge University Press, Cambridge, 1987

Tourrenc, P., Deruelle, N.: 1985, *Ann. Phys. Fr.* **10**, 241

Tyson, J.A., Giffard, R.P.: 1978, *Ann. Rev. Astron. Astroph.* **17**

van den Heuvel, E.P.J.: 1984a, *J. Astrophys. Astron.* **5**, 209

van den Heuvel, E.P.J.: 1984b, in *Proc. Workshop on Millisecond Pulsars*, NRAO, Greenbank, USA

Vinet, J.Y., Meers, B.J., Man, C.N., Brillet, A.: 1988, *Phys. Rev. D* **38**,

Wagoner, R., Will, C.M.: 1976, *Astrophys. J.* **210**, 764

Wagoner, R.: 1984, *Astrophys. J.* **278**, 345

Ward, H., Hough, J., Kerr, G.A., Mackenzie, N.L., Mangan, J.B., Meers, B.J., Newton, G.P., Robwertson, D.I., Robertson, N.A.: "The Glasgow Gravitational wave detector - Present progress and future plans", Proceedings of the INternational Symposium on Experimental Gravitational Physics, Guangzhou, China, August, 1987

Weber, J.: 1961, "General Relativity and Gravitational Waves"

Weiss, R.: 1972, "Quarterly Progress report of the Research Laboratory of Electronics of the MIT" **105**, 54

Zimmermann, M.: 1978, *Nature* **271**, 524

Neutron Star Masses

Tibor J. Herczeg
(University of Oklahoma)

Abstract and table of contents

A review of the literature is given to summarize the present evidence for masses of neutron stars in radio pulsars as well as in X-ray binaries. The material is arranged in the following sections:

Key words: binary stars -- apsidal rotation -- light time effect -- mass exchange and mass loss -- radial velocities -- general relativity, experimental tests of -- compact objects -- black holes -- neutron stars -- pulsars -- white dwarfs -- X-ray binaries.

Introduction

The following review of the neutron star masses is presented strictly from the observational point of view, trying to sum up the best evidence presently available while discussing binary radio pulsars as well as X-ray binaries. The importance of the neutron star masses for the theoretical understanding of highly condensed matter hardly needs to be further elucidated -- only one representative problem might be mentioned here, the connection between the maximum mass of a neutron star and various equations of state, at or above nuclear density. Furthermore, the problem of neutron star masses is quite inseparable from that of the evolution of these systems.

There is only one case known where the determination of neutron star masses can be termed a measurement as a physicist would understand it: that of the first binary pulsar PSR1913+16. In other cases we encounter estimates, at best, sometimes with unusually large uncertainty attached, sometimes hardly more than an educated guess.

There are important earlier surveys of this field of which the present discussion intends to be a sort of updating, however difficult may prove to come up to the standards set by these studies. In the first place we have to

C. İbanoğlu (ed.), Active Close Binaries, 693–727.

mention the review by J. Bahcall (1978), a rigorous treatment of the observational evidence for the best observed X-ray binaries, searching for "ultimate" lower and upper limits of the compact star masses. The results were presented in the form of widely opened intervals such as $0.5\,\mathcal{M}_o \leq \mathcal{M}_x \leq 1.8\,\mathcal{M}_o$ for SMC X-1 or $0.4\,\mathcal{M}_o \leq \mathcal{M}_x \leq 2.2\,\mathcal{M}_o$ for Her X-1. These wide bounds illustrate in a convincing form the difficulties and problems of obtaining truly reliable neutron star masses, but they are only of limited help for the physicist who would like to get "plausible" values of the masses. About the same time and also a little later P. Joss and S. Rappaport took up the problem, in several comprehensive review papers; see Rappaport and Joss, 1977 and 1983, Joss and Rappaport, 1984. The wide brackets have been only slightly reduced but most probable values of the pulsar masses have been added, derived from Monte Carlo analysis of error propagation. Their results are illustrated in Fig. 1 below, and can be summarized by saying that the "error bars" of the neutron star masses seem to overlap in the region $1.2 - 1.6\,\mathcal{M}_o$, with the only well defined binary pulsar mass right in the middle, at $1.4\,\mathcal{M}_o$, a value which happens to coincide with Chandrasekhar's limit for white dwarf masses. There was an early tendency to consider $1.4\,\mathcal{M}_o$ the "standard" neutron star mass. Recent evidence suggests that there may be deviations from this still important "benchmark" value; even phrases such as a canonical range of neutron star masses between $1.2\,\mathcal{M}_o$ and 1.8 or $2.00\,\mathcal{M}_o$, can be read occasionally. In this review we should like to pay particular attention to this question, especially to the "undermassive" objects; we hope to return to the possibly massive neutron stars in a follow-up paper.

Our aim is not so much to emphasize the still existing very large uncertainties but to identify likely values or at least narrower intervals for neutron star masses, compatible with plausible assumptions, spelled out carefully.

Methods of analysis

The main source of this topic of methodics, bringing a great wealth of information under the subtitle "Elements of the Analysis," is J. Bahcall's critical review (1978); it gave somewhat more space to the X-ray binaries, as PSR 1913+16 was for almost a decade the only known binary pulsar. Joss and Rappaport gave also short introductions to the methodics in their surveys; in particular their 1984 review article is comprehensive overview over the

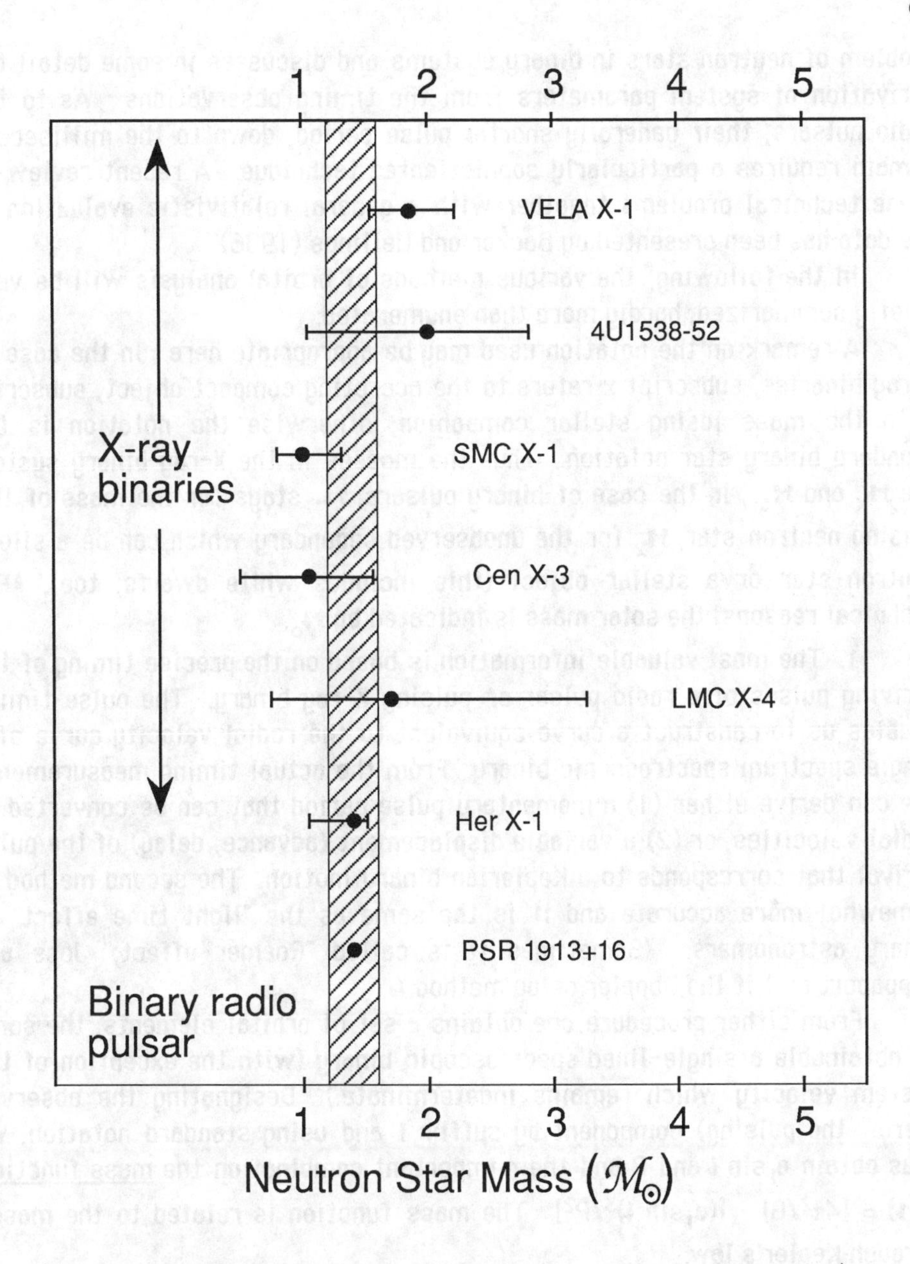

Fig. 1 Neutron star masses (after P. Joss and S. Rappaport) The System Vela X-1 is usually identified as 4U 0900-40.

problem of neutron stars in binary systems and discusses in some detail the derivation of system parameters from the timing observations. As to the radio pulsars, their generally shorter pulse period, down to the millisecond domain requires a particularly sophisticated technique. A recent review of these technical problems together with a general relativistic evaluation of the data has been presented by Backer and Hellings (1986).

In the following, the various methods of orbital analysis will be very briefly summarized, hardly more than enumerated.

A remark on the notation used may be appropriate here. In the case of X-ray binaries, subscript x refers to the accreting compact object, subscript s to the mass losing stellar companion; otherwise the notation is the standard binary star notation. Thus the masses in the X-ray binary system are $\mathcal{M}_x$ and $\mathcal{M}_s$. In the case of binary pulsars, $\mathcal{M}_1$ stays for the mass of the pulsing neutron star, $\mathcal{M}_2$ for the unobserved secondary which can be a silent neutron star or a stellar object (this includes white dwarfs, too). For technical reasons, the solar mass is indicated by $\mathcal{M}_o$.

1. The most valuable information is based on the precise timing of the arriving pulses of a radio pulsar or pulsing X-ray binary. The pulse timing enables us to construct a curve equivalent to the radial velocity curve of a single spectrum spectroscopic binary. From the actual timing measurements one can derive either (1) a momentary pulse period that can be converted to radial velocities, or (2) a variable displacement (advance, delay) of the pulse arrival that corresponds to a Keplerian binary motion. The second method is somewhat more accurate and it is the same as the "light time effect" of binary astronomers. (Sometimes it is called "Roemer effect;" Joss and Rappaport call it the Doppler delay method.)

From either procedure one obtains a set of orbital elements, the same as obtainable a single-lined spectroscopic binary (with the exception of the system velocity which remains indeterminate.) Designating the observed (here: the pulsing) component by suffix 1 and using standard notation, we thus obtain $a_1 \sin i$ and P and their important combination the mass function $f(\mathcal{M}) = [4\pi^2/G] \cdot [(a_1 \sin i)^3/P^2]$. The mass function is related to the masse through Kepler's law:

$$f(\mathcal{M}) = \mathcal{M}_2{}^3 \sin^3 i/(\mathcal{M}_1 + \mathcal{M}_2)^2 = \mathcal{M}_1 \sin^3 i/q(q+1)^2,$$ with $q = \mathcal{M}_1/\mathcal{M}_2$ the mass ratio.

This important relationship combines three unknown quantities, $\mathcal{M}_1$, $\mathcal{M}_2$

and the inclination angle i. The second form, with the mass ratio, has the advantage that the mass of the observed component can be at least estimated. Mass functions obtained through pulse timing are of very great accuracy and usually contribute but a negligible error to the treatment of the binary.

2. In the case of several X-ray binaries, particularly of those systems having an OB-type, massive and luminous stellar companion, an additional orbit, the spectroscopic one, can be derived. This orbit, in principle, would define the mass ratio by comparison of the amplitudes obtained from spectroscopic and pulse timing observations. Line distortion affects resulting from X-ray heating and possibly from presence of interstellar matter, in connection with the relatively low spectral resolution, make it difficult to obtain good radial velocities. In spite of the not too large formal errors claimed, 10-15%, the mass ratios derived the optical spectra should be considered useful estimates rather than measured quantities: in a typical case of a massive OB secondary, an error in K of 2-3 km/s may easily result in 0.1 $\mathcal{M}_o$, 0.15 $\mathcal{M}_o$, uncertainty in the mass of the neutron star.

3. A value of the mass ratio can also be obtained by accurate photometry of the "ellipsoidal" light variations show by the tidally distorted stellar component. The photometric effect is small, usually under 0.1 magn. and the X-ray heating certainly complicates the analysis of the light curve. The photometrically derived mass ratios are at best as reliable as the spectroscopic values. An example: from the ellipsoidal variations <u>and from the duration</u> of the X-ray eclipse, Avni and Bahcall predicted $15 < v_s \sin i < 25$ km/s (1974). The spectroscopic result turned out to be $v_s \sin i = 26 \pm 4$ km/s.

4. For point like masses (for a pair of pulsars) general relativity predicts, in contradiction to the classical two-body problem, that the apsidal line (major axis) of the orbit will not remain constant in the plane of orbit, but it will rotate at a (nearly constant) speed which depends on the sum of the two masses. The effect, first discussed by the mathematician T. Levi-Civita and almost simultaneously by Robertson (1938), is a generalization of Einstein's test of relativity in the case Sun + test particle (= Mercury). The relavistic periastron advance can be written in the form $\omega \propto (\mathcal{M}_1 + \mathcal{M}_2)^{2/3}$ with ω as the orbital element defining the periastron point, that is the position the apsidal line. Concerning the accuracy of pulsar orbits, it can be expected that for eccentric orbits, this particular relativistic effect will show comparatively early. It has been already observed in three systems.

If the secondary is an extended object, the classical advance of the periastron will be additionally observed. This is a perturbation of the motion due to the tidal (and to a lesser extent rotational) deformation of the components. The form of the dependence on the orbital elements and the masses was worked out by T. Cowling and by Th. Sterne in the late 1930's and is treated in several textbooks, see for instance Batten 1973. It is possible that even a low mass white dwarf (with larger radius) might be considered an extended object. In X-ray binaries, particularly in the massive systems, the classical effect would be strong and thus render the observed $\dot{\omega}$ unsuitable for mass determination.

General relativity imposes further constraints on the binary masses which can be used for mass determination. One of them is expressed in the form of a parameter γ and giving the amplitude of delays in the pulse arrival times, due to variations of the gravitational redshift and transverse Doppler effect as the pulsar moves in its orbit. The dependence of γ on the masses has the form $\gamma \propto \mathcal{M}_2 (\mathcal{M}_1 + 2\mathcal{M}_2)(\mathcal{M}_1 + \mathcal{M}_2)^{-4/3}$. This effect has been observed, as yet, only in the case of PSR 1913+16. (See, however, the note added on the last page of the list of references.)

For further literature, we may mention Weisberg and Taylor 1980, Taylor and Weisberg 1982, Taylor 1986, the review article by Will (1987), the textbook by Shapiro and Teukolsky (1982) and further references therein. The γ-constraint is discussed by Blandford and Teukolsky (1975, 1976), see also Haugan (1985).

5. For the X-ray binaries there is another possible constraint obtained by observing, if it occurs, the eclipse of the X-ray emitter behind the stellar companion. From the duration of the eclipse, a relationship including the inclination and the mass ratio can be derived; the duration D is usually expressed by the "eclipse half-angle" $\phi_{1/2} = (D/P)\cdot 180°$. If the orbit is circular and the occulting star spherical (undistorted), the simple geometric relation follows: $R^2 = a^2(\cos^2 i + \sin^2 i \cdot \sin^2 \phi_{1/2})$, a = orbital radius, R = stellar radius. A small eccentricity of the orbit, a rare occurrance among X-ray binaries, can be easily treated but it involves additional orbital elements. This simple equation gives good approximation even for tidally or rotationally slightly distorted stellar companions but in such cases, due to the sensitivity of the results to small changes in D, corrections have to be applied. A particularly important case is when the star fills or nearly fills its Roche

lobe. In this case the geometry of the system will depend on the mass ratio and a very useful table by Chanan *et al.* (1976) gives us the three-cornered relationship among the mass ratio q, i and $\phi_{1/2}$. Some X-ray binaries, such as Her X-1 with virtually no wind-driven component of the accretion, might exactly fill their respective Roche lobe and even the OB type secondaries in the massive systems are coming close to it. Then a correction factor usually designated by ß has to be applied to the R/a ratio following from the Roche geometry.

The X-ray eclipses, observed to occur for most of the massive X-ray binaries, are somewhat less helpful than expected earlier. They can show errors reaching 1° or more and in some cases they appear slightly variable with time. It is usually assumed that the eclipses occur behind a sharp photospheric boundary but there is evidence that this is not always the case, see the study of Cen X-3 by Clark *et al.* (1988).

6. The Roche configuration itself, even in cases it is only approximately realized, can mean a useful tool for us, particularly when we have to choose between conflicting models or reject a physically improbable one. Relative sizes of the components' Roche lobes (just like the position of the inner Lagrangian point), depend only on the mass ratio. There are several interpolating formulae for obtaining the "equivalent radius" of the Roche lobe of a component in a binary; in this paper it is obtained by interpolating in Z. Kopal's original table (1959). As an example we should like to mention the black hole candidate in the transient source A0620-00 where even a crude estimate of the Roche lobe radius seems to rule out masses for the compact component around 2 $\mathcal{M}_o$ (in favor of much larger masses, 4-5 $\mathcal{M}_o$).

We should like to mention that in all cases we are going to discuss, <u>synchronous rotation</u> of the secondary is <u>assumed</u>; in a Roche lobe filling configuration this condition is almost certainly fulfilled.

Finally we mention what can be called "<u>occasional treatments</u>," individual cases where a special consideration lends itself to make the solution easier. One should like to mention here attempts at detecting lateral orbital motion of pulsars by the aid of interstellar scintillation (A. Lyne), stability considerations in case of extremely short period (submillisecond) pulsars or, an observational <u>tour de force</u>, the observation of optical pulses interpreted as reprocessed X-ray pulses in the companions atmosphere, carried out by J. Middleditch, J. Nelson and others (Her X-1, 4U1626-67).

Binary and millisecond pulsars

For several years PSR 1913+16 was the only known pulsar in a binary system; the secondary was supposed to be another neutron star, not pulsing. Fortunate combination of orbital elements, long series of accurate observations and thorough theoretical studies made this 59 ms pulsar a crucial object in "experimental gravitation."

At present we know 6 binary pulsars of longer than a few ms period (>11 ms) and 8 millisecond pulsars, 6 of them in binary systems. Table 1 gives a summary of our present knowledge of these pulsars; for a clearer overview, systems with pulse periods over 11 ms are identified simply as "radio pulsars," meaning non-millisecond radio pulsars. The tabulated data are from recent reviews by Taylor, see for example Taylor 1987a, 1987b. References to newly discovered systems, PSP 0021-72 A and B, PSR 1820-11, are mentioned in the discussions later.

The extreme short period pulsars are almost certainly not very young objects -- according to the now generally accepted theory they are "resurrected" or "recycled" pulsars. This should be understood in the way, that these systems have been (some even vestigially are) in a phase of accreting matter from their respective companions. Thus they were temporarily accretion-driven systems, similar to X-ray binaries; the very short periods are the result of the spin-up of the accreting neutron stars, gaining material of higher specific angular inomentum. A review of the formation and evolution of ms-pulsars can be found in van den Heuvel, 1987.

The first ms-pulsar discovered, PSR 1937+21, proved a single object. The pathway of evolution through a phase of accretion in a binary system has been, however, verified in an almost dramatic manner by the later discovery of PSR 1957+20, a close system consisting of a ms-pulsar and an extremely low mass stellar companion, $\mathcal{M} \approx 0.02, 0.025\ \mathcal{M}_{o}$. There is a spectacular, comet-like nebula present, generated obviously through the interaction of the fast moving system (about 100 km/s) with the interstellar matter. The destruction of the companion, perhaps through intense, high energy particle radiation and tidal effects, becomes visible. By implication, the single millisecond pulsars are those which have entirely obliterated their respective companions; the coalescence of the two stars is a possibility also to be considered.

The stunning picture of the system PSR 1957+20 suggests that the process of the spin-up is not mass conserving. Nevertheless, the pulsar may

TABLE 1a. Binary Pulsars

PSR	P_{pulse} (ms)	P_{orb} (days)	e	f(m) ($m_\odot$)	Remarks
0655+64	196	1.03	<0.00005	0.0712	Secondary: cool wd
0820+02	865	1232.4	0.0119	0.0030	Secondary: hot wd?
1820-11	280	357.76	0.79462	0.0679	
1831-00	521	1.81	<0.005	0.00012	
1913+16	59	0.32	0.6171	0.1322	First bin. pulsar disc.
2303+46	1066	12.34	0.6584	0.2463	

TABLE 1b. Millisecond Pulsars

PSR	P_{pulse} (ms)	P_{orb} (days)	e	f(m) ($m_\odot$)	Remarks
1937+21	1.56	—	single	—	
1957+20	1.61	0.382	<0.001	5.2×10^{-6}	19th magn. star
1821-24	3.05	—	single	—	In M28
0021-72A	4.48	0.021	0.33	1.61×10^{-8}	In 47 Tuc
1855+09	5.4	12.33	0.00002	0.052	23rd magn. star?
1953+27	6.1	117.35	0.0003	0.027	
0021-72B	6.13	9 to 97	—	—	In 47 Tuc
1620-26	11.08	191.4	0.0253	0.0080	In M4

gain 0.2 M_o, 0.3 M_o in this phase of its evolution. Higher neutron star masses in these systems are quite possible and the proof of their existence desirable.

As to the "technical" aspects of mass determination, it is to be said that the observed "light time" curves, the Doppler delay curves, are (with the possible exception of PSP 1831-00) of very high accuracy, and the mass functions obtained from them very reliable, their uncertainty usually <1%. Spectroscopic determinations can seldom compete with this accuracy. Nevertheless, without the mass function being supplemented by other information, such as general relativistic constraints, the masses in the binary pulsars remain indeterminate.

Relativistic systems.

Among binary pulsars, we call "relativistic" systems those where the observed mass function can be supplemented by constraints following from the general relativity theory. Relativistic systems are still in the minority, mainly because the most noticeable effect, the relativistic apsidal motion, requires a markedly eccentric orbit while most binary pulsars -- and X-ray binaries, for that matter -- have circular or nearly circular orbits. The systems PSR1913+16, PSR2303+46, and PSR0021-72A show measurable rotation of the apsidal line and the long period system PSR1820-11 is expected to do it in the near future. PSR1913+16 is the only binary pulsar for the time being that shows all the orbital relativistic effects and thus became a most important test object for the relativity theory.

PSR1913+16. This is the first known binary pulsar discovered just about 15 years ago, see Hulse and Taylor, 1975. The relativistic apsidal motion, at a rate of $\dot{\omega} = 4\overset{\circ}{.}22\ yr^{-1}$, was found almost immediately and it leads to the constraint $M_1 + M_2 = 2.83\ M_o$ (Taylor *et al.* 1976). Thus the minimum separation of the components turns out to be 1.07 R_o (R_o = solar radius) and it is easy to show that the companion must be another compact object, judged from the masses, a neutron star, making the periastron advance a purely relativistic effect (see, for instance, Herczeg 1976). Several other relativistic constraints have been worked out by J. Taylor and his collaborators; the most spectacular among them is the "degradation" of the orbit (secular shortening of orbital period) due to loss of energy through gravitational radiation, where results from accurate pulse timing correspond exactly to the predictions of general relativity. There is virtually nothing to add to the exhaustive treatment of this star, it suffices to mention that the coefficient of the time dilation (γ) has also been found; this is still a unique case among binary pulsars. These three pieces of information, $f(\mathcal{M})$, $\dot{\omega}$ and γ are sufficient to solve the problem of the masses. J. Taylor gives the results of the most recent analysis (1987) as

$$\mathcal{M}_1 = 1.451 \pm 0.007\, \mathcal{M}_o, \quad \mathcal{M}_2 = 1.378 \pm 0.007\, \mathcal{M}_o$$

The corresponding value of the orbital inclination is $i = 46^\circ 9$. Two slightly weaker constraints, referring to sin i and the rate of change of the orbital period, verify that both masses have to be close to $1.4\,\mathcal{M}_o$; see the instructive diagram on p. 119 in Will 1987.

However accurate this mass determination is, it can only form the experimental foundation for models of neutron stars and their formation, if the mass of $1.4\,\mathcal{M}_o$ is characteristic for most or perhaps all neutron stars and can be made a "standard" value. As mentioned earlier, this is a question we wish to look at in this review.

PSR2303+46. The orbital eccentricity is 0.65838 and this pulsar is the second one where the relativistic apsidal advance has been observed. The first result, based on less than two months of observations, had $\dot{\omega} = 0^\circ 01092 \pm 0^\circ 0023\ \mathrm{yr}^{-1}$, leading to the preliminary value $\mathcal{M}_1 + \mathcal{M}_2 = 3.0 \pm 0.9\, \mathcal{M}_o$ (Stokes *et al*, 1985). Extension of the observations to more than two years resulted in the modified but obviously still not definitive value of $\dot{\omega} = 0^\circ 0092 \pm 0^\circ 0018\ \mathrm{yr}^{-1}$ (Taylor and Dewey, 1988). Since the sum of the masses depends on the $^3/_2$-power of $\dot{\omega}$, this relatively small change leads to $\mathcal{M}_1 + \mathcal{M}_2 = 2.3 \pm 0.6\, \mathcal{M}_o$, a value which is somewhat unexpectedly low. The system may contain a neutron star with the smallest known mass.

Although the last word is obviously not yet spoken in the case of $\dot{\omega}$ for PSR2303+46, taking for the time being the sum $\mathcal{M}_1 + \mathcal{M}_2 = 2.3\, \mathcal{M}_o$ for granted, we may obtain an interesting insight into the nature of the secondary component. Given the sum of masses, the mass function can be transcribed to an inequality defining a minimum mass for the secondary and consequently a maximum mass for the pulsar. Writing $(\mathcal{M}_2 \sin i)^3 = f(\mathcal{M}) \cdot (\mathcal{M}_1 + \mathcal{M}_2)^2 = X$, we at once obtain $\mathcal{M}_2 \geq X^{1/3}$. Since for PSR2303+46, $f(\mathcal{M}) = 0.2463\, \mathcal{M}_o$, we have

$$\mathcal{M}_2 \geq 1.09\, \mathcal{M}_2 \text{ and also}$$

$$\mathcal{M}_1 \leq 1.21\, \mathcal{M}_o.$$

To each value of the inclination angle, there is just one possible distribution of the masses between the components. Thus if

$$i = 90°, \mathcal{M}_1 = 1.21\,\mathcal{M}_o, \mathcal{M}_2 = 1.09\,\mathcal{M}_o,$$
$$i = 71°76, \mathcal{M}_1 = \mathcal{M}_2 = 1.15\,\mathcal{M}_o,$$
$$i = 60°, \mathcal{M}_1 = 1.04\,\mathcal{M}_o, \mathcal{M}_2 = 1.26\,\mathcal{M}_o.$$

If there is no preferential orientation in the orbital planes, the expectation that $\mathcal{M}_2 > \mathcal{M}_1$ is about 69%. The assumption that both components are low mass neutron stars appears justified. Only if the actual value of $\dot{\omega}$ is close to the upper bracket of the indicated 1 σ error, could we obtain pulsar masses around 1.4 $\mathcal{M}_o$.

Taylor and Dewey point out that the secondary can be, in principle, a solar type main sequence star yet they consider this possibility to be safely excluded since S. R. Kulkarni and R. J. Rand did not find any object brighter than the 26th magnitude at the pulsar's position, very well defined by radioastronomical observations.

PSR 0021-72A. This is an ms-pulsar discovered in the globular cluster 47 Tucanae by Ables *et al.* (1988). Its observed orbital elements are rather out of the ordinary:

$a_1 \sin i = 585$ km, P = 1824.3 s, $f(\mathcal{M}) = 1.61 \times 10^{-8} \mathcal{M}_o$ and, notably, e = 0.33. The extremely small mass function indicates that either the inclination is near zero degree or the mass-ratio << 1. Most importantly, a very rapid apsidal motion was found: $\dot{\omega} = 0°6$ day^{-1} (Ables, quoted in Wijers, 1989).

The observed apsidal motion leads to $\mathcal{M}_1 + \mathcal{M}_2 = 2.28\,\mathcal{M}_o$ if the rate of apsidal rotation can be identified with the purely relativistic value (without taking the second decimal digit too seriously). The interesting possibility arises that the pulsar mass is $\geq 2\,\mathcal{M}_o$. Using the two constraints: $f(\mathcal{M})$ and $\dot{\omega}_{rel}$ we can represent the solutions as function of a single parameter; in Fig. 2 this parameter is the mass ratio $q = \mathcal{M}_1/\mathcal{M}_2$. (In the figure $\mathcal{M}_x/\mathcal{M}_2$.) The corresponding values of i are also indicated and the representation suggests that for all inclinations $i \geq 3°$, the pulsar mass remains over 2 $\mathcal{M}_o$. The chance for i being in this range is to be expected higher than 99%.

Nevertheless, such an inference would not be justifiable; the inclination can be shown to be less than 1°.

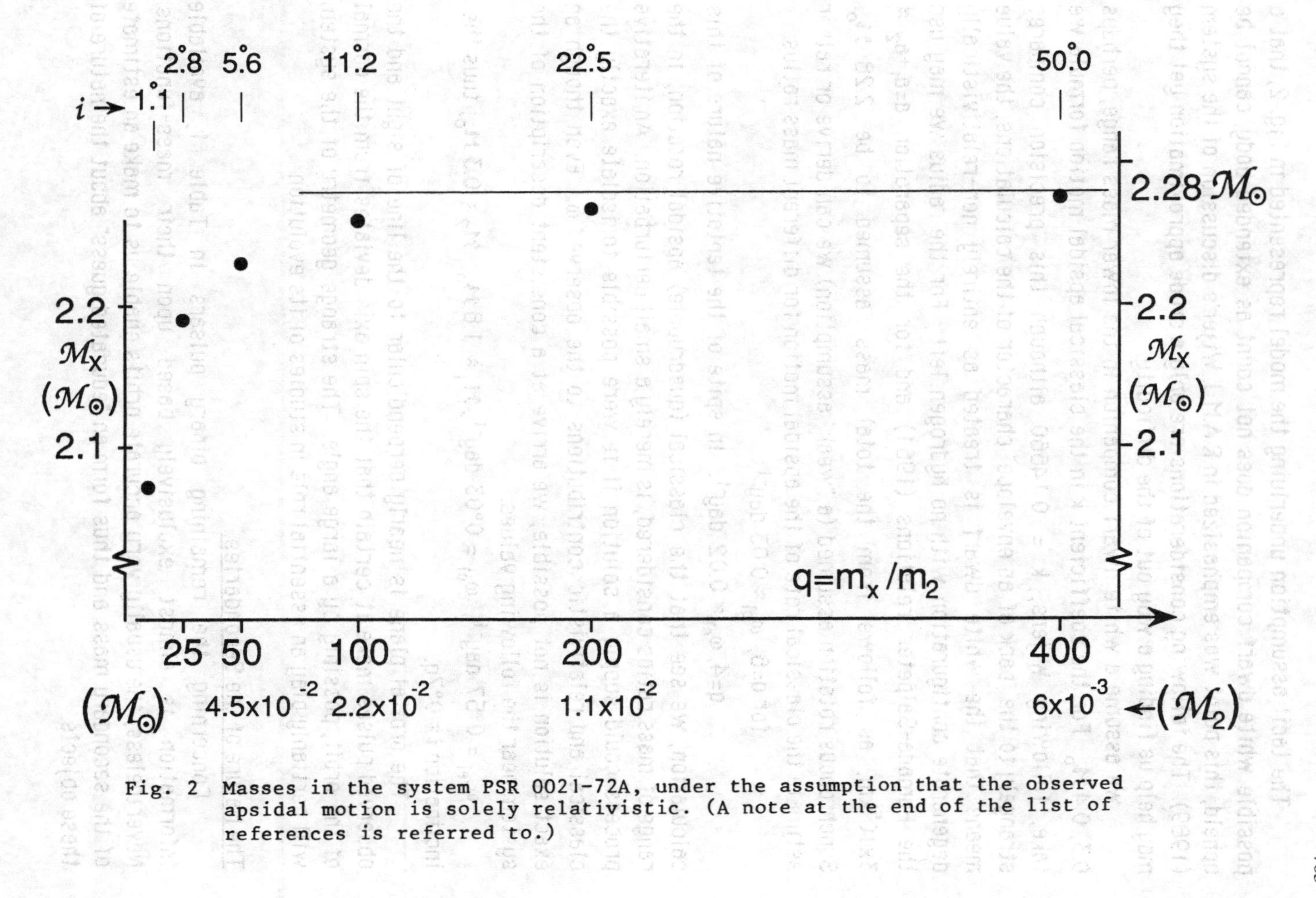

Fig. 2 Masses in the system PSR 0021-72A, under the assumption that the observed apsidal motion is solely relativistic. (A note at the end of the list of references is referred to.)

The tacit assumption underlying the model represented in Fig. 2, that a possible white dwarf companion does not count as extended body, cannot be upheld; this point was emphasized in R.A.M.J. Wijer's discussion of the system (1989). The following considerations are only a crude approximation yet they may help us finding a way out of the difficulty.

We assume a white dwarf companion in the lower mass range, perhaps 0.3, 0.4 $\mathcal{M}_o$. For the coefficient k in the classical apsidal motion formula we take, following Wijers, k = 0.14660 although this precision compares strangely to the "back of an envelope" character of the calculations; the value means that the white dwarf is treated as entirely non-relativistically degenerate configuration with no hydrogen left. For the radius we may use the Hamada-Salpeter relations (1961) and for the separation $a=a_1+a_2 \approx 3\times10^5$km, as follows from the total mass, assumed to be 2.28 $\mathcal{M}_o$. Synchronous rotation assumed (a "weak" assumption) we can derive or rather estimate the classical rate of the apsidal motion for different mass ratios:

for q=6, $\omega_{cl} \approx 0.03$ day^{-1}

q=4, $\omega_{cl} \approx 0.02$ day^{-1}. In spite of the tentative nature of this calculation, we see that the classical (quadrupole) apsidal rotation, in the range of mass ratios considered, is merely a small perturbation. An iterative process could supply a solution if it were possible to isolate exactly the classical and relativistic contributions to the observed ω. Even though an exact solution is not possible, we arrive at a consistent description of the system near the following values:

$\omega_{rel} = 0°57$ day^{-1}, $\omega_{cl} = 0°03$ day^{-1}, $\mathcal{M}_1 \approx 1.8\,\mathcal{M}_o$, $\mathcal{M}_2 \approx 0.3\,\mathcal{M}_o$; thus the inclination $i \approx 0°7$q.

The orbital plane is nearly perpendicular to the line of sight and the observed pulses make it certain that the spin axis deviates from the normal of the orbit, possibly by a large angle. The strange geometry of the system will certianly play an essential role in studies of its evolution.

The Nature of the Secondaries.

Concerning the remaining binary pulsars in Table 1, available information is almost exclusively based upon their mass-functions. Nevertheless, the usually very accurate orbits enable us to make an estimate of the secondary mass and thus form an "educated guess" about the nature of these objects.

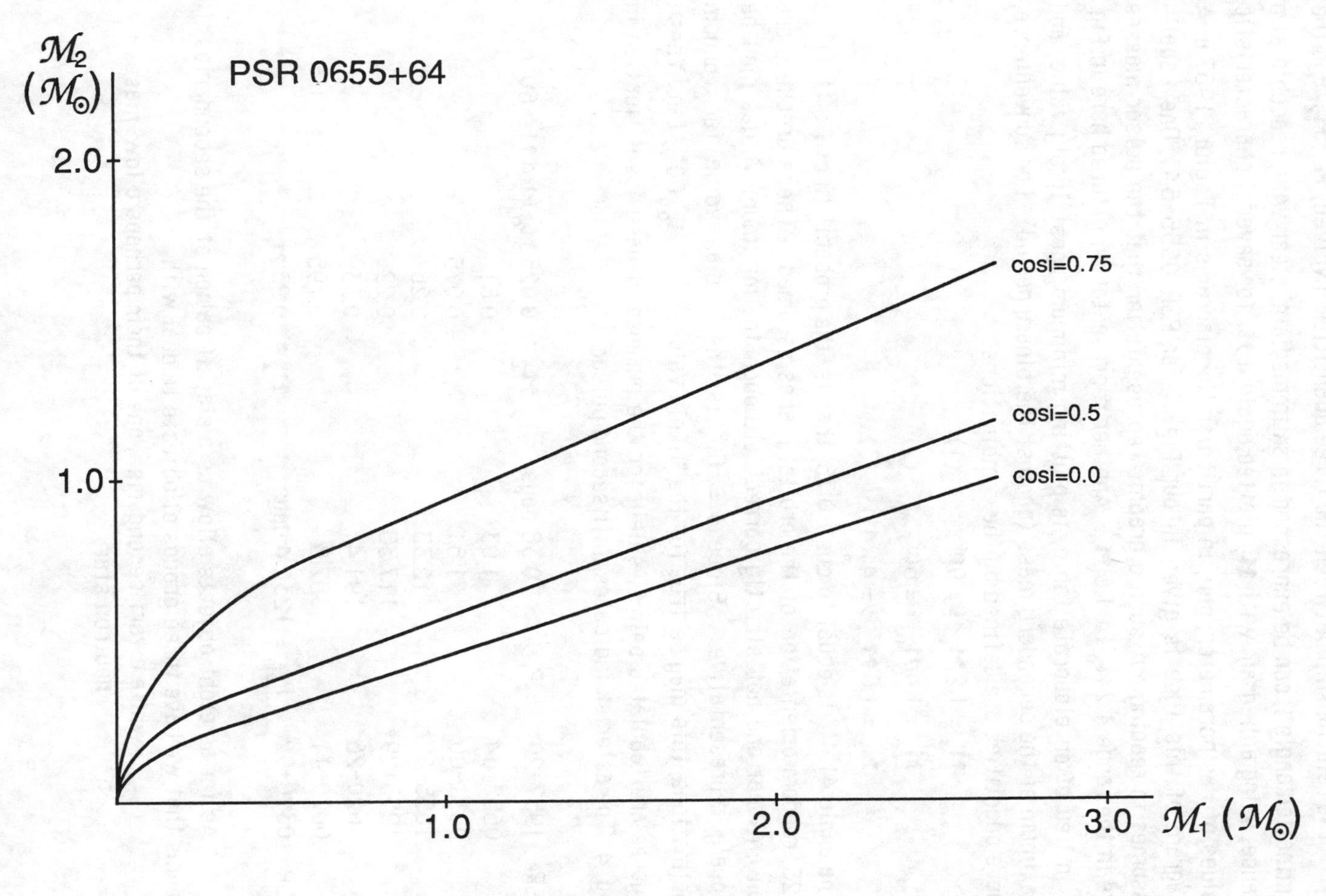

Fig. 3 Mass diagram for the 0.2 s binary pulsar 0655+64.

The relationship between the three quantities involved, $\mathcal{M}_x$, $\mathcal{M}_2$, and the inclination angle i, can be expressed in an immediate, clear way, readable at a glance, using a diagram with $\mathcal{M}_2$ plotted against $\mathcal{M}_1$ for several characteristic values of i as parameter; see, in particular the diagrams in Taylor, 1987.a A diagram of this type is given in our Fig. 3 for PSR 0655+64. One usually assumes in reading out such a graphic representation that the pulsar masses are in the range 1.2 $\mathcal{M}_o$ to 1.6 $\mathcal{M}_o$. Another representation offered here in Fig. 4, in lieu of an elaborate table, is plotting minimum, most likely value, and maximum of the secondary mass (against the binary period). For convenience, these quantities are defined by the combinations

$$\mathcal{M}_x = 1.2\,\mathcal{M}_o,\ i = 90°, \qquad (1)$$

$$\mathcal{M}_x = 1.4\,\mathcal{M}_o,\ i = 60°, \qquad (2)$$

$$\mathcal{M}_x = 1.6\,\mathcal{M}_o,\ i = 41.41°. \qquad (3)$$

(The angle 41.41° is equal to $\cos^{-1} 0.75$, that is, the probability of $i < 41.41°$ is 0.25, random orientation of the orbital planes assumed.) These brackets are conventional and not strictly correct although the minimum values may be closely representative. Since i = 60° is the median value for random inclinations, this may be true for the "likely values" (1.4 $\mathcal{M}_o$, 60°), too. These figures are tabulated below, solely for the purpose of easy identification in Fig. 4. The asterisk indicates millisecond pulsar.

PSR	1957+20*	P_{bin} =	0.38 days	$\mathcal{M}_2 \approx$	0.025 $\mathcal{M}_o$ (1.4 $\mathcal{M}_o$, 60°)
	0655+64		1.03		0.81
	0831-00		1.81		0.075
	1855+09*		12.33		0.28
	1953+29*		117.35		0.22
	1620-26*		191.4		0.33
	0820-11		357.8		0.795
PSR	0320+02	P_{bin} =	1232.4 days	$\mathcal{M}_2 \approx$	0.23 $\mathcal{M}_o$

As far as eight objects allow us a classification of the secondaries, it seems that we have three groups of objects to deal with:

2 white dwarf secondaries, one of them perhaps a low-mass neutron star;

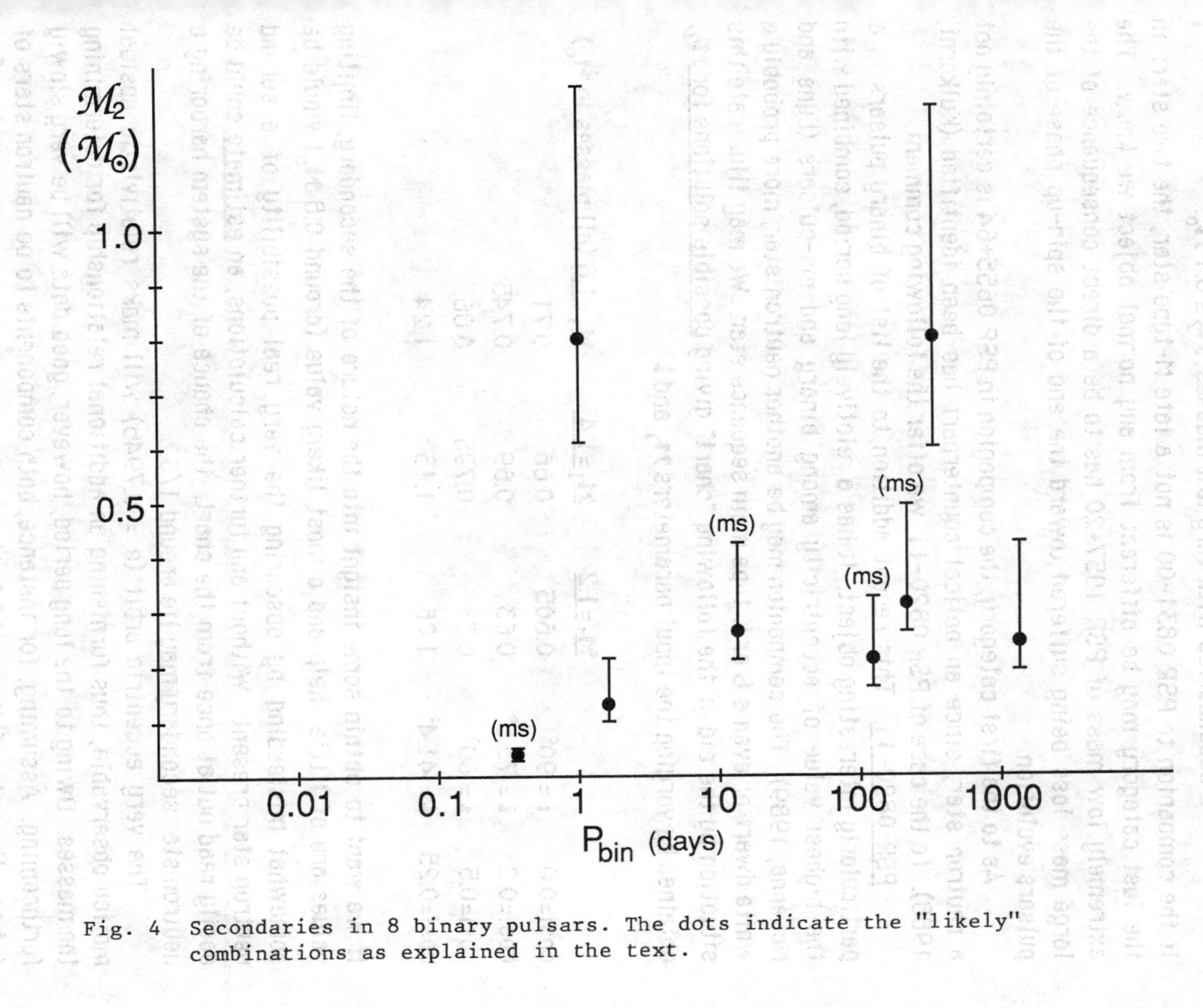

Fig. 4 Secondaries in 8 binary pulsars. The dots indicate the "likely" combinations as explained in the text.

4 secondaries with masses around 0.25–0.30 $\mathcal{M}_o$ (low mass white dwarfs or helium stars?);

2 secondaries with very low masses, $\mathcal{M}_2 \leq 0.1\ \mathcal{M}_o$.

If the companion to PSR 0831-00 is not a late M-type star, the two stars in the last category may be different from any normal object we know. The extremely low mass of PSR 1957+20 has to be a direct consequence of the large mass loss being suffered toward the end of the spin-up phase of the pulsars evolution.

As to the first category, the companion in PSR 0655+64 is certainly not a neutron star, since an optical counterpart has been identified (Kulkarni, 1988). To the case of PSR 0820-11, we offer the following comment.

PSR 0820-11. This recent addition to the list of binary pulsars is a particularly interesting object. It has a relatively long period, combined with the highest value of eccentricity among binary and ms-pulsars (Lyne and McKenna, 1989). The companion may be another neutron star, more probably a white dwarf or even a G or K type main sequence star. We may illustrate this situation by the aid of the following "chart" giving possible solutions for $\mathcal{M}_2$, obtained by varying the input parameters $\mathcal{M}_1$ and i.

		$\mathcal{M}_1=1.2$	$\mathcal{M}_1=1.4$	$\mathcal{M}_1=1.6$ (all masses in $\mathcal{M}_o$)
cosi=0.0	$i = 90°$	0.605	0.66	0.71
cosi=0.25	$i = 75.5°$	0.63	0.69	0.745
cosi=0.5	$i = 60°$	0.73	0.795	0.86
cosi=0.25	$i = 41.4°$	1.06	1.15	1.24

If we want to obtain some insight into the nature of the secondary, limiting values are of little help and a most likely value (around 0.8 $\mathcal{M}_o$) would be somewhat misleading by obscuring the very real possibility of a second neutron star present. Without any further calculations, an estimate could be easily read out at once from the chart: the chance of the system harboring a neutron star secondary may lie around 1/3.

The very eccentric orbit (e = 0.7946) will make relativistic apsidal motion observable, thus furnishing an additional relationship for determining the masses. Owing to the long period, however, good data will be very slowly forthcoming. Assuming, for instance, both components to be neutron stars of 1.4 $\mathcal{M}_o$, the rate of the apsidal advance turns out to be of the order of $\dot{\omega} \approx$

2×10^{-5}deg/yr. It would take about 100 years before the change in ω will reach the present uncertainty of ω, given as $\pm 0.002°$; a rather uncertain guess might be attempted earlier, perhaps already 10-15 years from now. On the other hand, the apsidal advance will be purely relativistic, the classical effect being entirely negligible: The relative orbit is of the order of 1 AU (slightly larger), thus even in the case of a solar type companion the critical factor $r_2^5 = (R_2/a)^5$ will remain under 10^{-10}.

X-Ray Binaries: The Better Determined Cases.

Most and quite possibly all point-like galactic X-ray sources with high X-ray luminosity ($\geq 10^{36}$erg/s) represent X-ray binaries. In these systems a stellar component is losing mass to an accreting compact object which is either a neutron star or in a few cases perhaps a black hole. The number of all known X-ray binaries may surpass hundred in the forseeable future; in their comprehensive review article of 1984, Joss and Rappaport listed over 20 pulsing sources which offer in most cases the best opportunity for a solution or at least a good estimate of the masses. The majority of the X-ray sources, however, do not show regular pulses and their spectroscopic study is particularly difficult. They remain rather weakly determined cases and in a later section we will present, for comparison, a short discussion of a few characteristic objects.

In the following section we are going to deal, in some detail, with the best studied, "classical" cases, treated already in J. Bahcall's early analysis. Particular attention will be paid to the systems Cen X-3, SMC X-1 and also Her X-1, studied *in extenso* since 1971. For the neutron stars in these systems one frequently finds in the literature masses proposed between 1.0 and 1.3 $\mathcal{M}_o$ and it would be interesting to ascertain the existence of neutron stars clearly less massive than the 1.4 $\mathcal{M}_o$; such an investigation seems to become more arduous and less convincing, the closer one looks at these systems.

On the other hand, several systems with a possible "massive" neutron star will be merely mentioned here, in the hope to return to them in a subsequent paper. To this group belong Vela X-1, LMC X-4, 4U1538-52 and also the somewhat neglected system Cyg X-2, certainly a peculiar low mass X-ray binary, where a thorough study by the Victoria (DAO) investigators reopened the case "for business." In this review, however, we are going to discuss in a short separate section those four X-ray binaries where the presence of a very massive compact object points toward a possible black hole in the system.

Hercules X-1. This extensively studied binary X-ray source with its unique, intricate structure and multiple periodicities ("the clockwork wonder," P. Boynton) was for some time considered the prototype of the low mass X-ray binaries, that is, an accreting compact object with a low mass (1-2 $\mathcal{M}_o$) stellar companion. Yet it is better to consider the system as a rare object, a class by itself, set apparently quite apart from the galactic bulge sources, also low mass X-ray binaries. (The faint peculiar variable and X-ray source V1729 Cyg was pointed out by astronomers at Sonneberg, Germany, as a possibly similar object.)

The stellar companion was identified as a variable star of puzzling nature already in the 1930's (HZ Her). We know today, after the UHURU discovery and optical identification, that both the large amplitude of the light variation, about 2 magnitudes, and the spectral type varying from B8 to F0 are results of the intense hemispheric X-ray heating of the stellar component. The orbital period is 1.70 days, the pulse period 1.24 sec and from the accurate X-ray orbit follows an amplitude K_x = 169.1 km/s and a mass function 0.85 $\mathcal{M}_o$ (see, for instance, Deeter *et al.* 1981). A particular feature of the X-ray light curve is a 35 day cycle: for 11 days the X-ray radiation remains strong then for 24 days it almost disappears. The generally accepted explanation is the precession of a thick accretion disk, periodically cutting off the X-rays directed toward us; the light curve shows that the X-ray source is radiating all the time and the X-ray heating is not cut off. This is an important circumstance since the precession axis is in all likelihood perpendicular to the accretion disk and thus to the orbital plane. This strongly suggests that the angle of orbital inclination might be close to 90°, possibly only a few degrees away from this value.

Without making use of the uncertain elements of the optical spectroscopy which has to cope with strongly distorted lines, one can derive surprisingly good masses for the system. This is due to a well defined eclipse half angle -- in fact, the only well defined eclipse angle among X-ray binaries. The UHURU determination was $\phi_{1/2}$ = 24.41° ± 1.06°; later by Deeter *et al.* (1981) an apparently much more accurate value was found, $\phi_{1/2}$ = 24.56° ± 0.03°. Since in the Her X-1 system the accretion is obviously driven by Roche lobe overflow, we may apply the tables by Channan *et al.* (1976) for

Roche configuration, to obtain a mass ratio for any given inclination. In this way a one-parameter set of the solution will result.

Another source of information is a remarkable observation by Middledich and Nelson (1976) who found regular optical pulses, X-ray pulses reprocessed in the atmosphere of the companion. They derived $i \approx 87°$ from these observations.

In order to illustrate the sensitivity of the solutions to the value of $\phi_{1/2}$, we may present the following table for possible pulsar masses:

	$i = 90°$	$85°$	$80°$	$75°$
$\phi_{1/2} = 24°$	1.64	1.41	0.95	0.55 ($\mathcal{M}_o$)
$\phi_{1/2} = 25°$	1.32	1.15	0.80	0.47 ($\mathcal{M}_o$)

In a cautious way one may state that the neutron star mass is between 0.8 and 1.6 $\mathcal{M}_o$. Nevertheless under the <u>assumption</u> that the limiting values indicated in the table, 24° and 25°, are really out of the question and $\phi_{1/2}$ is now known with the high accuracy indicated in the Middleditch-Nelson study, it is possible to establish the pulsar mass in Her X-1 with an accuracy almost comparable to that of the binary pulsar mass in PSR 1913+16. Taking $\phi_{1/2}$ = 24.5°, one can easily find the masses $\mathcal{M}_x = 1.28\,\mathcal{M}_o$ and $\mathcal{M}_x = 1.36\,\mathcal{M}_o$ for i = 85° and 87°, respectively. Expressed in the usual though not quite correct way $\mathcal{M}_x = 1.32 \pm 0.04\,\mathcal{M}_o$; it would be perhaps safer to write $\mathcal{M}_x = 1.3 \pm 0.1\,\mathcal{M}_o$.

From this value it follows at once (using $i = 86°$) $\mathcal{M}_s = 2.10 \pm 0.05$, q = 0.62 and for the Roche lobe filling radius $R_{RS} = 3.8\,R_o$.

Attempts at obtaining a usable spectroscopic orbit (see Hutchings *et al.*, 1985) resulted in $q \approx 0.5$, in marked deviation from the value above. Ellipsoidal variations are, of course, completely overwhelmed by the effects of X-ray heating. Both procedures for obtaining a viable mass ratio will regain their importance, if and when Her X-1 would enter another extended, several years long, "low state," presumably with a near complete stop of the accretion and X-ray emission. Existence of such states is well documented by studies in plate archives, both at Harvard and Sonneberg observatories.

<u>Centaurus X-3.</u> This famous object may be considered the prototype of the class of massive X-ray binaries, that is, binaries which consist of a massive OB star and a compact object, apparently the only homogeneous group

that can be easily identified as such and which represents the majority of well studied binary X-ray sources. The optical counterpart of Cen X-3 is a O7-star (main sequence?) as identified by Krzeminski (1974); the variable star determination is V779 Cen but the star is still frequently identified as "Krzeminski's star." The very accurate X-ray orbit yields the amplitude K_x = 415.1 km/s and the mass function $f(\mathcal{M})$ = 15.5 $\mathcal{M}_\odot$, a very large value. What can we say about the mass ratio?

The basic spectroscopic study is by Hutchings *et al.* (1979). The star proved a difficult object spectroscopically and the orbit is termed "highly undertain" in the Eighth Catalogue by Batten *et al.* Hutchings and his colleagues found $K_s = 24 \pm 3$ from the solution termed "CTIO All" and 26 ± 4 from combining all solutions. The spectroscopic work did not contradict earlier photometric attempts to derive the mass ratio from the observed ellipsoidal variations (as we have seen, Avni and Bahcall predicted between 15 and 25 km/s) but it hardly improved upon it. Probably a good way to illustrate the problem of the neutron star mass in Cen X-3 is to select a few representative values of K_s and combine them with selected values of the inclination angles i, as in the table below.

The mass of the neutron star (in $\mathcal{M}_\odot$):

	$i = 90°$	75°	60°
K_s=18 km/s	0.73	0.76	1.12
24 km/s	1.00	1.05	1.54
30 km/s	1.28	1.35	1.97

Since the system shows eclipses, orbital inclinations $i < 60°$ may perhaps be disregarded. Apart for higher angles of inclination and higher K values, the chart suggests a pulsar mass somewhat inferior to the value 1.4 $\mathcal{M}_\odot$. Nonetheless higher, less probable, values of K would suggest quite large masses for $\mathcal{M}_x$; taking for example K = 36 km/s, 2.5 σ away from the "best determination" we obtain, for the inclinations used above, the masses 1.59, 1.64 and 2.38 $\mathcal{M}_\odot$ respectively. The fit of the observation with K = 36 km/s is admittedly a poor one but as a plot would suggest it cannot be dismissed entirely. It is at any rate discouraging to note that a change in K by 2 km/s changes the pulsar mass by 0.1 $\mathcal{M}_\odot$. We cannot state after all <u>with certainty</u>

that the neutron star is less massive than the 1.4 $\mathcal{M}_o$ "standard value" and even limiting its mass to $0.8 < \mathcal{M}_x < 2.0\ \mathcal{M}_o$ cannot be objected to; Rappaport and Joss give similar limits.

The system exhibits regular eclipses but the eclipse half angle is uncertain by several degrees (it is around 35°). A Monte Carlo analysis of the data observed led Rappaport and Joss (1983) to the most probable value of 1.07 (+0.63, −0.57) $\mathcal{M}_o$. In a detailed study of the eclipses, Clark *et al.* (1988) found that egress and ingress are gradual, revealing important evidence of the atmospheric structure. They found a slightly smaller eclipse angle ($\phi_{1/2}$ = 32.9°) and suggested a mass $\mathcal{M}_x = 1.23 \pm 0.60\ \mathcal{M}_o$.

SMC X-1. This X-ray source of short pulse period (0.7 sec) is distinguished by its distance being reasonably well known (50 kpc); its X-ray luminosity turns out very high, near the Eddington limit. The pulsar mass has been given $0.8\ \mathcal{M}_o < \mathcal{M}_x < 1.8\ \mathcal{M}_o$ with the most probable value at 1.0 $\mathcal{M}_o$ (see, for instance, Rappaport and Joss, 1983). These values suggest that the compact component in SMCX-1 may be less massive than most other neutron stars, at any rate under the "standard" value of 1.4 $\mathcal{M}_o$.

The mass determination rests upon a very accurate X-ray orbit (Primini *et al.* 1977) and a relatively well determined spectroscopic orbit (Hutchings *et al.* 1977, *cf.* also Hutchings, 1977). The relevant data are: K_x = 299.5 km/s and $f(\mathcal{M})$ = 10.83 $\mathcal{M}_o$ from the SAS-3 X-ray observations and K_s = 19±2 km/s from the radial velocity curve; the spectral type is B1I. The duration of clearly discernible X-ray eclipses was characterized by the half eclipse angle $\phi_{1/2} \approx 28.2°$. Assuming the spectroscopic amplitude to be between 16 km/s and 22 km/s (1.5σ), we again obtain based on the mass function of the X-ray orbit, a tabulation of the masses as function of the inclination angle. (In parentheses the mass of the stellar companion; all masses in $\mathcal{M}_o$.)

	K_s=16km/s	K_s=19km/s	K_s=22km/s
i = 90°	0.64(12.0)	0.78(12.2)	0.92(12.5)
i = 70°	0.77(14.4)	0.94(14.8)	1.10(15.0)
i = 60°	0.99(18.7)	1.20(18.9)	1.41(19.2)
i = 50°	1.43(26.8)	1.73(27.2)	2.04(27.8)

The pulsar proves markedly "undermassive" only for inclinations between 90° and about 65°. Such high inclination values are not incompatible with the occurrence of eclipses, if the B-type supergiant, as generally assumed, nearly fills its Roche lobe. In this case we may make use of the observed eclipse angle, in connection with the table calculated by Chanan *et al.* (1976). In principle, K_x, K_s and $\phi_{1/2}$ could supply a complete solution, but of course the large uncertainties, in particular in K_s allow us only estimates of $\mathcal{M}_x$ resp. limiting values.

Taking $\phi_{1/2} = 28.2°$ at face value, we may arrive at a formal solution $i = 65° \pm 1.5°$, the uncertainty in the mass ratio being the main source of the error quoted. Taking $i = 65°$, we could insert into the table above, between $i = 60°$ and $i = 70°$, a line with the masses for the pulsar:

$$i = 65°, \quad \mathcal{M}_x = 0.86\,\mathcal{M}_o,\ 1.04\,\mathcal{M}_o,\ \text{resp. } 1.23\,\mathcal{M}_o.$$

On the other hand, if we allow for $\phi_{1/2}$ (probably pessimistically) the range from 25° to 30°, the values of i will scatter between 60° and 68° and we may only say that $0.8\,\mathcal{M}_o < \mathcal{M}_x < 1.4\,\mathcal{M}_o$, a very slight improvement. Nevertheless in both cases an "undermassive" neutron star is clearly suggested.

In order to illustrate the difficulties in obtaining more accurate data, we may try to verify the B-supergiant filling its Roche lobe. We will not quite succeed but at least one can show that the hypothesis is not contradicting the observations. Taking the solution $K_s = 19$ km/s, $i = 65°$, we easily arrive at $a = 27\,R_o$ and, for the Roche lobe radius, $R_s^* = 16.5\,R_o$. Turning to the photometric determination of the radius of the stellar component, it is easy to find a set of very plausible values (such as $A_V = 0.5$ mag., B.C. = -3.0, $T_{eff} = 32000$K) which then lead to $R_s = 16.5\,R_o$; the absolute bolometric magnitude is -8.8. The enforced accurate agreement is, of course, hardly convincing: changing A_V or B.C. by 0.1 mag., the resulting change in the radius will be about $0.7\,R_o$. In spite of the known distance, it is virtually impossible to come up in this way with any exact value for R_s.

Systems with possible black hole components

The mass-function is the main source of information in searching for possible black hole companions in binary X-ray sources. If the mass of the (otherwise invisible) accreting component as estimated from the mass function is likely to be larger, sometimes markedly so, than any currently proposed upper mass limit for neutron stars, consideration of a possible black hole in the system is almost inevitable. Additional pieces of information, in particular, the complete lack of regular X-ray pulses, in combination with erratic fluctuations of the X-ray intensity down to the millisecond level--quite different from the behaviour of millisecond pulsars--are of great importance, too.

Since pulses are not observed, all orbits refer to the stellar components and usually are of fair or poor quality. (Only Cyg X-1 has a "good" orbit, class b in the Catalogue of Batten *et al.*) Yet the mass functions are sufficiently reliable to show that already the minimum mass of the accreting component, corresponding to $i = 90°$, proves large enough to warrant a closer look at the system. Nevertheless, we have to bear in mind that the main problem is not so much the orbital inclination rather the assumption we have to make about the mass of the observed stellar component ($\mathcal{M}_s$). It is evident that these stars must have undergone mass loss, in some cases perhaps a substantial one, during the evolution of the system and the stellar components may be noticeably undermassive for their respective spectral types. Thus, out of caution, most cases are still considered not to be "open and shut."

In the early 1970's, several black hole companions were put forward for peculiar systems with massive but spectroscopically invisible secondary components. None of these proposals survived a closer scrutiny although β Lyrae is occasionally still mentioned as a weak candidate. Among the X-ray sources where the existence of an accreting compact component is evident from the beginning, four have been seriously considered harboring a black hole component: Cyg X-1, LMC X-3, the X-ray transient A0620-00, and more recently LMC X-1. Since we are interested mainly in neutron stars, we give only the basic data for these systems with a minimum of comments if it deems necessary. There are also review articles to be consulted such as Blandford 1987.

1. Cyg X-I, stellar companion O 9.7 Iab, $f(\mathcal{M}) = 0.24\,\mathcal{M}_o$; latest orbit: Gies and Bolton (1982).

$$\text{For } \mathcal{M}_s = 16\,\mathcal{M}_o,\ \min \mathcal{M}_x = 4.7\,\mathcal{M}_o\ (i=90°)$$
$$= 12\,\mathcal{M}_o \qquad = 3.95\,\mathcal{M}_o.$$

No eclipses have been observed and the mass of the invisible component may be higher than the above values. Studies of the ellipsoidal light variation, the rotational line broadening, the effective temperature and the surface gravity in connection with the estimated distance, all point toward the stellar component being, indeed, a massive (though probably somewhat undermassive) supergiant. The values quoted above for $\mathcal{M}_s$ are rather conservative estimates. If the orbital inclination is near 40°-45°, as estimated, even the ingenious alternative hypothesis of two neutron stars in a very close orbit, circling the O-supergiant, becomes very difficult to maintain (see Bahcall 1978).

2. LMC X-3, companion B3V, $f(\mathcal{M}) = 2.3 \pm 0.3\, \mathcal{M}_o$ (Cowley *et al.*, 1983).

$$\text{For } \mathcal{M}_s = 6.0\, \mathcal{M}_o, \quad \text{min. } \mathcal{M}_x = 9.95\, \mathcal{M}_o \ (i = 90°)$$
$$= 3.0\, \mathcal{M}_o \qquad\qquad = 5.5\, \mathcal{M}_o$$

Paczynski (1983) argued, under the assumption of near grazing eclipses of the X-ray source at the limb of a Roche-lobe filling stellar companion, that the mass of the compact object can be as high as $10 \pm 4\, \mathcal{M}_o$. On the other hand, Mazeh *et al.* (1986) mentioned the possibility that the star classified as B3V is really a low mass He-star ($\leq 1\, \mathcal{M}_o$) and the observed optical radiation is significantly enhanced by radiation from the accretion disk. In such a case the mass of the compact object could be as low as $2.5\, \mathcal{M}_o$ (a very massive neutron star?) The large mass-function in connection with $\mathcal{M}_x = 2.5\, \mathcal{M}_o$ however, tends to lead to implausible low masses for the alleged He-star of the order of $0.1\, \mathcal{M}_o$ or less. Only by selecting $f(\mathcal{M})$ as low as 1.8 or $1.9\, \mathcal{M}_o$ and assuming $i \approx 90°$ (yet we see no eclipses!) does it become possible to come up with masses around 0.35 or $0.4\, \mathcal{M}_o$ for a He-star component. The observed data favor a more realistic lower limit $\mathcal{M}_x = 3\, \mathcal{M}_o$ but one has to admit that until a closer spectroscopic study rules out the He-star hypothesis, LMC X-3 remains perhaps a less convincing case among the black hole candidates.

3. The transient source A 0620-00, the X-ray nova of 1975; companion K5V (Oke, 1977), $f(\mathcal{M}) = 3.18 \pm 0.16\, \mathcal{M}_o$ (McClintock and Remillard, 1986).

$$\text{For } \mathcal{M}_s = 0.6\, \mathcal{M}_o, \quad \text{min. } \mathcal{M}_x = 4.2\, \mathcal{M}_o \ (i = 90°)$$
$$= 0.5\, \mathcal{M}_o \qquad\qquad = 4.0\, \mathcal{M}_o.$$

The identification is quite certain owing to an optical outburst observed simultaneously; the variable star designation is V616 Mon. Masses well above 3.5 $\mathcal{M}_o$ for the compact star seem hard to avoid. An estimate of the Roche geometry strongly supports this conclusion. Taking $\mathcal{M}_s$ = 0.6 and $\mathcal{M}_x$ = 4.8 $\mathcal{M}_o$ or 6.0 $\mathcal{M}_o$ corresponding, in round figures, to inclination angles 70° and 60°, one easily obtains for the Roche lobe radius of the K-dwarf 0.75R_o and 0.76R_o, respectively. Comparing these values with the figures quoted in standard tables, around 0.73R_o, see for instance Schmidt-Kaler, 1982, or Allen, 1973, it is fair to say that in spite of all uncertainties the K-dwarf appears to be coextensive with its Roche lobe and it is quite possibly slightly smaller. (This may help to explain the intermittent, episodic flow of matter toward the compact object.) It seems important to notice that choosing a much lower mass, say 1.5 $\mathcal{M}_o$ or 2.0 $\mathcal{M}_o$, for the compact object, we invariably arrive at a geometry of the system where a normal size, even a somewhat undersized K-dwarf would be <u>substantially</u> larger than the corresponding Roche lobe.

4. LMC X-1, companion O7III, $f(\mathcal{M})$ = 0.14 $\mathcal{M}_o$ (Hutchings *et al.*, 1987).

For $\mathcal{M}_s$ = 25 $\mathcal{M}_o$, min. $\mathcal{M}_x$ = 5.0 $\mathcal{M}_o$ (i= 90°)
= 15 $\mathcal{M}_o$ = 3.65 $\mathcal{M}_o$.

The first proposed optical identification of the source (Hutchings *et al.* , 1983) was not uncontested due to the proximity of another candidate, a B5 supergiant. A recent study by Pakull and Angebault (1986) made the identification virtually certain, showing a remarkable, large He III region around LMC X-1, generated by the X-radiation of the source.

Only a reduction of the mass of the O7III-type stellar component to 20 - 25% of the value we may expect of this spectral type (≥30 $\mathcal{M}_o$) would allow the mass of the compact object to approach the range of the potentially most massive neutron stars. A mass ratio of considerable uncertainty, about $\mathcal{M}_s/\mathcal{M}_x \approx 2$, derived by Hutchings *et al.* in their second paper (based on the Doppler shift of some HeI and HeII emission lines) leads to implausibly low masses for the O-star, except for very low inclinations (i < 40°).

In summary, even the lowest estimated masses of the compact objects in these four systems seem to form a "separate group" in the range of 3 to 10 $\mathcal{M}_o$, clearly superior to neutron star masses in the range of, say, 1 to 2 $\mathcal{M}_o$. Individual cases may need further discussion yet should not we admit that at least in one case or the other, we are already dealing with systems containing a black hole?

<u>A few indeterminate cases.</u>

There are many important X-ray binary systems, where attempts to determine the masses of the components encounter more serious difficulties than was the case with the much studied "classical" systems such as Cen X-3, with a (relative!) wealth of data. A few such indeterminate systems will be briefly mentioned below, as an illustration of what can be learned in such cases; they are also quite important systems.

The much observed system <u>4U 1700-37</u> (HD153919) with a bright optical counterpart presents the difficulty that no X-ray pulses have been observed and instead of a well-determined X-ray orbit one has to resort to the spectroscopic orbit of the O6f stellar component. The most extensive study is by Hammerschlag-Hensberge (1978), giving a mass function $f(\mathcal{M}) = 0.0023 \pm 0.0005\ \mathcal{M}_o$; Bahcall estimates the error higher, to about 30%. Taking the mass function in the broad range of $0.0016\ \mathcal{M}_o$ to $0.0030\ \mathcal{M}_o$, we obtain for an assumed $\mathcal{M}_s = 20\ \mathcal{M}_o$ minimum values of the neutron star mass in the range $0.9\ \mathcal{M}_o$ to $1.1\ \mathcal{M}_o$. There are X-ray eclipses of apparently variable duration (Mason *et al.* 1976) and the inclination angle can easily reach 60-65°; the geometry of the system resembles the SMC X-1 system. The neutron star mass turns out to be between 1.0 and $1.3\ \mathcal{M}_o$ for $i = 65°$. These data indicate a mass possibly less than $1.4\ \mathcal{M}_o$ for the neutron star but this cannot be fully substantiated at this time. The O6-star mass is probably rather under- than over-estimated.

The early discovered X-ray source <u>4U 1907+09</u> became accessible for closer studies only during the last few years, after a difficult optical identification, discovery of an $8^d.38$-modulation of the X-ray flux identifiable with the orbital period, and finally after the discovery of slow ($437^s.5$) pulsations. For references to the extensive literature be referred to a recent spectroscopic study by van Kerkwijk *et al.*, 1989. The X-ray mass function is rather poorly defined: $f(\mathcal{M}) = 7.9 \pm 2.1\ \mathcal{M}_o$. Nevertheless, even crude estimates of the mass of the stellar component, spectrum "about B2," show the system to be a typical massive X-ray binary with minimum masses of the B-star component around 10-12 $\mathcal{M}_o$. The study quoted above makes the case for a supergiant stellar component, with strong stellar wind.

The X-ray source <u>in the globular cluster M15</u> (4U 2127+11) has been identified optically with the cluster member (designated AC211) by Aurière *et al.* (1884) and recently studied spectroscopically by Naylor *et al.* 1988,

Naylor and Charles 1989. Among several remarkable finds was the large negative radial velocity, about -150km/s relative to the motion of the cluster. The orbital data, in spite of their uncertainty, give some relevant information about the system. The period was found with some accuracy in the second paper 8.63±0.14 hrs. The HeIλ4471 line appears in emission and shows a radial velocity curve with an amplitude 36±16 km/s, and the motion it indicated has been tentatively ascribed to the compact component and its accretion disk. The large error of the amplitude (44%) makes the use of the mass function impractical for an analysis of the system. Nevertheless, a simple consideration avoiding the magnification of error by the third power may reveal the structure of this interesting binary.

Using the relatively well determined period, we can at once calculate the <u>relative</u> orbital velocity for a number of masses $\mathcal{M}_1 + \mathcal{M}_2$; we may also assume circular orbit. So we obtain for $\mathcal{M}_1 + \mathcal{M}_2 = 2\,\mathcal{M}_o$ (low mass model) v = 380 km/s, for $\mathcal{M}_1 + \mathcal{M}_2 = 20\,\mathcal{M}_o$ (high mass model). v = 800 km/s; these values are, of course, independent of the orbital inclination. Since sin 60° = 0.866, we may work with reasonable average values for the range $i = 90°$ to $i = 60°$, as we only want to obtain approximate data. Taking $K_x = v_x \sin i = 36$ km/s, ignoring for a moment the large errors, one easily finds the (approximate) values of the masses for these models:

low mass, $\mathcal{M}_1 = 1.8\,\mathcal{M}_o$, $\mathcal{M}_2 = 0.2\,\mathcal{M}_o$

high mass, $\mathcal{M}_1 = 19\,\mathcal{M}_o$, $\mathcal{M}_2 = 1\,\mathcal{M}_o$, without taking the decimal digits too seriously. The perhaps surprising result is that even in the high-mass system the accreting compact component is the massive one. There is no high mass secondary possible, at no value of $\mathcal{M}_1 + \mathcal{M}_2$; on the other hand, going from $2\,\mathcal{M}_o$ toward $20\,\mathcal{M}_o$, very soon we will reach the range of black hole masses for $\mathcal{M}_1$. This is not meant to propose a black hole companion for the system. The result of these simple estimates appears to be: the X-ray source in M 15 consists of a neutron star and a low mass, probably unevolved (captured) cluster member of about 0.2, 0.3 $\mathcal{M}_o$.

Taking the errors of K_x into consideration would not change this qualitative statement, neither would assuming that $i < 60°$. It is important to emphasize that these considerations rest with the assumption that the Doppler shift of the He-line reflects the orbital motion of the compact star.

The system <u>4U 1626-67</u> is of signal interest for the interpretation of low mass X-ray binaries. After its identification by McClintock *et al.* (1977) with a faint blue star of estimated absolute magnitude around +3 (the variable star designation is KZ TrA), the system might have appeared as a common member of the wide class of galactic bulge sources, were it not for two differences: its spectrum is much harder than those of the galactic bulge objects and it shows regular pulses of 7.68 sec period. Rather unexpectedly, the pulses did not show measurable shifts of the arrival time, due to orbital motion. This was interpreted as a sign of $a_x \sin i$ being very small, possibly under 0.1 lightsec, and S. Rappaport, P. Joss and others proposed the model of a highly compact binary, at that time an unusual new idea. A neutron star of 1 $\mathcal{M}_o$ and a presumably low mass, high radius white dwarf component were assumed, the latter filling its Roche lobe, and a separation of about 1 lightsec between the components. (For references to the rather extensive literature, see the 1984 review of Joss and Rappaport.) Optical pulses of about 2% contrast, detected by Ilovaisky *et al.* (1978), added a new feature to the picture.

New insight was gained into the structure of this extraordinary system by a set of extraordinary observations carried out by Middleditch *et al.* (1981). High speed optical photometry of the 18th mag. object revealed occasional sets of pulses at a second frequency, shifted by -0.40 mHz with respect to the main frequency of 130.26 mHz. The frequency shift of this radiation, reprocessed in the companion's atmosphere, can be interpreted as corresponding to the difference in the sidereal and synodic rate of rotation of the stellar companion, the latter being counted with respect to the neutron star as seen from the orbiting secondary. From these frequency shifts a binary orbital period of 2490s can be derived. Furthermore, the Fourier analysis allowed the detection of phase modulations due to orbital motion. Thus perhaps the system 4U 1626-67 should not be considered ill-defined at all as J. Middleditch and his colleagues were able to give a complete orbital solution albeit with very large errors attached to the elements. For the masses, this solution yields $\mathcal{M}_x = 1.8(+2.9/-1.3)\,\mathcal{M}_o$, $\mathcal{M}_s < 0.5\,\mathcal{M}_o$.

We just termed 4U 1626-67 an extraordinary system. It is still unique from the point of view of observation and analysis yet it can be a member in the class of galactic bulge sources, with some unusual features added. These sources also have to represent binary systems, most or perhaps all of them highly compact objects, as suggested by P. Joss and S. Rappaport and others working on 4U 1626-67.

It is sometimes assumed, like in earlier models for this system, that the neutron star in these low mass systems has a mass around 1 $\mathcal{M}_{\odot}$. Perhaps it should be emphasized that there is yet no direct evidence that neutron stars in these systems are themselves "low mass" objects. They can be, and probably are, of masses in a range around 1.4 $\mathcal{M}_{\odot}$, as in most of the better understood X-ray binaries.

The cases briefly considered up to now were not always really "indefinite" although much less complete than the "classic" cases of massive X-ray binaries or Her X-1, partly because some important piece of information, such as dopplershifted X-ray pulses, is missing partly because of large uncertainties in some data. A real contribution to the problem of neutron star masses is not possible in such cases. The question is here mainly to find out about the nature of the binary component, frequently by using "typical" values for assumed neutron stars, 1.4 $\mathcal{M}_{\odot}$ or 1.5 $\mathcal{M}_{\odot}$.

There are many systems -- and among them some of great importance -- where the information gained hitherto seems truly inadequate and only a set of sometimes debatable assumptions enables us to construct a model of the system. It goes without saying that many of these models are still controversial. Often it is a question of periodic modulations of the X-ray output. The identification with orbital motion is quite plausible although in most cases hardly provable. (Binarity is always assumed in case of strong sources) In this group we find the wondrous source Cyg X-3, furthermore GX 17+2 (see its discussion by Langmeier *et al.* 1986), or even the famous and still puzzling Geminga (see for example Leahy *et al.* 1986, or for the question of optical counterpart, Halpern *et al.* 1985).

Let us take a look at Cyg X-3. It starts out usually with deriving a characteristic dimension. From the observed 4.8 hours periodic modulation in the X-ray output an orbital separation of 10^{11} cm follows. Here the second assumption was made that Cyg X-3 is a low mass system. Indeed, P = 4.8 hrs and $\mathcal{M}_1 + \mathcal{M}_2 = 2\mathcal{M}_{\odot}$ leads to $a = 1.26 \times 10^{11}$ cm.

How arbitrary is declaring Cyg X-3 to be a low mass X-ray binary? There is an underlying hypothesis here, still somewhat arbitrary but not implausible. It is a question of considering some X-ray binaries as consisting of a neutron star and a Roche lobe filling lower main sequence star. This is a model adapted from the study of cataclysmic variables and it leads to an approximate relationship between the orbital period and the mean density of the secondary. In this case, the resulting value is close to the mean density of an M1 or M2 main sequence star. An M1 star having a mass about 0.5 $\mathcal{M}_\odot$, the total mass, neutron star + M-dwarf can will be near 2 $\mathcal{M}_\odot$. This is hardly an exact calculation but makes the consideration of a low mass system justifiable. Yet even if this simple exercise gives the dimensions of the system correctly, the model to be constructed is immensely more complicated if we want to explain all the exotic properties of Cyg X-3.

Conclusion

Our direct, observational knowledge about the masses of neutron stars is still very fragmentary. The first binary pulsar PSR 1913+16 is the only case where an accurate measurement almost free of assumptions has been possible, based on several detectable general relativistic effects. The masses of the components, both neutron stars, are close to 1.4 $\mathcal{M}_\odot$, sometimes considered the "standard" value; J. Taylor's recent figures are 1.45 $\mathcal{M}_\odot$ and 1.38 $\mathcal{M}_\odot$. For the other systems, binary pulsars as well as X-ray binaries, we can give only constraints for the masses of the compact objects, sometimes leading to quite wide intervals; in several cases, at least likely values can be added. The best solution for Her X-1 is remarkably narrowly limited: the formal result is a neutron star mass within about 0.05 $\mathcal{M}_\odot$ of 1.32 $\mathcal{M}_\odot$, it can reach 1.4 $\mathcal{M}_\odot$. The main assumption made here is that the latest measurement of the X-ray eclipse duration is accurate.

Although the value 1.4 $\mathcal{M}_\odot$ is compatible, within the error limits, with all determinations, there are indications that neutron star masses fall in a range of values. One can find systems with a likely value of neutron star mass distinctly smaller than 1.4 $\mathcal{M}_\odot$, notably SMC X-1 and the binary PSR 2303+46, the latter if the revised rate of apsidal advance is about correct. On the other hand, earlier studies suggested that a number of systems, not discussed here, among them Vela X-1, LMC X-4, Cyg X-2, may have neutron stars more massive than the 1.4 $\mathcal{M}_\odot$ standard. In the case of a "recycled" millisecond pulsar this actually should be the case; the pulsar PSR 0021-72A in the cluster 47 Tucanae yield, however, no unique results. To this question of massive neutron stars we should like to return in a subsequent paper.

References

Ables, J.G., Jacka, C.E., McConnell, D., Hamilton, P.A., McCulloch, P.M., Hall, P.J., 1988: I.A.U. Circular No. 4602.

Allen, C.W., 1973: *Astrophysical Quantities*, 3rd. ed., The Athlone Press, London.

Auriére, M., Le Févre, O., Terzan, A., 1984: *Astron. Astrophys.* **138**, 415.

Avni, Y. and Bahcall, J., 1974: *Astrophys. J.* **192**, L139.

Avni, Y. and Bahcall, J.N., 1975: *Astrophys. J.* **197**, 675.

Backer, D.C. and Hellings, R.W., 1986: *Ann. Rev. Astron. Astrophys.* **24**, pp. 537-575.

Bahcall, J.N., 1978: *Ann. Rev. Astron. Astrophys.* **16**, 241. See also an earlier assessment (1975) of selected systems in Bahcall, 1978a.

Bahcall, J.N., 1978a: in *Physics and Astrophyssics of Neutron Stars and Black Holes*, Scuola Internazionale "Enrico Fermi", LXV Corso (eds. R. Giacconi, R. Ruffini), Soc. It. di Fisica, Bologna.

Batten, A.H., 1973: Ch. 6 in *Binary and Multiple Systems of Stars*, Pergamon Press, Oxford, New York, etc.

Batten, A.H., Fletcher, J.M., and MacCarthy, D.G., 1989: *Publ. Dominion Astrophys. Obs.* **17**.

Blandford, R.D., 1987: Ch. 8 in *300 Years of Gravitation*, (S.W. Hawking and W. Israel, eds.), Cambridge Univ. Press, Cambridge, New York, etc.

Blandford, R. and Teukolsky, S.A., 1975: *Astrophys. J.* **198**, L27, see also *Astrophys. J.* **205**, 580 (1976).

Boynton, P.E., Deeter, J.E., Lamb, F.K., and Zylstra, G., 1986: *Astrophys. J.* **307**, 545.

Chanan, G.A., Middleditch, J., and Nelson, J.E., 1976: *Astrophys. J.* **208**, 512.

Cowley, A.P., Crampton, D., Hutchings, J.B., Remillard, R. and Penfold, J., 1983: *Astrophys J.* **272**, 118.

Crampton, D. and Hutchings, J.B., 1981: *Astrophys. J.* **251**, 604.

Deeter, J.E., Boynton, P.E., Lamb, F.K., and Zylstra, G., 1987: *Astrophys. J.* **314**, 634.

Gies, D.R. and Bolton, C.T., 1982: *Astrophys. J.* **260**, 240.

Halpern, J.P., Grindlay, J.E., Tytler, D., 1985: *Astrophys. J.* **296**, 190.

Hamada, T. and Salpter, E.E., 1961: *Astrophys. J.* **134**, 683.

Hammerschlag-Hensberge, G., 1978: *Astron. Astrophys.* **64**, 399.

Haugan, M.P., 1985: *Astrophys. J.* **296**, 1.

Herczeg, T.J., 1976: *Proc. 2nd SW Regional Conf.* (Lubbock, 1976), pp. 41-46.

Hulse, R.A. and Taylor, J.H., 1975: *Astrophys. J.* **195**, L51.

Hutchings, J.B., 1977: *Astrophys. J.* **217**, 537.

Hutchings, J.B., Cowley, A.P., Crampton, D., van Paradijs, J., White, N.E., 1979: *Astrophys. J.* **229**, 1079.

Hutchings, J.B., Crampton, D., Cowley, A.P., Osmer, P.S., 1977: *Astrophys. J.* **217**, 186.

Hutchings, J.B., Crampton, D., Cowley, A.P., Bianchi, L., and Thompson, I.B., 1987: *Astron. J.* **94**, 340.

Ilovaisky, S.A., Motch, Ch., Chevalier, C., 1978: *Astron. Astrophys.* **70**, L19.

Joss, P.C., Rappaport, S.A., 1984: *Ann. Rev. Astron. Astrophys.* **24**, pp. 537-592.

Kopal, Zdenek, 1959: *Close Binary Systems*, Wiley & Sons, New York.

Krzeminski, W., 1974: *Astrophys. J.* **192**, L138.

Kulkarni, S.R., 1988: *Astrophys. J.* **305**, L85.

Langmeier, A., Sztajno, M., Vacca, W.D., Truemper, J., Pietsch, W., 1986: in *Evolution of Galactic X-ray Binaries* (J. Truemper, W. Lewin, W. Brinkmann, eds.) Reidel Publ. Co., Dordrecht, Boston, etc.

Leahy, D.A., Damle, S.V., Naranan, S., 1986: *J. Astrophys. Astr.* **7**, 299.

Lyne, A.G., McKenna, J., 1989: *Nature* **340**, 367.

Mason, K.O., Branduarti, G., Sanford, P., 1976: *Astron. J.* **203**, L29.

Mazeh, T., van Paradijs, J., van den Heuvel, E.P.J., Savonije, G.J., 1966: *Astron. Astrophys.* **157**, 113.

McClintock, J.E., Rappaport, S., Nugent, J.J., Li, F.K., 1977: *Astrophys. J.* **216**, L15.

McClintock, J.A., Remillard, R.A., 1986: *Astrophys. J.* **308**, 110.

Middleditch, J., Nelson, J.E., 1976: *Astrophys. J.* **208**, 567.

Middleditch, J., Mason, K.O., Nelson, J.E., White, N.E., 1981: *Astrophys. J.* **244**, 1001.

Naylor, T., Charles, P.A., Drew, J.E., Hassall, B.J.M., 1988: *Monthly Notices R.A.S.* **233**, 285.

Naylor, T., Charles, P.A., 1989: *Monthly Notices R.A.S.* **236**, 1P.

Oke, J.B., 1977: *Astrophys. J.* **217**, 181.

Paczynski, B., 1983: *Astrophys. J.* **273**, L81.

Pakull, M.W. and Angebault, L.P., 1986: *Nature* **322**, 511.

Primini, F., Rappaport, S., Joss, P.C., 1977: *Astrophys. J* **217**, 543.

Rappaport, S.A. and Joss, P.C., 1977: *Ann. N.Y. Acad. Sci.* **302**, 460.

Rappaport, S. and Joss, P., 1983: in *Accretion-driven Stellar X-ray Sources*, pp. 1-39, (W. Lewin and E.P.J. van der Heuvel, eds.), Cambridge University Press, Cambridge.

Robertson, H.P., 1938: *Ann. of Math* **39**, 101.

Shapiro, S.L., Teukolsky, S.A., 1983: *Black Holes, White Dwarfs and Neutron Stars*, Wiley & Sons, New York, Chichaster, etc.

Schmidt-Kaler, T.H., 1982: in *Landolt-Börnstein*, New Series, Group VI, Vol. 2b, Springer Verlag, Berlin, Heidelberg, New York.

Stokes, J.H., Taylor, J.H., Weisberg, J.M., and Dewey, R.J., 1985: *Nature* **317**, 787.

Stokes, G.H., Taylor, J.H., and Dewey, R.J., 1985: *Astrophys. J.* **294**, L2.

Taylor, J.H., 1987a: in *The Origin and Evolution of Neutron Stars* (D.H. Helfand and J.-H. Huang, eds.), Dordrecht, Reidel, p. 383.

Taylor, J.H., 1987b: in *Proc. 13th Texas Symp. Relat. Astrophys.*, World Scientific, Singapore.

Taylor, J.H., 1987: *Proc. 13th Texas Symp. Relat. Astrophys.* (Chicago, 1986), pp. 476-477.

Taylor, J.H., Hulse, R.A., Fowler, L.A., Gallahorn, G.E., Rankin, J.M., 1976: *Astrophys. J.* **206**, L53.

Taylor, J.H., Weinberg, J.M., 1982: *Astrophys. J.* **253**, 908.

Taylor, J.H., Stinebring, D.R., 1986: *Ann. Rev. Astron. Astrophys.*, **24**, pp. 285-327.

Taylor, J.H. and Dewey, R.J., 1988: *Astrophys. J.* **332**, 770.

van den Heuvel, E.P.J., 1987: in *The Origin and Evolution of Neutron Stars* (D.H. Helfand and J.-H. Huang, eds.), Dordrecht, Reidel, p. 393.

van Kerkwijk, H.M., van Oijen, J.G.J., and van den Heuvel, E.P.J., 1989: *Astron. Astrophys.* **209**, 173.

Weinberg, J.M., Taylor, J.H., 1981: *Gen. Relat. Grav.* **13**, 1.

Wijers, R.A.M.J., 1989: *Astron. Astrophys.* **209**, L1.

Will, C.M., 1987: Ch. 5 in *300 Years of Gravitation*, (S.W. Hawking and W. Israel, eds.), Cambridge Univ. Press, Cambridge, New York, etc.

NOTE ADDED, JANUARY 1990

The lecture at the Kusadasi NATO Advanced Study Institute (19 September 1989) treated only the case of purely relativistic apsidal motion in connection with PSR 0021-72A. The discussion of a possible white dwarf companion was added later. In a recent letter to *Nature* (9 November 1989), Ables and colleagues reported in detail the unusually difficult observations and gave improved binary parameters, including the "Einstein delay" γ=510$\pm$30 μs. The derived pulsar mass is near 1.4 solar mass.

High Angular Resolution Studies of Binaries at Georgia State University

Donald J. Barry

Center for High Angular Resolution Astronomy
Georgia State University
Atlanta, Georgia 30303

ABSTRACT

KEYWORDS: speckle / interferometry / binary / spectroscopy / differential photometry / astrometry

CHARA, the Center for High Angular Resolution Astronomy at Georgia State University has conducted over 8,000 speckle-astrometric measurements of multiple stars. To complement the primary program of astrometric analysis, several other projects have been initiated to provide differential photometry, quadrant identification, spectroscopic orbital information, and "pure spectra" of binary components. The emphasis has been on real-time analysis of information, which has yielded two new algorithms: the "Fork" and "Directed Vector Autocorrelation" methods. A feasibility study has been completed for construction of a large array optical interferometer: if funded, construction could begin by late 1990.

ASTROMETRY

Since 1977, Georgia State University has been home to a growing program of binary star astrometry. The Center for High Angular Resolution Astronomy, directed by Hal McAlister, has compiled over 10,000 interferometric measurements by observers worldwide, including over 8,400 measurements and 225 discoveries of new binaries made by CHARA astronomers.

C. İbanoğlu (ed.), Active Close Binaries, 729–736.

The CHARA observing program, which will eventually include the entire Yale Bright Star Catalog, is thus a comprehensive survey of bright northern hemisphere stars. With 14 years of data accumulated, a number of stars in the observing program have completed orbits, and their fruit, orbital analyses yielding fundamental stellar parameters, are coming to harvest. [1]

Routine southern hemisphere observation commenced in April, 1989. The near absence of interferometric surveying below -30° declination has been noted [2]. It is expected that many new and interesting binaries may be found in this region, especially as several declination zones have not even received visual survey.

The current speckle camera system, described previously [3], has performed the majority of measurements at CHARA. An advantage of its design is the output of standard video images, which are archived as well as the vector autocorrelograms which are accumulated in real time at the telescope. This unreduced data has proven a rich lode for reanalysis by a new generation of hardware and software.

The impending acquisition of a PAPA camera [4] should extend observing to a magnitude limit of 14 or better, and permit further study of asteroids, and measurement of the Pluto-Charon system, before these objects have been scrutinized by HST.

PHOTOMETRY

Interferometry has classically dealt with pure astrometry, with little attention to photometry or even to true quadrant determination. In the classic case, the quadrant of Capella flipped several times during this century, until the definitive measurement by Bagnuolo and McAlister (1983) [5]. The vector autocorrelation procedure [6] has the disadvantage of yielding only crude Δm estimates, with no information as to the true quadrant. An emphasis at CHARA has been the development of rapid algorithms to establish quadrants, and to generate differential magnitudes when possible. For pairs which do not exceed 0.3 arcsecond separation, direct visual determination of quadrant is not possible, and catalogues of orbital motion, particularly those with poor coverage near periastron, may contain segments of data with inconsistent position angles, subject to the 180° position ambiguity of regular autocorrelation techniques. For example, an early orbit of McAlister 34 by Tokovinin [7] suggested a highly eccentric orbit which was compatible with later astrometry if one sense of position angle was chosen. The use of "Directed Vector Autocorrelation" (DVA) on archival data has shown a quadrant flip between epoch 1984 and 1988 which is incompatible with this early orbit, and which agrees well with a new orbit determined from the different assumption of quadrants. (see figure)

This DVA algorithm, which will be described in a forthcoming paper [8], is essentially an asymmetric vector autocorrelation, in which bright pixels which form the address-list for vector autocorrelation are weighted in the direction of the brighter pixel before the vector difference is accumulated. This algorithm, which has been implemented on a PC/286 and PC/386 architecture using Imaging Technologies PCvision frame digitization hardware, can process up to 15 video frames/second. An advantage in comparing results to archival autocorrelograms is that the symmetrized DVA is identical to the normal vector autocorrelogram.

A more sophisticated technique, which is still suited for real-time application with modest computational requirements, is the "Fork Algorithm" [9], also developed at CHARA, which samples quadruples of speckles in the image plane as "tines of a fork" aligned with the true separation and position angle of the binary. In this technique, each bright speckle in a frame serves as an independent estimate of Δm, and because there are many such estimates per frame, the algorithm outperforms such general purpose reconstruction methods as triple correlation and shift-and-add.

Both of these techniques operate in image, rather than Fourier space, and therefore do not involve the computational overhead of transforming each frame before processing. The image-space approach also avoids the ticklish matter of the "seeing correction" which is inherent in transform techniques. In addition, both algorithms are amenable to implementation in a pipeline arrangement, wherein a chain of processors perform subtasks in the analysis, permitting simultaneous astrometry, differential photometry, and quadrant determination to be performed at full video rates.

It has only been possible to analyze a small subset of archival data with these algorithms, due to the equipment limitations of devices intended for astrometric analysis only. In the case of Capella, however, photometric results suggested the opportunity to test models of stellar evolution for these co-evolved stars. Determination was made of the spectral types of the system [10]. The masses of the system were then calculated from orbital analysis combined with parallax data by van Altena [11]. These data suggested that the primary (G star) is at the beginning of the red giant branch, and not the core helium-burning phase as had been previously believed.

It is hoped that further measurements will yield more such pairs for which fundamental quantities can be derived from orbital data and compared with predictions for co-evolved stars.

SPECTROSCOPY

The CHARA spectroscopic program, coordinated by Ingemar Furenlid, consists of determination of spectroscopic orbits for as many interferometrically resolved systems as possible. The natural overlap between speckle interferometry and spectroscopic orbital characterization has yielded measures of geometric parallax for a number of systems, as well as direct measure of the empirical mass-luminosity relation. To date, such studies have been completed for α Aurigae, 17 Cephei, ϕ Cygni, χ Draconis, γ Persei, and 12 Persei. [12]

A recent initiative, suggested by CHARA member William Bagnuolo, has been to use fiber optics for dissection of a speckle image [13], and via either active control techniques, or passive statistical techniques, to accumulate many "time-sliced" spectra, which are then combined according to selection criteria which yield a net enrichment of the accumulated spectrum in light of the primary or secondary star. By manipulating "enriched" spectra vs. the joint spectrum, it will be possible to reconstruct the pure spectra of the components, and therefore to improve radial velocity measurements.

It is hoped that this will expand the ensemble of stars for which double-lined spectroscopic orbits can be determined, and that greater accuracy in abundance comparison of binaries may be obtained, particularly in evolved systems where abundance of Li and other light elements may yield clues to position on the evolutionary track. This technique may permit more accurate measurement of the Capella Li ratio, which was used to argue for assignment of the primary to the red-giant-branch phase of evolution as mentioned previously. Initial equipment development is underway, and first results are expected next spring.

Another project, under initial study, will involve the construction of a cost-effective 1.3m equivalent aperture spectroscopic telescope, at an estimated cost of only $50,000 U.S. Based on an idea by William Bagnuolo, light from several small-aperture (0.3m) mirrors will independently feed single fibers, which are merged into a linear "slit" in an inexpensive Ebert-Fastie spectrograph. Because of the efficiency of using single fibers, with no bundling loss, and because of the high efficiency coatings available only with small mirrors, a system of nine such mirrors can gather up to 50% more light than the equivalent collecting area larger-mirror using standard coatings and a fiber bundle. These equipment initiatives provide the additional benefit of experience with technology that is needed for the proposed CHARA array.

THE CHARA ARRAY

Since 1984, a primary goal of CHARA has been the design and implementation of a large optical array for very high spatial resolution astronomy. A grant by the National Science Foundation in 1986 has enabled a detailed feasibility study of such a facility to be conducted, with the result that a design now exists for an array which will provide a resolution a hundred times beyond that available via speckle interferometry at the largest single telescopes.

The CHARA array will consist of an arrangement of seven one-meter telescopes enclosed in a 400-meter circle. The alt-azimuth collecting telescopes will direct reduced, 20 cm diameter beams through evacuated light pipes to a central combining area, where 21 pair-wise measurements of fringe visibilities for the telescope combinations will permit measurement of stellar diameters and binary separations in a regime never before accessible.

As part of the feasibility study, a detailed study [14] was conducted of diffraction degradation of the propagated, atmosphere-distorted incipient beam. The reduction factor of five selected for the collecting telescopes was based on this analysis, and represented a compromise between increasing beam degradation, varying as the square of the reduction factor, and the inconvenience and expense of working with large beams.

Another area of concern was addressed by constructing a prototype of the system which will equalize path lengths between the various telescopes and the combining area. Computer control of a moving reflector assembly (optical trombone) was shown to be feasible; metrology was tested, and algorithms were investigated to acquire and maintain fringe lock in the system.

A detailed site survey was conducted, with meticulous emphasis on the seeing quality. CHARA member Wean Tsay, with Nat White of Lowell Observatory, constructed an automated, portable seeing monitor, by which several sites were analyzed. Microthermal measurements and results of the seeing monitor suggest that better seeing than might be expected is at hand at several attractive sites, provided some care is taken in the design of a telescope enclosure.

A detailed plan of implementation is now ready, and awaits only final funding. Three years thereafter, first light will occur, and a vast new pool of binaries can be explored. Nearly every spectroscopic binary will be resolvable with this facility, which will reach a resolution of 0.16 milliarcsecond. The Array's study of active spectroscopic binaries, such as Algol, which are associated with "wide" (0.1 arcsecond) companions with established orbits will further refine knowledge of masses and luminosities in these interesting systems. Performance will initially extend to magnitude 11, although adaptive optics under development will insert easily into the light path, and extend performance to magnitude 14.

CONCLUSION

A variety of novel instrumental initiatives promise a new era ahead in differential photometry and spectral analysis of binary systems. A decade's worth of patient measurement is now yielding a regular and bountiful harvest. As more speckle orbits are determined and integrated with spectroscopic and parallax information, meaningful discussions of a larger number of systems will be possible, and will present a useful confrontation of theory vs. experiment for models of stellar evolution, the mass-luminosity relation, and the cosmic distance scale.

The author would like to acknowledge the primary role of Hal McAlister in inspiring and supporting the work referred to in this summary. In addition, substantial credit must go to William Hartkopf, William Bagnuolo, Ingemar Furenlid, Douglas Gies, Jim Sowell, Edmund Dombrowski, Wean-Shun Tsay, Ali Al-Shukri, and the many CHARA collaborators who have made the projects referred to herein possible.

The program of binary star speckle interferometry is supported by the National Science Foundation through NSF grant No. AST 86-13095 and the Air Force Office of Scientific Research through AFOSR grant No. 86-0134. The CHARA array feasibility study was conducted under NSF grant No. AST 84-21304. We gratefully acknowledge this support.

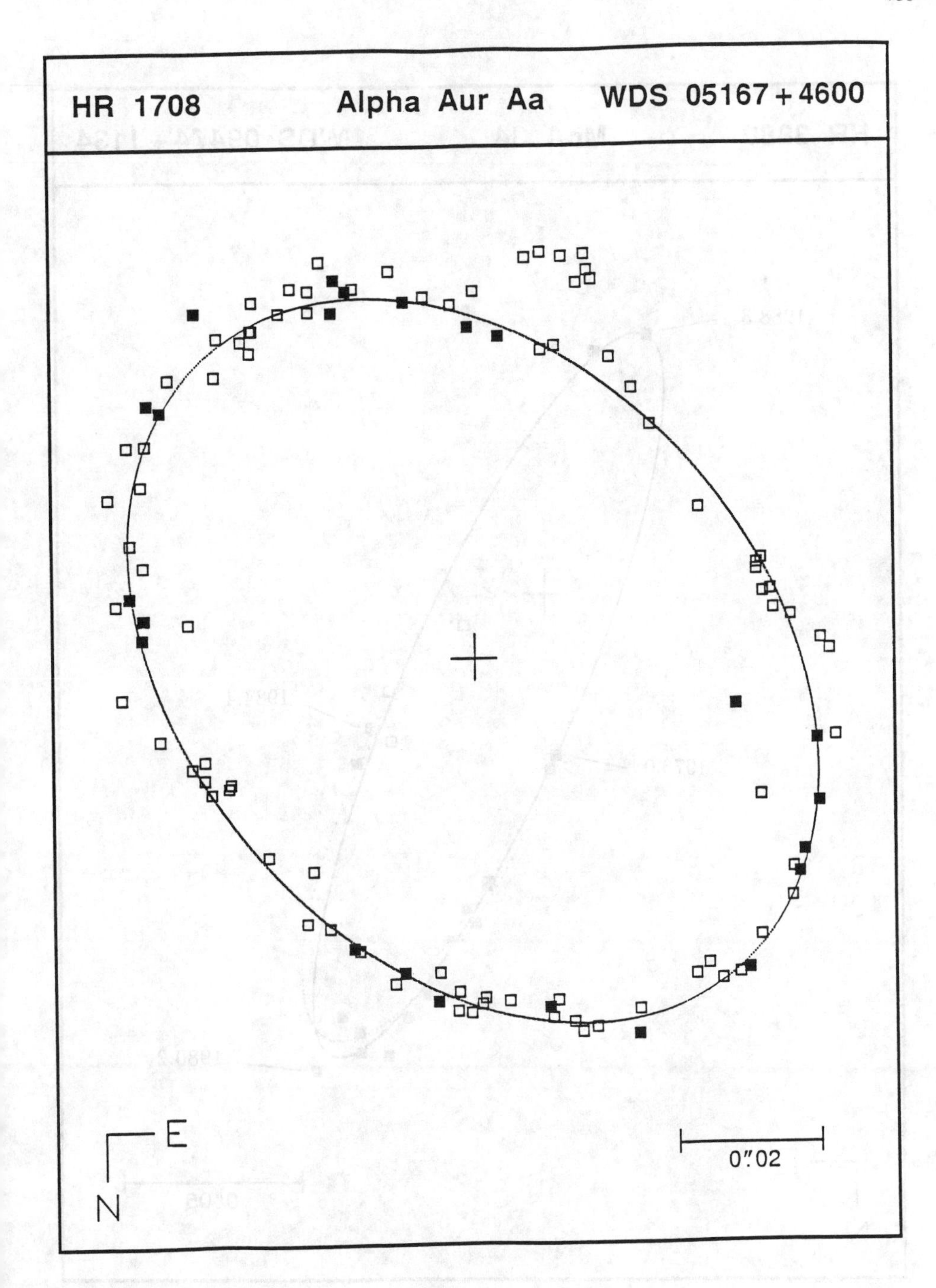
HR 1708
Alpha Aur Aa
WDS 05167+4600
E
N
0″.02

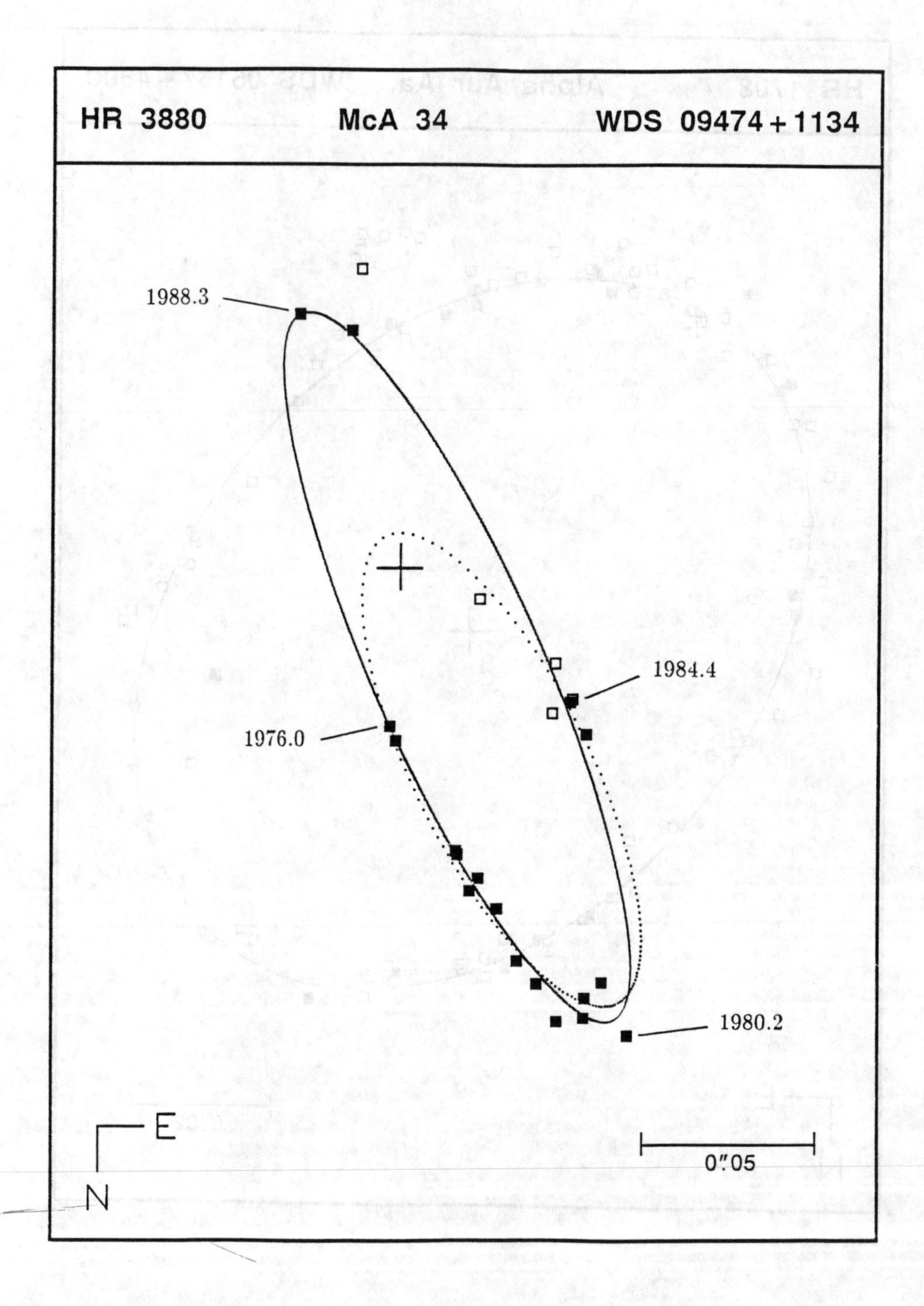
HR 3880
McA 34
WDS 09474+1134
1988.3
1984.4
1976.0
1980.2
E
N
0."05

EXTREME ULTRAVIOLET SPECTROSCOPY AS A PROBE OF ACTIVE CLOSE BINARIES

STUART BOWYER
Astronomy Department and
Center for Extreme Ultraviolet Astrophysics
University of California, Berkeley
Berkeley, California 94720
USA

ABSTRACT. Interactions between the members of close binary systems are expected on theoretical grounds to produce substantial amounts of extreme ultraviolet, or EUV, (λλ 100–1000 Å) radiation. The opacity of the interstellar medium is sufficiently low that a fair number of these systems should be observable. Spectroscopy of these sources should provide substantial insights into the character of these objects. Several spectrometers with capabilities in the EUV will be in operation in the early 1990s, including the Hopkins Ultraviolet Telescope (HUT), the Extreme Ultraviolet Explorer (EUVE), and the Orbiting Retrievable Far and Extreme Ultraviolet Spectrometer (ORFEUS). The spectroscopic capabilities of the EUVE satellite will be of special interest to workers in the field of active close binaries since NASA is planning an extensive multi-year guest observer program for this mission which will be carried out in a manner similar to that of the IUE program.

1. Introduction

The study of close binary systems extends well into the last century and is the cornerstone for the determination of masses of main sequence stars. However, in the past ten to twenty years, this field has received substantially increased attention and has expanded to impact a wide variety of different areas in astronomy. There are a variety of reasons for this increased interest; many of them are due to new discoveries from space, but this is certainly not the whole story. Some of the factors influencing this interest include:

a) the realization that one of the brightest X-ray sources in the sky was powered by interactions in a close binary system,
b) the discovery that many X-ray sources were binary systems,
c) the identification at optical wavelengths of the class of RS Cvn stars,
d) the discovery that RS Cvn stars were X-ray sources,
e) studies at optical wavelengths which showed that many parameters of these systems could be characterized through detailed spectroscopy, and
f) the discovery that cataclysmic variables were soft X-ray/EUV emitters.

A substantial amount of theoretical work has led to the conclusion that a considerable amount of the emission from these systems will be in the EUV. Hence, investigations of these objects in this band should be especially productive if these systems can be observed. Absorption of EUV emission by the interstellar medium was thought to be so extreme that this would not be possible; it is now recognized that the interstellar medium is quite inhomogeneous and

C. İbanoğlu (ed.), Active Close Binaries, 737–746.

is very tenuous in many directions. As a consequence, absorption by the interstellar medium will be minimal at least in these directions, and a substantial number of this class of objects should be observable.

At this time, we have only limited information on classes of objects that are EUV emitters because no all-sky surveys have yet been carried out. This situation will soon change. In 1990, the English Wide Field Camera (Pye et al. 1989) will carry out an all-sky survey from 60 to 200 Å, and in 1991 the Extreme Ultraviolet Explorer (EUVE) will provide an all-sky survey in several bands covering wavelengths from 70 to 750 Å (Malina and Bowyer 1989).

In the following, I will briefly review some of the limited observational results which indicate that at least some classes of active close binaries will be strong sources of EUV emission. I will then describe the capabilities of three instruments to be flown in the next three to five years which will provide the capability of carrying out detailed spectroscopy of sources in the EUV. The Extreme Ultraviolet Explorer, to be launched in 1991, is especially interesting in this regard, since all the spectroscopy on this mission will be carried out by guest observers.

2. Active Close Binaries as EUV Sources

A variety of classes of binary stars have been suggested as likely EUV-emitting objects. In the absence of surveys, these ideas are almost entirely conjectural. However, two classes of active binary stars have been detected in the EUV: cataclysmic variables and coronally active stars.

A. CATACLYSMIC VARIABLES

Cataclysmic variables are semi-detached binary stars; the donor star is a low-mass, late-type star that fills its Roche lobe and transfers matter to a white dwarf. The class has been divided into several subclasses depending on such parameters as the presence of a magnetic field in the white dwarf or other variables. Because the gravitational well of the white dwarf is very deep, it has long been theorized that accretion in these objects will lead to substantial X-ray emission. With accretion rates of 10^{-7} to 10^{-11} $M_\odot$/year, the total X-ray luminosity was expected to be in the range of 10^{32} to 10^{36} erg s^{-1}. At typical distances of 50 to 300 pc, these objects would be extremely bright X-ray sources and they would numerically dominate all other classes of X-ray objects. In reality, only a few of these objects have been detected in X-rays. A common explanation is that they are emitting most of their flux in the extreme ultraviolet, and a substantial amount of theoretical work has been carried out in support of this contention; the reader is referred to the articles by Lamb (1989), Raymond and Mauche (1989), and Wade (1989) for recent modeling efforts.

In addition to the intrinsic interest in understanding cataclysmic variables, these objects offer enormous possibilities for testing models of accretion disk physics. Accretion disks dominate the appearances and energetics of objects as diverse as protostars and active galactic nuclei, yet the physics of accretion is poorly understood. In the process of accretion, the gas in the disk must shed nearly all of its angular momentum. While this process can be modeled successfully by parameterizing the viscosity of the disk, the physical processes involved have been variously ascribed to convective turbulence (Cannizzo and Cameron 1988), or magnetic turbulence (Coroniti 1981), or to the interaction between a large-scale magnetic field and a disk wind (Blandford and Payne 1982). We await observational tests of these theories.

We have direct observational evidence that at least some members of this class will be observable in the EUV. SS Cygni was one of the first sources detected in the EUV when it was observed during outburst by the Berkeley EUV telescope on the Apollo-Soyuz Test Project (Margon et al. 1978). U Geminorum was discovered as a soft X-ray/EUV source by Mason et al. (1978), who detected the source at peak outburst and found it to be at least a factor of 100 times less intense during the subsequent quiescent optical phase. A substantial amount of work has subsequently been carried out on this source (see Patterson [1988] for a review), but the point to be made here is that it is certainly an EUV source.

VW Hyi is a dwarf nova that was monitored extensively by observers using the EXOSAT satellite. It was found to be an intense source of EUV radiation at wavelengths longer than 200 Å (van der Woerd and Heise 1987). This flux increased by two orders of magnitude during outburst and declined with V magnitude in the past outburst phase.

Polidan and Mauche (1989) used the EUV spectrometer on Voyager to search for five nearby CV's, listed in Table 1. It should be noted that these observations were at wavelengths longer than 600 Å where absorption by the interstellar medium is most severe.

TABLE 1 — Cataclysmic Variables Observed by Polidan and Mauche

Star	Type	V (mag)	d (pc)	N(H I) (cm^{-2})
SS Cyg	DN/U Gem	8–12	95	3.5×10^{19}
VW Hyi	DN/SU UMa	8–13	65	6×10^{17}
V3885 Sgr	NL/UX UMa	10	130	5.6×10^{19}
RW Sex	NL/UX UMa	10.5	150	8.9×10^{19}
IX Vel	NL/UX UMa	9.5	140	2.0×10^{19}

These authors were especially fortunate in observing SS Cyg in outburst and VW Hi in super outburst, but they detected no EUV emission at a limiting flux level of ~ 10^{-11} erg cm^{-2} s^{-1} $Å^{-1}$ from these objects. They detected no emission at ~ 10^{-12} erg cm^{-2} s^{-1} $Å^{-1}$ from the other three objects. We have insufficient knowledge to be overly surprised by a lack of detection at these intensity levels. It is important to note, however, that the spectrometer on EUVE will have *two to three orders of magnitude* more sensitivity over this waveband for continuum flux measurements, and more than an order of magnitude higher resolution, thereby further increasing its sensitivity to line emission.

VV Puppis is one of about 10 known AM Herculis stars that are one of the subclasses of the cataclysmic variable class. These systems contain strongly magnetized white dwarfs accreting from a binary companion with the white dwarf rotation locked to the orbital rotation. No accretion disk is present as the inflow material is channeled onto the pole of white dwarf by the magnetic field. Patterson et al. (1984) reported on *Einstein* observatory results on VV Puppis which showed that this star is an intense source of soft X-rays. These authors conclude it will certainly be an intense source of EUV emission. The optical and X-ray light curves indicate that the light emanates from a bright spot near one magnetic pole of the white dwarf. The ratio of soft to hard X-rays in VV Puppis is the highest for all of the AM Her objects, but this ratio is embarrassingly high for all these objects and is difficult to explain theoretically. Indeed, the problem is known as the "soft X-ray problem." Whether the problem is theoretical or observational in nature will only be determined by future work.

B. CORONALLY ACTIVE STARS

Chromospheric and coronal activity are thought to be interrelated, and studies of the sun have shown that spectroscopy in the EUV is especially productive in the study of these phenomena (Doschek 1989). This activity is enhanced in the case of interacting binaries such as the RS Cvn class of stars. Lemen et al. (1989) have reported on EXOSAT transmission grating studies for three nearby binary systems—Capella (G6III+F9III), σ^2CrB (F8V+GIV), and Procyon (F5IV/V+DF)—which show emission features in the EUV. Capella and σ^2CrB are both members of the RS Cvn class of binaries. They show emission features to 140 Å. These authors interpret the data in terms of two temperature coronal models. It seems almost inescapable that lower temperature enhanced chromospheric emission will be present in these systems. Procyron is a quite different source with strong EUV emission from 150 Å to the long wavelength limit of the EXOSAT instrumentation. The origin of this emission is not yet clear, but it may be a low-temperature ($\sim 0.6 \times 10^6$ K) corona.

V471 Tauri is a 12.5 hour eclipsing binary (DA+K2V) and has been classified as an RS Cvn object. Van Buren et al. (1980) identified the star as a soft X-ray/EUV source from HEAO-I data. They considered the K star to be the likely source of soft X-rays as did Young et al. (1983), who studied the object with the Einstein IPC detector. Jensen et al. (1986), using the EXOSAT satellite, established that **both** stars were soft X-ray/EUV sources. The flux from the DA is consistent with thermal emission at 35,000 K. The K star emission is describable in terms of a thermal plasma in the range 0.3–2 keV, but this result is much less certain. Sion et al. (1989) have made a case for the intriguing suggestion that this system is a pre-cataclysmic binary and may well represent a class of such objects that are bright in the EUV.

3. Prospects for the Future

Several missions are scheduled for deployment in space in the early 1990s with spectroscopic capabilities in the EUV. The Hopkins Ultraviolet Telescope (HUT), Davidsen et al. (1989), will be flown on the Space Shuttle as part of the ASTRO mission in the Spring of 1990, and at various times thereafter. HUT is a 0.9 meter telescope with a prime focus Rowland spectrograph with a resolution $\lambda / \Delta\lambda \sim 900$. The instrument is designed primarily for use at wavelengths above 900 Å. Its effective area is shown in Figure 1. There is a substantial effective area in the EUV in second order; the problem is that lines above 900 Å (which are likely to be stronger than the EUV lines because they will not be absorbed by the interstellar medium) will be difficult to separate from the EUV lines. This problem has been addressed by the addition of a thin aluminum filter which eliminates lines longward of 900 Å at the loss of some sensitivity. The effective area of the HUT telescope with this filter in place is shown in Figure 2.

The EUVE spectrometer will provide moderate resolution ($\lambda / \Delta\lambda \sim 200$) spectroscopy in the 80–760 Å band. The spectrometer design is based on the use of novel gratings (Hettrick and Bowyer 1983), which are placed in the converging beam of the mirror. The principle of this spectrometer is outlined in Figure 3. The gratings are large (10 × 30 cm) mechanically ruled plane gratings with a variable line-density that varies continuously from one end of the grating to correct first order aberration arising from the varying incidence angle at the grating. The spectrometer optical path thus has only three reflections achieving a high throughput. Each spectral channel uses one-sixth of the telescope aperture (the remaining half of the beam is

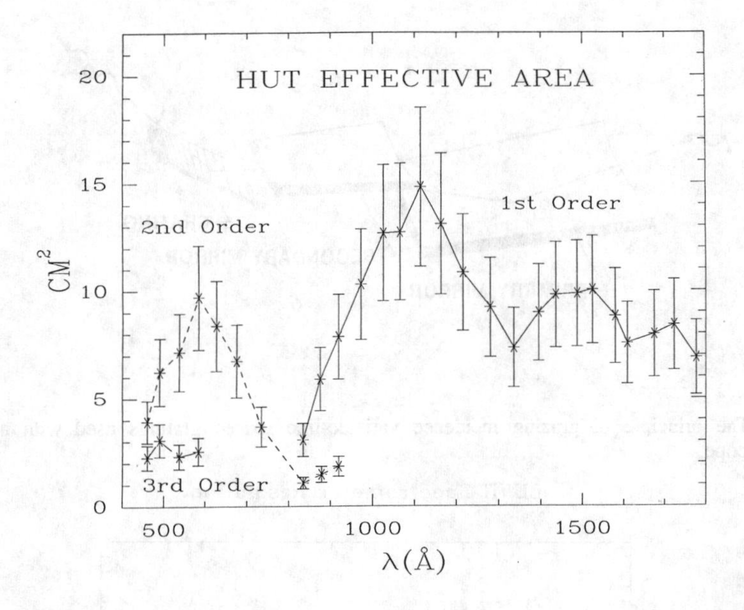

Figure 1 — Effective area of the HUT telescope in first, second, and third order, when observing with no order-separating filter in place.

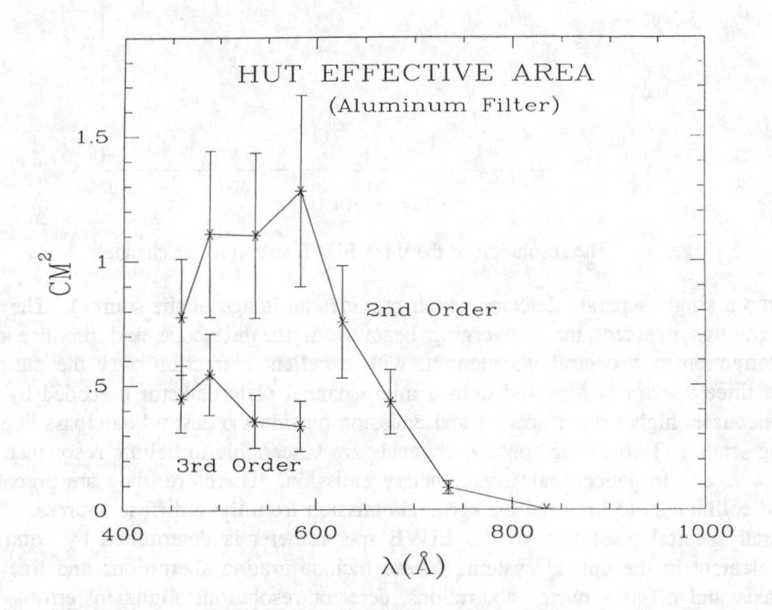

Figure 2 — Effective area of the HUT telescope in the EUV, when observing through the aluminum filter, which rejects first-order FUV light.

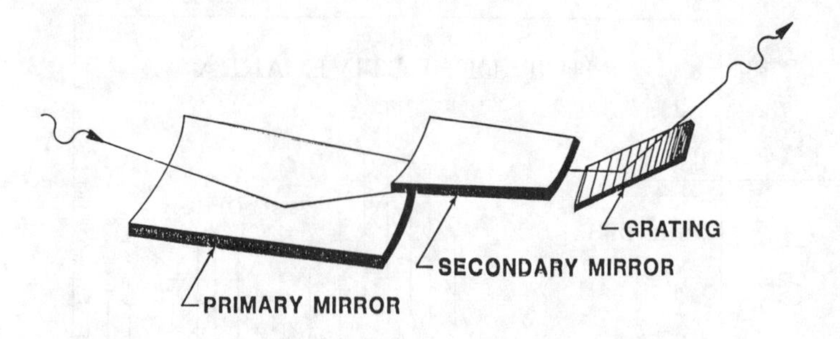

Figure 3 — The principle of grazing incidence variable-line spaced gratings used with a grazing incidence telescope.

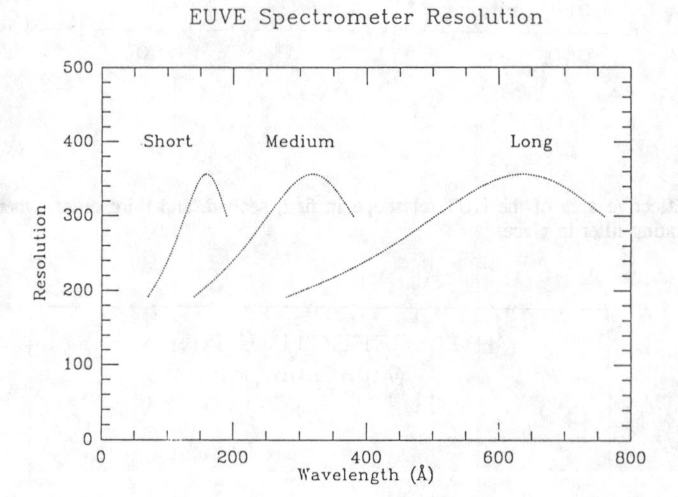

Figure 4 — The resolution of the three EUVE spectrometer channels.

focussed onto a single separate detector which provides an image of the source). The variable line-space gratings intercept the converging beam from the telescope and produce a quasi-stigmatic correction at a central wavelength, with excellent correction over the entire band. Each of the three spectra is focussed onto a microchannel plate detector preceded by a filter. The filter attenuates higher order spectra and emission outside the desired bandpass due to possible grating scatter. Two of the spectral channels are susceptible to helium resonance lines at 304 and 584 Å due to geocoronal/interplanetary emission. Therefore, they are precollimated by wire grid collimators to limit the background emission from these diffuse sources.

The overall spectral resolution of the EUVE spectrometer is determined by contributions from each element in the optical system. These include grating aberrations and line spacing errors, on-axis and off-axis mirror aberrations, detector resolution, alignment errors, and the in-orbit aspect knowledge uncertainty. Taking all these contributions into account, we show the expected in-flight resolution of each of the three spectrometer channels in Figure 4.

The overall sensitivity of each of the spectrometers is shown in Figure 5 for an integration time of one day, which is expected to be a typical observation. For reference, the spectrum of the hot white dwarf HZ 43 is shown for a similar integration period.

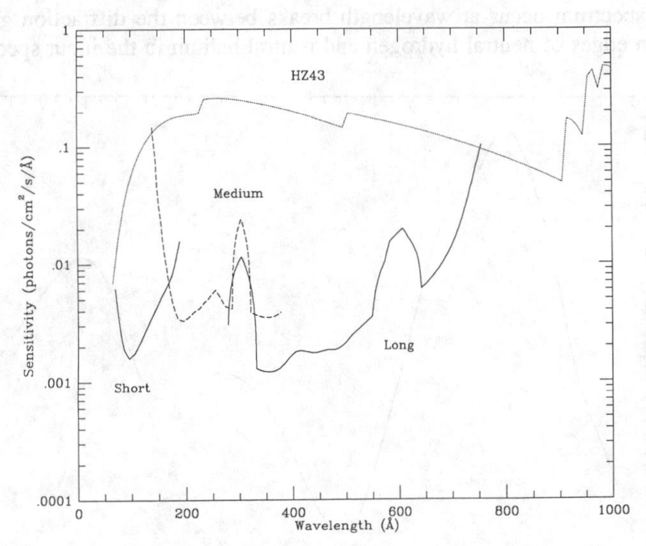

Figure 5 — The continuum level sensitivity of the three EUVE spectrometers for a typical one-day observing run. Data based on laboratory measurements of all components. The spectrum of HZ 43 is shown for comparison.

It is also of interest to discuss what will eventually be possible with higher resolution spectrometers in the EUV. The ORFEUS mission is a joint project of NASA and the BMFT in West Germany (Grewing et al. 1989). The Berkeley spectrograph (Hurwitz and Bowyer 1989) for this mission will obtain high resolution ($\lambda / \Delta\lambda > 7000$) spectra between 390 and 1200 Å. The Berkeley instrument incorporates a set of four novel, spherically figured, varied line-space gratings used in a geometry which is superficially similar to that of the classic Rowland mount. Unlike the conventional Rowland spectrograph, the groove spacing of each grating varies as a fourth-order function of position on the optic. Furthermore, the distances from the telescope image to the grating center, and from the grating center to the detector surface, are not equal to the values on the corresponding Rowland circle. The astigmatism and aberrations of this spectrograph are substantially lower than what would be achievable with a conventional Rowland spectrograph and toroidal, uniform line-space diffraction gratings.

The instrument contains four diffraction gratings, arranged so that each is illuminated by a ''wedge'' on the annular aperture of the telescope primary mirror. Each grating is used to study only one-fourth of the total 390–1200 Å bandpass, and each operates over a relatively small range of diffraction angle, allowing high diffraction efficiency to be maintained.

The telescope primary mirror will be coated with iridium, a stable coating which provides good normal incidence reflectivity across the instrument bandpass. The diffraction gratings will be coated with the same material. Preliminary measurements indicate that the gratings will achieve a groove efficiency of ~ 50% (Hurwitz et al. 1989) and that the quantum detection efficiency will be equal to that measured for the stable photocathode KBr by Siegmund et al. (1987). The overall effective area vs. wavelength is shown in Figure 6. To demonstrate the power of the instrument, in Figure 7 we show a simulated spectrum of the hot white dwarf HZ 43 as would be observed with this instrument for an integration time of 3600 s. Discontinuities in the spectrum occur at wavelength breaks between the diffraction gratings and also at the absorption edges of neutral hydrogen and neutral helium in the input spectrum.

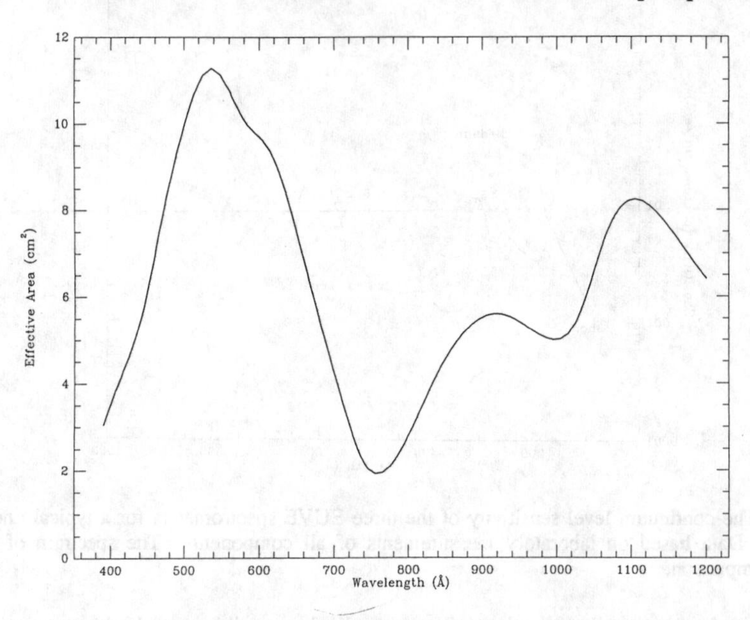

Figure 6 — Predicted effective area of the Berkeley ORFEUS spectrometer.

4. Conclusions

The study of binary stars has a long and illustrious position in astronomy. Much recent work in this field has emphasized active binary stars. Much of this work has built upon results from instruments in space observatories or ground-based observations used synergistically with data from space instrumentation.

The 1990s will provide astronomers with a variety of new tools to investigate the properties of these objects; spectroscopy in the EUV will be one of these tools. There are reasons to believe that EUV spectroscopy will be especially productive in this regard. NASA is promoting a strong Guest Observer Program for the EUVE in which resources will be allocated both to U.S. and foreign observers on a basis very similar to that used for the IUE observatory.

The science data center for this mission will be at the Center for Extreme Ultraviolet Astronomy at the University of California at Berkeley. Personal involvement at the Center will be encouraged, especially for first-time users, to allow familiarization with the data analysis programs which will be available. However, direct involvement at Berkeley, even for first-time

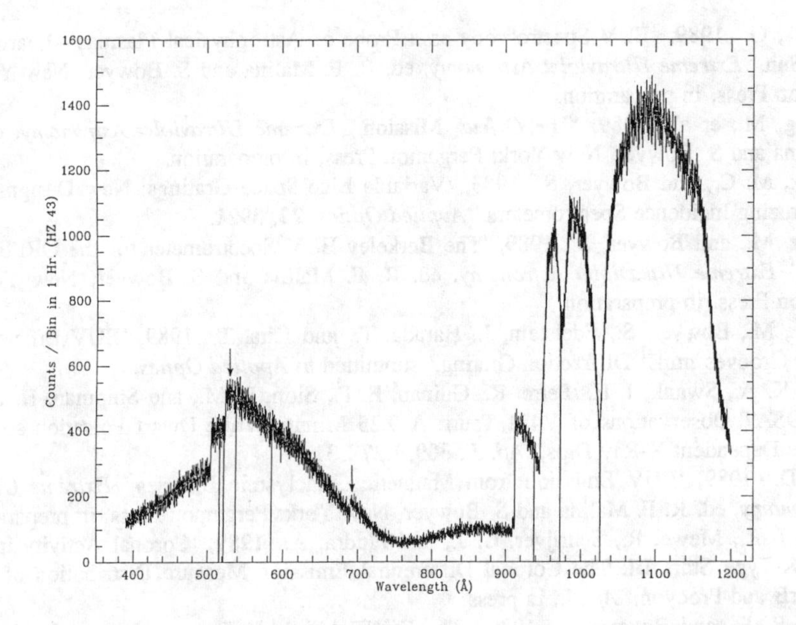

Figure 7 — Sample spectrum of hot white dwarf HZ43 as would be observed by the ORFEUS spectrometer. Counts per spectral bin for a 3600 s integration are plotted versus wavelength. Column density of H I assumed to be 2×10^{17} cm^{-2}; column of He I assumed to be 1/10 that of H I. Absorption edges of interstellar H I and He I are shown, as are strong resonance lines of these species. Because the spectral resolution changes discontinuously at grating crossover wavelengths, the simulated counts per spectral bin display those discontinuities (marked with *arrows*).

users, is by no means mandatory. Nonetheless, it is especially appropriate at a NATO Advanced Study Institute to point out that NATO country scientists who have observing programs on EUVE, or who intend to develop such programs, are eligible for NATO Collaborative Research Grants and NATO Science Fellowships which will support travel to Berkeley and related research costs.

I look forward to the new insights which will be available in the field of active binary stars in a conference such as this in 1995, which will be held, it is to be hoped, in conditions as ideal as they are at this conference.

Acknowledgements: This work was supported by NASA Contract NAS5-30180.

References

Blandford, R. D., and Payne, D. G. 1982, *MNRAS,* **199**, 883.
Cannizzo, J. K., and Cameron, A. G. W. 1988, *Ap. J.,* **330**, 327.
Coroniti, F. V. 1981, *Ap. J.,* **244**, 587.
Davidsen, A. F., Kimble, R. A., Durrance, S. T., Bowers, C. W., and Long, K. S. 1989, 'The EUV Capabilities of the Hopkins Ultraviolet Telescope,' *Extreme Ultraviolet Astronomy,* ed. R. F. Malina and S. Bowyer, New York: Pergamon Press, in preparation.

Doschek, G. 1989, 'EUV Spectroscopy as a Probe of Astrophysical Plasmas—Learning from the Sun,' *Extreme Ultraviolet Astronomy,* ed. R. F. Malina and S. Bowyer, New York: Pergamon Press, in preparation.

Grewing, M., et al. 1989, 'The *Orfeus* Mission,' *Extreme Ultraviolet Astronomy,* ed. R. F. Malina and S. Bowyer, New York: Pergamon Press, in preparation.

Hettrick, M. C., and Bowyer, S. 1983, 'Variable Line-Space Gratings: New Designs for Use in Grazing Incidence Spectrometers,' *Applied Optics,* **22**, 3921.

Hurwitz, M., and Bowyer, S. 1989, 'The Berkeley EUV Spectrometer for the ORFEUS Mission,' *Extreme Ultraviolet Astronomy,* ed. R. F. Malina and S. Bowyer, New York: Pergamon Press, in preparation.

Hurwitz, M., Bowyer, S., Edelstein, J., Harada, T., and Kita, T. 1989, 'EUV Efficiency of a 6000 Grooves mm^{-1} Diffraction Grating,' submitted to *Applied Optics*.

Jensen, K. A., Swank, J. H., Petre, R., Guinan, E. F., Sion, E. M., and Shipman, H. L. 1986, 'EXOSAT Observations of V471 Tauri: A 9.25 Minute White Dwarf Pulsation and Orbital Phase Dependent X-Ray Dips,' *Ap. J.,* **309**, L27-L31.

Lamb, D. 1989, 'EUV Emission from Magnetic Cataclysmic Binaries,' *Extreme Ultraviolet Astronomy,* ed. R. F. Malina and S. Bowyer, New York: Pergamon Press, in preparation.

Lemen, J. R., Mewe, R., Schrijver, C. J., and Fludra, A. 1989, 'Coronal Activity in F-, G-, and K-Type Stars III: The Coronal Differential Emission Measure Distribution of Capella, σ^2 CrB and Procyon,' *Ap. J.,* in press.

Malina, R. F., and Bowyer, S. 1989, 'The EUVE Mission,' *Extreme Ultraviolet Astronomy,* ed. R. F. Malina and S. Bowyer, New York: Pergamon Press, in preparation.

Margon, B., Szkody, P., Bowyer, S., Lampton, M., and Paresce, F. 1978, 'Extreme Ultraviolet Observations of Dwarf Novae from Apollo-Soyuz,' *Ap. J.,* **224**, 167.

Mason, K. O., Lampton, M., Charles, P., and Bowyer, S. 1978. 'The Discovery of Low-Energy X-Ray Emission from U Geminorum,' *Ap. J.,* **226**, L129-L132.

Patterson, J., Beuermann, K., Lamb, D. Q., Fabbiano, G., Raymond, J. C., Swank, J., and White, N. E. 1984, 'VV Puppis: The Soft X-Ray Machine,' *Ap. J.,* **279**, 785-797.

Patterson, J. 1988, 'Cataclysmic Variables as X-Ray Sources,' *Proceedings of "From EINSTEIN to AXAF,"* Columbia Astrophysics Laboratory Contribution Number 378, New York: Columbia University.

Polidan, R. S., and Mauche, C. W. 1989, *Ap. J.,* submitted.

Pye, J. P., Pounds, K. A., and Watson, M. G. 1989, 'The XUV Wide Field Camera for ROSAT,' *Extreme Ultraviolet Astronomy,* ed. R. F. Malina and S. Bowyer, New York: Pergamon Press, in preparation.

Raymond, J., and Mauche, C. 1989, 'The EUV Emission of Cataclysmic Variables,' *Extreme Ultraviolet Astronomy,* ed. R. F. Malina and S. Bowyer, New York: Pergamon Press, in preparation.

Siegmund, O. H. W., Lampton, M., Everman, E., and Hull, J. 1988, 'Soft X-ray and EUV Quantum Detection Efficiency of KBr-coated MCPs,' *Ap. Opt.,* **27**, 1568.

Sion, E. M., Bruhweiler, F. C., and Shipman, H. L. 1989, 'EUV Diagnostics of the Active K Dwarf — Hot White Dwarf Pre-Cataclysmic Binary U471 Tauri,' *Extreme Ultraviolet Astronomy,* ed. R. F. Malina and S. Bowyer, New York: Pergamon Press, in preparation.

Van Buren, D., Charles, P. A., and Mason, K. O. 1980, *Ap. J. (Letters),* **242**, L105.

Van der Woerd, H. and Heise., J. 1987, *Ap. J.,* **225**, 141.

Young, A., et al. 1983, *Ap. J.,* **267**, 655.

Wade, R. 1989, 'EUV Constraints on Boundary Layer Models for Cataclysmic Variables,' *Extreme Ultraviolet Astronomy,* ed. R. F. Malina and S. Bowyer, New York: Pergamon Press, in preparation.

MODELLING THE CORONAE AND CHROMOSPHERES OF RS CVn SYSTEMS BY THE ANALYSIS OF ULTRAVIOLET, X-RAY AND RADIO OBSERVATIONS

JEFFREY L. LINSKY[1]
Joint Institute for Laboratory Astrophysics
National Institute of Standards and Technology
and the University of Colorado
Boulder, Colorado 80309-0440
USA

'We all stood in the vast theatre of ancient *Ephesus*. – the stone-benched amphitheatre I mean – and had our picture taken. We looked as proper there as we would look anywhere, I suppose. We do not embellish the general desolation of the desert much. We add what dignity we can to a stately ruin with our green umbrellas and jackasses, but it is little. However, we mean well.' *Mark Twain*[2]

ABSTRACT. I will summarize the present status of chromospheric and coronal models of active stars in RS CVn systems. The first generation models, which assumed plane-parallel, one-component layers are now being supplanted by second generation models in which the covering fraction of the plage or flare plasma is estimated by a Doppler-imaging analysis. Third generation models should include the magnetic field geometry and the total energy balance in a self-consistent manner.

1 INTRODUCTION

Some ten years ago early observations of ultraviolet spectra with the IUE satellite, x-ray fluxes with the *Einstein* observatory, and radio flux measurements with single dish and interferometric arrays led to the first-generation models for the chromospheric and coronal regions of active stars in RS CVn systems. These first-generation models (e.g. Simon and Linsky 1980) assumed that the chromospheres are plane-parallel, one-component layers in hydrostatic equilibrium and that the coronae (e.g. Rosner, Tucker and Vaiana 1978; Swank

[1]Staff Member, Quantum Physics Division, National Institute of Standards and Technology
[2] *The Innocents Abroad*, ch 40 (1869)

C. İbanoğlu (ed.), Active Close Binaries, 747–759.

et al. 1981) may be explained by static flux tubes in which radiation balances thermal conduction from the hottest plasma located at the apex of the loop down to the chromosphere. In addition, the radio data were interpreted as gyrosynchrotron emission from mildly relativistic electrons in a region comparable in size with the binary star separation with pervasive $\sim$ 10 G magnetic fields (e.g. Brown and Crane 1978).

While each of these models succeeds in explaining a limited data set with a minimum of assumptions, all are inadequate for several reasons:

1. Each model explains only a small portion of the data and cannot explain data from the other wavelength regions.

2. Each model assumes a very simple geometry, although solar analogy and an increasing number of observations of RS CVn systems indicate that the geometry must be more complex.

3. The models for each atmospheric region are decoupled from each other with no boundary conditions or physical relations (e.g. energy and pressure balance) connecting the different regions.

4. The models do not explain the observed spatial and temporal variability in a self–consistent manner.

In this review, I will first describe the assumptions underlying the first–generation models and summarize the conclusions that have been obtained from these models. Unfortunately, these one–component models ignore the complex geometry that must be present and are therefore unrealistic. Second–generation models that begin to incorporate realistic geometries are now being built. I will describe these recent efforts, and suggest how third–generation models, which begin to incorporate energy balance, should be built.

2 THE ROLE OF MAGNETIC FIELDS

While magnetic field strengths and fluxes have not yet been measured for RS CVn systems, there are many reasons for believing that they play critical roles in controlling the atmospheric structure and energy balance in these systems. Both photometry (e.g. Rodonò *et al.* 1986) and Doppler imaging (e.g. Vogt and Penrod 1983) techniques clearly demonstrate the presence of dark starspots, which must be cooler than the surrounding photosphere on the basis of their redder colors and such spectral signatures as stronger TiO bands near optical minimum. Spots, which can cover up to 40% of the visible hemisphere, must be regions of strong magnetic fields. The close agreement between the size and location of spots on V711 Tauri in 1981.8 derived by the two complementary techniques (cf. Linsky 1988) indicates that both spot modelling techniques provide sensible estimates of the spot properties, although there remain uncertainties in both approaches (cf. Rodonò 1986). The presence of gyrosynchrotron radio emission, flares, and bright plage regions provide additional indirect evidence for large scale magnetic fields in these systems.

An important result of the direct measurements of magnetic fields in dwarf stars by the Robinson (1980) line broadening technique (see reviews by Saar 1990 and by Linsky 1985, 1989) is that nonspot photospheric field strengths appear to be proportional to $P_{gas}^{1/2}$, where

P_{gas} is the photospheric gas pressure. This relation is consistent with balance between the magnetic pressure in the flux tubes and the gas pressure of the surrounding magnetic plasma. This scaling law predicts nonspot field strengths of 400–800 G for photospheric gas pressures 4–16 times lower than for the Sun, which is expected for the photosphere of an early–K subgiant or giant. Such fields are very difficult to measure with present line broadening techniques especially for rapidly–rotating stars, although Gondoin, Giampapa and Bookbinder (1985) have measured a photospheric field strength of roughly 600 G for λ And (G8 III–IV+?). Given that the x-ray luminosities ($L_x \leq 10^{31.6}$, Majer *et al.* 1986) and surface fluxes ($f_x \leq 10^{7.6}$ erg cm^{-2} s^{-1}) for RS CVn systems approach the largest values measured for any stars, a situation called 'saturation' by Vilhu (1987), the filling factor for nonspot regions must exceed 0.5. The spot fields on these stars have never been measured, but likely exceed the nonspot fields by at least a factor of 2 to suppress convection sufficient to make the spots appear dark.

Given these estimates, I can make a guess about the total magnetic flux in the photosphere of a typical active star in an RS CVn system. I assume that the K0 IV star has a radius of $3R_\odot$, that 70% of the surface area is nonspot with a field strength of 800 G and a filling factor of 0.5, and that 30% of the surface area is spot with a field strength of 3x800 G = 2400 G and a filling factor of 1.0. The total magnetic flux of this typical active star is 1.5 10^{26} Maxwell. By comparison, the solar magnetic flux is roughly 5 10^{23} Maxwell, assuming that 1500 G fields cover 0.02 of the solar surface. Thus the magnetic flux of this typical RS CVn active star is 300 times larger than for the Sun and the magnetic energy density is about 400 times larger. The decrease in typical field strengths with distance above the photosphere is uncertain and should depend upon the geometry of the photospheric fields and the extent to which the field lines become twisted as a result of convective motions and differential rotation. Given that the spots are very large and the photospheric filling factors can approach unity, one might expect the field strength to decrease as $(R/R_\star)^{-2}$ or $(R/R_\star)^{-3}$. Thus typical coronal fields at $3R_\star$ should be in the range 40–110 G for the typical active star described above, and the field strengths above starspots could be in the range 100–270 G. The lower and upper ranges are for the two cited scaling laws. I will return to the magnetic field estimates later in the paper.

3 CHROMOSPHERIC MODELLING

3.1 First Generation: One–component models

The first generation of chromospheric models for stars in RS CVn systems assumed plane–parallel, one–component atmospheres in hydrostatic equilibrium. One technique for computing these atmospheres involves the solution of the radiative transfer and statistical equilibrium equations for such ions as Ca^+, Mg^+, Si^+, C^+, Si^{++} and H for an assumed distribution of temperature with column mass density. The temperature–mass distribution for these models (called *Tm models*) are modified by a trial and error approach until the computed line surface fluxes and profiles match the observed quantities. This technique is called *semi-empirical modelling.* I emphasize that the surface fluxes are derived from the observed fluxes assuming that the emitting region is the *entire visible hemisphere* of the star. Table 1 summarizes the results of the first–generation models that employ several different techniques. The first models computed in this way were reviewed by Linsky (1984).

A second technique, which I call the *DS technique*, involves computing densities at the top of the model (typically near 60,000 K) from such density–sensitive line ratios as F(CIII λ1909)/F(CIII λ1175), F(CIII λ1909)/F(SiIV λ1403), and F(SiIII λ1892)/F(CIII λ1909). These ratios are sensitive to electron density, because one or both lines are intersystem transitions with relatively small radiative de–excitation rates. Thus for densities higher than come critical value, typically in the range 10^8 to 10^{12} cm^{-3}, collisional de–excitation becomes more important that radiative de–excitation and the ratio depends on n_e^{-1}. Typically, the density at 60,000 K constrains the pressure at the top of a model, P_o (dynes cm^{-2}).

Table 1. First–generation Models for RS CVn Systems

System Modelled	*Spectral Features*	*Techniques*	P_o	*Reference*
λ And	CaII, MgII, Hα	Tm	0.9	Baliunas *et al.* (1979)
V711 Tau, UX Ari	MgII, SiII, SiIII, CII	Tm, DS	0.5	Simon & Linsky (1980)
UX Ari flare		DS	1.1	Simon *et al.* (1980)
λ And	Hα	Tm	0.4	Mullan & Cram (1982)
λ And flare		DS	5.5	Baliunas *et al.* (1984)
IM Peg flare	CII–IV, SiIII–IV, NV	EM	0.03	Buzasi *et al.* (1987)
V711 Tau, AR Lac	CII–IV, SiIII–IV, NV	EM, DS	0.3	Byrne *et al.* (1987)
II Peg quiet	CII–IV, SiIII–IV, NV	EM, DS	0.2	Byrne *et al.* (1987)
II Peg active	CII–IV, SiIII–IV, NV	EM, DS	1.8	Byrne *et al.* (1987)

The *EM technique* computes the emission measure, EM = $\int_{\Delta h} n_e^2$, from the observed stellar surface fluxes of effectively thin lines formed at temperatures above 10^4 K. The details of this technique can be found in Jordan and Brown (1981), and applications to RS CVn systems can be found in Byrne *et al.* (1987). In practice, EM(T_e) for the model is derived from the lower envelope of the individual EM(T_e) distributions of each line. Then the variation of pressure with temperature, P_e (T_e), is derived from

$$P_e^2 = P_o^2 + 2 \times 10^{-8} g_* \int_{T_e}^{T_o} EM dT,$$

where the subscript 'o' refers to the top of the model and g_* is the stellar surface gravity. P_o is typically derived from the density–sensitive line ratios. Once a model has been computed, the individual terms due to radiation, conduction, and wind losses may be balanced against mechanical heating, viz.

$$\Delta F_m = \Delta F_R + \Delta F_C + \Delta F_W.$$

These first–generation models provide a starting point for understanding chromospheres in RS CVn systems, since the models can be used to infer the radiative losses and other terms in the energy balance equation. The models summarized in Table 1, which assume that the *entire* visible hemisphere is an active region or a flare, show that active regions or flares have larger values of EM(T_e), consistent with the observed increased fluxes in flares by factors of 1.2–2.5 in the chromosphere and 2.4–5.5 in the transition region (Linsky *et al.*

1989). As a result, the pressures in the transition region (P_o) are also enhanced. However, the inferred surface fluxes for the active regions and flares depend upon the fraction of the stellar disk covered by this plasma. If the covering fractions (f) are small, then the surface fluxes, which are proportional to f^{-1}, become *very* large and realistic models will differ greatly from the quiet atmosphere. Are the covering fractions small or large?

3.2 Properties of Plages

Three techniques can, in principle, provide information on the location (longitude and latitude) and size of active regions and flares on the surfaces of stars – *eclipse light curves, rotational modulation*, and *Doppler imaging* (also called spectral imaging). I consider a flare to be a transient brightening on a star that persists for less than an orbital period, and an active region or plage to be a brightening that persists for at least several orbital periods. Monitoring campaigns that extend over several orbital periods are needed to determine whether an observed bright patch on a star is an active region or a flare. Little has been learned so far about active regions and flares in chromospheric emission lines from eclipse experiments in which one star of a binary covers and then uncovers the disk of another. In *rotational modulation* experiments the location of a bright patch on the stellar surface is inferred from the phase of a rapid increase in the flux of ultraviolet emission lines as the bright patch appears at one limb and the phase of the rapid decrease when the patch disappears at the other limb. The best example of a rotational modulation experiment is the observation of a bright plage on the surface of II Peg in October 1981 (Rodonò *et al.* 1987); these data yield a good estimate of the plage longitude, a less secure estimate that its latitude is nearly equitorial, and no reliable estimate of its size. This active region had essentially disappeared by February 1983 (Andrews *et al.* 1988).

The most reliable information on active regions in the chromospheres of RS CVn systems has come from using the *Doppler imaging* technique to study the MgII k line as a function of phase. As described by Neff (1987) and by Walter *et al.* (1987), this technique infers the location and size of a bright patch on the stellar surface from the changing wavelength displacement of an emission feature in a rotationally–broadened line profile as the plage first rotates into view (at maximum blueshift), then moves across the star (with increasing redshift), and finally rotates off the receding limb (at maximum redshift). The latitude of the plage can be inferred from the timing of its appearance and disappearance (provided the inclination of the rotational axis is known), and the size of the plage can be inferred from the broadening of its emission feature due to the different radial velocities of its leading and trailing edges.

Four observing campaigns to Doppler image the chromospheres of RS CVn systems with the IUE satellite have been described by Jim Neff earlier at this Workshop. Here I will use the results of these campaigns and mention some of their limitations. One problem is that the S/N in IUE high–resolution spectra of the MgII k line is typically 15 at peak line flux but only unity in the adjacent continuum. There are also artifacts in the data as processed with IUESIPS including ripples with scales of a few pixels. While these artifacts may be reduced and the S/N increased with the new data processing algorithms being tested by the IUE Project, the present assumption of Gaussian shapes for the line profiles of small patches on the stellar surface and for the quiescent emission from the whole stellar disk are the most complex representations warranted by the present data quality. We look forward to higher quality spectra in the MgII k lines and in higher–temperature lines when the GHRS on HST becomes operational. Another problem is the difficulty in telling whether a bright patch on the stellar surface is a flare or an active region from data sets that cover only one rotational period. Also, systematic streaming motions in the line of sight, if present, would lead to misplacement of the emitting regions on the stellar surface.

Table 2 summarizes the plage parameters derived from Doppler imaging analyses of AR Lac in February 1983 (Walter *et al.* 1987) and in September 1985 (Neff *et al.* 1989). Analyses of the September 1987 observations of AR Lac and HD 199178 and the September 1988 observations of EI Eri will be reported elsewhere. Despite the limitations in the data and the analysis technique, some important results have emerged from the Doppler imaging campaigns.

- The data are of sufficient quality to identify two discrete plages on the K star in October 1983 and three plages in September 1985. In addition, a flare was observed on the K star in October 1983 and on the G star in September 1985.
- The plages move across the star or disappear on time scales less than 2 years.
- During the second campaign the leading hemisphere of the K star was brighter in the MgII line than the trailing hemisphere, indicating the presence of many unresolved additional plages, and one hemisphere of the G star was unexpectedly dark.
- The large radial velocity variations of two plages in September 1985 with amplitudes 1.5 times larger than the equatorial rotational velocity indicate that the plages extend out to 0.5 radii *above* the limb. Near phase 0.1, Plage C, which extends 0.5 $R_\star$ above the limb of the K star, occults much of the G star and thus may be responsible for the appearance of the *dark* hemisphere of the G star observed in the K line. Future models should include geometrical extension.
- Since the plages are small, the emission line surface fluxes must be large.

Table 2. Plage Parameters for AR Lac

Plage	*Longitude*	*Latitude*	*Covering Fraction*	*Height* $R/R_\star$	*Reference*
1983 A	280°	0°	0.01	≤ 1.2	Walter *et al.* (1987)
1983 B	345°	0°	0.01	≤ 1.2	Walter *et al.* (1987)
1985 A	205°	0°	0.06	1.5	Neff *et al.* (1989)
1985 B	240°	±50°	0.02	1.0	Neff *et al.* (1989)
1985 C	130°	0°	0.09	1.5	Neff *et al.* (1989)

3.3 Second Generation: Two–component models

An essential component of what I call second–generation modelling is an accurate estimate of the surface area, or covering fractions, for each component of the atmosphere that contributes to the total emission from a star. One method is to derive the plage covering fraction in the chromosphere from a Doppler imaging analysis of the MgII k line, and to assume that the higher temperature layers cover the same area of the star. This assumption is valid for the upper chromosphere and transition region of the Sun, but it is not valid for the solar corona, where the magnetic structures expand greatly from the lower layers. Table 3 lists the surface fluxes (units 10^5 erg cm^{-2} s^{-1}) in several important lines for the K star (F_K) in each system and for the surface flux ratio of the plage to the K star (F_P/F_K). The plage surface fluxes assume that the individual plages on a star are equally bright. These data from Walter *et al.* (1987) and Neff *et al.* (1989) indicate that the plages are roughly 3–15 times brighter than the quiesecent K star in RS CVn systems with the brightness contrast with respect to the quiescent K star increasing with temperature.

Table 3. Surface Flux Ratios for Plages and Flares

Line	*AR Lac Oct 83*		*AR Lac Sept 85*		*V711 Tau flare*	
	F_K	F_P/F_K	F_K	F_P/F_K	F_K	F_F/F_K
NV 1240Å	0.8	11.3	0.8	7.6	0.5	800
CIV 1550Å	3.7	12.9	3.8	3.6	4.1	340
SiIV 1400Å	0.5	11.1	0.8	4.4	1.2	280
CII 1335Å	2.1	26.	1.6	2.8	3.1	210
SiII 1812Å	1.8	8.6	1.9	3.4	1.5	80
MgII 2800Å	36.	4.4	27.	4.0	47.	80

A second method is to estimate the covering fraction from the emission measure of such lines as CIV 1550Å and the electron density from density–sensitive line ratios. These two quantities yield the volume of the plage or flare plasma, which may be compared to the volume of the quiescent plasma in the same emission line to estimate the covering fraction. Linsky *et al.* (1989) used this approach to estimate a covering fraction of 0.005 for the 3 October 1981 flare on V711 Tau. The flare surface fluxes in Table 3 were computed by assuming this covering fraction. The F_F/F_K ratios in Table 3 are much larger than for the plages, but the same trend of increasing ratios with temperature is seen in the flare and plage data.

A third method is to assume that the area of the plage or flare is the same as the largest starspot on the visible hemisphere at that time, but there are major problems with this method. First, several spots may be on the visible hemisphere at the same time, which leads to ambiguity. Second, plages and flares need not be the same size as a spot or spot group. This is particularly important when spots cover as much as 40% of the visible hemisphere. For example, at the time of the 3 October 1981 flare on V711 Tau, a spot covering 18% of the visible hemisphere was located near the central meridian of the K star. If one assumes that the flare area was the same as the spot, then the surface flux ratios in Table 3 should be reduced by a factor of 36 and the deduced flare model will be very different.

The determination of plage or flare surface fluxes and emission measure distributions is useful for estimating the emitting volumes, total radiative losses over a broad range of temperature, typically $4.3 \leq \log T_e \leq 5.4$, and other terms in the energy balance. To proceed further to a detailed atmospheric model, one must assume a specific geometry – a plane–parallel slab or a magnetic flux tube. The latter has been more popular. For example, Byrne *et al.* (1987) have modelled the plage on II Peg in October 1981 by semi–circular loop of circular cross section with the cross sectional diameter 0.1 D, the footpoint separation. They estimate the emitting volume in the CIV and SiIV lines from the plage flux and the electron density derived from density–sensitive line ratios. They derive the loop footpoint separation by assuming that the loop is *filled* with plasma at the temperatures at which these ions are most abundant. These quantities are summarized in Table 4. Linsky *et al.* (1989) have assumed the same geometry for the 4 October 1981 flare on V711 Tau. Their results are also shown in Table 4 for comparison. Also shown in the table are the total radiative losses (L_R) and fluxes (F_R) for the temperature interval $4.3 \leq \log T_e \leq 5.4$.

Table 4. Loop Models for a Plage and a Flare

Physical Parameter	*II Peg Plage*	*V711 Tau Flare*
$\log N_e$ (at $\log T_e = 4.8$)	11.2	11.0
P_o (dynes cm^{-2})	1.3	0.8
V_{CIV} (cm^3)	$3.4\ 10^{29}$	$8\ 10^{29}$
V_{SiIV} (cm^3)	$4.0\ 10^{29}$	$36\ 10^{29}$
D (cm)	$6.7\ 10^9$	$5\ 10^{10}$
D/R_K	0.034	0.2
L_R (erg s^{-1})	$1.5\ 10^{28}$	$1.5\ 10^{32}$
F_R (erg cm^{-2} s^{-1})	$3.4\ 10^9$	$6\ 10^{11}$
A/A_K	$3.7\ 10^{-5}$	$1.0\ 10^{-3}$

The *filled loop model* just described is only one of many possible geometries. It may not be realistic as the loop or loops are more likely filled with hot coronal gas with the 10^5 K transition region material confined to a thin layer near the loop footpoints. Also, the loop cross section may not be constant along the length of the loop. If one assumes that the flaring loop on V711 Tau has the same area projected onto the stellar surface as the large spot located near the central meridian, then only 0.01% of the loop volume is filled with 10^5 K plasma. Linsky *et al.* (1989) noted that the MgII k line profile during the flare was asymmetric, and modelled it with a Gaussian for the global emission from the K star and a second Gaussian (for the flare) displaced 90 ± 30 km s^{-1} with respect to the K star. If this shift represents a systematic downflow onto the star, then the kinetic energy in this flow is comparable to the radiative luminosity in the temperature interval $4.3 \leq \log T_e \leq 5.4$. Clearly, the modelling of plages and flares in RS CVn systems is only beginning.

4 CORONAL MODELLING

4.1 First Generation: Static Loop Models

Rosner, Tucker and Vaiana (1978) proposed a model for the quiescent solar corona with magnetic loops as the basic structural element. Their model suggested by *Skylab* x-ray images (cf. Vaiana and Rosner 1978) assumes quasi–static loops in hydrostatic equilibrium for which thermal conduction from the hot loop apex and mechanical heating are balanced by radiation. They derived the now widely used relation $T_{max} \sim 1400\,(pL)^{1/3}$, where T_{max} is the temperature of most of the loop volume, p is the loop pressure, and L is the loop size. Recent x-ray solar images with excellent angular resolution (e.g. L. Golub, private communication) confirm that loops are the basic structural element of the solar corona and by implication stellar coronae. The magnetic character of the loops is confirmed by the correspondence between observed solar x-ray structures and coronal field lines calculated by extrapolating the measured photospheric field upwards assuming no electric currents (Poletto *et al.* 1975). Subsequent authors have extended the magnetic loop models to include extended geometries, systematic flows, and thermal instabilities (cf. review by Antiochos 1987). The discovery by Vilhu (1987) of an upper limit to the x-ray surface flux for the most active stars (including the RS CVn systems) suggests that saturation occurs when the loops fill the available coronal volume.

An important question is the coronal loop temperature(s) which can only be inferred from x-ray data. The first important measurements came from the *Einstein* Solid State Spectrometer, which had an energy resolution of 160 eV. Swank *et al.* (1981) showed that for seven RS CVn systems and Algol the SSS spectra are better fit by a two-temperature plasma with the lower temperatures in the range log T_1 = 6.6–6.9 and the higher temperatures in the range log T_2 = 7.4–8.0 than by a single-temperature plasma. The x-ray luminosity of the high temperature plasma was observed to have a larger range among the stars and as a function of time for a given star than the low temperature component. Since the introduction of additional thermal components did not significantly decrease the reduced χ^2 of the fits to their spectra, they concluded that only small amounts of plasma at intermediate temperatures are consistent with their data. The analysis of *EXOSAT* Transmission Grating Spectra of Capella and σ^2 CrB led Lemen *et al.* (1989) to the similiar conclusion that a bimodal temperature distribution with a dip in the EM(T_e) distribution at intermediate temperatures fit the observed spectra well.

Majer *et al.* (1986) arrived at a different conclusion from their analysis of *Einstein* IPC observations of 29 RS CVn systems. They found that for many of the nearby systems a two-temperature plasma model provides a significantly lower reduced χ^2 than a single-temperature plasma, but for the five systems observed by both the SSS and the IPC the two temperatures are significantly different. Since two-temperature models are a mathematically sufficient description of either the SSS or the IPC data but *not* both, they argue that the EM(T_e) distribution of the coronal plasma is continuous but appears bimodal due to the limited spatial and spectral resolution of the detector.

Irrespective of the shape of the deduced EM(T_e) distribution, the model of choice has been the static magnetic loop described by Rosner, Tucker and Vaiana (1978). Swank *et al.* (1981) proposed that each of the two thermal components of the plasma reside in loops with different characteristics. They argued that the plasma is in collisional ionization equilibrium on the basis that the observed fluxes did not vary significantly on time scales of hours. The unknown parameter for their loops was the gas pressure. If p = 1–10 dynes cm^{-2}, as is the case for solar active region loops and the top of the transition region for RS CVn systems (see Table 1), then the cool loops are small compared to a stellar radius and the hot loops are comparable in size to the binary separation. If, on the other hand, p $\geq$ 100 dynes cm^{-2}, as is the case for solar flares, then loops of both temperatures are small compared to a stellar radius and the high temperature loops cover only a small fraction of the stellar surface. Without additional information, the basic ambiguity about loop sizes cannot be resolved. The increase in cross sectional area of the loops with height ($\Gamma \equiv$

A_{top}/A_{base}) can be constrained by the shape of the EM(T_e) distribution. Schrijver, Lemen and Mewe (1989) state that a continuous distribution is consistent with the original Rosner, Tucker and Vaiana (1978) loop model in which $\Gamma = 1$, whereas a sharply–peaked EM(T_e) distribution is consistent with a $\Gamma \gg 1$ loop, which appears in x-rays to be isothermal. Schrijver *et al.* concluded that a double–peaked EM(T_e) distribution implies two distinct ensembles of loops each with $\Gamma \gg 1$. This rapid divergence with height is expected for magnetic field lines above a bipolar active region.

4.2 Second Generation: Extended Geometry

Observations of RS CVn systems during eclipses and throughout an orbital cycle have provided the missing information concerning the geometry of the x-ray emitting plasmas to resolve the basic ambiguity concerning the loop sizes. The first critical observations, obtained by Walter, Gibson and Basri (1983), consisted of *Einstein* IPC observations of the eclipsing system AR Lac. These data show that the leading hemisphere of the G2 IV star has a bright corona of small height ($h \leq R_\star$), because during the egress phase after primary eclipse the x-ray flux increases very rapidly. Also the x-ray light curve goes through a minimum at about the time of first contact of secondary eclipse, indicated that a component of the K0 IV star corona extends about 1 $R_\star$ above its photosphere. They concluded that both stars have coronae with small scale heights ($\sim 0.02\ R_\star$) and that the K star has an extended component near its equator on one hemisphere with a scale height of $\sim 1\ R_\star$. They speculated that this geometrically–extended component corresponds to the hotter spectral component and consists of magnetic loops that fill a volume comparable to the binary system. They identified the compact coronae of both stars with the cooler spectral component and proposed that these coronae consist of a large number of small magnetic loops.

While the geometry proposed by Walter, Gibson and Basri (1983) is consistent with the data, contemporaneous spectral information is needed to confirm it. This was provided by a continuous set of *EXOSAT* LE and ME observations of AR Lac over its two–day orbital period by White *et al.* (1987). Their low energy data show a deep minimum centered on primary eclipse but no dip in the higher energy data. These data *confirm* that the high–temperature plasma first seen in the SSS data is geometrically extended, whereas the low–temperature plasma is found mainly in two localized regions on the G star with small scale height. Table 5 summarizes the parameters of the loops in the three components of their model. *EXOSAT* observations of an eclipse in the Algol system (White *et al.* 1986) also provide evidence for hot, extended loops around the K0 IV star in this system. Future x-ray observations of eclipsing binary systems with both spectral and temporal resolution will provide far more complete models of the three–dimensional coronae of these stars.

Table 5. Coronal Model for AR Lac Based on *EXOSAT* Data

Component	*Star*	*Central Longitude*	*Width*	T (10^6 K)	$h/R_\star$	p (dynes cm^{-2})
A	G2 IV	3°	60°	7	0.07	~100
B	G2 IV	150°	60°	7	0.07	~100
C	K0 IV	225°	90°	15–30	≥1	~3

5 MODELLING OF THE RADIO CORONA

5.1 First Generation: One–component Models

RS CVn systems are powerful microwave sources during quiescence and especially during flaring episodes. A total of 66 systems have now been detected at 6 cm, and their properties are described in the extensive surveys by Morris and Mutel (1988) and by Drake, Simon and Linsky (1989). The radio emission mechanism is generally described as incoherent gyrosynchrotron emission of a power law distribution of relativistic electrons on the basis of the inferred brightness temperatures, circular polarization, and spectral shape (cf. Mutel *et al.* 1987; Dulk 1985). Drake, Simon and Linsky (1989) have most recently discussed correlations among the radio, ultraviolet, x-ray and optical properties of these systems.

5.2 Second Generation: Two–component Models

VLBI observations have provided critical insight into the geometry of the radio–emitting plasma and its emission mechanisms. Mutel *et al.* (1985) were able to resolve six systems at 6 cm, and two of the systems (UX Ari and Algol) were found to have both an unresolved core and a resolved halo. In each case the core size was smaller than an individual stellar diameter with inferred brightness temperatures, $T_B \geq 1\ 10^{10}$ K, whereas the halo sizes were comparable to the size of the binary system with $T_B = 4\text{–}8\ 10^8$ K. They interpreted the core flare source as gyrosynchrotron emission from relativistic electrons trapped in compact loops close to the surface of the active star, and the quiescent halo emission also as gyrosynchrotron emission from relativistic electrons but in much larger loops with smaller magnetic field strengths.

Drake *et al.* (1989) proposed as an alternative that the very low level quiescent radio emission may be thermal gyrosynchrotron emission from the *same* hot ($10^{7.7}$ K) electrons detected by the SSS located in the extended halo inferred from the VLBI observations. This model is consistent with all except the circular polarization data.

6 TOWARDS A UNIFIED PICTURE

In this review I have emphasized the importance of *spatial* information in order to model the atmospheres of RS CVn systems with a sensible geometry. Considerable progress is being made in this direction. Equally important is the need to tie together the various data sets (optical, ultraviolet, x-ray and radio) such that the *same* model explains *all* of the data. Only a few attempts have been made so far to tie together the various data sets. The Drake, Simon and Linsky (1989) model that uses the same hot electrons to explain the x-ray and radio data is an attempt along these lines. I believe that major advances in our understanding of the physical processes in RS CVn systems will emerge from attempts to tie together the various data sets using realistic geometries. When this happens we can proceed to third–generation models in which the energy balance and magnetic field structure are included in a self–consistent manner.

7 ACKNOWLEGEMENTS

This work is supported by NASA grants NAG5–82 to the University of Colorado and H–80531–B to the National Institute of Standards and Technology. I wish to thank the organizers of the NATO Advanced Study Institute *Active Close Binaries* in Kuşadası, Turkey and in particular Prof. C. İbanoğlu, for their kind support. Finally, I wish to thank Dr. A. Brown and D. Luttermoser for commenting on this manuscript.

8 REFERENCES

Andrews, A.D. *et al.* 1988, *Astr. Ap.* **204**, 177.

Antiochos, S.K. 1987, in *Cool Stars, Stellar Systems, and the Sun*, ed. J.L. Linsky and R.E. Stencel (Berlin: Springer–Verlag), p. 283.

Baliunas, S.L., Avrett, E.H., Hartmann, L. and Dupree, A.K. 1979, *Ap. J.* **233**, L129.

Baliunas, S.L., Guinan, E.F. and Dupree, A.K. 1984, *Ap. J.* **282**, 733.

Brown, R.L. and Crane, P.C. 1978, *A. J.* **83**, 1504.

Buzasi, D.L., Ramsey, L.W. and Huenemoerder, D.P. 1987, *Ap. J.* **322**, 353.

Byrne, P.B., Doyle, J.G., Brown, A., Linsky, J.L., and Rodonò, M. 1987, *Astr. Ap.* **180**, 172.

Drake, S.A., Simon, T. and Linsky, J.L. 1989, *Ap. J. Suppl.* **71**, 905.

Dulk, G.A. 1985, *Ann. Rev. Astr. Ap.* **23**, 169.

Gondoin, P., Giampapa, M.S., and Bookbinder, J.A. 1985, *Ap. J.* **297**. 710.

Jordan, C. and Brown, A. 1981, in *Solar Phenomena in Stars and Stellar Systems*, ed. R.M. Bonnet and A.K. Dupree (Dordrecht: D. Reidel), p. 199.

Lemen, J.R., Mewe, R., Schrijver, C.J. and Fuldra, A. 1989, *Ap. J.* **341**, 474.

Linsky, J.L. 1984, in *Cool Stars, Stellar Systems, and the Sun*, ed. S.L. Baliunas and L. Hartmann (Berlin: Springer–Verlag), p. 244.

Linsky, J.L. 1985, *Solar Phys.* **100**, 333.

Linsky, J.L. 1988, in *Multiwavelength Astrophysics*, ed. F. Córdova (Cambridge: Cambridge Univ. Press), p. 49.

Linsky, J.L. 1989, *Solar Phys.* **121**, 187.

Linsky, J.L., Neff, J.E., Brown, A., Gross, B.D., Simon, T., Andrews, A.D., Rodonò, M. and Feldman, P.A. 1989, *Astr. Ap.* **211**, 173.

Majer, P., Schmitt, J.H.M.M., Golub, L., Harnden, F.R. Jr. and Rosner, R. 1986, *Ap. J.* **300**, 360.

Morris, D.H. and Mutel, R.L. 1988, *A. J.* **95**, 204.

Mullan, D.J. and Cram, L.E. 1982, *Astr. Ap.* **108**, 251.

Mutel, R.L., Lestrade, J.–F., Preston, R.A. and Phillips, R.B. 1985, *Ap. J.* **289**, 262.

Mutel, R.L., Morris, D.H., Doiron, D.J. and Lestrade, J.–F. 1987, *A. J.* **93**, 1220.

Neff, J.E. 1987, Ph.D. Thesis, University of Colorado.

Neff, J.E., Walter, F.M., Rodonò, M. and Linsky, J.L. 1989. *Astr. Ap.* **215**, 79.

Poletto, G., Vaiana, G.S., Zombeck, M.V., Krieger, A.S., and Timothy, A.F. 1975, *Solar Phys.* **44**, 83.

Robinson, R.D. 1980, *Ap. J.* **239**, 961.

Rodonò, M. 1986, in *Cool Stars, Stellar Systems, and the Sun*, ed. M. Zeilik and

D.M. Gibson (Berlin: Springer–Verlag), p. 475.
Rodonò, M. *et al.* 1986, *Astr. Ap.* **165**, 135.
Rodonò, M. *et al.* 1987, *Astr. Ap.* **176**, 267.
Rosner, R., Tucker, W.H., and Vaiana, G.S. 1978, *Ap. J.* **220**, 643.
Saar, S.H. 1990, in Proceedings of IAU Symposium No. 138 *Solar Photosphere: Structure, Convection, and Magnetic Fields*, in press.
Schrijver, C.J., Lemen, J.R. and Mewe, R. 1989, *Ap. J.* **341**, 484.
Simon, T. and Linsky, J.L. 1980, *Ap. J.* **241**, 759.
Simon, T., Linsky, J.L., and Schiffer, F.H. III 1980, *Ap. J.* **239**, 911.
Swank, J.H., White, N.E., Holt, S.S. and Becker, R.H. 1981, *Ap. J.* **246**, 208.
Vaiana, G.S. and Rosner, R. 1978, *Ann. Rev. Astr. Ap.*, **16**, 393.
Vilhu, O. 1987, in *Cool Stars, Stellar Systems, and the Sun*, ed. J.L. Linsky and R.E. Stencel (Berlin: Springer–Verlag), p. 110.
Vogt, S.S. and Penrod, G.D. 1983, *Pub. Astr. Soc. Pacific* **95**, 565.
Walter, F.M., Gibson, D.M. and Basri, G.S. 1983, *Ap. J.* **267**, 665.
Walter, F.M. *et al.* 1987, *Astr. Ap.* **186**, 241.
White, N.E., Shafer, R.A., Parmar, A.N. and Culhane, J.L. 1987, in *Cool Stars, Stellar Systems, and the Sun*, ed. J.L. Linsky and R.E. Stencel (Berlin: Springer–Verlag), p. 521.
White, N.E. *et al.* 1986, *Ap. J.* **301**, 262.

MAGNETIC FLARES IN CLOSE BINARIES

JAN KUIJPERS
Sterrekundig Instituut
Rijksuniversiteit Utrecht
P.O. Box 80 000
3508 TA Utrecht
The Netherlands

ABSTRACT. Our present understanding of solar flares is reviewed and applied to close binaries, both with and without mass exchange, with a compact star and with both stars having convective envelopes. Although many details of the solar flare energy release are still not clear, in particular the processes of magnetic field reconnection and particle acceleration, it now appears that the global course of events in a flare is well understood and simple. In general magnetic flares arise when a dynamo is coupled electrodynamically to a magnetically dominated storage region. The dynamo converts kinetic (convection or accretion) energy into (electro)magnetic energy if the flow crossing time of the system is much less than the ohmic dissipation time. Magnetic energy can be transferred via a Poynting flux to a magnetically connected storage region. If the storage region is dynamically dominated by magnetic fields instead of by gas pressure or motion and if its Alfvén crossing time is much less than the flow crossing time, the magnetically dominated region evolves through a quasi-static series of force free fields. As the amount of stored energy becomes comparable to the energy of the original magnetic field structure before its distortion, the excess energy is released catastrophically. The release time is a number of Alfvén crossing times of the relaxing structure.

Eruptive phenomena have been observed in many close binaries, with and without compact objects. In fact various classes of close binaries satisfy the conditions for flare production and several of the observed explosions are probably magnetic flares. In these systems the dynamo is formed by the traditional convective envelope of a normal star, or the accretion disk around a normal star or compact companion, or more generally by the transferred mass as it falls in towards the companion and interacts with the magnetic field of the receiving star. The storage region is situated in the individual stellar coronae, in the common dilute envelope or in disk coronae, or in the magnetosphere of the compact object. As a particular example we show that magnetic flares occur in Low-Mass X-Ray Binaries and how they can explain the quasi-periodic oscillations (QPO) observed in X-rays.

1 Introduction

In this paper we review the theory of magnetic flares as we have come to understand from the study of solar flares. The main ingredients belong to the domain of plasma astrophysics and are not exclusive to the solar case. They can easily be applied to other objects with plasmas in relative motion, in particular to active close binaries. Common to all such systems is *a dilute plasma ("corona") which is powered electrodynamically by an adjacent dynamo.*

Observations of magnetic activity and flares in close binaries can be found in several

C. İbanoğlu (ed.), Active Close Binaries, 761–803.

papers in these proceedings, both in the presence or absence of compact objects. Further work on this subject can be found in IAU Coll. No. 104 on Solar and Stellar Flares [1,2].

In Sect. 2 I first discuss the conversion of kinetic into electromagnetic energy and the specific conditions required for the *appearance of magnetic flares.* A magnetic flare is roughly defined as a *release of magnetic energy in the form of motion, heating and particle acceleration on a time scale which is short in comparison to the typical time scale for changes on the bounding surface* (e.g. the flow crossing time). Clearly not all explosive events in close binaries are due to magnetic flaring and I therefore indicate more precisely in which observed eruptions magnetic flares may play a role.

In Sect. 3 I explain theoretical concepts which are important in the study of *energy release*: the use of lumped electric circuits, reconnection and particle acceleration, direct electric fields, electrostatic double layers and plasma turbulence, magnetic helicity and Woltjer-Taylor's theorem.

Finally in Sect. 4 a more detailed example is presented of magnetic flares in the magnetosphere of a compact object surrounded by an accretion disk.

2 Conversion of Kinetic into Electromagnetic Energy

There is an important difference between cosmic and laboratory plasmas: In cosmic plasmas the ultimate source of the magnetic activity is gravity, which causes confinement and generates motion (either indirectly by the intermediary of nuclear fusion in the stellar core which leads to a convective envelope, or directly as in the Keplerian motion in an accretion disk). In laboratory plasmas on the other hand the goal is to confine nuclear fusion by a strong magnetic field. Even so, plasma fusion has been plagued for decennia by instabilities, which it tries to suppress. In line with this tendency of laboratory plasmas towards *instability*, astrophysical plasmas show that magnetic activity and the occurrence of violent deviations from local thermodynamic equilibrium are the rule rather than exception. It is this violent interplay between magnetically stored energy and the kinetic energy of the generating source, which forms an important domain of study in plasma astrophysics and which is the subject of this paper. Conceptually plasma astrophysics is a rather simple branch of physics: Of the four forces in nature *only gravity and electromagnetism* play a role in most natural plasmas. Further, quantum effects can often be neglected since the energy per particle is usually much larger than the Fermi energy or the energy of a characteristic field quantum (exceptions are formed by the degenerate and solid state plasmas in the interiors of white dwarfs, neutron stars, and planets, and the strongly magnetized atmosphere of a neutron star). Finally in plasma astrophysics it is often an excellent approximation to describe the matter as consisting of pure hydrogen !

We consider astrophysical *plasmas where the flow crossing time is much less than the ohmic diffusion time*:

$$t_v \ll t_\eta . \tag{1}$$

Here these time scales are defined as

$$t_v \equiv \ell / \Delta v , \tag{2}$$

and

$$t_\eta \equiv \ell^2/\eta \tag{3}$$

where ℓ is the characteristic dimension of the structure, Δv the variation in velocity on the boundary over the length scale ℓ and η the electric resistivity. Although the conductivity (the inverse of the resistivity) of cosmic plasmas is rather poor (comparable to that of

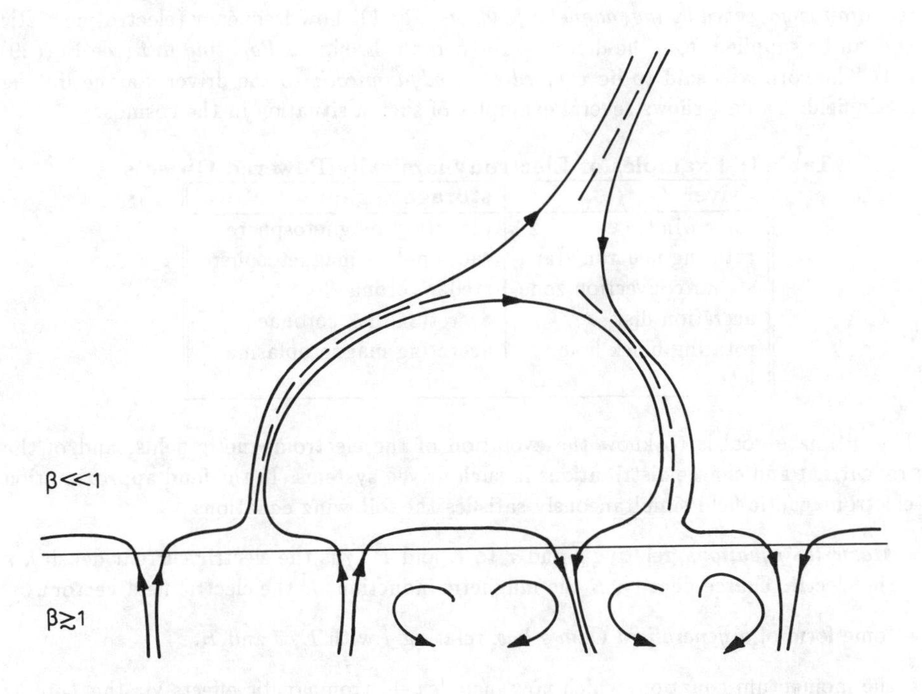

Figure 1: Electrodynamic coupling between magnetically connected regions. The dilute "corona", where the magnetic field dominates over gas pressure ($\beta \ll 1$), acts as a storage region and receives energy from a "driver" which is dominated by fluid motion and gas pressure ($\beta \geq 1$).

a metal conductor at room temperature) the length scale is usually extraordinarily large so that inequality (1) is amply satisfied. This is the reason why *to a first approximation astrophysical plasmas behave as an ideal plasma with infinite conductivity. For the energy release phase and in particular for flares it is however crucial to take into account the effects of a finite resistivity.* In the ideal plasma approximation then the gas is treated as a perfect conductor, there is no field diffusion or slip of the magnetic field with respect to the matter (in a direction perpendicular to the field) and the magnetic field is said to be frozen in. For the case of a convective envelope inequality (1) is usually rewritten in terms of the so-called dynamo number being much larger than unity.

We are interested in situations where magnetic fields extend from a spatial region with driving motions outwards into a dilute environment. The two regions differ in the relative importance of gas pressure p_{gas} and magnetic field pressure $p_{mag} \equiv B^2/(2\mu_0)$ ($\mu_0 = 4\pi \cdot 10^{-7}$ henry/m is the magnetic permeability of free space) as defined by the plasma bêta:

$$\beta \equiv \frac{p_{gas}}{p_{mag}}. \tag{4}$$

The proper dynamo or "driver" is characterized by an average plasma bêta larger than unity, $\beta \geq 1$, and is dynamically dominated by the fluid motion and the gas pressure forces. *The storage region or "corona" is characterized by bêta much less than unity, $\beta \ll 1$, and its evolution is governed by the magnetic field* (see Fig. 1). Low-frequency (electro)magnetic energy can be supplied from the driver to the corona through a *Poynting flux* (see Eq. (29) below). The corona is said to be *coupled electrodynamically* to the driver via the linking magnetic field. Table 1 shows several examples of such a situation in the cosmos.

Table 1: Examples of Electrodynamically Powered Objects

driver	**storage region**
solar wind	planetary magnetosphere
rotating neutron star	radio pulsar magnetosphere
stellar convection zone	stellar corona
accretion disk	accretion disk coronae
rotating black hole	accreting magnetoplasma
jets	cocoons

The ultimate goal is to know the evolution of the electromagnetic fields, and of the electric current and charge distributions in such driven systems. In the fluid approximation the electromagnetic field simultaneously satisfies the following equations:

- *Maxwell's equations*, relating $\vec{j}$ and $\vec{\tau}$ to $\vec{E}$ and $\vec{B}$ ($\vec{j}$ is the electric current density, τ the electric charge density, $\vec{B}$ the magnetic induction, $\vec{E}$ the electric field vector),
- some form of a *generalized Ohm's law*, relating $\vec{j}$ with $\vec{E}$, $\vec{v}$ and $\vec{B}$,
- the momentum equation, which now includes electromagnetic effects via the *Lorentz force* $\vec{j} \times \vec{B}$ and the *Coulomb force* $\tau\vec{E}$, and
- an energy equation, containing $\vec{j} \cdot \vec{E}$.

An important further simplification arises if the gas motions are non-relativistic: Since the ratio of the displacement current $\epsilon_0 \partial \vec{E}/\partial t$ to the material current in Ampère's law is of order v/c or smaller, the displacement current can be neglected. This has the important consequence that the magnetic fields are completely determined by the electric current distribution ($\nabla \cdot \vec{B} = 0$ and $\nabla \times \vec{B} = \mu_0 \vec{j}$) or vice versa. One can therefore choose to work with currents (or alternatively with magnetic fields) only. Taking the divergence of the reduced Ampére's equation one finds: $\nabla \cdot \vec{j} = 0$. In this approximation electric currents always close and do not cause important charge separations. In fact this allows one to consider geometrically simple current distributions as electric circuits (see Sect. 3). Similarly to the same order of accuracy the Coulomb force can be neglected in the momentum equation.

Finally in this approximation of non-relativistic motions the magnetic fields are independent of the frame of reference. Only *in this limit of both infinite conductivity and non-relativistic motion do the magnetic field lines acquire a frame-independent meaning, and can they be identified by the attached matter in which they are "frozen-in" and be followed in time.* This approximation is commonly called the *ideal* mhd (magnetohydrodynamic) limit.

There are two numbers that characterize the deviation of a magnetofluid from the ideal (zero resistivity and viscosity) state: the *Reynolds number*

$$\mathcal{R} \equiv \frac{\ell v}{\nu}, \tag{5}$$

where ν is the kinematic viscosity, and the *Lundqvist number*

$$\mathcal{S} \equiv \frac{\ell v_A \mu_0}{\eta}, \tag{6}$$

where $v_A \equiv B/\sqrt[2]{\mu_0 \rho}$ is the Alfvén speed (ρ is the mass density). The Lundqvist number can also be written as the ratio of the ohmic diffusion time Eq. (3) to the Alfvén crossing time Eq. (7). In cosmic plasmas of very different origin these numbers are usually many orders of magnitude larger than unity because of the large dimensions in comparison to the particle mean free path (high temperatures and small particle densities). A few examples are the solar corona with $\mathcal{S} = 2 \cdot 10^{13}$, the magnetosphere of an X-ray pulsar with $\mathcal{S} = 10^{14}$, and an accretion disk around a white dwarf with $\mathcal{R} = 10^{12}$ (based on the particle viscosity). Unfortunately the best numerical calculations available at present have Reynolds and Lundqvist numbers of order 10^4 at most. These calculations therefore represent only a small part of the physics; in particular the ratio of sizes of largest to smallest eddies in turbulence calculations is much too small in comparison with the real cosmic plasma. Important phenomena which occur in well-developed turbulence, such as self-organization, coherent structures and intermittency, may therefore be absent from the present computations while they are crucial in the appearance of the object to the observing astronomer. In this respect a present day Tokamak plasma with $\mathcal{S} > 10^9$ compares on the other hand favorably and is therefore important for the understanding of cosmic plasmas (although of course by stability design it has a small Reynolds number and is not suited to study fluid turbulence).

The nature of the interaction between driver and corona in the case of small resistivity (Eq. (1)) depends crucially on the *ratio of the Alfvén crossing time*

$$t_A \equiv \frac{\ell}{v_A} \tag{7}$$

to the flow crossing time, as we shall now show.

2.1 Emission of MHD Waves

If the flow crossing time is much less than the Alfvén crossing time

$$t_v \ll t_A, \tag{8}$$

the surrounding plasma does not have the time to relax and the moving object emits *non-stationary disturbances or mhd waves* (Fig. 2a). These disturbances consist of time-dependent electric current distributions which are closed in the mhd-approximation (Alternating Current (AC) circuits). The energy emitted in the form of mhd waves has been calculated for various cases both with and without magnetic connections of the (subalfvénically)

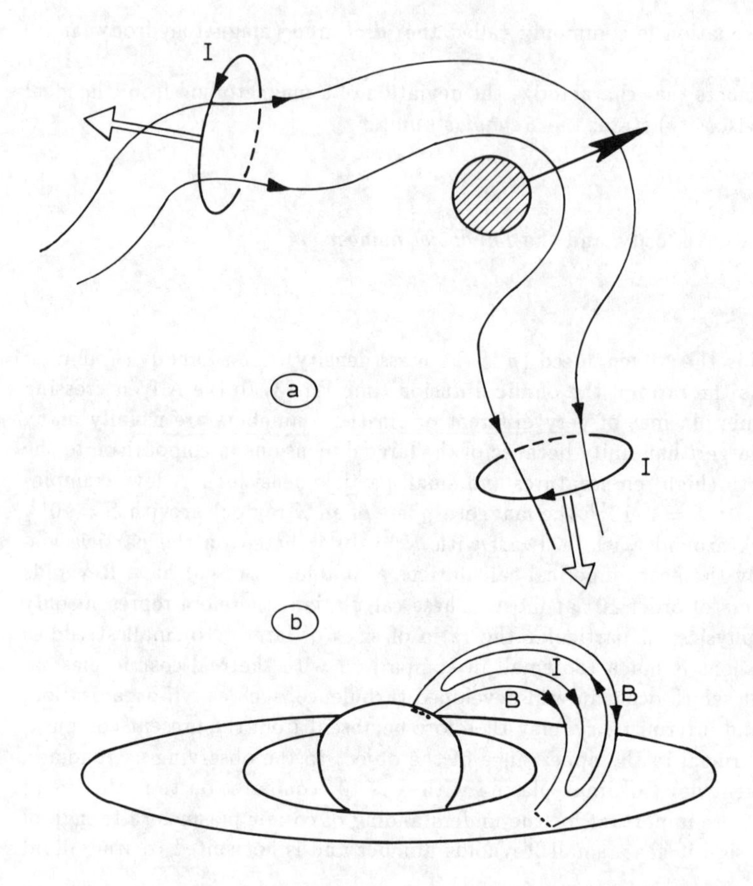

Figure 2: In (a) the moving conductor radiates mhd waves into the surrounding magnetoplasma as the flow crossing time is much less than the external Alfvén crossing time (Eq (8)). In (b) a magnetic link exists between a Keplerian disk and a central object and the reverse condition is assumed to be satisfied (Eq (10)); in this case a stationary current system may be possible (see Sect. 2.2.1).

moving object to the surrounding plasma [3-6]. For a perfectly conducting sphere of radius R moving with speed v perpendicular to the magnetic field of strength B in a homogeneous magnetoplasma with Alfvén speed v_A the power emitted is

$$P = \pi R^2 \frac{B^2}{2\mu_0} v (\frac{v}{v_A})^3. \tag{9}$$

2.2 Force Free Fields

In the opposite case when the Alfvén crossing time through the perturbed plasma structure is much smaller than the shearing time or flow crosssing time at the boundary:

$$t_A \ll t_v, \tag{10}$$

the coronal magnetic structure has ample time to continuously relax to a new magnetostatic equilibrium, which may however be unstable ! (see Fig. 3). At each moment such an equilibrium will have a specific distribution of *stationary electric currents*, which are closed in the mhd approximation (Direct Current (DC) circuits). Since in the coronal storage region $\beta \ll 1$, gas pressure perturbations can be easily balanced by Lorentz forces: electric currents only need to have a small angle with respect to the magnetic field. These equilibria are therefore to first order characterized by a current distribution that is everywhere aligned with the magnetic field direction. To a good approximation the field structures are free from Lorentz forces and called *force free fields* (FFF). It is important to realize that finite current systems can only locally be force free and must be bounded by a surface with non-vanishing forces [7]. This can be shown from the magnetic virial theorem.

For our driver/corona (or dynamo/store) system under condition (10) the coronal currents follow the magnetic field lines and the corona is approximately force free; the currents close perpendicular to the field lines in the driver situated in the dense subphotospheric layers which are not force free. For such systems the magnetic virial theorem can be used to set an upper limit to the amount of magnetic energy which can be stored in an equilibrium corona with given (magnetic) boundary conditions. The scalar virial theorem is derived by multiplication of the momentum equation for the magnetofluid with the position vector and subsequent integration over the volume V with bounding surface S (e.g. [8])

$$\int_V \left\{ \rho v^2 + \rho\Psi + 3p + \frac{B^2}{2\mu_0} \right\} d^3\vec{r} =$$

$$= \oint_S \left\{ \left(p + \frac{B^2}{2\mu_0} \right) \vec{r} \cdot d\vec{S} - \frac{(\vec{B}\cdot\vec{r})(\vec{B}\cdot d\vec{S})}{\mu_0} \right\} + \frac{1}{2}\frac{d^2 I}{dt^2}. \tag{11}$$

Here ρ is the density, v the velocity, Ψ the gravitational potential, p the pressure, $\vec{r}$ the radius vector and I the trace of the moment of inertia tensor of the matter inside the volume V. We apply Eq. (11) to a corona in equilibrium with the bounding surface consisting of two concentric spheres, one with radius going to infinity and the other above the photosphere in the dilute corona (see Fig. 4). In equilibrium the last term on the right-hand side vanishes. Further in a corona of finite energy the entire surface contribution from the sphere extending to inifinity vanishes. Next on the lower surface, which is still inside the low-bêta corona,

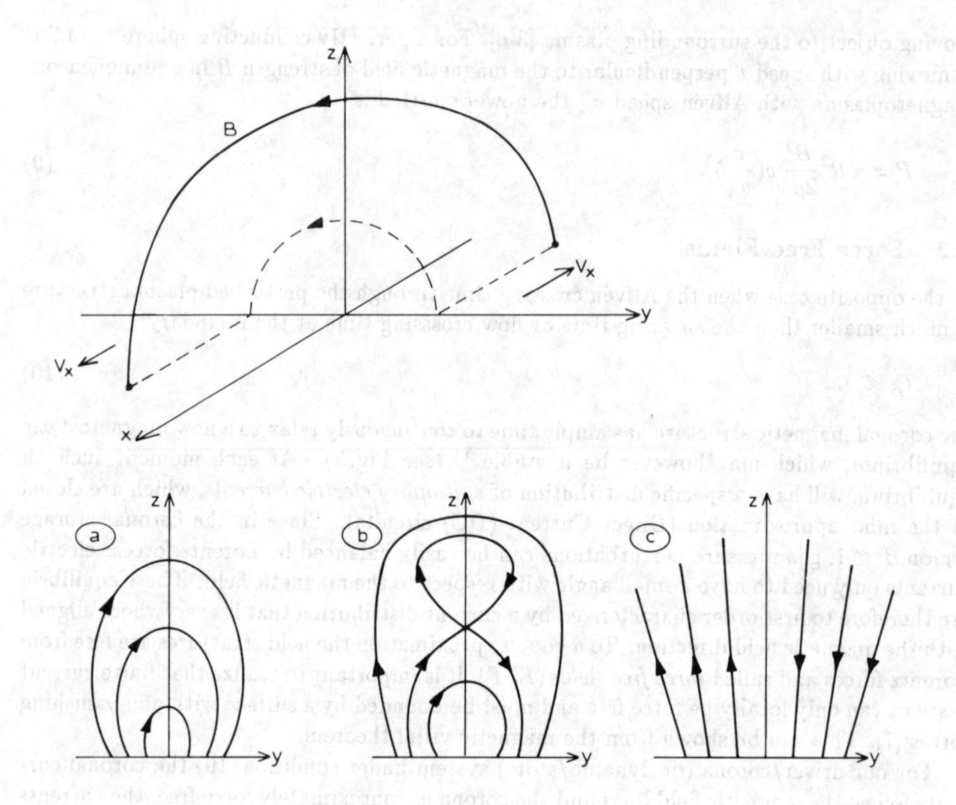

Figure 3: Shearing of a coronal magnetic structure ($z \geq 0$) by a relatively slow subphotospheric velocity pattern $\vec{v}$. The evolution of the projected field lines is sketched in (a)-(c): the asymptotic state (c) with open field lines and vanishing magnetic shear is reached only in the absence of resistivity; in reality resistive effects probably cause a flare- like transition to a different topology as sketched in (b).

the surface pressure term can be neglected in comparison with the magnetic term. Finally in the dilute, low-bêta corona only the total magnetic energy survives in the left- hand side. Eq. (11) then reduces to

$$W_M = \frac{2\pi R^3}{\mu_0}(< B_n^2 > - < B_t^2 >), \tag{12}$$

where W_M is the total magnetic energy inside the coronal volume V, the brackets denote surface averages and we have decomposed the magnetic field into components normal and tangential to the sphere with radius R. Each of the series of slowly evolving, approximately force free, coronal equilibria satisfies Eq. (12). In the absence of emerging or sinking magnetic flux the surface distribution of the normal field component is entirely determined by the frozen-in condition in the fluid moving just below and tangential to the sphere with radius R. Because of flux conservation the average value $< B_n^2 >$ can only increase if field flux is swept up into smaller areas. As the fluid below the boundary moves, magnetic energy

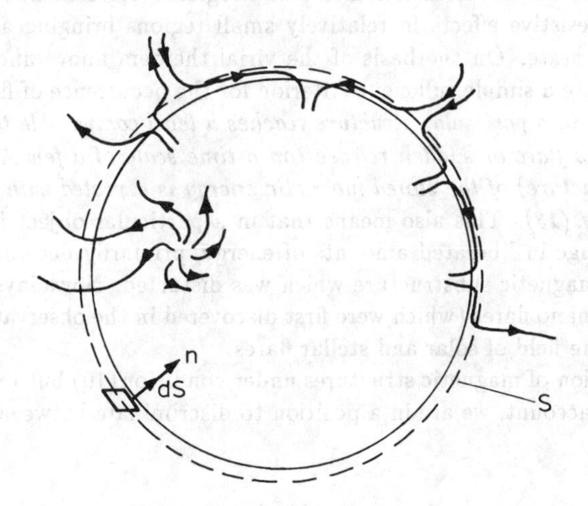

Figure 4: The figure shows the sphere of integration (dashed) used in Eq. (12) in a corona surrounding a spherical driver, some field lines (drawn) rooted in the driver and extending into the corona, and a surface element $d\vec{S}$ with normal vector $\vec{n}$ directed towards the center. The second sphere of integration lies at infinity.

is transported into the corona and stored. Eq. (12) then states that *the tangential magnetic field on average can never exceed the average normal field component on the bounding surface in an equilibrium situation* [9]. The second important conclusion from Eq. (12) is that *the maximum amount of magnetic energy which can be stored in a force free coronal structure of characteristic field strength B, area S and magnetic scale height H is of order*

$$W_M = SH\frac{B^2}{2\mu_0}. \tag{13}$$

Of course Eq. (13) is a rough estimate which neglects the details of the particular geometric field structure under consideration. It is however important for flares as we shall now explain. Up to now we have only considered ideal (non-resistive) mhd. As a result the field topology of the equilibria in a sequence does not change. If one includes resistive effects the field topology is no longer invariant. In particular resistive effects may allow the transition to a magnetic field structure of lower total energy and different topology but with the same flux distribution at the bounding surface (see Fig. 3b) [10]. The energy liberated during the transition would be observed as a magnetic flare. Such a sequence of events is indeed found from numerical resistive mhd calculations with slowly changing boundaries and for a Lundqvist number of 10^4 [11-13]: *First at a critical amount of shear the slowly evolving force free equilibrium becomes unstable and an ideal mhd instability appears on a fast (Alfvén crossing time) scale. The ideal mhd instability drives reconnection in a thin layer, creating a topologically separated gas blob or plasmoid. The plasmoid is ejected at the local Alfvén speed, while the stored magnetic energy is lowered by an amount comparable to the energy of the original potential field structure.* The details of the liberation of magnetic energy are still a matter of debate. In particular it is not clear if an ideal instability in the force free case appears first [14] which then drives reconnection in small regions. Agreement exists however on the importance of resistive effects in relatively small regions bringing about the transition to a lower energy state. On the basis of the virial theorem above and the numerical studies we can formulate a simple sufficient criterion for the occurrence of flares: *If the storage of magnetic energy in a particular structure reaches a level comparable to the original energy of the structure, a flare or sudden release (on a time scale of a few Alfvén crossing times of the coronal structure) of the stored magnetic energy is expected with total energy release of the order of Eq. (13).* This also means that in a particular object flares are expected to show a large range in liberated amounts of energy, primarily determined by the energy of the particular magnetic substructure which was distorted. Nowadays the concept of flares, mini flares and nano flares (which were first discovered in the observations by [15,16]) is quite common in the field of solar and stellar flares.

If we now consider the evolution of magnetic structures under condition (10) but taking a small non-zero resistivity into account, we are in a position to discriminate between two end results:

2.2.1 Stationary Circuits

If the currents in the force free corona generated by steady differential motion at the non-force free bounding surface remain so weak that the associated (by Ampère's law) field perturbation in the corona remains substantially less than the original field value, a stationary state may be possible (see Fig. 2b)

$$\delta B \ll B. \tag{14}$$

Apart from condition (14) a particular symmetric geometry and the existence of regions with sufficient resistivity to allow slip of the current system, are required (see Sect. 3.2).

2.2.2 Flares

If on the other hand the steady differential motion would distort the field strongly, so that

$$\delta B \leq B, \tag{15}$$

a sequence of flares is expected to occur. The characteristic time scale for the explosion would be of order a few Alfvén crossing times of the coronal magnetic structure (Eq. (7)). The repetition time of the flares on the other hand would be given by the flow crossing time (Eq. (2)).

The occurrence of magnetic flares is therefore probably a generic phenomenon in a great variety of cosmic objects where force free magnetic fields are anchored in a driver. In particular close binaries are rich in this respect. Fig. 5 shows how flares can occur in various

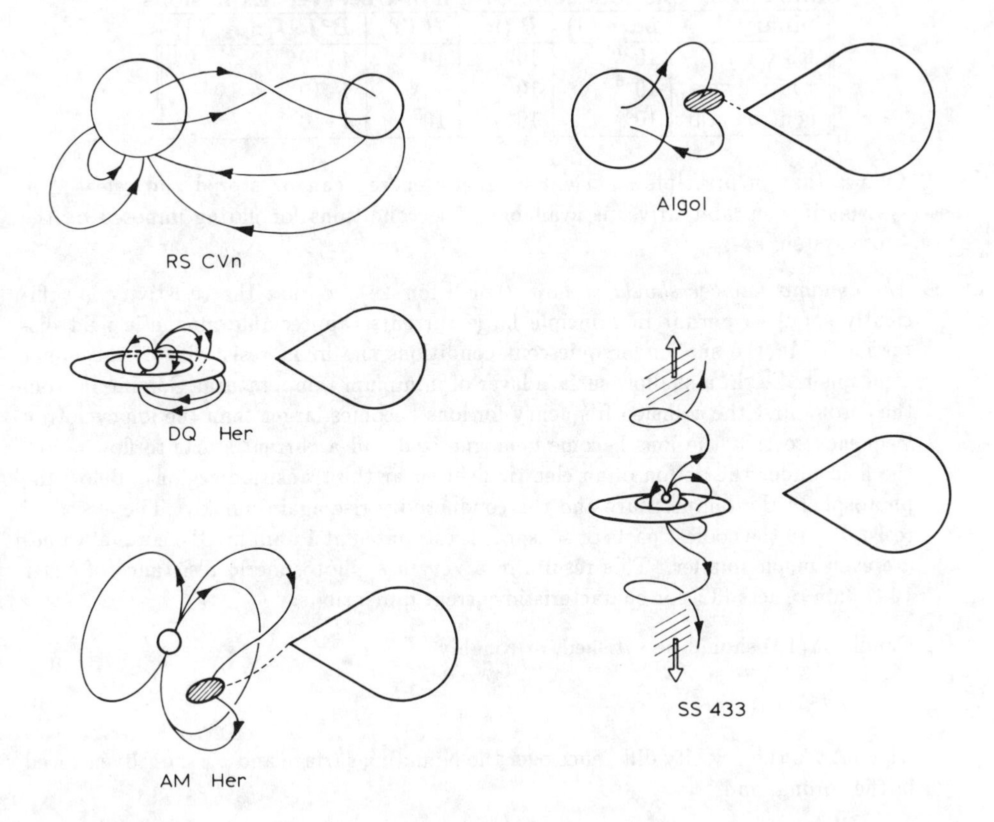

Figure 5: **The figure shows the possible generation of flares in various driver-corona systems, caused by, from top to bottom: differential rotation in RS CVn type binaries, mass exchange in magnetic Algol type binaries, accretion onto a white dwarf in a disk surrounding a magnetosphere in DQ Her type or accretion inside the magnetosphere in AM Her type binaries, and disk accretion onto a neutron star magnetosphere with the formation of jets and magnetic bullets as perhaps in SS 433.**

classes of close binaries: RS CVn stars with two interacting magnetospheres [17-22], Algol stars with transient mass exchange [23-25], contact binaries [26], cataclysmic variables with interactions between disk and magnetosphere of the white dwarf as in DQ Her systems or with interactions between infalling magnetized gas and magnetosphere as in AM Her systems [27,28], accreting magnetic neutron stars with the magnetic expulsion of plasmoids in jets [29]. The list of probable magnetic flaring objects can be extended to accretion disks around proto and T Tauri stars [28] and active galactic nuclei [30]. The observed amount of energy per explosion or "flare" in classic close binaries, cataclysmic variables and neutron star binaries (taken from [28]) is given in Table 2, together with characteristic dimensions of magnetic structures in these objects (put equal to the dimension of the stellar object) and the magnetic field strengths (based on estimate (13)) required to produce the observed explosion by magnetic means.

Table 2: Possible Magnetic Origin of Observed Explosions

binary	flare (J)	R (m)	B (T)	$B^2R^3/(2\mu_o)$ (J)
RS CVn	10^{27}	10^9	10^{-2}	$4 \cdot 10^{28}$
white dwarf	10^{26}	10^7	1	$4 \cdot 10^{26}$
neutron star	10^{33}	10^4	10^8	$4 \cdot 10^{33}$

It follows that in principle sufficient magnetic energy can be stored and released in these systems if a suitable driver is available. The conditions for flaring imposed on the driver-store system are:

1. *The dynamo number should be large* (condition (2)), so that the resistivity is sufficiently small to permit in principle large currents (appreciable magnetic field distortions). In the sun under quiescent conditions the main resistivity of the upper atmosphere lies in a shallow surface layer of minimum temperature: Coming in from the corona first the collision frequency for ions becomes larger than the ion cyclotron frequency so that the ions become demagnetized and a current starts to flow across the field under the action of an electric field (near the transition region). Below the photosphere the temperature and the conductivity rise again quickly. The electrical resistivity in the coolest parts of sunspots is estimated at 1 ohm m [31]; coronal values are even much smaller. This results in a very low photospheric resistance of order 10^{-8} ohm or less [32] for characteristic current dimensions.

2. Condition (10) should be satisfied, or roughly

$$\Delta v \ll v_A, \tag{16}$$

   where Δv is the velocity difference over the bounding surface and v_A the Alfvén speed in the corona, and

3. *The driver should be sufficiently powerful.* In the case of the sun the dynamic pressure in the photospheric layers is by far insufficient to act as a powerful driver [33,34], basically since $\rho(\Delta v)^2 \ll p_{gas} \approx B^2(2\mu_0)^{-1}$. Indeed
   the anchoring must occur in much deeper layers where the dynamic pressure of the convective eddies is relatively large; the forcing of the flux tubes is communicated to the surface layers by Maxwell stress in the form of Alfvénic motions. In the case of accretion the relative angular momentum of the driving gas should be much larger than the magnetic transport of angular momentum during the flare build-up, or roughly

$$\rho h(\Delta v)^2 \gg 2B^2 r/\mu_0, \tag{17}$$

where h is the height of the driver material along the field line at one footpoint, Δv and r are respectively the relative speed between both footpoints of the magnetic flux tube and their distance and we have put $\delta B \approx B$ (see Sect. 4).

Of course not all explosions in X-ray binaries are of magnetic origin: For the X-ray pulses of neutron star binaries with massive companions, for the type I bursts from X-ray pulsars and for the (recurrent) novae (nuclear fusion) there is no reason to invoque magnetic activity. On the other hand the Quasi-Periodic Oscillations in Low Mass X-Ray Binaries [35] and some of the eruptions in dwarf novae, polars and intermediate polars may be of the magnetic flaring type.

3 Energy Transport and Release

3.1 Lumped Astrophysical Circuits

Electric currents in cosmic plasmas generally occupy a large fraction of the volume of the particular magnetic structure. In contrast with current systems made of conducting wires the astrophysical magnetoplasma is not a lumped but a distributed circuit. Nevertheless the use of lumped circuits in nature is appealing because of their clarity by "lumping" the important physical properties of the system into a few coefficients of one simple ordinary differential equation. Early use of the circuit description can be found in [36] and its use has been quite common in magnetospheric current systems, e.g. the Birkeland currents [37,3,38] and more recently in solar physics [39]. However only in special circumstances can one construct a lumped circuit out of the actual current density field. The requirements are:

1. divergence-free current densities (the mhd approximation),
2. a simple topology of the current density lines, and
3. only changes on time scales larger than the Alfvén crossing time are considered so that the number and topological structure of the system of circuits remain constant in time.

Then one can proceed as follows: Integrate Faraday's law of induction along a closed curve of interest and eliminate the electric field with the usual form of Ohm's law in an isotropic plasma $\vec{E} = -\vec{v} \times \vec{B} + \vec{j}/\sigma$ ($\sigma \equiv 1/\eta$ is the conductivity) (or using the generalized Ohm's law to allow for a "battery field") to find

$$\frac{\partial \Phi}{\partial t} = -\oint \vec{E} \cdot d\vec{\ell} = \oint (\vec{v} \times \vec{B} - \vec{j}/\sigma) \cdot d\vec{\ell}. \tag{18}$$

Here Φ is the total magnetic flux contained by the closed curve. The first term on the right-hand side plays the role of an "electromotive force" $\mathcal{V} \equiv \oint (\vec{v} \times \vec{B}) \cdot d\vec{\ell}$. Eq. (18) describes quite generally the dynamo action in a moving magnetofluid, or how, for an arbitrary contour with given resistive properties, the current density and the contained magnetic flux change in response to the fluid motion.

Such behaviour is well-known from the example of the lumped *homopolar or unipolar dynamo* [7] in Fig. 6. In this case Eq. (18) can be further integrated over the current cross-section ΔS to give the standard result

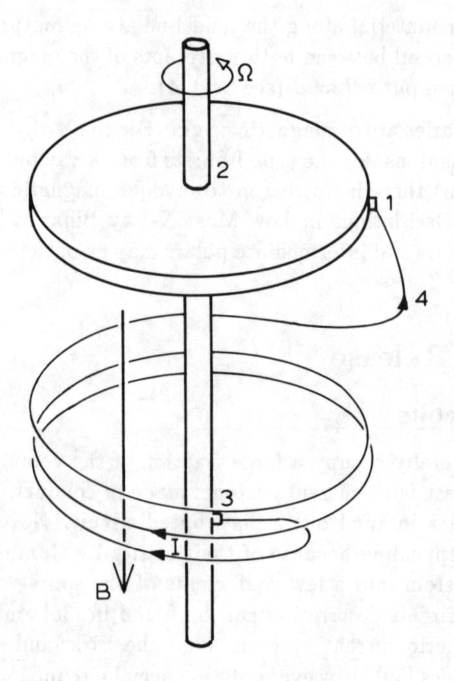

Figure 6: **The homopolar or unipolar dynamo. The rotating conducting disk is fixed in 2 to a conducting axis and connected in 1 and 3 to a coil with sliding contacts. The direction of rotation and the winding sense of the coil are chosen in such a way with respect to each other that a current starts flowing in the sense 1-2-3-4 which amplifies the magnetic field.**

$$\frac{d}{dt}(LI) = \mathcal{V} - IR, \tag{19}$$

where I is the total current at time t, L the total inductance of the circuit and R the total resistance. Note that we have kept the inductance inside the temporal differentiation to allow for a slowly varying wire length.

From Eqs. (18) and (19) it is clear that it only makes sense to make the identification $\Phi = LI$, $R = \oint(\vec{j}/\sigma) \cdot d\vec{\ell}\Delta S/I$ in an astrophysical plasma if the current is confined to a relatively thin tube, so that $I \approx j\Delta S$.

It is attractive to use a lumped circuit description because of its mathematical simplicity also under more general conditions. The use of a so-called "equivalent" electric circuit, e.g. for monochromatic Alfvén waves [40], can be useful to illustrate a specific property of the magnetoplasma under consideration. However *the resulting circuits then depend crucially on the chosen integration surface.* Therefore great care has to be taken in using the lumped circuit analogy, and for instance the character of the equivalent circuit may change abruptly in time when the resistivity in the actual plasma changes (see Sect. 3.4 below).

3.2 Steadily Rotating Loop

The concept of the unipolar dynamo has been applied to astrophysical plasmas in binaries: originally to the Io-Jupiter system [41,42], to cataclysmic variables [43] and to rotating black- holes with and without accretion disks [44-47]. In a steady state the electric current configuration must remain invariant during the differential motion of the system. For this a number of conditions have to be fulfilled:

1. The plasmas have to be highly conducting along a closed loop to allow for electric current generation by the moving material: Eq. (1) must be satisfied,
2. The differential motion between the field line footpoints must be subalfvénic: Eq. (10),
3. The field distortion must remain small: Eq. (14),
4. The original magnetic field distribution must possess a certain degree of symmetry: In the case of rotation the magnetic field must possess rotational symmetry, e.g. in Fig. 2b the undistorted magnetosphere around the central object must be symmetric with respect to the disk rotation axis, in Fig. 7 the topology of the field lines in the coronal loop must be invariant for rotation around the loop axis.

Let us first construct such a system for a magnetic loop structure, as depicted in Fig. 7. We warn the reader however that this steady situation is not representative for loops in the real stellar corona. The magnetic loop is assumed to consist of three portions:

- 0. Region number 0 below the photosphere, anchored in infinitely massive and perfectly conducting plasma which is rotating with prescribed angular speed Ω_0 so that the tube rotates subalfvénically,
- 1. A dilute region number 1 in the corona, which again contains perfectly conducting plasma apart from *two thin surface layers at the feet with finite resistivity,*

- 2. Region number 2 below the footpoint at the other end, with the same properties as region 0 except for its motion which we assume to vanish because of infinite inertia.

Then conditions 1, 2 and 4 of the previous paragraph are satisfied in our case provided a steady state exists. Assuming that a steady state is reached, let us now calculate the current and the associated field distortion and derive under what condition it remains small, as required for the steady state by condition 3 (see Eq. (14)). Since the magnetic field is twisted by the motion at one footpoint a current and a rotation will be set up in region 1 with angular speed Ω_1 lying in between the rotation speeds (Ω_0 and 0) of the subphotospheric colums. Throughout the approximately force free coronal part 1 the current flows along the field lines from footpoint 0 to the footpoint in 2 where it crosses the flux tube in the resistive layer, flows back over the tube surface to footpoint 0 where it closes through the resistive layer. The meridional projection of the current path is sketched in Fig. 7. Now outside the resistive layers we have $\vec{E} = -\vec{v} \times \vec{B}$, while inside approximately $\vec{E} = -\vec{v} \times \vec{B} + \vec{j}/\sigma$ is satisfied (for simplicity we neglect the Hall electric field $\vec{j} \times \vec{B}(ne)^{-1}$ which is allowed if the collision frequency of the charge carrier is much larger than its cyclotron frequency). In the steady state the electric potential jump across an individual flux tube is constant above and below the resistive layer. Going from region 0 to 1 and applying the above expressions for the electric fields, it follows that the coronal part 1 of the tube rotates slower than the driver 0. *The slip at footpoint 0 is associated with the current in the resistive layer at that location. Similarly the perpendicular current in the resistive layer at footpoint 2 takes care of the required slip of the rotating part 1 over the non-moving region 2.* The spatial distribution of the resistivity in the two layers in the radial direction now completely determines the stationary rotation speed of an individual nested flux tube. Using for simplicity two homogeneous slabs for the resistive layers, with uniform magnetic flux distribution inside the flux tube, one finds in the non-moving "laboratory" frame:

$$\mathcal{V}_0 = \mathcal{V}_1 = \mathcal{V}_2, \tag{20}$$

with

$$\mathcal{V}_0 = \Phi\Omega_0(2\pi)^{-1} - IR_0, \qquad \mathcal{V}_1 = \Phi\Omega_1(2\pi)^{-1}, \qquad \mathcal{V}_2 = IR_2. \tag{21}$$

Here $\mathcal{V}_i$ is the potential jump across the flux tube with flux Φ in region i ($i = 0,1,2$), I is the total current crossing the flux tube in each resistive layer at its outer radius and a formal total resistance is defined by $R_i \equiv (4\pi\sigma_i h_i)^{-1}$ for $i = 0,2$ with h_i the height of resistive layer i. One recognizes Eq. (20) as Kirchhoff's law for electric circuits. It is now easy to find the total current

$$I = \frac{\Omega_0\Phi}{2\pi(R_0 + R_2)} = \tag{22}$$

$$= \frac{(\Omega_0 - \Omega_1)\Phi}{2\pi R_0} = \tag{23}$$

$$= \frac{\Omega_1\Phi}{2\pi R_2}, \tag{24}$$

the rotation speed of the coronal part of the tube

$$\frac{\Omega_1}{\Omega_0} = \frac{R_2}{R_0 + R_2}, \tag{25}$$

and the power dissipated in the "load" R_2

$$I^2 R_2 = \left(\frac{\Omega_0 \Phi}{2\pi}\right)^2 \frac{R_2}{(R_0 + R_2)^2}. \tag{26}$$

From Eq. (22) and Ampère's law it follows that the magnetic field perturbation is of order

$$\frac{\delta B}{B} = \frac{\mu_0 \Omega_0 r}{4\pi(R_0 + R_2)}, \tag{27}$$

where r is the minor tube radius. The field perturbation remains relatively small (so as to satisfy condition 3 and to establish a steady state), if the total resistance of the resistive

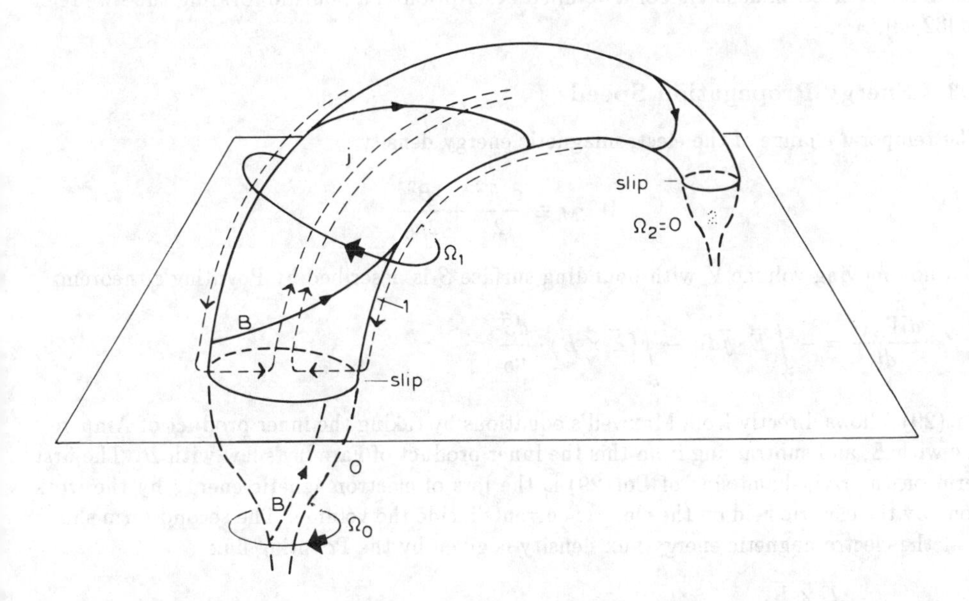

Figure 7: The figure shows a hypothetic stationary rotating coronal loop, which is forced to rotate by subphotospheric motion below one footpoint. Slip occurs between parts 0 and 1 (with the rotation rate decreasing from Ω_0 to Ω_1) and between 1 and 2 (decrease from Ω_1 to 0) where finite resistivity causes the existence of a current $\vec{j}$ (dashed lines) closing transversely to the magnetic field $\vec{B}$ (oblique drawn line). The current is distributed throughout the tube volume.

layers across the field is large enough. Therefore *to decide if a particular configuration, which obeys conditions 1,2 and 4, generates flares, one first calculates the relative field perturbation with Eq. (27): if the result is larger than unity a steady state cannot be reached and flares occur.*

From Eq. (25) it is seen that $\Omega_1 = \frac{1}{2}\Omega_0$ for $R_0 = R_2$, that $\Omega_1 = 0$ (no slip at footpoint 2) for $R_2 = 0$, and that $\Omega_1 = \Omega_0$ (no slip at footpoint 0) for $R_0 = 0$. In the last two cases the current across the field at the zero- resistance layer is formed by particle drifts in the discontinuous field at that footpoint.

Further from Eq. (26) follows the well-known result that the power dissipated at the load R_2 for fixed "internal" resistance R_0 maximizes if $R_0 = R_2$ (impedance matching).

Finally we mention that motion in a magnetoplasma in general is associated with a space charge distribution. In the coronal part of the tube one has approximately

$$\tau = \epsilon_0 \nabla \cdot \vec{E} = -\epsilon_0 \nabla \cdot (\vec{v} \times \vec{B}) = \frac{(\vec{\Omega} \times \vec{r}) \cdot \vec{j}}{c^2} - \frac{2\vec{\Omega} \cdot \vec{B}}{c^2 \mu_0} = \mathcal{O}\left(\frac{v v_D}{c^2} n_0 e\right), \tag{28}$$

where we have used solid body rotation $\vec{v} = \vec{\Omega} \times \vec{r}$ and we have written $j = n_0 e v_D$ (n_0 is the ambient electron density, v_D is the drift speed of the current carriers and $-e$ is the electron charge. Since the drift speed does not exceed the thermal electron speed, Eq. (28) shows that for non- relativistic motion this charge density is of order v/c or less relative to the ambient electron density. For a detailed description of a realistic rotating tube we refer to [32,48].

3.3 Energy Propagation Speed

The temporal change of the electromagnetic energy density

$$W_{EM} \equiv \frac{\epsilon_0 E^2}{2} + \frac{B^2}{2\mu_0}$$

in a non-moving volume V with bounding surface S is described by Poynting's theorem

$$\frac{dW_{EM}}{dt} = -\int_V \vec{E} \cdot \vec{j} dV - \oint_S (\vec{E} \times \vec{B}) \cdot \frac{d\vec{S}}{\mu_0}. \tag{29}$$

Eq. (29) follows directly from Maxwell's equations by taking the inner product of Ampère's law with $\vec{E}$, and subtracting from this the inner product of Faraday's law with $\vec{B}$. The first term on the right-hand side of Eq. (29) is the loss of electromagnetic energy by the work done by the electric field on the electric currents inside the volume. The second term shows that the electromagnetic energy flux density is given by the Poynting flux

$$F_{EM} = \frac{\vec{E} \times \vec{B}}{\mu_0}. \tag{30}$$

In the ideal mhd approximation, which is valid for low-frequency motion, the electric volume energy can be neglected in comparison with the magnetic energy: $W_{EM} \approx W_M \equiv \int_V B^2 (2\mu_0)^{-1} dV$ and we can use $\vec{E} = -\vec{v} \times \vec{B}$ to estimate the Poynting flux

$$| F_{EM} | = | \frac{B^2 \vec{v} - (\vec{B} \cdot \vec{v})\vec{B}}{\mu_0} | \approx \frac{B^2 v_\perp}{\mu_0}, \tag{31}$$

where $v_\perp$ is the velocity component perpendicular to the magnetic field.

For an Alfvén wave only a contribution of order $B < \delta B \delta v > / \mu_0 \approx v_A (\delta B)^2 / \mu_0$ survives after averaging over the wave period, where δB is the wave perturbation of the field (only in the direction perpendicular to the ambient field), δv is the oscillatory velocity amplitude and $\rho(\delta v)^2 \approx (\delta B)^2(2\mu_0)^{-1} \approx \rho(v_A \delta B/B)^2$. Note that the field perturbation becomes strong as the cross-field fluid velocity approaches the Alfvén speed. Clearly the propagation speed of low-frequency (in contrast with high-frequency photons) electromagnetic wave energy along the ambient field occurs typically at the Alfvén speed.

One can now also understand why it takes at least an Alfvén crossing time to set up or change a (stationary) circuit consisting of well-conducting magnetoplasma. The energy of a stationary circuit resides in the magnetic field, which fills a certain volume of plasma. If the currents change the fields change but since these are practically frozen in the plasma this will be established by Alfvénic motions. Similarly if one forces the sudden existence of a line current in a plasma, the associated magnetic field will be shielded from the surrounding plasma by the induction of a coaxial return current (initially an electric displacement current). The temporally changing magnetic field surrounding the current causes electric fields, which lead to (Alfvénic) motions in the plasma. The minimum time required for the magnetic field associated with the current to penetrate the surrounding plasma, is the Alfvén crossing time.

From Eq. (31) it also follows that magnetic energy in the form of a subphotospherically rising flux tube propagates with a velocity at most of the order of a (hybrid) Alfvén speed: The reason is that for a magnetic tube below the photosphere the rise speed v in surrounding gas with density ρ is limited by a balance between the aerodynamic drag per unit area perpendicular to the rising tube surface ρv^2 and the buoyancy lift $B^2(2\mu_0)^{-1}$.

From this follows the important conclusion that *subphotospherically stored magnetic energy can never appear in the corona at a rate substantially exceeding the characteristic magnetic energy density times surface area times minimum Alfvén speed.* For an individual rising solar flux tube of radius 500 km, field strength 0.1 T and Alfvén speed of 100 km/s an upper limit on the energy flow into the solar corona is typically $3 \cdot 10^{20}$ Watt; this is by far insufficient to explain the energy of 10^{25} J which is liberated during 10^3 s in a large solar flare. This estimate makes it plausible that the flare energy indeed has been accumulated in the corona on a time scale much longer than the individual flare release time. (Of course emerging magnetic flux may induce a global instability of the preexisting field structure and thereby trigger its energy release.)

A general expression for the energy flow from the driver into the corona (and vice versa) for the ideal mhd case is found from Eq. (29) by using $\vec{E} = -\vec{v} \times \vec{B}$

$$\frac{dW_M}{dt} = -\int_V (\vec{\jmath} \times \vec{B}) \cdot \vec{v} dV + \oint_S \frac{\vec{B} \cdot \vec{v} \vec{B} \cdot d\vec{S}}{\mu_0} - \oint_S \frac{B^2 \vec{v} \cdot d\vec{S}}{\mu_0}. \tag{32}$$

The first term on the right-hand side corresponds to the term in the same position in Eq. (29), and is the work done by the electric field on the current; in the ideal mhd case it consists entirely of the reversible work done by the Lorentz force inside the volume of integration; in the resistive case also an irreversible Ohmic dissipation term would appear. The next term shows how shear on the boundary causes transport of energy into the volume provided magnetic field extends through the boundary and provided the field has a component in the direction of the velocity along the boundary. The final term shows how the emergence of new magnetic flux through the boundary causes a change in electromagnetic energy. At a corona-driver interface characteristically all three terms are important, as both effects (shear and emergence) occur.

3.4 Sudden Bifurcation into Two Lumped Circuits

In connection to the example of the stationary rotating tube treated in Sect. 3.2 there appears to be a pittfall in deriving the effective energy propagation speed. The energy flow per unit area into the resistor at point 2 is of magnitude $F_{EM} \approx vB\delta B/\mu_0 \approx (vB/\delta B)(\delta B)^2/\mu_0$, where v is the rotation speed of the cylinder surface and B is the axial field component. It would seem that the energy propagation speed becomes arbitrarily large if $\delta B/B \downarrow 0$, which occurs if $R_2/R_0 \rightarrow \infty$ (see Eq. (27)). However in that case the Poynting flux goes to zero as is evident from the first expression (cf. also Eq. (26)). Apparently in this case it does not make sense to distinguish between a propagating and a non-propagating part in the total magnetic energy density $((\delta B)^2(2\mu_0)^{-1}$ and respectively $B^2(2\mu_0)^{-1}$. Physically this agrees with the absence of a propagating non-uniformity in the field.

What happens with the stationary tube of Sect. 3.2 if the resistance R_2 is suddenly increased strongly (and thereafter stays constant) in terms of lumped circuits requires careful consideration. First strictly the circuit approach only describes phenomena on a slow time scale larger than the Alfvén crossing time. Eqs.(20-27) can then be used to determine the initial and the end states. Compared to the initial configuration, in the final state the current has decreased because of the increased resistance R_2 (Eq.(22)), the tube rotation rate Ω_1 and the slip at footpoint 2 have increased (Eq.(26)), and the electric potential jump and electric field transverse to the tube axis have also increased (Eq.(21)). To get an idea about the transitory phase recourse must be taken to the full equations for the plasma. The situation is very similar to the sudden placement of a so-called double layer in a coaxial electric current as has been described in [49] and we refer to the left-hand part of Fig. 9 adapted from that paper, but now placed in the rotating coronal loop system of Fig. 7. The suddenly increased resistance R_2 causes an increased electric field at endpoint 2 initially when the current has not yet changed ($\vec{E} = \vec{j}/\sigma$). This temporal change of the electric field largely propagates back throughout the tube as an Alfvénic front, it causes a decrease in the magnetic field twist and the associated current, an increase in the rotation rate of part 1 ($\vec{E} = -\vec{v} \times \vec{B}$). As the information reaches endpoint 1, new feedback propagates to endpoint 2 until the final steady state is reached. It is interesting to note that the basic course of events can be modelled with circuits if one allows for the sudden bifurcation of the primary circuit at the instant of change in R_2 into two circuits: one with the same shape but already with the asymptotic values of new current, etc., the other circuit with the excess current and a continually diminishing shape with one end in 1 and the other in a retreating (from 2) Alfvén front (cf. the left-hand part of Fig. 9). Finally we note that the sudden introduction of a large resistance into a circuit of the kind of Sect. 3.2 only partly leads to enhanced dissipation of the current in that resistor (and in any case only during the transitory stage as it decreases asymptotically (see Eq. (26)) but that *part of the excess energy of the initial circuit is transformed into rotational energy of the plasma tube.*

3.5 Flare Circuits

Simple, in retrospect visionary, line current or circuit descriptions have been proposed for filamentary magnetic structures in the solar corona [50] and for the onset and energy release of solar flares [51,52]. The (two-dimensional) model is based on force balance on a line current running at altitude h above a plane conducting rigid surface and perpendicular to an ambient magnetic field anchored in the rigid surface, as sketched in Fig. 8. The ambient arcade shaped magnetic field of strength $B_0(h)$ exerts a downward Lorentz force on a current at the top of a field line at altitude h. The magnetically screening surface exerts an upward Lorentz force on the current. For simplicity we neglect the force of gravity on the filamentary current. Force balance per unit length of current then determines the magnitude of the current $I(h)$ at altitude h:

$$I(h)B_0(h) = \frac{\mu_0[I(h)]^2}{4\pi h}. \tag{33}$$

The left-hand side is the upward force on the current by the ambient field. The right-hand side is the repulsive force of the conducting boundary on the current filament and is obtained as follows. Because of the symmetrical geometry the magnetic field above the boundary surface is the same as that of a "virtual" antiparallel current placed at depth $-h$ below the surface. This so-called *mirror current* satisfies the condition that the normal magnetic field component at every point of the boundary is exactly opposite to the normal component from the original current, so that the net field of both currents does not penetrate the surface. From Eq. (33) the equilibrium current at altitude h is found to be

$$I(h) = 4\pi h B_0(h)/\mu_0.$$

The presence of a maximum in the curve $I(h)$ forms a local maximum to the amount of current which can be stored in stable equilibrium (see Fig. 8). This follows from a calculation of the restoring force per unit length on a current $I(h)$ which is displaced from its equilibrium position by an amount δh. From Eq. (33) the restoring force is $-\frac{\mu_0 \delta h I}{4\pi h}\frac{\partial I(h)}{\partial h}$. Therefore at small height the equilibrium is stable. As the current magnitude increases from a small value its stable equilibrium position shifts to larger height, untill it reaches its maximum value I_{max} at altitude h_0. For a slight increase in current magnitude the line current cannot remain in force balance and lifts off. In principle the current may reach a stable equilibrium position higher up by the presence of further maxima in the curve $I(h)$, which can be achieved for instance by the presence of a weak large-scale background field or by including the force of gravity as in [51].

Even if the real current distribution during a flare would not correspond to a line current but rather to a *spatially distributed* force free system, the model is of physical value. It *correctly represents two essential physical properties of distributed force free systems*: First, *as the current increases, it rises in altitude in accordance with the general behaviour of force free fields* which tend to rise and open out as the currents increase (cf. Fig. 3). Secondly the *characteristic magnetic energy content of the current per unit length at the point of instability* is $h_0^2(B_0(h_0))^2(2\mu_0)^{-1}$, which nicely agrees with the estimate (13) for the available flare energy in a general force free structure. *Therefore the macroscopic flare evolution is probably correctly represented by the line current model.* Also the stability

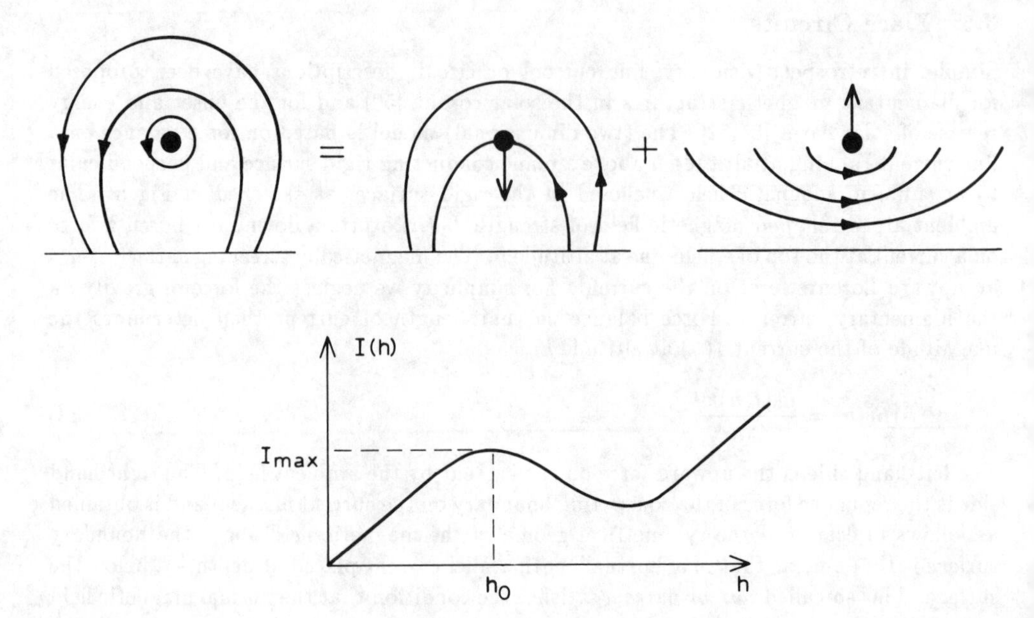

Figure 8: The top part shows the decomposition of the equilibrium of a line current into a downward Lorentz force on the current by the background field and an opposite Lorentz force arising from the shielding rigid wall. The current is located at the black dot and is directed towards the reader. For simplicity gravity is neglected. The lower half depicts the equilibrium current as a function of height h for an arbitrary background field.

calculations of line currents in more complex systems as the common corona of an RS CVn binary [22,53] are directly relevant to flares in these systems.

The presence of *singular current distributions* is expected to be quite common in a coronal field anchored in a convective dynamo. In Fig. 1 it is seen that the regions of contact between discrete flux tubes anchored at different places in the photosphere, contain singular current layers as the field orientation on both sides is unrelated. In fact the energy in these singular current layers (dashed in Fig. 1) again can be of the same order as the energy in a potential field with the same flux distribution at the boundary.

3.6 Reconnection and Particle Acceleration

Magnetic reconnection [54] is a term used for the topological change of the magnetic field structure in a nearly ideal plasma (where the field lines can be followed in time by the material motion) as a result of resistivity in small regions (see Fig. 10b,c). The process liberates part of the magnetic energy which is converted into motion by Lorentz forces of the reconnected state and into particle acceleration by electric fields in the reconnecting layer. *Reconnection allows a nearly ideal plasma to make the transition to a state of lower magnetic energy by finite resistivity in many (vanishingly small) regions.* A first description of steady reconnection is given in [55] and since then most research on reconnection has gone into finding stationary solutions [56]. However reconnection in the laboratory and in the cosmos may have little to do with these stationary models. In reality *reconnection is very time-dependent and leads to turbulent motions and turbulent magnetic fields in the plasma* [57,58]. Further progress in this area is badly needed.

One of the most conspicuous aspects of solar and stellar flares is *heating of plasma and acceleration of high-energy particles*, up to tens of MeV for electrons and GeV for ions in the solar case [59,60] and to much higher energies near compact objects. Radio observations of the high-energy electrons show that the acceleration often occurs in a multitude of spiky events [61,62,63]. The distinction between plasma heating and particle acceleration is based on a theoretical distinction between Maxwellian particle distributions with a well-defined local temperature and all other deviating distributions. In fact this distinction may not be very relevant: possibly the basic acceleration process in magnetic flares is universal while the result in terms of fractional heating versus acceleration depends on the importance of collisions at the accelerating site. Possible acceleration processes are [59,60]:

- *Stochastic Acceleration.* In cosmic plasmas, which are hot and dilute, collisions play a secondary role. The electron-ion collision frequency is of the order

$$\nu_{ei} \approx \frac{\omega_{pe}}{N_D}, \tag{34}$$

where $\omega_{pe} \equiv ne^2(\epsilon_0 m)^{-1/2}$ is the electron plasma frequency and

$$N_D \equiv n\lambda_D^3 = 3.3 \cdot 10^5 T^{3/2} n^{-1/2} \tag{35}$$

is the the Debye number, $\lambda_D \equiv v_{te}/\omega_{pe}$ the Debye length, $v_{te} \equiv (KT/m)^{1/2}$ the thermal electron speed, T the electron temperature in Kelvin and n the electron density m^{-3}. In most cosmic plasmas $N_D \gg 1$ and as a result waves and eigen oscillations, called *plasma turbulence*, can be easily excited. If in a particular plasma a particular wave mode is excited to a high level, particles can be accelerated. This occurs by some kind of resonance between the electric field of the particular wave and particles of the right charge and momentum. A simple example of such a resonance condition is that the Doppler shifted frequency of the wave in the frame moving with the projected particle speed along the magnetic field equals an integer (including zero and negative numbers) times the particle cyclotron frequency:

$$\omega^{\nu}(\vec{k}) - k_{\|} \cdot v_{\|} = N\omega_{c\alpha}, \tag{36}$$

for waves of mode ν with wave vector $\vec{k}$, wave frequency $\omega^{\nu}(\vec{k})$ and particle of kind α with speed $\vec{v}$ and cyclotron frequency $\omega_{c\alpha} \equiv q_{\alpha}B(\gamma m_{\alpha})^{-1}$, charge q_{α}, mass m_{α} and Lorentz factor γ. In that case the particle is kicked periodically in the same manner leading to a systematic energy transfer between particle and waves. Not surprisingly since in a magnetoplasma many wave modes exist, they correspond to as many different acceleration mechanisms. These processes are usually taken together under the name stochastic acceleration [63,64], as the waves often do not have specific phase relations and show a stochastic appearance. The main problem with these acceleration processes is to clarify how part of the flare energy is *first converted into wave turbulence of the specific wave mode and frequency interval.* For instance this problem exists for Langmuir wave turbulence which is very effective in accelerating mildly relativistic electrons, and for Alfvén turbulence of relatively high frequency which is very effective in accelerating ions to relativistic energies. Note however that Alfvén waves of low frequency are probably generated abundantly by the very process of bursty reconnection.

- *Direct Electric Fields.* Two possibilities are apparent: The first is by *electric field transverse to the magnetic field.* In the corona potential differences exist between the slowly moving magnetic field lines which are themselves equipotential lines to a first approximation. If a charged particle drifts along the electric field direction by the action of some other force it is accelerated. The process is known as *drift acceleration* [66]. Particular examples are acceleration near rotating compact objects and drift acceleration in quasi-perpendicular shocks [67,68]. The second acceleration possibility is by *electric field directed along the magnetic field.* Although for slow changes the individual magnetic field lines can be usually considered as electric equipotentials, this is not always true in the presence of currents. An electric circuit responds to local changes in the current geometry or in the plasma properties by generation of time-dependent inductive voltages. These electric fields although of a transient nature can be large because of the large length (and therefore of the inductance) of the coronal circuits. A special kind of parallel electric field acceleration is by so-called electrostatic double layers [49](see Sect. 3.7 below).

- *Diffusive Shock Acceleration.* For a strong shock the plasmas on both sides are in strong relative motion of approach. While this is true for the main body of the plasma some higher energy particles have a mean free path large enough that they can cross the shock transition. If they can be made to reflect many times, travelling back and forth between the approaching "mirrors", they undergo *first-order Fermi acceleration*, known as diffusive shock acceleration. The process has been elaborated for cosmic shocks as occur in supernova remnants and in shocks in (extra)galactic radio jets [69]. However this mechanism does not appear to be efficient for ion acceleration in solar flares as are inferred from gamma ray observations a few seconds after the flare onset.

3.7 Electrostatic Double Layers

Electrostatic double layers (DLs) have been proposed as the site of energy release in solar flares [70,71] and other energetic cosmic phenomena [38]. They have been observed in current carrying plasmas in the laboratory and in the earth magnetosphere [73-75]. A stationary DL can be considered as a coherent large-amplitude electrostatic plasma wave: The electric field is confined to the DL, which extends along the magnetic field over a typical distance of $10 - 10^3\lambda_{D0}$, and derives from a charge separation which is maintained self-consistently by the particle motion under the influence of the electric field (λ_{D0} is the Debye length in the surrounding plasma). Although stationary DLs are well studied theoretically, *the "explosive" type of DL is probably more common* under laboratory and cosmic conditions. DLs are of particular interest for magnetic cosmic explosions since in an electric plasma circuit they *accelerate both electrons and ions, and at the same time trigger the conversion of magnetic circuit energy into bulk motion* by allowing the magnetic field to unwind. How this occurs can be seen in Fig. 9, adapted from [49,76]. A DL is embedded in a cylindrical current tube shielded from its surroundings by a return surface current. The ideal mhd approximation is valid throughout the tube except for a thin region where the DL is located. After the establishment of the DL with central potential jump $2\mathcal{V}_0$ two Alfvénic fronts propagate away from the DL. The plasma through which the front has passed is set into rotation in agreement with the ideal mhd relation $\vec{E} = -\vec{v} \times \vec{B}$. The magnitude of the rotation speed and therefore of the electric field in the ideal mhd part is determined by continuity of the tangential electric field in a stationary state by the magnitude of the transverse electric field inside the DL. The latter is determined by the parallel potential jump $2\mathcal{V}_0$ and the radius of the cylinder (see Fig. 9 below). Part of the original total current, ηI passes through the DL, the rest is diverted in the Alfvénic front across the magnetic field where its Lorentz force accelerates the plasma up to the rotation rate Ω.

Let us assume that the angular rotation speed is uniform over the rotating parts and that the current is distributed uniformly over the tube cross section. It is then a simple matter to express the relative power put into particle acceleration as compared to motion in terms of η. The potential jump over the DL at radius r is

$$\mathcal{V}_{DL}(r) = 2\int_r^{r_0} r\Omega B_z dr = \Omega B_z(r_0^2 - r^2). \tag{37}$$

We define $\mathcal{V}_{DL}(0) \equiv 2\mathcal{V}_0$. Then the power dissipated over the DL in the form of particle acceleration is

$$\dot{W}_{DL} = \eta I \mathcal{V}_0, \tag{38}$$

the power put into rotational motion is

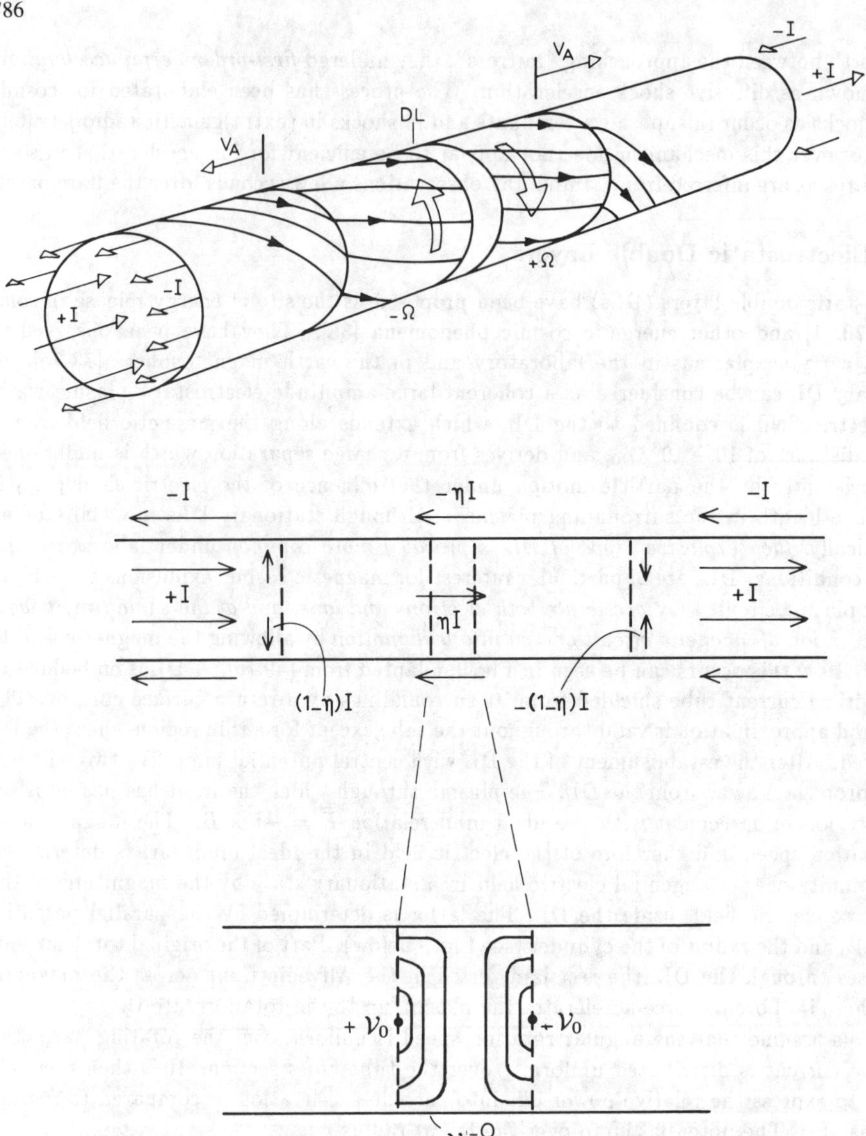

Figure 9: A cylindrical homogeneous current distribution shielded from its surroundings by a surface return current. After the appearance of a double layer inside the tube only a fraction η of the current passes through the DL. The rest of the current is diverted across the cylinder in Alfvénic fronts retreating from the DL at the Alfvén speed. The Lorentz force associated with the cross-field currents in the fronts accelerate the gas to a rotation speed Ω. Ideal mhd is valid throughout the tube except inside the DL. The potential jump from axis to cylinder surface is directly related to the rotation rate and must be equal to half the central potential jump along the DL.

$$\dot{W}_{kin} = \frac{\pi}{2}\rho r_0^4 \Omega^2 v_A = \frac{2\pi \mathcal{V}_0^2}{\mu_0 v_A}, \tag{39}$$

and the total power of converted current energy is

$$\dot{W}_{tot} = \frac{\mu_0}{8\pi}(1-\eta^2) I^2 v_A, \tag{40}$$

where we have used $B_\phi(r) = \frac{\mu_0 I r}{2\pi r_0^2}$ before the passage of the front and a fraction η times that value afterwards. Putting

$$\dot{W}_{tot} = \dot{W}_{DL} + \dot{W}_{kin}$$

and substituting Eqs. (38)-(40) one finds the potential jump

$$2\mathcal{V}_0 = \frac{\mu_0}{2\pi} I V_A (1-\eta) \tag{41}$$

and the relative power in particle acceleration over gas motion

$$\frac{\dot{W}_{DL}}{\dot{W}_{kin}} = \frac{2\eta}{1-\eta}. \tag{42}$$

The actual value of η is determined by the microscopic physics inside the DL. In particular since the DL has a current-voltage characteristic which initially rises monotonically, while the ensemble of pairs $(\mathcal{V}_{DL}, \eta I)$ permitted by physical consistency in the above model shows the reverse trend one unique value of η is chosen.

In the original proposal [71] the solar flare energy is released in one DL, which appears by a current evacuation process or instability when the drift speed of the electrons in the current exceeds the thermal speed. There are two main difficulties with this proposal. The first is that with the exception of the singular layers (cf. Sect. 3.5) it is highly unlikely that such high current densities occur in coronal circumstances, as this would require a constriction of the current cross section by a factor

$$\frac{\omega_{ce}}{\omega_{pe}} \frac{c/\omega_{pe}}{r/2} \frac{c}{v_{te}},$$

where r is the radius of the original current which is comparable to the scale height of the original magnetic stucture itself. The second difficulty is the following: In a large solar flare the total energy is 10^{25} J. If this is the energy content $\frac{1}{2}LI^2$ of a circuit with an "observed" length of $\ell = 10^7$ m and therefore an inductance of approximately $L = \mu_0 \ell = 4\pi$ Henry, one needs a current $I \approx 1.3 \cdot 10^{12}$ A. Since this energy is released in 10^3 s the power delivered by the circuit is $I\mathcal{V} \approx 10^{22}$ W. For the voltage drop of the DL one thus finds $\mathcal{V} \approx 8 \cdot 10^9$ Volt. Although GeV ions are produced in solar flares this is not typical for the energy of a flare accelerated electron which is 100 keV or less.

There is however an alternative route to the formation of DLs in coronal circuits which seems more realistic from an astrophysical point of view and which can be reconciled with many flare characteristics [77,28,78]. Both in the laboratory and in numerical experiments another way to produce DLs is by the *induction* of large voltage spikes (by a capacitor bank).

In the cosmos such capacitor banks are not commonly available; rather reconnection may do the same trick by locally modifying the field structure and thereby the current path. The global circuit reaction then is an inducted local voltage drop oppposing the forced change of current path. Out of this voltage drop a DL possibly develops. It is likely that this DL is of the *transient type* since it does not fit in with the boundary conditions of the preexisting plasma: On the low (high)-potential side it does not supply the half- Maxwellian of electrons (c.q. ions) in the ambient plasma travelling away from the DL and moreover produces an ion (c.q. electron) beam which may create anomalous resistivity in the surrounding plasma. In this way local reconnection may lead to the formation of one DL of limited extent (in directions transverse to the magnetic field), which itself triggers other transient DLs. Then the microscopic picture of *a flare appears as a multitude of transient DLs* distributed over the flaring magnetic structure with a small filling factor (in agreement with observations [79]), in which the magnetic energy is converted into acceleration of electrons and ions, the excitation of low- and high-frequency electrostatic plasma waves from the respective beam instabilities, into mass motions and mhd waves by the unwinding field and into coherent plasma radiation at radio frequencies observed on the earth.

3.8 Magnetic Helicity and Taylor's Hypothesis

With the use of Taylor's hypothesis [80,81] based on [82] one can calculate the energy which is liberated when a specific magnetic field structure relaxes by reconnection. *The hypothesis states that a magnetic field structure inside a magnetically closed volume V which relaxes under the only constraint that the the total magnetic helicity is conserved, relaxes to a linear force free field inside* V (a field with $\nabla \times \vec{B} = \alpha \vec{B}$ where α is constant in space). The magnitic helicity K inside V is defined by

$$K \equiv \int_V \vec{A} \cdot \vec{B} dV, \tag{43}$$

with $\vec{A}$ the vector potential corresponding to $\vec{B}$ ($\vec{B} = \nabla \times \vec{A}$). A magnetically closed volume is a volume formed by magnetic field lines (which are closed since $\nabla \cdot \vec{B} = 0$); consequently on its boundary with normal vector $\vec{n}$ the quantity $\vec{B} \cdot \vec{n}$ vanishes. Taylor's hypothesis has been successful in understanding the relaxation of a toroidal laboratory pinch. His idea has been further developed to stellar flares [83] and to slow motions at the boundary of a stellar corona [84-87]. We shall first illustrate the nature of magnetic helicity and then show what it has to do with reconnecting magnetic structures.

The helicity inside a volume V is a measure of the amount of *linkage* of the magnetic field lines within that (magnetically closed) volume. To see this consider two thin flux tubes which are linked as in Fig. 10a. Each tube i ($i = 1,2$) has a magnetic flux Φ_i, a volume V_i, a cross-section $\Delta\sigma_i$ and an arbitrary surface A_i spanning the loop i. Then the helicity of this structure (Eq. (43)) can be written as

$$K_{12} = \sum_{i=1,2} B_i \Delta\sigma_i \oint_{\ell_i} \vec{A} \cdot d\vec{\ell} = \sum_{i=1,2} \Phi_i \int_{A_i} \vec{B} \cdot d\vec{S} = 2\Phi_1\Phi_2, \tag{44}$$

showing that K is a measure of the *connectivity* of the magnetic field. Note that the vector potential is only defined up to a gradient of an arbitrary function χ so that the helicity

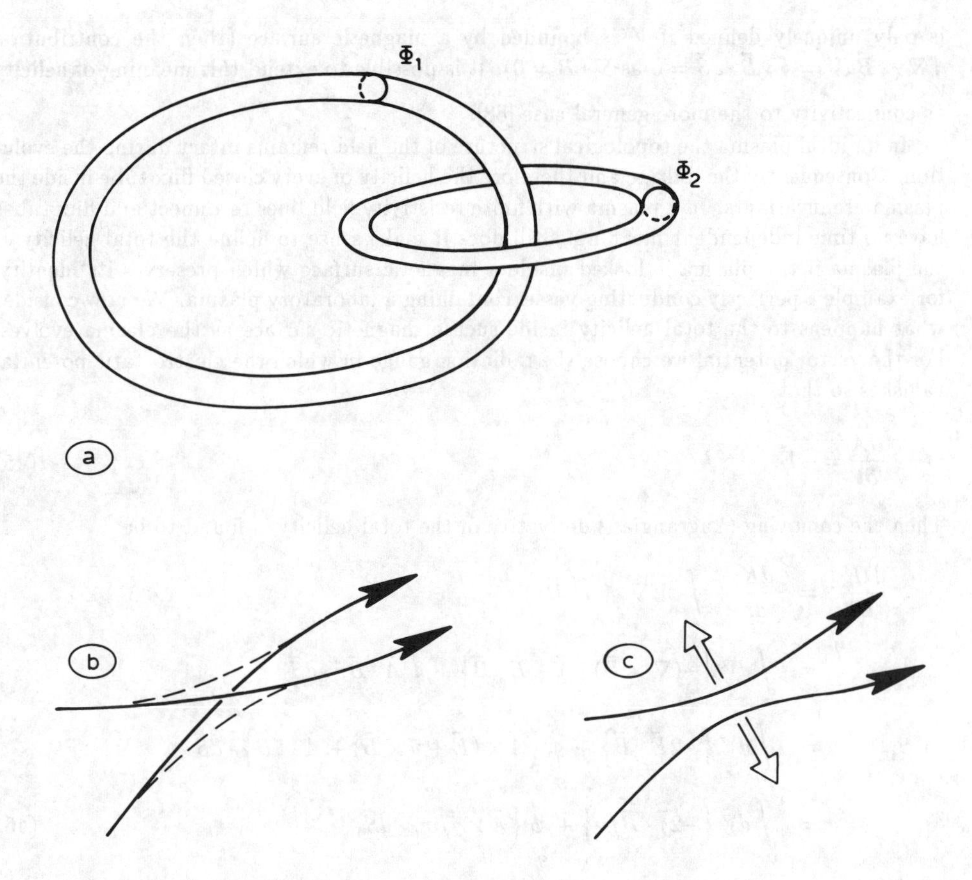

Figure 10: In (a) two flux tubes of flux Φ_1 and Φ_2 are linked. In (b) field reconnection occurs by the effect of resistivity in a small region, after which the further evolution in (c) is completely determined by ideal mhd.

is only uniquely defined if V is bounded by a magnetic surface (then the contribution $\int_V \nabla\chi \cdot \vec{B} dV = \oint_S \chi \vec{B} \cdot d\vec{S} = 0$ as $\nabla \cdot \vec{B} = 0$). It is possible to extend this meaning of helicity as connectivity to the more general case [88].

In an ideal plasma the topological structure of the field remains intact during the evolution. Consequently the linkage and therefore the helicity of every closed flux tube inside the plasma are invariants. In a plasma with finite resistivity field lines reconnect and flux tubes have no time independent meaning. Still does it make sense to define the total helicity of the plasma if the plasma is locked inside a magnetic surface which preserves its identity, for example a perfectly conducting vessel containing a laboratory plasma. We now consider what happens to the total helicity inside such a magnetic surface as the plasma evolves. For the vector potential we choose the radiation gauge in which the electrostatic potential vanishes so that

$$\frac{\partial \vec{A}}{\partial t} = -\vec{E}. \tag{45}$$

Then the comoving (Lagrangian) derivative of the total helicity is found to be

$$\begin{aligned}
\frac{DK}{Dt} &= \frac{dK}{dt} + \int_V dV \nabla \cdot (\vec{A} \cdot \vec{B} \vec{v}) \\
&= \int_V dV \left\{ -(\nabla \times \vec{E}) \cdot \vec{A} - \vec{E} \cdot \vec{B} \right\} + \oint_S \vec{A} \cdot \vec{B} \vec{v} \cdot d\vec{S} \\
&= \int_V dV \left\{ -2\vec{E} \cdot \vec{B} \right\} + \oint_S \left\{ \vec{A} \times (\vec{E} + \vec{v} \times \vec{B}) + \vec{A} \cdot \vec{v} \vec{B} \right\} \cdot d\vec{S} \\
&= \int_V dV \left\{ -2\vec{j} \cdot \vec{B}/\sigma \right\} + \oint_S \left\{ \vec{A} \times \vec{j}/\sigma \right\} \cdot d\vec{S},
\end{aligned} \tag{46}$$

where we have used Eq. (45), Faraday's equation, the definition for the vector potential, Ohm's law $\vec{j} = \sigma(\vec{E} + \vec{v} \times \vec{B})$ and finally $\vec{B} \cdot d\vec{S} = 0$ on a magnetic surface. Now suppose that reconnection occurs inside V in vanishingly small volumes, allowing the reconnection of neighbouring magnetic fields as sketched in Fig. 10b while the subsequent evolution after reconnection (Fig. 10c) is completely ideal. Then Eq. (46) shows that the total helicity of V is practically conserved during the reconnection processes.

It is now a straightforward mathematical operation [8] to show that the total magnetic energy in V under the only constraint of conserved total helicity is minimized for a linear force free field $\nabla \times \vec{B} = \alpha \vec{B}$ with α constant in space and having a value determined by the (invariant) total helicity. In the case of a coronal flaring magnetic structure any enveloping flux tube is anchored in the subphotospheric layers and the observable part does not form a magnetic surface. The concept of helicity can be extended however in the form of a gauge invariant relative helicity [85]. Therefore Taylor's hypothesis probably also applies to cosmic plasmas of high Lundqvist number in which relaxation occurs by reconnection in small volumes. It is then possible in principle *to calculate the free energy from the difference between the actual magnetic structure and a structure with a linear force free field having the same magnetic flux distribution at the boundary and the same total helicity.*

4 Flares from Disk Accretion onto Magnetospheres

Here we apply our understanding of magnetic flares to the specific case of Low-Mass X-Ray Binaries (LMXRB). These binaries contain a neutron star and a low-mass unevolved star filling its Roche lobe and spilling its outer envelope onto the compact companion in the form of a geometrically thin accretion disk without appreciable self-gravitation. Since the neutron star probably has a magnetic field of surface strength of 10^5 T or more, the interplay between magnetosphere and impinging plasma from the accretion disk is expected to lead to violent magnetic flares [29]. In particular we demonstrate that magnetic flaring provides a physical basis for the beat frequency model [89] proposed for the Quasi Periodic Oscillations (QPO) observed in LMXRB [35].

4.1 Definition of the Magnetospheric Radius

What determines the boundary between accreting material and magnetosphere ? For spherically symmetric accretion usually a so-called Alfvén radius is defined by a local balance between the ram pressure of the freely radially infalling gas and the magnetic pressure of the magnetosphere

$$\rho v_{ff}^2 = \frac{B^2}{2\mu_0}. \tag{47}$$

For thin disk accretion (see Fig. 11) often a magnetospheric radius r_m is defined by a condition similar to Eq.(47), now however with the Keplerian velocity v_K replacing the free-fall speed v_{ff}. Clearly for the same density and field strength the two values of the magnetospheric radius are not very different. However for the same value of the accretion rate and field strength the density for thin disk accretion is much higher and the magnetospheric radius therefore much smaller than for spherical accretion. Disk accretion onto a magnetosphere therefore allows for the powering by differential rotation of magnetic fields of much larger field strength than spherical accretion. To find out if this occurs in the form of flares rather than some smooth electrodynamic coupling, the concept of magnetospheric radius must be refined (see also [90]).

A more accurate definition of the radial distance within which the influence of the stellar field on the disk becomes important, is the radius r_p at which, coming from infinity, *equality appears first between disk gas pressure and external stellar field pressure.* Within this radius r_p what remains of the disk, is compressed in the vertical direction by the stellar magnetic rather than gravitational field (see Fig. 12). Therefore an inner ring exists in the disk in which the gas pressure is at any radius comparable to the stellar magnetic field pressure just outside. If magnetic connections could arise between star and gas in this ring, the disk gas would strongly power the magnetic link electrodynamically since the perturbing Keplerian ram pressure exceeds the magnetic pressure by a factor

$$\frac{\rho v_K^2 2\mu_0}{B^2} = \left(\frac{v_K}{v_s}\right)^2 \geq \left(\frac{R}{h(R)}\right)^2 \gg 1$$

for a thin disk (R is the radial distance of the link footpoint on the disk, $h(R)$ is the half thickness of the disk at that radius and v_s is the local sound speed inside the disk). The

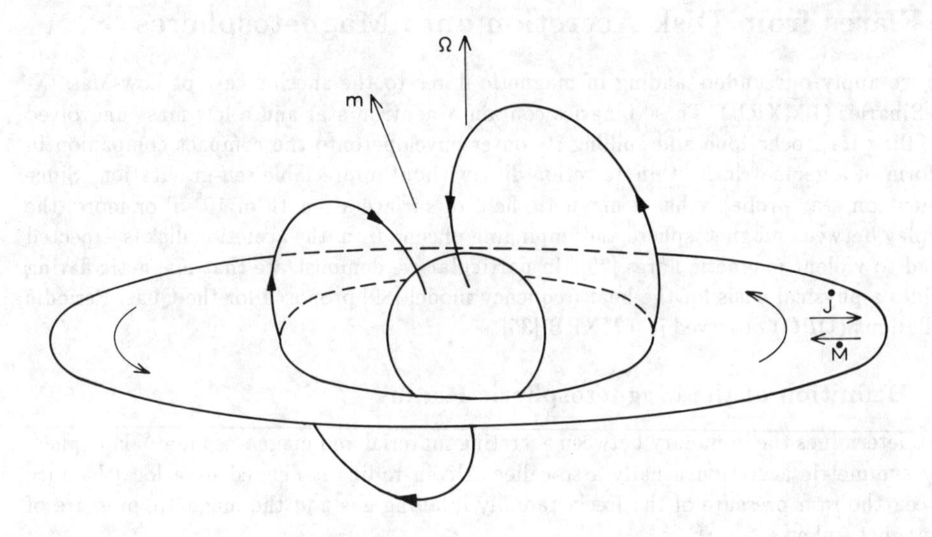

Figure 11: Disk accretion with accretion rate $\dot{\vec{M}}$ and outward rate of transport of angular momentum $\dot{\vec{J}}$ onto an oblique magnetic rotator with rotation $\vec{\Omega}$ and magnetic moment $\vec{m}$. If the magnetic moment is large enough the inner part of the disk is strongly modified by the magnetosphere.

first inequality follows from hydrostatic equilibrium inside the disk in the vertical direction and the vertical confinement by stellar gravity aided by magnetic compression. If the electrodynamic coupling leads to flares it is easy to estimate their energy. Using estimate (13) for the energy content of a flare, a dipolar dependence for the magnetic field of the central object and a scaling of the linked area $S(R) \propto R^2$ one finds that the flare energy depends on link radius as $W_M \propto R^{-3}$. Therefore we expect the largest flares to occur near compact objects with large field strength and for disk accretion with high accretion rate. Before we pursue the effects of such magnetospheric flares we first consider the magnetic properties of the accretion disk and the magnetic transport of angular momentum to show that flares do indeed occur.

4.2 Magnetic Disks

What are the magnetic properties of the disk? In many respects the disk resembles a two-dimensional star with the energy production by fusion replaced by gravitational accretion. The parts that have an effective temperature of several 10^3 K or more can be considered as well-conducting plasmas. To a first approach the disk therefore screens off the ambient stellar field and the magnetic interactions will be limited to relatively small regions of the disk. In the often used model of [91-93] the dominant interactions occur over a relatively small distance at about the magnetospheric radius r_m ($\ll r_p$) in the form of diffusion of stellar field into the disk and subsequent reconnection in a quasi-continuous manner. In this case the disk is assumed to have no magnetic field of its own. There are however two important physical effects which are neglected in this model. First, there is no reason

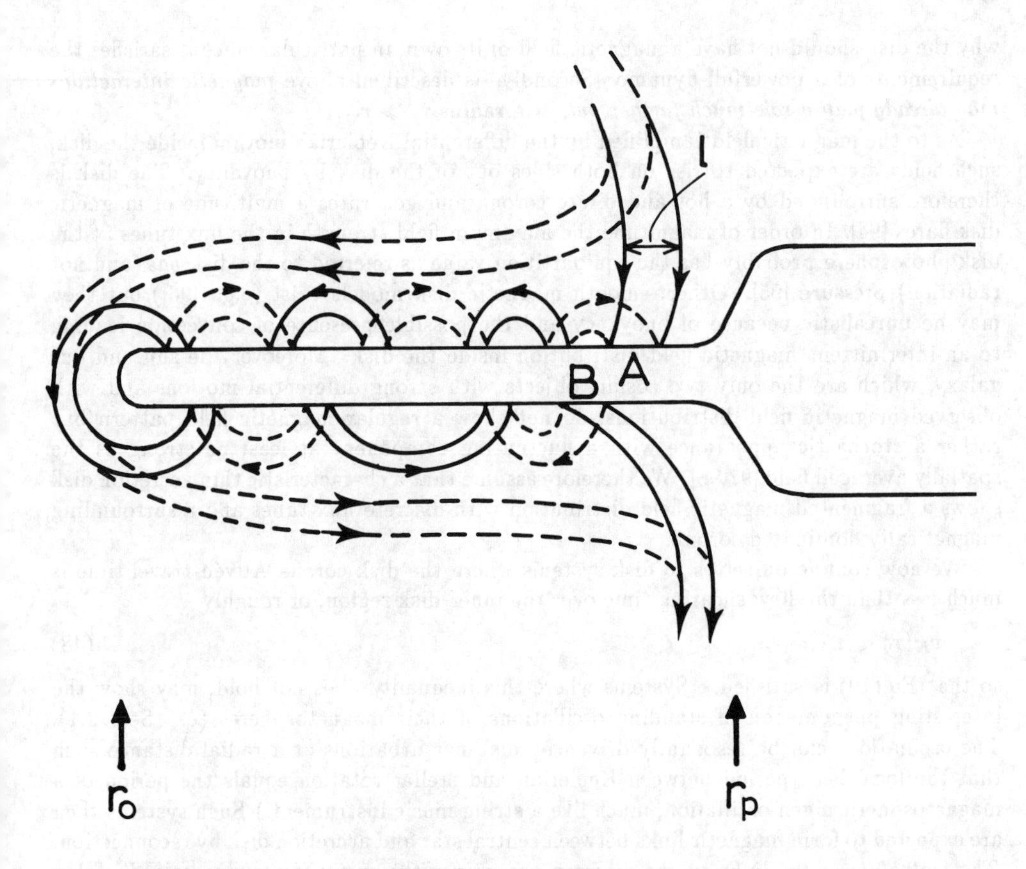

Figure 12: **Within the equal pressure radius r_p the disk is strongly modified by the ambient stellar field by vertical compression of the matter and by reconnection between stellar and disk field. The field lines are projected onto the meridional plane; the original field is dashed while the reconnected field is drawn. Note that only a small fraction of the stellar field (at A and B) is linked to the disk. Magnetic interactions break up the disk at an inner radius r_0 ($> r_m$).**

why the disk should not have a magnetic field of its own, in particular since it satisfies the requirements of a powerfull dynamo. Secondly, as described above *magnetic interactions may already play a role much further out*, at a radius $r_p \gg r_m$.

As to the magnetic fields amplified by the differential Keplerian motion inside the disk, such fields are expected to rise on both sides out of the disk by buoyancy. The disk is therefore surrounded by a hot and dilute corona and generates a multitude of magnetic disk flares [94]. In order of magnitude the maximum field strength in the flux tubes at the disk photosphere probably has the equipartition value as referred to the disk *gas* (and not radiation) pressure [95]. Other, smooth magnetic disk models exist (e.g. [96]) but they may be unrealistic because of buoyancy and the possible presence of convection leading to an intermittent magnetic field distribution inside the disk. Moreover the sun and our galaxy, which are the only two cosmic objects with strong differential motions and with observed magnetic field distributions, do not show a regular magnetic field pattern but rather a stochastic appearance with a fluctuating component at least as strong as the spatially averaged field [97,98]. We therefore assume that a characteristic thin accretion disk shows a fragmented magnetic field distribution with discrete flux tubes and a surrounding magnetically dominated corona.

We now confine ourselves to disk systems where the disk corona Alfvén travel time is much less than the flow shearing time over the inner disk region, or roughly

$$v_K(r) \ll v_A, \tag{48}$$

so that Eq.(10) is satisfied. (Systems where this inequality does not hold, may show the interesting phenomenon of standing oscillations of their magnetospheres (cf. Sect. 2.1). The oscillations can be resonantly driven by disk perturbations at a radial distance such that the local beat period between Keplerian and stellar rotation equals the period of a magnetospheric eigen oscillation, much like a string music instrument.) Such systems then are expected to form magnetic links between central star and accretion disk by reconnection. *The footpoints of the links on the disk can occur over the entire inner ring at radii below* r_p since there the characteristic disk field strength at the photospheric level is everywhere comparable to the ambient stellar field strength just outside the magnetically compressed disk. It should be realized however that the surface filling factor of linked flux is at any moment necessarily much less than unity on geometrical grounds (see Fig. 12).

4.3 Magnetic Transport of Angular Momentum

Magnetic fields are efficient transmitters of disk angular momentum in the radial direction. In this respect magnetic fields sticking out of the disk and linked to the star are much more efficient than fields entirely embedded inside the disk: The azimuthal magnetic stress per unit disk surface area from sheared field sticking out on both sides of the disk is $2B_zB_\phi/\mu_0$, and comparable to the azimuthal stress per unit cylinder surface area B_rB_ϕ/μ_0 from sheared field embedded horizontally inside the disk (assuming $B_r \approx B_z \approx B_\phi$). However in the former case (see Fig. 13) the column of gas subjected to the stress is of length $2h(R)$ compared to R in the latter case.

The magnetic rate of transport of angular momentum out of a column of unit disk surface area in a stellar link compares to the viscous transport from the same column in a Shakura-Sunyaev alpha-disk [99] as approximately

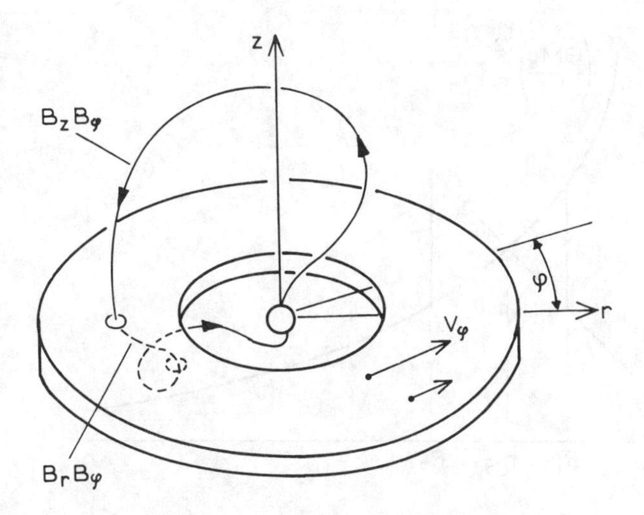

Figure 13: **Transport of angular momentum along magnetic field connecting a disk element to the central star. Angular momentum transport occurs by the stress $B_z B_\phi / \mu_0$ along field sticking out of the disk and by the stress $B_r B_\phi / \mu_0$ along field embedded inside the disk.**

$$\frac{B_z B_\phi R}{\mu_0 \alpha p_{gas}} \approx \frac{R}{\alpha h(R)} \gg 1, \tag{49}$$

where we have put $B_z \approx B_\phi$ (to be justified below) and $p_{gas} \approx B^2/(2\mu_0)^{-1}$. This proves that the *radial motion of disk matter inside links is dominated by magnetic stress.* We make a distinction between gas inside and gas outside the corotation radius, defined as the radius r_{co} where

$$v_K(r_{co}) \equiv \Omega_* r_{co}. \tag{50}$$

A magnetic link from the star to a disk element inside the corotation radius removes angular momentum from the disk since there the Keplerian rotation is faster than the stellar (see Fig. 14). Therefore inside the corotation radius a magnetic link strongly enhances the accretion rate. We can estimate the accretion speed from the magnetic torque on a column of unit disk area

$$2\rho h(R) \frac{d\{R^2 \Omega(R)\}}{dt} \approx -\frac{2RB_z B_\phi}{\mu_0}. \tag{51}$$

Substituting the Keplerian rotation speed for the actual rotation $\Omega(R) \approx \Omega_K(R) \equiv (GM_*/R^3)^{1/2}$ and putting $B_\phi \approx B_z$ we find from Eq. (51)

$$v_r \approx -\frac{2RB_z^2}{h\rho v_K(R)\mu_0} \approx -\frac{2B_z^2}{\rho v_s \mu_0} \approx -4v_s, \tag{52}$$

where we have used $h(R)/R = v_s/v_K$ and $B_z^2 = 2\mu_0 p_{gas}$. First we see that the assumption $\Omega(R) \approx \Omega_K(R)$ is a posteriori justifiable as for a thin disk the derived radial speed is

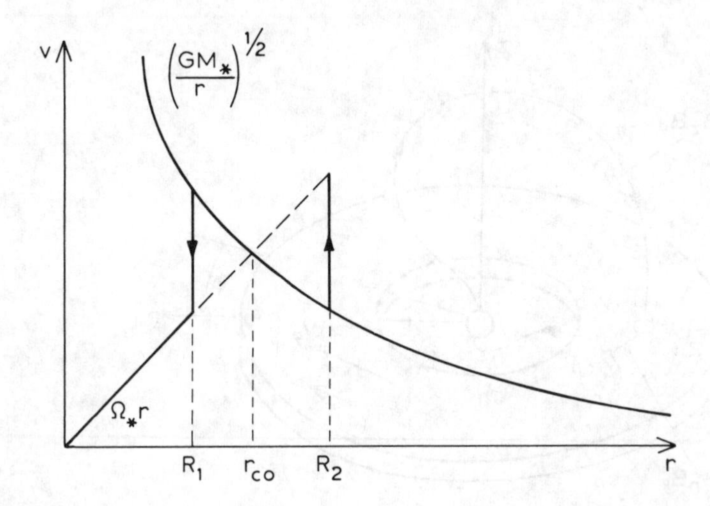

Figure 14: **The figure shows the corotation radius r_{co} where disk material with the Keplerian speed corotates with the star, $v_K(r_{co}) = \Omega_* r_{co}$. Disk material at $R_1 < r_{co}$ rotates faster than the star and falls in when brought into corotation, while disk material at $R_2 > r_{co}$ is accelerated away from the star when connected to the magnetosphere.**

still much less than the Keplerian speed. Secondly we find that the inward radial speed is much larger than the viscous inward radial speed for an alpha-disk, which is of order $\alpha v_s(R)h(R)/R = \alpha(v_s)^2/v_K$. (Alternatively this shows that magnetic links in an alpha-disk lead to local values of the parameter α of order $R/h(R)$, much larger than unity.) As a result in the inner ring of the disk ($r < r_p$ and $r < r_{co}$) *the disk breaks up into separate blobs* which spiral inwards much quicker than the outer viscous regions (see Fig. 15a).

Magnetic links can also end on the disk at a radial distance R larger than the corotation radius (as long as $R < r_p$). Such links however deposit stellar angular momentum into the disk (see Fig. 14) and they *prohibit rather than promote accretion.* Similar to the estimate in Eqs.(51) and (52) the disk material quickly acquires a supersonic radial speed, now in the outward direction, causing the formation of dissipative shocks in the surrounding matter. As in the picture of [91-93] the simultaneous existence of magnetic links between star and disk inside and outside the corotation radius leads to the possibility of equilibrium stellar rotation rates for stationary disk accretion.

We now justify that our approximation $B_\phi \approx B_z$ is characteristic for the magnetic field of the link at the disk footpoint. The azimuthal magnetic field component in a link B_ϕ only remains small compared to the vertical component B_z if the excess angular momentum in the gas with respect to the star (per unit area of order $2h\rho R\Omega_B(R)$)can be removed by magnetic stress $(2B_\phi B_z/\mu_0)$ in an interval much below the beat period between star and the footpoint of the link on the disk. Here the *beat frequency* is defined as

$$\Omega_B(R) \equiv \mid \Omega_K(R) - \Omega_* \mid . \tag{53}$$

It therefore directly follows that a necessary condition for the matter to reach corotation before the azimuthal field component reaches a value similar to the vertical field component, is

$$\frac{B_z^2}{2\mu_0} > \frac{1}{2}\rho v_s v_K \left(\frac{\Omega_B}{\Omega_K}\right)^2, \tag{54}$$

where we have used $h = v_s/\Omega_K$ and $\Omega_K = v_K/R$. From Eq.(54) we see that away from corotation (where $\Omega_B(R) \approx \Omega_K(R)$) the disk material never reaches corotation with the star for a small B_ϕ value since in the disk $B_z^2(2\mu_0)^{-1} \approx \rho v_s^2 \ll \rho v_s v_K$ for a thin disk.

Further by assumption the coronal Alfvén speed is large (Eq. (48)) so that the link evolves through a series of force-free equilibria. *Then the azimuthal field component can at most become equal to the vertical component after which a flare occurs and distortion of the newly reconnected field reappears.* We therefore conclude that the estimate $B_\phi \approx B_z$ is characteristic for the initial stage of a link as the matter spirals inwards. The link is periodically destroyed and reformed with different stellar field lines (see Fig. 15a).

Whether an individual gas blob remains spiralling in the plane of the disk as in Fig. 15a or starts flowing along the stellar field as in Fig. 15b, depends on the spatial dependence of the stellar field, which inside the inner disk radius is of course not limited to equipartition with the gas pressure. A simple estimate based on a dipolar dependence for the stellar field and conservation of flux through the infalling gas blob indicates that aligned infall towards a neutron star only occurs for very large surface fields (10^8 T or more) while for smaller fields the accretion remains in the disk plane [29]. In the latter case accretion of the different blobs occurs on an equatorial ring around the star and no X-ray pulses are expected to be observed as would be the case for aligned infall onto the poles. A definite answer to this problem however can only be given after a self-consistent solution of the thermodynamics of the radiation dominated inner region of accretion onto the neutron star.

4.4 A Flare Model for Quasi Periodic Oscillations

We now consider the energetics of the magnetic links. We have arrived at the following picture: The inner side of an accretion disk surrounding a magnetized compact stellar object largely excludes the stellar field and is compressed by it. As a result what remains of the disk has a gas pressure comparable to the external magnetic field pressure in a ring inside radius r_p, the radius of equal pressure between undisturbed disk and ambient stellar magnetic field. If the characteristic magnetic appearance of the disk consists of magnetic loops breaking out on both sides into a corona up to the local equipartition value ($B_{disk}^2(R) \approx 2\mu_0 p_{gas}(R) \approx B_*^2(R)$ for radial distances smaller than r_p) and if the coronal Alfvén crossing time is small (Eq. (48)), magnetic links between the disk and the star are expected to form on this ring $r < r_p$. Depending on the position of the corotation radius with respect to the ring two kinds of links with the star can be distinguished, those linking up to the disk inside and those outside the corotation radius, each with opposite directions of transport of angular momentum. *Links inside r_{co} promote accretion while links outside (temporarily) inhibit accretion of gas towards the star. The links are too weak to permit the matter to reach corotation with the star. As a result a given link undergoes a series of distortions, flares and reconnections. This occurs for both kinds of links: for links within the corotation radius the electromagnetic flare energy comes from the excess rotational energy of the disk material with respect to the star. For links ending on the disk*

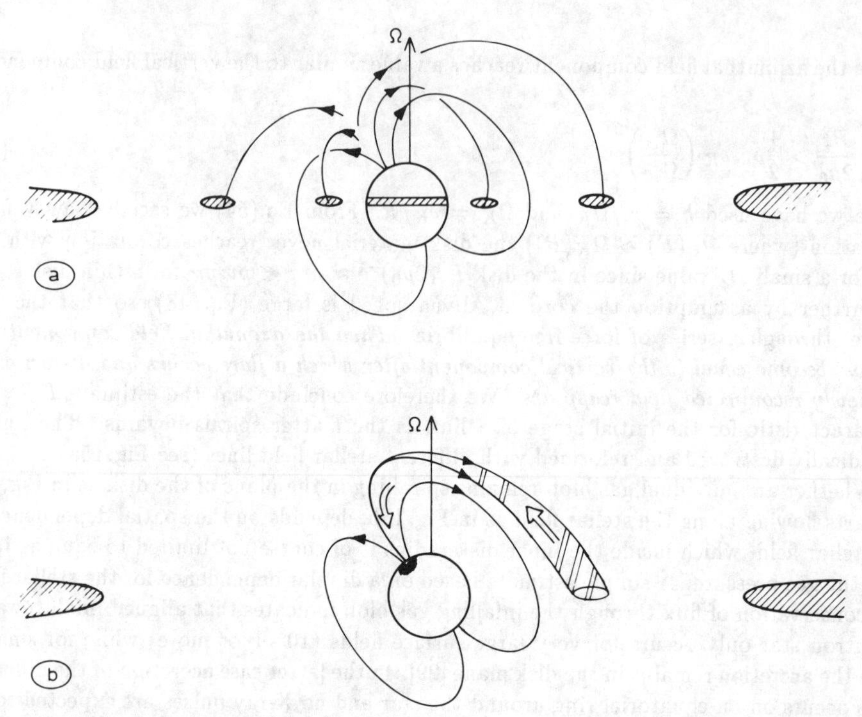

Figure 15: Accretion inside the magnetosphere of gas threaded by stellar magnetic field. In (a) the stellar field is relatively weak for given accretion rate and the gasblobs continue to spiral inwards in the disk plane, ultimately hitting the star along an equatorial belt. In (b) the stellar field is strong enough to guide the accreting material in the inner region along the field onto the magnetic poles. In (a) no sharp X- ray pulses are observed but only QPO; in (b) both X-ray pulses are observed from the guided accretion in the deep interior and QPO from the flares in the outer region.

outside the corotation radius on the other hand the flare energy derives from the excess rotation energy of the star.For the relatively weak surface stellar fields which are thought to occur in LMXRB (of order of 10^6 T or less) the accreting gas in links stays in the disk plane and eventually hits the star along an equatorial belt.

We first note that the flaring caused by a particular gas lump connected to the stellar field occurs in a more or less periodic manner. For an aligned magnetic rotator the flare period is of order $2\pi/\Omega_B(R)$ since the component B_ϕ approaches the value B_z after differential rotation over one radian. For the general case of an oblique rotator the asymmetry probably introducies a most conspicuous flare period at the beat period $\Omega_B(R)$ itself. Further the flares should be observable in X-rays and probably there is a preferred central distance for the best visible flares: flares which are too near the star may be obscured by surrounding material while flares too far from the star are less energetic (see below). *Going inwards from infinity towards the star the beat period first decreases from the stellar rotation period to zero at the corotation radius and then increases towards the high Keplerian rotation period near the stellar surface.* Since for a particular accreting system the equal pressure radius r_p decreases with accretion rate, the above effect would translate into a corresponding behaviour in the frequency of the flares as the observed X-ray luminosity (assumed to be a monotonic function of the accretion rate) increases. The low-frequency QPO (< 10 Hz) observed in several LMXRB [100], which show a weak anticorrelation between frequency and X-ray flux, would then correspond to flares occurring at relatively small accretion rates and originate from links outside the corotation radius. On the other hand the steep, positively correlated, branch in the observed QPO frequency versus flux diagram would correspond to flares at high accretion rates and links inside the corotation radius.

Finally we estimate the relative luminosity in these magnetic flares to show that indeed they can be comparable to the observed luminosity in QPO. In an axially symmetric (with respect to rotation axis) force-free magnetic link the quantity $B_\phi \tilde{r} sin\theta$ is constant along a field line ($\tilde{r}$ is the distance from a point on the field line to the stellar center and θ is the angle of the radius vector with the rotation axis). For a dipolar field ($sin^2(\theta)/\tilde{r}$ constant along a field line and $B_z \propto \tilde{r}^{-3}$) the main relative distortion of the field ($B_\phi/B_z \uparrow 1$) therefore occurs near the weak field at the disk footpoint. The energy in a flare from a link to the star is then of order (compare Eq. (13))

$$W_M = RS\frac{(B_z(R))^2}{2\mu_0}, \tag{55}$$

where R is the distance of the footpoint on the disk to the star and S is the tube cross-section. *This energy is far greater than from a flaring loop anchored with both feet in the disk* as the volume of the link is larger by a factor $R/h(R)$ (compare also [101]). Let us assume that the flaring links in a particular system can at any instant be approximated by a spherically symmetric thin shell of equatorial radius R, radial thickness at the equator ℓ and that the flaring rate is $\Omega_B^{-1}(R)$. For a stellar dipole field of strength B_* at the stellar pole we then find a *total flare luminosity* from one side of the disk

$$\begin{aligned} L_M &\approx \pi \ell R_*^2 \left(\frac{R_*}{R}\right)^4 \Omega_B(R)\frac{B_*^2}{4\mu_0} \\ &= 1.3\cdot 10^{28}\frac{\ell}{10^4\mathrm{m}}\left(\frac{R_*}{10^4\mathrm{m}}\right)^6\left(\frac{R}{10^5\mathrm{m}}\right)^{-4}\frac{\Omega_B}{200\mathrm{Hz}}\left(\frac{B_*}{10^6\mathrm{T}}\right)^2 \mathrm{Watt}, \end{aligned} \tag{56}$$

where R_* is the stellar radius. The observed luminosity in QPO is a few percent of the total source luminosity of 10^{31} Watt and can be reproduced by Eq. (56) for plausible conditions.

Finally the energy release of an individual flare from a gas blob of mass m at radius $R = r_p$ relates to the gravitational energy release at the moment of impact of the blob onto a (non-rotating) star as (use Eq. (55), pressure equality, thin disk assumption and Keplerian disk rotation)

$$\frac{W_M(R)}{W_G(R)} \approx \frac{h(R)R_*}{R^2}$$

and can easily be larger than 10^{-3}.

5 Conclusion

We have explained the basic ideas and developments in the field of astrophysics of magnetic flares. We have shown that the storage of electric currents and its subsequent disruption in the form of a flare, is probably a quite common phenomenon in magnetic atmospheres (coronae) which are anchored in a dynamo (driver) under a number of conditions. The most important conditions are that the resistivity is sufficiently small, that the flow crossing time at the boundary is much larger than the Alfvén travel time over the coronal structure, and that the driver is powerful enough to substantially distort the coronal structure. We have stressed the basic physics and shown how these concepts can be applied to close binaries, both of the classic type and with a compact companion. At present the global flare physics seem to be well understood in contrast with the microscopic processes of the energy conversion (reconnection, particle acceleration).

We have given a special example of magnetic flaring in disk accretion around a neutron star. We conclude that QPO may be the manifestation of magnetic flaring activity in LMXRB and can be understood if one combines the geometrical beat frequency model of [89] with the physics of magnetic flares.

Acknowledgement I like to thank Don Melrose and Bert van den Oord for stimulating discussions.

References

[1] *Solar and Stellar Flares, Proc. of the 104th Coll. of the IAU*, eds. B.M. Haisch and M. Rodonò, 1989, *Solar Phys.* **121**, Nos. 1/2.

[2] *Solar and Stellar Flares, Poster papers of IAU Coll. No. 104*, eds. B.M. Haisch and M. Rodonò, 1989, Catania Astrophysical Observatory, special publication, Catania, Italia.

[3] Neubauer, F.M.: 1980, *J. Geophys. Res.* **85**, 1171.

[4] Barnett, A. and Olbert, S.: 1986, *J. Geophys. Res.* **91**, 10117.

[5] Dobrowolny, M. and Veltri, P.: 1986, *Astron. Astrophys.* **167**, 179.

[6] Scheurwater, R. and Kuijpers, J.: 1988, *Astron. Astrophys.* **190**, 178.

[7] Roberts, P.H.: 1967, *An Introduction to Magnetohydrodynamics*, Longmans, Green and Co Ltd, London, Ch. 4.4, 1.1.

[8] Freidberg, J.P. 1987 *Ideal Magnetohydrodynamics,* Plenum Press, New York, p. 61, p. 477.
[9] Low, B.C.: 1986, *Astrophys. J.* **307**, 205.
[10] Aly, J.J.: 1985, *Astron. Astrophys.* **143**, 19.
[11] Mikic, Z., Barnes, D.C., Schnack, D.D.: 1988, *Astrophys. J.* **328**, 830.
[12] Biskamp, D.: 1989, in *Plasma Phenomena in the Solar Atmosphere,* eds. M.A. Dubois and D. Grésillon, Cargèse, France, in press.
[13] Forbes, T.G: 1990, *Solar Phys.* in press.
[14] Sakurai, T.: 1989, *Solar Phys.* **121**, 347.
[15] Schadee, A.: 1986, *Adv. Space Res.* **6**, No. 6, 41.
[16] Schadee, A., de Jager, C. and Svestka, Z.: 1983, *Solar Phys.* **89**, 287.
[17] Bahcall, J.N., Rosenbluth, M.N., Kulsrud, R.M.: 1973, *Nature Phys. Sci.* **243**, 27.
[18] DeCampli, W.M. and Baliunas, S.L.: 1979, *Astrophys. J.* **230**, 815.
[19] Simon, T., Linsky, J.L., and Schiffer III, F.H.: 1980, *Astrophys. J.* **239**, 911.
[20] Uchida, Y. and Sakurai, T.: 1983, in *Activity in Red-Dwarf Stars,* IAU Coll. No. 71, eds. P.B. Byrne and M. Rodonò, D.Reidel Publ. Cy., Dordrecht, Holland, p. 629.
[21] Kuijpers, J. and van der Hulst, J.M.: 1985, *Astron. Astrophys.* **149**, 343.
[22] van den Oord, G.H.J.: 1988, *Astron. Astrophys.* **207**, 101.
[23] White, N.E., Culhane, J.L., Parmar, A.N., Kellett, B.J., Kahn, S., van den Oord, G.H.J. and Kuijpers, J.: 1986, *Astrophys. J.* **301**, 262.
[24] van den Oord, G.H.J., Kuijpers, J., White, N.E., van der Hulst, J.M., and Culhane, J.L.: 1989, *Astron. Astrophys.* **209**, 296.
[25] van den Oord, G.H.J. and Mewe, R.: 1989, *Astron. Astrophys.* **213**, 245.
[26] Vilhu, O., Caillault, J.P., Neff, J., and Heise, J.: 1988, in *Activity in Cool Star Envelopes,* eds. O. Havnes, B.R. Pettersen, J.H.M.M. Schmitt and J.E. Solheim, Kluwer, Dordrecht, Holland, p. 179.
[27] Lamb, F.K., Aly, J., Cook, M., and Lamb, D.Q.: 1983, *Astrophys. J. (Letters)* **274**, L71.
[28] Kuijpers, J.: 1989, *Solar Phys.* **121**, 163.
[29] Aly, J.J. and Kuijpers, J.: 1989, *Astron. Astrophys.* in press.
[30] de Vries, M. and Kuijpers, J.: 1989, *A Magnetic Flare Model for X-Ray Variability in Active Galactic Nuclei*, 23rd ESLAB Symposium, ed. N.E. White, Bologna.
[31] Kovitya, P. and Cram, L.: 1983, *Solar Phys.* **84**, 45.
[32] Melrose, D.B. and McClymont, A.N.: 1987, *Solar Phys.* **113**, 241.
[33] McClymont, A.N. and Fisher, G.H.: 1989, *On the Mechanical Energy Available to Drive Solar Flares*, preprint.
[34] Melrose, D.B.: 1989, *The Energy Release in Solar Flares: a Model Involving Remote Energy Storage,* preprint.
[35] Lewin, W.H.G., van Paradijs, J. and van der Klis, M.: 1988, *Space Sci. Rev.* **46**, 273.
[36] Alfvén H. and Fälthammar, C.-G.: 1963, *Cosmical Electrodynamics*, 2nd ed., Oxford Univ. Press, London, p. 183.
[37] Akasofu, S.-I.: 1977, *Physics of Magnetospheric Substorms*, D. Reidel Publ. Cy., Dordrecht, Holland.
[38] Alfvén, H.: 1981, *Cosmic Plasma,* D. Reidel Publ. Cy., Dordrecht, Holland.
[39] Spicer, D.S.: 1982, *Space Sci. Rev.* **31**, 351.
[40] Scheurwater, R.S. and Kuperus, M.: 1988, *Astron. Astrophys.* **194**, 213.

[41] Goldreich, P. and Lynden-Bell, D.: 1969, *Astrophys. J.* **156**, 59.
[42] Piddington, J.H. and Drake, J.F.: 1968, *Nature* **217**, 935.
[43] Chanmugan, G. and Dulk, G.A.: 1983, in *Cataclysmic Variables and Related Objects,* IAU Coll. No. 72, eds. M. Livio and G. Shaviv.
[44] Blandford, R.D. and Znajek, R.L.: 1977, *Monthly Notices Roy. Astron. Soc.* **179**, 433.
[45] Lovelace, R.V.E.: 1976, *Nature* **262**, 649.
[46] Thorne, K.S., Price, R.H. and MacDonald, D.A.: 1986, *Black Holes: The Membrane Paradigm,* Yale Univ. Press, New Haven, USA.
[47] Phinney, E.S.: 1983, in *Astrophysical Jets,* eds. A. Ferrari and A.G. Pacholczyk, D. Reidel Publ. Cy, Dordrecht, Holland, p. 201.
[48] Boström, R.: 1973, *Astrophys. Space Sci.* **22**, 353.
[49] Raadu, M.A.: 1989, *Physics Reports* **178**, 25.
[50] Kuperus, M. and Raadu, M.A.: 1974, *Astron. Astrophys.* **31**, 189.
[51] van Tend, W. and Kuperus, M.: 1978, *Solar Phys.* **59**, 115.
[52] Kaastra, J.S.: 1985, *Solar Flares, An Electrodynamic Model,* Ph.D. Thesis, Utrecht University, The Netherlands, Ch.4 and 5.
[53] van den Oord, G.H.J.: 1987: *Stellar Flares,* Ph.D. Thesis, Utrecht University, The Netherlands, Ch.2.
[54] Kadomtsev, B.B.: 1987, *Physics Reports* **50**, 115.
[55] Petschek, H.E.: 1964, *AAS-NASA Symp. on the Physics of Solar Flares*, Washington, NASA SP-50, p. 425.
[56] Priest, E.R.: 1982, *Solar Magnetohydrodynamics*, D.Reidel Publ. Cy., Dordrecht, Holland.
[57] Dubois, M.A.: 1985, in *Magnetic Reconnection and Turbulence*, Cargèse Workshop, eds. M.A. Dubois, D. Grésillon and M.N. Bussac, Les Editions de Physique, Les Ulis, France, p. 213.
[58] Biskamp, D.: 1985, in *Magnetic Reconnection and Turbulence*, Cargèse Workshop, eds. M.A. Dubois, D. Grésillon and M.N. Bussac, Les Editions de Physique, Les Ulis, France, p. 19.
[59] Rieger, E.: 1989, *Solar Phys.* **121**, 323.
[60] Vlahos, L.: 1989, *Solar Phys.* **121**, 431.
[61] Slottje, C.: 1978, *Nature* **275**, 520.
[62] *Solar Radiophysics*, eds. D.J. McLean and N.R. Labrum, 1985, Cambridge Univ. Press, Cambridge, UK.
[63] Allaart, M.A.F. and van Nieuwkoop, J.: 1989, *Solar Phys.* submitted.
[64] Melrose, D.B.: 1986, *Instabilities in Space and Laboratory Plasmas,* Cambridge Univ. Press, Cambridge, UK.
[65] Melrose, D.B.: 1980, *Plasma Astrophysics,* Gordon and Breach Publ., New York.
[66] Stasiewicz, K. and Hultqvist, B.: 1989, *Evidence for the drift Acceleration of Auroral Electrons,* preprint.
[67] Holman, G.D. and Pesses, M.E.: 1983, *Astrophys. J.* **267**, 837.
[68] Webb, G.M., Axford, W.I. and Terasawa, T.: 1983, *Astrophys. J.* **270**, 537.
[69] Achterberg, A.: 1984, in *Particle Acceleration Processes, Shock Waves, Nucleosynthesis and Cosmic Rays*, eds. L. Koch-Miramond and M.A. Lee, Adv. Space Res. **4**, 193.
[70] Jacobsen, C. and Carlqvist, P.: 1964, *Icarus* **3**, 270.

[71] Alfvén, H. and Carlqvist, P.: 1967, *Solar Phys.* **1**, 220.
[72] Torvén, S., Lindberg, L. and Carpenter, R.T.: 1985, *Plasma Physics and Controlled Fusion* **27**, 143.
[73] Lindberg, L.: 1988, *Astrophys. Space Sci.* **144**, 3.
[74] Boström, R., Holback, B., Holmgren, G., Koskinen, H.: 1988, *Physica Scripta* **39**, 782.
[75] Hultqvist, B.: 1989, *Phil. Trans. Royal Soc. London A* **328**, 209.
[76] Raadu, M.A. and Rasmussen, J.J.: 1988, *Astrophys. Space Sci.***144**, 43.
[77] Hénoux, J.C.: 1987, in *Solar Maximum Analysis*, eds. V.E. Stepanov and V.N. Obridko, VNU Science Press, Utrecht, The Netherlands, p. 105.
[78] Kuijpers, J.: 1989, *Coherent Radiation from Electrostatic Double Layers,* in Plasma Phenomena in the Solar Atmosphere, 1989 Cargése Workshop, eds. M.A. Dubois and D. Grésillon, in press.
[79] Martens, P.C.H., van den Oord, G.H.J. and Hoyng, P.: 1985, *Solar Phys.* **96**, 253.
[80] Taylor, J.B.: 1974, *Phys. Rev. Lett.* **33**, 1139.
[81] Taylor, J.B.: 1986, *Rev. Modern Phys.* **58**, 741.
[82] Woltjer, L.:1958, *Proc. Nat. Acad. Sci.* **44**, 489.
[83] Norman, C.A. and Heyvaerts, J.: 1983, *Astron. Astrophys.* **124**, L1.
[84] Heyvaerts, J. and Priest, E.R.: 1984, *Astron. Astrophys.* **137**, 63.
[85] Berger, M.A. and Field, G.B.: 1984, *J. Fluid Mech.* **147**, 133.
[86] Berger, M.A.: 1984, *Geophys. Astrophys. Fluid Dynamics* **30**, 79.
[87] Berger, M.A.: 1988, *Astron. Astrophys.* **201**, 355.
[88] Field, G.: 1986, in *Magnetospheric Phenomena in Astrophysics,* eds. R.I. Epstein and W.C. Feldman, Am. Inst. of Physics, New York, p. 324.
[89] Alpar, M.A. and Shaham, J.: 1985, *Nature* **316**, 239.
[90] Vasyliunas, V.M.: 1979, *Space Sci. Rev.* **24**, 609.
[91] Ghosh, P. and Lamb, F.K.: 1978, *Astrophys. J. (Letters)* **223**, L83.
[92] Ghosh, P. and Lamb, F.K.: 1979, *Astrophys. J.* **232**, 259.
[93] Ghosh, P. and Lamb, F.K.: 1979, *Astrophys. J.* **234**, 296.
[94] Galeev, A.A., Rosner, R. and Vaiana, G.S.: 1979, *Astrophys. J.* **229**, 318.
[95] Coroniti, F.V.: 1985, in *Unstable Current Systems and Plasma Instabilities in Astrophysics,* IAU Symp. No. 107, eds. M.R. Kundu and G.D. Holman, D. Reidel Publ Cy., Dordrecht, Holland, p. 453.
[96] Lovelace, R.V.E., Wang, J.C.L. and Sulkanen, M.E.: 1987, *Astrophys. J.* **315**, 504.
[97] Parker, E.N.: 1979, *Cosmical Magnetic Fields*, Clarendon Press, Oxford, UK.
[98] Heiles, C.: 1987, in *Physical Processes in Interstellar Clouds*, eds. G.E. Morfill and M. Scholer, NATO ASI Series C, Vol. 210, p. 429.
[99] Shakura, N.I. and Sunyaev, R.A.: 1973, *Astron. Astrophys.* **24**, 337.
[100] Stella, L.: 1988, in *X-Ray Astronomy with Exosat*, eds. R. Pallavicini and N.E. White, Memoria della Societa Astronomica Italiana, **59**, 185.
[101] Barnes, C. and Sturrock, P.A.: 1972, *Astrophys. J.* **176**, 31.

THE EVOLUTION OF CHROMOSPHERIC STRUCTURE ON AR LACERTAE

JAMES E. NEFF
Lab. for Astronomy and Solar Physics
NASA Goddard Space Flight Center
Greenbelt, MD 20771, USA

ABSTRACT. I present preliminary results from the analysis of a series of ultraviolet spectra AR Lac obtained in 1987. These spectra are being used to image the spatial structure of the chromosphere and to study the dynamics of flaring regions. Comparing these results with previous results, I will discuss the evolution of the chromospheric structure from 1983 to 1987.

1. Introduction

Elsewhere in this volume (Paper 1) I describe the techniques by which *spatial resolution* can be obtained indirectly from *spectral resolution* for certain stars. In this paper I present an application of these "spectral imaging" techniques to ultraviolet spectra of the magnetically active system AR Lacertae. AR Lac (HD 210334; m_v=6.1) is a K0 IV+G2 IV detached, eclipsing binary system with an orbital period of 1.98 days. The ultraviolet emission line profiles from the two stars are broadened by the rapid rotation, but they are well separated at orbital quadrature phases by the high projected orbital velocity.

Walter *et al.* (1987) determined the size, location, and brightness of two plages on AR Lac using high-resolution ultraviolet spectra obtained in October 1983 with the International Ultraviolet Explorer (IUE) spacecraft. Their results were limited by the poor phase coverage of the data set, so a second-epoch data set was obtained with nearly full, uniform coverage of a single orbital cycle in September 1985. Neff (1987) and Neff *et al.* (1989) described the analysis of the 1985 data set (see also Paper 1). Based on the success of these two sets of observations, we observed AR Lac again in September 1987. In addition to full, uniform coverage with IUE throughout two orbital cycles, contemporaneous x-ray observations were obtained with the GINGA satellite. The system was also observed from ground-based observatories across the US, Europe, and the Soviet Union. In addition to a well-determined photometric light curve, we obtained high-resolution optical spectra and linear polarization measurements.

2. Observing Strategy

The goal of the September 1987 observing campaign was to observe the system continuously with IUE for at least two full orbital cycles ($\geq$ 4 days). We obtained a

C. İbanoğlu (ed.), Active Close Binaries, 805–808.

series of 34 high-resolution LWP spectra (2000–3200 Å) and 33 low-resolution SWP spectra (1200–2000 Å). This series provides uniform and complete coverage of orbital phase (same as rotational phase for this system). More importantly, observations of the same phase were obtained during subsequent orbits. Any spatially-induced variations must be periodic with the orbital phase and must therefore **repeat** on subsequent orbits. The LWP spectra are being used to map the spatial structure of the chromosphere, and the SWP spectra will be used to probe the radial structure of the chromosphere and transition region. The data analysis follows the procedures I outlined in Paper 1.

3. The Chromospheric Structure Of AR Lac In September 1987

The global stellar components are constrained only to be symmetric and to lie at the expected wavelength. Their flux and width are allowed to vary. In Figure 1 I show the total emission from the system and the uniform emission from each star. Notice that although the emission must arise uniformly from the stars, it is by no means constant. Both stars are roughly 50% brighter in their trailing hemisphere than in their leading hemisphere. The overall line widths (Figure 2) show the same effect. The structure that produces this variation must be very large-scale or approximately uniform across the disk, or else it would produce an asymmetric profile at some phases. This result is very firmly established, because spectra obtained at opposite quadratures, when the emission components are unblended, show the flux and width difference.

The features in Figures 1 and 2 that *repeat* are clearly spatially-induced. There are some noteworthy differences between the two orbits, showing that some of the features are transient. No major flares, however, were observed during this 4 day interval.

The final fits are not yet complete, but to illustrate the number, position, and motion of the asymmetric components, I plot in Figure 3 their velocity relative to the mean K star velocity.

4. Comparison With Previous Epochs

The 1983 data set included only 8 phases, nearly all clustered around the eclipses. The 1985 data set covered only 80% of a single cycle, completely omitting the secondary eclipse. This 1987 data set was the **first** obtained with the phase coverage necessary to conclusively demonstrate that the line profile variations are spatially-induced.

The global flux levels in 1987 are approximately the same as in 1985. The change in width is more pronounced in the 1987 spectra. The dashed lines in Figure 3 show the velocities of the three plage regions seen in the 1985 results. It is clear that the structure has changed. There is a strong qualitative similarity, however, and the 1987 image likely will contain the same three regions, but migrated slightly in longitude and perhaps latitude. No information is yet avalable on plage sizes and in September 1987.

5. Summary

Both stars show a pronounced difference in emission flux and line width between the leading and trailing hemisphere. About 50% more active regions likely are concen-

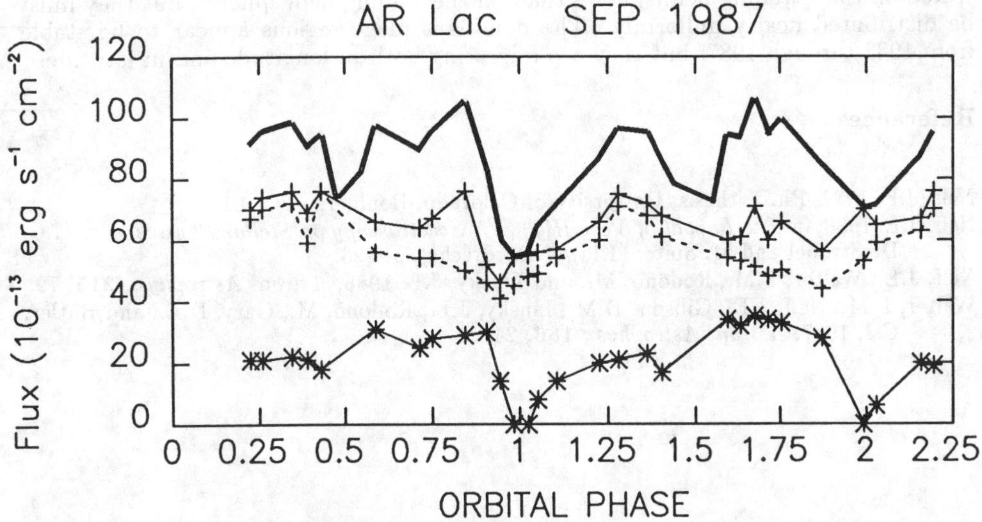

Figure 1: The integrated Mg II k flux in each component in September 1987 is plotted as a function of time, or equivalent orbital phase. The top curve is the total Mg II k emission from the system (observed flux corrected for interstellar absorption). The pluses indicate emission from the K star, and the asterisks indicate emission from the G star. The solid lines show the total emission from each star (including plages), while the dashed lines show only the emission from the global (uniform) components. The component fluxes are not shown during the secondary eclipse.

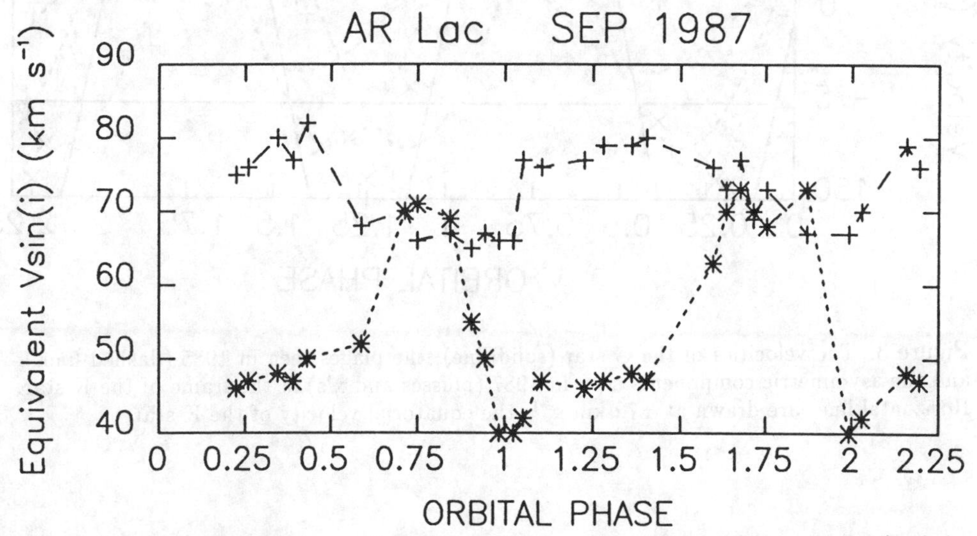

Figure 2: The measured line widths (with the instrumental profile deconvolved) of the global emission components from the K star (plusses) and the G star (asterisks).

trated on their trailing hemispheres than on the leading hemispheres, but they must be distributed nearly uniformly. The dominant plage regions appear to be stable from 1983 through 1987, but they are migrating both in longitude and in latitude.

References

Neff, J.E. 1987, Ph.D. thesis, University of Colorado, Boulder.

Neff, J.E. 1988, in *The Impact of Very High S/N Spectroscopy on Stellar Physics*, eds. G.C. De Strobel and M. Spite, (Kluwer:Dordrecht), p. 223.

Neff, J.E., Walter, F.M., Rodonò, M., and Linsky, J.L. 1988, *Astron. Astrophys.*, **215**, 79.

Walter, F.M., Neff, J.E., Gibson, D.M., Linsky, J.L., Rodonò, M., Gary, D.E., and Butler, C.J. 1987, *Astron. Astrophys.*, **186**, 241.

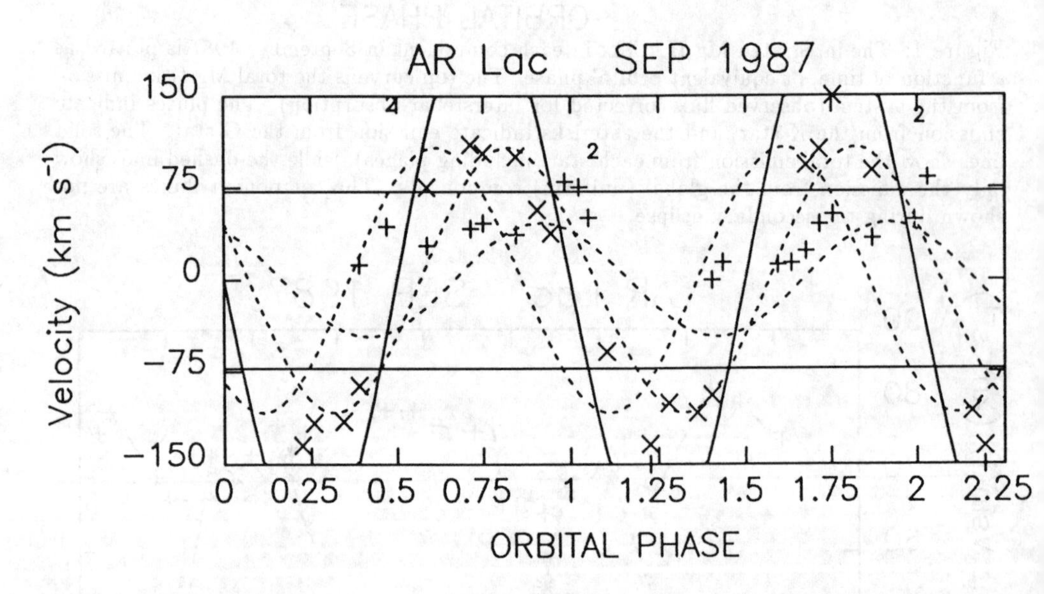

Figure 3: The velocities of the G star (solid line), the plages seen in 1985 (dashed lines), and the asymmetric components seen in 1987 (plusses and x's) in the frame of the K star. Horizontal lines are drawn at ± 72 km s^{-1}, the equatorial velocity of the K star.

SPATIAL RESOLUTION OF STELLAR ATMOSPHERES WITHIN ACTIVE CLOSE BINARIES

JAMES E. NEFF
Lab. for Astronomy and Solar Physics
NASA Goddard Space Flight Center
Greenbelt, MD 20771, USA

ABSTRACT. I discuss the general problem of mapping the spatial structure in the atmospheres of active, late-type stars. There are several major differences between the problems of imaging photospheres and of imaging chromospheres. Because of these differences, chromospheric spectral imaging must be based on a direct decomposition of the observed emission line profiles.

1. INTRODUCTION

Because of their vast distance from us, stars appear as mere points of light through even the largest telescopes. If stellar atmospheres were spatially homogeneous, plane-parallel emitters, there would be no need to spatially resolve them. However, the Sun—the only star that is close enough to be directly resolved—is covered with dark spots in its photosphere, and its chromosphere and corona consist of a non-uniform, inhomogeneous distribution of bright, magnetically confined regions (e.g. plages) and relatively faint, tenuous regions (e.g. coronal holes). Disk-averaged emission can serve as a probe only of the *average* atmosphere, which probably is very different from any of the atmospheric components that are actually present. In order to probe the characteristics of magnetically active regions, we must first somehow spatially resolve the atmosphere of the star.

Indicators of magnetic activity (e.g., photometric variability, Ca II H+K and Hα emission lines, ultraviolet emission lines) generally are independent of whether a star is a member of a close binary system. Global stellar properties–rotation, convection, age–are more directly linked with activity levels. Nevertheless, the most active stars tend to be close binary systems, primarily because of the conversion of orbital angular momentum into higher rotation rates. The close proximity of two active stars could conceivably lead to magnetic interactions between the stars. The preferred, non-spherical geometry of close binaries might induce a different structure and evolution of the active regions than that seen in single stars.

To go beyond simple correlations between indicators of total activity level and gross stellar properties is not a simple step. It is clear from the variability of these activity indicators that magnetic activity on other stars is confined in discrete regions, analogous to but larger and more energetic than those on the sun. It is critical that we study the discrete active regions themselves rather than the globally-averaged properties. Before this can be done, they must first be **spatially resolved**.

C. İbanoğlu (ed.), *Active Close Binaries*, 809–819.

The pioneering application to active stars of spectral imaging principles was by Vogt and Penrod (1983) and Vogt, Penrod, and Hatzes (1987). Their *Doppler imaging* technique can produce images of stellar photospheres using high-resolution, high S/N, visible absorption line profiles. In order to produce similar images of stellar *chromospheres*, the same basic principles can be exploited, but the techniques must be quite different (Neff 1988).

By combining spectral images of the photosphere and the chromosphere with other information about the structure of the corona (derived from radio and x-ray observations), it will be possible to develop a three-dimensional picture of stellar atmospheres. With such pictures, the size, location, and brightness of magnetic active regions can be studied as a function of time to determine stellar cycles and to measure differential rotation. It will then be possible to model the physical conditions within each atmospheric component separately.

In this talk, I discuss the general problem of producing images of rapidly rotating stars using observed spectral line profiles ("spectral imaging"). I will briefly review and compare these techniques and describe some recent results for both photospheres and chromospheres.

2. Basic Principles of Spectral Imaging

2.1 Direct Imaging Methods

Stars are too far away to be spatially resolved with conventional telescopes. A few close-binary systems have been resolved using VLBI (Lestrade *et al.* 1985), but the resolution limit typically is on the order of a stellar radius. Mutel *et al.* (1984) used VLBI to probe the spatial structure of the radio emitting plasma, which likely corresponds with the coronal structure, of several RS CVn stars. Elsewhere in this volume, Barry reports on the status of optical interferometry measurements of close-binary stars. Interferometry should eventually permit us to image directly *some* stellar photospheres and coronae, but it is not currently capable of resolving individual active regions or spots.

2.2 Photometry And Rotational Modulation

RS CVn and BY Dra stars exhibit visible-band variability that generally is believed to be due to the rotation of a star with a non-uniform photospheric brightness distribution, produced by regions analogous to sunspots. The phase of minimum light directly provides the central meridian longitude of the hemisphere of maximum spottedness. The amplitude of the variation is a measure of relative spotted area between hemispheres. Changes in the mean light level are diagnostics of total spot area. Using multi-color photometry, it is possible to constrain the temperature difference between the spotted and non-spotted atmosphere. Further details of the spatial distribution, size, and number of the spots can not be derived from photometry alone. The absence of photometric variation does not imply that the atmosphere is homogenous; it could be uniformly covered with spots. Only a highly non-uniform distribution of regions of greatly enhanced contrast will produce a variation in the disk-averaged light.

The same "rotational modulation" techniques can be applied to total flux measurements in any band and bandpass. For example, Rodonò *et al.* (1987) compared the modulation of ultraviolet emission line fluxes to the visible photometry for three

RS CVn stars. They saw in at least one case a clear anti-correlation between the photospheric and chromospheric variations, as would be expected if the variation is due to large spot/plage complexes analogous to those seen on the sun. Walter, Gibson, and Basri (1983) and White *et al.* (1987) used the rotational modulation of the total x-ray flux to probe the spatial structure of the corona of AR Lac. These techniques also can be applied to radio flux variability.

Photometric studies are based on a measure of intensity in a given bandpass versus time: $I_{\Delta\lambda}(t)$. Variations that are *periodic* with the stellar rotation (i.e. rotational modulation) are measures of the intensity as a function of rotational *phase*: $I_{\Delta\lambda}(\Phi_{\rm rot})$. This is the basic observational constraint of rotational modulation studies. If the relative brightness between the spotted and non-spotted atmosphere is known (e.g. from multi-color photometry to get the temperature difference or by assuming the maximum historical intensity represents the non-spotted intensity), then $I_{\Delta\lambda}(\Phi_{\rm rot})$ becomes $A(\Phi_{\rm rot})$, the spotted area (not the spot size!) as a function of phase. It is possible to go still a little further if you have very good phase resolution. In that case $A(\Phi_{\rm rot})$ is a crude measure of $A(\theta_p)$, where θ_p is the projected longitude on the star. Rotational modulation of total intensity can thus provide a **one-dimensional** map of stellar surface brightness.

In my judgement, that's about the maximum you can extract from the basic observational constraints. It will always be possible to match complex models to the measured constraints, but the models won't be any better constrained (with rotational modulation alone) than $A(\theta_p)$. If we wish to measure the location, size, shape, and distribution of the spotted areas themselves (in other words, to derive a two-dimensional map or image) then we need a new **dimension** of observational constraints. This new dimension can be provided by spectroscopy.

2.3 Spectroscopic Methods

A high-resolution spectroscopic measurement provides a new dimension of constraints: $I(\lambda, t)$. The rotational modulation of spectra therefore gives $I(\lambda, \theta_p)$, as described in the previous section. For most stars, that's all you can do, because there is no spatial information in the λ constraint. On the other hand, if the star is rotating very rapidly, then the wavelength on a line profile corresponds to emission from a specific locus on the stellar surface. This locus is essentially the distance, ρ, from the projected stellar rotation axis. Thus for rapidly-rotating stars, high-resolution spectra give $I(\rho, \Phi_{\rm rot})$. If the inclination is known and if the phase resolution is sufficient, then $I(\rho, \Phi_{\rm rot})$ can be mapped into $I(\theta, \phi)$, where θ stellar surface longitude and ϕ is the stellar surface latitude. $I(\theta, \phi)$ is an **image** of the star.

If the geometry is simple, the mapping is relatively straightforward. A discrete region of high contrast on the surface of a rapidly rotating star produces an effect on the line profile that moves from blue to red across the line profile as the star rotates. The amplitude of this velocity shift is $v_{rot} \sin i \cos \phi$ (where v_{rot} is the rotational velocity of the star, i is the inclination, and ϕ is the latitude of the region), if the region is on the stellar surface. When the region is on the central meridian, there is no velocity shift. By measuring the relative velocity of the feature with respect to the center of the line profile as a function of rotational phase, its longitude and latitude can be determined. If the region is spatially extended, the width of the spectral feature as a function of rotational phase can be used to determine the shape of the region on the stellar surface. Except for the case of a single, small region of very high contrast on an otherwise uniformly bright star, spectral images can not be derived by simply identifying the wavelength of "bumps and wiggles" on the line profiles.

3. Photospheric Imaging

3.1 Doppler Imaging

The "Doppler imaging" technique of Vogt *et al.* involves imposing a brightness distribution on the surface of a model star, synthesizing a predicted line profile, and then comparing the observed and predicted line profiles. The presumed brightness distribution is varied iteratively, using a maximum entropy algorithm, until a satisfactory match between the predicted and observed profiles is obtained.

Their model star is a spherical grid of roughly constant-area zones, up to 2500 covering the star. Each zone comprises a single element in an *image vector*, $\underline{I}$. The *response matrix*, $\mathbf{R}$, contains all of the known physics required to map the intensity of each element in their image vector into its resultant effect on an observed line profile. By stringing all of the observed line profiles end-to-end in a *data vector*, $\underline{D}$, they can compare a series of predicted line profiles with the observed profiles: $\underline{I}\mathbf{R}=\underline{D}$. Because the response matrix generally is not invertable, the image vector must be solved for iteratively.

There are several ways to guide this iteration. Vogt *et al.* use a maximum entropy method, while Rice, Wehlau, and Khokhlova (1989) and Piskunov, Tuominen, and Vilhu (1989) use the method developed by Goncharsky (Goncharsky *et al.* 1978, 1983; see also Khokhlova 1976). With perfect data, both techniques should yield the same result, but there appear to be different physical motivations guiding the iteration in each procedure. Unfortunately, even though the mathematical differences are clear, there is considerable argument over how to describe what these differences mean in physical terms. For example, should the resultant image be the "simplest"? the "smoothest"? have the minimum/maximum correlation between neighboring grid points? Which technique provides which of these qualities?

In any *inversion* problem like this (i.e., given the answer, what was the question?), any solution will be non-unique. That doesn't mean they need to be arbitrary or meaningless. Applying different types of physical guidance to the iteration can lead to different families of solutions. For example, one procedure might lead to a solution where the intensity in every grid box is completely uncorrelated with neighboring boxes and no coherent, large-scale structures are seen. A star looking like this **could** produce the observed profiles. A different procedure operating on the same data set might result in a solution with a large spot at the pole and a small coherent region near the equator. In order to decide which solution is better, we must understand the physical implications of each procedure. While I have no doubt that the various groups in this business do understand these *intrinsic* implications, I believe that more study needs to be made of the physical constraints on the final image that are *extrinsic* to the actual iteration.

First and foremost, of course, is the quality of the data and of the data reduction. Also, how do uncertainties in the assumed physical properties affect the final result? How sensitive is the image to uncertainties in spectral type, inclination, rotational velocity, magnetic field, etc.? Vogt *et al.* (1987) studied some of these issues, but the procedure itself renders a quantitative understanding virtually impossible. What we need is an image with error bars; not an easy thing to visualize or to present.

Short of this goal, it might still be possible to *parameterize* the solution. For example, a polar spot, or any longitudinally symmetric spot distribution, will affect the line profiles in the same way at all phases. Thus one could say that all observed phases provide a constraint on the polar spot. On the other hand, slightly incorrect physical assumptions might also affect the observed profile in the same way. Thus there is a great potential systematic error in the polar regions that doesn't exist in the equatorial regions. An equatorial spot would produce an unambiguous, moving signature on the line profile as the star rotates. There also is a trade-off between spot area and spot temperature. Vogt *et al.* allow each grid point in their image vector to have **any** non-zero brightness. Spectroscopy alone, however, really doesn't constrain the intensity, only the geometry. In order to get an image, they must decide upon some arbitrary intensity *cutoff* to define the spot boundaries. Doppler imaging using a single spectral line is not capable of a complete solution. Coupled with independent measure of spot temperature or area, it can provide the geometry.

3.2 Status

The first, and possibly the only, reliable photospheric image published in a refereed journal was by Vogt and Penrod (1983). This image shows a large spot at the pole of HR 1099, with a "finger" extending to lower latitudes, and a large spot (or group of spots) near the equator. Subsequent images derived by this group (Vogt 1988) indicate that these spot regions are long-lived, and the low-latitude spot shows a general poleward migration.

Vogt *et al.* have observed HD 26337, UX Ari, and HD 199178 since 1985 and V711 Tau since 1981 with the coudè spectrometer of the 3m Shane telescope at Lick Observatory. They use spectra with resolution of 0.13 Å and signal-to-noise of at least 400 obtained at many different rotational phases of a rapidly rotating star. In several cases, they acquired enough data to generate two independent images for a single observing season. Their images (see Vogt 1988) all tend to show slowly changing polar spots and a relatively small number of mid-latitude features, which tend to migrate poleward.

At least four other groups have been working on similar codes. Jankov and Foing (1987), for example, are working with a recently acquired data set of V711 Tau. An international Doppler imaging campaign of a single target is scheduled as part of the *Musicos* program late in 1989.

3.3 Future Improvements

Photospheric imaging is still in its developmental stages. The major factors holding back its application have been the inherent difficulty in acquiring and reducing high-quality data and the complexity of the method. The major weakness of the procedure is that it doesn't allow for easily parameterized solutions.

Simultaneous measurements of spot area or temperature are required to constrain the final image. A promising new procedure for measuring spot areas and temperatures is the measure the relative strength of the 7100 Å and 8860 Å TiO absorption bands. Saar and Neff (1990) show an example of this procedure. Simultaneous photometry can also help constrain these properties. With sufficient computing power, the Doppler imaging codes can be expanded to work on several lines simultaneously.

If lines of significantly different temperature sensitivities are observed, then Doppler imaging can directly probe the spot temperatures simultaneous with their geometry.

Still more study is needed to understand the physical implications of the various iteration schemes. To this end, three different groups are independently analyzing the same data set. A comparison of their results should shed light on the effect on the final image of their different assumptions. A code that would produce not just an intensity map but also an uncertainty map would be a big step forward.

4. Chromospheric Imaging

4.1 Contrast With Phospheric Imaging

While all spectral imaging techniques share the same fundamental principles, there are several noteworthy differences between visible Doppler imaging techniques (e.g., Vogt, Penrod, Hatzes 1987; Jankov and Foing 1987; Piskunov, Tuominen, and Vilhu 1989) and ultraviolet techniques (Neff 1987; Neff *et al.* 1989; Walter *et al.* 1987).

First and foremost, the physical properties of stellar *chromospheres* are not known. In fact, the goal of imaging is to determine these properties. The parameters needed to synthesize the intrinsic chromospheric profile therefore are not available. Current chromospheric model atmospheres typically represent only a global average, not the physically distinct regions that we expect to be present.

Second, in a given spectral line, most of the emission probably arises from discrete regions. This is emission above a very faint continuum. The more active the star, the brighter the discrete emission. By contrast, photospheric spots are dark regions viewed against a bright background. Further, the photospheric lines become shallower the more rapidly the star rotates, whereas the chromospheric lines get brighter and broader.

Because of these differences, chromospheric spectral imaging must be based on a direct *decomposition* of emission line profiles. While tedious and difficult to automate, this procedure is capable of providing spatial information using spectra of lower S/N. The resulting images will not have the fine detail provided by Doppler imaging. They are nevertheless a critical first step in deriving realistic model atmospheres.

4.2 Technique

The Mg II k (2795 Å) chromospheric emission line from a very rapidly rotating star provides a 1-dimensional map of the star's chromospheric brightness. For example, consider a star that is uniformly bright except for one discrete plage region. The plage will first come into view when it is on the approaching limb of the star. The emission from the plage will appear as a discrete emission feature on the blue-shifted wing of the line profile. As the star rotates and the plage transits the disk, this discrete feature will move redward across the profile. It should disappear from the profile roughly one half an orbital period later, as the plage rotates to the back side of the star. Using a series of spectra obtained at many different rotational phases, a 2-dimensional image can be built up to reveal the size, location, and relative brightness of the plage. Of course, real life is never so simple.

Emission arising from a **uniform** distribution on a stellar surface will produce a **symmetric** profile centered at the stellar velocity. Large-scale non-uniformities of the surface brightness distribution produce emission components that vary in wavelength due to the Doppler shift produced by the star's rotation. The net profile will be asymmetric, and a single-component fit to this profile will not lie at the stellar velocity. Therefore by fitting the observed profiles with a *symmetric component centered at the stellar velocity* plus additional components to account for the residual emission, the non-uniform distribution can be mapped. In practice, the minimum number of additional components required to match the observed profiles are determined iteratively, using an interactive multiple-component fitting routine. The primary constraints in this procedure are the positions of the uniform stellar components. Secondary constraints are the rough constancy of the stellar emission line widths and smooth point-to-point variation of the stellar flux.

4.3 Application to AR Lac in September 1985

To illustrate the power and the limitations of this technique, I summarize the analysis of a series of spectra of AR Lacertae (=HD 210334) obtained with the IUE satellite in September 1985 (these are fully described by Neff *et al.* 1989). Elsewhere in this volume I present the preliminary results for more recently obtained and far superior observations.

AR Lac is an eclipsing RS CVn binary system (K0 IV + G2 IV). The two stars rotate synchronously with the orbital period (1.98 days), yielding an equatorial velocity of 72 km s^{-1} for the K star and 39 km s^{-1} for the G star. Because the system is bright (m_v=6.1) and because both stars are active, the system is an ideal candidate for ultraviolet spectral imaging.

We observed AR Lac over 80% of a single orbital cycle. Eighteen high-resolution (R~20,000) spectra of the Mg II k (2795 Å) line and 18 low-resolution (6 Å) spectra covering the range 1150 to 1950 Å were obtained. The final multi-gaussian fits derived using the constraints discussed above are shown in Figure 1. The velocities relative to K star of these discrete emission components are shown in Figure 2. The components have been separated into three distinct groups, represented by different symbols, because their velocities can be well represented by the solid-body rotation velocities of three distinct locations on the K star (dashed lines). I will refer to the individual components as plages A, B, and C.

The longitudes of the plages are determined by the phase of the best fit solid-body (sinusoidal) rotation curves. The amplitude of the solid-body rotation curves are the product of the height above the photosphere, the stellar $v \sin i$, and the cosine of the latitude. The stellar $v \sin i$ is well known, but there is no way to distinguish *a priori* between height and latitude. However, amplitudes less than $v \sin i$ (e.g., plage B) indicate that the plage lies at high latitude. The minimum latitude can be determined by assuming that the plage lies on the photosphere. Radial velocity amplitudes greater than the stellar $v \sin i$ (e.g., plages A and C) imply that the plage must lie well above the photosphere, with the minimum height determined by assuming that the plage is equatorial. There is no way to distinguish between north and south latitudes, because the inclination of AR Lac is nearly 90°.

The radial velocity curves shown in Figure 2 provide the *locations* of the surface features. To produce a crude image of the stellar surfaces, we must first measure the plage *sizes*. I used the line widths of the plage emission components, after deconvolution of the instrumental profile, to determine their extension in longitude. I assume that all of the deconvolved component width is due to velocity smearing from a resolved region on the star. These measured sizes are upper limits, because there

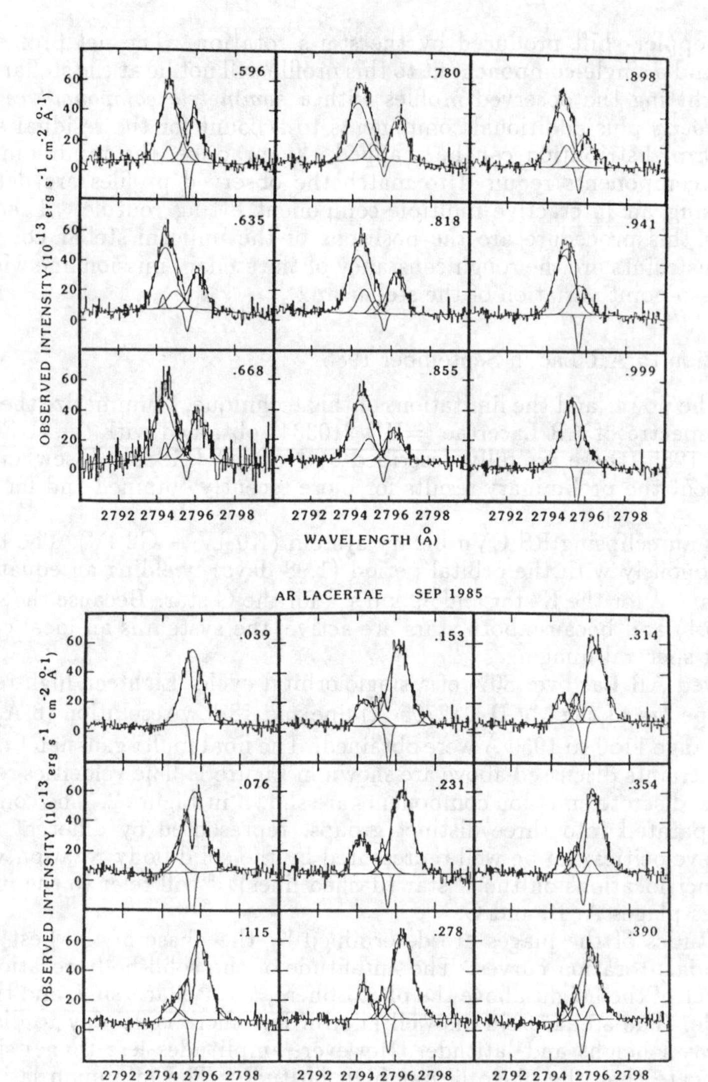

Figure 1: The final fits to the 18 IUE Mg II k spectra of AR Lac. The spectra in the top panel were obtained before the total eclipse of the G star, and those in the bottom panel afterwards. Symmetric (gaussian) components are centered on the expected positions of the K and G star (the K star is brighter) and on the interstellar absorption components. Additional components are added to match the phase-dependent asymmetries. The pre-flare G star profile was subtracted from the spectra between $\phi = 1.278$ and 1.390, to better display the emission from the flaring region.

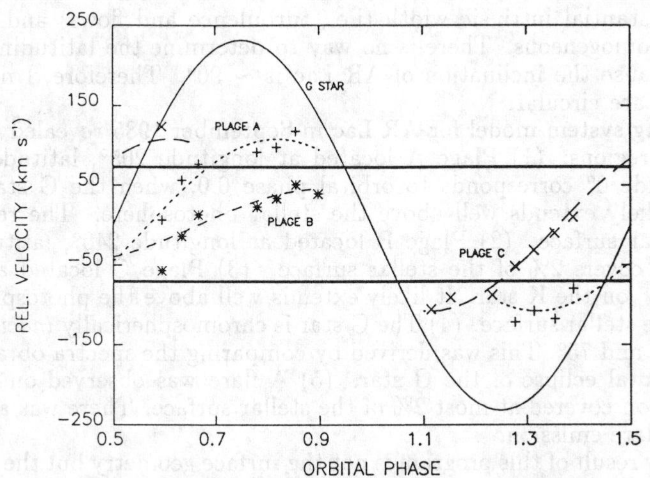

Figure 2: The velocities of the G star (solid line) and the additional emission components in Fig. 1 are shown here in the K star rest frame. Horizontal lines are drawn at ± 72 km s^{-1}, the equatorial rotational velocity of the K star. The emission components fall along 3 distinct solid-body rotation curves. The velocity curves are only plotted where the corresponding plage would be on the visible hemisphere.

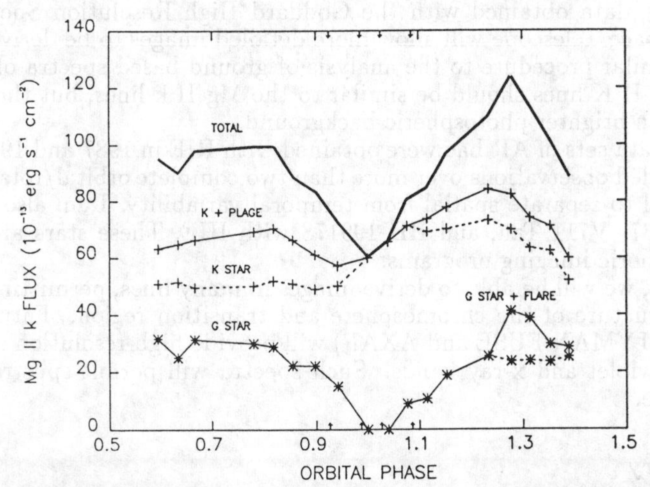

Figure 3: The integrated Mg II k flux in each emission component from Fig. 1 is plotted as a function of time, with $\phi > 1$ following total eclipse to indicate the time sequence of the observations. The top curve is the total Mg II k emission from the system (observed flux corrected for interstellar absorption). The plusses indicate emission from the K star, and the asterisks indicate emission from the G star. The solid lines show the total emission from each star (including plages and flare), while the dashed lines show only the emission from the global (uniform) components. The phases of 1st through 4th contacts are indicated by arrows.

could be a substantial intrinsic width (i.e., turbulence and flows) and the regions might not be homogeneous. There is no way to determine the latitudinal extent of the plages, because the inclination of AR Lac is $\sim 90°$. Therefore, I must assume that the plages are circular.

The resulting system model for AR Lac in September 1985 revealed five discrete chromospheric regions: (1) Plage A located at longitude 205°, latitude 0° on the K star (longitude 0° corresponds to orbital phase 0.0, when the G star is totally eclipsed). It likely extends well above the stellar photosphere. The region covers 6% of the stellar surface. (2) Plage B located at longitude 240°, latitude$\pm$50° on the K star. It covers 2% of the stellar surface. (3) Plage C located at longitude 130°, latitude 0° on the K star. It likely extends well above the photosphere, and it covers 9% of the stellar surface. (4) The G star is chromospherically *inactive* between longitudes 335° and 75°. This was derived by comparing the spectra obtained before and after the total eclipse of the G star. (5) A flare was observed on the G star. The flaring region covered at most 2% of the stellar surface. There was a systematic *redshift* of the flare emission.

The primary result of this program is not the surface geometry but the *surface flux* within each region. As shown in Figure 3, the global (uniform) flux from both stars varies, and the plage flux is not constant. All of this information can be presented coherently only as a series of images, as done by Neff *et al.* (1989).

4.4 Future Improvements

Higher quality data obtained with the Goddard High Resolution Spectrograph on the Hubble Space Telescope will allow more detailed images to be derived. I am also applying a similar procedure to the analysis of ground-based spectra of the Ca II K lines. The Ca II K lines should be similar to the Mg II k lines, but they are viewed against a much brighter photospheric background.

Superior data sets of AR Lac were obtained with IUE in 1987 and 1989. The 1987 data set included observations over more than two complete orbital/rotational cycles. This is critical to separate spatial from temporal variability. I am also studying the stars HD 26337, V711 Tau, and HD 199178 with IUE. These stars are included in most photospheric imaging programs.

With HST, we will be able to derive images in many lines, permitting us to probe the height structure of the chromosphere and transition region. Future spacecraft mission (e.g., LYMAN-FUSE and AXAF) will provide high-resolution spectra in the extreme ultraviolet and x-ray bands. Such spectra will permit spectral imaging of stellar coronae.

5. Summary

Very detailed images of stellar photospheres can be derived by synthesizing a line profile from an assumed brightness distribution and then varying this distribution until the synthetic profile matches the observed profile. Unfortunately, the same procedure can not be used to image stellar chromospheres. Nevertheless, a non-uniform chromospheric distribution manifests itself in the observed profiles. Because these non-uniformities are large scale and because the photospheric background is fainter in the ultraviolet, chromospheres can be imaged by a careful decomposition of the line profiles.

References

Goncharsky, A.V., Stepanov, V.V., Khokhlova, V.L, and Yagola, A.G. 1978, *Sov. Astron. Let.*, **3**, 147 (translated from *Pis'ma Astron. Zh.*, **3**, 278, 1977).

Goncharsky, A.V., Ryabchikova, T.A., Stepanov, V.V., Khokhlova, V.L., and Yagola, A.G. 1983, *Sov. Astron.*, **27**, 49 (translated from *Astron. Zh.*, **60**, 83, 1983).

Jankov, S., and Foing, B. 1987, in *Proc. Fifth Cambridge Workshop on Cool Stars, Stellar Systems, and the Sun*, eds. J.L. Linsky and R.E. Stencel, (Springer-Verlag:Berlin), p. 528.

Khokhlova, V.L. 1976, *Sov. Astron.*, **19**, 576 (translated from *Astron. Zh.*, **52**, 950, 1975).

Lestrade, J.-F., Mutel, R.L., Preston, R.A., and Phillips, R.B. 1985, in *Radio Stars*, eds. R.N., Hjellming and D.M. Gibson, (Reidel:Dordrecht), p. 275.

Mutel, R.M., Doiron, D.J., Lestrade, J.F., and Phillips, R.B. 1984, *Astrophys. J.*, **278**, 220.

Neff, J.E. 1987, Ph.D. thesis, University of Colorado, Boulder.

Neff, J.E. 1988, in *The Impact of Very High S/N Spectroscopy on Stellar Physics*, eds. G.C. De Strobel and M. Spite, (Kluwer:Dordrecht), p. 223.

Neff, J.E., Walter, F.M., Rodonò, M., and Linsky, J.L. 1989, *Astron. Astrophys.*, **215**, 79.

Piskunov, N.E., Tuominen, I., and Vilhu, O. 1989, *Astron. Astrophys.*, in press.

Rice, J.B., Wehlau, W.H., and Khokhlova, V.L. 1989, *Astron. Astrophys.*, **208** 179.

Rodonò, M., Byrne, P.B., Neff, J.E., Linsky, J.L, Simon, T., Butler, C.J., Catalano, S., Doyle, J.G., Andrews, A.D., and Gibson, D.M. 1987, *Astron. Astrophys.*, **176**, 267.

Saar, S.H., and Neff, J.E. 1990, in *Cool Stars, Stellar Systems, and the Sun*, ed. G. Wallerstein, ASP Conference Series, in press.

Vogt, S.S. 1988, in *The Impact of Very High S/N Spectroscopy on Stellar Physics*, eds. G.C. De Strobel and M. Spite, (Kluwer:Dordrecht), p. 253

Vogt, S.S., and Penrod, G.D. 1983, *Publ. Astr. Soc. Pacific*, **95**, 565.

Vogt, S.S., Penrod, G.D., and Hatzes, A.P. 1987, *Astrophys. J.*, **321**, 496.

Walter, F.M., Gibson, D.M., and Basri, G.S. 1983, *Astrophys. J.*, **267**, 665.

Walter, F.M., Neff, J.E., Gibson, D.M., Linsky, J.L., Rodonò, M., Gary, D.E., and Butler, C.J. 1987, *Astron. Astrophys.*, **186**, 241.

White, N.E., Shaefer, R., Parmer, A.N., Culhane, J.L. 1987, in *Proc. Fifth Cambridge Workshop on Cool Stars, Stellar Systems, and the Sun*, eds. J.L. Linsky and R.E. Stencel, (Springer-Verlag:Berlin), p. 521.

Information Limit Optimization Techniques applied to AB Doradus

by

Timothy Banks[1] and Edwin Budding[2]

1: Physics Department, Victoria University of Wellington, N.Z.

2: Carter Observatory, Wellington, New Zealand.

Abstract :

December 1984 UBVRI light curves of the rapidly rotating chromospherically active single star AB Doradus were modeled using circular starspots. Several different methods were employed to calculate the spot temperatures, resulting in a range of values slightly beyond 1000 Kelvin below the Photosphere. Grid searches varying the spot latitudes were performed at a range of inclinations in the expectation that an optimum combination of these parameters could be found. However the best fits, in a χ^2 sense, for each inclination tested were essentially the same, but when minimum spot area is included as an additional criterion an inclination of 70 degrees appears to be preferred, in reasonable agreement with literature values . A spot evolution sequence is discussed for this data. An analysis of Rucinski's (1983) 1982 data is also presented in support of a call for further accurate photometry of this system.

Introduction:

The question of whether certain stars might show the photmetric effects of non-uniform surface luminosity ("starspots") is an old one with rather a chequered history of scientific acceptability. In recent years, however, the existence has come to light of a certain class of variable star - the rapidly rotating cool stars known collectively as the RS CVn systems after their prototype RS Canem Venaticorum. They exhibit along with their low level of "wave-like" photometric variations, a whole range of additional phenomena including X-ray and UV excesses, radio "flares", chromospheric emission lines; all corroborating the general picture of highly enhanced electrodynamic activity of the kind that we see on the Sun (Tayler, 1989). The detailed study of these stars (see e.g Baliunas and Vaughan, 1986) has made it seem that it would have been surprising if photometric variations of the observed kind had not been found, and indeed, the whole business of modelling such variations by "starspots" of an assigned shape, temperature, and position has gained a considerable credibility; though care is still required in ensuring that inherent information limits present in any given data set are not surpassed by such models. In this paper we present an analytical approach to such problems, which we characterise as information limit optimization. Details of its methodology are given in Budding and Zeilik (1987).

C. İbanoğlu (ed.), Active Close Binaries, 821–830.

AB Doradus (HD 36705) is a rapidly rotating single star at a distance of 27 ± 7 parsecs (Innis et al., 1986) with a period of 0.51 day. Only some of the Pleiades K Dwarfs (Van Leeuwen and Alphenaar, 1983) rotate faster. This, in combination with a high Lithium abundance (Rucinski, 1985) and chromospheric activity (Hearnshaw, 1979) suggest that AB Doradus is a young star in the final stages of core contraction before Main Sequence. Since photoelectric observations of this star began in 1979 (Pakull, 1981) the light curve has varied quite dramatically over time scales of a month, with smaller variations evident within a few nights. Starspots are generally accepted as the cause of this behaviour, indeed such a volatile nature would be expected in such a rapidly rotating sun like star if the generator of stellar magnetic field is dynamo in nature (Weiss et al., 1984).

The 1984 data of Lloyd Evans:

A χ^2 optimising program based on the theoretical maculation wave work of Budding (1977) was used for the modelling of the UBVRI light curves obtained by Lloyd-Evans (1987), who used the 0.5m SAAO telescope in December 1984. This particular data set was chosen for detailed analysis because of its complete phase coverage obtained in as short an observational period as possible, and its high quality. The phases were calculated using Kubiak's (1985) ephemeris of HJD 244296.640 + 0.51339 E. A problem with a single star is that the angle of its rotational axis relative to us is uncertain, but as the effect of each spot lasts about 180 degrees it was initially assumed that AB Dor's inclination is 90 degrees. Rucinski's (1985) spectral type of K0-1V was adopted, resulting in a photospheric temperature of 5250 Kelvin (Allen, 1973, Hayes, 1978) being used. Al-Naimiy's (1981) tables provided suitable limb darkening values for each waveband.

The U band data was modeled with the spot/photosphere flux ratio (κ) set to nil. The strategy for the BVRI bands was to use the best U band fit's parameters optimising only the maximum light level of the star and the spot flux in order to determine a rough value of the latter, which was then used in a subsequent fit optimising the light level with the spots' longitudes and radii. Table 1 gives the results of these fits. As the observational wavelength increases so does the second spot's longitude, while the first's decreased. This apparent longitude variation is outside their formal error limits (for details of the calculation of the errors see Banks, 1989), and is presumably due to both the latitudes being set arbitrarily, or the model's simplification of circular spots.

The flux ratios obtained were used in an attempt to calculate the spots' mean temperature, assuming them both to be the same. Firstly it was assumed that both the spots and the photosphere behave as Black Bodies, but that the flux ratio is still effectively nil at the U Band. An average temperature of 4230 ± 70 Kelvin was obtained thus. When the V band flux was set to zero as in Zeilik et al. (1988), then values of 3840 ± 150 and 3710 ± 230 Kelvin were

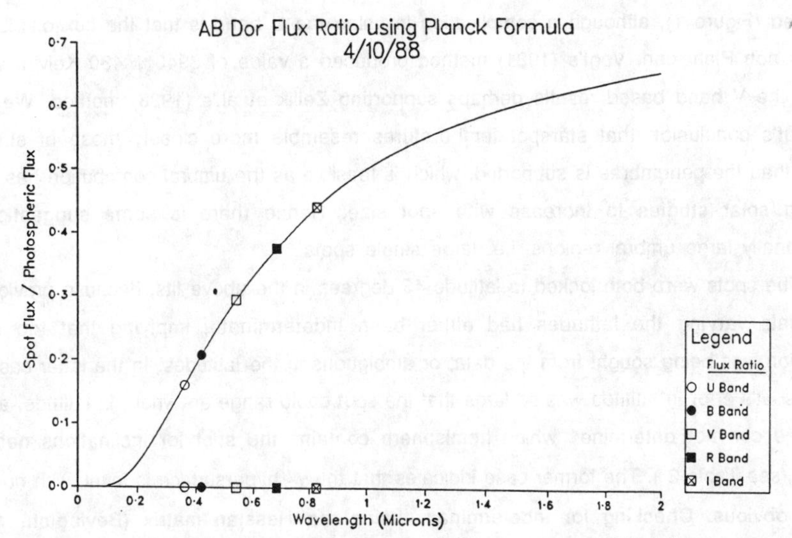

Figure One : Planckian Flux Ratio for a 4230 Kelvin Spot and a 5250 Kelvin Photosphere. The mean wavelength for the UBVRI filters are plotted.

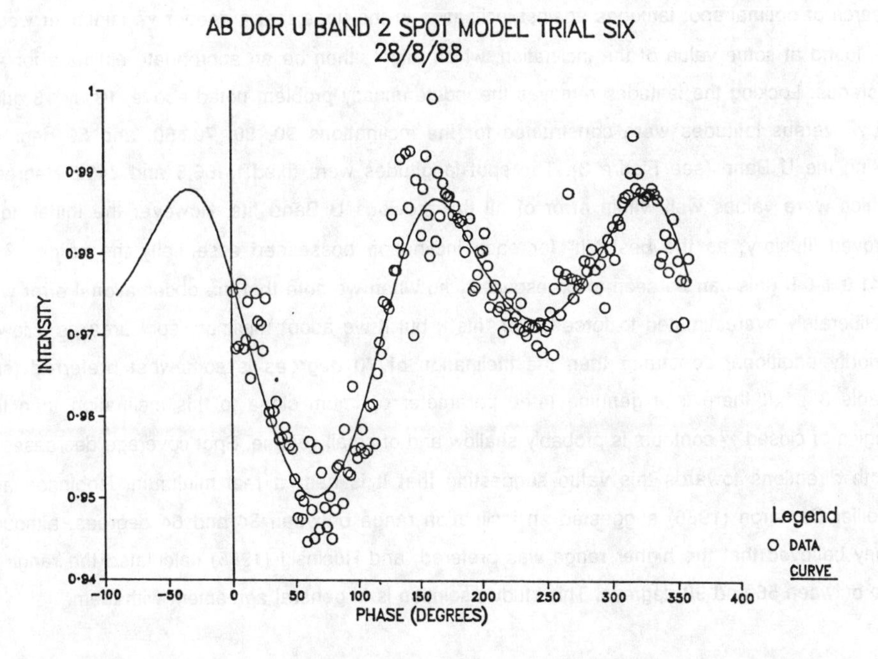

Figure Two : Dark Spot Model light curve plotted against the normalised U Band data. The spots are locked to latitude 45 degrees.

reached for the I and R bands respectively. However a systematic error would hereby be introduced (Figure 1), although a complication in using the U band is that the flux distribution is rather non Planckian. Vogt's (1981) method produced a value of 3940 ± 430 Kelvin, within error of the V band based results perhaps supporting Zeilik et al.'s (1988) method. We note that Vogt's conclusion that starspot temperatures resemble more closely those of sunspot umbrae than the penumbrae is supported, which is feasible as the umbral contribution has been noted in solar studies to increase with spot size. Hence there is some suggestion of proportionally large umbral regions, i.e. large single spots.

The spots were both locked to latitude 45 degrees in the above fits, because previous U band trials varying the latitudes had either been indeterminate, implying that too much information was being sought from the data, or ambiguous in the latitudes. In the latter case the second spot's error in latitude was so large that the spot could range anywhere in latitude, as our procedure can not determine which hemisphere contains the spot for inclinations near 90 degrees (see Table 2). The former case indicates that the χ^2 hypersurface is level, with no local minima obvious. Checking for indeterminacy using the Hessian matrix (Bevington, 1969, Budding and Najim, 1980) is an important part of our method, as it prevents overparameterising the data.

As the data had demonstrated its high information content it was then subjected to a grid search of optimal spot latitudes against inclination in the hope that a deeper χ^2 minimum would be found at some value of the inclination, which might then be an appropriate estimate for AB Doradus. Locking the latitudes removes the indeterminacy problem noted above. 18 by 18 grids of χ^2 versus latitudes were constructed for the inclinations 90, 80, 70, 60, and 50 degrees using the U Band (see Figure 3). The spot longitudes were fixed to 66.5 and 246.8 degrees, which were values well within error of all the previous U Band fits. However the initial hope proved illusiory, as the best "fit" for each inclination possessed essentially the same χ^2 of 541.0 ± 0.5 (this can be seen to be especially so when we note that the observational error was deliberately overestimated to force better "fits"), but if we adopt minimum spot area as a lower priority additional constraint then the inclination of 70 degrees is somewhat preferred (see Table 3). If there is a genuine three parameter optimum close to this inclination then the region of closed χ^2 contours is probably shallow and of small volume. Spot coverage decreases in both directions towards this value suggesting that it is near a real minimum. Robinson and Collier Cameron (1986) suggested an inclination range between 34 and 64 degrees, although they believed that the higher range was prefered, and Rucinski (1985) calculated the range to be between 56 and 90 degrees. This study's solution is in general agreement with them.

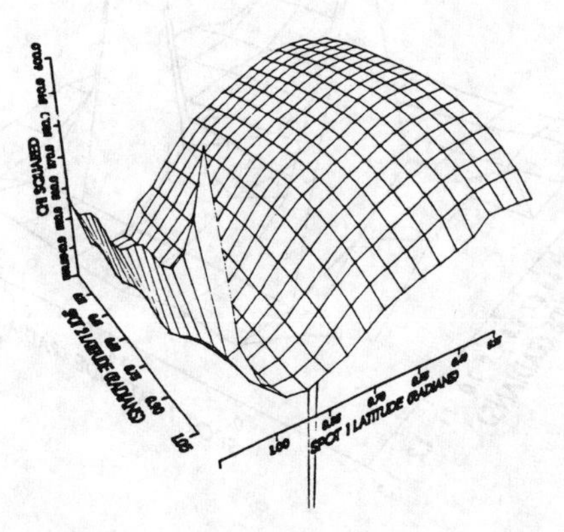

Figure Three : Contours of the Goodness of Fit versus the spot latitudes for the 90 degrees inclination. It can be seen that the first spot's latitude is tightly constrained to a valley, while the second spot is less constrained although the "thalweg" does decrease with the spot's latitude.

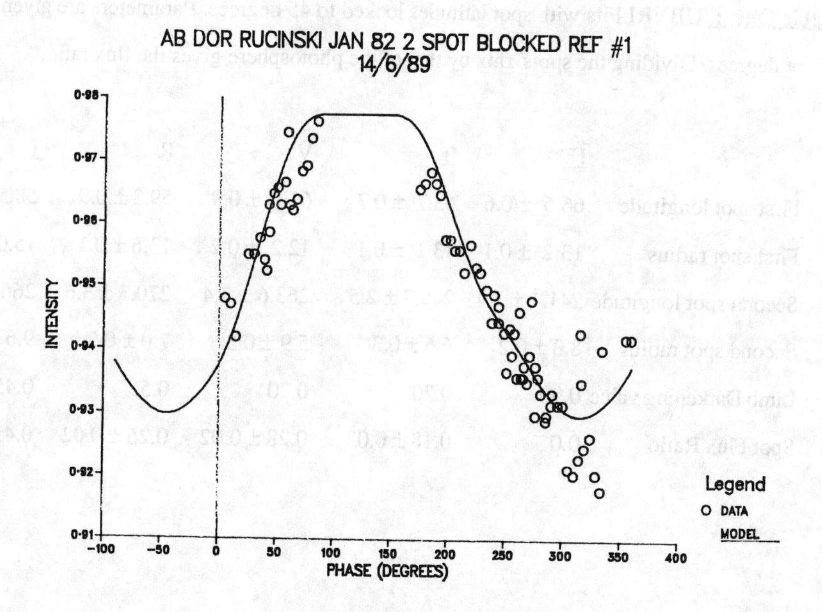

Figure Four : Two spot model fit to Rucinski's January 1982 data.

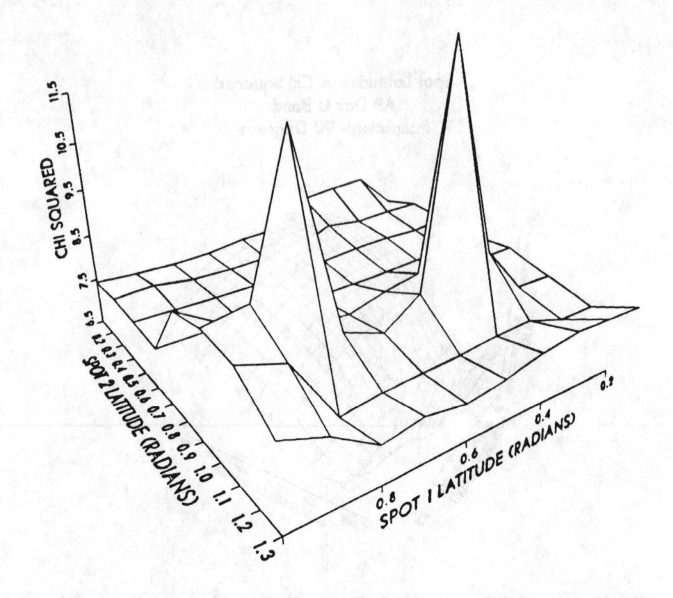

Figure Five : Contours of the goodness of Fit versus spot latitudes for Rucinksi's October 1981 data assuming an inclination of 90 degrees.

Table One : UBVRI Fits with spot latitudes locked to 45 degrees. Parameters are given in units of degrees. Dividing the spots' flux by that of the photosphere gives the flux ratio.

	U	B	V	R	I
First spot longitude	66.5 ± 0.6	63.7 ± 0.7	61.5 ± 0.9	59.3 ± 0.9	58.5 ± 1.1
First spot radius	13.2 ± 0.1	13.1 ± 0.1	12.2 ± 0.2	13.6 ± 0.1	15.0 ± 0.1
Second spot longitude	241.1 ± 1.4	249.2 ± 2.5	263.6 ± 3.4	270.3 ± 3.6	266.3 ± 2.9
Second spot radius	8.3 ± 0.2	6.6 ± 0.3	5.9 ± 0.3	7.0 ± 0.2	9.6 ± 0.2
Limb Darkening value	0.90	0.70	0.70	0.55	0.45
Spot Flux Ratio	0.0	0.18 ± 0.02	0.28 ± 0.02	0.26 ± 0.02	0.45 ± 0.02

Morphological Evolution of Spot Features :

The November 1984 and February 1985 light curves of Innis (1986) were examined in the hope that some evolution of the previously discussed spots could be seen. Innis had examined the data used in this study, and suggested that two spots were responsible for the light curve although he was uncertain of the second spot's existence (his spot D). The November data could be due to two spots centred on the approximate longitudes 77 and 310 degrees. Innis explained the light curve by a single spot at 90 ± 50 degrees but this would imply an extremely longitudinal distorted spot. The February light curve is possibly explained by spots about longitudes 54 and 234 degrees although Innis suggested a single spot at 126 ± 36 degrees.

We offer the following interpretation - in November both spots found for December's data existed, but the star was also spotted rather uniformly in longitude as well, lowering the the maximum "immaculate" intensity (Vogt, 1981). We justify this noting that the amplitude range was magnitude 6.88 to 6.93, while the maximum in the literaure is Pakull's of 6.77. In December this banding increased as the range then became 6.92 to 6.95. In February the variation is 6.95 to 6.86, and is attributed to the second spot moving further down in longitude and increasing in radius, its effect now merging with the other spot. Such behaviour would cause the maximum intensity to increase even if the banding remained constant. The first spot's longitude appears to have remained constant over this period, although it appears to have grown in slightly in radius after November, dropping the maximum light slightly. All of this discussion reinforces the need for frequent light curves as the spots on AB Doradus are altering very rapidly.

Good quality complete light curves obtained as quickly as possible are required - facts born out in an analysis of Rucinski's (1982) two light curves. Unfortunately Rucinski's data sets did not completely satisfy these requirements, leading to the analysis problems discussed below. Because of this, and more importantly the large amount of time between these data sets and those of Innis (1986) and Lloyd Evans (1987), our results can not be linked into a complete spot evolution sequence. We have already seen that the light curve variations are extremely rapid, indeed small changes can be seen in Lloyd Evans' data when it is sorted into the appropriate sections of the consecutive 9 night observational period, making such a comparsion over the two year gap between the data sets of dubious significance. However we present our results as evidence of the need for excellent data, and to demonstrate that spots could be responsible for these light curves too.

The January 1982 data was modeled using the 45 degrees spots - two spots were preferred to one by the χ^2 test at 5.1 compared to 3.6. The double spot fit's parameters were longitudes 343 ± 15 degrees and 255 ± 19 degrees, of radius 14 ± 2 and 13 ± 2 degrees respectively. This result involved locking the star's maximum intensity to an entirely arbitrary value as that crucial section of the light curve is missing, compromising the analysis. However Rucinski's October 1981 light curve was subjected to a successful latitude grid search at an inclination of 90 degrees, resulting in an optimum determinate solution of longitudes 270 ± 23 and 122 ± 6 degrees, latitudes 52 and 57 degrees, and spot radii 7.5 ± 6.1 and 0.9 ± 0.9 degrees. At other inclinations we were not successful in clearly resolving even a coherent two parameter solution, which was at least partially due to a large amount of scatter during the second spot's effect, perhaps due to this spot evolving substantially during Rucinski's observations. High latitudes, which appears to be a feature of chromospherically active stars (Budding and Zeilik, 1987), were preferred by the October 1981 for a 90 degree inclination. However if the inclination is actually near 70 degrees, as we have suggested above, then this circumstance would be no longer evident.

The new Automatic Photoelectric Telescopes could achieve all three light curve requirements (i.e. fast, accurate, and complete) and AB Doradus would make a good target for these instruments.

Conclusions:

In this paper we have applied some recently developed computer based optimization techniques to the peculiar, fast rotating dwarf AB Doradus. Our general conclusions are :-

(1) Dark circular regions (two in number) representing large aggregations of "starspots" can adequately account for the observed differential ("AC") photometric variation of AB Doradus. We also posit a general background of maculation effects, more or less uniformly distributed in longitude which accounts for the varying "DC" component of the light curve.

(2) We can find a value for the latitude of the main feature responsible for Lloyd-Evans's high quality 1984 light curve assuming that the rotational axis of AB Doradus is at 90 degrees to our line of sight, while the second's spot latitude can not be so well constrained being generally in a trough in the χ^2 hypersurface (see Figure 3). However at lower values of the inclination both spot latitudes were reasonably well constrained.

(3) Using total spot coverage as an extra selection criterion in addition to the goodness of fit, which were essentially the same over the range of inclinations tested, an inclination of about 70 degrees would be prefered. However, we acknowledge that the argument is not convincing by itself alone.

(4) We find evidence that features of a similar kind persist through the data of 1984 - 1985 (and similar such features could also explain the 1981 and 1982 lightcurves of Rucinski (1982)) and therefore light curves show some signs of an evolutionary trends in spot morphology.

(5) Uchida (1986) proposed the Active Longitude Belt model as an explanation of the

heavily spotted RS CVn binaries' activity, with its physical basis including flux tubes between the system's components. If AB Doradus, almost certainly a single star, exhibits such clustering into longitude bands on opposite sides of the star then such a mechanism could not be a consequence of an orbital effect. This would not be too surprising, as the Active Longitude Belt model has also been proposed for a rather slowly rotating single star - the Sun, indeed the concept comes froma solar physics background. Examination of all the published light curves for AB Doradus (see Innis et al., 1988) has been taken to show that all longitudes appear to be equally preferred, however spot modeling, which has previously been neglected for AB Doradus, might help to resolve the issue.

Finally, HD 36705 is highly active, it should be a good candidate for Vogt Imaging (Vogt and Penrod, 1981) and intensive (APT) photo-electric observations in the expectation that a definitive spot evolution sequence could be observed.

Table Two : Initial free latitudes B Band Spot Fit.

First spot longitude	62.4 ± 1.0 degrees
First spot latitude	71.2 ± 1.6 degrees
First spot radius	21.8 ± 3.0 degrees
Second spot longitude	247.7 ± 3.7 degrees
Second spot latitude	53.4 ± 74.8 degrees
Second spot radius	9.7 ± 12.3 degrees
Limb Darkening value	0.65

Table Three : Grid Search Optimum Solutions. All parameters are given in units of degrees. %H gives the percentage ("AC") spot coverage of a hemisphere.

Inclination	First Spot Latitude	First Spot Radius	Second Spot Latitude	Second Spot Radius	%H
90	53.9 ± 0.7	15.9 ± 0.1	61.9 ± 1.1	12.4 ± 0.1	5.02
80	33.8 ± 0.6	11.8 ± 0.1	29.8 ± 1.2	08.0 ± 0.1	2.51
70	17.8 ± 0.6	11.1 ± 0.1	21.8 ± 1.2	07.6 ± 0.1	2.23
60	13.8 ± 0.7	11.4 ± 0.1	09.7 ± 1.1	08.0 ± 0.1	2.40
50	08.0 ± 0.6	12.4 ± 0.1	05.7 ± 1.2	08.7 ± 0.1	2.83

Acknowledgements:

The authors would like to express their gratitude for the help extended by Mr Jim Gellen of the Research School of Earth Sciences (VUW), and to acknowledge many fruitful discussions with Mr Frank Andrews (Carter Observatory).

References :

Allen, C.W., 1973, *Astrophysical Quantities,* University of London Press.

Al-Naimiy, H., 1981, *Astron. Astrophys. Supp. Series,* **43**, 85.

Baliunas,S.L., and Vaughan, A.H., 1985, *A. Rev. Astr. Astrophys*., **23**, 379.

Banks, T., 1989, *Unpublished M.Sc Thesis*, Victoria University of Wellington, New Zealand.

Budding, E., 1977, *Astrophys. Spa. Sci*., **46**, 407.

Budding, E., and Najim, N.N., 1980, *Astrophys. Spa. Sci.,* **72**, 369.

Budding, E., and Zeilik, M., 1987, *Astrophys. J*., **319**, 827.

Hearnshaw, J, 1979, in *IAU Colloq. No. 79*, 371.

Innis, J.L., 1986, *Unpublished PhD. Thesis*, Monash University.

Innis, J.L., Thompson, K., and Coates, D.W., 1986, *Mon. Not. Roy. Astr. Soc.,* **223**, 183.

Innis, J.L., Thompson, K., Coates, D.W., and Lloyd-Evans, T., 1988, *Mon. Not. Roy. Astr. Soc*., **235**, 1411.

Kubiak, M., 1985, *Acta. Astron*., **35** (3-4), 369.

Lloyd-Evans, T., 1987, *South African Astron. Obs. Circ.,* **11**, 73.

Pakull, M.W., 1981, *Astron. Astrophys*., **104**, 33.

Robinson, R.D., and Collier Cameron, A., 1986, *Proc. astr. Soc. Aust.,* **6**, 308.

Rucinski, S.M., 1982, *Astron. Astrophys. Suppl. Ser.*,**52**,281.

Rucinski, S.M., 1985, *Mon. Not. Roy. Astr. Soc,* **215**, 591.

Stauffer, J.R., 1984, *Astrophys. J.,* **280**, 189.

Tayler, R.J., 1989, *Q. Jl. R. astr. Soc*., **30**(2), 1.

Uchida, X., 1986, *Astrophys. Spa. Sci.,* **118**, 127.

Van Leuuwen, F. , and Alphenaar, P. , 1982, in *Activity in Red Dwarfs*, IAU Colloq. 71, Eds - M.Rodono and P.B.Byrne, Reidel, Holland.

Vogt, S.S., 1981, *Astrophys. J.,* **250**, 327.

Vogt, S.S., and Penrod, G.D., 1982, in *Activity in Red Dwarfs,* IAU Colloq. 71, Eds - M. Rodono and P.B. Byrne, Reidel, Holland.

Weiss, N.O., Cattaneo, F., and Jones, C.A., 1984, *Geophys. Astrophys. Fluid Dynamics*, **30**, 305

Zeilik, M., Cox, D., De Blasi, C., Rhodes, M., and Budding, E., 1988, *Preprint submitted to Astrophys. J.*

THE COMPLEMENTARITY OF DOPPLER IMAGING AND STARSPOT MODELLING

E. Budding
Carter Observatory, Wellington

M. Zeilik
Institute for Astrophysics
University of New Mexico, Albuquerque

Abstract: Some essential formal similarities between photometric maculation wave modelling and spectrophotometric Doppler imaging are pointed out. This opens the way to a methodological approach to Doppler imaging on quite parallel lines to the authors' recent studies of stars which exhibit photometric maculation waves.

INTRODUCTION

In their pioneering paper on Doppler imaging techniques, Vogt and Penrod (1983) referred to the complementary roles of photometric starspot modelling and spectrophotometric line profile analysis, where use is made of the Doppler effect to determine the locations and sizes of spot–like features. The point has been recently confirmed in Fekel et al.'s (1987) detailed study of EI Eri by Doppler imaging.

The essential parallelism of the implied analytical tasks concerning these two procedures was already suggested in the case of close binary systems in chapters 4 and 5 of Kopal's (1959) well–known book. The "α–integrals" used to account for light variations in

C. İbanoğlu (ed.), Active Close Binaries, 831–843.

eclipsing binaries have a formally close counterpart in the "σ–integrals", which were used to express the light variation due to maculation effects (Budding, 1977). A procedural parallelism is also evident in the simulataneous approach recently presented by Strassmeier (1989). Our purpose here is to indicate that a formal parallelism between photometry and Doppler imaging in terms of σ or related integral representation is also present.

INTEGRATED EFFECT OF PROFILE FEATURES

Attention is drawn to Figure 1, where we set the axis of rotation of the star *OP* in the y, z plane of a 3–dimensional cartesian frame, such that the z axis coincides with the line of sight. The mean equatorial speed of rotation of the star we set to be v_o, which then projects into $\pm v_o \sin i$ in the line of sight at the points where the disk intersects the x–axis.

The concept of "Doppler Imaging" requires that v_o be sufficiently great to spread out a spectal line over a wavelength range much greater than the inherent natural broadening of the line (and yet not so wide that shallowness prevents a clear identification of the line.) In the absence of any particular "starspot" feature, a uniform photospheric source of intensity $I_u(\lambda)$ in a spectral line centered at a wavelength λ_o will then take the form of a rotationally broadened line, such that the intensity in the wavelength region λ to $\lambda + \delta\lambda$ can be associated with a strip of the star's photosphere having a radial velocity v to $v + \delta v$ relative to the average of the entire disk, corresponding to wavelength λ_o, where

$$\frac{\lambda - \lambda_o}{\lambda_o} = \frac{v}{c} \tag{1}$$

(c is the velocity of light).

In order to perform the necessary integration of surface flux we need to express the projected locus of the strip v = const. in the x,y plane. Now the radial velocity v, at the point x,y,z is given by the z component of the vector product of the angular velocity $\underline{\omega}$ and radius vector $\underline{r}$, i.e.

$$v = (\underline{\omega} \times \underline{r})_z \tag{2}$$

where we can expect that $\omega = \omega_o f(\varsigma^2)\hat{\varsigma}$, in general, $\hat{\varsigma}$ representing the direction of the polar axis in the y,z plane. Evaluating the required component we have, $v = \omega x \sin i$.

It is a convenient, and generally adopted, first approximation to regard $f(\varsigma^2)$ as a constant, in which case we have directly that

$$\frac{\lambda - \lambda_o}{\lambda_o} = \frac{\omega x \sin i}{c} \tag{3}$$

and the strips v=const correspond to the parallel strips x=const.

Let us return now to Figure 1 and consider the photospheric dark region S centered at x_o, y_o, z_o. This can be associated with a corresponding spectral feature $\sum$. The undisturbed absorption line profile $I_u(\lambda)$ becomes replaced with a new, featured profile $I_f(\lambda)$. Although we are allowing that it is only the region S which gives rise to the feature $\sum$, so that we may expect the shape of the rest of the line to remain undisturbed, in fact the continuum level I_c will itself be depressed due to the maculation light loss of S, so that I_f will actually refer to the reduced light level I_{cr}.

The photometric maculation effect $\Delta\ell_S$ from a feature such as S was already shown to correspond to an expression such as:

$$\Delta\ell_S = \ell_o(1 - \mathcal{H})\sigma \tag{4}$$

where ℓ_o is the "immaculate" light level, $\mathcal{H}$ is the "spot" to photospheric flux ratio and σ is the light loss function given as (Budding, 1977),

$$\sigma = \frac{3}{(3-u)}\left\{(1-u)\sigma^{o}_{o} + u\sigma^{o}_{1}\right\} \tag{5}$$

where σ^{m}_{n} is an integral of the form

$$\pi\sigma^{m}_{n} = \iint_{spot} x^{m}z^{n}dx\, dy \tag{6}$$

and u is the linear limb darkening coefficient.

Now the dependence of I_u on λ can be converted, through equation (3) into one on x, and we can write for the relative intensity of the line at some typical value x' along the profile,

$$I_u(x') = 2\int_{-1}^{1}\int_{o}^{\sqrt{1-x^2}} I'_x\,(y)dy\; r(x'-x)dx \tag{7}$$

where $r(x'-x)$ expresses the decrement of the suitably scaled intensity I'_x at the argument $(x'-x)$ associated with the inherent line formation process. We can express the equivalent width w_u of this feature by writing

$$w_u = \frac{1}{I_C}\int_{-\infty}^{\infty} I_C - I_u(x')dx' \tag{8}$$

or more fully

$$w_u = \frac{\int_{-\infty}^{\infty}\int_{-1}^{1}\int_{o}^{\sqrt{1-x^2}} I'_x\,(y)dy\;(1-r(x'-x))dx\;dx'}{\int_{-1}^{1}\int_{o}^{\sqrt{1-x^2}} I'_x\,(y)dy\;dx} \tag{9}$$

Clearly the inner integrals in (9) extend over the entire projected disk in the x, y plane. We may, for a convenient shorthand, apply the subscript D to the double (x, y) integral to denote an integration over the entire disk, S can be used in a similar way to denote the spot region, and $D-S$ the disk region excluding the spot. A similar procedure now follows for the equivalent width of the featured profile $I_f(\lambda)$, but here we introduce the quantity $\mathscr{H}_\lambda$, which gives the ratio of the surface flux in the region S to that of the surrounding photosphere, this time with particular reference to the spectral region of the line. For formal purposes we can take $\mathscr{H}_\lambda = 1$ except in the region S; there $\mathscr{H}_\lambda < 1$, and it is normally taken that $\mathscr{H}_\lambda$ is small and constant in S. We can then write

$$w_f = \frac{\iiint\limits_{D-S} I'_x (1-r)\, dx\, dy\, dx' + \mathscr{H}_\lambda \iiint\limits_{S} I'_x (1-r)\, dx\, dy\, dx'}{\iint\limits_{D-S} I'_x\, dx\, dy + \mathscr{H} \iint\limits_{S} I'_x\, dx\, dy} \tag{10}$$

where the function arguments have been omitted, for shorter writing.

Utilizing now the integration property that

$$\iint\limits_{D-S} I'\, dx\, dy + \mathscr{H} \iint\limits_{S} I'\, dx\, dy$$

$$= \iint\limits_{D} I'\, dx\, dy - (1 - \mathscr{H}) \iint\limits_{S} I'\, dx\, dy$$

$$= (1 - (1 - \mathscr{H})\,\sigma) \iint\limits_{D} I'\, dx\, dy \tag{11}$$

and noting that the outer integration (with respect to x') is independent of the limits of the inner areal integration, we can deduce that

$$w_f = w_u \frac{(1 - (1 - \mathscr{H}_\lambda)\sigma)}{(1 - (1 - \mathscr{H})\ \sigma)} \tag{12}$$

and infer that unless there is some particular reason why the flux ratio in the line, in moving from photosphere to spot, should be appreciably different from that in the surrounding continuum (–it appears to be a basic assumption of the original Doppler imaging procedure that $\mathscr{H}_\lambda \simeq \mathscr{H} \simeq 0$), then the equivalent width of a feature is not affected by the circumstance of maculation.

Note, though, that the relative variation of area of the line profile ΔA_Σ, due to the feature Σ, has a strictly parallel form to the corresponding expression for $\Delta \ell_S$ given in (4). i.e.,

$$\Delta A_\Sigma = A_o\ (1 - \mathscr{H}_\lambda)\,\sigma \tag{4a}$$

For the Sun we do know that certain strong absorption features are partially filled by chromospheric emission contributions, which correlate strongly with the maculation regions (—the "faculae"). Could not a corresponding situation (i.e., $\mathscr{H}_\lambda \neq \mathscr{H}$) be anticipated for certain lines in the spectrum of an "active" star?

The foregoing treatment, bringing up an equivalent width formulation, suggests a processing of narrow–band photometry (which measures equivalent width in terms of a proportional line index) comparable to the spot modelling teatment of broad–band photometry of the maculation effect as given by Budding and Zeilik (1987). (See also Budding and Marngus, 1980, for the comparable relationship of the eclipse effect problem.) Such procedures should also, in principle, be complementary at the same, or similar, source magnitude: S/N for simultaneous two channel narrow–band photometry having a comparable S/N to conventional (sequential) broad–band photometry.

At a more direct level, the areal effects of unambiguously spot related features in a set of line profile tracings, when they are all suitably scaled so that the unfeatured parts of the profile essentially overlap, could be modelled using exactly the same basic formulae as for the analysis of photometric maculation.

PROFILE FITTING

Vogt and Penrod (1983) addressed a somewhat different source situation, however, in that for certain suitably bright examples sufficient inherent S/N was present to enable profile fitting. Here would appear to be a distinct information advantage over maculation effect modelling, despite the implied significantly enhanced corresponding S/N for each single point in the light curve for the latter. In effect, the integration of photons that would give rise to such a well–defined photometric point for the latter case is bound to lose information, since a single point, no matter how precisely determined, can only yield a single parameter in principle; while the numerous less–well–defined points making up the line profile may be used to evaluate more than one parameter in a suitably well conditioned problem.

In order to express the form of the featured profile I_f, we depart from an expression similar to (7), but make use of the idea expressed in (11), i.e.,

$$I_f(x') = \int\int_{D-S} I'_x(y)\, dy\, r(x'-x)\, dx + \mathcal{H}_\lambda \int\int_S I'_x(y)\, dy\, r(x'-x)\, dx \tag{13}$$

As was previously remarked, the concept of Doppler imaging, as a practical procedure, on resolvable features in a line, has required that the Doppler effect be the predominating cause of line spreading, and much greater than the inherent broadening of the line as indicated by the function $r(x')$.

In this circumstance, we can replace $r(x')$ by the delta function ,$\delta(x')$ and so obtain

$$I_f(x') = \int_{D-S} I'_{x'}(y)\, dy + \mathcal{H}_\lambda \int_S I'_{x'}(y)\, dy \tag{14}$$

Clearly, in the range of abscissae excluding the region S the second integral is absent, and the first one reverts to the complete range of ordinates across the disk.

The form of the function I' can be spelled out without much difficulty according to the standard 'linear limb–darkening law' prescription. It is of the form

$$I' = \frac{3I'_o}{(3-u)}((1-u) + uz) \tag{15}$$

where u is the coefficient of limb–darkening, and $z = \sqrt{1-x^2-y^2}$ for the adopted spherical form of the photosphere.

In the profile region outside of $\sum$ we then find

$$I_f(x') = \frac{6}{(3-u)} I'_o \left\{ (1-u) \int_0^{\sqrt{1-x'^2}} dy + u \int_0^{\sqrt{1-x'^2}} \sqrt{1-x'^2-y^2}\, dy \right\} \tag{16}$$

$$= \frac{6}{(3-u)} I'_o \left\{ (1-u)\sqrt{1-x'^2} + \frac{\pi}{4} u (1-x'^2) \right\} \tag{17}$$

—the conventional "dish–shaped" profile, broadened by rotation.

In the profile region of $\sum$ we now have to define the range of integration of y. In the treatment of Budding (1977), the feature s is taken to have the form

$$\frac{(x_1-d)^2}{k'^2} + \frac{y_1^2}{k^2} = 1 \tag{18}$$

when referred to axes x_1, y_1, such that the centre of the elliptical outline of the spot lies on the x_1 axis at the distance d from the centre of the disk, with semi–major axis k and semi–minor axis k', where $k' = kz_0$, z_0 being the z–coordinate of the spot centre.

The rotation from the x, y system of Figure 1 to the x_1, y_1 system is actually the third rotation ψ about the z axis, discussed by Budding (1977), where the coordinates of spots in the plane perpendicular to the line of sight are related to the natural reference system of the star. In this way we are able to set quantities such as the longitude λ and latitude β of the spot centre. On this basis, we can determine that the positive rotation ψ turned in moving from the x. y to the x_1, y_1 system at phase ϕ and inclination i is given by

$$\cot \psi = \frac{\cos\,(\phi-\lambda)\,\cos\beta\,\cos i - \sin\beta\,\sin i}{\sin\,(\phi-\lambda)\,\cos\beta} \tag{19}$$

The line $x = x'$ in the x, y system transforms into the line

$$y_1 = mx_1 - c\,, \tag{20}$$

in the x_1, y_1 system, where $m = \cot\psi$ and $c = x' \;\; \mathrm{cosec}\psi$..

This will intersect the region S when the quadratic obtained by substituting (20) into (18), i.e.,

$$\frac{(mx_1 - c)^2}{k^2} + \frac{(x_1 - d)^2}{k'^2} - 1 = 0\,, \tag{21}$$

has two real roots. In these circumstances the roots in the x_1, y_1 system are given by

$$x_1^{(1,2)} = \frac{B \pm \sqrt{B^2 - AC}}{A}\,; \quad y_1^{(1,2)} = mx_1^{(1,2)} - c; \tag{22}$$

where $$A = \frac{m^2}{k^2} + \frac{1}{k'^2}\,;\quad B = \frac{mc}{k^2} + \frac{d}{k'^2}\,;\quad C = \frac{c^2}{k^2} + \frac{d^2}{k'^2} - 1\,. \tag{23}$$

Converting back to the x, y system then, the line $x' = \mathrm{const}$ intersects S at the points $y'_{1,2}$ where

$$y'_{1,2} = x_1^{(1,2)}\,\sin\psi + y_1^{(1,2)}\,\cos\psi\,. \tag{24}$$

We can now turn to the integral for I_f in the region containing Σ to find

$$I_f(x') = \frac{3}{(3-u)} I'_o \left\{ (1-u) \left[2\sqrt{1-x'^2} - (1-\mathscr{H}_\lambda) \int_{y'_1}^{y'_2} dy \right] \right.$$

$$\left. + u \left[\frac{\pi}{2}(1-x'^2) - (1-\mathscr{H}_\lambda) \int_{y'_1}^{y'_2} \sqrt{1-x'^2-y^2}\, dy \right] \right\}$$

$$= \frac{3}{(3-u)} I'_o \left\{ (1-u) \left[2\sqrt{1-x'^2} - (1-\mathscr{H}_\lambda)(y'_2 - y'_1) \right] \right.$$

$$\left. +u(1-x'^2) \left[\frac{\pi}{2} - \frac{(1-\mathscr{H}_\lambda)}{2} \left\{ \cos^{-1}\left(\frac{y'_1}{\sqrt{1-x'^2}} \right) - \cos^{-1}\left(\frac{y'_2}{\sqrt{1-x'^2}} \right) - y'_1 z'_1 + y'_2 z'_2 \right\} \right] \right\} \tag{25}$$

where the $y'_{1,2}$ are given by (24) and the $z'_{1,2}$ correspond to the z–coordinates for $x', y'_{1,2}$. Of course formula (25) will dovetail into formula (17) at either end of the profile region Σ.

OVERVIEW OF THE TECHNIQUES

With reference to the foregoing formulation for the fitting function to the featured line profile, we may note that its integral, in the x–direction, according to (11), involves the σ–integrals. Conversely, we could expect that some suitably weighted derivative of the photometric maculation wave function could imitate the profile function, insofar as the σ–integral's dependence on phase can parallel variation of the integration region S, at a given phase, with respect to its limits in x.

Essentially, the final curve fitting problem, whether one is dealing with a well–defined light curve or a well–defined line profile, would be amenable to quite the same methodology. In particular, the information content of a line profile may be specified by reference to the number of positive eigenvalues of the curvature Hessian derived at the optimum solution. Since a single profile scans effectively one complete hemisphere, one might expect the information content to be comparable to that of half a complete light curve on average, for a given level of probable error in the data set. With several profiles from different lines and at different phases one can expect to be able to resolve an increased number of parameters.

Also, as Vogt and Penrod (1983) point out, Doppler imaging can be profitably employed in somewhat different geometrical (non–eclipsing) circumstances than has often obtained in the light curve modelling of most (eclipsing) RS CVn stars.

Such matters are better clarified by actual experiment, but in any case, the theoretical information advantage of a single profile must lose practical significance as one moves to fainter stars, in the same way that a high S/N broad–band light curve of a certain star can be expected to yield intrinsically more information than that of another, say five magnitudes fainter. Budding and Zeilik (1987), and more recently Zeilik et al. (1988) considered what kind of parameter yield can be reasonably associated with broad–band photometry of a given S/N. For S/N values of around 100, a maculation wave can generally determine at least 3, sometimes 4 or 5 independent parameter values. More parameters can be specified, of course, but, apart from where independent evidence can fix values, such an overspecification can be associated with the "non–uniqueness" ambiguities referred to by Vogt and Penrod (1983). Thus the derivative curve of the broad–band photometric curve may well show various informative seeming "features". Whilst a number of parameters may be required to enable a fitting function to match these features, the real information content of the data cannot be increased by the operation of differentiation.

Finally we should like to stress the potentially useful role of narrow–band photometry to supplement both profile Doppler imaging and maculation wave photometry. In the former case it should monitor the extent of profile distortion from faculae as distinct from the dark spots. In the latter case it should allow the possibility of some confirmation that continuum and line distortions, associated with stellar "activity" are indeed correlated, as expected.

Figure 1.

A rotationally broadened but otherwise undisturbed spectral absorption line I_u (λ) is modified by the presence of a feature ∑, which maps the photospheric "spot" *S*. The resulting featured profile I_f (λ) is shown scaled to the same continuum level Ic, though actually there would also be a reduction of the continuum intensity to Ic_1 due to the photometric maculation effect of *S*. The photosphere is that of a star rotating with angular velocity $\underline{\omega}$ about the axis *OP* inclined at angle *i* to the line of sight.

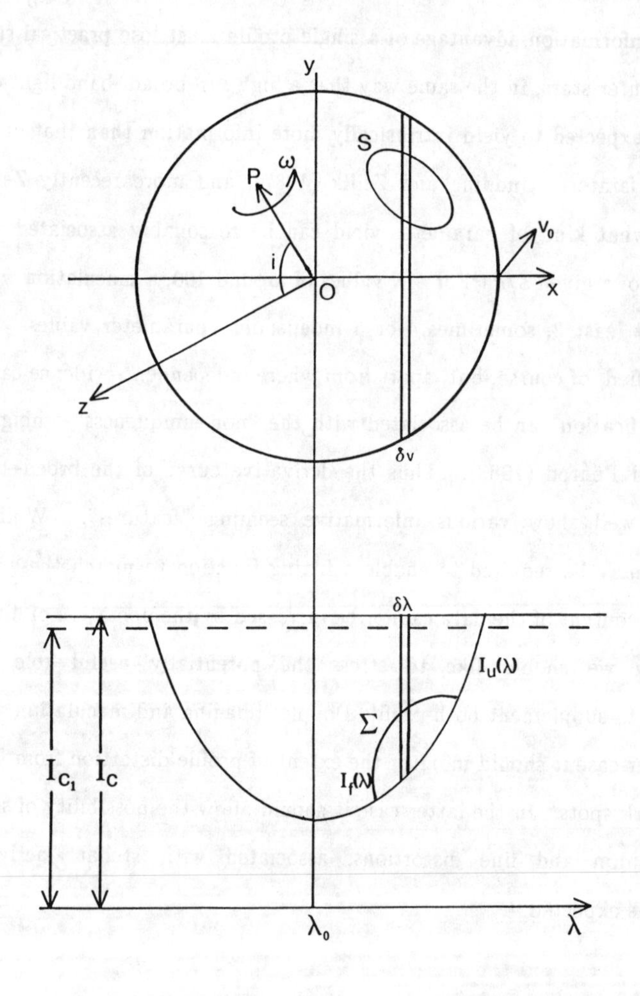

This work was supported in part by National Science Foundation grants NT–8722859 and AST–8903174 to MZ.

References

Budding, E.: 1977, Astrophysics and Space Sci. 48, 207.

Budding, E. and Zeilik, M.: 1987, Ap. J. 319, 827.

Fekel, F.C., Quigley, R., Gillies, K., Africano, J.L. : 1987, Astron, J. 94, 726

Kopal, A.: 1959, Close Binary Systems, Champman and Hall, Landau.

Vogt, S. S. and Penrod, G. D.: 1983, P.A.S.P. 95, 565.

Zeilik, M., DeBlasi, C., Rohodes, M. and Budding, E.: 1988, Astrophys.J. 332, Q93.

PERSPECTIVES FOR GROUND-BASED AND SPACE RESEARCH ON ACTIVE CLOSE BINARIES

Bernard H. FOING

ESA/ESTEC Space Science Dept, postbus 299, 2200 AG Noordwijk, Netherlands
and Institut d'Astrophysique Spatiale (IAS), BP 10, 91371 Verrières-le-Buisson, France

Summary. We review some aspects for diagnostic of magnetic activity phenomena in late-type stars and close binaries. This includes optical photometry and spectroscopy at various spectral and time resolution. Also time resolved spectroscopy in the UV, X-ray and radio range, obtained in the frame of multiwavelength multi-site observation campaigns is required for the observation of transient phenomena on those objects at short time scales. We review also briefly the prospects of some future space missions in which the European Space Agency is involved. We stress those missions that will address astrophysical problems relevant to the study of close active binaries.

Partly based on observations obtained at ESO, CFH, OHP observatories and with the IUE and EXOSAT satellites.

1. Magnetic activity in solar-like stars and close binaries

The presently available ground-based and space observations have already provided important results on stellar activity, such as averaged physical parameters of large spotted areas, plages and activity cycles. Further questions concern: 1) the physical differences between solar and stellar phenomena, 2) what is their variation on the H-R diagram, 3) the relation between magnetic fields, activity and global stellar parameters, 4) how they influence the evolution by affecting the rotation history and the mass loss through winds and mass ejections, and 5) how this can perturb their interstellar environment; 6) Surface structures; 7) what is the temporal behaviour of spots, plages active structures and of their collective properties that give rise to activity cyles; 8) how this is linked with internal properties, and coupled to the dynamo

C. İbanoğlu (ed.), Active Close Binaries, 845–857.

generation of magnetic fields.

The subsurface source of activity can be studied by separating the dependence on spectral type (convection zone depth), rotation, age and evolution. Different kinds of activity processes on other stars should be studied especially in the extreme cases of very thin convection zones (minimal expected dynamo) and of magnetic field covering nearly all of the stellar surface. Also, the effect of interconnecting magnetic fields in close binary systems should be considered.

Dwarf novae are close binary systems including a late-type (K-M) filling its Roche lobe and transferring matter, via an accretion disk onto a white dwarf. If the white dwarf has a magnetic field, the magnetosphere can inhibit the formation of a an accretion disk, and an accretion column is formed along the field lines. The variability occurs on various time scales from seconds to years, and affect all wavelengths, and can originate from the white dwarf, the accretion disk or from the accretion column in the case of polars. High speed photometry of these systems over long time allows to measure the history of the white dwarf spinning, phase jumps and dynamical effects in the column/disk or the orbit. Accretion disks may be the seat of a dynamo, and be surrounded by high temperature plasmas (cf Kuijpers in this conference). The problematics and required diagnostics of outbursts, flares may use similar observing methods than for solar-like activity phenomena. Optical, UV and X-ray time resolved spectroscopy allows to specify the energy budget, and to investigate different mechanisms proposed for the soft X-ray emission of these objects.

2. Spectral diagnostics

2.1. Photometric diagnostics

Since the bulk of photospheric radiation from late-type stars is emitted in the optical and infrared range, sufficient contrast is required to detect and quantify activity phenomena and flares against this background. Whereas solar optical white-light flares are rare events, broad-band optical enhancements are observed regularly in M-dwarf flare stars and sometimes in RSCVn stars, due to the faint background, especially in the U band. Thus, U-band enhancements up to few magnitudes are observed while I-band enhancements are only a few hundredths of magnitude. A very accurate photometry is also necessary to study the rotational modulation by active structures. On ground, the photometric precision is limited by the atmospheric scintillation for bright stars (up to 8th magnitude for 1m telescope) and by photodetection noise for fainter stars. For stars of 10th magnitude single-measurement are currently obtained with normalised precision in 10s of 5 10^{-3} for single-beam photometer and of 5 10^{-4} in two-beams photometry in good conditions. Thus photometric modulation lower than 1% are difficult to detect and characterize, and therefore the detection of small solar-type spots on stellar surfaces is beyond the capability of ground-based photometry. Systematic monitoring with space-borne instruments would bring a breakthrough in the measurement of activity phenomena that have escaped detection. They could reach a normalised precision in 10s of 5 10^{-5}, also adequate for asteroseismological studies.

During the impulsive phase of solar white-light flares, the optical continuum shows a strong Balmer discontinuity, and is generally flat during the decay phase. Rust (1986) attributes the impulsive spectrum to heating in the lower chromosphere producing H free-bound and free-free emission, while the decay phase emission is produced by the H^- continuum. The mechanism by which the strong soft X-rays could penetrate the upper photosphere , ionize H and produce sufficient H^- emission could be tested by detailed simultaneous optical and X-ray observations.

2.2. Optical spectroscopy

Optical spectroscopy at low resolution permits a more accurate investigation of the free-free, and the Paschen and Balmer bound-free continua. Flare spectra at medium resolution provide measurements of line fluxes. Balmer decrements may be used to diagnose typical temperatures and densities in active regions or during flares (Butler et al, 1987). The time behaviour of different line fluxes can be followed. Chromospheric line fluxes strengthen during flares and decay more slowly than the optical continuum. Some constraints on the temperature and density structure can be obtained from line fluxes, especially in the optical range from Balmer lines, He and higher excitation lines. The role of photoionisation and photoexcitation of these lines by soft X ray and XUV radiation should be estimated . The broadening and merging of higher Balmer lines dominated by Stark effect can be used to estimate electron densities in the chromosphere (Donati-Falchi et al, 1985). Measurement of line broadening of the lower Balmer lines, which are less affected by Stark effect, together with line shifts, provide information on the large-scale flows in plages or during flares.

2.3. Optical diagnostics at high spectral resolution

High spectral resolution allows measurements of chromospheric indicators, and also to measure magnetic fields through the Zeeman broadening of lines (Saar et al 1986, see also Foing in this conference) in the optical and the infrared. Polarisation measurements can be also used for the derivation of magnetic fields, e.g. using Zeeman-Doppler techniques (Donati et al 1988). Also methods of rotational modulation of spectroscopic changes give information on the distribution of surface structures on active stars by Doppler or spectral imaging (cf Vogt Penrod 1983, Vogt et al 1987, Jankov and Foing 1987, Neff, Foing in this conference). At high spectral resolution (20,000-100,000) and reasonable signal to noise (>50), optical spectroscopy allows the study of: i) broadening effects in the Balmer lines, He I and Ca II lines; ii) velocity fields associated with thermal, turbulent or directed motion; iii) and the time and velocity dependent signature of ejected components in emission or in absorption. Also high spectral resolution is required to separate blends by photospheric lines or molecular bands which are important for cool stars, and allows us to measure the core filling and activity or flare signature in photospheric lines. The observed chromospheric profiles can be compared with semi-empirical calculations using atmospheric models with different temperature stratifications. Those models , when consistent with the observations, can provide information on the depth of formation of each line or spectral feature, and also indicate the likely excitation and ionisation mechanisms. The presence of velocity gradients can be diagnosed from line shifts, asymetries and detailed line profiles but this diagnostic is crude because of the interplay between the plasma geometry in 3 dimensions, the dynamics, the NLTE radiative transfer effects and redistribution effects . However, it is hoped that a new generation of semi-empirical models which include the coupling between dynamics and radiation selfconsistently will shed light on the principal mechanisms at work that are responsible for atmospheric response to activity and flares.

2.4. Ultraviolet spectral diagnostics

There is an enormous literature on UV diagnostics of solar/stellar activity and flares (cf Feldman (1981), Dere and Mason (1981) and Dupree (1978) for these aspects). For collisionally excited resonance lines, the surface flux is related to the emission measure EM= $\int_{\Delta T} Ne^2 dh$ over the temperature range ΔT of line emission. For intersystem lines which are collisionally excited but depopulated by line radiation and collisions (proportional to the electron density), the observed line ratios provide a measure of electron densities and thus the emitting volume at the temperature of line formation . The presence in the UV spectrum of lines formed over a wide range of temperature and excited by different processes permits us to infer the distribution of emission measure with temperature and to constrain assumptions about the geometry, temperature distribution and electronic density stratification.

Optically thick chromospheric resonance lines and other transitions have been computed using various non-LTE codes in order to characterise the temperature and density distribution in active stars and flares. Rapid increases in UV line fluxes and continua have been detected during flares on dMe, RS CVn and other flare stars. In particular, UV continua as observed on the sun at 1600 Å and 2200 Å allow us to probe the temperature minimum region and middle photosphere, which play important roles in the energy balance of active regions and flares (cf Foing et al 1986a, 1986b).

Because of its limited sensitivity and operational constraints, IUE cannot observe fast phenomena such as stellar flares with time resolution less than 10 minutes . The Goddard High Resolution Spectrograph and the Faint Object Spectrograph on Space Telescope will improve the sensitivity, time resolution and spectral resolution significantly over that of IUE but with a decreased spectral range and scheduling flexibility. Also the large oversubscription will make it difficult to monitor flare stars or to monitor the variability associated with surface or circumstellar activity, and to coordinate multi-wavelength observations. Other instruments such as the ST Imaging Spectrograph or the LYMAN mission will extend the spectral range. With its 900-1200 Å prime spectral region and 30,000 resolution, LYMAN will observe the H Lyman series, the important O VI, S VI, C III lines at temperatures between $2\ 10^{5}$ and 10^{6} K, and include lines of the ions N I-III, P II-V, S III-VI, C I-IV. Useful density ratios are available for several of these ions. Also the coverage of the 100-912 Å range will extend the temperature range up to $2\ 10^{7}$K by observation of ionisation stages of Fe II-XXIV , and of the He I and He II Lyman series. Simultaneous coverage of the complete 1200-3200Å range is also required in a future successor to IUE. USSR instruments such as SUVT-170 planned for a 1995 launch with spectrometers covering the 1100-1900Å and 1900-3500Å ranges with resolution modes of 0.1, 3 and 30 Å, or the EUVITA instrument on SPECTRUM X planned for 1993 should be useful for flare studies.

These future satellites with their anticipated UV coverage, spectral resolution and temporal resolution of 10-100 s, should permit studies of active surface structures, thermal conduction, chromospheric evaporation, mass ejections, flare expansion, radiation and dynamics on a number of active close binaries and flare stars.

2.5.X-ray spectral diagnostics

X-ray photometry and spectroscopy can be used to study the 10^{6}-10^{8} K plasmas in stellar active regions and flares. Calculations of the emergent spectrum show the dominant role of emission lines for T< $2\ 10^{6}$ and of the bremmsstrahlung continuum at higher temperatures.

Low resolution spectroscopy (E/ΔE= 10-30), as was achieved with the Einstein Solid State Spectrometer, allows one to match an observed spectrum with a theoretical spectrum from a 2 temperature plasma folded with the instrumental response. This technique can also be applied for the analysis of low resolution spectra to be obtained with the JET-X instrument onboard SPECTRUM X , and later with higher throughput and time resolution by instruments on AXAF and by XMM/Focal Plane Imager. Moderate resolution spectroscopy (E/DE= 100-300) such as with the Objective Grating Spectrometer (OGS) on Einstein and Transmission Grating Spectrometer (TGS) on Exosat allowed to resolve spectral lines and to infer EM(T) distribution for the coronae on Capella and σ^2 CrB (Mewe et al, 1982). Also density-sensitive ratios from He, Be, C isoelectronic sequences allow us to infer densities and thus emitting volumes of the hot flaring plasma. With the enormous throughput of grating spectrometers on AXAF and XMM, moderate resolution with 100s time resolution of bright flares on dMe and RSCVn systems should be feasible (Linsky, 1987)

2.6. Radio observations

The usefulness of the radio spectral region for studying the hot thermal plasma and nonthermal electrons in coronal active region and flare plasmas is described in Kuijpers (1989) . The spectral, temporal and polarisation properties of different emission mechanisms, both coherent or incoherent are described in reviews by Kuijpers (1985), Dulk (1985), Melrose (1987).

Gyrosynchroton emission from mildly relativistic electrons in magnetic loops was invoked by Linsky and Gary (1983) for dMe stars and also by Mutel et al (1985) to explain VLBI observations of RSCVn systems. However radio emission during flares with brightness temperature > 10^{13} K (confirmed from spike rise times faster than 0.2 s in AD Leo, Lang et al 1985) and 100 % circular polarisation is explained as a coherent process such as electron cyclotron maser or plasma radiation. Recently Bastian and Bookbinder (1987) obtained the first dynamic spectra of flares on UV Cet analogous to radio bursts observed on the sun, thus inaugurating a new tool for radio flare studies.

3. Multiwavelength diagnostics

For solving the scientific questions concerning magnetic activity and flares, diagnostics are available in different wavelength ranges and at various spectral resolutions. These tools help to describe some of the existing processes at work during flares and to quantify some physical

properties of flare plasmas. In order to give a more complete description at different heights and temperature regimes in the flaring atmosphere, these diagnostics should be used simultaneously especially for flare or fast events (Foing et al 1988).
Coordinated multiwavelength campaigns involving future X-ray and UV satellites and ground based instrumentation with a good simultaneity and continuous coverage are needed to extend our understanding of stellar flares. Such campaigns will require special efforts for organisation, observations, calibration, data analysis and theoretical interpretation.

4. Space projects relevant for active close binary research

4.1 IUE

The international Ultraviolet Explorer is still offering outstanding opportunities for studying stellar magnetic activity, chromospheric/transition region structures and flares through diagnostics in the wavelength region 1200-3200 A. For stars, the 1200-3200 Å ultra-violet range covered by the IUE spectrograph is rich in spectral emission lines of species such as C I, O I , Si II, Fe II formed at 4000-6000 K at the base of the chromosphere, the Ly α line and the C II 1335Å at top of the chromosphere, and lines of Si III, C III, Si IV, O IV, C IV and N V formed in the transition region at 30,000-150,000 K.

4.2 HIPPARCOS

The Hipparcos mission aims at measuring the position of 120000 positions of stars brighter than 13, by covering simultaneously two 0.9x0.9 square degrees fields of view separated by 58 degrees, and using a modulating grid with step 8.2 micron through which the measured stars are scanned . Several groups where set to prepare the INPUT catalogue, to realise the data reduction either by the Northern Data Analysis or by the FAST consortia , and to achieve photometric analysis (TYCHO reduction) .

The Hipparcos satellite was put into space on 8 August 1989. Unfortunately, the solid apogee motor , the Mage-2 engine, which was supposed to place the satellite in a geostationary orbit, failed to ignite. The most likely cause, after current investigation is in one of the pyrotechnical elements. In the meantime the spacecraft has been placed in an elliptical orbit with a perigee of 525 km and an apogee of 36000 km, where it has started its scientific mission. The present orbit is stable over many years. The major lifetime limiting factor is the degradation of the solar arrays. With 3 additional ground stations for data recovery, the expected accuracy for the revised mission (if lifetime 18-30 months) is for the measurements of position (4-3 marcsec), parallaxes (6-4 marcsec) and proper motions (8-5 marcsec), while the nominal 30 months mission on geostationary orbit predicted a 2 marcsec outcome for all these values. A mission duration of less than 18 months do not allow decoupling contributions of parallaxes and proper motions, and thus the ability to survive long solar eclipse periods and the overall mission lifetime are critical.

4.3 HUBBLE SPACE TELESCOPE

The Hubble Space Telescope (HST) built by NASA , with a 15% contribution of ESA (Faint Object Camera, Solar arrays, support of the ST science institute and ST European Coordinating facility), is right now planned for launch on 26 March 1990 at Kennedy Space Center. The 2.4m HST telescope has a f/24 Ritchey-Chretien Cassegrain configuration that allows the use of several instruments with very different fields of view, wavelength and spectral resolution, and limiting magnitudes (cf table 1).

Table 1.: parameters of Hubble Space Telescope instruments

Instruments	Field of view	Wavelength (in nm)	Spect.Res.	Lim.Mag.
Wide field /planetary Camera	2.7x2.7 arcmin 1.2x1.2 arcmin	115-1100		9-28
Faint Object Camera	22"x22", $[11"]^2$ $[4"]^2$, 20"x0.1"	115-600		20-28
Faint Object Spectrograph		115-850	100-1000	9-26
High resolution Spectrograph		115-320	$10^3, 10^4, 10^5$	17
High Speed Photometer		115-700		0-24
Fine guidance sensor				

4.4 ROSAT

The launch of the ROSAT satellite, a joint german/UK/US mission, with a Delta II is foreseen in early 1990. This satellite will perform the first all-sky survey with imaging X-ray and XUV telescopes during the first six months, and pointed observations of selected targets with unprecedented angular/spectral resolution and sensitivity. The simultaneous operations of ROSAT, IUE and HST supply a unique opportunity to obtain simultaneous X-ray, XUV and UV observations for known X-ray sources.

4.5 ULYSSES

This ESA/NASA mission will be the first spacecraft to perform an exploration of the inner heliosphere covering the full range of heliographic latitudes, including the solar-polar regions. Specific topics to be adressed are: the three dimensional structure of the solar wind and heliospheric magnetic field, solar radio bursts and heliospheric plasma waves, solar hard X rays, the propagation and acceleration of solar energetic particles and galactic cosmic rays, the distribution of interplanetary/interstellar neutral gas and cosmic dust. Other aspects concern the science during Jupiter flyby, the study of cosmic gamma ray bursts and the search for gravitational waves. Observations of the solar heliospheric plasmas are useful for understanding similar physics in other astrophysical objects, and in particular in close binaries, active stellar environments or "asterospheres".

5 The SOLAR-TERRESTRIAL SCIENCE PROGRAMME

CLUSTER and SOHO are a joint venture between ESA and NASA as part of the STSP Programme, itself related to a broader Inter Agency coordination of missions of the next decade (from US, ESA, USSR, Japan,...) dedicated to the study of Solar-Terrestrial Science.

5.1 CLUSTER

The scientific objectives of this mission are also of interest for the plasma physics in other astrophysical contexts. They include the study of boundaries between plasmas, the processes for mass/momentum transfer , magnetic reconnection, plasma acceleration, MHD turbulence, shock waves, microstructures of the solar wind. The mission makes use of 4 identical spacecrafts to provide gradients on different time and length scales. The launch will be done by Ariane 5 in 1995/1996. On the 900 kg spacecraft, the 65 kg payload include magnetometers, field measurements, sounders, and analysers for electrons, ions and energetic particles.

5.2 SOHO: the Solar Heliospheric Observatory

This mission on a 3 axis stabilised satellite, is to be launched in 1995 in the L1 Sun-earth Lagrangian point. Its main objectives are 1) the study of the solar corona structure, heating and dynamics, and its expansion into the solar wind , and 2) the probing of the solar internal structure and dynamics from its core to the photosphere by helioseismological measurements and the monitoring of the solar irradiance variations.

5.2.1 SOHO Coronal/particle instruments

The coronal/particle instruments use remote sensing techniques (plasma diagnostics of densities, temperatures and flows from images and spectra) , or in-situ measurements (composition, charge, energy of particles) for studying structures and dynamics of the corona and heliosphere. The coronal/particle instruments include:

Table 2: Parameters of SOHO coronal/particle instruments

Instruments	Principle	Angular	Wavelength	Spectral res.
SUMER	Normal incidence spectrometer	1" res	50-160 nm	20000-40000
CDS	Grazing Incidence Spectrometer	2" res	17-60 nm	5000
EIT	Multi-Layer telescope Images	Full sun, 5" res	HeI, Fe IX, XII, XV	
UVCS	Coronograph Spectrometer	1.3-10 Ro	EUV lines	
LASCO	3 coronographs	1.1-30 Ro	optical	
SWAN	H cell spectrometer		Ly alpha	
CELIAS	Electrostatic, time of flight,SSD		ions (0.1-1000keV/e)	
COSTEP	Electrostatic, SSD (solid state detector)		protons, alpha, ions, electrons	
ERNE	SS, crystal scintillator, energy spectrum		ions (50-500MeV), electrons	

5.2.2 SOHO Helioseismology instruments

Seismology allows the investigation of internal properties of stars. On SOHO, this is aimed at by uninterrupted monitoring in intensity and velocity at very small amplitudes ($\delta I/I=10^{-6}$ or δV=1mm/s) for the detection and measurements of long period global oscillations giving at low order modes, information on the solar core. Measurements of intermediate or high degree modes investigate the convection zone properties. Also, solar irradiance variations give information on activity related phenomena, such as the blocking of energy by sunspots in the subphotosphere. These helioseismological investigations can give significant information on

the internal structure and rotation of the sun, as a paradigm for theories of stellar structure and evolution, and for models of dynamo and rotation.
The SOHO helioseismological instruments include (cf table 3) :

Table 3: Parameters of SOHO helioseismological instruments

Instruments	Principle	Signal	Angular information
GOLF	Na vapour resonance cell	velocity and B field	global degree 0-4
	Doppler shift measurements		
VIRGO	Global photometer	irradiance	global degree 0-4
	Active cavity radiometer	solar constant	
	Low resolution Imager	radiance on 12 pixels	global deg 0-7
MDI	Solar Oscillation Imager		degree l=4500
	Fourier tachometer		4 and 1.5 " resolution

6. XMM High- Throughput X-Ray Spectroscopy Mission

This misssion (the second cornerstone of ESA Horizon 2000 plan) is to be launched in 1998 on a highly eccentric 24 hours period orbit, to provide an X-ray observatory for 2 to 10 years. The science objectives include the study of : 1) large scale hot structures, 2) active galactic nuclei, 3) the hot interstellar medium, 4) the characteristics of degenerate stars (white dwarfs, neutron stars, black holes) and of the environment in close binary systems (time resolved X-spectroscopy of accretion and boundary layers), 5) the sample of X-binaries at distances until the limits of the Local group, 6) stellar coronae (with spectral and time resolution) in a huge sample of stars (allowing a thorough analysis of astrophysical parameters dependence) and 7) X-ray variability over time scales from minutes to years. The 3 collecting XMM-modules are made of 58 nested shells (of grazing-incidence Wolter I type) giving a field of view of 30 arcmin and a resolution of 30 arcsec. Instruments for the XMM mission have been selected by ESA in 1989.

Table 4: Main parameters of XMM instruments

FPI	Broad band spectro-imaging	CCD cameras	0.2-10 keV	$E/\Delta E$=50
TGS	Medium resolution spectroscopy	reflection grating+CCD	0.2-2.5keV	$E/\Delta E$=500
OM	Optical monitor		200-600nm	$M_V$lim=25

7. Requirements for future observations of active close binaries

For the diagnostics of enhanced activity and flares, we have seen that multi-band photometry over the whole range (radio, infrared, optical, ultraviolet, X ray) with time resolution as high as 0.1-1s can be necessary for the timing of the flare impulsive phase. Medium-resolution spectroscopy in the optical and ultraviolet with a time resolution of 10-30s is required to study line fluxes and continua diagnostics, for emission measure analysis, and to evaluate density-sensitive line ratios. High-resolution spectroscopy of line profiles provides insight into the plasma dynamics, and the radiative and excitation mechanisms .

All of these tools covering the full electromagnetic range must be used simultaneously with enough time resolution to study the plasma from active regions and flares over their complete temperature range and including nonthermal high energy components. The development of new instruments and satellite missions must take into account this multiwavelength coverage requirement. The need for simultaneous and continuous coverage requires the organisation of coordinated multiwavelength observing campaigns (Foing et al, 1988), and the use of networks of photometers and spectrometers around the globe such as the MUSICOS network (MUlti SIte COntinuous Spectroscopy) (Catala & Foing, 1988). An important ingredient is the knowledge of the magnetic field distribution and the indirect imaging of active structures and large scale motions, for studying build-up conditions for active regions and flares. Finally, a strong interaction between solar and stellar flare physics, together with atomic, plasma, radiative transfer physics is necessary . Those communities should jointly develop diagnostic methods, interpretative tools and theories for understanding the new observational results made possible with the next generation of space and ground-based instruments.

References

Bastian, Bookbinder: 1987, *Nature* 326, 678

Butler, C.J., Rodonò , M., Foing, B.H. and Haisch, B.M.: 1986, *Nature* 321, 679

Butler , C.J., Doyle, J.G., Foing, B.H. and Rodonò , M.: 1987, in *Tromso Midnight Sun Conference on Stellar activity,*

Butler, C.J., Rodonò , M. and Foing, B.H.: 1988, *Astron. Astroph. Lett.*. 206, L1

Catala, C., Foing, B.H., eds: 1988, *"1st MUSICOS Workshop on MUlti SIte COntinuous Spectroscopy"*

Dere and Mason :1981, in *"Solar Active Regions"*, ed F.Q. Orrall

Donati, J.F., Semel,M., Praderie,F.: 1988, in *"Ist MUSICOS Workshop on MUlti SIte COntinuous Spectroscopy"*, eds C. Catala and B.H. Foing, p.37

Donati-Falchi, A., Falciani, R., Smaldone, L.A.: 1985, *Astron. Astroph.* 152, 165

Dupree, A.K.: 1978, *Adv. Atomic and Molecular Physics* 14, 393

Dulk, G.A.: 1985, *Ann. Rev. Astron. Astroph.* 23, 169

Feldman: 1981, *Physica Scripta* 24, 681

Foing, B.H. et al: 1986a in *'The Lower Atmosphere of Solar Flares'* , NSO/SPO, ed D. Neidig, p.319

Foing, B.H., Bonnet, R.M., Bruner, M.: 1986b, *Astron. Astroph.* 162, 292

Foing, B.H., Butler, C.J., Haisch, B.M., Linsky, J.L., Rodonò , M. : 1988, in *"Coordination of Observational Projects"* eds. Jaschek,C. and Sterken, C., p.197

Foing, B.H.: 1990, NATO/ASI Kusadasi conference on *Active Close Binaries*

Jankov, S. and Foing, B.H.: 1987, in *"Cool Stars, Stellar Systems and the Sun"*, eds J.L.Linsky and R.E Stencel

Kuijpers, J.: 1985, in *"Radio Stars"*, eds Hjellming and Gibson (Dordrecht-Reidel), p. 185

Kuijpers, J.: 1989, *Solar Phys.* 121, 163-185

Lang, K. et al: 1985, in *"Cool Stars, Stellar Systems and the Sun"*, Santa Fe

Linsky, J.L. and Gary, D.E.: 1983, *Astrophys. J.* 274, 776

Linsky, J.L.: 1987, *Astro. Lett. and Comm.*, 26, 21

Melrose, D.B.: 1987, in *"Cool Stars, Stellar Systems and the Sun"*, eds J.L. Linsky and R.E Stencel, p. 83

Mewe, R. et al: 1982, *Astrophys. J.* 260, 233

Mutel, R., Lestrade, J.F., Preston, R.A. and Phillips, R.B.: 1985, *Astrophys. J.* 289, 262

Neff, J.: 1990, NATO/ASI Kusadasi conference on *Active Close Binaries*

Rust, D.: 1986, in *'The Lower Atmosphere of Solar Flares'* , NSO/SPO, ed D. Neidig

Saar, S.H., Linsky, J.L. and Beckers, J.M.: 1986, *Astrophys. J.* 302, 777

Vogt, S.S. and Penrod, G.D.: 1983, *Publ. Astr.Soc. Pacific,* 95, 565

Vogt, S.S., Penrod, G.D and Hatzes, A.P.: 1987, *Astrophys. J.* 321, 496

UNIFIED TREATMENT OF SATELLITE MOTION IN VARIOUS POTENTIAL FIELDS

B. ZAFIROPOULOS
Department of Physics, University of Patras, Greece
and
F. ZAFIROPOULOS
Department of Mathematics, University of Patras, Greece.

ABSTRACT. A general method for obtaining the perturbed elements of a satellite, moving in various forms of axi-symmetric potentials, is presented in this investigation. The first section contains various examples of axi-symmetric potentials. This section is followed by the explicit expressions for the disturbing accelerations. The procedure applied, in order to obtain the elements of the orbit, is outlined in the final part of this paper.

1. AXI-SYMMETRIC POTENTIALS

The figure of the Earth and other planets can be approximated by the form of a spheroid. All major planets in the solar system are known to have extended systems of rings. Disks and shells surrounding stars, binaries, or even galaxies are also known to exist.

The potentials of all the above mentioned celestial bodies possess an axis of symmetry. In what follows we present various forms of axi-symmetric potentials:

(i) Ring Potential. For a circular ring of radius A, the potential at a distance r greater than A, and co-latitude ϑ is (Taff, 1985; p.27)

$$U = -\frac{\mu}{r}\left[1 - \sum_{\nu=1}^{\infty} \frac{(-1)^{\nu}\,(2\nu-1)!!}{(2\nu)\,!!}\left(\frac{A}{r}\right)^{2\nu+1} P_{2\nu}(\cos\vartheta)\right], \qquad (1)$$

where $P_{2\nu}$ stand for Legendre polynomials, μ denotes the product of the ring's mass times the gravitational constant, and

$$(2\nu-1)!! = (2\nu-1)(2\nu-3)(2\nu-5)\ldots1. \qquad (2)$$

(ii) Disk Potential. The potential of a circular disk of radius A is given by the expression

C. İbanoğlu (ed.), *Active Close Binaries*, 859–863.

$$U = -\frac{2\mu}{r}\left[\frac{1}{2} - \sum_{\nu=1}^{\infty} \frac{(-1)^{\nu}(2\nu-1)!!}{(2\nu)!!}\left(\frac{A}{r}\right)^{2\nu} P_{2\nu}(\cos\vartheta)\right], \quad (r>A), \tag{3}$$

where the various symbols are defined as in the previous example (i).

(iii) Oblate Spheroid Potential. The gravitational potential of a homogeneous oblate spheroid at a point with distance r greater than the semi-major axes A is (MacMillan,1958;p.363)

$$U = -\frac{3\mu}{r}\left[\frac{1}{3} + \sum_{\nu=1}^{\infty} \frac{(-1)^{\nu}}{(2\nu+1)(2\nu+3)}\left(\frac{A\varepsilon}{r}\right)^{2\nu} P_{2\nu}(\cos\vartheta)\right], \tag{4}$$

where ε is the eccentricity of a meridian section. Similar expression is valid for the potential of a prolate spheroid.

(iv) Homogeneous Hemispherical Potential. For a hemisphere of radius A, the gravitational potential at distances $r>A$ is (cf., e.g., Mac-Millan, 1958; p.405, or Ramsey , 1981;p.160)

$$U = -\frac{\mu}{r}\left[1+3\sum_{\nu=0}^{\infty} (-1)^{\nu} \frac{(2\nu-1)!!}{(2\nu+4)!!}\left(\frac{A}{r}\right)^{2\nu+1} P_{2\nu+1}(\cos\vartheta)\right]. \tag{5}$$

(v) Hemispherical Shell Potential. The potential of a hemispherical surface of radius A has the form (Kellogg, 1954)

$$U = -\frac{\mu}{r}\left[1+\sum_{\nu=1}^{\infty} \frac{(-1)^{\nu+1}(2\nu-3)!!}{(2\nu)!!}\left(\frac{A}{r}\right)^{2\nu-1} P_{2\nu-1}(\cos\vartheta)\right], \tag{6}$$

where the symbols employed have their usual meaning.

(vi) Axi-symmetric Potential. All the potential presented by Equations (1) to (6) are given in the form of summations by means of spherical harmonics. These expressions, as well as some specific forms of potentials due to density distributions, can be represented by a general relation. After thorough examination of the above potentials we choose the following expression

$$U = -\mu \sum_{\nu=0}^{\infty} \frac{J_{\nu}A^{\nu}}{r^{\nu+1}} P_{\nu}(\cos\vartheta), \tag{7}$$

where the various coefficients J_{ν} can be estimated by comparing the last equation with any form of axi-symmetrical potentials given by (1) to (6).

2. EQUATIONS OF THE PROBLEM

We assume that a small satellite is moving around a celestial body with an axis of symmetry. The orbit of the satellite is defined by six independent parameters. These are the semi-major axis α, the eccentricity e, the longitude of the ascending node Ω, the inclination of the orbital plane ι, the argument of pericenter ω, and χ the mean anomaly at time $t=0$.

In order to estimate the orbit of the satellite we use the osculating elements and employ the Gauss-Lagrange form of variational equations. These relations are (cf., e.g., Smart, 1953).

$$\frac{d\alpha}{dt} = \frac{2\alpha^2}{rH}(er\sin\upsilon\, R + pS), \tag{8}$$

$$\frac{de}{dt} = \frac{1}{H}[p\sin\upsilon\, R + (p+r)\cos\upsilon\, S + erS], \tag{9}$$

$$\frac{d\Omega}{dt} = -\frac{r\sin u}{H\sin\iota} W, \tag{10}$$

$$\frac{d\iota}{dt} = \frac{r\cos u}{H} W, \tag{11}$$

$$\frac{d\omega}{dt} = \frac{1}{eH}[-p\cos\upsilon\, R + (p+r)\sin\upsilon\, S] - \cos\iota\frac{d\Omega}{dt}, \tag{12}$$

$$\frac{d\chi}{dt} = \frac{\sqrt{1-e^2}}{eH}[(p\cos\upsilon\, R - 2er)R - (p+r)\sin\upsilon\, S], \tag{13}$$

where R, S and W are the components of disturbing accelerations, H stands for the angular momentum of the satellite given by $H=\sqrt{\mu p}$, p denotes the semi-latus rectum($p=\alpha(1-e^2)$), r is the distance and u, υ are the true anomalies measured from the ascending node and the pericenter respectively.

By means of the disturbing potential

$$V = -\mu\sum_{\nu=1}^{\infty}\frac{J_\nu}{r^{\nu+1}}A^\nu P_\nu(\cos\vartheta), \tag{14}$$

derived from (7), and with the aid of spherical trigonometry, we finally obtain for the components R, S, and W, of the disturbing accelerations

$$R = \mu\sum_{\nu=1}^{\infty} J_{2\nu}\frac{A^{2\nu}}{r^{2\nu+2}}\sum_{j=0}^{\nu}\rho_{2\nu,j}\cos(2ju) +$$

$$+ \mu \sum_{\nu=0}^{\infty} J_{2\nu+1} \frac{A^{2\nu+1}}{r^{2\nu+3}} \sum_{j=0}^{\nu} \rho_{2\nu+1,j} \sin((2j+1)u), \tag{15}$$

$$S = - \mu \sin\iota \cos u \left\{ \sum_{\nu=1}^{\infty} J_{2\nu} \frac{A^{2\nu}}{r^{2\nu+2}} \sum_{j=0}^{\nu-1} \sigma_{2\nu,j} \sin((2j+1)u) + \right.$$

$$\left. + \sum_{\nu=0}^{\infty} J_{2\nu+1} \frac{A^{2\nu+1}}{r^{2\nu+3}} \sum_{j=0}^{\nu} \sigma_{2\nu+1,j} \cos(2ju) \right\}, \tag{16}$$

$$W = - \mu \cos\iota \ \{\text{same as in S}\}. \tag{17}$$

In all the above Equations (15) to (17) we have separated the various terms produced by the even spherical harmonics from those due to the odd harmonics. The symbols $\rho_{n,m}$ and $\sigma_{n,m}$ indicate functions of $\sin\iota$.

3. SOLUTION OF EQUATIONS

The procedure applied in order to obtain the perturbed elements of the orbit has as follows:

(i) change independent variable from time t to u by means of the relation

$$\frac{dt}{du} = \frac{r^2}{H}, \tag{18}$$

(ii) substitute for the components R, S, and W from Equations (15) to (17) respectively,

(iii) replace the radius vector $\vec{r}$ through the expression

$$r = \frac{p}{1+e\cos\upsilon}. \tag{19}$$

The orbital elements of the satellite have been presented in a compact form by means of summations and zero-order Hansen coefficients. These coefficients are functions of eccentricity, and analytical expressions can be found in Zafiropoulos and Kopal (1982).

The zero-order Hansen coefficients have been introduced into the relations for the perturbed elements by means of the equation

$$\left(\frac{r}{\alpha}\right)^{-n} = \frac{\sqrt{1-e^2}}{2} \sum_{k=0}^{n} (2-\delta_{o,k}) \, X_o^{-(n+2),k} \cos[k(u-\omega)], \tag{20}$$

where the symbol $\delta_{o,k}$ denotes the Kronecker's delta. Equation (20) represents finite summation.

The expressions for the perturbed elements are given in their most general form by means of summations and zero-order Hansen coefficients. All the elements are subject to periodic perturbations. With the exemption of the semi-major axis, all the rest elements include secular terms. Secular perturbations due to the second spherical harmonic do not appear in the elements for e and ι.

In order to reduce the length of the secular terms we have produced and employed two recurrence relations for zero-order Hansen coefficients

$$X_o^{n,m} - eX_o^{n,m+1} = \frac{(n+m+1)}{(n+1)} (1-e^2) X_o^{(n-1),m} \quad , \tag{21}$$

$$X_o^{n,m} - X_o^{n,m+2} = - \frac{2(m+1)}{(n+1)} \frac{(1-e^2)}{e} X_o^{(n-1),m+1} \quad . \tag{22}$$

The advantage of analytical methods is, that one can see the dependance of the orbital elements, on the various parameters of the problem. Combining analytical with numerical work may produce results of higher accuracy.

All the expressions produced for the elements can be easily adjusted to yield the perturbations for the various cases of axi-symmetric potentials presented in section 1.

A fuller account of the present investigation , which will include the proof of all the formulae employed, the elements obtained, and the tests applied, to check for the correctness of the results produced, will be given in a subsequent paper.

REFERENCES

Kellogg, O.D. : 1954, "Foundations of Potential Theory", Dover Publ., Inc., New York.

MacMillan, W.D.: 1958, "The Theory of the Potential", Dover Publ. Inc., New York.

Ramsey, A.S. : 1981, "Newtonian Attraction", Cambridge Univ. Press, Cambridge.

Smart, W.M. : 1953, "Celestial Mechanics", Longmans, London.

Taff, L.G.: 1985, "Celestial Mechanics", John Wiley and Sons, New York.

Zafiropoulos, B. and Kopal, Z.: 1982, Astrophys. Space Sci., 88, 355.

LIGHT CURVE VARIATIONS OF THE RS CVn-TYPE ECLIPSING BINARY ER VULPECULAE

Varol KESKİN
Ege University Observatory
Bornova, İzmir
Turkey

ABSTRACT. Photoelectric observations of the RS CVn-type double-lined spectroscopic and eclipsing binary ER Vul were obtained in B and V colours of Johnson's UBV system, between 1981 and 1986. The light curves show that the system is sometimes very active and sometimes very quiet and have changes in short time intervals. The wave-like distortion which is superimposed on the light curves has been obtained. Moreover, small-amplitude light fluctuations in the light curves are noticable.

1. INTRODUCTION

The classification of the RS CVn systems was made by Hall (1976). These systems are mostly Algol-type binaries which consist of nearly equal-mass components. The cooler ones are generally of K0 subgiants with strong H and K emission lines of CaII in their spectra and the hotter ones are between F4 and G9 with luminosity classes between IV and V. Some of these systems show strong radio and X-ray emission. Most of them also show H-alpha emission.

Northcott and Bakos (1956) stated that ER Vul is a double-lined spectroscopic binary. Its orbital period is short and components are of G0V and G5V. The first light curves of the system were obtained by Northcott and Bakos (1967) . H and K emission of Ca II were detected by Bond (1970) and Eggen (1978). Other light curves of the system were obtained by Al-Naimiy (1981) and also by Kadouri (1981). Existence of cool spots on one of the components were suggested.

2. OBSERVATIONS

The photoelectric observations of ER Vul have been made between 1981 and 1986 at the Ege University Observatory with the 48 cm Cassegrain telescope. The EMI 9781A photomultiplier and B and V filters were used during the observations. HD 200270 was used as the comparison star. HD 200425 was used as the check star. Twenty-four light curves, which contain 10 primary and 7 secondary minima, were obtained. We added these

C. İbanoğlu (ed.), Active Close Binaries, 865–871.

minima to the old ones and calculated the following new light elements :

$$\text{Min. I} = \text{J.D. Hel. } 24\ 46235.4395 + 0^{d}.69809472\ E$$
$$\pm 4 \qquad\qquad \pm 6$$

3. LIGHT CURVES, THEIR VARIATIONS AND CONCLUSIONS

The light variations due to the eclipses, in June and July 1986 (Figure 1) and August 1986 (Figure 2) are very less. There are short-term brightness variations at the outside eclipses. Maximum amplitude of the fluctuations is about $0^{m}.05$. The magnitude of the proximity effects has to be known in order to determine the shape of the wave-like distortion and the location of its minimum. These can be obtained from the geometrical and physical parameters of the system. The light curves which were obtained between 20 and 28 August 1984, are almost symmetric. So, we obtained a solution of the system using Wood's method (1972), with these light curves. Some determined parameters are :

$$i = 66^{\circ}.32 \pm 0^{\circ}.35 \quad , \quad q = m_2/m_1 = 0.98 \quad , \quad T_A = 6000\ \text{K (adopted)} ,$$

$$T_B = 5883 \pm 52\ \text{K} \quad , \quad r_A = 0.2713 \pm 0.0113 \ , \quad r_B = 0.2756 \pm 0.0326$$

These parameters are very different from the values obtained by Al-Naimiy (1981). The obtained synthetic light curve using these new parameters contains the proximity effects and represents almost all other light curves. The distortion effect can be obtained by subtracting the synthetic one from the one of the observed light curves. These are the results :
The distortions in the 1981, 1982, 1984, 1985, and 1986 light curves were plotted versus orbital phases in Figure 3, 4, 5 and 6. The distortion wave light curves, obtained in 8 - 16 July 1981, have an amplitude of $0^{m}.07$. However, the 19 - 21 August 1981 light curves, which were obtained approximately one month later, show no wave-like distortion. In 1984, 1985 and 1986 light curves, the short-term brightness variations are clearly seen. In the July 1981 light curve the distortion is regular and the wave minimum is located about the phase 0.05. In the 1984 light curve, there are two wave minima.

In Figure 7 and 8 the 20-28 August 1984 and 4-8 September 1985 light curves, which excluded the wave-like distortion, are shown together with the synthetic light curve in B and V colours. In these figures, at the top, the brightness differences between the observed and the computed are also plotted. The short-term light fluctations in the system are clearly seen.

Plotting wave minima phases versus years, we found that wave minimum migrates towards the decreasing phases and has a period of approximately 1.5 years (Figure 9). The mean brightness variation period is approximately 2.5 - 3 years. There are some regular changes in the orbital period of the system, but it is not certain that these changes are periodic. Brightness variations sometimes reach to $0^{m}.05$. The system is sometimes very quiet and shows no irregular variations.

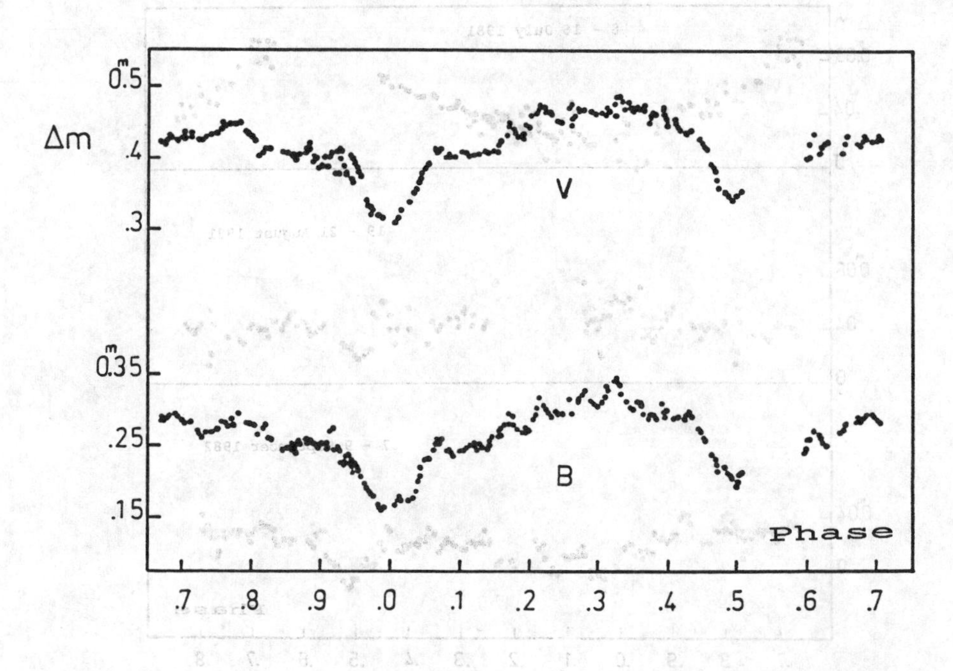

Figure 1. The B and V light curves of ER Vul obtained in June, July 1986.

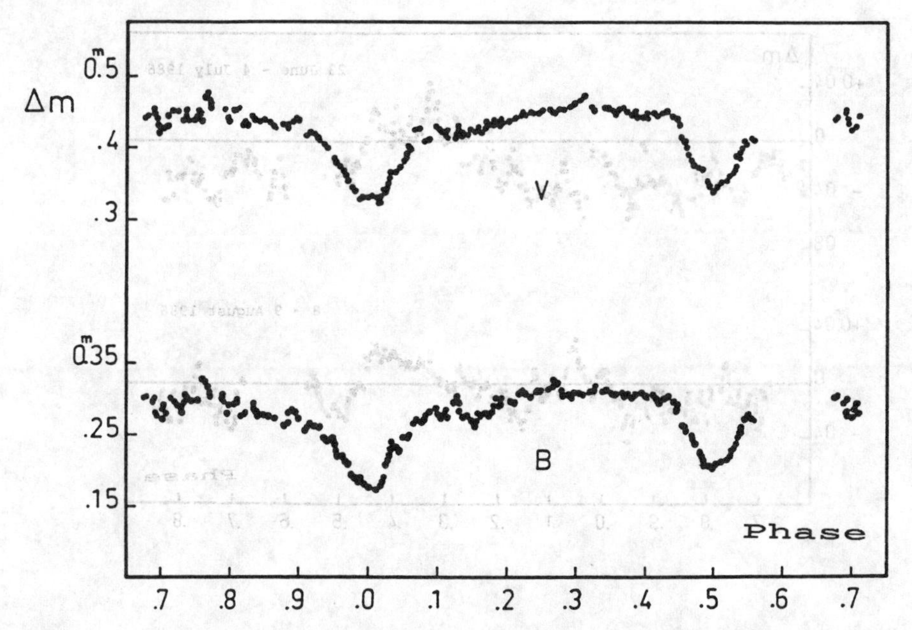

Figure 2. The light curves of ER Vul obtained in August 1986.

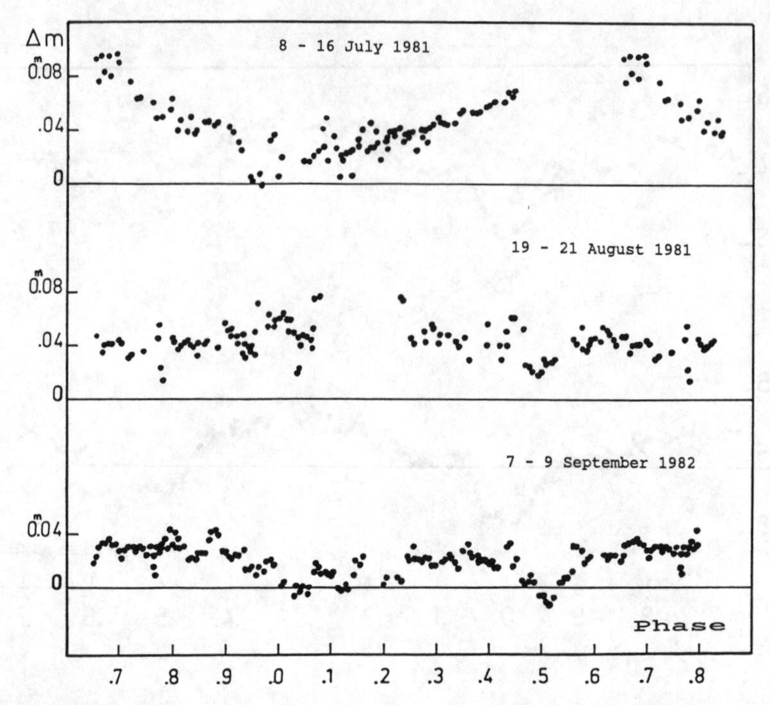

Figure 3. Distortions in the 1981 and 1982 light curves of ER Vul.

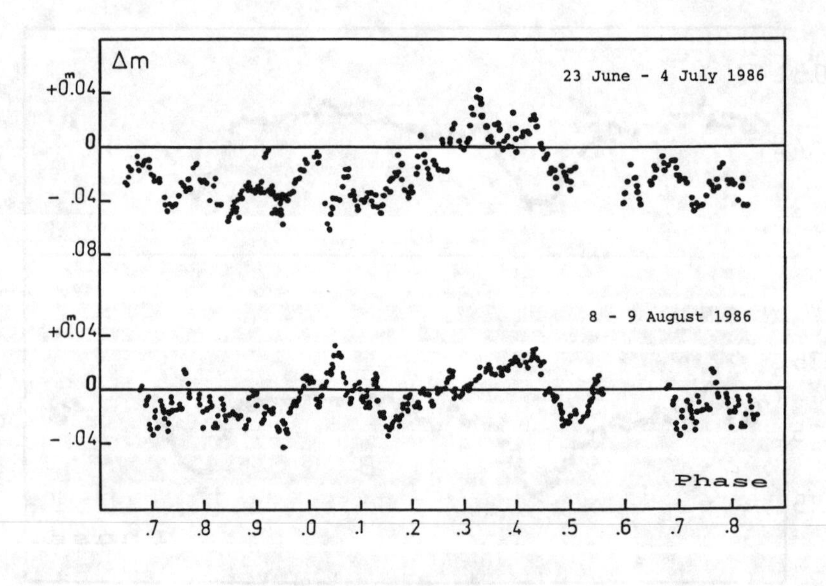

Figure 4. Distortions in the 1984 light curves of ER Vul.

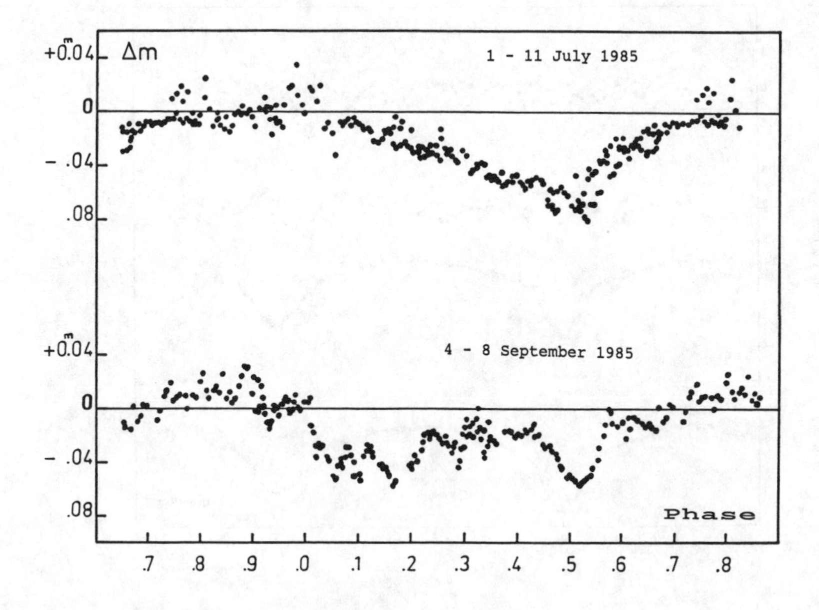

Figure 5. Distortions in the 1985 light curves of ER Vul.

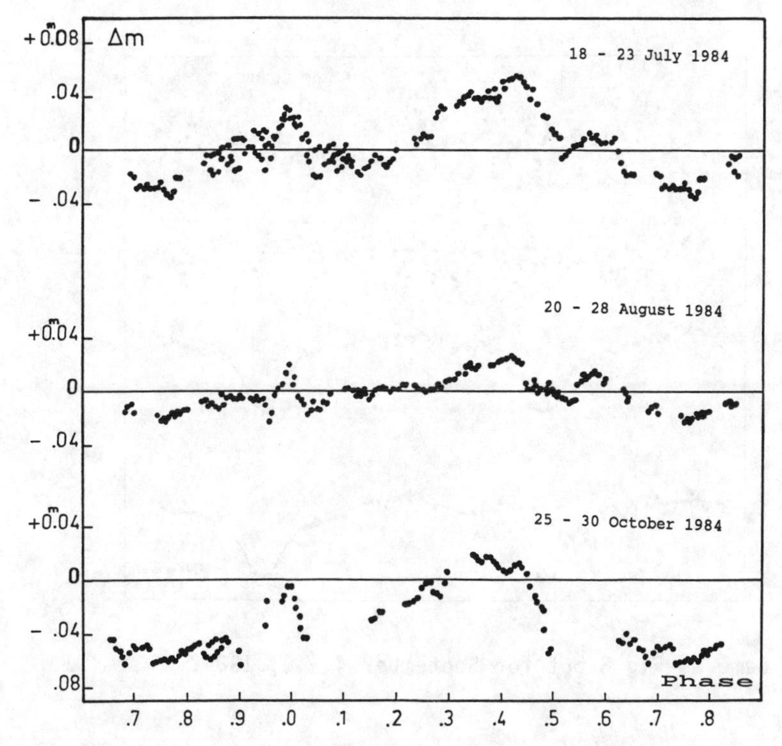

Figure 6. Distortions in the 1986 light curves of ER Vul.

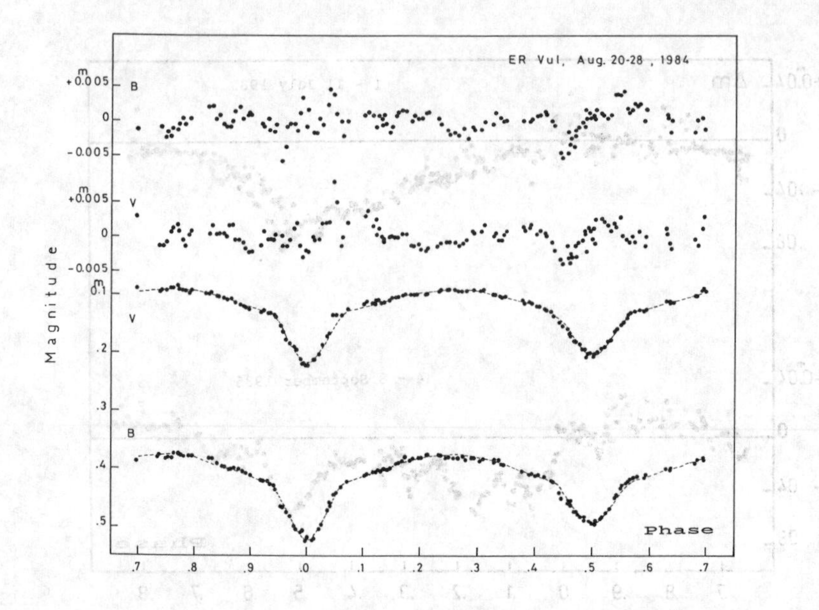

Figure 7. The observed light curves between 20 and 28 August 1984, excluded wave-like distortion, the computed curves and the brightness differences between observed and computed.

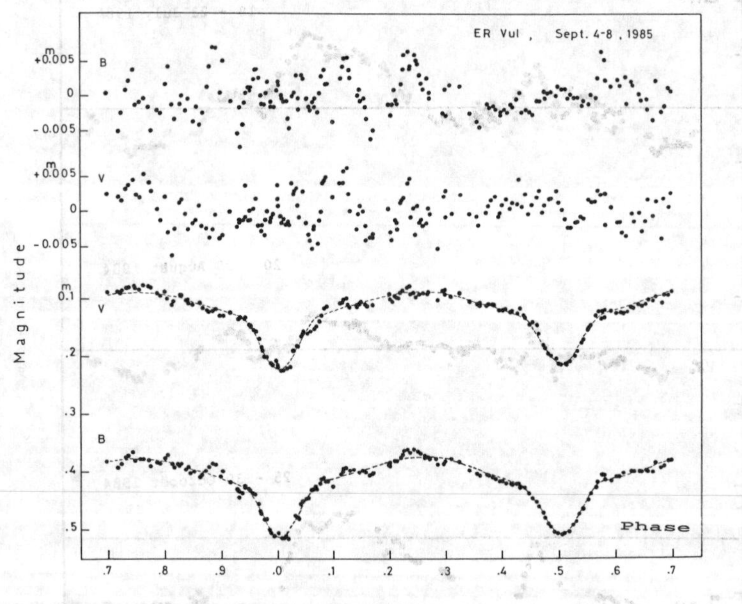

Figure 8. Same as Fig 8 but for September 4 - 8, 1984.

REFERENCES :

Al-Naimiy, H. M. K.: 1981, Astron. Astrophys. Suppl. Ser., **43**, 85.
Bond, H. E.: 1970, Publ. Astron. Soc. Pacific, **82**, 321.
Eggen, O. J.: 1978, Inf. Bull. Var. Stars, No.1426.
Hall, D. S.: 1976, Proc. of IAU Colloq., **29**, Part I, W.S. Fitch (ed.), Multiple Periodic Variable Stars, Dordrecht, Reidel, p.287.
Kadouri, T. H.: 1981, Inf. Bull. Var. Stars, No.2057.
Northcott, R. J. and Bakos, G. A.: 1956, Astron. J., **61**, 188.
Northcott, R. J. and Bakos, G. A.: 1967, Astron. J., 72, 89.
Wood, D. B.: 1972, Report X-110-72-473, Goddard Space Flight Center, Maryland.

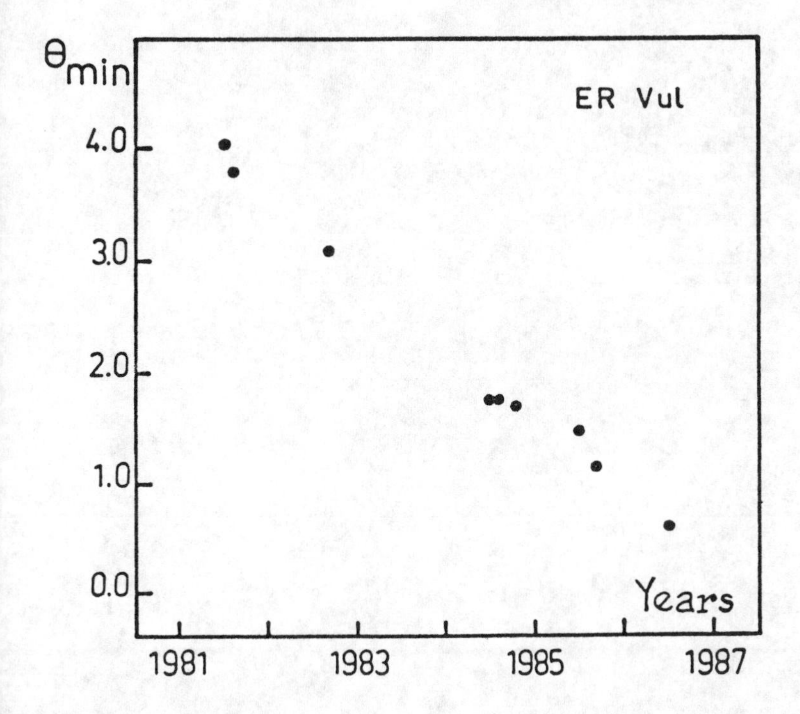

Figure 9. Migration of the wave minimum versus years.

SYNTHESIS METHODS FOR ECLIPSING-BINARY LIGHT CURVES

Paul B. Etzel
Department of Astronomy
San Diego State University
San Diego, California 92182-0334, USA

Kam-Ching Leung
Department of Physics and Astronomy
University of Nebraska
Lincoln, Nebraska 68588-0111, USA

ABSTRACT: A workshop was held to present overviews of two widely used synthesis models employed to solve the light curves of eclipsing binary stars; they are the Nelson-Davis-Etzel and Wilson-Devinney models. A brief description of each model is presented here, with references provided for more detailed information.

1. INTRODUCTION

In the early 1970s, sufficient computing power became available at the institutional level to encourage the development of synthesis methods for the solution of eclipsing-binary light curves. Today, similar computing power is available to scientists and students on their desk tops. As a consequence, the application of such synthesis methods has become wide spread. Many astronomers can now bring the full force of synthesis methods to bear upon the light curves that they obtain; in the past, their data were not fully explored as a result of computing restrictions.

Two widely-used synthesis models will be briefly discussed here. Many references are included so that the interested reader may pursue their uses in various applications. The first model is the very simple Nelson-Davis-Etzel model (Nelson and Davis 1972, Etzel 1981), which is intended to be used on simple, detached systems. Such systems usually provide the most reliable fundamental stellar properties because they do not suffer from complications introduced by proximity effects (i.e., ellipticity and reflection). The computing requirements of memory and time are rather modest. The second model is the more general Wilson-Devinney model (Wilson and Devinney 1971, Wilson 1979), which can be applied to the entire range of Roche-model configurations, from

C. İbanoğlu (ed.), Active Close Binaries, 873–879.

detached to overcontact. Thus, systems with very pronounced proximity effects become tractable problems. The computing requirements here are greater than for the former model.

2. THE NELSON-DAVIS-ETZEL (NDE) MODEL

The original definition of the NDE model was by Nelson and Davis (1972). Fundamental to the model is the use of the surface-brightness ratio to calculate the fractional luminosities of the spherical components. The effects of limb darkening are included by modelling the components with a series of concentric rings of varying intensity. Eclipse light losses are computed by numerical integrations based upon simple geometrical constructs. With typical ring sizes of five degrees, only 18 intensity values need be calculated for each star to reach computational accuracies of better that 0.00001 in intensity units.

Improvements made to the original model of Nelson and Davis are described by Etzel (1981) in a paper presented at the 1980 NATO ASI "Photometric and Spectroscopic Binary Systems" held in Maratea, Italy. Many of the participants at that meeting, and their students, have since used the NDE model, which is embodied in the EBOP (Eclipsing Binary Orbit Program) code. The most important changes made by Etzel were to: 1) replace the original Nelson and Davis treatment of ellipticity and reflection effects with perturbation terms derived by Binnendijk (1960) for biaxial ellipsoids; 2) make the ratio of the radii, k, a solution parameter along with the radius of the primary, r_p, since k is directly linked to the light ratio; 3) replace the orbital parameters e and ω with ecosω and esinω, since the latter parameters are sensitive to the eclipse displacements and durations, respectively; and 4) define the luminosity contribution of a third object in terms of a scaling factor. Perhaps more relevant to Section 4 of this paper, Etzel also introduced a more traditional differential corrections least-squares procedure to replace the less reliable sequential-optimization procedure employed by Nelson and Davis.

The first large-scale application of the NDE model was to seven detached systems by Popper and Etzel (1981). Of concern to this discussion was their comparison of the EBOP results to equivalent computer solutions with the widely used computer code WINK (Wood 1971, 1972, 1973 to 1978, Etzel and Wood 1982), which is based upon a triaxial-ellipsoid model that directly calculates the effects of reflection and ellipticity through the eclipses. Popper and Etzel's comparison of the results for the two codes demonstrates that EBOP is adequate for mildly distorted, detached systems, with the benefit of being 25 to 40 times faster than WINK. An interesting conclusion was that four of the seven systems investigated were found to have small but significant orbital eccentricities, which would have been ignored out of computational convenience just a decade previous to their investigation.

There have been other investigations made with the NDE model. Some of these will be mentioned only briefly here. Popper (1982) studied the B-type system DI Her (e=0.489) both spectroscopically and photometrically. Popper makes use of the observed ratio of absorption-line strengths for the components to place constraints upon the light ratio of the components, which is tied to k. Coates et al. (1983) applied EBOP to the RS CVn-type system HD 5303 (CF Tuc) to model the distortion wave produced by a large spot system on the outer hemisphere of the secondary star. The analyses of two light curves obtained at different epochs for the RS CVn-type system Z Her by Tümer et al. (1984) and Popper (1988) derived consistent geometrical elements for the stellar components; however, the radiative properties are much different, which can be attributed to an overall change in brightness of the system as a result of stellar activity. Andersen et al. (1985) determined the absolute dimensions of V760 Sco in their combined spectroscopic and photometric study. Of important note here, they have provided one of the most consistent comparisons between theory and observation for a system that exhibits apsidal motion. Woodward and Koch (1989) have recently improved the apsidal-motion characteristics of V523 Sgr from their photometric analysis with EBOP and make a strong appeal for a spectroscopic study of this neglected system.

3. THE WILSON-DEVINNEY (WD) MODEL

The WD model is perhaps the most widely used of all synthesis models for the solution of eclipsing-binary light curves. Its original definition was by Wilson and Devinney (1971). The WD model is embodied in the LC (Light Curve) and DC (Differential Correction) computer codes. A clear explanation of the various "modes" of operation for the DC code can be found in Leung and Wilson's (1977) article on V1010 Oph (a binary system discussed at this ASI). The various modes are used to impose different conditions unique to particular configurations of binary-star systems.

Of fundamental importance to the model is the definition of the stellar surfaces by a grid of typically thousands of points defined by any valid Roche equipotential from the mass ratio, q, and the direction cosines towards the points in question. The unnormalized luminosities of the components are defined by their fluxes integrated over the surfaces in the absence of reflection. The polar temperatures for each component are basic model parameters. The local intensity of each point is converted to a monochromatic flux in the direction of the observer by including the effects of limb darkening, gravity darkening, and reflection. The total observed flux is then the sum of the fluxes from those points that are above the horizon to the observer that are also not eclipsed. Wilson and Devinney (1971, 1972) first applied their model to the hot, detached system MR Cyg.

Wilson (1979) made major enhancements to the WD model, which were largely oriented towards a more generalized treatment of the problem. The model was extended to allow for the solution of the radial-velocity curves of the stellar components as well as the light curve of the system; simultaneous solutions of both types of data are allowed (although weighting such dissimilar data is not straightforward). In addition, the effects of orbital eccentricity and asynchronous stellar rotation are now included. The distortion upon the Roche surfaces by asynchronous rotation, led Wilson to propose a new morphological type of binary system called the double-contact binary. Such systems have both stars filling their limiting (critical) Roche lobes, yet if one component is rotating at a rate sufficiently greater than the synchronous value, the surfaces of the stars themselves do not touch.

Early on, it became obvious that the WD model would have its most interesting applications to systems with highly distorted components. Wilson and Biermann (1976) investigated the W UMa-type system TX Cnc to address the problem of temperatures of the components. They found that the light-curve could be explained by adiabatic coupling of the two stars' convective zones, which is a result of an overcontact configuration. Algol was the subject of an investigation by Wilson et al. (1972) in which simultaneous solutions were made on multicolor light curves. The rotation rates of the primaries of Algol-type systems have been pursued by Wilson and his collaborators. Van Hamme and Wilson (1986) investigated the systems U Sge and RY Per. While neither system is in the double-contact configuration, they discuss the importance of rotation in the context of mass-transfer activity. Wilson and Mukherjee (1988) found that for three Algol-type binaries with rapidly rotating primaries, the systems SW Cyg and AQ Peg appear to be in the double-contact configuration, while AW Peg is close to that configuration.

The grid-point nature of the WD model lends itself to the modelling of spotted stars. Various applications for both dark and bright spots were discussed at this ASI. An important improvement in the application of the WD model to spotted stars has been discussed by Kang and Wilson (1989) who allow for the least-squares adjustment of spot parameters. Each spot is described by its position, size, and temperature differential with respect to the undisturbed photosphere. Three RS CVn-type systems were investigated, with various numbers of spots allowed. The ability to systematically reproduce the migration wave at different epochs is encouraging.

4. SOLUTION CONVERGENCE AND TECHNIQUE

Unfortunately, synthesis models are not a panacea for the solution of the light curves of eclipsing binaries. The application of differential-corrections least squares does not insure strict convergence. Popper (1984) has addressed the degree to which parameters can be determined for various synthetic light curves subjected to random errors. The light curves were solved with the NDE model; however, the results are applicable to any synthesis model. Some parameters, such as limb-darkening coefficients, require very good observations in order to be adequately determined. Other parameters, such as orbital inclination and third light, can directly mimic each other in some situations. Popper finds that the formal mean errors of the derived parameters generally underestimate their true errors by a factor of three. De Landtsheer (1981) came to a similar conclusion concerning the derived errors of parameters in his analysis of the solution convergence of YZ Cas with both the WINK and WD codes. Rafert and Markworth (1986) have investigated the solution-convergence characteristics of the WD model. They propose testing solution space to verify that the minimum of the goodness-of-fit parameter (e.g., sum of the residuals squared, or mean error of one observation) found by the WD model is indeed the global minimum.

Various methods have been derived to improve solution convergence and reliability. Wilson and Biermann (1976) proposed the Method of Multiple Subsets (MMS) in recognition that some parameters are highly correlated in a general solution. They argue for splitting parameters into two or more subsets and solving for a general solution by a series of sequential (and repeated) solutions on the various subsets of parameters. Rafert and Markworth (1986) explored the convergence of CN And in inclination-log(q) space and addressed the MMS procedure.

Generally, the most difficult parameter to determine is the mass ratio, q. One should try to avoid a solution at a local minimum. Thus, a "q-search" method was introduced by Leung and Schneider (1978a) to locate the global minimum. Converging solutions are first obtained for several fixed values of q. The preliminary solution with the best goodness-of-fit parameter is then used to derive the final solution. The systems ZZ Aur and AX Vir (Leung, Zhai, Liu, and Yang 1985) have well determined values of q from this method. The systems BX and BB Peg (Leung, Zhai, and Zhang 1985) illustrate how easy it is to miss the global minima if the "q-search" is not carried out carefully. No unique solutions could be found for the badly behaved systems UW CMa (Leung and Schneider 1978b) and CQ Cep (Leung, Moffat, and Seggewiss 1983). These systems have very complex light curves that present multiple solutions as evident by similar goodness-of-fit measures for their respective ranges of q. Even though each system had multiple solutions, the mode (configuration) for each stayed the same for different values of q. However, when a system is very close to critical contact, such as ER Ori (Liu et al. 1988), the light curve can be represented by very similar sets of parameters with different configurations (modes 2, 4, and 5). Therefore, systems very close to being in critical contact should be explored with different configurations imposed.

When using synthesis models, the imposition of justifiable constraints to limit the number of free parameters is frequently desirable. Perhaps the most physically meaningful approach is to incorporate other independent data. The use of a spectroscopically determined light ratio, as demonstrated by Popper (1982), to constrain the light-curve solution, is an important technique. Frequently, a spectroscopically determined value of q may be available. Although desirable, such values need not be known very accurately for the light-curve solutions of very well-detached binaries. However, for the more complicated systems, where the effects of distortion of the components are more pronounced, spectroscopically determined values of q can become increasingly unreliable because of velocity-measurement limitations, stellar distortions, gas streams, etc., which has necessitated the various methods to constrain q in light-curve solutions as mentioned above.

Sometimes the selection of the basic model parameters may promote a lack of convergence. In this regard, the study by Hrivnak and Milone (1984) of the detached system AI Phe (e=0.173) is notable. A local minimum in e and ω was found when using the WD model; however, when using the NDE model, which employs ecosω and esinω, the true minimum was found. This situation is the result of the NDE parameters for orbital eccentricity being more directly linked to separable light-curve properties.

Other optimization techniques are being explored to find better the global minima in the hyperspace produced by synthesis models because of deficiencies inherent to the method of differential-corrections least squares. Kallrath and Linnell (1987) have argued for the replacement of the differential-corrections procedure with the simplex algorithm in conjunction with the WD model. Barone, Milano, and Russo (1990) have proposed a hyperspace optimization with the WD model to find the global minimum. Regardless of the algorithm used to solve for model parameters, there is no substitute for critical tests of the solutions by the investigator.

The authors wish to acknowledge the National Science Foundation for partial support under grants AST8822790 (Etzel) and INT8616425 (Leung).

REFERENCES

Andersen, J., Clausen, J. V., Nordström, B., and Popper, D. M. (1985). Astron. Astrophys. 151, 329.
Barone, F., Milano, L., and Russo, G. (1990). 'An Optimization Method for Solutions of Close Eclipsing Binaries,' (these proceedings).
Binnendijk, L. (1960). Properties of Double Stars, (Univ. of Penn., Philadelphia), pp. 288-326.
Coates, D. W., Halprin, L., Sartori, P.A., and Thompson, K. (1983). Mon. Not. R. Astr. Soc. 202, 427.
de Landtsheer, A. C. (1981). Astrophys. Space Sci. 80, 349.
Etzel, P. B. (1981). In Photometric and Spectroscopic Binary Systems, E. B. Carling and Z. Kopal eds., (Reidel, Dordrecht), p. 111.
Etzel, P. B., and Wood, D. B. (1982). WINK Status Report no. 10. (private circulation).
Hrivnak, B. J., and Milone, E. F. (1984). Astrophys. J. 282, 748.
Kallrath, J., and Linnell, A. P. (1987). Astrophys. J. 313, 346.
Kang, Y. W., and Wilson, R. E. (1989). Astron. J. 97, 848.

Leung, K.-C., Moffat, A. F. J., and Seggewiss, W. (1983). Astrophys. J. **265**, 961.
Leung, K.-C., and Schneider, D. P. (1978a). Astrophys. J. **222**, 917.
Leung, K.-C., and Schneider, D. P. (1978b). Astrophys. J. **222**, 924.
Leung, K.-C., and Wilson, R. E. (1977). Astrophys. J. **211**, 853.
Leung, K.-C., Zhai, D. S., and Zhang, Y. X. (1985). Astron. J. **90**, 515.
Liu, Q., Yang, Y., Leung, K.-C., Zhai, D., and Li, Y.-F. (1988). Astron. Astrophys. Suppl. Ser. **74**, 443.
Nelson, B., and Davis, W. (1972). Astrophys. J. **174**, 617.
Popper, D. M. (1982). Astrophys. J. **254**, 203.
Popper, D. M. (1984). Astron. J. **89**, 132.
Popper, D. M. (1988). Astron. J. **95**, 1242.
Popper, D. M., and Etzel, P. B. (1981). Astron. J. **86**, 102.
Rafert, J. B., and Markworth, N. L. (1986). Astron. J. **92**, 678.
Tümer, O., Ibanoglu, C., Tunca, Z, and Evren, S. (1984). Astrophys. Space Sci. **104**, 225.
Van Hamme, W., and Wilson, R. E. (1986). Astron. J. **92**, 1168.
Wilson, R. E. (1979). Astrophys. J. **234**, 1054.
Wilson, R. E., and Biermann, P. (1976). Astron. Astrophys. **48**, 349.
Wilson, R. E., de Luccia, M. R., Johnston, K., and Mango, S. (1972). Astrophys. J. **177**, 191.
Wilson, R. E., and Devinney, E. J. (1971). Astrophys. J. **166**, 605.
Wilson, R. E., and Devinney, E. J. (1972). Astrophys. J. **171**, 413.
Wilson, R. E., and Mukherjee, J. (1988). Astron. J. **96**, 747.
Wood, D. B. (1971). Astron. J. **76**, 701.
Wood, D. B. (1972). Goddard Space Flight Center Report X-110-72-473.
Wood, D. B. (1973 to 1978). WINK Status Reports no. 1 through no. 9. (private circulation).
Woodward, E. J., and Koch, R. H. (1989). Astron. J. **97**, 842.

TYPES OF CONTACT BINARY SYSTEMS AND WAYS TO FIND THEM

Kam-Ching Leung
Behlen Observatory
University of Nebraska
Lincoln, NE 68588-0111
U.S.A.

ABSTRACT. Generally, the designation of contact system among close binary systems means that the components are so close that they are physically touching each other or are joined together. They can be classified into four distinct groups based on the characteristics of their common atmospheres: late-type low mass systems (W UMa systems) with common convective atmospheres, early-type massive systems with common radiative atmospheres, mix- or intermediate-type systems consisting of a cooler component with a convective stellar atmosphere and a hotter companion with a radiative stellar atmosphere, and supergiant contact systems with very deep common convective atmospheres. In addition, there is one very peculiar group called double contact systems. In these systems, each of their components is being in contact with its respective critical equal potential surface but the components are not in physical contact with each other!

Some of the contact systems are believed to be at or very near zero-age, while most of the others are found to be evolved close binary systems.

Several ways of finding potential contact systems are suggested.

1. INTRODUCTION

1.1 Confusing Terminology

Historically, eclipsing binary stars have been divided <u>superficially</u> into three groups. They are classified traditionally according to the shape of their light variations or light-curves; namely the Algol type, the β Lyrae type, and the W Ursae Majoris (W UMa) type. Most unfortunately, these designations are very confusing in terms of modern astrophysics. The Algol system, the β Lyrae system, and the W UMa system are interacting binary systems with very specific physical properties, and are believed to be at

C. İbanoğlu (ed.), *Active Close Binaries*, 881–890.

very different stages of close binary evolution. A particular light-curve pattern/type does not (necessarily) tell us the type of physical system it belongs to (see discussion of Eggleton 1985). For example: contact systems are commonly found among systems with W UMa type light-curves as well as with β Lyrae type light-curves, and Algol systems are found frequently among systems with β Lyrae type light-curves, and sometimes among systems with Algol type light-curves. It is very important that our terminology should reflect the distinction between physical types (Algol, β Lyrae, and W UMa systems) and patterns of light-curves. Therefore, one should be very careful in applying the proper terminoloy accordingly.

1.2 Historical Note on Contact Binary Systems

The systems with W UMa type light-curves have the following characteristics: Short periods (about a quarter of a day), spectral type mostly of G To K, with continuous light variation in their light-curve with equal or almost equal minima, most of them showing a double-line (usually rather broad) spectra. It was believed that their components are joined together in the shape of a dumbbell. The term "contact binary" was first introduced by Kuiper (1941) and applied to the W UMa stars. In the same momentous papr on close binary systems, Kuiper also presented a very extensive discussion on characteristics and interpretation of contact binaries. At the time he assumed that the components of contact systems were main sequence stars. Kuiper proved that these stars could not simultanuously satisfy the mass-luminosity relation and Kepler's law unless the mass ratios of these systems had a value of unity.

In the earlier days, practically all the spectra taken for W UMa were of insufficent resolution with respect to phase and spectral dispersion because the periods are rather short, on the order of about a quarter of a day, and generally they were relatively faint. As a result, the radial velocity curves were not truly representative of orbital motion of the systems. Thus, the mass ratios estimated were unreliable. Since the light-curves of the W UMa stars showed minima of equal depths or almost equal depths, they easily led to the interpretation that these systems consisted of two identical or near identical stars in contact with each other. There was a common misconception that all the systems having W UMa type light-curves were W UMa systems and they were all contact systems!

In general, the configuration of a binary system is derived from analyses of its light-curve (i.e. photometric solution of the binary light-curve). The orbital parameters for a close binary (excluding visual binaries) are deduced from the radial velocity curve of the system. The absolute dimensions of a binary system are determined from the combined results of a photometric and spectroscopic analysis. As a result of improved spectroscopic observations (in terms of better phase resolution and spectral dispersion), the mass ratios of these systems were found to be mostly about a factor of 2 to 1 instead of unity. The question in most people's mind was whether all of the systems classified in catalogues as W UMa systems were really contact systems (e.g. the Variable Stars Catalogue, etc.). The major difficulty associated with this problem was that the classical methods of photometric analyses (e.g. Russell-Merrill, Kopal, etc.) for deriving the configurations of close binary stars were based on the assumption that the components (stars) were ellipsoidal in shape. Obviously, this cannot be the true shape of the stars for the very close systems. However, an alternative approach would have been formidable or prohibitively difficult without the current general availablility of fast computers. Usually, a system was "defined" as contact if the sum of the fractional radii of the components was 0,75 or larger. There was no reliable way to judge whether a system was in actual contact. However, if one thinks about this seriously, for a typical short-period system with a mass ratio of 2 to produce nearly equal eclipsing depths, it can mean only that there must be an energy exchange between the components. Thus, the stars have to be in contact.

With the beginning of the seventies, several groups around the world were developing more realistic models for deriving more accurate configurations for close binary systems, utilizing zero-velocity surface or Roche type geometry for stellar surfaces. Such methods of analysis were developed almost simultaneously by Lucy (1968), Hill and Hutchings (1970), Wilson and Devinney (1971), and Mochnachi and Doughty (1972), and by some others in later years. At present, the most flexible and most utilized computing codes are those developed by Wilson and Devinney in 1971, and being kept current with additional features by Wilson. Unfortunately, in most variable catalogues and journal papers, close binary systems have been classified very loosely as W UMa light-curves systems or the W UMa systems. Usually, they include any systems having light-curves with equal or nearly equal minima, as well as systems having significant differences in eclipse depths (though larger differences are reserved for B Lyrae light-curve systems). With the advance and general availability of fast computers, many systems with W UMa type light-curves are found to be having semidetached or even detached configurations! Therefore, one should be very careful in applying the term W UMa system. It should be reserved for the designation of late-type contact systems only.

2. TYPES OF CONTACT BINARY SYSTEMS

The term "contact system" in the modern sense means that the binary stellar surfaces of both components are in contact with the "same equal potential surface" (taking into account gravity and orbital rotation) which is loosely called "common Roche surface". That is, the two components are <u>physically</u> touching (critical/point-contact at the inner Lagrangian point) or joining (overcontact) with each other. The statistics show that some systems are at or near crictical contact, whereas most of the others are at a variety of degrees of overcontact. Contact systems can be classified into different types/groups according to their physical characteristics of their common envelopes. For single stars (excluding degenerated and exotic objects such as white dwarfs, neutron stars, black holes etc.), stellar atmospheres are defined by the mechanism of energy transportation in the outer layer of the star into radiative and convective atmospheres. Coincidentally, our sun is at about the dividing point with respect to its temperature or mass. Stars with earlier spectral-type have radiative atmospheres, and stars with spectra similar to or later than those of the sun have convective atmospheres. The convective atmospheres among the late-type giants and supergiants are believed to be extremely deep in comparison to convective atmospheres of mai sequence stars. A method to classify contact systems according to the atmospheric characteristics of their common envelopes was recently proposed by Leung (1988). Contact binaries could be classified into five different types as described in the following sections.

2.1 Late-type contact systems or W UMa systems

These systems have rather short periods of the order of about a quarter of day. They consist of late-type stars, mostly of G and K spectra, and common envelopes that are convective. They are low mass main sequence or near main sequence systems. They are also by far the most well-known and common types of contact systems. Most of the mass ratios are clustered around 2 to 1 which violates Kuiper's (1941) dilemma of mass ratio limited to unity. In the late sixties, Lucy (1968) had some success in constructing models of contact with convective common envelopes which allowed mass ratios different from unity. Binnendijk (1965) was the first to suggest that there are two subclasses of W UMa systems; A and W types (see extensive discussion by Rucinski 1985). A-type systems have their primary eclipses at transit, while W-type systems have their primary eclipses at occultation. Generally, the A-type systems have deeper-contact , slightly longer periods and earlier spectra. There is some observational evidence that the A-type systems may be evolved, but the evolutionary stage of the W-type systems are less certain. The interpretation of these two distinct types of late-type contacts remains a real challenge.

2.2 Early-type Contact Systems

These systems consist of early-type stars with spectra mostly of O and B. Since they consist of very hot stars, their contact common envelopes are believed to be radiative. They have longer periods (generally in terms of days) and very massive systems, compared with the W UMa systems. Most of these systems were discovered within about the last decade. One may say that this group is just the opposite of the late-tppe W UMa contact systems mentioned above. That is, their common envelopes are radiative instead of convective. There are about 15 members in this group that have been identified (see Leung 1988). At present, the number of early-type systems is much smaller than that of the late-type systems. The reason may be twofold: early type stars are fewer by the nature of the luminosity function of the galaxy, and their evolutionary life-time is far shorter for massive stars than for low mass stars. So far, there is no convincing evidence that there are two distinct subclasses of early-type contacts like the A and W types of the late-types systems. Unlike the late-type systems, some of the early-type systems have mass raios of unity or near unity. It is believed that there are distinct structural differences between early and late type contact systems.

2.3 Mix- Or Intermediate-type Contact Systems

Some members of this group have been called "contact systems with large temperature differences between the components". They generally consist of stars of intermediate spectral-types with a hotter component of spectral A (or F) and a cool companion of spectral G or later. This type of system can also be described as binary, consisting of early-type stars in contact with late-type stars. Their light-curves resemble B Lyrae type light-curves. With modern photometric analyses they were found to have contact configurations. They show huge temperature differences between the components (typically several thousand degrees) as a result of the large differences between the depths of the eclipses. According to their spectra, the systems consist of stars with radiative atmospheres in contact with stars with convective atmospheres. The common envelopes of such systems are most likely quite complicated, due to the huge temperature differences at the interface/neck. How can these systems maintain stablility under such huge temperature differentials at the interfaces! So far, there is no reliable theoretical model for contact systems of this type.

Some people seriously question the reliability of the contact solutions derived from modern analyses for these intermediate-type systems. At the moment no one is able to provide an answer for this tough problem. Certainly, more investigations of this group of "contact" systems are needed before any intelligent response can be put forward. At least one aspect of the photometric analyses should be thoroughly explored. So far, the contact solutions for these systems have been based mostly on mode 3 conditions (where the temperatures of the components are not a function of each other (see Leung and Wilson 1977 on modes of operation). It is suggested that a mode 1 contact solution (which ensures smooth temperature transition at the interface) should be attempted for each of these systems. If it is sucessful, a mode 1 solution, without large temperature discontinuity, could be a less controversial interpretation of these systems.

2.4 Supergiant Contact Systems

Recently, several late-type (G and K) binary stars (5 Cet, UU Cnc, PW Pup and possibly HD104901B), with periods of a hundred days or longer, were found to be contact or near contact systems (Leung 1988). The shape of the light-curves of these systems lies between that of W UMa and B Lyrae type light-curves except for their extremely long periodicity. Radial velocity curves are available for two of the systems (UU Cnc and 5 Cet). Boh of them are single-line systems. Their absolute dimensions (by means of photometric mass ratios) indicate that they consist of cool massive super-giant stars. Therefore, their common convective envelopes must be very deep or huge. Their extremely long periods and locations in the H-R diagram (for two of the systems) suggest that they may be the result of very advanced case B mass loss (i.e. very far from the terminal age sequence in a H-R diagram). Even for single stars, there are great diffi-culties associated with modeling the deep convective atmo-spheres of late-type giants and supergiants. It is very easy to see that modeling a common convective atmosphere for late-type supergiant contact systems would be an impossible task with our present knowledge of convection. No doubt, in the coming decade these systems will be a great challenge for the theorists.

2.5 Double Contact Systems

The term "double contact systems" was introduced by Wilson (1979, see also Wilson et. al. 1985). In general, most binary systems are in synchronized rotation. In the case of non-synchronized rotation, the rotation period of one or both components is different from the binary orbital period.

Usually it is the rotational period of one component that is faster than the orbital rotation as a result of gaining mass and angular momentum during the mass exchange phase of close binary evolution. Therefore, we have to include the distortion due to stellar rotation in calculating the equal potential surfaces of a binary system. As a result, we can define another critical equal potential (or zero-velocity) surface (beyond which a star would break apart because of rotational instability). Thus, in a binary system a star will have two different critical potential surfaces: the regular one mentioned in the introduction, and the one described above. A double contact system could mean that one of its components is contacting the regular critical potential surface, while the other is contacting its critical rotational potential surface. That is, both components are in contact with different critical surfaces. Thus, we have a double contact system! Wilson suggested several potential members belonging to this group. It is important to note that in a double contact system the components <u>are not in physical contact</u> with each other, whereas the components of the four types of contact systems described earlier <u>are in actual physical contact</u> with each other. In double contact systems, there are no common evelopes to seak of! Each component would have its own individual stellar atmosphere.

3. CONTACT BINARY SYSTEMS COME IN TWO KINDS; ZERO-AGE OR EVOLVED

Where do contact binary stars come from? Are they born as Siamese twins (contact), or are they evolved into contact? There are at least two ways to determine whether a system is at zero-age or being evolved. For the systems in clusters, the location of the components in the cluster H-R diagram will indicate the age of the systems. Lukily, there are about 4 or 5 systems which are members of open/galactic clusters. The second method is to compare the absolute dimensions (radii and masses) of the components with the radii and masses of zero-age main sequence stars. If the radius of the component(s) is larger than the radius of the zero-age star for the corresponding mass, the system is evolved. The observational data indicates that both kinds of contact systems exist. Most of the contact systems are found to be evolved contact systems. Only less than a handful are believed to be about zero-age. The evolved contacts outnumber the (near) zero-age contacts by at least a factor of ten. Most of the evolved contact systems are located near the main sequence. Therefore, they are mostly the result of case A mass loss of close binary evolution. Very few systems discovered are believed to be the result of case B mass loss. However, these statistics are subject to a very strong selection effect, since observers have a strong bias towards short-period systems. Short-period systems are found near the vicinity of the main sequence - case A mass loss systems.

The interior structures could be very different between the zero-age contacts and evolved contacts, even for the same spectral-type or mass. Modeling of contact systems is still at the pioneer stage. Lucy, and Shu and collaborators produced models of contact binary systems from fairly different basic physical assumptions. It is generally believed that these are preliminary models only, and that more realistic models may come only after a better understanding of the dynamical processes involved in the contact atmospheres.

4. SUGGESTIONS FOR FINDING CONTACT BINARIES

There are several ways to select candidates for contact binaries. Some of them are rather straightforward and easily selected from variable stars catalogues; systems located in or near the zero-age contact region of the "band-aid" zero-age contact of Leung and Schneider (1978), systems with both eclipses equal to or exceeding 0.75 of a magnitude, and systems with no inflection points at the shoulders of the light-curves (this method requires more experience!).

4.1 Period and Spectral-type With "Band-Aid Model"

The "Band-Aid Model" for critical contact systems of Leung and Schneider (1978) is the poor man's contact model which does not involve any high-power interior stellar structure! With zero-age mass and radius from published model stellar interiors from the literature one can make zero-age critical contact systems by pushing the stars together until they reach point contact (or critical contact). Then tape the stars in place with a band-aid. Thus, a contact system is made. The separation between the mass centers can be calculated from Roche geometry and the radii of the stars. The period of the point contact system can be calculated from Kepler's Law with the knowledge of the sum of the masses and the separation. Thus, one can calculate a period (critical) - mass relation for a given mass ratio (Leung and Schneider 1978 used mass ratio of 1.0 and 0.5). Unfortunately, one seldom has any idea about the mass of a close binary system, but the spectral-type of the system is often readily available. One translates mass into spectral-type through a mass spectral-type relation. As a result, one obtains a period - spectral-type relation for critical contact systems. Thus, with a system of a given spectral-type with a period shorter than the critical period, the system is overcontact and zero-age. In contrast, a longer period means the systems will be either detached or semidetached (if the components are zero-age). If the system is found to be in contact, then it has to be an evolved system because the radius (or radii) is (are) no longer zero-age. The region where the period of the system is shorter than the critical period is denoted as ZAC (zero-age contact) in Figure 9 in the paper by Leung and Schneider (1978). Therefore, any system that falls in or near the ZAC zone/region is very likely a good candidate for being a contact binary.

The attractiveness of the method is that it does not require any prior knowledge of a light-curve shape or radial velocity curve. Spectra and period are the only things needed. Usually, the period of a binary has been very accurately defined but the spectral-type, especially in general catalogues, is rather uncertain and could easily be off by two subtypes.

4.2 System With Both Eclipses Exceeding 0.75 Mag.

If we consider both stars/components with equal light, the maximum light lost has to be 0.5 for both eclipses or about 0.75 m (magnitude) for each eclipse. For ordinary/normal stars, one cannot obtain light lost more than 0.75 m for both minima. The only way to achieve this is to make the stars highly distorted. Therefore, in order to have both stars highly distorted is to make a contact system. Thus the systems one is looking for are: 1) systems with both eclipses deeper than 0.75 m, and 2) systems with one eclise of at least 0.75 m and the other even deeper.

In this method, one needs to search only for eclipsing variables in the catalogues and no knowledge of light-curve shape is necessary. One also has to keep in mind that some of the systems in the catalogues are not well known and their eclipse depths are not always reliable.

4.3 System With No Inflection Points

One may look at the inflection points in the shoulders of the light-curve as the points of external contact, θe. The phase of θe is very distinct in a light-curve where a system is detached. The curvature or the slope of the light-curve changes very abruptly from the out-of-eclipse portion to the eclipse portion for a detached system. That is, one could "eye-ball" the phase of θe without carrying out a photometric analysis. As the components get closer together, the phase of the θe or the point of inflection gets less noticeable. As the components become contact to each other, the inflection point starts to disappear. For an overcontact system the curvature of the light-curve changes continuously from maximum to minimum. There is no such thing as an inflection point or θe to speak of in such a light-curve. Therefore, if one cannot "eye-ball" the inflection point in the light-curve, the system must be a contact binary. This method requires the availability of good light-curves and extensive experience in light-curve analyses.

4.4 System With Very Large Rossiter Effect Or With Very Unequal Eclipse Duration

By definition a double contact system has one of its components filling the critical rotational surface. Usually this component rotates many times faster than the orbital period of the binary. This effect will be reflected in the radial velocity curve as the large Rossiter effect. This component is also highly distorted at the equator. It will cause large differences of eclipse duration between the primary and secondary minimum. In order to tell whether a system is double contact requires extensive analyses or the light-curve or the radial velocity curve, or both.

The author wishes to acknowledge the support of this work through a grant (INT8616425) from the National Science Foundation.

REFERENCES

Binnendijk, L. 1965, 3RD. IAU COLLOQUIUM ON VARIABLES ed. by W. Strohmeier, (Bamberg: Bamberg Obervatory), P.36.

Eggleton, P. P. 1985, INTERACTING BINARY STARS, ed. by J. E. Pringle and R. A. Wade (Cambridge: Cambridge University Press), p. 21.

ill, G. and Hutchings, J. B. 1970, Astrophys. J., 162, 265.

Kuiper, G.P.: 1941, Astrophys. J., 93, 133.

Leung, K. C. 1988, Critical Observations Vs Physical Models For Close Binary Stars ed. by K. C. Leung (New York: Gordon and Breach), p. 93.

Leung, K. C. and Schneider, D. P. 1978, Astrophys. J., 222, 917.

Leung, K. C. and Wilson, R. E. 1977, Astrophys. J., 211, 835.

Lucy, L. B. 1968, Astrophys. J., 151, 1123.

Monchnachi, S. W. and Doughty, N. A. 1972, M.N.R.A.S., 156, 243.

Rucinski, S. M. 1985, INTERACTING BINARY STARS, ed. by J. E. Pringle and R. A. Wade (Cambridge: Cambridge University Press), p. 85.

Wilson, R. E. 1979, Astrophys. J., 234, 1054.

Wilson, R. E. and Devinney, E. J. 1971, Astrophys. J., 166, 605.

Wilson, R. E., Van Hamme, W. and Pettera, L. E. 1985, Astrophys. J., 289, 748.

A PHOTOMETRIC ANALYSES OF ECLIPSING SYSTEMS IN NGC188

M. T. Edalati
Behlen Observatory & Department of Physics
University of Nebraska & University of Ferdowsi
Lincoln, NE 68588-0111 Mashhad
U.S.A. Iran

K. C. Leung
Behlen Observatory
University of Nebraska
Lincoln, NE 68588-0111
U.S.A.

ABSTRACT. NGC188 is one of the oldest open/galactic clusters known. This cluster is unique with respect to the large number of binary stars found. The latest survey (Kaluzny and Shara) reported 9 eclipsing variables (all short-period systems except one). We attempted to derive photometric solutions for the 8 short-period systems with the Wilson and Devinney method. Unfortunately, the phase coverage of the light curves is poor. We were able to obtain preliminary solutions to only 5 of the systems. Four of them were found to be semidetached systems and the the other is a contact system. The mass ratios, Mh/Mc, are less than unity for all 5 systems. There is some agreement between the photometric mass ratios and the spectroscopic mass ratios estimated by Baliunas and Guinan. It is concluded that both new and good quality photometric and spectroscopic measurements are urgently needed to reinvestigate these systems.

1. INTRODUCTION

Binary systems in star clusters are of great interest for us in respect to absolute dimensions of the components and binary evolution. With systems in a star cluster we are able to estimate the age of the binary systems from the cluster-age. There have been many studies on NGC 188 since it is one of the oldest open/galactic clusters known. The most recent variable stars survey for this cluster was carried out by Kaluzny and Shara (1987). They employed the 0.9 M telescope

C. İbanoğlu (ed.), Active Close Binaries, 891–896.

with a CCD detector at Kitt Peak Observatory. They discovered 5 new variables in addition to the 4 well-known short-period binaries (EP Cep, EQ Cep, ER Cep, and ES Cep). Unfortunately, with very limited telescope time at their disposal they were able to obtain only poor/marginal phase coverage of the light curves of the systems. We attempted to derive photometric solutions for the 8 short-period systems.

Baliunas and Guinan (1985) have been the only people who have attempted to estimate the spectroscopic mass ratios for the 4 well known systems. They, too, were greatly restricted by the very limited telescope time at their disposal. They obtained a few spectra with the MMT near the phase of radial velocity crossing, and near the phase of maximum separation. In cases where they were unable to determine the systems' radial velocities they adopted a value from the cluster radial velocity. Their method may not be the best way to estimate the mass ratios but it at least gives a preliminary value of the mass ratios of the systems.

2. ANALYSIS

The observations of Kaluzny and Shara (1987) were utilized for our photometric analyses of EQ Cep, ER Cep, ES CEp, V5, V6, V7, and V8. EP Cep was analyzed earlier by Nolin (1987). Generally, we started the computation assuming a detached configuration (mode 2, see Leung and Wilson 1977) and et the converging solution settle on a configuration via the Wilson and Devinney (1971) differential correction method. A q-search method (Leung and Schneider 1978, see also paper by Etzel and Leung at this meeting) was carried out for each of the systems in an attempt to locate the global minimum for the final/adopted solution.

2.1 EQ Cep

The system EQ Cep (P = 0.30690481 days, K0V, <B-V>o = 0.83) has an amplitude of a variation of 0.85 mag., and reaches its maximum brightness of 17.40 in blue. The difference in the depths of eclipses is about 0.1 mag. The phase coverage of Kaluzny and Shara's blue light curve is not good. In our analysis the Σ - q curve diagram (where Σ is the sum of squares of residuals and q = Mc/Mh is the mass ratio) is rather flat, with a slight depression at q = 2.22 or Mh/Mc = 0.45. The inclination, i, is about 83.5^{o}. The system was found to be semidetached where the low mass and hot component filled the critical/Roche surface. The agreement between the observed and computed light curves is satisfactory.

Baliunas and Guinan (1985) managed to secure two spectra at phases near the largest separation, but no measurement near the velocity crossing for EQ Cep. The estimate of the mass ratio Mh/Mc was 0.41 $\pm$ 0.12. We found the agreement between the photometric and spectroscopic values amazingly similar!

2.2 ER Cep

The system ER Cep (P = 0.28573676 days, G9V, <B-V>o = 0.76) has an amplitude of a variation of about 0.80 mag., and the difference in eclipse depth is close to 0.25 mag. Woeden et al. (1978) published a photometric analysis earlier. The phase coverage of the light curve is poor. Again, in our analysis the Σ - q diagram is very flat for this system. The slight depression in the diagram suggests a mass ratio q = Mc/Mh = 2.61, or Mh/Mc = 0.38. The inclination, i, is about 79.6°. There is a slight asymmetry in the light curve where Max I seems to be a little bit brighter. The computed light curve does not fit the Max I well, otherwise the overall fit is satisfactory. The system is found to be semidetached where the low mass and hot component filling the Roche surface. However, the system is about 4% to being in critical contact.

Baliunas and Guinan obtained two spectra near the phase of maximum separation, and one spectra near the radial velocity crossing. The spectroscopic mass ratio, Mh/Mc, estimated by Baliunas and Guinan, is about 0.65 $\pm$ 0.11. There is a significant discrepancy between the photometric and spectroscopic values.

2.3 ES Cep

The system ES Cep (P = 0.34245611 days, KOV, <B-V>o = 0.81) has an amplitude of a variation of about 0.50 mag., and the difference in eclipse depth is about 0.1 mag. The phase coverage is not good. In our analysis, the Σ - q diagram showed two minima with q at 0.3 and 3.5. The solution at q = 3.5 or Mh/Mc = 0.29 is slightly better. The inclination for this adopted solution is 73.7°. The configuration is semidetached , but it is only 2% to being in critical contact! Therefore, ES Cep is almost a marginal contact system.

Baliunas and Guinan obtained one spectra near the phase maximum separation and two spectra near the radial velocity crossing, and gave a mass ratio Mh/Mc = 0.11: $\pm$ 0.05. Again, there is a discrepancy between the photometric and spectroscopic values.

2.4 EP Cep

The system EP Cep (P = 0.28974188 days, KOV, <B-V>o = 0.83) has an amplitude of a variation of about 0.45 mag., and a difference in eclipse depth of about 0.15 mag. The phase coverage is fair for this system. The light curve of EP Cep was analyzed by Nolin (1987), and further work was done by us recently. Again, in our analysis the Σ - q diagram is very flat with a slight depression at about q of 1.8 or Mh/Mc of 0.55. The inclination is about 66.9^{o}. The configuration is semidetached with the low mass and hot component filling the Roche surface. The computed and the observed light curves are in good agreement.

Baluinas and Guinan obtained only one spectra near the phase of maximum separation and no observation near the radial velocity crossing. They suggested a spectroscopic mass ratio of Mh/Mc = 0.30: $\pm$ 0.30. There is a significant discrepancy between the photometric and spectroscopic values for the mass ratio.

2.5 V7 - NGC188

The new short-period eclipsing system (P = 0.3676(5) days) discovered by Kaluzny and Shara has an amplitude of a variation of about 0.50 mag. Unfortunately, there was no phase coverage at the assumed secondary minimum in Kaluzny and Shara's light curve! Also, neither color nor a spectral type has been estimated for this system. We assigned a (B-V)o from its period for this system according to the period-color relation for short-period systems published by Baliunas and Guinan (1987, Fig. 4) Thus, we adopted the temperature of 4900 K corresponding to a spectral type of about KOV. We also assumed that the temperature for the component eclipsed at the assumed primary minimum. In our analysis, the Σ- q diagram susggested a minimum at about q = 0.2. The final solution gave a value of q = 0.19. The system was found to be overcontact by 14%. The inclination is about 76.3^{o}. The computed light curve suggested that the seconday minimum at phase 0.5 is slightly deeper! Since there are no observations near the observed secondary minimum, we are unable to conclude whether the minima are reversed. The solution suggested the mass ratio of Mh/Mc = 0.19.

There is no published spectroscopic mass ratio for this newly discovered system.

2.6 V5, V6, V8 of NGC188

Very extensive attempts have been made in seeking photometric solutions for these systems, but no reliable solutions were found. Two of the systems (V6 and V8) have very small amplitudes of a variation of about 0.1 mag. We suspect that the period of V5 may be wrong or it may not be a double star system.

3. DISCUSSION

All the five systems which have photometric solutions are having values of mass ratio, Mh/Mc, less than one. Four of the systems having unequal minima are semidetached systems. Two of these four with small differences in eclipse depths are very close to being in critical contact (i.e. 2 and 4% to critical contact). In the case of V7-NGC188, the depth of the secondary minimum is not being observed. The photometric solution suggested a contact configuration for this system. In light of the poor coverage of its light curve, we conclude that it is premature to tell the true configuration of this newly discovered system.

Based on the fact that 1.) the light curves of Kaluzny and Shara indicated that the short-period eclipsing systems in NGC188 have unequal minima and 2.) subsequently found them to be semidetached systems, we conclude that these systems are not late-type contact W UMa systems. (Definition of W UMa system can be found in Leung's article on Types of Contact Systems in this meeting.) We strongly reccommend new and good quality photometric and spectroscopic observations for all systems in NGC188, especially for the five systems with photometric solutions mentioned above.

We understand that some of the light elements employed by Baliunas and Guinan were reliable (even though they were the best available at the time). Thus, there might be some small phase-shifts from the phases of maximum separation and radial velocity crossing. The small errors could lead to more uncertainty in the estimated spectroscopic mass ratios.

MTE very much appreciated the hospitality he received at the University of Nebraska during his one-year visit in 1988-89. We like to thank Otto Baugman who tried very hard to seek solutions for V5-NGC188. The research is partly supported by a grant from the NSF INT8616425.

REFERENCES

Baliunas, S. L., and Guinan, E. F.,: 1985, Astrophys. J., **294**, 207.
Kaluzny, J., and Shara, M. M.: 1987, Astrophys. J., 314, 585.
Leung, K. C., and Schneider, D. P.: 1978, Astrophys. J., **222**, 917.
Leung, K. C., and Wilson, R. E.: 1977, Astrophys. J., 211, 835.
Nolin, S. M.: 1987, Master Thesis, University of Nebraska
Wilson, R. E., Devinney, E. J.: 1971, Astrophys. J., 166, 605.

Woeden, S. P., Coleman, G. D., Rucinski, S. M., and Whelan, J. A. J.: 1978 M.N.R.A.S., 184, 33.

SUBJECT INDEX

Star Names

Constellation Designations

HR Numbers

HD Numbers

Durchmusterung Numbers

Other Designations